1st CCPS Asia-Pacific Conference on Process Safety 2013

Qingdao, China
4-5 September 2013

ISBN: 978-1-5108-1715-9

Printed from e-media with permission by:

Curran Associates, Inc.
57 Morehouse Lane
Red Hook, NY 12571

Some format issues inherent in the e-media version may also appear in this print version.

Printed by Curran Associates, Inc. (2015)

For permission requests, please contact AIChE
at the address below.

AIChE
120 Wall Street, FL 23
New York, NY 10005-4020

Phone: (800) 242-4363
Fax: (203) 775-5177

www.aiche.org

Additional copies of this publication are available from:

Curran Associates, Inc.
57 Morehouse Lane
Red Hook, NY 12571 USA
Phone: 845-758-0400
Fax: 845-758-2634
Email: curran@proceedings.com
Web: www.proceedings.com

TABLE OF CONTENTS

GENERAL PROGRAM

BEST PRACTICE MOC AND PSSR PROCESS 1
Mike Bearrow

IMPLEMENTING PSM IN INDIA - CHALLENGES AND OPPORTUNITIES 2
Ramesh Ramalingam, N V Subbarao, S Karthikeyan

MANAGING OPERATIONAL RISKS FOR EFFECTIVE BUSINESS RESULTS 13
Srinivasan Ramabhadran

BENCHMARKING - A GLOBAL AND ASIA-PACIFIC PERSPECTIVE AND COMPARISON OF PROCESS SAFETY MANAGEMENT PRACTICES 22
Louisa Nara, Cindie Pridy

SETTING MULTI-NATIONAL RISK TOLERANCE CRITERIA 39
Paul Baybutt

EXPERIMENTAL STUDY OF FLAME ACCELERATION IN CLOSED PIPE 54
Rafiziana M. Kasmani, Mimi H. Hassim, Miss Hasimawaty Mat Kiah

A CONSEQUENCE BASED ASSESSMENT METHOD/SOFTWARE MODEL TO FACILITATE THE SAFE LOCATION AND DESIGN OF OCCUPIED BUILDINGS ON OR IN CLOSE PROXIMITY TO PROCESS FACILITIES 65
Kehinde Shaba, Colin Hickey

AIR DISPERSION MODELING FOR LANDFILL GAS EMISSIONS AND FACILITY SITING 83
Tony Liu, Yi Liu, Morgan Reed, Eric Peterson

THE SIMULATION OF THE COMBUSTION REACTION OF GREASE WITH OXYGEN AT HIGH PRESSURE 91
Xu Ji, Dong Li, Lijing Zhang, Gang Tao

THE EFFICIENCY STUDY OF USING DIFFERENT EXTINGUISHERS ON TRICHLOROSILANE FIRES 104
Kuangyu Liao, Jaojia Horng, Kehun Liou, Yuanchun Liao, Jianhong Chen

DEVELOPING A FIRE MODEL FOR OFFSHORE QRA 119
Yongfu Xu

PROCESS HAZARD ANALYSIS FOR STEAM GENERATION AND COMBUSTION SYSTEM IN CAPITAL INVESTMENT PROJECT 134
Jianming Lu

INVESTIGATION ON MECHANICAL PROPERTIES AND STRENGTH EVALUATION FOR LARGE-SCALE CRUDE OIL TANK AFTER FIRE PROCESS 155
Wang Wenhe

RELIABILITY ANALYSIS OF TRUCK LOADING SYSTEM IN LNG TERMINAL BASED ON GO METHODOLOGY 165
Xi Chen, Chang-Hang Xu, Yuan Zhu, Guoming Chen, Guoliang Cao, Xiaoping Sun

RESEARCH ON MATHEMATICAL MODEL OF GAS DISPERSION AND ANALYSIS OF ENGINEERING APPLICABILITY 174
Jing Yuan, Yifei Meng

FIRE SAFETY MANAGEMENT BASED ON THE RISK ANALYSIS AND LAYER OF PROTECTION FOR THE CHEMICAL INDUSTRY PARK 184
Xuanya Liu, Qinlin Zhang, Xiaoyuan Xu

STUDY ON EMERGENCY MECHANISM FOR MAJOR PROCESS SAFETY ACCIDENTS IN PETROLEUM AND PETROCHEMICAL INDUSTRY 194
Shengli Chu, Zonghua Cheng, Fei Long

NUMERICAL SIMULATION OF THERMAL RADIATION OF TANK FIRE AND FIRE PREVENTION ASSESSMENT IN OIL PIPELINE STATION 204
Wenxing Feng, Chaopeng Wu

CCPS PROCESS SAFETY METRICS A GLOBAL APPROACH 215
Louisa Nara

BENCHMARKING RISK BASED PROCESS SAFETY EGG (RBPS TEMPLE + 2 UNIVERSES) AGAINST "THE ART OF WAR" 216
Long Zhang

DYNAMIC RISK ASSESSMENT OF WELL CONTROL SYSTEM ... 227
Faisal I. Khan, Vikram Garaniya
USING NEW TECHNOLOGY TO ENSURE CORPORATE STANDARDS COMPLIANCE AND LOWER OPERATIONAL RISK ... 238
Scott Lehmann
DETERMINE THE QUANTITATIVE INDIVIDUAL RISK CRITERIA FOR SINGLE ACCIDENT RISK IN FIXED CHEMICAL SITE USING THE BACKGROUND RISK METHOD ... 249
John Jianmin Zhao
DOMINO EFFECTS IN PROCESS HAZARD ANALYSIS ... 265
Paul Baybutt
RISK BASED DESIGN OF OVERPRESSURE PROTECTION SYSTEMS ... 280
Zhihe Wang
SAFETY CONDITION PROOF RESEARCH OF LARGE-SCALE LNG STORAGE TANK GROUP ... 290
Linghai Kong, Xiaoyuan Wang, Lianjun Tian
RISK BASED DECISION MAKING - APCI PROCESS SAFETY QRA AND RISK MANAGEMENT PROCESS INTRODUCTION ... 301
Xiaoming Zeng, Jenny Wang
A SYSTEMATIC FRAMEWORK FOR DESIGNING AN INHERENTLY HEALTHIER PROCESS DURING RESEARCH AND DEVELOPMENT STAGE ... 311
Mimi H. Hassim, Santha Pandian, Markku Hurme, Rafiziana M. Kasmani
AN INHERENT SAFETY IMPLEMENTING APPROACH FOR SAFETY PLANNING OF CHEMICAL INDUSTRIAL PARKS IN CHINA ... 320
Zhang Ming Guang
BEYOND PSM: INTEGRATING CULTURE AND LEADERSHIP INTO PROCESS INCIDENT PREVENTION ... 330
Ricky Yu
STATUS AND COUNTER MEASURES OF ENTERPRISE HSE AUDIT ... 339
Shaolin Qiu, Laibin Zhang, Apeng Ru
PROCESS SAFETY MANAGEMENT SYSTEM AUDITING ... 348
David A. Moore, Michael Hazzan
PROCESS SAFETY LEADERSHIP AND CULTURE - HOW TO MEASURE AND ENSURE IMPROVEMENT ... 361
Steve Arendt
IMPROVING PSM CULTURE THROUGH OPERATIONAL DISCIPLINE ... 382
Srinivasan Ramabhadran
DUST EXPLOSION HAZARDS ... 392
Ernest Cheam
MANAGING ELECTROSTATIC HAZARDS IN THE CHEMICAL INDUSTRY ... 402
Vahid Ebadat
DUST EXPLOSION RESEARCH DEVELOPMENT FOR COAL MINING SAFETY ISSUES ... 412
M Sam Mannan, Xiaodan Gao, Jiaojun Jiang
DUST EXPLOSIONS - THE NATURE OF THE PROBLEM AND PRACTICAL MEASURES FOR ITS CONTROL ... 423
Vahid Ebadat
INDUSTRY BEST PRACTICES AND EXPERIENCE WITH PROCESS SAFETY LEADING INDICATORS ... 440
Steve Arendt
RELATIONSHIP BETWEEN MINOR EVENTS AND PROCESS SAFETY MANAGEMENT ... 441
Wenhui Han
SAFETY CHALLENGES IN HARSH ENVIRONMENT: LESSONS FROM AN INDUSTRIAL WORKSHOP ... 450
Faisal I. Khan, Salim Ahmed, Ming Yang, Javad Hashemi
LESSONS LEARNT FOR SME'S PROCESS SAFETY MANAGEMENT ... 460
Jinsong Zhao, Johanna Suikkanen
ANALYSIS OF LEAKAGE AND EXPLOSION OF A NATURAL-GAS PIPELINE IN RESIDENTIAL AREAS ... 474
Zhongyu Wang, Lijing Zhang, Gang Tao, Shu Jiang

POSTER SESSION

QUANTITATIVE HAZOP MODELING AND OPTIMIZATION FOR LABORATORY OPERATION IN THE SEMICONDUCTOR INDUSTRY 483
Tianxing Cai

A METHOD BASED ON FUZZY CLUSTERING AND AHP FOR SAFETY CULTURE ASSESSMENT OF CHEMICAL ENTERPRISES 503
Bing Tong, Hua Zhang, Gang Tao, Lijing Zhang

THE NEURAL NETWORK CONTROL ALGORITHM RESEARCH OF STRAIGHT PULL SINGLE CRYSTAL FURNACE 513
Mingyu Li, Hua Meng

THE STUDY OF SUPPRESSION EFFECT OF POROUS MATERIAL FOR THE FLAME SPREADING IN THE TUBE 514
Fenglei Han

CFD BASED TWO-PHASE DISPERSION MODEL OF FLASHING LIQUEFIED JET AND EXPERIMENT VALIDATION 522
Zhu Yuan, Chen Guoming

SIMULATION AND FULL-SCALE EXPERIMENTS OF TUNNEL FIRE WITH OPENINGS ON ROOF 531
Yan Fu Wang, Jun Cheng Jiang

THE NUMERICAL ANALYSIS OF THE CYLINDRICAL VESSEL WALL RESPONSE UNDER DIFFERENT ENERGY LOAD 545
Wenling Guan, Jun Cheng Jiang, Xiaoli Wang

ANALYSIS OF MAJOR ACCIDENTS IN EUROPEAN CHEMICAL INDUSTRIES 554
Jingchun Fu

THE INVESTMENT OF INTEGRATE MODEL OF HSE MANAGEMENT ADOPTED AT THE HONGKONG-ZHUHAI-MACAO BRIDGE DURING CONSTRUCTION 555
Kun Liu

STUDY ON SAFETY EVALUATION TECHNOLOGY OF PLANT LAYOUT BASED ON DOMINO EFFECT 561
Yifei Meng, Dongfeng Zhao

APPLICATION RESEARCH ON INHERENT SAFETY METHOD IN CHEMICAL PROCESS SAFETY DESIGN 562
Wendi Yu

MODELLING AND SIMULATION OF MAJOR HAZARD INSTALLATION DYNAMIC CLASSIFICATIONS ON STORAGE TANK AREA OF WHARF USING LABVIEW 575
Xiao-Li Wang

THE EFFECT OF PARTICLE SIZE ON THE MINIMUM IGNITION TEMPERATURE OF FLOUR POWDER 586
Yue Shao, Fei You

A DISCUSSION OF THE GREY MODEL BASED METHODOLOGY IN WORK SAFETY TARGET FORECASTING AND SETTING 594
Huawen Zhang, Guohua Chen, Baikang Zhu

APPLICATION OF THE BOW-TIE RISK ANALYSIS METHOD ON METHANOL TANK LEAKAGE 618
Xin Wang, Lijing Zhang, Gang Tao, Hui Han

OPTIMIZATION ANALYSIS OF FIRE CONSEQUENCE OF BOG COMPRESSOR ROOM BASED ON FDS 628
Xiaoping Sun, Yuan Zhu, Guoming Chen, Changhang Xu, Xi Chen

DESIGN OF MONORTING AND EARLY-WARNING SYSTEM IN CHEMICAL INDUSTRY PARKS BASED ON RISK ASSESSMENT 637
Zhang Yong

STUDY ON EMERGENCY RESCUE DECISION MAKING BASED ON GIS FOR TYPICAL CHEMICAL ACCIDENTS 646
Shanshan Guo, Huihui Song, Gang Tao, Lijing Zhang

INVESTIGATION ON THE SAFETY PERFORMANCE OF THE ORGANIC INSULATION MATERIALS 655
Yanfei Chen

THERMAL STRESS ANALYSIS OF HYDROGENATION REACTOR BASED ON THE FINITE ELEMENT SOFTWARE 661
Xinmeng Sun, Wan Xu, Gang Tao, Lijing Zhang

A CORRELATION METHOD OF QUALITY PROBLEMS AND PROCESS SAFETY FOR IMPROVING SECURITY STATUS OF CHEMICAL PRODUCTION 671
Yaqi Gao, Bin Sun, Gang Tao, Lijing Zhang
A NEW COLLECTION SYSTEM OF RELIABILITY DATA FOR EQUIPMENT APPLIED TO MECHANICAL INTEGRITY MANAGEMENT 680
Hang Wang, Hai-Qing Wang, Yu-Tao Zhang
DYNAMIC RISK ANALYSIS TO IMPROVE PROCESS SAFETY USING BAYESIAN NETWORK 689
Jianfeng Zhu, Jinsong Zhao
A NEW SIL CONCEPTION AND ITS APPLICATION FOR EMERGENCY SYSTEM IN LNG TERMINALS 704
Jun-Fang Liu, Hai-Qing Wang
FEATURE EXTRACTION OF ACOUSTIC EMISSION FOR OFFSHORE PLATFORM STRUCTURAL HEALTH MONITORING 713
Guo-Rui Li, Chang-Hang Xu, Guo-Ming Chen
EXPERIMENTAL STUDY ON FAN WATER CURTAIN DILUTING WATER-INSOLUBLE HEAVY GAS IN OPEN SPACE 725
Xiaoyue Pi
OPERATIONAL PROCESS ON EMERGENCY RESPONSE SYSTEM OF CHEMICAL INDUSTRY PARK 740
Xue'E Zhao, Xin Cai, Hanyu Su, Hongqi Yao
INTEGRATED ALARM MANAGEMENT AND HUMAN FACTOR ANALYSIS FOR PIPELINE INDUSTRY 751
Yu-Tao Zhang
STUDY ON ADVANCED PETROCHEMICAL SAFETY TECHNOLOGY WITH OPTIMAL DESIGN OF FIRE & GAS SYSTEM(F&G) 760
Shengmei Shang, Haiqing Wang
EFFECT OF HEAT TREATMENTS ON STABILITY OF PENICILLIN G IN WASTE PENICILLIUM CHRYSOGENUM 769
Pu Wang, Huiling Liu, Huan Fu, Xiuwen Cheng, Bing Wang, Qinghua Cheng
THEORETICAL FLAMMABILITY DIAGRAMS 770
Tingguang Ma
STUDY ON SAFETY CAPACITY AND LAYOUT OF CHEMICAL INDUSTRIAL PARK BASED ON OVERALL REGIONAL RISK ANALYSIS 773
Yinting Chen, Mingguang Zhang
DESIGN OF A MANAGEMENT OF CHANGE PLATFORM BASED ON WEB AND RFID 784
Guoliang Cao, Changhang Xu, Huandi Shi, Lefeng Ren
FIRE CONSEQUENCE ANALYSIS FOR TWO PARALLEL-RUNNING PIPELINES 793
Tianming Ge, Lin Hong, Zhe Yao
OFFSHORE PLATFORM KITCHEN FACILITY CHANGE RISK 805
Zufeng Zhang
QUALITY CONTROL AND EVALUATION OF HAZOP ANALYSIS 809
Zhaohui Wang
THERMAL HAZARDS EVALUATION ON THREE KINDS OF INITIATORS IN SOLVENT UNDER ADIABATIC CONDITION 817
Minjun Peng, Jiayu Lv, Wanghua Chen, Liping Chen
THE EXPERIMENTAL STUDY OF EXPLOSION-PROOF PROPERTY OF HAN EXPLOSION SUPPRESSION MATERIAL 826
Tongshan Yin, Liu Yi
A TECHNICAL SOLUTION FOR THE OPTIMIZATION OF GAS DETECTORS' PLACEMENT IN HIGH-SULFUR CRUDE OIL PROCESSING INSTALLATIONS 836
Bo Zhang, Yan Jia, Lei Wang, Wei Xiong, Zhigang Wang
HIGH-SULFUR CRUDE OIL PROCESSING PROCESS H2S LEAKAGE GRADING WARNING AND PROTECTIVE MEASURES 846
Ming Chen, Dongfeng Zhao
THE RESPONSE TO THE LEAK ACCIDENTS OF LONG-DISTANCE NATURAL GAS PIPELINE 854
Cuiping Wang, Dongfeng Zhao, Fenglei Han
THE SAFETY DISTANCE STUDY OF THE LPG STORAGE FIELD BASED ON THE NUMERICAL SIMULATION OF VAPOR CLOUD EXPLOSION 863
Chao Zhang, Dongfeng Zhao, Yifei Meng, Yi Liu

THE BASIC CHARACTERISTICS AND HAZARD ASSESSMENT OF PARTICULAR RISK OF DUST EXPLOSION AS WELL AS SECURITY PREVENTION AND CONTROL COUNTERMEASURES ANALYSIS ... 876
Hao Ding

STUDY ON EMERGENCY MANAGEMENT CAPABILITY EVALUATION OF CHINESE OIL COMPANIES' OVERSEAS EMERGENCY BASED ON THE CATASTROPHE PROGRESSION METHOD ... 878
Hao Ding, Haodong Shi, Ming Liu

EVALUATION SYSTEM AND APPLICATION OF THE INTEGRAL SAFETY IN CHEMICAL INDUSTRY ZONE ... 888
Chen Chen, Xinmei Zhang

ARAMIS: POWERFUL METHODOLOGY IN QUANTITATIVE RISK ASSESSMENT OF CHEMICAL INDUSTRY ZONES ... 897
Xinmei Zhang, Chen Chen

LOSS FUNCTIONS AND THEIR APPLICATIONS IN LOSS-BASED PROCESS SAFETY MONITORING ... 906
Javad Hashemi, Faisal I. Khan, Salim Ahmed

THE RISK ANALYSIS AND CONTROL RESEARCH OF OIL SHALE PROCESSING ... 921
Jing Lan

IDENTIFICATION AND THERMOKINETICS OF AUTOCATALYTIC EXOTHERMIC DECOMPOSITION OF β-HMX ... 931
Shilong Bao

DEVELOPMENT AND IMPLEMENTATION OF SOFTWARE OF HAZOP ANALYSIS SEQUENCE DETERMINATION METHOD BASED ON THE INTERPRETIVE STRUCTURAL MODELING ... 940
Feng Wang

AN OVERVIEW OF MODERN CALORIMETRIC TECHNIQUES FOR ENERGETIC MATERIALS AND RUNAWAY REACTION STUDIES ... 952
J Zhou

RISK FACTOR IDENTIFICATION OF THE RISK EVALUATION FOR INDUSTRIAL PARK ALONG YANGTZE RIVE ... 963
Zhang Li, Bai Ping

FORMATION AND CHARACTERISTICS OF TULIP FLAME PROPAGATING THROUGH THE PREMIXED METHANE/AIR MIXTURE IN A DUCT ... 972
Dong-Liang Chen, Dong-Sheng Zhang, Kui-Sheng Wang, Jianfeng Yang, Chun-Ling Xin

FAULT TREE ANALYSIS OF NATURAL GAS PIPELINES FAILURE ... 981
Xinyong Liu, Y. N. Wang, Z. R. Wang

A PERFORMANCE EVALUATION SYSTEM OF SAFETY CULTURE CONSTRUCTION IN PETROLEUM ENTERPRISE BASED IN THE BALANCED SCORECARD ... 990
Daizhi Tang

STUDY ON ESTABLISHMENT AND ANALYSIS OF FAULT TREE MODEL OF BURIED NATURAL GAS PIPELINES FAILURE ... 998
X. Y. Liu, Y. N. Wang, Z. R. Wang

Author Index

Best Practice MOC and Pssr Process

Bearrow, M., OSyS, A Rolls-Royce company

Originally presented on: 9/4/2013 0:00:00 - 0:25:00

Management of Change (MOC) is still the most challenging element of OSHA's Process Safety Management (PSM) standard, the EPA's Risk Management Program (RMP) rule and now the US Department of Interior's Bureau or Safety and Environmental Enforcement's Safety and Environmental Management Systems (SEMS). Even though the PSM standard has been around since 1992 and industry has been managing changing forever, we still can get it wrong. The mindful or diligent efforts of many working in concert is necessary to ensure that change is identified, analyzed and executed in a quality way. Just when we get it right on paper and get the workforce up to speed, we have employee turnover, neglect and sometimes regulatory change. New actors and a constantly changing script make it hard to get change managed efficiently and effectively. Many high profile companies have addressed this challenge by creating standard processes and automating those processes across their manufacturing or production facilities. Two major processes that have been successfully automated are the MOC and PSSR processes. Technology has been used to automate each step of the MOC workflow; enabling the collection of change ideas, formalizing the analysis of the idea, approval and execution of the change. It can also help provide visibility of the health or current state of your MOC and PSSR processes. The use of software can enable the standardization of best practices. This paper discusses how the chemical process industry has decided what MOC and PSSR best practice is and how it should be automated.

1ST CCPS ASIA-PACIFIC CONFERENCE ON PROCESS SAFETY

Implementing PSM in India – *Challenges* and *Opportunities*

K Ramesh Ramalingam – DGM
rameshramalingamK@cholams.murugappa.com
N V Subbarao - Chief Executive
subbaraoNV@cholams.murugappa.com
S Karthikeyan - AGM
karthikeyan@cholams.murugappa.com

Chola MS
RISK SERVICES

Agenda

- ❖Status of Process Safety in India today
- ❖Challenges in implementing PSM in India
- ❖Opportunities
- ❖Conclusion

Introduction

- Spot light - Bhopal tragedy; Jaipur Fire; HPCL Vizag

Fig.1. Bhopal disaster

Fig.2. Jaipur IOC Fire

Fig.3. HPCL Vizag Fire

- last 5-7 years, the Indian companies have started adopting HSE at the design stage
- OISD is guiding for oil & gas sectors in India
- Indian Chemical Council has embarked on a journey of conducting PSM awareness session.

Chola MS RISK SERVICES

Process Safety Leadership

Challenges:

- Difference → Injury prevention v/s process safety
- Implement the PSM system where a well established safety management system.
- Form a PSM core committee → from various functions

Chola MS RISK SERVICES

PSI & PHA

Challenges:

Basic engineering / updated documents not in place – 40 to 50% of overall project time frame

Integrating safety during design stage

Lack of familiarity of project personnel on safety life cycle approach

Economics → leading to compromises on safety hardware.

Chola MS
RISK SERVICES

PSI & PHA (Conti…)

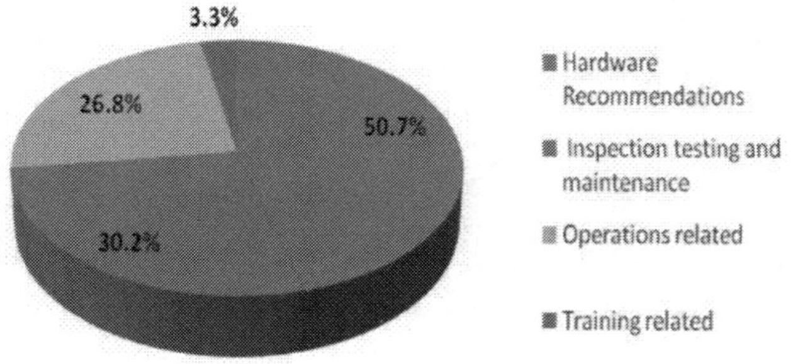

Figure 3. Analysis of 8 HAZOP studies carried out during operational stage to the process industries

Chola MS
RISK SERVICES

Asset Integrity

- Process Industries → Standards on inspection, testing and maintenance for its assets are not clearly defined.
- Understanding the Importance of carrying out proof testing of SIL

Figure 4. Importance of proof testing.

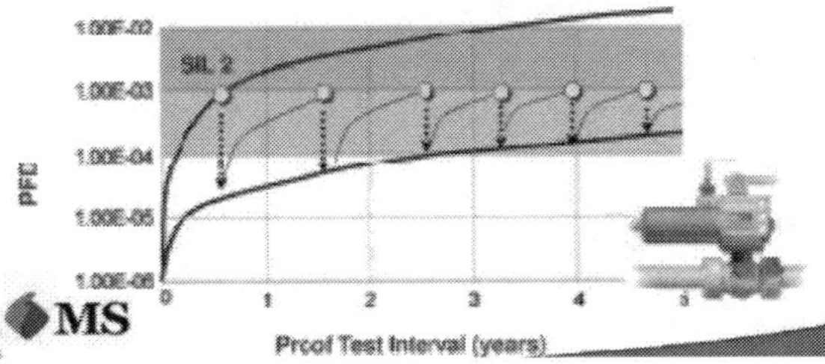

Asset Integrity (Conti…)

- Static equipment integrity program → neglected
- Lack of comprehensive maintenance guidelines leading to adopt risk based approach.
- Deriving asset integrity performance indicators

MOC

Challenges:

- Completion of MOC documentation post implementation of change
- Inadequate understanding of the terminology change across the organization
- Project activities are excluded from MOC process leading major layout and risk acceptability issues
- Use of improper hazard identification technique to identify hazards due to the proposed change
- Lack of participative approach by various functions
- No clear differentiation between a temporary change and a permanent change
- Documentation requirements are not fully understood and complied.

Chola MS RISK SERVICES

PSSR

Challenges:

- Absence of a formal approach to the PSSR – safety inspection
- Lack of full fledged and trained PSSR team members
- Does not comprehensively address the software aspects like operating procedure, training needs etc.,
- PSSR does not check into the implementation of PHA study recommendations related to the change.

Chola MS RISK SERVICES

PSSR (Conti...)

Inter relation with other PSM elements like process safety information, MOC, asset integrity, training etc are not comprehensively addressed.

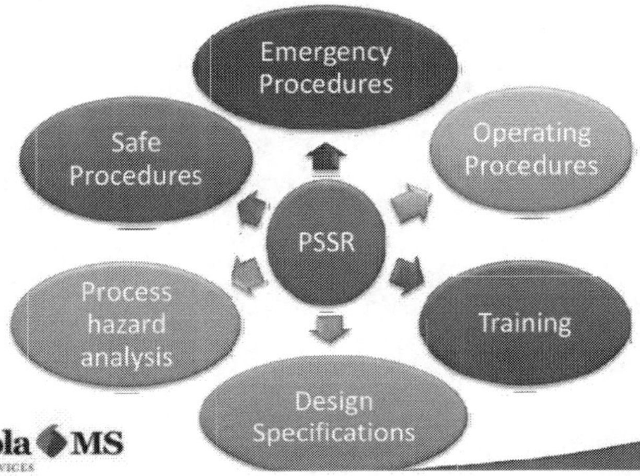

Chola MS RISK SERVICES

PSSR (Conti...)

Status of compliance of a third party PSSR audit for green field projects

- 96 recommendations in the report,
  - 55 High priority category
  - 41 Medium priority category

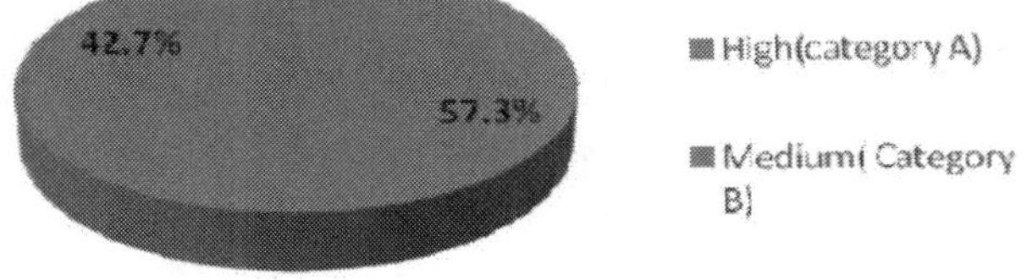

Figure 5. Status of compliance

Chola MS RISK SERVICES

Operating Procedures and Training

Challenges:

- Statutory requirements → Frequency of review is not clearly defined for process industries.
- Normal operating procedures → Not adequately address the steps involved in start up, shut down or temporary operations.
- Compliance to operating procedures→ for adopting innovative methods
- Training needs → Employee focus on injury prevention and process safety
- Qualified / accredited process safety trainers

Chola MS
RISK SERVICES

Contractor Safety

Challenges:

- Inadequate oversight procedures in monitoring the operations
- Contractor → manpower supply in nature
- Gaps in demand and supply mismatch and the resource requirements

Chola MS
RISK SERVICES

Incident Reporting, Investigation and Compliance Audits

Challenges:

- Participation of field level staff in reporting incidents
- Lack of motivation,
- Perception of being targeted if an incident is being reported,
- Absence of adequate feedback mechanisms
- Lack of understanding on what constitutes a process safety incident
- Transparency in sharing root cause between management team and field staffs

Chola MS
RISK SERVICES

Incident Reporting, Investigation and Compliance Audits (Conti...)

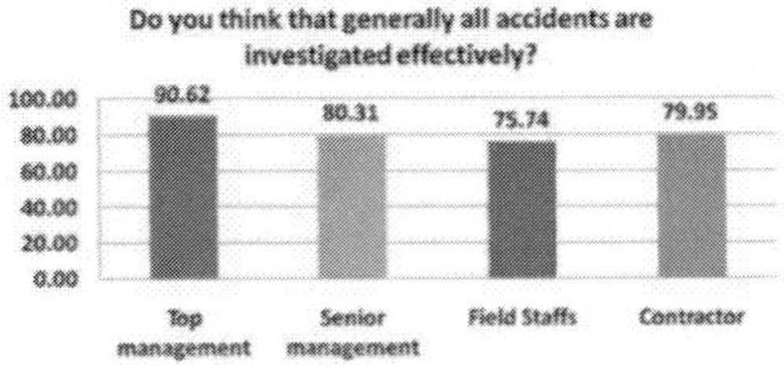

Figure 6. Graph derived from 21 facilities

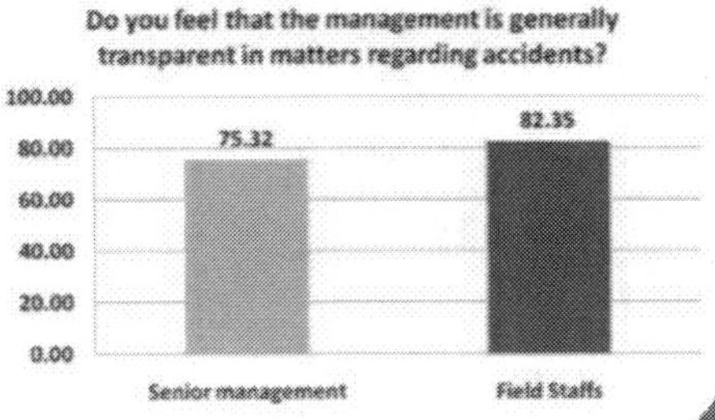

Figure 7. Graph derived from 9 facilities

Chola MS
RISK SERVICES

Opportunities

Industry Level (Sectoral) Initiatives

- Developing Process Safety Management system

Capacity Building Initiatives

- Organizations need to build internal capacities within their teams to derive long term benefits rather than relying on external support always.
- A national level →PSM awareness course and certificate courses.
- A robust system on incident reporting and sharing the learning.

Chola MS RISK SERVICES

Opportunities (Conti...)

Policy Level Initiatives

- A Process Safety Management System frame work shall be established.
- OHS frame work → a specific and focused frame work on PSM would go a long way in ensuring that the various agencies involved in enforcing the statutory requirements are made to participate in this mission.
- Establishing a nodal agency to provide technical and research support for implementing the PSM systems.

Chola MS RISK SERVICES

Conclusion

Implementing and sustaining PSM System in Indian context is a demonstration of the commitment by organizations towards enhancing the value for various stakeholders.

The challenges faced and the opportunities available present a positive view of benefits.

PSM is certainly the destination next for Indian industries.

Chola MS
RISK SERVICES

References

- *Oil Industry Safety Directorate (OISD) Standards and website www.oisd.nic.in*
- *Indian Chemical Manufacturers Association, website www.icma.org*
- *Indian Chemical Council, Website www.indianchemicalcouncil.com*
- *Directorate General of Factory Advice Services and Labour Institute (DGFASLI)*
- *Indian Institute of Chemical Engineers http://iiche.org.in*

Chola MS
RISK SERVICES

Thank You
CHOLAMANDALAM MS RISK SERVICES LTD.
PARRY HOUSE, 4TH FLOOR
NO.2, NSC BOSE ROAD, PARRYS, CHENNAI – 600 001
Phone: 044 – 3044 5400 Fax: 044 – 3044 5550
Chola MS Risk Offices:
Chennai | Mumbai | New Delhi | Gurgaon |
Kuwait & Singapore [Partner Offices]
inquiry@cholams.murugappa.com
www.cholarisk.com

1ST CCPS ASIA-PACIFIC CONFERENCE ON PROCESS SAFETY℠

MANAGING PROCESS SAFETY TO ENHANCE BUSINESS PERFORMANCE

Srinivasan Ramabhadran
DuPont
11-01 Harbour Front Tower One, Singapore
ramabhadran.srinivasan@dupont.com

Prepared for Presentation at
American Institute of Chemical Engineers
1st CCPS Asia-Pacific Conference on Process Safety
Qingdao, China
September 4 – 5, 2013

UNPUBLISHED

MANAGING PROCESS SAFETY TO ENHANCE BUSINESS PERFORMANCE

Srinivasan Ramabhadran
DuPont
11-01 Harbour Front Tower One, Singapore
ramabhadran.srinivasan@dupont.com

Keywords: process safety management, safety culture, performance improvement, management system

Abstract

The inherent nature of hazards and risks associated with operations and news about recurring process incidents continues to keep the spotlight on process and operational safety. Whilst most companies do have some sort of risk analysis, this is an area where many companies experience the biggest challenge. A good system for managing risks alone does not solve the problem. The need is for applying them consistently and effectively to get the desired results.

Many companies have embarked on operational transformation that has been triggered by a multitude of factors such as improving productivity, improving capacity through debottlenecking, changing workforce, etc. As a result the only thing that is constant in this industry is "Change". It has become inevitable for companies to manage risks through this operational transformation process.

This paper is intended to share the knowledge and experience from a practical perspective on how some of these challenges have been overcome by DuPont and how DuPont embarked on operational transformation by adopting a structured risk management approach. The emphasis will be on how process safety management can help in managing risks and avoid pains through the transformation process. The link between managing operational risks and strong and effective business results will be discussed and deliberated. It is a unique opportunity not only for operational leaders but also all professionals associated with process industries to gather and reinforce the belief that risks can be managed through a structured approach for achieving operational excellence.

1. Introduction

Many industrial companies today are either initiating or accelerating programmes designed to improve their process safety management (PSM) systems. These efforts are being driven by a number of factors, one of which is the prevention of large-scale, catastrophic industrial accidents. Recent incidents have shed new light on the need for continuous review and improvement of PSM – the explosion at the Chevron refinery in Pembrokeshire, UK and the explosion at the Marcoule nuclear site in France. These incidents come on the heels of the release of the US National Commission report on the Deepwater Horizon Oil Spill, in which it was highlighted that deepwater oil and gas exploration and production involves risks for which neither government nor industry are adequately prepared. Such incidents have led to not only incurred significant cost for the companies involved, but ultimately the erosion of their right to operate.

This is aligned with a wider macro-trend that is affecting industry across Asia – increased public and stakeholder accountability, as well as concern for corporate footprint, which DuPont defines as all injuries, incidents, waste, emissions, and utilisation of depletable forms of raw materials and energy. Indeed, the investment community and wider public are putting pressure on companies to improve both safety and environmental performance due to risks to financial health and corporate image, and, conversely, for the benefits associated with perceived positive performance.

In addition, another key driver for improvement of process safety is the impending impact on their operations of governmental regulations, especially in those cases where exploration and extraction is deemed "high-risk", such as offshore drilling. Indeed, governments are seeking to introduce prescriptive safety and pollution-prevention standards to supplement current regulatory requirements. Over time, this trend for increased oversight is likely to continue to increase.

In light of these trends, it is clear that PSM has become a strategic imperative for companies in hazardous industries across the Asia. While this does present a significant challenge, it also represents an opportunity: should proper initiative to enhance PSM be taken, owners and operators will be able to reinforce their right to operate, while realizing the benefits associated with improved performance - increased productivity, higher quality, reduced waste and lowered operational costs.

2. How DuPont Defines Process Safety Management

PSM is the application of management systems and controls (programmes, procedures, audits, evaluations) to a manufacturing or chemical process in a way that process hazards are identified, understood, and controlled. Within DuPont, PSM was initially developed to help prevent serious, process-related incidents, which might affect plant personnel, off-site communities, the environment, or result in significant property loss or loss of business. Its purpose was to establish a framework to help focus management efforts on this complex subject, to comprehensively describe the principles and essential features of PSM for use by sites in managing Process Safety and to define corporate business responsibilities and activities.

To do so, corporate leadership defined their vision for PSM: the "Goal is Zero". This programme focuses on the following themes:

- The goal is zero process-related injuries and incidents. The question is not if, but how this goal can be achieved, just as it is for personal safety.
- PSM is to be considered a business issue, not only a manufacturing or operational issue, requiring the contributions of a broad cross section of the organization to ensure success.

- All DuPont entities must therefore establish a sustainable continuous improvement process toward achieving the goal of zero PSM injuries and incidents that is embedded in business planning.

Once the vision was defined, all businesses and regional entities of DuPont were required to implement such a plan, one that included the following actions:

- Establish a senior executive as the "PSM Champion" to promote business/region accountability.
- Ensure that systems were in place to understand current process safety performance and for building value and awareness for achieving zero process incidents through specific goals and metrics.
- Ensure that a process for allocating resources to process safety was part of the business planning process.
- Apply PSM systems to all process operations, including lower hazard operations, with improved audit emphasis and metrics.
- Develop the necessary systems and training to further apply Inherently Safer principles to process and product development activities, including R&D, operations, and capital projects.
- Develop a programme to drive operational discipline – the deeply rooted dedication and commitment by every member of an organization to carry out each task the right way each time – at all sites.

The results of these efforts to drive continuous improvement have greatly impacted PSM performance in DuPont, allowing the company to realize an approximate 90% reduction in process safety incidents between 1995 and 2008. This was accompanied with a 20% increase in production and 19% decrease in energy use over the same period. However, the company has not yet reached its goal of zero, and thus will continue to review and upgrade the PSM performance.

Beyond the best practices highlighted above, DuPont created a model for PSM that involves the management and reduction of risk within three focus areas: facilities, technology and personnel. Underpinning these focus areas are 14 process elements, each of which serves as an avenue for enhanced risk management. Through PSM, businesses can capitalize upon the most effective group structures, documented procedures and human and technological capabilities. The PSM Programme includes specific requirements and recommended practices for 14 elements, depicted in Table 1.

Table 1: PSM Wheel

The behavioural elements and risk-based elements have a symbiotic relationship, ensuring maximum operational risk management.

This figure is typically called the PSM Wheel, with each of the 14 elements, such as Process Hazards Analysis, Operating Procedures and Safe Work Practices, and Personnel Training and Performance, arranged around the spokes of the wheel. Management leadership and commitment, necessary for implementing and maintaining strong PSM programmes, is shown at the centre of the PSM Wheel. Operational discipline is shown as the rim of the PSM Wheel, connecting all of the elements and translating the required managing systems into real results for preventing injuries and incidents.

The strength of this model lies in both its integrated nature and due to the single governance process. By integrating PSM principles into all business processes, ranging from capital projects planning and execution to performance management and training, companies are able to not only ensure that PSM is adopted within the corporate culture, but also to reap the soft benefits of adopting a structured, metrics-based methodologies: increased discipline and a proclivity to incorporate data-driven processes. Further, the rigorous protocols for PSM can certainly inform common corporate metrics, risk classification protocol, audit procedures and incident investigation.

3. The Human Factor

The PSM model can be applied to any hazardous process, including extra. However, many oil & gas companies tend to view the management of operational risk[1] only in relation to technology and facilities. While these are important foci, the human factor – behaviour – is largely sidelined. In contrast, DuPont believes that people are an essential factor that is interwoven through all elements of PSM.

Indeed, personnel must be given equal attention as facilities and technology, as is depicted in Table 1. This is largely done through the development of a strong safety culture and a high level of operational discipline. To achieve this, companies must strive to create an environment in which employees are empowered to maximize personal and collective operational discipline. This requires a sound behavioural management system that is conducive to the cultivation of operational discipline in such a manner. Certainly, people must be acclimated to the discipline that behavioral safety management instils, so they have the habit of following procedures. Unless employees see clear value in performing every procedure correctly – every time – there is a higher probability of shortcuts being taken, or lapses in judgment to occur. Once the appropriate context is created, lengths should be taken to ensure that each employee has an optimal level of knowledge to perform a task correctly, commitment to do the tasks properly, and awareness of the potential for problems.

[1] For the purposes of this paper, operational risk is defined as the portion of a business's total risk profile that arises from the execution of business functions. Essentially it is the risk of loss resulting from inadequate or failed internal processes or systems, human behaviour or an external event. Typical operational risks include employee health and safety, environmental damage, physical assets, business interruption, regulatory compliance, reputation, right to operate, employee morale and retention, as well as business value and market capitalization

The key to developing such a safety culture and cultivating operational discipline is, as Table 3 suggests, management leadership and commitment. Demonstrated leadership commitment – and engagement – is vital to creating a culture that perceives value in implementing and maintaining PSM improvements. This commitment should be codified in core values and treated as a long terms strategic issue by the company, with management ensuring it is treated as such, enterprise-wide. In other words, PSM should be given equal weight in management decisions as other business drivers, such as productivity, quality and profitability. While PSM cannot reasonably be everyone's constant focus, all levels of management should in some measure be seen demonstrating their commitment to PSM principles. That includes the discussion of PSM priorities in multiple communication channels, and visibly demonstrating personal commitment by providing resources, awarding staff and supporting initiatives.

4. Implementing Process Safety Management Effectively

Certainly, the implementation of PSM in any organization is preceded by integrating safety as part of the organizational culture and visibly demonstrating management commitment to safety. While a comprehensive programme is essential for achieving excellent results in preventing process incidents, it is only one of four key steps required to establish and maintain an effective PSM effort:

- Step 1 – Establishing the Safety Culture: As briefly noted above, a PSM programme will only be as effective as the underlying safety culture permits. If an organization has a weak safety culture, PSM efforts are doomed to conflicting priorities, such as production needs or lowering costs. A strong safety culture, based on a commitment to core values on safety, health, and environmental issues, and as evidenced by organizational policies, goals, metrics, and day-to-day decision-making, supports establishing safety systems where safety priorities are recognized, not as conflicting with other priorities, but rather as inherently necessary for completing any task the right way. DuPont's fundamental policy on safety was established by the company founder and has remained a core value for over 200 years, creating a strong overall safety culture.
- Step 2 – Providing Management Leadership and Commitment: As aforementioned, management leadership and commitment which is visible throughout the organization is required for continued excellence in PSM. Beginning with the formation and fostering of a strong safety culture, continuing with policies that set high expectations for excellent safety performance, and implemented in day-to-day decisions that provide resources and establish high priority for safety-related activities, the actions of all levels of management must support and reinforce strong PSM programmes and accountability.
- Step 3 – Implementing a Comprehensive PSM Programme: A comprehensive PSM programme includes integrated managing systems for various elements of process safety, such as Process Hazards Analysis, Incident Investigation, and Mechanical Integrity. A good starting point for developing a strong PSM programme is existing regulations, such as the OSHA PSM Rule and the EPA Risk Management Programme in the United States. Regulations should generally be considered as the starting point only, though, representing minimum essential levels of practice. Additional requirements should be

considered that go beyond regulations and better meet corporate needs for process safety and these requirements have been integrated in to the PSM model.

- Step 4 – Achieving Operating Excellence through Operational Discipline: No matter how comprehensive and well-designed PSM programmes may be, it is the day-to-day ability of everyone to practice operational discipline that successfully changes safety systems from concept to reality. Operational discipline not only helps prevent serious injuries and incidents, but also contributes to excellent business performance through increased operating excellence, including higher productivity, higher quality, reduced waste, and lower costs.

In addition, companies must be prepared to continuously upgrade and extend their PSM programme, especially in light of global trends that are altering the context in which companies operate, namely shifting workforce demographics, the slow global economic recovery, and rapidly evolving technology and modes of production. PSM should by no means be considered a static entity, and must have the capacity to develop and progress in concert with the business.

5. The "Other" Cultural Element – Implementing Process Safety Management in Diverse Cultures

One shifting context to which companies have found it particularly challenging to acclimate has been the need to implement PSM within facilities in different countries, especially following a merger or acquisition. For example, as DuPont began to globalize operations, the company faced significant challenges associated with the implementation of a comprehensive PSM programme into hazardous operations throughout its globe. As a result, implementation methodologies were adapted to accommodate cultural differences.

Indeed, when implementing a PSM programme within a different cultural context, the challenge is not to change the local culture, but to adjust the methodology to account for cultural differences while maintaining the basic elements required for the programme to be efficacious.

If the general cultural environment is evaluated early in the programme, an implementation plan can be developed that will enhance success. Drawing from the Geer Hofstede Cultural Dimensions Model (Hofstede 1980), culture was analysed based on five dimensions: Power Distance Index, Individualism, Masculinity, Uncertainty Avoidance Index and Long Term Orientation. DuPont first defined itself within the context of these criteria, and then defined cultures in which the company operates. The differences provide instruction as to how the implementation strategy can be adapted.

For example, ratings for the criterion "Individualism" can give a strong indication of how the implementation methodology should be changed to ensure maximum efficacy. In countries such as the United States, there is a high "Individualism" rating. That is to say that many workers in this country value network participation, in which individuals feel comfortable leveraging their individual skills and knowledge to benefit the corporation. Therefore, a network devoted to PSM would be quite useful in developing and ensuring compliance to a strong programme. In contrast, in China, there is a low "Individualism" rating. Individuals tend to prefer strong, directed leadership, with little deviation from formal reporting lines. Therefore, it is more useful to have corporate rules in place to which workers could be held accountable.

Furthermore, with the criterion "Uncertainty Avoidance" – defined as the low tolerance for uncertainty and ambiguity – the rating is generally quite high within companies in Latin America and the Middle East. That is to say that there is, generally, a strong acceptance and adherence to rules. In these cultures, people are more likely to readily accept the concept of PSM standards, or strict guidance from the company. Therefore, they are likely to effectively implement these requirements. In contrast, "Uncertainty Avoidance" is quite low in Western European countries for example. Thus, in general, there is little cultural resistance to ambiguity. If a rule or standard is developed, they are carefully crafted to describe only that which must be done. Sites are given considerable flexibility in determining how PSM is conducted. With this flexibility comes the need for a strong audit function to ensure compliance.

While the general observations for a particular culture may not fully apply in all situations, they can serve as valuable resources when devising implementation plans in different cultural contexts. It is generally much more effective to adapt methodologies to accommodate predominate cultural norms than to implement a strategy which may conflict with these norms.

6. Conclusion

Within hazardous industries, repetitive acts can lead to complacency, increasing operational risk. While workers understand the importance of safety, they tend to concentrate on unsafe conditions rather than unsafe acts. A greater dependence on technology may not provide the complete solution. Where incidents in the past have generally related to equipment or human failures, greater reliance on digital technology for process design and control introduces new complexity. As such, it is vital that companies across the Middle East improve process safety management to mitigate the high level of operational risk.

The goal now, and in the future, is to prevent serious injuries and catastrophic incidents – the goal must be zero. Learning from experience and going beyond the minimum essential practice defined by regulations are essential if zero is to be achieved. Seeking to understand and manage the hazards of our processes to help ensure a safe future, despite increasing complexity, will always be the challenge.

The task of implementing a process safety management system is indeed quite complex, as its mandate cuts across many functional areas of a company. However, when driven by a visible management commitment, it is possible to adopt methodologies that can be readily applied in each business process. Also essential to this process is the foundation of a strong, behaviour-based safety culture in which employees at all levels are empowered to feel ownership of not only their own safety, but also the safety of others

The DuPont PSM model offers a structured approach to manage these challenges. Through proper administration of PSM, it is possible to protect personnel from injury and prevent significant environmental harm, property damage, and business losses, while furthering the ability to achieve performance objectives.

7. Biography

Hofstede, Geert. “Motivation, Leadership, and Organization : Do American Theories Apply Abroad?” Organizational Dynamics. Summer 1980 42-63.

1ST CCPS ASIA-PACIFIC CONFERENCE ON PROCESS SAFETY℠

"Benchmarking – A Global and Asia-Pacific Perspective and Comparison of Process Safety Management Practices"

Scott Berger
Center for Chemical Process Safety (CCPS)
Scobt@aiche.org

Cindie Pridy
Phillip Townsend and Associates, Inc. (PTAI)
Cpridy@ptai.org

Prepared for Presentation at
American Institute of Chemical Engineers
1st CCPS Asia-Pacific Conference on Process Safety
Qingdao, China
September 4 – 5, 2013

UNPUBLISHED

"Benchmarking – A Global and Asia-Pacific Perspective and Comparison of Process Safety Management Practices"

Scott Berger
Center for Chemical Process Safety (CCPS)
Scobt@aiche.org

Cindie Pridy
Phillip Townsend and Associates, Inc. (PTAI)
Cpridy@ptai.org

Keywords: benchmarking, gap analysis, metric, performance gap, process safety management system, risk based process safety element, survey, Asia-Pacific

Abstract

The vision and recommendation of industry-wide process safety originated with the creation of the Center for Chemical Process Safety (CCPS®[SB1]) by AIChE following the December 3, 1984 Bhopal disaster, and has been refined and strengthened over time. Most recently, CCPS identified 20 process safety elements of Risk Based Process Safety (RBPS) to be used to drive continual improvement (CCPS, 2007). As organizations implement Process Safety Management (PSMS) systems through a structured and systematic approach there is a realization that the Process Safety journey includes the need to develop an understanding of the current status of existing PSMS, sustain Process Safety performance, and sustain continuous improvement. Benchmarking the process of PSMS practices, with reference to industry peers, is a very useful tool to help identify and address the gaps to best performers.

Viewed from a risk based perspective, in 2009 CCPS selected Phillip Townsend Associates (PTAI) to design a PSMS benchmark program. The CCPS benchmarking committee selected 6 of the 20 RBPS elements with the goal of seeking an understanding of the relationship between process safety outcomes and management practices. Benchmarking results of the global study provide participants element-specific quartile performance to the overall industry peers who participated in the study, as well as, specific Asia-Pacific peers. As part of the benchmarking process, the study develops and presents a gap analysis to senior management identifying key areas of improvement for implementing and evolving effective PSMS's.

This paper starts from a global benchmarking perspective. A detailed discussion of the PSMS benchmark, methodology, the value proposition to be gained, and the experience of participants in developing an understanding of their current status of existing PSMS is presented. Specific attention is on comparing the similarities and differences between the Asia-Pacific peer group and the overall, global industry peer group. Practices that will improve and sustain PSM performance from top performers are reviewed. The paper focuses on the journey to leverage from lessons learned through the enhancement of Process Safety Knowledge and adopting best practices. Continuous improvement and excellence can be achieved by focusing on the most important decisions and behavior changes that leadership can, closing the gap to the top performers, developing central and site specific action plans for improvement, and raising the internal level of all organizational units to the that of the top-performing internal site.

1. Introduction

Chemical production and the associated processing industries dealing with hazardous and toxic materials have a significant commercial and societal responsibility to eliminate and/or mitigate the loss of primary containment or fires related to process materials while also ensuring that adequate safety systems are in place. Process Safety Management Systems (PSMS's) differ from personnel safety management by focusing directly on the processes, equipment and technology involved in the root cause of incidents. The broader application of PSMS decision making and leadership behavior on a risk based perspective addresses loss of containment events of greater and lesser consequences, safety system challenges and failures that could led to failure, and operating discipline and management behaviors that ensure adequate layers of protection are being maintained. The focus shifts from the lagging indicators that track incidents to leading metrics used in a proactive manner to prevent incidents, adopt management practices supported by risk-based analysis, and continuous improvement as shown in Figure 1: Process Safety Indicator Pyramid (API RP 754, 2010).

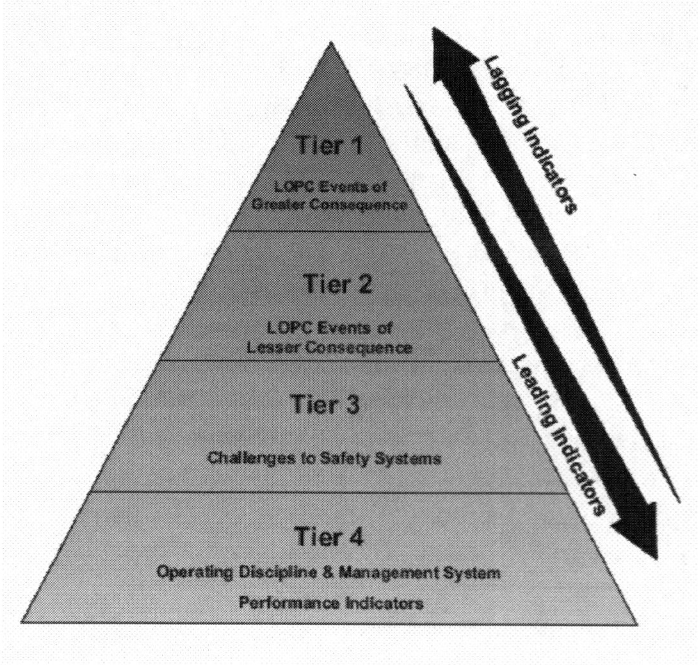

Figure 1: Process Safety Indicator Pyramid

Risk-based analysis is not a substitute for a standard/code based approach. Rather, RBPS is a method to evaluate and improve the design of PSMS within an organization in conjunction with current compliance practices. RBPS addresses three questions beyond standards-based codes;

"what can go wrong, how bad the result can be, and how often the incident might happen" (CCPS, pg.22).

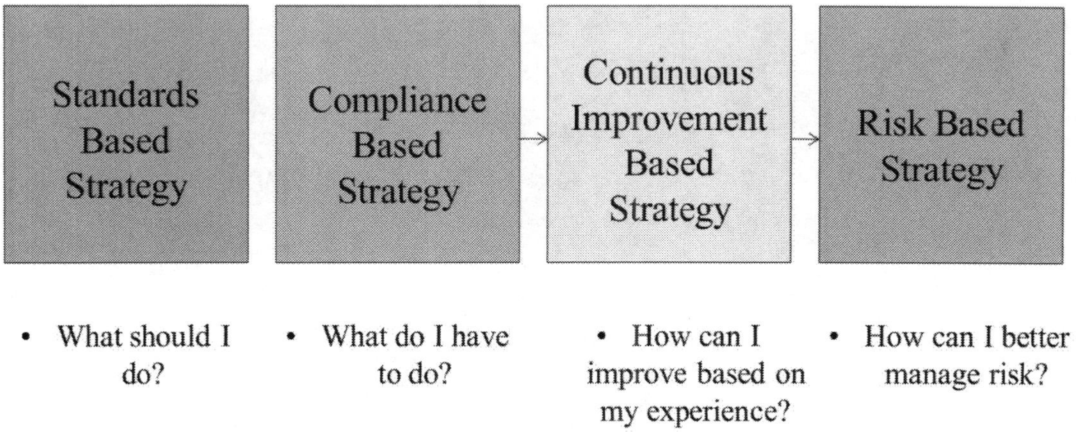

Figure 2: Evolution of RBPS and PSMS's

The vision and recommendation of industry-wide process safety originated with the setting up of Centre for Chemical Process Safety (CCPS) by AIChE as a result of the December 3, 1984 Bhopal incident killing more than 3000 people due to a toxic gas release. The need for continued focus on process safety was reinforced with the 2005 explosion at the BP refinery in Texas City, Texas. Since then, the Center for Chemical Process Safety (CCPS) has identified 20 process safety elements to be used as guidelines for a risk based safety process (CCPS, 2007). In 2009, CCPS selected Phillip Townsend Associates, Inc. (PTAI) to design a benchmarking program for PSMS's.

From a risk based perspective, the 20 risk-based process safety elements can be grouped into four categories: Commitment to Process Safety, Understanding Hazards and Risk, Managing Risk, and Systematic Learning from Experience. PTAI and the CCPS committee of experts designed the PSMS benchmark selecting six of the 20 elements: Process Safety Culture, Compliance with Standards, Hazard Identification and Risk Analysis, Asset Integrity and Reliability, Management of Change, and Measurement and Metrics (CCPS, 2007). The primary goal was to seek an understanding of the relationship between process safety outcomes and management practices.

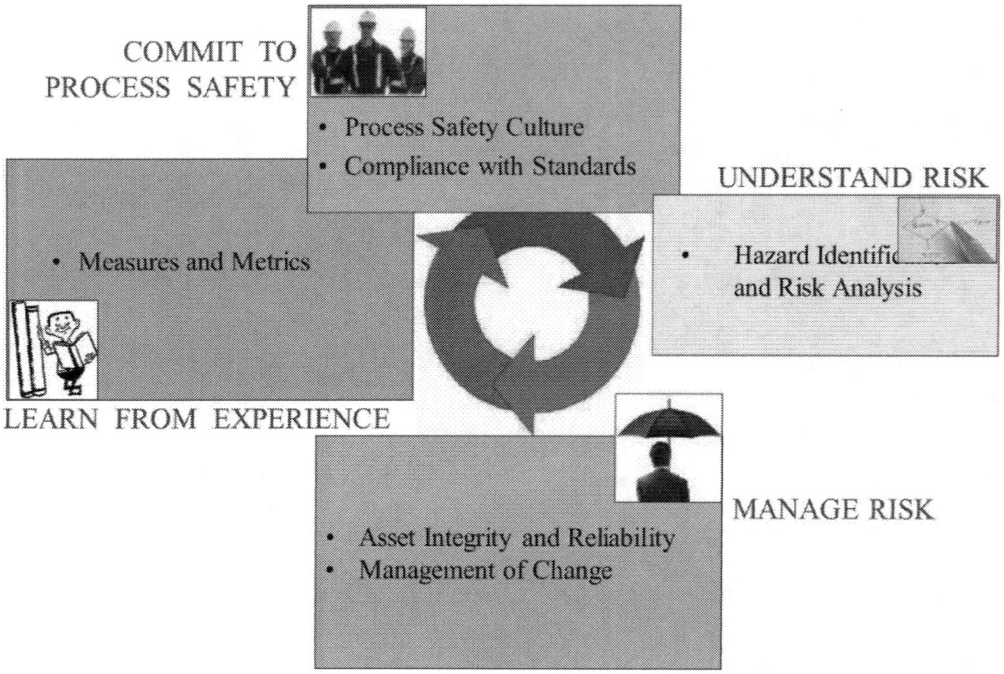

Figure 3: CCPS Risk-Based Elements and Benchmark Scope

This paper discusses the details of the PSMS benchmark, the value proposition to be gained, and the experience of participants in developing an understanding of their current status of existing PSMS's. Practices that will improve and sustain PSM performance from top performers are reviewed along with, the journey to leverage from lessons learned through the enhancement of Process Safety Knowledge and adopting best practices. Continuous improvement and excellence can be achieved by focusing on the most important decisions and behavior changes that leadership can make to close the gap to the top performers, develop central and site specific action plans for improvement, and raise the internal level of all organizational units to that of the top-performing internal site. The desired outcome is to take PSMS's and practices to the top quartile performance through a quantitative and qualitative benchmarking process of industry peers relative to PSMS practices.

2. Methodology

Benchmarking allows identification of how well a company is doing compared to its peers, what the industry top performance level is, how large the gap is to the top performers within the industry, and a stimulus for continuous improvement. PSMS Benchmarking seeks to understand the relationship between process safety outcomes and management process safety practices. The PSMS benchmarking program is a systematic study of process safety between companies that allows sites/companies to compare their management systems; with the goal of improving the design and implementation of management practices that impact process safety.

2.1 Value Proposition of Process Safety Benchmarking

PSMS benchmarking can elevate a company's process safety performance through four distinct steps:

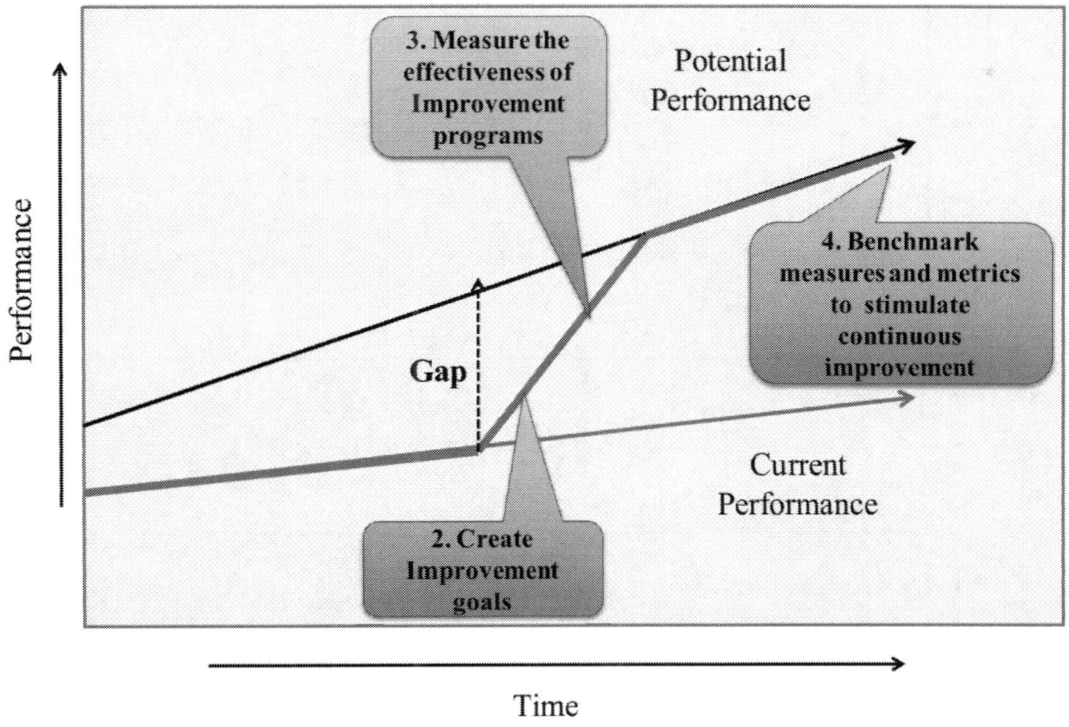

Figure 4: The Value Proposition PSMS Model

2.1.1 Identify performance gaps

The identification of a company's performance in relation to Performance Quartiles will highlight key drivers for performance enhancement versus its peers. These quantifiable results replace perceptions with a data-based reality check. In addition, these results identify if the enterprise policies, practices, and procedures are operating as intended and are achieving the desired results. The quantifiable methodology provides a rating for each question, an aggregated section rating, and an overall performance score aggregating all 6 sections.

Basis for Question Weighting

- $Question_{weight}$ ranges from 0 – 6.0
  - o Weighting based on 3 factors:
    - Significance to Process Safety
    - Breadth of Question
    - Effectiveness to Determine Performance
  - o Information only questions received no weighting

Figure 5: Question Rating Methodology

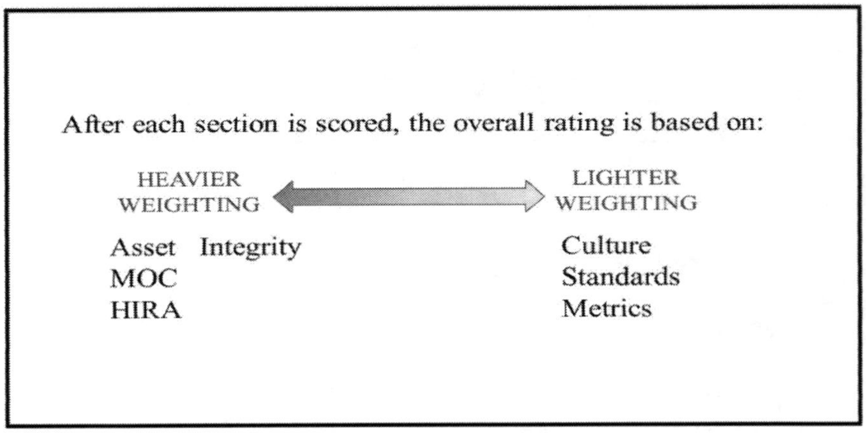

Figure 6: Bases for Overall Rating

2.1.2 Create performance enhancement goals

Data driven information provides a measure of the enterprise effectiveness – emphasizing strengths, weaknesses and potential for improvement over time. This helps focus attention on the critical aspects of an enterprise that create risk. It also underscores what improvements should be made – providing quantifiable performance enhancement goals.

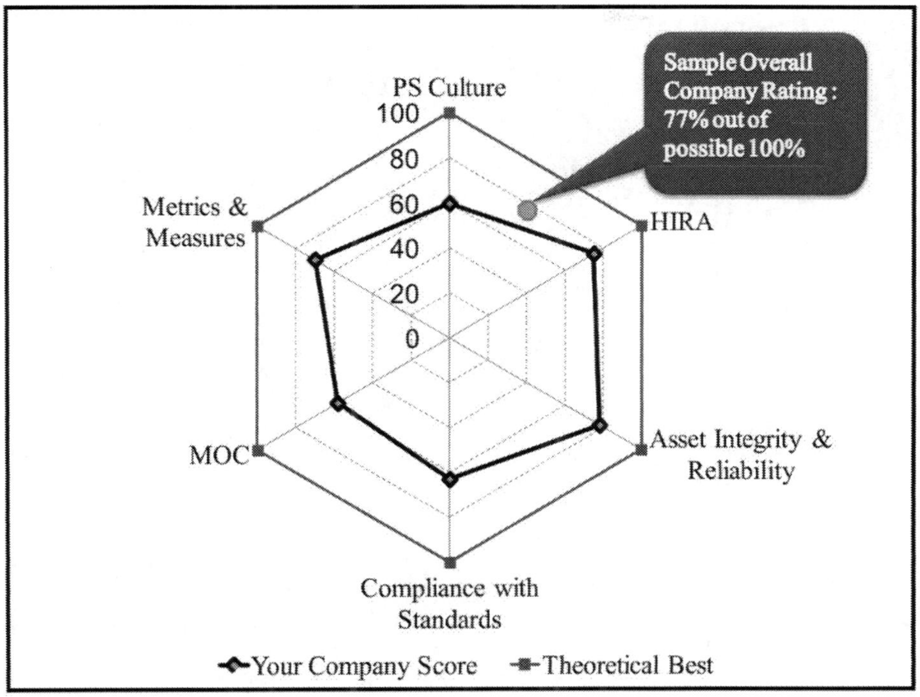

Figure 7: Participant Overall Benchmarking Score

2.1.3 Measure the effectiveness of improvement programs

PSMS benchmarking can be a starting point for goal setting and performance initiatives aimed at boosting step change improvements in process safety. The clear outcome is that this will add significant shareholder value through reduced risks of catastrophic incidents.

2.1.4 Maintain continuous improvement

The PSMS benchmark helps maintain a stimulus for continuous improvement. It provides the necessary motivation to ensure consistent top performance results.

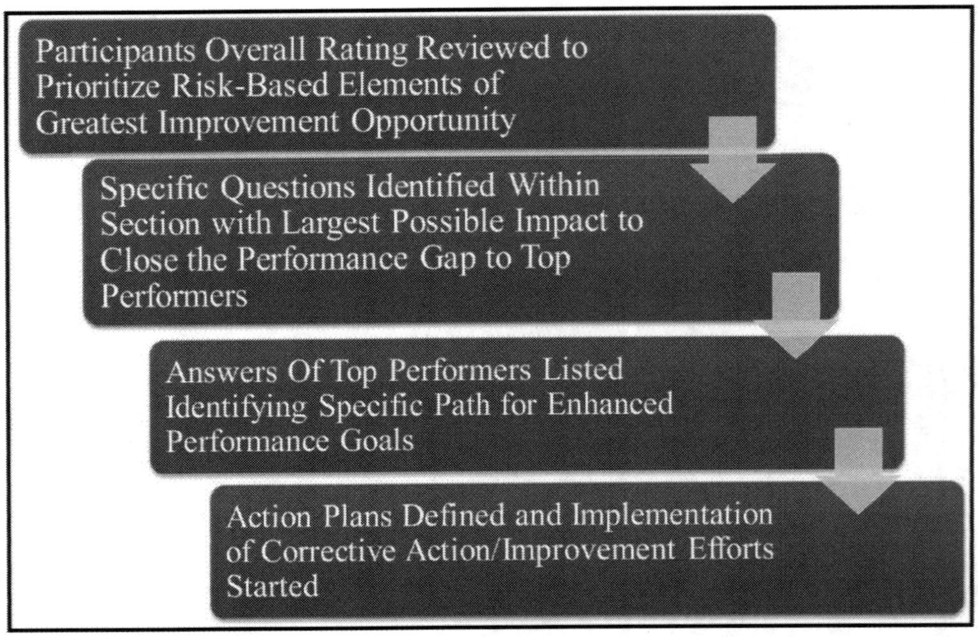

Figure 8: Sustaining Continuous Improvement

2.2 Benchmarking Process

A peer group is established by PTAI of sites/companies agreeing to compare their process safety programs and performance in an effort to identify best practices and drive improvement. Ensuring consistency and protection of each participant's interests are critical to a successful benchmark. A confidentiality agreement is completed between PTAI and participates, ensuring a confidential and unbiased third-party analysis of PSMS performance. The Data Collection Instrument (DCI), providing clear and consistent definitions for all participating sites/companies is made available through the use of a secure site over the internet. This ensures security of data and ease of use by participants as they are able to see all the survey questions and select their answers. A kickoff discussion to align key stakeholders at each company/site is critical.

Following submission of the data, PTAI validates the data on a company/site and industry basis. Validation is enhanced by an iterative process discussing and verifying the results reported via the on-line survey. Participation by key stakeholders and personnel during this process adds to the effectiveness and accuracy of data reported and provided a seamless voice to the sites between participants' PSMS Excellence function and PTAI.

Results are reported by statistical analysis of the data evaluating various drivers of process safety using a Gap Analysis comparing performance to the top quartile of the peer group. The quantitative and qualitative results are distributed to participants with numerical scores for each element along with an overall site/company specific score. Strengths, as well as the main opportunities for improvement to close the gap towards the top performers are identified. A discussion of the study results and gap opportunities is held with the executive management in an executive management summary, as well as, specific individual gap analysis presentations to the top Line and HSE management of each of the sites benchmarked.

Top Quartile Asset Integrity average rating 83% vs. Your Company 63%

Areas representing largest opportunities for 20% gap . . .

What percentage of time are your target inspection/testing intervals met?

Your Company: 75-90% of the time

Top performers: 98-99% of the time

%Δ Impact = 5.5%

How is metal thickness data from periodic inspections analyzed for trends?

Your Company: No formal analysis of trends

Top performers: Thickness measurement trends are analyzed, remaining life projections calculated, and adjustments made to inspection interval

%Δ Impact = 5.0%

Figure 9: Example of Asset Integrity Gap Analysis

2.3Benefits of PSMS Benchmarking

Management review is an integral part of moving from lessons learned to systematic improvement of PSMS. Reviews of the performance gaps can identify potential problems and address them before an incident occurs; identifying opportunities for improvement to move from current levels to the level of top performers. PSMS benchmarking helps an organization prioritize the allocation of process safety resources, both financial and human, to focus on the highest identified risk areas.

Among the benefits of PSMS benchmarking, there are certain ones that have been found to provide a big impact on an organization. There should be a marked increase in staff awareness of process safety issues. This awareness expands the issues from the Health & Safety group to Engineering and to Operations. Based on the organizational level of results discussion, it will potentially increase awareness from site management to impact those that are responsible for setting corporate strategy. A key learning from surveys within PTAI's database is the differences between the site scores and the perception of participants' PSMS Excellence functions and executive management. PSMS benchmarking creates a common understanding and language on process safety matters. Sometimes the opportunities for improvement can be very different from the perceived gaps in an organization. Identifying and gaining insight on perceived gaps versus actual gaps is greatly enhanced by an independent benchmark which effectively creates consistency across multiple sites within a company and with the others in the peer group.

- Internal Benchmark for participants with multiple sites
- Learning through the process of understanding the questionnaire
- Need for a detailed analysis of existing systems implementation
- Development of Plan Do Check Act (PDCA) cycles can result in collaboration between participants and PTAI/CCPS

Figure 10: The Value of Participation

3. Key Applications and Lessons Learned

Traditional process safety audits identify a list of lagging quantitative measures, while typical process safety surveys identify a list of qualitative issues. Providing leading, risk-based quantitative measures and translating qualitative issues into measurable actions is the strength of the PSMS benchmarking application. Identifying and defining actionable steps is a key issue for senior management decision making and allocation of resources. PSMS benchmarking results must be measurable and use a scoring methodology that will not only identify the performance gaps, but also quantify the impact of selected actions; enabling participants to have measurable action plans. Benchmarking is a systematic way to make site-to-site company comparisons and view a company's performance level in relation to that of top performers.

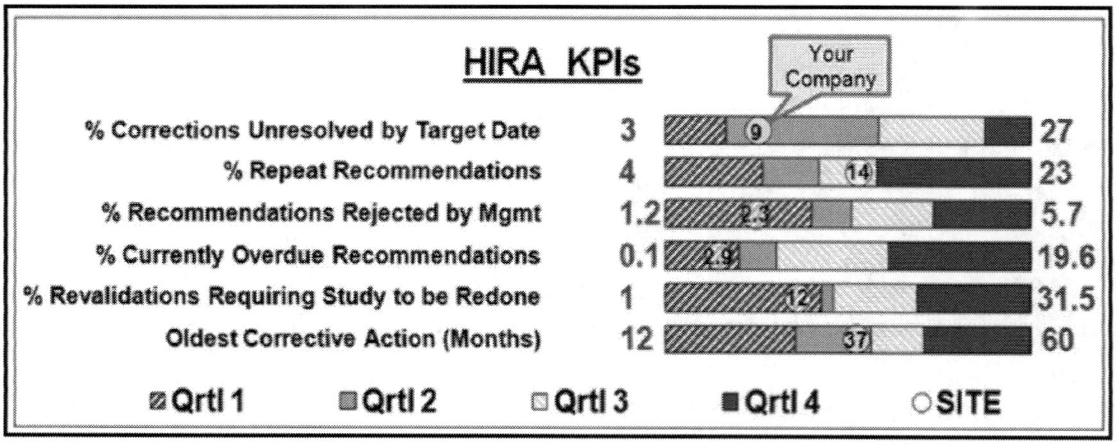

Figure 11: Quartile Placement within Each Section

The application of the benchmarking results to identify performance gaps provides meaningful insight for organizations as they develop goals and the associated improvement plans to reach those goals. Repeated benchmarking monitors the rate of improvement.

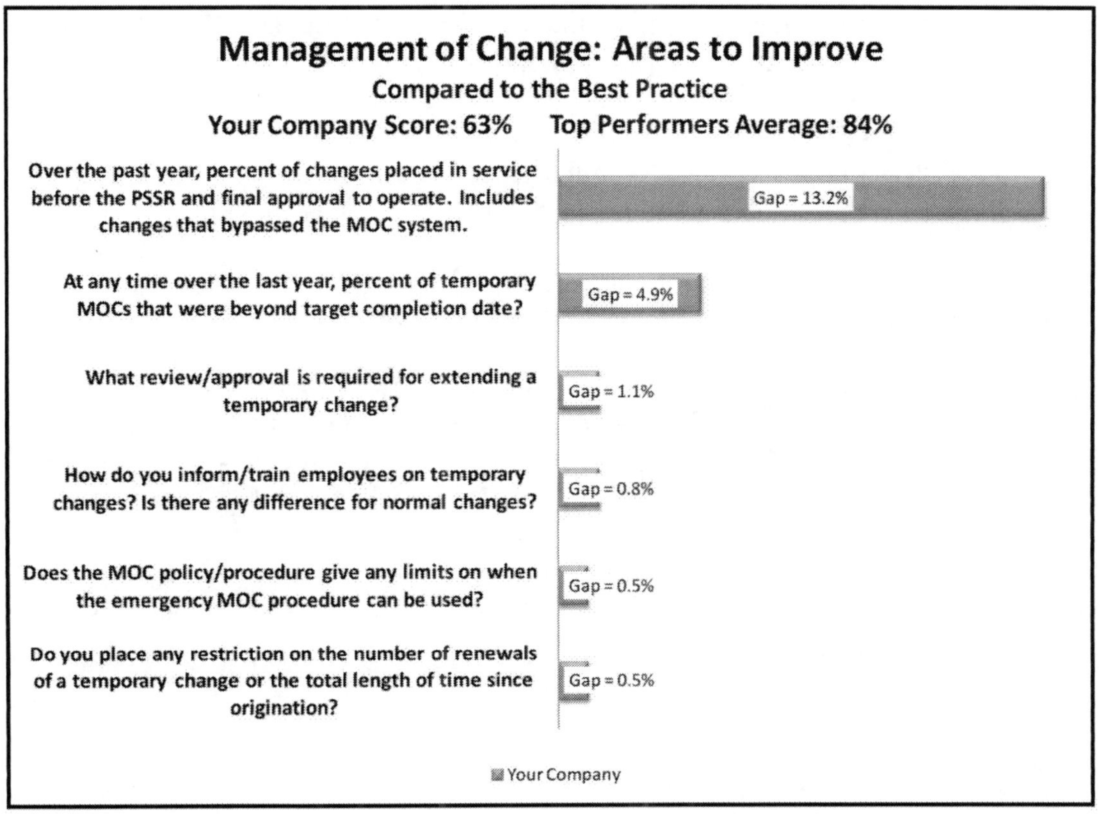

Figure 12: Example of Gap Analysis (sample data based on representative results)

One example of a lesson learned that applies to numerous organizations relates to Process Safety Culture. Findings from the Process Safety Culture section indicate the largest range of results compared to the other sections. The underlying cause of the low scores is a direct outcome of not monitoring the process safety culture through the use of an employee survey. Without a method of assessing the work force's values, attitudes and behaviors pertaining to process safety, an organization will have no meaningful basis for making and measuring improvements. Lack of process safety awareness and understanding as a key strategic driver for the organization signals a weakness in RBPS and effective layers of protection important to prevent any loss of containment scenarios.

4. Asia-Pacific Peer Group Key Learning's

PSMS Benchmark assessment and results specific to the Asia-Pacific peer group indentified three key interrelated observations and benefits: (a) internal benchmark of multiple sites within the same organization, (b) learning through the process of completing the benchmark questionnaire, and (c) the need for a detailed analysis of existing systems implementation.

4.1 Internal Benchmark of multiple sites within the same organization:

PSMS Benchmark surveys conducted for multiple sites within the same organization brought to the surface best practices followed at different sites beyond the existing standard and system requirements. Examples include; number of indicators tracked and analysis of metrics trended, reviews of Process Hazard Analysis reports, risk criteria used in HIRA studies (including probability and consequence of risk), Management of Change hazard review procedures and authorization, review and resolution of leading metric action items, and employee involvement through various PSMS practices. These findings provide learning internally among the sites and externally compared to the global peer group by identifying best practices to include in the corporate and site PSMS Excellence process.

4.2 Learning through the process of completing the questionnaire:

The benchmarking questionnaire itself generates opportunities for RBPS learning and enhanced PSMS knowledge. Understanding the "why" part of questions calls for discussions among various functions and responsible personnel within each of the 6 sections surveyed; reinforcing the understanding of PSMS's. The validation questions and iterative discussion between PTAI and participants enhances the quality of responses and engages the participation of different functions within the organizations PSMS function. Examples of validation questions that generated opportunities for enhanced learning through the process of completing the survey included:

Survey Question#	PSMS Element	Survey Question	Opportunity for Learning
HIRA42	Hazard Identification and Risk Analysis	Percentage of incident investigations in which HIRA was identified as the root cause? (% of total investigations)	If this number is 0%, is it due to inadequate investigation methods, poorly defined HIRA analysis, or correctly reported 0% of investigations identifying HIRA as the root cause?
AIR24	Asset Integrity and Reliability	How is the metal thickness data from the periodic inspections analyzed for trends?	If no formal analysis is conducted, how are remaining life cycles calculated and adjustments made to inspection intervals?
CS10	Compliance with Standards	How is conformance with standards measured/assessed?	If discovery of non-conformance is viewed as an incident and captured in the facility incident investigation system, why is it not included in the process safety audits?

PSM7	Process Safety Metrics and Measurements	Has the rate of leading metrics reported increased, decreased, or remained the same?	Has the rate of leading metrics collected increased due to increased emphasis on leading metrics or something else?

4.3 Need for a detailed analysis of existing PSMS:

The PSMS benchmark results within the Asia-Pacific peer group identified three major areas for further analysis to enhance the systems implementation for improved performance; Compliance with Standards, HIRA, and Management of Change. Overall section scores in all three of these segments are below the 50% level of best practices. Gap analysis indicated that assessment of compliance with standards for existing equipment, percentage of all HIRA team members meeting qualification requirements over the past three years, and the percentage of changes placed in service before PSSR are significant areas for further detailed analysis. One approach to leverage learning's is the implementation of the Plan, Do, Check, Act (PDCA) methodology for continuous improvement and 6-Sigma improvement teams. An example for Management of Change illustrates this process.

4.3.1: Identify best practices:

1. >90% explicitly identify facility and equipment changes, plant layout or equipment location changes, and process technology changes in their MOC program(s).

2. >95% explicitly include MOC procedures for origination of a change request, hazard review, authorization review, classification review, and pre-startup safety review.

3. 52% of provide specific formal hazard evaluation method tools to utilize during hazard reviews in MOC policy/procedures.

4.3.2: Define the PDCA Cycle in broad terms:

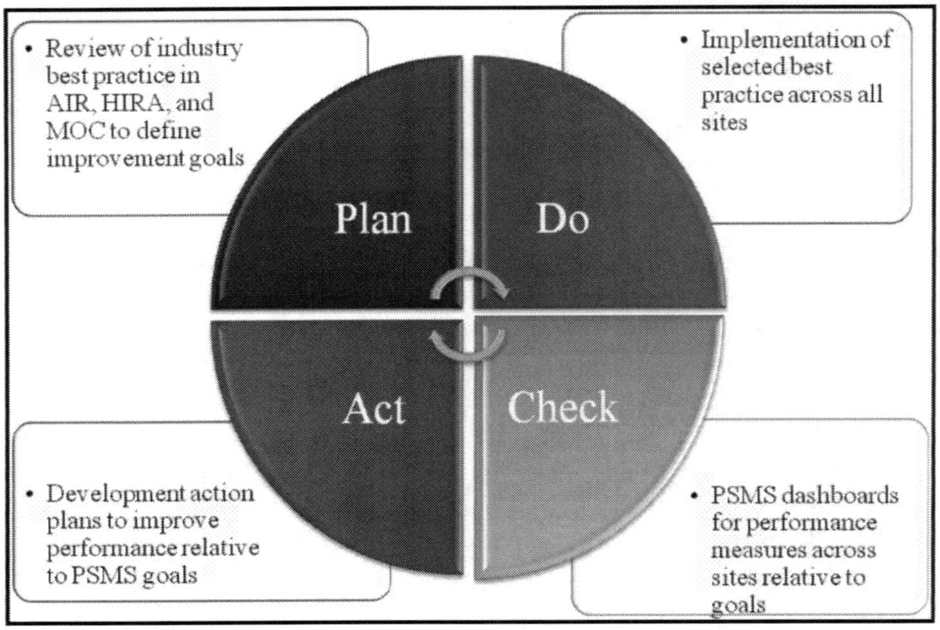

Figure 13: Integrating the PDCA Cycle for Continuous Improvement

4.3.3: Charter a 6-Sigma Improvement Team:

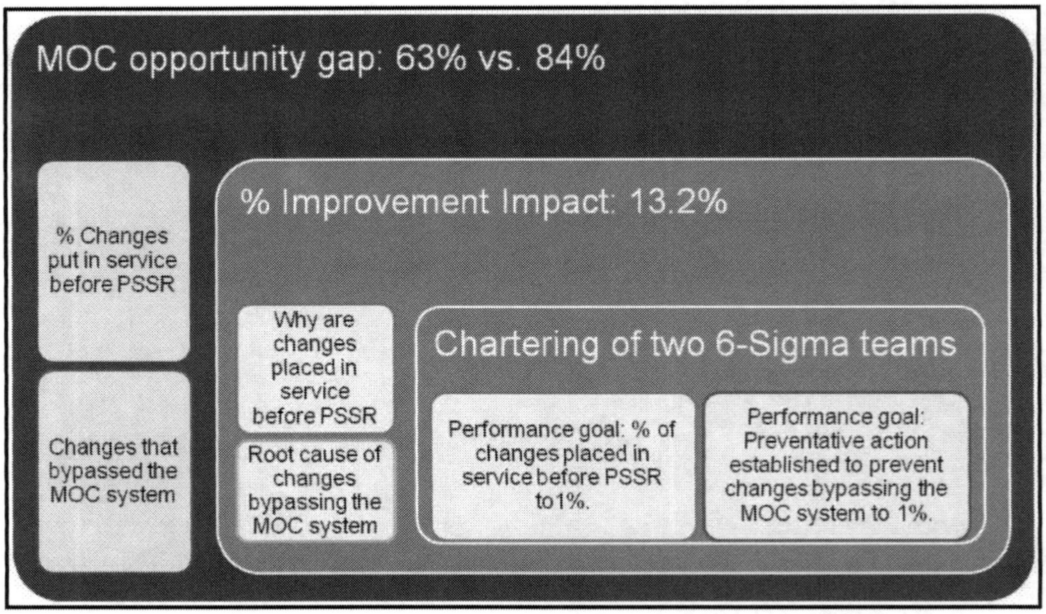

Figure 14: Leveraging Gap Analysis to PSMS Excellence

4. Key success factors:

The PSMS benchmark survey provides new insights to participants on the status of PSMS within their organization and in relation to global participants. The gap analysis converts results into measureable, actionable recommendations the organization can use to establish specific improvement opportunities based on the magnitude and impact of the gap.

The success of the PSMS benchmarking hinges on four key factors; management support of a systematic approach, the strength of the validation process, the ability to report results with a detailed gap analysis that could be translated into specific action plans, and the seamless voice of an organization's PSMS Excellence function and PTAI through strong communication and a systematic effort.

The survey data shows that the journey towards excellence PSMS's requires:

- Moving from a compliance model to a risk-based model with a comprehensive approach to managing process safety.
- Management review as an integral part of moving from lessons learned to systematic improvement of PSMS.
- PSMS benchmarking aids an organization's ability to prioritize the allocation of process safety resources, both financial and human, and focus on the highest identified risk areas.

5. Conclusion

PSMS is a mature, but continuously evolving concept with practices and procedures varying widely among the refining, chemical and chemical processing industry. Moving from a compliance model to a risk-based model requires a comprehensive approach to managing process safety. Substantial improvement opportunities exist for even the current top performers. Specifically, no site has made a perfect score on any of the risk-based elements of the PSMS benchmark.

The PSMS benchmarking program is a unique methodology to allow companies to identify how large the gap is to top performers and to "learn" how to prioritize resource allocations based on the greatest impact opportunity to improve their process safety management system. This shifts the understanding of current practices to the lessons learned that are measurable and actionable. The benchmarking process identifies top quartile management practices while highlighting not only a company's strengths, but also potential development opportunities. The program's scoring system provides a marker to track improvements and measure the gaps within the company sites and against external top performers.

Using a systematic PSMS benchmark drives lessons learned to strategic and organizational learning. Periodic assessment of an organization's PSMS, through repeat benchmarking at regular intervals, is a good value proposition to monitor trends of performance and to achieve pacesetter performance compared to industry peers.

6. References

[1] American Petroleum Institute, API RP 754, "Process Safety Performance Indicators for the Refining and Petrochemical Industries, First Edition," 2010.

[2] Center for Chemical Process Safety CCPS, "Guidelines for Risk Based Process Safety, 2007", John Wiley & Sons, Inc., Hoboken, NJ, ISBN 978-0-470-16569-0.

1ST CCPS ASIA-PACIFIC CONFERENCE ON PROCESS SAFETY℠

SETTING MULTINATIONAL RISK TOLERANCE CRITERIA

Dr. Paul Baybutt
Primatech Inc.
50 Northwoods Blvd., Columbus, Ohio 43235
paulb@primatech.com

Prepared for Presentation at
1st CCPS Asia-Pacific Conference on Process Safety
Qingdao, China
September 4-5, 2013

UNPUBLISHED

SETTING MULTINATIONAL RISK TOLERANCE CRITERIA

Dr. Paul Baybutt
Primatech Inc.
50 Northwoods Blvd., Columbus, Ohio 43235
paulb@primatech.com

Keywords: Risk tolerance criteria, risk analysis, individual risk, group risk, societal risk, process safety.

Abstract

Companies that operate in multiple locations around the world face the challenge of developing consistent risk tolerance criteria across their operations in various locations. However, the views of local stakeholders on the tolerability of the risks can vary significantly from one location to another. Indeed, the actual risks that people are exposed to and currently tolerate can vary significantly depending on local factors. The challenges in developing consistent criteria for such multiple locations are described and addressed in this paper. Regulators in several countries have developed numerical risk tolerance criteria and require that facilities meet them. The procedures used in the development of such criteria are described and form the basis for a method that is described in this paper for setting consistent criteria across national boundaries.

1. Introduction

Increasingly, numerical risk tolerance criteria are being used around the world to judge the tolerability of risks in the process industries [1]. Risk tolerance criteria express the level of risk that is tolerable, given the benefits provided, according to the collective standards of stakeholders in a facility. Their use recognizes that absolute safety cannot be guaranteed. Risk tolerance criteria express the maximum tolerable frequency of a specified level of harm to receptors such as people, property and the environment. They are used to determine if existing risks are tolerable by comparing estimates of the risks with the criteria to determine if there is a risk gap. If a gap exists, recommendations are developed for risk reduction measures to close the gap. For a valid comparison to be made, the type and form of risk estimates and risk tolerance criteria must match.

Risk tolerance criteria have been adopted in several nations, such as the United Kingdom (UK), the Netherlands, and Australia [1]. Companies that operate in different nations face the possibility of either being required to comply with different criteria in different nations or needing to develop criteria for nations where they are not currently required. Developing risk tolerance criteria for one location is challenging in its own right but, when multiple locations are involved, additional issues must be addressed. Also, compliance with risk criteria from multiple jurisdictions can be challenging, particularly when the criteria take different types and forms. This paper describes issues involved in setting and complying with multinational criteria.

2. Form and Type of Risk Tolerance Criteria

Risk tolerance criteria can be set for different types of receptors such as people (facility personnel, members of the public), the environment, property, etc. Where regulatory bodies have established numerical risk tolerance criteria in the process industries, typically, they have been for risks to people. This paper addresses only such criteria.

Both risk to individuals and risk to groups of people are important. *Individual risk* is the frequency at which an individual may experience a given level of harm as a result of exposure to one or more hazards and *group risk* is the relationship between frequency and the number of people in a given population experiencing a specified level of harm from the realization of one or more hazards [1]. Individual and group risk tolerance criteria specify limits to these risks for individuals and groups of individuals.

The term individual risk can be confusing. Sometimes, it is applied to the risk at a location within a facility regardless of whether an individual is actually present there. This can be viewed as the risk to a hypothetical person at the location, sometimes calculated for all times with no protection from the hazards. This particular risk measure is called *geographic* or *location* individual risk or *hypothetical* individual risk. Individual risk can also be determined for actual individuals and this measure is usually called just individual risk, hence the confusion.

Arguably, the preferred measure of individual risk is the actual risk to each individual determined by calculating individual risk at every geographical location where people are present and accounting for time spent at the location. Such a measure allows risk estimates to reflect variations in exposure durations, protections available, individual susceptibility to harm, etc. However, when performed as part of quantitative risk analysis (QRA), many calculations are needed to address the risk to each exposed individual and the resources required quickly become prohibitive and the analysis of the results becomes complex. Consequently, practitioners use the concept of a hypothetical person to provide a reasonable representation of people actually exposed to the risk, for example, a plant operator with a defined occupancy profile for plant areas during a typical working period. A number of hypothetical persons may need to be constructed to represent different categories of people with different exposures, protections, susceptibilities, etc. Note that this is not the same as the hypothetical person used to give meaning to location individual risk. Location risk is a theoretical construction that does not account for the presence of people and is not a measure of the risk actually experienced. However, it does represent the maximum possible risk that an individual could experience at a particular location.

Different types of individual risk measures can be used, for example, the maximum geographic individual risk and the maximum individual risk. The maximum geographic individual risk is the maximum value of the geographic individual risk. The maximum individual risk is the individual risk to the person exposed to the highest risk in an exposed population. It can be determined by calculating the individual risk at every location where people are actually present and selecting the maximum value. It is equivalent to the sum of the geographic individual risks at the locations where people are present for the maximum exposed person.

Individual risk is usually expressed as a single number, risk contour plot, or risk profile / transect [1] according to the type of individual risk measure used. It can be determined for different levels of harm, e.g. first-aid case, severe injury, and fatality. Often, the latter is used.

Historically, group risk has been called *societal risk*, although that is a misnomer as it does not relate to society at large. It was first used as a risk measure for the public, hence its name. It can also be used for employees at a facility. Group (societal) risk is often expressed as f-N or F-N curves [1], although F-N curves are more commonly used. f-N curves display the frequencies of all events that result in N casualties while F-N curves display the cumulative frequencies of all events that lead to N or more casualties (Figure 1). Both curves are functions of N, i.e. f-N = f(N) and F-N = F(N). Usually, both types of curve are plotted on a log-log scale since f, F, and N cover orders of magnitude. Although depicted as continuous lines, the curves really connect individual points that are plotted. When f-N or F-N curves are used for group (societal) risk tolerance criteria they are called limit lines [2]. They can be calculated using risk analysis techniques. Also, they can be constructed from historical data where available.

A shorthand notation is used to describe F-N limit lines using an anchor point and slope. For example, the triplet (10, 1×10^{-4}, -1) represents a limit line for which N = 10 at F = 1×10^{-4} with a slope -1, i.e. a straight line passing through the anchor point (10, 1×10^{-4}). The slope determines how quickly cumulative frequency falls off with increasing consequence. For slopes of -1, cumulative frequency decreases by an order of magnitude as N increases by an order of magnitude.

Group (societal) risk measures are a function of the number of casualties and this raises the issue of aversion to higher consequence casualties, called *risk aversion* or *scale aversion*. Early F-N limit lines used slopes of -1 in the belief that this was risk neutral (the higher the consequence, proportionately, the lower the frequency) and that consequence casualties should drop off at least as fast as the frequency. Risk aversion was introduced using slopes between - 1 and - 2 (the higher the consequence, the greater the reduction in frequency). However, it is now recognized that F-N limit lines with a slope of -1 actually incorporate a degree of risk aversion owing to the relationship between f-N and F-N limit lines [3]. Some risks may be viewed simply as unacceptable if their impacts are too high, for example, if an entire population is impacted.

Group (societal) risks usually cause much strong public reaction than individual risks. The number of casualties that occur at one time in catastrophic events cause them to be viewed as far more significant than the same number of casualties that may occur from a multiplicity of separate smaller events. A catastrophic event causes a devastating shock to the community where it occurs, and possibly to the entire nation, and even the world. Such events generate emotional responses.

Group (societal) risk can be expressed in simpler forms similar to individual risk, for example, the frequency of 10 fatalities occurring from a particular facility. Also, group (societal) risk can be reduced to a single number, or index. For example, the expectation value for casualties can be calculated by summing f(N) x N for all events at a facility. For fatalities, the expectation value is known as the *Probable Loss of Life (PLL)* or *average rate of death (ROD)*, or *average group (societal) risk*. Of course, as simpler group (societal) risk measures are used, information is lost about the distribution of group (societal) risk. For example, PLL values do not differentiate between low likelihood / high consequence events and high likelihood / low consequence events.

Integrals of the areas under F-N curves can also be used as measures of group (societal) risk. For example, the UK Health and Safety Executive (HSE) has developed a Scaled Risk Integral (SRI) as a measure of group (societal) risk [4]. It is used as a screening tool for land use planning in the vicinity of hazardous facilities.

3. Risk Tolerance Criteria and Decision Making

The need for risk tolerance criteria is driven by the necessity of deciding if process facilities pose unacceptable risks to people. Decisions on the tolerability of risk for a facility should involve people who are stakeholders in the facility, including owners, management, workers, other employees, the public, regulators, etc. Moreover, many factors are involved in the decisions, not just the assessed numerical value of risk.

An analysis of a variety of decisions made using quantitative risk analysis by the UK HSE concluded that individual risk was usually well below de manifestus values that had been proposed [5]. Considerations of group (societal) risk played a greater role, perhaps as a consequence. The UK HSE report identified over 40 different factors and expressed the view that the list was likely incomplete. At the time the decisions were made, the UK HSE risk-based decision, or tolerability of risk, framework was still evolving and it was not common practice to use numerical criteria in decision making. Thus, the retrospective UK HSE study may have taken a different perspective had the decisions been made with reference to formal numerical criteria, particularly with regard to the establishment of uniform group (societal) risk tolerance criteria which the report indicated were not deducible. However, the report's conclusion that factors other than those relating to numerical risk values need to be addressed will undoubtedly stand the test of time.

Actual decisions often reflect issues such as the level of risk being comparable to or less than other risks currently accepted, whether new risks are being created or existing risks are being increased, and lack of the means to reduce risk for a facility whose products are deemed necessary for society. Often risk is caused to some people as a result of the activity of others and the benefits of the activity usually extend beyond the people exposed to the risk to people who are not. Also, the nature of the benefits usually vary for those exposed and those not. Decisions involving such factors are inherently socio-political in nature although they must be informed technically and numerical risk tolerance criteria should play a key role.

Given the complexity of the decision-making process, the number and nature of the stakeholders involved, and its socio-political nature, the establishment of risk tolerance criteria should be the responsibility of a nation's government, institutions, legislative branch, and legal and regulatory authorities. However, many nations around the world have not established risk tolerance criteria for process facilities, for example, the United States of America. In such cases, when companies need criteria to comply with industry practices and standards, they are faced with the prospect of developing their own.

This process can be facilitated by a number of principles that have been advanced in the literature for setting risk tolerance criteria, for example, in the Royal Society study group report on risk assessment [6]. They include:

- The risk from a major hazard to an individual employee or member of the public should not be significant when compared with other risks to which a person is exposed in everyday life.

- The risk from any major hazard should, whenever reasonably practicable, be reduced. The comparison of risk against effort to reduce it in the judgement of reasonable practicability should be one of gross disproportionation and not equivalence.

- If the possible harm from an incident is high, the risk that the incident might actually occur should be made very low. This addresses the concept of scale aversion.

- For hazards that are involuntary and with little benefit to the people at risk, more stringent criteria should be applied.

Risk tolerance criteria are often used with the As Low As Reasonably Practicable (ALARP) principle [7, 8] which partitions the risks for processes into three regions (Figure 2):

1. Unacceptable risks that cannot be justified, except for activities entered into voluntarily in full knowledge of the risks, or under extraordinary circumstances.

2. Tolerable risks that are considered acceptable if further risk reduction is impractical, i.e. the benefit does not outweigh the impact.

3. Broadly acceptable, or negligible, risks that are so low as not to be of concern.

As risk decreases, the proportional benefit diminishes. The central element of the ALARP principle is that efforts to reduce risk should be continued until the incremental sacrifice is grossly disproportionate to the value of the incremental risk reduction achieved. Incremental sacrifice is defined in terms of cost, time, effort, or other expenditures of resources. The divisions between risk regions are used to set risk tolerance criteria. The division between the unacceptable and ALARP regions is used to define a minimum requirement (the de manifestus level) that must be achieved. The division between the ALARP region and the broadly acceptable region is used to define a target value (the *de minimis* level) that is the goal but may not be reached.

4. Issues Faced by Companies

Companies face challenges both in calculating risks for comparison with existing regulatory criteria and in developing criteria where they do not exist.

For nations where risk tolerance criteria have been established, the task facing companies is simplified to a certain extent as they must follow defined legal or regulatory requirements. However, the type and form and the values of criteria can vary from one nation to another. For example, the UK uses a hypothetical individual defined as a person who is in some fixed relation to the hazard [9], which is a form of actual individual risk, while the Netherlands uses geographic risk [1]. The UK addresses risk to both workers and the public while the Netherlands addresses risks only to the public [1]. Also, different jurisdictions may use different values for criteria. For example, maximum tolerable individual fatality risk in the UK has been set at 1×10^{-4} per year for the public while in Hong Kong it has been set at 1×10^{-5} per year [1].

When different types and forms of risk tolerance criteria are used, procedures employed for the calculation of risk estimates for comparison with the criteria will vary and companies must have access to the means to evaluate process risk appropriately. Obviously, it would be simpler for companies if the same methods could be used in all cases but that is not always likely to be possible.

In cases where the levels of tolerable risk are different from one nation to another, the promulgation of criteria by regulators provides justification for companies to operate so long as they comply with the criteria. It is when regulatory criteria are absent and companies must develop their own criteria that justification for criteria that vary by location becomes more difficult.

When companies are faced with the need to develop their own criteria, they must decide on the basis for doing so. Various possibilities exist including benchmarking with other companies, using prevailing risk levels from existing company facilities, comparing risks with other industries and everyday activities, and accounting for prevailing societal views on risk, if known. Other factors also need to be considered such as the feasibility of achieving risk levels, the costs involved, and the moral perspective that people should be exposed to the minimum risk possible.

When criteria must be developed for different nations, the additional issue arises as to whether the same criteria should be used for all locations where none currently exist, or whether local variations should be considered. The use of the same criteria for different nations may appear to be both equitable and the simplest approach. However, people in different nations around the world accept different levels of risk. People in an affluent society are likely to tolerate less risk than people who are impoverished economically. In the latter case, industrial risks are more readily accepted for the benefits offered.

While the case can be made that people should not be exposed to significantly different levels of risk in different nations, that does not reflect the current real-world situation. Consequently, if the same criteria are used, companies may experience difficulties in operating facilities in nations used to accepting higher risk levels. Indeed, it may prove very difficult to do so. A facility of the same design could be located in two different nations but pose significantly different risks owing to differences in operations, maintenance, and management because of differences in culture, human factors, and land-use planning. However, if a company sets different risk tolerance criteria for different nations, it is open to criticism that it is taking advantage of people in those nations with higher criteria. The criticism may have significant impacts on the company. Averse publicity may lead to a depressed stock price, boycotts of its products, and other difficulties.

5. Approach for Developing Criteria

The UK HSE's tolerability of risk (TOR) approach was developed over a number of years [9]. The approach has been subjected to extensive public consultation and has benefitted from public inquiries into disasters and the funding of research projects. It has been adopted as the basis for establishing criteria in other countries. It provides a rational and logical basis for establishing numerical risk tolerance criteria and is utilized in this paper.

Risk tolerance criteria cannot be set in a vacuum. There is no level of risk that is demonstrably tolerable except zero risk, which rarely can be achieved, without reference to other risk levels that have been tolerated. The UK HSE has set individual fatality risk tolerance criteria for facility personnel and the public with reference to the risk levels from, respectively, workplace accidents and non-work-related accidents, that actually have been tolerated. Risks to workers were surveyed in UK high-risk industries such as mineral extraction and construction. Such an analysis resulted in the determination of the maximum individual fatality risk that was ordinarily accepted by substantial groups of workers in any UK industry, i.e. the de manifestus risk tolerance criterion. The value chosen was exceeded only by deep-sea fishermen and relatively small sub-groups such as helicopter pilots, divers, and demolition workers [9, 10].
Improvements in social expectations, and the ALARP framework itself, likely will lead to lower levels in the future, although there have been no changes in the UK HSE criteria since they were first formulated.

The UK HSE took the view that the maximum level of fatality risk that should be tolerated for any single member of the public from any single hazardous facility should not be less than 10 times lower the value for workers. This level equated to the average annual risk of dying in a traffic accident in the UK, the implication being that if the public apparently has tolerated this level of risk from traffic accidents they will tolerate the same level of risk from industrial facilities.

The UK HSE set the same value for the de minimis individual fatality risk for both workers and the public. The value was predicated on the very small addition it would make to the ordinary risks of life. The value is not negligible but it was comparable to the risk of being electrocuted at home. Therefore, it was viewed as a level of risk which, provided there is a benefit to be gained, and proper precautions are taken, does not cause worry or the alteration of ordinary behavior in any way. Often, de minimis values are set two or three orders of magnitude lower than de manifestus values.

The UK HSE values for individual annual fatality risk from all hazards present at a single major industrial activity (i.e. a facility) are [9]:

de manifestus		de minimis	
Workers	Public	Workers	Public
1×10^{-3}	1×10^{-4}	1×10^{-6}	1×10^{-6}

People live in environments that expose them to a variety of background risks. In the UK, the risk of death is about one in 100 per year averaged over a lifetime [10]. The de minimis individual risk level of 1 x 10^{-6} is extremely small when compared to this background risk. Of course, the background risk has contributions from many sources and the UK HSE limits are intended to be applied only to the risk contribution from major hazard facilities. However, even individual human activities with risks that are routinely accepted, such as driving a car, exceed the UK HSE de minimis individual fatality risk level.

Group (societal) risk tolerance criteria are more difficult to develop in the same way as individual criteria. There are few F-N curves available using real data, particularly for high N events. Data can be used to establish points for lower N values but the curves must then be extrapolated which involves decisions on scale aversion.

The UK HSE has been considering the development of group (societal) risk tolerance criteria [11]. However, few F-N curves are available showing actual group (societal) risks that are tolerated. However, the UK HSE has suggested an anchor point for an F-N curve for a facility of a cumulative frequency of 1 in 5,000 per year at a value of 50 fatalities for N in a single event as intolerable risk for the public [9]. Independently, the UK HSE has endorsed a slope of -1 for a group (societal) risk tolerance limit line [12]. The UK Hazardous Installations Directorate (HID), part of the UK HSE, has adopted the limit line (50, 2 x 10^{-4}, -1) as a de manifestus risk level and a limit line two orders of magnitude below and parallel to this line, i.e. (50, 2 x 10^{-6}, -1), as a de minimis risk level [13] (Figure 3).

The UK HSE criteria are indicative and are intended to be guidelines to be interpreted with common sense. They are not intended to be rigid benchmarks to be complied with in all circumstances [9]. They are intended to apply broadly to industry and are to be averaged over working lifetimes.

The type and form of risk tolerance criteria developed by the UK HSE provide a good basis for nations and companies to use when setting their own criteria. However, numerical values must be determined using nation-specific data on workplace and non-workplace casualties. Of course data are needed on work and non-work-related accidents and such data may be unavailable, sparse, or unreliable. In such cases, conservative expert judgment can be used based on anecdotal evidence until more reliable data can be obtained, or the UK HSE values can be adopted on the basis of having received widespread acceptance in a developed country.

When national criteria do not exist, it is reasonable for companies to develop their own criteria for each location using the approach described above. However, a compromise approach that provides some uniformity to the criteria, but that allows for local differences, is to use the ALARP framework with the same set of de manifestus criteria for each location but with de minimis criteria that are reflective of local conditions. The de manifestus values are set conservatively. In time, de minimis criteria can be reduced further. In the meantime, the ALARP principle will move actual risks lower, but in a way that is feasible for each location.

The UK HSE has found that individual risk tolerance criteria "rarely bite" and that group (societal) criteria are more important in controlling risks [9]. This suggests that the value set for the de manifestus level of individual risk is not a major determinant for risk-based decisions and that attention should focus on setting suitable group (societal) criteria.

6. Other Issues in Settling and Using Criteria

Different people may have different vulnerabilities to exposure to hazards. Risk tolerance criteria should specify to whom they apply, such as an average individual or the most vulnerable individual. Definitions are also needed for such individuals.

Accident consequences entail not only fatalities. Often fatalities are accompanied by sub-lethal effects in the form of injuries, as well as harm to other receptors such as property and the environment. Also, in some cases, fatalities may be delayed rather than instantaneous. Typically, risk tolerance criteria used by regulators for the process industries do not address these issues.

Group (societal) risk scales with the population at risk [1]. Therefore, group (societal) criteria must be adjusted to reflect the size of each facility. Furthermore, besides group (societal) risk criteria for individual facilities, company-wide and even national criteria may also be needed. Group (societal) risk tolerance criteria should be set for both workers and the public but there are few precedents for doing so.

Ideally, group (societal) risk should be defined in terms of the total detriment that occurs rather than just fatalities or other casualties. Detriment affects large parts of the population that may not suffer mortality or physical injury but may be impacted in other ways. For example, it may be in the form of emotional pain and suffering or economic hardship.

New facilities may be judged using stricter criteria because technologies are expected to improve and more can be done to address risk reduction during the design of a new process than for an existing facility. The issue can be addressed by specifically setting lower criteria or using the ALARP principle to require further movement towards de minimis risk for new facilities. Of course, at some point the need to lower de minimis risk values may be reached.

The application of risk tolerance criteria must recognize the substantial uncertainties that are invariably present in risk estimates for hazardous facilities [7].

Usually, calculated risk it is averaged over a time period such as a year. Therefore, an individual who spends a short time during the year in a very high hazard area may face the same numerical risk as someone who works full time in a less hazardous area. Even though the average risk is the same, individuals may not want to be exposed to high risk levels even for a short duration. This issue has not yet been addressed in risk tolerance criteria although it has been addressed in the context of calculating risk [14].

Various risk analysis methods exist for the quantification of risk. The type and form of risk estimates must match those of risk tolerance criteria with which they are compared, and risk analysis methods must be chosen carefully. For analyses to be comparable for different locations, the risk analysis methods used must be comparable if similar risk tolerance criteria are used. Risk analysis requires a variety of data to be available and this can be a limiting factor from one location to another. Also, the calculations must be performed by competent personnel. The availability and qualifications of such personnel may vary by location.

7. Conclusions

Increasingly, risk tolerance criteria are being used to make judgments on the tolerability of risks for process facilities. Fundamentally, such criteria are numerical. Even qualitative means of judging risk tolerabilty, such as risk matrices, implicitly use embedded numerical criteria and must be calibrated using them.

Of course, companies must meet any legally-required or regulatory-mandated risk tolerance criteria in the jurisdictions within which they operate for facilities located in those jurisdictions. In cases where criteria have not been established for a nation, companies must develop their own if needed to comply with industry practices and standards. Two approaches are possible. A single set of criteria could be used around the world or criteria could be tailored to each location. The challenge of the first approach is that criteria that may be used by a multi-national company in one location may be too restrictive to be workable in another location. The challenge of the second approach is that companies may be criticized for exposing people to higher risks in one location than another. A possible compromise is to use the same de manifestus values but tailor de minimis values to each nation. The values can be set using the approach employed by the UK HSE in its tolerability of risk approach.

Generally, tolerance for risk is decreasing around the world and risk tolerance criteria must be reviewed and updated as needed. Undoubtedly, much higher risks were accepted in past times, for example in the UK during the industrial revolution. Views on the tolerability of risk evolve as living standards improve. Also, as population densities increase, public group (societal) risks increase. Land-use planning is very important in such cases [15].

8. References

[1] "Guidelines for Developing Quantitative Safety Risk Criteria", Center for Chemical Process Safety (CCPS) / American Institute of Chemical Engineers (AIChE), 2009.

[2] Cox, D.C., and Baybutt, P., "Limit Lines for Risk", Nuclear Technology, Vol. 57, pps. 320 - 330, June 1982.

[3] Ball, D. J., and Floyd, P. J., "Societal Risks", United Kingdom Health and Safety Executive report, 1998.

[4] Hirst, I. L., and Carter, D. A., A "worst-case methodology for obtaining a rough but rapid indication of the societal risk from a major hazard installation", J. of Hazardous Materials, A92, pps. 223 – 237, 2002.

[5] Quantified Risk Assessment: Its Input to Decision Making, HSE Books, 1989.

[6] Risk Assessment Report of a Royal Society Study Group, The Royal Society, London, 1983.

[7] ALARP Suite of Guidance, United Kingdom Health and Safety Executive, http://www.hse.gov.uk/risk/theory/alarp.htm.

[8] Baybutt, P., The ALARP principle in process safety, Process Safety Progress, accepted for publication.

[9] Reducing Risks, Protecting People, United Kingdom Health and Safety Executive Books, 2001.

[10] The Tolerability of Risk from Nuclear Power Stations. UK Health and Safety Executive, London, HMSO, 1988, Revised 1992.

[11] CD212, Consultative document on proposals for revised policies to address societal risk around onshore non-nuclear major hazard installations, United Kingdom Health and Safety Executive, 2007.

[12] Societal Risk and the Consideration of Technical and Policy Issues, Status Summary of 23 Issues, United Kingdom Health and Safety Executive, 2010, http://www.hse.gov.uk/societalrisk/technical-policy-issues.pdf.

[13] HID's Approach to ALARP Decisions, SPC/Permissioning/39, Annex, United Kingdom Health and Safety Executive, 2013, http://www.hse.gov.uk/foi/internalops/hid_circs/permissioning/spc_perm_39.htm.

[14] Lewis, D. K., Henselwood, F., Safety system impairment and the need to manage peak risk, Process Safety Progress, Vol. 31, Issue 4, pps. 373–376, December 2012.

[15] Risk Criteria for Land-Use Planning in the Vicinity of Major Industrial Hazards, HMSO, London, 1989.

Figure 1: F-N Limit Line

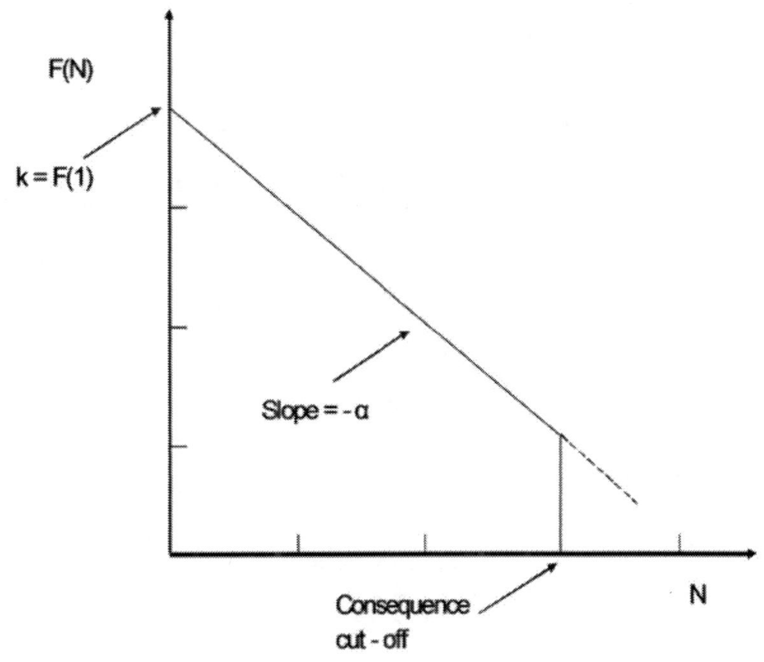

Figure 2: The ALARP Principle

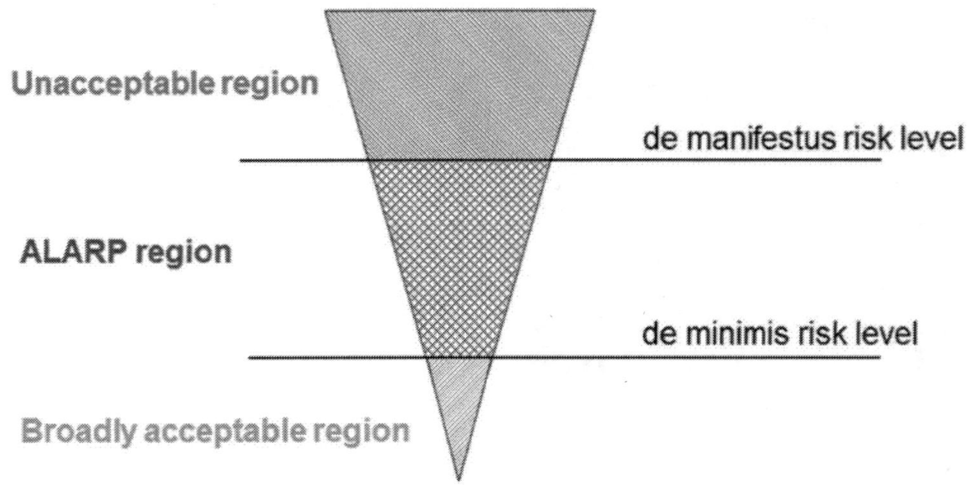

Figure 3: The ALARP Principle Used for F-N Limit Lines

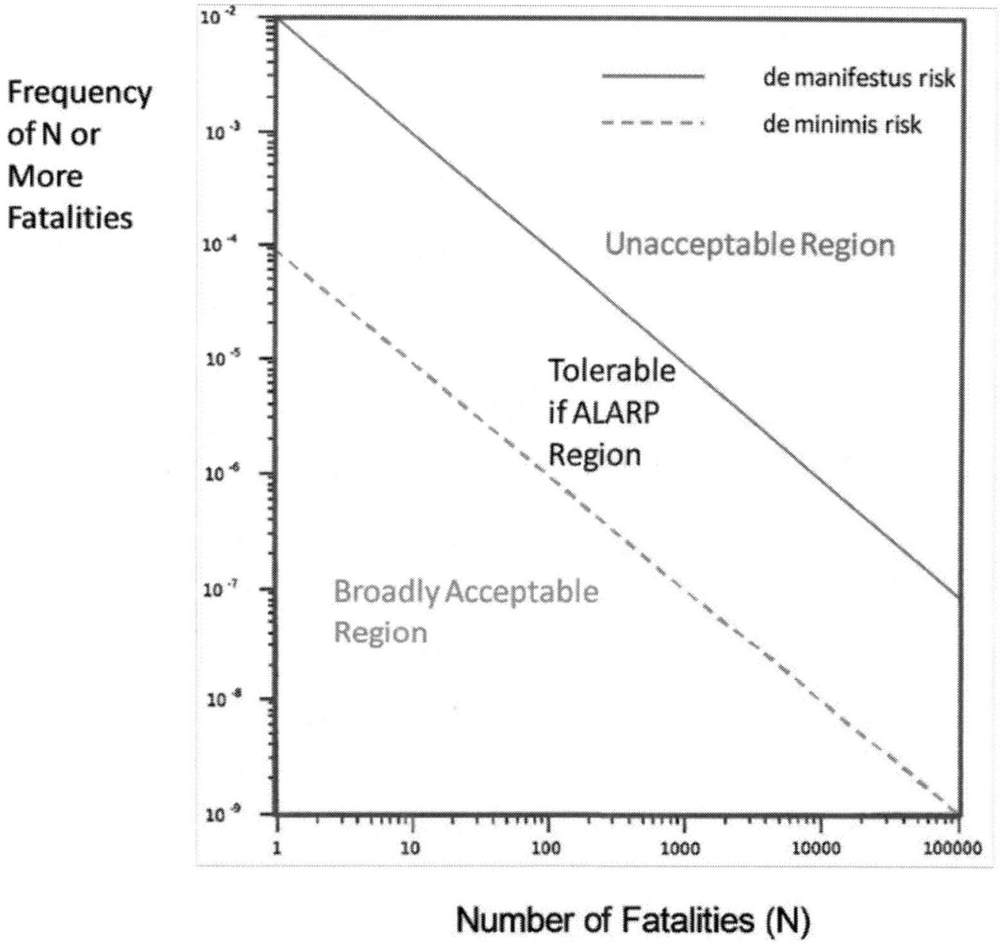

Experimental Study of Flame Acceleration in Closed Pipe

R.M.Kasmani
Faculty of Petroleum and Renewable Energy Engineering
Universiti Teknologi Malaysia, 81310 UTM, Johor, Malaysia
rafiziana@petroleum.utm.my

M.H.Mat Kiah
Faculty of Petroleum and Renewable Energy Engineering
Universiti Teknologi Malaysia, 81310 UTM, Johor, Malaysia
misshasimawaty@yahoo.com

Sina Sina Davazdah Emami
Faculty of Petroleum and Renewable Energy Engineering
Universiti Teknologi Malaysia, 81310 UTM, Johor, Malaysia
sinadavazdahemami@gmail.com

M.H.Hassim
Faculty of Chemical Engineering
Universiti Teknologi Malaysia, 81310 UTM, Johor, Malaysia
mimi@cheme.utm.my

Prepared for Presentation at
American Institute of Chemical Engineers
1st CCPS Asia-Pacific Conference on Process Safety
Qingdao, China
September 4 – 5, 2013

UNPUBLISHED

Experimental Study of Flame Acceleration in Closed Pipe

R.M.Kasmani
Faculty of Petroleum and Renewable Energy Engineering
Universiti Teknologi Malaysia, 81310 UTM, Johor, Malaysia
rafiziana@petroleum.utm.my

M.H.Mat Kiah
Faculty of Petroleum and Renewable Energy Engineering
Universiti Teknologi Malaysia, 81310 UTM, Johor, Malaysia

Sina Sina Davazdah Emami
Faculty of Petroleum and Renewable Energy Engineering
Universiti Teknologi Malaysia, 81310 UTM, Johor, Malaysia

M.H.Hassim
Faculty of Chemical Engineering
Universiti Teknologi Malaysia, 81310 UTM, Johor, Malaysia

Keywords: gas explosion, bending, equivalence ratio, flame acceleration, hydrogen/methane

Abstract

An experimental study has been done to investigate the flame acceleration in a closed pipe. A 90-degree bend pipe with L/D of 40 was used in this study. Ignition was affected at one end of the vessel while the other end was closed. Three different fuels–air mixtures (methane, methane/hydrogen and hydrogen) were used in this work. Only lean concentration of hydrogen/air, methane/air and hydrogen/methane/air mixtures were considered in this study i.e. 6% v/v. All mixtures were operated at ambient condition. The results show that bending is the critical part of pipeline and higher concentration of hydrogen can affect on maximum overpressure, flame speed and temperature rise of both pure hydrogen/air and methane-hydrogen/air mixtures. It was demonstrated that bending is the critical part of pipeline and even at lean concentration, hydrogen can affect significantly on maximum overpressure, flame speed and rate of pressure rise of both hydrogen/air and methane-hydrogen/air mixtures. This is due to bending acts similar to an obstacle, in which this mechanism can induce more turbulence, initiating combustion in an unburned pocket at the corner region and causing a high mass burning rate, which increases the flame speed. Further, at very lean concentration of hydrogen 0.5), methane/hydrogen/air and hydrogen/air mixtures gave about 9 to 11 times higher in explosion pressure compared to methane/air mixtures. This anticipates that the high reactivity and energy content of hydrogen gave significant effect in increasing the maximum explosion pressure, and this phenomenon is highlighted.

1. Introduction

The profiles of flame propagation and maximum pressure performed in pipe with of 90 degree bend were broadly studied by [1-3], using hydrogen-, methane-, propane- and ethylene –air mixture with variable concentration (lean and stochiometric). Phylaktou [1] observed that flame speed and overpressure increase in a factor of 5 of methane/air mixtures in a 90 degree bend tube compared to similar experiment carried out in straight pipes and this condition is similar to the effect using 20% blockage ratio baffle at the same position. This would be due to the ability of that 90 degree bend in pipe to provide an efficient way to create a turbulence yet enhance the flame speed, overpressure and shorten the DDT run-up distance for ethylene- and hydrogen –air mixture with stochiometric concentration[3]. To support the experimental observation on effect of 90 degree pipe on gas explosion, Chatrathi [2] observed that the highest flame speed and pressure attained at the bend using propane-and ethylene- air mixture with lean concentration due to fast burning rate. It is clearly shows that the presence of bend in pipe configuration plays an important role on creating turbulence as well as increase mass burning rate associated with combustion of normal fuel mixture.

Nevertheless, the effect of the obstruction on the explosion severity related to hythane fuel or so call hydrogen enrichment in methane is still unclear. Yet, hydrogen is commonly used to blend with methane as alternative fuel for power generation application for extending the flammability limit [4]. To be noted that hydrogen behaves intrinsically different with other hydrocarbon, i.e. hydrogen possess higher specific heat energy, higher diffusivity rate, wide range of flammability limit and such [5-6]. Thus blended hydrogen with hydrocarbon gas especially methane tends to create issues associated with reaction location and the flame stability [4, 7].

To date, there is sparse experimental studies have been carried out in micro scales [8-9], particularly in 90 degree bend [1,3, 10-11]. Current literature on the hydrogen-methane/air mixtures explosions in pipe line has been studied by Porowski and Teodorczyk [12] , however, the work focus on straight pipeline with blockage ratio 0.4 to 0.7. Hence, this study was carried out focusing on the flame propagation of hydrogen/air and hydrogen-methane/air mixtures explosion in 90-degree pipelines in order to investigate the physical and dynamic characteristics (e.g. maximum pressure, rate of pressure rise, temperature changes and flame speed) of this gas/air mixtures explosion.

2. Experimental and method

A series of tests were conducted to observe the flame behavior and pressure profile for different methane-air-hydrogen mixture concentrations. The test geometry consist of a horizontal steel pipe (length,=2.0 m, diameter=0.1m, volume=0.41m^3) with 90 degree bends (radius = 0.1 m) and added a further 1 m to the length of the pipe based on the centerline length of the segment. The pipe was made up of a number of segments ranging from 0.5 to 1 m in length, bolted together with a gasket seal in-between the connections and blind flanges at both ends. Refer to Figure 1 for the overall schematic of the experimental rig.

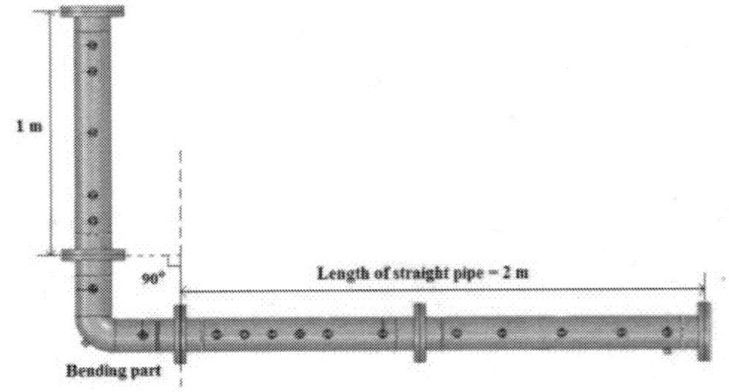

Figure 1. Schematic configuration of testing pipe

All tested methane-hydrogen-air mixtures were prepared using partial pressure method with different hydrogen and methane concentration initially at ambient condition. Only low concentrations of hydrogen, methane and hydrogen-methane mixtures were utilized in this study i.e. 6% v/v. The mixture was ignited at the center of one end of the pipe by means of a spark discharge. A 16 J ignition energy was used in all tests to ensure ignition is in near limit mixtures. The history of flame travel along the pipe was recorded by an axial array of mineral insulated, exposed junction, type K thermocouples. The time of flame arrival is detected as a distinct change in the gradient of the analogue output of the thermocouple and in this way, the average flame speed between any two thermocouples could be calculated. Flame speed was determined from the experiment by using flame arrival time on the mounted thermocouple with known distance from the spark plug. A 16-channel transient data recorder from National Instrument was used to record and process all the data. Each explosion was repeated at least three times for accuracy and reproducibility.

3. Result and discussion

3.1 Pressure-time histories of fuel/air mixtures during explosion development

This section discusses the severity of the hydrogen fuel could pose during the gas explosion. Taking methane-air mixtures as the basis, the overpressures of pure hydrogen and hydrogen enhancement in methane could give by one order of magnitude. Figure 2 shows the development of explosion pressure versus time for fuel mixtures of methane (CH_4), methane/hydrogen (CH_4 + H_2) and hydrogen (H_2) at 6%v/v concentration. At this lean concentration, the hydrogen/air mixture gave the highest explosion pressure (5.5 bars) compared to the methane/hydrogen/air mixture (4.5 bars) and the methane/air mixture (0.5 bars). Mixtures with hydrogen have high reactivity and energy content if compared to methane, which generates higher explosion pressures. The more reactive is the fuel, the higher is the pressure attained during the explosion. Higher hydrocarbon content indicates more reactive gases, contributing to a higher overpressure [13]. For the methane/air mixture, the highest pressure was obtained at t = 4.5 s because flame had a longer time to reach the end of the pipe due to its lower reactivity in comparison with the other two mixtures. For the methane/hydrogen mixture, the development of explosion pressure –time profile is sharper than for the other fuel mixtures. At t = 1.0 s, 2.5 s and 4.0 s, sharp peaks are clear compared with the other two fuel mixtures that show more constant pressure development profiles. On the basis of maximum overpressure of pure H_2/air mixtures, it can be said that the enhancement of H_2 concentration into methane-hydrogen/air leads to the dramatic increasing of maximum overpressure.

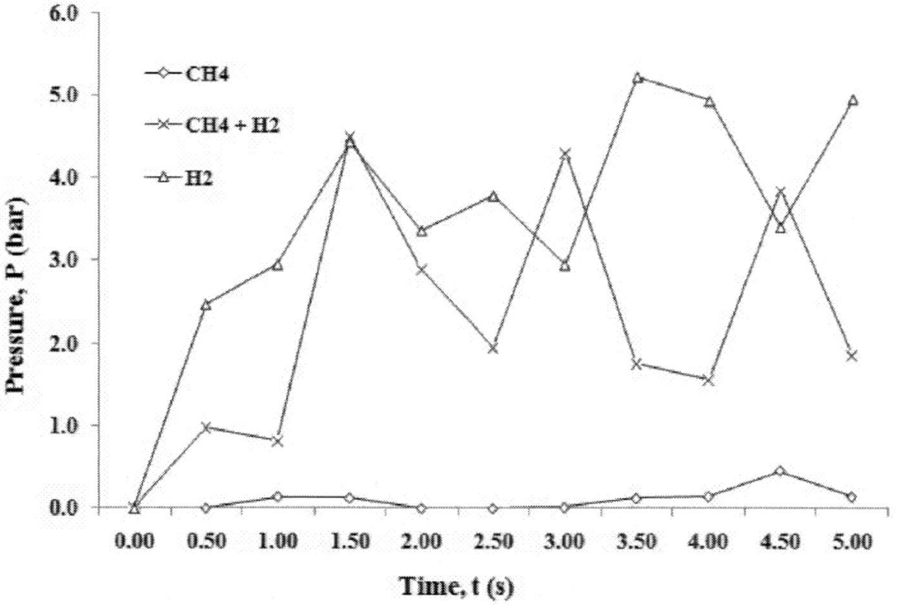

Figure 2 Pressure profile vs. time for methane, methane/hydrogen and hydrogen at lean concentration of 6% v/v.

3.2 Overpressure as a function of distance from ignition position at lean concentration (Φ = 0.5)

The explosion pressure as a function of distance from ignition position, x for fuel mixtures of methane, methane/hydrogen and hydrogen at $\Phi = 0.5$ (lean concentration) is shown in Figure 3. Figure 3 is clearly shown that similar pressure profiles were obtained for methane/hydrogen and hydrogen mixtures. For the methane mixture, the highest pressure was reached at a shorter distance from ignition (x = 1.26 m) compared to methane/hydrogen and hydrogen mixtures. At x = 2.79 m (the position of the bend), the maximum explosion pressures obtained for methane/hydrogen mixture (4.5 bars) and hydrogen mixture (5.5 bars) were due to the presence of the turbulence effect at that regime. These maximum pressures are higher by a factor of 9 to 11 than the value for the methane mixture (0.5 bar) at the same concentration. After the bending, the pressure decreased sharply before increasing at the end of the pipe due to the flame propagation being obstructed by the closed end as the closed end pipe acts the same as obstacles.

Dorofeev [14] obtained a high pressure (~30 bars) from a hydrogen/air explosion in a confined volume. He summarized that the reflections of pressure waves would induce auto-ignition in the mixture and cause localized explosions. Local explosions normally happen when there is an intense turbulent mixing of product and reactants.

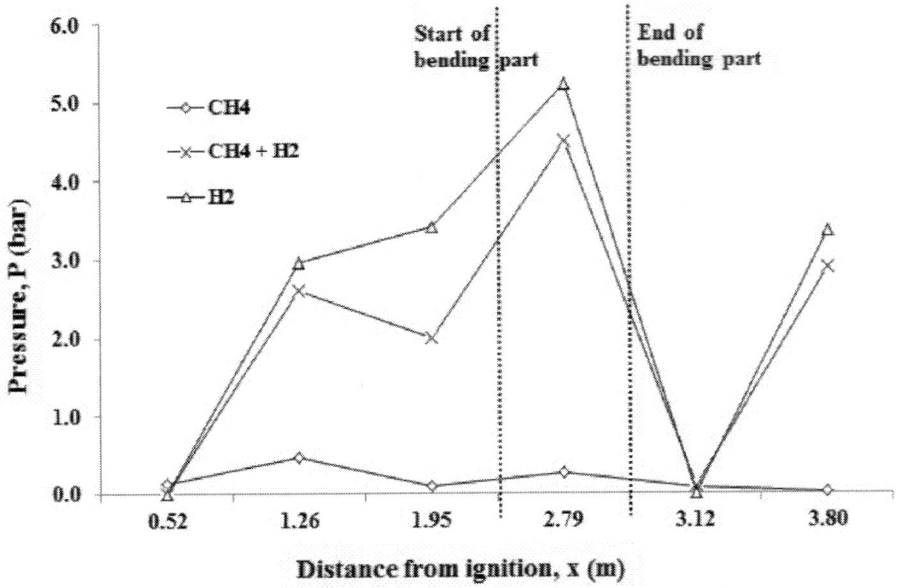

Figure 3 Explosion pressure as a function of distance from ignition for methane, methane/hydrogen and hydrogen/ air mixtures at lean concentration of 6% v/v

The reactivity of explosion on methane-hydrogen/air mixture has been studied in different vessels [15 - 20]. It was observed that by increasing the percentage of hydrogen in mixtures, the overpressure values increased slightly. It can be said that the additional of hydrogen content in methane-air mixtures could increase the burning velocity and thus, giving higher overpressure, especially at the bend area. Even though the result obtained in this study has a good agreement with previous works, yet, the value of overpressure increased dramatically in 90-degree bend.

3.3 Rate of pressure rise profile for methane, methane/hydrogen and hydrogen/ air mixures at lean concentration ($\Phi = 0.5$)

Figure 4 illustrated the rate of pressure rise for the studied fuel mixtures at lean concentration ($\Phi = 0.5$). For the recorded explosion time below 5.0 s, the highest rate of pressure rise was obtained for the methane/hydrogen mixture (7.0 bar s^{-1} at $t = 1.5$ s), compared to that for the hydrogen mixture (5.0 bar s^{-1} at $t = 0.5$ s) and the methane mixture (0.5 bar s^{-1} at $t = 4.5$ s). The sharp development of pressure profile versus time (as shown in Figure 2) for the methane/hydrogen mixture produced a higher maximum rate of pressure rise, dP/dt, due to the shorter flame travel time. It is expected that a higher rate of pressure rise will be obtained for stoichiometric and rich concentration mixtures of all three fuel/air mixtures tested.

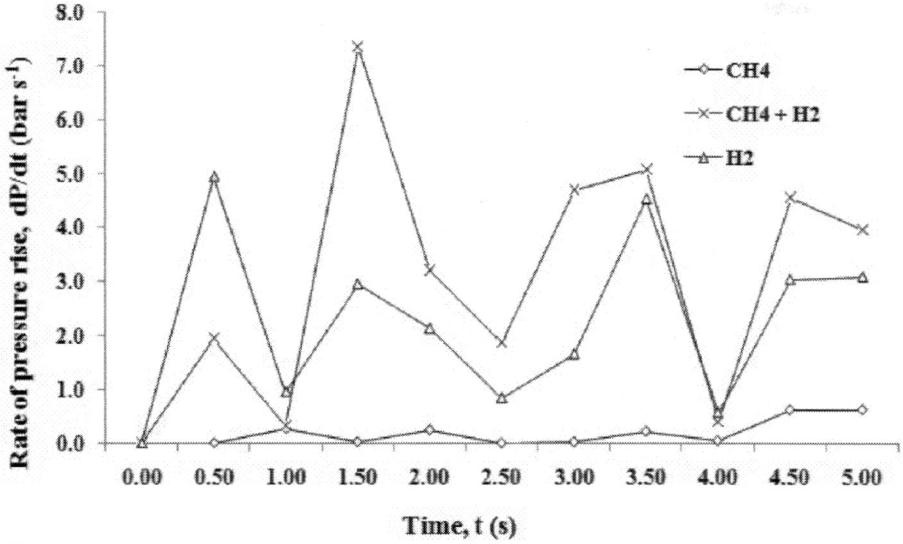

Figure 4 Rate of pressure rise vs. time for methane, methane/hydrogen and hydrogen at lean concentration of 6% v/v

3.4 Rate of pressure rise as a function of distance from ignition position at lean concentration (Φ = 0.5)

Figure 5 illustrates the rate of pressure rise as a function of distance from ignition position, x for a lean concentration ($\Phi = 0.5$) of methane, methane/hydrogen and hydrogen mixtures. Two sharp pressure peaks were observed at $x = 1.26$ m and $x = 3.12$ m for the methane/hydrogen and hydrogen mixtures, giving a maximum rate of pressure rise of 10.0 bar s^{-1}. The peaks show that the flame accelerates at its initial propagation and then decelerates before the bend. At the bend, flame speed is greatest because the obstruction caused by the bend triggers the turbulence effect, thereby enhancing the explosion pressure at that regime. The presence of hydrogen strongly promotes the increase in the maximum rate of pressure rise as reported by Cammarota et al. [7].

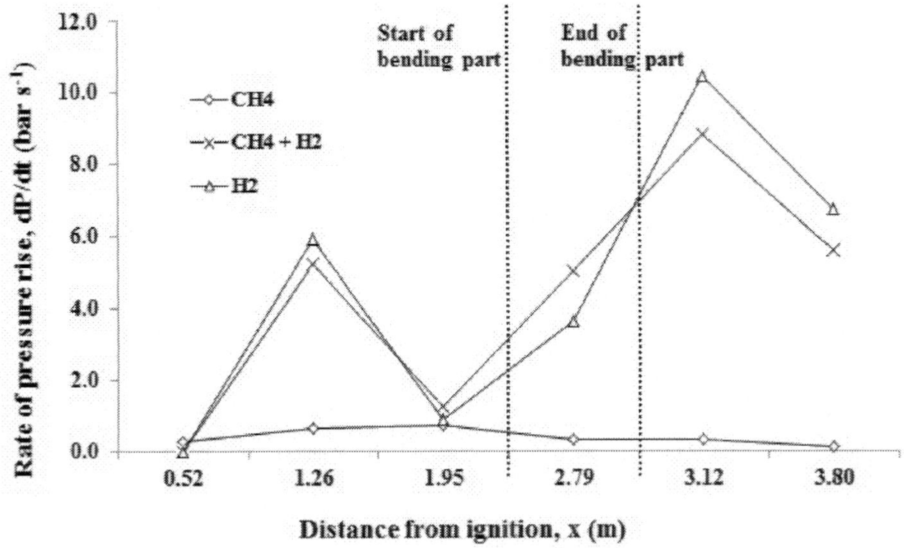

Figure 5 Rate of pressure rise vs. distance from ignition, x for methane,

methane/hydrogen and hydrogen at lean concentration of 6% v/v.

3.5 Flame speed as a function of distance form ignition position

Flame speed profile against the distance from ignition was shown in Fig. 5. A significant rate changes can be seen in flame acceleration along the centerline of the pipe due to different fuel reactivity [21]. The highest flame speed of 390 m/s was observed on H_2+ CH_4 mixtures, almost 26 times higher, as compared to the flame speed of a very lean concentration of CH_4 i.e. 15 m/s. By taking maximum flame speed for pure hydrogen as a basis, it can be seen that it is about 1.33 times increment of flame speed in additional of hydrogen into the mixtures. This confirms the previous results of Phylaktou et al. [1] and Zhu et al. [21]. They indicated that higher gas concentration increase the heat released by the reaction which it increase flame speed.

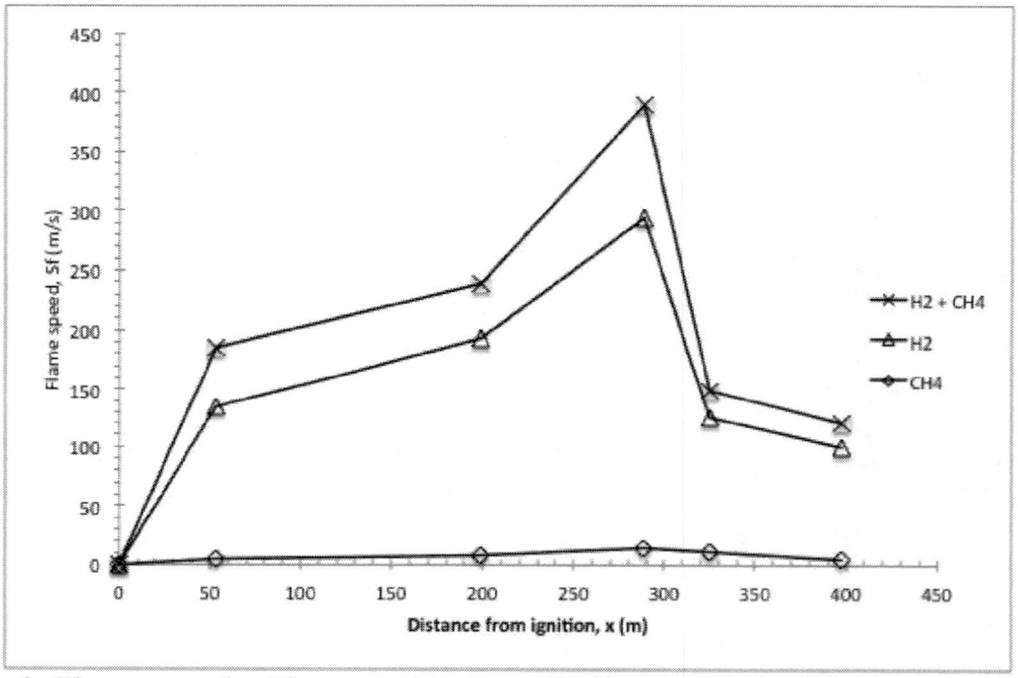

Figure 6 Flame speed, Sf as a function of distance from ignition, *x* for methane, methane/hydrogen and hydrogen at lean concentration of 6% v/v.

In other words, adding more hydrogen to mixtures resulted in significantly earlier times of flame arrival to 90 degree pipeline (Fig. 2 and 3), same as rig enclosure [8] and plenum chamber [22] giving higher flame speeds for both pure hydrogen/air and methane-hydrogen/air mixture as concluded by Hu et al. [23]. The acquiesced data showed that while the flame speed reached to the maximum value at the bend, it declined dramatically after the bend. The flame speed losses suffered at the bend are caused by both friction and momentum exchanges resulting from a change in the direction of flow. Interestingly, the recorded maximum flame speeds in this study are different in comparison with previous studies for pure hydrogen in 90-degree bend [3, 24]. This difference can be result of different experimental rig design.

Conclusion

This study was conducted to determine the maximum pressure, rate of pressure rise, flame speed and temperature changes of methane/air, hydrogen/air and methane-hydrogen/air mixtures in 90 degree bend pipeline in ambient pressure and temperature.

The findings showed that:

1. due to intense fluctuation of flame speed and pressure, it is observed that bending is the critical part in the pipeline. Accordingly, the potential risk of failure in pipeline becomes more probable at this part,
2. the results showed that methane-hydrogen/air and hydrogen/air mixtures gave about 11 times higher explosion pressure compared to the methane/air mixture. This indicates that the fuel type and its reactivity also affected pressure growth during explosions. The effect of hydrogen substitution to methane is dramatic, either in terms of maximum rate of pressure rise or overpressure, only for a lean hydrogen concentration i.e. 6 % v/v in the fuel, higher than 90%.
3. Even at lean concentration, the hydrogen enrichment into methane/air mixture can affect dramatically on maximum overpressure and flame speed.

References

[1] Phylaktou, H., M. Foley, and G. Andrews, *Explosion enhancement through a 90° curved bend.* Journal of loss prevention in the process industries, **6**(1): p. 21-29, 1993.

[2] Chatrathi, K., Going, J.E. and Grandestaff, B., Flame propagation in industrial scale piping. *Process Safety Progress,* 20**,** : p.286-294, 2001.

[3] Blanchard, R., et al., *Explosions in closed pipes containing baffles and 90 degree bends.* Journal of Loss Prevention in the Process Industries, **23**(2): p. 253-259, 2010.

[4] Di Sarli, V. and A.D. Benedetto, *Laminar burning velocity of hydrogen–methane/air premixed flames.* International Journal of Hydrogen Energy, **32**(5): p. 637-646, 2007.

[5] Raman, K. S., Laminar burning velocities of lean hydrogen-air mixtures. *Explosion Dynamics Laboratory Report.* Pasadena: Graduate aeronautical laboratories,California Institute of Technology, 1998.

[6] Sankaran, R. & Im, H. G., Effects of hydrogen addition on the markstein length and flammability limit of stretched methane/air premixed flames. *Combustion Science and Technology,* 178**:** p.1585-1611, 2006.

[7] Cammarota, F., et al., *Combined effects of initial pressure and turbulence on explosions of hydrogen-enriched methane/air mixtures.* Journal of Loss Prevention in the Process Industries, **22**(5): p. 607-613, 2009.

[8] Xiao, H., et al., *Experimental and numerical study on premixed hydrogen/air flame propagation in a horizontal rectangular closed duct.* International Journal of Hydrogen Energy, 2010. **35**(3): p. 1367-1376.

[9]. Zhian, H., et al., *Numerical simulation and study on the transmission law of flame and pressure wave of pipeline gas explosion.* Safety Science, 2012. **50**(4): p. 806-810.

[10] Guo, C., et al., *Cellular pattern evolution in gaseous detonation diffraction in a 90°-branched channel.* Combustion and Flame, 2007. **148**(3): p. 89-99.

[11]. Thomas, G., G. Oakley, and R. Bambrey, *An experimental study of flame acceleration and deflagration to detonation transition in representative process piping.* Process Safety and Environmental Protection, 2010. **88**(2): p. 75-90.

[12] Porowski, R. and Teodorczyk, A., Experimental study on DDT for hydrogen–methane–air mixtures in tube with obstacles. *Journal of Loss Prevention in the Process Industries,* 26**:** p. 374-379, 2013.

[13] Dennis, P. and Nolan, P. E., *Characteristics of Hydrocarbon Releases, Fires and Explosions.* Handbook of Fire & Explosion Protection Engineering Principles for Oil, Gas, Chemical, and Related Facilities, p. 41-61, 1996.

[14] Dorofeev, S. B., *Flame acceleration and explosion safety applications.* Proceedings of the Combustion Institute, 33(2):p. 2161-2175, 2011.

[15] Salzano, E., et al., *Explosion behavior of hydrogen–methane/air mixtures.* Journal of Loss Prevention in the Process Industries, 2012. **25**(3): p. 443-447.

[16]. Ilbas, M., et al., *Laminar-burning velocities of hydrogen–air and hydrogen–methane–air mixtures: An experimental study.* International Journal of Hydrogen Energy, 2006. **31**(12): p. 1768-1779.

[17] Lowesmith, B.J., et al., *Vented confined explosions involving methane/hydrogen mixtures.* International Journal of Hydrogen Energy, 2011. **36**(3): p. 2337-2343.

[18]. Lowesmith, B.J., G. Hankinson, and D.M. Johnson, *Vapour cloud explosions in a long congested region involving methane/hydrogen mixtures.* Process Safety and Environmental Protection, 2011. **89**(4): p. 234-247.

[19]. Wooley, R.M., et al., *Predictions of the consequences of natural gas-hydrogen explosions using a novel CFD approach*, in *Computer Aided Chemical Engineering*, B. Bertrand and J. Xavier, Editors. 2008, Elsevier. p. 919-924.

[20]. Lowesmith, B.J., et al., *Gas build-up in a domestic property following releases of methane/hydrogen mixtures.* International Journal of Hydrogen Energy, 2009. **34**(14): p. 5932-5939.

[21] Zhu, C., et al., *Effect of variation in gas distribution on explosion propagation characteristics in coal mines.* Mining Science and Technology (China), **20**(4): p. 516-519, 2010.

[22] Coppens, F.H.V. and A.A. Konnov, *The effects of enrichment by H2 on propagation speeds in adiabatic flat and cellular premixed flames of CH4+O2+CO2.* Fuel, 2008. **87**(13–14): p. 2866-2870.

[23]. Hu, E., et al., *Experimental and numerical study on laminar burning characteristics of premixed methane–hydrogen–air flames.* International Journal of Hydrogen Energy, 2009. **34**(11): p. 4876-4888.

[24]. Uchida, M., et al., *Pressure loading of detonation waves through 90-degree bend in high pressure H2–O2–N2 mixtures.* Proceedings of the Combustion Institute, 2011. **33**(2): p. 2327-2333.

1ST CCPS ASIA-PACIFIC CONFERENCE ON PROCESS SAFETY℠

A Consequence Based Assessment Method/Software Model to Facilitate the Safe Location and Design of Occupied Buildings on or In Close Proximity to Process Facilities

Kehinde Shaba
Det Norske Veritas Software
Palace House, London, UK
kehinde.shaba@dnv.com

Colin Hickey
Det Norske Veritas Software
Palace House, London, UK
colin.hickey@dnv.com

Prepared for Presentation at
American Institute of Chemical Engineers
1st CCPS Asia-Pacific Conference on Process Safety
Qingdao, China
September 4 – 5, 2013

UNPUBLISHED

A Consequence Based Assessment Method/Software Model to Facilitate the Safe Location and Design of Occupied Buildings on or In Close Proximity to Process Facilities

Kehinde Shaba
Det Norske Veritas Software
Palace House, London, UK
kehinde.shaba@dnv.com

Colin Hickey
Det Norske Veritas Software
Palace House, London, UK
colin.hickey@dnv.com

Keywords: Facility Siting, VCE modeling, Multi-Energy model, Baker Strehlow model, Consequence-based analysis methods, Occupied Building Assessment

Abstract

Ensuring the safe location and design of occupied buildings (permanent or temporary) is high on the agenda of key stakeholders – industry, regulators, and the workforce, not to mention members of the public who have a vested interest in the area. The significant amount of analytical work required in support of such an analysis represents a key barrier to conducting such studies in a detailed manner. It is often the case that a range of simplifications are utilised that can result in a less than realistic assessment.

This paper describes the implementation of two of the most widely used vapour cloud explosion (VCE) models, namely the TNO multi-energy (ME) and Baker-Strehlow-Tang (BST) into Phast, a general purpose onshore consequence analysis software package developed by DNV Software. This in direct response to the aforementioned challenges and the intention is to provide a robust platform by which a more detailed analysis of the VCE threats of process plant can be evaluated - without recourse for simplification which can be detrimental from a risk management perspective. The implementations of both models are quite detailed and include all parameters that have a bearing on the magnitude of a vapor cloud explosion. A simple case study of the tool as applied to a hypothetical occupied building siting challenge is presented and the value of using the tool discussed.

1. Introduction

Major accidents such as Texas city etc. have focussed attention on the need to ensure that occupied buildings are well protected from the often devastating impacts of vapour cloud explosions (VCE's) arising from process facilities. Ensuring the safe location and design of occupied buildings (permanent or temporary) is high on the agenda of key stakeholders – industry, regulators, and the workforce, not to mention members of the public who have a vested interest in the area.

In recent times, much has been done to develop and advance the knowledge gaps in this area. Notable projects include GAME (Eggen, 1988), GAMES (Mercx, et al., 1998), RIGOS (Van den Berg & Mos, 2005) to name a few. Yet, a key issue that remains is that the degree of analytical effort required in support of a detailed building evaluation can be significant due to the extensive time and cost required. This is largely due to the often complex, multi-system, large nature of process plant – there is "lots going on" and thus "lots to consider". It is clearly an unintended consequence of the prevailing process plant design ethos. Adding to the complexity is the intricate nature of the VCE accident sequence. A single loss of containment event can result in numerous outcomes – VCE or otherwise depending on factors such as the prevailing weather conditions, wind direction, degree of interaction with one or more regions of congestion or confinement and the location of ignition sources. Where a VCE outcome is realized, the possible impact to each building of interest has to be evaluated. It is often the case that most of the factors identified above are present in multiples e.g. a range of scenarios, weather conditions, obstructed regions, buildings of concern etc. The wide range of factors that need to be considered ("in isolation" or "as an interaction with other factors") also has a significant bearing on the level of effort required.

Various techniques are employed to make this task more manageable. The first is "screening exercises". Whilst there is much to be gained from making effective use of limited resources, screening is generally most useful and less harmful for screening hazards that clearly do not present VCE potential either intrinsically (e.g. degree of flammability) or by virtue of their context (no presence of a congested or confined region) - but can be dangerous when used beyond this point (i.e. in grey areas) largely because it becomes more subjective and thus less precise – whose standards of conservatism are being applied? How conservative is conservative enough? To be used effectively it also requires high levels of expertise and competence.

Another involves "scenario rationalisation". Much emphasis tends to be placed on establishing a representative set of scenarios that can be considered to be sufficiently reflective of the whole. Again, whilst this is useful, there is a danger that the selected scenarios might not be as representative as thought, leading to an underestimation of the threat level posed.

A third involves using "simplified models" which in some cases are then subject to further simplification. Much of the tediousness is made worse by the fact that the underlying physics of explosions is complex and this is reflected in the equally complex nature of the predictive models which are available. Computational fluid dynamic based models are generally considered to give the best approximations of this complex phenomenon. However, such models are largely considered impractical for day to day work due to the significant resource and time requirements needed and are generally only justifiable for highly uncertain issues. Additionally, the resource demands of this route is such that applying it for a plant wide assessment is generally not feasible. Simplified models are designed to solve this problem by allowing a robust level of characterization without the added expense. But then again there are varying degrees of simplicity and even the simplified models tend to require a wide range of inputs. The result is further simplification - it is common to place emphasis on the key parameters that are thought have a significant influence. An example is the implementation of the TNO multi energy model into the existing base configurations of Phast (versions 7.01 and below) which utilises the size of the confinement and blast curve number as the sole, key governing parameters. Additionally such models do not often examine the interaction of the flammable cloud with the region of confinement or confinement. Furthermore, such implementations tend to be conservative in their outlook and focus solely on the worst case outcome, a basis not always suitable for risk management. Together, these factors act as impediments to a realistic assessment of VCE potential to occupied buildings.

This paper describes a more detailed implementation of two of the most widely used vapour cloud explosion models, namely the TNO multi-energy (ME) (Van den Berg, 1985) and Baker-Strehlow-Tang (BST) (Baker, et al., 1996) into Phast, a general purpose onshore consequence analysis software package developed by DNV Software. This in direct response to the aforementioned challenges and the intention is to provide a robust platform by which full and detailed analysis of the VCE threats of process plant can be evaluated - without recourse for simplification which can be detrimental from a risk management perspective.

2. Background on Explosions

2.1 General

Prior to elaborating on the various methods by which the effects of a vapour cloud explosion can be estimated, it is useful to examine the nature of explosions in general.

An explosion is essentially a very rapid release of energy that results in a pressure discontinuity in the atmosphere. This discontinuity—called a "pressure wave"—initiates in the blast source and dissipates outwards. The energy associated with the wave is the primary source of destructive energy to objects in its path and is responsible for a large proportion of the damage that occurs from explosions on process facilities. Consequently, it is important to be able to characterize the nature of this wave as function of distance.

A pressure wave is typically characterized by three key components. First, the magnitude of the pressure it generates above ambient referred to as the "overpressure"; the maximum value is known as the "peak overpressure". Second, the duration of the wave, simply called the "blast duration". Third, the blast "impulse" which is defined as the change in momentum and is the force-time product. Together these are often referred to as "blast parameters" or "blast load" in the context of a particular building or location.

There are many types of explosions. Vapour cloud explosions tend to be the most significant for the bulk of process facilities. In some circumstances, explosions due to dust particles or pressure vessels can dominate.

2.2 *Vapour Cloud Explosions*

A wealth of knowledge has been accumulated on this area, such that there is an extensive range of material that elaborates to a great level of detail, the particulars of the science underpinning vapour cloud explosions and consequently how they can be modelled. Key references include (CCPS, 1994), (Lees, 1996) and (Crowl, 2003).The goal of the text that follows is to highlight the key aspects so as to provide sufficient contextual information to allow the uninitiated reader develop a better appreciation of the state of the knowledge without recourse to primary sources.

Various factors have a bearing on the potential for VCE's. Necessary preconditions include: a flammable vapour cloud (for example following a loss of containment event and sufficient mixing in air to produce a combustible mixture) and an ignition source. Other factors – some intrinsic to the material and others related to the environment also play a key role in determining the magnitude of the resulting overpressures produced. The two key external factors that influence the magnitude of a VCE are often referred to as the "degree of congestion" and the "degree of confinement" though it is not uncommon to find them described using disparate terminology.

A VCE occurs as a result of a flame propagating through a fuel-air mixture. Research into this area (TNO, 1997) has indicated that the magnitude of the resulting overpressure is directly related to the flame speed which is governed by the degree of turbulence realized. Turbulence is generated by the interaction of the flame with obstacles in its path. This interaction involves a vicious feedback cycle – the more obstacles faced, the higher the degree of turbulence and the increased acceleration of the flame front. With no turbulence, the cloud will burn as a flash fire without appreciable overpressures generated. High turbulence results in increased flame speeds and high overpressures. The opposite occurs with low turbulence. The level of obstacles within the cloud (i.e. obstacle density) is referred to as the "degree of congestion", whilst the location wherein the obstacles are situated is termed a "congested region" or "region of congestion".

As the cloud burns, it expands. This expansion allows for the rapid dissipation of energy. A vapour cloud allowed to expand in all 3 dimensions will lead to lower overpressures than confined ones (say beneath a platform or pipe-rack).The phrases "degree of confinement" or "degree of expansion" relates to the extent to which the clouds expansion is "unrestricted" or on the contrary "confined". Confinement is considered in one or all three dimensions (x, y, z), with consideration of intermediate positions. It is often the case that there are many opportunities for such areas on process plants. The number of congested/confined areas identified increases the calculation effort significantly.

Other factors such as material reactivity and not least the degree of "blanketing" between the flammable cloud and the region of confinement/congestion also have an influence.

2.3 *Key Vapour Cloud Explosion Models*

The most widely used simplified models to characterize—that is estimate the magnitude and duration of the resulting pressure wave as a function of distance from the explosion source—the nature of vapour cloud explosions are the TNO multi-energy (hereafter referred to as ME) (Van den Berg, 1985), the Baker-Strehlow-Tang (hereafter referred to as BST) (Baker, et al., 1996) and the Congestion Assessment Method (CAM) (Puttock, 1995). A fuller description/elaboration of each of the above models can be found in (CCPS, 1994), (Lees, 1996) and (Crowl, 2003).

The TNT equivalence model, also a simple model and once used extensively has now been shown to be limited in its ability to adequately characterize VCE's. Consequently, it is not generally recommended for use. For example API standard 752 (API, 2009) explicitly states that the model should not be used. Lees (1996) asserts that the "TNT equivalent method for vapour cloud explosion is obsolescent". Nevertheless, it still sees widespread application. This is in part driven by its simplicity – only a few inputs are required to the model. That it results in relatively conservative estimates of the peak overpressure (often used as a key indicator of explosion magnitude) is another factor – one that makes it especially appealing as an early screening tool. The sections that follow briefly outline the key premise and characteristics of the models that have been integrated into the Phast consequence package in a more robust manner, namely the ME and the BST models.

2.3.1 *TNO multi-energy model*

The TNO multi-energy model (Van Den Berg, 1985) is based on the supposition that the energy of an explosion is governed largely by the level of confinement and congestion (i.e. location specific factors) within which the vapour cloud is located and to a lesser extent by the intrinsic characteristics of the vapour cloud, most notably the flammable mass. By placing more emphasis on external factors, it represents a significant shift in thinking from the notions animating the TNT model.

The Multi Energy explosion model requires the identification of the obstructed region, volume blockage ratio, degree of expansion among other parameters as inputs. These are then utilised to calculate the overpressure generated with the level of congestion (input) provided. Alternatively, the model can use the "curve number" as input. The "Curve number" refers to different overpressure versus distance graphs. These graphs differ in range of severity from "Curve 1-weak explosion" to "Curve 10-severe/heavy explosion". Materials that may be released will be assigned a specific "curve number", i.e. strength of the explosion, depending on their properties such as their reactivity.

2.3.2 *Baker Strehlow Tang model*

The BST model (Baker, et al., 1996) also allows for the characterization of overpressure as a function of distance but uses the flame speed as the point of departure, the selection of which is based on three factors. First, the reactivity of the material of concern. Second, the potential for flame expansion within the region of the release (i.e. degree of confinement) and third, the obstacle density (degree of congestion).

In this model, the potential for expansion is characterized by the number of dimensions in which expansion is possible and includes one of four possibilities that sit on a spectrum from most confined (known as "1D", which reflects tunnels, corridors or similar geometries where expansion is only possible in one direction) to completely open ("3D" which reflects a situation where there is minimal restriction to expansion and the cloud can expand in all directions). Occupying the middle area is the "2D" and "2.5D" designations which aim to capture intermediate positions between these two extremes.

For the most part the ME and BST model are quite analogous, particularly in terms of their outlook and animating concepts. The ME model is known to give higher predictions of overpressure in the near field. The BST is also known to over predict, albeit in the far field.

3. The Phast Software Package and Recent Updates

3.1 The Software package "Phast"

Phast (acronym for Process Hazard Analysis Software Tool) is a consequence analysis software package developed by DNV Software for the modelling of the consequences associated with releases of flammable or toxic materials as would occur on process facilities.

Phast's modelling capabilities are numerous and diverse and allow for the evaluation of a wide range of hazard types in different situations. Its capabilities cover a wide range of release phases (gas, liquid and two-phase) and materials (various flammables and toxic compounds including LNG and CO_2) as pure components or in mixtures. In terms of release types, it can model unpressurised and pressurised releases as steady state or as time-dependent releases.

At its core is the Unified Dispersion Model (UDM) which enables complete and rigorous modelling of various release types accounting for transition to jet, dense and passive clouds; buoyancy effects, substrate interaction (ground effects); plume lift-off, capping at the mixing/inversion layer, droplet formation and rainout under diverse atmospheric conditions.

Phast is an integrated consequence analysis package that models all stages of an accidental release from:

- Discharge: including rainout, pool evaporation and spreading to
- Dispersion: from the pool and the discharge orifice as shown in Figure 1.
- Effects including toxicity, jet fires, fireballs, pool fires, flash fires and explosions

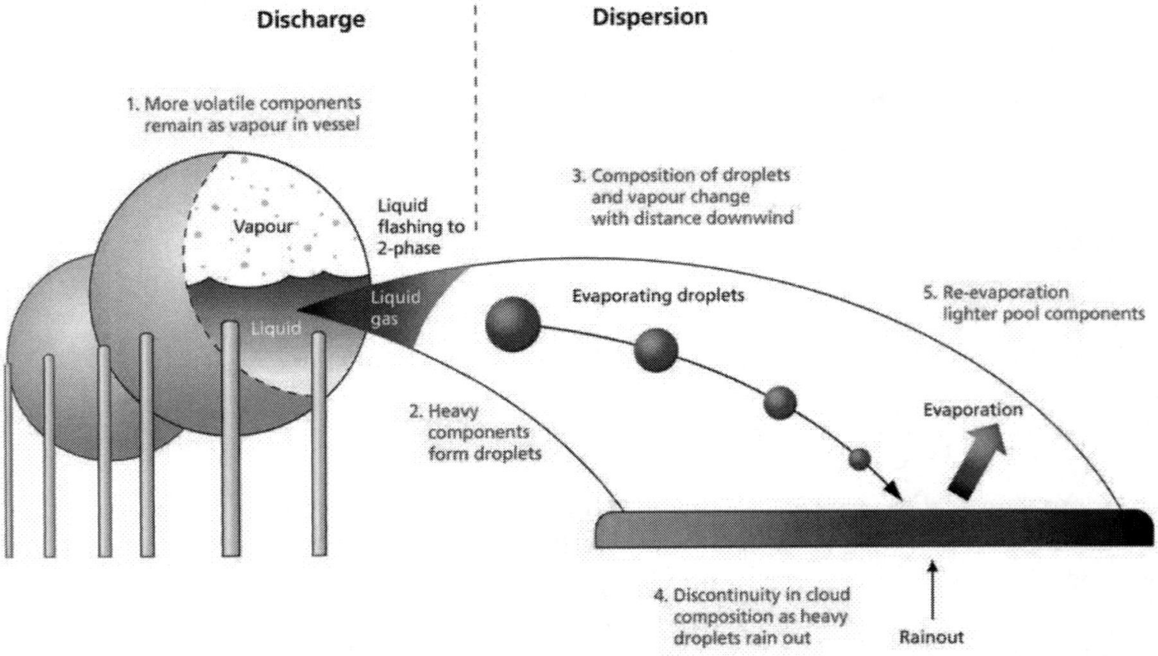

Figure 1: Initial stages of an accidental release as modelled by Phast

In terms of VCE modelling, Phast enables the key requirement namely the size and dimensions of the dispersing vapour cloud to be established.

The modelling takes into account various contextual factors that can influence the development of the hazard. More details of the technical capabilities of the Phast Software package can be found in body of technical reference documentation available as well as in papers by Witlox (2010) and Witlox & Oke (2008) amongst others.

3.2 Recent Updates

As indicated in the introductory section, the current base configurations of Phast (versions 7.01 and below) include what can effectively be referred to as a narrow implementation of both the ME and BST models. In the former, the blast curve number and the volume of confinement are considered to be key governing parameters to the exclusion of others such as degree of congestion. For the latter, the flame speed and confined volume are given primacy. Also, in both implementations, the interaction between the dispersion cloud and the region of congestion/confinement—a key element as the quantity of flammable mass located therein is a governing factor—is quite opaque and defined in a rather ambiguous manner. Additionally, the epicentre of the explosions was not located around the region of congestion but rather within the cloud – quite contrary to what would be expected in reality where the explosion centre will be located within the area of confinement. Finally, whilst the tool had the capability to evaluate numerous scenarios simultaneously and display the hazard range as contours on a location map, evaluating the impact to any one target was generally a tedious affair as this has to be done on an individual case-by-case basis – a herculean task where a large number of scenarios are being analysed.

The software package has now been updated to address the limitations highlighted above. Key innovations with regards to VCE modelling include: a fully featured and exhaustive implementation of the ME and BST explosion models as well as the inclusion "multi-scenario contouring capability". These updates are available as an add-on to the Phast base configuration for version 7.1 onwards. The implementation of both models into Phast is executed as part of the Obstructed Region Explosion Model (OREM) developed by DNV Software. The theory and validation of the OREM model is discussed in (Xu, 2010a) and (Xu, 2010b) respectively.

The validation work involved comparison of the results derived with the OREM model with those from seminal experimental studies such as BFEST3a, EMERGE and the Shell Deer Park Ethylene case (as reported in (Fitzgerald, 2001)), and the GAMES project (as reported by (Mercx, et al., 1998)). The GAMES study included verification versus three sub cases: a chemical plant case, a gas processing case and the LNG terminal case. Overall, the model showed good agreement with the experimental data in terms of its ability to predict the blast parameters. Better agreement was observed for the ME model than for the BST model.

Key aspects of these models are the definition of obstructed regions, consideration of the degree of overlap between the flammable cloud and the region so as to determine the explosive potential as well as consideration of critical separation distance between defined regions (features common to both models). The separation distance is important as it aids in the determination of how many explosion sources can be reasonably expected where a vapour cloud envelopes more than one region of obstruction. Finally, given the fact that the model will lend itself to the consideration of a wide range of scenarios, the software has been updated to allow for multi-scenario contouring - an efficient way of organizing and communicating the resulting hazard ranges especially when used to inform facility siting considerations as outlined in CIA (2010) (American Petroleum Institute (API), 2009) and (American Petroleum Institute (API), 2007)). All of these are discussed in the sections that follow.

3.2.1 Definition of "Obstructed Regions"

As noted above, "obstructed regions" (i.e. regions of congestion and/or confinement) are central to the potential for high strength VCE's and are thus a key element of both the ME and BST models. They are typically defined based on a review of the site layout diagrams/plot plans and the external boundaries of these areas are typically defined via "perimeter bounding boxes".

Whilst there are guidelines to facilitate this activity (see for example (Van den Berg & Mos, 2005) and (American Petroleum Institute (API), 2007), there remains a degree of subjectivity which can lead to some variation in the definition depending on the perspective of the person undertaking the task. Once defined, they are then approximated to distinct cuboid shaped volumes and input to the Phast model using the geographical information system (GIS).

Beyond the geometrical similarities in obstructed regions for both explosion models they both require a different set of parameters for their complete definition. As a consequence they are created differently within the Phast software package. Additionally, for each explosion model, two types of obstructed regions can be created: "Defined Strength" or "Calculated Strength" in the case of the ME model and "Defined Flame Speed" or "Calculated Flame Speed" in the case of the BST model. For the TNO multi energy model, key parameters for the "Calculated Strength" option include degree of expansion (or confinement), flame path length and volume blockage ratio (together used to determine the curve number that will apply). For the "Defined strength" option, the only inputs required are the curve number and the blockage ratio. For the BST model, the "Calculated Flame Speed" option requires degree of confinement, congestion level and flame reactivity are key (used to calculate the flame speed that will obtain). The flame speed and the blockage ratio are the only inputs required "Defined Flame Speed" option.

To sum, the following logic applies within Phast:

1) Obstructed regions are created as either ME or BST
2) For both model types, two types of obstructed regions can be created – "Defined" or "Calculated".
    a. For ME, these are referred to as "Defined Strength" or "Calculated Strength"
    b. For BST, these are called "Defined Flame Speed" or "Calculated Flame Speed"

A simple hypothetical case study has been developed to highlight how the tool will work from a practical standpoint. The objective of the study is to evaluate the side-on overpressure loads to three existing buildings of interest arising from a number of release events associated with a proposed installation of a gas compression facility. The loads are to be used as basis for consideration of possible mitigation strategies.

A range of pressurized ethane release scenarios associated with the proposed facility has been identified. These are located in and around the congested regions. For each scenario, three hole sizes have been considered – Small, Medium and Large. A single weather condition has been assumed. In terms of the wind rose, 8 directions have been accounted for.

Figure 2 illustrates the four obstructed regions (shown as pink boxes) defined for this study. The figure also shows the three occupied buildings of interest to this study.

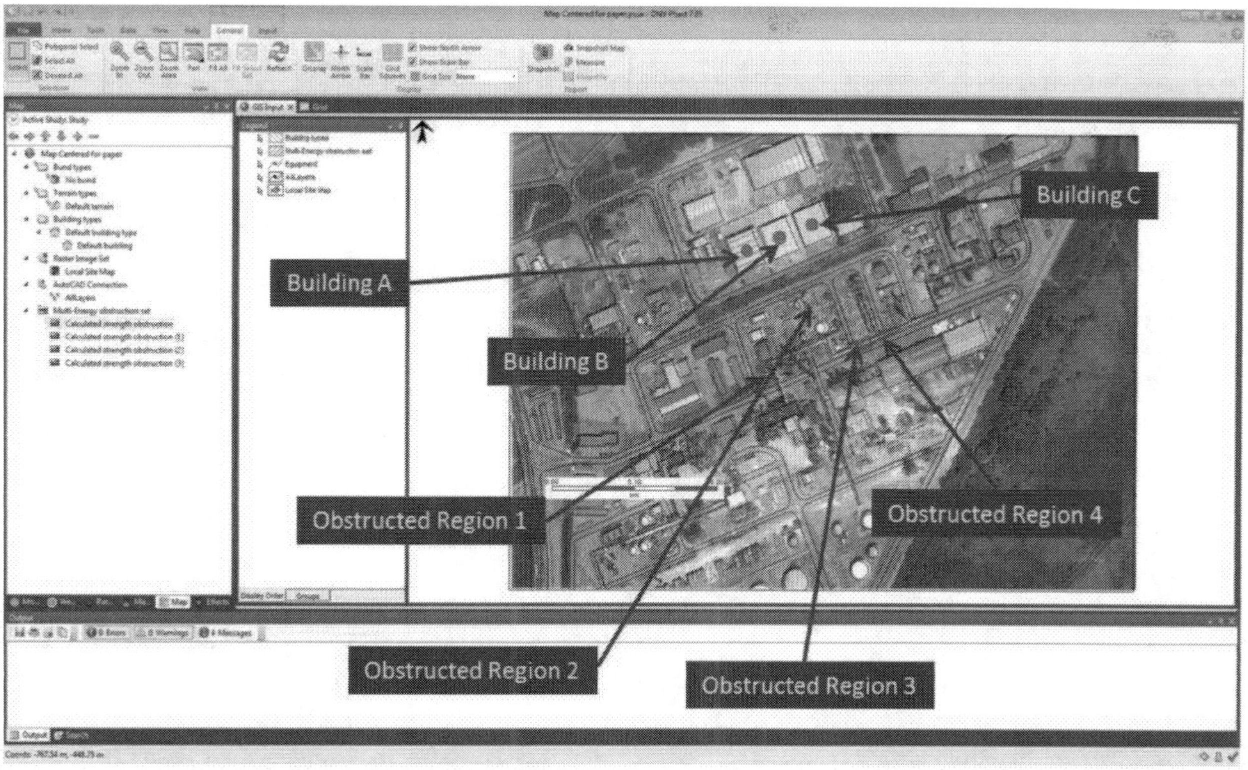

Figure 2: Graphical User Interface (GUI) representation of obstructed regions in Phast*.

***Annotations highlighting the positions of obstructed regions and buildings of interest have been added to improve clarity**

3.2.2 Consideration of flammable cloud/obstructed region overlap

It has been established that overlap of a flammable cloud with a region of confinement or congestion is a necessary prerequisite for a VCE of appreciable overpressure to occur, consequently it is important to evaluate the extent to which such blanketing occurs for the range of scenarios to be considered and obstructed regions defined on a facility.

A range of factors have an influence on this possibility. The relative distance between the release event and the obstructed region, the size of the resulting flammable cloud as it develops and disperses and the direction of travel of the cloud are key aspects to consider. The range of possibilities for a single dispersing flammable cloud travelling downwind with a single obstruction and building in its path is illustrated in Figure 3 for four representative time steps (i.e. snap shots) over the duration of the cloud. The situation becomes more complex, where multiple events, multiple obstructed regions and multiple wind directions are considered.

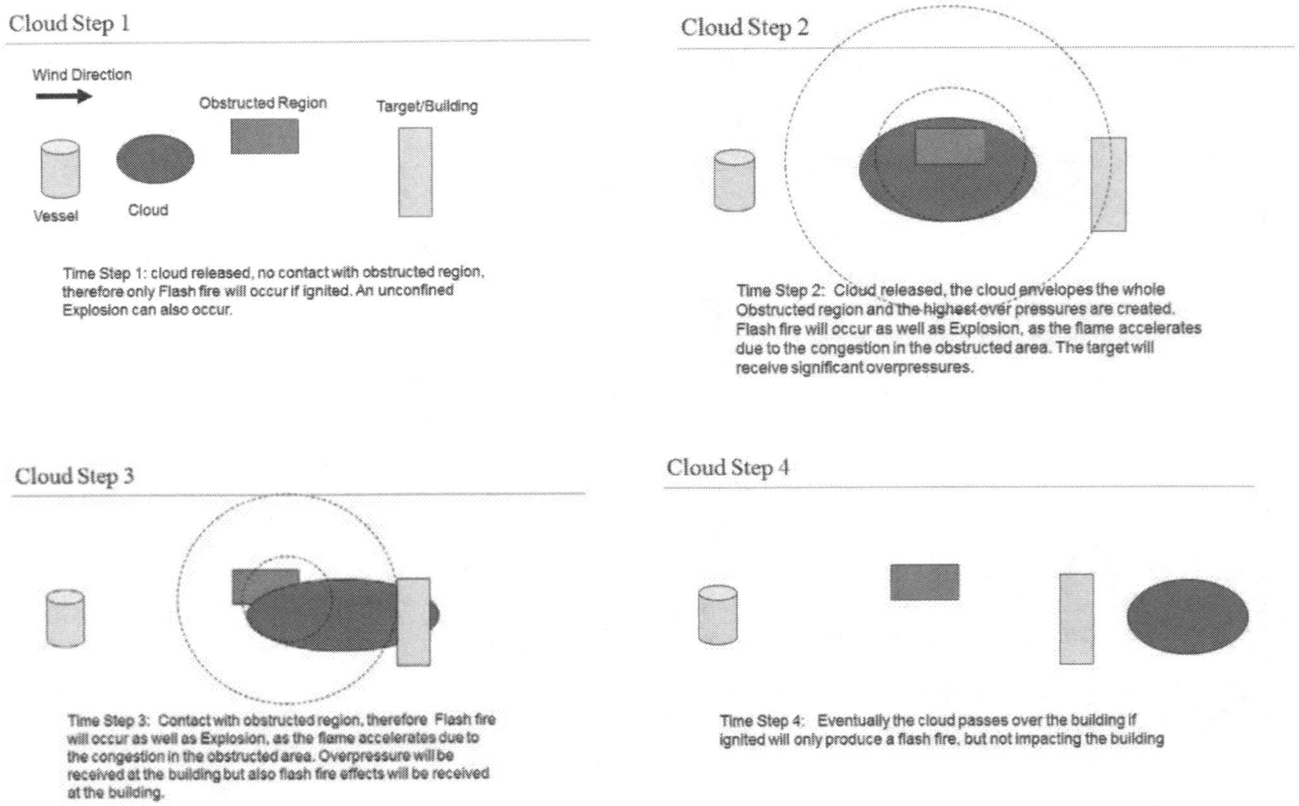

Figure 3: Possible cloud steps/outcomes as a flammable cloud disperses

Phast incorporates a sophisticated GIS system into which release locations and obstructed regions are defined. The system enables the model to examine in detail the degree of interaction between the resulting flammable clouds and the obstructed regions and thus determine the degree of fill to be used within the explosion calculations. It does so for all wind directions specified (four as a minimum; 18 as a maximum).

3.2.3 Critical Separation Distance

Under the right conditions, each identified obstructed region has the potential to act as an explosion source. However, where a dispersing flammable cloud overlaps more than one obstructed region there is the prospect for multiple explosion sources. The potential is governed by the "separation distance" between the regions. A large gap between multiple regions will mean that the flame front will decelerate resulting in multiple independent explosions. On the other hand, a small gap will result in a single explosion with the combined energy of all the sub-sources. The "critical separation distance" is the threshold criterion used to establish whether an explosion source should be considered as independent or part of a larger group.

There are divergent views on the specification of the critical separation distance. TNO (1997) suggests a critical separation distance equal to 10 obstacle diameters with an upper bound of 25m. This follows on from their definition of a congested region as one in which obstacles are located within 10 obstacles diameters of one another. Subsequent research into this area (specifically the work titled "Research into Guidance on Separation Distances, the "RIGOS" project, see Van den Berg et al, 2002) has indicated that this approach might not always be conservative particularly where the resulting overpressures are high. The report recommends the critical separation distance should be a function of the size of the maximum overpressure from the source explosion, referred to as a "critical separation ratio". Baker et al (1998) suggests than 5m should be adopted as the upper limit for defining obstructed regions are separate explosion sources.

The Phast model allows for the critical separation distance to be specified as a "ratio" or "fixed value". In the former case, users have to determine the ratio to apply based on a preliminary assessment of the maximum overpressures created within each region.

Based on the above, the following additional logic applies to how obstructed regions can be defined in the software tool:

1) Obstructed regions can be defined as being independent or as part of a group.
2) Where defined as part of a group, the critical separation criterion has to be specified to determine the number of potential sources within the group.
3) The criterion can be defined as affixed distance or a ratio.
4) If the criterion is not specified, all regions within that group are considered to be single source
5) Only obstructed regions defined within the same group are considered for combination potential. Regions in different groups or sets cannot be combined.
6) To avoid "double counting" resulting in possible overestimation of the explosion volume, overlap between obstructed regions within a set is not allowed. Violations are automatically checked within the software.

3.2.4 Multi-Scenario Hazard Contouring

Another key update is the capability to plot combined hazard contours to a defined threat level (e.g. as overpressure or impulse) for a range of scenarios on a facility layout diagram. This feature allows for clearer understanding and communication of hazards ranges and their impact footprint as well as makes for more efficient and comprehensive analysis - the full range of flammable release scenarios (as opposed to a limited subset) can be modelled simultaneously.

Table 1 below indicates the pressure calculation methods available to output the blast parameters (overpressure and impulse). This feature is not limited to explosion hazards only but can also be used to develop aggregate hazard contours for dispersing clouds (e.g. flammable or toxic) to a concentration of interest and thermal radiation from fires (jet and pool fires).

The side-on overpressure contours derived for the hypothetical study referred to in section 3.2.1 above is shown in Figure 4. Plotting the contours on a location map readily enables the overpressure loads buildings can potentially experience to be identified. The result for this hypothetical study indicates:

- The peak overpressure potentially experienced by the three buildings of interest will be of the order of 0.3 bar (this contour just marginally touches the periphery of these buildings.).
- The 0.2 bar overpressure contour overlaps most of the buildings suggesting that this is a more representative pressure loading.
- Due to the fact that the 0.3 bar contour just skirts the borders of the buildings, some further investigation of the results will be required to evaluate how sensitive this overpressure level is to key assumptions (e.g. size/degree of congestion of the obstructed region).
- Additionally, it is evident the blast loads to these buildings are driven by the obstructed congestion 1 (see Figure 2). Reducing the potential for congestion/confinement in this area represents a potential solution to mitigating the blast loads to the buildings of interest.

Table 1: Output options for blast parameters Overpressure and Impulse for ME and BST modelling options

Pressure Method	ME	BST
Side on Overpressure	Yes	Yes
Dynamic Overpressure	Yes	No
Side on + Dynamic Overpressure	Yes	No
Reflected Overpressure	Yes	Yes

A robust evaluation of explosion hazard potential to buildings requires a consideration of both the "overpressure level experienced as well as the duration" or the "overpressure and duration". Though not shown, the software model includes the capability to plot impulse contours for the range of overpressure methods available above.

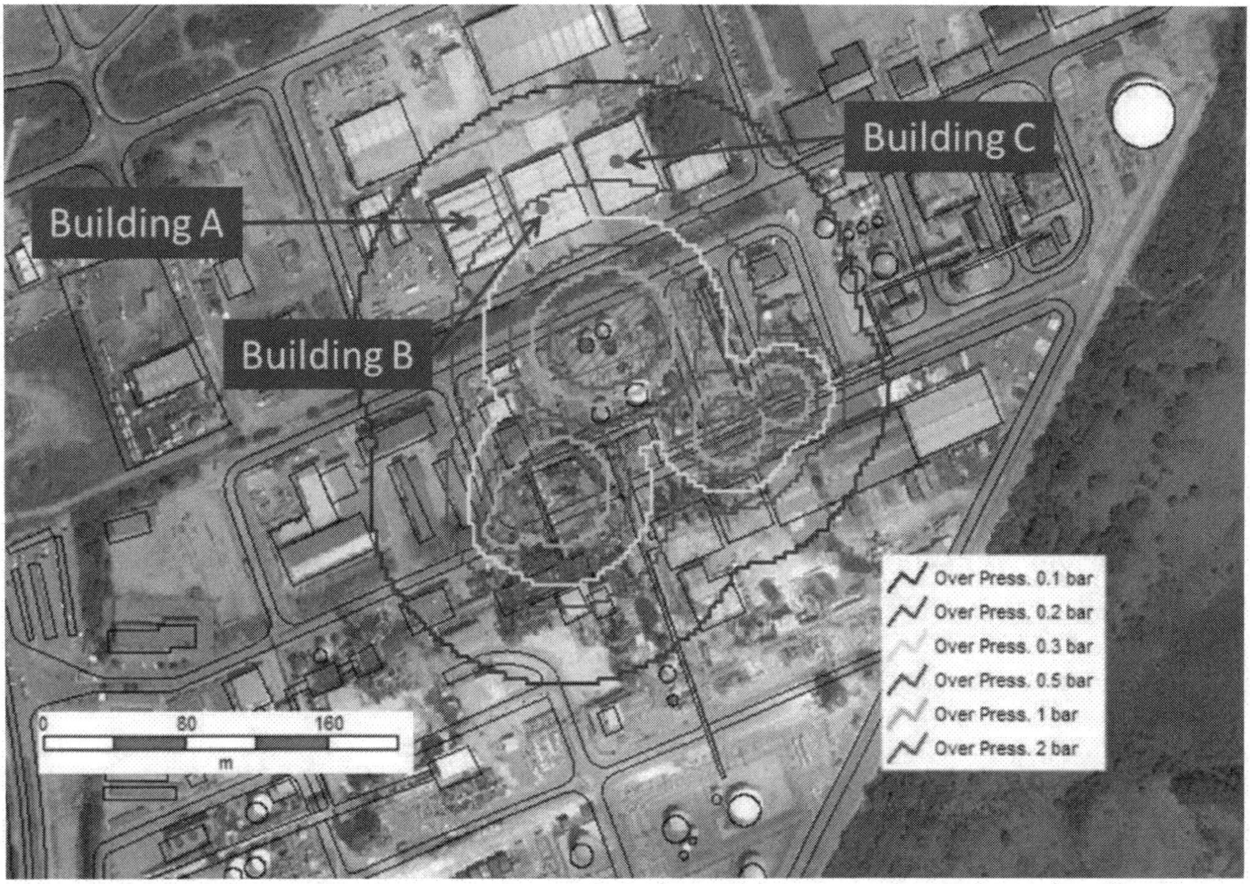

Figure 4: Side-on overpressure contour envelope for multiple scenarios and weather directions (representative result)*

***Annotations highlighting the positions of buildings of interest have been added to improve clarity.**

4. A new method to facilitate the "consequence-based" evaluation of VCE blast parameters to occupied buildings

Taken together, the above innovations allow for a new approach to the evaluation of blast loads to occupied buildings using the Phast Software package – one that is not necessarily limited by the often large number of events and obstructed regions that obtain on process facilities. Using this software model, the following simple three steps are required:

- Step 1: **Establish calculation basis**. Identify all release scenarios as well as areas of congestion/confinement. Select explosion model to be used (i.e. ME or BST). Input into the model together with parameters (e.g. separation distance between obstructed regions) required for their full definition.
- Step 2: **Undertake modelling.** Phast models each event in turn and examines the VCE potential and where found, determines the associated hazard ranges and blast loads as a function of downwind distance.

- Step 3: **View results and determine blast loads to buildings.** The results are presented on the location maps as contours for blast parameters of interest (e.g. Overpressure and Impulse).

It is good practice to examine the sensitivity of results to key input parameters/ assumptions, consequently, some iteration between steps 1 and 3 is also recommended.

The overall consequence of the above is that a more detailed evaluation of VCE blast hazards to all buildings on a facility resulting from all loss of containment events postulated can be realised simultaneously using this tool. This is particularly noteworthy given that such a task often represents a key challenge requiring significant amount of resources (cost and time).

Recognising that Phast is a consequence analysis tool, this advancement will facilitate the execution of building siting analyses that are predicated upon "consequence-based" evaluation methods. Other methods include "Simplified methods" (e.g. lookup tables that codify industry experience) and Risk-based methods (full consideration of both the likelihood and impact of an event e.g. using exceedance curves/hazard frequency contours). Safeti (Cavanagh, 2001; Worthington and Witlox, 2002), a risk assessment and consequence analysis package already incorporates this capability and is available for use in risk-based methods. The implementation of these VCE models into Safeti (versions 6.6 and above) is discussed by Cavanagh et al (2009).

5. Conclusion

Major accidents such as Texas city etc. have focussed attention on the need to ensure that occupied buildings are well protected from the often devastating impacts of vapour cloud explosions (VCE's) arising from process facilities. The large scale of such installations often implies that a substantial amount of analytical effort is required. Consequently, there is a need to make the effort much more manageable. This is usually achieved via some kind of screening activity or the use of much simpler approaches (e.g. using in-house or other standards). Whilst there is much to be gained from simplifying the analysis, there is a danger that if not executed properly, a less than adequate assessment of the hazard potential will be achieved. This is because such screening requires judgment and expertise ("good engineering judgement") and such skills are not widely available. Hence, it is desirable to have a software tool which allows for a substantive analysis to be undertaken in a rigorous, robust and resource (cost, time) efficient manner without the added risk simplification can introduce.

This paper has described and examined the recent advances in the Phast consequence model designed to address this challenge. It has demonstrated that increased rigour can be achieved in an effective manner for any scale study (both large and small) thus negating the need for screening/simplification activities. The advancements to the Phast software platform are a welcome addition to the range of software tools available to facilitate the evaluation of explosion risks to buildings and should help improve the robustness and overall quality of such evaluations resulting ultimately in more effective management of risks to occupied buildings.

6. References

1) American Petroleum Institute (API). (2007). *Management of Hazards Associated with Location of Process Plant Portable Buildings, API Recommended Practice RP753, 1st Ed (Reaffirmed January 2012).* American Petroleum Institute .
2) American Petroleum Institute (API). (2009). *Management of Hazards Associated with Location of Process Plant Buildings, API Recommended Practice RP752, 3rd Ed.* American Petroleum Institute.
3) Baker, Q., Doolittle, C., Fitzgerald, G., & Tang, M. (1998). Recent development in the Baker-Strehlow VCE analysis methodology. *Process Safety Progress*, 17(4), 297-301.
4) Baker, Q., Tang, M., Scheier, E., & Silva, G. (1996). Vapor cloud explosion analysis. *Process Safety Progress*, 15(2), 106-109.
5) Cavanagh, N. J. (2001). Calculating Risks. *Hydrocarbon Engineering (6).*
6) Cavanagh, N., Xu, Y., & Worthington, D. (2009). *A Software Model For Assessing Fatality Risk From Explosion Hazards Using The Multi Energy Method And Baker Strehlow Tang Approach. Hazards XXI SYMPOSIUM SERIES NO. 155* (pp. 647-656). IChemE.
7) Center for Chemical Process Safety (CCPS) . (1994). *Guidelines for Evaluating the Characteristics of Vapor Cloud Explosions, Flash Fires and BLEVEs.* New York: CCPS AICHE.
8) Center for Chemical Process Safety (CCPS). (1996). *Guidelines for Evaluating Process Plant Buildings for External Explosions and Fires.* New York: CCPS AICHE.
9) Chemical Industries Association (CIA). (2010). *Guidance for the location and design of occupied buildings on chemical manufacturing sites 3rd Edition.* London: Chemical Industries Association.
10) Crowl, D. A. (2003). *UNDERSTANDING EXPLOSIONS.* New York: CCPS AICHE.
11) Eggen, J. (1988). *GAME: development of guidance for the application of the multi-energy method (TNO Report PML 1995-C44 Prepared for the UK Health and Safety Executive: Research Report 202/1998).* Rijswijk: TNO Prins Maurits Laboratory.
12) Fitzgerald, G. (2001). A Comparison of Simple Vapor Cloud Explosion Model Explosion and Prediction methodologies. *Second Annual Symposium, Mary Kay O'Connor Process Safety Center, Texas A&M University.* Texas.
13) Lees, F. P. (1996). *Loss Prevention in the Process Industries: Hazard Identification, Assessment and Control. Vol. 2 (2nd ed).* Oxford: Butterworth-Heinemann Ltd.
14) Mercx, W., Van den berg, A., & Van Leeuwen, D. (1998). *Application of correlations to quantify the source strength of vapour cloud explosions in realistic situations. Final report for the project: 'GAMES'. TNO report PML 1998-C53.* Rijswijk: TNO Prins Maurits Laboratory.
15) Puttock, J. (1995). Fuel Gas Explosion Guidelines - the Congestion Assessment Method. *2nd European Conference on Major Hazards On- and Off-shore.* Manchester.
16) Tang, M., & Baker, Q. (1999). A New Set of Blast Curves for Vapour Cloud Explosions. *Process Safety Progress*(18), 235-240.
17) TNO (The Netherlands Organization of Applied Scientific Research). (1997). *Methods for the calculation of Physical effects due to releases of hazardous materials "Yellow Book" CPR 14E.* The Hague.

18) Van den Berg, A. (1985). The multi-energy method: a framework for vapor cloud explosion blast prediction. *Journal of Hazardous Materials*, (12) 1-10.
19) Van den Berg, A., & Mos, A. (2005). *Research to improve guidance on separation distance for the multi-energy method (RIGOS) (TNO Report PML 2002-C50 Prepared for the UK Health and Safety Executive: Research Report 369).* Rijswijk: TNO Prins Maurits Laboratory.
20) Witlox, H. W. (2010). Overview of consequence modelling in the hazard assessment package Phast. *In 6th AMS conference on applications of air pollution meteorology, Atlanta, USA, 17-21 January 2010.*
21) Witlox, H. W., & Oke, A. (2008). Verification and validation of consequence models for accidental releases of hazardous chemicals to the atmosphere. *IChemE Hazards XX symposium & workshop, Manchester, UK, 15-17 April 2008.*
22) Worthington, D., & Witlox, H. W. (2002). *SAFETI Risk Modelling Documentation – Impact of Toxic and Flammable Effects.* London: DNV Software.
23) Xu, Y. (2010a). *Obstructed Region Explosion Model (OREM): Theory.* London: DNV Software.
24) Xu, Y. (2010b). *Obstructed Region Explosion Model (OREM): Validation.* London: DNV Software.
25) Xu, Y., Worthington, D., & Oke, A. (2009). Correcting The Predictions By Baker-Strehlow-Tang (BST) Model. *Hazards XXI SYMPOSIUM SERIES NO. 155* (pp. 318-325). IChemE.

1ST CCPS ASIA-PACIFIC CONFERENCE ON PROCESS SAFETY℠

Air Dispersion Modeling for Landfill Gas Emissions and Facility Siting

Chaowei Liu
MMI Engineering
11490 Westheimer Rd, Suite 150, Houston, TX 77077
TLiu@mmiengineering.com

Yi Liu[1], Eric Peterson[2], and Morgan Reed[2]
1 Mary Kay O'Connor Process Safety Center
Texas A&M University, College Station, TX 77843
Yi.Liu1911@tamu.edu
2 MMI Engineering
11490 Westheimer Rd, Suite 150, Houston, TX 77077
EPeterson@MMIEngineering.com
MReed@MMIEngineering.com

Prepared for Presentation at
American Institute of Chemical Engineers
1st CCPS Asia-Pacific Conference on Process Safety
Qingdao, China
September 4 – 5, 2013

Air Dispersion Modeling for Landfill Gas Emissions and Facility Siting

Chaowei Liu
MMI Engineering
11490 Westheimer Rd, Suite 150, Houston, TX 77077
TLiu@mmiengineering.com

Yi Liu[1], Eric Peterson[2], and Morgan Reed[2]
1 Mary Kay O'Connor Process Safety Center
Texas A&M University, College Station, TX 77843
2 MMI Engineering
11490 Westheimer Rd, Suite 150, Houston, TX 77077

Keywords: Air Dispersion Modeling, Landfill Gas Emission, Facility Siting, CFD, AERMOD

Abstract

Air dispersion modeling studies how the air pollutants disperse in the ambient atmosphere using computer simulation. Dispersion models are used to estimate or predict the concentration of air pollutants or toxicants emitted from various sources such as industrial plants, road traffic vehicles or accidental chemical releases.

Landfill gas is a complex mixture of gases generated by anaerobic decomposition of biodegradable waste in landfills, which includes many hazardous air pollutants (HAP). In recent years there is increasing concern regarding the acute toxicity, chronic hazards, and risks associated with landfill gas emissions. A landfill site case study is demonstrated, with different screening and refined modeling software tools used for modeling landfill gas emissions and their resulting ambient air concentrations.

Facility siting assists decisions in managing the risk to on-site personnel located in new and existing buildings intended for occupancy, and air dispersion models can be applied to determine the consequences of accidental releases of hazardous or toxic materials. In this paper, analytical and computational fluid dynamics (CFD) software are used for a chemical gas plant permanent building location study.

1. Introduction

Air dispersion modeling studies how the air pollutants disperse in the ambient atmosphere using computer simulation. The dispersion and ground-level concentration of pollutants are determined by the complex interactions of many factors, including meteorological conditions at or near the site, the topography of the surrounding areas, physical characteristics of stack or other emission points as well as the physical and chemical characteristics of the pollutants [1]. Air dispersion models can provide future air pollution levels or to estimate air quality where no monitoring data are available.

Landfill gas is the natural by-product of the anaerobic decomposition of biodegradable waste in landfills. Landfill gas is a complex mixture of gases, which includes many hazardous air pollutants (HAP). Landfill gas migrates from the subsurface to the atmosphere via diffusion and advection mechanisms through the soil pores, fractures, gaps and defects in the cover materials, or it is collected and discharged via vent systems [2]. The emission may cause hazardous or toxic effects to the nearby personnel or residents. Therefore emission estimating is an important step in conducting risk evaluations, obtaining permits, demonstrating compliance with regulations, and designing emission control systems for landfills.

Air dispersion models are also used by process safety management for emergency planning of accidental chemical releases. Using creditable worst case scenarios and appropriate conservative meteorological conditions, dispersion modeling can provide an estimate for the location of impacted areas, ambient concentrations, and be used to determine the emergency plan in the event of an accidental release.

Significant advances in meteorological modeling research and applications have transpired, such as the development of sophisticated and now widely used advanced air dispersion modeling software. Some of these software are now being applied to air quality modeling applications and quantitative risk analysis. The advancement of computer software and hardware capability permits computational fluid dynamics (CFD) software to be economically used for air dispersion modeling with complex geometry and more complete weather conditions [3].

2. Landfill Gas Emission Study

This case study for landfill gas emission estimates the concentration for an unspecified hazardous air pollutant (HAP) within 1km range of a landfill site. The landfill is approximated by a rectangular area of length 490 m and width 230 m, where the highest concentration inside the landfill is 125 $\mu g/m^3$. The terrain within 1km is primarily a flat plain, with a small hill on the southeast side of the landfill site.

The EPA SCREEN3 program is a single source Gaussian plume model which is a good screening tool that provides maximum ground-level concentrations for point, area, flare, and volume sources. AERMOD is an EPA refined model designed to estimate average pollutant concentrations from multiple sources, at one or more receptor locations over the averaging period(s) of concern using actual hourly averaged meteorological data.

In SCREEN3, the landfill source release height is at the ground level (0 m), within a rectangular area of 490 m by 230 m. To account for surface roughness and its effect on the dissipation of atmospheric momentum to the surface, the dispersion coefficient is selected as rural for this landfill site. The receptor height used is at ground level (0 m). A simple flat terrain feature is selected to best mimic the topography, using the full meteorology category. The emission rate is reverse calculated for a limiting highest concentration of 125 μg/m^3 at the landfill property line. The result from SCREEN3 output is shown in Figure 1. From the output file, the worst case meteorology condition for the case study is wind speed 1 m/s with a Pasquill stability class F. Note that the concentration exponentially decays when it is dispersed away from the landfill, with a concentration at 1 km from the edge of the landfill of 18.27 μg/m^3.

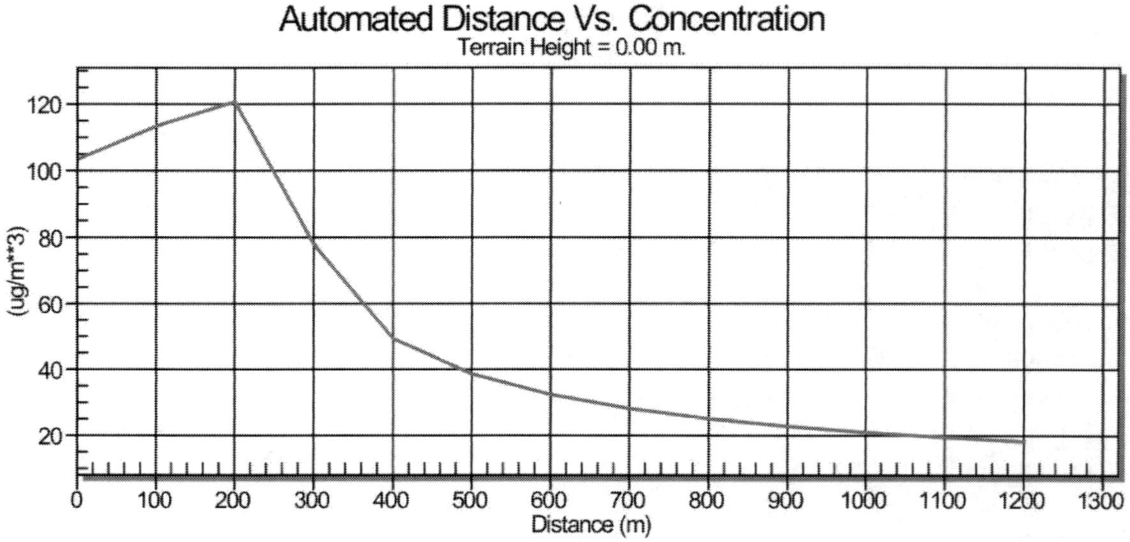

Figure 1 SCREEN3 results – distance vs. concentration

For the refined modeling, AERMOD is used to model the hazardous air pollutant concentration. The location of the same area source is defined. The HAP pollutant traced is 2.5 micron sized particulate material (PM2.5). The emission rate is also calculated so that the maximum concentration at the landfill property line is 125 μg/m^3. Representative yearly meteorology data from 1992 and the appropriate USGS topographic map are used for this study. A uniform receptor grid within 1km range of the landfill is used, with X- and Y-axis spacing of (50, 50) m. The average wind speed is 4.22 m/s based on the meteorological data, which is greater than the wind speed used in the SCREEN3 modeling. The final result for a 1 hour maximum concentration at 1km from the edge of the landfill is determined to be 13.55 μg/m^3. The comparison of calculated results for the two modeling software packages and different averaging time periods are depicted in Table 1. The 3 hours to annual averaging values for SCREEN3 are calculated based on a multiplying factor [4], while for AERMOD the values are read from the receptor output.

It can be seen that for this particular landfill site, the screening model result overestimates the concentration at a factor from 1.35 to 9.13. Based on the comparison, the upshot is that when meteorological and terrain data are available, AERMOD modeling should be performed for more accurate results at the specific landfill location.

Table 1 Concentration Comparison of SCREEN3 and AERMOD model

	SCREEN3	AERMOD	SCREEN3/AERMOD
Averaging Time	Conc. at 1 km from area edge (μg/m^3)	Maximum conc. at 1km from area edge (μg/m^3)	Ratio
1 Hour	18.27	13.55	1.35
3 Hours	16.44	8.26	1.99
8 Hours	12.79	5.39	2.37
24 Hours	7.31	2.24	3.26
Annual	1.46	0.16	9.13

3. Facility Siting Case Study

The facility site of interest is a chemical gas plant with two propane bullets on the west side, office building on the east side, and a storage building in between. There are also some pipeline racks and facility buildings on the north side of the plant. The distance between the propane bullets and the storage building is 10 m, and between the storage building and office building is 30 m. The propane inside the bullets is at temperature 100°F and pressure 250 psi$_g$. The tank has an outer diameter of 3 m, inner diameter of 2.97 m, length of 16 m, and the height of the bottom of tank is 0.5 m from the ground. The general plot plan is shown in Figure 2.

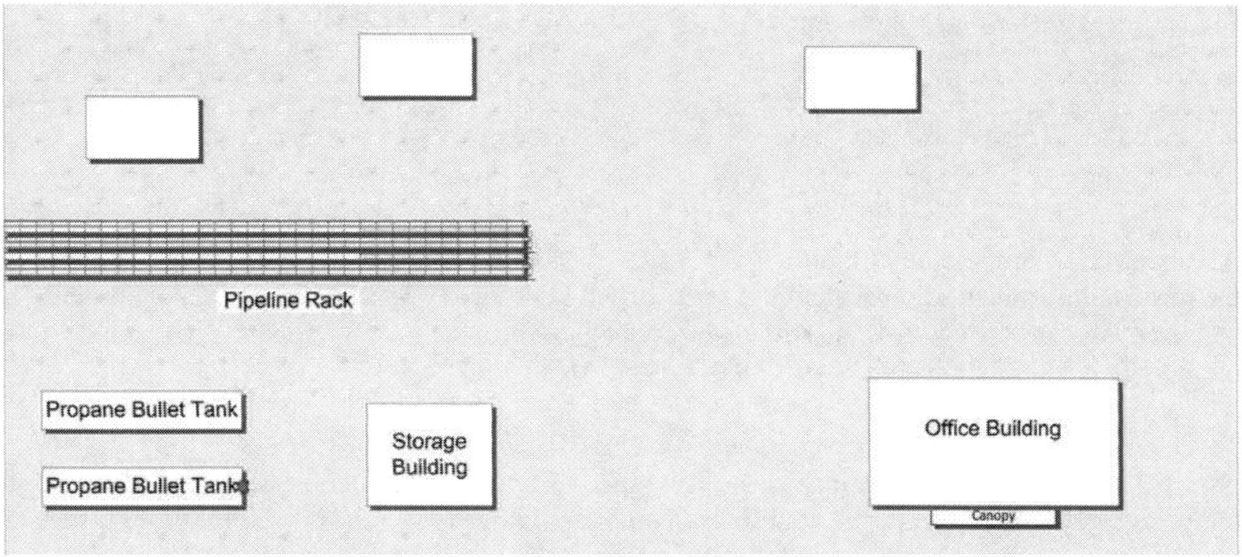

Figure 2 Chemical gas plant plot plan (red star is the release point)

The hazardous situation for the case study is an accidental release from the top of the bullet tank (shown as a red star in the plot plan) on the south side, producing a vapor plume that will expand and propagate. If an ignition source is present, a fire and/or explosion may occur which may affect the office building and its occupants. A hole diameter of 2 in (50 mm) and leak duration of 2 minutes before the blowdown system and emergency shutdown system start to response was assumed. A wind speed of 1 m/s, directed toward the office building with a Pasquill stability class of F will produce the maximum downwind plume size associated with a worst case scenario.

The dispersion of propane due to the leak on the tank is first calculated using DNV PHAST Software. It is modeled as a pressurized vessel with contents at a temperature 100°F and pressure 250 psi_g. The dispersion result at a height of 1 m is shown in Figure 3. The lower flammable limit (LFL) for propane is 0.02 (concentration by volume). It is seen that the building will be engulfed by the plume at 1/2 LFL within 25 seconds, while the south part of the building will be covered within an LFL concentration.

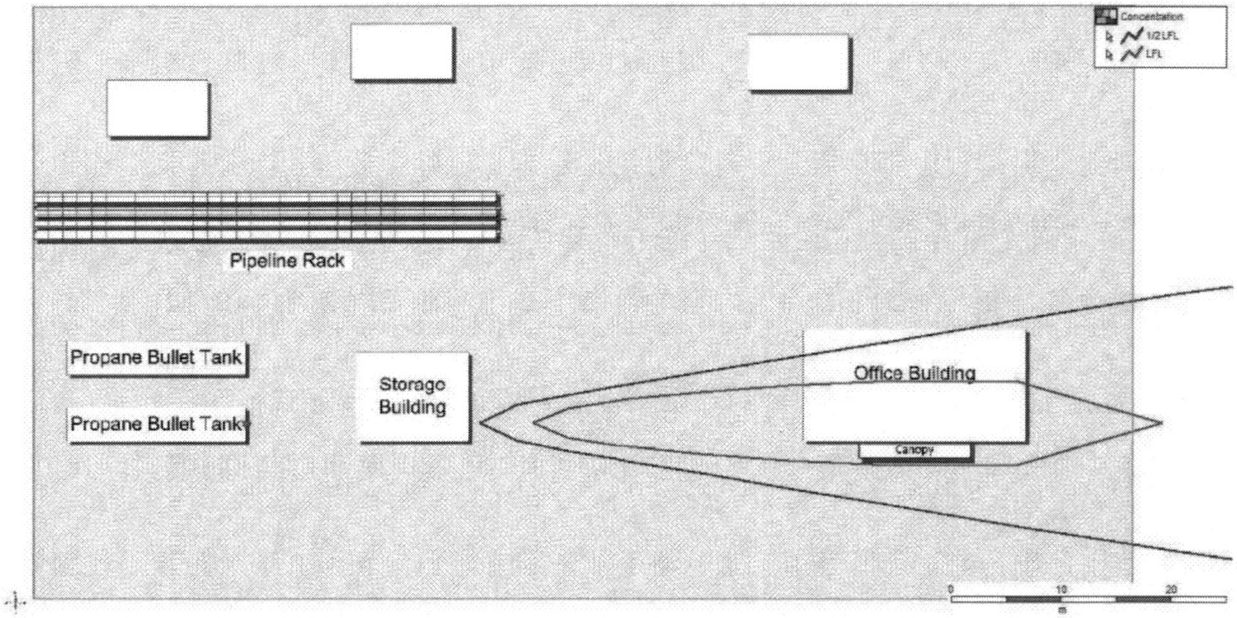

Figure 3 PHAST result of concentration for propane release from bullet tank

Alternatively, Computational Fluid Dynamics (CFD) simulation software is utilized to determine the maximum concentration around the same office building. Here we used two CFD simulation software, the explosion and dispersion modeling software FLACS from GexCon and flow modeling and simulation software FLUENT from ANSYS.

To calculate the release source characteristics, the flash program in FLACS [5] computes the physical properties of flashing releases of pressurized liquefied propane. The given orifice area, temperature and pressure of the liquid, are combined with the calculated the mass flow rate of released gas of approximately 50 kg/s. The released gas is mixed with air at the far field of the release as a jet leak for the calculated jet area of 2.77 m^2, mass flow rate of 174.4 kg/s and jet velocity of 32.3 m/s.

The result of maximum concentration around the office building at a height of 1.3 m is shown in Figure 4, at the release end time of 120 seconds. It is seen that the concentration at the west and south part of the building is above 0.05, while the concentration at the north and east part of the building is above 0.02 (the LFL).

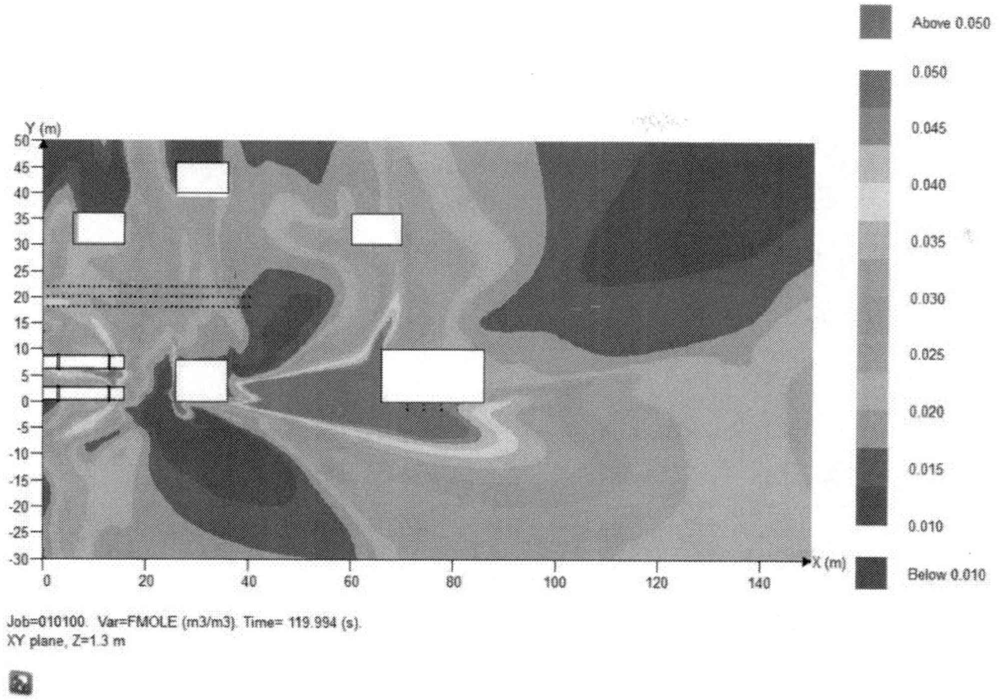

Figure 4 FLACS top view of propane concentration at Z=1.3 m

To verify the results from FLACS, the same input parameters were used to run the 3D air dispersion simulation in Fluent. The Fluent result is found to be very close to the FLACS result. The side view of the propane mass concentration profile is shown in Figure 5. It is seen that the propane concentration at west side of the office building just exceeds 0.045, while on top and east side of the building is above 0.03.

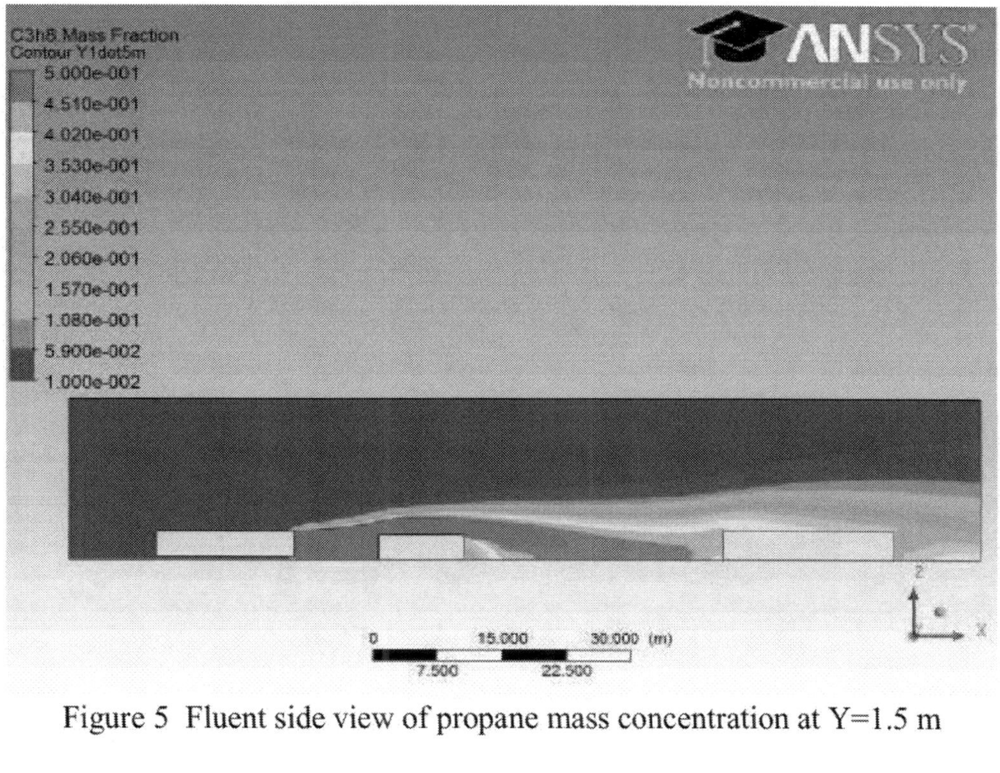

Figure 5 Fluent side view of propane mass concentration at Y=1.5 m

The results show that for a 2-minute release of propane from the pressurized tank, the maximum concentration around the office building will be above LFL, resulting in a fire and explosion hazard. When gas is detected, the HVAC systems should be shut-down to minimize any hydrocarbon intake and personnel inside the building should follow evacuation/escape routes as in the emergency response plan. The system should be shut-down as soon as possible as this vapor cloud could impact the office building in less than 30 seconds of the release initiation.

4. Conclusion

In this paper two type of air dispersion modeling related to process safety management are demonstrated. The first type of modeling is for landfill emission gas analysis. A hazardous air pollutant dispersed from a landfill site within 1km distance is modeled; the results from the screening software SCREEN3 and refined modeling software AERMOD are compared. It is shown that the screening results are more conservative as it uses the worst case weather conditions for the dispersion. If the terrain feature and meteorological data of the landfill site is available, the refined model will provide more accurate results.

The second type of modeling is an example used for a facility siting study. A propane bullet tank release in a chemical gas plant is studied using the CFD software FLACS and Fluent. CFD results show that building siting evaluation is important for the safety operation of the plant and safety of the onsite personnel. More detailed facility siting analysis such as CFD simulation studies can take into account the actual geometry within a facility and provide more accurate results about the dynamic dispersion behavior of the hazard or toxic vapors.

5. References

[1] Barratt, Rod (2001). Atmospheric dispersion modeling: an introduction to practical applications. Business and the Environment: Practitioner Series. London, UK: Earthscan.

[2] US Environmental Protection Agency. Guidance for Evaluating Landfill Gas Emissions from Closed or Abandoned Facilities (EPA/600/R-05/123a). September 2005.

[3] Irfan Shaikh, Morgan Reed and Eric Peterson. Use of CFD in Onshore Facility Explosion Siting Studies, 2012 International Symposium Proceedings, Mary Kay O'Connor Process Center, College Station, TX, Oct 23-25, 2012.

[4] US Environmental Protection Agency. Screening Procedures for Estimating the Air Quality Impact of Stationary Sources (EPA-454/R-92-019). October 1992.

[5] FLACS User Manual, FLACS Version 9.1.4. Gexcon AS, 2012.

The Simulation of The Combustion Reaction of Grease with Oxygen at High Pressure

Xu Ji
College of Urban Construction and Safety Engineering, Nanjing University of Technology
Xin Mofan Road 5#, Nanjing, China
jx_jixu@163.com

Dong Li
Sinopec Nanjing Engineering Construction Inc.
Ke jian road 1189#, Nanjing, China
361906915@qq.com

Lijing Zhang
Gang Tao
College of Urban Construction and Safety Engineering, Nanjing University of Technology
Xin Mofan Road 5#, Nanjing, China
zhanglj@njut.edu.cn
taogangs@163.com

Prepared for Presentation at
American Institute of Chemical Engineers
1st CCPS Asia-Pacific Conference on Process Safety
Qingdao, China
September 4-5, 2013

The Simulation of The Combustion Reaction of Grease with Oxygen at High Pressure

Xu Ji
College of Urban Construction and Safety Engineering, Nanjing University of Technology
Xin Mofan Road 5#, Nanjing, China
jx_jixu@163.com

Dong Li
Sinopec Nanjing Engineering Construction Inc.

Lijing Zhang
Gang Tao*
College of Urban Construction and Safety Engineering, Nanjing University of Technology

Keywords: Combustion; Cylinder; Grease; Simulation

Abstract

The grease in the cylinder could be oxidized, which can elevate the temperature. The high temperature could decrease the ignition point of grease, and then the combustion and explosion will occur. This paper presents the mechanism of the combustion reaction of grease with oxygen at high pressure. Fluent software is used in this paper to simulate the combustion reaction of grease with high-pressure oxygen. The simulation shows that grease is oxidized intensely by pure high-pressure oxygen in cylinder and the heat generates by combustion lead to higher pressure and temperature.

1. Introduction

Oxygen cylinder containing grease without stringent security checks before filling may result in cylinder explosion. Most oxygen cylinder explosion accidents are due to the grease enclosed in the cylinder. Grease filled into oxygen cylinder has following ways. Firstly, when replace or repair the parts of cylinder valve, the grease may enter into the cylinder. Secondly, other cylinders are used as oxygen cylinder, but not thoroughly cleaned with degreasing agent. Thirdly, cylinder without enough gas may cause grease fill into the cylinder. Fourthly, cylinder or cylinder valve contains grease.[1]

This paper presents the mechanism of the combustion reaction of grease with oxygen at high pressure. Fluent software is used to simulate the combustion reaction of grease with high-pressure oxygen.

2. The oxidation reaction mechanism of lubricant with oxygen at high pressure

Grease, especially those containing unsaturated fatty acids, can easily be oxidized and then release heat. Since pure oxygen has strong oxidation, it can promote combustible materials burning drastically. When grease is filled into oxygen cylinder by mistake, it would react with high-pressure oxygen and release heat. The reaction will instantaneously produce both high pressure and high temperature beyond the limit that the material can bear, and then the cylinder will burst. Briefly speaking, drastic combustion reaction of grease with high-pressure oxygen contributes to the explosion accident.[2]

Analyzing the oxygen cylinder accident in the filling process, the grease mixed into the oxygen cylinder in the filling process is almost lubricants. Therefore, lubricant is selected for the study.

The grease oxidation strength is proportional to concentration of oxygen and its oxidation rate is related to the temperature. When the temperature reaches about 60℃, the carbon precipitates lubricant has significantly oxidation. When the temperature is around 100~150℃, the oxidation rate is increased dramatically. Since the reaction is exothermic, it becomes the factor of elevated temperature. The existence of grease and the neutral tar also enhance the oxidation process. As the oxidation process of lubricant more in-depth, the presence of precipitate is more dangerous. Under atmospheric pressure, the auto ignition temperature scope of precipitate starts at 250 ℃~285 ℃.If the pressure in the compressed air system increases, the area will move to the lower temperature area. The possibility of spontaneous combustion may increase. When the increased partial pressure of the oxygen arrives at some points, the grease will self-ignite.[3]

Ignition energy of combustibles is inversely proportional to the oxygen pressure when in the high-pressure pure oxygen. Therefore, one of the main methods to reduce the risk of oxygen facility fire and explosion is to decrease the ignition point of combustible.[4]

- When the pressure is higher than 2.94MPa, the reaction of grease with oxygen is drastically. The grease would immediately self-ignite.[5]
- The ignition temperature of combustible decreases as the concentration and pressure of oxygen increases.

3. The simulation of grease combustion heat transfer in high-pressure oxygen

Based on the reaction mechanism of the lubricating oil with high-pressure oxygen, Fluent software is used to simulate grease combustion heat transfer in high-pressure oxygen. The Standard model and the Discrete-ordinates model (DO) are used in the simulation. The Species Transport model which based on the component transport equation is also established to solve reaction rate.[6]

Gambit is used to set grease block.[7,8]Considering the oxygen cylinder is symmetrical about the cylinder central axis, two-dimensional model is used to set the oxygen cylinder as rectangle. The rectangular is 1261mm in length and 202.8mm in width.[9] Diesel oil is selected as fuel in the simulation. In the simulation, the group of diesel oil is 15% -30% aromatic hydrocarbon and 70% -85% saturated hydrocarbons and olefin. It also contains a small amount of organic sulfur compounds and organic nitrogen compounds. Its liquids density is 0.9kg/m^3.The burning rate is 55.11kg/ (m^2 • h), heat of combustion is 42MJ/kg.

Making assumptions as follows:

- The grease completely attached to the bottom of the oxygen cylinder. The grease is at the center of bottom of the cylinder.
- The grease diameter is half of the bottom diameter, 101.4mm. The grease diameter is completely at the center line.
- Grease is cylindrical in the bottom of oxygen cylinder. The base diameter and height of the cylinder is 101.4mm and 50mm.

The rectangular is divided into two surfaces and it generated 30866 grids.

The Gambit Geometric model is shown in Figure 3-1.

Fig. 3-1 Gambit Geometric model

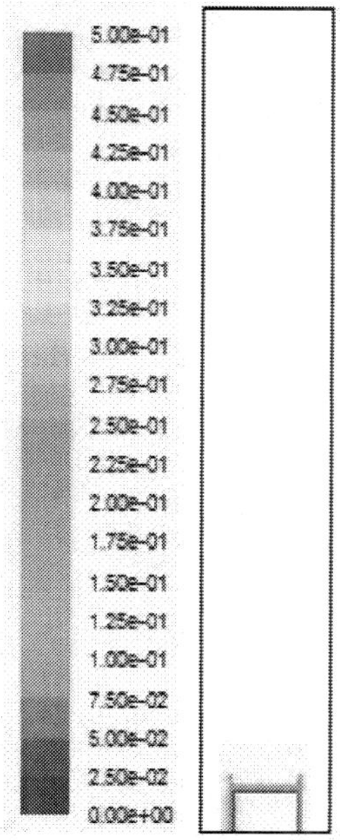

Time Step Size is set to 0.001s. Number of Time Steps is set to 2000. Max Iterations per Time Step is set to 100. File-Write-Autosave Data File Frequency is set to 1. It means that it will iterative 2×10^5 times to simulate the condition of grease burning in high-pressure oxygen. Simulation data is saved every 0.001s.

3.1 Grease mass fraction contour lines

Take 0.1s as the time interval. Grease mass fraction contour lines of each time period are shown in the Figure 3-2, Figure 3-3, Figure 3-4, Figure 3-5, Figure 3-6, Figure 3-7, Figure 3-8, Figure 3-9 .

Fig. 3-2 0.001s Grease mass fraction contour lines

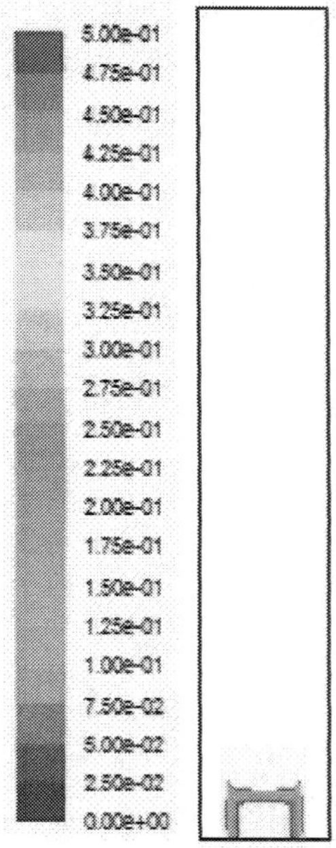

Fig. 3-3 0.1s Grease mass fraction contour lines

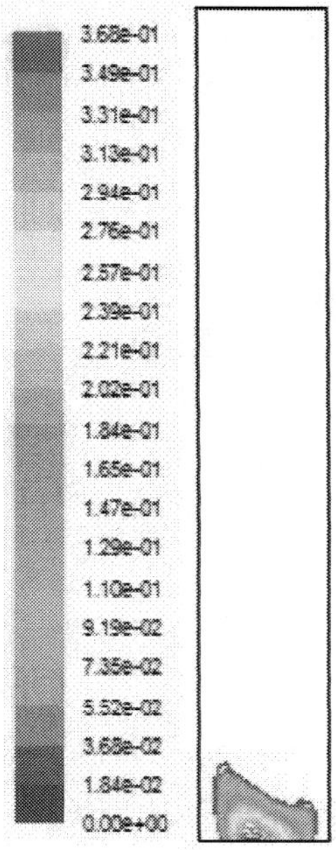

Fig. 3-4 0.2s Grease mass fraction contour lines

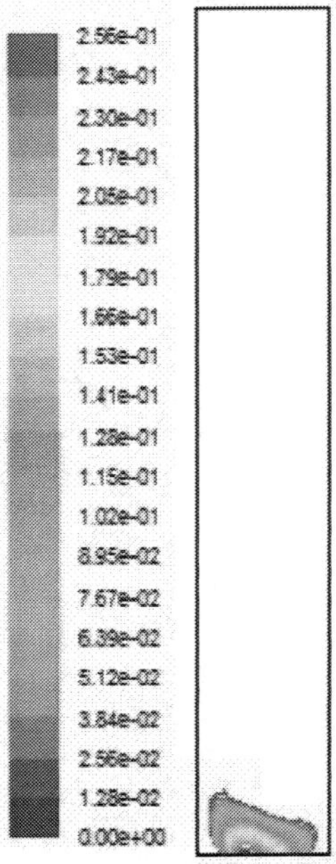

Fig. 3-5 0.3s Grease mass fraction contour lines

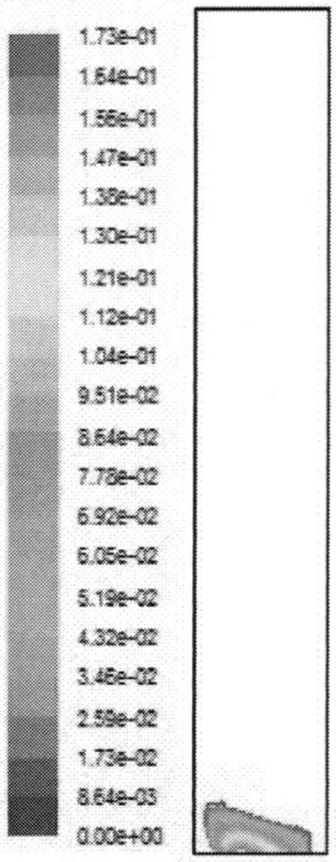

Fig. 3-6 0.4s Grease mass fraction contour lines

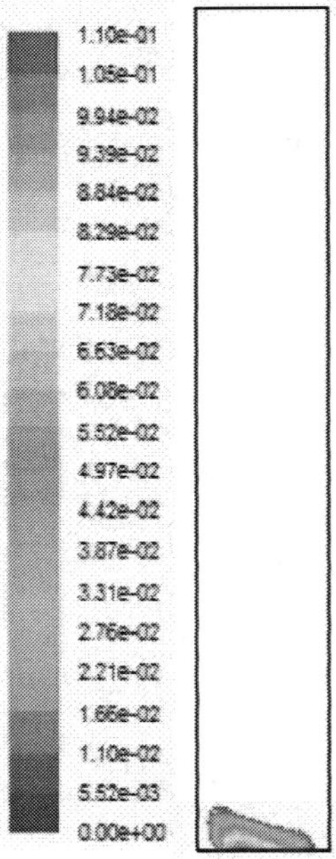

Fig. 3-7 0.5s Grease mass fraction contour lines

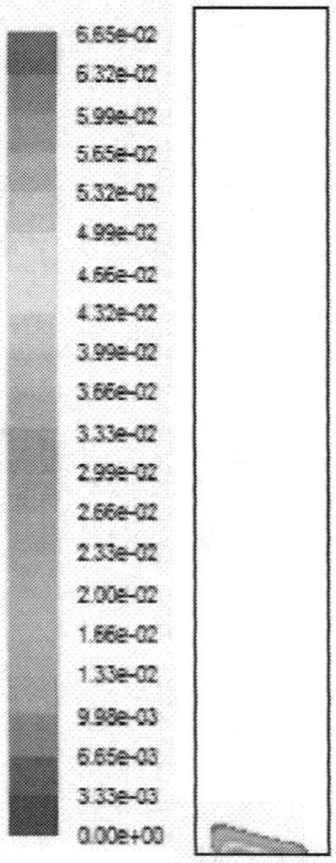

Fig. 3-8 0.6s Grease mass fraction contour lines

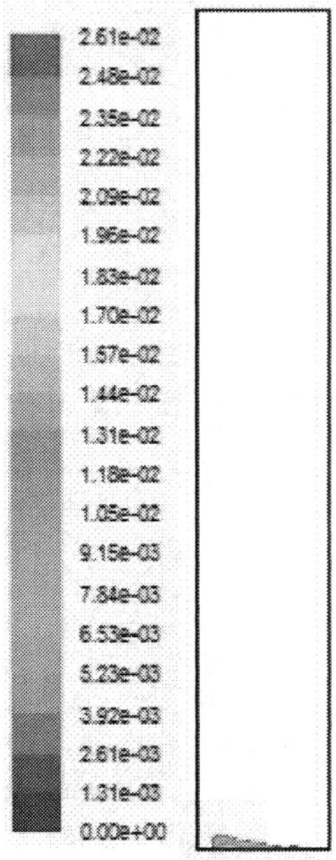

Fig. 3-9 0.65s Grease mass fraction contour lines

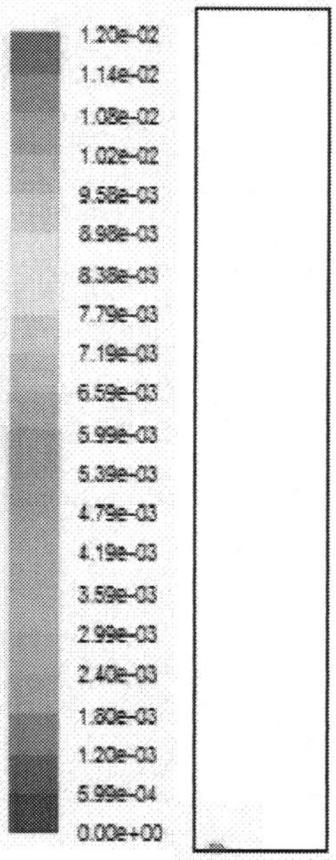

A series of grease mass fraction contour lines can be seen that the grease produced a layer of vapor on its surface before combustion. The fuel vapor is mixed with oxygen and heated to ignition, and then combustion flame is formed. In this process, grease reacts with oxygen rapidly. After 0.65s, grease involved in the reaction is almost completely exhausted.

3.2 Temperature contour lines

Take 0.1s as the time interval. Temperature contour lines of each time period are shown in the Figure 3-10, Figure 3-11, Figure 3-12, Figure 3-13, Figure 3-14 Figure 3-15, Figure 3-16, Figure 3-17. The unit is K.

Fig. 3-10 0.001s temperature contour lines

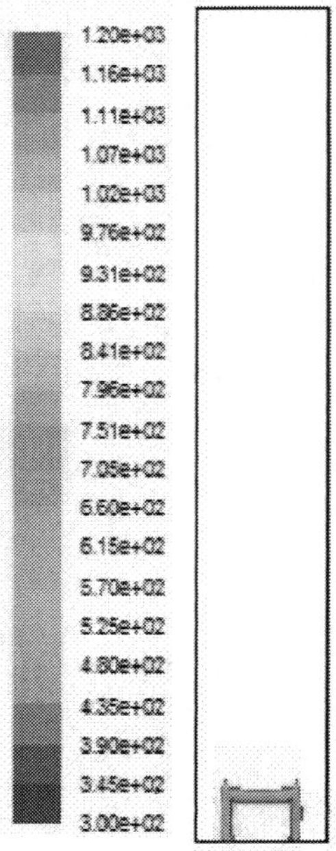

Fig. 3-11 0.1s temperature contour lines

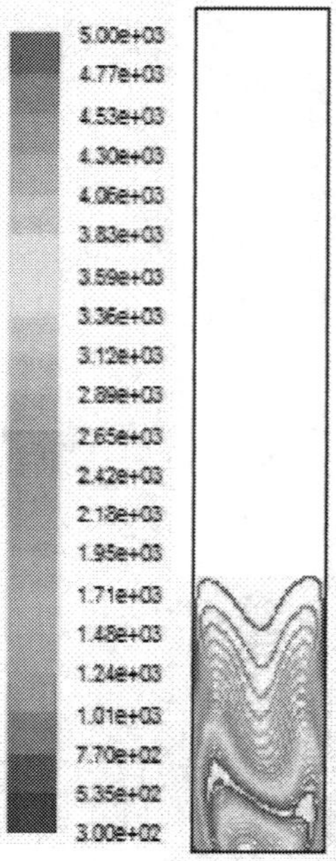

Fig. 3-12 0.2s temperature contour lines

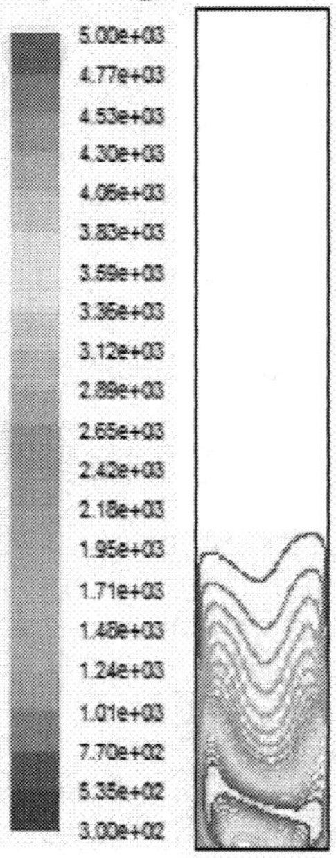

Fig. 3-13 0.3s temperature contour lines

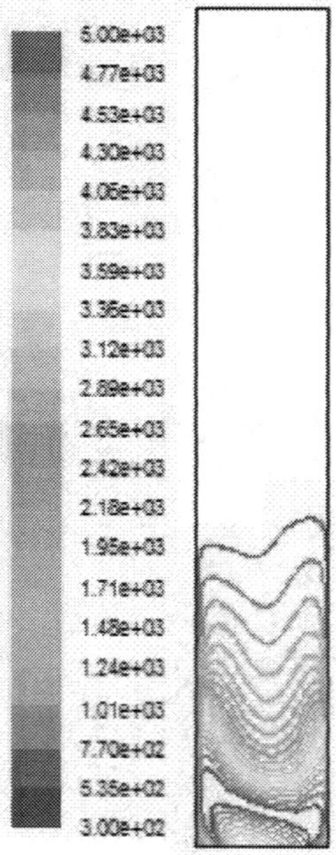

Fig. 3-14 0.4s temperature contour lines

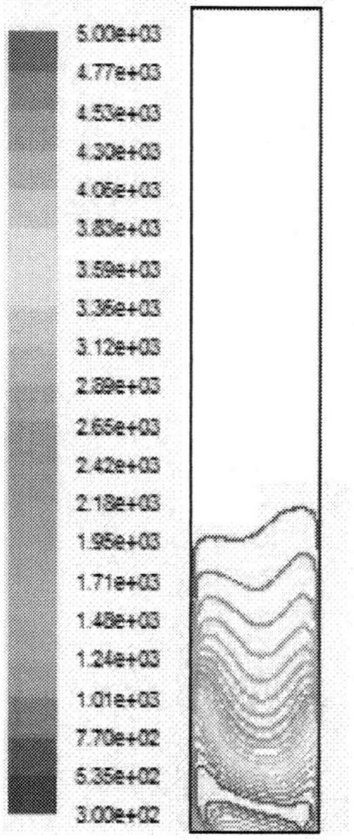

Fig. 3-15 0.5s temperature contour lines

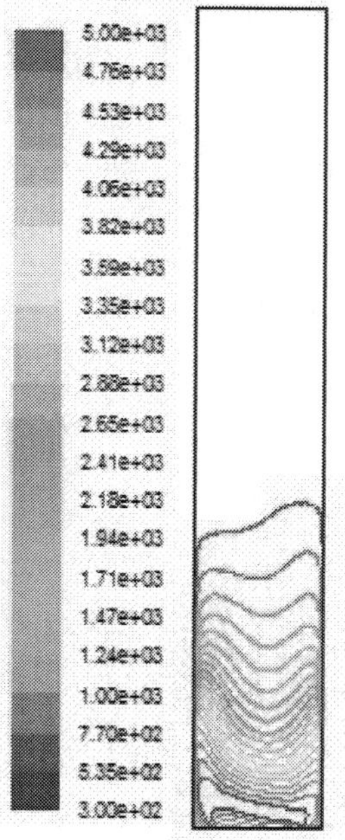

Fig. 3-16 0.6s temperature contour lines

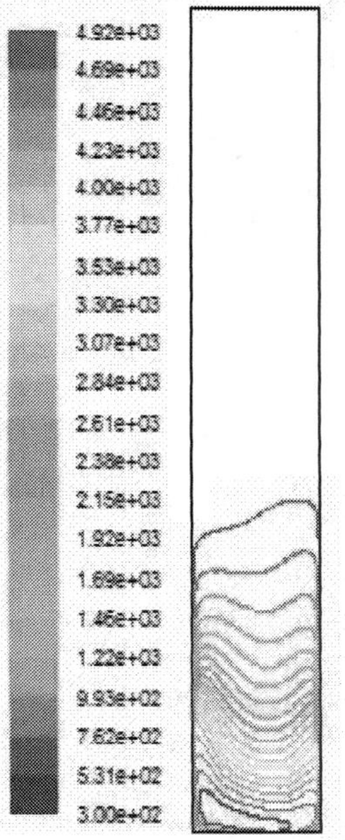

Fig. 3-17 0.65s temperature contour lines

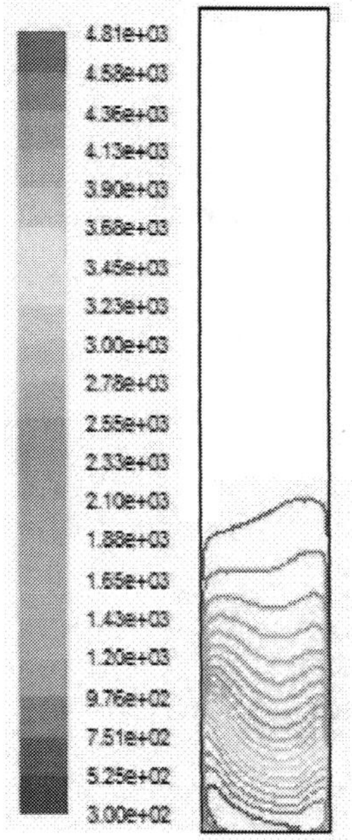

A series of temperature contour lines can be seen that grease reacts with high-pressure oxygen drastically. The highest temperature of grease is 5000K because of the adiabatic environment in simulation. At 0.2s, 0.3s, 0.4s, the reaction keeps approximate steady-state heat transfer. After 0.65s grease reaction completely. Then the temperature of combustion decreased slightly.

3.3 Pressure contour lines

Take 0.1s as the time interval. Pressure contour lines of each time period are shown in the Figure 3-18, Figure 3-19, Figure 3-20, Figure 3-21, Figure 3-22 Figure 3-23, Figure 3-24, Figure 3-25. The unit is pa.

Fig. 3-18 0.001s pressure contour lines

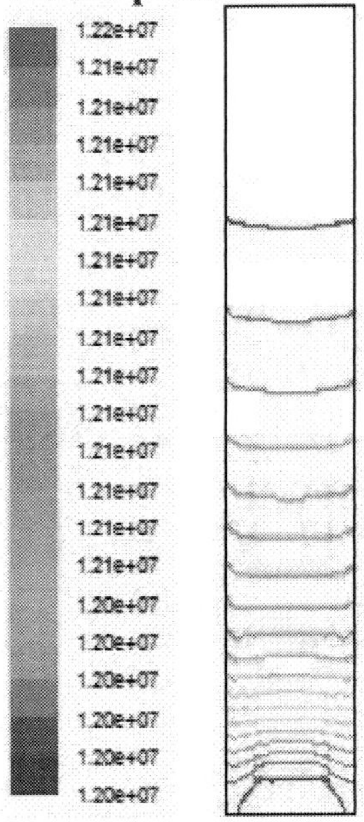

Fig. 3-19 0.1s pressure contour lines

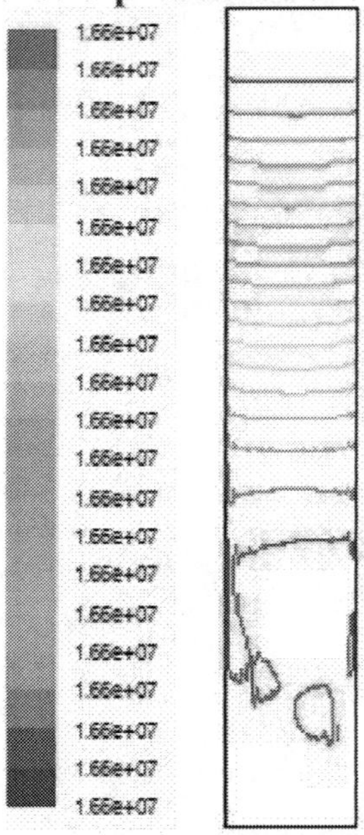

Fig. 3-20 0.2s pressure contour lines

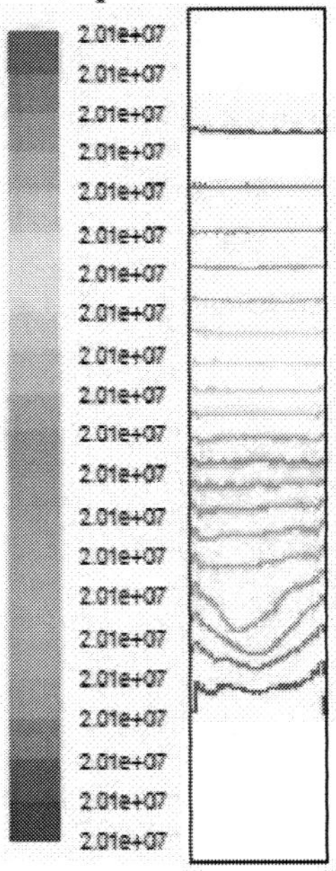

Fig. 3-21 0.3s pressure contour lines

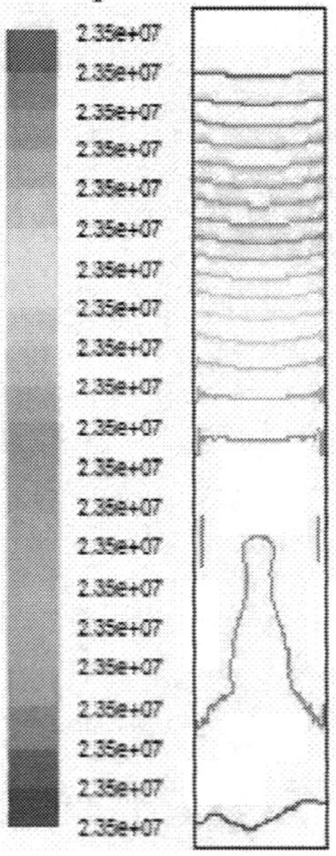

Fig. 3-22 0.4s pressure contour lines

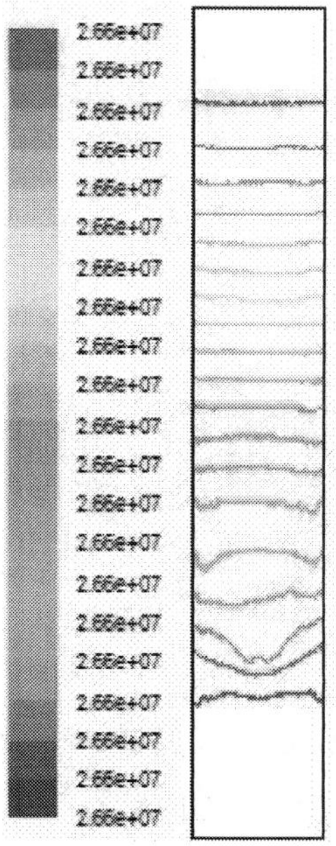

Fig. 3-23 0.5s pressure contour lines

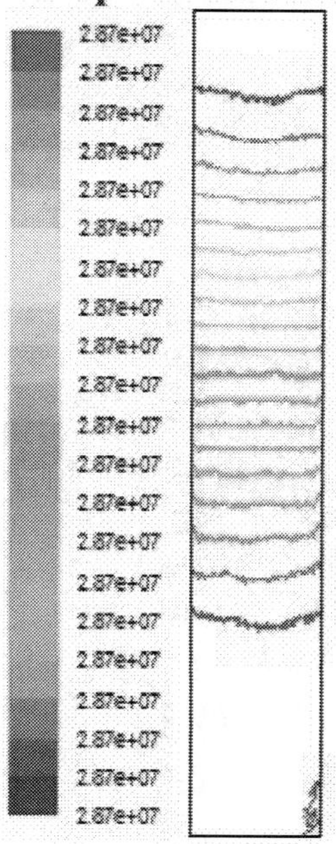

Fig. 3-24 0.6s pressure contour lines

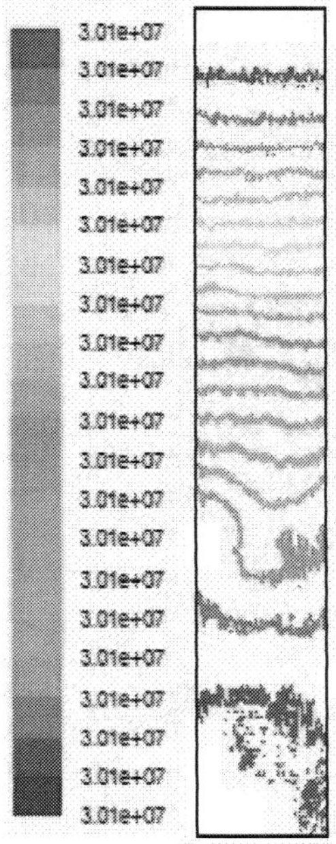

Fig. 3-25 0.65s pressure contour lines

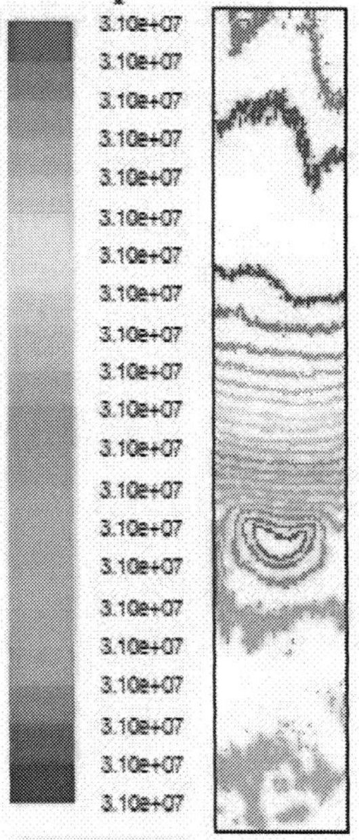

The above contour lines can be seen that at the initial stage of the combustion, heat generated by combustion transferred slowly. The pressures applied on the cylinder are the same at any moment. For gas, the convection heat transfer coefficient increases with the increase of pressure.[10] At the same time, heat transfer will increase the oxygen pressure in the cylinder. The pressure of cylinder increases to 31MPa in a short time within 0.65s. When the grease reacts completely, the pressure growth rate decreases.

The simulation shows that grease is oxidized intensely by pure high-pressure oxygen in cylinder. And the heat generates by combustion leads to higher pressure and temperature.

In detail, the grease produces a layer of vapor on its surface before combustion. Then the grease is ignited and flame is formed. Then the process is approaching the steady-state heat transfer. When the reaction finished, the temperature of combustion decreased slightly. At the initial stage of the combustion, heat generated by combustion transferred slowly.

4. Conclusion

- The grease reacts with oxygen at high pressure will release heat energy, which can elevate the temperature of the cylinder. At the same time, the increased pressure and temperature in the cylinder will decrease the ignition point of grease. When the pressure of the cylinder increases at some points that may make the grease self-ignited, the grease will self-ignite. And then the grease will instantaneously produces both high pressure and high temperature beyond the limit that the cylinder can bear, and then the rupture of cylinder will occur.
- Fluent software is used to simulate the combustion reaction of grease with high-pressure oxygen. The simulation shows that heat generated by combustion transferred slowly at the initial stage of the combustion.
- The results of simulation show the pressure is less than the pressure limit of cylinders. There are two reasons: firstly, the combustion process which simulated is heat transfer process, but the actual situation is combustion and explosion. The actual process of reaction is more rapidly. Secondly, grease simulated in this paper is at the bottom of the cylinder. But in practice, the grease attached to the inner wall of the cylinder and the bottom of the cylinder. Therefore, the grease exposes to high pressure oxygen with the largest surface area. The reaction accumulates large amounts of heat in a short time and causes pressure to a higher value.

5. References

[1]Zhang Liang, Tao Gang, Zhang Lijing, et.al., "Reaction of Oil with High Pressure Oxygen Measured by PTG," Journal of Safety Science and Technology, Volume 7, NO. 1, 2011

[2]Han Suqin, "Why Oxygen Cylinder Can Not Containing Grease," Safety, Volume 25, NO. 2, 2004

[3]Je-Lueng Shie, Jyh-Ping Lin, Ching-Yuan Chang, et.al., "Oxidative Thermal Treatment of Oil Sludge at Low Heating Rates," Energy & Fuels, Volume 18, NO. 5, 2004

[4]Zhang Feng, Li Wei, Shi Peiyao, "Risk Analysis and Preventive Measures of Burning Explosion in Air Separation Unit and Relevant Facilities," Safety Technology, Volume 1, 2008

[5]Zhuang Shengqiang, "Analysis of Oxygen Cylinder Explosion Accident of Jiangyin Entey-Exit Inspection and Quarantine Bureau," Cryogenic Technology, NO. 1, 1996

[6]Fan Weicheng, Wan Yuepeng, "The Model and Calculation of Flow and combustion," Press of University of Science and Technology of China, Hefei, 1992.

[7]Han Zhanzhong, Wang Jing, Lan Xiaoping, "The Simulation Examples and Application of Fluent Fluid Engineering," Beijing Institute of Technology Press, Beijing, 2005

[8]Lv Xinrong, Li Qing, "Calculation of Flamelet Structure of Turbulent Combustion in Diesel Engine", Journal of Dalian Maritime University, Volume 31, NO. 2, 2005

[9]Liu Mingan, Pan Keyu, Chen Shuo, "New Model for Computing Flame Radiation Heat Transfer in Cylinder of Diesel Engin," Journal of Combustion Science and Technology, Volume 6, NO. 1, 2000

[10]Gu Shouyan, Zhang Luwei, An Zhiqiang, "Effect of Pressure on Air Forced Convection Transmit Heat Coefficient," Liaoning Chemical Industry. Volume 38, NO. 10, 2009

The efficiency study of using different extinguishers on Trichlorosilane fires

Liao Kuangyu
+886-5-5342601-4450
No.123, Sec. 3, Daxue Rd., Douliu City, Yunlin County 640, Taiwan (R.O.C.)
liaoky@gmail.com

Horng Jaojia, Liou Kehun, Liao Yuanchun, Chen Jianhong
+886-5-5342601-4450
No.123, Sec. 3, Daxue Rd., Douliu City, Yunlin County 640, Taiwan (R.O.C.)
horngjj@gmail.com, soul0801@gmail.com, steveycliao@gmail.com, cch.hugo@gmail.com

Prepared for Presentation at
American Institute of Chemical Engineers
1st CCPS Asia-Pacific Conference on Process Safety
Qingdao, China
September 4 – 5, 2013

The efficiency study of using different extinguishers on Trichlorosilane fires

Liao Kuangyu
+886-5-5342601-4450
No.123, Sec. 3, Daxue Rd., Douliu City, Yunlin County 640, Taiwan (R.O.C.)
liaoky@gmail.com

Horng Jaojia, Liou Kehun, Liao Yuanchun, Chen Jianhong
+886-5-5342601-4450

No.123, Sec. 3, Daxue Rd., Douliu City, Yunlin County 640, Taiwan (R.O.C.)
horngjj@gmail.com, soul0801@gmail.com, steveycliao@gmail.com, cch.hugo@gmail.com

Keywords: Trichlorosilane (TCS), water-reactive, extinguishers

Abstract

In recent years with rapid development of petro-chemical and high-tech technology, a variety of chemicals have proliferated all over the industries. The unique properties of chemicals not only improved the product yields and productivity but also brought the disastrous consequences to the extremely, complex, and high output manufacture sites.

This study was investigating TCS (Trichlorosilane, $SiHCl_3$) which abundantly used in semiconductor and solar energy industries as the principal resource to generate ultrapure silicon. The flammability property of TCS has caused numerous fires, such as the TCS truck overturned in Chengdu, China; solar manufacturer TCS leakage in Pingtung, Taiwan (on 2011). Many concerns aroused among the relevant agencies on how to extinguish TCS fires and reduce related hazards.

TCS ignites in air and reacts with moisture to generate silica oxides, hydrogen gas, chlorine and hydrochloric acid in vapor or droplets. Therefore, those hazardous, flammable, and corrosive gases may cause severe respiratory tract and health injury. The study investigated the effectiveness of the different fire extinguisher agents (dry powder, CO_2, Halon, water, and foams) on putting out fires involving a small leakage of 3 kg TCS. Results showed that agents of dry powder, CO_2 and Halon could not put out the fire but even caused other hazards. Water and foams could accelerate the degradation of TCS, however thick foams could put out fires successfully as some emergency response guides suggested. Thus, fundamental researches on TCS burning characteristics should be studied through appropriate field extinguishing tests in order to establish effective measures.

1. Introduction

Trichlrosoilane (TCS) was used abundantly in petrochemical industry in the past. TCS has become one of the important raw materials in the semiconductor, solar panel, opt-electronic manufacturers and is also the principal source of ultrapure silicon. TCS was categorized as Hazard Class 4.3 with subsidiary class 3 and 8 where NFPA 3-4-2-W [1]. It was dangerous when wet. However, several incidents were caused by its hazard properties. TCS and Alkyl Silicon Chlorides are all acutely toxic, water-reactive and corrosive substances, and when reacting with air releases several dangerous gases and byproducts [2]. TCS could also react with moisture or water, a dangerous hydrogen gas and hydrochloride acid [3]. Those hazardous, flammable, and corrosive gases may cause severe respiratory tract injury and health problems. Evacuation is mandatory not only for manufacture plants but also nearby communities in massive TCS leakage incidents. However, proper emergency response measures need to be implemented to reduce or eliminate the hazards of TCS exposures. In the fires, TCS might form chlorine, hydrochloride acid, and silica oxides in vapor, small particles and droplets. To stop the TCS degradation, and/or to segregate the dangerous products are important response measures in those leakage or fire incidents. This study was to find the better extinguisher agent(s) and/or strategy to achieve those goals.

2. Literature Review

The TCS dangerous properties were assembled and analyzed. Several special properties of TCS degradation with water or under fire were then discussed. Several incidents were discussed with the fundamental knowledge and response measures. Then, those measures were linked to the types of fire extinguisher. Finally, several popular guides and instructions on responding to TCS fires were collected and analyzed.

2.1 TCS Physical and Chemical Properties

TCS ($SiHCl_3$) is a colorless liquid with volatility, with irritating acid odor in gas form (Boiling point at 32℃ and Flash point at -27.8℃). Flammable and water-reactive TCS reacts vigorously with vapor in the air. Reacting with water forms corrosive gas (such as hydrochloric acid) and hydrogen, in another hand, produces massive hydrogen chloride gas and chlorine upon burning.

Table 1. TCS Physical and Chemical Properties [4][5]

Boiling point	32℃	Melting point	-128.2℃
Autoignition temperature	104℃	Flash point	-27.8℃
Flammable Limits in Air	1.2 – 90.3 %	Specific Gravity	4.7(air=1)
Specific Gravity	1.33(H_2O=1)	Vapor Pressure	500mmHg@20℃

2.2 TCS Hydrolysis and Burning Equation

1. TCS reacting with moisture/water

When exposure to the air, the corrosive TCS reacts with vapor in the air and the equation as (1):

$SiHCl_3+2H_2O \rightarrow SiO_2+3HCl+H_2$................................ [Eq. 1]

Knowing from equation (1), TCS reacts with vapor and forms derivatives such as Silica (SiO_2), hydrogen (H_2), hydrochloric acid (HCl) those are hazardous to personnel and equipment. TCS reacts violently with water then be reduced by hydrogen, the reduction equation as (2):

$3SiHCl_3+3H_2O \rightarrow H_2SiO_3+H_2+HCl$.......................... [Eq. 2]

The reduction reaction, equation (2), forms derivatives such as dehydrate Silicic Acid (H_2SiO_3), hydrogen (H_2), hydrochloric acid (HCl). Explosion may occur as hydrogen gas mixing with oxygen and ignition source(s).

2. TCS reacting with air or oxygen while igniting:

$SiHCl_3+O_2 \rightarrow SiO_2+HCl+Cl_2$.................................... [Eq. 3]

$5SiHCl_3+6O_2 \rightarrow 5SiO_2+HCl+7Cl_2+2H_2O$................. ..... [Eq. 4]

Learning from the above two equations that the derivatives may contain Silica (SiO_2), hydrochloric acid (HCl) and chlorine (Cl_2) after burning or oxidizing reactions [3].

2.3 TCS Major Accidents in the past

TCS major accidents occurred worldwide as table 2, which showed that TCS leakage or ignition will release massive acidic substances that are harmful to personnel, working staff and residents nearby. The emergency response to TCS leakage or fire should be defensive such as suppressing the burning, covering the surface and reapplying water and foam extinguisher agents.

Table 2. Major Accidents [6][7] [8] [9][10]

Time	Description	Reaction	Casualties
April 1974.	Calumet Harbor, Chicago, USA A Silicon Tetrachloride leakage occurred between the connections of storage tank and the flexible hose formed massive acid mist in volume of 1600m lengths, 400m widths and 450m heights.	Emergency response Personnel took wrong response by covering the runoff liquid in the overflow protection dike with fuel and covering the leak with sand and accelerated cement. The incident was brought under control after two days and the entire tank of liquid was transferred.	Over 14,000 people were evacuated.

July 1989.	Freeland City, Michigan, USA The freight train derailed accident caused TCS leakage and was caught on fire.	Extinguished the fire with Sodium carbonate after burning for 3 days but accelerated the fire. Fire lasted for 6 days and 2000gallons of liquid was isolated.	Nearly 1,000 people were evacuated, 11 people injured.
June 1994.	Gainesville City, Florida, USA. A result of the transfer hose fractured on the trailer, TCS flowed back and spilled.	The spill caught on fire by extinguished it with water and fire lasted for 3 hours. The acid gas clouds spread for 3miles and aqueous foam was applied to reduce the amount of acid gas produced by TCS hydrolysis reaction.	No injury.
August 1997.	Midland City, Michigan, USA. TCS leakage caused by the corroded overhead pipeline.	Fire department was called to deal with the incident.	No injury.
October, 1998	Morse Lake, Washington DC, USA. Fractured pipeline caused around 34,000 pounds of high-temperature, high-pressure silicon tetrachloride and TCS mixtures leakage.	Fire department was called to deal with the incident.	Two dead.
December 2008.	Hubei province, China. TCS leakage at Yi-Chang glass materials factory during equipment cleaning process.	Fire department was called to deal with the incident.	No injury.
January, 2011	Taiwan, one solar panel company. During production process, the fluidized bed reactor was overheated and the flang gaskets were over loaded then caused the fire involving TCS.	Applied dry powder extinguisher for cooling down and suppressing the oxidizing reaction.	No injury.

2.4 Main Composition of Conventional Fire Extinguishers

Since TCS reacts violently with water and produces hydrogen gas, an idealized extinguishing should includes two inhibition mechanisms of avoiding recombination of radicals and HO_2 destruction as Yong et al suggested in their study of the chemistry of hydrogen flame extinguishment.[11] The types of fire extinguishers that commonly find as shown in chart 3. Each class of fire extinguisher usually designates the fuel types involve in the fire for certain classes. Ordinary combustibles fires are designated as Class A which the dry powder of fire extinguisher could cover the fuel and put the fire out.
As flammable or combustible gas or liquid as fuel source, the fires will be designated Class B. Those fires could be put out by AFFF (Aqueous film forming foams) by segregating fuel and oxygen using foam type of extinguishing agents. While the electrical fires designated as Class C are those involving electrical equipment that could be fought off by gas agent extinguisher, such as CO_2. Fires involving certain flammable or combustible metals are designated as Class D. Specific types of dry powder agents for specific fuel/metal could only fight these fires.

Table 3. Conventional fire extinguishers

Type	Composition	Fire Classes
Dry Chemical (Powder)	$NH_4H_2PO_4$	Class A,B and C
	$NaHCO_3$	Class B and C
	NaCl(90%)	Class D (ex. Sodium metal)
Carbon Dioxide	CO_2	Class B and C
Halon	HFC-227ea (Heptafluoropropane)	Class A, B and C
Foam	Surfactant Foam Agent	Class A and B

2.5 Suggestions for Fire Involving TCS

According to the suggestions from TCS's Material Safety Data Sheet (MSDS), Emergency Response Guidebook (ERG), ASTM MNL33 Manual, other related regulations and references, dry powder and carbon dioxide extinguishers are suggested to fight the small TCS fires. However, the suggestions for fighting great fire by applying 6 percent AFFF of medium-expansion foam type extinguisher were quite ambiguous. The objectives of suppression matter are to cool and control the fire (i.e., prevent it from spreading) and to extinguish the fire. A variety of fire suppression methods are available to protect a facility. [12] The fact that foam type extinguisher could cover the fire effectively but could not stop the generation of derivatives. The suggestions gave by many sources for emergency response as shown in table 4. The study tested and discussed the uses of several types of extinguisher for small TCS fires to evaluate their effectiveness.

Table 4. Suggested extinguishing agents

Source	Suggestions for small fire	Suggestions for great fire	Note
MSDS (PRAXAIR) [3]	1. Use copious quantities of water to react with the TCS fire. 2. Apply 6% in water foam to suppress the release of Hydrochloric acid.		
MSDS (SIGMA-ALDRICH) [4]	Dry powder agent is recommended	Dry powder agent is recommended	No suggestions were given to different scale of fire.
International Chemical Safety Cards[13]	Do not use water; AFFF foam, dry powder and carbon dioxide extinguishers are recommended.	Do not use water; AFFF foam, dry powder and carbon dioxide extinguishers are recommended.	No suggestions were given for different scale of fire.

ASTM MNL33 Manual[3]	Dry chemical powder, carbon dioxide and low or medium-expansion foam are suggested.	Massive corrosive vapors were released after applying the low or medium-expansion foam agent. Water can be used to suppress the vapor releasing. Water can be used to keep the vessels cool till the fire is extinguished but never be allowed to enter a vessel containing TCS and should keep far away from the vessels that are on fire.	
2012 Emergency Response Guidebook[14]	Use sand, dry powder and soda ash as extinguishing agent.	Dry powder, soda ash and lime (calcium oxide) can excite the silicon tetrachloride type of fires and make them worse to release great amount of explosive hydrogen gas.	

3. Methods

The study investigated the effectiveness of the different fire extinguisher agents (dry powder, CO_2, Halon, water, and foams) on putting out fires involving a small leakage of 3 kg of TCS.

3.1 Equipments and Materials

1. Type 304 Stainless steel drum: diameter 29cm
2. TCS (Trichlorosilane) 250 Kg: 99.9999%, 4N; Provider: GREAT CHANCE TECHNOLOGY Co., Ltd.
3. 47 Liters of nitrogen gas cylinders: 99.9999%, 4N; Provider: TAIWAN STANDARD GAS Co., Ltd.
4. 20 pounds of ABC-Dry Chemical Extinguishers: Main component: $NH_4H_2PO_4$; 2sets Provider: CHENG DEH FIRE PROTECTION INDUSTRIAL CORP, Model: ABC3015
5. 20 pounds of Carbon Dioxide Fire Extinguishers: Main component: CO_2; 2sets Provider: CHENG DEH FIRE PROTECTION INDUSTRIAL CORP, Model: C015
6. 5 pounds of Halon Extinguisher: Main component: HFC-227ea; 4sets Provider: CHENG DEH FIRE PROTECTION INDUSTRIAL CORP, Model: NAF-5P
7. Infrared Thermography: NEC R-300; Provider: CHING HSING COMPUTER-TECH Ltd.

3.2 Testing

Prepare 3 containers with 3kgs TCS liquid in each and ignite them. Apply dry powder, carbon dioxide and Halon extinguishers to 3 containers respectively for observing the extinguishing effectiveness after complete combustion for at least 1 minute. Maximum 2 extinguishers are allowed for each container and the process shown as figure 1.

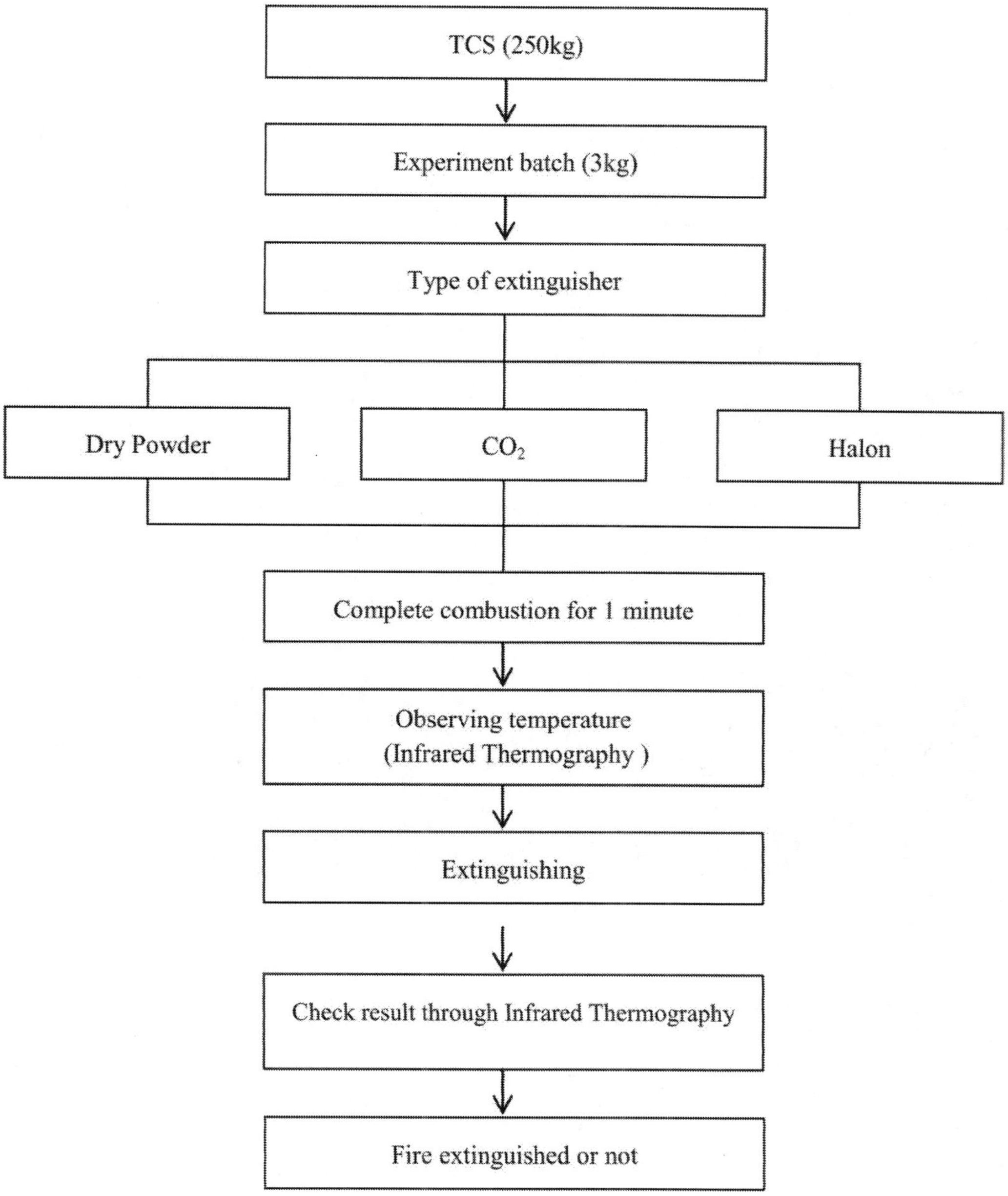

Fig 1. Extinguishing process flow of TCS fire

4. Results and Discussion

The effectiveness of applying extinguisher agents of dry powder, CO_2, and Halon to small fires of 3 kg TCS was evaluated by their temperature changes by Infrared thermography and flame conditions.

4.1 The Effectiveness of Dry Powder Extinguisher to TCS Fire

Observing through the Infrared Thermography at the beginning of burning, the flame temperature was higher than 500 Celsius (Our maximum detection temperature) after ignition for 1 minute. The experiment showed that dry powder could not effectively cover the fire but suppress the flame outward. The applying location on the central of fire where temperature of flame decreased to 50 Celsius rapidly but the outer zone was much higher and smoked. The temperature drop could be explained by dry powder application. The study showed that TCS evaporation (ER) and burning rate (BR) positively correlated to the burning surface area (Chen et al., 2008)[15]. Therefore, as the extinguisher application increased the contact area between burning TCS and air, the flame was pushed to outer regions with higher temperature. Once application of dry powder was stopped, the central temperature recovered rapidly. Thus, the application of dry powder extinguisher failed to put out the fire and enlarge fire prorogation instead. The experiment showed that applying dry powder to a TCS fire resulted in larger fire as shown in Figure 3.

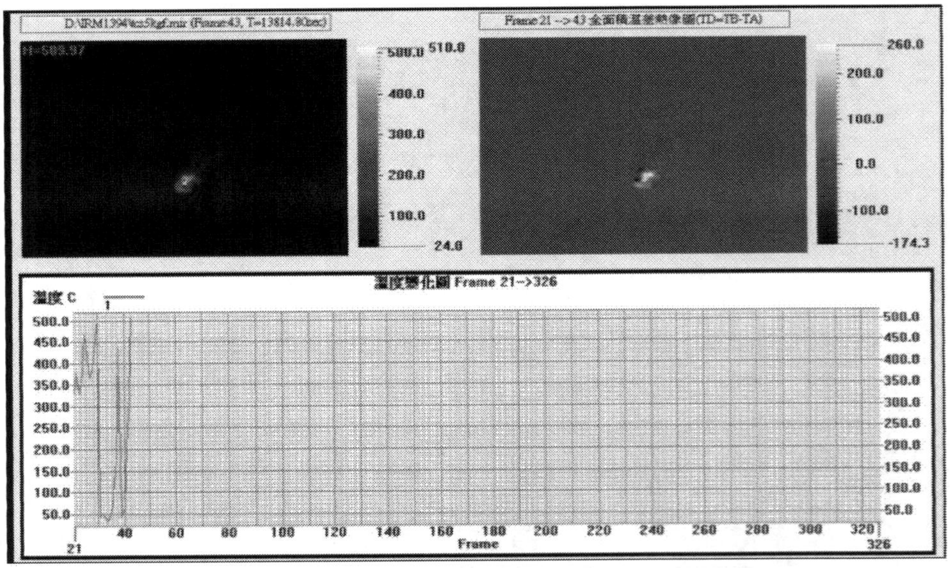

Fig 2. Temperature variation of dry powder fighting TCS fire

Fig 3. Dry powder extinguisher (a) complete combustion for 1 minute, (b) dry powder extinguisher applying, (c) applying 2nd extinguisher, (d) TCS fire kept burning.

4.2 The Effectiveness of CO_2 Extinguisher to TCS Fire

Since TCS reacts rapidly with air or water (vapor) and forms hydrogen chloride, hydrogen, silicic acid or silica (SiO_{2}), segregating TCS from air by CO_2 extinguisher could be the way to put out the fire. However, it resulted in more smoke than TCS burning alone and more derivatives generated as shown in Fig. 4 (a). The center flame temperature was lower than ambiance but the flame sustained and larger smoke was released. TCS was still burning after applying 2 sets of CO_2 extinguisher as Fig. 4 (b). Thus, CO_2 extinguisher failed to extinguish TCS fire.

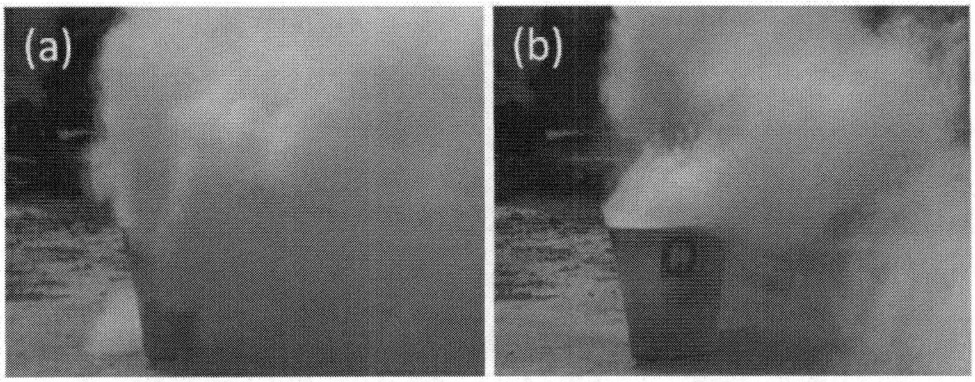

Fig 4. As CO_2 the First extinguisher (a) applying (b) smoke still released after applying.

4.3 Halon Extinguisher to TCS Fire

Brownish smoke was released from the stainless steel drum while Halon extinguisher was applying as Figure 5. The brownish smoke might be the formation of fluorine as Halon decomposed. The flam temperature was observed at above 500C. According to the reference, the major composition of Halon, heptafluoropropane (HFC-227ea), will start its self-decomposition ($CF_3CH_2CF_3 \rightarrow CF_3CH{=}CF_2 + HF$) when temperature reaches 750 Celsius[16] [17]. The flame was lesser developed and in lower temperature when applied second Halon extinguisher and white smoke was produced. However, the TCS fire still sustained.

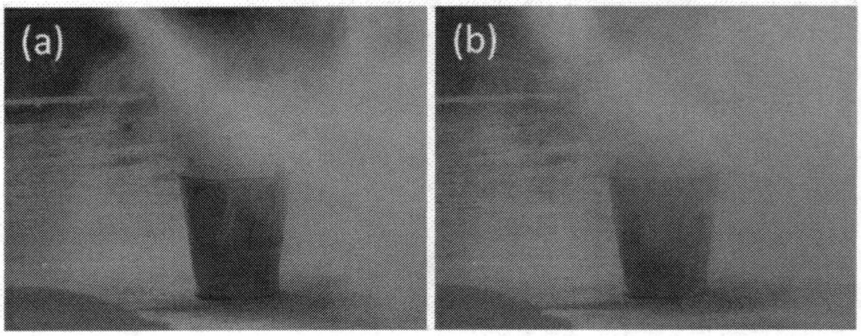

Fig 5. Halon extinguisher (a) applying (b) brownish smoke releasing after applying

4.4 Discussion

As previous discussed, 6 percent AFFF medium-expansion foam was suggested for fighting off the large TCS fires in many guides. One study showed that heat input and foam expansion ratio contributed by evaporation rate, energy balance, and temperature distribution. The energy balance was dominated by foam heat absorption and evaporation[18]. The main mechanisms for fire fighting by foams were heat absorption and oxygen segregation. Hence, establishing a relatively thick blanket of foam (15cm minimum) over the entire flame area could extinguish the TCS fire as shown in Fig. 6. Litmus paper tested to be acidic on remaining solution. Even so, the fire fighting mechanism of water is different than foam due to TCS's reactivity with water.

Water is a frequently used fire-extinguishing agent. The liquid-agents appear to be better thermally effective suppressants than the gas and water is considered the 'best' one [19]. However, extinguishing TCS fire with water caused explosion of hydrogen gas and massive white smoke of hydrochloride acid releasing as shown in Figure 7. The phenomena of hydrolysis reaction formed flammable hydrogen and corrosive hydrochloric acid that were resulted from burning, explosion, and smoking. TCS fire could be extinguished after applying water for 10 more minutes continuously. It indicated that the hydrolysis reaction was completed by large amount of water and the TCS fire was thus put out. TCS fire extinguished by AFFF was due to the duo mechanisms—segregation to avoid hydrogen gas explosion and TCS hydrolysis as hydrochloride acid dissolved in "large" amount of water that applied. However, further researches are required.

Fig 6. Foam extinguisher (a) foam agent given (b) two drums at the right with foam thickness of 5cm and 10cm were not working.

Fig. 7. Water extinguisher (a) applying water but accelerated the fire and formed large amount of smoke. (b) Still burning

5. Conclusion and Suggestions

The study gathered suggestions from all the chemical hazards emergency response data for effective extinguishments against small amount of TCS leakage. Dry powder, CO_2 and Halon were applied to 3kg of TCS spill fire and all failed to put of the fires.

5.1 Conclusion

1. The application of dry powder extinguisher failed to put out the fire and to enlarge fire prorogation instead. Observed from Infrared Thermography, the temperature of outer flame was higher than inner and the inner temperature rose back to 500 Celsius rapidly once the dry powder application stopped.
2. The effect of segregation from contacting with air by CO_2 extinguisher could be the way to extinguish the TCS fire, however, experiment results confirmed failed.
3. Halon extinguisher is available on the market but none of the reference and data mentioned about the effectiveness of Halon against TCS fire. The results of experiment showed that Halon is not suitable for TCS leakage fire as the fluorinated gas was formed would be hazards to human body.
4. Foam and water agents were also discussed. Results showed that 6 percent AFFF solution in water of medium-expansion foam was required to accumulate higher than 15cm of blanket to allow the extinguishing to occur. Water accelerated the TCS hydrolysis therefore reduced the combustion period.

5.2 Suggestion

1. The Emergency Response Guidebook instructed to apply dry powder and CO_2 extinguisher for small fire involving TCS, nevertheless, none of the references indicate the results. This experiment proved dry powder; CO_2 and Halon were all failed.
2. The agents to apply for TCS fire are varied or even contradictory among references, which make it difficult to set up fire fighting system. Further experiments should be done based on TCS burning characteristics in order to establish the effective fire fighting system for the plants.

6. References

1. NOAA, CAMEO Chemicals version 2.4. (Available from: URL:http:// http://cameochemicals.noaa.gov/chemical/1638).
2. Cruice, M. J. ,1988, Flammability studies on monosilane and disilane in air, Hazards ,Research Corporation, Report # 6469
3. John T. Higgins, Wilmuth E. Hubbs et al. Manual on Cholorosilance Emergency Response Guideline, American Society for Testing and Materials (ASTM) ,1998.
4. Praxair, Inc. 2009. Trichlorosilane Materials safety Data sheet.
5. SIGMA-ALDRICH, Inc. 2012. Trichlorosilane Materials safety Data sheet.
6. NTSB. 1991. Derailment of CSX Transportation Inc Freight Train and Hazardous Materials Release Near Freeland, Michigan on July 22, 1989. NTSB Report RAR-91/04.
7. EPA completes chemical safety audit at PCR Inc. in Gainesville, February, 1995.
8. Dow corning. Responsible Care Report. 1999.
9. Washington stage department of labor and industries, Violations alleged in fatal Moses Lake silicon plant explosion, 1999,April 14.
10. Xin hua net, Hubei Yichang CSG company spill occurred trichlorosilane, December 10, 2009.(Available from: URL:http://news.qq.com/a/20081210/002881.htm)
11. Yong Jiang, Jiangtao An , Rong Qiu , Yong Hu , Ning Zhu, Improved understanding of fire suppression mechanism with an idealized extinguishing agent, International Journal of Thermal Sciences 64 (2013) 22-28.
12. Dennis P. Nolan, Handbook of Fire and Explosion Protection Engineering Principles, Elsevier Inc, 2010.
13. Centers for Disease Control and Prevention, International Chemical Safety Cards. (Available from: URL:http://www.cdc.gov/niosh/ipcsnswh/nswh0591.html)
14. Department of Transportation Pipeline and Hazardous Materials Safety Administration (US). Emergency Response Guidebook, 2012.
15. Chen Cheng-Jen, Tsai Peng-Chi, Li Chia-Wei, Trichlorosilane of Emergency Operating Procedures and Implementation of Field Testing Program, Southern Taiwan Science Park Administration , 2008.
16. Tan Ling-hua, Li Qin-hua, Mao Ai-qin, Gao Fei, Pan Ren-ming, Li Feng-sheng, Thermal Decomposition Properties and Mechanism of Hexafluoropropane, Vol. 34 No.5 Journal of Nanjing University of Science and Technology , Nanjing University of Science and Technology, 2010.

17. Tan Ling-hua, Li Qin-hua, Mao Ai-qin, Gao Fei, Pan Ren-ming, Li Feng-sheng, Study on Thermal Decomposition Properties of Hexafluoropropane Clean Gaseous Fire-Extinguishing Agent, Vol. 30 No.7 Spectroscopy and Spectral Analysis, Nanjing University of Science and Technology, 2010.
18. Christopher F.Boyd, The behavior of a fire-protection foam exposed to radiant heating, US. Nuclear Regulatory Commission, 1998.
19. C.T. Avedisian, C.Presser, J.Stiehl, R.E.Cavicchi, A simple method to assess the quenching effectiveness of fire suppressants, Fire Safety Journal 45 (2010) 206–210.

Developing a fire model for offshore QRA

Yongfu Xu & David Worthington
DNV Software
Palace House
3 Cathedral Street
London SE1 9DE

yongfu.xu@dnv.com
david.worthington@dnv.com

Prepared for Presentation at
1st CCPS Asia-Pacific Conference on Process Safety
Qingdao, China
September 4-5, 2013

UNPUBLISHED

Developing a fire model for offshore QRA

Yongfu Xu & David Worthington
DNV Software
Palace House
3 Cathedral Street
London SE1 9DE

yongfu.xu@dnv.com
david.worthington@dnv.com

Keywords: Offshore QRA, flammable effect, pool fire, jet fire, fire hazard, escalation modeling, Modfire.

Abstract

The complex and densely packed nature of offshore facilities presents a series of challenges in terms of the potential for fire hazards to develop as well as the ability to model them. In the former, the key issue is the increased potential for escalation, i.e. hazards originating in one area could pose a risk on equipment and people in other areas, and the difficulties to evacuate workers. With the later, the challenge is being able to understand the extent to which the geometry impacts on the development of the hazard and account for this in the modeling. Modfire, a new empirical fire model, has been developed specifically for fires in offshore modules for offshore QRA. It extends the application of simple fire models developed for pool fires and jet fires for onshore QRA to the offshore environment. Using discharge results as the input, Modfire predicts flame shapes with consideration of the interaction between a fire and surrounding boundaries, without the need of huge computing power. The predicted fire characteristics are dependent on release conditions, boundary geometries and ventilation. This paper presents the modeling approach of Modfire and validation work undertaken against CFD predictions.

1. Introduction

An offshore installation normally houses complex facilities for a range of production processes. Space limitation leads to closely packed equipment and pipework and, subsequently, the likelihood of escalation from a relatively small fire is much higher compared to onshore installations and it is far more difficult to evacuate workers when an accident occurs. For instance a fire escalated to riser failure during the Piper Alpha incident in 1988 and caused major loss of life and loss of the platform.

CFD models are the most sophisticated tool available to predict fire dynamics in offshore installations. They solve the fundamental equations of combustion with explicit account for the influence of geometry (e.g. walls, decks, large equipment items) on the development of a fire. However, due to the high demand on resources (e.g. cost, manpower and time) and the large number of scenarios to be considered in a QRA, CFD based analysis becomes impractical for Quantitative Risk Assessment (QRA). Instead, simple fire models developed for onshore scenarios, such as the jet fire model by Chamberlain[①] and the pool fire models reviewed by Mudan[②], are normally used. Whilst the simple models offer quick solutions to very practical problems, they have the obvious drawback that they cannot take account of the platform geometry.

Modfire is a new empirical fire model developed specifically for offshore QRA. It extends the application of simple fire models to the offshore environment. Using discharge results as the input, Modfire predicts flame shapes with consideration of the interaction between fires and boundaries, without the disadvantages which have prevented CFD models for being used routinely for QRA. Modfire predicts the spread of the initial fire into wider areas. Fire characteristics predicted are dependent on release conditions, material properties, boundary geometries and ventilation. The model also considers escalation of fires through boundaries or by causing further process leaks. This paper presents the modeling approach of Modfire and validation work undertaken.

2. Modeling approach

2.1 Background

The flames of hydrocarbon fire consist of high-temperature combustion products with temperatures up to 1600K. Heat is transferred between hot gases of the flame to objects in the surrounding area through convection, conduction and radiation. Heat convection and conduction are important for calculating heat load on facilities engulfed by fire. At some meters away from the fire, heat radiation is normally predominant in heat transfer[③]. Very near to the flame, radiation is usually high and close to fatality levels for personnel risk. The objectives of this fire model are to predict:

- flame shape of fires to assess whether equipment or walls and decks are engulfed by fires and to enable the prediction of subsequent escalation by fire.
- radiation away from fires for the assessment of flammable risk to workers, process equipment and boundaries

Temperature of the hot gases from a fire fluctuates due to turbulence caused by jet momentum, thermal buoyancy and wind. The radiated heat depends on a wide range of parameters, such as fuel property, release and environmental conditions, and it is difficult to make accurate predictions. In models for QRA, the concept of 'solid-flame' is normally used. Heat from the combustion process of a fire is considered to be emitted from the visible flame surface with constant surface emissive power. So once the flame shape is known, the radiation at targets can be estimated as:

$$q = VF * SEP * \tau \qquad (1)$$

Where q is the radiation flux at a target (kW/m^2), VF is the view factor of the target to the flame, SEP is surface emissive power of the flame (kW/m^2) and τ is the atmospheric transmissivity.

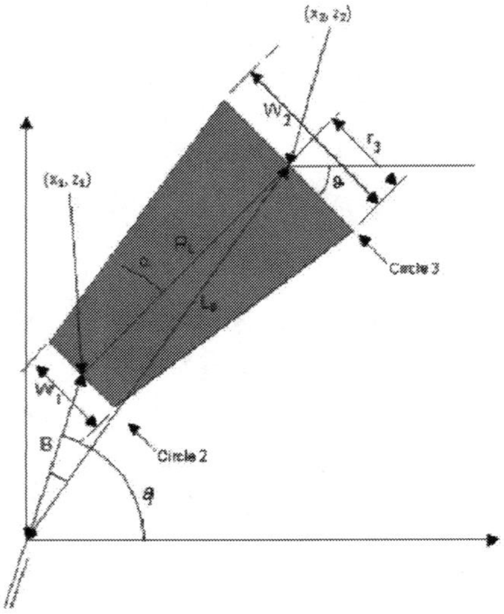

Figure 1 'Solid-flame' model for jet fires

Even though simple, models based on the concept of 'solid-flame' have proved to produce satisfactory results for risk assessment, e.g. the jet fire model by Chamberlain① and tilted cylinder flames for pool fires②. These models are implemented in Phast & Safeti, software tools for risk assessment. Figure 1 illustrates the flame shape for jet fires of the Chamberlain model. The flame has a cone shape with its dimensions correlated to the discharge and weather conditions. The flames of pool fires are represented by tilted cylinders.

Surface emissive power (SEP) of the flame surface is assumed to be constant and can be calculated as:

$$SEP = \frac{F_s m H_{COMB}}{A} \qquad (2)$$

Where F_s is the fraction of heat radiated from the surface of the flame, m is the fuel burning rate [kg/s], H_{COMB} is the heat of combustion of the fuel mixture [kJ/kg] and A is the total visible surface area of the flame.

Simple 'solid-flame' models for jet fires and pool fires are primarily developed for fire hazards in relatively open spaces with good ventilation. They have no consideration of the interaction between fires and the surrounding obstacles, and this interaction is an important characteristic of offshore or dense onshore installations because of the closely packed equipment and pipework. Therefore for Offshore QRA, the fire model needs to reasonably predict the interaction between flames and the surrounding boundaries, and subsequent spreading of fire into wider areas through openings and connections. Modfire is a new fire model developed specifically for offshore QRA with these capabilities.

2.2 The Modfire approach

In Modfire, the concept of 'solid-flame' has been extended to allow a flame to change shape when interacting with boundaries and to spread into other areas when reaching openings. Depending on complexity of the offshore installation concerned and the release conditions, the flame of a fire could spread into wider areas so that the fire cannot be accurately represented using a single cone, cylinder or sphere. In Modfire, this kind of fire is represented a combination of sub-flames of various shapes, i.e. cone, cylinder & cuboid.

Without considering the wind effect, Figure 2 illustrates the flame shapes of a pool fire on a deck of four scenarios:

- Scenario (a): there is no upper deck so the flame shape remains like a cylinder, which tilts when wind is blowing.
- Scenario (b): there is an upper deck, but it is not reached by the flame, so the fire still has a cylinder shape as for Scenario (a).
- Scenario (c): the pool fire impinges the upper deck and spreads outwards beneath it. The top section of the flame is represented by a cone and the low part by a cylinder in Modfire.
- Scenario (d): the flame has spread to the edge of the upper deck and goes upwards after the edge. Apart from the cone and cylinder flames as shown in Scenario (c), a cuboid flame may be needed to represent the hot layer near the upper deck depending on shape of the upper deck and position of the fire, the escaped flame above the upper deck is represented by cylinder flames which tilt according to wind..

The following algorithm is followed to determine the flame shape due to the interaction with boundaries and spreading into openings:

- Flame shape of a fire is determined in an iterative searching process starting at a relatively small volume. Sub-flames are gradually introduced when a fire gets into new spaces when the volume is increased in the searching process.

- Volume of the fire is defined by the fire of the same release rate in the open at a wind speed of 3m/s, i.e. volume of the undisturbed fire of the same release conditions. This wind speed is selected by taking into consideration of generally slow air movement inside offshore installation. For an area to be classified as 'open area' according to IP15 of Institute of Petroleum[4], air velocity inside is defined as rarely below 0.5 m/s and frequently above 2m/s. Air movements in sheltered areas are less than that in open areas.

  Volumes of the undisturbed fire are estimated using the Chamberlain model for jet releases and the model by Cook et al[5] for pool fires, which uses the correlations overview by Mudan.

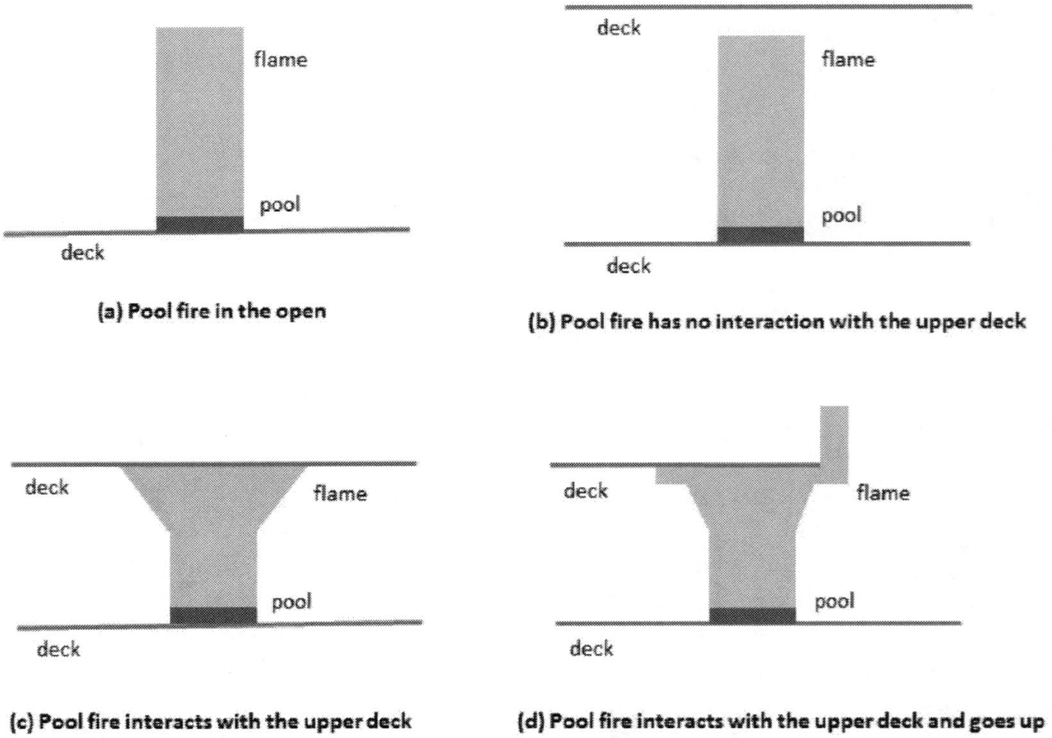

Figure 2 Flame representation for pool fires of four scenarios

- When ventilation in a module is limited, the flame volume inside is capped to the volume of the undisturbed fire at a transitional fuel rate as given by Lowesmith et al[6] as

  $$m_{tr} = m_{vent} / r \qquad (3)$$

  Where r is the mass ratio of air to fuel required for stoichiometric burning, m_{vent} is the mass flowrate of air available for the fire and m_{tran} is the transitional rate at which the fire is controlled by ventilation available, instead of the release rate.

- The fraction of emitted radiation is estimated using the correlations by Chamberlain for jet fires and by Mudan for pool fires.

- Surface emissive power (SEP) of a fire is calculated using the visible surface area of the predicted flames and equation (2). Overlapped flame surfaces between sub-flames are excluded to prevent over-estimation of flame surface area. When there is no interaction between a fire and any boundary, the surface emissive power predicted by Modfire would be the same as that of undisturbed fires.

- Radiation at targets is the sum of all sub-flames which are visible to them, i.e. no boundary in between.

3 Model validation

Even though the 'solid-flame' concept has been used with success to predict radiation from pool fires and jet fires, Modfire is an extension of the concept and needs to be validated to be used with confidence. Validation was carried out by comparing Modfire predictions against the results of CFD codes. The validation includes qualitative comparison of the predicted flame shapes and quantitative comparison of radiation predicted for pool fires and jet fires, i.e. against the results of FDS for cases with a two-deck setup as shown in Figure 3 and the results of KFX on a medium-size offshore platform shown later.

FDS is a computational fluid dynamics CFD) model developed by National Institute of Standard and Technology (NIST) of USA for fire-driven fluid flows[7]. It is an open-source CFD code and has been widely used all over the world and validated by the developer and the users. Kameleon FireEx KFX® 2010 is a CFD code from ComputIT for fire simulations and has been used widely to model fires for risk assessments.

For the simple test cases shown in **Figure 3**, a fire is initiated between two decks from a pool or a jet release in the middle of the domain. All sides of the computation domain are open. When the fire is large, it would impinge on the upper deck, spread outwards in all directions and go upwards after reaching the edge of the upper deck. The central vertical plane is assumed to be symmetric to reduce computing time using the mirror option available in FDS.

The pool fire is located on the lower deck and is modeled with a given heat release rate equivalent to the pool burning rate calculated by Modfire. Vertical jet releases are positioned upwards at the center with the required mass release rate. No wind speed is specified for the cases, heat transfer between flame and upper deck has been ignored.

Temperature contours on the symmetry plane are used to qualitatively assess the flame shapes predicted by Modfire. Temperature contours by FDS are representative snapshots selected from the animated results over a period of 180s. For quantitative validation, predicted radiations are compared on six transects as shown by the red lines in Figure 3. The radiation values predicted by FDS are the averaged values at positions over the last 120s of the 180s period simulated. Location of these transects and observer orientations used to calculate the radiation are given in Table 1. Radiations predicted by FDS are the values as measured by radiometers with the specified orientations. Planar observers are placed at the same orientations in Modfire predictions for a direct comparison against the FDS results.

Table 1 Location and observer orientation at transects for radiation predictions

Transect Name	*Location*	*Observer orientation*
Y2	Y=2 & Z=0	Facing the upper deck
Y3	Y=3 & Z=0	Facing the upper deck
Y5	Y=5 & Z=0	Facing the upper deck
X3	X=3 & Y=8	Facing into the direction of -Y
X8	X=8 & Y =8	Facing the flame central line
Z2	Y=5 & Z=2	Facing into the direction of -Y

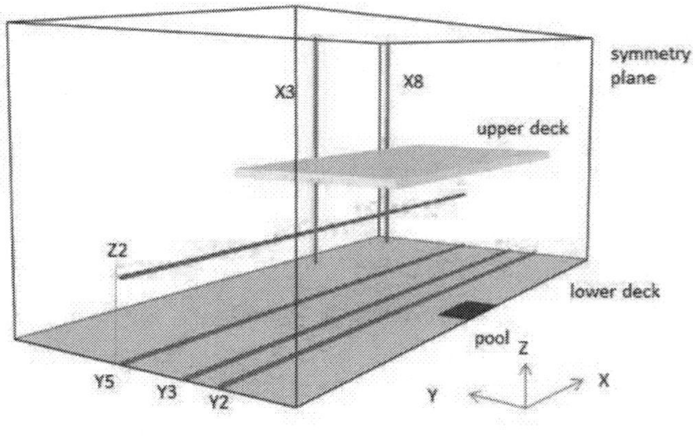

Figure 3 Setup of simple test cases to validate Modfire

3.1 Simple test cases: pool fires

Table 2 gives the input data of the pool fire test cases. The pool has a square shape in FDS models with a specified HRRPUA (heat release rate per unit area) calculated using the pool burning rates.

Table 2 Input data for the test cases of pool fire

Case name	Material	Pool area (m^2)	Pool burning rate (kg/s m^2)	HRRPUA (kW/ m^2)
P1	n-Pentane	1	0.106	4812
P2	n-Pentane	4	0.106	4812
P3	n-Pentane	9	0.106	4812
P4	n-Pentane	16	0.106	4812

Figure 4 shows the predicted results for the pool fire with a pool area of $1m^2$. In this case, the fire is quite small and the flame has just reach the upper deck without much spreading. Therefore the flame by Modfire is close to a cylinder confined between the decks and this is confirmed by FDS as shown by the temperature contours. Modfire does not predict the hot layer under the upper deck, but temperature of the layer is still relatively low as shown by the FDS results. Results of the two models agree well in trends and Modfire has slight over-predictions on all transects.

When the pool size has been increased to 9 m^2, the fire has spread beneath the upper deck to the edge as indicated by both FDS and Modfire as shown in **Figure 5**. At this point, radiation predictions by Modfire are still in good agreement with FDS results with slight over-predictions along transects on the lower deck, i.e. Y2, Y3 & Y5. However, a region of under-prediction appears on the vertical transects at the height about 2m above the lower deck.

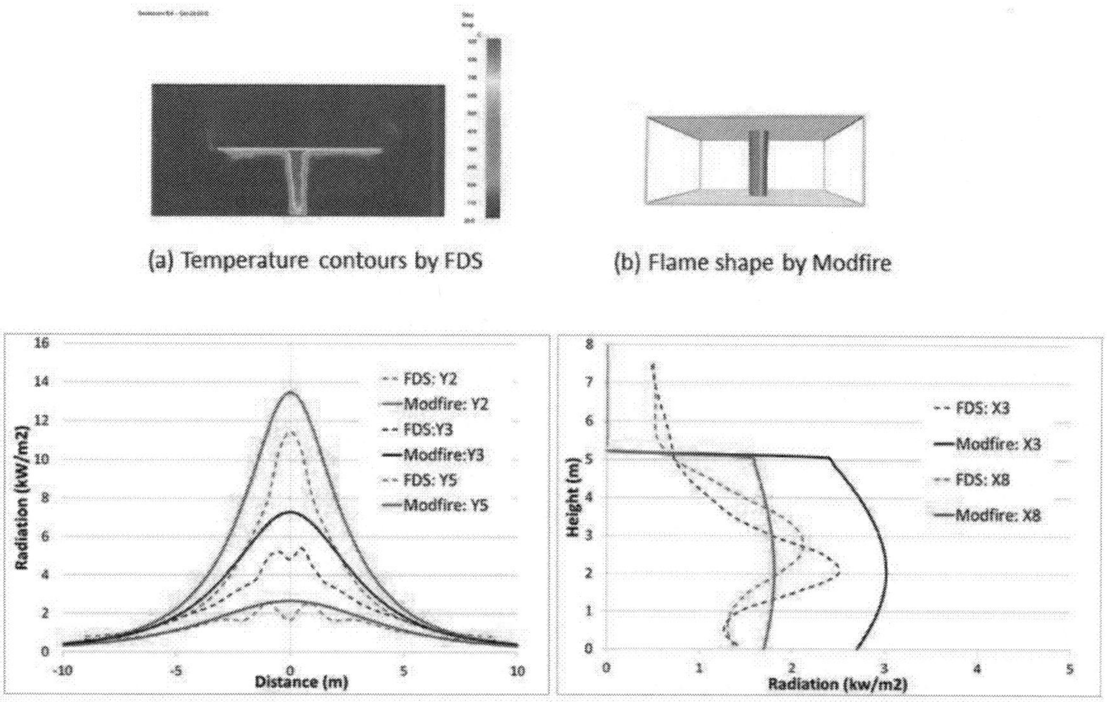

(a) Temperature contours by FDS (b) Flame shape by Modfire

(c) Radiations on horizontal transects (d) Radiations on vertical transects

Figure 4 Modeling results of tset case P1 for pool fire, i.e. a pool area of 1 m^2

When the pool size is increased to 16 m^2, fire is escaping the region between the two decks and going upwards as predicted by both models as shown in **Figure 6**. In Modfire the flame of the escaped fire is represented by two cylinders at each side. Modfire starts to under-predict in the region very close to the flame on transect X2, i.e. near to the centre. Despite the under-prediction seems large, but the radiations predicted in the region are higher than 60 kW/m^2, which would have caused 100% fatality if the Probit method is used for flammable risk[⑧]. So under-predictions at this level on the lower deck are unlikely to cause problems in risk calculation.

There are also under-predictions on the vertical transects. Flames predicted by FDS can reach low positions occasionally due to turbulence and unsteadiness of the burning process, as illustrated in Figure 7 for the pool fire case P4. Modfire is too simple to predict these fluctuations and this should have contributed to the under-predictions. On the other hand, as shown in the validation for the Chamberlain model[⑨], the discrepancy between predicted and measured radiations can be ± 40% at some locations, and this is consistent with results by Modfire.

In QRA, radiation is usually calculated using optimised observer orientation, i.e. the orientation that gives the highest radiation for planer observers, to represent the worst case by radiation. Figure 7 compares the predicted radiations with optimised observer orientation on transect Z2, which locates at the height where Modfire produces the highest under-predictions as shown in **Figure 5** and **Figure 6**. The under-predictions are almost eliminated if optimised observer orientation is used in the case.

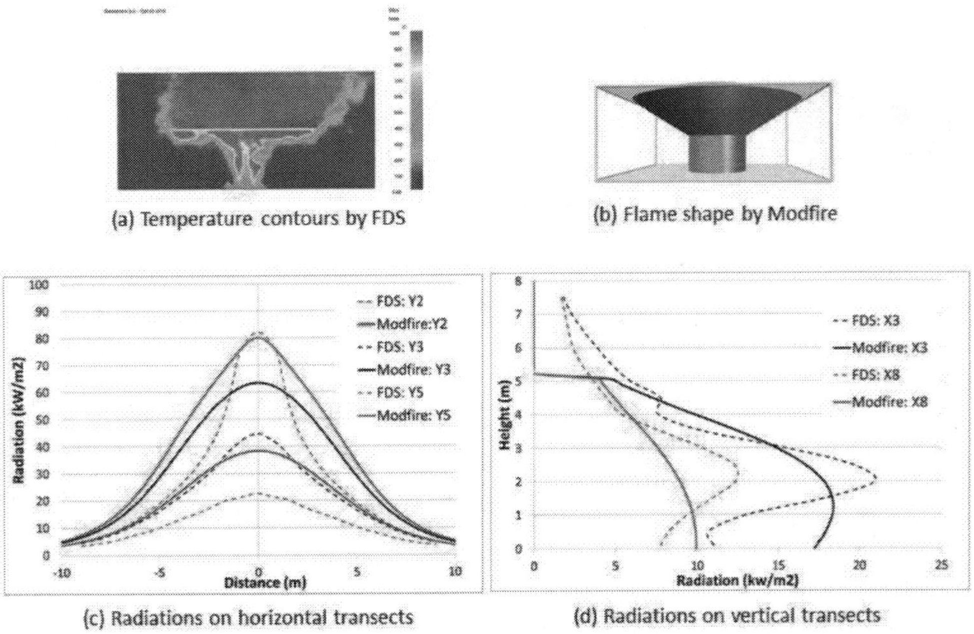

Figure 5 Modeling results of the test case P3 of pool fire, i.e. a pool area of 9 m^2

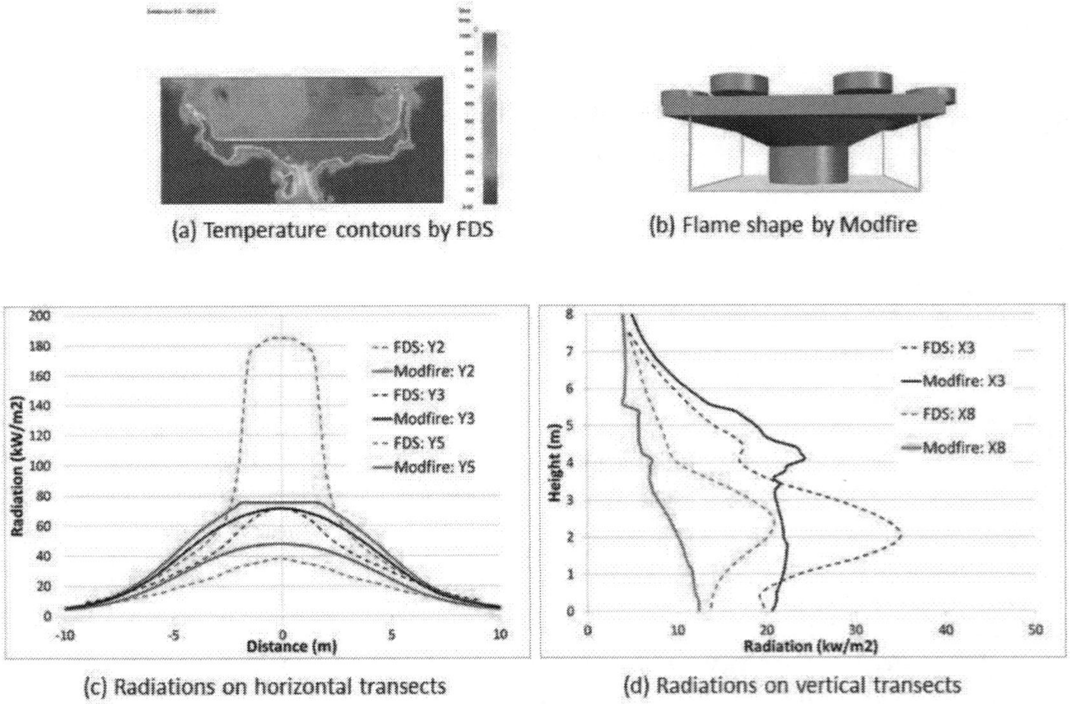

Figure 6 Modeling results of test case P4 of pool fire, i.e. a pool area of 16 m^2

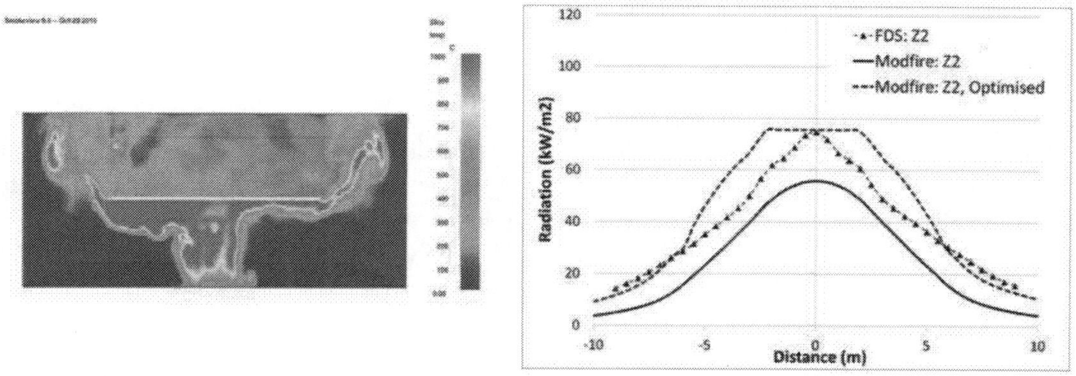

Figure 7 Predicted radiations on the horizontal transect Z2 for the pool fire of 16m^2.

3.2 Simple test cases: Jet fires

Table 3 gives the input data of the jet fire test cases. The jet release is positioned upwards in the middle of the computational domain with a square shape in the FDS model. The jet release conditions are the identified release scenarios of an offshore QRA.

Table 3 Input data for test cases of jetfires

Case	Material	Release rate (kg/s)	Jet velocity (m/s)
J1	methane	1	400
J2	methane	2	400
J3	methane	4	400

Because of the momentum of jet releases, there is usually an uplift of the jet fire away from the release point. The lift-off distance is calculated as in the Chamberlain model. In Modfire, the lift-off distance is capped to the half distance between the release point and the upper deck to reflect the restriction of upper boundaries on fires. **Figure 8** shows the predictions of a jet fire at a release rate of 4 kg/s. At this rate, the fire has spread outwards beyond the edge of the upper deck, which is confirmed by FDS simulations. The flame predicted by Modfire is represented by a cone sub-flame and a cuboid sub-flame below the upper deck as a hot layer and cylinder sub-flames for the escaped fire around the upper deck.

The predicted radiation on the lower deck is satisfactory with slight under-predictions in the far regions from the centre. That indicates the hot gas may have travelled further in the horizontal direction after reaching the edge of the upper deck, instead of going upwards immediately as assumed in Modfire. The lift-off of the jet flame in Modfire has produced no fire between the lower deck and the lift-off point as shown in **Figure** 8(b), this should have contributed to the under-prediction of radiations between the decks on the vertical transects X3 & X8. The under-predictions can be addressed using optimised observer orientation as demonstrated in Figure 7 for a pool fire.

For the jet fire shown in **Figure 8**, FDS has predicted large entrainment of high-temperature combustion products into the region behind the upper deck. Modfire has not included this kind of phenomena. Instead, Modfire has predicted larger flames with increased heights. Even so, radiation values predicted on the transects are still very reasonable.

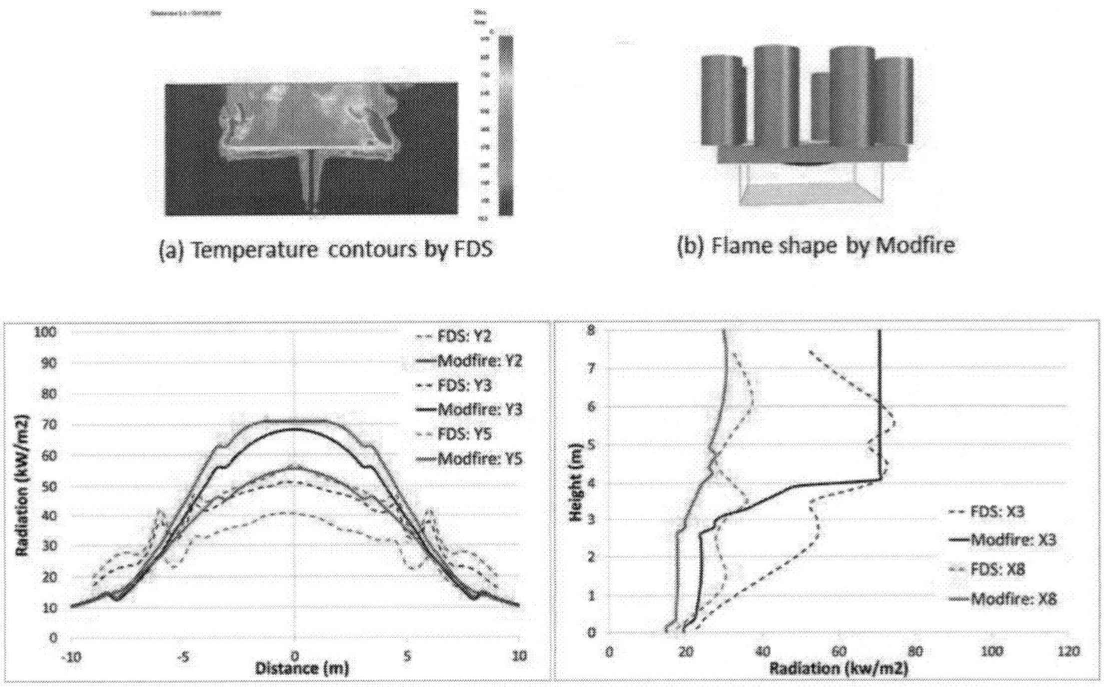

(a) Temperature contours by FDS

(b) Flame shape by Modfire

(c) Radiations on horizontal transects

(d) Radiations on vertical transects

Figure 8 Modelling results of test case J3 of jet fire, i.e. at a release rate of 4kg/s.

3.3 An offshore platform

Offshore installations are usually complex and packed with facilities. **Figure 9** shows a medium-size platform for oil production. Modfire has been used to model several release scenarios in the production area, i.e. the left section of the platform shown **Figure 9**. In a complex situation as this, a network of cuboids has been created to represent the platform in Modfire. Boundaries, i.e. walls & decks, are defined as a property of the cuboids. A fire starting inside the production area will spread into connected cuboids if there is no boundary to stop it. **Figure 10** shows the cuboids generated to represent the platform and the flames of a pool fire.

The predictions of Modfire are compared against the results of KFX. As the results shown above against FDS predictions, Modfire is predicting the spreading of fire within the platform with good accuracy. **Figure 11** shows the predicted radiations on two horizontal transects at the south side of the platform of a pool fire case. The flat section of Modfire results indicates the locations are on fire.

In a simulation for QRA with Modfire, properties of the cuboids are updated according to the consequences they have experienced. Boundaries are being impaired due to explosion and fires, and the fire development history is predicted using the updated geometry as it changes with time.

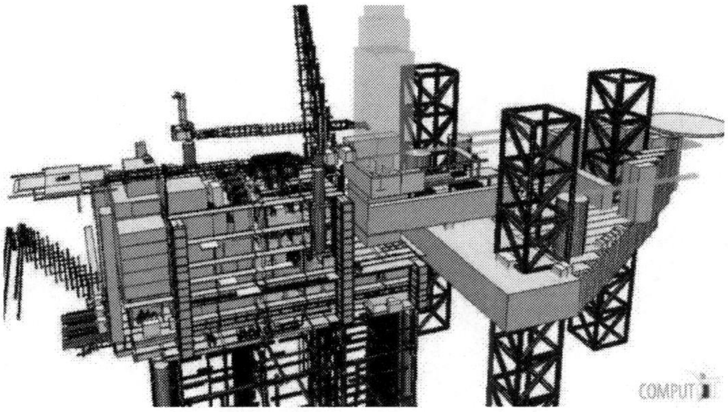

Figure 9 A medium-size offshore platform for oil production

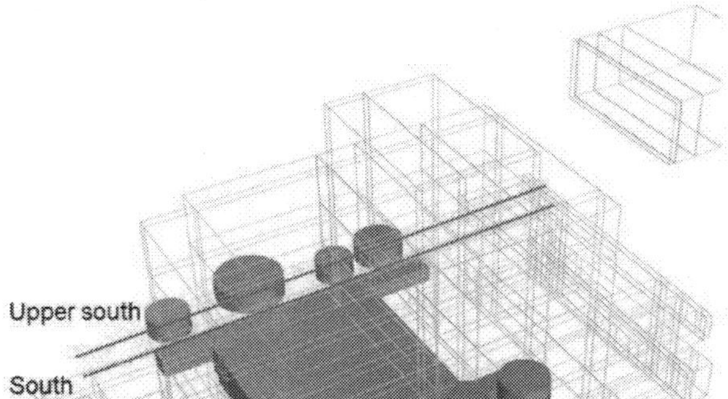

Figure 10 Cuboids for Modfire predictions and the predicted flames

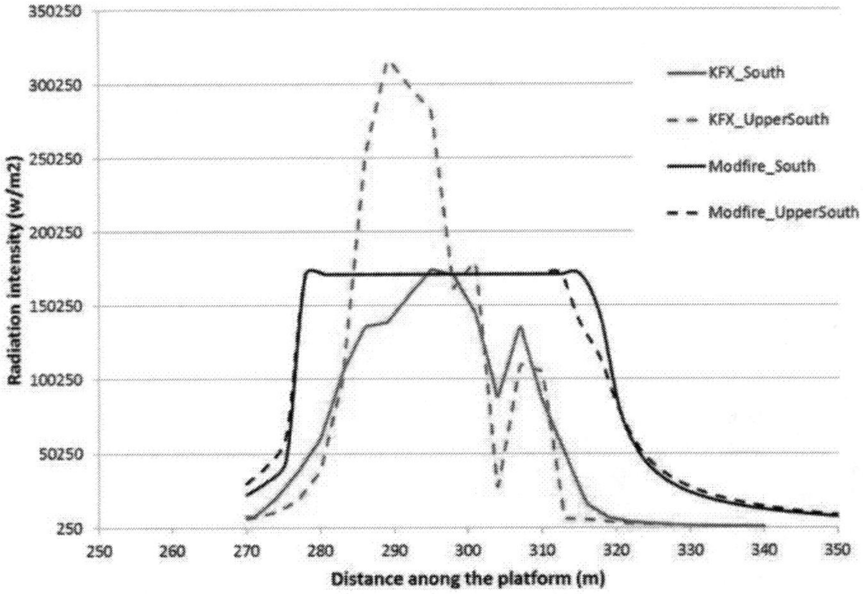

Figure 11 Comparing the predicted radiations on two transects along the platform

4 Conclusions

A new method has been developed to model fire hazards and risk in the offshore environments. This method is an extension of the 'solid-flame' concept which has been used successfully for the QRA of onshore installations. Predictions of this new method have been compared against well validated CFD fire models using results for a range of test cases:

- Pool fires with four pool diameters within a two-deck setup
- Jet fires at three release rates within a two-deck setup
- Pool fires and jet fires on a medium-size platform.

In all of these test cases, predictions of Modfire have been compared against the CFD results qualitatively for the predicted flame shapes and quantitatively for radiation predictions. The predicted radiation is generally in good agreement with CFD results. However, there are under-predictions at some locations when using a fixed plane orientation. This is a weakness of the solid flame approach in that the flame geometry is idealized and does not fluctuate with time. In QRA, radiations are usually calculated with optimised observer orientation to detect the worse-case scenarios. With this option, the under-predictions by Modfire using fixed orientations are largely addressed and overcome this limitation of the solid flame approach.

In practice, offshore hydrocarbon fires can vary widely in characteristics due to the variations in production process and in design of the installations. Further validation is needed to test the model for more scenarios, such as multiphase and multi-component releases, wind effects and horizontal fires.

References

① Chamberlain, G. A., "Developments in design methods for predicting thermal radiation from flares", Chem. Eng. Res. Des., Vol. 65, pp 299-309, (July 1987)

② Mudan, K.S., "Fire hazard calculations for large open hydrocarbon fires", Section 3-11 in "SFPE handbook of fire protection engineering", Second Edition, National Fire Protection Association, Quincy, MA (1995)

③ Van den Bosch C. J. H., and Wetterings R. A. P. M. (Eds), "Methods for the calculation of physical effects 'Yellow Book' CPR 14E (Part 2)", 3rd ed., Sdu Uitgevers, The Hague, pp 6.17-6.138, (1997)

④ Institute of Petroleum, "Area Classification Code for Petroleum Industry, Part 15 of Institute of Petroleum Model Code of Safe Practice in the Petroleum Industry", JohnWiley and Sons, March 1990.

⑤ Cook, J., Bahrami, Z., and Whitehouse, R.J., "A comprehensive program for calculation of flame radiation levels", Proceeding of First Int. Conf. on Loss of Containment, London, UK, 1989. J. Loss Prev. Process Ind. 3, pp 150-155 (1990)

⑥ Lowesmith, B.J., Hankinson, G., Acton, M.R. and Chamberlain, G., "An Overview of the Natural of Hydrocarbon Jet fire Hazards in the Oil and Gas Industry and a Simple Approach to Assessing the Hazards", Trans IChemE, Part B, Process Safety and Environmental Protection, 85(B3): pp 207–220, 2007.

⑦ McGrattan, K., McDermott, R., Hostikka, S. and Floyd, J., " Fire Dynamics Simulator (version 5), User's guide" National Institute of Standard and Technology, US Department of Commerce. 2010

⑧ Committee for Prevention of Disasters, "Guidelines for Quantitative Risk Assessment", CPR 18E, The Purple Book, Den Haag, 1999.

⑨ DNV Software, "JFSH (Jet Fire) Model Theory", Palace House, 3 Cathedral Street, London SE1 9DE, UK, 2012. http://www.dnv.com/software

1ST CCPS ASIA-PACIFIC CONFERENCE ON PROCESS SAFETY℠

Process Hazard Analysis for Steam Generation and Combustion System in Capital Investment Project

Jianming Lu
Asia Pacific Regional Engineering, Du Pont China Holding Co., Ltd.

6F, Building No. 11
No.399, Keyuan Road, Zhangjiang, Pudong, Shanghai, China
Jerry.Lu@dupont.com

Prepared for Presentation at
American Institute of Chemical Engineers
1st CCPS Asia-Pacific Conference on Process Safety
Qingdao, China
September 4 – 5, 2013

UNPUBLISHED

Process Hazard Analysis for Steam Generation and Combustion System in Capital Investment Project

Jianming Lu

Asia Pacific Regional Engineering, Du Pont China Holding Co., Ltd.

6F, Building No. 11

No.399, Keyuan Road, Zhangjiang, Pudong, Shanghai, China

Jerry.Lu@dupont.com

Keywords: Steam, Combustion, Boiler, PSM, PHA

Abstract

Steam generation is an integral part of many manufacturing facility as well as other types of facilities such as lab and office building. Normally combustion of gas, liquid or solid types of fuel is involved in the steam generation equipment, i.e. boiler. There are certain process hazards in the steam generation and combustion process, e.g. flammability of the gas and liquid fuel and combustible dust hazard for pulverized coal, high energy stored in the hot water and steam.

Although in many cases steam generation and combustion system is not covered by PSM (Process Safety Management) related regulations and standards, e.g. US OSHA's 29CFR 1910.119 and AQ/T3034-2010 in China, some companies elect to establish PSM requirements for this type of process as there can be high consequence if hazardous events happen, PHA (Process Hazard Analysis) is an important element in PSM system to identify, evaluate the process hazards and develop the recommendations to reduce the risk to acceptable level.

However, PHA for this kind of process is often overlooked in real practice, the reasons as follow, the boiler is equipment from vendor as a whole package, the user just needs to provide water, fuel supply, steam piping and ventilation equipment in designated location, the potential hazards and hazardous events that can originate in the boiler are not well understood and analyzed. Besides, for lab and office facility where the boilers are used, usually PSM system is not in place.

It is critical to conduct the PHA in the capital investment project to identify all the safeguards needed to control the hazards and define design requirements in equipment specification before placing order to vendor, delay in carrying out those activities may causes difficulties in modification of system design or the modification can be impossible.

PHA normally consists of the following elements, i.e. hazard identification, consequence analysis, hazard evaluation, facility siting review, interlock evaluation, inherently safer process review and human factor analysis. This paper will provide a summary on how these reviews are conducted for the steam generation and combustion system with some examples.

1. Introduction

Although it has been a long history since boiler equipment was firstly used in industry revolution in 18th century, safety incidents related with the steam generation and combustion process still happens despite of technology improvement, following statistics is from The National Board of Boiler and Pressure Vessel Inspectors [1], on the steam heating boiler incidents happened between 1995 to 1998 in US, which has caused human injuries, fatalities and severe property damages.

Table 1. Steam Heating Boiler Incidents in US (1995-1998)

Failure Mechanism	Accidents	Injuries	Fatalities
Safety Valve	26	0	0
Low Water Condition	2708	2	0
Limit controls	218	2	0
Improper Installation	94	0	0
Improper Repair	75	0	0
Faulty Design or Fabrication	99	0	0
Operator Error or Poor Maintenance	746	25	0
Burner Failure	196	28	2
Totals	3532	57	2

There are also boiler safety incidents happened in China, following statistics is from the bulletin No.122[1994] issued by labor department [2] on the boiler incidents happened in China in 1992 & 1993.

Table 2. Boiler Incidents in China (1992-1993)

Incident Statistics	1992	1993
Explosion Incidents	24	40
Severe Incidents	406	391
No. of Fatalities	19	37
No. of Severe Injured Persons	42	61

In the incidents happened in 1993, 75% of the incidents are classified as caused by lack of water supply and low water quality. In all the 40 explosion incidents, 32 is due to using boiler from the vendor without boiler manufacturing license, 7 is caused by adding water right after stop of water supply, one is due to failure of safety devices.

It can be concluded from the above statistics that the direct causes of these boiler incidents are quite diverse, including faulty design, fabrication and installation, equipment and safety device failure, operation and maintenance errors. Thus a process safety management system is needed to control the process hazards within the process.

There are some PSM regulations and guidelines worldwide, e.g. US OSHA has issued the PSM regulation 29CFR 1910.119 "Process safety management of highly hazardous chemicals" and there is AQ/T3034-2010 "Guidelines for process safety management of petrochemical corporations" in China. Although 29CFR 1910.119 may not be applicable to the combustion system in the boiler equipment [3], and AQ/T3034 is non-mandatory and only applicable to chemical industry, many companies decide to apply the PSM requirements for the steam generation and combustion system due to the high consequence if the process hazards are not well controlled.

PHA is one element in the PSM system, it is defined as the application of organized, methodical approaches to identify, evaluate, and control the hazards associated with the process facilities.

Following activities are usually conducted in the PHA, i.e. hazard identification, consequence analysis, hazard evaluation, facility siting review, interlock evaluation, inherently safer process review and human factor analysis.

The purpose of this paper is to provide some suggestions on what to be considered in the above PHA activities for steam generation and combustion process.

2. PHA process overview

2.1 Purpose

A Process Hazard Analysis is the systematic, comprehensive study of a process, using recognized methods of analysis, the purpose is to:

- Identify hazards and hazardous events in the process;
- Identify all the ways in which each hazardous event can occur;
- Evaluate the likelihood of each event occurrence and the seriousness of consequences; i.e. risk assessment.
- Based upon the risk assessment, develop practical recommendations, where necessary, to reduce the risk through the elimination of process hazards (i.e., make the process inherently safer) or through the addition of appropriate safeguards to decrease event frequency or consequences.

2.2 PHA in the Whole Life Cycle of Facility

PHA needs to be considered in the whole life cycle of a process facility, following are some types of PHAs can be conducted in different phases.

Table 3. PHAs in the Whole Life Cycle of Process Facility

Facility Type	Phase of Life Cycle	Types of PHA
New Facility(Capital Investment)	Front-End Loading (Before Project Approval)	Screening and Preauthorization PHA
	Production Engineering and Design (After Project Approval)	Detailed PHA
Existing Facility	At Startup of Operation	Baseline PHA
	Ongoing Operation	Cyclic PHA
	Management of Changes	Some activities of PHAs, e.g. hazard review.
	Incident Investigation	Some activities of PHAs, e.g. hazard review.
	Facility being Mothballed	PHA for Mothballing Operations
	Facility being Dismantled	PHA for Dismantling Operations

For project PHAs, the purpose of screening PHA is to identify acute hazards in the process and to develop broad recommended changes in scope to significantly reduce these hazards (including consideration of the use of inherently safer technology) in the basic data development period. Purpose of preauthorization PHA is to confirm all process hazards have been identified, and determine if current project scope is adequate to control all the hazards before the project is authorized. The purpose of detailed PHA is to carry out all the activities mentioned in section 2.1 when the detailed design is almost ready. For the process that involving boiler which normally is a package equipment from vendor, it is critical to have effective and complete project PHAs conducted, otherwise, it may be difficult or impossible to make modification of system design after the project is completed.

2.3 PHA versus RAGAGEP & Regulation Compliance

There are many industrial codes and standards, normally called as RAGAGEPs (Recognized And Generally Accepted Good Engineering Practices) related with steam generation and combustion system, e.g. NFPA 85, ASME Boiler and Pressure Vessel Code, there are also some relevant regulations, such as China Steam Boiler Safety Technical Supervision Administration Regulation, etc. These codes, standards and regulations have set very detailed prescriptive requirements in boiler design, installation, operation and maintenance, most of them are safety related. One may argue that a compliance check with those RAGAGEPs and regulations is enough to ensure safe operation of the process and PHA may not be necessary, but from risk control perspective, PHA is needed beyond compliance with RAGAGEPs and regulations, that is due to:

- Those RAGAGEPs and regulations only address the requirements for the boiler itself, while the boiler is only one piece of equipment in the whole process with other peripheral equipment and piping, actually every process is unique, need to systematic analyze the whole process to identify all the hazardous event scenarios and conduct further analysis.
- Since some codes and standards are mandatory according to local regulation, and some companies define compliance with certain RAGAGEPs as corporate policies even jurisdictional authority do not declare those RAGAGEPs are mandatory requirements. So following the RAGAGEPs is the minimum requirements while PHA is risk based decision making process, each company may have different criteria of risk tolerance levels, in some cases PHA team may conclude to take additional safeguards beyond those prescriptive requirements in RAGAGEPs and regulations to reduce risk to acceptable level according to the risk criteria. PHA team may also define more detailed requirements on specific safeguards required by RAGAGEPs, e.g. some interlocks are required to prevent hazardous conditions in boiler design, integrity level of the interlocks can be determined by risk based interlock evaluation methods, this will be further discussion in section 3.9, the integrity level requirement for the interlock with same function may be different in specific applications if the process design and operation methods are different.

3. Specific Considerations for Steam Generation and Combustion Process PHA

3.1 Scope of PHA

The process boundaries of the PHA shall be clearly defined to ensure no sub process and operation mode is missed, the whole process needs to be separated into different parts. For a typical steam generation and combustion process, following sub processes need to be considered,

- Fuel storage and supply outside the boiler package
- The burner and furnace
- Combustion air supply system
- Combustion product vent system (flue stack and draft fans)
- Water storage, treatment, supply system
- Water/steam side of the boiler (water drum, steam drum, tube bank, economizer, superheater, etc.)

Usually the steam distribution system is covered by a separate PHA but PHA team may also decide to review this part of process together with the steam generation process.

The operation modes to be reviewed include startup, normal operation, abnormal operation, shutdown and maintenance.

3.2 Process Technology information needed before the PHA

Sufficient and accurate process technology information is needed to ensure the effectiveness of the PHA, the process technology essentially contains following information of

- Hazard of the process
- Process design basis
- Equipment design basis

These information need to be assessed by the PHA team before PHA review is started, the team should correct minor errors as they are found, or document the need for updates and corrections as action items for follow-up outside of the PHA.

For steam generation and combustion system, the following need to be considered for each part of the process technology information,

3.2.1 Hazard of the process

The hazard of the process is inherent property of the materials (i.e. raw materials, intermediates, byproducts, products, etc) in the process and/or the process conditions (including startup, normal operation, abnormal operation, shutdown and maintenance) which can cause human injury, irreversible health damage and environmental pollution.

Some hazards of process need to be considered for the steam generation and combustion system,

- Dissolved oxygen in the water which can cause pitting corrosion and loss of containment.

- Hardness of water which can cause scale buildup in the heating surface which can then cause overheating in local area, and loss of containment of hot steam and water.
- Water treatment chemicals, which is potentially corrosive, toxic or reactive with other chemicals.
- Sulfur containing fuel such as coal and heavy fuel oil, which can be turned into sulfuric acid cause corrosion of equipment and potential loss of containment of hot water and steam. Sulfur dioxide in the flue gas also is air pollutant.
- Corrosion due to water with very low conductivity, erosion caused by water at high flow rate, these all can cause loss of containment in long term.
- Flammability of the fuel which can cause fire and explosion.
- Coal powder is combustible dust which can cause dust fire and explosion.
- High temperature of water and steam which can cause burn injury upon loss of containment.
- High energy of superheated water and steam mixture in the boiler drum, which can cause BLEVE (Boiling Liquid Expanding Vapor Explosion) after loss of containment. This can happen in fire tube boilers.
- Steam is also an asphixiant.

3.2.2 Process Design Basis

Following information need to be prepared before PHA,

- A process flow diagram containing the basic process parts such as burner fuel trains, combustion air supply, induced draft fans, water supply and treatment equipment, superheater, economizer and flue gas treatment equipment.
- The standard operating conditions including the normal pressures, temperatures and flows for steam, all fuels, feedwater, and combustion air.
- Safe steam and feed water system operating limits, safe fuel and combustion air operating limits.
- Consequence of deviations above the maximum and below the minimum operating limits needs to be documented if there is safety, health and environmental impact.
- Process description including functional description of steam generator firing rate controls, burner management system functions, alarm and interlocks, description of fuel supply systems and fuel trains and the combustion air supply system, detailed functional description of boiler water level indication and control, alarms and interlocks.
- The potential contamination sources and the impact on the steam generation equipment.

3.2.3 Equipment Design Basis

Following information need to be prepared before PHA,

- Piping and Instrument Diagrams (P&ID's),
- Electrical classification including drawings and reports that establish classifications
- Pipe, valve, and gasket codes
- Safety Instrumented Systems (SIS) technical information, including
    - Safety interlock electrical schematics and settings
    - SIS software configuration
    - Safety integrity level determination and PFD calculations
- Relief system design and design basis
- Ventilation system design
- Fire protection system design.
- Bonding and grounding system design.

Besides, there are many RAGAGEPs and regulations related with steam generation and combustion system such as the following, need to have documents to verify that the RAGAGEPs and regulations are followed per local authority and corporate policy requirements,

- Steam Boiler Safety Technical Supervision Administration Regulation
- Boiler House Safety Management Rules
- Safety Technical Regulation for Oil and Gas Burners, TSG ZB001-2008
- Boiler House Design Requirements, GB50041-2008
- Boiler Air Pollutant Emission Standard, GB13271-2001
- Industrial Boiler Water Quality Specification, GB/T1567-2008
- Industrial Boiler Installation and Acceptance Requirements, GB50273-1998;
- Boiler Periodical Inspection Rules

Some international standards and codes are also good references, e.g.

- NFPA 85 - Boiler and Combustion Systems Hazards Code
- ASME Boiler and Pressure Vessel Code
- The ASME Code for Power Piping, B31.1
- The ASME Code for Process Piping, B31.3

It need to be noted that NFPA 85 has set very detailed requirements on BMS (Burner Management System), i.e. the control system dedicated to ensuring combustion unit safety, providing assistance to operators in the starting and stopping of fuel preparation and burning equipment, and preventing misoperation of and damage to fuel preparation and burning equipment. It monitors operating parameters and includes the safety interlock system, fuel trip system, flame monitoring and tripping system, ignition subsystem, and main burner subsystems. The standard requires that the BMS shall be independent from other logic systems, e.g. firing rate, combustion fuel and air modulating control. The purpose is to ensure the reliability of the safety interlocks and safety critical functions.

There are no requirement on the independence of BMS in local codes such as TSG ZB001-2008 and Steam Boiler Safety Technical Supervision Administration Regulation, but from a risk control perspective, some companies have decided to follow the NFPA 85 requirements of separation of BMS from other control logics.

Besides, most of the boilers belong to special equipment according to local regulation, and there are pressure vessels, pressure pipe in the steam generation and distribution system, need to make sure the special equipment documentation requirements are followed per the regulation.

3.3 Hazard Identification

Hazard identification need to be done in the initial stages of the PHA. Hazards are inherent to the specific chemicals used (e.g., flammability, toxicity, reactivity, and corrosiveness) and/or their conditions of use (e.g., high pressure or temperature) in the process under review. Specific hazards can result in different kinds of undesirable hazardous events such as explosion, fire, other major energy release, or loss of containment of toxic or other harmful materials, which are capable of resulting in consequences such as serious injury to humans, irreversible human health effects, or significant damage to the environment. The results of hazard identification are used during the PHA to help focus the discussion and need to be included in the PHA report and in communication of the hazards to the affected personnel.

Normally hazard identification can be conducted by reviewing relevant incident investigation reports, previous PHA reports or checklist reviewing, etc.

Following is a summary of hazard identification for example,

Table 4. Hazard Identification of A Typical Boiler with Gas Burner

Sub Process	Hazard	Hazardous Events	Consequence
Gas supply system outside the boiler furnace	Flammability of fuel gas	Fire and explosion after loss of containment of the fuel gas	Serious human injury or fatality, equipment and facility damage.
Boiler burner and furnace	Flammability of fuel gas	Explosion inside the boiler furnace after ignition of flammable gas /air mixture pre-existed in furnace or over pressurization when fuel rich condition happens with no loss of flame, with subsequent return of air flow leading to rapid combustion of accumulated fuel gas.	Serious human injury or fatality, equipment and facility damage due to released heat and flame, the explosion blast wave and flying broken equipment pieces and secondary fire, explosion.
Boiler burner and furnace	High temperature	Furnace pulsation or pressurization conditions caused by abnormal process condition, e.g. insufficient induced draft.	Skin burn due to exposure to hot combustion product and/or flame, potential eye burn while furnace flame condition check.
Water storage, treatment, supply system	Water treatment chemical hazardous property, may be toxic, corrosive, reactive with other chemicals	1. Human exposure to the water treatment chemical after loss of containment. 2. Water pollution due to chemical loss of containment. 3. Energy release due to	Human injury and environmental damage, the extent of consequence depends on property of the chemicals.

		unexpected chemical reaction.	
Water/steam side of the boiler(water drum, steam drum, tube, etc.)	High energy of water and steam	Low water level in fire tube boiler: Uncovering furnace tubes leads to overheating and then BLEVE (Boiling Liquid Expanding Vapor Explosion) due to large amount of super heated water flashed into steam. Low water level in water tube boilers: Uncovering tubes leads to tube failure and loss of containment of water and steam and pressure loss prior to drum failure. For both fire tube and water tube boiler: Overpressure and shell/steam drum rupture (shell/steam drum material reaches its ultimate strength limit rather than small leaks) due to overfiring or blocked steam outlet, loss of containment of hot water and steam.	Serious human injury or fatality, due to expose to high temperature steam, blast wave of explosion and flying broken equipment pieces. Thermal burn injury Serious injury or fatality if spilled by hot water or steam, and/or hit by the broken pieces of equipment.
Flue stack and draft fan	Environment impact of incomplete combustion product	Release of incomplete combustion products during fuel rich burning condition.	Air pollution
Steam piping system	High temperature of steam and condensate	Loss of containment of steam and condensate or contact with poor, non-insulated hot pipe and equipment.	Thermal burn, could be serious especially when superheated steam is released, which is invisible.

Other types of boiler may have other types of hazards and hazardous event, e.g. environment pollution if loss of containment of combustible liquid fuel oil, coal dust fire and explosion for boiler using pulverized coal.

3.4 Consequence Analysis

Consequence analysis (CA) is an activity to understand the outcome of hazardous event scenarios assuming complete loss of all engineering and administrative controls. Consequences are analyzed independently of the event's probability. This evaluation includes estimating release amounts and conditions, evaluating consequences and affected areas, and determining the resulting acute safety, health and/or environmental effects.

The purpose of consequence analysis is to help the PHA team understand the type, severity, and number of potential injuries; possible adverse impact on facilities and structures; and significant environmental effects, at both on-site and off-site locations, of unmitigated events.

Quantitative consequence analysis can be conducted by using modeling tools to understand the safety impact to human beings and equipment, building to certain existing criteria, in some cases, quantitative CA is needed per jurisdictional requirement or the consequence is high, e.g. there is off-site impact or impact multiple persons and buildings. Qualitative CA can also be considered if the consequence is relatively low. In some cases for fire and explosion scenarios, there may be no well-developed technology to accurately quantify the consequences and qualitative analysis may be better. Qualitative CA can be done by comparing the event scenarios with other documented accidents, e.g. as shown in Figure 1, to understand the consequence or experienced consequence analyst can be consulted as they understand the boundaries and limitations of some modeling tools and may be able to give best estimation of the consequence.

Figure 1 Explosion of A Fire Tube Boiler due to Low Water Level

For the steam generation and combustion system, consequence of the following scenarios needs to be analyzed,

- Loss of containment of flammable gas/vapor into open area or confined space e.g. boiler furnace, boiler room, which can cause fire and/or explosion.
- Combustible dust fire and explosion.
- BLEVE of fire tube boiler
- Thermal burn due to exposure to the flame when there is pressurization in the furnace.
- Thermal burn due to exposure to the released steam and hot water.

Quantitative CA can be done for some scenarios such as the boiler BLEVE, indoor or in vessel explosion of flammable vapor/air mixture, following is the result of modeling result which use the isentropic expansion energy for estimation of the explosion energy from the vapor phase expansion during BLEVE of boiler drum, which gives overpressure and impulse at the distance away from the center of explosion.

Figure 2 Over Pressure and Impulse Vs. Distance From a Hypothetic Boiler BLEVE

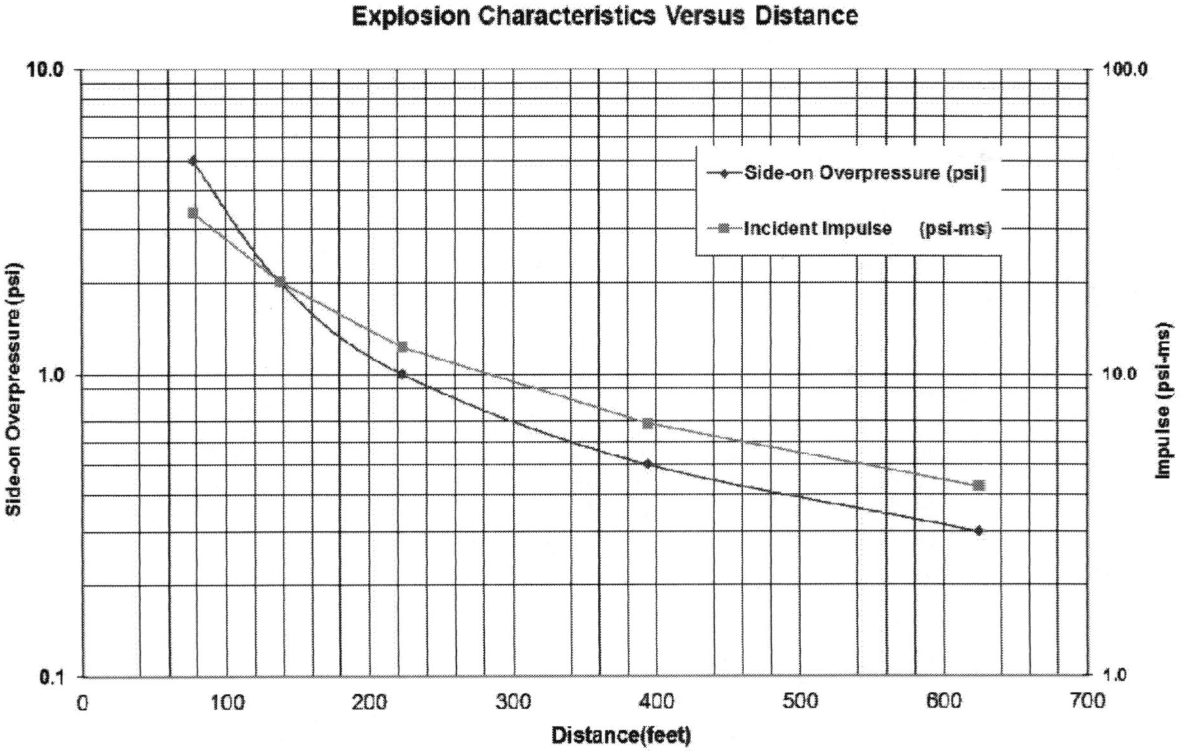

The level of damage of facility and injury of human beings at specific distance from the center of explosion can be estimated against certain criteria such as Table B-1 and B-2 in the CCPS book "Guidelines for Evaluating the Characteristics of Vapor Cloud Explosions, Flash Fires, and BLEVEs" [4]

It need to be noted that the fragments of the boiler after the BLEVE may also hit the personnel or nearby building on its' path of travel, cause severe human injury or building damage and if the boiler is located indoor, the wall of the boiler room may impede the travel of the explosion overpressure and reduce the impact to the nearby buildings.

For the scenario of flammable gas release indoor, NFPA 68-2007 provides a methodology for estimating the vent area requirements from vapor explosions in low strength enclosures(<1.5 psig maximum pressure) where there are few internal obstructions and the length to "Diameter" of the enclosure is very low, and only applicable for medium to low reactivity hydrocarbons, e.g. methane and propane. Section 7.2.2 of the standard provides the formula for calculation. The formula can be used to estimate the damage level of the building by the explosion inside the room, normally where the boiler is located, caused by ignition of air and flammable fuel gas mixture.

$$Av = C \times (As)/P_{red}^{1/2} \qquad \text{[Eq. 1]}$$

Av is the vent area of the enclosure, C is the venting parameter which is specific for the flammable gas used, As is the internal surface area of enclosure, normally these parameters are known for the installation, the Pred, i.e. the maximum pressure developed in a vented enclosure during a vented deflagration, can be calculated. The damage level of building can be estimated based on the value of Pred calculated.

3.5 Hazard Evaluation

The hazard evaluation is the core process of the PHA, the purpose is to identify the potential routes to the hazardous events of concern in thorough and systematic ways, and assess the risk by evaluating the consequence of the event considering failure of all the engineering and administrative controls, and the frequency of happening of the consequence, i.e., the combination of the various initiating events, the effectiveness of the existing or planned lines of defense.

There is different methodologies that can be used for hazard evaluation, e.g. Hazard and Operability Analysis (HAZOP), Failure Mode and Effect Analysis (FMEA), What If/Checklist, etc. Following is a brief comparison between the methods on the specific advantages and disadvantages.

Table 5. Comparison Between Different Hazard Evaluation method

Method	Advantages	Disadvantages
Whatif /Checklist	Combine the power of brainstorming of team members and the structural review of specific checklists.	No inherent structure of what if review, rely on PHA facilitator and team members' skill and knowledge.
HAZOP	Structured systematic method focusing on deviations from the normal operation conditions	1. Need experienced facilitator and team members to avoid missing or mis-interpretation of any questions. 2. Need large amount of time, team member may get "brain dead" after long time discussion. 3. Focus on steady state of process, may miss reviewing on intransient periods.
FMEA	Structured systematic method focusing on all the potential failure modes of components in the system, mainly each part of equipment.	If the method is used alone, may need to expand the types of components, e.g. including operator and each material, or may miss some hazardous scenarios.

For steam generation and combustion system, it may be appropriate for the use of structured what if/checklist method, i.e. combining the strength of HAZOP, FMEA, whatif/checklist methods. It can be conducted in following steps.

1. Ask HAZOP type questions regarding to deviation from normal process conditions for different part of process, line by line and equipment by equipment, e.g.
    - What if fuel pressure in fuel train is high/low?

- What if flow rate in the fuel train is high/low?
- What if combustion air flow rate is high/low? etc.
- What if ratio of fuel/combustion air flow rate is high/low?
- What if drum water level high?
- What if drum water level low?

  …

2. Ask FMEA type questions focusing on failure of components in the system, e. g.
   - What if the boiler drum level transmitter gives false high/low level reading?
   - What if the safety shutoff valve on the fuel train leak through/ fails opened/ fails closed, etc?
   - What if the fuel pressure regulator stuck fully opened?
   - What if the fuel pressure regulator stuck fully closed?
   - What if loss of induced draft fan?

     …

3. Let the team members to ask classic what if type questions by creative thinking, this can find questions that are not discovered in the HAZOP and FMEA type review and can help the team to identify potential causes of hazardous events that may originate outside of the process itself. e.g.
   - What if the flue stack flow is blocked or obstructed?
   - What if there is fuel gas leak in the boiler house, and the leaked gas is sucked into the burner during light off?
   - What if the drain valve of the boiler drum is left open?
   - What if loss of electricity to the combustion air blower?

     …

4. Use checklists to develop questions that are not asked in the above steps, there is checklist which focuses on specific area, such as these on the subject of loss of containment and some relevant codes and standards can also be used as checklist, e.g. those listed in section 3.2.3.

Following is an example what if question and relevant analysis.

Table 6. Example of What if Question and Analysis

What if	Consequence	Existing Safeguards	Frequency Level	Consequence Level	Risk Level	Recommendation
The online boiler feed water pump stop working due to mechanical	Loss of water supply to the fire tube boiler, leading to low water level, the furnace tubes are overheated and could fails eventually, BLEVE happens when large amount of	1. Site has feed water pump preventive maintenance procedure. 2. Operator continuously attended in the area. Low water alarm set point	1	4	III	The PHA team feels that the risk is adequatel

failure?	superheated water flashes to steam due to rapid loss of pressure, multiple serious injuries and fatalities can happen if people are exposed to the missiles of broken equipment pieces, blast wave, high temperature steam.	in BPCS with about 30 minutes to uncover main furnace tube; Two level indications are provided; There is low water level response procedure and the operators are trained on the procedure. 3. Two independent low level interlocks devices hard wired to close fuel safety shutoff valves and open gas bleed valve. 4. Weekly partial blow down test is done and annually full functional test are done on the low level alarm and interlocks. 5. There is maintenance procedure to inspect the devices of the low level interlocks.				y mitigated by the existing safeguards, no further action required.

3.6 Facility siting

The facility siting analysis is intended to assess the risks posed to occupants of buildings on a site by process-related events originating outside those buildings. Recommendations may be generated to control the risk to acceptable level. For sites having multiple process units, this analysis is typically integrated into a single, site-wide facility siting study, which may be revalidated periodically. If a new steam generation and combustion system will be installed, need to evaluate the impact of the new system on existing buildings in the site, this include

1. Hazard Screening - identify the hazards, hazardous events that can impact the buildings, this can be some items selected from the results of hazard identification as listed in section 3.3.
2. Building Screening- review the design, function, and occupancy of site buildings, the buildings with critical functions such as fire pump house, emergency command center also need to be included although normally they are unoccupied.
3. Consequence screening based on site-specific modeling or certain spacing criteria. For the explosion scenarios, some modeling tools can provide overpressure and impulse data at certain distance from the origin of the explosion. With the building structure information and data of overpressure and impulse received by the building, damage to the buildings can be estimated by using of specific software. BEAST (Building Estimation And Screening Tool) is one of these kinds of software, it is developed by Baker Engineering and Risk consultants, Inc. for Explosion Research Cooperative. There is some spacing criteria can also be used as reference, e.g. those listed in API RP753.

Following is an example that using BEAST to estimate the impact to the building and occupants in the case of hypothetic boiler BLEVE, the relation of distance to overpressure and impulse is shown in Figure 2.

Table 7 Building Damage and Occupant Injury Estimates- Fire tube Boiler BLEVE

Building Name	**Structure Name**	**Damage Level**	**Over Pressure (psi (g))**	**Impulse (psi-ms)**	**Building Occupant Vulnerability**
Building A	1. Steel Frame, Metal Siding/Roof	2A - Light Moderate	0.84	10.73	0.1
Building B	3. Steel Frame, Unreinforced CMU Infill	2B - Heavy Moderate	3.58	28.38	0.5
Building C	1. Steel Frame, Metal Siding/Roof	2A - Light Moderate	1.02	12.50	0.1
Building D	1. Steel Frame, Metal Siding/Roof	1 - Minor	0.59	7.92	0.01

There are specific definitions for each building damage level in BEAST application. One way to estimate the level of consequence of people get injured from the scenario is by multiplying building occupant numbers with the vulnerability, i.e. probability of serious or fatal injury.

4. Risk screening based on qualitative or quantitative risk assessment, according to pre-set risk criteria, i.e. evaluate the risk level by considering both the consequence level and frequency level, certain recommendations may be generated to control the risk level.

Of course, some local standards also need to be considered in the design and siting of the boiler house, e.g. Boiler House Design Requirements GB50041-2008, these standards set the mandatory requirements to be followed, this review shall be done before the facility siting analysis, the aim of the recommendations resulted from the facility siting review is to further reduce the risk to acceptable level.

3.7 Inherently Safer Process Review

Processes can be made inherently safer through the elimination of hazards, rather than control of hazards, the use of ISP technology is an approach to the safe design and operation of facilities.

ISP technology need to be considered in the whole life cycle of a facility, but the best time to consider it is during the early stage of capital project for new facility, as there are more options of process design in the conceptual design period. During the PHAs, ISP evaluation can be incorporated into What if/ Checklist review, by ask questions of whether there are choices to eliminate or greatly reduce the hazards in the process, e.g. what could we do to reduce the amount of hazardous material in the process? What could we consider to run the process at lower temperature and pressure, etc. There is also specific ISP review checklist can be used such as that in the CCPS book "Inherently Safer Chemical Process- A Life Cycle Approach" [5].

For steam generation and combustion system, one may argue that the process technology is well developed and there may not be opportunities for application of ISP options, but still there are some can be applied, e.g. boiler can be classified as water tube boiler and fire tube boiler based on hot combustion air and water flow path design, water tube boilers are designed to flow water through the heat transfer tubes, instead of combustion air flow in the tubes as of fire tube design, because of the smaller diameter of the pressurized components, water tube are better able to contain the higher pressure than the fire tube units, while the fire tube boilers are susceptible to BLEVE, thus the water tube boiler design is an inherently safer option comparing with the fire tube boiler, if it can meet process needs.

If gases are chosen as fuel for the boiler and vent stack is installed for the furnace, although pre purge and post purge by forced ventilation is normally designed to remove residue fuel gas remained in the furnace to prevent explosion at light off period, gas lighter than air such as natural gas would be much more easily dispersed from the stack comparing gas heavier than air such as LPG.

Besides, other heating method can be considered instead of combustion heating system, the author once has worked in a capital project PHA for the process involving an oil burner to provide heat to the HTF (Heat Transfer Fluid). After evaluating the hazards in the combustion system, project team finally selected electrical heating as heating method.

3.8 Human Factor Analysis

Many process safety incidents are caused by human errors directly, for example, as shown in the Table 1, i.e. the statistics of boiler incidents happened in US (1995-1998), human error are the direct cause of about 26% of the incidents, including improper installation, improper repair, operator error or poor maintenance. Similarly, as mentioned in the section 1, 7 of 40 (i.e. 18%) boiler explosion incidents happened in China in 1993 is caused directly by operator error, which is adding water right after stop of water supply. NFPA 85, 2011 version section A 4.3.1 also mentions that "Statistics indicate that human error is a contributing factor in the majority of explosions."

But putting all the blames on the field operator and mechanics doesn't not help to define the safeguards to effectively prevent the reoccurrence of the incidents. One important part of the PHA is human factor analysis, the focus of the analysis is to identify and avoid situations where human error is likely, both in the process and in the maintenance of the equipment and systems associated with the process, some error likely conditions including,

- Deficient procedures or procedural violations
- Inadequate, inoperative, or misleading instrumentation
- Poor layout or design of controls
- Poor task design (e.g., excessive mental tasks or extended periods of uneventful diligence)
- Poor communications.
- Conflicting priorities (e.g. safety, production, cost, etc.)
- Circumstances resulting in unusual or excessive fatigue of the work force

For steam generation and combustion system, following items need to be emphasized during human factor analysis,

- Normal startup and shutdown procedure,
- Normal operation and emergency operation procedure,
- Emergency shutdown procedure,
- Startup following emergency shutdown,
- Inspection and maintenance procedure

Human factor need to be considered during the field observation, using the hazards evaluation methodology to identify hazardous events, and while considering the lines of defense. A human factor review checklist can be used to identify and evaluate the human factor issues, such as the checklist in the book "Human Factors Methods for Improving Performance in the Process Industries" [6]. Besides, there are relevant requirements in the standards that can be taken as reference during human factor review, e.g. In NFPA 85, instructions on operation are provided for different types of boilers, section 4.4 describe the requirements on maintenance, inspection, training and section 4.5 list basic operation requirements. Requirements on operation and maintenance procedures are also described in local standards such as Steam Boiler Safety Technical Supervision Administration Regulation, Boiler House Safety Management Rules.

3.9 Interlock Evaluation

Interlocks are one type of safeguards for hazardous scenarios. For boilers, specific interlocks are required in the codes and standards to prevent hazardous events due to loss of control of fuel flow, combustion air and water level in boiler drum.

NFPA85 2011 version, section 4.11 describe the requirements of Burner Management System Logic, in section A 4.11, it provide alternative that meets an appropriate safety integrity level (SIL) for safety instrumented systems (SIS) based on the methods of the ISA 84 standard series.

There are other similar RAGAGEPs such as IEC 61511 call for specific actions to be taken as part of the hazards evaluation activity with respect to the evaluation of interlocks as instrumented layers of protection against process hazardous events. China SAWS also issued a draft standard named "Guidelines for layer of protection analysis of chemical enterprises" in December 2012.

LOPA (Layer of Protection Analysis) is one of the methods for SIL determination, which is a semi quantitative method, LOPA uses order of magnitude categories for initiating event frequency, consequence severity, and the likelihood of failure of independent protection layers (IPLs) to approximate the risk of a scenario. LOPA is an analysis tool that typically builds on the information developed during a qualitative hazard evaluation, such as what if, HAZOP, FMEA analysis.

LOPA only apply to the event of single cause and consequence pair, i.e. a chained events starting from the single initiating event, the failure of each IPL (Independent Protection Layer) in the time order to the final consequence. For other types of events, QRA method such as event tree analysis should be used.

In addition to an initiating event and a consequence, a scenario may also include enabling event(s) that have to occur or be present for the initiating event to result in the consequence. For example, an initiating event might be fuel gas supply pressure regulator stuck open which would result in explosion and over pressurization inside the furnace due to fuel rich condition, the undesired consequence of injury or fatality only happens when the enabling event of "operator presents in the nearby area" exists.

There are different types of IPLs, e.g. basic process control system(BPCS), alarm and human intervention, safety instrumented functions(SIFs, which is called safety interlock previously), physical protection and mitigation devices such pressure relief valve and rupture disc, dike. A SIF is functionally independent of the BPCS. The failure rate of IPL is called Probability of Failure on Demand (PFD),

Based on the hazardous scenario developed in hazard evaluation, the initiating event, IPLs and consequence are identified, as the chain of events marked in bold line in figure 3.

Figure 3 Hazardous Event Analyzed By LOPA Method

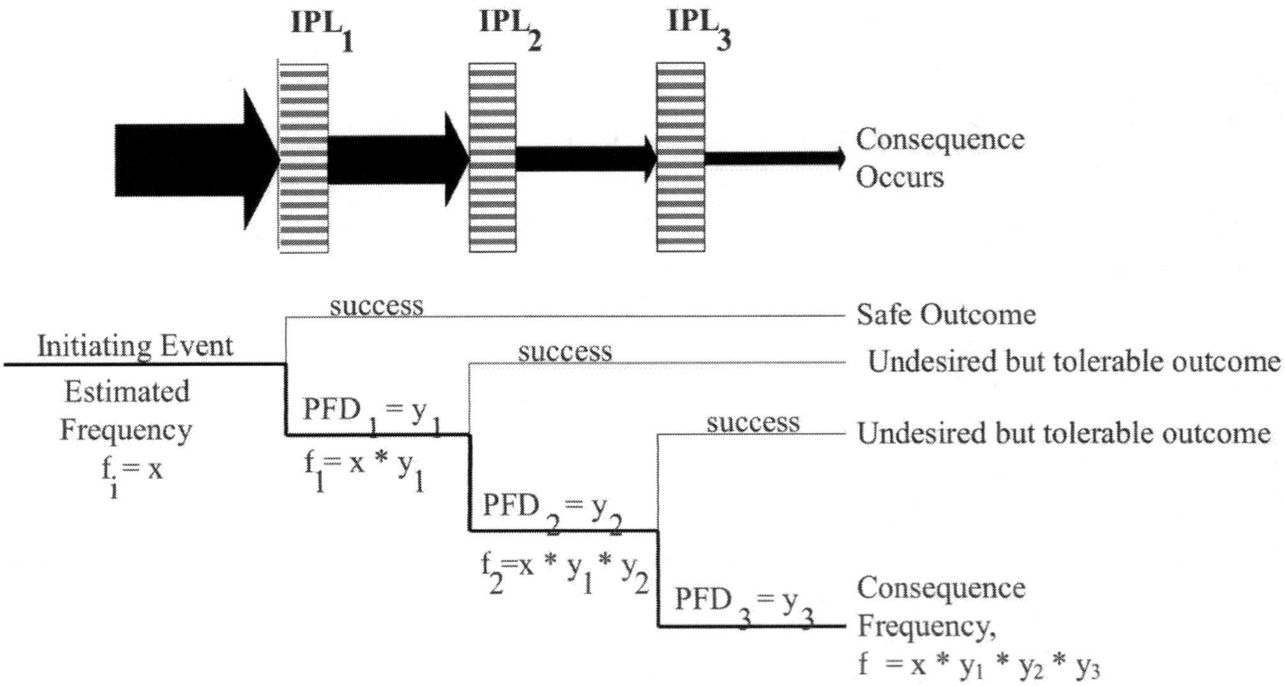

The tolerable frequency for the consequence is determined by the risk criteria of each company. The frequency of the different types of initiating events and PFDs of IPLs can be found in industry statistics and some standards such as IEC 61511-3, table F.4 and F.3, some companies have defined their own guideline on these data.

For the scenario involving interlocks as safeguard for the final consequence, the PFD of the interlock can be determined when tolerable frequency for the consequence, initiating event frequency and PFDs of other IPLs for the consequence are known. For SIFs, the safety integrity level (SIL) is then determined based on the PFD, which can be referred to Table 3 & 4 of IEC

61511-1. It needs to be noted that same SIL requires different PFD under low demand condition and high demand condition. For the interlock function that is designed in the BPCS, according to section 3.2.3 of IEC 61511-1, it does not perform any safety instrumented functions with a claimed SIL ≥ 1 and common mode failure need to be considered when other BPCS functions are claimed to IPL, such as alarm and human intervention.

Following is an example LOPA scenario on fire tube boiler low water level interlock, which is based on the example what if scenario in Table 6.

Table 8 Example LOPA scenario on Fire Tube Boiler Low Water Level Interlock

Scenario Description	The online boiler feed water pump stop working due to mechanical failure, loss of water supply to the fire tube boiler, leading to low water level, the furnace tubes are overheated and could fails eventually, BLEVE happens when large amount of superheated water flashes to steam due to rapid loss of pressure, multiple serious injuries and fatalities can happen if people are exposed to the missiles of broken equipment pieces, blast wave, high temperature steam.
Consequence category	Level 4
Event frequency goal	0.00001/year (per company guideline)
Initiating event frequency	0.1/ year (mechanical failure of feed water pump)
Enabling event probability	1 (Operator continuously attended in the area.)
Likelihood of significant injury	1
Additional mitigation IPL	In this case, it is BPCS alarm and operator response, 0.1/ year
Targeted Safety Interlock PFD	0.001/ year for low water level interlock

Since one interlock may be safeguard for different initiating events leading to same consequence, for example, other initiating events for low water level conditions could be closing the valves on feed water pipe or leaving boiler drain valve opened by mistake, feed water pipe catastrophic leak, loss of electricity supply to feed water pump etc. Target PFD of the interlock need to be analyzed for each scenario and the lowest PFD value is selected as the target PFD for the interlock.

4 Conclusion

In the past several years, the author has been involved in some capital investment projects related with steam generation and combustion process. The type of facilities where the process is installed includes manufacturing sites, lab and office building. The PHA activities introduced in the above sections have been performed in these project PHAs, the precautions and benefits of these activities, as considered by the author are as follow,

- In the early stage of the project, starting the hazard identification, inherently safer process review, consequence analysis and facility siting would help the PHA team to understand the hazards involved in the process, the extent of safety and environmental impact and the options in design to eliminate or reduce the hazards.

- While prepared for a detailed hazard evaluation, process technology information including hazard of process, process design basis and equipment design basis need to be prepared in advance, the emphasis is to identify the RAGAGEPs and regulations that need to be followed, this is not only for regulatory compliance purpose, as these RAGAGEPs and regulations also set very detailed prescriptive requirements related with process safety.
- Need to consider all the operation modes and sub processes in the hazard evaluation, human error is an important cause of the process safety incidents related with boilers, identifying the potential human error scenarios and the countermeasures shall be included in the review.
- Through interlock evaluation, the integrity of interlocks can be identified to meet the risk reduction target.
- The safeguards and recommendations identified by the PHA frequently are related with other elements of PSM system such as mechanical integrity and quality assurance program, operation procedure, employee training, emergency preparedness, etc. Operation team can start to establish the PSM system at the project phase based on the PHA results.

The capital project PHA activities usually takes significant time and efforts, but it is well worthy of doing so, as early planning and establishing the incident prevention and mitigation system is the most cost effective way for safe operation.

5 References

[1] The National Board of Boiler and Pressure Vessel Inspectors
https://www.nationalboard.org

[2] Bulletin of Boiler and Pressure Vessel Incidents in China, 1993, issued by Department of Labor, No.122, 1994

[3] USA OSHA, Standard Interpretation,
http://www.osha.gov/pls/oshaweb/owadisp.show_document?p_table=INTERPRETATIONS&p_id=21933

[4] Center for Chemical Process Safety, Guidelines for Evaluating the Characteristics of Vapor Cloud Explosions, Flash Fires, and BLEVEs, American Institute of Chemical Engineers, 1994

[5] Center for Chemical Process Safety, Inherently Safer Chemical Processes A Life Cycle Approach, Second Edition, John Wiley & Sons, 2009

[6] Center for Chemical Process Safety, Human Factors Methods for Improving Performance in the Process Industries, John Wiley & Sons, 2007

1ST CCPS ASIA-PACIFIC CONFERENCE ON PROCESS SAFETY

Investigation on Mechanical Properties and Strength Evaluation for Large-scale Crude Oil Tank after Fire Exposure

Wen-he Wang
College of Civil Engineering, Central South University
Central South University, Yuhua District, Changsha, 410083, P.R.C.
College of safety engineering,Chongqing University of Science & Technology
Chongqing University of science and technology, HuXi Town, ShaPingBa District, Chongqing, 401331, P.R.C.
wangwenhe518@163.com or whwang@cqust.edu.cn

Zhi-sheng Xu
College of Civil Engineering, Central South University
Central South University, Yuhua District, Changsha, 410083, P.R.C.
zhshxu@mail.csu.edu.cn

Jun Yi
College of safety engineering,Chongqing University of Science & Technology
Chongqing University of science and technology, HuXi Town, ShaPingBa District, Chongqing, 401331, P.R.C.
Yijun2322@yahoo.com.cn

Dong Liang
Guangdong Provincial Key Laboratory of Fire Science and Technology
Sun Yat-Sen University, Xingang Xi Road, Guangzhou, 510275, P. R. C
liangdong@mail.sysu.edu.cn

Prepared for Presentation at
American Institute of Chemical Engineers
1st Asia-Pacific Conference on Process Safety
Qingdao, China
September 3-5, 2013

UNPUBLISHED

Investigation on Mechanical Properties and Strength Evaluation for Large-scale Crude Oil Tank after Fire Exposure

Wen-he Wang
College of Civil Engineering, Central South University
Central South University, Yuhua District, Changsha, 410083, P.R.C.
College of safety engineering,Chongqing University of Science & Technology
Chongqing University of science and technology, HuXi Town, ShaPingBa District, Chongqing, 401331, P.R.C.
wangwenhe518@163.com or whwang@cqust.edu.cn

Zhi-sheng Xu
College of Civil Engineering, Central South University
Central South University, Yuhua District, Changsha, 410083, P.R.C.
zhshxu@mail.csu.edu.cn

Jun Yi
College of safety engineering,Chongqing University of Science & Technology
Chongqing University of science and technology, HuXi Town, ShaPingBa District, Chongqing, 401331, P.R.C.
Yijun2322@yahoo.com.cn

Dong Liang
Guangdong Provincial Key Laboratory of Fire Science and Technology
Sun Yat-Sen University, Xingang Xi Road, Guangzhou, 510275, P. R. C
liangdong@mail.sysu.edu.cn

Keywords: Large-scale storage tank; Crude oil; Fire exposure; Strength evaluation

Abstract

In recent years, the demand of the crude oil is increasing in the world, and the oil storage tanks are also developing larger and larger. Higher requirements of safety for storage tank, especially safety evaluation of the oil tanks in fire environment, was proposed because the oil tank volume is large, as well the oil is volatile, flowing, inflammable and explosive easily. In the paper, the fire process was simulated by the heat treatment for the key position, and the relationship between mechanical property and heating temperature of large tank after fire was obtained. The strength evaluation for large-scale crude oil storage tank after fire exposure was implemented and the result showed that the strength for large crude oil tank was satisfied with requirement.

1. Introduction

In recent years, with the rapid growth of globle economy, the demand of the crude oil is increasing in various industries, and the oil storage tanks are also developing larger and larger[1].

Higher requirements of safety for storage tank, especially safety evaluation of the oil tanks in fire environment, was proposed because the oil tank volume is large, as well the oil is volatile, flowing, inflammable and explosive easily.

Large-scale oil tanks are generally rest on the basis of reinforced concrete ring beam in the form of non-anchor. Equal strength design are used for oil tank wall thickness mostly, the stress from junction of tank bottom and the bottom wall plate is very complicated when it is subjected to hydrostatic pressure. From the research status at home and abroad[2-7], the studies about mechanical properties of material after fire exposure focused on building materials mostly, especially the relationship of fire temperature and burning time about the steel and concrete structure after fire exposure,and there is less studies on the structural strength, the relationship of strength and burning temperature,fracture assessment with defects of oil tanks after fire exposure.Three level assessment method recommended by API579 is a qualitative evaluation method for equipment after fire exposure, but the current research about the specific evaluation method and process for oil tanks exposed in fire is less[8].

Heat treat process can be used for simulating key position of the tank materials after fire process, and the material exposed in fire is cooled with water to room temperature[9-10]. In this paper, mechanical properties of SPV490Q steel with weld exposed in different fire temperatures are studied, the relationship about yield strength, tensile strength and different fire temperatures of SPV490Q steels are obtained.meanwhile, large-scale oil storage tank strength is evaluated in accordance with standards.

2. Parameters of large-scale crude oil storage tanks

Take the $10\times10^4m^3$ large-scale crude oil storage tanks in Zhoushan oil reserve base for example, the cylindrical tanks top is two-disc external floating roof structure, its wall plate and edge of the bottom wall are T-shaped plate welded structure. Tanks are based on reinforced concrete ring beam.,the relative overhang length between the tank bottom edge of the plate and the tank wall is 100mm. The base of tanks are concrete ring beam and backfill sand. The structure of tank is shown as Figure 1.

Seen from Figure 1, the inner diameter of tank is 80m, total height 21.8m and designed liquid height 20.2m.Corrosion allowance of tank wall is 1mm, tank bottom edge of plate material is SPV490Q and its thickness is 21mm, widest width is 1980mm; the center plate material is Q235-B and its thickness is 11mm, the tank wall panels welded together by the 9 circles, the tank material and geometric parameters is list in Table 1.

Figure 1:10,0000 cubic meters external floating roof oil tank

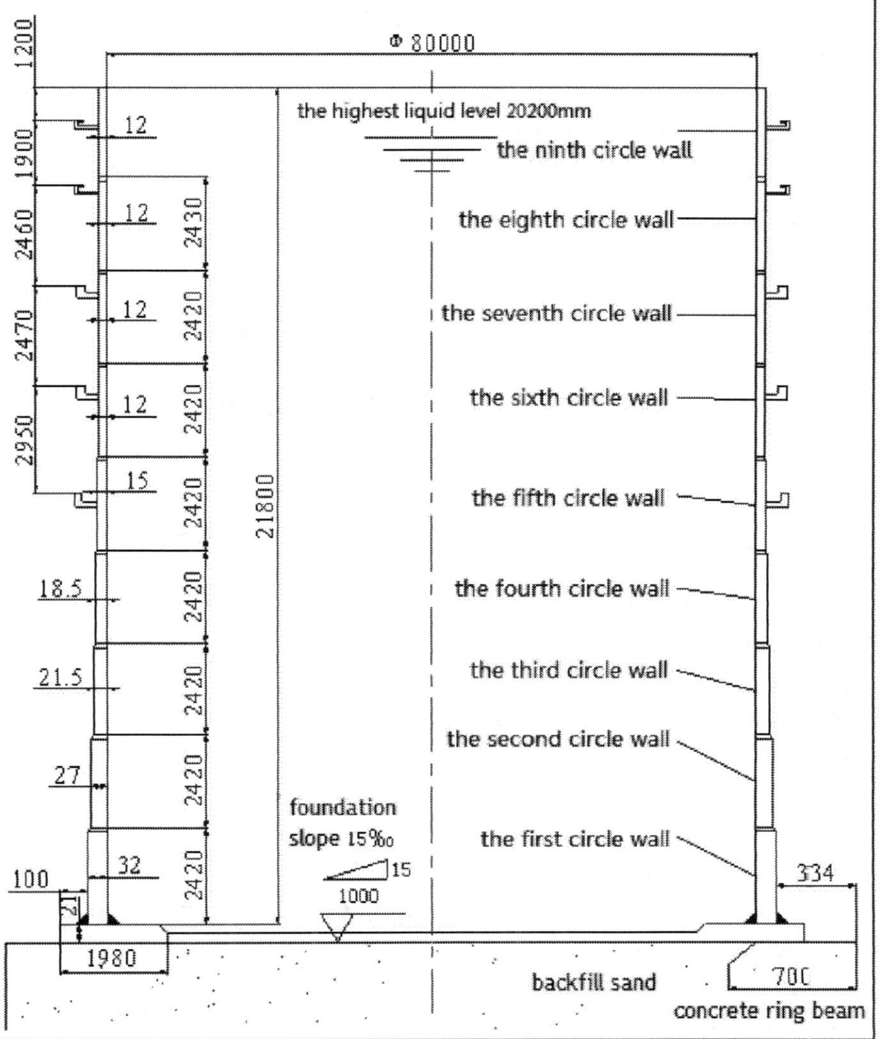

Table.1 Geometric parameters of tank wall

Tank wall layers	Materials	Wall thickness/mm	Heigth/mm
9	Q235B	12	2 430
8	Q235B	12	2 430
7	16MnR	12	2 420
6	SPV490Q	12	2 420
5	SPV490Q	15	2 420
4	SPV490Q	18.5	2 420
3	SPV490Q	21.5	2 420
2	SPV490Q	27	2 420
1	SPV490Q	32	2 420

3. Determining key positions of large-scale crude oil storage tanks

According to10×10^{4}m^3 floating roof tanks stress measurement by Chen Difei , Liu Yuming and Chen Zhiping et al [2,3],and finite element strength analysis of large crude oil storage tanks by Chen Zhiping, Zhou Lifang and Hui hu et al[4-6], the results indicated that it.Joints of the bottom internal surface and the inner surface of the tank wall is the large fillet weld where the stress is biggest.The maximum hoop stress on the outer surface of the tank wall located in the second layer near the first layer of wall plate.

Refering to these conclusions[2-6], the strength evaluation of the large-scale crude oil storage tanks include two parts,strength evaluation of tanks wall and tank bottom large fillet welds strength evaluation after fire exposure.

4. Stress distribution of large-scale crude oil storage tanks

On the condition of Hydrostatic test,the deformation and stress distribution on the bottom and wall of 10×10^{4}m^3 large-scale crude oil storage tanks can be expressed as follows[5].

1) Large fillet welds stress distribution. Due to the liquid pressure and different settlement of the base, the portion of the tank wall close to the bottom occurred deformation in the direction off the ground. Internal surface and external surface of the tank bottom is under radial tensile stress and radial compressive stress respectively, joints of the bottom internal surface and the inner surface of the tank wall is the large fillet weld where the stress is biggest and the maximum stress value is 520. 6MPa.

2) Hoop stress distribution on the outer surface of the tank wall. The maximum hoop stress on the outer surface of the tank wall located in the second layer near the first layer of wall plate. 6 layers under the lower part of the tank wall subjected to larger hoop stress and the maximum value is 258. 8MPa.

5. Experiment of large-scale crude oil storage tank after fire exposure

5. 1 Test sample and procedure

Test sample material is SPV490Q steel and its size is 360 mm *304 mm* 12 mm,which are cut from the site of welding(width is 10mm), its chemical composition as shown in table 2.

Table.2 Chemical compositions of SPV490Q steel (wt%)

Steel Type	C	Si	V	P	S	Mn	Ni
SPV490Q	0.14	0.26	0.04	0.007	0.003	1.43	0.02

According to combustion temperature while tank burning and a method for determining heat exposure region from API579 [7], and the heat exposure region whthin V level is not enough to cause damage to equipment. Steel heating temperature is adopted respectively at room temperature, 750 ℃, 800 ℃ and 850 ℃,and the heat holding time is one hour.Samples were heated to the specified temperature and heat holding time in the heating furnace, they were took out from furnance and cooled to room temperature by water. The test samples after heat treatment were processed into tension specimen which shape and size as shown in figure 2.

Figure 2: Dimension of tension specimen

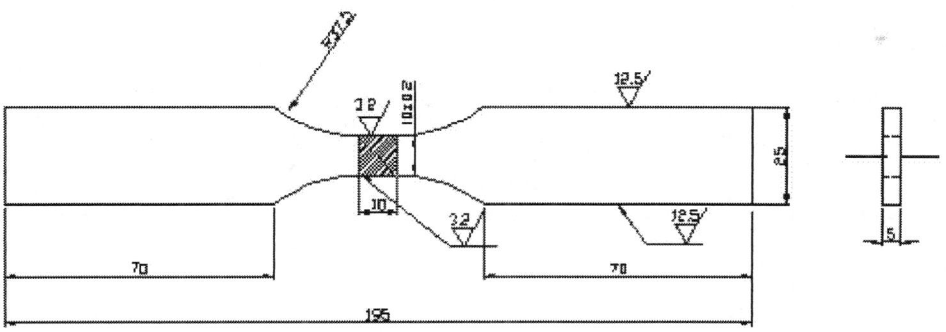

5. 2 Experiment results

In order to obtain the material mechanical properties including the yield strength, tensile strength,etc.Samples after fire exposure in deferent temperature point were test by tensile testing machine ,and its test results are shown in table 3.

Table.3 The mechanical properties of tensile test after fire exposure

Number	Heating treatment temperature/℃	Yield strength /MPa	Tensile strength /MPa	Elongation /%	Reduction of area/%
T01	room temperature	540	651	20	42
T02	600	501	622	25	46
T03	700	430	565	25	54
T04		471	685	15	33
T05	750	495	728	20	35
T06		510	750	17	30
T07	800	462	714	17	32
T08	850	421	635	27	60
T09	room temperature	551	655	22	36
T10	600	495	627	23	52
T11	700	435	629	24	49
T12	750	504	693	17	36
T13	800	475	705	23	41
T14	850	416	647	25	58
T15	room temperature	532	661	26	43
T16	600	513	634	18	44
T17	700	451	636	22	59
T18	750	482	689	18	32
T19	800	473	721	16	35
T20	850	431	631	31	58

6 Storage tank strength evaluation after fire exposure

The tank stress can be analyzed and evaluated by the JB4732-95 about analysis design standard of steel pressure vessel [10]. The design of the steel stress intensity should be a minimum of the following values: a) 1/2.6 of minimum tensile strength value at room temperature. (b) the yield strength at room temperature or 1/1.5 of 0.2% minimum yield strength value. (c) the design temperature yield strength or 1/1.5 of 0.2% minimum yield strength value. So the allowable stress of SPV490Q material after fire exposure can be calculated and are shown in table 4.

Table.4 The allowable stress of SVP490Q with weld after fire exposure

Items	Mean value of yield strength /MPa	Variance of yield strength /MPa	Mean value of extension strength /MPa	Variance of extension strength/MPa	Mean value of allowable stress /MPa
room temperature	541	9.5	656	5.0	252
600°C	503	9.2	628	6.0	241
700°C	439	11.0	627	9.6	242
750°C	492	11.1	689	4.0	265
800°C	470	7.0	713	8.0	274
850°C	422	7.6	638	8.3	245
900°C	959	14.8	1036	19.9	398

6.1 Tank wall plate strength evaluation after fire exposure

On the basis of hydraulic test and finite element simulation results[8-10],the strength of large-scale storage tank can be analysed and evaluated. The primary membrane stress *Re* are situated in the maximum stress of wall plate is 258.8MPa, and there is a combination effect of primary general membrane stress and bending stress in the location. According to documents [10] , the primary membrane stress *Re* should be less than $1.5S_m$, so the evaluation results of stress are checked by documents is qualified in the location and are shown in table 5.

Table.5 The strength evaluation of tank wall after fire exposure

Items	Mean value of allowable stress /MPa	The maximum equivalent stress of tank wall /MPa	Mean value of allowable limited stress /Mpa (Calculated)	Results of stress checked
room temperature	252	258.8	378	Qualified
600°C	241	258.8	361.5	Qualified
700°C	242	258.8	363	Qualified
750°C	265	258.8	397.5	Qualified
800°C	274	258.8	411	Qualified
850°C	245	258.8	367.5	Qualified
900°C	398	258.8	597	Qualified

6.2 Large fillet weld strength evaluation of tank after fire exposure

On the buttom and wall plate junction , the stress of the large fillet weld area belongs to the combination of primary and secondary stress, the allowable combination stress should be controlled within $3S_m$ according to JB4732-95 standards[10]. The maximum stress σ_{max} of the large fillet weld area is 520.6MPa which is less than $3S_m$, so the stress are checked is qualified in the location and are shown in table 6.

Table.6 Strength evaluation of weld between wall and bottom of tank after fire exposure

Items	Mean value of allowable limited stress /MPa	The maximum stress of the large fillet weld /MPa	Mean value of allowable limited stress /Mpa (Calculated)	Results of stress checked
room temperature	252	520.6	756	Qualified
600℃	241	520.6	723	Qualified
700℃	242	520.6	726	Qualified
750℃	265	520.6	795	Qualified
800℃	274	520.6	822	Qualified
850℃	245	520.6	735	Qualified
900℃	398	520.6	1194	Qualified

7 Conclusions

1) Crude oil storage tank wall subjected to axial stress, radial stress which are relatively smaller and mainly distributed in the first circle. The axial stress, hoop stress of edge plate and three-dimensional stress of center plate are little.

2) The maximum hoop stress on the outer surface of the tank wall located in the second layer near the first layer of wall plate. 6 layers under the lower part of the tank wall subjected to larger hoop stress and the maximum value is 258. 8MPa.Tank large fillet weld is the most dangerous position, its maximum stress can reach 520. 6MPa.

3) Based on the finite element analysis results under the condition of large storage tank Hydrostatic test,and material mechanical performance test results of large-scale crude oil storage tank, strength of tank wall plate and large fillet weld of large-scale crude oil storage tank are checked, the evaluation results show that the large-scale oil storage tanks meet the strength conditions.

8 Acknowledgements

This work is supported by National Natural Science Foundation of China (51254001),the Project (2012M511418) supported by China Postdoctoral Science Foundation and Central South University Postdoctoral Science Foundation; Project (CK2010B17,CK2013Z03) supported by the Fund of Chongqing University of Science & Technology; Project(2013A02) supported by the opening Fund of Guangdong Provincial Key Laboratory of Fire Science and Technology.

9. References

[1] Tongzhu Wu. Summary of Development on Large Scale Verticle Oil Storage Tank [J]. Petro-Chemical Equipment Technology, 2004,25(3) : 63-66.

[2] Difei Chen, Xuetong Li. Stress Testing and Analysis of Crude Oil tank with 100,000m3 Oil [J] . Oil & Gas Storage and Transportation, 1988, 7(6): 29-37.

[3] Yuming Liu, Zhiping Chen, Xiaobin Wu, et al.Stress Testing and Analysis of Crude Oil tank with 100,000m3 Oil [J]. China Petroleum Machinery, 2001, 29(3): 26-28.

[4] Zhiping Chen, Fei Wang, Jianmin Shen, et al. Stress Analysis on Large Unanchored Oil Storage Tanks [J]. Journal of Mechanical Engineering, 2006, 42(11): 206-211.

[5] Lifang Zhou. Study on Thermal and Mechanical Response of Oil Tank on Fire [D] . Nan Jing: Nanjing University of Technology, 2008.

[6] Hu Hui, Hutang Song, Yunlong Wu, et al. Finite Element Intensity Analysis for Large Oil Tank [J]. Oil & Gas Storage and Transportation, 2004, 23(12): 21-25.

[7] Yuntao Yao, Yang Du. Study on Mode and Behavior of Oil Tank on Fire [J]. Natural Gas and Oil, 2000, 18(3): 20-31.

[8] API579, Fitness For Service[S].

[9] GB/T228-2002, Metallic Materials-Tensile testing at room temperature[S].

[10] JB4732-95, Steel Pressure Vessel-Analysis and Design Criteria[S].

1ST CCPS ASIA-PACIFIC CONFERENCE ON PROCESS SAFETY℠

Reliability Analysis of Truck Loading System in LNG Terminal Based on GO Methodology

Xi Chen
College of Mechanical and Electronic Engineering
China University of Petroleum
chenxi0831@126.com

Changhang Xu
Yuan Zhu
Guoming Chen
Guoliang Cao
Xiaoping Sun
College of Mechanical and Electronic Engineering
China University of Petroleum
changhangxu@126.com
zhy3323@163.com
offshore@126.com
824145817@qq.com
sunxiaoping87@163.com

Prepared for Presentation at
American Institute of Chemical Engineers
1st CCPS Asia-Pacific Conference on Process Safety
Qingdao, China
September 4 – 5, 2013

Reliability Analysis of Truck Loading System in LNG Terminal Based on GO Methodology

Xi Chen
College of Mechanical and Electronic Engineering
China University of Petroleum
chenxi0831@126.com

Changhang Xu
Yuan Zhu
Guoming Chen
Guoliang Cao
Xiaoping Sun
College of Mechanical and Electronic Engineering
China University of Petroleum

Keywords: goal oriented (GO) methodology; liquefied natural gas (LNG); terminal; reliability; Truck loading system

Abstract

Liquefied natural gas (LNG) has the characteristics of low temperature, volatile, flammability and it is easy to explode. LNG terminal is the hub of LNG transportation and its security issues have become increasingly prominent. Truck loading system is the main process system and its security issues are top priority in LNG process system. GO methodology is a success-oriented method for system reliability analysis, which are commonly used in multi-state and timing sequence system analysis especially in actual logistics and production process. In this paper, GO methodology was used in the truck loading system of LNG terminal. The GO chart was established and both qualitative and quantitative analyses were carried out. The reliability of the system is 0.960393 and the mean time between failures is 209 days which are totally consistent with the real situation This paper verified the feasibility of GO methodology applied to the reliability analysis of LNG terminal process system. The results of the reliability analyses are important for the formulation of safety regulations of the terminals.

1. Introduction

Liquefied natural gas (LNG) for its environmental impact, energy efficiency, security and other aspects of notable features, has become an important energy in the economic development of 21st century energy [1,2]. As demand increases, China's coastal cities such as Shenzhen, Putian, Qingdao, Dalian and other places have established a number of LNG terminals [3]. LNG is a liquid mixture consisting mainly of methane, and it is initially intensely cold (-260°F) and about 1.5 times the density of ambient air [4,5]. Currently, China's LNG industry is in a boom period, and truck loading system is one of the main ways to exporting. It is obviously necessary to have an analysis of its security.

GO methodology is a success-oriented method for system reliability analysis, which is proposed by the U.S. Kaman Sciences Corporation in the mid-1960s. Lately, it is improved by American Electric Power Institute to developing a universal GO methodology procedure for solving complex system reliability problems. In the late 1980s, the Japanese ship institute developed GO-FLOW methodology to make it more suitable for the multi-state and timing sequence system [6,7]. In the 1990s, a Tsinghua University professor Shen Zupei improved GO theory and algorithms and brought it to China [8-12]. Currently, GO methodology is mainly used in the field of machinery, supply chains, power systems and nuclear power plant systems for safety analysis. The petroleum industry institution firstly used GO methodology in the reliability analysis and supply computation of natural gas pipeline transport system, and it marks the beginning of application of GO methodology in China petroleum industry [12].

In this paper, reliability analysis for truck loading system of a certain LNG terminal is made. Based on the analysis of truck loading system process, GO chart was established, and both qualitative and quantitative analyses were carried out associated with the probability formula algorithm. This paper verified the feasibility of the GO methodology applied to the reliability analysis of LNG terminal process system. The results of the reliability analysis are important for the formulation of safety regulations of the terminals.

2. Methodology

The basic principle of GO methodology is to translate system schematics, flowcharts, drawings or other system information into GO program through system analysis. And then you may take quantitative calculation based on the probability formula algorithm. The system is usually made up of units. GO program includes two aspects, which are operator units and signal flow. Operator units represent the unit function and logical relationship between unit input and output. Signal flow represents the association between units and connects operator units to establish GO program. The main steps for reliability analysis based on GO methodology are as follows.

1. Step one is system analysis which means that the logical relationship of each unit should be given and the input and output boundary of system is determined.
2. Step two is the establishment of GO program which means system schematics, flowcharts, drawings or other system information should be translated into GO program.
3. Step three is data inputting which means the data of operator characters should be determined.
4. Step four is calculation which means the results of quantitative analysis based on the probability formula algorithm are acquired.

There are seven reliability characteristics that can indicate system reliability. λ represents a system failure rate. f represents the average failure frequency. μ represents a system maintenance rate. T_U represents mean time between failures of a system. T_D represents mean time to repair of a system. T_C represents mean the total cycle time of a system. A represents the success probability of a system.

The followings are operators and calculation formula using in this paper:

- <u>Type 1</u> - Two state unit

The unit state is other success or failure, and it has an input signal and an output signal. When the unit is in the state of success, it means the operator is in the state of working. When the unit is in the state of failure, it means the operator is not working. When the signal is independent, the calculation formula is:

$$P_R = P_S \cdot P_C \qquad \text{[Eq. 1]}$$

$$\lambda_R = \lambda_S + \lambda_C \qquad \text{[Eq. 2]}$$

Where P_R is the successful state probability of output signal, P_S the successful state of operator, P_C the successful state probability of input signal, λ_R the failure rate of output signal, λ_S the failure rate of operator, λ_C the failure rate of input signal.

- Type 2 - Or gate

Or gate is a logical symbol which represent the relationship between M input signals and the output signal. As long as there is one input signal which is working, the output signal is working. When the signal is independent, the calculation formula is:

$$P_R = 1 - \prod_{i=1}^{M}(1-P_{Si}) = 1-(1-P_{S1})\cdot(1-P_{S2})\cdots(1-P_{SM}) \qquad \text{[Eq. 3]}$$

$$\mu_R = \sum_{i=1}^{M}\mu_{Si} = \mu_{S1} + \mu_{S2} + \cdots + \mu_{SM} \qquad \text{[Eq. 4]}$$

Where P_{Si} is the successful state of i operator, μ_R the repair rate of the output signal, μ_{Si} the repair rate of i input signal.

- Type 5 - Single signal generator

There is no input signal and the only output signal is the input signal of the system. the calculation formula is:

$$P_R = P_C \qquad \text{[Eq. 5]}$$

$$\lambda_R = \lambda_C \qquad \text{[Eq. 6]}$$

Usually, there are two combinations of components that needing a correlation in repairable system. One is tandem structure of shutdown related, the other is parallel structure of redundancy design. When you are in the calculation of GO methodology, the first thing you should take into account is to analysis the combination of component in the system and replaced it with the equivalent element. The methods are as follows:

- Tandem structure of shutdown related— because of the tandem structure if one of the operator is shutdown, there will be no output signal. The tandem structure of M operators can be replaced by one equivalent unit, and the calculation formula is:

$$P_R = \sum_{i=1}^{M} P_i \qquad \text{[Eq. 7]}$$

$$\lambda_R = \sum_{i=1}^{M} \lambda_i \qquad \text{[Eq. 8]}$$

Where P_R is the successful state probability of the equivalent unit, λ_R failure rate of the equivalent unit.

- parallel structure of redundancy design – the first step is to calculating equivalent units of each branch with the above mentioned, and replaced each branch with one equivalent unit using or gate calculation formula.

3. Example

3.1. Analysis of process and GO chart

Fig.1 shows the schematic of process system for truck loading system. The pipelines of this system include the liquid line of LNG, the vapor line of boil off gas (BOG) and other secondary pipelines including pressure relief line, cold circular line and nitrogen line. The process of truck loading system is usually with pressure loading. LNG is pumping into the liquid line before going into the truck. When LNG is pumping into the truck, BOG is pumping out of the truck through vapor line. And then it goes into the BOG recovery processing system. Fig.2. shows the GO chart of truck loading system, and the first number in GO operator is the type of the operator, the second number is the numbering of the operator, and the numbers on the signal line is the numbering of signals flow. There are 18 operators in the GO chart, and the necessary data of operators is given in Table 1.

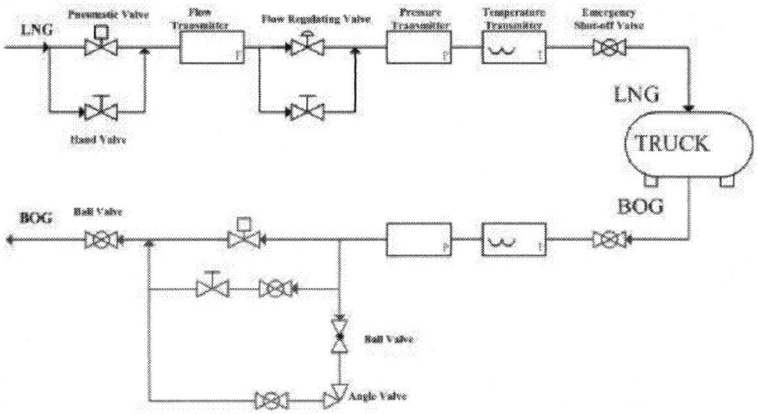

Fig.1. Picture of schematic of process system for truck loading

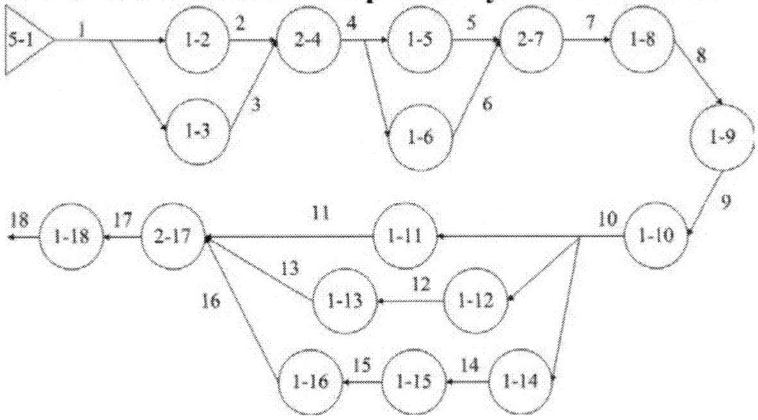

Fig.2. Chart of GO methodology for the truck loading system

Table 1. Data of operator characters

No.	Type	Component	$\lambda(a^{-1})$	$\mu(h)$
1	5	LNG source	0.1	10.0
2,11	1	Pneumatic valve A,B	0.38	8.0
3,6,13	1	Hand valve A,B,C	0.38	8.0
4,7,17	2	Or gate	...	...

5	1	Flow control valve	0.43	5.0
8,10	1	Emergency shut-off valve A,B	0.38	8.0
9	1	Tuck system	0.1	12.0
12,14,16,18	1	Ball valve A,B,C,D	0.21	8.0
15	1	Angle valve A	0.42	8.0

3.2. Quantitative analysis

As mentioned above, operator 8,9,10 are tandem structure of shutdown related and replaced by an equivalent repairable unit which is a Type 1 operator 10*. The same calculation is done to operator 12,13 and 14,15,16,and replaced by Type 1 operator 13*and 16*. Then their equivalent units are calculated ahead by the formal formulas and the results are listed in Table 2.

Table 2. Equivalent reliability parameters of tandem structure

Signal No.	Unit	Reliability	$\lambda(a^{-1})$	μ(h)
8,9,10	10*	0.9752	0.86	42.4684
12,13	13*	0.9801	0.59	29.0582
14,15,16	16*	0.970299	0.84	27.4419

There are three combinations of parallel structure of redundancy design, and need to be replaced by equivalent units. The calculation formulas are mentioned above, and the results are listed in Table 3.

Table 3. Equivalent reliability parameters of parallel structure

Signal No.	Unit	Reliability	$\lambda(a^{-1})$	μ(h)
2,3	3*	0.9999	0.219	4.0
5,6	6*	0.9999	0.3504	2.5
11,13*,16*	16**	0.999994	0.0069	7.6075

Fig.3 shows the simplified GO chart for the truck loading system with equivalent units. There are only six operators in the simplified GO chart and five of them are Type 1 operator and the remaining one is a Type 5 operator. Based on the probability formula algorithm, the reliability characteristics were calculating along the signal flow, and the results are showed in Table 4.

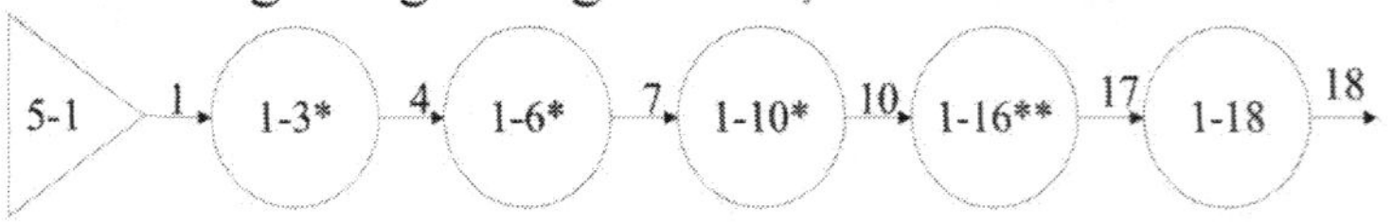

Fig.3. Chart of GO methodology for the truck loading system

Table 4. Calculations based on the probability formula algorithm

Signal No.	Reliability	$\lambda(a^{-1})$	μ(h)
1	0.995	0.1	10
4	0.9949	0.319	140.7681
7	0.9948	0.6694	68.4047
10	0.9701	1.5294	176.5389
17	0.970094	1.5363	175.7813
18	0.960393	1.7463	206.8751

Finally, the exact success probability of truck loading system is obtained according to the formulas mentioned above, and the calculation results are shown in Table 5.

Table 5 Results of quantitative analysis based on the probability formula algorithm

Parameters	results
A	0.960393
f /(a^{-1})	1.6771
μ /(a^{-1})	42.34584
λ /(a^{-1})	1.7463
T_U /(h)	5016.3202
T_D /(h)	206.8751
T_C /(h)	5223.1953

Under the conditions of continuous production, the probability of success of the truck loading system is 0.960393, the failure rate is 1.6771 times per year, the maintenance rate is 42.34584 times per year, the mean time between failure is 5016.3202 hours, the mean time to repair is 206.8751 hours, the mean cycle time is 5223.1953 hours(about 217 days). In reality, LNG terminal safety inspection cycle is generally 4-6 months. It is easy to see the results of quantitative analyses are totally consistent with real situation. The results can be used as a critical value for the truck loading system, and provide a reference for the engineering, scientific management and decision-making in LNG terminal.

3.3. Qualitative analysis

Qualitative analysis of truck loading system was based on GO chart. The method of qualitative analysis could be described as follows: First, assume that there are M operators in the system in addition to logical operators. Then, assume that there is only one operator that is failed of working while other M-1 operators are in the state of success. If the system is failed of working, it is certain that the only failed operator is a first-order system cut-set. Assume the M operators respectively, and then get all first-order system cut-set. It is similar to get the second-order system cut-set. First, assume that there are only two operators that are failed of working while other M-2 operators are in the state of success. If the system is failed of working, then the only two operators are a second-order system cut-set. Assume the combination of every two operators except for the operators that are in the first-order system cutest, and then get all the second-order system cut-set. Followed by analogy, it is easy to calculate all the cut-set of the system. It's worth noting that when you calculating a higher order there is no need for calculating when the higher order combination containing a lower order cut-set.

Table 6. Analytical result of the minimum cut-set

No.	Phase	Operator No.	Operators	probability of failure
1	1	1	LNG source	0.005
2	1	8	Emergency shut-off valve A	0.01
3	1	9	Tuck system	0.005
4	1	10	Emergency shut-off valve B	0.01
5	1	18	Ball valve D	0.01
6	2	2,3	Pneumatic valve A, Hand valve A	0.0001

7	2	5,6	Flow control valve, Hand valve B	0.0001
8	3	11,12,14	Pneumatic valve B, Ball valve A, B	0.000001
9	3	11,12,15	Pneumatic valve B, Ball valve A, Angle valve A	0.000001
10	3	11,12,16	Pneumatic valve B, Ball valve A,C	0.000001
11	3	11,13,14	Pneumatic valve B, Hand valve C, Ball valve B	0.000001
12	3	11,13,15	Pneumatic valve B, Hand valve C, Angle valve A	0.000001
13	3	11,13,16	Pneumatic valve B, Hand valve C, Ball valve C	0.000001

Table 6 shows the analytical result of the minimum cut-set. After calculating, we got 13 cut-set of the system including 5 first order cut-set, 2 second order cut-set, and 6 third order cut-set. The maximum probably occur that would lead to the system failing are emergency shut-off valve A, emergency shut-off valve B, and ball valve D with the same probability of 0.01. Therefore, making redundant design for the key equipments or strengthening supervision and inspection in order to reduce the probability of failure, can improve the overall system security level. The total probability of system failure is 0.040206, which means the probability of system success is 0.959794. That is very close to the quantitative analysis which is 0.960393. The results of qualitative analysis prove the correctness of the quantitative calculation indirectly.

4. Conclusion

On the basis of previous research, this paper is a first attempt to use GO methodology into a new area of LNG terminal. Reliability analyses of both quantitative and qualitative analysis were made after the process analysis of truck loading system. Since the construction of LNG terminal in China is in its infancy, it is important to make security and reliability analysis of LNG terminal production system. This paper is an attempt to promote GO methodology in petrochemical industry for risk analysis. LNG terminal truck loading system and other process system needs a more detailed and in-depth analysis.

Acknowledgement

This research is supported by "National Key Technology Research and Development Program of the Ministry of Science and Technology of China" (2011BAK03B08), "the Shandong Provincial Natural Science Foundation, China" (ZR2011EL048), "the Fundamental Research Funds for the Central Universities" (12CX04060A).

References

[1] Xueqiang ZH, Haifeng L, "Fluent based numerical simulation of LNG release and diffusion," *Journal of Jimei University: Natural Science*, Volume 16, No. 4, 2011

[2] Arun Parihar, Charles Vergara, J. Keith Clutter, "Methodology for consequence analysis of LNG releases at deepwater port facilities," *Safety Science*, Volume 49, 2011

[3] Hairong W, Xiaoqian M, "Improved quantitative risk-assessment method for major accidents of LNG gas station," *Journal of Safety and Environment*, Volume 6, No. 5, 2006

[4] Jingjing W, Xiaoqian M, "Catastrophe in sudden boosting of LNG tank pressure under thermal attack," *Journal of Safety and Environment*, Volume 6, No. 3, 2006

[5] D. W . Hissong, "Keys to modeling LNG spills on water," *Journal of Hazardous Materials*, Volume 140, 2007

[6] Matsuoka T, Kobayashi M, "GO-FLOW: A new reliability analysis methodology," *Nuclear Science and Engineering*, Volume 83, No. 1, 1988

[7] Matsuoka T, Kobayashi M, "GO-FLOW methodology: a reliability analysis of the emergency core cooling system of a marine reactor under accident conditions," *Nuclear Engineering and Design*, Volume 84, No. 3, 1989

[8] Zupei S, Jia G, "GO methodology and improved quantification of system reliability," *Journal of Tsinghua University：Science and Technology*, Volume 39, No. 6, 1999

[9] Zupei S, Xiangrui H, "Application of GO Methodology for reliability analysis of repairable system," *Nuclear Power Engineering*, Volume 21, No. 5, 2000

[10] Zupei S, Tao Zh, "Exact algorithm for complex system reliability using the GO methodology," *Journal of Tsinghua University：Science and Technology*, Volume 42, No. 5, 2002

[11] Zupei S, Xiangrui H, "Principle and Application of GO Methodology," *Beijing: Tsinghua University Press*, 2004

[12] Dongyu S, Zupei S. "The application of GO approach in the reliability analysis in natural gas pipe system," *Technology Supervision In Petroleum Industry*, Volume 16, No. 11, 2000

Research on Mathematical Model of Gas Dispersion and Analysis of Engineering Applicability

Jing Yuan
Mechanical and electrical engineering college, China University of Petroleum
Shandong Qingdao, China
forever10123@163.com

Yifei Meng
Chemical engineering collage, China University of Petroleum
Shandong Qingdao, China
Myf213@163.com

Prepared for Presentation at
American Institute of Chemical Engineers
1st CCPS Asia-Pacific Conference on Process Safety
Qingdao, China
September 4 – 5, 2013

Research on Mathematical Model of Gas Dispersion and Analysis of Engineering Applicability

Jing Yuan
Mechanical and electrical engineering college, China University of Petroleum
Shandong Qingdao, China
forever10123@163.com

Yifei Meng
Chemical engineering collage, China University of Petroleum

Keywords: gas diffusion, mathematical model, DEGADIS, UDM, FEM3

Abstract

The leakage and dispersion of hazardous substance in petroleum and petrochemical companies often lead to the significant fire or explosion accidents. This paper gave a simple general review over the currently used methods of numerical simulations for gas dispersion including the basic principle, merits and demerits, and the range of engineering applicability. In addition, by means of the widely used softwares, such as DNV-SAFETI, ALOHA, FLUENT, this paper made a result contrastive analysis about the numerical simulation, it pointed out that the 3D simulations always have the higher accuracy on the results. However, due to the simply calculation, short time-costing and easy understood on the consequences, the simplistic empirical simulations were generally accepted and had a better engineering applicability. Last of all, the paper also put forward some recommendations for this field.

1. Introduction

With the development of modern chemical industry, the increasing devices, larger equipment, and more and more complex pipelines, the leakage of flammable, explosive, toxic and hazardous chemicals often occurs in the storage, transportation, production and other process. Once the hazardous chemicals are discharged out, it will spread rapidly and expand the area of possible hazards, then it often leading to fires, explosions and other serious accidents of poisonings, causing huge damage and loss to the surrounding environment, personnel, equipment. To reduce damage and loss of chemical accidents, it is particularly important to simulate of the possible consequences of leakage accident, establish and improve the emergency response system for prompt, timely and effective response to incidents[1].

To research the diffusion of substances after leaking, the main means include field trials, physical modeling and numerical simulation, all of which are based on theoretical calculation and analysis[2-4]. Given that the numerical simulation is cost, short-consuming and highly reproducible, such studies have been widely used. Numerical simulation of diffusion is generally divided into the neutral gas and heavy gas, and the gas cloud formed by hazardous chemicals in most leakage accident is heavy gas, the diffusion harm of which is far more than that of general positively buoyant or neutrally buoyant. After then, a number of mathematical models based on

different mechanisms and assumptions from the large-scale field trials and the fluid model have appeared.
According to the principles and the complexity, the modeling can be divided into the nomogram model, intermediate model, computational fluid dynamics model, etc. In order to promote research in the field of further development, this paper have summarized the numerical simulation models at home and abroad, and analyzed on the advantages, disadvantages and scope of their applications.

2. Model overview and contrast

2.1 Nomogram model

Nomogram model, which is also called phenomenological models or empirical model, is based on the measurements in field and laboratory experiments. In the nomogram model, the diffusion behavior of heavy gas cloud is described by the column line chart or a simple equation, which can only illustrate a axis concentration of smoke at a certain distance. Typical representatives of such models are the Gaussian model, and britter & McQuaid's BM model[5].
Nomogram model can only shows a rough comparison of the predicted values because of fewer considerations, such as ignoring atmospheric stability, surface roughness, so it should only do the screening model for everyday applications.

2.2 Intermediate model

Intermediate model is a kind of model referring to its complexity which is intermediate among the nomogram model, three-dimensional computational fluid dynamics model and Lagrangian particle model. It includes boxes and similar model, uniform or normal plume model, GM plume model, integral jet model and shallow model.
Typical intermediate model contains Cox&Carpenter dense gas dispersion model (box model)[6], DEGADIS model[7], SLAB (shallow model)[8], UDM model (embedded into SAFET)[9], TWODEE model and DISPLAY2 model.

2.3 Computational fluid dynamics model

Computational fluid dynamics model(CFDM) are three-dimensional model. CFDM is depended on the solution of partial differential equations on time and the three-dimensional coordinates or Navier-based -Stokes (NS) equations, which reflect the conservation principle of the mass, energy, momentum and substance (component). Typical three-dimensional CFD models have FEM3 model[10], and MDPG model.
Essentially, this model can be applied to any release conditions, terrain or weather conditions. But NS equation model is facing the thorny issue of mathematical closure assumptions, and strictly speaking, this assumption has no solid physical basis to follow, which will inevitably affect the universality of simulation.
This paper will present the detailed analysis of the advantages, disadvantages and scope of commonly used gaseous diffusion field Gaussian model, sutton model, B & M model, Cox&Carpenter dense gas dispersion model, SLAB model, DEGADIS model, UDM model and FEM3 model. Detailed information can be seen in Table 1.

Table 1 Comparison of Several Common Gas Diffusion Models

Name	Formula	Advantages	Disadvantages	Applicable Scope
Gaussian Model	$C(x,y,z)=\frac{Q}{2\pi u\sigma_y\sigma_z}\exp\left[(-\frac{y^2}{2\sigma_y^2})+(-\frac{(z-H)^2}{2\sigma_z^2})\right]$ Plume Model PuffModel $C(x,y,z)=\frac{m}{(2\pi)^{3/2}\sigma_x\sigma_y\sigma_z}\exp\left[-\frac{1}{2}(\frac{x^2}{\sigma_x^2}+\frac{y^2}{\sigma_y^2}+\frac{z^2}{\sigma_z^2})\right]$ u: Average wind speed m/s; $\sigma_x\cdot\sigma_y\cdot\sigma_z$: Diffusion coefficient in x、y、z directions m; H is height of effective source m; Q: Source intensity kg/s; C(x,y,z) means the intensity in one point, kg/m^3.	Early proposed; many experimental values; proficient calculate methods; simple model; easy calculation.	Ignorance of the affection of gravity to diffusion; low accuracy of simulation.	Applicable to calculation of neutral gas; Plume Model and Puff Model applicable to the stimulation of instantaneous and continuous leakage respectively.
Sutton Model	$C(x,y,z)=\frac{Qe^{-\frac{y^2}{C_y^2x^{2-n}}}}{\pi uC_yC_z}\left[e^{-\frac{(z-h)^2}{C_z^2x^{2-n}}}+e^{-\frac{(z+h)^2}{C_z^2x^{2-n}}}\right]$ u: Average wind speed m/s; C_y、C_z: Diffusion coefficients relating with meteorological conditions m; h: is height of effective source, m; Q: Source intensity, kg/s; C(x,y,z): Gas concentration, with volume fraction being expressed by percentage.	Usage of the Statistical theory of Turbulent Diffusion to deal with turbulent diffusion.	Ignorance of the affection of gravity ; low precision of calculation with large difference with the experimental values.	Applicable to rough research of neutral gas.
Cox&CarpenterDense Gas Diffusion Model(Box Model)	$\frac{dR}{dt}=\left[kgh(\rho_2-1)\right]^{1/2}$ R: The conformation radius of instantaneous leaking clouds, m; t: time, s; k: experimental values, 1 generally; ρ_2: gas density, g/m^3.	Empirical Model, trivial calculation.	Ignorance of the affections of statistical wind, topography, obstacle, etc.	Applicable to the calculation of toxic gas diffusion in flat terrain.

Note: Do not add page numbers. Do not refer to page numbers when referencing different portions of the paper

Model Name	Formula	Advantages	Disadvantages	Applicable Scope
B&M Model	Continuous Leakage: $\frac{C_m}{C_0} = f_i \left\{ \frac{x}{(V_{i0})^{1/3}}, \frac{g_0' V_{i0}^{1/3}}{u^2} \right\}$ Instantaneous Leakage: $\frac{C_m}{C_0} = f_c \left\{ \frac{x}{(C_{c0}/u)^{1/2}}, \frac{g_0' V_{c0}^{1/2}}{u^{5/2}} \right\}$ C_m、C_0: Average and initial concentration of the gas cloud in the cross section, kg/m^3; V_{c0}、V_{i0}: Volume flow of the initial gas cloud released by continuous and instantaneous heavy gas, respectively; g_0': Initial converted gravity term; f_c、f_i: Generalized dimensionless function. In the equation above, the first term is dimensionless distance, the second term is Richardson number of the source. In the B&M Model, the horizontal axis is source Richardson number, the vertical axis means dimensionless distance and depict the experimental values of C_m/C_0 in the chart.	Empirical Model, convenient application, trivial calculation.	Low accuracy, limited application.	Only used as benchmark screening model; applicable to continuous or instantaneous leakage and dispersion, and unreasonable when used in two-phase fluid or complex terrain conditions.
SLAB Model	---	Divided into Steady Plume Model and Transient Fluid Model; easily and fast applicable; accurate results available; applicable to evaporation of liquid pool on the ground, horizontal jet above ground and instantaneous volume source.	Ignorance of the complex conditions with buildings existence and terrain change; the variance of concentration in the vertical direction is excluded.	Good forecast results are available in different atmospheric environment; applicable to prediction of the variance of edge layer.

Model Name	Formula	Advantages	Disadvantages	Applicable Scope
DEGADIS Model	Diffusion parameters $S_y(x)$, $S_z(x)$, characteristic widths b(x), density p(x) and temperatures in different distance of the downwind should satisfy the conservation of total mass, the conservation of leaking constituents and the conservation of energy; relative equations are not showed here.	The affections of gravity to diffusion and admixture ; gathering effect of gas due to wind; dealing with area, not just point source; dependence of the time; these factors above have been taken into consideration.	Only the vapor cloud leaking and diffusing on smooth surfaces have been taken into consideration, with the variant diffusion due to terrain, buildings, gas tanks being ignored. Cannot deal with the evaporation on the water surface; steady atmospheric environment.	This model has synthesize the diffusion results, and the simulation accuracies are high when used to deal with high-risk high density gas and leakage of aerosol.
UDM Model	---	Could be used to simulate the diffusion on the ground/ two-phase release under high pressure; diffusion forms of many kinds; consideration of comprehensive factors.	Low accuracy of the simulation results of low speed wind and sudden change of new-field concentration. Complication	Applicable to various kinds of diffusion conditions, easy to use with being embedded in software. Fail to fully consider the buildings and change of terrain.
FEM3 Model	---	3 dimensional finite element calculation model; deal with diffusion under complicated conditions through solving unsteady continuity equations,etc; high accuracy calculation.	Complicated calculation	Applicable to continuous or limited time leakage of heavy gas source; could be used to simulate the diffusion of gas under complex terrain conditions, without limiting the leakage time and amount.

3. Comparison among typical engineering application models

In this paper, with the help of DNV SAFETI(UDM model), ALOHA(DEGADIS model) and FLUENT softwares, the consequences of LNG diffusion in different terrains and chlorine diffusion under different wind speed are simulated and analyzed.

3.1 LNG diffusion in different terrains

3.1.1 Scenario introduction

LNG tank storage, 100000m3. Provided that methane accounting for around 98%, the remaining is ethane. The style of leak was continuous and will last 20 minutes with a hole of 4 inch. Storage temperature was 161.4oC, pressure was 1.013bar. D=80m, H=21.8m, reserves was about 32910t. The atmosphere parameters included relative humidity , atmosphere temperature and wind speed, 70%, 10 oC and 6m/s repectively.

3.1.2 Comparison of results

3.1.2.1 Diffusion in open space

1. Using the DEGADIS model、UDM model and FEM3 model to simulate the scenario separately. The results are shown in Figure 1.

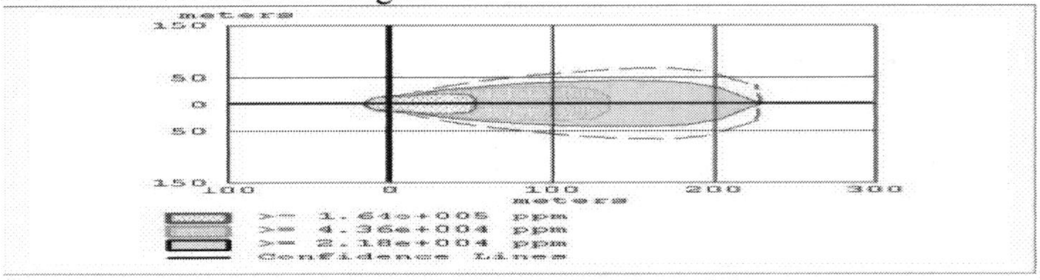

(a) Result of DEGADIS model

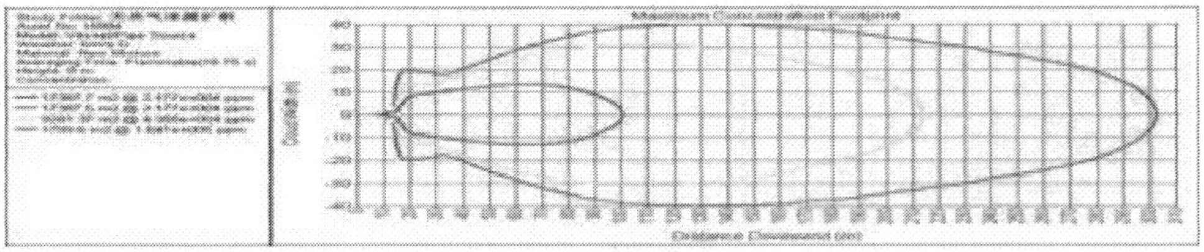

(b) Result of UDM model

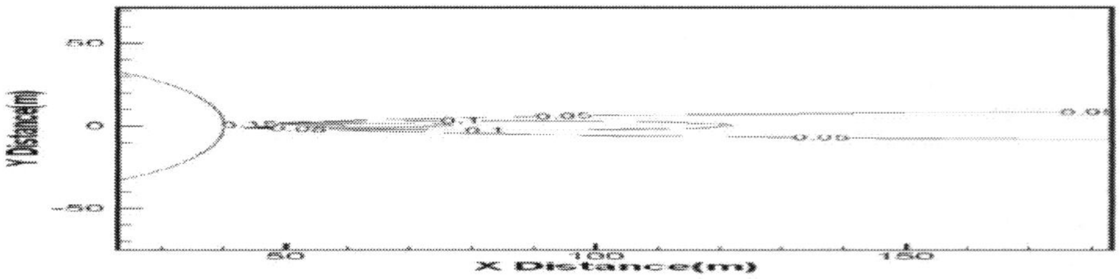

(c) Result of FEM3 model

Figure 1 The results of three different calculation models in open space (DEGADIS,UDM,FEM3)

2 Results analysis

(a) Tendency of simulated results are consistent.

(b) Compared with DEGADIS model, the simulated results of UDM model and Fluent predicted high range of diffusion. In terms of high-concentration region, DEGADIS models and FEM3 model had a high degree of consistency;

(c) Aimed at cloud width, fluent result was limited in the scope of 5m, far less than other two models. In terms of low-concentration region, DEGADIS models and UDM model had a high degree of consistency.

3.1.2.2 Diffusion in industrial area

1. Using the DEGADIS model、UDM model and FEM3 model to simulate the scenario separately. The results are shown in Figure 2.

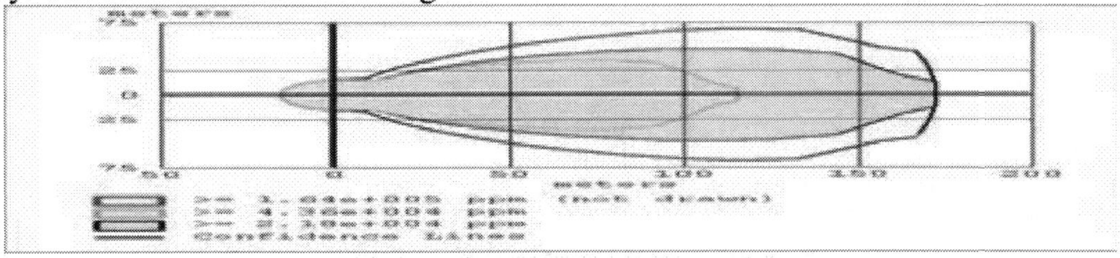

(a) Result of DEGADIS model

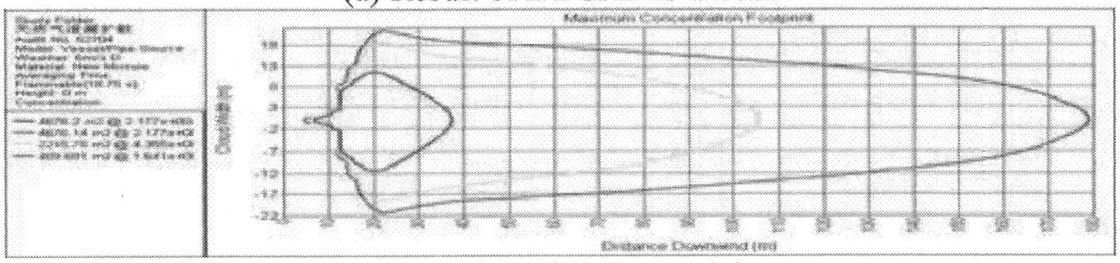

(b) Result of UDM model

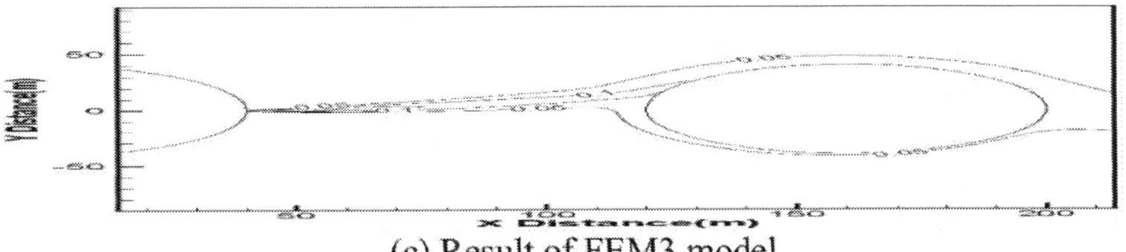

(c) Result of FEM3 model

Figure 2 The results of three different calculation models in industrial area (DEGADIS,UDM,FEM3)

2 Results analysis

(a) There is no huge difference about results between open space and industrial area except slightly decrease in the downwind distance with DEGADIS model. UDM model showed that gas had a slower diffusion rate in industrial area, and had a relatively accurate estimation range. In FEM3 model, barriers were taken into consideration, diffusion range of cloud width had close relationship with barriers which can express the real diffusion condition.

(b) Under the condition of industrial area , DEGADIS model couldn't calculate the region of high-concentration, on the contrast, the results of UDM and DEGADIS models about far field were similar to each other.

3.2 Chlorine diffusion under different wind speed

3.2.1 Scenario introduction

Chlorine tank storage which was horizontal type contained 200kg material. The style of leak was valve leak. And lasted 6 minutes with a hole of 1.5 cm. Parameters about the tank: volum=800L, D=0.8m, B=1.65m. The atmosphere parameters included relative humidity , atmosphere temperature, 65% & 28oC. Atmospheric stability was D.

3.2.2 Comparison of results

3.2.2.1 Diffusion under 5m/s

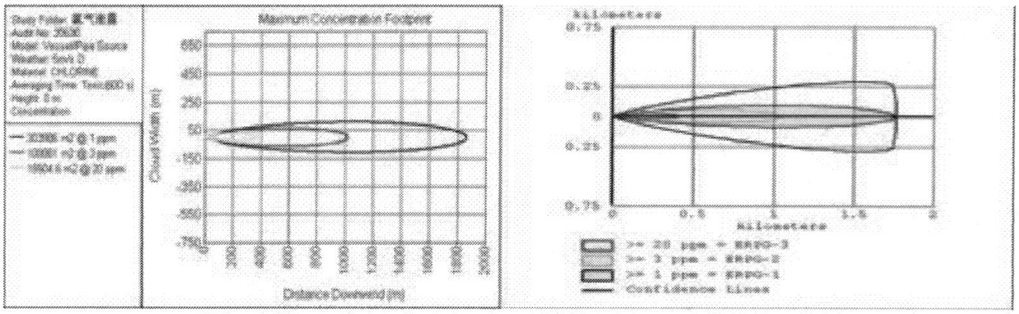

Fig.3 Results of chlorine diffusion under 5m/s

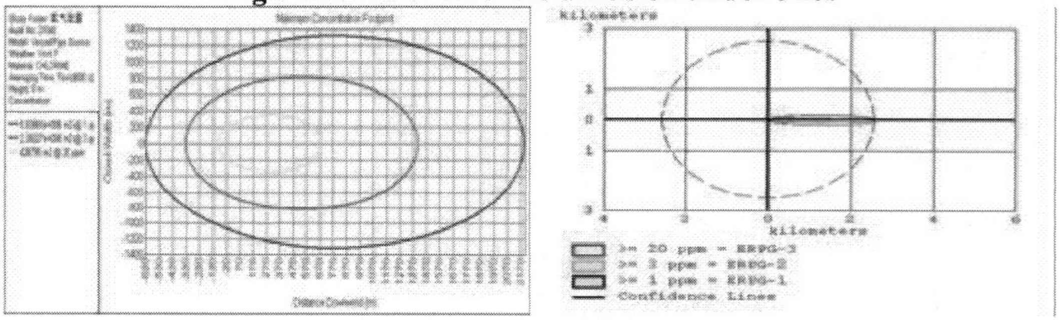

Fig.4 Results of chlorine diffusion under 1m/s

Figure3 and Figure4 show that:

(1) DEGADIS models and UDM model had a high degree of consistency under the condition of 5m/s.

(2) DEGADIS model couldn't calculate the region of upwind and the scope of diffusion is far less than the scope of UDM model.

4. Conclusion

Nomogram model and intermediate model (which are enbedded in DNV-SAFETI and ALOHA softwares) are mainly used in conventional risk assessment and emergency response systems. These models can be considered as ideal models for engineering application as the simple principle, less time-consuming and easy-getting data. Otherwise, these models should be improved continuously, but because of the over-simple principle, the improvement potential is limited.

Compared to nomogram model and intermediate model, computational fluid dynamics simulation model has a higher accuracy, which can take both environmental factors and obstacles into consideration, however, this kind of simulative model is less used in engineering application but more in science research field because it needs lots of time to modeling and calculation.

At present, complex terrains, obstacles, change of external environment, multi-component, low wind speed and something with chemical reaction conditions have little influence on the diffusion results, which will be the main direction of future models of gas cloud diffusion.

5. References (examples given to show format)

[1] HASEGAWAK, OHNOS, SEKIYAM, et al. On the safety measures against poisonous substance accidents and the management of poisonous substances [C]//HUANG Ping, WANG Yajun, LI Shengcai, et al. Progress in safety science and technology, Vol. VI. Beijing: Science Press, 2006: 1287-1292.

[2] VENARTJ E, "Flixborough: A final footnote", *Journal of Loss Prevention in the Process Industries*, Volume 20, NO. 4/5/6, 2007

[3] NIELSENM, OTTS, JORGENSENHE, et al, "Field experiments with dispersion of pressure liquefied ammonia", *Journal of Hazardous Materials*, Volume 56, NO. 1/2, 1997

[4] MERONEYRN, "Guidelines for fluid modeling of dense gas cloud dis-persion".*Journal of Hazardous Materials*, Volume 17, NO. 1, 1987

[5] BRITTERRE, MCQUAIDJ, "Workbook on the dispersion of dense gases". *Sheffield, UK: Research and Laboratory Services Division(HSE)*, 1988

[6] Wei Lijun, "Numerical Simulation of Heavy Gas Dispersion", *BeiJing: BeiJing University of Chemical Technology*, 2000.

[7] GUINNUPD. Dispersion model for elevated dense gas jet chemical releases (Ooms/DEGADIS) (for microcomputers) [R]. United States: Environmental Protection Agency, Research Triangle Park, NC. Office of Air Quality Planning and Standards, 1988.

[8] Jiang Chuansheng, Ding Hui, Liu Guoliang, "Comparison and Analysis of Wind Tunnel Test Dataand Dispersion Model Prediction for Accidental Continuous Release of Dense Gases", *China Safety Science Journal*, NO.13, 2003

[9] Liu Moshan. Simulator Investigation of the Sulfur-containing Natural Gas Pipeline Leak Consequence in Sichuan and Chongqing[D]. BeiJing:China University of Geosciences, 2010.

[10] CHANST, ERMAKDL, MORRISLK, "FEM3model simulations of selected Thorney Island phase I trials". *Journal of Hazardous Materials*, Volume 16, 1987

Fire Safety Management Based On the Risk Analysis and Layer of Protection for the Chemical Industry Park

Xuanya, Liu
Qinlin Zhang
Xiaoyuan Xu

Tianjin Fire Research Institute of MPS
Weijin Road No.110 Nankai district of Tianjin, China
liuxuanya@tfri.com.cn
zhangqinglin@tfri.com.cn
xuxiaoyuan@tfri.com.cn

Prepared for Presentation at
American Institute of Chemical Engineers
2013 Spring Meeting
9th Global Congress on Process Safety
San Antonio, Texas
April 28 – May 1, 2013

Fire Safety Management Based On the Risk Analysis and Layer of Protection for the Chemical Industry Park

Xuanya, Liu
Qinlin Zhang
Xiaoyuan Xu

Tianjin Fire Research Institute of MPS
Weijin Road No.110 Nankai district of Tianjin, China
liuxuanya@tfri.com.cn
zhangqinglin@tfri.com.cn
xuxiaoyuan@tfri.com.cn

Keywords: fire safety management; risk analysis; layer of protection; chemical industry park

Abstract

According to the status and characteristics of chemical industry park planning and fire safety management in China, the dynamic fire risk analysis method is put forward by the dynamic hazard identification and consequences of fire disaster evolution analysis. And the layer of protection analysis method for chemical industry park is established to evaluate its fire safety management. Based on the fire risk analysis and safety layer of protection analysis, it is proposed to build a multi-level protection, regional cooperation fire safety management model and emergency fire safety management system for chemical industrial park in the paper. That is, during the whole process of the park industrial positioning, planning and construction, enterprises settled and running in the park, the concept of fire safety planning, design and construction management will be carried out throughout the process of chemical industrial park management. From park planning to enterprises running management, the implementation of fire safety management based on dynamic risk monitoring on the whole process is help to improve the large industrial park overall fire prevention and management level..

1. Introduction

Large oil & gas storage and transportation, oil refining and ethylene refining and fine chemical products processing are all the mainstay of Chemical Industry Park, including the chemical raw materials, warehousing, production and processing as a whole. Through the integration and optimization of chemical industrial structure and scale of industrial agglomeration effect, it can improve the industry intensification, reducing the external circulation of dangerous chemicals, which reflects the current major trends in the economic high degree cycle. However, with the development of large-scale chemical industry park construction booming, not only promote local economic development, but also posed a serious challenge of fire to the local fire safety management of chemical industrial park. In the event of leakage, fire, a large number of major hazard of aggregation, explosion often lead to large-scale disasters and large-area chain of environmental pollution and other secondary disasters, and a serious threat to the safety of many enterprises within the park surrounding the production and lives of the residents and property. In recent years, due to lack of scientific safety planning, safety supervision weak link, some counties and municipal industrial zone, a series of major disasters, causing heavy casualties and economic losses. As for the gas blowout accident in Kai County, Chongqing of China on December 23, 2003, caused 243 deaths and economic losses of nearly billion[1]. On April 15, 2004 in Chongqing, the Tianyuan Chemical Plant chlorine leakage accident occurred in the repair process had occurred many times explosion, causing more than a dozen casualties and tens of thousands of people had to be evacuated. As for the PetroChina Jilin Petrochemical Company Dual benzene plant "11.13" fire and explosion accident was not only caused heavy casualties and property losses, but also triggered a large area of the Songhua River water pollution [2]. On January 07, 2010, the PetroChina Lanzhou Petrochemical Company 316 tank leakage caused fire and explosion accidents, resulting in around eleven storage tanks comic deflagration, causing 12 casualties [3]. July 16, 2010 PetroChina Dalian International Logistics Co Ltd pipeline explosion fire caused a large oil spill which formed nearly 60,000 square meters of floor flowing fire burned area, and caused the tank District No. T103 crude oil tank fire, resulting in a large area of the sea water near Dalian polluted [4].

The scientific planning and safety management of chemical industry park, especially the fire safety management, is very important to protect the safety and healthy development of the park. Well, because the chemical enterprises in the park are highly specialized and the production processes are complex, it is particularly important to carry out scientific management of fire safety. At present, the chemical industrial parks fire safety managements in China are mainly the post-disposal managements which include the investigation of different forms of fire hazards, blitz inspection and temporary administrative means of fire safety to supervise fire prevention measures, with reference to the relevant standards of chemical industrial fire safety. These measures curbed petrochemical major fire accidents to some extent. However, due to the lack of long-term, dynamic and comprehensive evaluation of fire safety regulatory mechanism, the unannounced fire safety inspection and fire hazard investigation can only correct some cosmetic problems. For the chemical industrial park enterprises planning, construction of the industrial park, as well as the dynamic fire risk management, relevant firefighting force settings, accident safety precautions, emergency management plans and fire risk control effectiveness and other aspects, the current fire safety management has been relatively absent. How to build a scientific fire safety management mode based on the intrinsically safety, whole process and dynamic, multi-level security protection, according to the whole fire risk characteristics of the chemical enterprises in the park. It is very important to carry on multi-level security monitoring long-term, dynamic evaluation of fire safety management during the park planning and construction to the daily supervision of the whole process, which to control the fire risk chemical parks, and to improve safety and fire prevention, emergency response level of technological capability.

2. Overall fire risk characteristics of Chemical Park

According to the products features, petrochemical enterprises can be divided into the oil and gas storage and transportation business, the oil refining companies with the midstream and downstream industry chain, as well as fine chemical companies with petroleum products as raw material. Well , the chemical industry park in China can also be divided into the following four types: (1) Petrochemical Park with the refining, petrochemical ethylene-based demonstrably on the downstream industry chain model; (2) which does not have significant chemical industry chain process with small and medium enterprises (including fine chemical enterprise, small and medium sized chemical enterprises with the cities relocation) gathering area; (3) with chemical raw materials (including oil, gas and other chemicals) warehousing, transportation and processing of raw materials-based chemical storage Harbor Park; (4) coal chemical-based chemical industrial park. For the different types of chemical companies, the characters of the fire hazard form and distribution are different. Storage Petrochemical enterprises are involved in large number of oil & gas storage and transportation. The main fire hazards in this kind of petrochemical enterprises include: a large amount of flammable explosive oil and gas storage tanks, oil and gas pipelines and transportation. The main fire & explosion accident types are: tank lightning and static fire, tank and pipeline leakage fire, explosion, fire ignited, the impact fire, boiling liquid evaporate vapor explosion (BLEVE), vapor cloud deflagration, physical explosion, explosion of energetic materials and so on, mainly in the form of pool fire, jet fire, and fire flows. While for the oil refining enterprise with industry chain, there is not only a large amount of flammable liquid, superheated liquids and other dangerous chemicals, but also some high temperature and pressure, presence of hydrogen and other complex high-risk chemical production equipment and processes, in which the form of fire & explosion and hazards evolution and spread process are relatively complicated. While the main hazardous materials in the fine chemical enterprises not only include the oil refinery products, but also other dangerous chemicals materials and technological process, which is possible to burn and explosion, such as material handling, crushing, mixing and distillation, drying technology and equipment, etc. The main casualty devices (including storage tanks and pipelines) leakage fire, explosion, technology and equipment out of control fires, explosions. Typical types of disasters and accidents include explosion, jet fire or pool fire, flowing fire and so on caused by combustible gas or steam leakage from the device. Since the park hazardous chemical species, quantity and type of different production processes, the different types of chemical parks fire risk characteristics are also different. For the petrochemical and coal chemical industrial parks with complex industrial process chain, which reserve a large number of both flammable liquids, liquefied hydrocarbons, high volatile coal and other dangerous chemicals, but also has a high temperature, high pressure, hydrogenation process and other high-risk chemical process, and its major hazards of fire relatively concentrated, the dissemination process of fire, explosion and disaster evolution is relatively complex, mainly as a device or process unit uncontrolled leakage caused by fire, explosion accidents. For the petroleum chemical storage parks with chemical raw materials (including oil, gas and other chemicals) storage and transportation, such as large-scale strategic oil storage, warehousing and freight chemical materials base, which usually do not involve in complex chemical process, but have a large number of oil and gas distribution park tanks, pipelines and materials, transportation lines, densely distributed sources of danger, its disasters type to leak oil and gas pipelines or storage tanks are usually fire and explosion accidents. Since petrochemical process units within the park (including storage tanks and pipelines) densely distributed, flammable materials inside the park single tank volume and tank farm size is increasing, and in the event of fire, explosion accidents it can easily produce a domino effect, causing a large area of fire, explosion and other chain reaction. The chemical companies aggregated chemical industrial parks, because of the relatively small production scale, relatively independent enterprises in the park, although its chemical production process and product categories are more complicated, such a fire hazard zone distribution is relatively dispersed, spacing of the enterprises in the park which make the domino effect disasters and accidents risk are relatively low.

According to the different types of chemical park fire hazard characteristics and distribution of fire in the industrial park, the overall chemical park fire risk arises primarily depend on the park fire hazard types, status, quantities, enterprise layout and the security protection levels. Major hazard for the chemical park accidents caused by regulatory requirements and safety rules, the chemical industrial park main safety layers are usually shown in Figure 1. The first layer of protection in the park is the intrinsically safe process plant protection level; the second is control system layer of the process unit in park; the third is the monitoring and human intervention protective layer of process unit in the park; the fourth is the fixed fire safety protection layer of the process unit in the park; the fifth is the security protective layer of the process unit structure in the park; the sixth is the park accident emergency disposal and regional isolation protective layer. For the initial accident caused by fire incidents, when the basic process plant control and protection layer, emergency protective intervention, safety devices and safety of fixed fire protection systems have been in the case of failure, the initial event after accumulating rapidly developed into a large-scale fire, explosions and chain disasters, and many major fire and explosion accidents are often caused by the cumulative effect of multiple initial event occurred.

Figure 1: Chemical Industrial Park security protection layer strategy

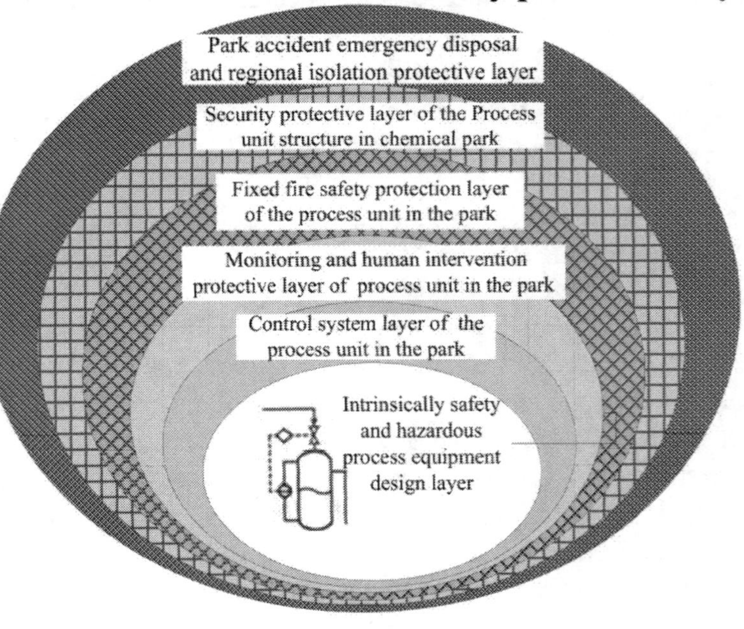

Taken into account of the thermal effect model of the fire and the consequences of fire risk analysis, the mainly evaluation basses of accident consequences are the calculation data of the consequences of the accident jet fire, pool fire, flash fire, fire ball fire and other types of thermal radiation distribution characteristics and so on. Considering the park chain disaster and accident probability of the development of uncertainties, for different fire scenarios to form a chain reaction of the park overall fire risk calculation model by the following formula [5]:

$$R=\prod \lambda \sum_{i=1}^{n} g(s_i)p(s=s_i) \qquad \text{[Eq. 1]}$$

Formula (1), n is the number of fire scenarios considered, g (s_i) for the fire scenario i may cause thermal radiation intensity function, P (s = s_i) is the probability of fire scenario i, λ is uncertain dynamic factor. Combination of different fire scenarios Park numerical simulation analysis, the model can be calculated through the park all possible fire scenarios radiant heat load distribution probabilistic risk curve, using the results of the risk analysis within the park can be a source of danger for the degree of harm, prevention focus decision-making and risk control measures guidance.

3. Protective layer based dynamic fire safety management mode

Fire safety management of chemical industrial park as an important part of safety management is mainly to prevent the risk of accidents, to carry out emergency response, reducing the impact of disasters and other accidents task. At present, fire safety of chemical industrial park in China mainly relying on the public security fire brigades and full-time fire brigades of the enterprises in the park, for the small and medium sized chemical companies aggregated chemical industrial park, it is mainly relied on the local public security fire brigade Park (Station). For the installations and processes with major hazard in the chemical industrial park, people usually conduct security settings in the park or enterprise planning and design to realize intrinsically safe control, industrial critical warning disposal, security guards, physical protection and other aspects of security settings. The firefighting forces are often taken as a passive disposition forces emergency disasters disposal in the security protection and physical protection uncontrollable circumstances. The existing chemical industrial fire safety management is mostly confined to fire-fighting facilities operation and management, as well as various types of fire investigation and fire safety awareness campaigns and other content, is still unable to deal with the accident prevention and in early fire accidents to play in a role. Usually in the firefighting force emergency intervention, the initial event fires in the park have already evaluated from the signs of small-scale fires to large fire and difficult to control. From Lanzhou Petrochemical "1.7" and Dalian "7.16" major fire explosion accidents, from the initial combustible gas leaks, plumbing accidents overpressure signs, the incidents developed to large-scale fire and explosion accidents, it is often experienced the evolution for a long time, but during the period the chemical safety protection measures are mostly in a stale or missing status. Therefore, improving the level of fire safety management in the chemical industrial park, according to the fire risk characteristics, it is important to make the fire safety precautions disposal action sooner in the process of firefighting, which the firefighting warning and disposal intervene at the same time of process and equipment warning. It can effectively improve the firefighting abilities against fire risk to and fire safety management.

According to the overall fire risk characteristics and different development stages of chemical industrial park, based on different levels of protection requirements, the dynamic fire safety management model diagram is shown in Figure 2. The whole process of fire safety management of chemical industrial park mainly includes three stages of firefighting regional planning, construction management and operation of fire safety dynamically management, corresponding to the different stages of industrial planning and design, security planning and enterprises construction to managements. Mainly through the regional fire risk analysis, urban areas firefighting planning, chemical industrial park fire safety planning and firefighting settings and other means, from the beginning of the park planning and construction to the park running and operation on the whole process, according to the park overall fire risk characteristics in different levels of security protection processes and apparatus that implement effective fire control and emergency response, in order to improve the industrial park technology, intrinsically safety. The park dynamic fire safety management measures and procedures are shown in Figure 3. Key parts of the park for the fire, through the establishment of spray/ mist, inert gases and other fire alarm protection devices, and with the technology, the device's security warning measures for linkage, to raise awareness of key parts of the initial safety warning accident disposal capacity. In addition, to strengthen monitoring and early warning fire park accident disposal techniques at the same time, also through GIS, GPS, RS and video, communication networks and other information technology tools to strengthen key positions and dynamic monitoring of process equipment and prepare corresponding positions emergency disposal plan, to improve the park staff emergency disposal levels. When the emergency disposal and safety isolating protective layer of chemical industrial park cannot control the spreading of disasters and accidents, then based on the overall risk prediction and analysis in the park, through cross-regional fire emergency response mode, start serious accidents regional emergency response mechanism to conduct joint controlling of the accidents disasters.

Figure 2: Chemical Industry Park dynamic Firefighting safety management model diagram

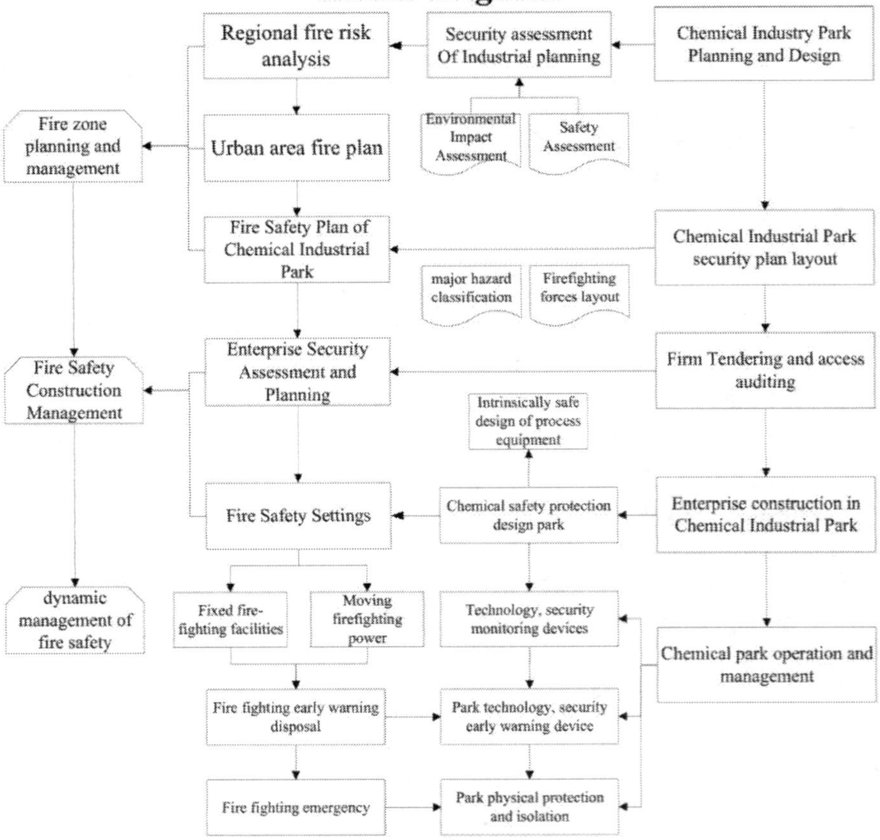

Considered the whole process of the fire accidents in the petrochemical plant, the usual safety protection measures mainly include four types and stages. Firstly, during the plant planning and design stage, the intrinsic safety protection is the most important safety measures which control the plant process and prevent process accidents and devices deviation. The second is the safety monitoring and supervision measures through which to detect the leakage dangerous materials and potential accidents signals in the dangerous equipment and instrument areas to provide the prewarning signals and process safety disposal measures. The third is accident disposal technical measures such as firefighting techniques which to control the fire accidents spreading and development till to eliminate the accidents. The fourth is the hazard controlling and safety evacuation measures which to limit the imperiling range of accidents hazard and protect the safety of life. The four stages of safety protection measures just compose four layers of safety protection in the whole plant.

According to the different layers of safety protection measures in the petrochemical plant, the emergency management systems should realize different objectives and responding stages. For the first layer of intrinsic safety protection, the emergency management should focus on the routine safety monitoring and management of the process and equipment. Once the process or equipment occurring failure, the emergency management triggers the second startup program to carry out prewarning and primary safety disposal. In the third stage of layer protection, the emergency response should focus on the accidents evaluation, emergency resources management and command decision through emergency response platform. Petrochemical enterprises in the event of fire & explosion accidents, it need the police, fire safety supervision departments and other organizations to be unified to carry on the deployment of emergency resources, coordination, linkage to carry out emergency rescue under emergency command mechanism. Due to the complex causes of the petrochemical spills, fires, explosions, failure of devices, process control and process defects such as device technology factors, including human error, mismanagement and other human factors, there is some chance of probability and sudden characteristics on the occurrence of the accidents. Existing emergency managements are mainly depending on the monitoring to carry on the emergence deployment in accordance with the emergency settings plans, which is often lack of evolution forecasting and early warning on accidents process and development. Moreover the emergence disposal plans are mostly determined by on-site decision to make contingency plans. The fire emergency management is required to analyze risk of the fire hazard and predict the fire hazard evolution based on the fire risk prediction, which can determine the possible disasters mode in the region model, the accident evolution trends, affecting regional and extent of risk, accident probability and other related factors. And according to the risk prediction and analysis results, emergency measures, emergency response resources and equipment, emergency routes and other content related to contingency plans should be set up. Risk prediction based of the emergency management models focus on the different stages and objectives, such as accident prevention, risk prediction, early warning and emergency response decision. And it is important to deploy the prevention and early warning measures before the accident or on the initial stages, by which to carry on the continuous monitoring on the position of the hazard technology process, equipment and operations about the abnormal temperature, pressure, level changes and component concentration parameters on-line monitoring and analysis, in order to achieve early warning of potential risks and disposal, control of the hidden dangers in the bud stage, and avoid disasters and accidents happening.

Figure 3: Park dynamic fire safety management measures and procedures

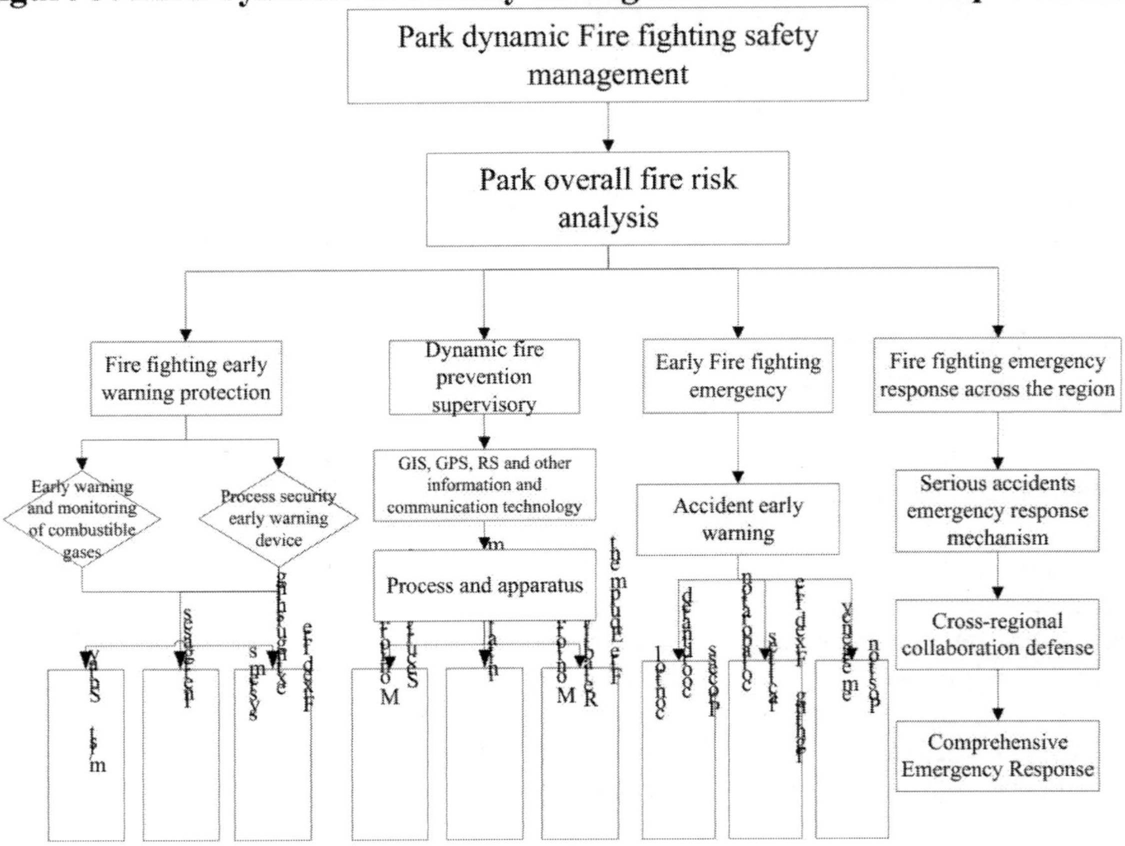

4. Conclusion

In this paper, based on the overall fire risk and fire safety management status analysis of the chemical industrial park in China, the dynamic model of fire safety management based on the protective layer was constructed. And the dynamic fire safety management measures and procedures were also analyzed. According to the overall fire risk characteristics, based on different levels of industrial park security protection layer, it is important for the prevention of the disasters and accidents occurrence in the park to incorporate fire safety planning, firefighting monitoring and early warning, accident early disposal etc., into the dynamic fire safety management mode of chemical industrial park in order to achieve integration of firefighting security management.

For the operation risk and safety protection levels in the chemical industrial park, it is proposed to establish fire safety management measures and procedures with firefighting early warning protection, dynamic fire prevention supervisory, fire and early emergency disposal, inter-regional linkage dynamic fire emergency, which make monitoring and early warning signs of the fire, security monitoring and emergency management as the core, make the fire safety management disposal sooner, which can effectively improve the efficiency of the park accident prevention and fire safety technology management level.

5. References

[1] Fan Hongjun., "From Chongqing Kai County "12 • 23" gas blowout accidents talk about disposal and prevention of casualties" *Chemical Safety and Environment*, NO. 6, 2004,13-14.

[2] Jiang Juncheng., "Accident investigation and analysis techniques" Beijing, Chemical Industry Press, 2009.

[3] State Administration of Production Safety Supervision and Management, "About the China National Petroleum Corporation, Lanzhou Petrochemical Company "1.7" explosive fire notification", Beijing: State Administration of Production Safety Supervision and Management, 2010.

[4] State Administration of Production Safety Supervision and Management, Ministry of public security, "About PetroChina Dalian International Logistics Co Ltd "7 • 16" pipeline explosion fire of the briefing", No.122. Beijing, 2010.

[5] Liu Xuanya, Zhang Qinglin, Yixing B. "Study of the petroleum storage tank zone fire risk assessment method based on the fire scenario analysis: Proceedings of the 3rd International Conference on Risk Analysis and Crisis Response", Laredo, Texas, USA, 2011. ATLANTIS PRESS

1ST CCPS ASIA-PACIFIC CONFERENCE ON PROCESS SAFETY℠

Study on Emergency Mechanism for Major Process Safety Accidents in Petroleum and Petrochemical Industry

Yan Lunjiang
China University of Petroleum (Beijing)
China University of Petroleum-Beijing, 18 Fuxue Road, Changping, Beijing China
chushengli@cnpc.com.cn

Zhang Laibin
Chu Shengli
Cheng Zonghua
China University of Petroleum (Beijing)
CNPC Research Institute of Safety & Environment Technology
China University of Petroleum (Beijing)
China University of Petroleum-Beijing, 18 Fuxue Road, Changping, Beijing China
Xishatun Bridge West, Shahe Town, Changping District, Beijing, China
China University of Petroleum-Beijing, 18 Fuxue Road, Changping, Beijing China
zhanglb@cup.edu.cn
chushengli@cnpc.com.cn
chengzonghua@cnpc.com.cn

Prepared for Presentation at
American Institute of Chemical Engineers
1st CCPS Asia-Pacific Conference on Process Safety
Qingdao, China
September 4 – 5, 2013

Study on Emergency Mechanism for Major Process Safety Accidents in Petroleum and Petrochemical Industry

Yan Lunjiang
China University of Petroleum (Beijing)
China University of Petroleum-Beijing, 18 Fuxue Road, Changping, Beijing China
chushengli@cnpc.com.cn

Zhang Laibin
Chu Shengli
Cheng Zonghua
China University of Petroleum (Beijing)
CNPC Research Institute of Safety & Environment Technology
China University of Petroleum (Beijing)

Keywords: Process safety accident; Emergency mechanism; Emergency tactics

Abstract

Three features of major process safety accidents in petroleum and petrochemical industry, such as strongly professional, high accident harm and liability to secondary disaster have been summarized in the paper. Combined the accidents characteristics and practical situation, it is presented five key emergency tactics for these accidents, adequate risk identification, professional emergency organization, plentiful supply of emergency materials, efficient joint-action mechanism of government & enterprise and right-hand guidance of public opinions. Finally main work points of constructing perfect risk identification mechanism, training professional emergency team, establishing complete emergency conservation system, building efficient joint-action mechanism and practical emergency tactics for social crisis have been analyzed in detail.

1. Introduction

Petroleum and petrochemical industry has a large probability of accident and high level difficulty of safety protection because of the energy concentration, flammable objects and complicated working process. Some of the key process equipment or process contains enormous energy risk that can lead to severe accidents. "1.7" fire accident in Lanzhou, "4.26" platform explosion in the Gulf of Mexico and "7.16" tubing explosion in Dalian in 2010 are three typical process accidents. Compared with common safety incidents of single or a limited number of people in industrial production, the process safety incidents of petroleum and petrochemical industry have greater energy. It can easily lead to a chain reaction, induce secondary environmental pollution incidents and cause huge losses to the enterprise. Limited by economic and technological conditions, as well as by human factors and uncontrollable environmental factors, petroleum and petrochemical industry still can't completely eliminate the process safety incidents. When an accident occurs, emergency action for the accident is the key to control the situation, to reduce losses and casualties. We must pay high attention to it and continue to improve the emergency response system and enhance the level of emergency in order to better guarantee safety.

The State clearly requires to establish an emergency management system according to the overall framework of "one case, three systems" in "Eleventh Five-Year" period. "One case, three systems" refers to emergency plans, emergency system, law and mechanisms. Among them, the emergency mechanism refers to a series of emergency measures and institutional arrangements to deal with emergency incidents. Judging from the emergency procedures, emergency response mechanism should at least include four main parts: preventive mechanisms, early warning mechanisms, response mechanisms and recovery mechanisms [1]. For the major process safety incidents are usually accompanied with large-scale accidents, multi-chain reaction, lots of derivative accidents, therefore, the starting and running of the emergency mechanism is not a purely technical operation. It is a very complex emergency management process when faced with emergency. In order to better play a crucial role in an emergency, it requires in-depth research and constantly advancement and improvement.

2. Analysis of features and emergency tactics of major process safety accidents

2.1 The characteristics of major process safety accidents

Suddenness, anomalism and loss are the common features of safety accidents. Suddenness refers to that accidents are always unexpected, unpredictable and the occurring process is very quickly. Anomalism refers to that the normal security incident is caused by the deviation from process or normal conditions, and that can lead to suspending of production and business activities of enterprises. Loss refers to that the result of an accident is always accompanied with economic loss or personal injury.

Except for the three characteristics as above, major process accidents also have the following three important features:

The first is highly specialized. The process of major process safety accidents are generally very complex and highly specialized [2]. For example, a typical chemical plant tends to place a large number of devices and a variety of different materials within a small area. The situations after the accident are usually very complicated.

The second is huge accident energy. The vast majority materials in petroleum and petrochemical production process are flammable and explosive. If the accident occurs, it will be a large-scale accident for the material usually has a large number and enormous energy.

The third is that it easy to form a chain reaction to induce environmental pollution incidents. There are lots of normal flowing flammable materials, toxic and hazardous substances in petroleum and petrochemical production process, which may easy to form a chain of fire explosion. If the materials are spilt in one accident, the large quantity of fire protection water accompanied with fire protection will be prone to environmental pollution incidents.

2.2 The emergency tactics analysis of major process safety accidents

According to the accident characteristics as mentioned as above and by combining with the significant process safety accident cases at home and abroad, five significant process safety accidents emergency tactics are listing as the following:

Firstly, fully risk identification. In emergency work, the most important role of risk identification is to guide emergency preparedness. Only the risks were fully identified, the preparations can be well made, and accident situations would be controlled more quickly and effectively. If the risk identification is not completely when the accident occurs, powerless emergency work will lead to disaster. There are many such accident cases in history, such as "Indian Pa Boer Accident", "Japan's Fukushima nuclear power plant accident". So, the primary jobs to the major safety process accidents are deep risk identification.

Secondly, professional and powerful emergency guarantee. Major process safety accidents are different from general safety accidents. It usually occurs in the area with intensive devices, complex processes and intricate materials. Carrying out emergency work requires in-depth understanding of production processes, device conditions and professional emergency equipment. Therefore, the emergency work of major process safety accidents should mainly rely on professional emergency team and provide strong guarantee for the accident's emergency work by building professional emergency team. For example, China National Petroleum Corporation (hereinafter referred to as CNPC) and China Petroleum & Chemical Corporation (hereinafter referred to as SINOPEC) have constructed professional emergency response team which with full-time staff and professional equipment. It provides specialized emergency support and plays a key role in major process safety emergency accidents.

Thirdly, adequate emergency resources. Major process safety accidents usually have huge energy, serious consequences, and a long emergency time. It always requires mobilizing a lot of emergency resources. So, well-stocked emergency resources are very important to the response work of major process safety accidents.

Fourthly, efficient linkage between enterprise and local government. Major process safety accidents usually have three features large scale, long time and wide range. In order to better reduce casualties and mobilize a great number of materials to carry out a large-scale emergency operation, it needs enterprises to work with local government closely.

Fifthly, correctly guide of public opinion. Enterprises need to initiatively cooperate with local government, media and well monitor public opinion to avoid mass incidents to protect corporation reputation. The well accident information releasing work can create a good social environment for the emergency of accidents.

3. Analysis on key points of the emergency mechanisms of major process safety production

3.1 Establishing sound risk identification and assessment mechanism of process safety accidents.

Petroleum and petrochemical industry must pay high attention to risk identification work. From safety management perspective, enterprises need through the risk identification to check out accident risks and assess damage degree to put forward effective preventive measures to prevent accidents as much as possible. From the perspective of the emergency, the identification and analysis of process accidents can guide enterprises to take the necessary and sufficient emergency preparedness work after fully understanding the evolution of accidents [3]. Risk identification and evaluation analysis are suitable for such stages as the process equipment research and development, design and operation. In the research and development phase, researchers can learn the hazard of process envisaged line or test equipment through process hazard analysis, in order to timely take the necessary safety measures to prevent accidents in the development process. In the design phase, based on the results of the process hazard analysis, the designer can improve design scheme to enhance the safety of process system. In the operational phase, the change parts need hazard analysis. In addition, when the factory goes into operation, it needs to re-confirm previous hazard analysis at regular intervals, which are also known as hazard analysis of running factory.

The Dow Chemical Company recommends using layer risk method, which entire analysis process is divided into four layers, to analysis production process [4]. The first layer is making process hazard analysis for all facilities by using the method of fire and explosion index, chemical exposure index, the protective layer target index, the active chemical - process hazard analysis checklist and other methods. The second layer is making additional risk inspection for specific unit of the facility by using the method of identify cause and effect in pairs, the protective layer analysis, blast evaluation, HAZOP and other methods [5]. The third layer is making enhanced risk inspection for the target process. It is primarily quantitative risk assessment. The fourth layer is to select a few scenarios for quantitative risk assessment and according to the analysis of the results of the combination and accident frequency to assure how to choose. It mainly focused on high-risk activities.

Recent years, on one hand CNPC has carried out system combing and screening for the original equipment by the means of risk management, on the other hand CNPC has carried out risk identification for proposed, under construction and new equipment installation system by using the HAZOP method. It effectively promoted the level of essential safety and risk identification of the process.

3.2 Cultivating professional emergency team to continuously enhance professional emergency response capability

In 2010, the General Office of the State Council clearly required to strengthen the construction of grass-roots emergency team in the "Opinion on Strengthening the Construction of Emergency Response Team", and required the chemicals production units, mines, non-coal mines to build its own system of emergency rescue teams. Because of the emergency particularity of major process safety accidents, cultivating professional emergency team is the foundation to guarantee efficient emergency and ensure that "Be ready to assemble at the first call and be capable of fighting and winning".

Judging from the building work of professional emergency team, firstly, professional emergency team must try to approach the production site that need protection so that they can reach the scene within the fastest time to carry out emergency operations. The second is to establish and continuously improve accident drilling system and carefully carry out accident drilling work, so that professional emergency team can better understand the scene and production personnel are more proficient in performing their emergency duties. The third is that the better enterprises can improve the professional emergency drill results by establishing the digital field model and with the interactive training system and the using of interactive training system.

At present, CNPC has constructed well control, pipeline emergency maintenance, professional firefighting and maritime emergency rescue four full-time emergency teams, which respectively undertake the four emergency missions of oil and gas wells blowout, pipeline leak, fire and explosion and maritime oil spill and personnel search & rescue. There are two main kinds of construction modes of professional emergency team. The first is point construction, which is setting up emergency teams in different points of high-risk areas (or affiliated companies). Well control, professional firefighting and marine emergency team have taken this pattern. The second is the linear deployment, which is setting up emergency teams along the linear risk areas. Pipeline maintenance emergency team covered all long-distance pipelines through this construction mode. The emergency response team equipped with professional equipment and supplies and got a very good emergency response capacity by specialized training and simulation exercises in normal times. It also played an extremely important role in the emergency rescue of CNPC's major process safety accidents.

3.3 Establishing sound emergency resources reserve system, and constantly improving the level of resources reserve

Petroleum and petrochemical enterprises should focus on process safety accident risks. Study and formulate emergency resources with security systems to form a good security system. At present, the issues as following should be focused on.

The first is setting allocation standards of the emergency resources (mainly goods and materials). This requires solving many very complex problems and the key point is how much and what to distribute. For the "what" question, it needs to closely combine the risk identification results and response capability assessment results to research the same type of risk resources by class and finally formulate outfit schemes. For the "how much" question, it needs to take a different scheme. If the price is not high, what can be fully directly equipped must be equipped. If the price is particularly expensive, it needs to consider regional linkage.

The second is to establish regional linkage reserve system for emergency resources. For the complexity of major process safety incidents, for some expensive or easy failure emergency supplies, the cost of separate individual business unit is so high that they need to consider regional linkage reserve system [6]. However, relevant supporting systems and coordination work need to keep timely in order to play a better role in emergency accidents.

The third is to establish advanced emergency resource management system. An integrated emergency resource management system was established to manage emergency resources in one uniform standard by information technology, network technology, logistics technology etc. It includes not only the company's own emergency resources, but also the same industry or the emergency resources nearby to provide emergent supporting. Establishing an informatization system of the mobilization of emergency resource can quickly coordinate and mobilize resources to the site of the incident.

There are two ways in CNPC's emergency resource deployment, one is allocating appropriate emergency resources according to enterprise's risk, the other one is allocating appropriate emergency resources according to the level of emergency response team. For example, pipe emergency maintenance team is consisted of emergency maintenance center, emergency team and maintenance teams three parts. Based on different function of the three parts, China Petroleum Pipeline Company has set up different standards of emergency resources deployment. The standard not only considering the maintenance needs of pipeline body, but also allocating appropriate emergency resources for different geographical and climatic regions which occurred oil spill accidents. CNPC's HSE management systems and emergency management system have emergency resource management modules. When subordinate enterprises reporting and updating information in it, the headquarters can learn the newest information and unified deploy the emergency resources at any time.

3.4 Promoting the construction of sound regional emergency linkage mechanism and enhancing the ability of linkage response to major accidents

The operating units with major process safety accidents should actively promote regional emergency linkage mechanism among enterprises or between enterprise and government, or between enterprise and society. Because the consequences of large-scale major process safety incidents are very serious, however the probability of occurrence is relatively lower. Therefore, an enterprise cannot reserve enough scale emergency response force to respond to all scale major process safety accidents. After an accident, in order to ensure the efficiency of emergency rescue, enterprise should prepare in advance and construct regional emergency linkage mechanism.

From theoretical perspective, the region of emergency linkage with self-organizing features is a complex organization system. First of all, the emergency linkage area is located in an open environment. It needs material, energy and information exchange with the outside world. When emergency occurs, the main body of regional emergency management needs to obtain relevant information from the outside world. It needs to mobilize large emergency equipment, and also needs assistance of specialized rescue teams and volunteers [7]. Secondly, the internal sub-region of emergency management linkage region has competition and cooperation as well as sub-regions. But cooperation and competition is essentially a non-linear. As a result, in the emergency management linkage region, imbalance develops between each subsystem and the system is in the equilibrium state. Finally, the emergency management linkage region has the negative and positive feedback in regulating material circulation and energy flow. The complex interactions within the system may generate synergies to promote the orderly development of regional emergencies linkage mechanism [8].

As major process safety incident emergency of petroleum and petrochemical enterprises, three things are most important to the regional emergency linkage mechanism [9]. The first thing is to keep communication and coordination with government to afford assurance for emergency accidents when carrying out emergency warning, personnel evacuation, work coordination, requesting support departments to provide protection. The second thing is to establish resource mechanism linkage of emergency resources to make full use of social emergency resources to provide strong support for the emergency work. The third thing is to establish and form an emergency assistance mechanism in the same industry to improve the emergency response capabilities for major process safety accidents.

The regional emergency linkage of Dalian "7.16" pipeline explosion incident is a typical case. Not only the government organized a large number of fire emergency power for fire emergency accident, maritime emergency teams of CNPC and CNOOC were respectively dispatched to participate in oil spill contingency and a large number of local fishermen boats were organized to carry out oil spill collection. Currently, our country's three major oil companies have established an emergency linkage mechanism of information resource sharing in major safety incident. CNPC's enterprises and local governments have built a wide range of corporate linkage. The mechanism provides an important guarantee for major process safety accidents.

3.5 Establishing and improving information warning, news releasing, crisis response plans, and enhancing public crisis response capability

With the rapid development of information technology, especially the widely used network significantly weakens the role of gatekeepers in the emerging media information dissemination. Information is easily distorted, the negative news will bring huge damage for corporate's reputation. Petroleum and petrochemical companies need to establish crisis response mechanism. The first is to establish the appropriate organizations to formulate emergency plans for news release and public crisis; The second is to actively carry out monitoring public opinion, not only paying attention to traditional media, but also to QQ group, Weibo and other new ways of communication to keep abreast of trends in public opinion; The third is to establish a unified news release channels, speaking with one voice, and publish information organized and targeted; The fourth is to timely open accident's information, to reduce communication links to do a good job of the information guidance; The fifth is do everything possible to avoid stimulating public sentiment[10].

When Gulf of Mexico drilling platform accident occurred in 2010, a web project described in detail of the progress of the accident was established in BP's official website. It included timely reporting emergency disposal progress, instructing emergency methods, publishing accident photos and videos data and reporting rescue details about the states which affected by the accident, such as Alabama, Florida, Louisiana and Mississippi. In order to introduce to the outside world the open attitude of the company, BP also had bought the search results about oil spill in Google, Yahoo search engines. It easy for people to understand that BP's efforts in dealing with Gulf of Mexico oil spill incident, meanwhile people also can easy find some links with key information, such as how claims and so on. The main purpose of BP was to show its transparency in dealing with this incident and make efforts to alleviate the criticism of public and media on them. Finally BP gained a good effect in improving the environment of public opinion, mitigating malicious speculation and hype and successful saving an increasingly damaged company reputation.

4. Conclusions

Petroleum and petrochemical major process safety incidents have strong professionalism, a large scale and can easily cause secondary accidents, etc. which are full of serious challenges for the accident emergency work. Petroleum and petrochemical enterprises should enhance accidents risk identification and assessment, build and strengthen professional emergency teams, promote emergency resource reserves, establish emergency system, improve public opinion monitoring and crisis response capability and constantly improve the emergency level of major process safety incident to offer a safe assurance for safe and stable development of enterprises.

5. References

[1] Jiang Naibing. "The Research on Emergency Incidents Mechanism". *Public Administration and Law*, Volume 10, 2003

[2] Bai Yongzhong,Yu Anfeng,Dang Wenyi,Wu Zhifeng "Current Development and Advice of Process Safety System". *Safety Health and Environment*. Volume 9, NO. 1,2009

[3] Wang Lihong, Liu Yi, Zhuang Xu. "Discussion about Process Safety Management from Fire and Explosion of Largest Refinery in USA". *Safety Health and Environment*. Volume 9, NO. 5,2009

[4] Zhang Guangwen, Bai Yongzhong, Li Qi, Wan Gujun. "Process Hazards Analysis——Core Part of Process Safety Management". *Safety Health and Environment* , Volume 9, NO. 9,2009

[5] Wan Gujun , Li Qi, Zhang Guangwen, Bai Yongzhong. "Implementing HAZOP to Control Major Accidents". *Safety Health and Environment*. Volume 11, NO.9, 2009

[6] Wu Xiaotao. "Research on the Connotation and Construction of Regional Emergency Response Mechanism". *Journal of Management*. Volume 24, NO.1, 2011

[7] Xie Daiyin , Chen Lan. "Strengthening Emergency Material Support Improving Disaster Response Time" . *Science &Technology Review*, Volume 26, NO. 18,2008

[8] Liu Yajing. "Research on Coordination Mechanism for Dealing with Cross-regional Public Crisis". *Journal of Hunan International Economics University*. Volume 11, NO. 1, 2011

[9] Sun Yuanming, "Research on Construction of Emergency Response Mechanism and Platform for Domestic Urban Incidents". *Journal of Chongqing University of Posts and Telecommunications (Social Science)*, Volume 19, NO. 1,2007

[10] Lin Aijun. "Government crisis management and public opinion guidance in the new media environment". *China Emergency Management*. NO. 3, 2011

Numerical Simulation of Thermal Radiation of Tank Fire and Fire Prevention Assessment in Oil Pipeline Station

Wenxing Feng
Changqing Branch of PetroChina Pipeline Company
Guandao Base,Yinchuan, Ningxia, 750000
wxfeng@petrochina.com.cn

Chaopeng Wu
University of Science and Technology of China
No.96, Rd.Jinzhai, Hefei, Anhui, 230026
cpwu@ustc.edu.cn

Shuxin Li, Zhongning Wen, Yi Ren, Lianjun Zhang
Changqing Branch of PetroChina Pipeline Company
Guandao Base,Yinchuan, Ningxia, 750000
376093516@qq.com

Haishan Zhang
HSE Department of PetroChina Pipeline Company
No.408, Rd.Jianshe, Langfang, Hebei, 065000
28644097@qq.com

Prepared for Presentation at
American Institute of Chemical Engineers
1st CCPS Asia-Pacific Conference on Process Safety
Qingdao, China
September 4 – 5, 2013

Numerical Simulation of Thermal Radiation of Tank Fire and Fire Prevention Assessment in Oil Pipeline Station

Wenxing Feng
Changqing Branch of PetroChina Pipeline Company
Guandao Base,Yinchuan, Ningxia, 750000
wxfeng@petrochina.com.cn

Chaopeng Wu
University of Science and Technology of China

Shuxin Li, Zhongning Wen, Yi Ren, Zhang Lianjun
Changqing Branch of PetroChina Pipeline Company

Haishan Zhang
HSE Department of PetroChina Pipeline Company

Keywords: tank fire; thermal radiation; water cooling system; pipeline station; numerical simulation

Abstract

Tank fire is one of the most frequent accidents in oil pipeline transportation station. The hazard of tank fire to people, building, process piping, and other facilities, is mainly from the thermal radiation. It is necessary and meaningful to study the distribution of the thermal radiation of tank fire for emergency response, preventing and reducing loss.

To research tank fire in the pipeline station, computer fluid dynamics (CFD) technology, a useful tool, is introduced to carry out the numerical simulations of tank pool fire and dike pool fire. The fire heat release rate of gasoline and diesel is set by combustion properties. Three kinds of fires, including disk pool fire and ring pool fire on the top of a tank, pool fire in the dike, are set in the six simulation scenarios. A three-dimensional station model is built in the CFD software package and used to research the thermal radiation distribution under the different conditions.

By the results of simulation, the effects of the thermal radiation to the facilities and people are analyzed. It concludes that the water cooling system has great function of decreasing the thermal radiation and protecting the nearby tanks and facilities away from the fires. Without water protection, the disk fire or ring fire can destroy or damage the nearby intensively. The dike pool fire can involve the nearby facilities into the catastrophic consequence and environmental wind can not change the thermal radiation distribution significantly. On basis of simulation, the fire safety risk is evaluated and the countermeasure and tactics is proposed to enhance fire prevention in a scientific way.

1. Introduction

Tank fire is one of the most frequent and terrible accidents in oil pipeline transportation station. The hazard of the tank fire to people, building, process piping, and other facilities, is mainly from thermal radiation. It is necessary and meaningful to study the distribution of thermal radiation of tank fire for emergency response, preventing and reducing loss.

Since a few tank fires occurred in the last years causing plenty of loss, PetroChina Pipeline company (PPC) develops a research program to evaluate the safety of the oil pipeline station[1]. One branch study aims to assess the safety of tank fire although the tanks in PPC are designed and built according to the mandatory codes. It includes assessing the safety distance, the consequence of tank fire and the function of water cooling system.

Fortunately, many experimental and theoretical studies of thermal radiation from tank fires are carried out in the past years and many inexpensive and effective methods are developed in a rational way. PPC selects the compute fluid dynamics (CFD) technology, a numerical simulation method, instead of the costly and unpractical real-scale experiments.

Here, Fire Dynamics Simulator (FDS), a professional fire simulator, is applied to simulate the tank fire in pipeline transportation station. FDS is a large-eddy simulation (LES) code for low-speed flows, with an emphasis on smoke and heat transport from fires[2]. It is developed by National Institute of Standards and Technology (NIST) of U.S.A. It is widely used to simulate the industrial and civil fires and is confirmed to provide a credible result. In the research, a real oil pipeline station is introduced as a case to be simulated and be studied.

2. Case and simulation scenarios description

2.1 Case and simulation inputs

The station has a tank area of 270m length and 230m width and has 5 main facilities including station control building, fire pumps room, power station room, valves room and pumps room as Figure 1. The tank area has 8 production tanks. Tank 1, 2, 3 and 4 is of 24m height and 40m diameter. Tank 5, 6, 7 and 8 is of 20m height and 28m diameter. The distance between tanks, d1, d2, d3 and d4, is 24m, 19m, 24m and 23m, respectively. Simulation inputs of parameters such as thermal conductivity, density and specific heat of facilities and tanks are list according to the materials property as Table 1[3,4].

Figure1: Layout of the oil pipeline station

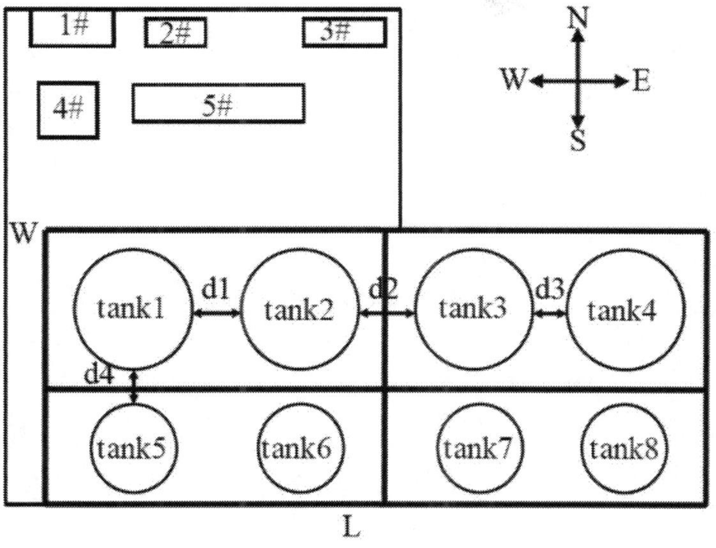

Table 1. Facilities and material parameters for simulation

No.	Facilities	Materials	Thermal conductivity ($W \cdot m^{-1} \cdot K^{-1}$)	Density ($kg \cdot m^3$)	Specific heat ($kJ \cdot kg^{-1} \cdot K^{-1}$)
1	station control building	concrete	1.2	2200	0.88
2	fire pumps room	concrete	1.2	2200	0.88
3	power station room	concrete	1.2	2200	0.88
4	valves room	steel	45.8	7850	0.46
5	pumps room	concrete	1.2	2200	0.88
6	Tanks(1~8)	steel	45.8	7850	0.46

The intensity of thermal radiation depends on the scale and intensity of the fire. As a parameter describing intensity of fire, heat release rate (HRR), in unit of kW/m^2, is defined as [Eq. 1]. χ is fuel burning efficiency; m_f is mass burning rate, estimating combustion velocity in fire; ΔH is heat of combustion. In simulation, the parameters are assigned the values as Table 2[4,5,6].

$$HRR=\chi \cdot m_f \cdot \Delta H \quad [Eq. 1]$$

Table 2. Values assigned to HRR parameters

	Gasoline	Diesel
χ	0.85	0.68
m_f ($kg \cdot m^{-2} \cdot s^{-1}$)	0.083	0.062
ΔH ($kJ \cdot kg^{-1}$)	46500	42500

2.2 Simulation scenarios

The 3D station model in FDS software is established as Figure 2. 6 scenarios are set with different fire source, fire type and the status of water cooling system as Table 3.

Scenario 1 is a disk pool fire on the top of dome roof tank 3 with diesel, as the water cooling systems of fire tank (tank 3) and nearby tanks (tank 2, tank 4 and tank 7) are ineffective. Scenario 2 is same as scenario 1 but the water cooling systems are effective. They are usual scenarios when the dome roof tank gets fires. Scenario 3 is a ring pool fire on the top of floater tank 2 with gasoline, as the water cooling systems of fire tank (tank 2) and nearby tanks (tank 1, tank 3 and tank 6) are ineffective. Scenario 4 is same as scenario 3 but the water cooling system are effective. They are frequent fire scenarios after the oil leaks from the seals of the floater tank. Scenario 5 is a big pool fire in the dike of tank 1 and tank 2 and is a representative condition of leaking from tank 1 and/or tank 2. Scenario 6 is same as scenario 5 but the environmental wind speed from the west is 2.6 m/s.

The purpose of scenario 1, 2, 3 and 4 is to study the thermal radiation (heat flux) distribution of a disk fire or a ring fire on the top of tank and the protection function of the water cooling system. The purpose of scenario 5 and 6 is to study the thermal radiation (heat flux) distribution of a dike pool fire and the influence of wind speed to the fire consequence.

Figure 2: 3D station model built in FDS software

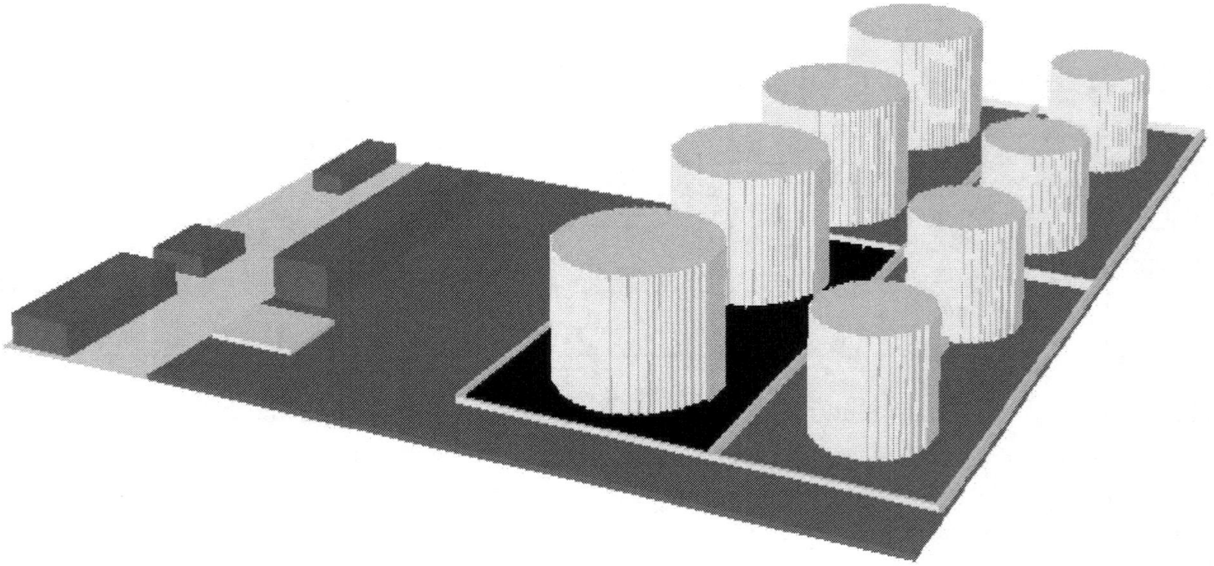

Table 3. Six simulation fires scenarios

Scenario No.	Fire source	Fire type	Fire area (m^2)	Oil	Wind speed (m/s)	Water cooling system /flow rate($L \cdot min^{-1} \cdot m^{-2}$)
1	top of tank 3	disk pool fire	1257	diesel	0	off
2						on / 2.0
3	top of tank 2	ring pool fire	175	gasoline	0	off
4						on / 2.0
5	dike of tank 1 and 2	dike pool fire	6618	gasoline	0	off
6					2.6	

2.3 Damage criteria of thermal radiation

To evaluate the heat hazard of the fires to the facilities and people, the damage criteria of thermal radiation is used in the study showed in Table 3[7]. Actually, the time, for which thermal radiation continue, is another important factor which decides whether the equipment will be destroyed or not. In the study, it is assumed that fuel is enough to burn for a long time.

Table 4. Equipment status to different heat flux

Heat flux (kW/m^2)	Equipment status
35.0	Facilities will be destroyed; exposed to radiation for 30 minutes, steel structure will be broken or collapsed. People will be killed.
25.0	Exposed to radiation for 30 minutes, steel structure will be deformed obviously.
12.5	Facilities will be damaged. Wood is ignited by flame. Plastic will be melted.

3. Simulation results

The spatial distribution of the thermal radiation of the fire scenarios are get from the numerical simulation. To assess the fire influence on the nearby tanks and facilities, the maximum heat flux is collected and shown as Table 5.

Table 5. Simulation results of 6 scenarios

Scenario No.	Maximum heat flux of nearby tank wall (kW/m^2)	Maximum heat flux of facilities (kW/m^2)
1	Tank 2: 25 Tank 4: 15 Tank 7: 20	—
2	Tank 2: 5 Tank 4: 3 Tank 7: 3	—
3	Tank 1: 10 Tank 3: 12 Tank 6: 10	—
4	Tank 1: 3 Tank 3: 5 Tank 6: 3	—
5	Tank 3:>35 Tank 5:>35 Tank 6:>35 Tank 7:25	1#:10 2#:6 3#:6 4#:13 5#:13
6	Tank 3:>35 Tank 5:>35 Tank 6:>35 Tank 7: 25	1#:10 2#:6 3#:6 4#:13 5#:13

The 3D flame model of disk pool fire in scenario 1 is shown as Figure 3 and the heat flux distribution of scenario 1 is shown as Figure 4. The 3D model of water cooling system in scenario 2 is shown as Figure 5 and the heat flux distribution of scenario 2 is shown as Figure 6. In scenario 1, the maximum heat flux of the nearby tank wall is $25W/m^2$(tank 2) and is high risky. In scenario 2, with the water cooling system protection, the maximum heat flux of the nearby tank wall is $5W/m^2$(tank 2) and is in little danger. It indicates the water cooling system can decrease the radiation heat flux greatly and protect the nearby tanks effectively.

Figure 3: 3D flame model of disk pool fire in scenario 1

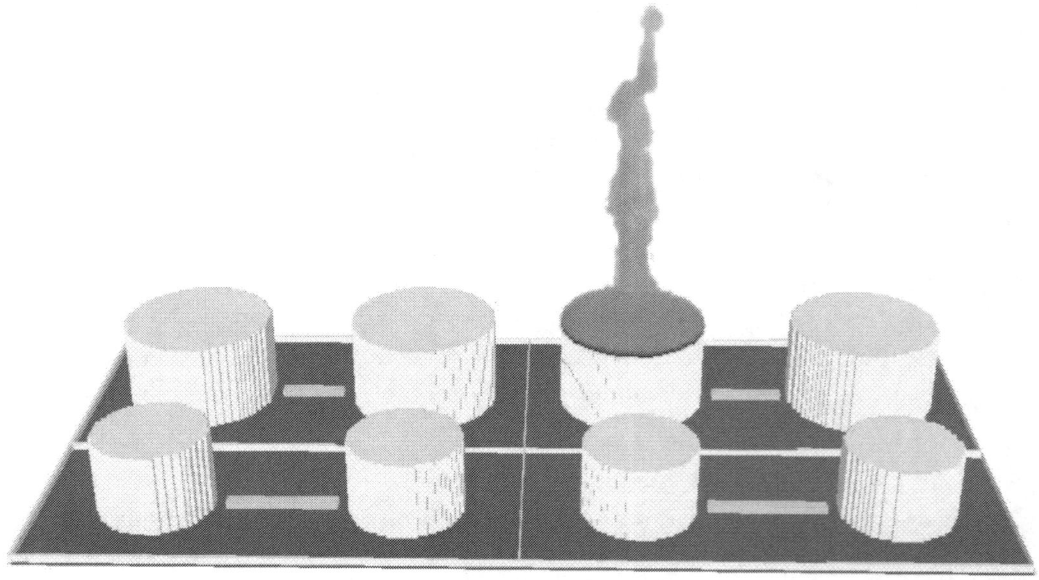

Figure 4: Thermal radiation distribution of scenario 1 (top view)

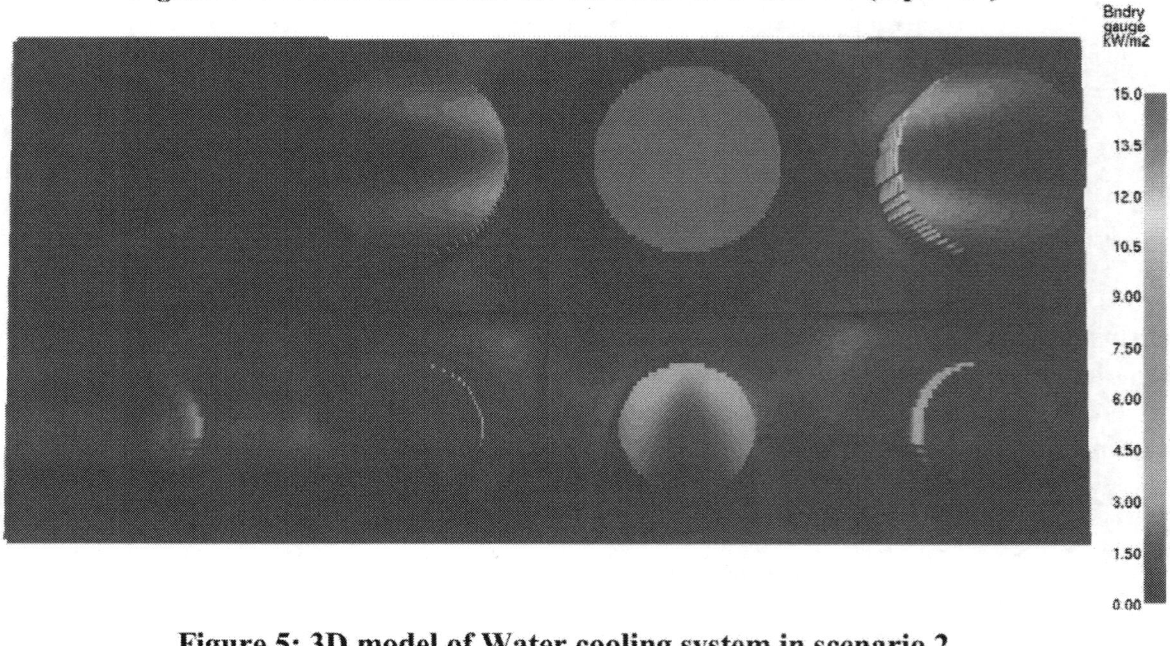

Figure 5: 3D model of Water cooling system in scenario 2

Figure 6: Thermal radiation distribution of scenario 2 (top view)

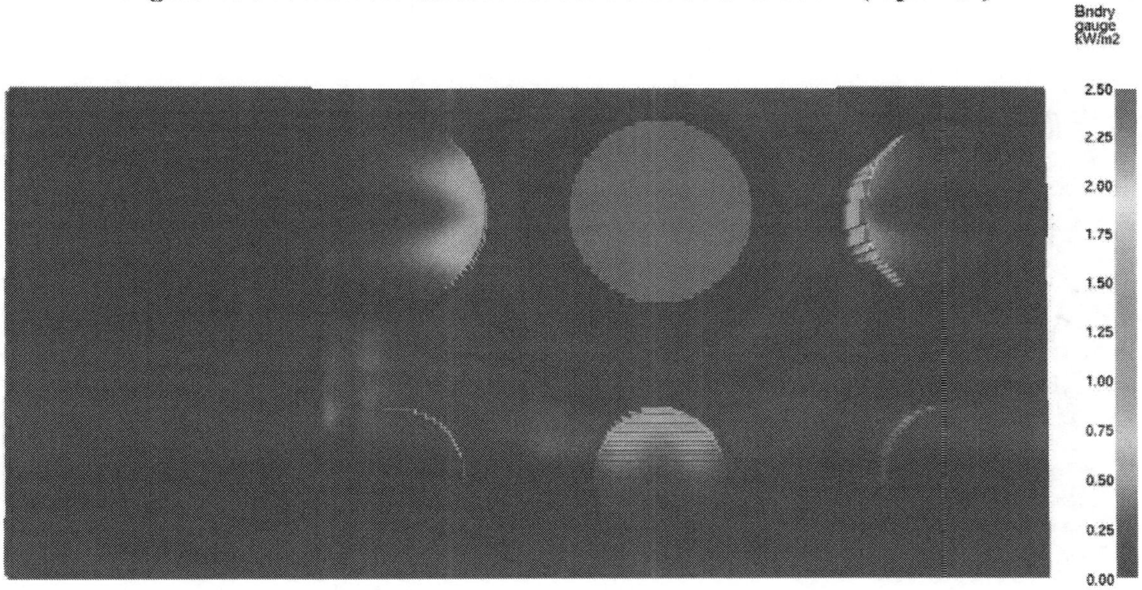

The heat flux distribution of scenario 3 and 4 is shown as Figure 7 and Figure 8. In scenario 3, the maximum heat flux of the nearby tank wall is $12W/m^2$(tank 3) and is high risky. In scenario 4, with the water cooling system protection, the maximum heat flux of he nearby tank wall is $5W/m^2$(tank 3) and is in little danger. It indicates great protection of the water cooling system as the above. It also shows the ring pool fire is less dangerous than the disk pool fire since it has much less fire area.

Figure 7: Thermal radiation distribution of scenario 3 (top view)

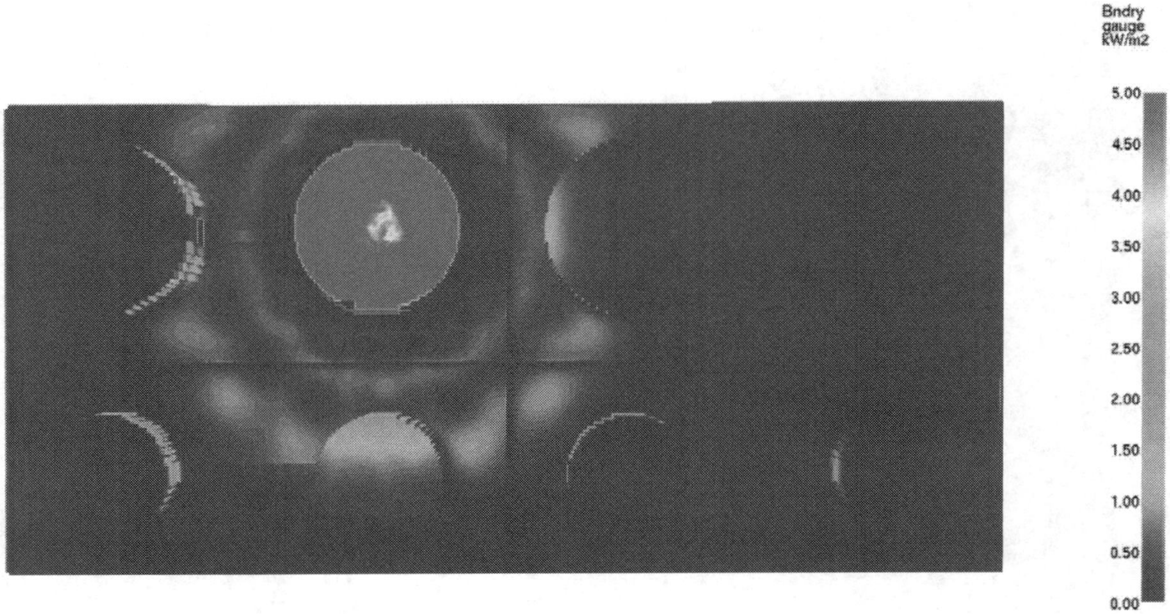

Figure 8: Thermal radiation distribution of scenario 4 (top view)

The heat flux distribution of scenario 5 is shown as Figure 9 and the vertical velocity distribution in scenario 6 is shown as Figure 10. In scenario 5, the maximum wall heat flux of tank 3, tank 5 and tank 6 is more than 35.0kW/m^2 and can lead catastrophic consequence. The maximum wall heat flux of tank 7 is 25kW/m^2 and is in high danger. Facilities such as valve room and pump room face a heat flux of more than 12.5 kW/m^2, that can cause great damage and the secondary fire. In scenario 6, the vertical velocity of the fire flame driven by fire is about 60m/s, much more than the wind speed of 2.6m/s, because the buoyancy of the fire is more powerful than the wind. so the fire flame has very little offset that do not influence the heat flux distribution significantly.

Figure 9: Thermal radiation distribution of scenario 5

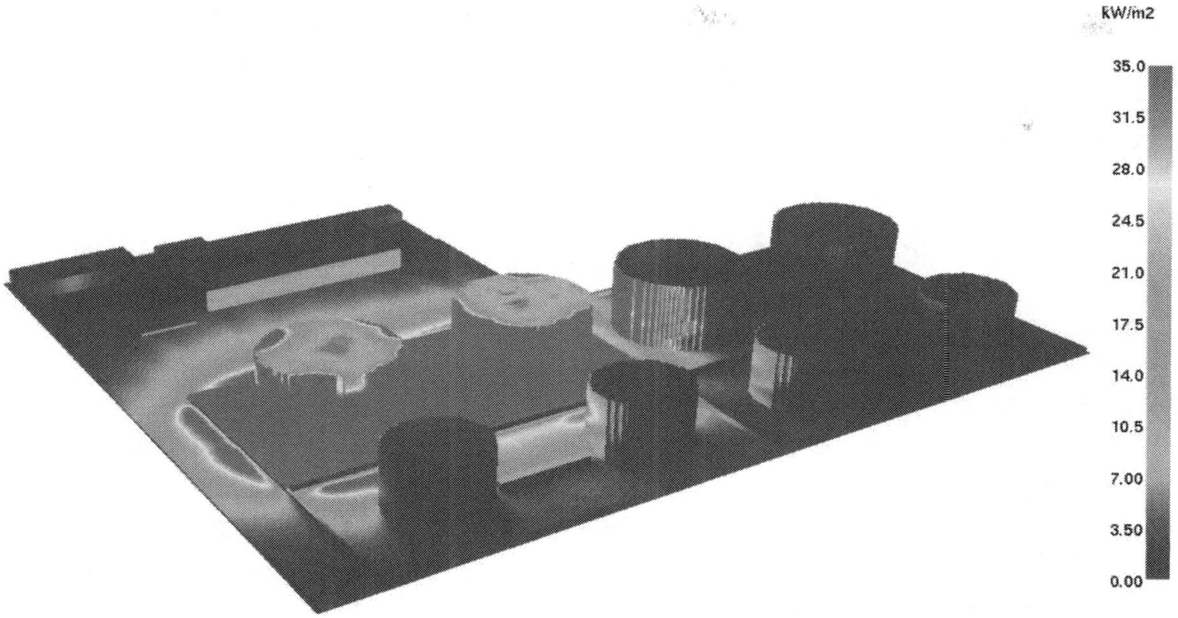

Figure 10: Vertical velocity distribution of fire flame in scenario 6

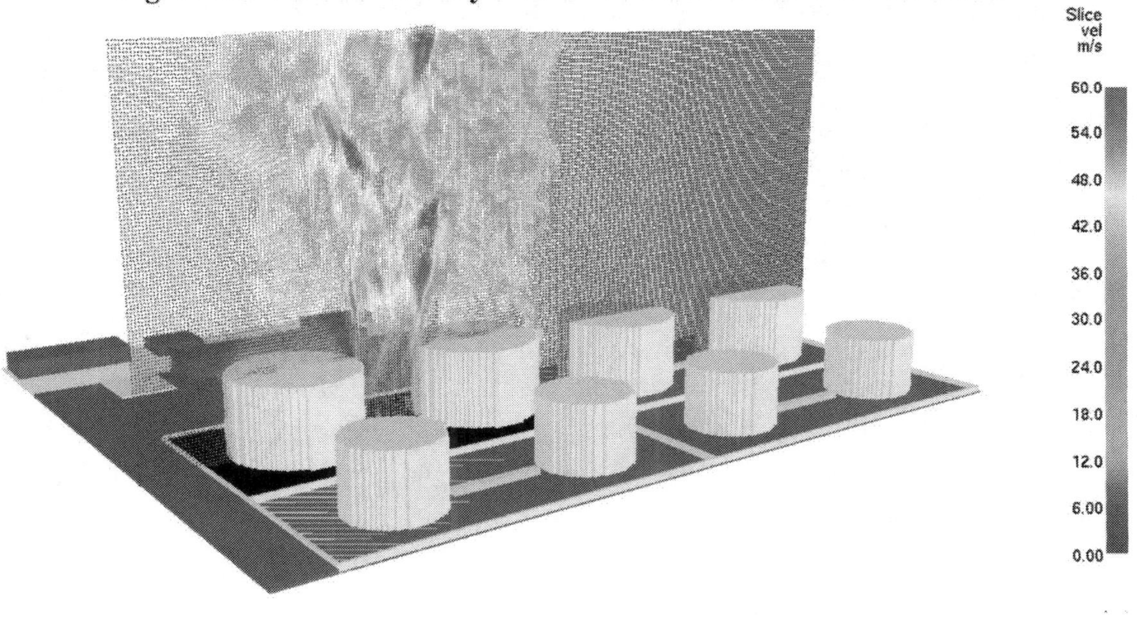

4. Conclusions

In the study, a numerical simulation method is used to study the thermal radiation of tank fire in oil pipeline transportation station. Six fire scenarios including disk fire, ring fire on the top of a tank and dike pool fire are set to be simulated and analyzed. The following are conclusions.

(a) The ring fire is less dangerous than the disk fire because it has much less fire area and burning heat. Even though, both of them can destroy or damage the nearby tanks strongly without the water cooling system protection.

(b) The water cooling system can play a great role on decreasing the thermal radiation and protecting the nearby tanks when tank fires occur. It is very important to maintain it and keep it in a right way in fire prevention of oil pipeline station. When the disk fire or ring fire on the top of a tank happens, the active and offensive measures should be taken to reduce and eliminate the fires.

(c) The dike pool fire is catastrophic. It can involve the nearby facilities and tanks into the horrible consequence. When it happens, the conservative and defensive measures should be taken to control the fires and loss.

5. References

[1] Feng Wenxing, Shui Biyuan, etc., "Application of Quantitative Risk Assessment at Stations of Oil Product Pipelines," *Oil and Gas Storage and Transportation*(in Chinese), Volume 28, NO. 10, 2009

[2] "Fire Dynamics Simulator (Version 5) Technical Reference Guide," 2010, National Institute of Standards and Technology, Maryland, U.S.A

[3] "Fundamental of Heat Transfer," 1999, University of Science and Technology of China, Hefei, China.

[4] M. Muñoz, J. Arnaldos, J. Casal, E. Planas, "Analysis of the Geometric and Radiative Characteristics of Hydrocarbon Pool Fires," *Combustion and Flame,* Volume 139, NO. 3, 2004

[5] Yi Liang, Huo Ran, Zhang Jingyan, etc., "Characteristics of Heat Release Rate of Diesel Oil Pool Fire," *Journal of Combustion Science and Technology*(In Chinese), Volume 12, NO. 2, 2006

[6] Zhuang Lei, Chen Guoqing, Sun Zhiyou, etc., "On the Damage Study of the Thermal Radiation of the Large Oil-Tank Fire Accidents," *Journal of Safety and Environment y*(In Chinese), Volume 8, NO. 4, 2008

[7] "Guidelines for Quantitative Risk Assessments", 2005, Committee for the Prevention of Disasters, Netherlands

CCPS Process Safety Metrics A Global Approach

Nara, L., AIChE/CCPS

Originally presented on: 9/5/2013 0:00:00 - 0:25:00

The use of global metrics is essential for companies and industries who strive to understand their process safety incidents and improve performance. These metrics are an important component in the comparison and evaluation of site to site, or unit to unit, performance within a company, and to help determine where focus is needed and resources should be allocated. In addition to internal comparisons, global process safety metrics can assist companies (and industries) in benchmarking their performance with others or in comparison to an industry average or goal.

In 2007, CCPS published "Process Safety Leading and Lagging Metrics…You Don't Know What You Don't Measure" (available for free download at http://www.aiche.org/ccps/resources/tools/process-safety-metrics). The definitions established in the document were a key input to the creation of a new ANSI / API standard (ANSI/API RP 754) which was finalized and released in April 2010, with CCPS support. In 2011, CCPS updated their metrics publication with the intent of aligning CCPS and API documents. CCPS has gone further in 2013 to develop the use of process safety leading and lagging indicators by publishing the results of a member company survey "Process Safety Leading Indicators Industry Survey" (also available to download). This publication provides an update on the chemical, petroleum and other process industries in the use, direction and effectiveness of leading indicators to help drive performance improvement and the reduction in the number and severity of process safety incidents.

This presentation will discuss the new CCPS metrics document and ongoing work towards global metrics.

1ST CCPS ASIA-PACIFIC CONFERENCE ON PROCESS SAFETY℠

Benchmarking Risk Based Process Safety Egg (RBPS Temple + 2 Universes) against *The Art of War*

Steven (Long) Zhang
ConocoPhillips China
TEDA, Tianjin, P.R. China
Steven.Zhang@conocophillips.com

Prepared for Presentation at
American Institute of Chemical Engineers
1st CCPS Asia-Pacific Conference on Process Safety
Qingdao, China
September 4 – 5, 2013

Benchmarking Risk Based Process Safety Egg (RBPS Temple + 2 Universes) against *The Art of War*

Steven (Long) Zhang
ConocoPhillips China
TEDA, Tianjin, P. R. China
Steven.Zhang@conocophillips.com

Thus the reason the farsighted ruler and his superior commander conquer the enemy at every move, and achieve successes far beyond the reach of the common crowd, is foreknowledge.

- Sun Tzu *The Art Of Warfare*
Roger T. Ames's translation

It (culture) is in fact a prison unless one knows that there is a key to unlock it.

- Edward T. Hall *The Silent Language*

Keywords: Process safety, *The Art of War*, RBPS, Espionage, Traditional Chinese Medicine (TCM)

Abstract

Within this paper, some attempts to benchmark Risk Based Process Safety Egg (RBPS Temple + 2 Universes) against Sun Tzu's *The Art of War* are made in terms of the strategic assessment approach Sun Tzu called 5 criteria, i.e. the Tao, the weather, the terrain, the leadership and the discipline, so that the results of the war against catastrophic process safety accidents could be foretold. And furthermore, the role and purpose of the industry wide process safety metrics campaign is examined in light of 'Espionage' and Traditional Chinese Medical diagnosis.

1. Introduction

Within the paper presented at 9th Global Congress on Process Safety, the author identified 2 new elements /universes, i.e. RBPS Core Value and Process Safety Cultural Communication for RBPS temple or 4 pillars, and recommended a new model i.e. RBPS egg to the process industry and beyond. [1] And another idea the author advised is: process safety culture and management should be inherently of *The Art of War* against catastrophic process safety accidents so that what might have been taken for granted in process safety culture and management for such a long period of time could be reexamined and reflected in light of Sun Tzu's wisdom and insight.

The benchmarking results show that RBPS egg model is very similar, or even could be said being exactly the same, with the one of *The Art of War:* RBPS Core Value - Sun Tzu's Tao; Commit to Process Safety - Weather & Leadership; Understand Hazards & Risk - Terrain; Manage Risk - Discipline; and Learn from Experience - Espionage.

And furthermore, the other universe i.e. Process Safety Cultural Communication is also examined and the implications of culture as a prison and core value as the key explored. At the end, based on the benchmarking results, the frequently mentioned "Plant, Procedure and People" are characterized with new implications, and a few of recommendations made.

2. Culture is a Prison - when RBPS Egg meets *The Art of War*

2.1 RBPS Core Value - the Tao

Sun Tzu defined the Tao like this (Roger T. Ames' Translation): what brings the thinking of the people in line with their superiors. Hence, you can send them to their deaths or let them alive, and they have no misgivings one way or the other. [2] And for surety, we need not our operation workforce to go to death, but to achieve a highly reliable conduct of operations (COO) and operational discipline (OD), especially of critical operation activities. [3] And when facing issue of human factors and challenges for a highly reliable conduct of operations and operational discipline, Sun Tzu's solution is Tao, which we might call core value today.

And 2500 year later when globalization is still deepening and our global village increasingly smaller, the author crafted a universal Tao, i.e. RBPS Core Value, whose definition is like this: There is no power, authority or governance on process safety except on behalf of process safety expertise, which the author characterized as the divine essence of the highest strategic insights and vision of process safety management or the Tao for process safety culture, is really an ultimate solution to the power inequality induced organizational failure in context of process safety strategic communication and decision making, and without which the subordinates usually stay in an 'ethical' dilemma where process safety expertise and the superiors' chain of command compete for chairing all the decision making and behavior determining. [4]

Conclusion: a correct Core Value should be the ultimate solution to the issue of human factors, for highly reliable conduct of operations and operational discipline, and overall sphere of process safety management.

2.2 *Understand Hazard & Risk - the Terrain*

Sun Tzu defined the Terrain like this (Roger T. Ames' Translation): Terrain refers to the fall of the land, proximate distances, difficulty of passage, the degree of openness, and the viability of the land for deploying troops. [5]

From above we could see that understanding the actual specifics of the Terrain is very critical for waging warfare, so does *2nd pillar Understand Hazards & Risk* pose to modern process safety management. And there is rationale that not only a general knowledge and expertise of Terrain is necessary but also the actual specifics of some Terrain should be acquired and understood. And the following 2 elements within this pillar function similarly as Terrain: *Process Knowledge Management* and *Hazard Identification and Risk Analysis.* [6]

Conclusion: metaphysically, *2nd pillar Understanding Hazards & Risk* is the same with Terrain.

2.3 *Manage Risk - the Discipline*

Sun Tzu defined the Discipline like this (Roger T. Ames' Translation): And regulation (Discipline) entails organizational effectiveness, a chain of command, and a structure for logistical support. [7]

And the purpose of the 3 critical aspects at the tactical level that were highlighted is to reliably 'manage risk' during warfare. And when we look back at the following 10 elements of the *3rd pillar Manage Risk,* we may be aware that there might be some underlying interrelationship between the 3 critical aspects and the 10 elements of *3rd pillar.* [8]

Operating Procedure;
Safety Work Practices;
Asset Integrity & Reliability;
Contractor Management;
Training and Performance Assurance;
Management of Change;
Operational Readiness;
Conduct of Operations; and
Emergency Management

And the author advises that each of the 10 elements of 3rd pillar should be continuously assessed and improved in light of the 3 critical aspects of Sun Tzu's Discipline, among which 'a chain of command' might be completed with RBPS Core Value so that the 'ethical' dilemma for subordinates could be eliminated and the performance of 'manage risk' improved revolutionarily through inherently safer design of new type of chain of command, i.e. chain of command directly and primarily from RBPS Core Value. [9]

And the 2 elements of Contractor Management and Management of Change might be more special and deserves a separate topic in the future.

Conclusion: At tactical level, 3rd pillar and its 10 elements should be continuously assessed and improved in light of the 3 critical aspects, i.e. organizational effectiveness, a chain of command and logistical support.

2.4 Commit to Process Safety - the Weather & the Leadership

Sun Tzu defined the Weather (Climate) and Leadership (command) like this [10] (Roger T. Ames' Translation): Climate is light and shadow, heat and cold, and the rotation of the seasons. Command (Leadership) is a matter of wisdom, integrity, humanity, courage, and discipline.

And this time we have to disintegrate *1st pillar Commit to Process Safety* [11] so that the subgroups of the 5 elements below could find their counterparts within *The Art of War*.

Process Safety Culture;
Compliance with Standards;
Stakeholder Outreach;
Process Safety Competency; and
Workforce Involvement

The element Process Safety Culture might be absorbed into the Universe Process Safety Cultural Communication, which is to be examined within section 3.1.

Standard & Outreach – Climate/Weather: Metaphorically, we might understand within process industry what the 'weather' or 'climate' within process safety management relates to could be external public communication with stakeholders, and the 2 elements of *Compliance with Standards* and *Stakeholder Outreach* function together as interacting with 'climate', and manage the image, reputation and regulatory environment, whose impact usually proves to be profound and at the strategic level so that the consistent and stable expertise of climate prediction and strategic decision making should be heavily counted on.

Workforce & Competency – Leadership: Pursuing the Tao, Sun Tzu's made the job spec for leadership as a matter of wisdom, integrity, humanity, courage and discipline. And in other words, the ultimate purpose of the leadership, Sun Tzu intended, is to lead the organization to pursue the Tao so that "you can send them to their deaths or let them alive, and they have no misgivings one way or the other." i.e. highly reliable conduct of operations and operational discipline. And in the author's opinion, the 6 characters of the "leadership" are actually of the organization's competency of pursuance and conveyance of the Tao, which could be detailed as below: [12]

Competency in embracing RBPS Core Value;
Competency in upward Process Safety Cultural Communication;
Competency in downward Process Safety Cultural Communication; and
Competency in process safety leadership planning, assessment and intervention

At this stage, we might be aware the *Element Process Safety Competency* of the 1st pillar should be rewritten to include RBPS Core Value and Process Safety Cultural Communication. And some people might argue, however, that within this element the knowledge learning should remain. The author would say, however, value and communication based should be more inherently safer, and only through value and communication could knowledge based be reliably maintained.

Conclusion: the *element Process Safety Competency* might be rewritten to include RBPS Core Value and Process Safety Cultural Communication so that the 2 elements of *Workforce Involvement* and *Process Safety Competency* could function equally as Sun Tzu's Leadership.

2.5 Learn from Experience - Espionage or Traditional Chinese Medical Diagnosis

Espionage: From perspective of Sun Tzu, war could be strategically assessed in terms of 5 critical criteria so that war results foretold and ascertained. One question, however, has to be answered is how the actual and accurate data of the 5 criteria be foreknown so that the strategic assessment be reliably carried out? The answer is 'espionage' or 'spy', of which Sun Tzu specifically assigned the final chapter to explain and make rationale within *The Art of War*. [13]

Rationally, the espionage to an enemy should be carried out as deep as possible, the deeper the better, and more comprehensive and valuable the collected intelligence the better. When retrospect, however, we might be aware that the current industry wide process safety metrics campaign is in fact of the 'espionage' regarding the actual process safety culture and management, not of any enemy, but of the organization ourselves. So the question becomes how 'shy' we would be when we face process safety metrics of our organizations, i.e. could we touch or even dig deep within ourselves the topics such as RBPS Core Value, Process Safety Cultural Communication, leadership competency, 'Weather/Climate', 'Terrain', and 'Discipline'. And as the author emphasized within the paper at 9th Global Congress on Process Safety, all these 4 pillars including 'espionage' are of process safety cultural communication, whose ultimate purpose is to convey the message of the actual core value of the organizations, rather than finding flaw in it.

Traditional Chinese Medicine (TCM): within TCM, there are 4 diagnostic methods, i.e. inspection, listening & olfaction, inquiry, and pulse-taking & palpation, all of which function as diagnosis i.e. TCM 'process safety' metrics, audit and incident investigations etc [14], or 'espionage' so that symptoms (syndromes) and roots (diseases) could be identified and strategically addressed in wise manner. [15] There was an old Chinese proverb saying: a doctor with a disease shall not diagnose and cure himself. In another words, an organization with cultural issues neither shall nor are able to diagnose and cure itself.

Conclusion - culture is a prison: Anthropologist Edward T. Hall says within *The Silent language*: It (culture) is in fact a prison unless one knows that there is a key to unlock it. [16] So what is 'culture' in context of process safety? The 4 pillars! And the author goes further saying: RBPS Core Value is the key to the prison [17], and Process Safety Cultural Communication the key's brightness. Should the key rusted, the prison could not be opened and substantively the organization living within might starve to death. And Sun Tzu actually also characterized that 'key and its brightness' as similar: Only the sagacious ruler is able to employ spies; only the most humane and just commander is able to put them into service; only the most sensitive and alert person can get truth out of spies.[18]

3. Philosophy of Process Safety – recommendations for a new architecture

3.1 Plant, Procedure & People – Philosophy of Process Safety

In light of the implications of Weather/Climate, we may be aware our 'plant' is not a pure engineering product, but rather a social process influenced artifact. [19] So Weather/ Climate and Terrain become the 2 dimensions of 'plant'. And similarly, 'people' also acquires some new implications, i.e. 'people' means leadership, means 4 dimensions of process safety competency, and 'people' does not simply mean people. And 'procedure' implies the pillar 'manage risk'.

And traditionally, following procedure might have been overemphasized in order to pursue operational discipline, but in fact should need to be addressed first at higher strategic level, i.e. level of RBPS Core Value and Process Safety Communication. So we could conclude 'Philosophy of Process Safety' as below:

RBPS Core Value is the 'worldview' for process safety;
Process Safety Cultural Communication is the 'methodology'.

In other words, the harmonization between People, Plant and Procedure only could be acquired through embracing the Process Safety Worldview (RBPS Core Value) and practicing the Process Safety Methodology (Process Safety Cultural Communication).

<table>
<tr><td colspan="2">RBPS Egg</td><td>The Art of War</td><td>Philosophy of Process Safety</td><td></td></tr>
<tr><td colspan="4">RBPS Core Value - Tao - RBPS Worldview</td><td>Worldview</td></tr>
<tr><td rowspan="4">Commit to Process Safety</td><td>Process Safety Competency</td><td rowspan="2">Leadership</td><td rowspan="2">People</td><td rowspan="6">Process Safety Cultural Communication (Process Safety Culture) RBPS Methodology</td></tr>
<tr><td>Workforce Involvement</td></tr>
<tr><td>Compliance with Standards</td><td rowspan="2">Weather/Climate</td><td rowspan="3">Plant</td></tr>
<tr><td>Stakeholder Outreach</td></tr>
<tr><td>Understand Hazards & Risk</td><td></td><td>Terrain</td></tr>
<tr><td>Manage Risk</td><td></td><td>Discipline</td><td>Procedure</td></tr>
</table>

Learn from Experience		Espionage	Engineering & Humanities	

Contemporary technology philosopher Carl Mitcham categorized philosophy of technology as engineering philosophy of technology and humanities philosophy of technology. [20] And the author borrows this idea saying: regarding process safety, there should also be 2 traditions i.e. engineering philosophy of process safety and humanities philosophy of process safety, both of which could help us succeed in 'espionage', far beyond where we have been.

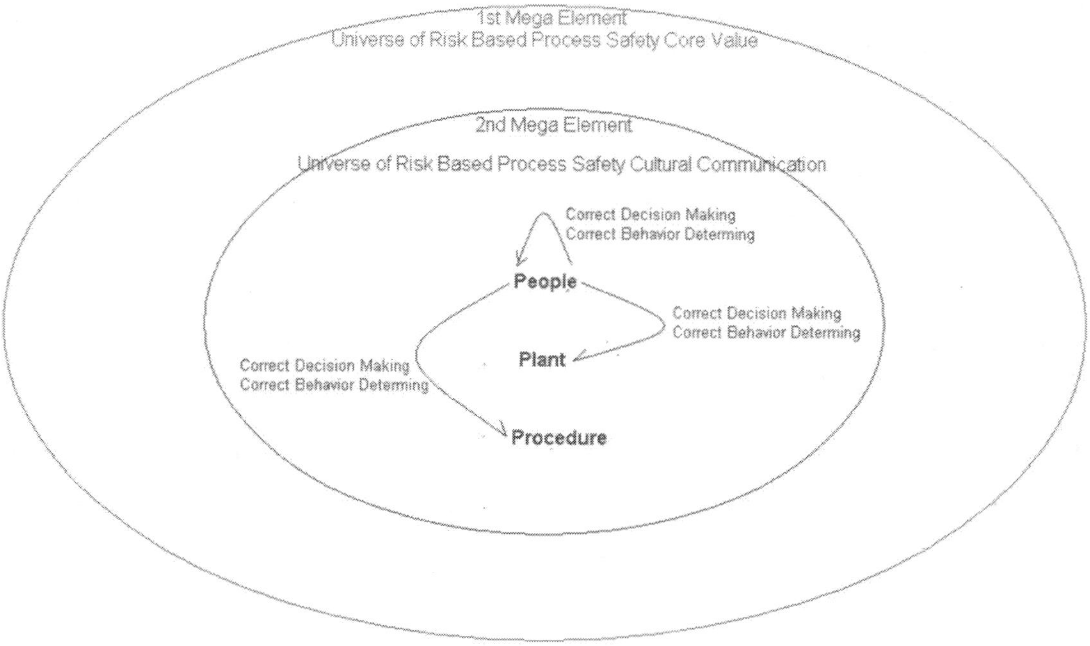

3.2 Seven Mega Elements – recommendations for a living architecture

Upon the benchmarking results, a newly adjusted architecture of RBPS Egg, which consists of the following 7 mega elements, might be presented to process industry and beyond:

RBPS Core Value;
Process Safety Cultural Communication;

Leadership;
Weather/Climate;
Terrain;
Discipline; and
Espionage

And furthermore, when the element of *Process Safety Competency* is rewritten and the mega element Leadership is characterized with the implications of the 4 dimensions of new process safety competency, it's not difficult among the last 5 mega elements to recognize the fundamental independency of Leadership, upon which all the rest 4 mega elements strategically depend to some extent, more or less.

And the author makes the following recommendations with regards the benchmarking results:

1. The main gap is: no correct leadership within RBPS Egg as within *The Art of War.* The element *Process Safety Competency* might be rewritten to include the new 4 dimensions of process safety competency.

2. Only Tao (RBPS Core Value) is what brings about a highly reliable conduct of operations and operational discipline, rather than all relatively tactical addressing of the issue of human error [21], which might turn out not to be so substantive as RBPS Core Value, and which is only of the symptoms not of the root.

3. Process Safety Culture is a prison, except we know there is a key to unlock it. And Process Safety Cultural Communication is the key's brightness, and is to resurrect all the elements, especially those 5 new pillars which would function collectively as organs so that the RBPS Egg would become 'biologically' alive.

4. Except with stable RBPS Core Value and Process Safety Cultural Communication, MOC for 'plant, procedure and people' is bound to get problem, sooner or later, here or there.

5. RBPS Core Value and Process Safety Cultural Communication mandate a new type of chain of command, a shorter chain of command, directly from RBPS Core Value, rather than from superiors. And with this new type of chain of command, line management [22], performance contract [23] and Management by Objectives (MBO) [24] would acquire new meaning, implications, and dimensions.

6. The future of the current industry wide process safety metrics campaign might turn out not to be as brighter as universally expected, except that the new 'Leadership' with those 4 dimensions of the new process safety competency is pursued.

4. Conclusion

The experience of benchmarking RBPS Egg against *The Art of War* turns out to be fascinating, and the results shows that RBPS Egg is surprisingly similar, or even could be called exactly being the same, with *The Art of War,* except one main gap, i.e. lack of correct 'Leadership' within the Egg, all of which means that we are very near to the Art of Process Safety or Philosophy of Process Safety, where 'plant, procedure and people' are characterized with new implications and a few of recommendations attempting to correct the prevailing misunderstanding and misconception across the industry and beyond made.

5. References

[1] Zhang, Long, *An Upgraded Definition of Risk Based Process Safety Core Value and Its Impact in Light of Anthropology & Intercultural Communication on the Future Evolution of Process Safety Culture & Management*, 2013 AIChE Spring Meeting and 9th Global Congress on Process Safety, San Antonio, April 2013

[2] Sun-Tzu and Ames, Roger T., *Sun-tzu: The Art of Warfare*, Ballatime Books, 1993, p.103

[3] Center for Chemical Process Safety, *Conduct of Operations and Operational Discipline*, John Wiley & Sons, Inc., 2011, pp. xxiii & xxv.

[4] Zhang, Long, *An Upgraded Definition of Risk Based Process Safety Core Value and Its Impact in Light of Anthropology & Intercultural Communication on the Future Evolution of Process Safety Culture & Management*, 2013 AIChE Spring Meeting and 9th Global Congress on Process Safety, San Antonio, April 2013

[5] Sun-Tzu and Ames, Roger T., *Sun-tzu: The Art of Warfare*, Ballatime Books, 1993, p.103

[6] Center for Chemical Process Safety, *Guidelines for Risk Based Process Safety*, John Wiley & Sons, Inc., 2007, pp.169-244.

[7] Sun-Tzu and Ames, Roger T., *Sun-tzu: The Art of Warfare*, Ballatime Books, 1993, p.103

[8] Center for Chemical Process Safety, *Guidelines for Risk Based Process Safety*, John Wiley & Sons, Inc., 2007, pp.243-547.

[9] Zhang, Long, *An Upgraded Definition of Risk Based Process Safety Core Value and Its Impact in Light of Anthropology & Intercultural Communication on the Future Evolution of Process Safety Culture & Management*, 2013 AIChE Spring Meeting and 9th Global Congress on Process Safety, San Antonio, April 2013

[10] Sun-Tzu and Ames, Roger T., *Sun-tzu: The Art of Warfare*, Ballatime Books, 1993, p.103

[11] Center for Chemical Process Safety, *Guidelines for Risk Based Process Safety*, John Wiley & Sons, Inc., 2007, pp.37-165.

[12] Zhang, Long, *An Upgraded Definition of Risk Based Process Safety Core Value and Its Impact in Light of Anthropology & Intercultural Communication on the Future Evolution of Process Safety Culture & Management*, 2013 AIChE Spring Meeting and 9th Global Congress on Process Safety, San Antonio, April 2013

[13] Sun-Tzu and Ames, Roger T., *Sun-tzu: The Art of Warfare*, Ballatime Books, 1993, pp.169-171.

[14] Center for Chemical Process Safety, *Guidelines for Risk Based Process Safety*, John Wiley & Sons, Inc., 2007, pp.549-647.

[15] Wang, Lufen and Li, Zhaoguo and Bo, Bai, *Diagnostics of Traditional Chinese Medicine*, Shanghai University of Traditional Chinese Medicine Press, 2002, pp.1-7.

[16] Hall, Edward T., *The Silent Language*, Anchor Books, 1990, p.182

[17] Zhang, Long, *An Upgraded Definition of Risk Based Process Safety Core Value and Its Impact in Light of Anthropology & Intercultural Communication on the Future Evolution of Process Safety Culture & Management*, 2013 AIChE Spring Meeting and 9th Global Congress on Process Safety, San Antonio, April 2013

[18] Sun-Tzu and Ames, Roger T., *Sun-tzu: The Art of Warfare*, Ballatime Books, 1993, p.170

[19] Bucciarelli, Louis L., *An ethnographic perspective on engineering design*, Design Studies, Volume 9, Issue 3, Jul 1988, pp.159-168.

[20] Mitcham, Carl, *Thinking through Technology: The Path between Engineering and Philosophy*, University of Chicago Press, 1994, p.17

[21] Center for Chemical Process Safety, *Guidelines for Preventing Human Error in Process Safety*, John Wiley & Sons, Inc., 2010

[22] B.P. U.S. Refineries Independent Safety Review Panel, *The Report of the B.P. U.S. Refineries Independent Safety Review Panel*, 2007, p.252

[23] B.P. U.S. Refineries Independent Safety Review Panel, *The Report of the B.P. U.S. Refineries Independent Safety Review Panel*, 2007, p.244

[24] Hofstede, Geert H. and Hofstede, Gert Jan. and Minkov, Michael, *Cultures and Organizations: Software of Mind: Intercultural Cooperation and Its Importance for Survival*, McGrawHill, 2010, p.335

Dynamic Risk Assessment of Well Control System

Faisal Khan
Offshore Safety and Risk Management Group
Australian Maritime College
University of Tasmania, Launceston, TAS 7250, Australia
Email: faisal.khan@amc.edu.au

Vikram Garaniya
Offshore Safety and Risk Management Group
Australian Maritime College
University of Tasmania, Launceston, TAS 7250, Australia
Email: v.garaniya@amc.edu.au

Prepared for Presentation at
American Institute of Chemical Engineers
1st CCPS Asia-Pacific Conference on Process Safety
Qingdao, China
September 4 – 5, 2013

Dynamic Risk Assessment of Well Control System

Faisal Khan and Vikram Garaniya
Offshore Safety and Risk Management Group
Australian Maritime College
University of Tasmania, Launceston, TAS 7250, Australia

Keywords: Process Safety, Well Control , Dynamic Risk Assessment, Accident Precursors

Abstract

The recent offshore drilling accident such as Deepwater Horizon drilling rig explosion in April 2010; Vermillion 380 oil rig explosion in September 2010, and Chevron Nigeria limited oil rig explosion in January 2012 have raised serious concerns about the safety of drilling operations. The major reasons for such accidents are the lack of understanding of how such events unfold or how the early warning signs of a potential major accident could be integrated into a risk assessment and management framework. A well blowout is among the most undesired and feared accidents during drilling operations. The dynamic nature of a blowout accident is a result of rapidly changing physical parameters and time-dependent failures of safety barriers. To better predict and prevent such accident, one needs real time risk assessment of a well control system. This paper briefly recapitulates the application of dynamic risk assessment to well control systems using early warnings usually known as accident precursors. Drilling operation and well control systems are modeled using Bow-tie (BT) approach. The BT model represents a potential accident scenario and associated consequences. Further, a real time predictive model for the failure probabilities of key elements of well control systems is developed and subsequently used in conducting the risk assessment of drilling operations. The present work provides both a real time assessment of well control systems and a basis to make timely safety critical decisions. The risk assessment methodology is demonstrated for a well blowout accident. This case study highlights the applicability and advantages of the proposed method in drilling operations.

1. Introduction

Safety assessment involves identification of potential hazards associated with an operation and evaluation of the likelihood of the occurrences of such hazards. This approach is known as Quantitative Risk Assessment (QRA). It also identifies the possible consequences resulting from these hazards and quantifies the associated likelihoods. QRA was first developed in the nuclear industry in early 70s; since then it has been steadily improving. Khan and Abbasi (1998) have presented a review of available techniques to conduct QRA in process industries. Khan and Abbasi (2002) have developed a criterion for identification of the most credible accident scenario used for risk assessment. Bucci et al. (2008) have recently presented an approach to construct dynamic event trees and fault trees and addressed the concerns with the traditional event tree and fault tree methodology. Although QRA has been proven as an effective method in the industry, it lacks an important element of risk profile that is time dependency; in other words, it lacks of learning from the previous accidents in process industry. In recent years researchers have been working on developing a more dynamic approach than QRA. Cepin et al. (2002) developed a method which represents an extension of the conventional fault tree with time requirements; Boudali et al. (2005) defined discrete-time Bayesian Network formalism and demonstrated its capabilities from a modeling and analysis point of view. Other notable contributions in the area of dynamic risk assessment include Bier and Mosleh (1990), Meel and Seider, (2006) Meel et al. (2007), Kalantarnia (2008, 2009) Khakzad et al. (2012, 2013a,b,c), Rathnayaka et al (2012, 2013)

The QRA concept has been successfully applied to many operations to increase safety and reduce loss and damage by asking the following questions:

- What can go wrong?
- How can it go wrong?
- How severe will the consequence be?
- How likely is such situation to occur?

Dynamic risk assessment is a novel approach for the process safety assessment and management. It enables the real time failure analysis of any operation. Dynamic risk assessment is a new approach to process safety assessment and management. It enables the real time failure analysis of an operation. In dynamic risk assessment, Bayes' theorem and joint probability distributions are used to develop a predictive failure model of an operation. As the operation generates accident precursors such as mishaps, incidents and near misses, the accident occurrence probability is updated. The updated probability of the accident is then used to revise the associated risk profile. In subsequent section details of the dynamic assessment framework is presented.

2. Dynamic Risk Assessment

Dynamic risk assessment is a new approach for QRA. In this approach, real time data in form of incidents and near misses (referred to as accident precursor data), are collected from the system and used to update the risk profile (Bier and Mosleh, 1990). Dynamic risk assessment enables the real time estimation of the risk of an operation. The algorithm for dynamic risk assessment is shown in Figure 1.

Dynamic risk assessment starts with defining the system. Since the dynamic risk assessment uses accident precursor information, it is important to identify these events and analyze them to identify the accident sequence process. Each event is assigned a specific severity level. The key step to an effective dynamic risk assessment is to model all possible accident scenarios. This serves as a basis to perform quantitative risk analysis. The combination of a fault tree and an event tree, which is referred as a Bow-tie diagram, is used to develop the accident sequence process, characterizing its cause-consequence relationships. The results obtained through Bow-tie analyses are known as prior failure probabilities, which are used along with the consequences to estimate the prior risk. Consequences are assessed in equivalent dollar values considering the event's impact on assets, people, environment, and the company reputation.

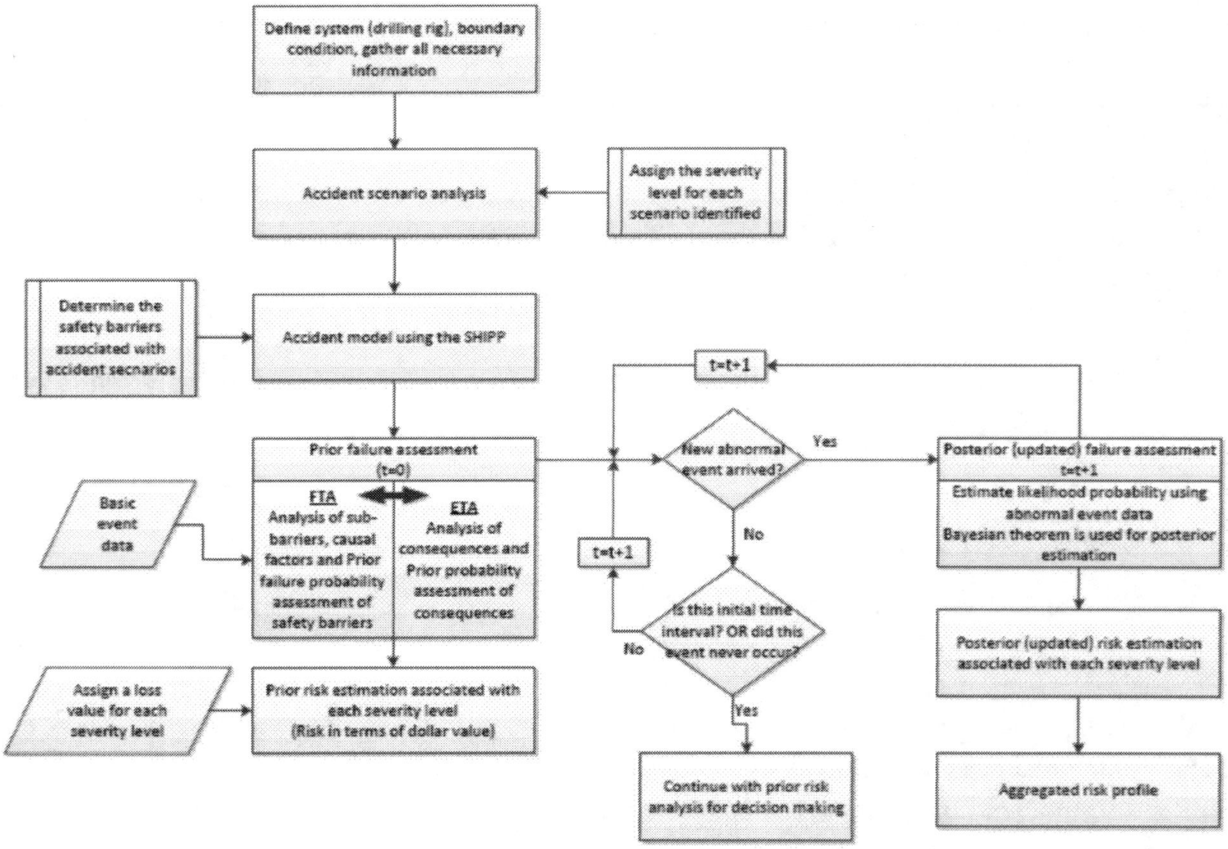

Figure 1. Dynamic Risk Assessment Framework (Rathnayaka et al., 2013)

To obtain the updated failure probabilities of the safety barriers and the resulting consequences, respective likelihood functions are developed using accident precursor data obtained from the operation. The prior probabilities along with the likelihood functions are then used via Bayes' theorem to obtain posterior (updated) probabilities. This updating process is performed for each time interval as long as accident precursors are generated through the operation. If a particular accident precursor has not been observed in the system for a while, the prior probability of the related safety barrier is not updated and remains unchanged till a new observation is made. However, it is worth noting that for Poisson processes, the prior probabilities are updated when time passes even if no accident precursor is observed (Khakzad et al., 2012). Having the updated probabilities and the severity of consequences, the updated risk can be estimated.

2.1. Bayes' Theorem:

Bayes' theorem is a tool for reasoning under uncertainty:

$$f(\theta|X) = \frac{\pi(\theta)l(X|\theta)}{\int \pi(\theta)l(X|\theta)} \qquad 1$$

In dynamic risk assessment, Bayes' theorem is used to update the probabilities of failures, likelihood of consequences, and the risk. The parameters in Equation 1 are:

- $\pi(\theta)$ - Prior distribution: a probability distribution function representing the prior knowledge available for a process. Priors can be derived from historical data (objective), defined by experts (subjective), or be uninformative showing that there is no initial data available for the system.

- $l(X|\theta)$ - Likelihood function: the probability of observing the value of X given the prior value of θ.
- $f(\theta|X)$ - Posterior distribution: a probability distribution function representing the state of knowledge after observing the data.

The term in the denominator of Equation 1 is the normalizing factor, interested readers may see Meel and Seider (2006) Kalanatarnia et al. (2008, 2009), and Khakzad et al. (2012).

The predictive model is developed to estimate the number of accident precursors in the next time interval. To obtain the predictive model, Bayes' theorem and joint probability distributions are used (Bier and Yi, 1995; Vose, 2000; Meel and Sieder, 2006):

$$p(y_{N+1}|X) = \int_0^\infty p(y_{N+1}, \lambda|X)\, d\lambda \quad (2)$$

$$= \int_0^\infty p(y_{N+1}|\lambda, X)\, p(\lambda|X)\, d\lambda \quad (3)$$

where y_{N+1} is the number of precursors in the next time interval, X is the data, and λ is the rate of occurrence of the accident precursor.

2.2. Consequence Analysis:

To estimate the potential damage caused by an accident, a consequence analysis is conducted. A generalized algorithm used for this purpose is shown in Figure 2.

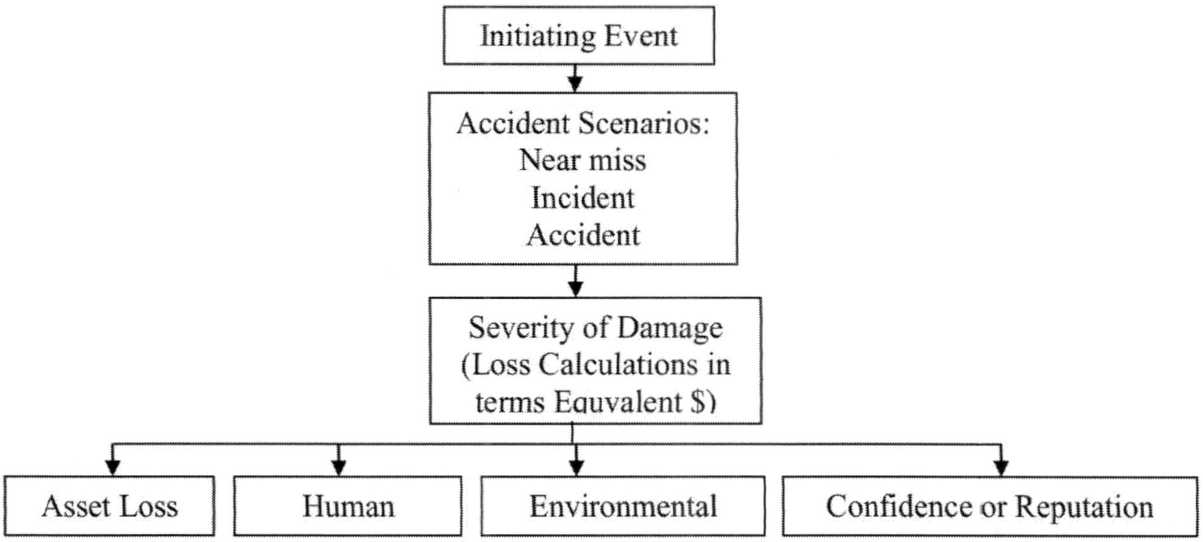

Figure 2. Algorithm for consequence analysis

Consequences can be represented in quantitative terms such as number of injuries and fatalities or dollar value, though some consequences such as reputation loss are hard to do so. There are many approaches to calculate the equivalent dollar values of damages in each category using mathematical models [Reference]. However, selecting a specific model depends on the scenario under study (Bier and Yi, 1995; Khan and Abbasi, 2002; Khan and Amyotte, 2007). The results of this analysis can be directly used to derive the risk profile.

3. Well Control Modeling

A well control process includes monitoring, kick detection, modeling the kick's escalation into a blowout, and the initiation of mechanisms to prevent or control the kick and blowout. A blowout is an uncontrolled flow of hydrocarbons (e.g., gas and condensate) to the surrounding environment as the final outcome of a kick. A kick is an unwanted influx of formation fluids into the wellbore as a result of loss of well control. It is often caused when pressure of formation fluids, i.e. pore pressure (P_P), exceeds the pressure exerted by the column of drilling fluid on the bottom of the wellbore, i.e. bottom hole pressure (*BHP*). A kick can result in a blowout if it is not detected in a timely manner and properly prevented. Well control operation comprising of technical, managerial and organizational measures to maintain well integrity. It is achieved by reducing the risk of blowout through kick prevention, kick detection, blowout prevention and kill operation (Khakzad et al., 2013c).

Recently, a complex series of human errors and mechanical failures resulted in a blowout in the Macondo Well on April 20, 2010, leaving deaths, injuries and a significant amount of hydrocarbon spill (BP, 2010; DHSG, 2011; Chief Counsel's Report, 2011). The fire and explosions that followed the blowout finally caused the Deepwater Horizon drilling rig to sink. According to the report provided by the BP incident investigation team (BP, 2010; DHSG, 2011; Chief Counsel's Report, 2011), a chain of events was to blame for the loss of well control. Of these events, poor cementing caused a kick to occur while failure to notice the kick indications such as changes in the flow rate and the wellbore pressure resulted in the kick not being detected until it flowed up into the riser. The failure of the blowout preventer (BOP) to close in the well escalated the kick into a blowout. Since the kill lines as well as the choke lines of the BOP were damaged, it was also impossible to perform a kill procedure to re-establish the well control (BP, 2010; DHSG, 2011; Chief Counsel's Report, 2011; Khakzad et al., 2013c).

The risk of blowouts cannot be eliminated, however can be reduced through preventive and mitigative measures though priority should be given to preventive measures. Safety measures and critical decision-making are often done considering risk analysis. To conduct the well control risk analysis, the blowout probability is estimated using Bow-tie approach and real time data obtained from the site. Below a brief description of well control risk assessment with is presented via an example.

4. Dynamic Risk Assessment of Well Control System: Bow-tie approach

Bow-tie is an effective graphical method commonly used for process accident risk analysis (Khakzad et al., 2012, 2013a, c). Focused on an undesired event as the pivot node, Bow-tie applies both a fault tree and event tree to determine the potential causes and consequences of the undesired event, respectively. That is, the undesired event is the top event of the fault tree and the initiating event of the event tree. Figure 3 illustrates a typical Bow-tie in which components such as basic events (BE), top event or critical (CE), safety barriers (SB), are connected to represent different accident scenarios and associated consequences level or outcome events(OE).

The probability of the top event is calculated using the fault tree while the safety barrier failure probabilities are analyzed using the event tree. For example, the probability of the consequence OE3 in Figure 3 is estimated as P(OE3) = P(CE) P(SB1) P(SB2).

Fault Tree: Kick detection	Event Tree: Blowout prevention			
	Critical Event (CE)	[c]SB_1	SB_2	Outcome events (OE)

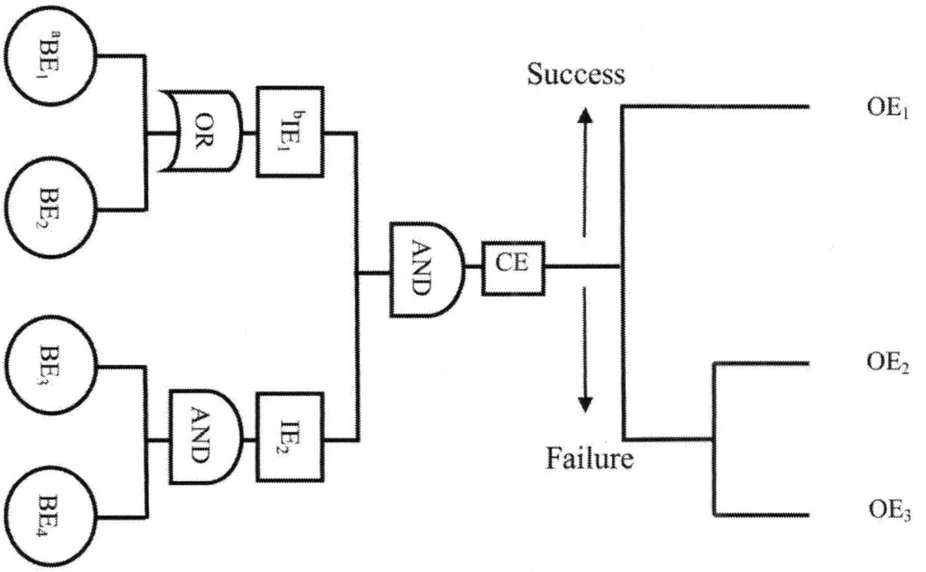

Figure 3: A typical "Bow-tie" diagram
([a]BE-Basic Events; [b]IE-Intermediate Events; and [c]SB-Safety Barrier)

Bow-tie is not widely recognized in the context of dynamic risk analysis where information about frequent accident precursors (e.g., kicks) could effectively have been used to update the estimated risk of major accidents (e.g., blowouts). To consider new information in the context of dynamic risk analysis, the Bow-tie is used along with Bayes' theorem (Khakzad et al., 2013a).

4.1 Blowout risk analysis using bow-tie approach

A kick is the first undesired event in a series of events leading to a blowout. Fault tree is used to assess the kick occurrence probability (Khakzad et al., 2013c). In the case of an overbalanced drilling scenario, a decrease in wellbore pressure will create a negative differential pressure (i.e., $\Delta PL < 0$), thus causing influx of hydrocarbon into the wellbore and initiating a kick. Considering the basic event probabilities presented in Khakzad et al. (2013c), the probability of the kick is calculated as 1.22 E-02. Following the kick occurrence, another fault tree is used as to estimate the failure probability of the kick detection. A variety of drilling parameters, operational parameters and mud characteristics are used to detect a kick occurrence. Using these probabilities, the failure probability of the kick detection is calculated as 8.60 E-06.

In addition to the kick detection, a high-pressure 4-stage Blowout Preventer (BOP) stack including two pipe rams, a blind/shear ram and an annular preventer is considered as the second safety barrier. It is to be noted that even if the kick is detected and the BOP is activated, the blowout may not be prevented unless the casing is strong enough to hold the kick inside the wellbore. The event tree is then used to analyze the failure sequence of safety barriers devised for blowout prevention. Considering the kick as the initiating event and the kick detection and the BOP as the first and second safety barriers, respectively, the probabilities of the possible consequences are estimated and used in the risk assessment. The Bow-tie model for the blowout is shown in Figure 4. It may be seen that Bow-tie comprises of a fault tree associated with the kick generation, a fault tree associated with the kick detection, and an event tree to investigate the propagation of the kick to an atmospheric (surface) blowout. Using the probabilities reported in Khakzad et al. (2013c), the probability of the blowout is estimated 2.55 E-06 (Figure 5). The probability of the blowout is then updated considering new information. The updated probabilities of the kick and the blowout are plotted in Figure 5 (Khakzad et al., 2013c).

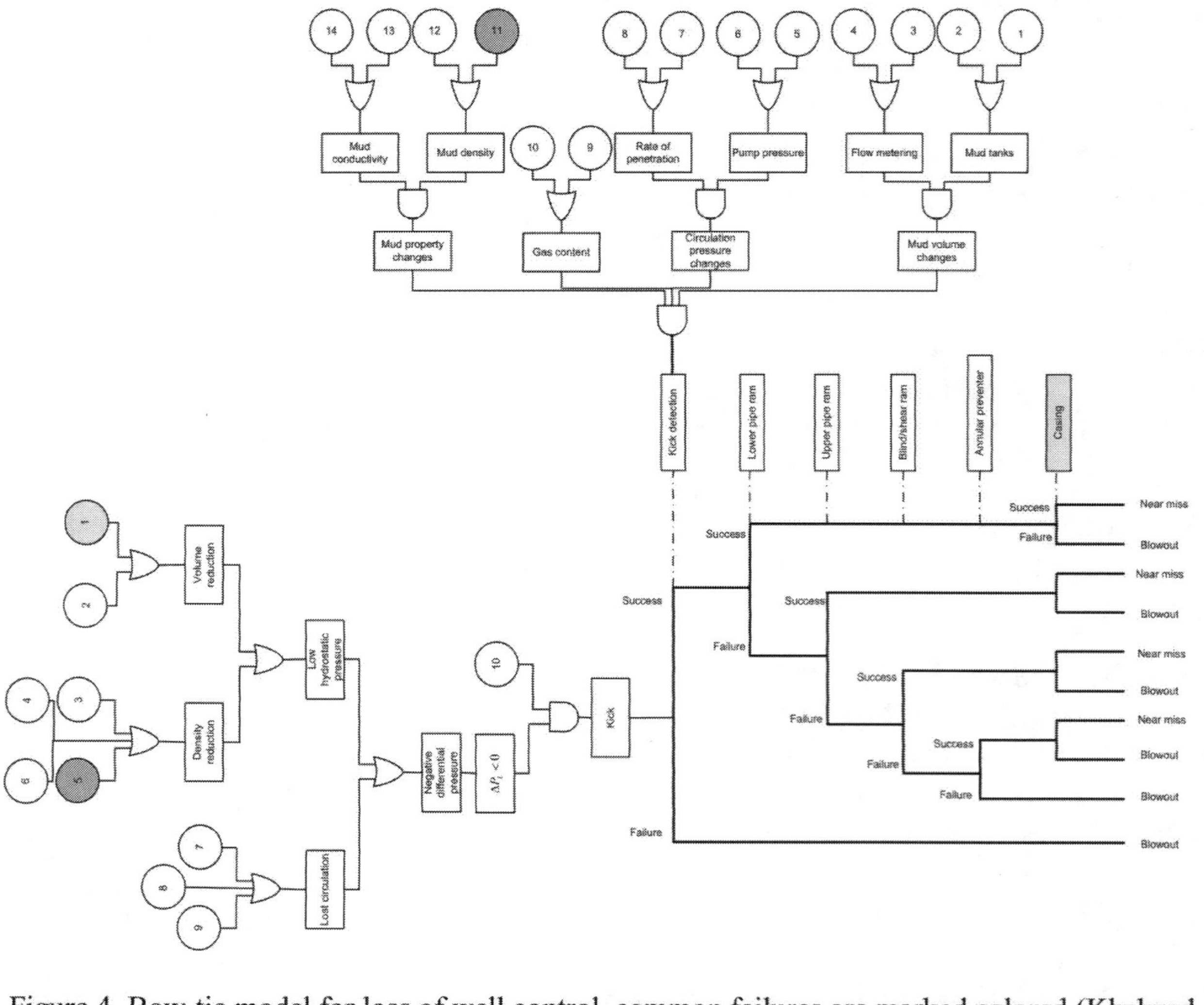

Figure 4. Bow-tie model for loss of well control, common failures are marked colored (Khakzad et al, 2013c).

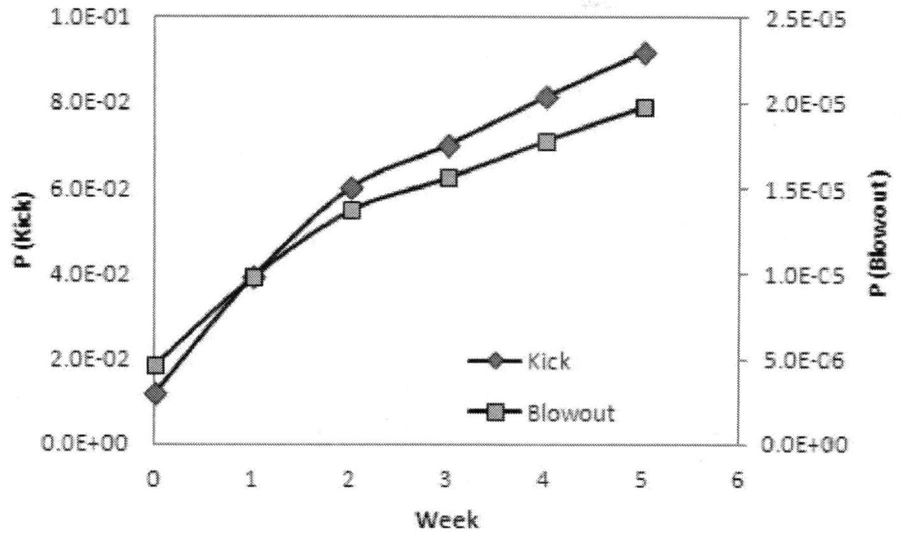

Figure 5. Updated probabilities for kick and blowout (Khakzad et al, 2013c)

Bow-tie has inherent limitations; it is unable to account the common-cause failures and conditional dependencies among the basic events and the safety barriers (Khakzad et al., 2012, 2013a,c). For example, consider a case where the same density meter is used to measure the mud density both when the mud is mixed and pumped into the wellbore and when the mud is brought back to the surface to be tested for kick signs. In this scenario, mud density meter becomes a common cause of failure for both the kick occurrence and the failure of the kick detection. In other words, the initiating event, kick, and the first safety barrier, the kick detection, become conditionally dependent. This dependency cannot be captured in a traditional Bow-tie.

As mentioned earlier, risk is updated by revising the probability of an accident scenario. The probability of accident scenario updating is performed in terms of posterior probability of event x_i given new evidence. The posterior probabilities of the basic events given a blowout, i.e., $P(x_i \mid Consequence = Blowout)$, are estimated and reported elsewhere (Khakzad et al., 2013c). The most probable configuration of the events is determined as: 1) the failure of the casing led to annular losses, which in turn caused a negative differential pressure. 2) Due to the presence of an effective hydrocarbon reservoir, this negative pressure caused an influx of the reservoir fluids into the well, resulting in a kick. 3) Although the kick was detected and the BOP was accordingly activated and functioned properly, the casing failed to hold the influx in the well and thus a blowout occurred (Khakzad et al., 2013c).

5. Conclusions

This article discusses the application of a dynamic risk assessment methodology to a well control system. To this end, first the fundamentals of dynamic risk assessment methodology were briefly reviewed. . A Bow-tie model was developed for risk assessment of the kick and the blowout in an overbalanced drilling operation. The failure probability assessment is performed using a fault tree analysis, while the consequences assessment is conducted using an event tree analysis. Having the prior probabilities, the prior probabilities of the kick and the blowout are calculated using a Bow-tie approach. To capture the dynamic behavior and to reduce the uncertainty of the model, the posterior probabilities are obtained through Bayesian updating. The posterior probabilities provide comprehensive understanding about the performance of safety barriers, occurrence likelihoods of consequences, and the risk associated with an accident scenario. The accuracy of the analysis depends strongly on the precision and availability of precursor data, and therefore a comprehensive monitoring and reporting mechanism is recommended. It is observed that if the proposed dynamic risk assessment methodology is implemented properly at any well, the blowout accident could be avoided or at least controlled with less severe outcomes.

6. References

1. BP (2010). Deepwater Horizon. Accident Investigation Report. Available online at www.bp.com/liveassets/bp_internet/globalbp/globalbp_uk_english/gom_response/STAGING/local_assets/downloads_pdfs/Deepwater_Horizon_Accident_Investigation_Report.pd (last checked July 17, 2013).

2. DHSG. (2011). Deepwater Horizon study group final report on the investigation of the Macondo well blowout. Deepwater Horizon Study Group, Centre for Catastrophic Risk Management, University of California, Berkeley, Available online at http://ccrm.berkeley.edu/pdfs_papers/bea_pdfs/dhsgfinalreport-march2011-tag.pdf (last checked July 17, 2013).

3. Chief Counsel's Report. (2011). Macondo: The Gulf oil disaster. The Chief Counsel's Report, National Commission on the BP Deepwater Horizon oil spill and offshore drilling investigation report. USA. National Commission on the BP Deepwater Horizon Oil Spill and Offshore Drilling. Available online at www.oilspillcommission.gov/sites/default/files/documents/C21462-408_CCR_for_web_0.pdf (last checked July 17, 2013).

4. Bier, V. M., and Mosleh, A. (1990). The Analysis of Accident Precursors and Near Misses: Implications for Risk Assessment and Risk Management, *Reliability Engineering and System Safety,* **27**, 91-101.
5. Bier, V.M., and Yi, W. (1995). The performance of precursor-based estimators for rare event frequencies, *Reliability Engineering and System Safety,* 50, 241-251.
6. Boudali, H., and Dugan, J.B., (2005). A Discrete-Time Bayesian Network Reliability Modeling and Aanalysis Framework, Reliability *Engineering & System Safety*, 87(3), 337-349.
7. Bucci, P., Kirschenbaum, J., Mangan, A.L., Aldemir, T., Smith, C. and Wood, W. (2008). Construction of Event-tree/Fault-tree Models from a Markov Approach to Dynamic System Reliability, *Reliability Engineering & System Safety*, 93(11), 1616-1627.
8. Cepin, M., and Mavko, B., (2002). A Dynamic Fault Tree, *Reliability Engineering & System Safety*, 75(1), 83-91.
9. Kalantarnia, M., Khan, F., and Hawboldt, K., (2008). Dynamic Failure Assessment Using Bayesian Theory, *58th CSChE conference*, October, Ottawa, 2008.
10. Kalantarnia, M., Khan, F., and Hawboldt, K., (2009). Risk Assessment and Management

using Accident Precursors Modeling in Offshore Process Operation, *28th International Conference on Ocean, Offshore and Arctic Engineering,* May, Hawaii, 2009.

11. Khakzad, N., Khan, F., Amyotte, P. (2013a) Dynamic safety analysis of process systems by mapping bow-tie into Bayesian network. *Process Safety and Environmental Protection*, 91. 46 – 53.

12. Khakzad, N., Khan, F., and Amyotte, P. (2012) Dynamic risk analysis using bow-tie approach. *Reliability Engineering and System Safety*,104, 36 – 44.

13. Khakzad, N., Khan, F., and Amyotte, P. (2013b) Risk based design of process systems using discrete-time Bayesian networks. *Reliability analysis and System Safety*, 109, 5 – 17

14. Khakzad, N., Khan, F., and Amyotte, P. (2013c) Quantitative risk analysis of offshore drilling operations: A Bayesian approach. *Safety Science*, 57, 108 –117

15. Khan, F. and Abbasi, S.A., (2002). A criterion for developing credible accident scenarios for risk assessment, *Journal of Loss Prevention in the Process Industries*, 467-475.

16. Khan, F. and Amyotte, P., (2007). Modeling of British Petroleum Refinery Accident, *Journal of Loss Prevention in the Process Industries*, 20, 387–395.

17. Khan, F., and Abbasi, S., (1998). Techniques and Methodologies for Risk Analysis in Chemical Process Industries, *Journal of Loss Prevention in Process Industries*, 11, 261-277.

18. Meel, A. and Seider, W. D. (2006). Plant Specific Dynamic Failure Assessment using Bayesian Theory. *Chemical engineering science*, 61, 7036-7056.

19. Meel, A., O'Neill, L.M., Levin, J.H., Seider, W.D., Oktem, U., and Keren, N. (2007). Operational Risk Assessment of Chemical Industries by Exploiting Accident Database. *Journal of Loss Prevention,* 20, 113-127.

20. Rathnayaka, S., Khan, F., and Amyotte, P. (2012). Accident modeling approach for safety assessment in an LNG processing facility. *Journal of Loss Prevention in the Process Industry*, 25, 414 – 423.

21. Rathnayaka, S., Khan, F.I., and Amyotte, P. (2013) Accident modeling and risk assessment framework for safety critical decision making: Application to deepwater drilling operation. Journal of Risk and Reliability , DOI: 10.1177/1748006X12472158

22. Vose, D. (2000). Risk Analysis, A Quantitative Approach. John Wiley & Sons Ltd. 146-149.

1ST CCPS ASIA-PACIFIC CONFERENCE ON PROCESS SAFETY

Using New Technology to Ensure Corporate Standards Compliance and Lower Operational Risk

Jodie Crawford
Regional Product Manager
Petrotechnics Singapore
80 Raffles Place
UOB Plaza 1, Level 35
Singapore, 048624
Jodie.Crawford@petrotechnics.com

Scott Lehmann,
Vice President Product Management & Product Marketing
Petrotechnics Ltd.
Petrotechnics Ltd, Craigshaw Road
Aberdeen, AB12 3QH, UK
Scott.lehmann@petrotechnics.com

Prepared for Presentation at
American Institute of Chemical Engineers
1st CCPS Asia-Pacific Conference on Process Safety
Qingdao, China
September 4 – 5, 2013

Using New Technology to Ensure Corporate Standards Compliance and Lower Operational Risk

Jodie Crawford
Regional Product Manager
Petrotechnics Singapore
Jodie.Crawford@petrotechnics.com

Keywords: Compliance, Asset Integrity Risk, Barrier Management, Risk Management, New Technology

Abstract

In hazardous industries, recent events such as the Macondo disaster in the Gulf of Mexico have demonstrated the enormous cost of mismanaging operational risk. This has resulted in a stronger drive for corporate governance and assurance processes to ensure that people at the front line managing risk are making the right decisions. This paper will address how new technology can be used to systematize a standardized approach of safe working practices across the organization and lower operational risk. Through this standardization we can generate critical operational risk related data and the tools to proactively manage a more holistic operational risk by linking Asset Integrity Risk to Cumulative Workload Risk.

1. Introduction

Organisations in hazardous industries have one overriding objective; deliver performance in a responsible and sustainable manner. Stakeholder expectations of operational performance have never been higher as the consequences of getting it wrong are significant. From reputational damage to substantial fines and penalties to serious threats to the financial viability of the organisation, it's understandable that balancing performance against operational risk management is now a board level challenge.

Implementing a robust Process Safety Management (PSM) program significantly reduces the likelihood of a major event happening and can mitigate some of the consequences where the occurrence of these hazards remain significant. As a result, most focus on proper design and traditional PSM approaches. In addition best efforts are made to improve operational performance and strengthen Operational Discipline initiatives supported by new policies, procedures, better rulebooks and further training. The expectation is that all safety related near misses and incidents will be substantially reduced and behavioural compliance will follow suit, reducing the likelihood of unwanted outcomes.

Yet despite these efforts, which are often time-consuming and complex, improvement is often illusory or unsustainable. Take for example, the occurrence of a significant incident prompting major changes that lead to a complete re-write of standards and procedures. As a result, the HSE management standards and practices, the very systems developed to support safe and sustainable operations typically become so complicated as to restrict their utility to people in the Frontline- the people who need them the most.

In hazardous industries, recent events such as the Macondo disaster in the Gulf of Mexico have demonstrated the enormous cost of mismanaging operational risk. This has resulted in a stronger drive for corporate governance and assurance processes to ensure that people at the front line managing risk are making the right decisions. This paper will address how new technology can be used to systematize a standardized approach of safe working practices across the organization and lower operational risk. Through this standardization we can generate critical operational risk related data and the tools to proactively manage a more holistic operational risk by linking Asset Integrity Risk to Cumulative Workload Risk.

2. The Operational Context & Single Biggest Area of Operational Risk

At its most basic level our frontline operational environments can be characterised by three things; the Plant, the People and the Interaction between the People and the Plant. The Interaction between the People and the Plant represents millions of man hours per year for the typical organization and thousands of people every day routinely intervening in plant operations to do repair, maintenance, inspection and construction work.

Not only does this type of work expose people directly to the surrounding process hazards but the work they do and the equipment being used can introduce new hazards to those doing the work, to other workers and to the plant itself. PSM approaches recommend creating multiple barriers to prevent unplanned releases but when we examine typical Work Execution (WE) operations, we see that often these barriers meant to keep the plant safe are deliberately removed; safety devices inhibited and new hazards are introduced to the workplace in the course of carrying out work. For example, to access, replace or repair equipment it is often necessary to:

- Break containment of hydrocarbon bearing systems,
- Inhibit or take out of service, safety related devices such as alarms and shutdown systems in order to perform work without shutting down the plant,
- Access equipment which often involves working at height or having to perform heavy lifts over live plant

Such scenarios are daily occurrences on most plants.

Fig 1. The Interaction of People with the Plant = Operating Risk

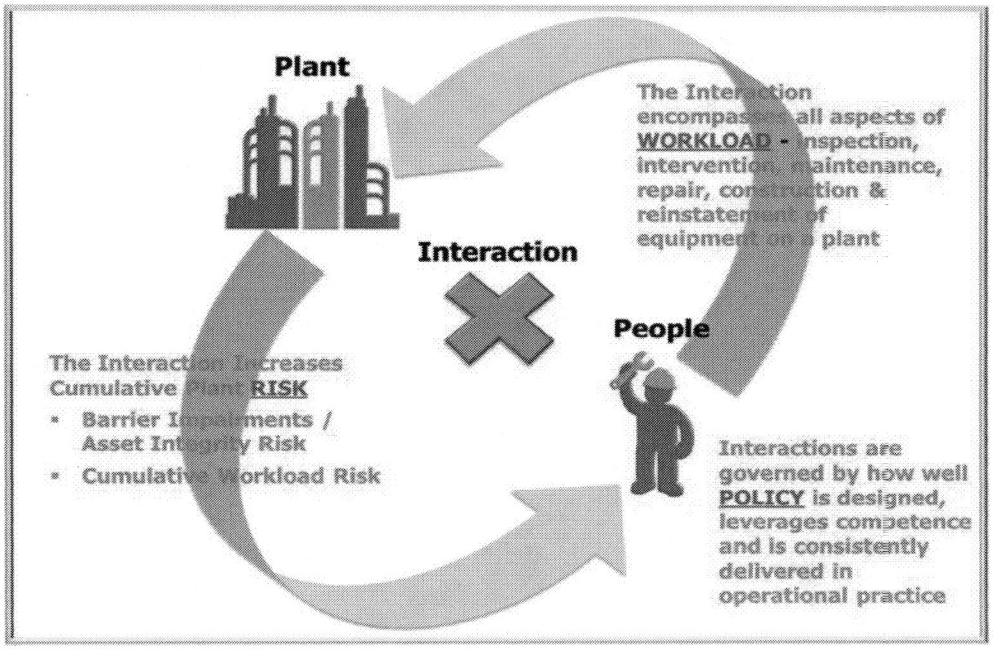

However, the status of the barriers (equipment, people and processes) put into place as safeguards and controls prevent the realisation of a major accident hazard is typically managed in isolation from the cumulative workload risk as a result of the interventions on the plant at any given time. The lack of a link between asset integrity risk and cumulative workload risk especially within a given area of the plant can be critical. Nancy Leveson, professor of Aeronautics and Astronautics at MIT in her testimony before the United States Senate Committee on Energy and Natural Resources on May 17, 2011 said in reference to confusion between occupational and system safety…

> *Most industries separate these very different problems. Occupational safety focuses on controlling injuries to employees at work by changing individual behavior. System safety puts an emphasis on designing the system, including the engineered and operational components, to prevent hazardous system states and thus losses. Confusion between these two very different problems and solutions can lead to overemphasis on only one type of*

safety, usually occupational or personal safety, while thinking that the other types of accidents or losses will also be prevented—which they will not.[2]

Barrier impairments do not exist in isolation of the on-going intervention work that being performed around them and in fact are sometimes caused by the intervention work itself. Without an understanding of how the cumulative workload in a given area of the plant is directly impacting the barrier status, is what we currently understand about the health of our barriers correct? And from the workload side, do we understand in simple terms what the impact of the barrier impairments should have on the scheduled work within a specific area of the plant? Are we simply managing a part of the operational risk equation? Or are we managing two critical parts separately?

Fig 2. Asset Integrity Risk & Cumulative Workload Risk Impact Each Other

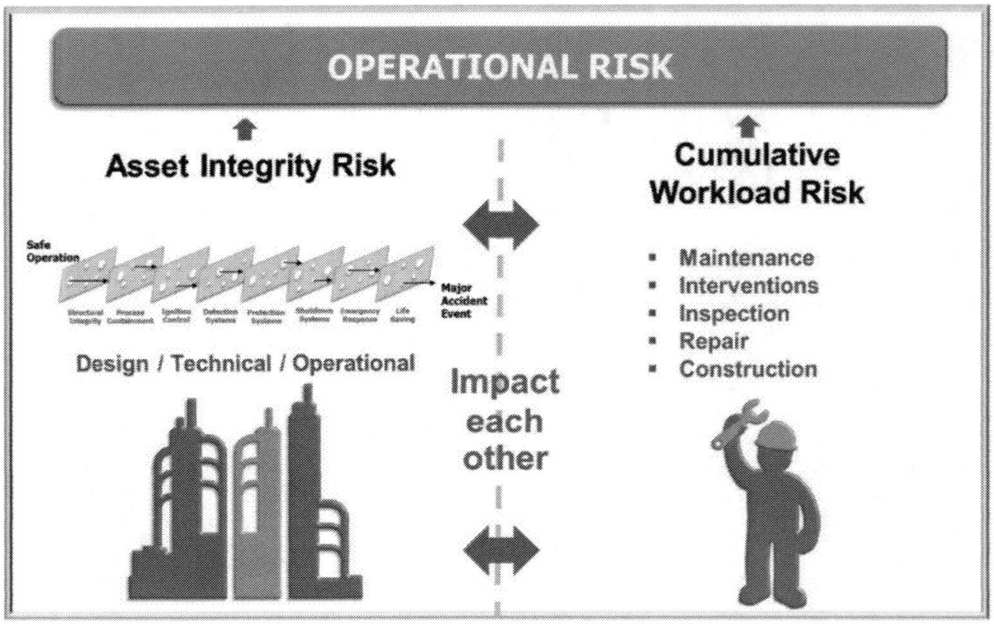

With each day typically presenting different hazards and risks and with inadequate tools to support the decision making process, the quantity and frequency of incidents is not surprising. Most global corporations have a strong need to standardize their practices in this area. The next few sections will examine why this standardization is important and why it is very difficult to implement.

3. Why is Standardization so important within an organization? Why is it so hard to achieve?

Shareholders demand that their executives control and direct company activities in accordance with regulatory laws and with their best interests in mind. Typically, this is achieved through the creation of layers of governance which must be adhered to at all levels of the organization. In the context of managing frontline hazardous work, this translates into common policies, standards and practices for managing specific hazards - hazards which should be the same in all places with similar activities.

A great many obstacles make it very difficult for large organizations to successfully align behaviors at the front line with global corporate policy and often the result is a simplification of the requirements to such general terms that their effectiveness is often reduced. Some of these barriers are created by the organization management structure itself. Autonomous business units are not designed to readily accept dictates from the corporate center. National Regulations may not align with corporate policy requiring you to do more or less than policy dictates. If a region has a long established industry with normalized and deeply embedded practices, it will be difficult to change. Sometimes, even an apparently contradictory cultural practice, 'it's just the way we do things around here', which leads to apparent improved efficiencies may inadvertently cause leadership to reward rather than address procedural violations.

Andrew Hopkins, a well-renowned Associate Member of the National Research Centre for OHS Regulation, emphasizes the importance of this in one of his recent OHS papers, Risk Management and Rule Compliance Decision Making in Hazardous Industries (2011) "Clearly, rulemaking by itself is of little value in achieving safe operation. What is required is compliance with those rules. This in turn depends on adequate enforcement or accountability mechanisms. In other words, the rule-compliance approach requires both rule-making and rule-enforcement."[1]

4. Using Technology to Systematize Corporate Standards & Improve Compliance

Following the plan, ensuring the execution of work is carried out accordingly to policy and consistently across the plant or organization is critical to the safety and success of operations. Typically organizations have assurance over the PSM approaches and plant design that include policies, standards, practices, systems, data, KPI's/metrics and feedback. In this way, they have confidence and feedback that their policies are being followed and they have the means through the use of the right technology, the collection of data, to measure status and progress of their performance.

However, as WE takes place and the PSM barriers meant to keep the plant operationally safe are temporarily disabled to carry out the critical maintenance work, there is often ambiguity around the compliance of work being carried out according to policy. Although there are some policies and standards to guide the frontline workers, they are limited often to paper based systems with no data, no metrics and as a result, no feedback loop of any consequence. The rules and guiding principles in these limited systems are provided in a one way and static format that absolutely ensures divergence in the way they are delivered in operational practice even within the same plant. Without systems we do not have data and without data or feedback we cannot measure performance thus severely limiting management's effectiveness in driving improvements in operational performance and operational discipline.

By capitalizing on new technology such as, Operational Performance & Predictive Risk software platforms, we can embed policies, rules and regulations into operational workflows and automatically generate consistent and reliable operational and risk related data. We can ensure all the task risk assessments take into account process hazards relevant to the location and equipment being worked on and are communicated to work crews. Frontline workers can be prompted by the system to be aware of process and hazard safety considerations and control mechanisms relevant to the area of work.

Through their scalable and enterprise nature, these systems can be deployed across a single plant, multiple assets, and even the enterprise. The standardized and harmonized implementation cultivates a realistic and pragmatic approach to working across the plant and yet retains the flexibility to support regional, local and cultural contexts. The frontline workforce can flourish under the guidance of a tool that provides operational decision support, raises risk management capability and competencies. Over time this can manage out variances and different ways of working. It can nullify practices that can foster subconscious bias toward inappropriate behaviors leading to unpredictable work execution, compliance failure notices, and inaccurate planning. It can reduce high levels of shutdown and incidents, while generating and sustaining improved culture, discipline and behavior for operational excellence and enhanced safety.

The diagram below shows how an enterprise deployment of the platform would allow rules to be set in a cascading hierarchy. Under such a hierarchy, the rules set at a corporate level would be common for the whole hierarchy. Those set at a regional level would contain all the common corporate rules but allow each region to add their own specific rules for flexibility, maximum efficiency and effectiveness. At the plant level, rules can be set which are specific to the plant but include all corporate and regional rules. This hierarchy can even continue down to separate parts of the plant.

Fig 3. Embedding Policy into Practice

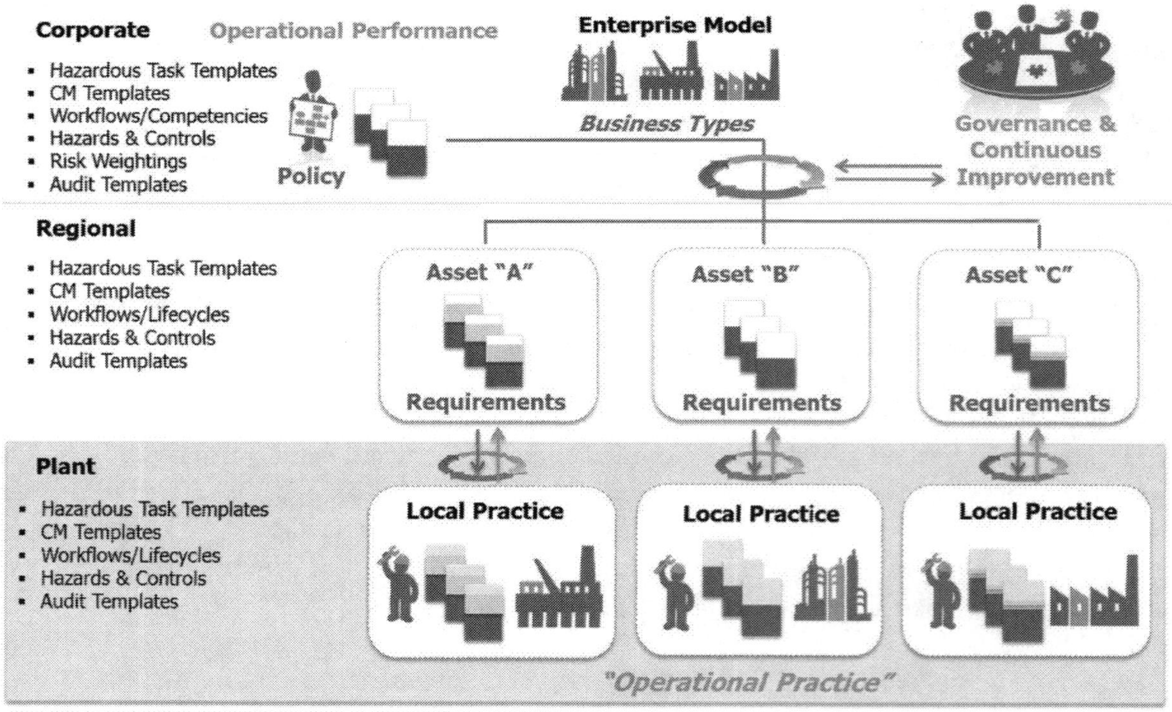

As Hopkins pointed out rule making itself isn't sufficient, so the system ensures that rules are intelligently embedded within operational workflows. In this way rules are placed within the most pragmatic and effective place to facilitate successful compliance and discipline. Organizations and senior management can be confident and assured that its strategic intent is translated efficiently and effectively into operational practice.

5. Improving the Way We Manage Operational Risk

Effective Operational Risk Management is an important and highly specialized skill that demands careful balance between risk appetites, risk assessment, risk response decisions and risk awareness whenever operational work is executed. All too often it is a daunting task to provide the safest integrated response to the cumulative risks arising across multiple tasks in multiple locations as well as addressing conflicting production demands and dynamic plant conditions where barriers may be impaired. It is inevitable that the necessary skills for effective risk management are constrained by entrenched local pressures and cultures, as well as the misleading bias that arises from an absence of incidents.

With the additional pressure of getting more work done, we often leave it up to individual competency to interpret how operational risk is identified and managed, leading to poor quality risk assessments being approved. Furthermore frontline supervisors and workers lack robust tools to guide them to make the appropriate decisions. In other words, decisions are being made without the full information, decisions which could lead to catastrophic failure. As is always the case when there is no systematized process, the variance in how workload is managed against risk doesn't just vary from facility to facility but from frontline supervisor to frontline supervisor.

As we systematize our frontline work processes in a consistent manner across our organization we automatically generate data from scheduled and executed frontline operational work and processes. This provides us with visibility on how work is to be carried out, how it is actually carried out, how the current status of barrier impairments may impact it and the degree to which the frontline is executing work processes in line with the organization's policies, procedures and improvement initiatives.

Operational Performance and Predictive risk platforms provide an organization with Operational Decision Support to enable the management of the cumulative effect of multiple jobs happening in the same proximity as impaired barriers along with other factors and other issues. They provide the following capabilities that can be used to lower operational risk:

- Automated capture of critical barrier and workload related operational risk data
- Real-time Visualization of Barrier Impairments & Simultaneous Work Activity across your plant(s) - in Time, in Space (Area, Plant etc.)
- Common view of Risk to all Disciplines in time & space
- Operational Decision Support Tools to proactively manage human performance and activity workload against risk

- Prioritise work to minimise risk
- Better utilise resources to complete work efficiently

It is not unusual for 20 or more independent jobs to be in progress simultaneously within a defined area. While these individual tasks may have been independently risk assessed and authorised, there are dependencies and interactions between the jobs that need to be taken into consideration in order to reduce the potential overall risk of the combined jobs. The cumulative risk effect of multiple jobs coupled with any human factor or environmental hazards and how they impact or are impacted by the barrier impairments in a specific location of the plant are seldom if ever assessed on a day to day basis. Work schedules are assessed for clashes where two or more jobs may interfere with each another but we do not equip our plant supervisors with the right tools and techniques to properly assess these types of risks.

In the diagram below, the barrier impairments and the concurrent tasks are visualized to provide a view of the risk in time by area and location of the plant.

Fig 4. Visualization of the barrier impairments and concurrent tasks by location of the plant.

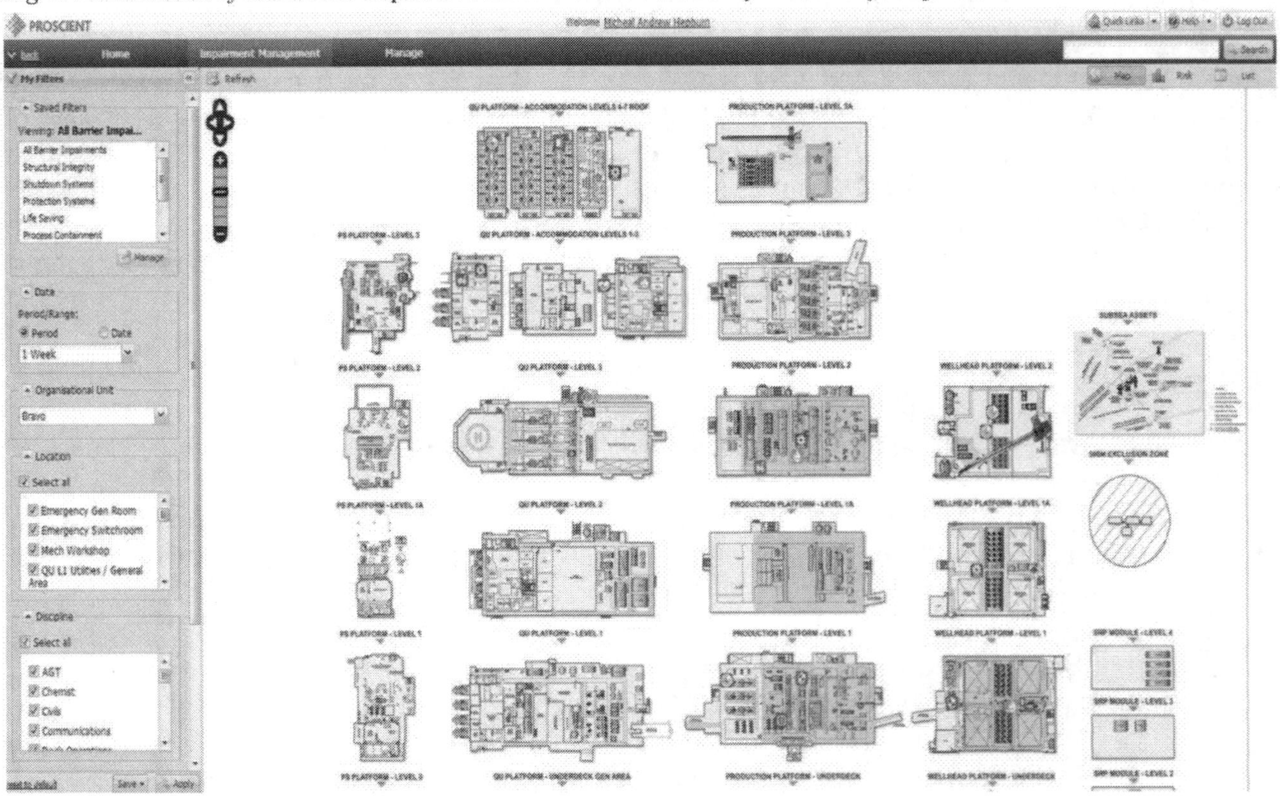

In the following diagram, frontline supervisors can quickly ascertain the status of the barriers, the cumulative risk of the scheduled work and proactively reschedule work to improve the risk profile in a location in the plant.

Fig 5. Visualization of the Status and Plant Cumulative Risk enables work to be rescheduled to improve the risk profile of an asset

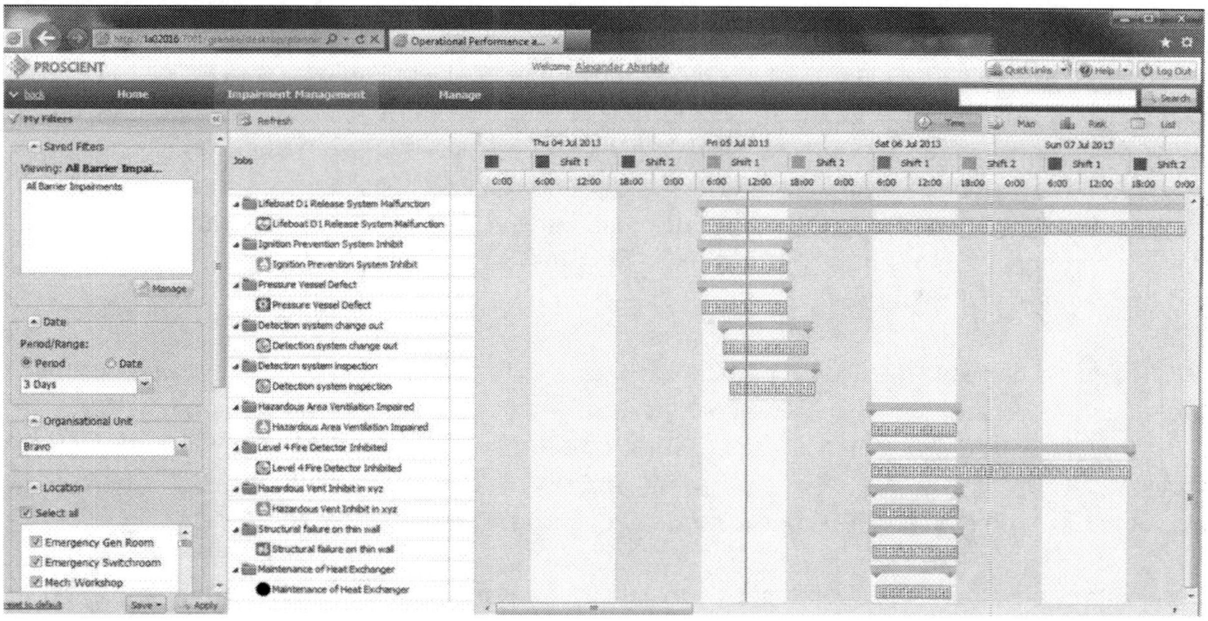

While visualization of risk is important the ability to understand what is driving it is critical to improving Operational Decision Support and ensuring Compliance with Corporate Standards. In the diagram below, we can see the status of the barriers within a specific location of the plant and what jobs or tasks are impacting the risk profile of the barriers and what specific plant conditions are impacting it as well. With this knowledge work can either be rescheduled, more effective controls put into place or at a minimum frontline management can more effectively communicate the hazards and the risks associated with executing the work.

Fig 6. Drill down: How do specific Tasks impact barriers & risk?

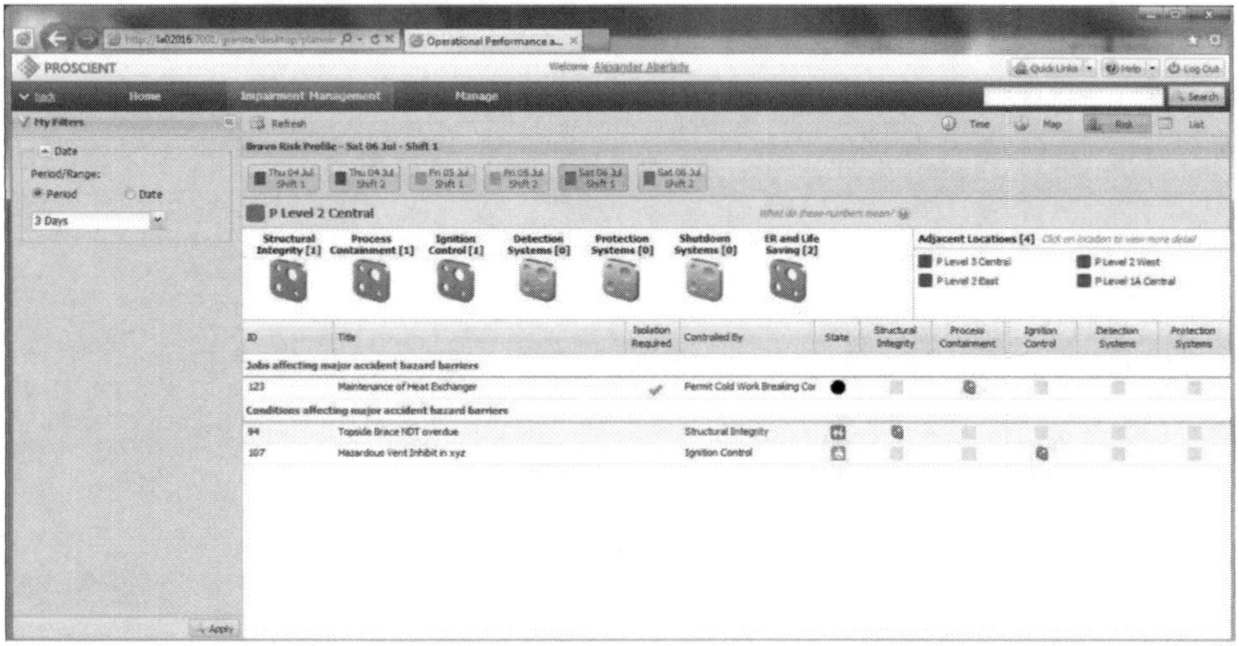

6. Summary & Conclusions

New technology in the form of Operational Performance & Predictive Risk software platforms can support the systematization of a standard approach of safe working practices across the organization. Corporate policies can be configured as the baseline templates to be rolled out in a standardized and harmonized way to achieve the desired level of conformance at the regional and local levels yet retaining the flexibility to support regional, local and cultural contexts. The frontline workforce can flourish under the guidance of a tool that continually raises risk management capabilities and competencies by providing a systemic approach to managing workload against risk according to policy.

Over time this can start to improve practices that can foster inappropriate behaviors leading to unpredictable work execution, compliance failure notices, and inaccurate planning. It can also have tangible benefits across the organization in terms of reducing unplanned shutdowns and incidents, while generating and sustaining improved culture, discipline and behavior for operational excellence and enhanced safety.

Furthermore this standardization drives consistent and reliable data generation including critical operational risk related data. With the tools to dynamically capture, visualize and manage operational risk in our plants in order to get more work done safely. By understanding the status of barrier impairments on our plants and the impact of the current or planned workload on those barrier impairments we can link asset integrity risk with work risk. Consistent data generation can provide us with the risk profiles of the compounded risk of doing work against a backdrop of dynamic plant conditions in specific areas of the plant thus improving frontline decision making.

The ability to provide a real time view of frontline activities across an organisation's global operations is no longer a pipe dream. Technology can empower and enable the workforce and management to proactively identify, track, monitor and manage the health of their barriers and the impact of workload in real time to ensure and improve overall plant health integrity, safety and efficiency. Organisations can start to move from a "find and fix" mode of operating to one of "predict and prevent". With line of sight provided by new technologies, organisations now have the ability to see and understand the relationships between workload and risk and can manage the organization to a more optimized and less risky performance

References

[1] HOPKINS, ANDREW Risk Management and Rule Compliance Decision Making in Hazardous Industries, Working Paper 72 OHS Regulation 2010, p.3

[2] LEVESON, NANCY, US Senate Testimony on Energy and Natural Resources, 2011, P.3

1ST CCPS ASIA-PACIFIC CONFERENCE ON PROCESS SAFETY℠

Determine the Quantitative Individual Risk Criteria for Single Accident Risk in Fixed Chemical site using the Background Risk Method

John Zhao
Shegle Risk Consulting Co. Ltd. (Shanghai)
Room 404, 1st Building 2742 Lane PuDong Avenue Shanghai
jmzhao@shegle.com

Zhu Xiaodong
Yantai Wanhua Polyurethanes Co. Ltd

Zhang Bing
Shegle Risk Consulting Co. Ltd. (Shanghai)
Room 404, 1st Building 2742 Lane PuDong Avenue Shanghai
Bing.zhang@shegle.com

Ren Jing
Shegle Risk Consulting Co. Ltd. (Shanghai)
Room 404, 1st Building 2742 Lane PuDong Avenue Shanghai
Jing.ren@shegle.com

2013 Spring Meeting
9th Global Congress on Process Safety
San Antonio, Texas
April 28 – May 1, 2013

UNPUBLISHED

Determine the Quantitative Individual Risk Criteria for Single Accident Risk in Fixed Chemical site using the Background Risk Method

John Zhao
Shegle Risk Consulting Co. Ltd. (Shanghai)
Room 404, 1st Building 2742 Lane PuDong Avenue Shanghai
jmzhao@shegle.com

Zhu Xiaodong, Zhang Bing, Ren Jing
Yantai Wanhua Polyurethanes Co. Ltd, Shegle Risk Consulting Co. Ltd. (Shanghai)

Keywords: Accident Risk Criteria, Layers of Protection Analysis, LOPA, Background Risk Method

Abstract

LOPA (Layers of Protection Analysis) is more popular in the process industry, especially in determination of SIL (Safety Integrity Level) for SIS (Safety Instrument System). When using LOPA, the Tolerable Risk Criteria for single risk scenario has to be pre-defined. This paper presents a simple method to define the Tolerable Risk Criteria.

1. Quantified risk criteria

How safe is safe, it is a difficult question to be answered. The core to answer this question is to determine the risk acceptance criteria and it is influenced by many factors, such as: the development level of the society, people's risk perception and some specific characters of the industry[i]。. The risk matrix used in HAZOP is a kind of qualitative risk criteria, as shown in figure 1. As for the quantitative risk criteria, it is not obviously mentioned, is hidden or in another word, implied in the qualitative risk matrix[ii]。. In recent years, LOPA (Layer of Protection Analysis) is getting more attention as a fairly simple risk analysis tool than other quantitative tools, especially in the area of determining the SIL (Safe Integrity Level) of SIS (Safety Instrument System). LOPA needs a quantified risk criteria which specifies which risk is acceptable and which risk is tolerable[iii]。 , as shown in figure 2. The equation for SIL determination in LOPA is:

$$PFD_{avg} = \frac{F_{tolerable}}{F_{demand}}$$

The quantitative risk criteria used in LOPA is normally the individual risk that specifically related to one accident scenario, this is different with the individual risk data we normally see. One well-known example is, 0.001 is the maximum individual tolerable risk criteria for employee, raised by UK Health & Safety Executive (HSE) in its famous book *Reducing risks - protecting people*[iv]. This value is the aggregated risk that the employee exposed to, not the specific risk caused by a specific accident scenario. The risk criteria have different meanings as implied at different levels of a company in the process industry, they are interrelated as demonstrated in figure 3.

Figure 1. Qualitative risk matrix

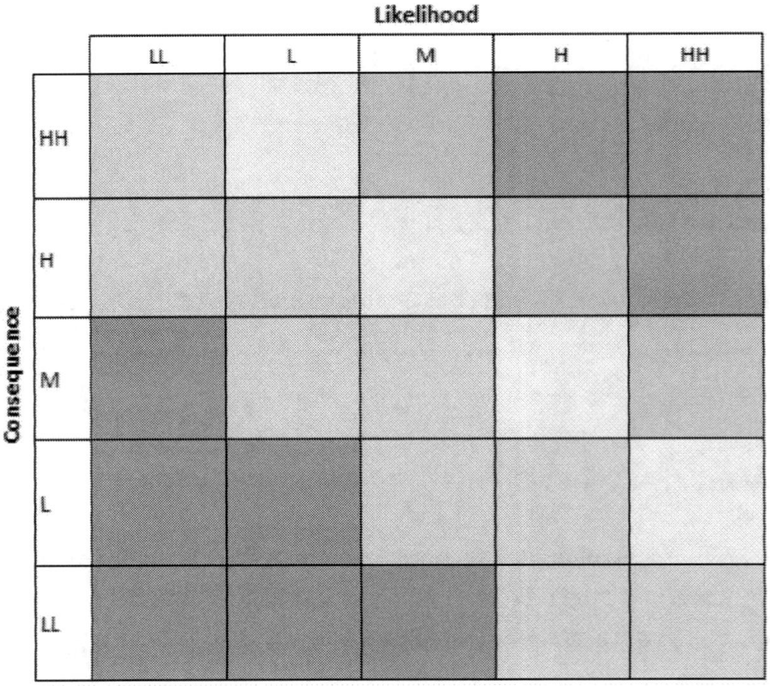

Figure 2. Quantitative risk criteria is determining SILv

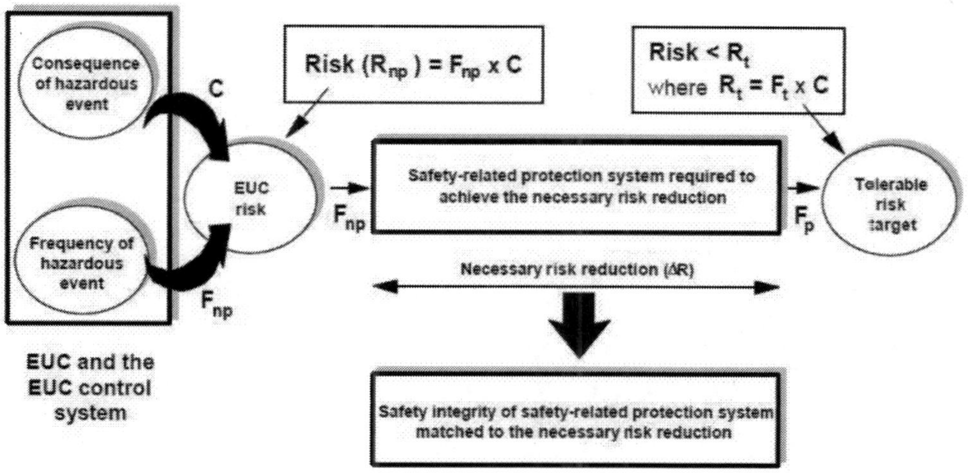

Figure 3. Relation of different risk criteriavi

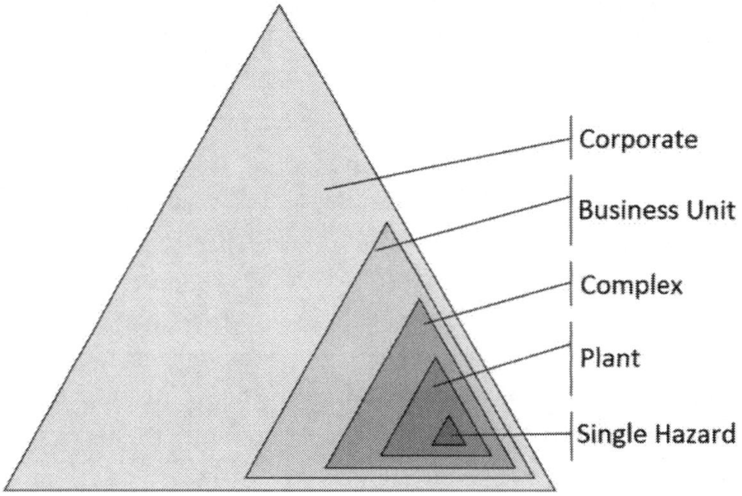

2. Background Risk Method (BRM)

This paper raises a method to determine the specific individual risk criteria for single accident scenario. The basic principle is shown in figure 4. The risks that people face can be divided into different levels and different risk sources, reversely, if we want to know the risk criteria for each individual risk scenario (source), it can be derived from the value of the overall risks, all risks people face in their daily worked related or non work related life.

Figure 4 . Relation of different risk in another perspective

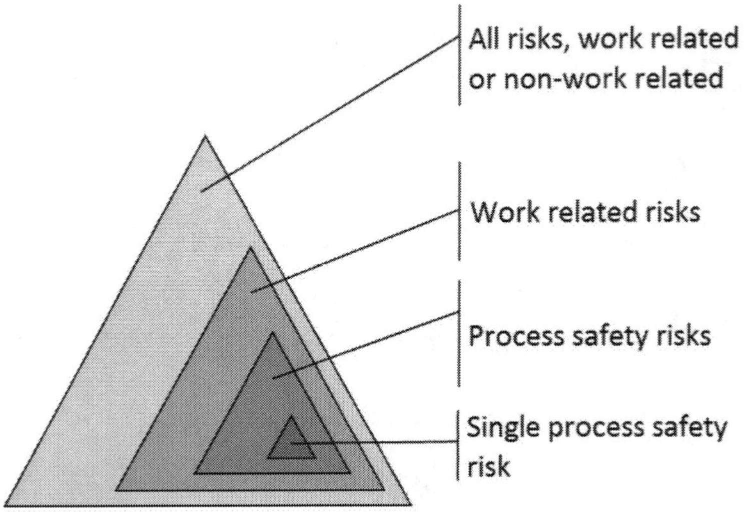

There are several steps in determining the individual risk criteria for single accident risk as shown in figure 5. The first step is to determine the value of background risk, that is the risk that people face in their daily worked related or non work related life. Then, imposed risks can be determined by determining the imposed risk factor, the imposed risk is work related risk that people have to bear or can be tolerated due to the expectation of some benefit. The third step is to determine the risk category factor, normally, the accident risks come from 3 categories in the process industries, say, process safety related, personal safety related and traffic related. In different industries or different companies, the distribution of risks in these 3 categories may be different and can be quantified by statistical analysis of accident data, in another word, the risk category factor can be determined. In the category of process safety risk, some risks can be quantified by the model of protection layers but some are not, the examples are mechanical integrity induced accidents, such as, total rupture or partial rupture risk of pressure vessel or pressure pipe due to manufacturing defects rather than process deviation such as overpressure. Normally these risks are deemed as the residual risk from those can be reduced by process controls such as overpressure protections or over fill protections. If we call these mechanical integrity related risk unquantifiable risk, then these unquantifiable risks have to be deducted from the process safety risks, then we get the quantifiable process safety risks that can be reduced or controlled by process means. Each process plant has its unique risk distribution, for example, the process safety risk for the liquid phase bulk polymerization process of polypropylene is different from the gas phase process, though they are all polyethylene processes. To count and numerate each specific process safety related risks in the specific process plant and quantify them is the last step.

Figure 5. Steps of BRM to determine the risk criteria for single accident risk

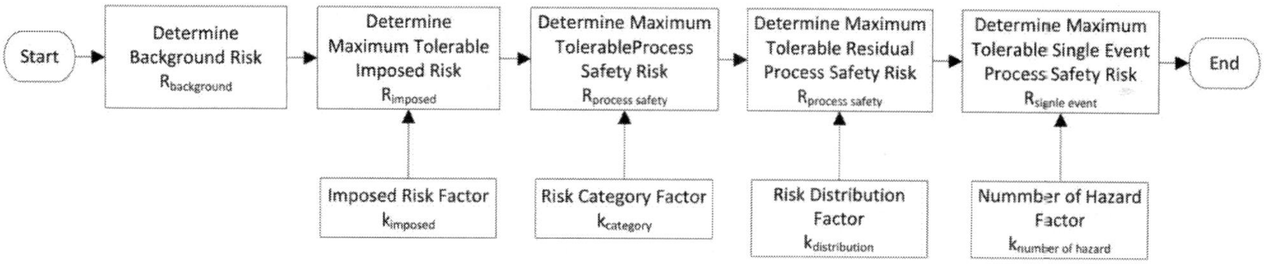

2.1 Determine the background risk

The first step is to determine the value of background risk, the risk that people face in their work related or non work related life. A good data source for this calculation is the life table that the insurance companies use for their business. From the life table, we get the data of Force of Mortality (FoM) which represents the instantaneous rate of mortality at a certain age measured on an annualized basis. It is identical in concept to failure rate, in process safety and reliability. The FoM can be regarded as the value of background risk.

Table 1. 1990-1993 life table for Chinalife life insurance (non pension business, male)[vii]

Age	Survival	Mortality	Age	Survival	Mortality	Age	Survival	Mortality	Age	Survival	Mortality	Age	Survival	Mortality
0	1,000,000		22	979,084	1,027	44	949724	2,584	66	768225	17,022	88	145745	26,465
1	996,963	3,037	23	978,075	1,009	45	946893	2,831	67	749942	18,283	89	121473	24,272
2	994,813	2,150	24	977094	981	46	943791	3,102	68	730349	19,593	90	99580	21,893
3	993,210	1,603	25	976144	950	47	940393	3,398	69	709410	20,939	91	80182	19,398
4	991,968	1,242	26	975222	922	48	936670	3,723	70	687094	22,316	92	63325	16,857
5	990,976	992	27	974320	902	49	932594	4,076	71	663386	23,708	93	48979	14,346
6	990,163	813	28	973428	892	50	928133	4,461	72	638287	25,099	94	37041	11,938
7	989,480	683	29	972535	893	51	923251	4,882	73	611815	26,472	95	27344	9,697
8	988,893	587	30	971627	908	52	917911	5,340	74	584010	27,805	96	19668	7,676
9	988,379	514	31	970692	935	53	912075	5,836	75	554939	29,071	97	13760	5,908
10	987,916	463	32	969714	978	54	905699	6,376	76	524694	30,245	98	9345	4,415
11	987,484	432	33	968682	1,032	55	898738	6,961	77	493399	31,295	99	6148	3,197
12	987,058	426	34	967582	1,100	56	891144	7,594	78	461211	32,188	100	3911	2,237
13	986,606	452	35	966400	1,182	57	882867	8,277	79	428319	32,892	101	2400	1,511
14	986,097	509	36	965123	1,277	58	873853	9,014	80	394946	33,373	102	1418	982
15	985,502	595	37	963737	1,386	59	864047	9,806	81	361348	33,598	103	805	613
16	984,806	696	38	962229	1,508	60	853391	10,656	82	327810	33,538	104	438	367
17	984,007	799	39	960583	1,646	61	841825	11,566	83	294241	33,569	105	228	210
18	983,114	893	40	958785	1,798	62	829288	12,537	84	262166	32,075			
19	982,150	964	41	956819	1,966	63	815720	13,568	85	230719	31,447			
20	981,140	1,010	42	954666	2,153	64	801060	14,660	86	200629	30,090			
21	980,111	1,029	43	952308	2,358	65	785247	15,813	87	172210	28,419			

Figure 6. Force of Mortality (FoM) from age 0-105 years

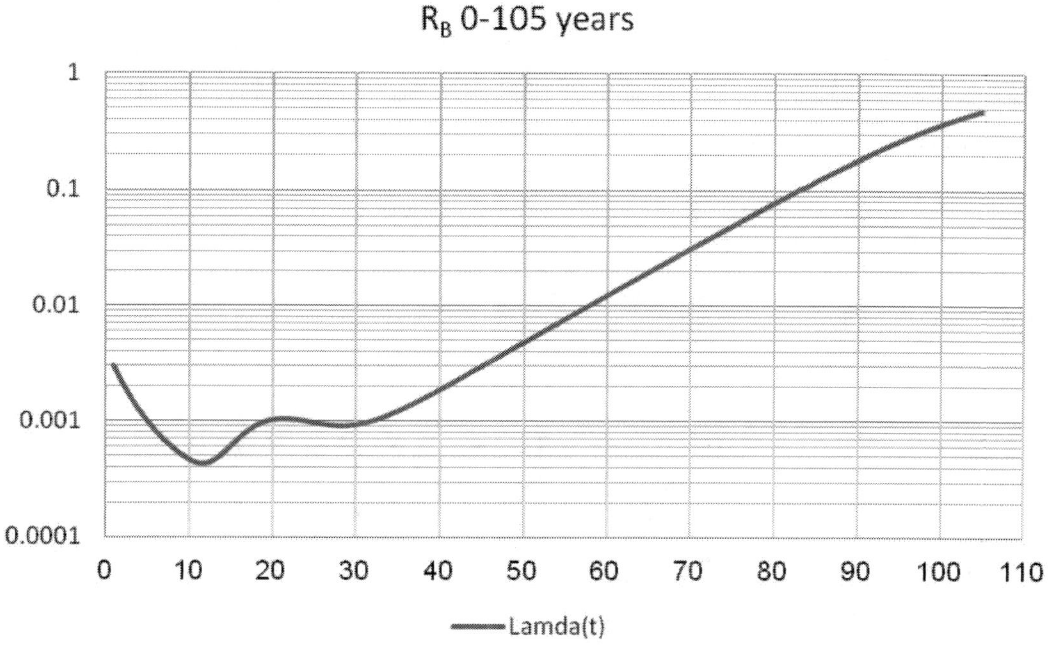

From figure 6, we can see that the FoM is lowest at the age of about 12, it is about 0.0004. When the age is between 20 and 30, the FoM is about 0.001. The FoM increases with the increase of age.

For the background risk, there are several options:

- The average FoM between age 20 – 30, it is 0.0011;
- The average FoM between age 20 – 60, it is 0.0035;
- The average FoM between 30 – 60, it is 0.0043;
- The average age of employee for a company is 35, then take the FoM of age 35 as the background risk, which is 0.0012;
- ……

Here in this paper, we chose 0.001 as the valve of background risk ($R_{background}$).

2.2 Determine the Imposed Risk Factor

Someone raised a way to evaluate the imposed risk to average human life, and then decide if the imposed risk is tolerable or acceptable to the society or individuals. Form table 1, we can calculate the average human life span is 74.12 years. Suppose the human work life is from 20 to 60, and in this period, they are exposed to the background risk and also exposed to the imposed risk ($R_{work\ related}$) which is work related, consider the impact to the average human life span from different values of $R_{work\ related}$, we get the data in table 2.

Table 2. Relation between imposed risk and average life span

$R_{work\ related}$	0.1	0.05	0.01	0.005	0.001	0.0005	0.0001	0.00005
$\triangle E(T)$	-44.54	-35.49	-11.89	-6.43	-1.36	-0.69	-0.14	-0.07

$R_{work\ related}$: Imposed risk from work, fatality/year;

$\triangle E(T)$: the reduction of average life span due to imposed risk from work, year;

$E(T)$: Average (expected) life span

Figure 7. Relation between imposed risk and average life span

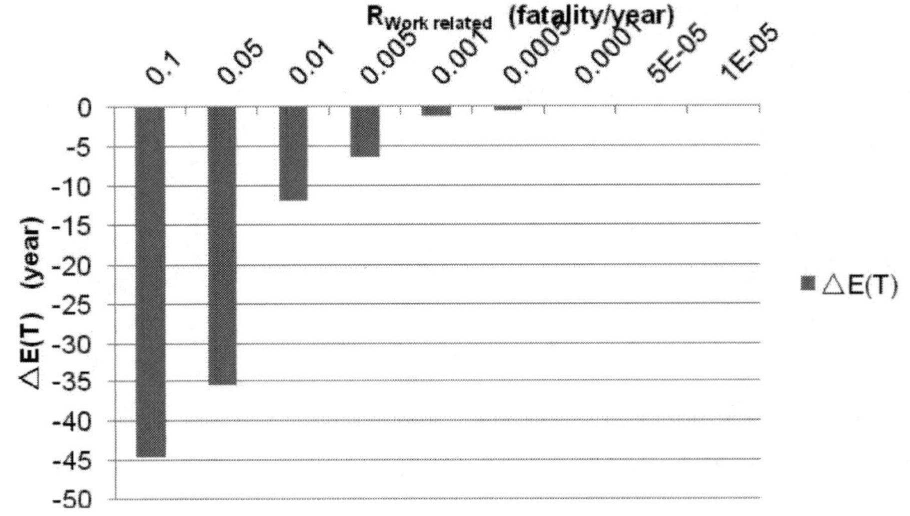

Suppose the imposed risk has the following relationship with the background risk:

$$R_{work\ related} = R_{imposed} = k_{imposed} \times R_{background}$$

$R_{work\ related}$: work related risk, fatality/year;

$R_{imposed}$: imposed risk;

$k_{imposed}$: imposed risk factor;

$R_{background}$: background risk;

From table 2, if the work related risk is 0.01, the average life span reduces 11.89 years, this is a fairly big impact, there is no sufficient reasons to bear this kind of risk. If the work related risk is 0.001, the average life span reduces 1.36 years, if taking into consideration that the employee get benefit from their job and voluntary to bear some risks, this level of risk seems tolerable. The same risk of 0.001, if we consider the general public, the general public does not get benefit directly from the company's business, if expose them with the same level of risk as the employee, it seems not defendable. Generally speaking, the risk exposed to the general public shall be a level of magnitude lower than that of the employee. In summary, as to the employee:

The maximum value of $R_{work\ related}$ can be 0.001 for employee, the value of $k_{imposed}$ is 1 accordingly. Some company strives to target zero accident, they may consider to lower the value of $R_{work\ related}$ to as low as 0.0001, the value of $k_{imposed}$ is 0.1 accordingly.

$$0.1 \leq k_{imposed\ for\ employee} \leq 1$$

As to the general public:

$$k_{imposed\ for\ public} \leq 0.1$$

2.3 Determine the Risk Category Factor

The risks that employee face in their work can be divided into 3 categories, as shown in figure 8:

- Traffic related risk: risk caused by traffic accident;
- Personal safety risk: risk due to trips, slips and falls from height, etc, these are normally categorized as personal safety accident;
- Process safety risk: risk due to the fire, explosion and toxication caused by chemical release, these are normally categorized as process safety accident;

Figure 8. 3 categories of work related risks

If not considering other risks such as occupational health risk, then:

$$R_{work\ related} = R_{driving} + R_{personal\ safety} + R_{process\ safety}$$

Define the Risk Category Factor as:

$$k_{category\ driving} = \frac{R_{driving}}{R_{work\ related}}$$

$$k_{category\ personal\ safety} = \frac{R_{personal\ safety}}{R_{work\ related}}$$

$$k_{\text{category process safety}} = \frac{R_{\text{process safety}}}{R_{\text{work related}}}$$

$k_{\text{category driving}}$: Risk Category Factor – driving, percentage of driving risk in the work related risk;

$k_{\text{category personal safety}}$: Risk Category Factor – personal safety, percentage of personal safety risk in the work related risk;

$k_{\text{category process safety}}$: Risk Category Factor – process safety, percentage of process safety risk in the work related risk;

Easy to know:

$$k_{\text{category driving}} + k_{\text{category personal safety}} + k_{\text{category process safety}} = 1$$

Considering a site operator in a chemical plant, if his job involves not so much with vehicles, the risks he may be exposed to is mainly from personal safe and process safety risks, generally, the following equation is true:

$$k_{\text{category personal safety}} = k_{\text{category process safety}} = 0.5$$

Of course, the specific value of $k_{\text{category personal safety}}$ and $k_{\text{category process safety}}$ depends on the specific statistic data for that specific company.

2.4 Determine the Risk Distribution Factor & Number of Risk Factor

The employee may be exposed to multiple hazard or risks in a chemical plant, as shown in figure 9, the employee face the risks from the reactor, the tower and the storage tank. Suppose the 3 hazards are independent, then the risk the employee face can be expressed as:

$$R_{\text{process safety}} = R_{\text{reactor}} + R_{\text{column}} + R_{\text{tank}}$$

R_{reactor} : risk from the reactor;

R_{column} : risk from the tower;

R_{tank} : risk from the storage tank

The more general expression is:

$$R_{\text{process safety}} = \sum_{i=1}^{n} R_i = R_1 + R_2 + \cdots + R_n$$

R_i : risk from the ith hazard;

The process safety hazards in a chemical plant can be numerous. Even in a QRA (Quantitative Risk Assessment) study, only some big risks are listed and calculated[viii]. In a LOPA study, it seldom needs to evaluate all the hazards, normally it evaluate the major hazards one by one, for each single major hazard, it compares the risk of the hazard with a predefined risk criteria to see if the risk is tolerable or generally acceptable[ix].

In order to determine to maximum tolerable risk for single process safety hazard, we can sort the process safety risks in order, consider the number of risks ($N_{R_i \geq 10^{-4}}$) which are higher than 1×10^{-4}, number of risks ($N_{R_i \geq 10^{-5}}$) which are higher than 1×10^{-5}, number of risks ($N_{R_i \geq 10^{-6}}$) which are higher than 1×10^{-6}.

According to the analysis in 2.1, 2.2 and 2.3, if:

$$R_{background} = 0.001$$

$$R_{work\ related} = k_{imposed} \times R_{background} = 1 \times 0.001 = 0.001$$

$$R_{process\ safety} = k_{category\ process\ safety} \times R_{work\ related} = 0.5 \times 0.001 = 0.0005$$

If want the process safety risk to meet the risk criteria of 0.0005, need:

$$N_{R_i \geq 10^{-4}} \leq 5$$

$$N_{R_i \geq 10^{-5}} \leq 50$$

$$N_{R_i \geq 10^{-6}} \leq 500$$

$$N_{R_i \geq 10^{-7}} \leq 5000$$

To have a precise numeration of the risks at the magnitude of 1×10^{-6}, 1×10^{-7} or even lower, it is too onerous. The good way is to have a rough estimation.

To consider the number of risks at the magnitude of 1×10^{-6}, let look at some of the practices in the industries or of the government. UK HSE recommends the failure rate of a pressure vessel failure (full rupture) [x] as 3.5×10^{-6}, China State Administration of Work Safety recommends the failure rate of a pressure vessel (full rupture) [xi] as 6×10^{-6}. If we make some simplification, define a concept of critical vessels, those vessels that have a serious consequence if the vessels fail, say, operating pressure is very high, or the content inside is highly hazardous, etc. Suppose the failure rate of the critical vessel is 6×10^{-6}, if we count the number of critical vessels ($N_{critical\ vessels}$), then, the risk contribution from the critical vessels is $N_{critical\ vessels} \times 6 \times 10^{-6}$.

$$N_{R_i \geq 10^{-5}} \leq \frac{0.0005 - N_{critical\ vessels} \times 6 \times 10^{-6}}{1 \times 10^{-5}}$$

Value of $N_{critical\ vessels}$ is plant or process specific. One further work needs to be done is to have typical processes evaluated for $N_{critical\ vessels}$.

Suppose the number of critical vessels in a chemical plant is 50, then:

$$k_{distribution} = \frac{0.0005 - N_{critical\ vessels} \times 6 \times 10^{-6}}{0.0005} = 0.4$$

Easy to know:

$$R_{process\ safety} = R_{R_i \geq 10^{-6}} + R_{R_i < 10^{-6}}$$

$$R_{R_i \geq 10^{-6}} = R_{critical\ vessels} + R_{others \geq 10^{-6}}$$

$$R_{others \geq 10^{-6}} = N_{others \geq 10^{-6}} \times R_{others\ single\ scenario}$$

In the equation listed above, $R_{\text{process safety}}$ and $R_{\text{critical vessels}}$ is known, if $R_{R_i<10^{-6}}$ can be discounted, or they are covered by the conservative estimation of $R_{\text{critical vessels}}$, then $R_{\text{others}>10^{-6}}$ is easy to get. If $N_{\text{others}>10^{-6}}$ (the number of risks other than the critical vessels at the magnitude of 1×10^{-6} and higher) is known, the maximum tolerable value for these risks can be calculated.

Define:

$$k_{\text{distribution}} = \frac{R_{\text{process safety}} - N_{\text{critical vessels}} \times 6 \times 10^{-6}}{R_{\text{process safety}}}$$

In the above example, suppose the number of critical vessels is 50, then:

$$k_{\text{distribution}} = \frac{0.0005 - N_{\text{critical vessels}} \times 6 \times 10^{-6}}{0.0005} = 0.4$$

Normally, in HAZOP study, the residual failure of pressure vessel or pressure pipe due to purely mechanical integrity reasons is not counted or recorded in the HAZOP worksheet. There are several options for $N_{\text{others}>10^{-6}}$, one way is to search the HAZOP worksheet and count all the scenario that have a consequence of fatality, the number of such scenario is $N_{\text{others}>10^{-6}}$. Another way might be to count the number of safety interlocks, as the value of $N_{\text{others}>10^{-6}}$.

Define:

$$k_{\text{number of hazard}} = \frac{1}{\text{number of safety interlocks}}$$

If the number of safety interlocks in a plant is 30, then:

$$k_{\text{number of hazard}} = \frac{1}{30} = 0.033$$

Figure 9. Multiple hazards[xii]

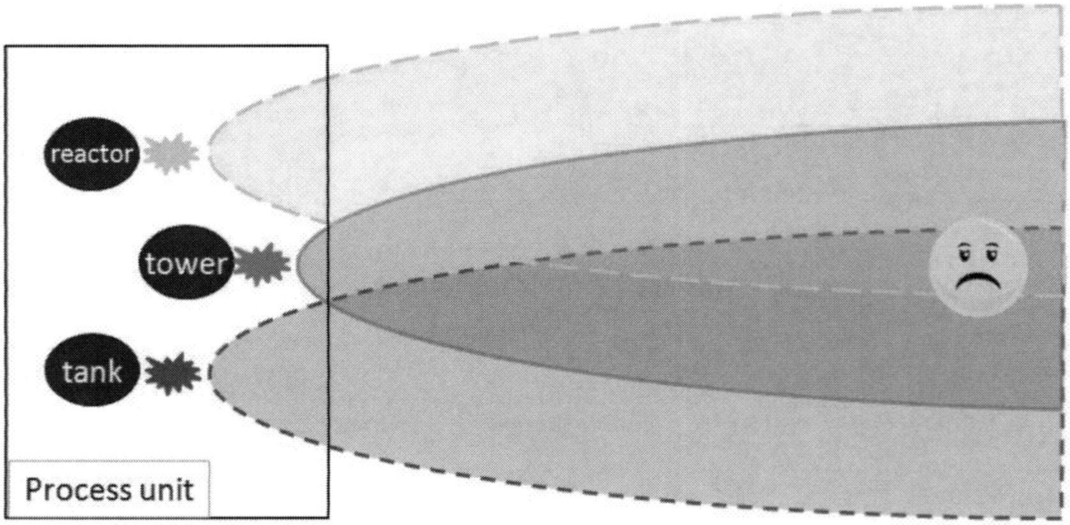

2.5 Determine the maximum tolerable individual risk criteria for single hazard

To determine the maximum tolerable individual risk criteria for single hazard in a chemical plant, needs to follow the following steps. The first step is to determine the background risk, if we chose 0.001, then:

$$R_{background} = 0.001$$

The second step is to determine the Imposed Risk Factor, as to employee, the range for the imposed risk factor is $k_{imposed\ for\ employee} \leq 1$, if $k_{imposed\ for\ employee} = 1$, then:

$$R_{imposed} = k_{imposed} \times R_{background} = 0.001$$

The third step, determine the Risk Category Factor, if $k_{category\ process\ safety} = 0.5$, then:

$$R_{process\ safety} = k_{category\ process\ safety} \times R_{work\ related} = 0.0005$$

The 4th step, determine the Risk Distribution Factor, if the number of critical vessels $N_{critical\ vessels} = 30$, then:

$$k_{distribution} = \frac{R_{process\ safety} - N_{critical\ vessels} \times 6 \times 10^{-6}}{R_{process\ safety}}$$

$$= \frac{0.0005 - 30 \times 6 \times 10^{-6}}{0.0005}$$

$$= 0.64$$

$$R_{residual\ process\ safety} = k_{distribution} \times R_{process\ safety} = 0.00032$$

The 5th step, determine the Number of Hazard Factor, if the number of safety interlock is 30, then:

$$k_{number\ of hazard} = \frac{1}{30} = 0.033$$

$$R_{max\ tolerable\ for\ signle\ scenario} = k_{number\ of hazard} \times R_{residual\ process\ safety}$$

$$= 0.033 \times 0.00032$$

$$= 0.000011$$

3. Conclusion

Using the Background Risk Method (BRM) to determine the risk criteria for single hazard is simple, logical and easy to be understood by employee and management. This method avoids the discussion of Value of life which is sensitive, it is easy for employee and management.

This paper provides a general framework for application of the Background Risk Method (BRM). It is suitable for large corporation in the process industry to develop their internal guidelines at the corporate level, to provide guidance for their business units or subdivisions to define their specific risk criteria which are tailored to their business size, process characteristics, etc.

[ii] How to develop a risk matrix for SIL determination
[iii] IEC61511
[iv] HSE, *Reducing risks - protecting people*, Paragraph 132, 2001
[v] IEC, *IEC61508 Part5 – Examples of Methods for the determination of safety integrity levels*, Figure C.1, 1998
[vi] CCPS, *Guidelines for Developing Quantitative Safety Risk Criteria*, FIGURE 4.4, 2009
[vii] 卢仿先，张琳，《寿险精算数学》，中国财政经济出版社，2006 年
[viii] CCPS, Guideline for CPQRA
[ix] IEC61511
[x] HSE failure rate database
[xi] 国家安全生产监督管理总局，《化工企业定量风险评价导则（征求意见稿）》，第 26 页，http://www.chinasafety.gov.cn/newpage/Contents/Channel_5826/2009/1221/80687/files_founder_169812096/2696328152.doc
[xii] CCPS, *Layer of Protection Analysis: Simplified Risk Assessment*, Figure7.2, 2001

1ST CCPS ASIA-PACIFIC CONFERENCE ON PROCESS SAFETY℠

THE TREATMENT OF DOMINO EFFECTS IN PROCESS HAZARDS ANALYSIS

Dr. Paul Baybutt/Principal Author
Primatech Inc.
50 Northwoods Blvd., Columbus, Ohio 43235
paulb@primatech.com

Prepared for Presentation at
1st CCPS Asia-Pacific Conference on Process Safety
Qingdao, China
September 4-5, 2013

THE TREATMENT OF DOMINO EFFECTS IN PROCESS HAZARDS ANALYSIS

Dr. Paul Baybutt/Principal Author
Primatech Inc.
50 Northwoods Blvd., Columbus, Ohio 43235
paulb@primatech.com

Keywords: Domino effect, domino event, process hazard analysis.

Abstract

Domino effects produce hazard scenarios that involve escalating consequences as a chain of linked events propagates throughout and beyond the process where they originate. Such scenarios have been the subject of accident survey studies and quantitative risk analyses but their identification in process hazards analysis (PHA) has received less attention. This paper discusses the nature and types of domino effects and a procedure is provided to address them in PHA. The role of separate domino effects studies is described, insights into domino effects that can assist PHA practitioners are provided, and some pitfalls in addressing domino effects are identified. The identification of domino effects in PHA is illustrated using an actual accident that occurred in a refinery.

1. Introduction

Domino effect is defined by the Center for Chemical Process Safety (CCPS) as, "the triggering of secondary events, such as toxic releases, by a primary event, such as an explosion, such that the result is an increase in consequences or area of an effect zone" (CCPS, 2000[14]). The effect zone is the area over which the effect equals or exceeds the level of concern. Each event may spawn consequences in multiple process locations. The term domino effect derives its name from an analogy with a row of dominoes, each standing on end, all of which fall when one is pushed. Sometimes, the term *knock-on effect* is used instead of domino effect but some practitioners use knock-on effect to describe situations where an event occurs indirectly, and possibly distinct temporally, from the preceding event.

Variations occur in the definitions of domino effect and other related terms (Reniers, 2010[13]). As used in this paper, the term *domino effect* refers to the phenomenon of a hazard scenario in which a linked sequence or chain of events occurs wherein the first, or primary, consequence of the scenario produces further cascading consequences in succession, often escalating, branching, and extending spatially. These multiple consequences (primary, secondary, and higher-order) occur serially as a direct result of one initiating event. Thus, one hazard is realized and triggers the realization of one or more additional hazards of the same or different types, and so on in what can become a complex sequence of events. A key aspect of many definitions is the requirement for escalating consequences. *Domino event* means any of the consequence events that occur in the causal chain of domino effects. *Domino scenario* means a hazard scenario that involves domino events. *Super event* means the set of domino events in a domino scenario.

Domino effects have played a role in numerous incidents dating back at least to the 1940's (Kourniotis et al., 2000[11], Abdolhamidzadeh et al., 2011[2]). Their consequences are amongst the greatest that can occur in process facilities and their importance has been recognized (Khan and Abbasi, 1998[10], Gledhill and Lines, 1998[8]). Domino scenarios often involve multiple fatalities and have destroyed entire facilities (Abdolhamidzadeh et al., 2011[2]). Domino effects have been shown in case studies to increase equipment failure frequencies by up to a factor of three (Abdolhamidzadeh et al., 2010[1]) (considering only primary domino events). Thus, neglecting domino effects will underestimate risk.

While approaches have been developed to address domino effects using Quantitative Risk Analysis (QRA) (CCPS, 2000[14], Abdolhamidzadeh et al., 2010[1], Cozzani, 2005[4], Landucci et al., 2009[12], Kadri et al., 2011[8], Kadri et al., 2012[9]), less attention has been paid to treating them in process hazards analysis PHA (CCPS, 2008[16]). Today, PHA is performed on most hazardous facilities. However, QRA is less common. Consequently, it is essential that PHA address possible domino effects so that these potentially significant contributors to process risk can be evaluated and appropriate risk management actions taken.

Separate domino studies are valuable but usually they are performed for assumed major releases of flammable materials from processes. As is true of many QRA studies, they are event-driven rather than scenario-driven. Consequently, they may not address more subtle initiating events that can be identified with PHA. Also, such domino studies primarily look outwards from the source of the domino effect to the target or receptor. PHA looks outwards from the source to the target but also inwards from the target back to the source which provides a valuable check to help identify domino scenarios as completely as possible.

2. Nature of Domino Effects

Sometimes, domino effects have been defined to mean situations in which a major incident on one site gives rise to consequential incidents on a neighboring site (HSE, 1984[21], Reniers, 2010[13]). Thus, a distinction can be made between intra-site domino effects that occur at the same site and inter-site domino effects that occur between separate but proximate sites. As might be expected, domino effects are more common at the same site than between different sites (Gledhill and Lines, 1998[8]). The Seveso Directive specifically requires the consideration of domino effects wherein an incident in one hazardous establishment impacts a neighboring hazardous establishment where the risk or consequences of a major incident may be increased because of the geographical position and the proximity of the establishments and their inventories of hazardous materials (CD, 2012[18]). The term domino effects, as used herein, refers to escalation events that occur either within or outside a particular facility site. Also, the effects may occur within a unit or process.

Domino effects can be categorized as direct or indirect (Gledhill and Lines, 1998[8]). Direct effects occur when a domino event results in an immediate escalation of the scenario consequences. Indirect effects occur when a domino event causes an impact that leads to escalation. Such impacts include:

- Loss of process control systems, e.g. basic process control system (BPCS), safety instrumented system (SIS).
- Loss of equipment, e.g. pumps, compressors.
- Loss of equipment supports, e.g. piping hangers, vessel foundations.
- Loss of mitigation systems, e.g. disabled fire protection system.
- Loss of utilities, e.g. electric power, cooling water, instrument air.
- Loss of actions by people, e.g. through incapacitation, evacuation, or the inability to access a process area, for example, inaccessible firewater monitors or isolation valves owing to the presence of fire.

The distinction between direct and indirect domino effects influences the strategies used to control them. Protective measures against direct domino effects include, for example, spacing, barriers, and hardening. Protective measures against indirect domino effects include, for example, diversity, redundancy and separation.

Some hazard scenarios involve a simple escalation of their consequences wherein the consequences steadily worsen, such as a fire that increases in size owing to an increase in the amount of flammable material released. Such scenarios do not involve domino effects, which, by their nature, extend the scenario consequences to other parts of the process, other processes, or other neighboring sites.

3. Types of Domino Effects

There are several ways in which domino effects can occur:

- Fire

Domino effects may result from *thermal radiation* or *flame impingement*. Fire may cause escalation by inducing the mechanical failure of equipment that contains hazardous materials, or it may heat the contents of process equipment producing a release. For example, a jet fire from a tank may impinge on an adjacent tank and cause its rupture, or heat radiation from a pool fire resulting from a tank leak may cause the overpressurization failure of an adjacent tank.

- Explosion

Domino effects may result from *overpressure, or blast wave*, and from *projectiles, or missiles*. Primary projectiles may result from fragmentation of the equipment within which the explosion occurs, separation and rocketing of the equipment, or ejection of an object from the equipment. The projectiles absorb energy from the explosion. Secondary projectiles result from the explosion blast wave imparting energy to objects in its path.

Explosions can release energy physically, or chemically. Physical explosions include a pressure burst from the rupture of a system under pressure, such as an air receiver, a compressor, a boiler, or a process pressure vessel; and a rapid phase transition involving a change of state of a substance, such as a steam explosion resulting from the instantaneous vaporization of water to steam on contact with molten metal. Chemical explosions can be condensed phase when explosive material is present in the form of a liquid or solid, such as ammonium nitrate, or vapor or gas phase when explosive material is a vapor or gas, such as methane.

- Toxic material release

A toxic gas or vapor cloud may result in the incapacitation or degradation in the performance of personnel, require their evacuation, or prevent their access to a process area resulting in the escalation of the consequences of the incident. For example, a release of a toxic material may prevent an operator from activating the emergency shutdown for a process resulting in a runaway reaction that destroys the reactor building and causes further casualties.

Some practitioners exclude toxic release as a type of domino effect because they do not produce direct effects. However, they can make important risk contributions through their indirect effects and they are included in this paper as a type of domino effect.

- Other hazardous material release

Domino effects may be produced by the inadvertent release of other hazardous materials through their impact on the process or its personnel. Usually, PHA focuses on the major hazards posed by flammable, explosive, reactive, and toxic materials and seeks to identify possible causes of their release from containment in the process. Typically, PHA teams focus on direct causes of releases such as leaks from piping, valves left open, and loss of electric power. However, there are also indirect causes from domino effects. For example, a leak of corrosive material may cause a failure of another piece of process equipment and the release of process materials.

Similarly, a leak from oxygen cylinders into a process area may increase the likelihood of fires, and a leak from a nitrogen supply system may incapacitate operators resulting in loss of process control and a release. A release of a cryogenic liquid may cause low temperature embrittlement of process piping or vessels resulting in their failure and producing a release. Such indirect causes of hazard scenarios may be missed by a PHA team unless they are considered explicitly. Other hazardous materials that may produce such scenarios include corrosives, oxidizers, asphyxiants, cryogenics, steam, oil, and fuel gas. Of course, such materials are specific to each process.

- Physical effects

Domino effects may be produced through the impact of various physical effects on the process or its personnel. For example, a high-pressure gas jet may impinge on process piping or vessels resulting in their failure owing to low temperature embrittlement. Hose whipping may incapacitate people or damage equipment. Failure of rotating equipment may generate projectiles. Process materials may be released in these and other such events.

Domino effects may occur over very short or longer time periods. For example, the effects of projectiles and blast overpressures will be immediate while the effects of thermal radiation and toxic gas dispersion will take longer to be realized. The distinction is important because more prevention and mitigation measures are possible when sufficient time is available. The concept of a "local" domino effect has been proposed wherein the domino events occur within the same process unit but separated by a time period so that mitigative actions may be possible (Abdolhamidzadeh et al., 2011[2]). Certainly, domino scenarios that occur within a process unit should be included in the scope of a PHA, but those that may not permit mitigation should not be excluded from consideration as it is likely they will have the higher risk.

4. Consideration of Domino Effects in PHA

Historically, to the extent that domino effects have been addressed in PHA, usually they have been considered as external events, along with natural and human-induced events and utility and service failures that originate outside the process (Baybutt, 2013[3]). Thus, when considering hazard scenarios for a process section, PHA teams address possible domino effects from external fires, explosions, etc. This approach looks inwards and focuses on contributors to scenarios for the process section. It is possible that PHA teams may not recognize that the domino effect is part of a chain and that further domino effects may be possible. Indeed, the domino event for the process section may be a tertiary or higher-order event in a domino scenario. Even if the domino effect for the process section is second order, fundamentally, it is a consequence of an initiating event in another process section rather than an initiating event for the process section where it occurs. Therefore, it should be recorded in the PHA worksheet as a consequence for the other process section, not in the worksheet for the process section where the domino event occurs.

A further problem with the inward-looking approach is that the PHA team may tend to focus on large direct domino effects that clearly will adversely impact the process section under consideration. They may miss lesser and indirect domino effects. Furthermore, the approach encourages PHA teams to focus attention on possible releases of process materials from a specific piece of equipment. In actuality, releases may occur from multiple pieces of equipment as a result of the same domino effect. The consequences of the domino scenario may be underestimated if they are considered individually, rather than as a super event, owing to possible synergistic effects. However, the inward view provides an important check to ensure that domino effects are not missed.

It is preferable to begin the identification of domino scenarios with their initiating event in an approach that looks outwards. PHA teams should brainstorm possible initiating events for domino sequences for each process section considered in a study and examine how their consequences may extend beyond immediate impacts. Potential direct and indirect domino effects of each type should be considered. Successive domino events should be traced throughout process, and beyond, to the extent feasible. It is preferable to consider indirect domino events first. If the team starts with direct domino events they may be distracted by the magnitude of their immediate consequences and may not return to consider the indirect domino events.

PHA teams must be careful to consider hazard scenarios with immediate consequences of low severity that may result in domino scenarios with consequences of high severity. For example, the rupture of a small diameter pipe carrying a flammable material may result in a small jet fire with immediate consequences of low severity. However, if the jet fire impinges on a larger diameter pipe that ruptures and produces a larger jet fire that impinges on a storage tank, a Boiling Liquid Expanding Vapor Explosion (BLEVE) may result producing much greater consequences. The initial pipe rupture event may not receive a high PHA risk ranking if only its immediate consequences are considered. Moreover, if rupture of the large pipe is not considered credible by other means, the BLEVE may be overlooked if domino effects are not addressed.

As each process section is considered, it is valuable to inquire of the team if any domino effects could impact it as they may have been missed when other process sections were considered. However, they should be traced back to their origin and recorded in the worksheet for the process section where the domino scenario originates.

Various types of fires and explosions should be considered. Pool fires result from the ignition of spilled flammable liquids. Jet fires occur when high pressure flammable material is ignited as it is released from containment. Torch fires are produced at lower pressures. Flash fires are produced by the ignition of a cloud of flammable vapor with the flame velocity less than the sonic velocity. Fireballs result from the sudden release of a flammable gas or volatile liquid stored under pressure coupled with immediate ignition. Container fires happen when flammable materials in storage containers are ignited. Each type of fire can result in different domino effects.

Explosions may be confined, unconfined, BLEVE, dust, or reactive chemical in nature. Confined explosions occur within the process containment or within a building, or when a flammable release is subject to confinement by process structures. Unconfined explosions occur in the open. BLEVE's occur when a vessel containing a flammable liquid above its atmospheric boiling point is ruptured by exposure to a pool fire or a jet fire resulting in flashing of the liquid and ignition. Thus, BLEVE's are actually secondary or higher-order domino events. Dust explosions result from the rapid combustion of fine solid particles. Reactive chemicals may produce not only explosions but also fires, over-pressurization of process equipment, and the generation of toxic gases or other hazardous materials. Several types of reactive chemical hazards exist. They include uncontrolled decomposition, re-arrangement, or polymerization of a single unstable compound; self-reacting materials, such as organic peroxides and monomers; loss of control of chemical reactions; and chemical interactions from the inadvertent mixing of two or more chemicals leading to an uncontrolled chemical reaction.

The following procedure is suggested for the identification of domino events in PHA. As scenarios are identified for each section of the process:

1. Brainstorm scenarios that may result in domino effects in other process sections.

2. Consider effects on the current process section from scenarios that may originate in other process sections.

3. In all cases, consider indirect and then direct effects. Use checklists, as appropriate, to facilitate team brainstorming. Checklists can be developed using lessons learned from analyses of the historical record of domino scenarios [2, 3].

4. Record scenarios in the process section where they originate.

An alternative approach, which may be merited by the importance of domino effects, is to conduct a PHA first without searching for domino effects and then to make another pass through the process and the PHA worksheets looking specifically for domino effects using the approach described above. This alternative approach allows the team to focus exclusively on domino effects. However, it may encounter resistance from the team who may be reluctant to extend the study, particularly if some domino scenarios were identified in the first pass.

A considerable amount of information is needed to analyze domino effects formally using quantitative methods. For the purpose of PHA, information is needed on plant layout, separation distances, and the vulnerability of people and equipment. The vulnerability of receptors owing to their proximity to sources of domino effects is the key issue.

5. Separate Domino Effect Studies

PHA is not the place to explore all the possible paths and branches of domino scenarios. At best, PHA may be able to identify effects clearly up to secondary events and the possibility of higher-order ones. However, the exploration of higher-order events would sidetrack the team from their primary responsibility of identifying initiators of domino scenarios. Consequently, domino effects also should be addressed in separate studies or in facility siting studies (CCPS, 2003[15]). Traditional analytical techniques also are limited in their ability to analyze complex domino scenarios. A Monte Carlo simulation method has been proposed to address this difficulty (Abdolhamidzadeh et al., 2010[1]).

Recent efforts in facility siting have focused attention on the effects of fires, explosions and toxic releases on the occupants of buildings (CCPS, 2012[17]). However, the occurrence of domino events emphasizes the importance of also continuing to address the appropriate separation of equipment in processes.

PHA studies that include domino effects should be conducted before separate studies. If a domino study is performed first, the PHA team will feel less need to search for domino scenarios because they may believe that the separate study adequately addressed the domino effects. Also, the domino study for the process will benefit from prior identification of domino scenarios. However, the PHA study may need to be updated after the domino study is performed so that it provides a complete record of all hazard scenarios that are known for the process. If the domino study is performed first, it will have to be reviewed after the PHA study to see if any important domino scenarios have been missed.

6. Insights Into Domino Effects

The potential for domino effects depends on numerous factors (Gledhill and Lines, 1998[8]). For example, the spreading of fires depends on the pathway available, presence of combustible materials, fire mitigation systems such as firewalls and dikes / bunds, meteorological conditions such as wind direction, and ignition delays. Prevention of domino effects is best addressed during the initial design of processes when it is more likely that spacing of equipment, segregation of processes, plant topography, and robust safety systems can be employed (Gledhill and Lines, 1998[8]). Strategies for the prevention of domino effects have been addressed in the literature (Cozzani et al., 2007[5], Cozzani et al., 2009[6]).

Fires and explosions have been found to be the most frequent causes of domino effects [6]. Pool fires and vapor cloud explosions historically have been the dominant causes of primary unit-to-unit domino events (Abdolhamidzadeh et al., 2011[2]). Domino effects have most frequently originated in storage areas, process areas and transportation (Darbra, 2010[7]). Explosion-to-fire, fire-to-explosion, and fire-to-fire were the most frequent domino sequences (Darbra, 2010[7]). Data on the length of the domino event chains for unit-to-unit domino effects indicates that about half progress up to a second domino event and about half progress to a third domino event or further (Abdolhamidzadeh et al., 2011[2]). Minor jet fires caused by releases from small diameter pipes or valves, and pool fires caused by leaks from seals, are the more likely events that may cause scenario escalation. Such insights from the analysis of actual domino effects can be used by PHA teams to help focus on those domino scenarios that are likely to be important.

Impacts on adjacent sites will depend on the type of domino effect. For example, explosions may well cause impacts off-site but pool fires are less likely to do so.

7. Pitfalls in Addressing Domino Effects

Practitioners must try and avoid numerous pitfalls that can be encountered in the analysis of domino effects.

Chains of events may be difficult to identify or dismissed as not credible. Congested layouts present many possibilities for domino scenarios. Usually, domino scenarios affect multiple process sections and their occurrence may be identified when analyzing each of the process sections. However, there is only one domino scenario and it should be recorded only once to avoid double counting. Also, double counting of consequences must be avoided. For example, if people are killed by the primary event, they cannot be killed again by a secondary event, or if property is damaged by the primary event but destroyed by a secondary event, the consequences are not additive. Domino events in a domino sequence are not independent. For example, injuries to personnel in a primary event may result in fatalities in a secondary event owing to their inability to evacuate the area.

The summation of primary, secondary and any higher-order consequences of domino events taken individually may result in an underestimate of the actual consequences because synergistic physical effects may contribute when the domino events occur near simultaneously. For example, thermal radiation from two fires occurring together will produce greater consequences than if the two fires are considered individually owing to reinforcement of the fire effects that results in impacts on people and the process that would otherwise not occur. However, the proximate occurrence of domino events does not necessarily mean there are synergistic effects. For example, the occurrence of two successive explosions does not produce synergistic physical effects.

The likelihoods of some scenarios may also be underestimated if domino effects are not considered. For example, a PHA may identify two similar hazards scenarios that result in a fire at each of two identical tanks. The likelihoods of each scenario are the same and depend on the events that make up the scenarios. However, if the possible domino effect of a fire in one tank causing a fire in the other tank is not considered, the likelihood of fire at each tank will be underestimated.

Fires and explosions may result in not only further breaches of process containment, and the additional consequences involved, but also the destruction or impairment of protective systems, possibly causing further domino events. Fires may result in domino effects from thermal radiation or flame impingement. However, fires may also produce toxic combustion products or toxic fumes due to thermal decomposition. The US Occupational Safety and Health Administration's process safety management regulations are silent on whether such secondary consequences should be addressed (CFR, 1992[19]). However, they may result in an indirect domino effect by incapacitating personnel.

8. Example of Domino Effects

On February 16, 2007, liquid propane was released from piping in the Propane DeAsphalting (PDA) unit at Valero's McKee Refinery near Sunray, Texas, resulting in a major fire that injured three employees and a contractor, caused extensive equipment damage, and required the evacuation and total shutdown of the refinery (CSB, 2008[20]). A vapor cloud of propane traveled downwind towards the boiler house, where likely it ignited and the fire flashed back to the leak source.

The rapidly expanding fire prevented operators from closing manual isolation valves or reaching local pump controls that might otherwise have been used to control the propane discharge. A high-pressure propane jet fire impinged on a non-fireproofed steel pipe rack support column which buckled, collapsing the rack and causing multiple pipe failures. Liquid petroleum products discharged from the damaged pipes and significantly increased the size and duration of the fire contributing to its rapid spread and damage to surrounding equipment. The refinery's emergency response team attempted to activate stationary fire water monitors, but high and shifting winds and the rapid growth of the fire hampered their efforts.

Three one-ton chlorine containers used to provide a biocide for water in an adjacent cooling tower were exposed to radiant heating from the fire and vented leading to the release of more than 2.5 tons of highly toxic chlorine. Responders were unable to enter the refinery to extinguish the fire owing to the chlorine release and leaks of sulfuric acid that made entry too hazardous.

Radiant heat from the PDA fire blistered the paint on a butane storage sphere. Fortunately, the wind at the time of the fire tended to move the flames away from the sphere. However, heat from the fire kept responders from reaching the manual firewater deluge valves for the sphere which were located too close to the PDA unit and prevented the responders from opening the valves and establishing a protective flow of water over the surface of the sphere.

The Chemical Safety Board (CSB) report on the incident found fault with the refinery's PHA (CSB, 2008[20]). According to the report, the PHA did not identify and address hazards posed by fire exposure to neighboring equipment, including the chlorine containers and the butane storage sphere.

This incident illustrates several domino effects:

- Operators were prevented from controlling the propane discharge (indirect).
- Impingement of the propane jet fire on the pipe rack support (indirect).
- Escalation of the fire caused by the discharge of petroleum products from damaged pipes in the collapsed pipe rack caused further damage to surrounding equipment (direct).
- Responders were unable to activate fire water monitors (indirect).
- Chlorine containers were exposed to radiant heating from the fire and vented (direct).
- The fire brigade could not enter the refinery owing to the chlorine release and sulfuric acid leaks (indirect).
- Radiant heat from the PDA fire blistered the paint on butane storage sphere (direct).
- Responders were unable to reach the manual firewater deluge valves for the butane sphere (indirect).

Thus, there were a total of eight domino effects in this incident. Two of the three direct domino effects involved thermal radiation, while the third involved flame impingement.
Four of the five indirect domino effects involved inaccessible mitigation systems, two owing to the presence of fire and one owing to the presence of hazardous materials. The fifth indirect domino effect involved flame impingement.

The method by which domino effects can be identified and addressed in PHA can be illustrated using this incident. The analysis involves these steps:

1. The team identifies an initiating event that causes a release of propane from the PDA unit.

2. The team considers possible scenarios that may result in a fire or explosion.

3. The team identifies the immediate consequence of a fire owing to ignition of the propane by the boiler house and determines its possible effects on receptors of concern, such as people, property, etc.

4. The team identifies any safeguards against the release, the ensuing fire, and its consequences. This step is repeated for each domino effect.

5. The team considers whether the fire can produce any direct or indirect domino effects.

6. In discussing the impact of the fire on controls and safeguards, the team identifies the possible inaccessibility of equipment used to control the release (first domino effect).

7. No direct domino effects on other parts of the process are identified as separation distances are judged adequate. However, the team observes the lack of fireproofing on pipe rack supports. A check reveals that industry spacing standards have been met but the team is still concerned. A recommendation is made to perform some calculations which reveal that the supports may be engulfed in a serious fire (second domino effect).

8. The team considers the impact of exposing the pipe rack supports to fire and realize that the rack may collapse causing pipes to fail and further hydrocarbons to be released enlarging the fire and damaging surrounding equipment (third domino effect).

9. The team notes the presence of fire water monitors but on examining their locations realizes they will be inaccessible (fourth domino effect).

10. On considering whether the escalating fire can produce any further direct or indirect domino effects, the team realizes that chlorine containers may be exposed to radiant heating from the fire and could vent (fifth domino effect) and, consequently, emergency responders may be prevented from fighting the fire (sixth domino effect).

11. The team also realizes that the butane storage sphere may be exposed to radiant heat and could BLEVE (seventh domino effect).

12. On considering the availability of the firewater deluge system, the team realizes that the manual valves required for its operation would be unreachable (eighth domino effect).

During or subsequent to this analysis, the team will consider recommendations to prevent or mitigate the domino effects such that if a suitable recommendation is made and implemented, no further domino effects may occur. For example, the installation of remotely operable isolation valves may have prevented the first domino effect and fireproofing of pipe rack supports may have prevented the second one. Either protective measure may have terminated the scenario in its early stages.

In all likelihood, few PHA teams would follow a domino scenario through so many steps. Most likely, they would reach a point where further escalation is viewed as incredible. Also, it is probable that they would rely on recommendations that prevent further escalation as part of the justification for not proceeding further in the event chain. The recommendations indeed may reduce the risk below tolerable levels but they do not necessarily completely eliminate the risk of higher-order events.

This incident provides an example of PHA being expected to identify deficiencies in good engineering practices for processes. Arguably, a number of the process aspects that allowed this complex domino scenario to occur could have been identified without the need to perform PHA. Processes that are presented for PHA should comply with government regulations, industry standards, and company practices so that PHA can focus on correcting unusual situations that are not addressed by regulations, standards and practices.

9. Conclusions

Domino scenarios should be addressed in PHA because of their potentially catastrophic consequences. They can be addressed best using an approach that combines the identification of domino scenarios that may originate within each process section with the consideration of the impacts on the process section from domino scenarios that may originate in other process sections. Both indirect and direct domino effects from all important domino sources should be considered. An understanding of the characteristics of domino scenarios from actual accidents can help to guide their identification in PHA.

PHA can help to identify scenarios that may be missed by separate quantitative studies of domino effects. Usually, such studies are event-based and focus on domino scenarios with primary events that have major consequences rather than lesser consequences. Both types of primary domino events can be identified using PHA. However, PHA can provide only an indication of possible domino scenarios. Their complexity and importance warrants detailed analysis using more quantitative methods but their identification in PHA is an important precursor to such studies.

10. References

[1] Abdolhamidzadeh, B., Abbasi, T., Rashtchian, D., & Abbasi, S.A., 2010, "A new method for assessing domino effect in chemical process industry", J. of Hazardous Materials, 182, 416 - 426.

[2] Abdolhamidzadeh, B., Abbasi, T., Rashtchian, D. & Abbasi S. A., 2011, "Domino effect in process-industry accidents: An inventory of past events and identification of some patterns", J. of Loss Prevention in the Process Industries, 24 (5), 575 - 593 .

[3] Baybutt, P., 2013, "Analytical Methods in Process Safety Management and System Safety Engineering – Process Hazards Analysis", in Handbook of Loss Prevention Engineering, Haight, J. M. (ed), Wiley-VCH.

[4] Cozzani, V., Gubinelli, G., Antonioni, G., Spadoni, G. & Zanelli S, 2005, "The assessment of risk caused by domino effect in quantitative area risk analysis", J. of Hazardous Materials, 127, 14 – 30.

[5] Cozzani, V., Tugnoli, A. & Salzano, E., 2007, "Prevention of domino effect: From active and passive strategies to inherently safer design", J. of Hazardous Materials, 139(2), 209- 219.

[6] Cozzani, V., Tugnoli, A., & Salzano, E., 2009, "The development of an inherent safety approach to the prevention of domino accidents", Accident Analysis and Prevention, 41 (6), 1216 - 1227.

[7] Darbra, R., Palacios, A. & Casal, J., 2010, "Domino effect in chemical accidents: Main features and accident sequences." J. of Hazardous Materials, 183, 565–573.

[8] Gledhill, J. & Lines, I., 1998, "Development of methods to assess the significance of domino effects from major hazard sites", CRR 183, Health and Safety Executive. Kadri, F., Châtelet, E., & Chen, G., 2011, "Method for quantitative assessment of domino effect caused by overpressure", in Advances in Safety, Reliability and Risk Management, Guedes Soares, C. (Ed.), 1360 - 1367, CRC Press.

[9] Kadri, F., Chatelet, E., & Chen, G., 2012, "Quantitative risk assessment of domino effect caused by heat load", International J. of Systems Control, 3, 42 - 48.

[10] Khan, F. I. & Abbasi S. A, 1998, "Models for domino effect analysis in chemical process industries", Proc. Safety Prog. 17(2), 107 - 123.

[11] Kourniotis, S.P., Kiranoudis, C.T. & Markatos, N.C., 2000, "Statistical analysis of domino chemical accidents", J. of Hazardous Materials, 71, 239 - 252.

[12] Landucci, G., Gubinelli, G., Antonioni, G. & Cozzani V., 2009, "The assessment of the damage probability of storage tanks in domino events triggered by fire", Accident Analysis and Prevention, 41(6), 1206 - 1215.

[13] Reniers, G. L. L., 2010, "Multi-Plant Safety and Security Management" in the Chemical and Process Industries, Wiley-VCH.

[14] Center for Chemical Process Safety (CCPS), "Guidelines for Chemical Process Quantitative Risk Analysis", 2000, 2nd Edition, Center for Chemical Process Safety / American Institute of Chemical Engineers, New York, NY.

[15] Center for Chemical Process Safety (CCPS), "Guidelines for Facility Siting and Layout", 2003, Center for Chemical Process Safety / American Institute of Chemical Engineers, New York, NY.

[16] Center for Chemical Process Safety (CCPS), "Guidelines for Hazard Evaluation Procedures", 2008, 3rd Edition, Center for Chemical Process Safety / American Institute of Chemical Engineers.

[17] Center for Chemical Process Safety (CCPS), "Guidelines for Evaluating Process Plant Buildings for External Explosions, Fires and Toxic Releases" 2012, 2nd Edition, Center for Chemical Process Safety / American Institute of Chemical Engineers, New York, NY.

[18] CD, 2012, Council Directive 2012/18/EU of 4 July 2012 on the control of major-accident hazards involving dangerous substances, amending and subsequently repealing Council Directive 96/82/EC, http://eur-lex.europa.eu/LexUriServ/LexUriServ.do?uri=OJ:L:2012:197:0001:0037:EN:PDF.

[19] CFR, 1992, Federal Register, Volume 57, Number 36, pages 6355 to 6417, Final Rule, Process Safety Management (PSM) of Highly Hazardous Chemicals, 29 CFR 1910.119.

[20] CSB, 2008, Investigation Report, LPG Fire at Valero – Mckee Refinery Report No. 2007-05-I-TX, U.S. Chemical Safety and Hazard Investigation Board, Washington, DC.

[21] HSE, 1984, Advisory Committee on Major Hazards, Third Report: The Control of Major Hazards, HSE Books.

1ST CCPS ASIA-PACIFIC CONFERENCE ON PROCESS SAFETY℠

Risk Based Design of Overpressure Protection Systems

Zhihe Wang
Tony Downes
Binhui Yu
Lindsay Steve
Hong Xiao

430 Li Bing Road, Zhang Jiang Hi-Tech Park
Pudong New Area, Shanghai 201203, China
Zhihe.Wang@Honeywell.com
Anthony.Downes@Honeywell.com
Steve.Lindsay@Honeywell.com
Jane.Xiao1@Honeywell.com
Ben.Yu@Honeywell.com

Prepared for Presentation at
American Institute of Chemical Engineers
2013 1st Global Congress on Process Safety
Qingdao, China
Sep 4 – 5, 2013

Risk Based Design of Overpressure Protection Systems

Zhihe Wang
Tony Downes
Binhui Yu
Lindsay Steve
Hong Xiao

430 Li Bing Road, Zhang Jiang Hi-Tech Park
Pudong New Area, Shanghai 201203, China
Zhihe.Wang@Honeywell.com
Anthony.Downes@Honeywell.com
Steve.Lindsay@Honeywell.com
Jane.Xiao1@Honeywell.com
Ben.Yu@Honeywell.com

Keywords: Process Hazard Analysis (PHA), Layer of Protection Analysis (LOPA), Safety Instrumented Systems (SIS), Overpressure Protection.

Abstract

"When Honeywell International's Performance Materials and Technologies group builds a new facility, the project applies risk based work processes to Safety Instrumented Systems (SIS) design and application, including a multi-step PHA and LOPA process that recommends Safety Instrumented Functions to deal with overpressure scenarios, etc.. Other applicable IPLs are included in the LOPA risk evaluation as well. Honeywell's Automation and Controls group is assisting with the design of these systems. Using examples, this paper will illustrate how hazards can be identified and then managed using high reliability safeguards. It will also illustrate the importance of using "standard" loops by providing several practical examples, to simplify designs as well as proof-testing over the life of the plant."

1. Introduction

In the chemical industry, an important safety issue is the prevention of loss of containment due to vessel or pipeline overpressure situations. Loss of containment could result in serious impacts to personnel and the environment, if flammable and/or toxic materials are released to the atmosphere. Loss of containment could also result in economic impact due to production unit replacement/repair costs and production losses. Let's assume some potentially hazardous scenarios were identified and addressed during the Process Hazard Analysis (PHA). The consequence from those scenarios was assessed qualitatively or quantitatively and some were found to require additional safeguards. A risk based analysis methodology was used to ensure the correct identification and assessment risk of overpressure protection systems, especially for Safety Instrumented Systems (SIS) design.

As is common in the industry, Honeywell uses a multi-step PHA approach to manage risks, beginning with coarse-scale reviews early in a project and progressing through HAZOP and

LOPA as the documentation becomes finalized. PHA is adopted to address credible accident scenarios. Results of such studies include scenarios with consequences that may be impractical to eliminate through facility designs. Rather than designing facility for a high level of protection against the most severe credible events, a risk-based approach provides a practical and a cost effective approach to maximize safety using Layer of Protection Analysis (LOPA) methodology according to catastrophic consequences following PHA output.

The benefits of a risk based approach to design overpressure protection system include reasonable frequency, reasonable consequence and safeguard credit, action priority, and consensus from all involved discipline personnel.

This paper outlines the basic steps to identify hazard, consequence and safeguard using some overpressure scenarios as examples. It also illustrates steps how to design SIS using LOPA using several cases as examples. This industry-standard methodology helps establish practical overpressure protection systems to minimize risk while achieving cost effective designs.

2. Risk Based Analysis Work Process

Risk is the likelihood of occurrence or exposure to the hazard and which results in harm to people, property or the environment as a result. PHA is a proven effective way to identify, mitigate and manage risk from chemical processes. It is recommended that PHA be deployed first to identify all potential causes and consequence of fire, explosions, releases of toxic or flammable chemicals, prioritize each scenarios risk and identify scenarios for further study using advanced techniques such as Layer of Protection Analysis (LOPA)

A LOPA is typically applied after the PHA report is issued as part of the PHA recommendation resolution process using the scenarios identified by the PHA team. LOPA is a more detailed approach and should not be applied to all scenarios. The amount of effort required to implement LOPA may be excessive for lower consequence/lower risk scenarios.

One fundamental principle is required to be met for an overpressure protection design to qualify as "Risk Based". The principle is that the overpressure protection is designed in a way such that it reduces affected personnel risk to an acceptable level. Employing this fundamental principle ensures that risk mitigation resources to be optimized. The acceptable level risk generally is defined using national and international standards and regulations, corporate policies, and discussion and agreement between interested parties such as the community, local jurisdiction and insurance companies supported by good engineering practices.

3 LOPA Work Process

LOPA is one of the most widely used semi-quantitative methods of analyzing and documenting safeguards. An important outcome of LOPA is identification of Independent Protection Layers (IPLs) essential for required risk reduction. The required Safety Integrity Level (SIL) of each IPL is also determined during LOPA. SIL defines the target performance level of an IPL in terms of a range of average probability of failure on demand. In the case of overpressure protection safeguards, the SIL is determined by the overpressure scenario with the highest risk. The assigned SIL value will close the overpressure scenario risk gap. Other applicable LOPA IPLs are included during the LOPA risk evaluation as well.

3.1 Consequences assessment

In Honeywell, the LOPA methodology will take account of the four consequences category of receptors or affected populations as below:

- Local personnel working on the immediate process equipment that are assigned to that work station
- Remote personnel that work away from the process equipment
- Community
- Environmental

The LOPA team will assess the worst credible case result assuming that passive safeguards relevant to the hazard scenario are functioning, but that active safeguards are not functioning. Then assign the severity of consequences for the four categories of potential receptors or potentially affected populations. For example, a potential scenario might include a chemical release in large enough quantity that it could cause a serious injury to local and/or remote personnel and adverse public reaction in the community.

A quantitative description should include quantity of material, release rate, modeling information, siting factors (source relative to receptors), assumptions and consequence action Level. Threat zones should be estimated as a distance around the source, using computer dispersion models and/or other consequence models as required. The threat zones, representing the areas downwind of the release that may be at risk; Modeling, if applicable, shall be developed for each of the identified receptor scenarios, including off-site, unless results of a higher degree are applied. Compare the threat zone radial distance to the distance between the source and potential on-site, as well as off-site, receptors to determine severity level to determine the consequence severity from loss of containment. In determining the severity level, it shall require evaluation of factors such as the expected number of personnel in the area during the scenario, the amount or rate of the release, the likely meteorology as it effects dispersion of a toxic cloud, or the distance from the source as it affects overpressure from an explosion or thermal load from a fire. Consequence Analysis techniques can be used to quantify the impact at a specified location. Passive safeguards in the process are assumed to be completely effective, having a negligible likelihood of failure to modify the severity of an event during a catastrophe. Examples of passive safeguards include buildings, enclosures, dikes (bunds), etc. that is effective for the hazard scenario under consideration and as otherwise documented in this procedure.

3.2 Likelihood Assessment

The qualitative levels of likelihood are intended to be significantly different from each other. The likelihood will be estimated on the basis of industry published data plus Honeywell experience. The likelihood of an Initiating (hazardous) event should be estimated first, before considering the impact of any existing IPLs. Company specific data, when available, should be employed to establish the likelihood of occurrence of hazardous events. Otherwise industry generic data can be used.

The first step is to determine the Base Likelihood of the initiating event. The Base Likelihood level is determined by the highest likelihood (lowest number in our system) among the individual descriptions.

In the second step for the hazard scenario under consideration, all passive and active safeguards are considered. Apply the credits identified for each additional safeguard (layer) which acts to prevent the scenario from occurring or mitigates the effects. More independent safeguards reduce the likelihood. Unless noted, the active safeguards are understood to be typical plant controls.

3.3 Risk Ranking

After determination of severity and likelihood, risk for each scenario is determined using risk matrix. The final risk ranking is determined at the intersection of the Likelihood and Severity. The Risk Matrix is similar to many other companies in our industry. If the risk is not low enough, the PHA team will recommend additional safeguards or perhaps more detailed study using LOPA. For risk rankings of that are low enough, the residual risk is acceptable. Further modifications are not necessary. This does not mean that recommendations cannot be made. Typical recommendations at this level would address operability issues or conformity with pre-existing methods. If recommendations are made, a response is required, but implementation is not necessary for risk reduction purposes. A simple example might be when a PHA team recommends an alarm for operations reasons – a reminder to the operator to swing to a new filter after a certain number of hours. If this is not a safety concern then it may not be necessary to add the alarm. It is well known in the industry that too many "operations" alarms can lead to alarm flooding in some cases.

When selecting recommendations the team should consider the sequence of increasing consequence according the following sequential order of means to identify/prevent/mitigate the hazard scenario.

- Inherently Safer, so that the hazard or consequences become insignificant.
- Operator response with no alarm, e.g., checklist SOP
- Alarm layer of protection/Operator response
- Engineered controls (e.g. interlocks, SIFs, etc)
- Relief protection vented to a safe location

3.4 Safeguard Validation

The risk can be reduced to an acceptable level by implementing safeguard. Each company defines its tolerable risk level for safety, environmental and commercial hazards. Each safeguard reduces the risk by a certain order of magnitude. No single safeguards cannot eliminate risk and protect a facility and its personnel against harm or mitigate the consequence from overpressure scenario. For this reason, safeguards exist in protective "layers":

- Procedures (Operating or auditing procedure with documented readings, or alarms with operator response when the alarm is independent of the cause of a possible problem)
- Basic process controls system (BPCS),
- Auto shutdown system external to BPCS, for example, a Safety Instrumented System (SIS)
- Mechanical relief devices,
- External response measures (e.g. dike, building) which prevent or mitigate a hazardous event.

If one IPL fails, successive layers will be available to take the process to a safe state. If one of the protection layers is a safety instrumented function (SIF), the risk reduction allocated to it determines its safety integrity level (SIL). As the number of protection layers and their reliabilities increase, the risk reduces. The required SIL is determined by the calculated

Probability of Failure on Demand (PFD) and Risk Reduction Factor (RRF). When the hazards identification and risk assessment phase concludes that a SIS is required, the level of risk reduction afforded by the SIS and the target SIL has to be assigned.

IPLs must meet the following criteria in order to be classified as such:

- Specificity
  - Designed to completely mitigate/reduce risk under consideration
  - Must be able to perform function under normal, abnormal, and design conditions
  - When that is achieved by mitigating or reducing the consequences of a specific hazard scenario, to take any credit, the severity level must be reduced by one or more.
- Independence
  - Independent from all other layers, so that failure of one layer does not cause the failure of another
  - Separation from the cause (initiating event)
- Availability
  - Must reduce the risk by at least one order of magnitude (probability to fail on demand)
  - The IPL must be available at least the percent of time credited
- Verifiable
  - An IPL must be designed to allow for regular, periodic validation of the protective function in order to ensure that the specified level of risk reduction is being achieved
  - Allows auditability of the assumptions made that are necessary for risk reduction value used to be achieved

3.4.1 Restriction/Assumption on Safeguard Credit

The BPCS is suitable for a single reduction credit for a given scenario except when the Initiating Event is contained in the BPCS. In those cases, any safeguards involving the BPCS shall not count as a reduction credit. This requirement is applicable when the Initiating Event failure includes any component of the control loop including the primary device, transmitter, I/O card, transducer or final control element, as well as human errors such as an incorrect set point.

In cases when operating procedures as safeguards are taken credit, for a full likelihood reduction credit, operating or auditing procedures must include a mechanism for verifying that critical parameters were measured or observed, such as a log sheet of recorded process variables or a log of specific items inspected.

4 SIL determination using LOPA

Various methodologies are used for assignment of target SILs. The determination must involve people with the relevant expertise and experience. Methodologies used for determining SIL include Risk Matrix, Fault Tree Analysis, Layer of Protection Analysis (LOPA) and Markov Analysis. This paper will illustrate SIL determination by LOPA methodologies. LOPA defines risk as a function of severity and likelihood. The LOPA assessment team assesses the level of risk for each scenario, and then determines what reliability (in terms of SIL-rating for Safety Instrumented Systems) is needed for the proposed safeguards. The necessary risk reduction is achieved by a combination of all the safety protective features.

The basic risk reduction process is shown as sketch 4-1. First, Non-SIS IPLs (such as pressure relief valves) have been applied to reduce the likelihood while passive IPLs (such as dike) are applied to reduce the consequence. A SIS reduces risk by reducing the likelihood, and the SIL is the order of magnitude by which the likelihood is reduced. A SIL-1rated SIS reduces the likelihood by a factor of ten, a SIL-2 reduces the likelihood by a factor of 100, and a SIL-3 reduces the likelihood by a factor of 1000. If the LOPA identifies that the risk prior to application of the SIS is an order of magnitude higher than desired, then an order of magnitude reduction in likelihood is required so the SIL requirement is 1. If at the risk prior to application of the SIS 100 order of magnitude of likelihood reduction is needed then the required SIL is 2.

Sketch 4-1 typical risk reduction process

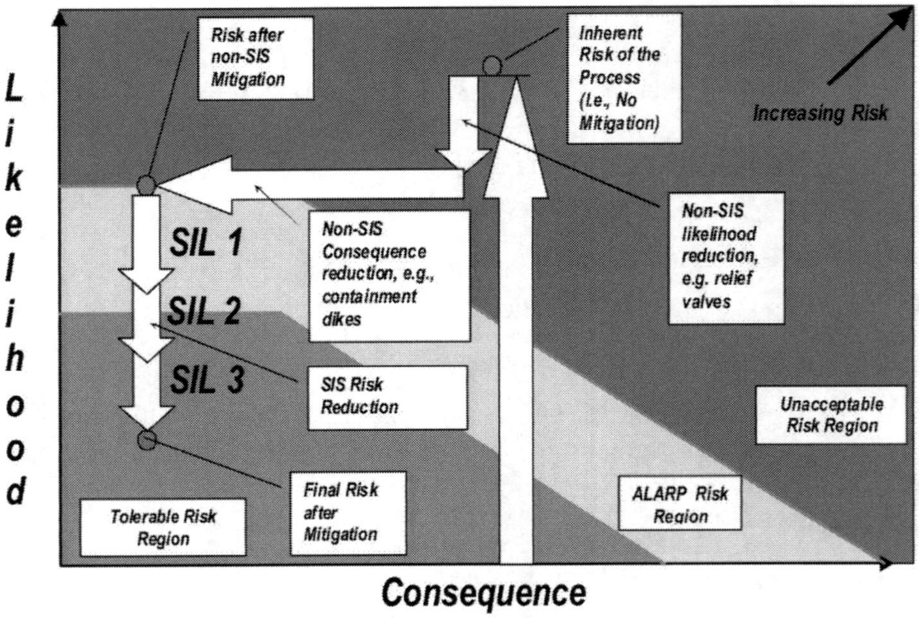

4.1 SIS design using Honeywell SIL certified equipments

Honeywell's Automation and Controls group makes SIL certified equipment which has been reviewed and tested by a third-party agency according to IEC 61508 and IEC 61511. Honeywell's IEC 61511 and IEC 61508 SIL certified solution provides the optimal level of safety and process integration while still maintaining functional safety separation as mandated by those standards. Through Experion® operational integration, all systems are unified into one operationally integrated architecture, providing a unique opportunity to improve safety, process availability and efficiency. Honeywell's Safety Manager® (shown as sketch 4-2), part of the Experion® Process Knowledge System, enhances the safety, reliability and efficiency of critical processes. Safety Manager® complies with the following international standards:

- For burner management: NFPA 85, 86, EN50156
- For emergency shutdown and other critical applications:
- IEC 61508, IEC61511, ISA S84.01, DIN V 19250, UL, FM, ATEX
- For fire and gas: EN54-2, NFPA 72

As a standard loop, Safety Manager® provides the basis for safety unification, reducing risks and installed costs, and improving safety while increasing uptime. Also, using a Safety Manager® as the logic solver of the SIS helps in ease of integration with the plant BPCS (Honeywell Experion® system). This allows for seamless integration in terms of alarm reporting, Sequence of event reporting, display on a common HMI and communication between the Experion®

controllers and the Safety Manager®. The Safety Manager® does not use information from Experion® controllers for any executive action but informs Experion® controllers if any SIF does execute so that Experion® controllers can initiate a smooth shutdown or other required actions. Other advantages of using the Safety Manager® (especially in redundant mode) are online changes to the code at a later date and higher availability in the event of a failure of any of its modules.

Sketch 4-2 Honeywell Safety Manager® Loop

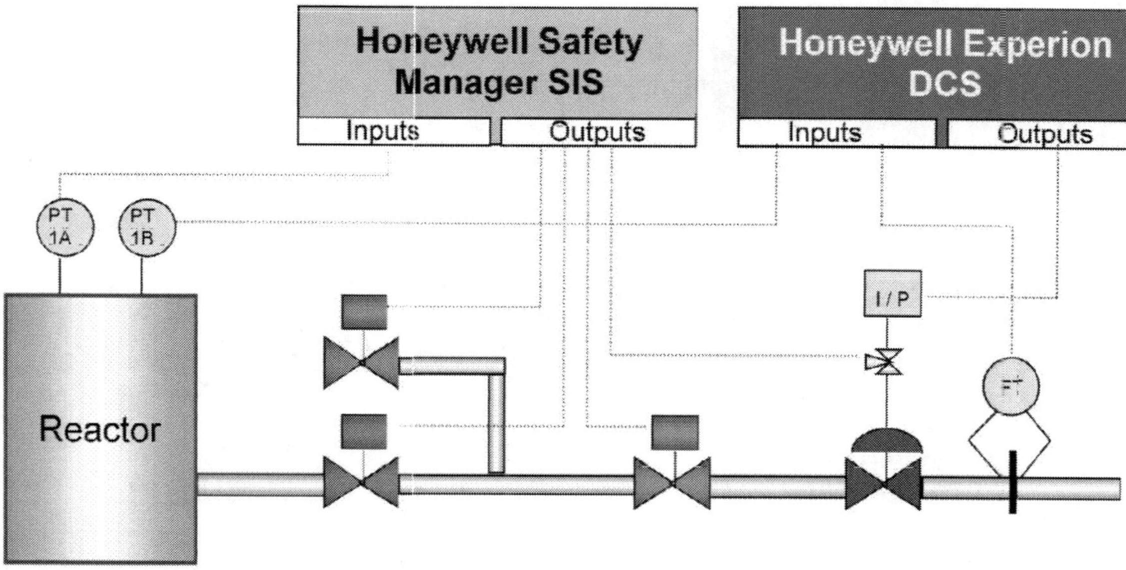

5 Case Studies

Below are example cases that require SIS.

5.1 Possible runaway reaction

Scenario: Reactant A reacts with catalyst in reactor in an exothermic reaction. Agitation is required for this reaction to ensure that (unreacted) Reactant A does not accumulate. Pressure relief devices can often be used as one of the Independent Protection Layers (IPL).

SIS Example: Reactor

Cause: What if the agitator of reactor stops during normal operation?

Consequence: Reaction stops due to poor mixing. If Reactant A continues to be fed into the reactor, it can accumulate, and a hazardous condition could result. The pressure relief IPL provides protection but it might not be enough for all cases.

Solution/Recommendation:

1) Require low current alarm for the agitator of reactor and operator to take prompt action and/or provide a DCS-based interlock to stop the feed of Reactant A
2) Apply an independent SIS to prevent overpressure. Use low, low speed of agitator interlock to stop and prevent continuing feed of Reactant A into the reactor. The required SIL-rating would be set by the LOPA team depending on the details of the scenario.

5.2 High Flow Steam Supply System

Scenario: Steam heat is connected to water bath via a temperature control valve to heat up a reactant B ton cylinder for vaporization. All such BPCS control loops have fault modes that could possibly result in overheat of the ton reactant B cylinder.

SIS Example: Reactant B Cylinder Vaporization

Cause: What if temperature control valve fails or sticks open (one possible BPCS control loop failure)?

Consequence: Overheat the cylinder creating higher reactant B vaporization rate and thus higher pressure of Reactant B. The maximum water bath temperature is controlled to 38 °C, but could theoretically reach 100 °C. The fusible plug on the cylinder is designed to melt at 70 °C, which would release reactant B， leading to a hazardous condition. Note that the fusible plugs are provided by the suppliers of reactant B as standard safeguards in order to protect the cylinder in case it is ever exposed to a fire as it is being transported.

Solution/Recommendation:

1) Provide a high temperature alarm external to BPCS that requires the operator to manually close the warm water block valve. The temperature measurement will be located in the water line before it comes into contact with the cylinder.
2) Consider adding a thermostatic valve (mechanical, no electricity involved) diverts hot water from the heating system to the basin on high temperature.
3) Provide a SIL-rated interlock using high water temperature input to close steam supply valve, all independent of the BPCS. Again, the required SIL-rating would be determined by the LOPA.

5.3 Thermal Oxidizer with quench tank

Scenario: The Thermal Oxidizer receives waste gas from the process units, and treats them by thermal oxidization, then sends them to downstream scrubber. Loss of cooling water can lead to high temperature and overpressure.

SIS Example: Thermal Oxidizer with quench tank

Cause: What if there is a loss of lose cooling water for the cooling medium circulation stream (BPCS control loop failure)

Consequence: Temperature increases at the Quench tank and could cause high temperature and the possibility of eventual damage of the Quench Tank and the vapor line to the downstream Absorber which is made of Fiber-Reinforced Plastic (FRP). This could eventually lead to possible release of toxic materials.

Solution/Recommendation:

1) Apply SIS to mitigate high temperature. Using high high temperature alarm and interlock to shutoff thermal oxidizer. Again, the SIL-rating that is required would be determined during the LOPA.

6 Conclusions

A risk based design approach provides a practical and a cost effective approach to maximize safety using LOPA methodology to analyze selected scenarios identified during the PHA. The examples and associated discussion above prove LOPA is a practical and a cost effective approach to maximize safety. Meanwhile, this paper provides guidance on how to mitigate overpressure scenarios with SIS using LOPA. This approach has many potential applications beyond the examples discussed in this paper. In addition, it also illustrate that Honeywell Safety Manager® can provides a basis for safety unification, reducing risks and installed costs, and improving safety while increasing uptime. Using a Honeywell Safety Manager® as the logic solver of the SIS helps in ease of integration with the plant BPCS as well.

7 References

[1] IEC 61508 Functional safety of electrical/electronic/programmable electronic safety-related systems, Edition 2.0, 2010.

[2] IEC 61511 Functional safety–Safety instrumented systems for the process industry sector, first edition, 2003.

[3] Application of Safety Instrumented Systems for the Process Industries, ISA Standard, ISA-S84.01-1996, ISA TR-84.02 and S84.01-2004.

[4] Scott A. Tipler, "Practical Examples of Alternate Overpressure Protection Systems An Owner's Perspective", CCPS 9th Global Congress on Process Safety, San Antonio, April 2013.

[5] Prophet, N. and G. Gorski, "The Importance of a Team-Based Approach to Overpressure Scenario Development", CCPS 9th Global Congress on Process Safety, San Antonio, April 2013.

[6] API RP 581 Risk-Based Inspection Technology.

Safety condition proof research of large-scale LNG storage tank group

Kong Linghai1 , Tian Lianjun1 , Wang Xiaoyuan2
(1 Guangdong Zhuhai Golden Bay LNG Co., Ltd. Zhuhai 519090 , Guangdong)
tianlianjun@gdzhlng.com

(2 Qingdao Oasis Environment & Safety Technology Co.,Ltd. , Qingdao 266580 , Shandong)
Key words:large LNG storage tank ; separation distance ; NFPA 59A-2009 ; quantitative risk; ; risk criteria

Abstract

With the growing demand for LNG, Zhuhai LNG terminal project plans to construct a large LNG storage tank group. In order to ensure its safety, the rationality of layout design is analyzed respectively from storage tank fire separation and tank farm quantitative risk analysis.According to NFPA 59A-2009 standard, using PHAST software to simulate accident to determine storage tank separation distance, it is considered that the current tank spacing should be further adjusted according to scope of accident; DNV SAFETI software is applied to conduct quantitative risk analysis for tank top pool fire and pipeline leakage jet fire, and receives personal and social risk chart for tank farm, it is thought that tank layout planning is feasible though compare with risk standard, also can take safety measures to make risk level as low as possible considering factors of costs and benefits.

1 Introduction

Natural gas as a kind of high quality, efficient and clean energy is widely used[1], its social demand is surging. Zhuhai LNG receiving terminal project plans on completing LNG tank group of 16-27 million cubic volumes. Since the design size is huge, beyond the scope of conventional design specification, safe condition argument should be carried out for tank farm to ensure its safety. The demonstration includes safety and rationalanalysis of tank fire separation and tank farm quantitative risk. On the one hand, it's able to provide effective basic for planning and design of large LNG tank group. On the other hand, it also can drastically ensure the safety of LNG terminal and surrounding area.

2 LNG storage tank farm overview

Guang dong zhuhai LNG (Liquefied Natural Gas, referred LNG) receiving terminal plans on constructing tank farm, which selects full containment concrete storage tank in tank farm with better safety ,technical and economic performance and comprehensive performance[2].

15 storage tanks are planned on constructing in LNG tank farm, symmetrically divided in two rows, the first row consists of 7 tanks, and second row consists of 8 tanks. LNG tube rows are arranged in the middle of two rows of storage tanks, all tanks input / output lines are fixed vertically along the tank wall and are connected with the intermediate tube row, specific layout planning and the tank spacing is shown in figure 1, which No.1,2,3 are 16×104m3 tanks, No.4,5,6 are 20×104m3 tanks, the rest are 27×104m3 tanks, inner tank diameters respectively are 80m, 84m and 90m, the concrete outer tank diameter respectively are 82.4m, 86.4m and 92.4m, the outer tank diameter respectively are 84m, 88m, 94m[3~4]. LNG storage tank operating temperature is - 162 ℃, the operating pressure is 0.025 MPa; Discharge input pipeline diameter is 40 "(1016 mm), the pressure is 0.4 Mpa.

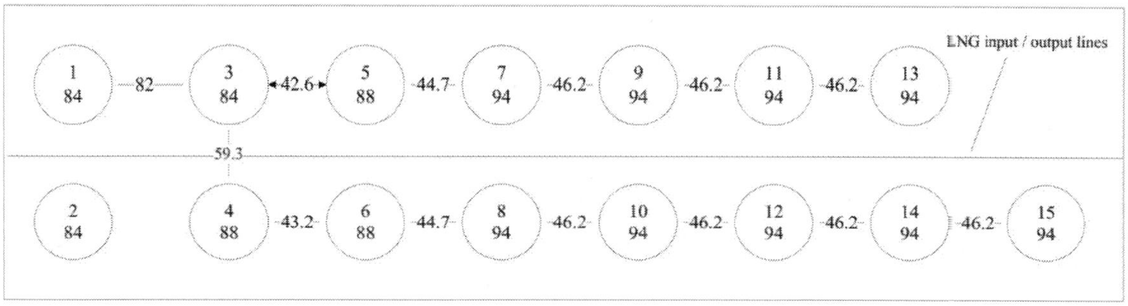

Fig.1 Sketch map of LNG tank layout

3 Analysis of LNG storage tank fire separation

3.1 Method for determiningtank fire separation

3.1.1 Basis for determiningtank fire separation

The determination of LNG storage tank fire separation is mainly on the basis of the relevant domestic and international execution and the related specification on LNG station. In general, the main criteria are used at home and abroad: National Fire Protection Association standard, NFPA 59A-2009 "LNG production, storage and shipment "; European standard, EN1473-2007 "LNG equipment and installation: continental device design ";Japanese standardand norm, high pressure gas Security Law "provisions of liquefied natural gas equipment (Note 1)";Chinese standards and norms, GB50183-2004 "code for fire protection design of petroleum and natural gas engineering (chapter 10 LNG station) ";GB/T 20368-2006 "LNG production, storage and shipment", GB/T 22724-2008 "Installation and equipment for liquefied natural gas onshore design", etc[5].

NFPA 59A standard for determination of safety distance only aims at two types of risk that are the flame thermal radiation damage caused by the largest flame (pool fire) of storage tank or impounding area andLNG vapor cloud spread. EN1473-2007standardnot approves only "safe distance", but requires to do risk assessment to evaluate whether the LNG site design is acceptable [6]. Japanese standard for determination of safety distance is based on the intensity of the shock wave blasting; this idea is extended from LPG safety regulations.Domestic standards, GB50183-2004 is suitable for determination of tank safety distance that capacity is less than 100000 m3 storage tank [7], GB/T 20368-2006 and GB/T 22724-2008 respectivelyequal to NFPA 59-2001 and EN1473-1997[8], domestic standards almost refer to foreign standards.

3.1.2 Content for determiningtank fire separation

At present, NFPA 59A is widely used on the international in the process of production, storage andtransportation and installation. Therefore, the paper refers tothe content of NFPA 59A-2009 that the latest version of standard to determine storage tank fire separation.

Section 5.3.4.3 of NFPA 59A-2009 specifies the minimum tank fire spacing. The standard rules tank fire spacing should not be less than 1/4 of the sum of adjacent tank diameter for storage tank thattank volume is more than 265 m3[9].

Section 5.3.4.3 of NFPA 59A-2009 rules tank fire spacing shall be determined in accordance with the influence of tank pool fire accident for full containment or double containment tank, and according to one hundred percent of tank volume to calculate (impounding area of full containment or double containment tank is itself). When this accident scene is simulated, 30 kW/m2 should be as thermal radiation threshold, which is the maximum radiation does limit value that concrete buildings can reinforce near fire at the property line in the impoundment area. Meanwhile, the standard sets forth in either active of passive measures should be considered to reduce accident consequences [9].

3.2 Tank fire separation analysis

According to the standard content of NPFA 59A-2009, PHAST of DNV which simulates accident consequence isselected to determine tank fire separation. PHAST is dedicated to safety calculation in petrochemical and natural gas fields [10], and the software adopts solid flame model that NFPA 59A-2009recommends, whichfurthestmake a best fit between computation and experimental data.

3.2.1 Simulation for accident consequence

According to the actual circumstance in tank farm, No.1 to 3 tanks have been in the stage of construction, and the paper should pay much attention to the rest of 12 tanks. Thus, tank roof pool accident is as accident scene; respectively simulate accident consequence of20×104m3 and 27×104m3 tanksto determine accident influent. The scene composition and various input parameters are as ①tank volume :20×104m3, 27×104m3;②liquidpooldiameter:84m, 90m;③wind speed: 4.5m/s④atmospheric stability :D;

The result of tank roof pool fire accident is predicted withthermal emission standard as 30kw/m2 that tank can withstand, Forecast result can be seen in tabale1 and fig.2-3. As can be seen from the result analysis, the accident influence scope of 20×104m3 and27×104m3 storage tanks respectively are 165.4m and 174.6m.

Table1. Tank roof pool fire accident forecast results

Research object	thermal radiation value(kW/m2)	distance of thermal radiation (m)	area of affection(m2)
tank of$20\times10^4m^3$	30	165.4	44353.76
tank of$27\times10^4m^3$	30	174.6	49892.94

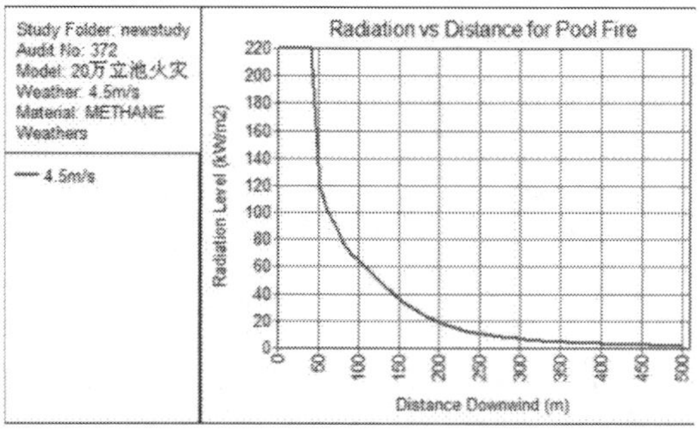

Fig.2 Radiation vs Distance for pool fire of 20×104m3 storage tank

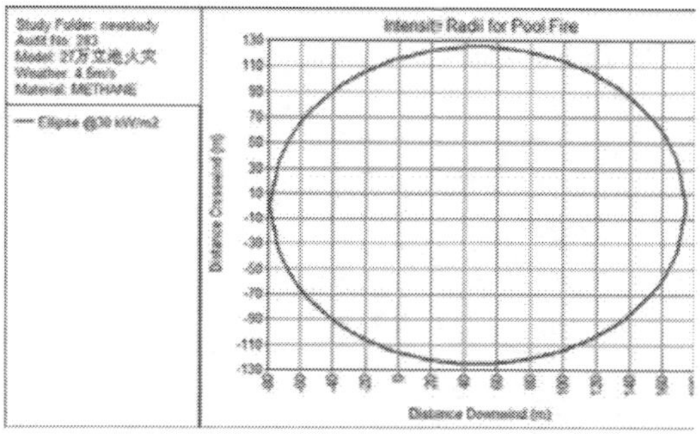

Fig.3 Intersity Radii for pool fire of 27×104m3 storage tank

3.2.2 Safety compensating coefficient

According to the requirement for tank fire separation of NFPA 59A-2009,positive and passive measures should be taken into consideration on the reduction function of accident influence scope treated tank roof pool fire as accident scenario to determine accident effect. So in order to reflect the weaken effect of passive or positive measures, safety compensating coefficient system should be established, which the size of safety compensating coefficient depends on the control of positive or passive measures on potential dangerous situation .Because NFPA 59A - 2009 is not clear the quantitative relation between the active or passive measures and fire separation spacing. Therefore, Dow chemical for security compensation coefficient values is mainly use for referencein the establishment of this system. According to the actual circumstance of tank farm, the compensation coefficient of LNG tank area is determined, the detail process and consequence can be seen in table 2[11].

Table2 . LNG storage tank farm safety compensation coefficient

1、Process control safety compensation coefficient

Item	Range of compensation coefficient	Using compensation coefficient	Item	Range of Compensation coefficient	Using compensation coefficient
1)emergency powersupply	0.98	0.98	6) inert gas shielding	0.94~0.96	

2)coolinginstallation	0.97~0.99		7) operating instruction	0.91~0.99	0.95
3) Explosion suppression device	0.84~0.98		8) chemical activity inspection	0.91~0.98	0.98
4)emergency cut-off device	0.96~0.99	0.96	9) other process hazard analysis	0.91~0.98	0.96
5) computer control	0.93~0.99	0.93			

C1value：0.78

2、physical isolation security compensation coefficient

Item	Range of Compensation coefficient	Using compensation coefficient	Item	Range of Compensation coefficient	Using compensation coefficient
1) remote-controlled valve	0.96~0.98	0.98	3) drain-off system	0.91~0.97	
2) Discharge/emptying device	0.96~0.98	0.98	4) interlocking device	0.98	0.98
C2 value:0.94					

3、Fire prevention measures to safety compensation coefficient

Item	Range of Compensation coefficient	Using compensation coefficient	Item	Range of Compensation coefficient	Using compensation coefficient
1)Leak detection device	0.94~0.98	0.94	6) water curtain	0.97~0.98	
2)steel structure	0.95~0.98	0.98	7) foam installation to fight fire	0.92~0.97	0.94
3)Fire water supply system	0.94~0.97	0.94	8) Portable fire extinguisher/spray gun	0.93~0.98	0.95
4)Special fire fighting system	0.91		9) cable protection	0.94~0.98	0.98
5)water sprinkler system	0.74~0.97	0.97			
C3 value :0.75					
Safety compensating coefficient C= C1×C2×C3=0.54					

Notion 1: Functions of fire prevention measures fit for the relevant standard specification

3.2.3 Fire separation distance analysis

Combined accident consequences with safety compensating coefficient that is established, the range of 20×104m3storagetank roof pool fire narrows to 165.4×0.54=89.3mwith the safety measures; the range of 20×104m3storagetank roof pool fire narrows to 174.6×0.54=94.3m. But distance of tank roof pool fire is from the center of the liquid pool, the fire separation distance of tank 20×104m3 and 27×104m3 can be respectively 45.3m and 47.3m.(the tank diameter respectively is 88m and 94m)considered the effect of storage tank diameter for accident.

When two adjacent tanks contain different volumes, the fire separation spacing should determined by the larger one. Therefore, all fire separation spacing acquired and compared with the layout planning, the result is showed in Table 4. Tank fire spacing in the same row should be adjusted by the content of Table 3. For the tank fire spacing about adjacent row, the planning space is 59.3m;the tank fire spacing is respectively45.3m, 47.3mbased on accident consequence. So the planning space should ensure the effect of tank roof pool fire on adjacent tank, enterprises can depend on adjust actual situation adjust adjacent row spacing.

Table3. Tank fire separation compared to the result analysis

tank fire spacing in the same row				
Adjacent tank type	**16/20**	**20/20**	**20/27**	**27/27**
Based on accident consequences (m)	45.3	45.3	47.3	47.3
Layout planning (m)	42.6	43.2	44.7	46.2
Relevant suggestion	spacing increases 2.7m	spacing increases 2.1m	spacing increases2.6m	spacing increases 1.3m

4 Quantitative risk analysis of LNG tank farm

4.1 Risk assessment standards

Individual risk is commonly used risk contour to indicate, the standard is established combined historical statistics of petrochemical industry with foreign authorities enforced standards.Literature[12] statisticsindustrial injury accident mortality in 10 developing countries is 1.3×10-4; China petroleum, China petrochemical in the reports showed that enterprise accident mortality from 2004 to 2008 respectivelywere2.4×10-5, 1.1×10-5. Foreign authority enforces acceptable value ofmaximum risk is 1×10-5or 1×10-4. American standard ISA - S84.01 recommendsindividual risk is 10-6~10-4[13]. Therefore, the paper recommends maximum acceptable risk is 1×10-,4, negligible risk is 1×10-6, individual risk between 10-6 to 10-4should use ALARP (As Low As Reasonable Practice) principle[14], which is as low as possible under the reasonably practicable condition.

Social risk uses F/N curve to indicate [15], the standard also learns from foreign social risk criteria for analysis. Table 5 is international and regional social risk acceptance criteria. National standards are compared, and select the most stringent standardas social risk of LNG receiving terminal. Dutch standard is most stringent combined with social risk limit and slope. Therefore, the Dutch standard is selected as social risk standard of LNG receiving terminal.

Table4 . International and regional social risk acceptance criteria[16~17]

Country	Social risk higher limit	Social risk lower limit	Slope	Application object
Hong Kong	10-3	10-5	-1	Dangerous device
UK	10-2	10-4	-1	Dangerous device
Denmark	10-2	10-4	-2	Dangerous device
Netherlands	10-3	10-5	-2	Dangerous device

4.2 Risk identification of tank farm

According to the hazard analysis of LNG tank farm, the main causes of fire and explosion accidents are leakage of tanks or attached pipelines [18]. The import or export pipelines are easily damaged when they suffer impacts from outside, vibration or the inherent defects of materials. The Purple Book published by Dutch CPR recommends the probability of failures about different failure patterns. There are two different kinds of failure patterns, one is piperupture, the other is pipe leakage which means that the materials flow from a hole whose effective diameter is 10% of its nominal diameter, but less than 50mm. As for those pipe that the diameter are more than 150mm, the probability of rupture is 1×10-7m-1y-1, and the leakage probability is 5×10-7 m-1y-1。

The main fire types of LNG leaking are jet fire and pool fire. But taking the drainage design of pipes into consideration, it is not easy to form a liquid pool for the LNG will be introduced into collection pool, so jet fire is the significant kind of accidents. This paper chose the roof pool fire, rupture completely of LNG unloading pipeline and jet fire as the main accident models.

4.3 Quantitative risk analysis

SAFETI(DNV) is a versatile calculation software which is used to quantitative risk assessment. This software can calculate the scope and level of fire and explosion accidents of any kind of petrochemical plants. Results are displayed in graphic forms so as to assess the individual and social risks. In conclusion, SAFETI was selected to analysis the quantitative risk of LNG tank farm.

4.3.1 The input parameters

According to the results of risk identification, we input detailed parameters to calculate the quantitative risks of specialized occasions. On the basis of local conditions, we chose the environment data as following:①Atmospheric temperature：22.9℃；②Relative humidity：77%；③Wind speed：4.5m/s；④Atmospheric stability：D；⑤Solar radiation：1kW/m2；⑥Ground roughness：0.1m。

Table5 . The input parameters of Quantitative risk analysis

Accident pattern		Fire type	Phase of material	Mass of leakage (kg)	Frequency (m-1y-1)	Operating temperature (°C)	Operating pressure (MPa)
Roof failure		Pool fire	Liquid	——	1×10-8	-162	0.025
Unloading pipeline	Totally rupture	Jet fire	Liquid	53760	1×10-7	-162	0.4
	Leak from hole	Jet fire	Liquid	1302	5×10-7	-162	0.4

4.3.2 Results of risk calculation

On the basis of accident scenes, probability and environment data, this paper calculated the risk level of LNG tank farm. The individual and social risk contours were shown in Figure 4 and Figure 5. In figure 4, the innermost is 10-5 individual risk contour line.

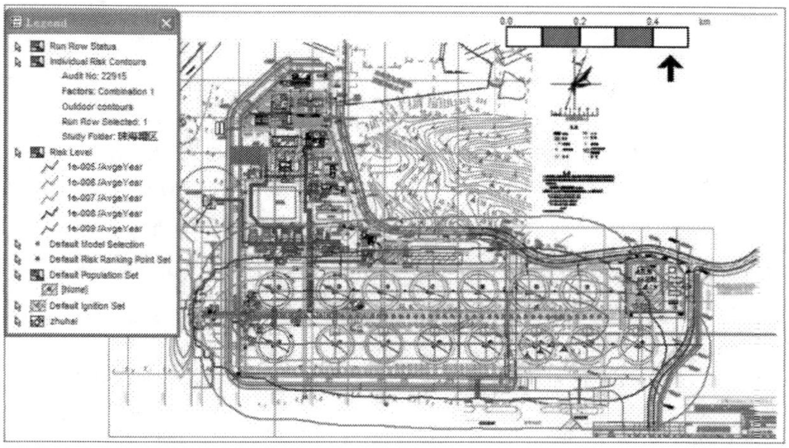

Fig.4 Individual risk contours of LNG tank farm

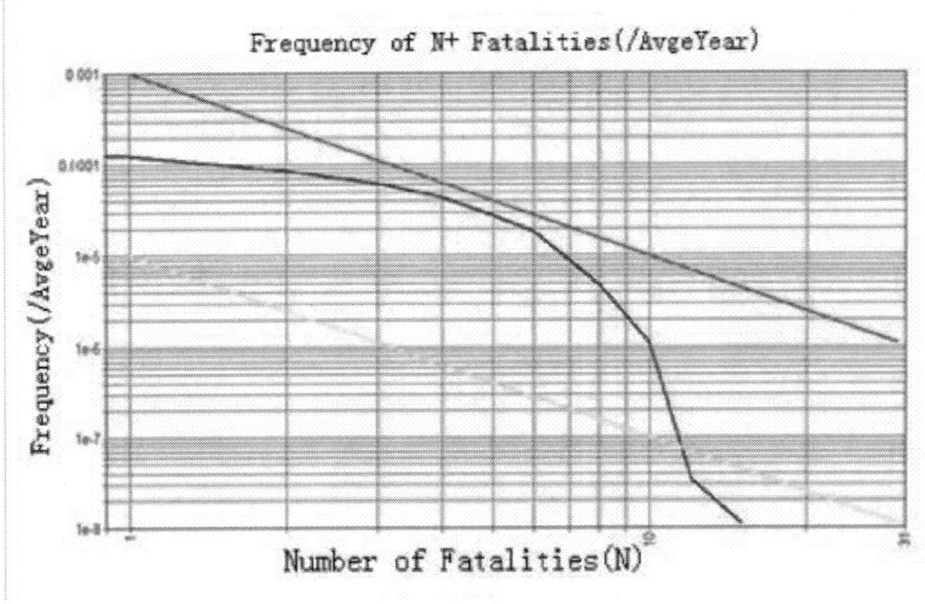

Fig. 5 F/N curve of LNG tank farm

4.3.3 Consequence analysis

Based on therisk criteria, the analysis of calculation results are as follows:

1) The frequency of 10-4/y is the upper limit of individual risk, there wasn't a 10-4 contour in figure 4 which means the individual risk level of tank farm was acceptable.

2) The covered area between 10-6/y and 10-4/y stays in the ALARP region. According to the simulation results and the correspondence relationship between venture capital and benefits, the managers could adopt more safety measures to low the risk level of tank farm.

3) The frequency of 10-6/y is the lower limit of individual risk. The individual risk of the area beyond 10-6/y was completely acceptable.

4) Based on the F/N curve, the effects of tank farm on station were also acceptable. The social risk curve lay between the upper limit and lower limit.According to ALARP, tooking the venture capital and benefits into consideration, the managers could take more risk-reduced measures to meet the principle of "as low as reasonable possible".

5 Conclusion

Take the Zhuhai LNG tank group for example: this paper demonstrated the safety condition of large LNG storage tank group. Through the security analysis of fire separation and QRA of tanks, the paper concluded that the design of Zhuhai LNG storage tanks was feasible. But we should still pay more attention to the following aspects:

The fire separations between the adjacent tanks in the same row need increase in accordance with the analysis above.

The individual risk of tank farm stayed in the region of ALARP, managers should adopt more measures to increase the safety conditions.

The social risk of tank group lay in the acceptable area. Considering the composite factors of costs and benefits, the risk level of this tank group was as low as reasonable possible.

Reference

[1] LuoXiaoling, Qi Changyong, Cheng Huanxin. Large-scale liquid natural gas storage tank development research[J]. Machinery Design & Manufacture, 2009, (9):255-257

[2] Liu Li. The design technology of receiving station[J]. Petroleum and Chemical Construction.2005,27(4):8-11

[3] Young-myung Yang, Ji-hoon Kim, Heung-seokSeo. Development of the World's Largest above-ground Full Containment LNG Storage Tank[A]. 23rd World Gas Conference[C], Amsterdam, 2006

[4] Young-myung Yang. Latest tank design for the world's largest 270,000m3 full containment LNG tank[A]. International Gas Union Research Conference[C]. Seoul, 2011

[5] WU Ce- yu , DANG Wen- yi , LIU Xiao- long. Study on the safe distance for siting of LNG receiving terminal[J]. Journal of Safety Science and Technology.2012, 8(6): 167-172

[6] HUANGPU Li-xia, GUO Kai-hua. Current issues anddevelopment for LNG plant safety standards[J]. CIESC Journal.2009, 60:11-15

[7] GB 50183-2004 , Fire prevention design code of petroleum and gas[S].Beijing: China Planning Press. 2004

[8] GB/T 20368-2006 , The production, storage and transport of LNG [S].Beijing: General Administration of Quality Supervision, Inspection and Quarantine of the People's Republic of China, 2006

[9] NFPA 59A (2009).Standard for the Production, Storage, and Handling of Liquefied Natural

[10] ZHU Bo-ling ,YU Xiao-chun. Analysis of the Dangerous Area Caused by Hole Leak andDispersion in LNG Tank[J]. Safety and Environmental Engineering.2009,16 (3) : 104-108

[11] Wang Xian-zheng, Zhao Tie-chui. Safety Assessment[M]. Beijing: China Coal Industry Press, 2004

[12] Luo Yun, Huang Yi. China Strategy of Safety production-five factors of safetyproduction [M].Beijing: Chemical Industry Press, 2005

[13] Ma Ke-wei. Research on quantitative area risk assessment based ondomino effect model[D].Zhejiang University of Technology. 2010

[14] CHEN Xiao-dong. Study on theAppropriate Risk Capacity of Chemical Industry Park[J]. China Safety Science Journal.2009,19(5):132-137

[15] Wu Zhi-feng, Dang Wen-yi, Yu An-Feng.Study on QRA routine of petrochemical company[J]. Safety Environment &Health, 2010,10(10): 35-38

[16] Tang Qiu-xiang. Study on Regional Quantitative Assessment Method forPetrochemical Industry[D]. Qingdao, China University of Petroleum(East China), 2010

[17] GAO Jian-ming, WANG Xi-kui, ZENGMing-rong.The research development and indication of the acceptablestandard of individual risk and social risk[J]. Journal of Safety Science and Technology, 2007,3（3）：29-34

[18] Liu Ruo-zhu. Research on risk Assessmentand controlor LNG storagesystem[J]. Shanghai. East China University of Science and Technology, 2011

[19] Advisory Council on Dangerous Substance. Guideline for quantitative risk assessment CPR 18E, 2005

[20] Liang Tao, Chen Guo-hua, Zhang Rui-hua. The application of SAFETI in assessment of storage tank accidents [J].Oil & Gas Storage and Transportation, 2006,25(2): 53-58

1ST CCPS ASIA-PACIFIC CONFERENCE ON PROCESS SAFETY℠

Risk Based Decision Making APCI Process Safety QRA and Risk Management Process Introduction

Xiaoming Zeng
Air Products and Chemicals, Inc.
No. 88, Lane 887, Zu Chong Zhi Rd, Zhanghjiang Hi-Tech Park, Shanghai, 201203
zengx@airproducts.com

Jenny Wang
Air Products and Chemicals, Inc.
(address same as above)

Prepared for Presentation at
American Institute of Chemical Engineers

1st CCPS Asia-Pacific Congress on Process Safety
Qingdao, China
September 4 - 5, 2013

Risk Based Decision Making

APCI Process Safety QRA and

Risk Management Process Introduction

Xiaoming Zeng
Air Products and Chemicals, Inc.
No. 88, Lane 887, Zu Chong Zhi Rd, Zhanghjiang Hi-Tech Park, Shanghai, 201203
zengx@airproducts.com

Jenny Wang
Air Products and Chemicals, Inc.

Keywords: risk based, decision making, process safety, quantified risk analysis, QRA, APCI, risk criteria, risk management process, LOPA, risk tolerance

Abstract

Risk Based Process Safety (RBPS) has become a trend for process safety improvement in many chemical organizations. With reference to the CCPS publication, APCI representative introduces the overview on Why, What and How risk based decision making is practiced in the corporation. The corporate Risk Management Process, Quantified Risk Analysis, Risk Analysis methods and tools are presented for experience sharing in the conference. The presentation include some contents on local regulation developments in terms of risk based decision, like China's risk target – how it compares to other countries and organizations. APCI risk management process is one of the four pillars that support risk based process safety in the corporate – culture, management system, work process and technical competency.

1. Introduction

Process Safety awareness is improving in Asia Pacific region esp. China in the recent decade. With more and more public concerns on chemical industry for safety and environmental protection, manufacturers in this region are seeking all efforts to improve their process safety performance as critical means of sustainability growth. They are not only satisfied with meeting the minimum legal requirements. However, the companies have to balance with cost and be economically competitive. Air Products and Chemicals, Inc. has many years of experience on process safety risk analysis and management process. Risk based Process Safety Management has become an integrated part of the investment and business activities.

Air Products and Chemicals, Inc. (APCI) is an internationally recognized supplier of industrial gases and related industrial process equipment, intermediate and specialty chemicals, and environmental and energy system in the variety of medical and specialty gases. The company also designs, manufactures, and operates cryogenic and non-cryogenic equipment for air separation, natural gas liquefaction, hydrogen purification, and other gas processing applications.

Air Products' chemical businesses produce specialty chemicals such as polyvinyl alcohol, surfactants; performance additives for polyurethane, epoxy, and other applications; and polyurethane and industrial chemical intermediates.

2. Learning from BP Texas incident

A massive explosion kills 15 and injures 180 at the BP Texas City refinery in 2005. OSHA fined BP totaled over $21 million – majority labeled egregious willful fines. CSB published an animated video of the incident. Afterwards, CSB issued emergency recommendation to API.

When we look at the Technical Causes published by CSB, we may list a few key areas for lesson learning:

- Operators did not fully understand Raffinate Splitter Tower operation and did not recognize:
    - Material fed to column but did not exit
    - Level exceeded safe limits; level device beyond effective range
- Level instrumentation in blow-down tank not working; not repaired prior to startup
- Startup procedures not fully followed
    - Critical valve not opened during startup
- Temporary work trailers were located 100 to 150 feet away from the vent stack. Workers who were in and around these trailers were killed or injured when the explosions occurred.
- *Hazard/Risk Assessments* – Siting of trailers and safety systems not properly evaluated
- Atmospheric venting hazard - BP chose not to change design
- And many more …

We may ask ourselves questions to assess how good we perform on process safety:
Are we taking chances that would not have been tolerated before?
Is our risk assessment process effective? Are we addressing our risks on a timely basis?
If the safety of an activity is in question, where dose the burden of proof lie? Do we require proof of safety to permit the activity to continue… or proof of danger to prevent the activity?
Do our employees feel empowered to share their safety concerns freely with management? Do we "shoot the messenger" if we disagree with the message?
Do we frequently see business and customer pressures at odds with safety requirements?

Besides the culture impact, do we often feel the difficulty to make decisions based on limited understanding of risks?

3. Risk Based Process Safety (RBPS)[1] Overview

3.1 What is risk?

Risk is defined as

$$\text{Risk} = f(\text{HAZARD, CONSEQUENCE, FREQUENCY}) \quad [\text{Eq.1}]$$

When we talked about risk, we need to describe how bad the things can go wrong and how likely the bad consequence can happen.

3.2 Why are we doing Risk Based Process Safety Decision Making?

After an initial surge of activity, process safety management activities appear to have stagnated within many organizations. Incident investigations continue to identify inadequate management system performance as a key contributor to the incident. And audits reveal a history of repeat findings indicating chronic problems whose symptoms are fixed again and again without effectively addressing the technical and cultural root causes.

This problem can be reflected with following findings:

- Lack thorough understanding RAGAGEP (recognized and generally accepted good engineering practices) and inconsistent interpreting and applying them.
- Management systems are overemphasized while the technical aspects of process safety, which actually control the hazards and manage risk, are neglected.
- Management does not understand or apply risk-based decision processes

A risk-based approach reduces resources to managing lower-risks, thereby freeing up resources for higher-risks. It would help to handle increasing demand from Continuous Improvement activities and business pressure. It addresses the increasing public awareness and the Process Safety professionals' workload concern.

The relation of risk and cost can be illustrated by following correlation graph:

Figure 1

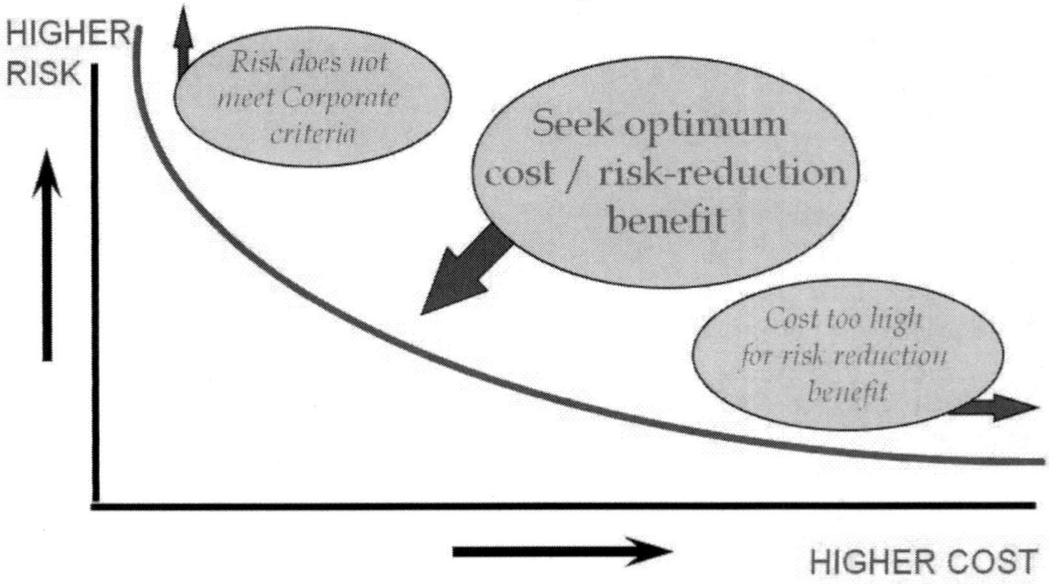

This may serve as consideration for setting up corporate risk criteria. From views of major injury frequencies, the criteria can be explained as an example below:

Figure 2

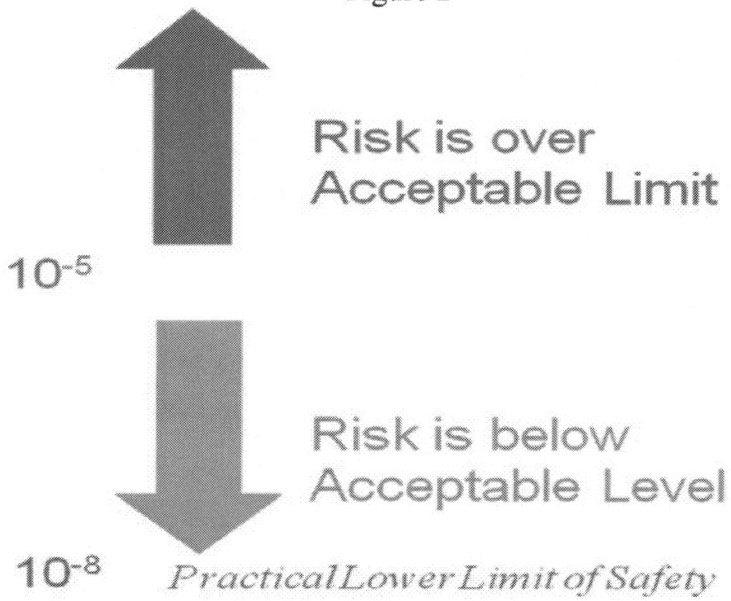

Evolution of Process Safety and Accident/Loss Prevention Strategies shows the industry is seeing organizational behavior changes from passive to active and rational:

Figure 3

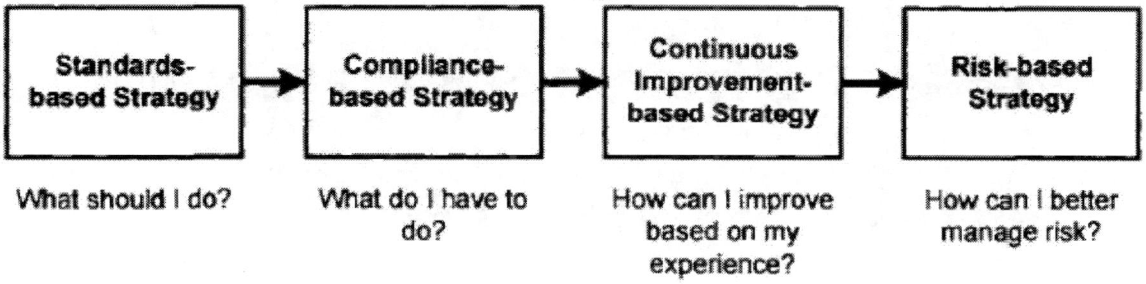

China is a typical world factory. It gains economic growth but in the same time sacrifices with many major process safety incidents and environmental pollutions. It is not a surprise to see process safety regulations are being developed. For example, the China PSM regulation[2] was implemented on May 1, 2011. It includes 12 elements and requires PHA and audit at 3-year basis. National Regulation[3] on major hazardous chemicals requires PSM documentation, quantitative risk assessment and some specific requirements such as emergency shutoff devices shall be installed for those toxic gases, liquids and flammable gases systems. Emergency spill treatment systems shall be installed for toxic gases. SIS shall be considered for high risk processes. Highly toxic materials shall be monitored with CCTV and connected with Public Security office. Etc.

China first setup the process safety criteria for public risk and individual risk:

Table 1 Individual Risk Criteria

Key targets and sensitive populations	Risk Criteria (/year)
Schools, hospitals, nurseries, Military setups, government offices, Stadiums, open markets, transportation stations...	$<3\times10^{-7}$
Residential areas, hotels, Malls, office buildings, recreations, parks...	$<1\times10^{-6}$

Social risk:

Upper curve: $FN=e^{-3}$ [Eq. 2]
Lower curve: $FN=e^{-5}$ [Eq. 3]

Figure 4 Social Risk Criteria

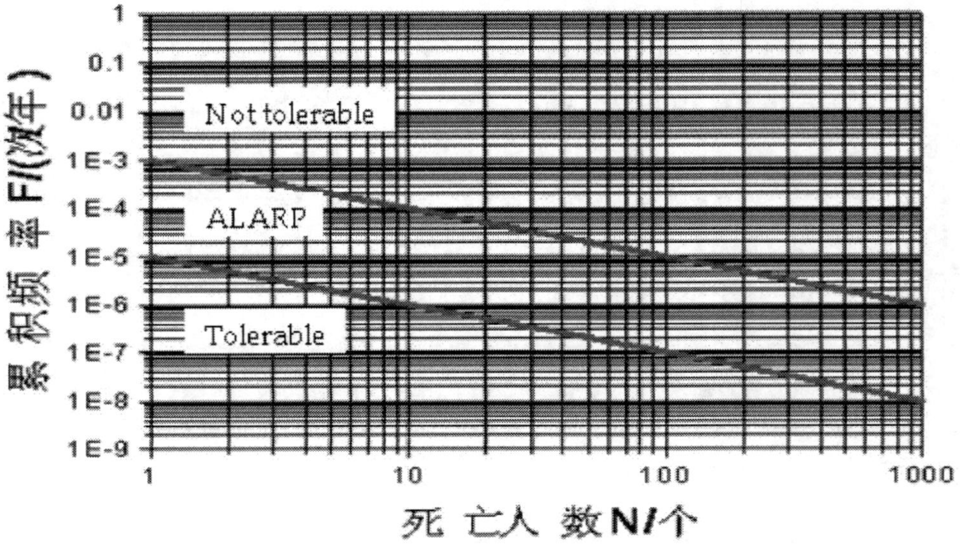

3.3 *What can we do Risk Based Process Safety Decision Making?*

This approach encourages companies whether novices or veterans on PSM to:

- Evolve their accident prevention from a compliance-based to a risk-based strategy.
- Integrate the process safety into business processes.
- Decrease unnecessary PSM work, based on risk judgments. Perform PSM activities more efficiently.

3.4 *How can we do Risk Based Process Safety Decision Making?*

The hazards and level of risk associated with the facilities or operations should be the primary consideration in the design and improvement of PSM activities. The demand for the activity, resources needed, and embedded safety culture also influence design and improvement decisions.

The main emphasis is to put just enough energy into each activity to meet the anticipated needs for that activity. Risk-based process safety criteria shall be based on the understanding of risk, resource demand and culture impact. It is necessary to measure the effectiveness of performance and efficiency. The CCPS guideline book[1] describes the 4 accident prevention pillars – Commit, understand risk, manage risk and learn from experience.

4. APCI Risk Based Process Safety (RBPS) Practices

4.1 Risk Management Process

In APCI, the so called Tiered Risk Review Work Process is triggered by various operational and business activities, internal and external events. Process Safety risk as related with public risk, on-site risk and transportation risk along with other EH&S risks are managed within the risk management process system. Activities of any potential deviation from corporate criteria have to be hazard identified, risk assessed and then reviewed in tiered management groups according to the magnitude of risks. For example, if a functional / business level review is triggered, the process safety engineer (risk analyst) shall call a review meeting when he / she finished risk assessment and quality reviewed by peers. The review team includes process safety functional leaders, operations and/or other functional leaders and business leaders. The process safety engineer will present the methodology of risk assessment and get agreement of quality analysis meaning adequacy of analytical methods and tools and data processing. The team also review the risk reduction recommendations and determine if the presented risks are tolerable and manageable before and after recommendations. Decision will be made at the end of the review meeting to continue the business activity/project, halt the business activity/project, or elevate the issue to the EHS Risk Council if no practical or agreed recommendations are able to bring the risk down to corporate criteria. If the risk review is elevated, the business leader will present the case to the Corporate Risk Council for final decision. The risk council is consist of (mainly) Global VPs of EHS (Risk Council Chair), Supply Chain, Engineering & Manufacturing, Operations, legal, technical fellow and Chief Risk Officer. The outcome of risk council review can be withdraw of the business proposal.

4.2 Process Safety Culture

APCI imbeds process safety in the design and operations. The engineering standards and EH&S policies are based on risk assessment result to meet corporate risk criteria. The performance is measured by Process Safety leading and legging KPIs. The employees including business leaders are trained for process hazards and risk management process. The organization has commitment to process safety and learn from process safety incidents and near-misses which happen within the company and the industry.

4.3 Process Safety Management System

The process safety requirements are put in deferent levels of document system from top policy level down to instructions level.

Figure 5

Process Safety Standards and Procedures

These documents include: policies, procedures, work instructions, compliance documents, and engineering standards. These documents are housed in several repositories.

 	**Corporate Policies and Practices (Tier 1 Policies)** particular interest to process safety are: 25-0803 Process Safety 25-0804 Hazard Reviews 25-0807 EH&S Corporate Risk Management Policy
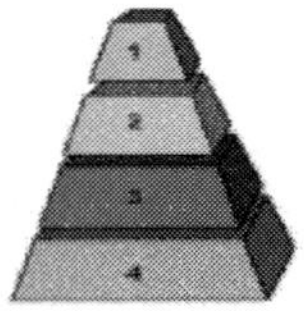 	**EH&S Management System** The EH&S Management System is a four-tiered collection of documents which define the company's EH&S policies, values, system elements, and objectives, along with standards and procedures necessary to comply with those policies and meet the stated objectives.
Engineering Standards	**Engineering Standards** The engineering standards set contains many standards related to process safety in three tiers: The tier 2 global standards (also known as the Basic Engineering Practices). The tier 3 global standards, and The tier 4 regional engineering standards.
 	**Process Safety Analysis Methods Manual (PSAMs)**
	**Process Safety Work Instructions** Process Safety Work Instructions provide guidance related to the "work process" (*who* does *what*, *when*, and general guidance on *how* and *why*) for various process safety tasks.
Global EH&S Manual	**Global EH&S Manual** **Section 9** of the manual covers Process Safety Management
 	**R&D EH&S Manual** Section 9 of the manual covers Process Safety Management
 	**Quality Management System (QMS)** **Engineering Procedures and Work Instructions** Several Tier II Engineering Procedures (EP's) in Engineering's Quality Management System are relevant to process safety.

4.4 Process Safety Technical Competency

APCI process safety technical competency covers capability to assess the following risks:

- stationary site public and employee risk
- LOX tank risk

- overpressure risk from vapor cloud explosion or pressure vessel failure (including cause of reactive chemistry)
- Toxic, asphyxiation and fire exposure risk from process vents and emergency relief vents
- risk from neighboring plants
- transportation public risk
- pipeline public risk

Figure 6 Spectrum of tools for risk-based decision making [4]

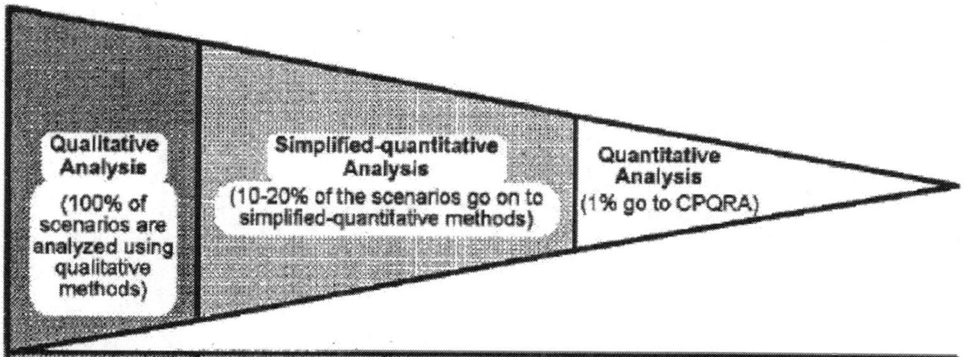

Techniques:	HAZOP What-if/Checklist FMEA	Quantified FMEA F&EI CEI	LOPA	Rough estimate with event tree	Event tree Fault tree HRA
Applicability to simple issues:	Good	Good	Good	Excessive	Excessive
Applicability to complex issues:	Poor	Poor	Fair	Fair	Good

Corporate process safety have developed series of methods (PSAMs) to support QRAs for purpose of consistency among risk analysts and keeping most update with the industry knowledge.

- Reference Values for Use in Consequence Analysis
- vent analysis for toxic, O2, asphyxiant, thermal radiation and enclosure releases
- LOX release
- analysis of explosions
- vulnerability calculation
- analysis of common mode failure
- human reliability assessment
- occupied bldg siting
- loss of containment frequencies
- population determination
- weather distribution
- transportation risk
- pipeline risk
- ignition probabilities
- Etc.

The standard material physical & toxic data, equipment and instrument failure rates are configured in tools database such as dispersion model software, Fault Tree Analysis and Layer of Protection Analysis programs.

5. Conclusion

APCI as a corporate member of CCPS has committed to implement RBPS in depth and wider range, more elements than PSM regulation are being adopted. The organization is continuing to improve understanding of RBPS at all levels, improve management efficiency, improve risk analysis technology and improve tools accuracy. In order to meet business fast pace, the Process Safety risk review process and risk analytical techniques are deem to be tiered approach: screening methods are simple, quicker and more conservative while in-depth quantitative methods are sophisticated, time consuming but less conservative.

Process Safety is manageable and science and technology based, like the saying "If we always do what we've always done, we'll always get what we've always got." - Stephen Covey (author of the 7 habits).

6. References

[1] CCPS Guidelines for Risk Based Process Safety

[2] *AQT 3034-2010 Guidelines for Process Safety Management of Petrochemical Corporations* (China PSM)

[3] *Measures on the safety supervision of hazardous chemical key dangerous sources*

[4] CCPS Layer of Protection Analysis

A Systematic Guideline for Designing an Inherently Healthier Process during Research and Development Stage

Santha Pandian[a]
Mimi Haryani Hassim[a]
Rafiziana M. Kasmani[a]
Markku Hurme[b]
[a]Universiti Teknologi Malaysia, Department of Chemical Engineering, 81310, UTM Johor Bahru, Johor, MALAYSIA
[b]Aalto University, School of Biotechnology and Chemical Technology, P.O. Box 16100, FIN-00076 Aalto, FINLAND

Corresponding email address: mimi@cheme.utm.my
emenda_27@hotmail.com
markku.hurme@iki.fi
rafiziana@petroleum.utm.my

Keywords: inherent occupational health; inherent safety; computer aided tool; process design; R&D.

Abstract

The number of people dying due to occupational related diseases is increasing each year. Therefore the effort to reduce, if not eliminate, the causes is very much in need. One of the factors that lead to this incident is due to exposure to harmful workplace conditions and chemicals especially in chemical process industries. In this paper, a systematic guideline for designing an inherently healthier chemical process during research and development stage is introduced. The proposed framework provides a guideline for the user to design a process with reduced, if not eliminated, health hazards that are commonly arise in a chemical plant.

1. Introduction

Occupational health is the promotion and maintenance of the highest degree of physical, mental and social well-being of workers in all occupations by preventing departures from health, controlling risks and the adaptation of work to people, and people to their jobs [1]. It is not widely introduced in chemical process design industry as compared to medical field. Recently, the awareness of occupational health issues has been gradually increased in chemical and petrochemical industry. This is because it is reported that there are more than two million people worldwide die due to work related diseases each year [2]. As compared to process safety, occupational health has always received much poor attention especially in chemical process design due to its more complex underlying principle.

Hassim [3] introduced inherent occupational health - a concept to prevent occupational health hazards that have potential to harm workers health by eliminating the use of hazardous chemicals, process conditions and operating procedures. Basically, this concept is derived from inherent safety, which was introduced by Trevor Kletz in 1970s. The ideology behind inherent safety is, when a process is said to be safe, it should be due to its nature and not due to add-on of safety devices [4].

In principle worker does not create hazards if the guidelines and the working environment are properly designed [5]. Therefore, it is always good to prevent the hazards by designing a safer workplace from the start. Also inherently safer plants are usually cheaper than the conventional ones because they do not need much additional protective equipment. Table 1 shows the six main principles defining inherent safety, which was formalized by Kletz (1991). The idea behind these principles is to reduce or eliminate hazards by modifying the design of the process itself.

There are many researches conducted to incorporate inherent safety in chemical process design, notably those by Edwards and Lawrence [6] (Prototype Index of Inherent Safety), Heikkila et.al [7] (Inherent Safety Index), Palaniappan et.al [8] (i-Safe), Gupta and Edwards [9] (Simple Graphical Method), Khan et.al [10] (Integrated Inherent Safety Index), and Leong and Shariff [11] (Inherent Safety Index Module).

The assessment of inherent occupational health and process design based on this concept however, are much more lacking compared to inherent safety. This is because work-related illnesses are typically as a result of long-term exposure; hence more difficult to be evaluated. Nevertheless, Hassim and Hurme [12] introduced an index called the Inherent Occupational Health Index (IOHI). This index is used to measure the health hazard level of a chemical processes during the research and development (R&D) stage.

Later, there was an effort by Pandian et.al [13] to develop a computer aided tool based on the IOHI method. The tool poses several advantages over the manual IHO method – it is more user friendly and swift in getting the results. Besides, having a computer aided tool can minimize the chances of human errors. This paper presents computer aided approach for desigining an inherently healthier process the R&D stage, which is demonstrated by the case study of methyl methacrylate production.

Table 1. General principles of inherent safety [5]

Principle	Definition
Intensification	Reduction of the inventories of hazardous materials
Substitution	Change of hazardous chemicals
Attenuation	Reduction of the volumes of hazardous materials required in the process. Reduction of operation hazards by changing the processing conditions to lower temperatures, pressures or flows
Limitation of effects	The facilities must be designed in order to minimize the effects of hazardous
Simplification	Avoidance of complexities such as multi-product or multi-unit operations, or congested pipe or unit settings
Error tolerance	Making equipment robust, processes that can bear upsets, reactors able to withstand unwanted reactions,etc.

2. Computer aided IOHI tool for occupational health assessment

As discussed in Section 1, Hassim and Hurme [12] developed the IOHI as a simple approach to develop an inherently healthier process. This method was designed carefully so that only those data available in the R&D stage is required for the health hazard assessment.
In the IOHI, there are two index which needs to be calculated; 1) Index for Physical and Process Hazards (I_{PPH}) that represents the possibility for workers being exposed to chemicals and some physical hazards and 2) Index for Health Hazards (I_{HH}) that characterizes the impacts on health due to the exposure. The IOHI for each process route is calculated by summing up these indexes.

$$\mathbf{I_{IOHI} = I_{PPH} + I_{HH}} \quad \text{[Eq. 1]}$$

The method is comprehensive since it was developed by considering all the problems of the previous methods. The only drawback is that it requires manual intervention from the users. It is very time-intensive as during the R&D stage, there is possibility that hundreds of routes need to be screened within a limited time. This feature is definitely unattractive to the industries and therefore Pandian et al. (2012) extended the work by Hassim and Hurme [12] by developing a computer aided tool based on the IOHI method. Besides due to the rapid growth of technology nowadays, such computer aided process engineering (CAPE) tool is very much in need.

4. Development of systematic framework for designing an inherently healthier process

The work by Pandian et al. [13] is mainly for assessing the level of inherent occupational health hazards during the R&D of a chemical process route. In this paper, a systematic framework for designing an inherently healthier process is proposed (presented in Figure 1), which requires an input on the hazard assessment results from the tool proposed by Pandian et al. [13]. The systematic design approach is proposed to reduce the health hazards in a chemical process based on the very limited data in the R&D stage.
As mentioned earlier, this design tool is used in parallel with the computer aided assessment tool. From Figure 1, it can be seen that the user will start to perform the assessment by specifying the operating parameters (i.e. temperature and pressure) as well as the chemical substances. Based on the input data, the user is now able to calculate the physical and process hazards subindex as well as the health hazards subindex automatically with the aid of the assessment tool. After the subindexes are calculated, the assessment tool will then proceed to calculate the IOHI. Once the IOHI value is calculated, the results are compared with the hazard level benchmark. If the hazard level is indicated as 'safe' or 'moderately safe', then the user can proceed with the design. Otherwise ('moderately hazardous' or 'hazardous' level), the user needs to modify the design and re-calculate the health hazards once again. This iteration process will continue until the lowest reasonably and practically achievable index value is reached. Of course ideally, the process improvement should continue until the hazard level reaches 'safe' or 'moderately safe' status. However this is not always practically feasible; hence the best aspiration is to have the index value which is as low as reasonably practicable (ALARP). To get a better understanding

on the proposed framework, a case study on the methyl methacrylate (MMA) process is performed. This is the same case study used by Pandian et al. [13] so that the assessment results from their work can be used as an input data for this particular case study presented in this paper.

Figure 1: A systematic guideline for designing an inherently healthier chemical process

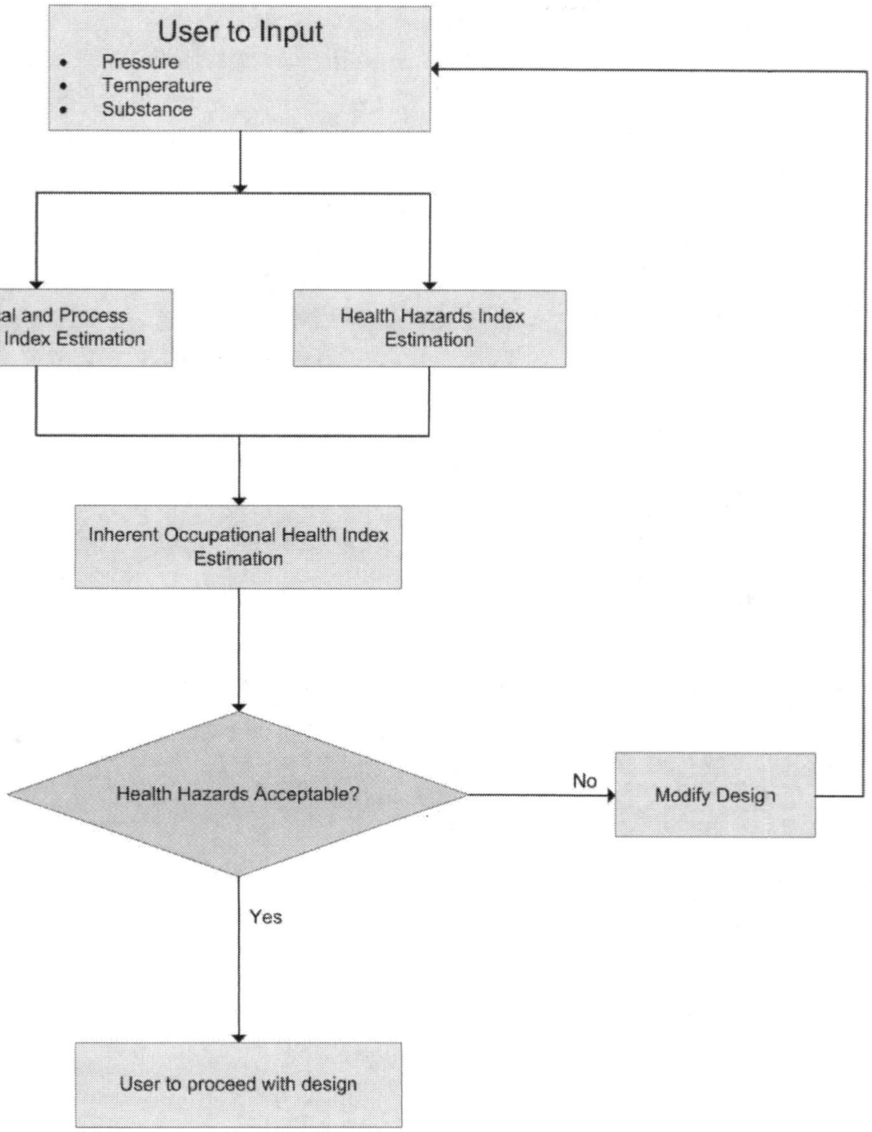

5. Case study

Based on the results obtained from a case study performed by Pandian et.al [13] on MMA, an inherently healthier process design is demonstrated in this section using the proposed guideline. There are six alternative routes for MMA manufacture. Below are the six synthesis routes:

1. Acetone cyanohydrin based route (ACH)
2. Ethylene via propionaldehyde based route (C2/PA)
3. Ethylene via methyl propionate based route (C2/MP)
4. Propylene based route (C3)
5. Isobutylene based route (i-C4)
6. Tertiary butyl alcohol based route (TBA)

As we can see from Table 2 (based on the study by Pandian et al. [13], all the six routes are assigned with 'moderately hazardous' status based on the calculated IOHI value. Therefore, the framework suggests modifying the process design until 'safe' or 'moderately safe' is achieved, or at least a reduced hazard index value is achieved. Details on the health hazard level benchmark are in Hassim and Hurme [12].

In order to illustrate the proposed guideline, only one route is selected. In this case C2/MP route is taken as an example. To lower down the hazard index value, one of the options is to modify the reaction conditions to a more moderate condition i.e. temperature and pressure. This can be achieved by using catalyst in the second and the third reaction steps of the C2/MP based route. Based on the article by Nagai [14], the reaction temperature of these particular steps can be lowered down to 180°C from 470°C (for the second step) and from 350°C (for the third step) with the help of Lewis acid catalyst. The IOHI value is recalculated using the proposed computer aided tool (refer to Figure 2) based on this new operating condition. With the new reaction temperature in both steps, the I_{PPH} value has reduced from 25 to 23 which makes the total IOHI value of the C2/MP route to reduce from 40 to 38. However, the synthesis route is still under the status of 'moderately hazardous'. The user may need to repeat the same procedure until 'moderately safe' or 'safe' is achieved. However as mentioned earlier, this is not always achievable; it depends on the feasibility of the modifications made, both from the technology and the costs aspects.

Table 2. The Inherent Occupational Health Index (IOHI) values for MMA [13]

Route	No. of steps	I_{PPH}	Average I_{PPH}	I_{HH}	Average I_{HH}	Average I_{IOHI}	Status
ACH	3	21	7.00	23	7.67	14.67	Moderately hazardous
C2/PA	4	29	7.25	28	7.00	14.25	Moderately hazardous
C2/MP	3	25	8.33	15	5.00	13.33	Moderately hazardous
C3	4	31	7.75	26	6.50	14.25	Moderately hazardous
i-C4	3	23	7.67	20	6.67	14.33	Moderately hazardous
TBA	3	21	7.00	20	6.67	13.67	Moderately hazardous

Figure 2: Snapshot of systematic computer-aided IOHI tool

Process Route: Ethylene based via methyl propionate — Reaction Step: 1

Chemical Reaction: C_2H_4 + CO+ CH_3OH → Methyl propionate — Reaction Phase: Liquid

Operating Conditions (Process)

Factor	Condition	Penalty
Temperature (°C), I_T	100	1
Pressure (bar), I_P	15	1

Mode of Process	I_{MP}
Continuous	1

Physical Property (Material)								Health Property (Material)					
Substance	Material State	I_{MS}	L & G: Boiling Point (°C)	S: Type of solid	I_V	Material Corrosivity	I_C	R-Phrase (R)	Selected R-Phrase	I_H	V: EL(ppm)	S: EL(mg/m³)	I_{EL}
Ethylene	Gas	1	-104	0	3	Carbon Steel	0	R12 R67	R67	[illegible]	250	163.0	1
Carbon Monoxide	Gas	1	-192	0	3	Carbon Steel	0	R12 R23 R33 R48 R61	R61	[illegible]	25	29	2
Methyl Alcohol	Liquid	2	64.5	0	1	Stainless Steel	1	R11 R23/24/25 R39 R39/23/24/25	R39/23/24/25	[illegible]	200	260	1
Methyl Propionate	Liquid	2	79	0	1	Stainless Steel	1	R11 R20	R20	[illegible]	-	-	0
Maximum Penalty Selected		2			3		1			[illegible]			2
Total Subroute I_{OHI}		15											

L: Liquid, G: Gas, S: Solid, V: Vapor, EL: Exposure limit value

Process Route: Ethylene based via methyl propionate — Reaction Step: 2

Chemical Reaction: CH_3OH + O_2 → Methylal — Reaction Phase: Gas

Operating Conditions (Process)

Factor	Condition	Penalty
Temperature (°C), I_T	180	2
Pressure (bar), I_P	4.5	0

Mode of Process	I_{MP}
Continuous	1

Physical Property (Material)								Health Property (Material)					
Substance	Material State	I_{MS}	L & G: Boiling Point (°C)	S: Type of solid	I_V	Material Corrosivity	I_C	R-Phrase (R)	Selected R-Phrase	I_H	V: EL(ppm)	S: EL(mg/m³)	I_{EL}
Methyl Alcohol	Liquid	2	64.5	0	1	Stainless Steel	1	R11 R23/24/25 R39 R39/23/24/25	R39/23/24/25	2	200	260	1
Methylal	Liquid	2	41.5	0	2	N/A		R11 R36/37/38 R43	R43	3	1000	3100	1
Maximum Penalty Selected		2			2		1			3			1
Total Subroute I_{OHI}		11											

L: Liquid, G: Gas, S: Solid, V: Vapor, EL: Exposure limit value

Process Route: Ethylene based via methyl propionate — Reaction Step: 3

Chemical Reaction: Methyl Propionate + Methylal → MMA + CH_3OH — Reaction Phase: Gas

Operating Conditions (Process)

Factor	Condition	Penalty
Temperature (°C), I_T	180	2
Pressure (bar), I_P	1	0

Mode of Process	I_{MP}
Continuous	1

Physical Property (Material)								Health Property (Material)					
Substance	Material State	I_{MS}	L & G: Boiling Point (°C)	S: Type of solid	I_V	Material Corrosivity	I_C	R-Phrase (R)	Selected R-Phrase	I_H	V: EL(ppm)	S: EL(mg/m³)	I_{EL}
Methyl Propionate	Liquid	2	79	0	1	Stainless Steel	1	R11 R20	R20	2	-	-	0
Methylal	Liquid	2	41.5	0	2	N/A		R11 R36/37/38 R43	R43	3	1000	3100	1
Methyl Methacrylate	Liquid	2	100	0	1	Stainless Steel	1	R11 R36/38	R36/38		100	410	2
Methyl Alcohol	Liquid	2	64.5	0	1	Stainless Steel	1	R11 R23/24/25 R39 R39/23/24/25	R39/23/24/25	2	200	260	1
Maximum Penalty Selected		2			2		1			3			2
Total Subroute I_{OHI}		12											

L: Liquid, G: Gas, S: Solid, V: Vapor, EL: Exposure limit value

6. Conclusion

In this paper, a systematic guideline for designing an inherently healthier process during the R&D stage is proposed. The systematic guideline is able to help the user to decide on the best chemical process route with lowest index value i.e. inherently healthier. With both the assessment tool and the design framework, the user can now get the assessment result fast and more efficiently. Also, the results obtained can be claimed to be more reliable now as human errors can be minimized with the presence of the computer aided tool.

7. References

[1] ILO/WHO., Joint ILO/WHO Committee on Industrial Hygiene: Report of the First Meeting, ILO, Geneva, 1950

[2] Eijkemans G., WHO/ILO joint effort on occupational health and safety in Africa. In international Occupational Hygiene Association (IOHA) 6th international scientific conference, Pilanesberg, 2005

[3] Hassim MH., Inherent Occupational Health Assessment in Chemical Process Development and Design, PhD Thesis, Aalto University School of Science and Technology, Espoo, Finland, 2010

[4] Kletz TA., Preventing catastrophic accidents, Chemical Engineering, 1976; 83(8), 124-128

[5] Kletz TA., Plant Design for Safety: A User Friendly Approach, Hemisphere Publishing Corporation, New York, 1991

[6] Edwards DW, Lawrance D., Assessing the inherent safety of chemical process routes: Is there a relation between plant cost and inherent safety?, Process Safety and Environmental Protection, 1993;71(B4), 252-258

[7] Heikkila AM, Hurme M, Jarvelainen M., Safety considerations in process synthesis, Computers and Chemical Engineering, 1996; 20(A), 115-120

[8] Palaniappan C, Srinivasan R, Halim I., Selection of inherently safer process routes: A case study, Chemical Engineering and Processing, 2002;43 (2004) 647–653

[9] Gupta JP, Edward DW., A simple graphical methodology for simple measuring inherent safety. Journal of Hazardous Materials, 2003; 104 (2003), 15-30.

[10] Khan FI, Amyotte PR., Integrated inherent safety index (I2SI): a tool for inherent safety evaluation, Process Safety Prog, 2004; 23(2): 136–148.

[11] Leong CT, Shariff AM., Inherent safety index module (ISIM) to assess inherent safety level during preliminary design stage, Process Safety and Environmental Protection, 2007; 86 (2008) 113-119

[12] Hassim MH, Hurme M., Inherent Occupational Health Assessment During Process Research and Development Stage, Journal of Loss Prevention in the Process Industries, 2010; 23(1) 127-138

[13] Pandian S, Hassim MH, Hurme M., Computer Aided of Occupationally Healthier Processes During Research and Development Stage, Journal of Loss Prevention in the Process Industries, 2013; 26(1) 705-713

[14] Nagai K., New developments in the production of methyl methacrylate, Applied Catalysis A: General, 2001; 221(2001) 367-377

An Inherent Safety Implementing Approach for Safety Planning of Chemical Industrial Parks in China

Zhang Mingguang
Nanjing University of Technology
No.200 Zhongshan North Road, Nanjing, China
Mingguang_zhang@163.com

Ahmed Mebarki[a],Chenyinting[b],Qian chengjiang[b],Wang rui[b]
[a] University Paris-Est Marne-la-Vallée [b]Nanjing University of Technology
[a] Marne la Vallée Cedex 2 [b]No.200 Zhongshan North Road,Nanjing, China

Prepared for Presentation at
American Institute of Chemical Engineers
1st CCPS Asia-Pacific Conference on Process Safety
Qingdao, China
September 4 – 5, 2013

UNPUBLISHED

An Inherent Safety Implementing Approach for Safety Planning of Chemical Industrial Parks in China

Zhang Mingguang
Nanjing University of Technology
No.200 Zhongshan North Road,Nanjing, China
Mingguang_zhang@163.com

Ahmed Mebarki[a],Chenyinting[b],Qian chengjiang[b],wangrui[b]
[a] University Paris-Est Marne-la-Vallée [b]Nanjing University of Technology
[a] Marne la Vallée Cedex 2 [b]No.200 Zhongshan North Road,Nanjing, China

Keywords: inherent safety; safety planning; chemical industrial park.

Abstract

In the last two decades chemical industrial parks have developed greatly in China. Due to lack of sufficient safety planning technologies, many problems have been brought about to dramatically cause hazards,which could not be amended or may cost too much money to make up.Although there are some related technologies in foreign countries,some factors affected their application in China,such as lack of foudamental data, difference of management mechanism and equipment quality.In this paper, statistics of main accidents in Chinese chemical industrial parks were carried out to study accident process mechanism and risk characters, from which impact parameters were picked up. Based on the validation of the information about the impact parameters, the lifecycle of a chemical industry park was divided into several stages.Safety planning targets of each stage were figured out in the view of inherent safety theory.Considering the current state of chemical industrial parks in China, quantified indice for former targets and achieving strategy were dicussed, then models were built to assess the quantified indices.

1. Introduction

As far as 2011, more than 1200 chemical parks had set up in China, nicluding 15 national parks, 112 provincal parks, 305 municipal parks and 829 county-level parks[1].Although chemical parks contribute much to economy, many past accidents in the chemical parks have caused daramatic casualty and loss,even great social panic,which indicate obvious lack of safety planning in the rapid development of the chemiacal parks. To overcome such difficulities,the State Administration of Work Safety Department have pointed some important concepts and targets of safety planning[2].Some Chinese researchers advised to apply relevant foreign experience on chemical park's safety planning,but they may not be put into use because of Chinese special environment(such as lack of foundamental datas,different administrative mechanism),and a systematic approach is still missing[3].Inherent safey theory was proposed by Trevor Kletz in 1977[4], it is an advanced idea facing lifecycle of chemical process. It has supported lots of safer design work on chemical process industry[5,6] ,especailly on chemical installation, production technology and layout assessment. Chemical park is centralizaiton and integration of different chemical plants, thus a systematic approach is proposed to achieve safety planning of chemical parks in China.

2. Content of safety planning for Chinese chemical Parks

2.1 Satatistics of accidents in Chinese chemical Parks

To discover main targets of safety planning, statistics of accidents (1991-2013) in chemical parks was carried out[7], finally 32 accidents were gathered and studied.Most of them are collected from website of Chinese State Administration of Work Safety, and have relatively detailed description. Among these accidents, three main initial cause are equipment failure (7 cases), human error(15 cases) and technology out of control(5 cases). 15 accidents (46.8%) caused by human error indicate poor tolerance of the plant system. 22 accidents (68%) involved domino accident chains. Domino effcet always aggravated accident's consequence, which implies that the failures in layout design and process tehchnology design.

Table 1. Statistic result of accidents in chemical parks in China

Initial cause	Number of accidents	Existence of domino effect
Equipmen failure	7	5
Human error	15	13
Technology out of control	5	4

2.2 Targets of inherent safety design in the lifecycle of chemical park

2.2.1 Inherent safety theory

Inherent safety aims for the elimination, or the reasonably reduction of the hazards in a system. As for a chemical park, the key idea of the inherent safety approach is achieving relatively inherently safer status,which could be effectively implemented in the early stages of lifecycle[8]. Although hazards can not be completely eliminated in the chemical process industries, a wide range of opportunities and improvements which lead to inherently safer systems are possible.Commonly, six guidewords (minimization, substitution, attenuation, simplification, and limitation of effects) are widely accepted in inherent safer design of process industry.

2.2.2 Content of safety planning in chemical park's lifecycle

In China, chemical parks usually are divided into 4 levels: national, provincial, municipal and county-level. Considering economic income, local government has enough enthusiasm to support the construction of chemical parks in their administrative district, even ignore some safety need not directly implied in industrial practice and simple guidelines .According lifecycle theory and common developing procedure of Chinese chemical park [9, 10], four stages exist in the lifecycle of a chemical park, safety planning content of different stages are listed as follows:

1) Feasibility study stage
   - Location of a chemical park

     The location of the chemical park must be fixed in the administrative district, under comprehensive consideration of many impacting factors, such as economy, environment, geography and transportation. For safety, the interactive influence with sensitive targets around the park must be taken into account.Sensitive targets are the units which may reslut in accidents to cause damage of the park and which could be destroyed in the accidents of the chemical park.
   - Chemical industrial chain decision

     To meet the need of centralizaion and integration production, types of chemical production could be chosen to make up industrial chain, producing better economic effect.
2) Initial construction stage
   - Land-use decision:

     The whole park consists of many land parcels, each land parcel will be used for a fixed function,thus a chemical park could be divided into different functional areas.They are plants area, storage area, pbulic works area (water,electricity, gas, public pipes), administrative area and living area.Method to position different functional regions in the park is a mutiple targets decision problem, for which safety target is sessential, especially the boundary between different funcitonal regions and design of buffer zones.
   - Layout design

     Layout means spatial arrangement of units (chemical plant, building, and warehouse) in one functional area. It plays an important role to indicate inherent safety extent of a functional area.Taking plants area for example, quantified principles can effectively guide the layout of different factories to reduce the possibility of domino accidents and magnitude of possible consequence. If we can finish inherently safer design in early stages, better safety performance will be achieved, but may not be carried out in detail before the preliminary design of equipment.
3) Mature operating stage.
   - Safey access criteria for factory

     Besides choosing chemical reaction types to achieve centralizaion and integration in a park, elimilating outdate technology and choosing safer technology, minimizing quantity of hazardous materials in storage or transportation process must be taken before a factory is permitted to enter the park. Safety access criteria are critical standard to evaluate the following risk to undertake after entrance of a factory. In this stage, accessing and exiting of factories may happen at the same time, it enables choosing factory in the sight of safety.

- Safey capacity of the park

In recent years, the price of land has been arising rapidly.Making full use of the land means obvious economic profit. In China, the current law of safety distance in chemical industry usually is half –quantitative values base on past industrial experience, can't protect nearby plant when accident happens, that can be observed from the accident statistic of widely existing domino phenomenon. Creating risk criteria to stand for it is a hotspot in relevant research work [11].

- Emergency rescue planning

It contains emergency resources, plans and organization. Quantity threshold and layout optimazation could be calculated after the layout of factories is fixed.

4) Declining stage.

- Waste management

After factories move out of the park, waste hazardous materials and demolished plants should be reported to administration department to decide the treating way.

2.2.3 Targets of inherent safety design

As discussed above, safety planning has eight main issues in overall lifecycle, but not all of them involve opportunities to implement inherent safety design. The goal of the present work is to bring inherent safety concepts into design of these issues by means of a systematic approach. This requires evaluation of potential applicability of applying guidewords to possible issues in early design stage.

The targets of possible issues are explained in Table2.

Table 2. Safety targets of inherent safety design in the lifecycle

Lifecycle stage	Content of Safety planning	Safety targets
Feasibility study stage	Location of the park	1.Aviod or limit interactive damage between chemical park and surrounding sensitive targets
Initial construction stage	Land-use decision	1.Limit affected areas in the accidents
	Layout design	1.Avoid domino accidents among units in different systems 2.Limitation of domino accdents among units in same system
Mature operating stage	Safety access criteria for factory	1. simplification of hazardous materials flow(relative coefficient to the main type of industrial chain) 2. Minimization of hazardous materials(quantity and distribution)
	Safety capacity of park	1. minimization of hazardous materials and energy 2. attenuation of process

3. Implementing strategy and assessment models

3.1 Location of the park

Actually, safety isn't the first item considered in deciding location of the park.Our job is picking up an optional plan from several potential positions and land area. In this stage, preliminary layout of the park haven't been designed, only little information about types of chemical reaction, magnitude of the production. We have to assume the main installations of the reaction from practical experience of existing similar project. Through investigation, we can collect the information about nearby sensitive targets, here they are places with many people or lots of wealth. If an option has less potential hazard to nearby targets, it means inherently safer.

Therefore, the potential hazard to nearby targets can be calculated by the following equation:

$$C_i = \sum_{j=1}^{n} \left(R_{ijeffect} / R_{ijreal} \right) \quad \text{[Eq. 1]}$$

$$C_{total} = \sum_{i=1}^{m} C_i \quad \text{[Eq. 2]}$$

Where C_{total} is the potential damage capacity of the park; m is the total number of installarions in the park; n is the total number of sensitive targets; i stands for one installation; j stands for one sensitive target; $R_{ijeffect}$ is the maximum damage radius in potential accident scenarios of installation i ; R_{ijreal} is the distance from installation i to target j.

3.2 Land-use decision

In this period, we can get more details about installations in each special function area. To limit affected areas in the accidents, the focus is making surrounding other functional areas less affected by one functional area.To distinguish different functional areas, buffer zone like road or green land are widely used, therefore probability of domino effect cross two functional areas is ignored. An assuption of the most dangerous situation is made here: All the units are placed at the edge of the functional area it belongs to, death radius of each installation was calculated to finish drawing death circle. The circle may cover different function areas, multiply the area by density of people, finally we can get overall number of affected people. Obviously, less people we get, the plan is safer.

Therefore, the potential affected area can be calculated by the following equation:

$$N_i = \sum_{j=1}^{n} \left(P_{density} \times S_{area} \right)_j \quad \text{[Eq. 3]}$$

$$N_{total} = \sum_{i=1}^{m} N_i \quad \text{[Eq. 4]}$$

Where N_{total} is the potential number of death people ; m is the total number of functional areas in the park; n is the total number of units able to cause accidents in a functional area; i stands for one functional area; j stands for one unit; S_{area} is the amount of affected area caused by unit j. $P_{density}$ is the people density of S_{area}; Analysis of $P_{density}$ and S_{area} could be assisted by GIS technology.

3.3 Layout design

Domino effect is a kind of very important phenomenon in this problem [12, 13]. Two kinds of domino effect must be treated as equal.One is domino effect among units in different systems, blast, heat radiation or faragment produced in primary accident destroy another chenical pant to cause secondary accident.In Cozzani's work [14], and escalation radius was suggested as critical value to evaluate the propogation of the former. The other is domino effect among units in the same system. The plants may be linked by pipes, named "process linkage domino effect" and simple probit models were built in reference [15]. Then overall risk colud be calculated as coupling of primary accident risk and domino accident risk.

Therefore, the overall societal risk of a layout can be calculated by the following equation:

$$R = \sum_{j=1}^{n} \left(R_{Pj} + R_{j2d}\right) \quad \text{[Eq. 5]}$$

Where n is the total number of units in this layout; j stands for one units; R_{pj} is the risk of primary accident ; R_{j2d} is the comprehensive risk of two kinds of domino effect.It's noted here the risk disscussed above is represented by Potential Loss of Life (PLL).

3.4 Safety access criteria for factory

Safety accessing mechanism is very complex, in fact it relates to management behaviour of the government.At present, there's only one qualitative rule in China:the chemical reaction can't be outdate technology listed in a directory announced by the central government. It can't Some parameters should added:a) attitude of around people.Different group of people have different tolerance limits to the same risk.Their safety need can be realized throug statistics.b) the quantitiy of risk it adds to the existing chemical park. Commonly people hardly care the size of land the factory occupies, thus a risk index like individual risk was selected.

Therefore, the safety access criteria canbe expressed with this equation:

$$R_{edge} \leq R_{c-individual} \quad \text{[Eq. 6]}$$

Where R_{edge} is the maximum individual risk at the edge of the factory; $R_{c-individual}$ is the risk accepted value of around people,result from risk perception investigation. It's a dynamic parameter,any safety improvement or past accident may influence it.

3.5 Safety capacity of park

For the whole chemical industrial park, safety capactiy is a subjective opinion to control risk in the park.After some factories are constructed and put into use, the factory, daministrative department and the public could communicate to reach agreement of an accepted safety stadard. A sample factory is choosen to represent a status of accepted risk. Safety management organization, quality of equipment, human resource and other related information are able to be collected in details for exact risk calculation. An averge value of total PLL on unit squares is valid. This value is used for future decision in the following development of the park. This work need continuous cooperation.

Therefore, the safety capacity of the park can be calculated by the following equation:

$$PLL_{average} = \frac{PLL_{cf}}{S_{cf}} \quad [Eq. 7]$$

$$PLL_{total} = PLL_{average} \times S_p \quad [Eq. 8]$$

Where PLL_{total} is the safety capacity of the park; PLL_{averge} is the total PLL on unit square meters of the sample factory; PLL_{cf} is the total PLL of the sample factory; S_{cf} is the total area of the sample factory (in square meters). S_p is the total area of the park.

4. Conclusion

A systematic analysis of safty planning about chemical industrial park was carried out to achieve the main items. Based on the practise in China, exact content of the item were explained. In the view of inherent safety, five items of the content were chosen, and exact planning targets were demonstrated. Finally, implementing approach and evluate models were established.

1. Not all of the safety planning content have opportunities to implement inherently safer design.Inherent safety is part of the safety planning, effect of added safety measures can't be ignored.But inherently safer design can reduce the entire cost for safety.That's a very important characterstic of inheren safety theory.
2. Assessment models of inherent safety design targets are simple and rough in this work, afterall safety planning has only some degrees of freedom to perform, many designs for other purpose come first in practical experience of chemical park management. More influence factors sholud be dicussed in future work. Ideas of these models mostly derive from concept of risk, but to some extent these models have better prospect than risk calculation because risk calculation need many detailed information which may not be available at a certain satge.
3. Nowdays safety planning relates to public opinion in China, technology of quantifying the public attitude is essential for the sustainable development of chemical park.More attention should be paid on risk tolerance research.

5. Acknowledgements

The author gratefully acknowledges the financial support of the National Natural Science Foundation of China (No. 71001051) and the Natural Science Foundation of Jiangsu Province, China (BK 2012824).

6. References

[1] Wu Zong-zhi, Wei Li-jun, WANG Ru-jun, et al," Methods and applications of safety planning for chemical industrial parks," Journal of Safety Science and Technology, 8 (2012), 46-51

[2] Li C.G, Liu Y.J, Liu J, et al, Study on Safety Planning for Chemical Industrial Park Base on Gross Risk Quantity. China Safety Science Journal, 19(2009), 116-121.

[3]LI B, L, Zhao D.F, Meng YF, Study on Security access of enterprises in chemical parks. Journal of Safety Science and Technology, 07(2011), 159-163.

[4] Kletz, T.A., Inherently safer design—its scope and future. Trans IChemE, Part B: Process Safety and Environmental Protection 81 (2003), 401.

[5]Kletz, T. Process Plant; A Handbook for Inherently Safer Design. (Second edition) Taylor and Francis, Bristol PA, 2010

[6]Khan, F.I., Amyotte, P.R., Inherent safety in offshore oil and gas activities: a review of the present status and future directions. Journal of Loss Prevention inthe Process Industries, 15 (2002), 279–289.

[7]CHEN Yinting, ZHANG Mingguang, GUO Peijieb, JIANG Juncheng、Investigation and analysis of historical Domino effects statistic[J]. Procedia Engineering, 2012.45 (10): 152 – 158.

[8] Hurme, M., Rahman, M., Implementing inherent safety throughout process lifecycle. Journal of Loss Prevention in the Process Industries, 18(2005), 238-244.

[9] Leong, C.T., Shariff, A.M., Inherent safety index module (ISIM) to assess inherent safety level during preliminary design stage. Process Safety and Environmental Protection, 86(2008), 113-119.

[10] Shariff, A.M., Leong, C.T., Zaini, D., Using process stream index (PSI) to assess inherent safety level during preliminary design stage. Safety Science, 50(2012), 1098-1103.

[11] Rahman, M., Heikkilä, A.M., Hurme, M., Comparison of inherent safety indices in process concept evaluation. Journal of Loss Prevention in the Process Industries, 18(2005), 327-334.

[12]Tugnoli, A, Khan, F.I., Amyotte, P.R, Cozzani, V., Safety assessment in plant layout design using indexing approach: Implementing inherent safety perspective. Part 1 - Guideword applicability and method description.Journal of Hazardous Materials, 60(2008), 100-109.

[13]Tugnoli, A, Khan, F.I., Amyotte, P.R, Cozzani, V, Safety assessment in plant layout design using indexing approach: Implementing inherent safety perspective. Part 2-Domino Hazard Index and case study.Journal of Hazardous Materials, 60(2008), 110-121.

[14] Cozzani, V. Tuqnoli, A.Salzano, E.Prevention of domino effect: From active and passivestrategies to inherently safer design.Journal of Hazardous Materials, 139(2007), 209-219.

[15] Zhang Mingguang, Jiang Juncheng. An improved probit method for assessment of domino effect to chemical continuous processing equipment caused by overpressure. Journal of Hazardous Materials,2008, 158(2-3):280-286.

1ST CCPS ASIA-PACIFIC CONFERENCE ON PROCESS SAFETY℠

Beyond Process Safety Management: Integrating Culture and Leadership into Process Incident Prevention

Ricky Yu
BST
41 Science Park Road
The Gemini
Singapore Science Park 2, #04-03C
Singapore 117610, Singapore
bstap@bstsolutions.com

Prepared for Presentation at
American Institute of Chemical Engineers
1st CCPS Asia-Pacific Conference on Process Safety
Qingdao, China
September 4 – 5, 2013

UNPUBLISHED

Beyond Process Safety Management: Integrating Culture and Leadership into Process Incident Prevention

Ricky Yu
BST
41 Science Park Road
The Gemini
Singapore Science Park 2, #04-03C
Singapore 117610, Singapore
bstap@bstsolutions.com

Keywords: process safety, catastrophic event prevention, organizational safety, process incident prevention, organizational culture, leadership

Abstract

A comprehensive approach to process incident prevention is needed to take organizations beyond common practices in process safety management and ensure that the systems designed to avert disaster are fully supported and properly implemented at all levels of the company. A Comprehensive Process Incident Prevention approach builds on the foundation of process safety management to create a safety process that integrates culture with leadership characteristics critical to catastrophic event prevention. Essential to the success of this approach—or any process designed to mitigate serious accidents—is the development of a strong safety culture. Culture guides people's actions even when no one is watching. It is the driving force that takes a workforce good at safety to one that excels at it. This paper examines key organizational characteristics indicative of culture that can objectively measure catastrophic event prevention programs and predict the effectiveness of these programs. Building an effective safety culture starts with leaders. Leaders need to recognize that safety is not an isolated part of the organization, but an integral part of everything the organization does. The paper outlines seven areas leaders can examine in their own company to begin to apply comprehensive approach to process safety. The paper also explores the four major components or areas of focus that all organizations successful in process safety do effectively every day.

1. Introduction

Prominent incidents such as Oxy's Piper Alpha explosion, the Esso Longford gas explosion, and the BP Deepwater Horizon spill have given rise to new rules and practices in the implementation of process safety management. Despite this attention, major incidents continue to occur. Research has shown that alongside the proximate causes of each incident, there also existed an underlying fabric of systems, mechanisms, and culture that allowed risk to persist and processes to be ignored. The conclusion is that traditional approaches to process safety management are not sufficient for addressing the precursors that lead to catastrophic events.

A comprehensive, organization-wide approach is needed to ensure that safety systems are properly executed and people are fully supported and trained to recognize and speak up about weak signals and minor events that could lead to major incidents. A comprehensive approach requires strong leadership to build a robust safety culture that will cultivate an environment in which everyone is working together to prevent catastrophic events.

2. Creating Comprehensive Process Incident Prevention

To advance beyond the level of safety achieved through process safety management systems alone, organizations should adopt a Comprehensive Process Incident Prevention (CPIP) approach that builds on the foundation of process safety management to create a safety process that integrates culture with leadership characteristics critical to catastrophic event prevention. This approach includes four major components: Anticipation, Inquiry, Execution, and Resilience. Both individually and collectively, these four components build a strong safety culture that supports effective technical and management safety systems designed to prevent catastrophic events.

2.1 Anticipation

Anticipation refers to having an organization with the systems and culture to be sensitive to the recognition of "weak signals" that may be indicative of increased risk of catastrophic events.

An organization that is strong in Anticipation will have mechanisms to capture information from a variety of sources. Process deviations, unusual maintenance requests, and even frontline workers detecting differences in sounds can potentially be meaningful early indicators of change to exposure. If there are not systems to seek out and capture this information it can be easily lost, along with the opportunity to prevent a disaster.

Once captured, the various types of information that can be early indicators must be accessible and structured in a way that allows analysis of patterns across various types of data. In many organizations there is a wealth of information collected, but different types of information cannot be integrated and effectively used.

Even when systems are in place, it is incumbent upon the organization to ensure that the systems function as intended. Individuals will not come forward with critical information if they believe that nothing will happen if they do, or, even worse, that such an effort on their part will be perceived as a nuisance. It is important for leaders to encourage individuals to be alert for weak signals and to reinforce employee vigilance.

Finally, the information gathered must be used effectively. This requires having people with appropriate skills who are given encouragement and attention. Early warning indicators will not always lead to actual findings of increased exposure. By its very nature, sensitivity to weak signals will result in numerous "false positives." It is easy for the organization to begin to discount early warnings in the manner of the boy who cried wolf. Organizations with strong cultures have leaders who visibly value the search for early warnings and reinforce the analysis of these indicators, even when this does not result in identification of a serious risk. These leaders understand that supporting the detection and investigation of many false positives is working if it results in just one catastrophic event avoided.

2.2 Inquiry

Inquiry involves making effective use of information to analyze, understand, and plan mitigation of risks. Traditional PSM includes a number of elements (e.g., process-hazard analysis, pre-startup safety review, management of change) designed to evaluate and plan for control of hazards and risks. However there are common (but often undetected) cultural characteristics that can undermine the effectiveness of these efforts and leave the organization vulnerable.

When we examine what we knew before an accident or injury occurred, we often see that we had all the information we needed to prevent it but failed to realize it at the time. Why are the right choices so clear in hindsight, but clouded in the moment? A rich scientific literature in cognitive psychology helps explain this seemingly irrational phenomenon: Human beings tend to make inaccurate judgments about future probabilities in predictable ways. These tendencies toward faulty judgments are called cognitive biases, and they can seriously undermine process safety management.

Cognitive bias refers to the tendency we all have to rely on intuitive rather than analytical thinking in order to process information efficiently. Our knowledge and experience allow us to reach conclusions and make decisions quickly and efficiently in many circumstances. However this can also trap us in poor decisions. Recency bias is a common example—we tend to overemphasize our recent experience in assessing data and situations in favor of long-standing and proven analysis. Confirmation bias is another example—we tend to give more weight to data that supports our intuition and beliefs than to data that refutes them.

Where cognitive bias becomes particularly problematic for catastrophic event prevention is that our methods for systematic hazard analysis can become predisposed by our biases and can cause us to underestimate, or miss entirely, potential failure scenarios. In addition, the routine operational decision making that occurs day to day can be influenced by cognitive bias, leading to unintended increases in risk.

While awareness of cognitive bias helps to counteract its outcomes, the best way to guard against the insidious effects of cognitive bias is to have an organizational culture that combats it. There are specific leadership practices that promote a culture in which the effect of cognitive bias is minimized (e.g., encouraging employees to voice dissenting opinions). There are also specific skills involved in asking the right question in the right way to get the right data. Organizations that want to prevent the problems that can arise from cognitive biases should promote and measure the use these leadership practices and skills.

2.3 Execution

Excellent hazard identification and assessment, as well as hazard control efforts, can be undermined if the programs and practices are not followed as intended. While many organizations use periodic audits to provide a check on implementation, the key to ensuring consistent and ongoing activity is managers who systematically monitor, reinforce, and verify effective program execution.

Monitoring involves regularly acquiring information on what subordinates are doing, how they are progressing toward achieving goals, and what issues or problems they may be encountering. This is not micromanaging; rather it is ensuring that leaders have sufficient information to recognize good performance, provide support when subordinates need it, and offer corrective feedback when subordinates fail to fulfill their responsibilities.

A leader's monitoring behavior may take many forms. Depending on the situation and the leader's particular role, these may include: observing the worksite, informal conversations with frontline workers, periodic meetings with subordinates to review progress, use of written progress reports, review of appropriate metrics, and so on.

Reinforcement involves providing feedback that recognizes good performance. This communicates the importance and priority of the catastrophic event prevention activities and maintains focus on consistent execution. Effective reinforcement is based on effective monitoring, which provides the leader with specific data on which to base reinforcement, avoiding the vague and ineffective "good job" type of feedback to subordinates.

Verification is similar to monitoring, but where monitoring is focused on the performance of subordinates, verification is focused on activities and programs. Ensuring that audit findings are resolved in a timely manner is an example of verification, and leaders who do this effectively are more likely to have an organization in which consistent execution is valued as part of the culture.

2.4 Resilience

Upset conditions occur from time to time in any system. Resilience refers to the organization's ability to react in ways that prevent upset conditions from becoming catastrophic events, and then learning from the experience. This has a major influence on ultimate results. Even where automated control systems are designed to handle upset conditions, it is important that workers understand when and how to intervene. It is also important that they know how to intervene and are not only able but also willing to make appropriate interventions early. An organization that is strong in resilience is more likely to prevent a small process disruption from becoming a major incident.

One requirement for strong resilience is knowledge; that is, do people at various levels have a broad enough understanding of the operation so that they can make good judgments in case of emergency. Some organizations approach this by using extensive sets of rules and procedures. This approach is intended to ensure consistency and avoid reliance on technical knowledge at the operating level. However two problems arise from this approach. First, the range of possibilities that must be planned for can produce such a mountain of procedures and rules that it becomes impractical, if not impossible, for anyone to know them all or to employ the right one for the given situation. Second, this approach assumes that all possibilities (with all variations in every scenario) can be identified in advance—something that is unlikely to be true. The alternative is to develop an organization with people who are knowledgeable and who are able to make sound judgments based on the information at hand.

Another requirement for resilience is willingness, and this relates directly to culture. Simply put, people are less likely to take action on their own initiative if they are not confident that the organization will support them. Perceptions of the culture's support for resilience are formed over a long period and are based on many small actions taken and not taken by leaders. An organization desiring strong catastrophic event prevention will be sensitive to this and will create the culture that supports resilience.

3. Developing Organizational Characteristics that Support Safety

An important starting point for organizations wishing to develop a CPIP approach is to assess the impacts their culture has on existing technical and management systems. There are key organizational characteristics indicative of culture that can be measured to track the progress of process safety programs and predict how effective they are at mitigating catastrophic events. These characteristics are:

1. Procedural Justice—Measures the extent to which the individual worker perceives fairness in the supervisor's decision-making process.
2. Leader-Member Exchange—Measures the relationship between an employee and supervisor. Particularly, the level of trust between the two.
3. Management Credibility—Measures the employee's perception that what management says is consistent with what management does.
4. Perceived Organizational Support—Measures employees' perceptions of how well the organization cares about, values, and supports them.
5. Workgroup Relations—Measures the perceptions employees have of their relationships with coworkers. To what degree do they help each other, listen and show respect, and follow through on commitments made?
6. Teamwork—Measures the extent to which the employee perceives that working with team members is an effective way to get things done.
7. Safety Climate—Measures employee perceptions of how well the organization cares about safety performance improvement.
8. Upward Communication—Measures the extent to which communication about safety flows freely upward through the organization.
9. Approaching Others—Measures the extent to which employees feel free to speak to one another about safety concerns.

In addition to measuring these key organizational characteristics, it is important to assess the extent to which Anticipation, Inquiry, Execution, and Resilience are present and consistent in the organization. This assessment will help leaders understand the strengths of the organization and identify improvement opportunities for creating a cultural foundation that optimizes technical and management systems. It will also highlight key leadership practices that will raise the company's effectiveness in each of these nine characteristics, strengthen culture, and improve overall performance.

Changing leadership practices to ones that support a strong safety culture cannot be done through training alone. While training is an important component, creating a culture that complements the CPIP approach requires a process that clarifies the need for change, elucidates for each individual the leadership practices he or she must focus on, provides structured reinforcement to help establish the new culture, and measures progress for corrective action.

4. The Role of Culture in Process Safety Management

Leaders in process safety management (PSM) have recognized that the systems designed to control and mitigate major events are only as effective as the culture in which they operate. Culture shapes behavior and behavior is how systems are implemented.

Culture is often described as "the way we do things around here" or "the unwritten rules." Culture arises from shared values and beliefs held within the organization, which lead to shared norms of behavior. These behavioral norms are reinforced over time as they lead to successful outcomes and positive consequences. A common example relates to operating procedures. It is not uncommon to find that over a period of time, operators and their supervisors deviate from the letter of the written procedure, finding ways to do their jobs more quickly and efficiently (at least in their perception). They are praised, reinforcing the belief that efficiency and speed are valued. During this period they experience no adverse consequences (e.g., injuries or incidents), bolstering the belief that their practice is sound. In this situation, leaders would probably say that operating procedures are important and operation discipline is valued, but shortcuts become the norm.

Because all PSM systems are ultimately implemented through the behavior of engineers, managers, supervisors, frontline workers, and others within the organization, the behavioral norms created by the culture have a direct impact on the effectiveness of the PSM systems. For example, in one organization decisions were made, supported by a weak safety culture, that allowed a proven rule to slide for months, resulting in a dangerous incident. In this organization a transfer hose was replaced every month according to the procedures that were in place based on prior hazard assessments. When no replacement was available, the rule was ignored for two months until a new hose could be obtained. Since no incident occurred during that two-month period, the replacement cycle was informally changed to every other month and later to every quarter. This eventually led to a chemical-release accident when the hose failed—not due to poor hazard assessment or a mechanical integrity program that was flawed on paper, but because various individuals allowed the program to slip based on their perception of risk derived from recent experience.

Incidents like this can be avoided if leaders are willing to commit to making safety "who we are" as an organization. In other words, safety must become embedded in every part of the organization's culture. The culture of an organization and the organization's leadership are inextricably linked. Leaders drive change in the organization and they create and shape culture by how they talk about safety, approach opportunities, what they focus on, ignore, choose to emphasize, what they delegate to others, and how they provide feedback. A leader who never asks about the resolution of safety audit items but requests frequent updates on installation of new production equipment helps create a culture that values production but may take safety shortcuts to achieve it.

5. How Comprehensive Is Your Process Safety Management System?

Virtually all catastrophic events in man-made systems are related to technical failures made possible by organizational failures. This explains why catastrophic events continue to occur despite widespread implementation of sophisticated technical and management systems. In thinking about a comprehensive approach to process safety, leaders can ask themselves the following questions. The answers can reveal a lot about where the organization is in terms of process safety and where it needs to go to become truly successful at preventing tragedies.

1. Do people throughout the organization understand the difference between personal safety and prevention of catastrophic events? If not, they may mistake good performance in personal safety with good control of the potential for catastrophic events.
2. Do we have the right technical and management systems in place, and do they get implemented as intended (and how are they monitored)? Ultimately what happens day to day is more important than the design intent of these systems.
3. Does our culture support consistent and rigorous use of safety-technical and management systems? Culture determines the way things are really done in the organization, and if the culture does not support key systems the systems are bound to break down.
4. Do leaders act in ways that promote identification of exposure and reduction of risk? There are key leadership behaviors that can help ensure effective hazard recognition, evaluation, and control. If leaders are not aware of these and actively practicing them, technical and management systems are likely to be compromised.
5. Do our "consequence management" systems support the activities critical to prevention of catastrophic events? The way performance is assessed, promotions awarded, and recognition and rewards distributed are all examples of how key systems influence perceptions of what is truly important to succeed in an organization.
6. Do we have the right skills available for supporting all aspects of safety? Do we assign roles and responsibilities in a way that ensures clarity, alignment,

coordination, and communication? Prevention of catastrophic events is generally a multifunctional effort. This requires careful consideration of how the various participants interact and collaborate.

7. Are the metrics in place to detect changes in exposure and ensure focus on key processes and procedures? If the only metrics in use for safety are those addressing personal safety, the organization will be unable to detect trends in catastrophic event potential.

6. Conclusion

The Comprehensive Process Safety Prevention approach integrates organizational culture and leadership with workers, systems, and processes to achieve unparalleled results in process safety management. Through a focused initiative, individual leaders will understand their role in catastrophic event prevention and learn to cultivate worker commitment toward that objective. With knowledge of the key practices that support Anticipation, Inquiry, Execution, and Resilience, along with the appropriate measurement tools, leaders can assess where their company currently is and chart a course to achieving their process safety goals.

7. References

[1] Duncan, C., *The Zero Index: A Path to Sustainable Safety Excellence.* Ojai, CA: Safety in Action Press, 2012

[2] Erickson, J.A., "The Relationship Between Corporate Culture and Safety Performance," *Professional Safety*, May 1997

[3] Groover, D., "Taking Ownership of the Injury-Free Culture: 8 Questions Every Leader Needs to Ask & Answer," *Perspectives,* Fourth Quarter 2007

[4] Krause, T.R., *Leading with Safety*. Hoboken: Wiley Interscience, 2005

[5] Krause, T.R., "Influencing the Behavior of Senior Leaders in EHS," *Professional Safety*, June 2004

Status and Countermeasures of Enterprise HSE Audit

Qiu Shaolin （ Deputy Chief Engineer ）[1]
Department of Safety Environmental Protection & Energy-Saving for CNPC1
Zhang Laibin(principal)[2]
China university of petroleum(Beijing)
Ru Apeng(Engineer)[3]
CNPC Research Institute of Safety & Environment Technology Zip code 100007

Key words: HSE management system, HSE audit, continuous improvement

Abstract

Combined with the practice of HSE management system building and implementation of China National Petroleum Corporation (CNPC), this paper systematically expounds the meaning, significance and role of HSE management system audit as well as the relationship between HSE audit and safety inspection. Besides it analyzes the main problems and causes existing in enterprise HSE audit at present. Finally, it proposes specific countermeasures to strengthen HSE audit from the following aspects: improving the awareness of the importance of HSE audit, improving HSE audit system, preparing operable audit checklist, establishing an auditor team with high professional quality, constantly optimizing the ways and methods of HSE system audit and strengthening the application of audit results.

Introduction

HSE management system is a common scientific management method in international oil and gas industry. Its core purpose is to control enterprise HSE risks. It follows the principle of PDCA, and has the characteristics of achieving self-continuous improvement. Carrying out HSE audit is an important part of establishing and running enterprise HSE management system, and an effective tool of promoting the continuous improvement of HSE management. This paper analyzes the status and problems of HSE audit, elaborates on internal audit and proposes targeted countermeasures.

1. Implications of audit

1.1 Definition of audit

HSE management system audit is a systematic, independent and documented process which obtains and evaluates audit evidence objectively to determine compliance with applicable audit criteria. It is an important way to assure the effectiveness and continuous improvement of enterprise HSE management system.

1.2 Classification of audit

Audit can be divided into two types based on different audit purposes: internal audit and external audit. Internal audit is also called "first-party audit" which is performed by the enterprises themselves, or in the name of the enterprises for management review and other internal purposes. External audit includes "second-party audit" and "third-party audit". Among them, second-party audit is conducted by the enterprise's relevant parties or others in the name of the relevant parties, for instance, the contractor or supplier audits by the enterprises, etc. Third-party audit is performed by an independent, impartial, and competent external certification body in accordance with the applicable audit procedures. If the audit results are consistent with specific laws, regulations and standards, the certification body will issue certification certificate to the organization. Therefore, the third-party audit is also called "certification audit".

Audit can be divided into centralized audit and rolling audit based on different audit arrangements: centralized audit refers a one-time audit within a certain period of time which covers all departments, units and elements of the HES system. Rolling audit is also called sectional audit which generally carries out an audit on one or several departments or elements on a quarterly or monthly basis to ensure that all departments and elements are covered at least once within a year.

Audit can be divided into all-element audit and special audit based on different audit contents: all-element audit is an audit covering all elements of the HSE system. Special audit is an audit taking one or several special operations, activities, system elements as the audit theme, such as those for process safety, behavior safety, fire management, well control safety, traffic safety, and work permit.

2 Significance and role of HSE audit

2.1 HSE audit is a common practice in international oil and petrochemical enterprises

Currently, "Management must be reviewed" has become a common practice in large international oil companies. HSE audit is an important aspect to establish and maintain HSE management system, a reliable way to evaluate the effect of HSE management system implementation and an effective tool to improve the enterprise HSE management. Audit can verify the actual implementation of HSE management system in enterprises, and find out the advantages and disadvantages and improvement, so as to achieve the goal of HSE management system integrating into all production and business activities, serving safety production and continuous improvement.

2.2 HSE audit is an important part of HSE system construction

HSE system construction is the main line of safety and environmental protection work which is gradually established by CNPC after a long time practice. HSE audit is an important part of HSE system construction. Promoting HSE audit and strengthening performance management is an important way to ensure the continuous improvement of HSE management system. At present, the safety culture of CNPC as a whole is still at strict supervision stage. HSE audit is an effective safeguard measure during the strictly regulated stage, an important starting point to strengthen safety production management and an important carrier to promote the enterprise management level.

2.3 HSE audit is an important tool and starting point to promote the implementation of all works

CNPC began to implement HSE headquarters audit in 2007 and formed a HSE headquarters audit mechanism of full-coverage-twice-a-year in 2012. Through establishing audit teams of different professionals and combining with the annual key work deployment, the headquarters conducts comprehensive and systematic reviews on the operation of the HSE management system of the subsidiaries which play an effective promotion, publicity and guiding role. It timely publicizes and promotes the HSE management concepts and tools, improves the awareness of the leadership, creates a full participation atmosphere, and promotes the key task implementation.

2.4 HSE audit is an important guarantee to promote continuous improvement

In HSE audit, the audit team follows the principles of objectivity and fairness. Through a combination of audit and guidance, the audit team helps enterprises find out the weaknesses in their HSE management and analyze the reasons of the problems, and puts forward practical recommendations or corrective measures, which effectively promote the standard operation and continuous improvement of the enterprise HSE management system.

3. Relationship between HSE audit and safety inspection

In the establishing and running process of enterprise HSE management system, both HSE audit and safety inspection are the management tools and methods to find out the problems or weaknesses in HSE management, so as to promote the HSE performance. There are some similarities and dissimilarities between the two methods. HSE audit is a scientific, rigorous tool for improving management system, which intends for reviewing the continuing suitability, adequacy and effectiveness of the whole HSE management system. It looks for reasons from management and defects from the system. Safety inspection, however, is a professional, standard method for improving certain state, and focuses on certain production scene or particular businesses to look for hidden dangers and problems.

The following table shows the relationship between HSE audit and safety inspection:

Items compared	HSE audit	Safety inspection
Basis	HSE management system standards	Technical standards, parameters
	Enterprise system and documents	Enterprise system
	National and industry standards and codes	
Object	Functional departments, production scene	Production scene
Contact	Management, functional departments	
	Employee, contractor	Employee, contractor
Way	Interviews, consulting, field observation	Scoring, assessment
Timing	Implementing according to the planned arrangements (active)	Implementing according to the relevant requirements (passive)
Personnel	Auditors with qualification	Managers or professionals
Method	Safety observation and communication	Judgment, comparison
	Tracking, analysis	Giving lectures
Treatment methods	Cause analysis, continuous improvement	Taking the matter on its merits, immediate rectification

4 The problems in HSE audit

4.1 Consciousness

Managers do not realize the purpose and significance of establishing HSE management system nor the important role of carrying out HSE audit. Therefore, managers at all levels don't pay due attention to HSE audit, and their involvement is inadequate. They simply cope with the requirements of the higher authorities, and never really want to seriously perform HSE audit from the ideology and practical action which lead to the momentum and ethos of "solidly going in for formalism, seriously going through the motions".

4.2 Planning and preparedness

4.2.1 The scope of the audit is incomplete

Enterprises do not carry out timely and standardized HSE audit. When conducting audit planning, instead of covering all the internal functional departments and top management, they merely review the subordinate units, or just part of the subordinate units. Meanwhile, they do not assign audit elements in accordance with the duties and authorities and related business process of the functional departments which result in arbitrary audit and neglecting important HSE management activities of the relevant departments. Therefore they cannot conduct systematic assessment and diagnosis of HSE management system running at all levels from the leadership, management to operation, nor give full play to the positive role of HSE audit for management improvement.

4.2.2 Ways for audit organizing are single

Audit arrangements are relatively simple and unreasonable. Regardless of the sizes, geographical distributions and professionals of different organizations, enterprises only adopt a centralized audit way. Because of the time, effort, organizations, and other reasons, audit quality is greatly reduced.

4.3 Auditors

4.3.1 The insufficiency of auditor training

Before each audit, there is no centralized targeted training for auditors, which leads to the unclarity of the understanding of HSE management system standards, audit implementation requirements, as well as the considerations during the audit process.

4.3.2 Uneven quality of the auditors

The auditors' level and capability are uneven. Some auditors lack audit experience. They do not know audit standards well and their audit level and skills need improving. Moreover the auditors still mainly come from the safety and environmental protection departments. Some auditors lack management or technical knowledge. Therefore they cannot find out the deep-seated problems and management risks in HSE system running which affect the quality and depth of the audit.

4.3.3 The instability of the auditor teams

There is no stable experienced auditor team. Almost every audit is performed by "new hand". The large personnel changes cannot guarantee the consistency, continuity and quality of the audit.

4.4 Audit checklist

4.4.1 Checklist content lacking pertinence and operability

Some audit checklists mechanically apply the requirements of HSE system standards. Their contents do not conform to the HSE responsibilities and operation characteristics of various departments and units; some checklists only include audit contents, lacking sampling requirements, audit objects and methods. Therefore the checklists are too hollow to be operable in field audit.

4.4.2 The checklist lacking flexibility

When using audit checklists, the auditors do not conduct reasonable and effective tracking to potential problems or clues found in field audit. They perform dogmatic audit according to the checklists, lacking appropriate modifications and flexibility.

4.5 Audit implementation

4.5.1 Audit focus is not prominent

Key audit objects are not recognized. When auditing various departments and units, there is a lack of interviews to managers and key post staff. Instead of directly auditing those who carry paramount responsibilities, audits are only restricted to the staff who are in charge of document and safety.

4.5.2 Audit depth is not enough

During audit, auditors only point out management problems of documents, records or identifications, such as "no signature", "no number", etc., and they rarely come to real, deep-seated problems of design, production and system management and improvement. This does not actively promote system improvement. Meanwhile, the auditors mechanically apply the clauses of the standard, using personal experience as audit criteria. Therefore audit conclusions are not reasonable and convincing.

4.5.3 The audit approach is too simple, and does not pay enough attention to the effectiveness

At present, the main audit method is to review documents and records. Employee interviews and field observation and communication are inadequately used. Enterprises ignore the validity diagnosis of the system running and concerns of the actual effect of system running. They only pay attention to the completeness of documents and records, lacking analysis and study on related information and data of the records, which lead to the fact that HSE audit cannot promote HSE management.

4.6 Audit findings and improvement

4.6.1 The audit conclusions are not accurate and systematic

The description of audit findings are not specific and accurate, lacking traceability and hard for auditees to accept. Meanwhile, instead of systematic arrangement and analysis of the audit findings, the overall audit conclusion merely lists the problems. It does not clarify the main problems and weaknesses of system operation, and its improvement suggestions are not targeted as well.

4.6.2 Rectification is not complete, just taking the matter on its merits

Some enterprises' audit rectifications are incomplete, which confine themselves merely to facts as they are, and do not infer other things from one fact. The enterprises do not analyze the causes of the problems, nor develop and implement effective corrective measures. They fail to avoid similar problems recurring. In addition, they do not stipulate the person responsible and completion deadlines of the rectifications, nor verify the validity. They try first to make their mistake sound less serious and then to reduce it to nothing at all.

5 Countermeasures for strengthening HSE audit

5.1 Improving the awareness of the importance of HSE audit

Staff at all levels should establish an idea of system management. More over they should unequivocally understand the positive role of HSE audit to enhancing HSE management performance, and improve the awareness of the importance of HSE audit. In safety production, using system thinking of closed-loop management of PDCA to analyze and solve the problem from organizing, responsibilities, processes, systems, resources and capabilities to promote the continuous improvement of comprehensive management level of the enterprises.

5.2 Improving HSE audit system

Combined with actual situation, enterprises should improve HSE audit system, define the workflow of HSE audit, and follow the principles of overall planning, objectivity and fairness, standard procedures, concise operation and scientific sampling to give full play to the systematic diagnosis, decision-making and guidance on HSE system running.

5.3 Preparing operable audit checklist

Audit checklist is a guarantee of audit quality. Combined with the characteristics of professional management requirements and risks, enterprises should prepare targeted, highly operable, adaptive-to- professional HSE audit checklist, making the checklist both a guidance document for HSE system construction of the grassroots units, and audit instructions for auditors. Enterprises should make the grassroots know what HSE system means, and the auditor understand what to audit and how. They should have a definite object in view, give prominence to the key points, and avoid the phenomenon that daily management is a kind of practice, and audit is another.

5.4 Establishing an auditor team with high professional quality

The level of the auditors largely influences and determines HSE audit quality and effectiveness. In order to establish an auditor team with outstanding comprehensive ability and high level of professional work, enterprises can divert professional managers and technical experts who know business well, and can audit, analyze, and speak out. Through centralized training and routine audit, the understanding of audit standards and the use of audit approaches and techniques can be strengthened, and the auditors' coaching skills, training skills and team-work skills can be fostered. Meanwhile, enterprises should ensure the relative stability of the auditor team and the continuity of the audit work.

5.5 Constantly optimizing HSE system audit ways and methods

As for audit organizing ways and methods, combined with the actual situations, enterprises should conduct HSE audit work which is a combination of total-element audit and special audit and a complementation of centralized audit and rolling audit. Instead of taking audit frequency as the main purpose, enterprises should pay more attention to the effectiveness of the audit.

5.6 Strengthening the application of the audit results

Enterprises should timely count, collect, sort and analyze the problems identified in audit so as to find out the short board and weaknesses of HSE management, and to provide scientific and rational information for management and decision-making; at the same time, linking audit results with performance assessment, and when appropriate, integrating audit results into the safety and environmental protection liability statement, can maximize the positive role of HSE audit.

Conclusion Currently, in HSE management system building, "no measure no progress, no audit no lasting" have become consensus which highlights the importance and necessity of HSE audit at present. Therefore, enterprises should make use of PDCA to conduct HSE audit, starting from the ideological understanding, earnestly carrying out the audit planning and preparation, focusing on the implementation of the audit processes and procedures, enlarging the talent pool as well as strengthening problem rectifications. Through improving each link of the audit process, enterprises should further enhance the effectiveness of HSE audit, which can play a greater role for the continuous improvement of HSE management system.

References

[1] China National Petroleum Corporation Enterprise Standard Health, Safety and Environment Management System Part 1: Specification

[2] HSE & Energy Conservation Department of China National Petroleum Corporation, A Textbook of HSE Management System Audit, Petroleum Industry Press, 2012.

[3] Chen Hui, On the Differences between Internal Audit and Inspection, Safety, Health and Environment, 2004,4 (9) .

Process Safety Management System Auditing

David A. Moore
President & CEO AcuTech Consulting Group
dmoore@acutech-consulting.com

Michael J. Hazzan
Technical Manager, AcuTech Consulting Group
mhazzan@acutech-consulting.com

1919 Gallows Rd. Suite 900
Vienna, Virginia, 22182

Prepared for Presentation at
1st CCPS Asia-Pacific Conference on Process Safety
Qingdao, China
September 4-5, 2013

UNPUBLISHED

Process Safety Management System Auditing

David A. Moore
President & CEO AcuTech Consulting Group
dmoore@acutech-consulting.com

Michael J. Hazzan
Technical Manager, AcuTech Consulting Group
mhazzan@acutech-consulting.com

1919 Gallows Rd. Suite 900
Vienna, Virginia, 22182

Keywords: Process safety management, PSM, Auditing, Management System, Metrics

Abstract

Process safety management system auditing is becoming more of a business critical function, and as PSM evolves, it is receiving more attention. Auditing is a key opportunity to validate the process and improve performance to meet expectations for success. Auditing is more effective if aligned to organizational design and goals. Many auditing processes provide an incomplete approach, without consideration of the management system design or organizational goals.

Considerations such as management model, elements of the process, if the program is regulatory or performance driven, multiple performance objectives and key performance indicators should be key considerations when developing PSM management systems and the associated audit process.

1. Introduction

Process Safety Management (PSM) is a "blend of engineering and management skills focused on preventing catastrophic accidents, particularly explosions, fire and toxic releases associated with the use of chemicals and petroleum products" (CCPS, 2010). The term was first used following the Flixborough accident in the UK and popularized following the Bhopal accident in 1984. PSM is recognized throughout all levels of the enterprise from the operating level to the boardroom. Thanks to the efforts of numerous organizations and individuals, there is a growing global recognition of the necessity and value of process safety.

Given the numerous catastrophic and highly publicized events (BP, others), the need for improvement in process safety is paramount. This fact was made clear in the Baker Panel Report of the BP Texas City accident [1]. A critical aspect of any management system is the need to review performance and to then make corrections to the way it is designed or being implemented to improve results. Corporate governance can oversee improvements in process safety.

Process safety management system (PSMS) auditing is becoming a business critical function, receiving more attention than previously as PSM evolves. The expectation for success in PSM is high and auditing is a key opportunity to validate the process and improve performance. Most auditing processes address the PSM system from an incomplete approach without consideration of the actual design of the management system or goals of the organization. Auditing can be more effective if aligned to organizational design and goals.

Many companies are only now facing the rationalization of global PSM performance vs. only regulatory compliance in the United States to the OSHA PSM Standard [2]. Global companies realize the need for uniform approaches to managing PSM and assessing its performance. A notable exception is the American Chemistry Council's (ACC) Responsible Care Management System (RCMS) [3].

This paper will outline an approach for defining corporate objectives, understanding options available, defining a PSM program, defining performance objectives and KPIs, designing an audit program, and then executing the audit process. This approach ensures that the enterprise-wide process is defined and managed for responsible management of process hazards.

2. Defining the Process Safety Management System

The ISO 14001 Standard defines a management system as "that part of the overall management system that includes organizational structure, planning activities, responsibilities, practices, procedures, processes, and resources for developing, implementing, achieving, reviewing, and maintaining the environmental policy." According to the CCPS Guidelines for Auditing Process Safety Management Systems [4], process safety management systems are "comprehensive sets of policies, procedures, and practices designed to ensure that barriers to episodic and potential process safety incidents are in place, in use, and effective". All modern management systems typically follow the Plan-Do-Check-Act (PDCA) model used in quality and environmental management systems as defined by Deming and as recommended in ISO 19011[5]. The PDCA cycle, shown in Figure 1, starts with the need to plan the activities. The author's experience is that many corporations do not have a clearly defined PSM management system for global operations and then an aligned governance function including auditing that optimizes the process. Often the audit function is vague and unplanned, leading to unsystematic reviews and a lost opportunity to achieve enterprise wide insights and conformance.

Figure 1. Example Plan-Do-Check-Act Approach to Management Systems

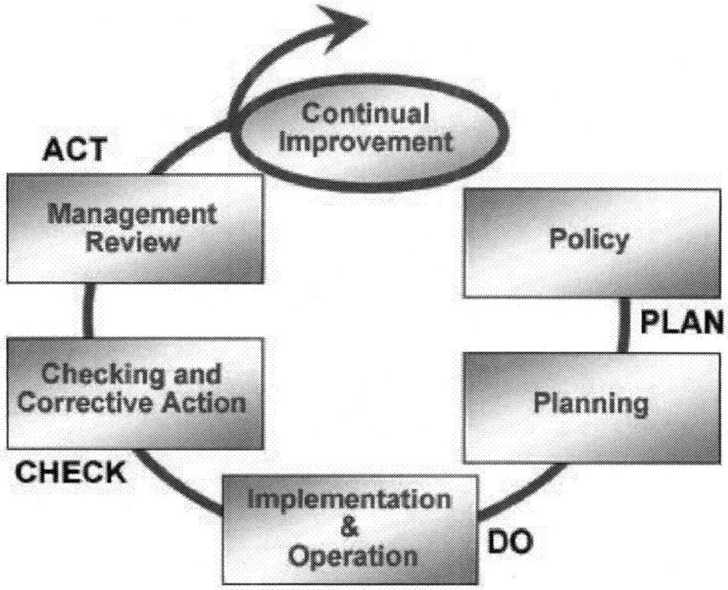

The choice of PSM management system defines the underlying opportunities for managing process safety risks. PSM management systems and the audit process that reviews them need to be defined from the fundamentals, including the following considerations:

- What management model will be used?
- What processes and hazards should the program be applied (e.g. what chemicals)?
- Which elements will the process comprise?
- Is the program regulatory driven or performance driven?
- Are there multiple performance objectives to be met?
- What key performance indicators will be used to measure performance?

By addressing these considerations, the auditing process can be defined to assure compliance to the required system.

Recent incidents and developments have shown that the core PSM elements introduced by OSHA in the PSM Standard do not fully include all of the relevant issues [6]. As a result, one improved choice is the Risk Based Process Safety (RBPS) model described in the AIChE CCPS

"Guidelines for Risk Based Process Safety", and outlined in Figure 2. For those who have existing programs, changing all of the elements may be problematic. Others may implement some but not all of the elements, or to supplement it with their own elements. Other models may include proprietary models, other regulatory models, or minimum efforts such as the U.S. OSHA PSM Standard 14 elements.

If the above 20 elements are already being addressed, the author recommends extending the scope beyond these factors. They represent the technical aspects of the program and they do not necessarily give a full indication of the health of the process safety program. Other areas that could be examined include:

1. Management policies and procedures to support the PSM effort
2. Funding and budget vs. actual expenditure
3. Staffing and resource allocation
4. Organization
5. Progress vs. Key Performance Indicators (KPIs) review
6. Incident statistics
7. Integration with other EH&S elements and departments
8. Senior management attention to the PSM program

Figure 2: AIChE CCPS RBPS Elements

Management Commitment	Asset Integrity and Reliability
Process Safety Culture	Contractor Management
Compliance With Standards	Training and Performance Assurance
Process Safety Competency	Management of Change
Workforce Involvement	Operational Readiness
Stakeholder Outreach	Conduct of Operations
Process Knowledge Management	Goals, Objectives and Plans
Hazard Identification and Risk Assessment	Emergency Management
Operating Procedures	Incident Investigation
Safe Work Practices	Measurement and Metrics
Auditing	Management Review and Continuous

	Improvement

3. Defining the Process Safety Management Audit Objectives

ISO 19011 Clause 5 provides guidance for those who need to establish and maintain an ongoing set of audits for an organization. The standard utilizes the Plan-Do-Check-Act cycle to define the audit program. Some of the key actions addressed are:

- Establishing the objectives and extent of the audit program;
- Establishing the responsibilities, resources, and procedures;
- Ensuring the implementation of the audit program;
- Monitoring and reviewing the audit program to improve its efficiency and effectiveness; and
- Ensuring that appropriate program records are maintained.

Section 5.2 details the requirements for developing the program objectives and extent. This is key to correcting the issues previously mentioned, "Objectives should be established for an audit programme to direct the planning and conduct of audits. These objectives can be based on consideration of:

a) Management priorities;
b) Commercial intentions;
c) Management system requirements;
d) Statutory, regulatory and contractual requirements;
e) Need for supplier evaluation;
f) Customer requirements;
g) Needs of other interested parties; and
h) Risks to the organization."

For PSM, these objectives and priorities need to be carefully considered in light of operating requirements, goals of the organization, risk tolerance, risk governance requirements, legal requirements, and other considerations. The day-today working of the PSM management system (its' programs, objectives, and content) must be aligned with the auditing program.

Some potential objectives of a PSM audit program include the following, which can address process safety risks and ensure corporate governance:

- Improving process safety performance
- Identifying process safety risks and issues
- Improving understanding of PSM requirements

- Assessing and influencing PSM culture
- Achieving compliance to regulatory requirements
- Sharing best practices
- Due diligence as part of merger or acquisition.

Most companies fail to define the objectives and scope. This leads to an unfocused audit program, creating issues that range from lost opportunities to address risk to failures in effort and scope, and can lead to serious consequences.

4. Defining the Process Safety Management Audit Scope

The scope of a PSM audit program is critical to its success. A PSM audit examines management system design, followed by an evaluation of management system implementation. The design of the management system must be understood and then evaluated to determine if the system, when functioning as intended, will meet the applicable criteria. Management has to decide on the scope of the audit initiative including the number of elements to be examined, the depth of the study, the frequency of the audits, and other factors.

ISO 19011 outlines the Extent of an audit program in Section 5:

"The extent of an audit programme can vary and will be influenced by the size, nature and complexity of the organization to be audited, as well as by the following:

a) The scope, objective and duration of each audit to be conducted;
b) The frequency of audits to be conducted;
c) The number, importance, complexity, similarity and locations of the activities to be audited;
d) Standards, statutory, regulatory and contractual requirements and other audit criteria;
e) The need for accreditation or registration/certification;
f) Conclusions of previous audits or results of a previous audit programme review;
g) Any language, cultural and social issues;
h) The concerns of interested parties;
i) Significant changes to an organization or its operations."

Clearly the PSM audit function is receiving more respect and attention that when it was first introduced to industry. Industry has found that PSM is a very complex issue that deserves a thorough audit at least on par with other corporate governance functions such as finance, environment, and business operations.

Management is responsible for establishing the overall policies that will control the audit activity. Also, management should establish and implement the appropriate management-system procedures for the PSM audit program. From the CCPS Audit Guidelines, "a typical PSM audit procedure should address the following topics:

- Selecting facilities for PSM audits
- Establishing frequency of PSM audits
- Planning and conducting audits, including scheduling
- Determining training and qualifications of auditors, including lead auditors
- Selecting and determining audit teams and assignment of the lead auditor
- Developing and maintaining the audit protocol
- Selecting focus units/processes and sampling guidance
- Documenting audits
- Following up on audit findings
- Determining format and content of audit reports
- Distributing and retaining audit reports
- Communicating audit results to the employees
- Providing access to employees of audit results
- Certifying audits (certification required by some process safety regulations)

Figure 3 illustrates the audit process as a Plan-Do-Check-Act model. [7]

In addition, many companies have recognized the importance of using qualified independent auditors, i.e. second- or-third party auditors, in order to avoid conflicts of interest and help ensure a non-biased review. The Safety and Environmental Management System (SEMS) requirements enacted by U.S. Bureau of Safety and Environmental Enforcement (BSEE) for the offshore oil and gas industry specifically incorporates this requirement.

Figure 3. Audit Program Flowchart

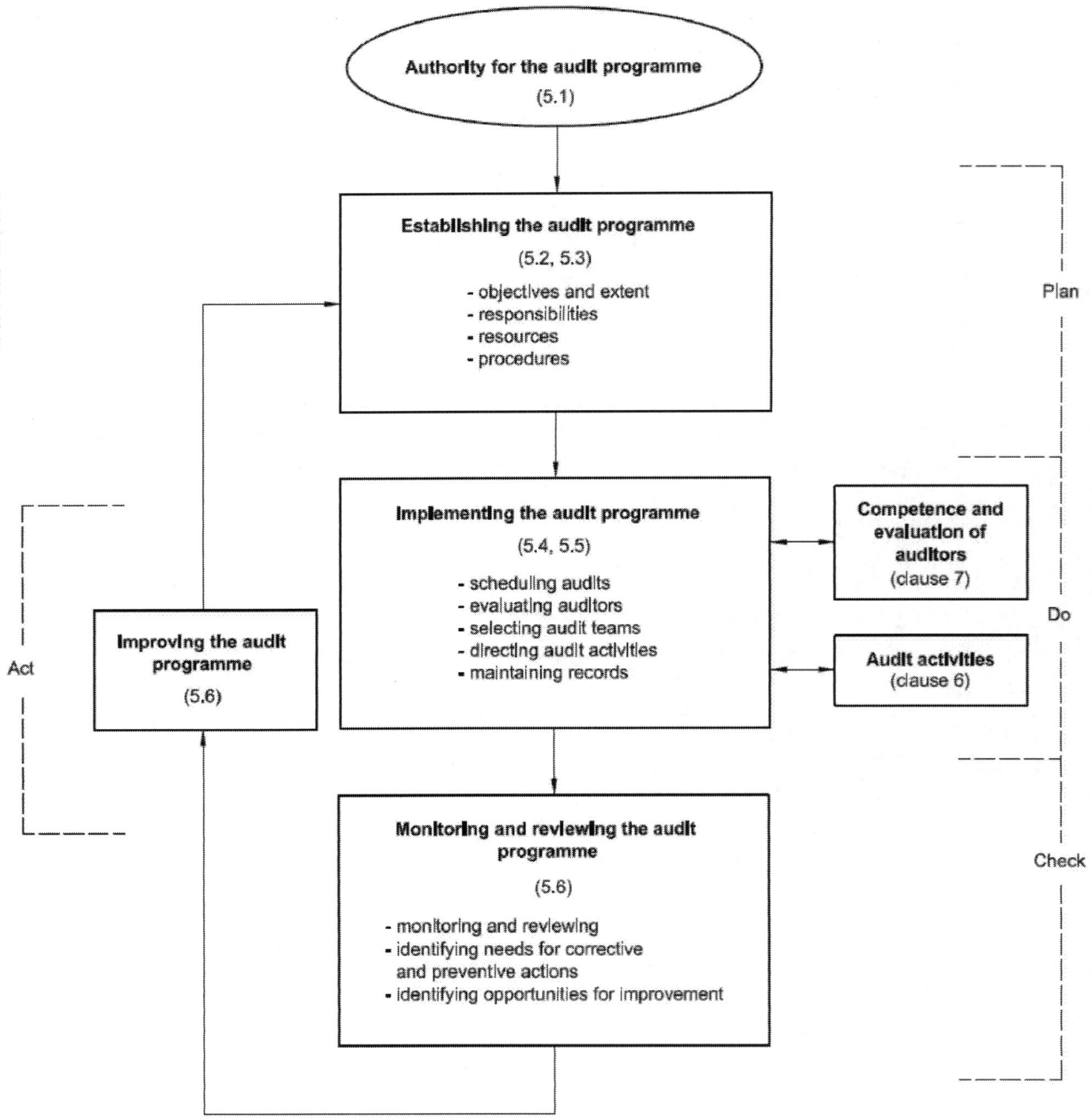

5. PSM Audit Protocol

PSM audits benefit from a defined protocol. Many auditors utilize their judgment and experience, focusing on issues that they recognize. This may lead to an imbalanced and incomplete result. Defining a protocol of minimum questions or issues tailored to the expectations of the enterprise for PSM will result in a more complete appraisal. The CCPS "Guidelines for Auditing Process Safety Management Systems" includes a complete protocol based on the elements of the CCPS Risk-Based Process Safety guidelines. This should be modified to suit the management systems being audited to ensure alignment between the PSM goals, the audit approach, and the report to management on discrepancies to the PSM management system.

6. PSM Metrics

PSM metrics is a topic that is only recently been gaining traction, again following the recommendations of the CSB for the BP Texas City accident. The CCPS Audit Guidelines recommend that metrics can provide useful input for determining the scope of PSM audits. For example, PSM program area deficiencies can be used to determine the scope of an audit. CCPS and others (CEFIC [8], HSE [9], OECD [10]) have developed a set of metrics or guidance for measuring PSM programs The CCPS "Guidelines for Risk Based Process Safety" and "Guidelines for Process Safety Metrics" [11] contain additional guidance for use and auditing of PSM metrics.

During the planning of the audit, metrics should be used to define the focus area, maximizing limited time and ability to sample. Also, auditors should review metrics to gain valuable insight to the status of the program implementation as well as various leading and lagging indicators of performance. The audit may lead to enforcement or modification of the metrics that are collected.

7. PSM Audit Scoring Systems

While there is no universal audit process or scoring system for PSM, it is good to have a scoring system for the audit results. The company needs to choose the design of an audit scoring system. The author recommends both a score for each question in a formal protocol and a system to summarize this into an overall score. Such a system is illustrated in Figure 4. This enables relative measurement against other elements as well as benchmarking at least within the same ranking system, such as across the enterprise. The design of the scoring system must be able to objectively show results and allow for progress to be recorded in a tangible way forward.

Figure 4. AcuTech PSM System Element Ranking Levels

Level	Criteria
5.0 Superior (90-100)	Highest level of effectiveness and completeness, Global Standard as benchmarked against best practice
4.0 Advanced (80-90)	Program exceeds the minimum level of compliance in all requirements; best practices applied
3.0 Minimal (60-80)	Meets minimum intention in all programs requirements
2.0 Incomplete (50-60)	A program exists, but it is lacking completeness or effectiveness in one or more areas
1.0 Inadequate (< 50)	No program exists or one or more are incomplete or ineffective

Figure 5 shows how this system could be used to track the status of development of PSM across all operating facilities of an enterprise and to provide management with an executive dashboard.

Figure 5. Example of Using the System

		Facilities					
Element	Elem. Wt.	A	B	C	D	E	F
Applicability	2	85.0	75.0	50.0	50.0	50.0	50.0
PSM Culture	2	79.6	94.4	94.4	88.9	88.9	88.9
Compliance w/Stds	1	25.0	25.0	62.5	100.0	100.0	81.6
PS Competence	1	90.6	83.0	88.7	84.0	91.5	81.1
Workforce Involvement	2	88.5	88.5	86.5	44.2	44.2	50.0
Stakeholder Outreach	1	76.9	100.0	100.0	100.0	100.0	100.0
Process Safety Information	3	88.0	87.2	88.0	90.2	88.0	72.8
Process Hazard Analysis	3	88.6	85.7	70.0	72.9	85.7	71.4
Operating Procedures	3	79.8	86.9	96.4	96.4	86.9	82.1
Safe Work Practices	2	54.0	74.0	74.0	100.0	94.0	94.0
Asset Integrity	3	88.4	81.5	95.8	98.4	75.3	80.0
Contractors	2	90.3	95.2	79.0	95.2	95.2	90.3
Training	2	84.6	89.7	85.9	92.3	89.7	88.5
Management of Change	3	67.0	73.6	100.0	97.2	50.0	87.5
Pre-Startup Safety Review	2	64.3	67.9	64.3	71.4	21.4	82.1
Conduct of Operations	1	80.8	80.8	80.8	80.8	80.8	69.2
Emergency Response	2	95.4	98.0	93.4	96.7	96.7	97.4
Incident Investigation	1	85.5	93.4	98.7	89.5	90.8	76.3
Metrics	1	64.6	73.2	81.7	56.1	61.0	61.0
Audits	1	68.0	72.0	70.0	80.0	72.0	76.0
Management Review	2	62.5	82.5	80.0	70.0	80.0	70.0
AVERAGE	40.0	78.4	82.6	83.7	84.3	76.9	78.7

8. Conclusion

Carefully designing the PSM audit program to reflect the key expectations and objectives of the PSM management system and to drive future performance will pay dividends. While the audit team will do their best, without a clear strategy and forward planning, it is a missed opportunity. In lieu of an exception only approach where the audit team gives findings and recommendations, quantitative measurement of PSM efforts against goals and specifications for the management system will drive improvements in performance. Expanding the scope of the PSM audit to include all of the now recognized necessary elements and program support issues will help ensure the fullest picture of PSM effectiveness.

9. References

[1] The Report of the BP U.S. Refineries Independent Safety Review Panel, 2007

[2] Occupational Safety and Health Administration (OSHA) 29 CFR §1910.119, Process Safety Management of Highly Hazardous Chemicals, Explosives and Blasting Agents; Final Rule, Washington, DC, February 24, 1992

[3] American Chemistry Council, RCMS® Technical Specification, RC101.02, March 9, 2005, American Chemistry Council, RCMS® Technical Specification Implementation Guidance and Interpretations, RC101.02, January 25, 2004, American Chemistry Council, RCMS® Technical Specification Implementation Guidance and Interpretations Appendices, RC101.02, January 25, 2004

[4] Center for Chemical Process Safety (CCPS), Guidelines for Auditing Process Safety Management Systems, American Institute of Chemical Engineers, New York, 2007, CCPS, 2011

[5] ISO 19011:2002, Guidelines on Quality and/or Environmental Management Systems Auditing. International Standard, International Organization for Standardization, Geneva, Switzerland, 2002

[6] Investigation Report, Refinery Explosion and Fire, BP Texas City Refinery, US Chemical Safety and Hazard Investigation Board, March 20, 2007

[7] ISO 19011:2002, Guidelines on Quality and/or Environmental Management Systems Auditing. International Standard, International Organization for Standardization, Geneva, Switzerland, 2002

[8] CEFIC - Guidance on Process Safety Performance Indicators, 2nd Edition, 2011

[9] Developing Process Safety Indicators; A Step-by-Step Guide for Chemical and Major Hazard Industries, UK Health & Safety executive, HSG254, 2006

[10] Organisation for Economic Co-operation and Development (OECD), Guidance on Developing Safety Performance Indicators, 2008

[11] Center for Chemical Process Safety (CCPS), Guidelines for Process Safety Metrics, American Institute of Chemical Engineers, New York, 2007, CCPS, 2011

1ST CCPS ASIA-PACIFIC CONFERENCE ON PROCESS SAFETY℠

Process Safety Leadership and Culture – How to Measure and Ensure Improvement

Steve Arendt, PE, Vice President
sarendt@absconsulting.com

Rick Curtis, Senior Risk Consultant
RCurtis@absconsulting.com

Eddy Schedule, Manager Training Programs
WSchedule@absconsulting.com

ABS Group
16855 Northchase Drive
Houston, TX 77060

Prepared for Presentation at
American Institute of Chemical Engineers
1st Asia-Pacific Conference on Process Safety
Qingdao, China
September 4-5, 2013

UNPUBLISHED

Process Safety Leadership and Culture – How to Measure and Ensure Improvement

Steve Arendt, PE, Vice President
sarendt@absconsulting.com

Rick Curtis, Senior Risk Consultant
RCurtis@absconsulting.com

Eddy Schedule, Manager Training Programs
WSchedule@absconsulting.com

ABS Group
16855 Northchase Drive
Houston, TX 77060

Keywords: PSM, Process Safety Management, RBPS, Risk Based Process Safety, Culture, Leadership, Operational Discipline, Conduct of Operations

Abstract

Process safety leadership and culture are recognized as key factors for effective process safety management implementation. In spite of this, major accidents investigations continue find these lapses as significant contributors. Leadership and process safety culture assessment and improvement in a multi-cultural environment is particularly difficult. This paper addresses a method for baseline evaluation of culture/leadership and ongoing near-real-time measurement of improvement based upon the CCPS essential features of a good process safety culture as published in *Guidelines for Risk Based Process Safety.*

1. Introduction

Companies have been taught many times that organizational factors have been important contributors to major accidents. Some of those organizational characteristics have to do with not having a proper safety culture, failing to exhibit strong leadership to support the culture, and not creating the consistent operational discipline at all organizational levels. One theme common to all three of these aspects has been the failure of companies to learn from experience – either from their own or from others. The following are some characteristics of a "process safety learning-disabled company:"

- Dysfunctional safety culture
- People hide things and kill messengers
- Fail to question; procedures not followed without accountability
- Mixed/improper safety/production messages
- Complacency, low trust, silo mentality
- Misplaced safety ownership, invisible/ineffective leadership

- Superficial causal analysis of problems
- Things don't get fixed – or things don't stay fixed
- No company memory

In order to address these learning disabilities, companies should strive to improve operational discipline, leadership, and eventually their culture. The following sections describe examples of each of these aspects and how to improve them.

2. Relationship among Culture, Leadership and Operational Discipline

Process safety culture has been a key organizational factor in several major process safety incidents. Leadership has been a focal point of several industry group promoting performance improvement CCPS's 20 elements for success and Vision 2020 and RBPS Culture and Conduct of Operations elements.[1,2] The UK Process Safety Leadership Group's report following the Buncefield accident.[3] Subsequent tools provided by the OECD and the UKPIA.[4,5] In the U.S., the API/AFPM have been focusing on Process Safety Leadership as one of its assessment areas. [6] Finally, lack of operating discipline, as a product of a Conduct of Operations program, has been a sore point for many companies whose incident causes increasingly involve "people did not follow the procedures."[7]

How Are These Aspects Connected?

Culture is the tendency in all of us – and our organization - to want to do the right thing in the right way at the right time, ALL the time – even when/if no one is looking. Leadership is an essential feature of a good culture. Operational discipline (or the lack thereof) is a behavioral result of your culture and leadership. A ConOps program is a system designed to create and manage operational discipline.

3. Safety Culture

The Center for Chemical Process Safety made culture an official safety management system (SMS) element for the first time when it published its ***RBPS Guidelines***.[1] CCPS safety culture working group and ABS Consulting evaluated major organizational accidents and prepared a Safety Culture Awareness tool, which has been widely distributed via CCPA's web page. Subsequently, Process Safety Culture was defined as an element in the ***RBPS Guidelines*** wherein it created structure for a culture management practice and laid out the "Twelve Essential Features of a Good Culture."

Table 1. CCPS Process Safety Culture – Essential Features

1	Establish safety as a core value
2	Provide strong leadership
3	Establish and enforce high standards of performance
4	Formalize the safety culture emphasis/approach
5	Maintain a sense of vulnerability
6	Empower individuals to successfully fulfill their safety responsibilities

7	Defer to expertise
8	Ensure open and effective communications
9	Establish a questioning/learning environment
10	Foster mutual trust
11	Provide timely response to safety issues and concerns
12	Provide continuous monitoring of performance

Other organizations have followed suit with establishing their own industry structures for culture. For example, the following table lays out the framework for the U.S. offshore industry recently published by the U.S. Bureau of Safety and Environmental Enforcement (BSEE).[7]

Table 2. BSEE Robust Safety Culture Characteristics

1	Leadership safety values and actions
2	Work processes
3	Personal accountability
4	Respectful work environment
5	Inquiring attitude
6	Environment for raising concerns
7	Effective safety communication
8	Continuous learning
9	Problem identification and resolution

As one could expect, the more of these frameworks created by various groups – almost with respect to which industry they come from – the more these frameworks tend to cover the aspects. Ultimately, it will not matter which framework you follow, but that you excel in the aspects of any one of them. When this doesn't happen and a poor culture persists, here are some lessons the authors have learned about why and what needs to be done.

- If you have poor culture, marked by mistrust or needs large improvement, the worst thing you can do is too just start "talking" about it at the top
- The "top" needs to first start "behaving" better to address culture weaknesses
- Then, the talk will build up from the bottom
- If you survey, do it anonymous and voluntary; you should commit to sharing the results – quickly
- Any education/training, etc. should extend to ALL of the workforce, including contractors
- BUILD OWNERSHIP

It has been said that management's role in nurturing culture involved "enabling the environment" – not so much "manufacturing" the culture. Management:

- Sets the standards
- Sets the tone
- Provides the resources

- Provides reinforcement

MIT Professor Dr. Edgar Schein has stated "...one could argue that the only thing of real importance that leaders do is to create and manage culture..." The authors of this paper agree with the spirit of Dr. Schein, but prefer to think of Management as Enablers and that the culture is advance via strong leadership at ALL organizational levels – not just the top management.

4. Process Safety Leadership

So, given the above and the recognition by organization that the most practical way to advance culture is by improving leadership, what does it mean to exhibit good process safety leadership traits? The following table lays out the traits that the authors can complied by winnowing through myriad leadership reference and applying it to a Process Safety context.

Table 3. ABS Group Process Safety Leadership Attributes

1	Ethical	10	Communication
2	Committed	11	Accountability
3	Vision	12	Connectivity
4	Competent	13	Develop others
5	Courage	14	Empowering
6	Emotional intelligence	15	Change navigator
7	Sense of vulnerability	16	Responsive
8	Influence	17	Consistent
9	Visibility		

1. Ethical Behavior: demonstrates organizational core values in all process safety decisions and follows the correct procedures in process safety management no matter what the current conditions or pressures are to deviate.

2. Commitment: the ability of a process safety leader to stay focused and not deviate from the requirements of process safety no matter what other influences or pressures are being applied either externally or internally.

3. Vision: the ability of a leader to see the present process safety culture as it is and formulates a future plan that grows the culture out of and improves upon the present.

4. PS Competence: to be an effective leader in process safety, a leader must have a high level of experience and competence in process safety management systems.

5. Courage: the process safety leader must be able to say what needs to be said and the fortitude to stop any unsafe process or work practice that could adversely affect people, the environment and or the plant.

6. Emotional Intelligence: the ability of the leader to control their emotions and recognize other's emotional reactions in stressful or conflicting situations in regards to achieving positive process safety outcomes or decisions.

7. Sense of Vulnerability: leads and drives others to maintain a high awareness of process hazards and their potential consequences and is constantly vigilant for indications of system weaknesses that might foreshadow more significant safety events.

8. Influence: in process safety leadership, a leader must be fully aware of how their actions or words compel others behaviors, opinions and actions toward or away from process safety. To have a positive impact on the process safety culture, they must use their level of influence to promote best practices and practice what they say at all times.

9. Visibility: takes the initiative to get out into the process areas to observe and get feedback from the workforce. Sets the example in leading from the front in process safety and displays the organization's core values during every event or task.

10. Communication: a process safety leader must ensure healthy and open communication channels exist both vertically and horizontally within the organization. Vertical communications are two way – leaders listen as well as speak. Horizontal communications ensure that all workers have the information needed for safe operation.

11. Accountability: the acceptance of personal responsibility in ensuring process safety management systems are in place and being followed by the workforce.

12. Connectivity: the understanding that process safety culture is not just about procedures or requirements. It's about developing relationships within the entire organization to aid the process safety leader in driving a learning organization focused on sharing process safety knowledge.

13. Develop Others: ability to grow and expand the skills of the workforce in process safety to develop future process safety leaders and create a learning organization.

14. Empowering: demonstrate the ability to be the catalyst in workforce involvement and encourage employees to provide input openly to support process safety culture and learning opportunities

15. Change Navigator: the process safety leader must be an expert in management of change and also have the ability to lead others through required changes and recognize the various reactions to change people may experience.

16. Responsive: listens to process safety concerns and follows up on issues with corrective action.

17. Consistency: displays fairness and consistent application of process safety management systems no matter what the current production climate may be.

Even with a good framework that is energized and enabled by organizationally penetrating leadership, sustaining a great culture over a long period to withstand all types of stressors requires maintaining Operational Discipline.

5. Operational Discipline and Conduct of Operations

Operational discipline involves people in an organization routinely following the policies, rules, and procedures prescribed in order to meet organizational expectations from top-to-bottom in the chain of command forever. A Conduct of Operations element is a PSM program element containing the organized management practices intended by the company to create and sustain operational discipline. CCPS published its ***ConOps Guidelines*** which contain the following people-related aspects influencing leadership and culture.[8]

Table 4. People Aspects of ConOps

Clear Authority/Accountability
- Accountability is assigned
- Workers understand the lines of authority
- Technical and other support personnel understand the lines of authority
- Accountability flows upward as well as downward

Communications
- Structured messages in a standard format
- Written instructions or procedures
- Key parts of the message repeated back to the sender
- Structured protocols, checklists, and logs to supplement written instructions
- Pre-job briefing for the entire work crew and a field walk down by people with knowledge/experience
- Important information reinforced via multiple communication channels

Compliance with Policies and Procedures
- Personnel who interrupt work due to an anomaly are congratulated on their attention to detail rather than chastised
- Improvements in procedures, policies, or practices are subject to a formal change management (MOC) system
- Management ensures that improved outcomes are not based on shortcuts or unauthorized modifications

Intolerance of Deviations
- Management sets the example
- When standards are changed, the reasons are made clear and the rationale is sound
- There is a process for authorizing exceptions, and it is strictly followed
- Workers hold peers accountable for adhering to policies, procedures, and standards
- When standards are violated, individuals who should have been enforcing the standards, who authorized the activity, and who actively participated are all held accountable

Supervision/Support
- Embraces the ConOps system
- Maintains a predictable balance between taking corrective action and taking disciplinary action
- A sense among all leaders that they are ultimately responsible for everything that their organization does and fails to do
- Supervisors find ways to make workers successful by leveraging their strengths and compensating for their weaknesses
- Management supports supervisors' role in enforcing the OD concepts

6. How to Measure Culture, Leadership and Operational Discipline

Process safety culture is hard to measure and more difficult to change. There are few direct indicators of process safety culture, and because of its nature, it cannot be evaluated very frequently. Leadership and operational discipline are essential attributes of sustaining a healthy process safety culture. So, how do you know if better culture or operational discipline is needed? What evidence would lead you to believe that you need better operational discipline and that you need a ConOps element? Typical ways to get a handle on process safety culture are:

Employee surveys – Surveys are the most frequently used method. Typically, a company will prepare an anonymous survey (20-70 questions, shorter is better) for both hourly employees and management. The content of the survey historically has been focused more on occupational safety issues, but recently they have been adapted to address process safety issues. Survey questions are developed to see how employees "feel" about important process safety-related matters. Respondents are given a choice of five answers to gauge the strength of their feelings about the issues – strongly agree, agree, neutral, disagree, strongly disagree. Questions and results are normally placed in categories relating to the process safety issues of concern (Process Safety Reporting, Commitment to Process Safety, Supervision, Procedures and Equipment, Employee Involvement, Process Safety Training, and Safety Processes).

The following are a few examples of typical process safety survey questions:

Typical Process Safety Culture Survey Questions
Plant management is sincerely interested in feedback on process safety from the employees.
My co-workers routinely report all process safety issues - including near misses - to our supervisor.
Process safety is #1 at this company – we do not cut corners in order to maintain or improve production
When an incident occurs, management focuses on correcting the issue rather than on "blaming someone."
Our supervisor never asks us to perform an unsafe operation or task.
My coworkers never "work around" a safety problem, rather than report it
Overall, I am satisfied with the process safety training we receive
Safety systems are not left disabled, but are restored as rapidly as possible

Some difficulties with surveys are (1) that they should be voluntary, which may lead to insufficient participation to achieve statistical validity and (2) surveys cannot be repeated very often or else workers will become accustomed to it and can tend to feed back "what you want to hear" rather than what they are really feeling.

Interviews – another way to elicit process safety culture insights is through limited, representative, but targeted, interviews of company personnel. These interviews may last from 15 minutes to an hour. Interviews may address the same types of issues as surveys; however, the questioner can go into more detail on a topic if the answer warrants. A disadvantage of interviews is that they are very time-consuming and resource intensive and the results are more difficult to pull together in a consistent framework for analysis.

Work observations – Process safety culture issues that deal with the tendency for employees to not following procedures, safe work practices, etc. can be identified via workplace observations or peer-audits. These can very effective, but are difficult to conduct in a consistent fashion using a large number of observers. The biggest limitation is that they are difficult to do without the person being observed knowing that they are being watched/evaluated. If they know, you are unlikely to get the "real" information about how the worker behaves "without anyone looking."

Process safety leading indicator metrics – More companies are using leading indicators of process safety as a window into process safety culture. For example, the rate of reporting of near-misses, the rate of close-out of action items, or the completeness of training compliance can be used to gage a company's leadership in process safety. Metrics are good because they can be refreshed frequently, but they are usually very "indirect" measurement of process safety issues.

Depending upon the situation, we typically use a combination of these means, anchored by some variation on a culture survey.

Process Safety Performance Assurance Review

ABS Consulting has devised an approach for connecting process safety culture survey results to process safety outcomes. This process safety **Performance Assurance Review** approach (a) categorizes the culture survey results and maps them to the 12 essential features of a good process safety culture, (b) categorizes the results from a review of recent and historical process safety performance at a plant (e.g., current PSM or EHS audit results) and maps these results to the same 12 essential features. The "weighted outcome" of the mapping of contributions of both the survey results and the PSM audit results to the 12 essential features are totaled and the most significant process safety culture issues are identified for the plant/company.

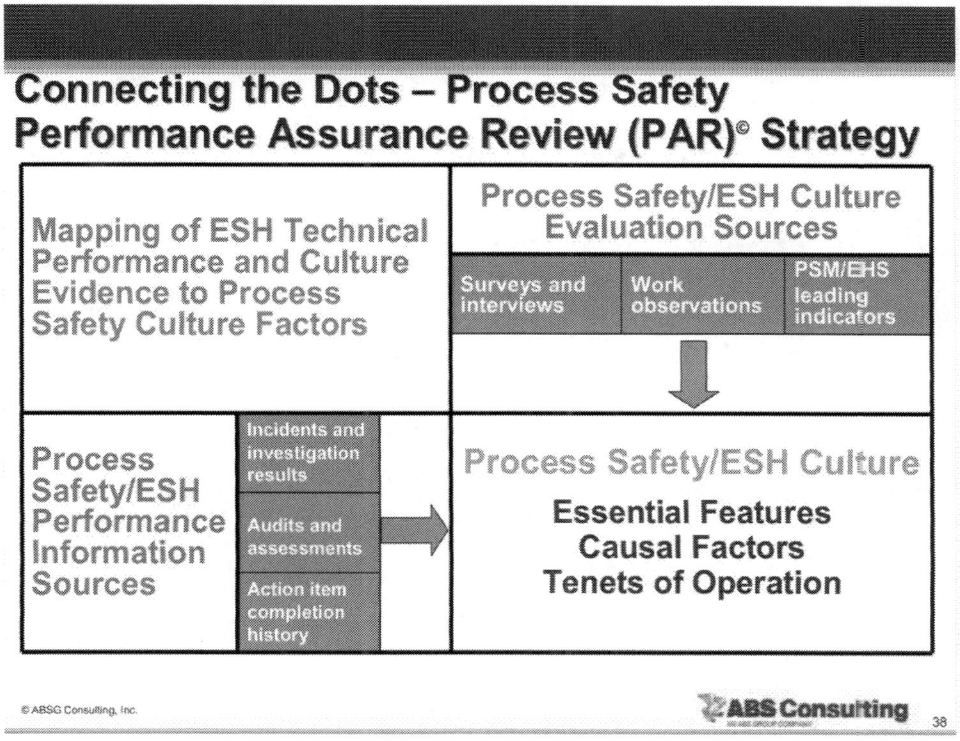

Categorizing the Process Safety Culture Survey Results

The results of the process safety culture survey are categorized into the 12 essential features of a good process safety culture. The survey results for the positive responses (Strongly Agree plus Agree) are ranked as shown in the example below.

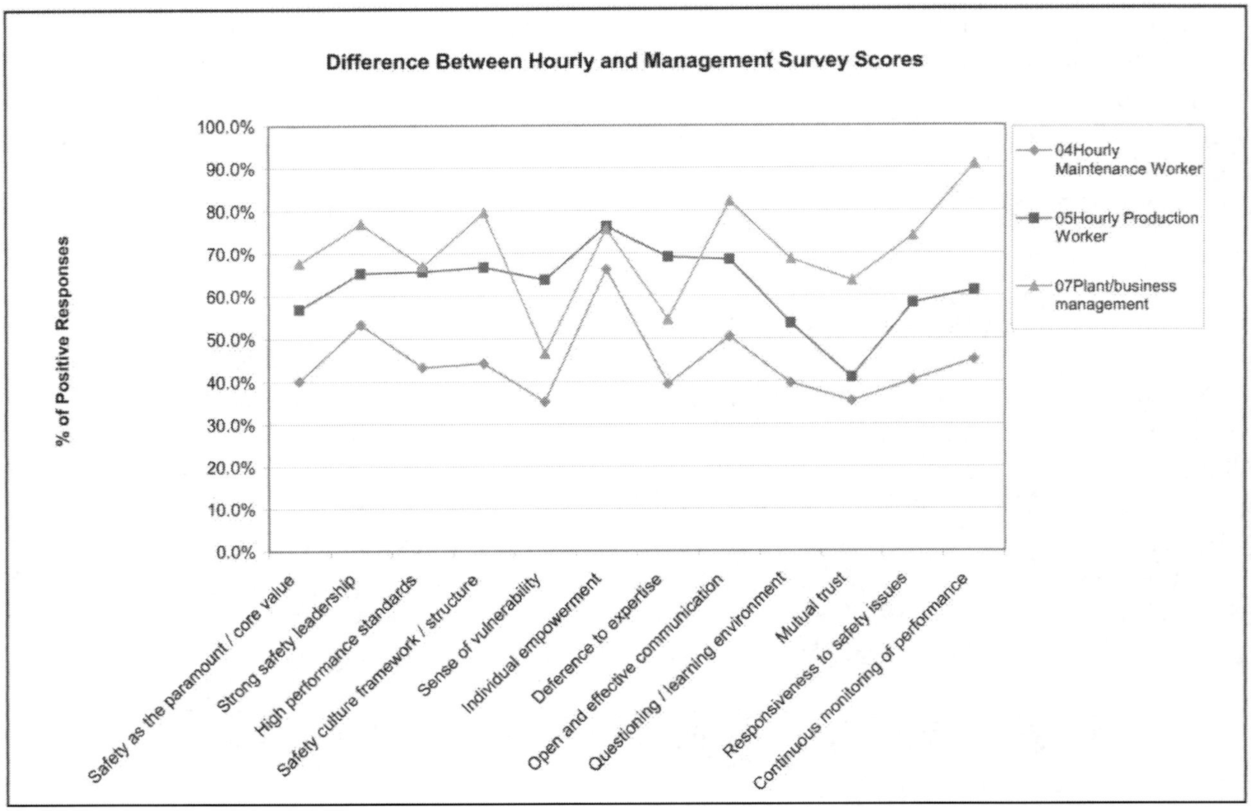

Also shown is a typical difference one sees in survey responses between demographic groups.

Categorizing the Process Safety Performance Review Results

Technical findings are given varying degrees of "risk significance" based upon (a) the type of finding (e.g., imminent integrity issue vs. improper training records) and (b) the number of items contributing to the finding (e.g., 93 problem situations found for one category of finding vs. 8 situations found for another).

- Containment integrity issue allowed to exist
- Unsafe work practice
- Action item not completed or late
- Safety hazard situation is allowed to exist
- Inadequate inspection, testing, maintenance
- Inadequate management practice/ system / procedure
- Inadequate hazard, risk, root cause analysis
- Inadequate monitoring or auditing
- Inadequate training
- Inadequate recordkeeping and documentation
- Inadequate communication or signage/placards

Increasing Risk Significance ↑

Each of these categories is mapped to the 12 essential features of a good process safety culture. This mapping shows the relationship between the existence of the apparent underlying culture problem and how it may have results in the evident process safety outcome. For example, a plant that has (1) known, imminent apparent safety hazard situations that have not been fixed in a long time, (2) a large number of action items not completed, (3) large deficient equipment and inspection backlogs, and (4) poor operating discipline indicated by a lack of following safety work practices indicates that the plant has the culture disease symptom of "Normalization of Deviance – Lack of Setting and Follow Process Safety Standards and Rules".

Connecting Process Safety Performance to Culture Results

The survey results and the technical results are both weighted and mapped to the same 12 essential features. The weighted contributions from survey results that also have contributions from the technical results are summed to give the overall score for the essential feature. Features that have high survey results but do NOT have high scores for process safety evidence are reduced in significance. Features that have significant survey scores that also have significant process safety evidence scores are enhanced in importance.

Safety Culture Essential Features:						Safety as the paramount value	Strong safety leadership	High performance standards	Safety culture framework / structure	Sense of vulnerability	Individual empowerment	Deference to expertise	Open and effective communication	Questioning/ learning environment	Mutual trust	Responsiveness to safety issues	Continuous monitoring of performance
Scores from Survey Results of Dataset : 01						71%	73%	73%	71%	68%	76%	62%	68%	55%	70%	76%	66%
Safety Performance Deficiency	Safety Cultural Performance Ranking					4	4	4	4	4	4	6	4	6	4	4	4
Plant 1 Dataset	Audit Findings	Consequence Level	Severity Adjustment Factor (1-4)	Risk Rating	Risk Category	EHS Technical Performance / Culture Evidence to Process Safety Culture Factors											
Containment integrity issue allowed to exist	0	50	0.0	0	0	0.0	0.0	0.0		0.0	0.0	0.0		0.0		0.0	0.0
Unsafe work practice	0	25	0.0	0	0	0.0	0.0	0.0	0.0	0.0	0.0	0.0		0.0	0.0		0.0
Action item not completed, late, or chronic	0	25	0.0	0	0	0.0	0.0	0.0		0.0			0.0	0.0		0.0	0.0
Safety hazard situation is allowed to exist	0	20	0.0	0	0	0.0	0.0	0.0		0.0	0.0	0.0		0.0		0.0	0.0
Inadequate maintenance, inspection, testing	4	20	2.0	40	6	0.5	2.4			0.5		0.7		0.7			2.4
Inadequate management practice / procedures / system	2	15	3.5	53	25	2.0	10.0		10.0	2.0			2.0				2.0
Inadequate hazard, risk, or RCA review / analysis	9	15	3.5	53	25	2.0	10.0		2.0	10.0	2.0	15.0		15.0			2.0
Inadequate monitoring or auditing	1	10	1.5	15	4	0.3	1.6	0.3	1.6	0.3	0.3			2.4	0.3		1.6
Inadequate training	3	10	3.0	30	6	0.5	2.4		0.5	0.5	2.4	0.7		0.7	2.4		
Inadequate recordkeeping and documentation	1	5	1.0	5	1	0.1	0.4									0.1	0.1
Inadequate communication or signage	1	5	2.5	13	1	0.1	0.4				0.4		0.4	0.1	0.1		
UPDATE COLORS				Priority Guide:		3	2		2	2	3	1	3	1	3		3
21						5.4	27.2	0.3	14.1	13.3	5.1	18.4	2.4	19.0	2.8	0.1	8.1
						7	3	11	4	5	8	2	10	1	9	12	6

The process safety performance assurance review (PAR) seeks to find technical evidence consisting of process safety outcomes (e.g., findings and improvement areas) that seem to support the opinions and feelings of the respondents from the process safety culture survey. The results of this comparison showed which process safety culture problems indicated by "opinions" in the survey were "proven out" by the company's recent historical performance in the field. This way, a company can have greater confidence that the potential culture issues uncovered from the survey are, in fact, real – and supported by visible evidence in the field. The following figure gives an example of the "survey and technical evidence" for a typical essential feature.

Evidence of 9 - Lack of a Questioning/Learning Environment

Technical Evidence

- 21 open PHA recommendations, 14 overdue, all 21 would become overdue within 2 weeks of the review
- All actions from investigations were not recorded in tracking tool, some were overdue, others closed without complete resolution of the concern

Culture Evidence

- 59. Contractors at my site are adequately trained in safety – 36% agree
- 63. The process safety training that workers receive at my company is of high quality – 48% agree
- 60. Newly hired workers are adequately trained in safety – 50% agree
- From interviews, training materials are out of date (not reflecting current equipment) and training received is inconsistent, including lack of adequate mentoring of new employees

The following figure illustrates the results from a typical PAR, indicating the top six process safety culture issues that need to be addressed by the company.

Process Safety Cultural Causal Factor	Ranking
9. Lack of a questioning/learning environment	1
12. No performance monitoring/pursuit of improvement	1
5. Lack of sense of vulnerability	2
7. No deference to expertise/burden of proof on safety	2
11. Non-responsiveness to safety concerns	3
3. Normalization of deviance	3
2. Lack of strong process safety leadership	3
1. Process safety is not a core value	4
4. No formalization of a "culture process"	4
8. Ineffective communications	4
10. Lack of trust – unsafe reporting environment	4
6. Lack of personal responsibility for process safety	5

In this table the process safety features are characterized by their negative representation – thus, called cultural causal factors, similar to causal factors identified in a root cause analysis. The cultural causal factors are listed in five strata – the more significant culture issues are nearer the top of the table. From this table it is apparent that this company is challenged with several process safety culture factors – the primary ones being the lack of a questioning/learning environment and a lack of monitoring of process safety performance/pursuit of continuous improvement.

No this company can move forward in improving process safety with better insights into not only the physical and management system problems that need to be fixed, BUT ALSO the underlying cultural issues that have contributed to these problems existing for so long. Addressing ALL of these issues will give the company the best chance at sustaining its process safety improvement for a longer time.

Process Safety Culture Evaluation Case Studies

Over the past six years, process safety performance and culture reviews have been conducted for 30+ companies in the oil, chemical, pharmaceutical, consumer products industries, involving tens of 1,000's of employees, 100+ facilities, domestic U.S. and international and old companies and new companies

All of these studies included an evaluation of safety culture – most also involved mapping to process safety performance outcomes. The following table summarizes the frequency at which the 12 essential features of a good process safety culture were found lacking at the company or facility – company names not included.

Process Safety Culture Problem	Company Problem Location								
	A	B	C	D	E	F	G	H	J
Process safety is NOT a core value	6	3	12	6	5	10	10	9	8
Lack of strong PS leadership	7	7	11	7	6	11	11	10	1
Normalization of deviance	1	1	4	2	1	2	7	1	2
No formalization of culture process		12	10		12	12	12	12	12
Lack of sense of vulnerability	5	2	5	4	8	7	8	5	7
Lack of personal responsibility for safety			8			9	9	8	9
Not listening to technical experts		10	6		9	8	1	6	4
Ineffective communications		11	2		11	6	6	11	10
Lack of a questioning/learning environment	4	5	7	5	3	3	2	4	5
Lack of trust – unsafe reporting environment		6	1		10	5	5	7	6
Non-responsiveness to safety concerns	3	8	3	3	7	1	3	3	3
No performance monitoring or pursuit of improvement	2	9	9	1	2	4	4	2	11

Some observations can be made from these combined results:

- The top three cultural problems were way above all the others
- Surprising that "culture foundation issues" were so low – core value and strong leadership

- Two of the 10 companies had process safety culture problems that were not high in the other 8 cases
- Seven out of 10 of the companies had undergone significant organizational change in this decade
- No direct information on cultural root causes – research continues as to how these companies got to the point where they are
- Even without having robust cultural root cause information, it is possible to heal culture disease

7. How to Improve Process Safety/HSE Culture

Once you have determined the areas of process safety/HSE culture weakness, it is time to begin the process of improvement. Usually, culture improvement activities will take many months or years to have a deeply rooted effect. Regardless of the size of the plant or company, the following activities and concerns are important to creating a last-lasting culture change.

- Education
- Planning
- Workforce involvement

Education is given to all layers in a plant/company. Typically, three main audiences are addressed with approximate initial awareness training durations given:

- 1 to 1½-days for plant management or corporate executives
- 4 to 6-hours for supervisors, engineers, and team leaders
- 1- to 2-hours for all employees (including contractors) for awareness

The following are some example topics used for reach layer of training.

Management and Supervision Course Topics and Activities

Addresses organizational culture as a key element in the successful implementation of both process safety and conventional safety programs. Attendees learn how to:

- Assess current plant culture
- Determine changes that need to be made
- Establish an environment to foster change
- Measure the effect of culture change
- Discuss suggestions for ways to correct process safety cultural problems

Management courses should include how to design and implement a management system for evaluating and nurturing plant process safety culture. Courses should revolve around facilitated workshop exercises; however, the number and timing of such activities will be limited in the shorter duration course.

Employee and Contractor Awareness Level Course Topics

This training focuses primarily on:

- Examples of accidents that occurred due to process safety culture problems
- 12 essential features of a good process safety culture
- Taking personal responsibility for evolving your process safety/HSE culture
- Understanding potential historical root causes for culture problems
- Soliciting ideas for improving culture

Culture Workshop Ideas

Such training, whether for executives or hourly workers, is more effective is the delegates are fully engaged and thinking about culture change. Their opinions matter and it is important that we discover their ideas for culture change. Interactive, facilitated workshops provide an effective means for dialogue, soliciting ideas, engagement, and learning. These workshops are most effective if done using small groups that consider some of the following issues:

1. What can you do to assess the culture in your company and/or your facility?
2. What can you do to identify which of the process safety culture features will have the greatest impact in your company; which is the weakest in your company?
3. Pick one of the culture areas and develop an action plan to address the following issue at your company/site.

    - Develop a culture improvement plan for your company/site. Be as specific as possible.
    - Decide which culture elements you are going to address
    - Decide who in management is needed to support your efforts
    - Develop a plan for the next month, the next six month, the next year, and the next three years.
    - Determine what metrics will be used to monitor progress

Identify Culture Improvement Options

The following steps should be addressed in culture workshops at all levels of the plant/company.

1. <u>Identify possible culture weakness root causes.</u>

    - Brainstorm what has happened in the past that could be in the memory of the work force
    - Look at culture survey/interview results by feature.
    - What is going on the past year/recently that could affect this feature?

2. <u>Identify existing activities that could be used or amended to help correct the cultural issues.</u>

- What things are going on now in the plant/company that could improve or adversely affect this issue?
  - Capital expansion/operations contraction
  - Employee welfare
  - Workforce changes
  - Implementation of new safety programs (e.g., Behavior Based Safety)

3. Look at the PSM/HSE audit finding resolution actions.

- See how they could be done in a "culture-smart way
- Avoid initiative overload and unfunded mandates that could adversely affect culture efforts

4. Identify additional actions to fix the culture issues.

- See the list of potential remedial actions in the next section as a starting point.
- Have small groups during culture awareness training develop company and personal culture action plans using this list.

5. Develop an overall, integrated list of activities to help deal with the culture challenges.

- Work the plan as you would any process safety/HSE action plan
- Monitor implementation
- Develop culture improvement metrics

Remedies for Culture Weaknesses

Following the above approach will require that you eventually address what actions you should take to improve culture. You will follow an interactive process that engages all levels of the plant and company. At each occasion early on, you will seek people's opinions about the root cause of culture weaknesses, factors that contributed to the problems, conditions that the company faces in the near-term as it pursues culture improvements.

Normally, a plant or company faces one or more culture issues at the same time. These issues should be addressed together to avoid adverse interactions of improvement actions and to increase efficiency. The following tables give some examples for each culture essential feature of possible remedies that have been used by many companies to address culture weaknesses.

Table 5. Process Safety/HSE Is NOT a Core Value

• Establish a process safety policy • Communicate the Business Case for Process Safety to senior executives; communicate the process safety policy regularly • Hold training courses/workshops on the policy • Reinforce policy via company newsletters and meetings • Ask the workforce for feedback on whether they consider process safety to be a core value and whether management is acting that way • Demonstrate that process safety is a core value by making decisions and taking visible action in support of the

policy • Celebrate workforce actions that demonstrate adherence to process safety being a core value • Encourage senior executives to participate in process safety functions/committees/symposia with other respected executives • Widely share the CCPS Process Safety Beacon • Share significant incidents from similar facilities and relate the broader consequences of a process safety incident • Help people at all levels in the organization to understand the importance of process safety • Appeal to the business case for process safety, the humanitarian case for process safety, and the legal case for process safety • Find a process safety champion high in the organization that is willing to try to influence other senior leadership • Provide ideas/input for annual reports and other visionary documents that are periodically published

Table 6. Lack of Strong Process Safety/HSE Leadership

- Establish a clearly defined process safety leader at a level that can influence operational vice-presidents and facility managers
- Clearly define process safety responsibilities
- Establish process safety element owners at the company level and at the site level
- Include process items in annual objectives for line management leadership positions
- Provide short talking points for company executives to mention during town hall meetings
- Periodically publish a short company process safety newsletter that includes notes from executives
- Provide process safety topics to be included in the company annual report
- Fund specific process safety activities
- Participate in process safety activities (e.g., management reviews)
- Process safety significance is reinforced throughout the chain of command
- Take decisions and action on the side of process safety
- Attend process safety training courses/workshops
- Address process safety issues in every company/work group meeting
- Discuss process safety during day-to-day conversations
- Visit workers and discuss process safety issues
- Investigate lapses in process safety performance
- Participate in external venues to promote process safety as a core value
- Participate in industry groups that support process safety
- Establish a process safety management review system

Table 7. Normalization of Deviance – Not Maintaining High Standards of Performance

- Create reasonable, high standards of performance and communicate expectations
- Create a dialogue within the company on the importance of operational discipline
- Promote accountability throughout the organization
- Institute a constructive discipline policy
- Promote a safe reporting environment; if necessary, create an anonymous reporting system
- Create metrics and communicate results
- Consider a zero-tolerance policy for severe or willful transgressions and for chronic failures. Take strong action when necessary to protect company performance standards
- Appropriately communicate both positive and negative consequence situations to significant breaches
- Highlight examples throughout the chain of command – not just at the hourly worker level
- Institute some fashion of pay-at-risk for performance failures and reward for performance successes

Table 8. Culture Process Not Formalized

- Develop a vision statement for process safety culture
- Ensure that process safety is included in annual objectives for facility leadership
- Provide "elevator speeches" for senior leadership so that they can speak frequently and consistently about the importance of process safety

- Establish clear documented accountabilities for process safety
- Establish and enforce high standards of performance for activities that affect process safety results
- Implement a policy of zero tolerance for willful violations of process safety policies and procedures
- Document how the organization approaches evaluating and nurturing process safety culture
- Consider establishing a Process Safety Culture element is the PSM system using the CCPS RBPS framework
- Create a process safety culture element with the company process safety management system
- Institute process safety culture evaluations
- Provide process safety culture training
- Highlight process safety culture as an evaluation area in audits, incident investigations, etc.
- Write articles and newsletter items on improving company process safety culture
- Make process safety performance and culture a part of every worker, supervision, and management annual evaluations

Table 9. Lack of Sense of Vulnerability

- Create institutional memory about past company accidents; communicate reminders
- Investigate and communicate lessons learned from recent accidents throughout the company
- Communicate lessons from external accidents
- Provide hazard/risk awareness training to all new employees; provide regular refreshers
- Provide root cause analysis training to all employees
- Institute a burden-of-proof in favor of pessimistic action to preserve safety rather than a presumption of safety
- Adopt incident investigation policy to look at what could have happened instead of just what did happen
- Relate on-the-job safety events to off-the-job situations
- Publish the CCPS Process Safety Beacon internally

Table 10. Lack of Empowerment and Personal Responsibility for Safety

- Clearly define accountability for process safety systems; who is responsible for what
- Effectively communicate expectations by training employees in process safety policies and procedures
- Implement a policy of zero tolerance for willful violation of process safety policies and procedures
- Be certain that management response to acceptable and unacceptable performance of process safety responsibilities is timely, consistent, and fair
- Establish annual process safety objectives/goals throughout the organization
- Institute an off-the-job safety program
- Employ an "anyone can and everyone should shut it down" policy
- Provide awareness training on workplace hazards/risks and accident statistics
- Consider adopting an employee-driven behavior-based safety program including peer observations
- Promote highlighting of employee safety concerns through formal and non-traditional means
- Celebrate employee safety decision/action successes
- Establish a Conduct of Operations element that encourages rigorous implementation of process safety policies and procedures

Table 11. Not Listening to Technical Experts

- Identify the technical experts for various functions and types of operating equipment
- Define necessary technical disciplines that need to be involved in specific process safety activities (e.g., types of MOCs) and use them
- Train the individuals identified as technical experts to fill their technology expertise weaknesses
- Establish a network of technical experts within the company so that experts can share ideas and learn from experts at other sites
- Encourage technical experts to attend technical symposia in their area of expertise
- Involve management in defining expertise that is valuable/respected/needed
- Enhance the employee participation element so that management is more likely to naturally involve experts
- During discussions about critical issues, periodically ask management if they are using all of the available expertise to resolve the issue
- Expand MOC to non-traditional change types (e.g., organizational, policy, budget changes)

- Formalize the process for making risk decisions; itemize types; consider procedures and requirements
- Ensure SMEs are included in the decision making loop; seek out SME input
- Develop a process safety training competency matrix for all levels of the organization
- Consider designating a facility or company process safety "authority"

Table 12. Ineffective Communications

- Encourage senior management in the facility to be in the plant frequently to be available to listen and respond to employee concerns. Remember to seek input from professional staff as well as hourly employees
- Provide a mechanism for anonymous input to management so that those that have fear of reprisal have an avenue to provide input
- Do not shoot the messenger
- Praise "bad-news" messengers in visible, public ways
- Provide frequent status of lengthy projects that are important to employees so they are aware that the project is progressing
- Help management include process safety messages in their periodic newsletters or other communications
- Establish process safety committees that include a vertical slice of the organization
- Hold regular process safety management reviews
- Provide communications training to everyone
- Solicit workforce opinions on effective communication means and frequency
- Create newsletters, increase safety meeting frequency and effectiveness
- Emphasize communication throughout chain of command
- Develop/improve communications response discipline for phone mail, email, and web responses

Table 13. Lack of a Questioning/Learning Environment

- Widely circulate the CCPS Process Safety Beacon
- Develop and distribute summaries of incident reports that include what happened, lessons learned, and how the lessons learned might apply locally
- Employ a "high potential incident" practice of communicating notable incidents across the company and industry
- Modify the incident investigation system to more fully address "what could have happened" instead of only the actual incident consequences
- Conduct table top drills with operating teams to discuss response to operating problems and incident scenarios
- Review key hazard scenarios with highest potential consequences from PHAs with operating and technical teams
- For outside incidents with lessons learned that have serious potential local consequences, require documented follow-up to ensure similar conditions do not exist or are well managed locally
- Conduct hazard awareness training for operating teams and for technical teams
- State a policy from the top that it is okay/required for people to appropriately question issues
- Hold training and workshops
- Hold meetings specifically to question issues
- Demonstrate the questioning attitude in designed situations and then walk the talk
- In company meetings, invite people to question senior management on issues – model positive approaches to the workforce
- Institute a formal hazard evaluation system
- Consider using peer feedback of work practices

Table 14. Lack of Mutual Trust

- Design a system that ensures response to incidents is consistent
- Establish an employee committee to help address consistency
- Institute blameless, root cause analysis system. Incidents result in minimal consequences unless willful policy violations occurred.
- Create an anonymous safety issue reporting system

- Communicate the need for reporting of all incidents
- Conduct refresher training about the progressive discipline system
- Include an elapsed time at which previous violations no longer impact future discipline. For example, incidents more than two years old do not impact the progressive discipline system
- Make examples of supervisors and management who do things right – and wrong
- Establish/re-invigorate an employee/management safety committee
- Include workforce representatives on an accident investigation committee
- Provide human factors/behavioral safety training

Table 15. Non-Responsiveness to Safety Concerns

- Relate the safety concerns of the company to actual incidents within the company or to incidents outside the company
- Present the business case for process safety to company executives and site management
- Ensure senior management understands the results of realistic scenario consequence modeling
- Communicate to senior management the OSHA fines from serious incidents
- and the possibility of personal criminal litigation
- Periodically share the learnings from incidents within your own company or from similar processes
- Publish the CCPS Process Safety Beacon internally
- Develop a strong near miss identification system for potential process safety incidents
- Do a survey of all process safety-related action items; determine status and dues dates
- Determine roles and responsibilities and schedules for completion
- Develop a unified action item tracking system to stay on top of process safety items
- Develop an improvement strategy and supply resources to reduce backlog; communicate progress to workforce
- Determine performance standard on completion of action items
- Develop a management system to deal with process safety issues and action item life-cycle
- Create process safety issue metrics and monitor metrics during regular management reviews
- Incorporate safety concern responsiveness into regular management reviews
- Provide a means for employee input of safety concerns
- Provide awareness and detailed training as needed
- Include metric in employee accountability and performance reviews

Table 16. No Performance Monitoring or Pursuit of Improvement

- Include process safety topics in annual objectives for line management
- Start monitoring a few site-wide easily obtained but meaningful metrics; start small
- Ask each operating unit to monitor a few easily obtained but meaningful metrics; develop appropriate safety metrics scorecards and communicate broadly
- Establish a process safety management review system
- Form a site-wide process safety committee from several disciplines to promote PS improvement
- Ask senior leadership to promote process safety improvement in town hall meetings
- Establish a business case for process safety for your company
- Discuss with senior executives the consequences of a "Texas City type" event to your company
- Consider putting "pay at risk" based on performance to process safety objectives and metrics
- (Re)establish continuous improvement as a core value and communicate broadly
- Develop strategy to incorporate learnings from audits, hazard reviews, incidents into the work life
- Hold management reviews to spotlight good performance and focus on weaknesses
- Make improvement a part of the accountability and performance review system
- Create ways to solicit and accept employee suggestions

8. Conclusions

Many companies have overall safety policies or visions that embody some sort of "pursuit of zero accidents". Why can't we have a similar for vision for pursuing "Perfect Process Safety."

- A culture based on proper ownership of HSE empowered by visionary leadership
- Risk-informed sensitivity that guides everything
- Effective, fit-for-purpose management systems
- PS practices embraced and followed with good operational discipline at ALL levels
- Learning from ALL sources – internal, external and outside industry group
- Well-formed/visible performance pyramid; metrics at every level that drive intended behaviors
- Goals and actual performance that improves

"Perfect Process Safety" is a worthy and valuable stretch goal for all. It can only be effectively pursued by highly reliable organizations that embody effective learning patterns and have a growing culture. Sustainable process safety cannot not allow learning to just evaporate or "retire". Companies that focus on improving process safety leadership and operational discipline have the best chance.

9. References

[1] Guidelines for Risk Based Process Safety, Written by ABS Consulting, AIChE Center for Chemical Process Safety, John Wiley & Sons, New York, NY, April 2007.

[2] A Management Approach for Creating a Strong Process Safety Culture – 20 Elements for Success, AIChE Center for Chemical Process Safety, New York, 2007.

[3] Principles of Process Safety Leadership, Process Safety Leadership Group, UKHSE, July 2009

[4] Successful Leadership Self-Assessment Module, UK Petroleum Industry Association Ltd., November 2010

[5] Corporate Governance for Process Safety - Guidance for Senior Leaders in High Hazard Industries, OECD Environment, Health and Safety, Chemical Accidents Programme, June 2012

[6] Process Safety Assessment Program, API/AFPM, January 2012

[7] Final Safety Culture Policy Statement, Bureau of Safety and Environmental Enforcement, May 9, 2013

[8] Guidelines for Conduct of Operations and Operational Discipline, Written by ABS Consulting, American Institute of Chemical Engineers, Center for Chemical Process Safety, John Wiley & Sons, New York, NY, March 2011.

1ST CCPS
ASIA-PACIFIC
CONFERENCE
ON PROCESS SAFETY℠

IMPROVING PSM CULTURE THROUGH OPERATIONAL DISCIPLINE

Srinivasan Ramabhadran
DuPont
11-01 Harbour Front Tower One, Singapore
ramabhadran.srinivasan@dupont.com

Prepared for Presentation at
American Institute of Chemical Engineers
1st CCPS Asia-Pacific Conference on Process Safety
Qingdao, China
September 4 – 5, 2013

UNPUBLISHED

IMPROVING PSM CULTURE THROUGH OPERATIONAL DISCIPLINE

Srinivasan Ramabhadran
DuPont
11-01 Harbour Front Tower One, Singapore
ramabhadran.srinivasan@dupont.com

Keywords: process safety management, safety culture, performance improvement, management system, operational discipline

Abstract

There's an old saying: "Strategy is everything. Execution is everything else." In the case of Process Safety Management (PSM), many companies have a strategy and a plan. It's when the rubber meets the road – implementation – that failure occurs. A poorly implemented PSM program can result in increased incidents that threaten the well-being of employees, companies and the communities where they operate.

DuPont defines operational discipline as "the deeply rooted dedication and commitment by every member of an organization to carry out each task, the right way, each time." A high level of Operational Discipline (OD) is integral part of cultivating a strong PSM culture and ensuring optimum performance and overall operational excellence.

Fortunately, PSM culture can be enhanced through operational discipline. When done properly, operational discipline can result in rapid, substantial and sustainable improvements in day-to-day PSM.

DuPont's paper will answer the following:

- What is operational discipline?
- DuPont defines operational discipline as "The deeply rooted dedication and commitment by every member of an organization to carry out each task, the right way, every time." With DuPont's cumulative working experience, we have identified 14 success characteristics that are the foundation of world-class operational discipline, cover leadership,
- How will operational discipline enhance my existing Process Safety Management culture?
- Effective implementation of PSM does not happen only by designing good PSM systems; effective implementation requires a continuous, consistent and persistent application of these good systems in day to day operations. The principle of operational discipline is recognized as a key success factor in driving effective PSM implementation. Organizations are increasingly realizing that operational discipline can result in rapid, substantial and sustainable PSM improvements.
- How do I embed operational discipline into my managing processes to make the culture sustainable?

- The DuPont approach to achieving operational discipline addresses change at organizational and individual levels. We believe top-down cultural improvement is driven by leadership and is focused on everyone in the organization. By involving individuals, a learning environment is created that is innovative, agile and continuously. So as to implement world-class PSM and make the culture sustainable.

1. Introduction

Even a moderately-sized facility performs thousands of PSM related tasks and manages a vast amount of data round the clock every day of the year. Aging facilities require ongoing maintenance and renewal, which, of necessity, entails the updating of equipment and introduction of new materials. All such changes to processes – from new products to plant improvements - must be evaluated. This requires a tremendous attention to detail and operational discipline.

At DuPont, our PSM managing system is fully integrated into our operational strategy. It is not considered as a separate safety priority, but as an integral approach to safe and efficient operations. Successful PSM implementation requires both a strong safety culture and strong PSM technical execution.

We are well aware of the leading operational risk factors. Among the oft cited primary challenge of "insufficient time and inadequate resources", the list of key risk factors also includes the perceived competing pressures of production, quality, cost and safety; the lack of strong visible leadership commitment to safety; an incorrect identification and/or evaluation of operational risks; the failure to manage process safety risks differently to workplace safety hazards; ineffective audits and a failure to comply with procedure. Given the multitude of risk factors, it is thus vital that companies apply a rigorous PSM management system that places great emphasis on cultivating a high level of operational discipline.

2. Defining Operational discipline: Where to Start?

DuPont defines operational discipline as the deeply rooted dedication and commitment by every member of an organization to carry out each task, the right way, each time. A review of some of the operating practices of the most hazardous industries, including airline transportation and nuclear power generation, shows that operational discipline is built into their daily functions through check sheets, step-by-step sign-off's and independent verification of critical steps. However, in most DuPont manufacturing operations, there is no reliance on direct supervisory oversight or direct secondary observation in order to ensure employees perform tasks correctly; checklists and sign-offs are often reserved for the most critical tasks. Notably, even checklists, sign-offs and other observation approaches rely upon workers' doing the job right, following required procedures and without taking short cuts. Instead, DuPont depends on a healthy operational discipline culture.

An increasing number of firms share DuPont's perspective vis-a-vis the need for a strong operational discipline culture as a means to protect employees and the environments they operate in as well as manage quality and reliability issues, poor process optimization and higher costs.

Utilizing DuPont's cumulative work experience, we have identified a methodology to effectively implement operational discipline (see box 1) providing step-by-step guidance for firms to continuously and effectively improve their overall PSM standards.

After confirming the desire to strengthen operational discipline, firms should perform an honest, comprehensive assessment of the current state of their PSM. This establishes a baseline sense of what tasks are performed throughout a business's functional operations — both quantitatively and qualitatively — as there is often a disconnect between hands-on employee's awareness of procedures and that of their superiors. Furthermore, firms must have a sense of what training, re-training, documentation, compliance, and enforcement mechanisms exist in relation to these procedures before they can begin to amend or strengthen them.

Box 1: 8 STEP OPERATIONAL DISCIPLINE IMPLEMENTATION METHODOLOGY

Step 1: Be convinced improved OD is needed
Step 2: Assess the current state
Step 3: Develop a comprehensive roll-out plan
Step 4: Focus on the "It"
Step 5: Establish priorities
Step 6: Strengthen discipline processes
Step 7: Implement
Step 8: Audit, audit, audit

As with every reformative initiative, especially those that affect the organization's fundamental culture, detailed and comprehensive planning is essential, and firms should have a sense of when, where, and how they will implement these changes to OD. Building a strong operational discipline culture in an organization has been described by some as Management 101 – in other words, it is good, fundamental management practice. As such, its implementation plan should be kept as simple as possible, with its central considerations revolving around management consistency, exhaustive two-way communication and a prompt response to questions or concerns. However, even firms committed to improving their OD still face the difficult task of determining what they should strategically change first — rendering the fourth step in the methodology, "focusing on the it", a crucial one.

In DuPont's definition of OD – *everyone does it right every time* – "it" refers to the performance of day-to-day tasks. In order to standardize a particular process, the "it" is usually detailed in instructions or procedures. These procedures should be the codified best practices that represent the organization's complete knowledge and experience. Furthermore, they should convey in writing how a particular task is to be performed each and every time.

Unfortunately, organizations seeking to improve their operational discipline often focus their first efforts on too diffuse of an initial target, thinking that "fixing" a lax safety culture will more quickly ameliorate PSM weaknesses. Instead, firms should **begin** by examining the quality of their procedures, both from the perspective of content and the worker's subsequent interactions with them. To wit, if management instructs its employees to follow procedures that do not represent the best knowledge and experience of the facility, they place that employee in an untenable position: obey the bad procedure to dubious ends; or to ignore the procedure altogether and conduct the task in their own, "better", fashion, with potential ramifications for failure to comply with management's instructions. This not only introduces a higher probability of an adverse event occurring, but management will invariably suffer a loss of credibility and loyalty in the eyes of employees forced into a comprising decision process.

For this reason, effective OD is predicated on a robust process of technical review and employee involvement, combined with talented procedure writers and a document management and control system. Management should not insist upon compliance to procedures until the latter are of sufficiently high quality to merit compliance. Put otherwise, if management seeks to develop a culture where workers follow procedures every time, they must first invest in and confirm that the procedures are indeed helpful, accurate and owned by the users.

3. The Importance of Quality Procedures for OD

A number of recent incidents highlight the catastrophic potential of poorly constructed operational procedures and/or the failure to adhere to them. The notorious Deepwater Horizon incident in the Gulf of Mexico in April of 2010 was one of the most costly and damaging accidents in history; a later investigation implicated insufficient procedural specificity across a range of tasks — including basic communications — and poor employee familiarity with established procedures. Not only did BP face unprecedented financial liabilities and disrupted operations, but worker fatalities and widespread environmental damage held grave ethical and public relations consequences for the firm.

However, the case of nuclear submarines in the United States Navy provides an example of how improving procedural standards can mitigate and allow proper management of operational risks in inherently dangerous industries. Between 1915 and 1963, the U.S. Navy lost roughly one submarine every three years to noncombat causes, with mounting casualties and pressures to contain costs of lost vessels. In response to the loss of the USS Thresher in 1963, the SUBSAFE system was developed to set procedural priorities and establish a mindset of strict insistence on verification and compliance. The attached graphic highlights the major tenets of the SUBSAFE system.

An analysis of these 12 principles (see box 2) indicates that 7 of them (58%) specifically speak to the need for a strong, written procedure program, supplemented by review and training. Another four speak to the establishment of a supportive culture of operational discipline and accountability. The remaining item describes the important role of top management commitment, reinforcing that sterling leadership is required for each of the 12 principles to be realized.

Box 2: PRINCIPLES OF THE NAVAL "SUBSAFE" SYSTEM

- Top management commitment to safety
- Clear and written safety requirements
- Education, not just training
- Regular rewriting of requirements
- Separation of powers and assignment of responsibilities
- Emphasis on rigor, technical compliance, and work discipline
- Documentation capturing what is done and why it is done
- Participatory audit approach and requirements for objective quality evidence
- Program based on written procedures, not personality-driven
- Continual certification of a facility
- Accountability and accompanying responsibility
- Special efforts to be vigilant against complacency

4. Prioritizing Procedures

As the SUBSAFE system highlights, written procedure programs are crucial to improving OD. However, all procedures are not created equal. Employees understand that to operate a facility or perform their tasks safely, some procedures and regulations are more important than others. Thus, to strengthen operational discipline, a focus on compliance with the most important procedures should take precedence. The following graphic (Box 3) lists the groups of procedures in order of their impact on OD.

Box 3: Prioritization of Procedures
1. Management Procedures or Managing Systems
2. Safety Procedures
3. Operating Procedures
4. Maintenance/Repair Procedures
5. Testing or Quality Procedures
6. Emergency Response Procedures

First, management must set the proper example for building OD, establishing and then following appropriate managing systems. Whenever possible, these systems should be documented and accessible to those employees that need to reference them regularly. Unfortunately, these are often the most neglected procedures, as managers mistakenly consider their documentation to be less crucial than other procedures on the list.

After managing systems, safety procedures are the next important. Safety is an obvious, highly-visible priority, and it is imperative that all employees contribute to the efforts to maintain the highest of standards. Having excellent safety procedures builds a strong culture of operational discipline, while failure to comply with them graphically undermines safety and OD culture. Taking this approach, DuPont and many other companies compile and publish a list of the most critical safety rules. The list becomes a line in the sand, and there is zero tolerance for willful violation.

The content of the safety rules themselves will vary from company to company and site to site, depending on culture and risk profiles: management's complete commitment to seeing them followed without exception should be unwavering, however. Other operational and maintenance procedures can be equally critical and equally valuable in an organization's drive for full compliance and greater operational discipline. Shutdown and startup procedures, interlock test procedures, relief valve test procedures, and personal protection equipment compliance can be given high priority and no-tolerance status.

All workers should know the organization's inviolable rules - without any doubts. The rules should be highly visible, with framed copies in conference rooms, control rooms, maintenance shops and laboratories. Contractors should also be fully aware of the rules and should be expected to comply. Every possible effort should be made to avoid the situation where an employee claims, usually during an incident investigation, that he or she was not apprised of violating an inviolable standard.

5. Operational Excellence - A Result of Effective Operational Discipline

The state of operational effectiveness is the product of the quality of the management standards and the operational discipline. This requires that firms must be vigilant about establishing a culture of OD across all levels of their operational and management hierarchy: weakness in either case will result in ineffective operations and safety incidents. As is obvious, high quality, high compliance procedures form the bedrock of this efficacy. DuPont has observed eleven best practices for establishing such effective procedures, each explained in detail below:

5.1. A procedures "Standard" describes what is expected and by whom.

A "procedure for procedures" will not only establish clear expectations and the basis for monitoring progress (via metrics and audits), it will also establish procedures as a top management priority that will be supported with resources, time and money.

5.2. Every procedure has a single person owner.

Procedure owners own the quality of the content, the review processes and can provide instructor led training.

5.3. Procedures are well written and easily followed. Visuals and other technology tools are employed.

Procedure writing is a skill, and many organizations have dedicated procedure writers who have been internally trained. Some elect to hire professional writers, while others will include procedure writing as part of their other duties.

In either approach, the quality of the writing is of paramount importance, as poorly prepared procedures will not be readily followed. Additionally, creative use of educational tools including color, pictures, diagrams and even embedded videos aid in the communication and explanation of procedural content. Placement of laminated versions of procedures where they are most often utilized is another creative approach to encourage their use and value.

5.4. Content is consistent with core documents and balances comprehensiveness with user needs. A consistent format is used universally with appropriate procedure headers.

There must be clear accountability in the development of a procedure for its accuracy. In particular, operating procedures must be consistent with technology documentation. All procedures must also be consistent with corporate requirements and all applicable regulations.

It is easy to over-burden procedures with interesting, adjacent information that explains the reasons for its existence. Though the latter is important, over-contextualization too often confuses the reader or simply obfuscates the pertinent information. Finding the right balance between these two needs should be influenced by the criticality of the procedure.

The ability to manage a procedure is facilitated when a common header format is used for all of a facility's procedures. The title, the procedure's owner, the location of the master copy, importance/risk mitigation classification (see #5 below), dates of its last revision and its next required revision, and also an appropriate procedure number should be included. An optional procedure revision history can also prove useful.

5.5. Every procedure is classified as to importance and risk mitigation.

Allocating appropriate resource support to a procedure is made possible by classifying all procedures by their importance and criticality. Several factors are feasible, though the most straightforward is simply a brief consequence analysis: "What is the worst thing that could happen if this procedure were not followed correctly?"

Some firms integrate their Risk Registry or Risk Classification System (Category I to V, for example) into the way they classify procedure importance, while others simply use a simple Tier I, Tier II and Tier III approach.

5.6. All procedures are routinely reviewed by an appropriate cross section of the organization.

Reviewing procedures for accuracy and completeness must be an important ritual in every organization, the frequency of which is predicated on the criticality of the procedure, e.g. once every 12 months for critical procedures, every 6 years for the least critical. That said, proactive changes to a corporate standard, a process or equipment that may render an existing procedure inaccurate must trigger an immediate procedure review.

Essential to any review process is adequate time and use of the "right" people. Procedure owners should plan three months in advance of a procedure expiration date so that the quality of the review is not compromised.

5.7. A verification strategy is applied to all procedures

As cited previously, the U.S. Navy SUBSAFE motto became "Trust but Verify." The same approach should be employed for all procedures, consistent with the procedures criticality. Trust that employees will perform the procedure correctly but just in case, verify.

There is a wide range of verification techniques available to organizations. On one end of the spectrum lies duplicate review of every step with initials, such as what occurs in the cockpit of commercial airliners before every flight. This may be appropriate for your most critical procedures.

On the other end of the spectrum may be little or no verification, assuming, of course, that the individual is trained and competent to perform the task. The principle is that appropriate verification strategies should be identified and applied to all procedures. An appropriate action is determined by the procedure's criticality classification.

5.8. Content is maintained up-to-date via a robust Management of Change Process (MOC).

Failure to identify and mitigate new risks introduced via facility, technology and even organizational change are often the cause of serious, even catastrophic incidents. MOC processes are intended to methodically review changes so that incidents are prevented. Codification of the proper way to implement and execute desired changes must include procedure reviews and revisions plus the corresponding appropriate training.

5.9. Initial and refresher training is provided on a regular basis. Written, field, and supervisory assessments are performed.

It goes without saying that accurate, up-to-date procedures are of little, real value if their users are not knowledgeable and familiar with the content. Training must be done initially, and periodically thereafter to ensure that knowledge is not lost. For critical procedures, refresher training may be required annually. Other less critical procedures should be less frequent.
It is critical that managers have full confidence in each individual's demonstrated ability to follow procedures. Job Cycle Checks, conducted by peers and/or supervisors are an effective way of demonstrating acquired skills. Other more formal assessment methods should also be employed as appropriate.

5.10. Procedure revision documents are controlled without impacting broad access to the procedures.

Document control is essential. Many facilities utilize systems and standards such as ISO 9000 to bring structure to the document management process. Hopefully, the rigors of these approaches do not get in the way of an equally important priority – procedure access. Auditing of both document controls and ease of access, including how much time it takes for users to access key procedures will allow managers to judge whether an appropriate balance between these two priorities has been achieved.

5.11. Continuous improvement is driven by metrics and audits.

If the procedures standard is properly prepared, it will be very straightforward for an organization to collect appropriate metrics (overdue procedure reviews, overdue procedure training, overdue MOC procedure updates, etc) as well compare actual performance vs. expectations via proactive auditing. Continuous improvement requires data (metrics and audit results) that illustrate where improvement opportunities can be found.

6. The Next Steps: What Comes After Procedures?

Although operational discipline can and should be a positive element in an organization's culture, it is important to make a clear statement regarding the individual consequences for failure to follow procedures. To encourage cooperation and adherence to operational discipline principles, many facilities use a process called progressive discipline that establishes negative consequences for failing to comply as expected.
Typical steps in a progressive discipline system are, depending on the number of infractions or severity, 1) verbal warning, 2) written warning, 3) probation, 4) suspension with or without pay, and 5) termination.
The discipline system must universally apply to all employees within the hierarchy, from shop floor to managers and superintendents. Although the human resources function is typically the administrator of the process, it is most effective if line management responsible for its broad communication and implementation.
Communication is also key for the broader implementation of a strengthened OD culture. Managers should solicit feedback from employees in order to determine what progress is being made, and where there might be difficulties or misunderstandings. Clear accountability for key tasks should be assigned, timelines established, and deadlines met.

Similarly, the final step of the implementation process revisits the assessment and audit step to trigger the next round of planning and improvement, as the process of amendment should be continual and reflexive. While the assessment and audit tools are obviously customizable to the situation — and may be different than the ones used for establishing initial baselines — it is important to be able to determine whether true progress is being made by a comparison with prior results. In a trusting and mature culture, assessing the level of operational discipline is important for all members of the organization, regardless of their roles. Defensiveness on the part of individuals or work groups or the failure to report the "whole truth" can usually be avoided if management effectively communicates the critical importance of a strong OD culture and the changes that are being made.

7. Conclusion

In an organization where a strong operational discipline culture exists, there will be a deeply rooted dedication and commitment by every member of an organization to carry out each task the "right way, every time". When an organization needs to create or strengthen its OD culture, it must undertake a careful analysis of its current PSM benchmarks work to craft pertinent and sustainable procedures, and be prepared to reassesses the efficacy of the procedures it has implemented. As with all cultural changes, the implementation of these new patterns of behavior can be a laborious process, and continuous effort by management will be required to sustain the gains. However, coupled with proper leadership commitment and a culture of safety, operational discipline provides a crucial component of process safety management, and with it serves to protect the employees, environment, and assets involved in a firm's day-to-day operations.

1ST CCPS ASIA-PACIFIC CONFERENCE ON PROCESS SAFETY℠

Dust Explosion Hazards

Ernest Cheam
SABIC
Singapore
ernest.cheam@sabic-ip.com

Larry Bowler
SABIC
Mt Vernon, USA
Larry.Bowler@sabic-ip.com

Prepared for Presentation at
Center for Chemical Process Safety
1st CCPS Asia-Pacific Conference on Process Safety
Qingdao, China
September 4-5, 2013

UNPUBLISHED

Dust Explosion Hazards

Ernest Cheam
SABIC
Singapore
Ernest.Cheam@sabic-ip.com

Larry Bowler
SABIC
Mt Vernon, USA
Larry.Bowler@sabic-ip.com

Keywords: dust explosion, SABIC, combustible powder, case study

Abstract

Overview of dust explosion basics, hazards present in Innovative Plastics Compounding plants in SABIC, and existing Powder Handling policy and standards to control the hazard. Two case studies of actual past powder explosions, and lessons learned is also discussed.

1. Introduction

Presentation presents overview of explosion hazards of combustible powders, in use at SABIC Innovative Plastics Compounding plants, and basic background knowledge, and preventative and protective measures which are applicable.

Two actual powder explosion incidents case studies are discussed and lessons learnt shared.

2. Overview

DUST EXPLOSIONS ACCORDING TO INDUSTRY

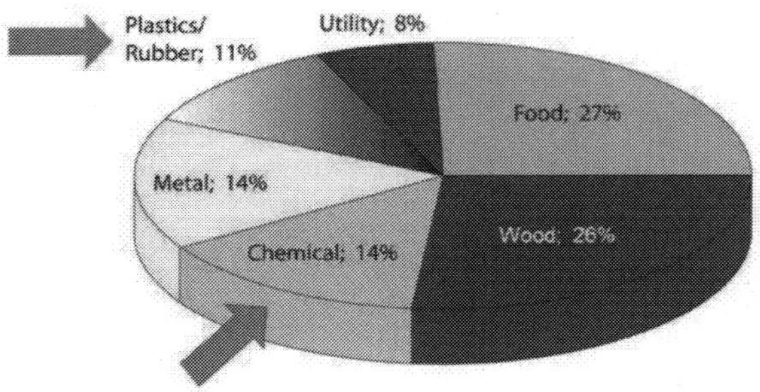

CLASSIFICATION OF DUST FLAMMABILITY

Class	Response	Std. Example
1	No Ignition	NaCl, table salt
2	Ignition but not sustained	Tartaric acid
3	Ignition sustained but does not propagate	D+Lactose
4	Propagation with no open flame	Tobacco
5	Propagation with open flame	Sulfur
6	Explosive propagation	Black Powder

Most concern

EXPLOSION CONDITIONS

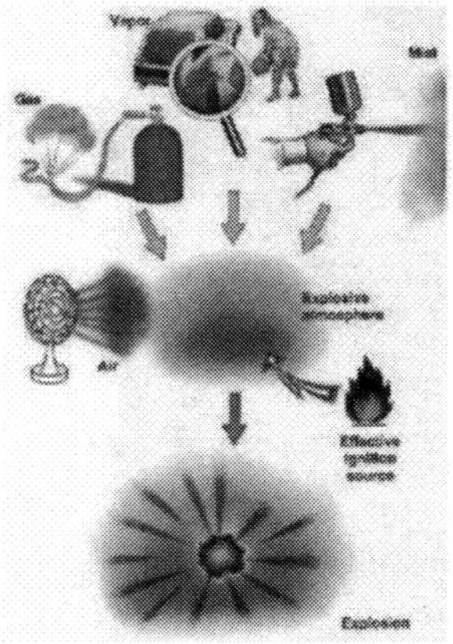

- Dust explosions are comparable to gas explosions as long as dust is fine
- Gas, vapour or dust needs to be in right concentration range between LEL (or MEC) and UEL
- Ignition source or "auto-ignition" conditions need to be present

DUST EXPLOSION PENTAGON REPRESENTING EXPLOSION REQUIREMENTS

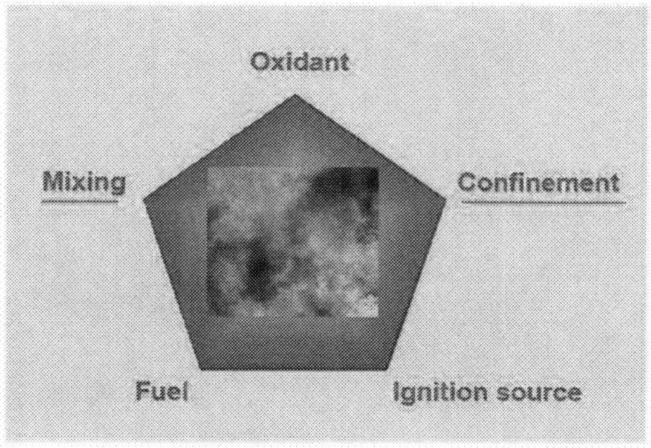

Two additional conditions beyond the traditional "Fire Triangle"

No. 9

JUDGMENT OF POWDER PROPERTIES

- **Explosibility** →
  - Minimum Explosive Concentration (MEC)
  - Limit O2 (below Explosion) Concentration (LOC)
- **Ignitability** →
  - Minimum Ignition Energy (MIE)
  - Minimum Ignition Temperature (MIT)
- **Explosion Effect** →
  - Maximum Explosion Pressure (Pmax)
  - Max Rate Pressure Rise (Kst)

No. 10

MINIMUM EXPLOSIVE CONCENTRATION (MEC OR LEL)

- Minimum Explosive Concentration (MEC) or Lower explosion limit (LEL) for dusts
  - ➢ usually 30 - 100 $g.m^{-3}$
  - ➢ visually dense cloud
- UEL not measurable nor useful
  - ➢ several $kg.m^{-3}$
- In contrast, OEL or PEL, in context of Industrial Hygiene, is in the range of 10^{-2} - 10^{-3} $g.m^{-3}$, << MEC

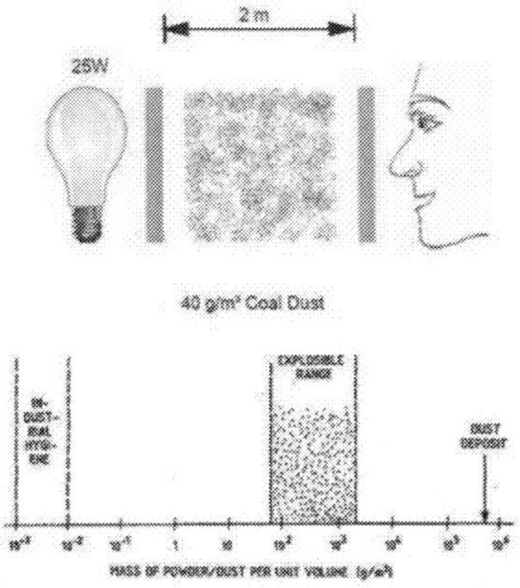

No. 11

MINIMUM IGNITION ENERGY OF DUST VS GASES

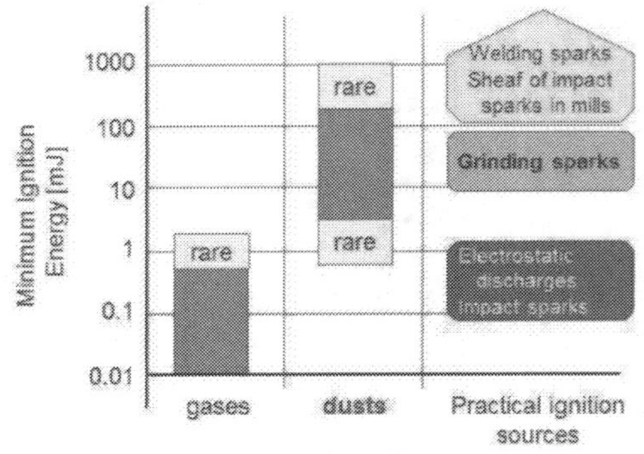

Comparison between the minimum ignition energy of gases, dusts and practical sources of ignition

No.12

EXPLOSION IN A CLOSED VESSEL

- Combustion creates hot gases that expand, or in a closed vessel the pressure will increase
- P_{max} is volume independent, P_{max} for dust explosions is normally in range of 6~10 bar
- Rate of pressure rise is volume dependent:

  $K_{st} = (dP/dt)_{max} . V^{1/3}$
- K_{st} the Deflagration Index for dusts, provides an indication of severity of dust explosion, the larger the value, the more severe the explosion
- Kst is key parameter used for sizing explosion vent area.

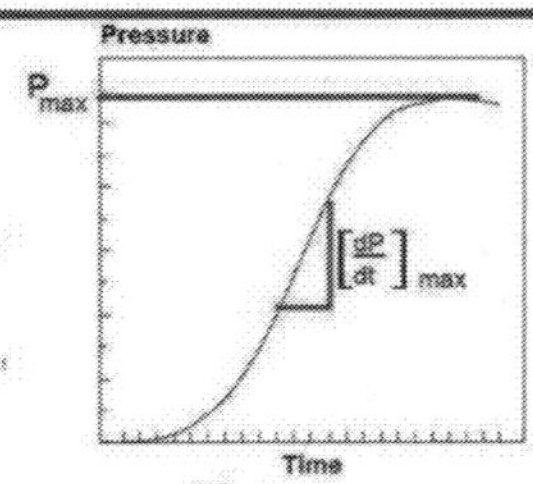

20-liter test sphere

No.13

EXPLOSION CONTROL MEASURES

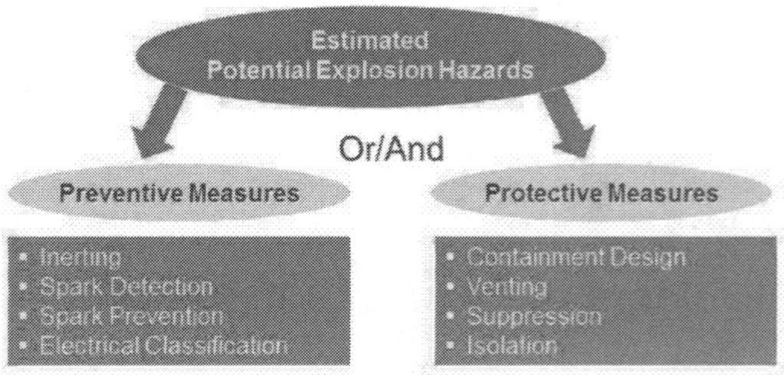

No.14

PREVENTIVE MEASURES

- Inerting
  - Typically with Nitrogen dilution to bring atmosphere below LOC (applicability depending on powder MIE)
- Ignition Prevention:
  - Static Electricity
  - Use conductive material of construction
  - Grounding / bonding
  - Limiting gravity drop height, rate
  - Hot work permitting program, Safe Work Practices
  - Electrical Classification
  - Mechanical Friction
  - Hot surfaces

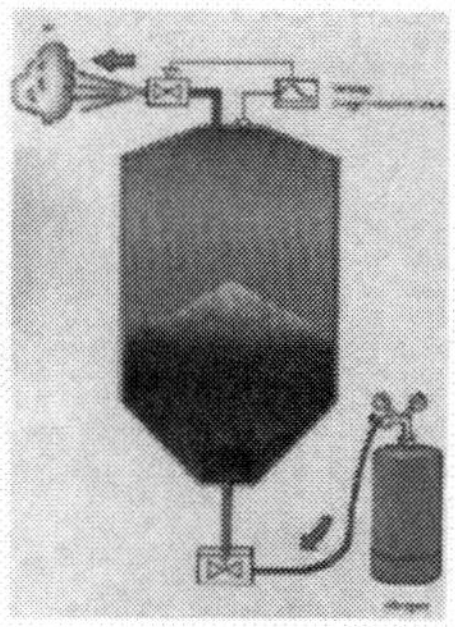

Inerting

No. 15

PROTECTIVE MEASURES

- Containment Design
  - Explosion Pressure Proof Design (no permanent deformation)
  - Explosion Pressure Shock Resistant Design (can be permanent deformation after the explosion)
  - Only practical for smaller vessels
- Venting
  - Generally follow NFPA 68 or VDI 3671. Vent to safe location.
- Suppression
  - Used when venting is not possible
- Isolation
  - Isolation Valves whenever there are large diameter piping between vessels

Containment

Venting

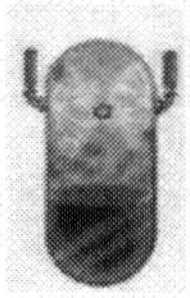

Suppression

Isolation

No. 16

SECONDARY DUST EXPLOSIONS

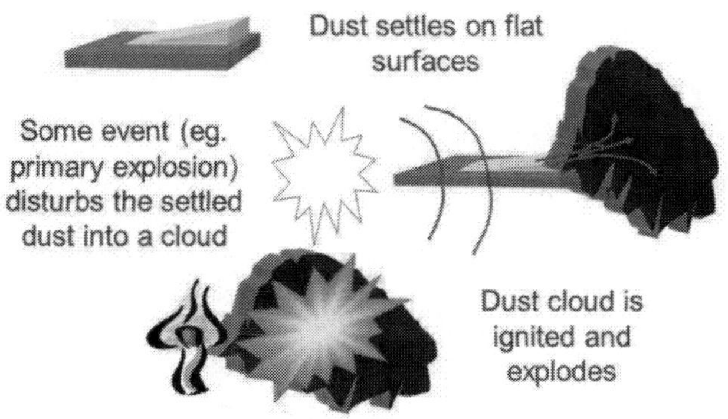

Secondary Explosions Often More Destructive Than Primary Explosion

No. 17

HOUSEKEEPING IS IMPORTANT!

❑ How to minimize dust accumulation

- Connecting equipment exhaust ducts to a suitable collector.
- Operating equipment under a slight negative pressure to prevent dust leakage outside of equipment.
- Minimizing surfaces and ledges in buildings where dust might accumulate.
- Repair process equipment leaks immediately with a vigilant maintenance activity.

❑ How to proper remove dust

- Vacuum cleaning (assure that electrical equipment is suitably classified for atmosphere).
- Use soft push brooms for light accumulation.
- Continuous suction at processes where large quantities of dust are liberated at frequent intervals.

3. Case Study 1

RAIL CAR EXPLOSION

An explosion and subsequent fire occurred on rail car while loading ABS.

Top of rail-car after explosion

Damage to shed

Damage to structure and piping

No. 20

RAIL CAR EXPLOSION

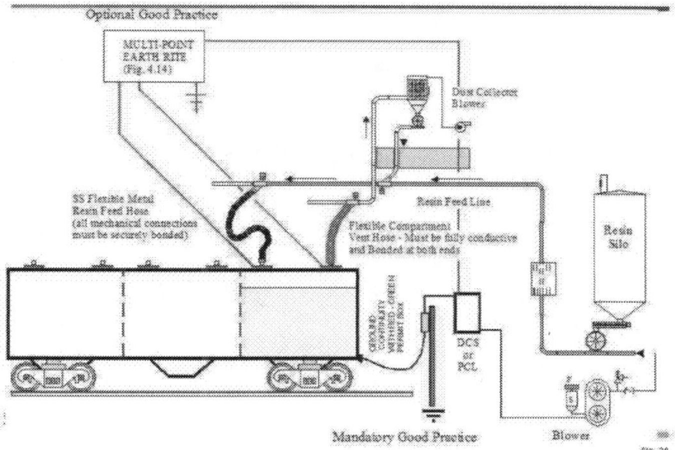

No. 21

RAIL CAR EXPLOSION

- Root Cause
  - Inadequate grounding of the vent hose along with improper (non-conductive) hose material
- Contributing Factors
  - MOC process was not followed for the hose change.
  - Lack of awareness of the static hazards associated with this type of transfer hose.
- Lessons learnt
  - Everyone needs to be aware of explosion risk of combustible powders.
  - Proper MOC review! Even for A seeming trivial change
  - Proper grounding and bonding of all equipment and correct material selection is vital to prevent the buildup of electro-static charges.

No. 22

4. Case Study 2

DUST COLLECTOR EXPLOSION

A fire broke out at an extruder line running an ABS grade. Fire propagated to main Dust Collector via dust collection ducting. The main dust collector located on the roof of building exploded causing severe damage to surrounding facilities. All staff were evacuated safely and there was no injury.

Dust Collector Destroyed By Dust Explosion

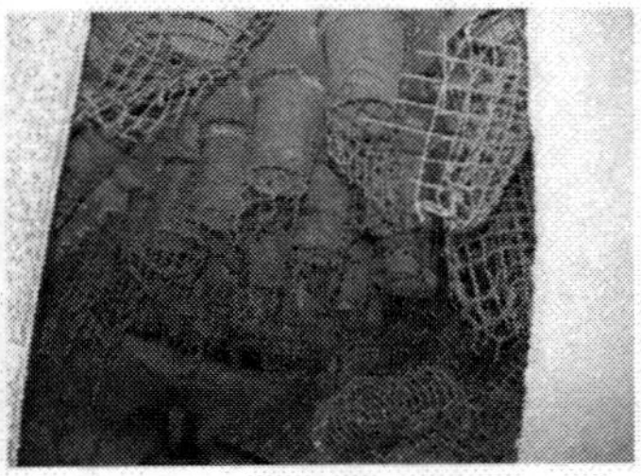

Damage Inside Dust Collector

DUST COLLECTOR EXPLOSION

- Sequence of Events:
  - Die fire started, caused by die being clogged, and not noticed by operator.
  - Over time, fire developed larger, and some embers were drawn into the dust collection system at the throat hopper.
  - Due the dust dusting having powder deposits, fire/embers propagated to main dust collector, eventually causing a dust explosion.
- Contributing Factors:
  - Lack of operator or system response to high temperature alarm which allowed this situation to escalate.

LESSONS LEARNT

- Have routine inspection to keep dust collection ducting free from powder deposits, re-design higher velocity if necessary.
- Previous design of dust collector was inadequate (too weak, explosion vent insufficient), replaced with higher specification
- Provide interlock to shut down dust collector blower in case of fire, to avoid drawing air to dust collector in case of fire.
- Operators trained to watch out for fire around extruder, and to turn off dust extraction and fume extraction to prevent fire and embers being drawn in.

5. Summary

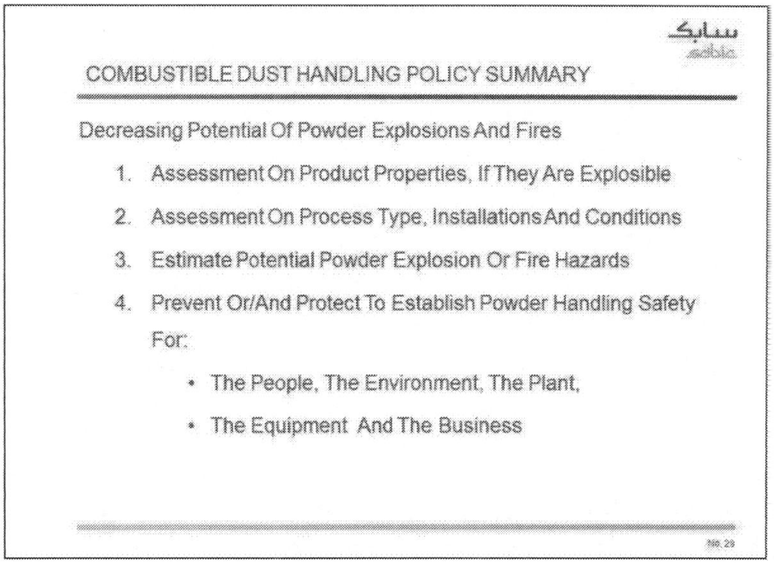
COMBUSTIBLE DUST HANDLING POLICY SUMMARY
Decreasing Potential Of Powder Explosions And Fires
1. Assessment On Product Properties, If They Are Explosible
2. Assessment On Process Type, Installations And Conditions
3. Estimate Potential Powder Explosion Or Fire Hazards
4. Prevent Or/And Protect To Establish Powder Handling Safety For:
• The People, The Environment, The Plant,
• The Equipment And The Business

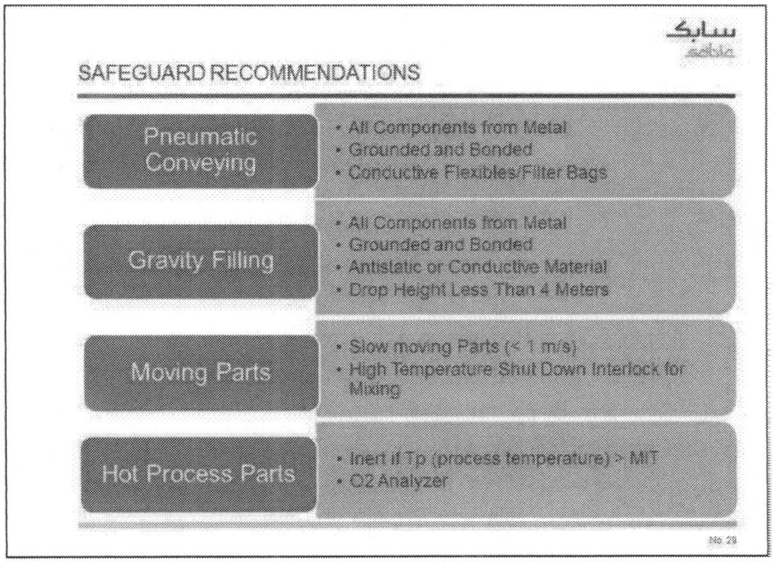
SAFEGUARD RECOMMENDATIONS
Pneumatic Conveying
• All Components from Metal
• Grounded and Bonded
• Conductive Flexibles/Filter Bags
Gravity Filling
• All Components from Metal
• Grounded and Bonded
• Antistatic or Conductive Material
• Drop Height Less Than 4 Meters
Moving Parts
• Slow moving Parts (< 1 m/s)
• High Temperature Shut Down Interlock for Mixing
Hot Process Parts
• Inert if Tp (process temperature) > MIT
• O2 Analyzer

MANAGING ELECTROSTATIC HAZARDS IN THE CHEMICAL INDUSTRY

Vahid Ebadat, Ph.D.
Chilworth Technology, Inc. (A DEKRA Company)
113 Campus Drive, Princeton, New Jersey 08540, USA
vahid.ebadat@dekra.com

ABSTRACT

Many processes and operations in a chemical plant involve handling and processing of liquids and powders, most of which could, under certain conditions, be flammable. Flammable gas, vapor, and dust cloud atmospheres can be ignited if a sufficiently energetic ignition source is present. It is noteworthy that most flammable gas/vapor atmospheres and some dust clouds can be ignited by very small energy ignition sources. One potential ignition source that, in many cases, could be an inherent part of the process/operation is an electrostatic discharge. Electrostatic discharges result from the generation and accumulation of electrostatic charges. Electrostatic charge generation most commonly occurs whenever two materials, solids and/or liquids, make and then break contact with each other. The accumulation of the electrostatic charge can result in electrostatic discharges. This suggests that there are many occasions during the handling, processing, and packaging of flammable liquids and powders in a chemical plant when flash fire and explosion hazards exist.

This paper discusses the practical measures that can be considered to prevent/control electrostatic ignition hazards that are associated with liquid and powder handling operations in the chemical industry.

INTRODUCTION

Most electrostatic hazards in process industries arise when unwanted or uncontrolled charging occurs. Electrostatic charge generation usually occurs whenever any two materials come into contact causing electrons to transfer from one contacting material to another. Under such conditions the surface that acquires an excess of electrons becomes negatively charged and the ~~other~~ surface that loses electrons becomes positively charged. If the two contacting bodies are rapidly separated ~~sufficiently quickly~~, and one or both bodies remain isolated from ground or are insulating in nature, they will retain their excess charges. The magnitude as well as the polarity of the generated charge on a surface is often influenced by many factors such as surface contamination, impurities, and adsorbed materials, as well as the chemical and physical properties of the solids. Higher electrostatic charges are also expected when the contacting surfaces also rub against each other because of the increase in contact area between the two surfaces.

Typical operations in a chemical processing plant that can give rise to electrostatic charging include liquid transfer through hoses and pipes, mixing, settling of two-phase mixtures, filtration, drying, grinding, blending, sieving, powder transfer, and packaging operations.

It should be noted that electrostatic charge generation itself is not necessarily hazardous. The accumulation of the generated charge; however, can be hazardous if the charge levels are sufficient to cause the electrical breakdown of air (discharge) in the presence of a flammable atmosphere. In order for an ignition to occur the static discharge energy has to be equal to or greater than the minimum ignition energy (MIE) of the flammable atmosphere.

This paper discusses electrostatic hazards that may be associated with some common operations involving liquids and powders in a chemical plant and makes practical suggestions for the control of these hazards [Ref 1, 2].

ELECTROSTATIC AND FLAMMABILITY CHARACTERISTICS OF MATERIALS AND EQUIPMENT

Adequate precautions must be taken to avoid ignition of flammable vapor, dust cloud, and or hybrid (mixtures of gaseous, liquid vapor or powder) atmospheres during processing and handling operations in the plant. The first step in achieving this is to obtain or determine the relevant electrostatic and flammability properties of the liquids and powders as well as the materials of construction of the process equipment. The characteristics of interest typically include those listed in Table 1.

Table 1 - Some Electrostatic and Flammability Characteristics

Property	Standard Test Method	Description
Minimum Ignition Energy (MIE)	ASTM E2019[Ref 3]	The MIE test determines the lowest electric-spark energy that is capable of igniting a dust cloud at its optimum ignitable concentration. The test is used primarily to assess the susceptibility of the flammable dust cloud atmosphere to ignition by electrostatic discharges (sparks).
Electrostatic Volume Resistivity	General accordance with ASTM D257[Ref 4]	Volume Resistivity is the primary criterion for classifying materials (powders) as low, moderately, or highly insulating. Insulating materials have a propensity to retain electrostatic charge and can produce hazardous electrostatic discharges even if they are in contact with grounded conductive surfaces.
Liquid Conductivity	General accordance with CLC/TR 50404[Ref 5] and ASTM D2624[Ref 6]	Conductivity is the primary criterion for classifying liquids as low, medium, or high conductivity. When liquids flow through pipes and hoses or are otherwise processed, they may become electrostatically charged. In many cases, the charge on high or medium-conductivity liquids can be controlled or eliminated by handling such liquids in grounded conductive vessels, equipment, and piping. However, non-conductive liquids may remain charged even when handled in grounded conductive equipment.
Electrostatic Chargeability	General Accordance with ASTM D257[Ref 4]	Chargeability is a measure of the propensity of materials (powder and liquids) to become charged when flowing through piping or when processed. The test data can be used to develop appropriate material-handling guidelines from an electrostatic-hazards point of view.
Electrical resistance of floors	Using a "Megger" Megohmmeter and "floor" electrode	This is an on-site measurement that determines the ability of the flooring to dissipate to ground the charges from any (conductive) object that may be placed on that floor. This measurement should be conducted particularly on flooring in locations where vapor or dust cloud atmospheres are likely to be present; for example, in manual powder handling and in liquid-dispensing areas.
Electrical resistance of footwear	Using a "Megger" Megohmmeter and electrodes	This is an on-site measurement that determines the ability of an operator's footwear to dissipate to ground the electrostatic charges on his/her body when standing on sufficiently conductive flooring. If the combination of the resistance of operators' footwear and the flooring on which they stand/work is greater than $1x10^8$ ohms to ground, operators can become electrostatically charged and give rise to spark discharges when touching a conductive object.

GENERAL PRECAUTIONS FOR CONSIDERATION THOUIGHOUT THE CHEMICAL FACILITY [Ref 1]

The following precautions should be implemented throughout the plant where flammable vapors or dust clouds may exist. It should be noted that - depending on the actual electrostatic and flammability properties of the materials (liquids, powders, solids) and the nature of the operations - different and/or additional precautions may be required for ensuring safety from electrostatic ignition hazards.

Conductive (Metal) Items of Equipment

For fixed conductive installations; the conductive objects should be grounded by a direct path to the ground or by bonding it to another conductive object that is already connected to the ground. Where the system is all metal, resistance in continuous ground paths should not exceed 10 Ohms.

For portable conductive equipment; to prevent the accumulation of static electricity on conductive equipment, the total resistance of the ground path to earth should not be high enough to prevent the dissipation of charges that are otherwise likely to be present. Where the portable equipment is all metal, resistance in continuous ground paths should not exceed 10 Ohms; however a resistance of 1 megaohm (10^6 ohms) or less is generally considered adequate when grounding is provided through non-metal systems such as conductive rubber wheels on metal carts.

Grounding connections should be routed to protect them from accidental damage and their purpose should be known to operators. Furthermore, periodic inspection and testing of bonding and grounding systems is especially important. Proper inspection and testing ensures that the likelihood for an electrostatic charge to accumulate is eliminated particularly when bonding and grounding installations that are prone to corrosion, movement, or insulating surface coatings.

Footwear

Where a person could come into contact with a flammable-vapor or combustible-dust-cloud atmosphere having minimum ignition energy up to 30 mJ, it is recommended that he/she be grounded. Personnel can be considered grounded if their resistance-to-ground is less than $1x10^8$ ohms. Grounding of personnel may be achieved by the use of suitable static-dissipative or conducting footwear and floors. For static-dissipative footwear to be effective, the resistance of the floor/surface on which the operator is standing should also be less than $1x10^8$ ohms. Additionally, where a risk of exposure to power supplies cannot be excluded, the footwear resistance should not be below $5 \times 10^4 \Omega$. Alternatively, if practical, grounding of personnel may be achieved through the use of grounding wrist straps.

Floor

The flooring should not raise the total resistance between the person and ground above $10^8 \Omega$. Deposits of insulating contaminants on the floor, such as high resistivity powder, paint, or other floor coating should be removed.

If the flammable mixture is confined to a limited area, e.g. near the manway of a vessel containing a flammable liquid or in front of a powder dump hopper, it may be sufficient to render only a small area of the floor conductive. This may be achieved by providing a grounded metal plate or conductive mat, with grounding wire, for the operator wearing static dissipative footwear to stand on.

CONTROLLING ELECTROSTATIC HAZARDS ASSOCIATED WITH SOME SPECIFIC EQUIPMENT AND OPERATIONS

Containers for powders and liquids

Containers and packages of various sizes, shapes, and materials of construction are often utilized in the chemical industry for the transport and storage of flammable liquids and powders. Additionally, when handling solvent wet or dry powders it is convenient to use an inner liner (often plastic) in the container.

Flammable vapor and dust cloud atmospheres that are often present during filling and emptying of containers can be ignited if a sufficiently energetic ignition source is present. One potential ignition source that, in many cases, could be an inherent part of the filling and emptying of containers and packages is an electrostatic discharge. Electrostatic charge generation most commonly occurs due to the process of contact and separation which takes place between the powder particles or the liquid stream and the filling equipment upstream of the container and between the powder or liquid and the internal surfaces of the container during emptying. The accumulation of the electrostatic charge on the container and/or its contents can result in electrostatic discharges and these discharges may be incendive to the flammable atmosphere generated during the activities. This section will discuss the practical approaches for eliminating/controlling electrostatically initiated flash fires and explosions. The risk will be present when containers and packages, including flexible intermediate bulk containers (FIBCs), metal and fiberboard containers, plastic containers, plastic liners for use in various containers, are being used for flammable liquid and powder transfer operations

Conductive (metal and fiberboard) Containers [Ref 7]

It should be noted that for electrostatic hazard assessment purposes, containers constructed from uncoated and unpainted fiberboard material are considered to be sufficiently conductive and should be treated like metal containers.

Conductive containers and any other conductive object in the vicinity, including the fill pipe and personnel, if present, should be electrically grounded.

Insulating (plastic) Containers for Liquids [Ref 8, 9] - *Containers fabricated from materials having volume resistivity > 10^9 ohm.m and/or surface resistivity > 10^{11} ohm/square*

Insulating containers can pose an electrostatic ignition hazard when used for flammable liquids or in locations where flammable vapor atmospheres might exist. Static charge can be developed when liquids are dispensed into or from the container. Plastic containers may also become charged due to handling and rubbing against other surfaces by personnel. Therefore, the use of containers made from ordinary plastics or other electrically-insulating materials for flammable liquids or in locations where flammable atmospheres having a minimum ignition energy (MIE) of 4mJ (millijoules) or less is not recommended due to the risk of ignition from "brush"-type electrostatic discharges. Instead, the use of containers made from metal or static dissipative or conductive plastics should be considered.

"Brush"-type discharges are produced when a charged insulating surface is approached by a grounded conductor having a relatively large radius of curvature (> 1 mm), i.e., a rounded or blunt electrode. Since charge mobility on insulating surfaces is impeded by the resistivity of the material, only charge from a limited surface area participates in a brush discharge. Also, charge cannot flow to form one distinct discharge channel and instead a network of smaller channels is formed. The term brush discharge refers to the appearance of this network of discharge channels, which resemble a brush when observed under low light conditions. Brush discharges are known empirically to exhibit effective energies of as much as about four (4) millijoules (mJ) [Ref 10].

Additionally, it would be prudent to implement measures to limit the development of electrostatic charge when low conductivity liquids are being handled. Such measures may include: (1) limiting the rate at which the container is filled; and (2) use of a grounded metal dip tube that extends sufficiently close to the bottom of the container. Dispensing, handling, and processing of flammable liquids should be performed in a suitable ventilated enclosure in order to minimize the spread of the flammable vapor.

Insulating (plastic) Containers for Powders [Ref 7] - *Containers fabricated from materials having volume resistivity > 10^9 ohm.m and/or surface resistivity > 10^{11} ohm/square*

If the contents of the container are solvent-wet a flammable atmosphere can be present, even if the solvent content is as little as 0.5% by weight. It is therefore generally suggested that insulating containers not be used for solvent wet powders. For dry powders electrical grounding of any conductive section/equipment that might be associated with the container should provide adequate protection, if it is established, through on-site measurement or laboratory testing, that the rate of electrostatic charge input with the powder (powder chargeability) is insufficient to give rise to propagating brush discharges from the inside surfaces of the container. Propagating brush discharges are highly energetic electrostatic discharges capable of igniting many flammable atmospheres, including gases, vapors, and dust clouds. These discharges -- exhibiting effective energies of as much as one to two Joules -- derive their energy from the formation of a double layer charge on both sides of a surface, such as a fabric, film, or coating [Ref 10].

If the powder is conductive, it should be in contact with an electrical ground at all times during filling and emptying.

Insulating (plastic) Liners for use in Containers [Ref 7]

When handling solvent wet or dry powders in drums, it is often convenient to use an inner liner. However, insulating liners can become electrostatically charged during filling, emptying, or if the liner is removed from the container. When filling the container with a highly charged insulating powder, there is a possibility of a propagating brush discharge from the liner. Additionally, the liner may insulate its contents from ground, even when the outer container is conductive and grounded. It is therefore suggested that electrically grounded static dissipative or conductive liners with a surface resistivity less than 10^{11} ohm per square be used.

Flexible Intermediate Bulk Containers (FIBCs) - *FIBC Types and Selection Criterion*

There are currently four types of FIBCs; namely Types A, B, C and D as defined by IEC 61340-4-4 [Ref 11]. IEC 61340-4-4 specifies requirements for FIBCs between 0.25m^3 and 3m^3 in volume, intended for use in hazardous explosive atmospheres created by FIBC contents or that may exist outside the FIBC.

According to the IEC 61340-4-4 an FIBC intended for use in the presence of a flammable material or in a hazardous explosive atmosphere shall not itself produce incendiary discharges. The absence of incendiary discharges shall be verified for at least the smallest and largest sizes of FIBC of a particular design when tested as manufactured, prior to usage. Type qualification certificates shall be supported by a test report and unless otherwise specified or agreed by interested parties, a type qualification certificate for a FIBC design shall be valid for a period of 3 years from date of issue.

Table 1 provides the features of the various FIBC types and the laboratory testing requirements for each FIBC [Ref 12].

Table 1 - Features and Testing Requirements for Various FIBC Types

FIBC Type	Features	Electrostatic Testing Requirements
Type A	- No protection from static discharge - May be used for non-combustible solids or solids with MIE >1000mJ	None
Type B	- Provides protection from Propagating Brush discharge - Use for combustible solids with MIE >3mJ in the absence of flammable vapors and gasses	✓ Breakdown Voltage ✓ Propagating Brush Discharges on liner/fabric combination
Type C	- Provides protection from Propagating Brush discharges and Brush discharges - Uses inter-connected conductive threads or conductive coatings to dissipate the charges - Must be grounded during filling and emptying - Use when transporting combustible solids with MIE <3mJ and/or when loading and unloading in the presence of flammable vapors and gasses	✓ Resistance-to-Groundable Point ✓ "Surface Resistivity" and/or "Breakdown Voltage" on liner
Type D	- Provides protection from Propagating Brush discharges and Brush discharges - Incorporates interwoven specialist antistatic threads and/or specialist antistatic coatings - Does not require grounding - Use when transporting combustible solids with MIE <3mJ and/or when loading and unloading in the presence of flammable vapors and gasses with MIE >0.14mJ	✓ Full-scale discharge Incendivity test ✓ "Surface Resistivity" and/or "Breakdown Voltage" on liner

Manual Additions to Vessels Containing Flammable Liquids [Ref 13]

Many chemical operations involve the manual addition of bulk solids and powders, as well as small quantities of liquid additives, to vessels containing flammable liquids. These are typically introduced through open manways or hatches located on the top of vessels. Several methods are used to transfer solids to vessels, including paper and plastic bags, metal, plastic and fiber drums, pneumatic conveyors, and flexible intermediate bulk containers (FIBCs). The contents of bags and smaller containers are often poured manually into a vessel by operators. The contents of larger drums and FIBCs are poured using a hoist or lift.

The introduction of ingredients through an open manway to vessels containing flammable liquids can pose a significant electrostatic hazard unless suitable precautions are taken. The pouring of powders and liquids can develop hazardous levels of static charge on the operator, bag or container, lifting equipment, and any chute that may be used for the conveying of the material to the receiving vessel. Flammable atmospheres include vapors evolved from liquid in the receiving vessel, dust clouds evolved in the area of the manway or hatch, and hybrid mixtures of vapors and dusts.

The following suggestions are offered to control the potential electrostatic ignition hazard associated with manual additions to vessels where flammable vapors or explosible dusts may be evolved:

1. In general, the loading of ingredients directly into a vessel containing a flammable liquid is not recommended. Instead, suitable intermediate transfer equipment such as a flap gate or air lock, rotary valve, or bag dump hopper should be used. The objective of such a system is to minimize the potential fire and explosion hazard by: (a) isolating potential ignition sources, such as electrostatic charges developed by solids handling, from the flammable atmosphere within the vessel; and (b) keeping the flammable atmosphere and inert gas (if used) within the vessel.

2. All metal equipment associated with manual additions to vessels where a flammable vapor or explosible dust may be evolved should be electrically grounded.

3. All personnel loading ingredients into vessels who may be exposed to flammable atmospheres should be electrically grounded. It is also suggested that such personnel should also be protected from possible flash fire by wearing long-sleeve flame-resistant outer clothing, with hand, face, and head protection.

4. If a flammable-vapor atmosphere is present in the vicinity of the vessel's manway, items made from plastics should not be used to introduce ingredients to the vessel, due to the risk of ignition from brush-type electrostatic discharges. Items of concern include plastic bags, scoops, chutes, funnels, and drum liners. Instead, only containers and items made from metal, or static-dissipative or conductive materials should be used. Such containers and items should be grounded during use.

Blending and Mixing (liquids, or liquid-solid)

Many chemical processes involve the blending or mixing of liquids or liquid-solid mixtures.
Electrostatic hazards may arise during blending and mixing as a consequence of static charging of materials due to continuous contact at liquid-liquid and liquid-solid interfaces. Static charging will be pronounced in electrically-insulating liquids, especially those containing solids, immiscible components, or multiple phases. Static charge can accumulate on insulating liquids even when they are being mixed in a grounded conductive vessel, and can accumulate on isolated conductors. Potential electrostatic ignition sources include: (1) spark-type electrostatic discharges from isolated conductors; and (2) discharges from charged liquid surfaces.

The following general suggestions are offered to control the potential electrostatic ignition hazards associated with blending and mixing operations when flammable liquids and vapors may be processed, present, or evolved:

1. All metal components of blenders and mixers should be suitably grounded such that the resistance-to-ground is less than ten (10) ohms. Items that should be grounded include agitators, baffles,

flanges, valves, lids, manway covers, and the vessel itself. Grounding should be checked regularly by measurement.

2. All personnel who may be exposed to a flammable atmosphere should be suitably grounded such that their resistance-to-ground is less than 1×10^8 ohms.

3. Virtually all types of blending and mixing operations -- particularly those involving electrically-non-conductive liquids -- are characterized by the development of high levels of electrostatic charge. Consequently, using control of ignition sources as a basis of safety for blending and mixing operations where flammable vapors are evolved may prove difficult. Discharges from the surface of charged liquids may have energies that exceed the ignition energy of flammable-vapor atmospheres. Therefore, the preferred basis of safety for blending and mixing operations where flammable vapors are evolved is inerting.

4. For blending and mixing vessels having an internal electrically-insulating lining or coating such as glass or plastic, a path to ground should be provided to allow for dissipation of electrostatic charge from the vessel contents. A suitable pathway will reduce the accumulation of electrostatic charge on the liquid and thereby minimize the probability of ignition from an electrostatic discharge from the liquid surface. A pathway to electrical ground is typically provided by: (a) a tantalum patch below the liquid level; (b) a conductive bottom discharge valve; or (c) a grounded dip leg.

Solids Blending

1. During blending and mixing operations, there is a risk of ignition from electrostatic discharges from accumulating bulk solids or powders. Thus, inerting and/or explosion protection measures should be considered when blending explosible bulk solids or powders having a minimum ignition energy (MIE) of 25 millijoules (mJ) or less and a volume resistivity of 1×10^{10} ohm-meters or greater. These same precautions should be considered when the bulk solid or powder is wet with a flammable liquid or a flammable gas or vapor atmosphere may otherwise be involved.

2. Metal vessels and the metal chimes of fiberboard drums into which bulk solids or powders are transferred after blending and mixing operations should be electrically grounded. Fiberboard typically has sufficient "conductivity" to be static-dissipative.

3. Only static-dissipative plastic liners -- having a surface resistivity of less than 1×10^{11} ohms per square -- should be used in vessels, drums, or other containers.

CONCLUDING REMARKS

The subject of static electricity has come a long way from being considered "black magic". As a direct result of many years of scientific research and practical industrial experience, one can now conduct an electrostatic hazard assessment by:

- Determining the magnitude and polarity of the electrostatic charge that can be generated in any process/operation;
- Investigating the ability of materials to retain electrostatic charge;

- Determining the igniting power of discharges resulting from electrostatic charge generation and accumulation; and
- Determining the equivalent electrostatic spark energy (Minimum Ignition Energy) that is required to ignite a flammable gas, vapor, dust or hybrid atmosphere.

An electrostatic ignition hazard can exist if the discharge energy is equal to or greater than the Minimum Ignition Energy of the flammable atmosphere.

Depending on the specific process/operation conditions safety may be achieved by preventing the charge from being generated in the first place, elimination the charge as it is being generated, preventing the accumulation of the charge, avoiding the flammable atmosphere, or considering a suitable explosion protection. Finally, any measure or combination of measures that are selected for ensuring safety from electrostatic ignition hazards must be recognized, understood, and maintained by facility personnel.

REFERENCES

1. National Fire Protection Association (NFPA), "Recommended Practice on Static Electricity", NFPA 77 (2007).
2. Vahid Ebadat, "Minimize Electrostatic Hazards During Process Operations", Chemical Engineering Progress (CEP), June 2013
3. ASTM E 2019, "Standard Test Method for Minimum Ignition Energy of a Dust Cloud in Air", ASTM International, West Conshohocken, PA, USA
4. ASTM D 257, "Standard Test Methods for DC Resistance or Conductance of Insulating Materials", ASTM International, West Conshohocken, PA, USA
5. CLC/TR 50404, "Electrostatics Code of Practice for the Avoidance of Hazards Due to Static Electricity", British Standards Institute, London, UK
6. ASTM D 2624, "Standard Test Methods for Electrical Conductivity of Aviation and Distillate Fuels", ASTM International, West Conshohocken, PA, USA
7. Vahid Ebadat, "Control of Electrostatic Hazards – Containers for Powders", Safety Watch 2005, No. 1, Chilworth Technology, Inc., 113 Campus Drive, Princeton, New Jersey 08540, USA, Safety-usa@chilworthglobal.com
8. Chantell Laing, "Case Study - Use of Non-Conductive Containers with Flammable Liquids, Safety Watch 2003, No. 1, Chilworth Technology, Inc., 113 Campus Drive, Princeton, New Jersey 08540, USA, Safety-usa@chilworthglobal.com
9. Vahid Ebadat and Paul Cartwright, "Electrostatic Hazards from Small Non-Conducting Containers for Flammable Liquids", Proceedings of the 27th Annual Loss Prevention Symposium, American Institute of Chemical Engineers, Houston, Texas, USA, March 29 – April 1, 1993
10. Handbook of Chemical Health and Safety, Chapter 36, American Chemical Society, Oxford University Press, 2001
11. IEC 61340-4-4, Electrostatic Classification of FIBCs, 2nd Edition, 2012-01
12. Vahid Ebadat, "Selection and Safe Use of Flexible Intermediate Bulk Containers (FIBCs), World FIBC Guide 2013/2014, MCI Media Ltd., UK
13. Vahid Ebadat, "Electrostatic Hazards in The Chemical Industry", Proceedings of the 28th Annual Loss Prevention Symposium, American Institute of Chemical Engineers, Atlanta, Georgia, USA, April 17 - 21, 1994

1ST CCPS ASIA-PACIFIC CONFERENCE ON PROCESS SAFETY℠

Dust Explosion Research Development for Coal Mining Safety Issues

Jiaojun Jiang
Mary Kay O'Connor Process Safety Center, Artie McFerrin Department of Chemical Engineering, Texas A&M University, College Station, TX 77843, USA
jiaojunjiang@tamu.edu

Xiaodan Gao
Mary Kay O'Connor Process Safety Center, Artie McFerrin Department of Chemical Engineering, Texas A&M University, College Station, TX 77843, USA
xiaodan_gao08@mail.che.tamu.edu

M. Sam Mannan*
(Presenter/Principal author)
Mary Kay O'Connor Process Safety Center, Artie McFerrin Department of Chemical Engineering, Texas A&M University, College Station, TX 77843, USA
mannan@tamu.edu

Prepared for Presentation at
1St CCPS Asia-Pacific Conference on Process Safety
Qingdao, China
September 4 – 5, 2013

UNPUBLISHED

Dust Explosion Research Development for Coal Mining Safety Issues

Jiaojun Jiang, Xiaodan Gao, M. Sam Mannan*

Mary Kay O'Connor Process Safety Center, Artie McFerrin Department of Chemical Engineering, Texas A&M University, College Station, TX 77843, USA

Keywords: Coal Mining Explosion, Dust Explosion, Hybrid Mixture Explosion

Abstract

China, U.S. and India are the three largest coal producers and consumers in the world. Statistics in recent years show that the number of incidents, death tolls and fatality rates per million tons are still high, especially in China and India. These incidents have attracted the attention of the public and the media for only a short period after the occurrence of the incident. Unfortunately, this type of attention does not help to prevent the next tragedy. The much more important thing is to learn the lessons from these tragedies by doing a complete study of the underlying hazards in the coal mining process. Coal mining incidents are generally caused by explosion, fire, haulage, ground fall/bump, and inundation. With regard to explosion, a lot of attention has been paid only to pure dust or pure gas explosions during the past few decades. However, several disasters with massive loss of lives and properties were caused by the hybrid mixture explosions with methane and coal dust together. Research on hybrid mixture explosions rather than pure gas or dust explosions can rarely be found.

This paper discusses the present dust explosion research developments from the Mary Kay O'Connor Process Safety Center (MKOPSC) and how to apply the research methodologies to coal mining safety issues. Specifically, studies on hybrid mixtures show that the presence of a small amount of flammable gas can significantly increase the explosion index of dust as well as the maximum explosion pressure. Furthermore, the lower flammable limits of the hybrid mixture are lower than each of the components' limits. The research methodology and results can be used for risk assessment and prevention in the coal mining industry.

1. Introduction

1.1 Coal is an important energy source

On April 5, 2010, a massive coal dust explosion occurred at the Upper Big Branch Mine-South (UBB). Twenty-nine people were killed and two were injured in the incident. It was the largest coal mining disaster in the United States since 1970 [1]. This disaster along with others has not lowered the pace of coal mine production, because coal is a main energy source which plays a vital role in delivering electricity across the global economy, in the creation of steel and concrete, and in providing energy for transport [2].

In 2011, over 6,185 million tons (Mt) of hard coal and 1,042 Mt of brown coal/lignite were produced worldwide [3]. China is the world's largest coal producer, followed by the U.S. and India. The economy and energy demand have been growing rapidly in China over the last 30 years, and especially in the last decade. In 2011, coal consumption in China grew more than 9%, continuing its upward trend for the twelfth consecutive year [4]. Coal supplies the vast majority of China's energy needs. The government has a program on energy saving, and has made massive investments in alternative energy sources and new technology on efficiency improvements [5]. However, coal will still remain the major energy source, and the production is expected to be continuously increasing in the foreseeable future.

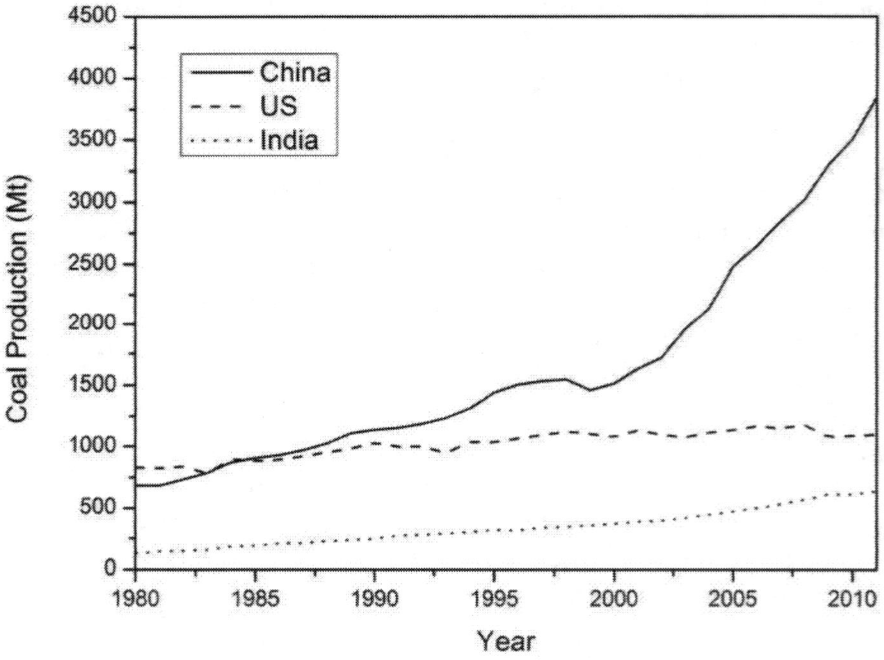

Figure 1 Coal mine production historic statistical data in China, U.S. and India [6]

As a developed country, over 90% of U.S. coal consumption is used for power plants which generated 42% of its electricity in 2011 [7]. As shown in Fig. 1, the production of coal grew slowly in the last 30 years compared with China. At the same time, U.S. coal exports climbed to 10% in 2011, which is the highest level in the last two decades. Although the combination of strong price competition with natural gas/shale gas, and increased use of renewable technologies, the absolute amount of U.S. coal production is expected to increase to meet growing demands for power in the absence of new policies to limit or reduce emissions of carbon dioxide [7].

India, as another developing country similar to China, was the third largest coal producer in 2011. Its coal production has continuously increased in the last 30 years. The country depends heavily on coal as a source of energy. India is likely to be the second largest consumer of coal, surpassing the U.S., in the next five years according to the International Energy Agency (IEA) [3].

1.2 Safety and health in coal mining industry

Due to the continuous growth of coal demand and the inherent risk of coal mining, every country including China, U.S. and India have made many efforts to improve coal mining safety. Unfortunately, similar or even worse incidents like UBB still occur.

Table 1 Fatalities and fatal injury rates from coal mining incidents

	China [8]			US [9]			India [10]		
Year	Production/Mt	Fatalities	Fatal injury rate/per Mt	Production/Mt	Fatalities	Fatal injury rate/per Mt	Production/Mt	Fatalities	Fatal injury rate/per Mt
2000	1314	5798	4.4110	1074	38	0.0354	335	144	0.4300
2001	1459	5670	3.8869	1128	42	0.0372	344	141	0.4100
2002	1380	6995	5.0681	1094	28	0.0256	359	97	0.2700
2003	1668	6702	4.0191	1072	30	0.0280	377	113	0.3000
2004	1956	6027	3.0807	1112	28	0.0252	381	99	0.2600
2005	2190	5491	2.5070	1132	23	0.0203	403	117	0.2900
2006	2381	4895	2.0563	1161	47	0.0405	406	138	0.3400
2007	2524	3786	1.5002	1146	34	0.0297	416	79	0.1900
2008	2716	3092	1.1384	1172	30	0.0256	417	100	0.2400
2009	2960	2631	0.8889	1179	18	0.0153	526	96	0.1825
2010	3200	2433	0.7603	1086	48	0.0442	572	117	0.2045
2011	3498	1973	0.5640	1095	21	0.0192	589	112	0.1900
2012	3650	1365	0.3740	1017	19	0.0187	558	116	0.2079

Statistics shows that 26 major incidents with more than 100 fatalities happened in China from 1949 to 2009 [11]. As indicated in Table 1, the fatalities and fatal rate per million tons decreased from 2000 to 2012. 2002 was the worst year in the last decade with 6,995 people killed and a 5.07 fatality rate per million tons. While in 2012, a milestone was indicated by Chinese government, and the fatalities dropped to 1,365 with a 0.3740 fatality rate which is the lowest in the past decade. In general, China has made many significant improvements regarding coal mine safety since 2000, and these improvements definitely have helped save many miners' lives. But the fatal injury rate is still 20 times higher than in the U.S. in 2012.

The annual coal production coupled with the fatalities and the fatal injury rate of the U.S. from 2000 to 2012 is also presented in Table 1. The Mine Safety and Health Administration (MSHA) have established comprehensive standards, compliance and enforcement procedures for coal mining industry. Eighteen people who were killed in 2009 led to a 0.0153 fatal injury rate which made this year the safest year in the past decade in coal mining in the U.S.. However, the UBB incident in 2010 killed 29 miners which made this year the worst one in the last decade. This incident also alerted researchers to the inherent risk in coal mining industry. Disasters could happen even with a good record of safety. Continuous attention and improvement should be made.

Table 1 also includes the data from India. The annual coal production kept increasing in last decade. The fatalities and fatal injury rate decreased from 2000 to 2012. It fluctuated around 0.30 until 2006, then around 0.20 until 2012. In general, the coal mining safety in India is better than China in the last decade. But the fatal injury rate is also 10 times higher than in the U.S. A lot of work needs to be done in India as well.

1.3 Causation analysis for coal mining incidents

Coal mining incidents capture the attention of the public and the media only for a short period of time after occurrence of the incident. Unfortunately, this type of attention does not help prevent the next tragedy. The much more important thing is to learn the lessons from these tragedies by doing a complete study of the underlying hazards in the coal mining process. Coal mining incidents are generally caused by explosion, fire, haulage, ground fall/bump, inundation and others [12]. Explosions that occur in mines involving coal dust usually also involve methane. Methane coexists with coal and is released while the coal is being extracted. The released methane itself can be highly explosive when mixed with air and reaches a concentration of 5.0-15.7% in volume [13]. A methane explosion can suspend coal dust in the air, which may lead to a second explosion. Sometimes, the methane and coal dust show up and are ignited together which causes a massive explosion. There have been many incidents where mining incidents have taken place in China, U.S. and India that involves coal and methane.

Table 2 Twenty-six major coal mining incidents in China from 1949 to 2009 [11]

Time	Place	Type incident	Fatalities
1950	Henan	Gas explosion	187
1954	Inner Mongolia	Gas and coal dust explosion	104
1960	Shanxi	Coal dust explosion	684
1960	Sichuan	Coal and gas outburst	125
1960	Henan	Gas and coal dust explosion	187
1960	Sichuan	Gas and coal dust explosion	124
1961	Liaoning	Fire	110
1968	Shandong	Coal dust explosion	108

1969	Shandong	Coal dust explosion	115
1975	Shaanxi	Gas and coal dust explosion	101
1977	Jiangxi	Gas explosion	114
1981	Henan	Gas and coal dust explosion	133
1991	Shanxi	Gas and coal dust explosion	147
1996	Shanxi	Gas and coal dust explosion	114
2000	Guizhou	Gas explosion	162
2002	Heilongjiang	Gas explosion	124
2004	Henan	Gas explosion	148
2004	Shaanxi	Gas explosion	166
2005	Liaoning	Gas explosion	214
2005	Guangdong	Inundation	121
2005	Heilongjiang	Coal dust explosion	171
2005	Hebei	Gas explosion	108
2007	Shandong	Inundation	172
2007	Shanxi	Gas explosion	105
2008	Shanxi	Inundation	276
2009	Heilongjiang	Gas explosion	108

The detailed types of the 26 major incidents in China from 1949 to 2009 are represented in Table 2. Twelve of them happened since 2000 even with the improvements China has made. Explosion is the major cause for these incidents. In these 26 incidents, specifically, seven of them were caused by hybrid gas and coal dust explosions.

Table 3 Number of underground coal mining worker fatalities by type of disaster in U.S., 1900 through 2008 [12]

Type incident	Number of Events	Number of Fatalities
Explosion	420	10390
Fire	35	727
Haulage	21	145
Ground fall/Bump	14	92
Inundation	7	62
Other	17	199

The number of fatalities by type of disaster in the U.S. from 1900 to 2008 is indicated in Table 3. In 420 disasters caused by explosions, 10,390 people lost their lives. This indicates that explosion is undoubtedly the major reason for most incidents.

Table 4 Hazard-wise probability of at least one major incident per year in Indian coal mines [14]

Hazard Category	Probability of at least one major accident
Roof and side fall	0.76
Explosion	0.32
Inundation	0.27
In shaft	0.17
Miscellaneous	0.20

The type of incidents and their probabilities in India are represented in Table 4. It shows that explosion is a major hazard which can cause at least one major incident with the probability of 0.32 each year.

Historical data shows that explosion was the major hazard in the coal mining industry and can still cause disasters with massive loss of lives and properties. Research with regards to explosion, especially gas and coal dust explosion, must be done to fully understand the hazards, to develop safety measures, and to prevent similar incidents in the future.

1.4 Hybrid mixture explosion

When an explosion involves dust and gas together, it is referred to a hybrid mixture explosion [15]. Aside from the mining industry, hybrid mixtures are also encountered in industries such as paint factories (pigments and solvents), pharmaceutical industries or grain elevators (small grains and fermentation gases). Unlike pure dust or gas explosions, which have been widely studied in past decades, only minimal research focuses on hybrid mixtures, and data concerning hybrid mixtures can rarely be found [16-18]. Because of this, hybrid mixture is one of the three most important research areas for dust explosion [19].

MKOPSC has a 36L vessel that can be used to evaluate the explosibility characteristics of dust samples and hybrid mixtures. With this equipment, one area of research MKOSPC is focused on is to determine the hazard class for the analyzed hybrid mixture samples (K_{St} values) as well as the Lower Flammable Limits (LFL). K_{St} is an explosion index which used to characterize the severity of the explosion. LFL represents the lowest concentration of material in air that can give rise to flame propagation upon ignition. These results are valuable for conducting a hazard assessment of process/operations that involve hybrid mixture such as coal mining industry.

A cornstarch and methane mixture was studied in this paper as an example to show the method developed at MKOPSC. This method can be applied to coal mining safety research.

2. Method

The experiments have been performed in a 36L apparatus. It was calibrated and the results agreed well with a standard 20L apparatus. For pure dust, the tests were performed based on the standard test procedure ASTM E1515. After the dust sample had been loaded into the container, the nozzle and igniter were installed in the chamber. Then the flange was bolted on and the chamber was partially evacuated to 10.3psi in order to start the ignition at 14.7psi. The air reservoir was pressurized with compressed air to 314.7psi. At the start of the test, the fast valve was triggered to open and lasted for 50ms. The chemical igniter was activated 25ms after the fast valve closed. The total ignition delay time is 75ms. In order to test hybrid mixtures, the partial pressure method was adopted to prepare methane-air mixtures inside the chamber prior to test. The chamber was evacuated to a target pressure, and then pure methane was fed until 10.3psi. The value of the target pressure is determined by different methane concentrations. The air from the reservoir with the methane in the chamber generated the final test mixture at 14.7psi.

The explosion behavior of the cornstarch/methane/air mixtures has been performed in the conditions close to minimum concentration values of both cornstarch (0 up to 200g/m3) and methane (0 up to 9.52v/v %).

3. Results and discussions

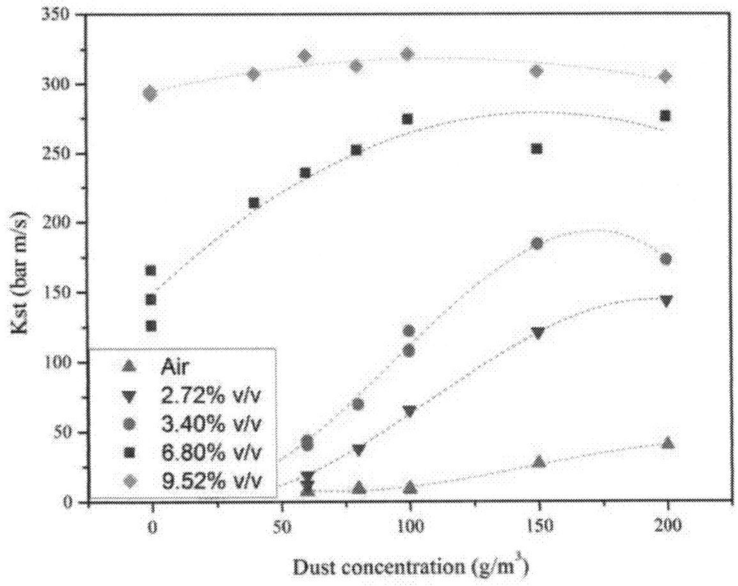

Figure 2 K_{St} as a function of cornstarch concentration at different values of the methane concentration

Figure 2 represents in detail the K_{St} values for the studied system. When the methane concentration is lower than its LFL (2.72v/v % and 3.40v/v %), the K_{St} values dramatically increase with the increase of cornstarch content. When methane concentration is equal to 6.80v/v %, the K_{St} will increase to 270bar·m/s as dust concentration increased to 100g/m^3. Then it appears to remain steady or even decrease as dust content increases. When the methane concentration is increased to 9.52v/v %, the K_{St} appears to be nearly independent of the dust content. Similarly, at cornstarch concentrations below the MEC (100g/m^3), it is not able to be ignited, but it could be easily ignited along with methane. The presence of methane has significant influence on the deflagration index (K_{St}).

At a dust concentration above MEC, the deflagration index will increase with the presence of methane, but the impact of methane will be less. Particularly, at a cornstarch concentration equal to MEC, the K_{St} increases from 10 to 340bar·m/s when the methane content increases from 0 to 9.5v/v %.

These phenomena are also found in other systems such as niacin/methane systems as reported by Garcia [16], and cork/methane systems as reported by Pilão [20]. They can be explained by the oxygen competition between the gas and dust during the explosion. For high dust concentrations, the deflagration index is imposed by the oxygen content and the increase of gas content also has little impact on the deflagration index due to the same reason.

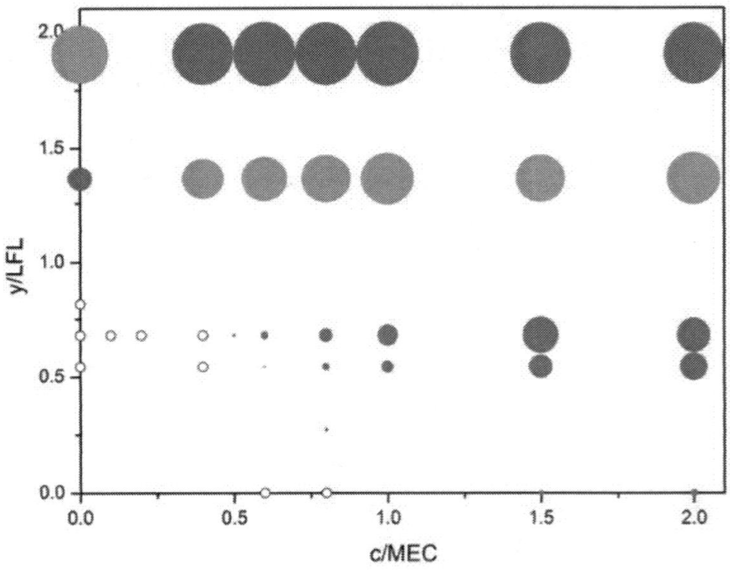

Figure 3 K_{St} in the 2-dimensional plane methane content/cornstarch concentration

In order to rationalize all the data, a map of the explosion behavior of the methane/cornstarch mixtures has been developed as shown in Fig. 3. The x-axis is the dust concentration (c/MEC), and y-axis is the methane content (y/LFL) respectively. In the figure, the deflagration indexes are represented by circles in which diameter is proportional to the value of K_{St}.

The hazard of the dust deflagration is divided into three classes by NFPA 68. Red circles refer to a K_{St} value higher than 300, which is located in class St-3. Yellow circles are in the range of 200 to 300 and are located in class St-2. Green circles are less than 200 and are in class St-1. The empty circles represent experiments where no explosion occurred. Compared to the pure dust in class St-1, the cornstarch can easily be transferred into St-2 and St-3 by adding methane. The effects of gas cannot be neglected, and the result shows that the hybrid mixture is much more dangerous than pure gas or dust.

4. Conclusions

As an important energy source in China, U.S. and India, coal production is expected to be continuously increasing in the foreseeable future. Although a lot of efforts have been made, coal mining incidents which can cause massive loss of lives and properties still occur, especially in China and India. Statistics show the hazard of explosion was the main reason for previous disasters. Several of those explosions were caused by coal and methane together as a hybrid mixture. Research on this hybrid mixture is required.

The Mary Kay O'Connor Process Safety Center has conducted significant research on hybrid mixtures. The experiment with the cornstarch and methane mixture shows that the presence of a small amount of flammable gas can significantly increase the explosion index of dust as well as the maximum explosion pressure. Furthermore, the lower flammable limits of the hybrid mixture are lower than that of each component.

The research methodology employed in this paper can be extended to other hybrid mixtures, including coal dust/methane. These results (K_{St}, LFL) are valuable for conducting a hazard assessment of the coal mining industry to develop proper prevention measures in order to avoid similar incidents from happening in the future.

5. References

[1] Fatal Underground Mine Explosion, Report of Investigation, http://www.msha.gov/Fatals/2010/UBB/FTL10c0331noappx.pdf (accessed July 2, 2013)

[2] "Coal – Energy for Sustainable Development", http://www.worldcoal.org/coal-energy-access (accessed July 2, 2013)

[3] "World Energy Outlook 2012", *International Energy Agency*, 2012

[4] "China Consumes Nearly as much Coal as the Rest of the World Combined," *U.S. Energy Information Administration*, Jan. 29, 2013

[5] Yardley, J., "Chinese Dam Projects Criticized for Their Human Costs," *The New York Times*, Nov. 19, 2007

[6] International Energy Statistics Database, *U.S. Energy Information Administration*

[7] "Annual Energy Outlook 2012," *U.S. Energy Information Administration*, 2012

[8] State Administration of Coal Mine Safety, http://media.chinasafety.gov.cn:8090/iSystem/shigumain.jsp (accessed July 2, 2013)

[9] "Coal Mine Fatalgrams and Investigation Reports," Mine Safety and Health Administration, http://www.msha.gov/fatals/fabc.htm (accessed July 3, 2013)

[10] Chakrabort, S.R.B., "Safety and Health Performing in Indian Coal Mining," http://www.sari-energy.org (accessed July 6, 2013)

[11] Wu, L., Jiang, Z., Cheng, W., Zuo, X., Lv, D., Yao, Y., "Major Accident Analysis and Prevention of Coal Mines in China from the Year of 1949 to 2009," *Mining Science and Technology (China)*, Volume 21, 2011

[12] Brnich, M.J., Kroalski-Trakofker, K.M., "Underground Coal Mine Disasters 1900 -2010: Events, Responses, and Look to the Future," http://www.cdc.gov/mmwr/preview/mmwrhtml/mm5751a3.htm (accessed July 6, 2013)

[13] Li, Z., Gong, M., Sun, E., Wu, J., Zhou, Y., "Effect of Low Temperature on the Flammability Limits of Methane/Nitrogen Mixtures," *Energy*, Volume 36, Issue 9, Sept. 2011

[14] Mandal, A., Sengupta, D., "The Analysis of Fatal Accidents in Indian Coal Mines," Indian Statistical Institute," http://faculty.franklin.uga.edu/amandal/sites/faculty.franklin.uga.edu.amandal/files/Fatal_accidents_in_Indian_Coal_Mines.pdf (accessed July 7, 2013)

[15] Cashdollar, K.L., "20-L Explosibility Test Chamber for Dusts and Gases," *Review of Scientific Instruments*, Volume 20, No. 4, 1985

[16] Garcia Agreda, A., Di Benedetto, A., Russo, P., Salzano, E., Sanchirico, R., "Dust/gas Mixtures Explosion Regimes," *Powder Technology*, Volume 205 No. 1, 2011

[17] Denkevits, A., "Hydrogen/dust Explosion Hazard in ITER: Effect of Nitrogen Dilution on Explosion Behavior of Hydrogen/tungsten Dust/air mixtures," *Fusion Engineering and Design*, Volume 85, No. 7, 2010

[18] Garcia Agreda, A., "Study of Hybrid Mixture Explosions," *Università degli studi di Napoli Federico II*, 2010

[19] Eckhoff, R., "Current Status and Expected Future Trends in Dust Explosion Research," *Journal of Loss Prevention in the Process Industries*, Volume 18, No.4, 2005

[20] Pilão, R., Ramalho, E., Pinho, C., "Explosibility of Cork Dust in Methane/air Mixtures," *Journal of Loss Prevention in the Process Industries*, Volume 19, No. 1, 2006

DUST EXPLOSIONS

Understanding of the Hazard and Practical Measures for its Control

Vahid Ebadat, Ph.D.
Chilworth Technology, Inc. (A DEKRA Company)
113 Campus Drive, Princeton, New Jersey 08540, USA
vahid.ebadat@dekra.com
www.chilworth.com

KEYWORDS

Dust Explosion, Explosion Testing, Electrostatic Hazards, Explosion Prevention, Explosion Protection, Ignition Sensitivity, Explosion Severity, Explosion Isolation, Inert Gas

INTRODUCTION

The processing industry has experienced many serious incidents caused by explosions of what might be normally thought of as harmless materials such as plastics, rubber, foodstuffs, and wood dust. Table 1 lists some of the more serious dust explosions in the United States alone.

Table 1 Some serious dust explosions in the United States

Year	Company	Location	Dust	Injuries/Fatalities
1999	Ford Motor Company	Dearborn, Michigan	coal dust	6 fatalities, 14 injuries
2002	Rouse Polymerics International, Inc.	Vicksburg, Mississippi	Rubber dust	5 fatalities, 7 injuries
2003	West Pharmaceutical Services	Kinston, North Carolina	Polyethylene dust	6 fatalities, 38 injuries
2003	CTA Acoustics, Inc.	Corbin, Kentucky	phenol formaldehyde resin dust	7 fatalities, 42 injuries
2003	Hayes Lemmerz International, Inc.	Huntington, Indiana	Aluminum dust	1 fatality, 2 injuries
2008	Imperial Sugar	Savannah, Georgia	Sugar dust	14 fatalities, more than 60 injuries

A dust cloud explosion occurs when a number of conditions simultaneously exist in one location. These conditions are:

- Dust must be combustible

  The first stage of a dust explosion hazard assessment is to determine whether the dust will explode when dispersed as a cloud. The combustibility of a dust cloud can be assessed by conducting an

"Explosion Classification Test" on a representative sample of the dust. In this laboratory test the observation of flame propagation or pressure development in a closed vessel determines whether or not a suspended dust is capable of initiating and sustaining an explosion.

The particle size is a property which influences the explosibility of the dust cloud. Generally the particle must be smaller than about 0.5mm in diameter (500 microns); however, irregular shaped particle of larger sizes can also participate in an explosion. The finer the particles, the greater the surface area per unit mass, and thus the more explosible a given dust is likely to be. When the cloud is composed of a series of particle sizes ranging from fine to coarse, the fines play a more prominent role in the ignition and explosion propagation throughout the dust cloud. The presence of dusts should be anticipated in the process stream, regardless of the starting particle size of the material, due to physical size reduction caused by mechanical forces.

The moisture content of a product will also affect the explosibility, or explosion risk. A dry dust usually contains less than 5% moisture. Dry dusts of small particle size will be more easily ignited and produce more severe explosions when compared to dusts containing moisture. Moisture within or on the particle surface reduces both the ease of ignition and the severity of the explosion.

Table 2 lists some examples of combustible dusts from various industries [ref 1].

Table 2 Examples of Combustible Dusts

Agricultural Products	Cottonseed	Soybean dust	**Chemical Dusts**	Epoxy resin
Egg white	Garlic powder	Spice dust	Adipic acid	Melamine resin
Milk, powdered	Gluten	Spice powder	Anthraquinone	Melamine, molded
Milk, nonfat, dry	Grass dust	Sugar (10x)	Ascorbic acid	(phenol-cellulose)
Soy flour	Green coffee	Sunflower	Calcium acetate	Melamine, molded
Starch, corn	Hops (malted)	Sunflower seed dust	Calcium stearate	(wood flour and
Starch, rice	Lemon peel dust	Tea	Carboxy-methylcellulose	mineral filled phenol-
Starch, wheat	Lemon pulp	Tobacco blend	Dextrin	formaldehyde)
Sugar	Linseed	Tomato	Lactose	(poly) Methyl acrylate
Sugar, milk	Locust bean gum	Walnut dust	Lead stearate	(poly) Methyl acrylate,
Sugar, beet	Malt	Wheat flour	Methyl-cellulose	emulsion polymer
Tapioca	Oat flour	Wheat grain dust	Paraformaldehyde	Phenolic resin
Whey	Oat grain dust	Wheat starch	Sodium ascorbate	(poly) Propylene
Wood flour	Olive pellets	Xanthan gum	Sodium stearate	Terpene-phenol resin
	Onion powder		Sulfur	Urea-formaldehyde/
Agricultural Dusts	Parsley (dehydrated)	**Carbonaceous Dusts**		cellulose, molded
Alfalfa	Peach	Charcoal, activated	**Metal Dusts**	(poly) Vinyl acetate/
Apple	Peanut meal and skins	Charcoal, wood	Aluminum	ethylene copolymer
Beet root	Peat	Coal, bituminous	Bronze	(poly) Vinyl alcohol
Carrageen	Potato	Coke, petroleum	Iron carbonyl	(poly) Vinyl butyral
Carrot	Potato flour	Lampblack	Magnesium	(poly) Vinyl chloride/
Cocoa bean dust	Potato starch	Lignite	Zinc	ethylene/vinyl
Cocoa powder	Raw yucca seed dust	Peat, 22%H_2O		acetylene suspension
Coconut shell dust	Rice dust	Soot, pine	**Plastic Dusts**	copolymer
Coffee dust	Rice flour	Cellulose	(poly) Acrylamide	(poly) Vinyl chloride/
Corn meal	Rice starch	Cellulose pulp	(poly) Acrylonitrile	vinyl acetylene
Cornstarch	Rye flour	Cork	(poly) Ethylene	emulsion
Cotton	Semolina	Corn	(low-pressure process)	copolymer

- Dust must be airborne

The dust must be suspended and intimately mixed with a suitable oxidizing gas atmosphere such as the oxygen in air to allow oxidation to occur at the particle surface. Because of the availability of oxygen around each particle, the combustion process will be very rapid, giving rise to a dust explosion and pressure effects.

- Dust concentrations must be within explosible range

  An explosion cannot propagate if the concentration of the dust cloud is below the "Minimum Explosible Concentration" (MEC). The sensitivity of the dust cloud to ignition and also the resulting explosion violence increase as the dust cloud concentration is increased above the minimum explosible concentration until an optimum concentration is reached. This would cause the highest explosion violence. At dust cloud concentrations exceeding the optimum concentration, the explosion violence decreases and can even reach a point at which propagation will not occur because the atmosphere is too fuel rich.

- The atmosphere in which the dust cloud is present must be capable of supporting combustion

  A dust cloud explosion will only occur if there is a sufficient amount of oxidant available. In practice the oxygen in air is the most common oxidant. Other oxidants include Chlorine, Fluorine, Nitrous Oxide, Nitric Oxide, and Nitrogen Tetroxide.

- An ignition source with sufficient energy to initiate flame propagation must be present

  The ignition sources that have been found to be the cause of the majority of explosions in dust handling/processing plants include (this is not an exhaustive list) welding and cutting, heating and sparks generated by mechanical failure, sparks generated by mechanical impacts, flames and burning materials, self-heating, electrostatic discharges, electrical sparks, and others.

A systematic approach to identifying dust cloud explosion hazards and taking measures to ensure safety usually involves the following steps:

- Determining the dust cloud's ignition sensitivity and explosion severity characteristics through appropriate laboratory tests on representative dust samples
- Identifying areas of the facility where combustible dust cloud atmospheres could exist under normal and abnormal operating conditions
- Identifying potential ignition sources that could exist under normal and/or abnormal conditions;
- Preventing the formation of explosible dust clouds in the plant and reducing the extent and duration of any clouds that may form
- Implementing an effective housekeeping procedure
- Taking measures to eliminate or control all likely ignition sources
- Taking measures to protect against the consequences of dust cloud explosions. Explosion protection measures include explosion relief venting, explosion suppression, explosion containment, along with explosion isolation. Alternatively, where practical, one could consider the application of inert gas purging or blanketing to prevent the combustion process.

Management personnel in facilities processing and handling combustible powders need to be aware of the potential for fires and explosions and how industry best practice and applicable codes and standards can help manage the risk. Taking measures against potential dust explosions as described above is referred to as establishing "a Basis of Safety".

This paper shall provide practical information on how to assess and manage dust explosion hazards in the processing industry.

APPLICABLE UNITED STATES CODES AND STANDARDS

According to the US Chemical Safety and Hazard Investigation Board (CSB) [ref 2] three areas very often require management efforts in order to control dust explosion hazards. These are:

1. Understanding and use of National Fire Protection Association (NFPA) Codes and Standards

   Codes and Standards are an important method used by industry, legislators and professionals to track and maintain a continuously expanding and changing experience-based information database in support of Best Industry Practice(s). In the United States, industry must comply with OSHA's "General Duty Clause" (employers must provide a safe place to work free from recognized hazards likely to cause injury or death) as well as specific safety directives. In order to comply, it is critical that industries handling and processing explosible dusts be fully aware of Best Industry Practices as described in pertinent dust Codes and Standards, and follow these recommendations.

   Relevant NFPA codes and standards include:

   - NFPA 61, "Standard for the Prevention of Fires and Dust Explosions in Agricultural and Food Products Facilities" [ref 3]
   - NFPA 68, "Standard on Explosion Protection by Deflagration Venting" [ref 4]
   - NFPA 69, "Standard on Explosion Prevention Systems" [ref 5]
   - NFPA 77, "Recommended Practice on Static Electricity" [ref 6]
   - NFPA 484, "Standard for Combustible Metals" [ref 7]
   - NFPA 499, " Recommended Practice for the Classification of Combustible Dusts and of Hazardous (Classified) Locations for Electrical Installations in Chemical Process Areas" [ref 8]
   - NFPA 654, "Standard for the Prevention of Fire and Dust Explosions from the Manufacturing, Processing and Handling of Combustible Particulate Solids" [ref 9]
   - NFPA 655, "Standard for Prevention of Sulfur Fires and Explosions" [ref 10]
   - NFPA 664, "Standard for the Prevention of Fires and Explosions in Wood Processing and Woodworking Facilities" [ref 11]

   The majority of the above codes and standards are cited in the International Fire Code under dust explosion hazard management and are requirements of that code.

2. Use of Material Safety Data Sheets (MSDS) to Identify Combustible Dust Explosion Risks

   Quantitative combustible dust fire and explosion properties are not specifically and clearly required to be included on Material Safety Data Sheets by the existing OSHA Hazard Communication Standard (HCS) or the American National Standards Institute consensus standard for MSDS format and preparation, ANSI Z400.1. As a result, when explosivity information is included, it is often in the form of less than adequate qualitative statements such as "Powder may form Explosive Dust/Air Mixtures". Qualitative statements give no hint as to the conditions required to create an explosion hazard, or the relative violence of the resulting deflagration. Good information on the hazardous properties of materials is critical for an adequate process hazards analysis. Therefore, all pertinent fire and explosion, electrostatic, and thermal instability information should be included in the MSDS in order to adequately understand and control the hazards associated with combustible dusts. The Chemical Safety Board report identified the deficiencies in the content of the MSDS sheet with regard to failure to provide adequate information concerning combustibility properties to the end-user. Since the publication of the report OSHA has revised the Hazard Communication Standard to address these deficiencies [ref 12]. The revised standard has renamed the MSDS as a Safety Data Sheet (SDS) and added requirements to address the deficiencies including adding combustible dust to the definition of a "hazardous chemical", requiring that a hazard evaluation be conducted to determine if a combustible dust hazard exists and specifying a format for presentation of the pertinent information that must be provided in the SDS. Since there is a two year compliance period users should use good judgment with regard to the information that is provided on the current MSDS sheet and seek the services of the product manufacturer or an expert if there are any questions.

3. Awareness and training to ensure that operating personnel and those responsible for safety and fire prevention oversight are knowledgeable in the subject of dust explosion hazards and can apply existing codes and standards to prevent dust explosions.

Table 3 lists some United States codes and standards related to process safety and their equivalent European and Chinese documents.

Table 3 Some United States and equivalent European and Chinese Codes and Standards

US Standard		European Standard		Chinese Standard	
NFPA 30	Flammable and Combustible Liquids Code [ref 13]	**EN 1127-1**	Explosive atmospheres – Explosion prevention and protection - Part 1: Basic concepts and methodology		中华人民共和国国务院令 第591号自2011年12月1日起施行 《危险化学品安全管理条例》
NFPA 51B	Standard for Fire Prevention During Welding, Cutting, and Other Hot Work [ref 14]			AQ 3022	化学品生产单位动火作业安全规范
NFPA 70	National Electrical Code (NEC 500) [ref 15]		Country Specific	GB13869 GB 19517	用电安全导则 国家电气设备安全技术规范
NFPA 77	Recommended Practice on Static Electricity [ref 6]	**CLC/TR 50404**	Electrostatics – Code of practice for the avoidance of hazards due to static electricity, CENELEC	GB 12158	防止静电事故通用导则 General Guideline for Preventing Electrostatic Accidents
NFPA 68	Guide for Venting of Deflagrations [ref 4]	**EN 14491** **EN 14797** **EN 14494**	Dust explosion venting protective systems Explosion venting devices. Gas explosion venting protecting systems	GB/T 15605	粉尘爆炸泄压指南
NFPA 484	Standard for Combustible Metals [ref 7]	**X**	Generally, Directive ATEX 99/92/EC. There are no specific European standards for this	GB 17269-2003	铝镁粉加工粉尘防爆安全规程
NFPA 497	Recommended Practice for the Classification of Flammable Liquids, Gases, or Vapors and of Hazardous (Classified) Locations for Electrical Installations in Chemical Process Areas [ref 16]	**EN 60079-10-1** **EN 60079-14** **EN 60079-17**	Explosive atmospheres – Part 10-1: Classification of areas. Explosive gas atmospheres Explosive atmospheres Electrical installations design, selection and erection. Explosive atmospheres. Electrical installations inspection and maintenance.	GB 3836.14-2000	爆炸性环境用防爆电气设备 第14部分- 危险场所分类
NFPA 499	Recommended Practice for the Classification of Combustible Dusts and of Hazardous (Classified) Locations for Electrical Installations in Chemical Process Areas [ref 8]	**EN 60079-10-2** **EN 60079-14** **EN 60079-17**	Explosive atmospheres. Part 10-2 Classification of areas. Combustible dust atmospheres. Explosive atmospheres. Electrical installations design, selection and erection. Explosive atmospheres. Electrical installations inspection and maintenance	GB12476.1-2000 GB12476.2-2010	可燃性粉尘环境用电气设备 第1部分 用外壳和限制表面温度保护的电气设备 可燃性粉尘环境用电气设备 第2部分：选型和安装
NFPA 654	Standard for the Prevention of Fire and Dust Explosions from the Manufacturing, Processing, and Handling of Combustible Particulate Solids [ref 9]	**EN 1127-1**	Explosive atmospheres – Explosion prevention and protection - Part 1: Basic concepts and methodology	GB 15577	粉尘防爆安全规程

LABORATORY TESTING

As was mentioned in the *introduction* section, the first step in the assessment and control of dust cloud explosion hazards and selecting the most suitable basis of safety involves the determination of the explosibility, ignitibility, thermal instability, and electrostatic properties of the dust(s) that are being handled / processed in the facility. The explosion characteristics of powders normally fall within one of two groups, "likelihood of an ignition (ignition sensitivity)" and "consequences of an explosion (explosion severity)".

- Tests to Determine the "Likelihood of an Ignition"

  Table 4 contains information on the laboratory tests along with the applicable United Sates standard methods that are commonly utilized to determine the sensitivity of a powder or dust cloud to ignition. Depending on the nature of the process/operation some or all of these tests may need to be considered.

Table 4 Laboratory Tests to Determine the Likelihood of Ignition of a Powder or Dust Cloud

Test	Standard Test Method	Description
Explosion Classification (Screening)	ASTM E1226, Standard Test Method for Explosibility of Dust Clouds [ref 17]	This test determines whether a dust cloud will explode when exposed to a sufficiently energetic ignition source. The test results in a powder being classified as either "explosible" or "non-explosible". This test, commonly referred to as a Go / No Go test, answers the question "Can this dust explode?"
Minimum Ignition Energy (MIE)	ASTM E2019, Standard Test Method for Minimum Ignition Energy of a Dust Cloud in Air [ref 18]	MIE test determines the lowest electric energy that is capable of igniting a dust cloud at its optimum ignitable concentration. The test is used primarily to assess the susceptibility of dust clouds to ignition by electrostatic discharges (sparks).
Minimum Ignition Temperature of a Dust Cloud (MIT Cloud)	ASTM E1491, Standard Test Method for Minimum Auto-ignition Temperature of Dust Clouds [ref 19]	This test determines the lowest temperature capable of igniting a dust dispersed in the form of a cloud. The MIT_{cloud} is used in evaluating the ignition sensitivity of dusts to such ignition sources as heated environments, hot surfaces, electrical devices, and friction sparks.
Minimum Ignition Temperature of a Dust Layer (MIT Layer)	ASTM E2021, Standard Test Method for Hot Surface Ignition Temperature of Dust Layers [ref 20]	This test determines the lowest surface temperature capable of igniting a dust layer of standard thickness (5 to 12.7 mm). The MIT_{layer} is used in evaluating the ignition sensitivity of powders to ignition by hot surfaces.
Self-Heating	J.A. Abbott, "Prevention of Fires and Explosions in Dryers", Institute of Chemical Engineers, 1990 [ref 21]	Ignition of bulk powders can occur by a process of self-heating when the temperature of the powder is raised to a level at which the heat liberated by the exothermic oxidation or decomposition reaction is sufficient to exceed the heat losses and to produce a "runaway" increase in temperature. The minimum onset temperature for self-ignition of a powder depends mainly on the nature of the powder and on its dimensions. If these variables are predictable, the onset temperature for self-ignition and also the induction time to self-ignition can be reliably assessed by appropriate small-scale laboratory tests: • Bulk Powder Test: Simulating bulk powder in IBCs, bags, and the bottom of hoppers • Aerated Powder Test: Simulating fluidized bed processing • Powder Layer Test: Simulating powder deposits on dryer walls/surfaces and tray drying • Basket Test: Simulating large-scale storage or transport conditions

- Tests for ensuring safety through the avoidance/control of flammable atmospheres or explosion protection

Table 5 contains information on the laboratory tests along with the applicable standard methods that are commonly utilized to obtain the required information when ensuring safety from dust cloud explosions through the avoidance/control of dust clouds or the oxidant or protecting people and facilities against the consequences of an explosion through relief venting, suppression or containment. Depending on factors such as the nature of the process/operation, explosion properties of the dust, and the company culture, one of these tests will need to be considered.

Table 5 Tests for Taking Measures to Prevent an Explosion or Protect against its Consequences

Test	Standard Test Method	Description
Minimum Explosible Concentration - MEC	ASTM E1515, Standard Test Method for Minimum Explosible Concentration of Combustible Dusts [ref 22]	MEC test determines the lowest concentration of a dust cloud in air that can give rise to flame propagation upon ignition. This test helps to answer the question "How easily can an explosible dust cloud be formed?"
Limiting Oxidant Concentration - LOC	EN 14034-4, Determination of the Limiting Oxygen Concentration of Dust Clouds [ref 23]	LOC test determines the minimum concentration of oxygen (displaced by an inert gas such as nitrogen or carbon dioxide) capable of supporting combustion. An atmosphere having an oxygen concentration below the LOC is not capable of supporting a dust cloud explosion. The LOC test is used to study explosion prevention or severity reduction involving the use of inert gases and to set oxygen concentration alarms or interlocks for inerted vessels.
Maximum Explosion Pressure, Maximum Rate of Pressure Rise, Deflagration Index (K_{st})	ASTM E1226,Standard Test Method for Explosibility of Dust Clouds [ref 24]	The maximum explosion pressure and maximum rate of pressure rise are measured and used to calculate the Deflagration Index (K_{st}) value of the dust cloud and to classify a material's explosion severity. These data can be used for the purpose of designing dust explosion protection measures such as explosion relief venting, suppression, and containment and also as a predictor of the potential for equipment and plant damage in the event of an explosion.

- Electrostatic Properties

Table 6 contains information on the applicable standard test methods that may be considered to obtain information on the magnitude and the polarity of the electrostatic charge that is generated on powder particles during processing and handling operations as well as the length of time that it takes for the charge to relax to ground once the particles come into contact with a grounded conductive surface.

Table 6 Tests for Determining Electrostatic Properties of Powder

Test	Standard Test Method	Description
Electrostatic Volume Resistivity	General Accordance with ASTM D257, Standard Test Methods for DC Resistance or Conductance of Insulating Materials [ref 25]	Volume Resistivity is used for classifying powders as conductive, semi-conductive, or insulating. Insulating powders have a propensity to retain electrostatic charge and can produce hazardous electrostatic discharges.
Electrostatic Chargeability	General Accordance with ASTM D257, Standard Test Methods for DC Resistance or Conductance of Insulating Materials [ref 25]	Electrostatic Chargeability is a measurement of the propensity of powder particles to become charged when flowing through conveyances or when handled in containers. This test provides data that can be used to develop appropriate material handling guidelines from an electrostatic hazards point of view.

Table 7 specifies the type of data that might be required to assess dust explosion hazards associated with some common unit operations in powder handling/processing facilities.

Table 7 Suggested Combustible Dust Testing for Assessing Risks Associated with Specific Unit Operations [ref 26]

Unit Operation	Explosion Screening[1]	MIE (mJ)	MIT – Cloud (°C)	MIT – Layer (°C)	Explosion Severity – Kst (bar.m/s)	LOC[2] (%)	MEC (g/m^3)	Volume Resistivity[3] (Ω.m)	Chargeability[4] (C/Kg)	Self-Heating (°C)
Manual Handling / Pouring	X	X						X	X	
Sieving / Screening	X	X						X	X	
Tumble / Double Cone Blending	X	X			X	X		X	X	
Ribbon Blending	X	X	X	X				X	X	
Milling	X	X	X	X	X	X		X	X	X
Jet Milling	X	X			X	X		X	X	
Spray, Fluidized Bed, Tumble Drying	X	X			X			X	X	X
Tray Drying	X	X								X
Pneumatic Conveying	X	X					X	X	X	
Screw Conveying	X	X	X					X	X	
Transfer to Hopper / Bin / Tote / Container	X	X			X			X	X	
Dust Collector and Exhaust Ventilation	X	X			X		X	X	X	

Full form of abbreviations used in Table 7:

- *MIE - Minimum Ignition Energy*
- *MIT- Cloud - Minimum Ignition Temperature of a Dust Cloud*
- *MIT-Layer - Minimum Ignition Temperature of a Dust Layer*
- *K_{st} - Deflagration Index*
- *LOC - Limiting Oxidant Concentration*
- *MEC - Minimum Explosible Concentration*

Notes on Table 7:

1. *Explosibility Screening test is only conducted if the combustibility of the powder/dust (as being present in the process/facility) is not yet established. If the powder is found to be non-combustible, other tests in the table may not be required.*
2. *LOC is determined if the basis of safety is inert gas blanketing.*
3. *Volume Resistivity should be considered if the Minimum Ignition Energy is less than 25mJ.*
4. *Chargeability should be considered if the Minimum Ignition Energy is less than 25mJ.*

Table 8 provides both the United States and equivalent European references for the laboratory tests sites above.

Table 8 United States and equivalent European references for the laboratory tests

TEST	US METHOD	EUROPEAN METHOD
Explosibility Screening	ASTM E1226	Kuhner Operating Instructions
Minimum Ignition Energy - Dust Cloud) (MIE)	ASTM E2019	IEC 61241-2-3 (Part 2, Section 3) / BS EN 13821: 2002
Minimum Ignition Temperature - Dust Cloud (MIT Cloud)	ASTM E1491	IEC 61241-2-1 (1994), Method B, BS EN 50281-2-1: 1999 PART 2-1: Method B
Minimum Ignition Temperature - Dust Layer (MIT Layer)	ASTM E2021	IEC 1241-2-1: 1994 BS EN 50281-2-1, METHOD A
Minimum Explosible Concentration (MEC)	ASTM E1515	BE EN 14034-3: 2006
Limiting Oxygen Concentration (LOC)	WK1680 ISO/6184/1 Kuhner Operating Instructions	BS EN 14034-4: 2004 Kuhner Operating Instructions
Explosion Severity (Kst)	ASTM E1226	PMAX: EN 14034-1 DP/DT: EN 14034-2
Powder Resistivity / Chargeability	ASTM D257	BS 5958: Part 1: 1991 and BS 7506: Part 2: 1996

BASIS OF SAFETY FROM DUST CLOUD EXPLOSIONS

Managing the risk of potential dust cloud explosions could include taking measures to avoid an explosion (explosion prevention) or designing facilities and equipment so that in the event of an explosion people and facilities are protected (explosion protection). Selection of explosion prevention and/or protection measures is usually based on:

- The availability of information on the sensitivity of the powder(s) to ignition and the resulting explosion severity
- The nature of the processes and operations
- The level of personnel's knowledge and appreciation regarding the consequences of a potential dust explosion and adherence to the preventive measures
- The environmental effects of a dust explosion
- The business interruptions resulting from a dust explosion
- Safety Culture: The desire to protect personnel and property

Explosion Prevention Measures

The risk of an explosion is removed when one of the following measures is taken:

- Controlling the spread of an explosible dust cloud atmosphere beyond the immediate source of release through the use of correctly designed and utilized local exhaust ventilation (LEV). Local exhaust ventilation is designed to intercept the dust at the source of release, dilute the dust cloud atmosphere and direct it into a system where the dust is recovered from the air stream.

- Reducing the oxidant concentration below a level that could no longer support combustion (Limiting Oxidant Concentration, LOC), by adding an inert gas. Nitrogen gas is perhaps the most commonly used inert gas. Other inert gases include carbon dioxide, argon, helium, steam, and flue gas (waste gas from on-site processes). Oxidant can also be removed by working under vacuum (safe vacuum pressure should be obtained by testing).

  Limiting Oxidant Concentration (LOC) for combustion is dependent on the nature of the dust and type of inert gas used. Once the LOC of the dust has been determined for the inert gas that will be used, the inert gas needs to be introduced into the vessel. Successful inert gas blanketing will only be possible if the entire volume of the vessel is inerted and the inert atmosphere is maintained at all times, even when the vessel is opened during the addition of solids and/or liquids to the vessel.

- All ignition sources capable of igniting the dust cloud are eliminated. Elimination of ignition sources involves:

  <u>Control of Heat Sources</u>

  - In heat processing (e.g. during drying), the temperature should be maintained below the self heating temperature (for solids)
  - Preventing the overloading of processing equipment (grinders, feeders, conveyors, blenders etc.). Internal buildup will BOTH reduce heat loss from material AND increase operating temperature above "normal". Consider the installation of overload protection devices on drive motors

- Isolation or shielding of hot surfaces
- Prevention/removal of dust accumulations on hot surfaces
- Use of approved electrical equipment (correct temperature rating)

Control of Friction/Impact Sparks

In any operation where friction/impact sparks could be expected, the measures that may be considered for avoiding a dust cloud ignition include:

- Preventing overheating due to misalignment, loose objects, belt-slip/rubbing etc. by regular inspection and maintenance of equipment
- Preventing foreign material from entering the system when such foreign material presents an ignition hazard. Consider use of screens, electromagnets, pneumatic separators, etc. This is especially important at milling and grinding feed locations.
- Controlling hot work operations by a "hot work permit" system in accordance with NFPA 51B, Standard for Fire Prevention During Welding, Cutting and Other Hot Work
- Preventing the formation of dust clouds and dust deposits

Electrical Equipment and Instruments

Incorrectly specified electrical equipment is a potential ignition source for flammable gases, vapors and dusts by causing electrical sparks and also causing hot surfaces. Therefore, in facilities handling flammable materials, electrical equipment must be suitable for the environment in which it is to be used. In order to determine the type of equipment, it is necessary to define hazardous (classified) locations, where ignitable atmospheres could occur.

Article 500 of the National Electrical Code® (NEC®) / NFPA 70 contains guidelines that specify area classifications and determines the type and design of equipment and wiring methods that are permitted to be used in classified areas. The intent of Article 500 is to prevent electrical equipment from providing a means of ignition for an ignitable atmosphere. Additional guidance on the extent of classified areas for dust/air atmospheres is provided in NFPA 499 "Recommended Practice for the Classification of Combustible Dusts and of Hazardous (Classified) Locations for Electrical Installations in Chemical Process Areas".

Electrostatic Discharges [ref 6]

General precautions for controlling electrostatic ignition sources include:

- Bonding and grounding Spark discharges can be avoided by electrically grounding conductive items such as metal equipment, fiberboard drums, low resistivity powders, and people.

- Restricting the use of insulating materials Where there could be high surface charging processes, non-conductive materials should not be used, unless the breakdown voltage across the material is less than 4 kV. Examples of non-conductive objects include plastic pipes, flexible connectors, containers, bags, coatings and liners.

Explosion Protection Measures

If formation of an explosible dust cloud atmosphere cannot be prevented and all sources of ignition cannot be reasonably eliminated or excluded, then the possibility of a dust cloud explosion exists. Under such conditions, explosion protection measures should be taken to protect people and minimize damage to facilities. It should be noted that explosion protection measures should be considered in addition to taking all reasonable steps to reduce the possibility of formation/spread of dust clouds and to exclude potential ignition sources. Explosion Protection measures include:

- Explosion Containment

- Explosion containment involves constructing the equipment to withstand the maximum explosion pressure resulting from the deflagration (propagation of a combustion zone at a velocity that is less than the speed of sound in the un-burned atmosphere) of the dust/air mixture that is present in the equipment Explosion Suppression

  Explosion Suppression relies on early detection of an explosion and the rapid injection of a suitable flame suppressant before the explosion pressure reaches a level that could damage the equipment. A typical explosion pressure at the moment of detection is 0.035 to 0.1 barg. Suppressant extinguishes the flame within 0.08 seconds. An explosion suppression system normally includes explosion detector, control unit, suppressor, and a suitable suppressant. After activation, the system will have attained a "safe" reduced pressure well below the unsuppressed explosion and prevent enclosure rupture.

- Explosion Relief Venting

  The principle of explosion relief venting is that a dust explosion in a vessel causes vent(s) of sufficient area to open rapidly and relieve hot gases and (burning) dust to a safe location and prevent over-pressurization of the protected enclosure. Explosion relief venting has the advantage of being relatively inexpensive compared to explosion containment and suppression methods and is simple to install in many cases. Explosion relief venting is not suitable; however, for toxic materials, some of which will be released to the atmosphere in the event of explosion. Additionally, the venting of explosion products (flame, heat and materials) to the inside a building is not usually acceptable unless a vent duct can be installed to direct the products of combustion, from the vent, to a safe location outside the building.

- Explosion Isolation

  Regardless of what type of explosion protection measure is considered, the dust cloud explosion should be prevented from propagating from the location where it originates to other locations in the plant. This is referred to as "explosion isolation". Dust explosions can propagate through ducts, pipes, chutes, conveyors, etc. The first step in isolating an explosion is to avoid unnecessary connections. If this is not possible, barriers should be created in the path of the explosion.

There are two types of barriers that could be considered for isolating a dust explosion:

- Mechanical Barriers Explosion propagation is prevented by some type of physical barrier. Mechanical barriers could include rotary valves that have a sufficient number of blades to form a barrier, screw feeders that are modified to continuously contain a plug of material, and fast acting shutoff valves.

- Chemical Barriers The flame front or pressure wave is detected and a suitable suppressant is injected to extinguish the flame. Although chemical barriers extinguish the flame, they cannot prevent the explosion pressure from propagating. Downstream process equipment should therefore be able to withstand the resulting "suppressed" pressure.

CONCLUDING REMARKS

Elimination/control of dust cloud explosion hazards requires:

- Understanding of the explosion characteristics of the dust(s),
- Identification of locations where combustible dust cloud atmospheres are or could be present during normal and abnormal operating conditions,
- Identification of potential ignition sources that could be present under normal and abnormal conditions,
- Taking measures to eliminate and/or minimize the occurrence of dust explosions and protect people and facilities against their consequences,

The measures intended to prevent dust explosions must be recognized and understood and systems must be in place to control the hazards. Operators should be aware that signs of overheating, excessive vibration, or noises indicating mechanical malfunction need prompt attention before a small smoldering clump of dust could lead to a serious explosion. Protective safeguards such as explosion relief vents and suppression systems must be appropriately designed and maintained with clear, safe discharges. A routine inspection and testing program should be created for safety critical explosion protective systems.

REFERENCES

1. Combustible Dust; Does your company or firm process any of these products or materials in powdered form? Occupational Safety and Health Administration, OSHA, US Department of Labor
2. United States Chemical Safety Board (CSB) Investigation Report: Combustible Dust Hazard Study, November 9th, 2006, 2175 K Street NW, Suite 400, Washington, DC 20037-1809
3. National Fire Protection Association (NFPA) 61, "Standard for the Prevention of Fires and Dust Explosions in Agricultural and Food Products Facilities", 1 Batterymarch Park, P.O. Box 9101, Quincy, MA 02169, USA
4. National Fire Protection Association (NFPA) 68, "Standard on Explosion protection by Deflagration Venting, 1 Batterymarch Park, P.O. Box 9101, Quincy, MA 02169, USA
5. National Fire Protection Association (NFPA) 69, "Standard on Explosion Prevention Systems", 1 Batterymarch Park, P.O. Box 9101, Quincy, MA 02169, USA
6. National Fire Protection Association (NFPA) 77, "Recommended Practice on Static Electricity", 1 Batterymarch Park, P.O. Box 9101, Quincy, MA 02169, USA

7. National Fire Protection Association (NFPA) 484, "Standard for Combustible Metals", 1 Batterymarch Park, P.O. Box 9101, Quincy, MA 02169, USA
8. National Fire Protection Association (NFPA) 499, " Recommended Practice for the Classification of Combustible Dusts and of Hazardous (Classified) Locations for Electrical Installations in Chemical Process Areas", 1 Batterymarch Park, P.O. Box 9101, Quincy, MA 02169, USA
9. National Fire Protection Association (NFPA) 654, "Standard for the Prevention of Fire and Dust Explosions from the Manufacturing, Processing and Handling of Combustible Particulate Solids", 1 Batterymarch Park, P.O. Box 9101, Quincy, MA 02169, USA
10. National Fire Protection Association (NFPA) 655, "Standard for Prevention of Sulfur Fires and Explosions", 1 Batterymarch Park, P.O. Box 9101, Quincy, MA 02169, USA
11. National Fire Protection Association (NFPA) 664, "Standard for the Prevention of Fires and Explosions in Wood Processing and Woodworking Facilities", 1 Batterymarch Park, P.O. Box 9101, Quincy, MA 02169, USA
12. Code of Federal Regulations 29CFR 1910.1200, Occupational Safety and Health Standards, Toxic and Hazardous Substances, Occupational Safety and Health Administration, OSHA, US Department of Labor
13. National Fire Protection Association (NFPA) 30, "Flammable and Combustible Liquids Code", 1 Batterymarch Park, P.O. Box 9101, Quincy, MA 02169, USA
14. National Fire Protection Association (NFPA) 51B, "Standard for Fire Prevention During Welding, Cutting, and Other Hot Work", 1 Batterymarch Park, P.O. Box 9101, Quincy, MA 02169, USA
15. National Fire Protection Association (NFPA) 70, "National Electrical Code (NEC 500)", 1 Batterymarch Park, P.O. Box 9101, Quincy, MA 02169, USA
16. National Fire Protection Association (NFPA) 497, "Recommended Practice for the Classification of Flammable Liquids, Gases, or Vapors and of Hazardous (Classified) Locations for Electrical Installations in Chemical Process Areas", 1 Batterymarch Park, P.O. Box 9101, Quincy, MA 02169, USA
17. American Society for Testing and Materials (ASTM) E1226, Standard Test Method for Explosibility of Dust Clouds, ASTM International, 100 Barr Harbor Drive, PO Box C700, West Conshohocken, Pennsylvania, 19428-2959 USA
18. American Society for Testing and Materials (ASTM) E2019, Standard Test Method for Minimum Ignition Energy of a Dust Cloud in Air, ASTM International, 100 Barr Harbor Drive, PO Box C700, West Conshohocken, Pennsylvania, 19428-2959 USA
19. American Society for Testing and Materials (ASTM) E1491, Standard Test Method for Minimum Auto-ignition Temperature of Dust Clouds, ASTM International, 100 Barr Harbor Drive, PO Box C700, West Conshohocken, Pennsylvania, 19428-2959 USA
20. American Society for Testing and Materials (ASTM) E2021, Standard Test Method for Hot Surface Ignition Temperature of Dust Layers, ASTM International, 100 Barr Harbor Drive, PO Box C700, West Conshohocken, Pennsylvania, 19428-2959 USA
21. Prevention of Fire & Explosions in Dryers", J. A. Abbott (Technical Editor), 2nd Edition, 1990, Institution of Chemical Engineers, Rugby U.K.
22. American Society for Testing and Materials (ASTM) E1515, Standard Test Method for Minimum Explosible Concentration of Combustible Dusts, ASTM International, 100 Barr Harbor Drive, PO Box C700, West Conshohocken, Pennsylvania, 19428-2959 USA
23. EN 14034-4, Determination of the Limiting Oxygen Concentration of Dust Clouds
24. American Society for Testing and Materials (ASTM) E1226, Standard Test Method for Explosibility of Dust Clouds, ASTM International, 100 Barr Harbor Drive, PO Box C700, West Conshohocken, Pennsylvania, 19428-2959 USA

25. American Society for Testing and Materials (ASTM) D257, Standard Test Methods for DC Resistance or Conductance of Insulating Materials, ASTM International, 100 Barr Harbor Drive, PO Box C700, West Conshohocken, Pennsylvania, 19428-2959 USA
26. Vahid Ebadat, Dust Explosions in the Food Industry, Materials Handling and Logistics Magazine, August 2011

ADDITIONAL REFERENCES

Please refer to the following references for more detail on the topics that have been discussed in this paper:

1. Center for Chemical Process Safety (2005), "Guidelines for Safe Handling of Powders and Bulk Solids", American Institute of Chemical Engineers, New York, New York
2. Barton, John (2002), "Dust Explosion Prevention and Protection, A Practical Guide", Institution of Chemical Engineers, Rugby, Warwickshire, UK
3. Eckhoff, Rolf K (1997), "Dust Explosion in the Process Industries", 3rd Edition, Butterworth-Heinemann Linacre House, Jordan Hill, Oxford
4. Buschart, Richard J. (1991), "Electrical and Instrumentation Safety for Chemical Processes", Van Nostrand Reinhold, New York, NY
5. Bartknecht, Wolfgang (1989), "Dust Explosions Course: Prevention, Protection", Springer-Verlag, Germany
6. Bodurtha, Frank T. (1980), "Industrial Explosion Prevention and Protection", McGraw-Hill Book Company, New York, NY

Industry Best Practices and Experience With Process Safety Leading Indicators
Arendt, S., ABS Consulting
Originally presented on: 9/5/2013 0:00:00 - 0:25:00

Industry Best Practices and Experience with Process Safety Leading Indicators

CCPS has been a leader in the development and deployment of process safety metrics since the mid-1990's. Several international groups in the process industries have developed guidance on how to develop/deploy metrics and numerous examples of leading indicators have been proposed. Many companies have adopted process safety metrics and are beginning to generate implementation lessons. Nonetheless, numerous major accident investigations have listed inadequate line-of-sight on effective process safety metrics as causal factors. This paper will cover the following issues:

- Brief history on process safety metrics development
- Comparison of globally effective approaches for metrics deployment
- Benchmarking of company process safety metrics
- Discussion of indicator action limits
- Lessons learned

The presentation will compares the effective practices of several companies. It will contrast the benefit and limitations of lagging/leading indicator strategies. The paper will give some specific examples of where monitoring of specific process safety leading indicators have been credited with preventing/reducing the seriousness of process safety incidents.

Relationship between Minor Events and Process Safety Management

Wenhui Han
Dushanzi Petrochemical Company,
Xinjiang, China
hanwenhui@petrochina.com.cn

Prepared for Presentation at
American Institute of Chemical Engineers
1st CCPS Asia-Pacific Conference on Process Safety
Qingdao, China
September 4 – 5, 2013

UNPUBLISHED

Relationship between Minor Events and Process Safety Management

Wenhui Han
Dushanzi Petrochemical Company,
Xinjiang, China
hanwenhui@petrochina.com.cn

Keywords: Minor events, accident survey, PSM elements

Abstract

The survey, analysis, disposal and learning of minor events from the past experience play an important role in verifying and improving process safety performance, which can also prevent the major accidents from happening. By means of collecting and analyzing the cases of minor events and accidents, this paper concludes the occurrence rule of different kinds of accidents. Meanwhile, it also devotes efforts to find and demonstrate the relationship between the primary causes and the elements of PSM.Some improvement approaches are also put forward.

1 Introduction

In the 1980 s, many major accidents occurred in the world continuously, such as India Bhopal's methyl isocyanate leaked in 1984 which caused the deaths of one thousand people at least, the spherical tank explosion which killed more than 500 people in Mexico in 1984 and Texas PE large explosion which killed 23 people and caused major property damage in 1989 and so on. These accidents have played a huge role in the deepening development and promoting improvement of petrochemical industry safety work, so the governments, the public and the industry began to reflect on how to prevent catastrophic accidents, and had carried out much useful and fruitful research work. U.S. OHSA issued the process safety management (PSM) in 1992 including 14 factors which is used by most of the countries later. Dushanzi Petrochemical Company of CNPC in 1997 compiled the SYT6230-1997 PSM in the field of oil and gas processing. In 2010, China enacted AQ/T 3034-2010 PSM implementation guideline of chemical enterprises.

2 Status of PSM

The application of PSM popularized rapidly in oil chemical industry of our country in recent years, and gradually received the recognition from the enterprises. Through the implementation of PSM management system, we finally reach the goal of safe and running of petrochemical industry smoothly.

China has got some achievements and reduced the frequency of accidents since the United States OHSA enacted PSM in 1992, but the government still remain to face grim safety situation. Big explosion of BP refinery in Texas caused 15 deaths on March 23, 2005; The tank explosion in a Lanzhou petrochemical enterprise killed 6 people on January 7, 2010; BP drilling rig explosion in gulf of Mexico led to 11 workers death and serious environmental pollution on April 20, 2010; On July 16, 2010, the Dalian oil tank exploded. A series of accidents remind us that our PSM is

insufficient especially on running period. How to judge the real situation of our management system and what is the shortage are the two big problems that we need to solve urgently.
"Accident is a kind of precious resource" is true, but it is meaningful only for the first accident and also should be established on the basis of drawing lessons fully. Frequent repeated accidents should not be regarded as resources, but rather a pure loss and tragedy! There are many precursor events before major accidents. For example, a glass level gauge of a spherical tank in a company unit has occurred 11 congealing events successively in just one month in winter based on the accident statistics of this article. For such minor incident, there is not an effective investigation and improvement. Again for instance, a device alone has caused 4 events of dropping from patrol stairs causing serious damage from the record, and also before the explosion of BP refinery in Texas in 2005, the raffinate fractionating column and F - 20 blow-down tank had happened 19 incidents (hydrocarbons leakage, steam release or fire) from March 1991 to February 2005 from the document. None of the above incidents has been thorough investigated and analyzed. So, the occurrence of serious accidents is inevitable according to the incident pyramid theory.

3 Incident/accident cases

3.1 Accident statistics

This accident statistics collects 1714 accident cases in total. The specific distribution is shown in figure 1.

figure-1 incident/accident cases distribution in 10 years

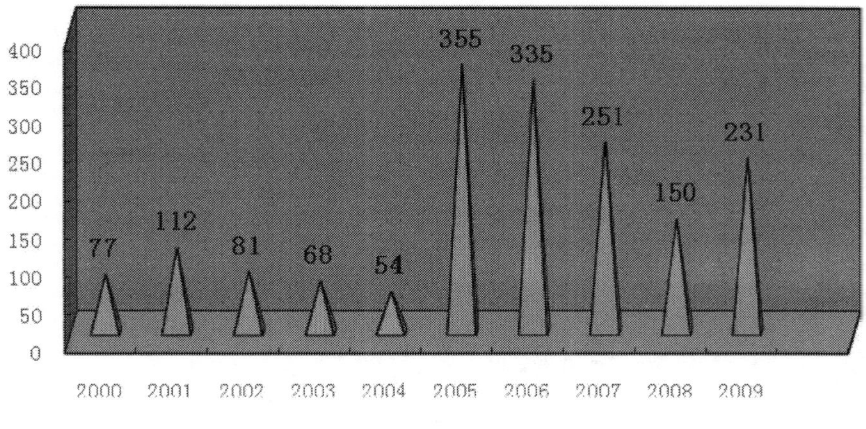

Because the company encourages the grassroots units and staff of each branch device to report the minor incidents and near misses actively since 2005. The high statistics number of accidents/incidents does not mean its safety management is poor, but it shows that the unit is serious to the management and on the other hand it provides accurate basis for accident/incident statistical analysis. The reason why we do it is to make everyone learn lessons seriously and prevent major accidents. At the same time, we need to realize deeply that all kinds of accidents are the precious wealth, and make full use of accident resources and improve the relevance of safety work to avoid the similar accident to happen.
Thus accident/incident reporting mechanism is a key step, so that we can share accident/incident statistics which gives us a certain reference and warning at the same time, and it can clear the direction we need to improve. Through careful investigation, analysis and sharing every

accident/incident, controlling the minor accident/incident which is more economic and effective can control major accidents indirectly based on the accident pyramid theory.

BP Texas explosion on March 23, 2005 causing 15 people death drew one conclusion in the report "safety performance mainly focuses on indicators after the accidents rather than the indicators before the accidents, or at least related and minor high frequent events before the accidents. When indicators after the accidents such as event number of loss of working hours and recordable injury frequency are declining, we feel that the safety situation has improved. Concentration on the indicators before the accidents is more likely to master accurate site operation safety situation such as loss of material, overflow, ignition and operation deviation and so on, many of which are not official reported or rarely investigated." We can't avoid the happening of the tragedy if we lose the best improvement opportunity.

3.2 Accident/incident and PSM element

After the US chemical safety investigated the BP refinery in Texas explosion accident on March 23, 2005 in the United States and BP itself, they found that almost all the PSM system elements exist shortcomings and loopholes through detailed analysis of the accident, which eventually led to major accidents. This study analyzes and ratiocinates the root causes of the above 1714 accidents/incidents involving almost all the basic PSM elements. 11 elements of which account for a large proportion which related data and message is shown in figure 2.

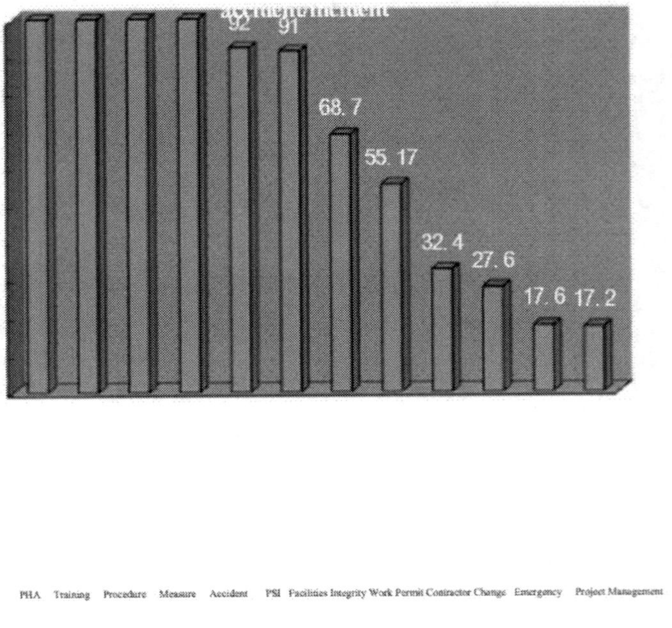

3.2.1 Hazard identification and risk evaluation elements

PSM emphasizes the risk management as the core and takes the initiative to prevention. Statistical analysis found that the causes of 100% of the 1714 cases are related to the risk assessment element. They lacked the necessary PHA or conduct PHA incomplete and didn't understand process safety. Though conducting PHA to BP Texas units every five years according to OHSA, it's inefficient. Many accidents investigation showed that no HAZOP study or inefficient HAZOP study is a main reason for the accident as Trevor Kletz said. Designs of not matching with the specification brought us a painful lesson.

Countermeasures: We have to provide the training about the risk evaluation method to the personnel targeted and implement a comprehensive PHA. The new and renewal projects have to conduct the PHA by drawing lessons from American OHSA and other multinational companies. The personnel conduct the risk identification effectively every five years and take corresponding measures. At the same time, take full advantage of the results of the analysis to improve process safety information, technical indicators, operating rules, shutdown scheme and emergency scheme etc.

3.2.2 Regulations, procedures, operation cards

Enterprises have made great progress in the procedures, writing operation cards, operability and effectiveness through continuously learning the successful experience from international well-known enterprises (such as TOT). But statistical analysis found that reasons of 100% of the accidents are those procedures and operation cards are insufficient, the operability is not strong or operation risk identification is not deep enough. For instance, operation procedures after modification are not timely revised or lack applicability and operability, and most of the companies/units still keep the attitude that the production is more important than safety or safety is subject to the production. For example, in the BP Texas explosion accident, operators on night duty fed too much material to the tower making the liquid level exceed 100%, but the level requires to be set at 50% according to the operation procedures. Meanwhile, operators on day duty set the level of fractionating column back to 50% just before the accident. Such phenomena widespread exist for a long time.

The root cause of inefficient rules and procedures is not in-depth hazard identification and risk analysis. What needed is keeping improvement by strengthening the recognition, using evaluation methods such as JHA and JCA to ensure that the whole process is controlled. What's more, the regulations not only write on the paper, but also should take into action which is more important. Good operation is taking the PSM (process safety management) schemes into action. Each member of an organization needs to complete each task in the correct way all the time.

3.2.3 Ability, training and awareness elements

Companies organize every profession to hold skills contest every year, and many units can always get good grades showing the effectiveness of our training work indeed. But the statistical analysis of the accidents found that training is inadequate, staff's consciousness, abilities, skills and behaviors are not good pushing us to think about the running effect of the element, as a major element of management system, is also unsatisfactory such as guidance of the leadership, lack of targeted training, lax management, lack of effective training evaluation system, etc. Take a poisoning of N2/H2S as an example, rescuers rescued people blindly without wearing protective equipments causing more casualties as a result which showed that our training and staff safety consciousness existed great space to improve.

The Times once pointed out: "the most successful companies in the future will be those companies/enterprises which base on the learning organization". We need to strengthen technical training and the comprehensive safety training, and ensure the effectiveness by conducting periodic review and evaluation. Establish training matrix according to the need of devices and employees for constantly supplement and perfecting the training content and quantity. Perfect the employment standards and attach more importance to the evaluation and inspection on the sense of responsibility for key positions and operating staff to constantly enhance its awareness. Choosing qualified managers and operators is one of the key factors to avoid accidents.

3.2.4 Supervision, measuring and audit management element

The company/enterprise will conduct a lot of inspection quarterly, monthly and holidays every year, and the device also has its own daily inspection and the superior department will arrange related audit regularly. But reasons of 100% accidents in the statistical analysis are related with our supervision, measuring and audit management which is not comprehensive, detailed and thorough. As a result, we can't judge correctly PSM of enterprises which leads to the accident.
Countermeasures: Investigate risk (hazard) persistently. Encourage employees' enthusiasm. Improve their self-management. Strengthen independent inspection and diagnosis management work. At the same time, we must make a unified risk standard through scientific scheme and all kinds of quality inspections and audits to improve its effectiveness. Only by constantly improving all elements of PSM system can we achieve the closed-loop virtuous circle of PDCA.

3.2.5 Accident management element

Statistical analysis found that 92.8% of the accidents and accident management element are directly related. Reasons are the analyses are not enough thorough, measures are not proper and lacking follow-up and implementation. What's more, the accident study and communicate is not thorough and the definition and row level of accident is defected making a lot of accidents/incidents not being detailed investigated and analyzed and the accident experience not being shared fully and effectively. In addition, now the examination and appraisal system is against experience sharing of accidents letting the accidents repeat continuously rather than decline.
Countermeasures: Review and analyze the effectiveness of PSM system based on the accidents/incidents as an opportunity. Make full use of resources and attach great importance to the minor and attempted accident management and encourage the reporting of accidents. We thoroughly analyze the short board, management system and the insufficiency for minor accidents truly achieving "the wise use lessons which patients in exchange for with blood to avoid bleeding again". At the same time, improve and promote the enterprise's PSM system by other accidents to avoid the similar accident happening in this enterprise. The goal is to learn seriously, do preventively and prevent major accidents.

3.2.6 Work permit management

Both in normal production and during inspection and maintenance, work permit is an important link of controlling accidents. But over the last decade, accounted for about 55.17% of the total number of accidents are related with work permit. Work permit procedures and management exist loopholes, and some habitual violate regulations and wrong operations abound such as job ticket issued by someone else, privately modifying operation ticket content, altering operation time, not confirming the scene, not analyzing effectively operation risk and using unqualified facilities and protective equipments fully demonstrating loopholes in the procedure of inspection, supervision and audit management. Some wrong operations exposed currently in the process of production and equipment maintenance we are used to have not been realized and timely corrected which led to the accident. So work permit management is still and will always be the focus of our safety management.

Countermeasures: Strengthen dependency management. Improve the linear safety management mechanism. Implement safety production responsibility system of "the one who is in charge is responsible ". Create a good atmosphere of improving both safety and production. Increase risk identification PHA at the same time. Make effective and operational procedures. Strengthen inspection, supervision, measuring and audit management ensuring the operation of the element effectively.

3.2.7 Mechanical integrity

Statistical analysis found that 68.8% of the accidents are related to equipment and facilities Integrity. Some high risky (hazardous) equipment, facilities exist for a long time, so we can't guarantee the safety which consequence usually is very severe. Take a case of BP refinery explosion in Texas in 2005 for example, the independent liquid level alarm in a tower had not been effectively tested and maintained and wasn't be repaired when shutdown making the operation staff lose another monitor and judge measure.

Based on "bath curve", the period of high equipment failures is mainly concentrated in the old factories, and the second is the new device just putting-in-service which facilities are all in the reconciliation and adjustment period. So the initial and the last stage is the peak period of equipment accidents. For instance, an old coking unit's bottom manhole leaks leading to a fire accident in 2009. Reasons are those workshop equipment identification work is not meticulous, basic data is not comprehensive and not clearing about manhole material; in 2008 the raw water in single tower stripper of a new unit overpressures caused the damage accident of the tank. The reason is that upstream old device container of R-301's position control valve LV-303 inner leaked excessively. Although the device had been twice repaired since September 2007, equipment and instrument workshop both didn't calibrate, test and maintain the valve resulting hydrocarbon C3, C4 material in liquid directly entered into the atmospheric storage tank causing a serious accident.

Countermeasures: Improving the integrity of the facilities not only need the management after putting-in-service, but also include the design, procurement, supervision, verification, construction installation, acceptance inspection, daily inspection, maintenance procedures and change management in the whole life cycle. Take some examples, strengthen the quality assurance of equipment; increase pre-startup safety review (PSSR); conduct the reliability analysis of equipment maintenance, risk based inspection (RBI) and reliability centered maintenance (RCM) and so on. Comprehensively improving equipment management and ensuring the equipment reliable and in good condition is the key and also the center of the safety work.

3.2.8 Contractors and suppliers element

Contractors and suppliers management element exist problems in 32.4% of accidents which includes the selection management of contractors and suppliers, performance management and the whole procurement chain indemnification. Poor, low effective and difficult regulation of contractors and suppliers is one of the main factors of serious accidents. Such as the explosion accident killing 13 people in 2006, when the contractors carried out anti-corrosive and coating work on the top of a 100000 cubic meters crude oil tank, the operator violated to replace the antirust paint thinner without taking any forced ventilation under the condition of disqualified electrical wiring and lighting lamps and poor regulation causing such malignant accident occurrence frequently and repeatedly.

Countermeasures: Require the contractors to follow the same safety management mode as the owner, which has become an international practice. The contractors' safety performance is one of the important factors to consider of bidding and calibration qualification assessment, so we should be strict to the access management of contractors and the suppliers especially the subcontractors and the affiliated parties with strengthening and evaluating its performance on a regular basis. Strengthen the implement and compliance checks strictly of the relevant state laws and regulations in construction process, and carry out territorial management principles positively. Operators of contractors should be trained in accordance with the contract. Facilities and equipment of the contractors need to be evaluated and work needs to be conducted JHA/JSA. Form a joint investigation team to investigate the accident of contractors. Share fully the experience of accident timely. Establish appropriate safety and emergency scheme management system, operation scheme and emergency scheme with strict supervision.

3.2.9 Change management element

Change management element includes staff, facilities and technology. Accident statistics found that about 27.6% of all accidents are related to change management. Mainly for the design scheme change, the shutdown scheme change, the maintenance construction scheme change, organization staff change, process scheme change and so on, we did not take effective risk identification, inform relevant staff timely and reliably and update information description in time after the change. In 2002, for example, a vent line of four outlet of a compressor exploded. It's because the reconstruction design change did not conduct effective PHA and specification compliance audit adding unreasonable material classification causing a tragedy of overpressure explosion. Take a device fire accident in 2008 as a case, one of the reasons is the lack of file management after the staff change leading to much work not effectively continuation and transmission. Again as in 2005 BP refinery in Texas explosion accident, the manager on duty directly put the work to a deputy because of some things at home which lacks effective change management procedure under the condition that the deputy lacks experience and ability of independent organizing shutdown eventually leading to a severe accident.

Main countermeasures: Strengthen controlled management of all kinds of changes. All kinds of process schemes, equipment, operations and maintenance scheme change and staff change should be managed as a significant risk, conducted PHA and inspected, supervised and audited on a regular basis as the important part of safety management.

3.2.10 Missing full PSI(process safety information)

File and its controlling factors account for 91%. Much data loses and MSDS, process and equipment design data is inadequate. Just completed construction projects also exist the problem that technical information is not accurate and complete and the file or record of accidents and abnormal conditions (deviation) are defected seriously. Due to the long period of lacking induction and summary, document management is not formed. So experience can't be shared and inherited fully and effectively. We only improve the level of personal not the overall level which is more important.

Countermeasure is to implement controlled system and document comprehensively, unswervingly and thoroughly. Supplement and improve the basic data of the old device and ensure all data and information comprehensive and integral of the new device.

3.2.11 Emergency management element

Accounted for about 17.6% of the total number of accidents are related to inadequate emergency preparedness or improper emergency response. Some international statistics show that reliable emergency rescue can reduce accident losses to one-tenth of no effective emergency rescue. For example, a device had an accident of acute benzene poisoning in 2009 which contingency scheme in advance was poor applicability and didn't conduct centralized special emergency drill and training. Operators have not yet fully understood the use of anti-virus production and lack knowledge of the danger, protective measures and accidental release measures of benzene leading to the accident.

Therefore, only by effective hazard identification, risk assessment and making, training and drilling targeted contingency scheme can we reduce or eliminate the consequence of an accident effectively.

Other elements of PSM system are also involved in accident analysis in varying degrees, and each element is linked together. PSM system is an organic whole. Only by changing and enhancing the company's safety culture can we guarantee each element effective, reliable and keep comprehensive PDCA virtuous circle which can express our respect and care for the previous victims. We need to learn lessons from every event and accident to constantly upgrade the company's safety performance and the safety of device, and ultimately achieve the goal of zero accidents.

4 Conclusions

1 Treat every tiny event and accident seriously. Learn from the accidents of other enterprises for reference to find out and make up the deficiency and loopholes existing in the enterprise management elements. Strengthen accident management to demonstrate the leader's commitment to safety management and improve employees' knowledge and confidence on workplace safety;

2 Hazard identification and risk assessment is the precondition of PSM and a way of realizing management in advance. Improve safety essentially, consummate the operation procedures and improve the operability through effective risk assessment;

3 Human elements are the key to do a good job. Promoting company staffs' safety training (including consciousness, ideas, skills, and habits) is the only way to prevent accidents and the important guarantee of the system application;

4 Design reasonably and accurately. Choose good contractors and suppliers. Guarantee the high quality of construction installation and maintenance. Ensure the procurement chain which is the guarantee of realizing safe production. Safety must conduct from construction quality;

5 To strengthen quality inspection, supervision, measuring and audit is the link to ensure that all the elements of PSM operation normally;

6 Establishing a good corporate culture is the core to make sure all PSM process elements operation effectively.

5. References

[1] BP Texas explosion accident investigation report on March 23, 2005.

[2] Accident analysis and countermeasures since Petro China chemical sector lists.

[3] 29 CFR 1910.119 high hazardous chemicals PSM promulgated by US OHSA

Safety Challenges in Harsh Environments: Lessons Learned

Faisal Khan[a,b]
fikhan@mun.ca

Salim Ahmed[a]
sahmed@mun.ca

Ming Yang[a]
myang@mun.ca

S. Javad Hashemi[a]
j.hashemi@mun.ca

Susan Caines[a]
susan.caines@mun.ca

Dan Oldford[c]
doldford@eagle.org

(a) Safety and Risk Engineering Group, Faculty of Engineering & Applied Science, Memorial University, St. John's, Canada
(b) Australian Maritime College, University of Tasmania, Launceston, TAS 7450, Australia
(c) ABS Harsh Environment Technology Center, 240 Prince Philip Dr. St. John's, NL. Canada

Prepared for Presentation at
American Institute of Chemical Engineers
1st CCPS Asia-Pacific Conference on Process Safety
Qingdao, China, September 4-5, 2013

UNPUBLISHED

Safety Challenges in Harsh Environments: Lessons Learned

Faisal Khan[a,b]
fikhan@mun.ca

Salim Ahmed[a]
sahmed@mun.ca

Ming Yang[a]
myang@mun.ca

S. Javad Hashemi[a]
j.hashemi@mun.ca

Susan Caines[a]
susan.caines@mun.ca

Dan Oldford[c]
doldford@eagle.org

(a) Safety and Risk Engineering Group, Faculty of Engineering & Applied Science, Memorial University, St. John's, Canada
(b) Australian Maritime College, University of Tasmania, Launceston, TAS 7450, Australia
(c) ABS Harsh Environment Technology Center, 240 Prince Philip Dr. St. John's, NL. Canada

Keywords: harsh environment, process safety, winterization, human factor, risk analysis

Abstract

Development of natural resources in harsh environments presents significant technical and logistical challenges. An industrial workshop on "safety and integrity management of operations in harsh environments" was organized by the safety and risk engineering group at Memorial University to bring together industrial practitioners, regulatory authorities, and research and development institutions to identify the safety and integrity challenges in harsh environments, share experience and develop a roadmap for desired solutions. This article summarizes the lessons learned from the workshop on safety issues in harsh environments. The workshop identified that there are safety challenges regarding construction and operation including a lack of detailed standards, optimization with respect to winterization, and data scarcity. The remoteness of operations in harsh environments is an additional challenge. Finally, the human factors add another set of challenges that arise from the physical and psychological behavior of plant personnel in harsh and remote environments.

1. Introduction

Harsh environments refer to those climatic conditions which are very difficult for people to work in and for process plants to be operated. This definition includes very warm to extremely cold climates. Operations in harsh conditions have become a necessity and industrial operations are being pushed more and more into difficult environments. In recent years, the development of natural resources in the Arctic and sub-Arctic regions has created huge challenges; however, this has become one of the priorities for the Canadian oil and gas industry. The Arctic and sub-Arctic regions have some of the most harsh environmental conditions in the world with unique features such as ice and permafrost, complex degradation rates, unpredictable climate changes, and high uncertainty due to lack of knowledge and data. Oil and gas, mining, utility, construction, and shipping companies need to know how their operations can safely progress in these challenging conditions.

On March 19-20, approximately 120 researchers and practitioners came together for a workshop on "safety and integrity management of operations in harsh environments". The participants were from major oil and gas companies, regulatory bodies, and academia from different parts of the world. This workshop was organized by the safety and risk engineering group (SREG), a research division of the Faculty of Engineering and Applied Science at Memorial University. The objectives of the workshop were to identify and prioritize the risks and challenges for operations and exploration in harsh environments, strategize a plan for solutions, and disseminate knowledge and advances in safety and asset integrity in harsh environments.

The workshop included eight technical sessions and panel discussions on important topics in the area of safety and asset integrity in harsh environments. In total, 20 keynotes and technical presentations were delivered. This article summarizes the lessons shared by industry practitioners and academics on safety challenges in harsh environments. It further identifies fundamental research activities necessary to address practical safety issues in the Arctic regions.

The workshop identifies that there are safety challenges regarding process and occupational safety, risk assessment with scarce data, winterization as well as human factors. The following sections provide details about the challenges identified by the workshop participants.

2. Process and occupational safety in harsh environments

2.1 Process safety vs. occupational safety

Core elements of process and occupational safety are the same and many aspects have overlaps; however, with different foci. It is a good practice to differentiate between process safety and occupational safety, especially for harsh environments. Process safety focuses on prevention and mitigation of process hazards in processing facilities which may result in release of chemicals or energy. Occupational safety mainly relates to personal safety and the focus is on prevention and mitigation of hazards which could result in health issues (e.g. slips, trips, and falls). In harsh environments, the relationship between these two areas of safety is stronger and one may lead to another. In harsh environments, there is less known information and different practical aspects; the link between them requires special consideration.

Process related incidents may not be prevented by the same methods used for occupational safety related incidents. Industries usually report the occupational near misses. To prevent process related incidents, industry should also focus on process related near misses. Different risks exist in each of these two areas of safety. Therefore, each one needs a different approach and tool for risk analysis. These risk analysis tools will help better assessment of occupational and process safety risks and thus help in designing strategies to prevent incident occurrence and limit consequences.

2.2 Process safety best practices

Five important best practices in process safety are identified as:

- Risk management
- Inherently safer design
- Human error and human factors
- Safety management systems
- Safety culture

Harsh environments add an extra degree of complexity in the assessment and management of risk. This is due to more unknowns in extreme conditions which are discussed in more detail in the next section. Risk perception is as important as likelihood and consequences of failure, particularly for harsh environments. How companies and public perceive risk and how regulators perceive risk are all important. A common understanding of risk by different stakeholders will help to take better decisions. More importantly, it will also help to minimize false perception of risk.

Implementation of inherently safer design (ISD) in offshore facilities should be considered. Unmanned platforms, relocation of accommodation modules and lifeboats, compact heat exchangers, and subsea manifolding to minimize the number of risers and wellhead platforms are some of the established idea; however, still not widely practiced. Minimization (intensification), substitution, moderation (attenuation), and simplification are four fundamentals of inherently safer design in offshore facilities. ISD is also of great importance for human safer design for operations in extreme conditions.

In terms of safety culture, following issues need more consideration in harsh environments:

- Sharing lessons learned and learning from one another is a good safety culture which should be taken seriously. Any safety issue should be shared among operating companies while respecting company policy and reputation.
- Training workers to refuse unsafe work is fundamental to a successful safety culture. If people believe that something is unsafe or if they feel is unsafe, they are not required to do the work. Saying 'no' to the unsafe task must be supported by management; employees must know that management support is available.
- Safety culture is not just an engineering issue; engineers and managers need also to understand it from non-engineering viewpoints such as sociology and psychology.

2.3 Design, construction, and operation safety

Safety in harsh environments can be studied in three main aspects; design aspects particularly the effect of winterization and enclosures on fire and explosion risk, construction aspects with focus on protection during winter fabrication, and operational aspects considering equipment and human factors in cold temperatures.

Harsh environments often require enclosed process modules. Enclosing people and equipment to protect them from extreme climate needs special design consideration for fire and explosion safety. The risk contribution could be from several factors including module size, HVAC (heating, ventilation, and air conditioning), ignition source isolation efficiency, and relief wall configuration. When enclosing harsh environment facilities, splitting large offshore facilities modules into smaller modules will reduce fire frequency and blast loads due to less potential ignition sources and gas accumulation in lesser volume. Smaller modules are more rapidly filled with smoke (bad for escape), but at the same time favor to extinguish a fire rapidly as there is less air to burn (provided HVAC is shut down). Smaller modules also reduce potential loss of life (PLL) by separating personnel from accidents.

Regarding operational and construction aspects, several challenges are identified. Protection during winter fabrication is of great importance as freezing of standing water is a hazard in cold weather. The message is when you are in cold situations, you should be aware of what are there in the pipes and equipment and should protect those from freezing. To provide protection against cold, insulation is one of mots preferred approach. However, insulation introduces a complexity by having possibility of corrosion under insulation. Effect of ice buildup and snow on escape, evacuation, and rescue (EER) surfaces is another major factor in cold situations. In such conditions, access to safety equipment (emergency phones, lifeboats, rescue boats) may be restricted. Ice falling on personnel from structures and slippery surfaces are two other potential hazards. Wind, temperature, high humidity, rain, sleet, snow, and icing may affect availability and operability. Air, sea, and ground transportations may become very difficult during extreme conditions.

3. Risk assessment with scarce information

Risk and safety are more critical in harsh Arctic regions with limited infrastructure. Operations that experience widely different harsh conditions due to seasonal variations pose additional challenges in their design. This makes risk assessment for operations in harsh environments even more important. Several other challenges are also encountered.

Lack of information and data was identified as the main impediment in the assessment of risks of operations. Choosing appropriate data that can best represent the conditions in harsh environments is challenging. Potential sources of information and data include:

- Expert experience and knowledge
- Input from local residents in the Arctic and sub-Arctic regions
- Data and information shared across industries that have operations in harsh environments

Methodologies for recording and analyzing process related near misses and unsafe conditions should be developed to systematically gather and share information.

The determination of an acceptable risk level may pose the biggest challenge for safety practitioners, industry leaders, and other stakeholders. The traditional risk assessment methods will continue to play an important role in assessing the risks of exploration, development and operations in harsh environments. Conventional ways to determine acceptable risk level, such as "As Low As Reasonably Practicable (ALARP)" criterion, will still be useful.

The effects of remoteness of operations and harsh conditions on the physical ability and the cognitive behavior of operators are yet to be studied. Qualitative risk assessment methods may be better used to evaluate risk of human factors in a data-scarce environment. A qualitative approach provides a proper tool for risk assessment in the absence of data when a setting is unknown. It also helps to organize available information in a transparent manner. However, quantitative risk assessment may be preferred for process systems with well-known structures. Insurance industry may have effective tools in conducting risk assessment with scarce information. Further investigation is needed to identify suitable approaches to assess risk with scarce data.

4. Optimum winterization

In harsh cold environments, many vessel and process systems will be operated at or close to their design limits. Winterization aims to adapt conventional installation designs and their operations for continued safe operations. Winterization is traditionally related to issues that include de-icing, ice effects mitigation, heat tracing, protection of operating condition, and prevention of ice accretion. When dealing with winterization hazards, the majority of these issues should be identified in the design phase of a project, if possible. The winterization processes, although practiced, need to be optimized. Major challenges and their potential solutions are presented as follows.

When looking at winterization, some of the factors to consider are; area of operations, environmental extremes, duration of activities (length of operating season), exposure of personnel, walking surfaces/deck camber, and emergencies and equipment. Lifeboats and lifesaving equipment, firefighting systems, escape and egress routes, machinery, helidecks, piping systems/drains, pressure regulation and relief systems, deck machinery (hydraulics), accommodations and navigation equipment are some of the critical focus areas for winterization.

Avoiding over or under winterization and thus balancing the cost of winterization and its extent is a challenging task. For example, insulation is usually installed at a considerable cost and the proper level of insulation needs to be applied at a reasonable cost. The main factors that may need to be considered in deciding the extent of winterization are ambient temperature, humidity, and wind velocity. Most guidelines for vessels' winterization use ambient temperature as a design criterion. For example, the "ABS Guide for Vessels Operating in Low Temperature Environments (Low Temperature Guide)" specifies the design service temperature (DST) and the minimum anticipated temperature (MAT) [1]. The DST is taken as the lowest mean daily average temperature in the area of operation for data taken over at least a 20-year period. The MAT is determined as the DST minus 20 °C. When the extent of winterization is dependent on the DST and MAT, this may lead to over-winterization and add unnecessary costs because the minimum temperature may only occur for a very short period or at a very low frequency. This may not cause system failures or impair safety [2].

Moreover, the few guidelines for winterization that have been published offer some prescriptive requirements, guidance, and suggestions based on the available knowledge. Although the prescriptive approach is easy to follow and implement, it may cause over or under winterization problems and does not drive changes for performance improvement. A potential solution is to move towards a performance-based approach. The performance-based approach will be able to advise on which winterization technologies should be applied to operations in the Arctic, and to inform the reliability of these technologies. A risk-based approach is one way forward in this direction.

There is no clear interpretation of winterization standards. For example, more scientific justifications are needed to support the use of DST as design criterion for winterization.

5. Human factors in harsh environments

5.1. Human factor challenges in harsh environments

In harsh environments, the first mistake could be the last mistake. Learning from experiences is very important to make analytical and predictive decisions in harsh environments. Recent studies show significant difference between relative human error probabilities in cold and normal conditions and the importance of the effect of cold temperature on human performance [3]. Emotions, behaviors, culture, and rotation shifts have impacts on human reliability; these also pose challenges for quantitative risk analysis involving human factors. Complexity, uncertainty, and time available for response are additional factors in human response which are affected by harsh environments. To minimize these effects, safety barriers should be selected based on harsh cold environmental conditions.

Daily rhythm of human physical conditions will be affected by working in a harsh environment. Fatigue is a common and expected condition for offshore people and has many symptoms such as sleepiness, irritability, and lack of motivation as well as many long term issues like increased susceptibility to illness, depression, headaches, and loss of appetite and digestive problems. Restricted flexibility and range of motions is another challenge in a cold environment. Therefore, ergonomic design is important for harsh environment operations. Ergonomic design and consideration are required also for specific clothing design. Even the instructions on equipment needs to consider human factors.

Temporary enclosures for people working in harsh environments, such as confined space beside equipment, introduce new hazards. These hazards should be addressed because they were not considered in the original design of the facility. Modification by crew or operators for easier use of equipment is a frequently observed problem. As a result, design should prioritize easy use of equipment in extreme conditions to avoid operator's modification later during operations.

5.2. Current practices

Currently, there are standards and works regarding human and health aspects in harsh environments which include:

- ABS 2010, Guide for vessels operating in low temperature environments [1]
- ISO 19906:2010, Petroleum and natural gas industries - arctic offshore structures [4]
- NORSOK S-002, Working environment [5]
- OGP Report Number 398, Health aspects of work in extreme climates [6]
- Barents 2020, Assessment of international standards for safe exploration, production and transportation of oil and gas in the Barents sea [7]

There is an on-going work to develop a new ISO/CSA standard using expertise from countries with interests and experience from work in Arctic conditions, including the following Arctic specific aspects:

- Work in cold climate: health aspects of work in extreme climate
- Winterization issues to improve health and working conditions in cold climate
- Special ergonomic, chemical, noise and vibration challenges in these conditions
- Stress management: work situations in extreme climates
- Training and competence: working in the Arctic

5.3. Potential solutions in human factor engineering for harsh environments

Following main issues and some potential solutions regarding human factor engineering in harsh environments are identified. However, there is still a need for more studies on a number of factors:

- Role of age: retirement age for people working in harsh environment and offshore facilities could be different. However, people should not be excluded because of age. Instead, medical examinations and fit-for-work analysis based on the age can be used to approve people for working in harsh environments. Review of the retirement age for harsh environments is required.
- Clearances in worksites: appropriate clearances should be included in designs for harsh environment regarding mobility and flexibility of people.
- Anthropometry: there is a need for special clothing and heating (individual fitting instead of free size clothing). Anthropometry is more than clothing design it includes design of interfaces, clearances, access etc.
- Physiology: motor skills, reaction time, speed of movement, strength (e.g. willingness to remove the gloves in cold temperature to use a tool), physical workload, and capacity are all important and require careful consideration.
- Sensory: the issues such as UV reflection from snow, effect of fog on visibility, effect of vibration and noise from harsh environment need further study.

- Cognition: effects of harsh environment on the lack of attention, correct decision making, and tasks requiring rapid responses need better understanding.

Human reliability is on the soft side of the quantitative risk analysis (QRA) spectrum. QRA methods for process and mechanical equipment are accepted by industry; however, for human beings there is more resistance. QRA is required for industry to think in a new way and make more informed decisions. If a model produces an unacceptable result, this could be a point for thinking why this unexpected number is observed. The value of the models is because of the questions raised. Therefore, simulation and modeling can be taken into account to obtain human reliability under extreme conditions.

6. Conclusions

Development of natural resources in the Arctic and sub-Arctic regions is facing huge challenges. Although some of the most important safety aspects have been identified in this article, these important issues require concerted efforts from all entities. The workshop on "safety and integrity management of operations in harsh environments" was an initial step to capture the main safety and asset integrity research areas by recognizing the practical challenges shared by industrial practitioners. The safety aspects mentioned in this article are identified by industry as being critical to the development of Arctic resources. Through sustained practical research efforts focused on the critical issues identified in this work, advancements in knowledge, technology, and training are expected. To benefit from the vast natural resources in the Arctic and sub-Arctic regions, there is still a need for similar efforts to overcome existing barriers to development.

7. Acknowledgment

The authors would like to extend their sincere gratitude to the participants and the members of the organizing committee who contributed their valuable time, experience, and insights to the workshop. Contributions made by individuals from industry have been instrumental in guiding and shaping the outcomes of this workshop; their participation is most appreciated. The authors would like to acknowledge the financial support from Faculty of Engineering and Applied Science at Memorial University, ABS, VALE, Wood Group PSN, RDC, and NSERC.

8. References

[1] ABS, *Guide for vessels operating in low temperature environments*, Houston, USA: American Bureau of Shipping, May 2010.

[2] Yang, M., Khan, F., Lye, L,. Sulistiyono, H., Dolny, J., and Oldford, D., "Risk-based winterization for vessels operating in Arctic environments," *Journal of Ship Production and Design*, 2013.

[3] Deacon, T., Amyotte, P. R., Khan, F., and MacKinnon, S., "A framework for human error analysis of offshore evacuations," *Safety Science*, vol. 51, no. 1, pp. 319–327, Jan. 2013.

[4] ISO 19906, *Petroleum and natural gas industries - Arctic offshore structures*. International Organization for Standardization, 2010.

[5] NORSOK Standard S-002, *Working environment*, Standards Norway, August 2004.

[6] OGP Report Number 398, *Health aspects of work in extreme climates: a guide for oil and gas industry managers and supervisors*. London: International Association of Oil & Gas Producers, 2008.

[7] Barents 2020: Phase 4 final report, *Assessment of international standards for safe exploration, production and transportation of oil and gas in the Barents sea*. .

1ST CCPS ASIA-PACIFIC CONFERENCE ON PROCESS SAFETY℠

Lessons Learned for SME's Process Safety Management

Jinsong Zhao
Responsible Production and APELL Center (UNEP), Department of Chemical Engineering, Tsinghua University, Beijing, 100084, China
jinsongzhao@tsinghua.edu.cn

Johanna Suikkanen
Division of Technology, Industry and Economics, United Nations Environment Programme (UNEP), 15 rue de Milan, Paris, France
apell@unep.fr

Prepared for Presentation at
American Institute of Chemical Engineers
2013
The 1st CCPS Asia-Pacific Conference on Process Safety
Qingdao, China
Sep 4 – Sep 5, 2013

UNPUBLISHED

Lessons Learned for SME's Process Safety Management

Jinsong Zhao
Responsible Production and APELL Center (UNEP), Department of Chemical Engineering, Tsinghua University, Beijing, 100084, China
jinsongzhao@tsinghua.edu.cn

Johanna Suikkanen
Division of Technology, Industry and Economics, United Nations Environment Programme (UNEP), 15 rue de Milan, Paris, France
apell@unep.fr

Keywords: SME; lesson learning; process safety; APELL

Abstract

To learn from accidents, accident reporting and publication are the two major approaches for the whole chemical industry. However, most of these accidents are not well-known and well-published in China so that lessons have not well been learned and a number of major accidents and environmental emergencies still occur every year in China. In the Major Accident Information website of State Administration of Work Safety (SAWS), there are only 20 accident investigation reports published from April 2006 to July 2013 with preliminary causal analysis. Moreover, among the 26 accident cases listed on the SAWS's website, about 80% accidents occurred in SMEs. Some common causes of those accidents are analyzed in this paper for helping SMEs to learn lessons. Compliance with this regulatory framework requires substantial resources and may therefore appear too complex to be efficiently implemented by small and medium sized enterprises (SMEs) in the chemical sector. To help solve this problem, UNEP's Awareness and Preparedness for Emergency at Local Level (APELL) program is introduced in this paper and suggested as a coordinated loss prevention approach targeted specifically at SMEs which, regardless of handling hazardous chemicals in their daily operations, may not have the knowledge or capacity to efficiently implement PSM. By introducing PSM to SMEs in a more manageable way, relevant steps can be progressively implemented by companies towards full compliance with the current regulatory framework, contributing to increased safety in chemical industry parks in China.

1. Introduction

Since the USA OSHA's process safety management (PSM) standard and Europe's Seveso Directive were promulgated more than twenty years ago, major chemical accidents still deprive hundreds of people of their lives every year worldwide. People keep questioning why these accidents have not been eliminated, why people have been making the same mistakes and why lessons have not been learned. In fact, since the series of chemical disasters in the 1980s, i.e. Bhopal's methyl isocyanate release (1984), Mexico City's LPG terminal explosion (1984), Piper Alpha offshore platform explosion (1988), Pasadena's polyethylene plant explosion(1989), there has been a significant impulse for increased research and efforts in the field of safety. However, the effectiveness of learning from reported incidents can often be questioned. Jacobsson, Ek and Akselsson (2011) proposed a 6-step method for evaluating learning from incidents. Based on their preliminary study of six companies, they found that the levels of learning from incidents were mostly level 0 or level 1 which means essentially no learning or learning only at the specific place in the plant where the incident occurred with limited documentation and organizational learning. None of the companies had a very structured way of dealing with their incident reports on an aggregated basis so that the general patterns of underlying causes could not be derived. Besides individual learning, multilevel learning including organization learning, sector learning and authority learning is indispensable for effective learning. However, the effectiveness of learning from incidents can often be questioned. In many cases the learning process stops at the reporting step. Hovden, Storseth and Tinmannsvik (2011) presented a set of learning criteria that support and contribute to multilevel learning processes. By implication of their multilevel approach, technological, organizational and regulatory changes in the aftermath of an accident were included.

With the fast economic development in China, the production scale of many chemicals such as synthetic ammonia, sulphuric acid and sodium carbonate has surpassed most of other countries in the world. The living standard of the Chinese population has increased to a historically high level. Unfortunately, public safety and health are being threatened by chemical accidents resulting in societal stability issues. These kinds of accidents occur at chemical, pharmaceutical, refining facilities, at chemical, crude oil transfer pipelines and at highways, waterways during hazardous chemical transport. Since about 70% of chemical facilities in China are located near rivers, lakes and seas, these accidents have resulted in environmental disasters. However, most of these accidents are not well-known and well-published in China, therefore lessons have not been learned, and a number of major accidents and environmental emergencies still occur every year in China. Even though lesson learning has been emphasized after every chemical accident, the effectiveness of lesson learning has hardly been studied in China. In this paper, we will present an overview of lesson learning in China and identify the areas in need of further research.

2. Lesson learning in China

Learning has been a core value in the Chinese philosophy. Confucius ever said "In a group of three people, there is always something I can learn from". One of the key points in the Chinese policy on reforming and opening is to encourage Chinese people to learn advanced culture and technologies from other countries. In the area of chemical safety management, China has also learned many lessons from developed countries and international organizations.

About one decade ago, China ranked No. 1 in the world not only in terms of population but also of the total death toll from industrial accidents (Table 1). In the chemical industry, according to Liu (2003), 416 chemical accidents were recorded in 2000 with 1092 deaths and 2156 injuries while 564 accidents occurred in 2001 with 785 deaths and 1421 injuries. In the year 2000, only 47 people died of chemical accidents in the United States. Therefore, there was a huge gap in loss prevention in the chemical industry between China and the United States (Liu, 2003).

Table 1 Comparison of death tolls and rates of industrial accidents between China and some other countries (Liu, Zhong, Xing, 2005)

Country	Death	Death rate ($1/10^5$)	Year
United States	5900	4.0	2001
Japan	1790	2.79	2001
Germany	1155	7.0	2001
Canada	882	7.1	2000
France	730	4.4	2000
Australia	198	2.0	2002
United Kingdom	210	0.8	2001
Austria	122	4.5	2001
Denmark	50	2.0	2001
China	14,924	14.14	2002
Russia	4370	15.0	2001
Brazil	2503	11.5	2000
South Korea	1298	6.16	2001
Italy	1155	7.0	2000
Ukraine	1321	9.2	1999
Mexico	1502	12.0	2001
Argentina	915	18.6	2000
Malaysia	858	10.8	2002
Thailand	616	11.3	2000
Poland	510	5.0	2001

Learning from the tragic incidents, the Chinese government enacted its first Work Safety Law in 2002. Many key articles in the law such as the duties of employers and employees, inspections, investigations and recordkeeping and so on were enacted by referring to the Occupational Health and Safety Act was promulgated in the U.S.A. in 1970.

In order to prevent major accidents, the International Labor Organization released its Prevention of Major Industrial Accidents Convention (C174) in 1993. Seven years later, to fight against major accidents, the Chinese national standard on Identification of Major Hazard Installations (GB 18218) was released in 2000 and then amended in 2009. The definition of major hazard installation in GB 18218 is the same as that in C174.

The global epidemic of severe acute respiratory syndrome (SARS) during the first half of 2003 resulted in over 8000 cases with more than 800 deaths (Tang, Chan, 2004). The "SARS" exposed the weaknesses of original emergency response system in China. On November 13, 2005, a petrochemical plant exploded in Jilin Province, China, and approximately 100 tons of toxic substances including benzene, aniline and nitrobenzene spilled into the Songhua River with waste fire-fighting water. Drinking water supply to the downstream city HarBin was terminated for four consecutive days. The impact of this Songhua River pollution emergency event was huge. The Head of the then State Environmental Protection Administration resigned. Lessons learned from these two major emergency events and the natural gas well blowout accident described below catalyzed the enaction of Emergency Event Response Law of China in 2007. According to this law, emergency response plan has to be made by each chemical plant and each level of government.

The first published chemical accident in China with detailed causal analysis took place in Kaixian of Chongqing: on December 23rd, 2003, an overflow suddenly broke out from the bottom of a natural gas well in Kaixian of Chongqing, China, and eventually resulted in uncontrollable release of a high concentration of natural gas and hydrogen sulfide. The region that was mainly affected was a circular area around the well with a radius of 5 km. This disaster caused 243 deaths, 1242 hospitalizations and 65,000 evacuations. The direct economic loss was up to 64,323,100 Yuan. Dogs, ducks, rabbits and other domestic animals were killed. Even fish did not escape. The Xiaoyan village was the most affected area from this accident. Over 90% of the deaths occurred in Xiaoyan. Most of the families living 300–500m away from the well did not survive . The detail of this accident is described and analyzed by Li, Zhang, Wang and Liu, 2009 together with lessons learned, which will help prevent future similar accidents.

One lesson learned from the above accident is demonstrated by a new State Administration of Work Safety (SAWS) of China standard named 'Specification for ignition time of out of control on wellhead of natural gas well involving hydrogen sulfide' enacted in 2008. Based on this new standard, an out of control natural gas from a well has to be ignited within 15 minutes after a blowout if there is any resident not evacuated 500 meters away from the well, or the concentration of hydrogen sulfide exceeds 100ppm in a nearby unsafeguarded community which is less than 500 meters away.

Since process safety has been widely recognized as one of the keys to major accident prevention in the chemical industry, SAWS released its first Process Safety Management (PSM) regulation AQ/T 3034-2010 (SAWS, 2010). It resembles OSHA's PSM standard which contains requirements for the management of hazards associated with processes using highly hazardous chemicals (US Department of Labour, 1992). Figure 1 briefly describes the twelve elements which form the basis of the SAWS's standard.

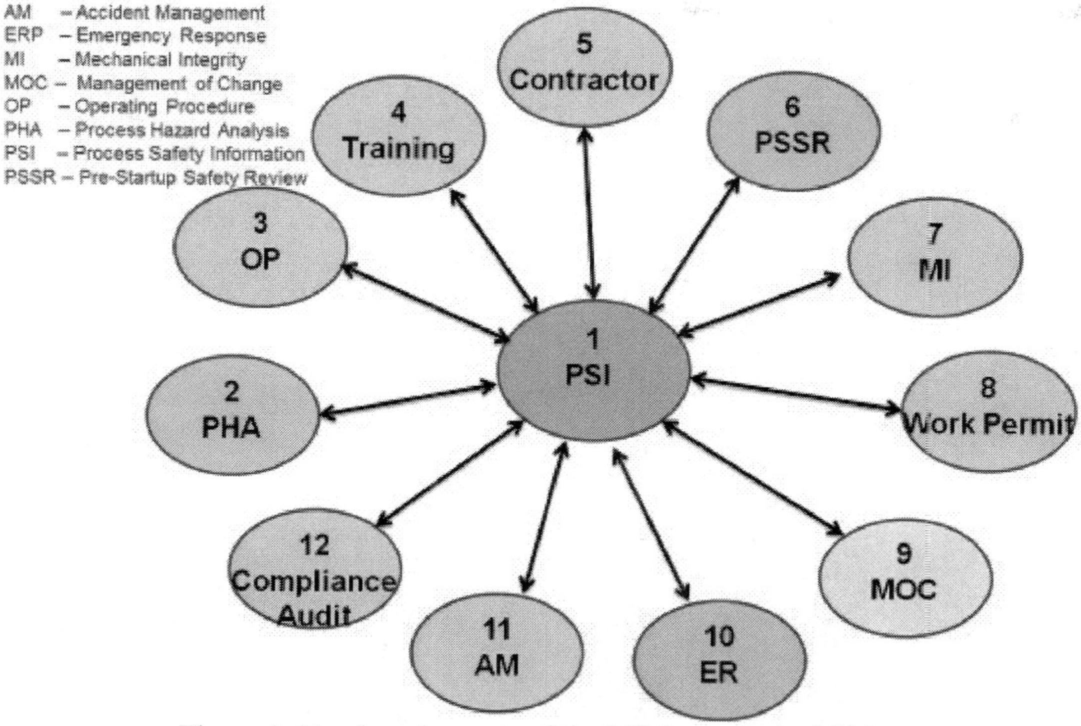

Figure 1. Twelve elements of the PSM standard of China

Development of industrial ecosystems is an effective method for recycling, reuse and resource conservation. In such as system, waste from one plant can be used as a resource by another plant. Since the establishment of the agro-chemical ecosystem in Kalundborg, Denmank, many industrial ecosystems have been established worldwide (Ding, Hua, 2012). In addition to the above environmental benefit, concentration of chemical companies into a specified zone, work safety supervision and administration can also be facilitated. Therefore, China has adopted a policy to relocate small and medium enterprises (SMEs) into chemical industrial parks rather than having them sporadically located throughout the country. According to statistics of China Petroleum and Chemical Industry Federation (CPCIF) (www.cpcia.org.cn/english/parks), there are 1,568 national and provincial development zones, among which 357 are primarily engaged in the chemical or petrochemical industry.

Chemical industry parks in China usually host many companies with significantly varying sizes and capabilities of safety management. The operations at chemical industry parks range from production, storage, utilization and transport of chemicals. These chemicals include raw feed, intermediates or final products, most of which are flammable, explosive and toxic. In the event of an accident, it would very likely lead to a domino effect, causing damages which might be extensive. Therefore, safety management of chemical industry parks needs to be regulated, including land use planning, risk assessment, safety supervision, and emergency rescue and response. In order to improve the risk management capability of all stakeholders, a pilot project on responsible production was implemented in China through the support of the United Nations' Environment Programme (UNEP) (Zhao, Joas et. al., 2013).

Due to the work safety law enforcement and other lesson learned by the governments, enterprises and academic organizations, safety performance in the chemical industry has been significantly improved in China over the past one decade. From Figure 2, it can be seen that the number of chemical accidents and death toll have dropped consecutively from 2004 to 2012. Please note that chemical accidents caused in the transport sector or in the mining industry are excluded in those statistics. Over the past 9 years, the number of chemical accidents decreased 77% while the fatalities decreased 66%. In 2012, 44 chemical accidents occurred with a 99 death toll.

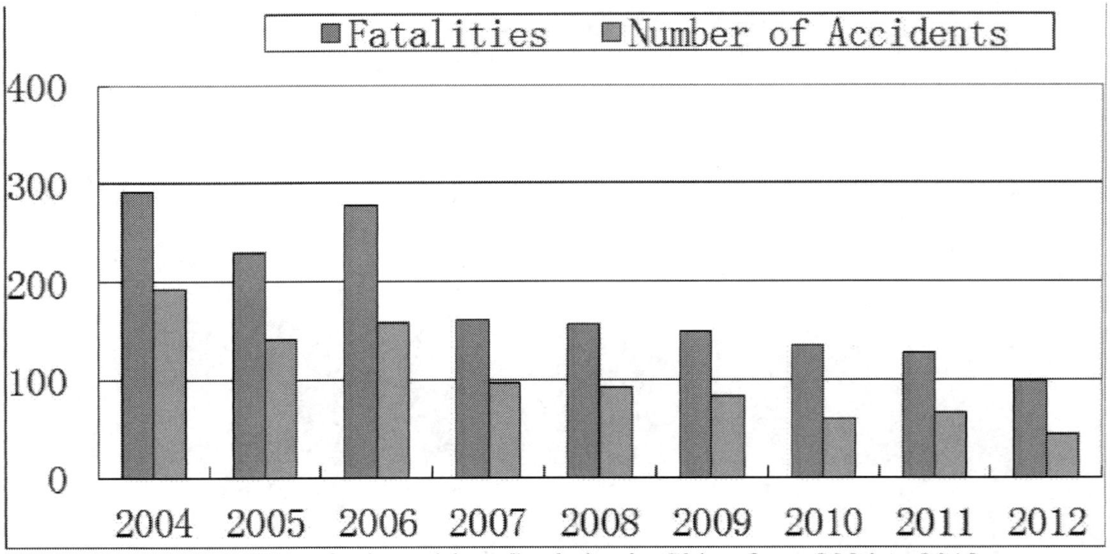

Figure 2. Chemical Accident Statistics in China from 2004 to 2012

3. Analysis of accident reports

There can be no doubt that learning from accidents has been one of the important factors which contributed to the improvement in chemical accident prevention observed over the past half century worldwide. To better learn from accidents in the chemical industry, two approaches other than incident investigation are useful: accident reporting and publication. For preventing future accidents, accident reports have been gradually published by SAWS and the Ministry of Environmental Protection (MEP) of China either through websites or through book publications.

In 2011, about 100 selected environmental accident reports was compiled and published by MEP in a book through the China Environmental Science Press. The MEP accident reports focus on emergency responses and the lesson learned section is very helpful for emergency responses to similar accidents in the future. However, these lessons or expert comments are hardly helpful for accident prevention.

In the Major Accident Information (MAI) website (http://www.chinasafety.gov.cn/newpage/wxhxp/wxhxp2013_jdys.htm) of State Administration of Work Safety (SAWS), there are only 26 accident cases from April 2006 up to July 29, 2013 (Table 2). Only 20 cases contain preliminary causal analysis. Among those 26 accidents, 20 accidents occurred in SMEs. For the 20 SME accidents, causal analysis is not available in 3 cases. For the 17 SME accidents with causal analysis, PSM elements process hazard analysis, emergency response plan, and training are the top three causes. 14 accidents were caused by poor process hazard analysis (PHA), 12 accidents were caused by unprepared emergency response plan (ERP) while 11 accidents were caused by inadequate training or no training (Figure 3).

Table 2 Accident causal analysis information extracted from the MAI website

Case # Date	PSI	PHA	OP	Training	CM	PSSR	MI	WP	MOC	ERP	SME	D/I
Case 1 08/05/2006	1	1	1	1		1				1	Yes	22/29
Case 2 11/02/2006	Causal analysis not available										No	13/0
Case 3 03/03/2007	Causal analysis not available										Yes	4/1
Case 4 05/15/2007	Crude oil pipeline broken by criminals										No	0/0
Case 5 05/28/2007			1	1						1	No	5/80
Case 6 07/24/2007	1	1				1	1				Yes	9/1
Case 7 09/12/2007		1		1				1		1	Yes	6/3
Case 8 12/11/2007		1		1		1				1	Yes	8/5
Case 9 06/29/2008		1		1		1				1	Yes	6/29
Case 10 09/18/2008	Causal analysis not available										Yes	20/60
Case 11 08/01/2010	1				1			1			Yes	13/120
Case 12 07/13/2011		1		1						1	Yes	3/1
Case 13 07/25/2011	Causal analysis not available										No	0/0
Case 14 08/09/2011		1		1			1			1	Yes	3/2
Case 15 08/09/2011	Causal analysis not available										Yes	3/1
Case 16 02/24/2012				1					1	1	Yes	3/0
Case 17 03/13/2012		1		1			1		1	1	Yes	29/46
Case 18 03/23/2012		1	1					1		1	Yes	3/0
Case 19 09/14/2012		1	1	1		1				1	Yes	3/7
Case 20 09/14/2012		1		1			1			1	Yes	3/2
Case 21 09/14/2012		1		1						1	Yes	3/4
Case 22 04/26/2013		1				1	1	1			Yes	7/0
Case 23 05/20/2013	Illegal production										Yes	3/1

Case 24 05/20/2013	Causal analysis not available										No	3/0
Case 25 06/08/2013		1			1			1			No	4/0
Case 26 07/09/2013		1							1		Yes	1/5
Subtotal of SMEs	3	14	3	11	1	6	5	4	3	12	20	152/317

Note: CM=contractor management; WP=work permit

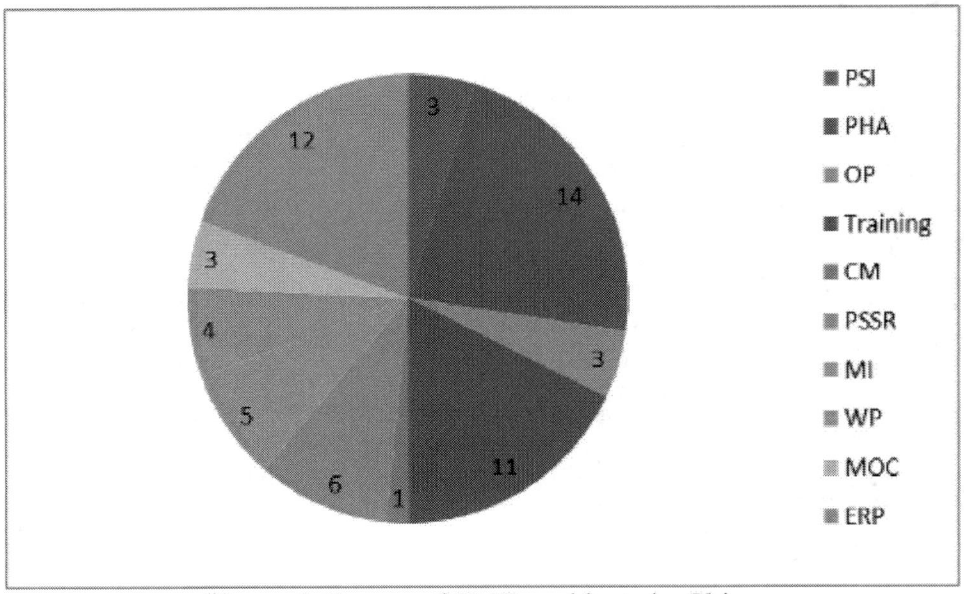

Figure 3 Causes of SME accidents in China

Blair (2004) conducted a simple statistical study on the U.S.A. Chemical Safety Board (CSB) data of 21 chemical accident investigations. The top five ranked causes were identified as "maintenance procedures", "process hazard analysis (PHA)", "engineering design and review", "management change", and "operation procedures". Yang and Dinh et. al.(2011) presented the following 10 common lessons through the study of 10 historical incidents:

1. Carefully review the materials of construction
2. Periodically inspect and maintain equipment to ensure adequate equipment reliability
3. Develop effective worker training program
4. Manage the design of hazardous installations to minimize the impact of an explosion
5. Install plants far away from residential areas
6. Design and effectively implement a good emergency response plan
7. Periodically conduct process hazard analysis
8. Take proactive actions to use lessons learned to prevent potential hazards
9. Periodically inspect safeguards and protection system to ensure their functionality when needed
10. Prevent accidents by designing and effectively using operating procedures that include standard industrial best practices

From the above lessons learned(Blair, 2004; Yang and Dinh et.al., 2011), it can be found that the SME accidents could have been prevented by the Chinese SMEs if they took proactive actions to improve their process designs and procedures by periodically reviewing the collected lessons learned. However, the analysis of the incident reports and the following implementation of appropriate measures and improvements are often ineffective and the full lessons are therefore seldom learned. (Jocobsson, Ek, Akselson, 2011). For example, on March 23rd 2005, the isomerization unit at the BP Texas City refinery exploded. 15 workers were killed and more than 170 people were injured. This Baker Panel (The BP US Refineries Independent Safety Review Panel) did a systematic and thorough investigation through which it was found that personal safety had overshadowed process safety. However, the lessons about the importance of process safety had been widely available in numerous accident analysis reports before. BP just had not really learned those lessons because of a kind of organizational "learning disability" associated with issues regarding safety management, cost cutting, reward structure, decentralization and leadership (Hopkins, 2008). What should be paid attention is that among the 20 SME accidents, 5 hydrogen sulfide poisoning accidents are quite similar. Work procedure in constrained space was not established so that more people got poisoned when they tried to rescue one suffering from hydrogen sulfide. The people involved did not even have basic knowledge of hydrogen sulfide toxicity. The accident reports have on the website of SAWS for a few years. However, the chemical companies just did not learn the lessons that are easily accessible. Therefore, the organizational "learning disability" of Chinese SMEs is very severe. This problem can be gradually solved by attending conferences, seminars, trainings, continuing educational courses, or by simply browsing relevant accident reports on the SAWS website.

By learning from past incidents operating companies can improve reliability, reduce risk and improve financial performance. In order for this to work effectively, operating companies must be willing to share experience by reporting accidents and near misses to the chemical process safety community at large. However, few chemical accidents reports are currently available for sharing lessons in China. One of the obstacles to an effective and proactive lesson learning effort is the prevalence of blocking access to accident investigation reports from the public or disclosing the accident information superficially. A special lunchtime workshop was held as part of the 2nd meeting of the Hazards Asia Pacific Symposium in Malaysia on 16 April, 2013, in partnership with the Major Accident Hazards Bureau of the European Commission, as well as IChemE's Loss Prevention Bulletin. The workshop's topic was 'Safety is not proprietary – why we need to share accident reports'. The authors here once again ask the chemical companies and relevant government agencies to share their accident reports.

4. Recommendations

Rabino (1994) believed that the future development of technology (nuclear power, genetic technology, etc.) depended heavily on public acceptance and that public perception would have a strong impact on the field's progress. In the same manner, public perception of the risk related to the chemical industry will affect the industry's development in China. Furthermore, Salihoglu and Karaer (2004) found that the more developed that a country was, the greater its population's concern was for their environment and the greater their demand was for risk controls. For example, in June 2007, a chemical project involved in producing p-xylene in Xiamen City was suspended and finally relocated as a result of public demonstrations in opposition to the project (Yuan, 2007). This event indicates that public attitudes toward the chemical industry have already become an important factor in investment decisions and, thus, industrialists and government regulators will need to account for public opinion during their decision-making processes (Huang, 2013).

Awareness and Preparedness for Emergency at Local Level (APELL) is a programme developed by UNEP in 1986 in conjunction with governments and industry with the purpose of minimizing the occurrence and harmful effects of technological accidents and environmental emergencies. The strategy of the APELL approach is to identify and create awareness of risks in an industrialized community, to initiate measures for risk reduction and mitigation, and to develop coordinated preparedness between the industry, the local authorities and the local population (Figure 4).

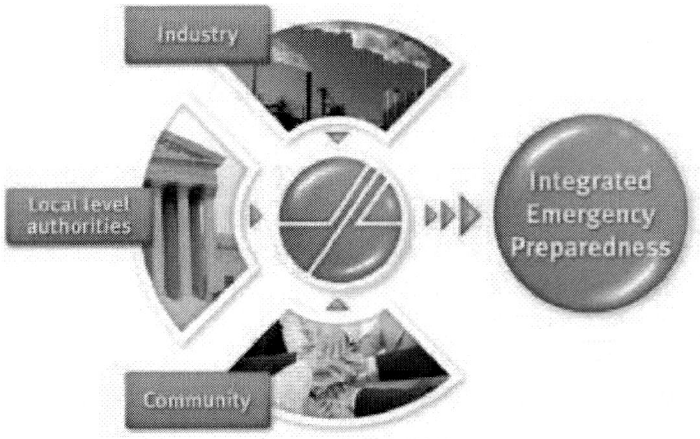

Figure 4. APELL

The APELL process consists of ten steps which are described in Figure 5.

Figure 5. Steps of APELL

During the last twenty 27 years, the APELL programme has been introduced in more than 30 countries such as Argentina, Thailand, India and China and implemented in over 80 communities. The APELL programme was quite successful in the chemical park in Bahia Blanca, Argentina in a sense that the industry's credibility was improved and the community got ready for emergency events through risk communications guided by the APELL. An APELL pilot project sponsored by Dow Chemical was implemented in ZhangJiaGang chemical park in Jiangsu province in China from 2008 to 2010. UNEP experts including the first author of this paper provided trainings on risk identification technologies, APELL methodologies, and emergency drills to the stakeholders in the chemical park. The comprehensive emergency response plan of the chemical park was evaluated by the UNEP experts and 14 recommendations were given to improve the plan. One recommendation, for example, was to establish an emergency warning system in the nearby community. A joint emergency drill which involved four neighboring companies was observed by the UNEP experts and 17 recommendations were given to the chemical park's management for highlighting the weakest points in the emergency response process. That was probably the first joint drill in China without any script. An APELL website in Chinese is going to be created soon at Tsinghua University. Hopefully the Chinese SMEs will get used to using this website for learning lessons and developing sustainably.

In reality, compliance with regulatory frameworks such as the PSM standard requires substantial resources and may therefore appear too complex to be efficiently implemented by SMEs in the chemical sector. A light tool "Responsible Production (RP) approach for Chemical Hazards Management along the Value-Chain" was established by UNEP and recommended by the authors (Zhao, Joas et. al., 2013) as a start point for risk reduction in SMEs. By introducing PSM to SMEs in a more manageable way, relevant steps can be progressively implemented by companies towards full compliance with the current regulatory framework, contributing to increased safety in the SMEs in China.

Other than APELL and RP, the following recommendations are given to SMEs and government agencies that intend to help SMEs in accident prevention:

1) To enhance lesson learning, a better chemical accident reporting and information sharing system has to be developed and deployed to achieve more timely and accurate reporting of accidents, as well as updated accident investigation and emergency response procedures. Although trust is easily eroded and difficult to rebuild, some relevant measures should be taken immediately, especially in regions where a lack of published environmental information made people distrust local agencies. Governing agencies should promulgate more information about every chemical accident to the public in a timely manner. When people are treated with fairness and honesty, and their right to make their own decisions is respected, they are less likely to overestimate small hazards and will support policy actively.

2) Full accident reports should be published in a dedicated website that is accessible for the public. A benzene pollution accident that occurred in Chongqing in November 2005 was caused by a reactor explosion which was triggered by excessive addition of hydrogen peroxide into the reactor for removing sulfur-containing impurities in benzene. This accident report is available in the MEP's environmental emergency cases book published in 2011 by the China Environmental Science Press. An oil pipeline explosion accident occurred in July 2010 in the storage tank farm of the PetroChina's Dalian Subsidiary. The pipeline explosion accident resulted in severe sea pollution. According the accident investigation report, it was also caused by excessive addition of hydrogen peroxide into the crude oil pipeline for removing hydrogen sulfide from the crude oil. If the lessons learned from the 2005's

benzene pollution accident were accessible before the 2010's oil pipeline explosion, the latter accident could have been prevented.

3) Although some officials have the incentive of enforcing regulations in a developing country like China, they can still be less effective than their counterparts in developed countries. Limited regulatory capacity and limited accountability in developing countries are among the major reasons why regulatory efforts are ineffective. For one, regulatory agencies in China are usually short of resources, especially lack of trained professionals, to fulfill their duties. Limited regulatory capacity and limited accountability both can be seen in the recent accidents in China. For example, the major reason why the Guanidine Nitrate reactor exploded in Hebei province in February 2012 is that the operating temperature of the heating oil was raised about 40 ℃ without going through management of change procedures. In other words, the regulator failed to enforce the rules in advance. Regulators can avoid such accidents if they fully realize the importance of operational requirements for safety.

Regulations should be modified to require reporting accidents not leading to deaths and/or severe injuries. Lessons could still be learned in those minor accidents. Causal analysis should go to the process safety management (PSM)'s element level. Root causes on PSM elements should be examined thoroughly and systematically. Lessons learned should cover best practices on both PSM and emergency management.

5 . Conclusion

Lesson learning is not only difficult for SMEs which have limited human resources and expertises, but also for giant organizations such as BP and PetroChina. Incident investigation is regulated by the law. However, how to effectively and systematically learn lessons from incident investigation reports is not specified. Lesson learning doesn't require high quality of investigation report, but also high self-learning capability. Learning is not completed until a relatively permanent change of behavior including process design or procedure is verified. Lesson learning should not be constrained within one organization. It should be encouraged and facilitated across industries and countries. All of these things can be realized over one night. Priorities for Chinese SMEs should be put on improving their capabilities in process hazard analysis and emergency response and providing all necessary trainings to their employees. For sustainable development, SMEs should also practice APELL and responsible production.

Acknowledgement

The authors gratefully acknowledge financial supports from the National High Technology Research and Development Program of China (863 Program, Grant No. 2013AA040702) and the National Basic Research Program (973 Program, Grant No. 2012CB720500)

References

Blair, A. (2004). Management system failures identified in incidents investigated by the U.S. Chemical Safety and Hazard Investigation Board. Process Safety Progress, 23(4): 232-236

Ding Jiansheng, Hua Weiqi (2012). Featured chemical industrial parks in China: History, current status and outlook. Resources, Conservation and Recycling. 63:43-53

Hopkins Andrew (2008). Failure to learn. The BP Texas City Refinery Disaster. CCH Australia Limited.

Hovden Jan, Storseth Fred, Tinmannsvik Ranveig Kviseth(2011). Multilevel learning from accidents – case studies in transport. Safety Science, 49: 98-106

Huang Lei, Ban Jie, Sun Kai, Han Yuting, Yuan Zengwei, Bi Jun (2013). The influence of public perception on risk acceptance of the chemical industry and the assistance for risk communication. Safety Science. 51(1)；232-240

Jacobsson Anders, Ek Asa, Akselsson Roland (2011). Method for evaluating learning from incidents using the idea of "Level of Learning", Journal of Loss Prevention in the Process Industry, 24:333-343

Li Jianfeng, Zhang Bin, Wang Yang, Liu Mao(2009). The unfolding of '12.23' Kaixian blowout accident in China. Safety Science. 47(8):1107-1117

Liu Qiang (2004). Implement the urgent notice to carry out safety management of dangerous chemical conscientiously. China Chlor-Alkali, No.7：1-5

Liu Tiemin, Zhong Maohua, Xing Juanjuan (2005). Industrial accidents: Challenges for China's economic and social development. Safety Science. 43(8):503-522.

Salihoglu, G., Karaer, F.(2004). Ecological risk assessment and the problem formulation for Lake Uluabata ramsar state in Turkey. Environmental Management 33, 899–910.

State Administration of Work Safety (SAWS) of China (2010). Guidelines for Process Safety Management of Petrochemical Corporations (AQ/T2034-2010).

Tang J. W., Chan R.C.W.(2004). Severe acute respiratory syndrome (SARS) in intensive care units (ICUs): Limiting the risk to healthcare workers. Current Anaesthesia & Critical Care. 15(3): 143-155

US Department of Labour (1992). Process safety management of highly hazardous chemicals, OSHA 29 CFR1910.119. Available at: http://www.osha.gov/pls/oshaweb/owadisp.show_document? p_table=STANDARDS&p_id=9760. (Last assessed: July 29th, 2012)

Yang Xiaole, Dinh Linh T.T., Castellanos Diana, Amado Carmen H. Osoria, Ng Dedy and Mannan M. Sam. (2011). Common lessons learned from analysis of multiple case histories. Progress Safety Progress, 30(2):143-147

Yuan, Y., 2007. Xiamen PX Event. Sanlian Life Weekly. http://news.sina.com.cn/c/2007-09-27/165713986641.shtml (Last accessed on March 27th, 2013)

Zhao Jinsong, Joas Reinhard, Abel Jochen, Marques Tomas, Suikkanen Johanna (2013). Process Safety Challenges for SMEs in China. Journal of Loss Prevention in the Process Industries,26（5）：880-886

1ST CCPS ASIA-PACIFIC CONFERENCE ON PROCESS SAFETY℠

Analysis of Leakage and Explosion of a Natural-Gas Pipeline in Residential Areas

Zhongyu Wang
College of Urban Construction and Safety Engineering, Nanjing University of Technology
Xin Mofan Road 5# Nanjing China
wzy315219@njut.edu.cn

Lijing Zhang
Gang Tao
Shu Jiang
College of Urban Construction and Safety Engineering, Nanjing University of Technology
Xin Mofan Road 5# Nanjing China
zhanglj@njut.edu.cn
taogangs@163.com
shujiang@njut.edu.cn

Prepared for Presentation at
American Institute of Chemical Engineers
1st CCPS Asia-Pacific Conference on Process Safety
Qingdao, China
September 4 – 5, 2013

UNPUBLISHED

Analysis of Leakage and Explosion of a Natural-Gas Pipeline in Residential Areas

Zhongyu Wang
College of Urban Construction and Safety Engineering, Nanjing University of Technology
Xin Mofan Road 5# Nanjing China
wzy315219@njut.edu.cn

Lijing Zhang*
Gang Tao
Shu Jiang
College of Urban Construction and Safety Engineering, Nanjing University of Technology
Xin Mofan Road 5# Nanjing China
zhanglj@njut.edu.cn
taogangs@163.com
shujiang@njut.edu.cn

Keywords: Explosion, TNO Multi-Energy model, TNT Equivalency, Buried pipeline

Abstract

As an actual flammable gas explosion occurred in a residential building, the purpose of the paper is to analyze and evaluate the process of the accident. Based on the accident consequence, the explosive overpressure and gas leakage volume are calculated by TNO Multi-Energy model and TNT Equivalency respectively. In addition, the leakage model is applied to estimate the fracture degree of the pipeline. Comprehensive analyzing and evaluating on the process of the explosion and the situation of pipeline leakage are presented in the paper. The analysis results are very close to the actual situation.

1. Introduction of the accident

In the evening of January 4th, 2013, a flammable gas explosion occurred in a residential building in Nanjing, China. Unfortunately, nine people were injured in varying degrees due to the blast overpressure. On the site of the accident, the ceiling of the bedroom was fallen and the enclosed balcony with the window frames was totally destroyed as shown in Figure 1. There was a buried nature-gas pipeline near the house and several burning traces were found along the pipeline by the accident investigation as shown in Figure 2. Based on the analysis above, it can be concluded that the leakage might be caused by the rupture of the buried pipeline. Finally, the leakage gas resulted in the explosion.

* Corresponding author. Tel.: +86 13851810118.

Figure 1: The site of the accident

Figure 2: Burning traces along the wall

2. Estimation of overpressure by TNO Multi-Energy model

For vapor cloud explosions, the TNO Multi-Energy model is often used to determine the overpressure. The procedure of using the multi-energy model for this incident is as follows.

2.1 Determine the extent of the cloud

By the accident investigation, the total volume of the accident room is 98 m^3.

2.2 Estimate the combustion energy of equivalent fuel-air

The combustion energy is determined by Eq. (1) [1].

$$E = V \cdot E_0, \tag{1}$$

Where

E (J) is the combustion energy,

V (m^3) is the confined volume, and

E_0 (J/ m^3) is the heat of combustion of hydrocarbon.

The E_0 value of methane is $3.5 \times 10^6 J/m^3$. Thus, the combustion energy is

$$E = (98m^3)(3.5 \times 10^6 J/m^3) = 3.43 \times 10^8 J.$$

2.3 Estimate the strength of the blast

By the practical application experience, the strength of a strong blast is divided into ten levels. Each number represents the different strength of the blast [2, 3]. This explosion in the building belongs to deflagration in confined space and the strength can be defined as level 6.

2.4 Calculate the scaled distance [1]

$$R' = \frac{R}{(E/P_0)^{1/3}}, \tag{2}$$

Where

E (J) is the combustion energy,
P_0 (101,000Pa) is ambient pressure,
R (m) is the distance from the explosion center, and
R'(Dimensionless) is the scaled distance.

By the accident investigation, the cloud was ignited by the refrigerator which was located between the bedroom and the north room as illustrated in Figure 3. Taking the balcony as the analysis object, which is 5.5m away from the explosion center, the scaled distance can be computed by Eq. (2):

$$R' = \frac{R}{(E/P_0)^{1/3}} = \frac{5.5m}{[(343\times10^6 J)/(101000Pa)]^{1/3}} = 0.37$$

Figure 3: The pipeline and the accident house

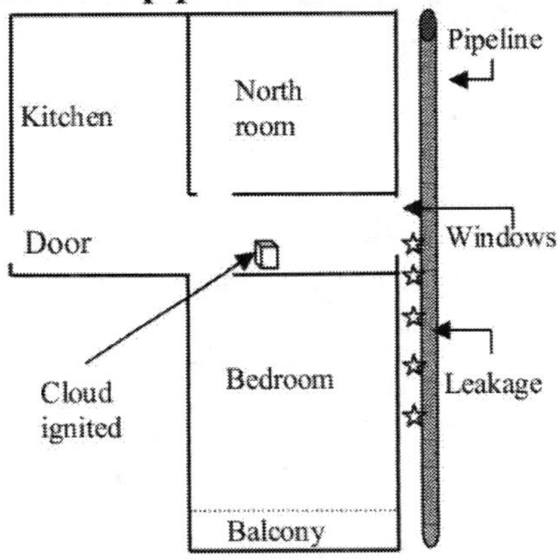

2.5 Calculate the scaled overpressure

The relationship of the scaled distance and the scaled overpressure is given by Eq. (3) [4].

$$P' = aR'^{b} , \qquad (3)$$

Where
P' (Dimensionless) is the scaled overpressure, and
a and b depend on the strength of the blast and R'.

The level is 6 and R' is 0.37. From Table 1, $a = 5\times10^{-1}$ and $b = 0$. Substituting into Eq. (3), the scaled overpressure can be obtained:

$$P' = 0.5$$

Table 1. Fitted equations for scaled overpressure

Level	Interval for R'	a	b	Interval for R'	a	b
1	0.23~0.6	10^{-2}	0	0.6~7	6.40×10^{-3}	-0.97
2	0.23~0.7	2×10^{-2}	0	0.7~12	1.32×10^{-2}	-0.98

3	0.23~0.6	5×10^{-2}	0	0.6~30	6.05×10^{-2}	-0.99
4	0.23~0.5	10^{-1}	0	0.5~70	6.44×10^{-2}	-0.99
5	0.23~0.6	2×10^{-1}	0	0.6~90	1.17×10^{-1}	-0.99
6	0.23~0.6	5×10^{-1}	0	0.6~100	3.01×10^{-1}	-1.11
7	0.23~0.5	1	0	0.5~100	4.06×10^{-1}	-1.20
8	0.23~0.5	2	0	0.5~1	4.76×10^{-1}	-2.08
	1~2	4.67×10^{-1}	-1.58	2~100	3.18×10^{-1}	-1.13
9	0.2~0.35	5	0	0.35~1	4.87×10^{-1}	-2.03
	1~2	2×10^{-1}	-1.58	2~100	3.18×10^{-1}	-1.13
10	0.23~1	4.41×10^{-1}	-2.39	1~2	4.67×10^{-1}	-1.58
	2~100	3.18×10^{-1}	-1.13	-	-	-

2.6 Calculate the practical overpressure [1]

The practical overpressure can be calculated by Eq. (4).

$$P = P' \cdot P_0 \,, \tag{4}$$

From Eq. (4), the overpressure at the enclosed balcony is obtained:

$$P = 0.5 \times (101kPa) = 50.5kPa$$

3. Estimation of TNT equivalency and gas leakage volume

TNT equivalency is based on the assumption of equivalence between the flammable material and TNT (trinitrotoluene). The approach is based on the assumption that an exploding fuel mass behaves like exploding TNT on an equivalent energy basis [5].

3.1 The theoretical calculation of TNT equivalent in explosive limits

The upper and lower flammability limits of methane are 5% and 15% in air. Some researches show that the maximum pressure occurs at the stoichiometric ratio which is 10% methane in air [6]. The theoretical calculation of TNT equivalent should consider the stoichiometric ratio concentration, upper and lower flammability limits respectively.

The equivalent mass of TNT is estimated by the following equation [1].

$$m_{TNT} = \frac{\sigma m \Delta H_C}{E_{TNT}} \,, \tag{5}$$

Where

m_{TNT} (mass) is the equivalent mass of TNT,

σ is the empirical explosion efficiency,

m (mass) is the mass of hydrocarbon,

ΔH_C is the energy of explosion of the flammable gas, and

E_{TNT} is the energy of explosion of TNT.

A typical value for the energy of explosion of TNT is $4.52\times10^3 kJ/kg$, $\Delta H_C = 5\times10^4 kJ/kg$ and $\sigma = 0.04$. The total volume of the explosive mixtures is $98m^3$ and the density of methane is $0.72kg/m^3$. It is assumed that the concentration distribution of the mixtures is homogeneous.

If the explosion occurred at 5% methane, the mass $m = (5\%)(98m^3)(0.72kg/m^3) = 3.53kg$, the equivalent mass of TNT is determined by Eq. (5):

$$m_{TNT} = \frac{\sigma m \Delta H_C}{E_{TNT}} = \frac{(0.04)(3.53\text{kg})(5\times10^4 kJ/kg)}{4.52\times10^3 kJ/kg} = 1.56\text{kg}$$

If the explosion occurred at 15% methane, the m_{TNT} value is 4.68kg.
If the explosion occurred at 10% methane, the m_{TNT} value is 3.12kg.

3.2 The actual value calculation of TNT equivalent in explosive limits

Through large numbers of experiments, the relationship of the overpressure and TNT equivalent is given by Eq. (6) [7].

$$P = 7.1\times10^5\left(\frac{R}{m_{TNT}^{1/3}}\right)^{-2.09}, \tag{6}$$

Where
m_{TNT} (mass) is the equivalent mass of TNT,
P (Pa) is overpressure, and
R (m) is the distance from the explosion center.

Based on the calculation of section 2, the overpressure at the balcony is 50.5kPa. The distance from the explosion center to the enclosed balcony is 5.5m. Substituting into Eq. (6), the actual value of TNT equivalent is

$$m_{TNT} = 3.65kg$$

Substituting this m_{TNT} value into Eq. (5), we obtain $m = 8.25kg$. Then the volume of explosive methane in the accident room is $V = m/\rho = (8.25kg)/(0.72kg/m^3) = 11.46m^3$. Therefore, when the explosion occurred, the methane concentration is $(11.46/98) = 11.7\%$ which is between the stoichiometric ratio and upper explosive limit as shown in Figure 4.

By the accident investigation, the gas released from the ruptured buried pipeline could diffuse into the accident room through the enclosed balcony and the small windows. Considering the effect of wind, it is assumed that 10% of the leakage gas diffuses into the house. So the total volume of the leakage gas is 114.6 m^3.

Figure 4: TNT equivalent at different methane volume% in air

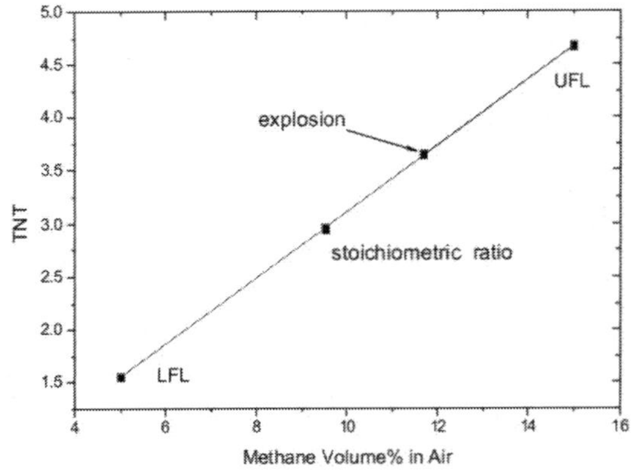

4. Estimation of the leakage area of the pipeline

The leakage area of the pipeline can be estimated utilizing the leakage model. The release process can be regarded as adiabatic process and nature-gas is ideal-gas. Through the Bernoulli equation, the volume flow rate of the gas is given by[8]

$$Q = C_D A P_1 \sigma \sqrt{\frac{M}{RT}}, \tag{7}$$

Where

- Q (m^3/s) is the volume flow rate of the gas,
- C_D (0.9) is leakage coefficient,
- A (m^2) is leakage area,
- P_1 (181,800Pa) is the internal pressure of the pipelines,
- σ is correction coefficient,
- M ($0.017 kg/mol$) is the molecular weight,
- T_1 ($283.15K$) is pipe temperature, and
- R is $8.314\ J/(mol \cdot K)$.

As the household's description, he left home at 19:40 and the explosion occurred at 20:30. Time interval was approximate 50 minutes. Therefore, the volume flow rate of the leakage gas is obtained: $Q = (114.6\ m^3)/(3000s) = 0.038 m^3/s$.

Here, σ is the coefficient of the gas leakage flow rate which depends on the type of the gas flow. If $(P_2/P_1) > [2/(k+1)]^{k/k-1}$ (The k value of methane is 1.3 and P_2 is the external pressure of the pipelines), the gas is subsonic flow. Otherwise, the gas is supersonic flow. As the pipeline is buried pipeline, the external pressure $P_2 \geq P_0$ (P_0 is atmospheric pressure). Then three results are obtained, which are $(P_2/P_1) \geq 0.56$, $[2/(k+1)]^{k/k-1}$=0.54 and $(P_2/P_1) > [2/(k+1)]^{k/k-1}$. Therefore, it can be concluded that the gas is subsonic flow .Then the coefficient is given by Eq. (8).

$$\sigma = \sqrt{\frac{2k}{k-1}[(\frac{P_2}{P_1})^{2/k} - (\frac{P_2}{P_1})^{k+1/k}]} \quad (8)$$

If the resistance of soil isn't taken into account, the external pressure P_2 is equal to the atmospheric pressure P_0. From Eq. (8), the σ value is 0.67. Thus the rupture area of the pipeline can be obtained from Eq. (7).

$$A_s = 1.29 \times 10^{-4} m^2$$

However, the pipeline is buried and the gas diffuses into soil first. The resistance of soil pore leads to the alleviation of diffusion rate and decrease of the amount of releasing gas. Thus, the actual leakage area may be larger than the simulation results.

Figure 5: The ruptured pipeline

After the buried pipeline was excavated as shown in Figure 5, the actual rupture area was confirmed to be 2×10^{-4} m^2, demonstrating that the actual rupture area A_a is 1.55 times as large as the area of the simulation A_s.

$$A_a : A_s = (2\times10^{-4}) : (1.29\times10^{-4}) = 1.55$$

When transforming into relationship of pressure ratio, the actual external pressure of the buried pipeline is 1.6 times the large of the atmospheric pressure owing to the existence of soil.

$$P'_2 = 1.6P_0$$

Therefore, the atmospheric pressure P_0 can not represent the external pressure of the buried pipeline and the influence of soil pressure should be considered in similar accident investigations of buried pipeline.

5. Conclusion and Discussion

The explosion overpressure, TNT Equivalency, the volume of leakage gas and the pipeline fracture degree are calculated in the paper. By calculating, the overpressure at the balcony is 50.5kPa. The total volume of leakage gas is 114.6m^3. When the explosion occurred, the methane concentration is 11.7% which is between the stoichiometric ratio and upper explosive limit. In addition, the actual external pressure of the buried pipeline is 1.6 times the large of the atmospheric pressure owing to the existence of soil. So the influence of soil pressure should be considered in the relevant gas release accident. The analysis results are very close to the actual situation. Those data provide a basis for restoring the process of the accident. Meanwhile, this can be referenced for similar accident investigation.

6. Acknowledgements

Financial support for this work was provided by Jiangsu ordinary university postgraduate research and innovation project (Grant No. CXLX13_442).

7. References

[1] Crowl, D. A, Louvar, J. A.. Chemical Process Safety: Fundamentals with Applications, 3rd ed. Pearson Education, Inc, Boston, MA, 2011.

[2] Mercx, W. P. M., et. al. "Application of correlations to quantify the source strength of vapor cloud explosions in realistic situations. Final report for the project: 'GAMES'." TNO Report PML 1998-C53.

[3] Zhang Ruihua, Chen Guohua, Zhang Hui. Engineering Applications of TNO Multi-Energy Method in VCE Simulation Assessment [J]. Journal of South China University of Technology, 2006, 34(5).

[4] Alonso, Fernando Díaz, et. al. "Characteristic overpressure–impulse–distance curves for vapor cloud explosions using the TNO Multi-Energy model." Journal of hazardous materials 137.2 (2006): 734-741.

[5] Dobashi, Ritsu, et. al. "Consequence analysis of blast wave from accidental gas explosions." Proceedings of the Combustion Institute 33.2 (2011): 2295-2301.

[6] Mashuga, Chad V., Daniel A. Crowl. "Application of the flammability diagram for evaluation of fire and explosion hazards of flammable vapors." Process Safety Progress 17.3 (1998): 176-183.

[7] Sun Li, Zhou Qin, Zhang Li. Study on the Failure Result of the Indoor Gas Leakage [J]. Guangdong Chemical Industry, 2009(9): 9.

[8] Liu Fu-jun, Ling Zhang-wei, Kong Shuai. Research on the Real-time Calculation of the Gas Leakage and Diffusion and Hazards Prediction of Fire and Explosion for Nature Gas Pipeline [J]. Pressure vessel Technology, 2010 (010): 21-29.

1ST CCPS ASIA-PACIFIC CONFERENCE ON PROCESS SAFETY℠

Quantitative HAZOP Model and Optimization for Laboratory Operation in the Semiconductor Industry

Tianxing Cai
Dan F. Smith Department of Chemical Engineering
Lamar University, Beaumont, TX 77710, USA
tcai@lamar.edu

Prepared for Presentation at
American Institute of Chemical Engineers
1st CCPS Asia-Pacific Conference on Process Safety
Qingdao, China
Sep 4 – 5, 2013

UNPUBLISHED

Quantitative HAZOP Model and Optimization for Laboratory Operation in the Semiconductor Industry

Tianxing Cai
Dan F. Smith Department of Chemical Engineering
Lamar University, Beaumont, TX 77710, USA
tcai@lamar.edu

Keywords: HAZOP, Quantitative Model, Optimization, Laboratory Operation, Semiconductor Industry, Laboratory Hazards, Operation Debottleneck

Abstract

Unlike the semiconductor manufacturing process with fully automatic equipment, the laboratory operation in the semiconductor industry will still involve multiple types of manual handling for chemical analysis, reliability test, material characterization and failure analysis. The test operators or lab engineers have different levels of exposure to a lot of chemicals and physical hazards in multiple phases. Compared with the traditional qualitative HAZOP analysis, this paper will describe our developed quantitative HAZOP (Hazard and Operability Study) model to identify the hazards and assess the risks of laboratory operations. The quantitative description will be deployed in the ways of manpower, machine, materials, method and environment. The analysis output can be regarded as the aid to optimize safer laboratory design, operator training and real-time fault diagnosis. This will help to debottleneck and achieve operation safety and occupational health in the semiconductor industry.

1. Introduction

Unlike the semiconductor manufacturing process with fully automatic equipment, the laboratory operation in the semiconductor industry will still involve multiple types of manual handling for chemical analysis, reliability test, material characterization and failure analysis. These operations will involve the handling of advanced and complex equipment or instruments such as Gas Chromatography (GC), Gas Chromatography-Mass Spectrometry (GC-MS), Fourier Transformed Infra Red Spectroscopy (FT-IR), Ultraviolet Visible Spectroscopy (UV/Visible), Liquid Chromatography (LC), Thermal and Elemental Analysis (TEA).

Furthermore, due to rapid changes in this industry, manufacturing processes and their associated hazards may change completely every few years. These changes make hazard assessments more difficult to complete and require that they be conducted more often. Common hazards may include exposure to solvents, acid and caustic solutions, toxic metals, and radiation. The above Table 1 has listed the typical hazardous chemicals with their roles in the manufacturing process and potential health problems linked to the exposure. These hazardous chemicals have included, but not limited to acetone, arsenic, arsine, benzene, cadmium, hydrochloric acid, lead, methyl chloroform, toluene and trichloroethylene. The test operators or lab engineers have different levels exposure to a lot of chemicals and physical hazards in multiple phases.

Table 1:Hazardous Chemicals in the Semiconductor Industry

Chemical name	Role in manufacturing process	Health problems linked to exposure
Acetone	Chemical-mechanical polishing of silicon wafers	Nose, throat, lung, and eye irritation, damage to the skin, confusion, unconsciousness, possible coma
Arsenic	Increases conductivity of semiconductor material	Nausea, delirium, vomiting, dyspepsia, diarrhea, decrease in erythrocyte and leukocyte production, abnormal heart rhythm, blood vessel damage, extensive tissue damage to nerves, stomach, intestine, and skin, known human carcinogen for lung cancer
Arsine	Chemical vapor deposition	Headache, malaise, weakness, vertigo, dyspnea, nausea, abdominal and back pain, jaundice, peripheral neuropathy, anemia
Benzene	Photoelectrochemical etching	Damage to bone marrow, anemia, excessive bleeding, immune system effects, increased chance of infection, reproductive effects, known human carcinogen for leukemia
Cadmium	Creates "holes" in silicon lattice to create effect of positive charge	Damage to lungs, renal dysfunction, immediate hepatic injury, bone defects, hypertension, reproductive toxicity, teratogenicity, known human carcinogen for lung and prostate cancer
Hydrochloric acid	Photoelectrochemical etching	Highly corrosive, severe eye and skin burns, conjunctivitis, dermatitis, respiratory irritation
Lead	Electroplated soldering	Damage to renal, reproductive, and immune systems, spontaneous abortion, premature birth, low birth weight, learning deficits in children, anemia, memory effects, dementia, decreased reaction time, decreased mental ability
Methyl chloroform	Washing	Headache, central nervous system depression, poor equilibrium, eye, nose, throat, and skin irritation, cardiac arrhythmia
Toluene	Chemical vapor deposition	Weakness, confusion, memory loss, nausea, permanent damage to brain, speech, vision, and hearing problems, loss of muscle control, poor balance, neurological problems and retardation of growth in children, suspected human carcinogen for lung and liver cancer
Trichloroethylene	Washing	Irritation of skin, eyes, and respiratory tract, dizziness, drowsiness, speech and hearing impairment, kidney disease, blood disorders, stroke, diabetes, suspected human carcinogen for renal cancer

Meanwhile, The continuous fast development and innovations in the semiconductor industry have expedited the increase of the semiconductor industry employment, which has further invoked an unavoidable concern of occupational safety to the semiconductor manufacturing operation. Semiconductor industry revenue in 2012 is expected to reach $323.2 billion, up a slight 3.3 percent from last year's revenue of $312.8 billion, according to an IHS iSuppli Global Manufacturing Market Tracker report from information and analysis provider IHS.

Based on the professional estimation, the growth in the semiconductor industry from 2013 to 2015 will average between a more encouraging 6.6 to 7.9 percent, as shown in the figure below, with total semiconductor revenue by 2015 rising to some $397.7 billion. In the coming several years, the strong semiconductor growth can be seen. Figure 2 and Figure 3 have also shown the same increasing trend for material and equipment forecast.

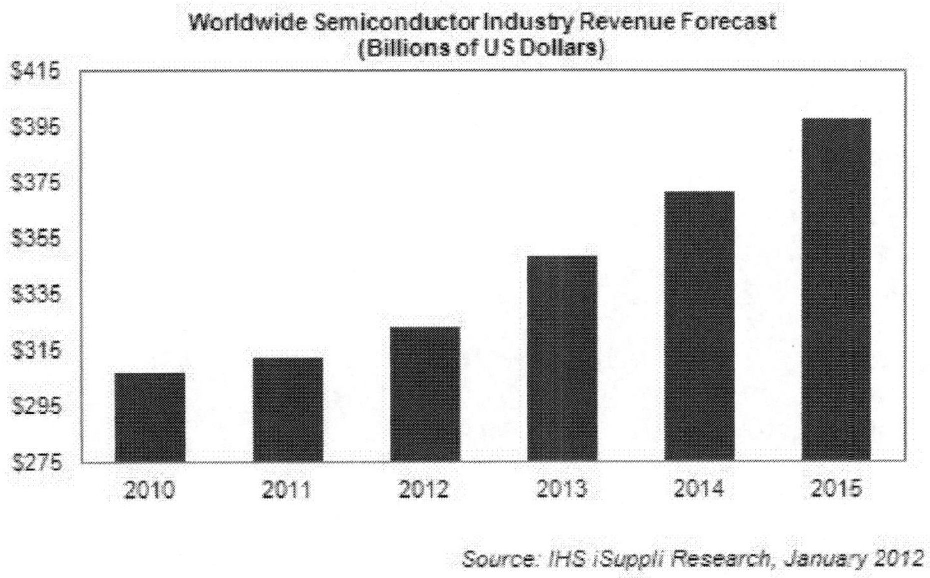

Figure 1: Worldwide Semiconductor Industry Revenue Forecast

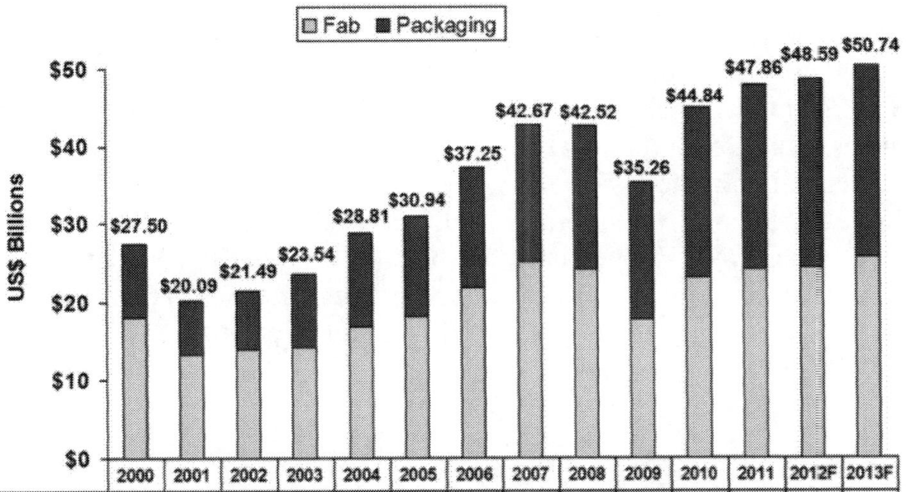

	2000	2001	2002	2003	2004	2005	2006	2007	2008	2009	2010	2011	2012F	2013F
■ Packaging	9.44	6.84	7.54	9.27	11.95	12.77	15.54	17.59	18.32	17.51	21.80	23.67	24.24	25.02
□ Fab	18.06	13.25	13.95	14.27	16.86	18.17	21.70	25.08	24.19	17.75	23.05	24.19	24.35	25.72

Totals may not add due to rounding.

Figure 2: Worldwide Semiconductor Industry Material Forecast By Segment. (Source: SEMI Materials Market Data Subscription August 2012).

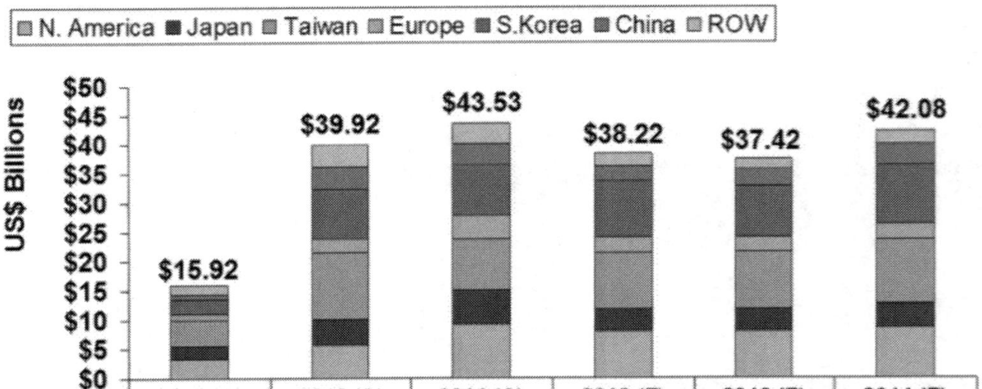

	2009 (A)	2010 (A)	2011 (A)	2012 (F)	2013 (F)	2014 (F)
ROW	1.44	3.84	3.41	2.12	1.85	2.18
China	0.94	3.68	3.65	2.58	2.86	3.70
S.Korea	2.60	8.63	8.66	9.59	8.63	9.98
Europe	0.97	2.33	4.22	2.68	2.60	2.79
Taiwan	4.35	11.25	8.52	9.60	9.80	10.88
Japan	2.23	4.44	5.81	3.72	3.78	4.01
N. America	3.39	5.75	9.26	7.95	7.90	8.55

Totals may not add due to rounding

Source: SEMI December 2012

Figure 3: Worldwide Semiconductor Industry Equipment Installation By Segment. (Source: SEMI December 2012).

Coupled with the fast development, the semiconductor employment has globally stretched. In USA, recently on Dec 11th, 2012, the Semiconductor Industry Association (SIA) announced that U.S. semiconductor industry jobs stretch across almost every region of the country and into the majority of states, according to an analysis of government data. California leads all states with 47,100 semiconductor jobs, followed by Texas (28,800), Oregon (23,400), Arizona (18,800), Massachusetts (10,100) and New York (7,600). According to BLS, the U.S. semiconductor industry's workforce grew by 3.7 percent over the previous year. In comparison, jobs throughout the broader U.S. economy increased by 1.2 percent over the same time period. Figure 4 has shown national wide employment distribution of semiconductor industry by state.

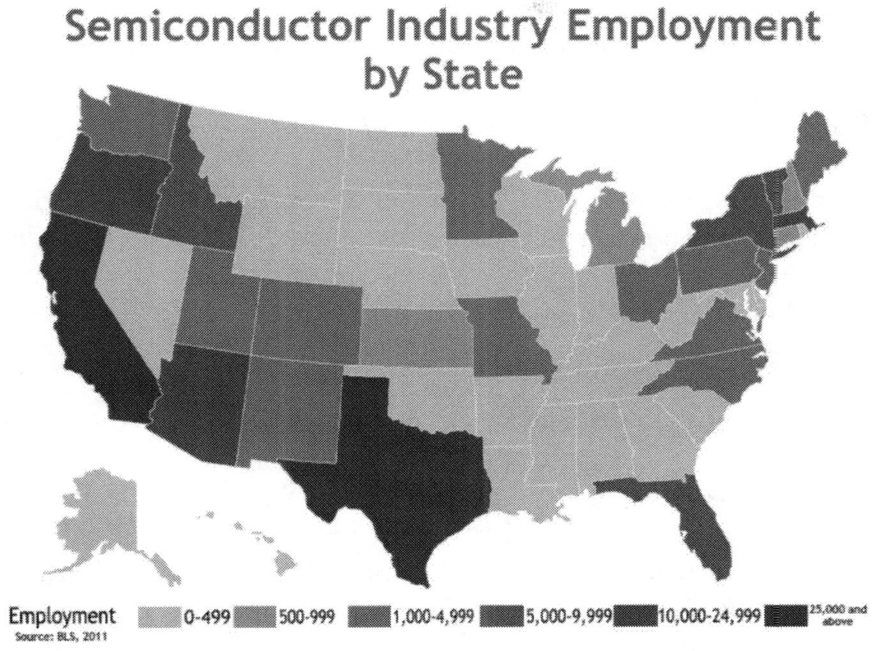

Figure 4: Semiconductor Industry Employment by State

In order to mitigate the potential risk during the chemical process operation, HAZOP study is always conducted. A hazard and operability study (HAZOP) is a structured and systematic examination of a planned or existing process or operation in order to identify and evaluate problems that may represent risks to personnel or equipment, or prevent efficient operation. The HAZOP technique was initially developed to analyze chemical process systems, but has later been extended to other types of systems and also to complex operations and to software systems. A HAZOP is a qualitative technique based on guide-words and is carried out by a multi-disciplinary team (HAZOP team) during a set of meetings.

This paper will firstly describe the straightforward HAZOP process to identify the hazards and assess the risks of laboratory operations. The analysis will be deployed in the ways of manpower, machine, materials, method and environment(4M1E). Beyond the traditional qualitative HAZOP analysis, this paper will further describe our developed quantitative HAZOP (Hazard and Operability Study) model to identify the hazards and assess the risks of laboratory operations. The quantitative description will also cover the above mentioned five aspects of 4M1E. The analysis output can be regarded as the aid to optimize safer laboratory design, operator training and real-time fault diagnosis. This will help to debottleneck and achieve operation safety and occupational health in the semiconductor industry. The identified quantitative index will also be used to reflect the comprehensive safety performance of the laboratory operation.

2. General Methodology

Several tools and methodologies are adopted to deploy the HAZOP Analysis and debottleneck for laboratory operation in the semiconductor industry. This has included Laboratory Operation Flow Diagram (LOFD), HAZOP Meeting, Parameters and Guide Words. The quantitative

HAZOP metrics have been defined for the elements which will be used to quantify the final model to identify the hazards and assess the risks of laboratory operations.

2.1 Laboratory Operation Flow Diagram (LOFD)

The method applies to existing or planned laboratory operation for which design information is available. Similar with the commonly used process flow diagram, this commonly includes a laboratory operation flow diagram, which is examined in small sections, such as individual items of laboratory test instruments, glassware or chemicals between them. For each of these, a operation design intention is specified. For example, in a failure analysis laboratory, a standard solution of aqua regia (the combination of HNO_3 and HCl with the volume ratio of 1:3) may have been used. The intention to remove gold layer in the integrated circuit so that the further visual inspection can be conducted to check the failure mode. The HAZOP team then determines what are the possible significant deviations from each intention, feasible causes and likely consequences. It can then be decided whether existing, designed safeguards are sufficient, or whether additional actions are necessary to reduce risk to an acceptable level. Besides the operation safety, HAZOP can also give the detailed instruction on the opinion whether the potential deviation will affect the quality of experiment or test analysis result.

In order to quantify a test item according to the LOFD, a variable of Document Definition Rate of LOFD, R_{DD} has been defined to quantitatively describe the detailed level of Standard Operation Procedure (SOP). It equals to the ratio of the number of standard operation procedures which have been defined by the documents of working instructions, N_{DDP} to the total number of standard operation procedures, N_{TP}. This can be illustrated by below formulation.

$$R_{DD} = \frac{N_{DDP}}{N_{TP}}$$

2.2 HAZOP Meeting

When HAZOP meetings were proceeded by the involved laboratory engineers, technicians and operators, they were generally scheduled for one to two hours per day. For a medium-sized laboratory where the total number of test items to be considered is 600 (items of equipment and pipes or other transfers between them), about 20 such meetings would be needed.

In order to quantify the effectiveness of HAZOP meeting, a variable of Action Implementation Rate of HAZOP Meeting, R_M has been defined to quantitatively describe the status of in time action implementation for the action plans decided from HAZOP meetings. It equals to the ratio of the number of implemented action items, N_{IA} which need to be completed by the verification time to the total number of improvement action which need to be completed by the verification time, N_{TA}. This can be illustrated by below formulation.

$$R_M = \frac{N_{IA}}{N_{TA}}$$

2.3 Parameters and Guide Words

The key feature is to select appropriate parameters which will apply to the design intention. The general words used in the laboratory operation are usually solution concentration, solution temperature, sampling volume, and instrument detection limit. In the above example, it can be seen that variations in these parameters could constitute deviations from the design intention of experiments. In order to identify deviations, a set of guide words is applied to each step of the experiment process. The current standard Guide Words are as follows:

Table 2: Guide Words for HAZOP Study of Laboratory Operation in the Semiconductor Industry

Guide Word	Meaning
NO OR NOT	Complete negation of the design intent
MORE	Quantitative increase
LESS	Quantitative decrease
AS WELL AS	Qualitative modification/increase
PART OF	Qualitative modification/decrease
REVERSE	Logical opposite of the design intent
OTHER THAN	Complete substitution
EARLY	Relative to the clock time
LATE	Relative to the clock time
BEFORE	Relating to order or sequence
AFTER	Relating to order or sequence

In order to quantify the completeness of guide word application, a variable of Guide Word Application Rate R_{GW} has been defined to quantitatively describe the status whether the guide words have been considered for each stand operation procedure. It equals to the ratio of the total items which has considered the impact of the situation of the guide word, N_{GV} to the product of number of total standard operation procedures and the number guide word items, which is 11. This can be illustrated by below formulation.

$$R_{GW} = \frac{N_{GW}}{11 \times N_{TA}}$$

2.4 HAZOP Software

Various software programs are now available to assist in HAZOP meetings. Even though they are always used in the chemical industry, the principles will be quite applicable to laboratory operation in the semiconductor industry.

The software can also have the capability and capacity to count and calculate the above mentioned variables and parameters automatically. If the software have the auto email function, it will be better and more effective to guarantee the conduct of HAZOP for laboratory operations.

2.5 *HAZOP Team*

HAZOP is normally carried out by a team of people with roles as follows

Table 3: Team Members for HAZOP Study of Laboratory Operation in the Semiconductor Industry

Name	Position	Role
HAZOP Group Leader	Lab Manager	someone experienced in HAZOP and laboratory operation, to ensure that the method is followed carefully
Recorder	Lab Engineer 1	someone to ensure that problems are documented and recommendations passed on
Designer	Lab Engineer 2	someone to explain any design details or provide further information
User	Lab Operator	someone to consider it in use and question its operability, and the effect of deviations
Specialist	Lab Engineer 3	someone with relevant technical knowledge
Maintainer	Lab Technician	someone concerned with maintenance of the process.

The above mentioned mention parameters can also be calculated for each responsible person respectively by below formulations in order to appraise the performance.

$$R_{DD,i} = \frac{N_{DD,i}}{N_{TP,i}}$$

$$R_{M,i} = \frac{N_{IA,i}}{N_{TA,i}}$$

$$R_{GW,i} = \frac{N_{GW,i}}{11 \times N_{TA,i}}$$

where the subscript *i* represent the *i-th* person who is responsible for that item.

3. HAZOP Management System

Based on the above defined variables and parameters, the performance change trend can be plotted in a three dimensional space. The x-axis, y-axis and z-axis represent the rate of R_{DD}, R_M and R_{GW} which change within the domain between 0 and 1. One of the example has been shown by below Figure 5.

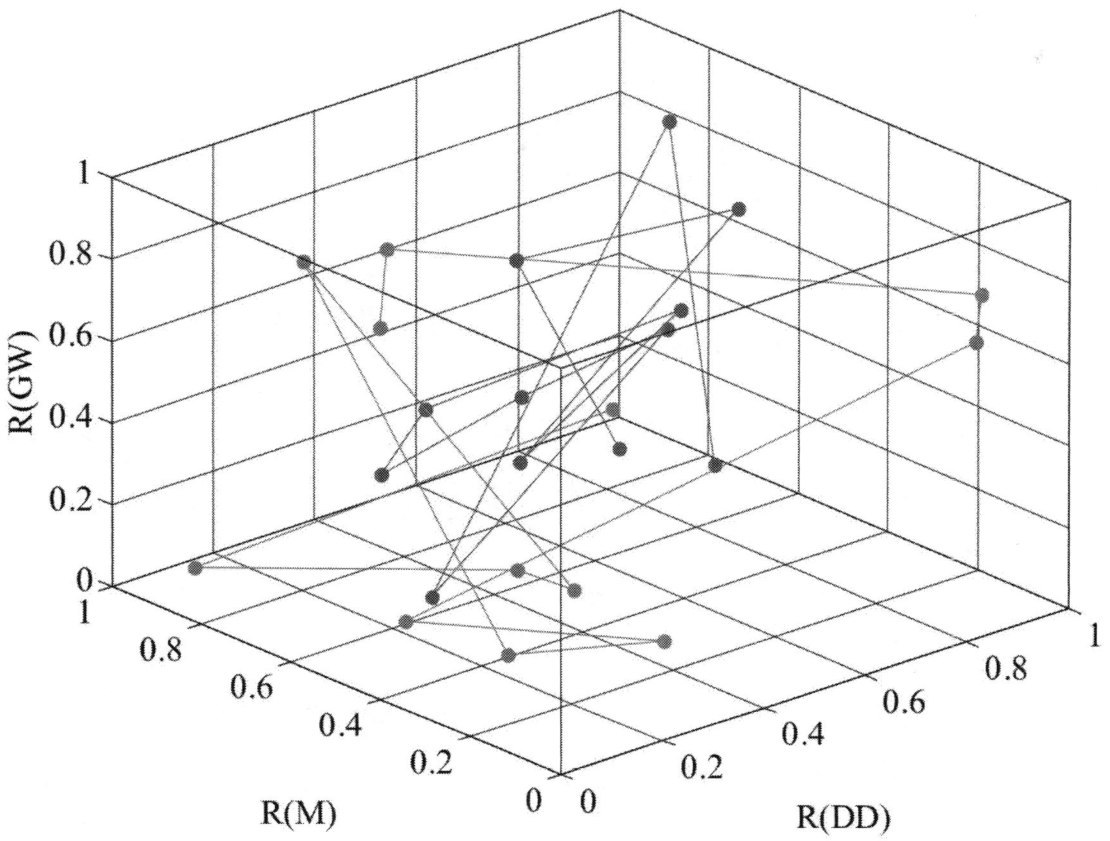

Figure 5: Demo of Performance Change Trend by 3D Plotting

For the purpose of effective implementation of the HAZOP management policy and solid application of introduced general methodology, the laboratory HAZOP system should be set up. As what has been illustrated by Figure 6, the system has coupled general management policy (operation specification and work instructions), environment control (temperature and humidity), operation safety (chemical storage, hazards assessment, chemical identification and MSDS), equipment management (PM, calibration, MSA), chemical control (data analysis and shelf life time control), training and certification (orientation, OJT, regular qualification). The whole system has covered multiple layers of analysis factors. The first layer will include manpower, machine, materials, method and environment.

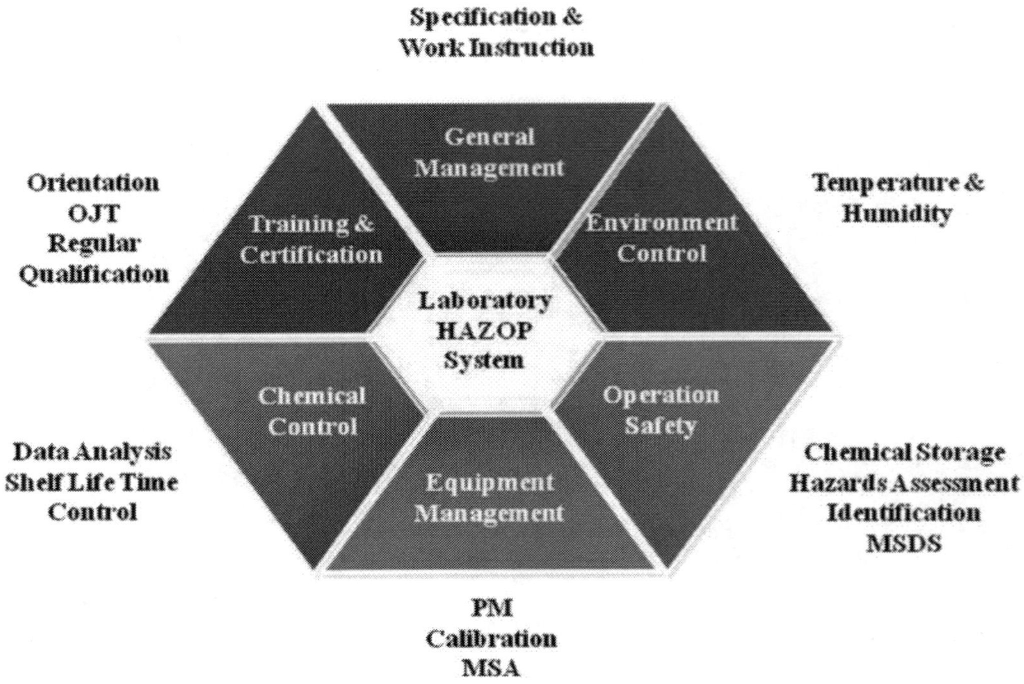

Figure 6: HAZOP Management System for Laboratory Operation in the Semiconductor Industry

Figure 7 shows that these factors can be expanded with below fishbone diagrams so that the second layer of factors will be included into HAZOP matrix system: for example, sample collection and sample preparation will be considered in the material part; training, parallax, reproducibility, practice and ergonomics will be considered in the manpower part; samples, test methods, workmanship and standards will be considered in the method part; discriminations, repeatability, bias, linearity and calibration will be considered in the instrument part; vibration, lighting, temperature and humidity will be considered in the method part.

Next, these practices can form a closed PDCA (Plan, Do, Check and Act) loop for continuous improvement of HAZOP activity, as shown in Figure 8. The close loop include four phases: plan, do, check and act. In the first phase of planning, the procedures and schedules to conduct the study should be identified and the input data should be collected. Then, the deviation should be identified by the application of key words and the causes and consequences should also be determined in the second phase. Next, the improvement actions need to be followed up. It is better to restudy the necessity of current HAZOP system. After the verification of the above items, the study report needs to be generated. Then the activity of HAZOP study should be evaluated to confirm whether the current scope, objective and responsibility need to be updated. Furthermore, the team should decide whether they need more team members to cover a larger scope.

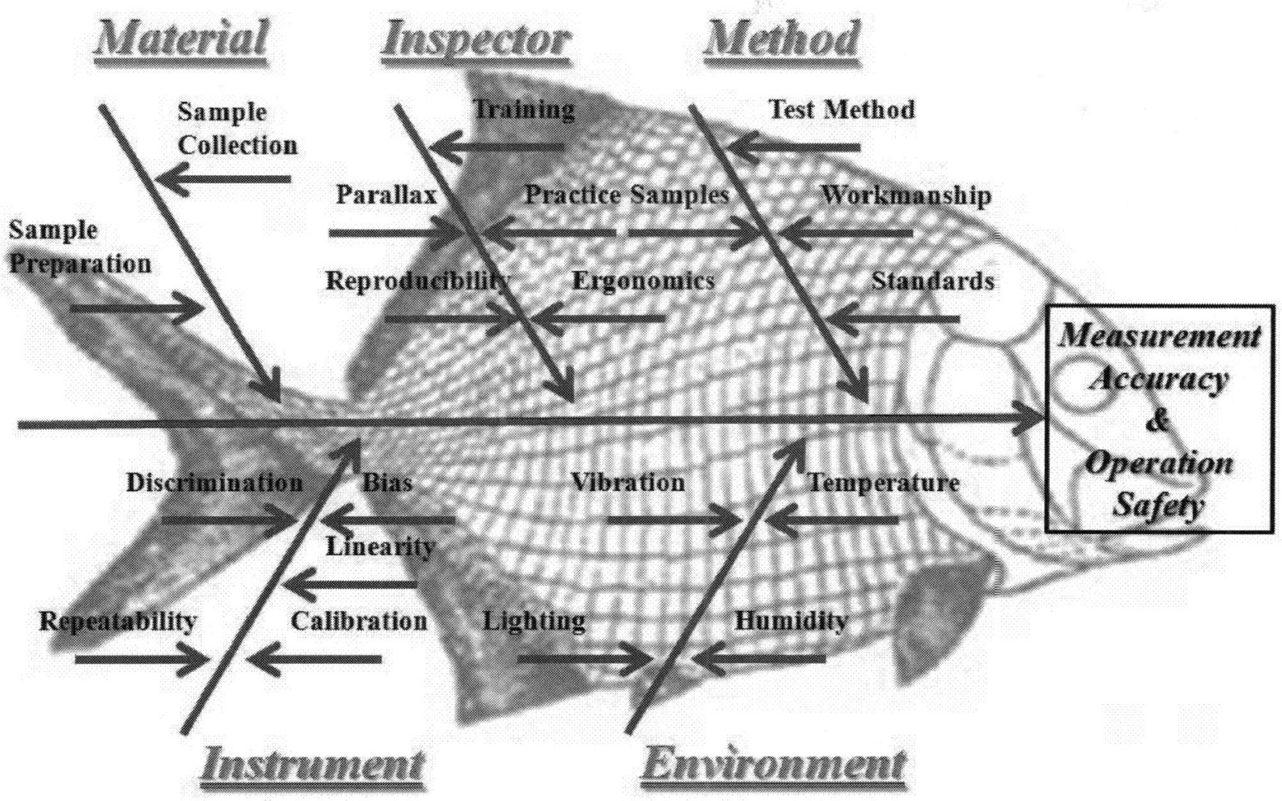

Figure 7: Fish Bone Diagram for HAZAP Study for Laboratory Operation in the Semiconductor Industry

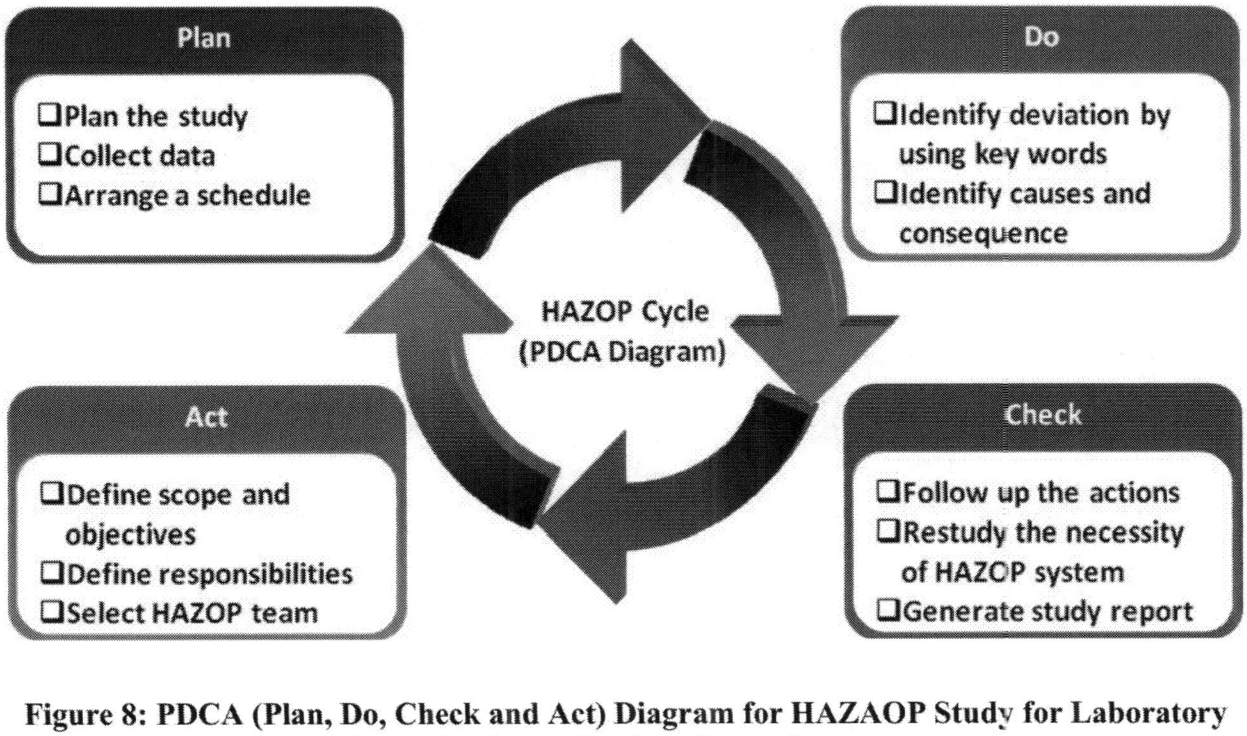

Figure 8: PDCA (Plan, Do, Check and Act) Diagram for HAZAOP Study for Laboratory Operation in the Semiconductor Industry

The analysis output can be regarded as the aid to safer laboratory design, operator training and real-time fault diagnosis. This will help to debottleneck and achieve operation safety and occupational health in the semiconductor industry.

4. Application Illustration

To demonstrate the efficacy of the developed systematic methodology, one case study has been conducted as the illustration of the methodology application in the real practice of plating bath monitoring of wafer bumping in the semiconductor industry. Wafer bumping is an advanced packaging technique that attaches solder spheres or gold bumps to the bond pads on a selected die. Gold bumping is generally used for Tape Automated Bond (TAB) assembly or Anisotropic Conductive Film (ACF) flip-chip assembly. Solder bumping is primarily used for reflow flip-chip assembly which unites the components face down, directly with the substrate or board through conductive bumps on the bond pads. These bumps deliver the electrical, mechanical and thermal interconnection which provides direct contact between the chip package and the device. Another process is called stud-bumping. Unlike gold and solder bumping, this process can be done at the die level. Wafer bumping also known as flip chip has traditionally been used for high-end niche applications. However in the last few years it has gone mainstream to include a wider range of consumer applications geared towards both the performance driven market and the form factor driven market.

Wafer bumping provides the electrical and mechanical connection between a die and substrate, which is one of the most critical elements of any flip chip package structure. Predominantly lead-tin and lead-free solders at the present time, these connections - or bumps - must exhibit superior adhesion to the die, minimal resistance, and result in high assembly yields. Solder bumps are formed by using either thin film metal deposition or ball loading techniques. Currently the advanced semiconductor assembly and test company such as Amkor offers state-of-the-art capability in electroplated solder technologies including the electroplating capability to support fine pitch bumping and copper pillar structures.

Wafer bumping utilizes thin film deposition, photolithography and plating techniques to place solder or gold onto IC (Integrated Circuit) lead pads. The wafer solder bump process is to produce the solder bumping of the wafer pad. Then it is going to melt the bumping by heat in assembly. This technology can substantially reduce IC size and have the advantages such as high density, low sensitivity, low cost and effective heat dissipation. This bumping process is suitable for flip chip assembly for LCD, memories, microprocessors, and microwave RF ICs applications. The process includes sputter UBM (under Bump Metallurgy), Photolithography, Plating and Etching etc., This bumping is suitable to apply on TAB, Flip Chip. The detailed process flow chart has been illustrated in below flow chart of Figure 9.

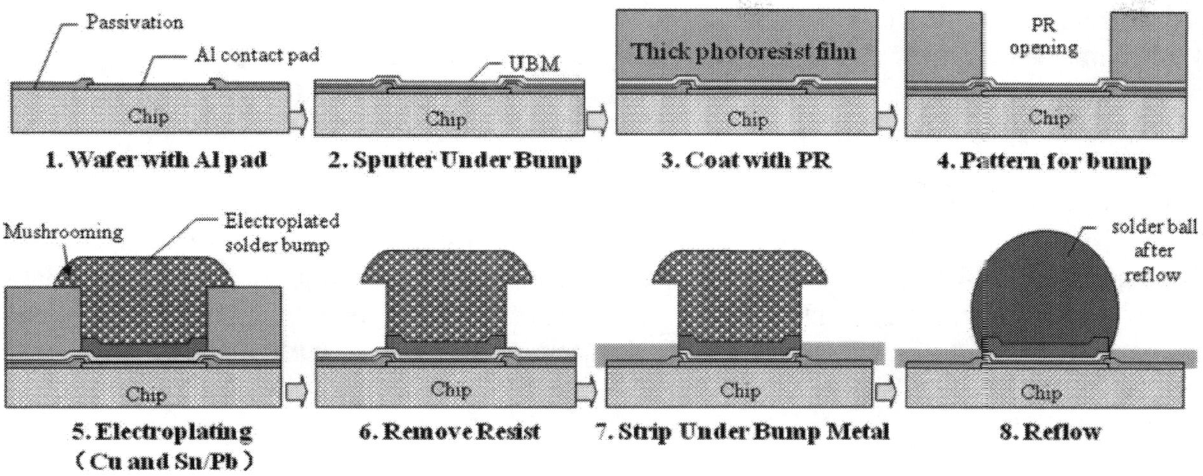

Figure 9: Process Flow Chart for Wafer Bumping Process

One of the key test item for wafer bumping operation is the electroplating bath monitoring and composition control of solder materials and additives. Plating solutions must be routinely analyzed in order to maintain the recommended bath formulation and to preempt the occurrence of problems related to improper levels of bath constituents. Contaminant levels in the solutions must also be monitored. Manufacturers of plating systems establish optimum specifications to ensure maximum solution efficiency and uniformity of deposits. The various factors that cause the concentrations of bath constituents to deviate from their optimum values are as follows: 1. drag-out; 2. solution evaporation; 3. chemical decomposition; and 4. unequal anode and cathode efficiencies. Regular analysis of plating solution becomes the responsibility of process engineering and quality control group.

The techniques employed for the quantitative analysis of plating solutions are classified as volumetric (titrimetric), gravimetric, and instrumental. Volumetric and gravimetric methods are also known as "wet" methods. The analyst must select the method that is best suited and most cost effective for a particular application. The wet methods outlined here are simple, accurate, and rapid enough for practically all plating process control. They require only the common analytical equipment found in the laboratory, and the instructions are sufficiently detailed for an average technician to follow without any difficulty. The determination of small amounts of impurities and uncommon metals should be referred to a competent laboratory, as a high degree of skill and chemical knowledge are required for the determination of these constituents. Instrumental methods differ from wet methods in that they measure a physical property related to the composition of a substance, whereas wet methods rely on chemical reactions. The selection of an instrument for the analysis of plating solutions is a difficult task. Analysts must decide if the cost is justified and if the analytical instrument is capable of analyzing for the required substances with a high degree of accuracy and precision. Instruments coupled to computers can automatically sample, analyze, and record results. Mathematical errors are minimized and sample measurements are more reproducible than with wet methods. Instrumental methods are also extremely rapid when compared with wet methods. Analytical instruments frequently used in the analysis of plating solutions can be categorized as spectroscopic, photometric, chromatographic, and electro-analytical. Spectroscopic methods (flame photometry, emission

spectrometry, X-ray fluorescence, mass spectrometry, and inductively coupled plasma) are based on the emission of light. Photometric methods (spectrophotometry, colorimetry, and atomic absorption) are based on the absorption of light. Chromatographic methods (ion chromatography) involve the separation of substances for subsequent identification. Electro-analytical methods (potentiometry, conductometry, polarography, amperometry, and electrogravimetry) involve an electric current in the course of the analysis. The instrumental methods, comprehensively reviewed below, are most applicable to plating environments.
Here, the manual titration procedures to determine copper concentration in the sample plating bath is used as illustration because the manual titration will have more potential safety concerns compared with the instrument analysis methods. Below is the procedure of the test.

PROCEDURE:

1. Preparation of EDTA (Ethylenediaminetetraacetic Acid) (0.01M). Dry EDTA for 2 hours at 80°C. GMW = 372.2., or check the reagent bottle. Dissolve needed grams in a 500 ml vol. flask. Store in a conditioned polyethylene container. EDTA solutions attack glass and deteriorate after 1 or 2 days.
2. Preparation of known copper solution. Place 0.2 g to 0.25 g of pure Cu. wire (or shot) (weighed on an analytical balance) in a 100 ml beaker. Add 15 ml conc. HNO3 and 1 ml conc. H2SO4. Place a watch glass over the beaker. Boil until dense white fumes of SO3 cease. Transfer carefully to a 200 ml vol. flask. Remember to rinse the watch glass. Dilute to the mark.
3. Preparation of unknown copper solution. Weigh 0.5 to 0.6 grams of unknown into a 200 ml vol. flask. Add 30 ml of 1:5 nitric acid. Heat until all particles are dissolved. Dilute to the mark.
4. Titration with EDTA - murexide indicator. Do this step for 4 samples of known and 4 samples of unknown.

KNOWN: Place 10 ml of Cu known solution (step 2) in 250 ml Erlenmeyer flask. Add pH buffer drop-wise until the solution turns completely blue and the precipitate first formed dissolves. Add about 100 ml. Add more buffer if solution becomes cloudy. Add 1 tablet of murexide indicator and titrate with EDTA until color changes from yellow (green) to deep violet. If the volume of EDTA is less than 10 ml, repeat using 25 ml liquid of Soln. 2.
UNKNOWN: Place 25 ml of Cu unknown solution (step 3) in 250 ml Erlenmeyer flask. Add pH buffer drop-wise until the solution turns completely blue and the precipitate first formed dissolves. Add 1 tablet of murexide indicator and titrate with EDTA until color changes from yellow (green) to deep violet. If the vol. of EDTA solution is less than 10 ml. use a larger liquid of Soln. 3.

Remarks: The copper-murexide complex is pure yellow, but the color at the starting point of the titration is greenish depending on the amount of copper present. The green is the mixture of yellow (indicator complex) and blue (copper-tetramine complex). As the titration proceeds the tetramine complex is transformed to the less intensely colored EDTA complex and the yellow of the indicator complex becomes permanent.
CALCULATION: Calculate the molarity of the EDTA. Calculate percent copper in the sample using the value of the Molarity of the EDTA found. Include the spread sheet for the calculation.

Based on the available procedures to determine the copper concentration in the sample solution of plating bath, HAZOP factors can be expanded so that the second layer of factors will be included into HAZOP matrix system: for example, sample collection and sample preparation will be considered in the material part; training, parallax, reproducibility, practice and ergonomics will be considered in the manpower part; samples, test methods, workmanship and standards will be considered in the method part; discriminations, repeatability, bias, linearity and calibration will be considered in the instrument part; lighting, temperature and humidity will be considered in the method part. These factors are listed as the column. Then, the HAZOP guide words will be listed as the row. The related information of the deviation stored in LOD(Laboratory Operation Database) and HPD(HAZOP Project Database) can be selected into the blank spaces from the drop-down list box in the above expanded table. Next, the data record of regular monitor results will be proceeded into data analysis to validate the impact of the above deviation . The data analysis result will couple with the expanded HAZOP table to quantitatively identify the potential deviation of operation safety and test accuracy. The data collections of two groups have been applied to calculated the above defined performance parameters and the change trend has been shown in Figure 10 through Figure 13. The performance of each group can be seen clearly and the comparison can also be made for benchmark. Therefore the metrics can be used to identify the weakness of HAZOP implementation and improvement direction: the target should let the performance point in the 3D space to be as close as possible to the corner point of (1,1,1).

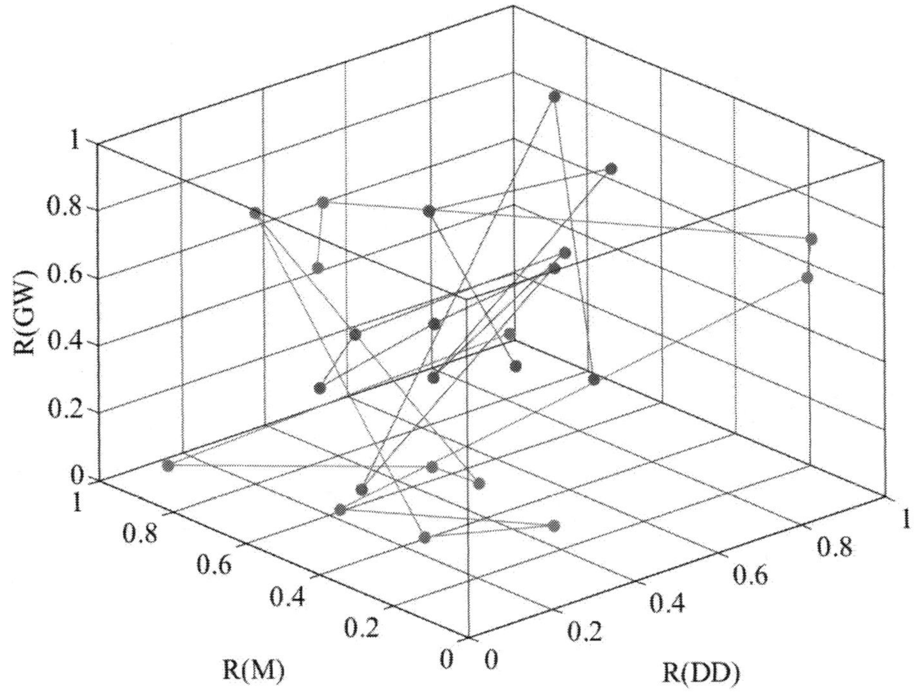

Figure 10: Performance Change Trend by 3D Plotting for Group One (Red Dot) and Group Two (Blue Dot)

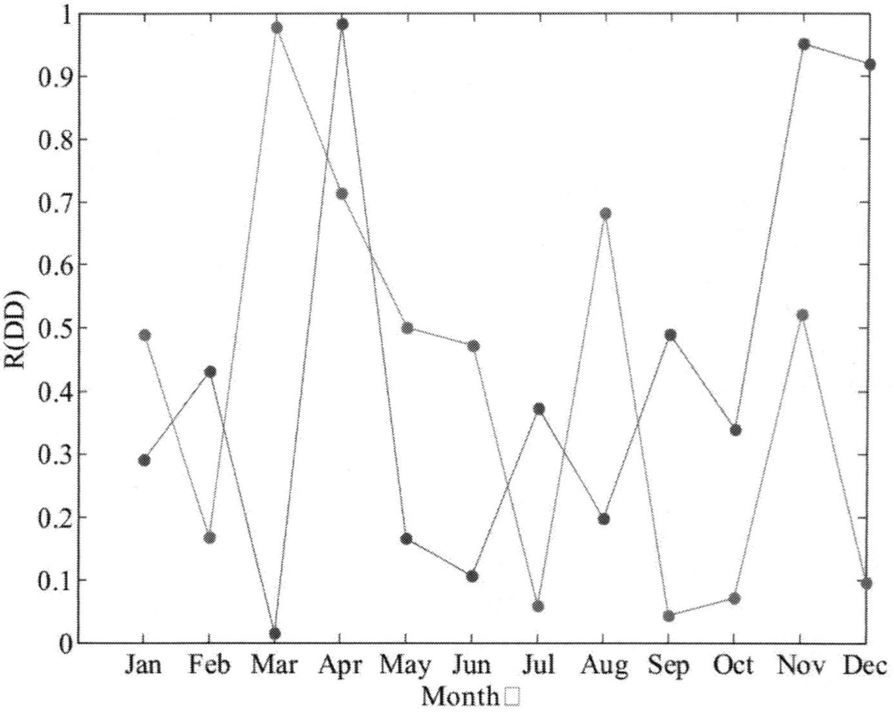

Figure 11: Performance Change Trend of R_{DD} for Group One (Red Dot) and Group Two (Blue Dot)

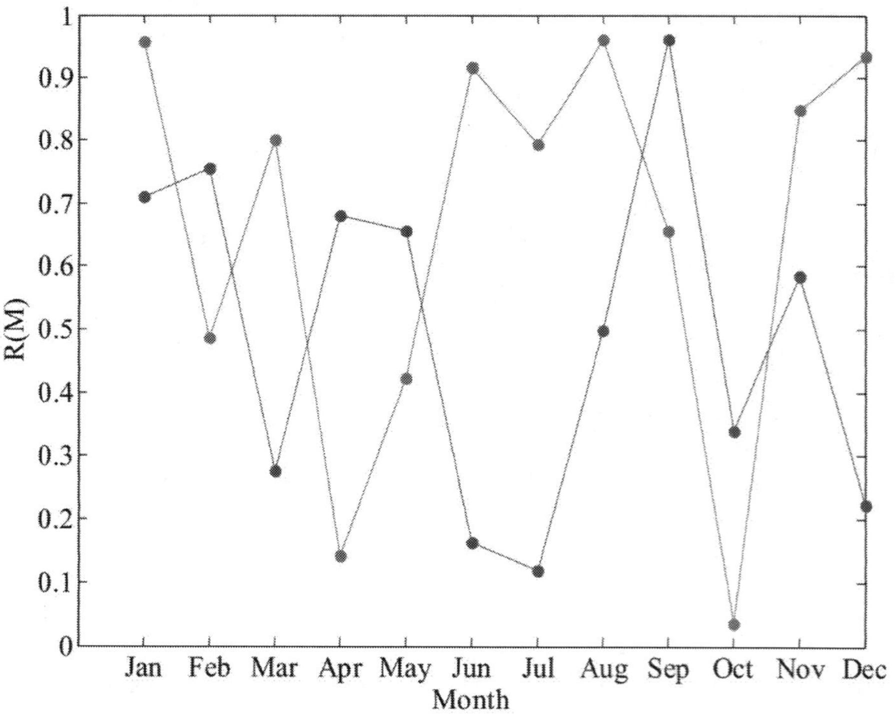

Figure 12: Performance Change Trend of R_M for Group One (Red Dot) and Group Two (Blue Dot)

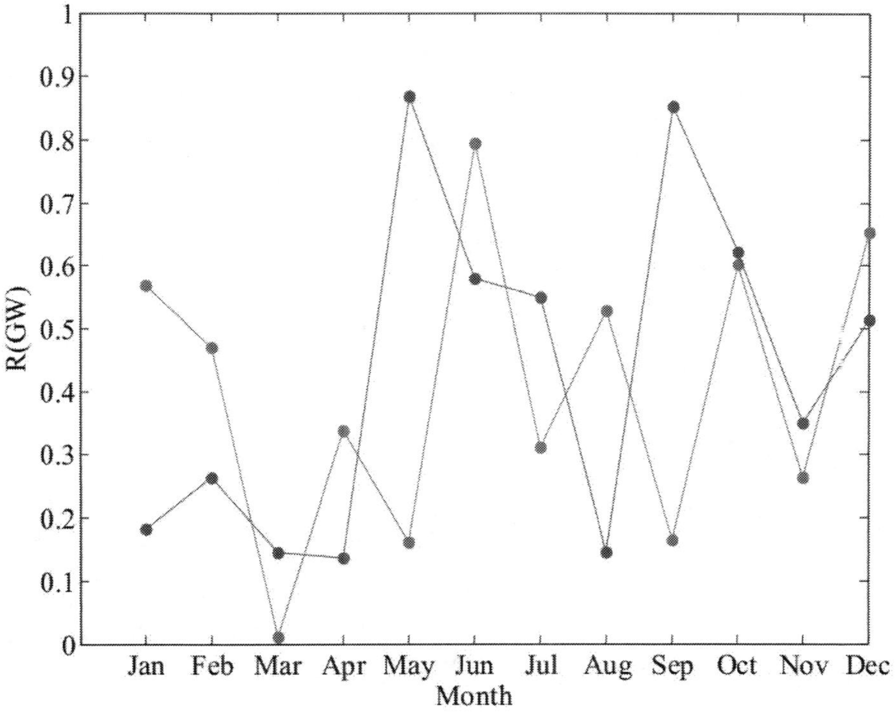

Figure 13: Performance Change Trend of R_{GW} for Group One (Red Dot) and Group Two (Blue Dot)

5. Conclusion

The laboratory operation in the semiconductor industry will have more possibilities of manual handling for chemical analysis, reliability test, material characterization and failure analysis. Therefore higher attention should be paid to the laboratory operations. As illustrated in this paper, the methodology of HAZOP study procedure can be applied to identify the potential deviations in both experiment accuracy and operation hazards. Multiple layers of factor consideration by the application of 4M1E (Man, Machine, Material, Method and Environment) will help the brainstorm to give the aid to safer laboratory design, operator training and real-time fault diagnosis. Our developed quantitative HAZOP (Hazard and Operability Study) model can be used as the metrics to effectively identify the hazards and assess the risks of laboratory operations. The quantitative description can be deployed in the ways of manpower, machine, materials, method and environment. The analysis output can be regarded as the aid to debottleneck and achieve operation safety and occupational health in the laboratory operation of semiconductor industry.

6. Reference

[1] Chepesiuk, R., “Where the Chips Fall: Environmental Health in the Semiconductor Industry.” Environmental Health Perspectives, Volume 107, pp.452-457, 1999

[2] Swann, C. D., & Preston, M. L., "Twenty-five years of HAZOPs", Journal of Loss Prevention in the Process Industries, Volume 8, No 6, pp.349-353, 1995

[3] Amkor Technology Wafer Bumping Service Introduction, http://www.amkor.com /go/turnkey-services/wafer-bumping (accessed Jan. 5, 2013)

[4] Nolan, D.P., "Application of HAZOP and What-If Safety Reviews to the Petroleum, Petrochemical and Chemical Industries". William Andrew Publishing/Noyes, 1994

[5] Lawley, H. G., "Operability studies and hazard analysis", Chemical Engineering Progress, Volume 70, No 4, pp.45, AIChE,1974

[6] Chemical Industries Association., "A Guide to Hazard and Operability Studies", 1977

[7] IHS iSuppli Research., "Semiconductor Industry Revenue to Ensure Slow Growth in 2012",http://www.isuppli.com/Semiconductor-Value-Chain/News/Pages/Semiconductor-Industry-Revenue-to-Endure-Slow-Growth-in-2012.aspx (accessed Jan. 5, 2013)

[8] U.S. Bureau of Labor Statistics (BLS) Data Report,2011

[9] Semiconductor Industry Association, http://www.sia-online.org/(accessed Jan. 5, 2013)

[10] Bolmen, Richard A., "Semiconductor Safety Handbook: Safety and Health in the Semiconductor Industry", Noyes Publications, 1998

[11] SEMI, http://www.semi.org

A Method based on Fuzzy Clustering and AHP for Safety Culture Assessment of Chemical Enterprises

Bing Tong
College of Urban Construction and Safety Engineering, Nanjing University of Technology
Xin Mofan road 5# Nanjing, China
tb1989111@126.com

Hua Zhang
Tangshan Caofeidian Shihua Crude Oil & Terminal Co., Ltd
Tangshan, China
zhanghua69@sina.cn

Gang Tao
Lijing Zhang
College of Urban Construction and Safety Engineering, Nanjing University of Technology
Xin Mofan road 5# Nanjing, China
taogangs@163.com
zhanglj@njut.edu.cn

Prepared for Presentation at
American Institute of Chemical Engineers
1st CCPS Asia-Pacific Conference on Process Safety
Qingdao, China
September 4 – 5, 2013

UNPUBLISHED

A Method based on Fuzzy Clustering and AHP for Safety Culture Assessment of Chemical Enterprises

Bing Tong
College of Urban Construction and Safety Engineering, Nanjing University of Technology
Xin Mofan road 5# Nanjing, China
tb1989111@126.com

Hua Zhang
Tangshan Caofeidian Shihua Crude Oil & Terminal Co., Ltd
Tangshan, China
zhanghua69@sina.cn

Gang Tao*
Lijing Zhang
College of Urban Construction and Safety Engineering, Nanjing University of Technology
Xin Mofan road 5# Nanjing, China
taogangs@163.com
zhanglj@njut.edu.cn

Keywords: Fuzzy clustering, Safety culture assessment, Chemical enterprises, Analytic hierarchy process

Abstract

Safety culture is playing an increasingly influential role in the development of chemical enterprises. However, it seems that safety culture is difficult to be measured in quantitative methods. Fuzzy mathematics is applied to assessing the safety culture of chemical enterprises to solve the above problem in this paper. Six chemical enterprises are chosen to analyze the levels of their safety culture in Nanjing, China. Firstly, a comprehensive and reliable safety culture assessment index system for chemical enterprises is established including six safety culture factors. Then, the weights of six indexes are determined by applying analytic hierarchy process (AHP). Furthermore, a dynamic clustering figure for six chemical enterprises is obtained and the standard levels of safety culture are confirmed by using fuzzy clustering and MATLAB. Finally, the Generalized Hamming distance equation is used to compare the scores of safety culture of six chemical enterprises with the standard scores, and the levels of six chemical enterprises' safety culture are confirmed.

1. Introduction

* Corresponding author. Tel.: +86 13451919654.

Safety culture is a great spiritual prop for enterprises and the soul of safety management, especially for chemical enterprises. The chemical enterprises have been known to present a significant risk. Indeed, with the increasing development of chemical enterprises, many incidents including fire, explosion and poisoning are also following, and many of which have led to significant numbers of fatalities, injuries and damage. The frequent accidents make us realize that it isn't only technical problems, but also the quality of individuals. So the Center for Chemical Process Safety (CCPS) defines the safety culture as "Safety culture is how the organization behaves when no one is watching" [1]. Since safety culture is important in improving the security of chemical enterprises, it is very necessary to propose effective and accurate methods for assessing the safety culture of chemical enterprises.

Fuzzy clustering is a method which introduces fuzzy mathematics into clustering, and the purpose of it is to classify the similar samples into the same class and different samples into separate class. Compared with traditional clustering methods which classify samples into a certain class rigidly, the results of fuzzy clustering are dynamic and the final results can be obtained by analyzing the actual situation [2]. As a systematic analysis method, the AHP was put forward by operations researcher T.L. Saaty in the middle of 1970s and it is adopted to confirm the weight of indexes in this paper.

2. Safety culture assessment index system

In recent years, some researchers have studied the content of safety culture [3]. Some index systems for safety culture are also put forward, such as the Integrated Safety Culture Model (ISCM) [4]. Because of the different safety culture in different industry, the safety culture for chemical enterprises is proposed according to the special features of chemical enterprises. Figure 1 shows an index system with a kind of target chart which contains six factors and every factor stands for one kind of safety culture in chemical enterprises, that is, "safety material culture, safety institution culture, safety behavior culture, safety concept culture, safety culture's openness and safety culture's persistence".

In the chemical enterprises, safety material culture is the hardware of an enterprise and the basis of safety culture. Safety institution culture is made up of laws and regulations formulated by government departments and enterprises. Safety behavior culture is how staffs of enterprise behave to improve the security. Safety concept culture is the staffs' cognition degree to safety culture. Safety culture's openness includes the changes affecting the safety of enterprise in the external environment and the measures to deal with changes taken by enterprise. Safety culture's persistence is that the safety culture of enterprise is to be constantly learning and improving. The construction of safety culture is just like shooting. The edges of the culture target may be shot at the beginning, such as U_1, U_2, or U_3. However, it is not a bad thing. Thanks to the continuous exercises and improvements, the target of "Safety culture's persistence" will be shot at last.

Figure1: Safety culture target chart

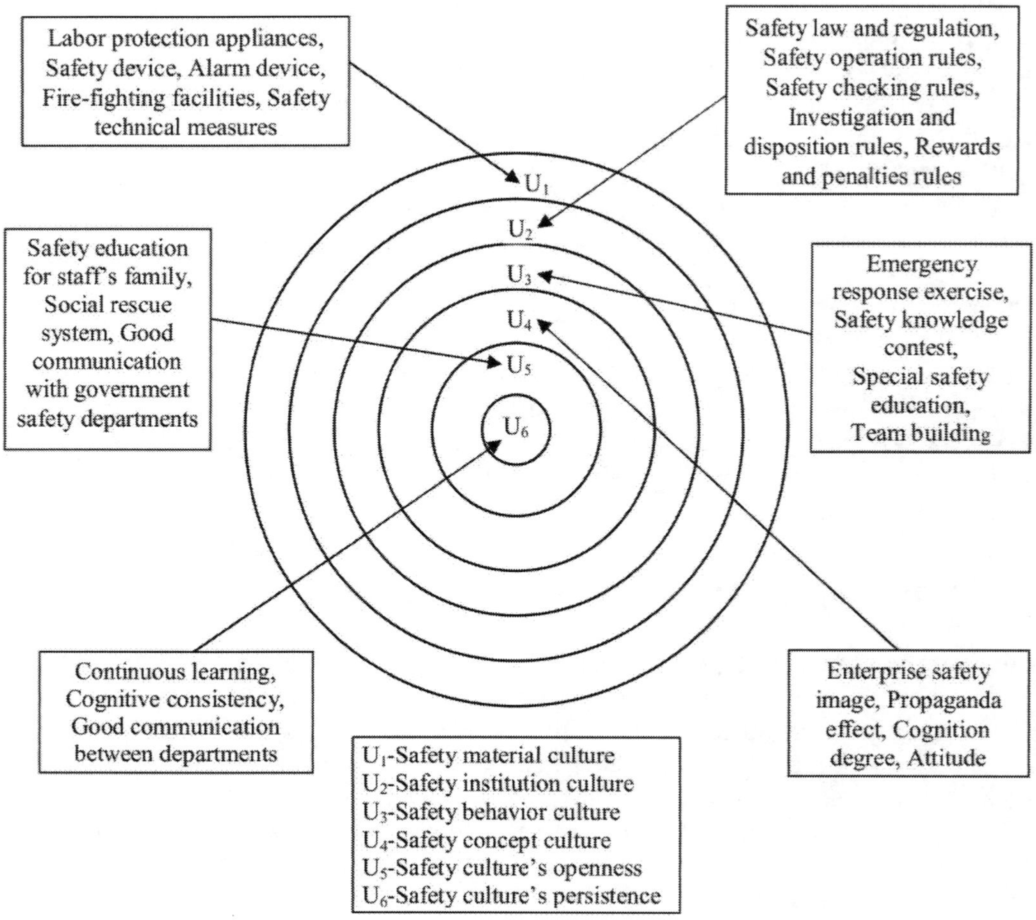

3. Weight of indexes

The AHP is used to confirm the weight of six indexes for safety culture. The AHP is a method which decomposes complex problems into various factors by comparing one with the other to determine the relative importance degree of factors. The key of AHP is constructing a judgment matrix by "1-9 scale method" [5].

3.1 Construct a judgment matrix

Table 1. 1-9 scale method

Scale value	Meaning
1	Two elements are equally important
3	One element is a little more important than the other

5	One element is obviously more important than the other
7	One element is strongly more important than the other
9	One element is extremely more important than the other
2,4,6,8	The comparison result is in the middle of the above results
Reciprocal	If the importance ratio of i and j is a_{ij}, the importance ratio of j and i is $a_{ji}=1/a_{ij}$

Table 1 shows the 1-9 scale method. For example, if m is obviously more important than n, a_{mn} =5 and a_{nm} =1/5. In this way, on the basis of intercomparsion among six safety culture factors, fifteen safety experts give the following result which is listed in Table 2.

Table 2. The comparison result

Index	U_1	U_2	U_3	U_4	U_5	U_6
U_1	1	3	4	4	5	5
U_2	1/3	1	1/2	3	4	4
U_3	1/4	2	1	1	3	4
U_4	1/4	1/3	1	1	2	4
U_5	1/5	1/4	1/3	1/2	1	3
U_6	1/5	1/4	1/4	1/4	1/3	1

Turning Table 2 into the form of matrix, a judgment matrix $J_{6\times6}$ is obtained.

$$J_{6\times6}=\begin{Bmatrix} 1 & 3 & 4 & 4 & 5 & 5 \\ 1/3 & 1 & 1/2 & 3 & 4 & 4 \\ 1/4 & 2 & 1 & 1 & 3 & 4 \\ 1/4 & 1/3 & 1 & 1 & 2 & 4 \\ 1/5 & 1/4 & 1/3 & 1/2 & 1 & 3 \\ 1/5 & 1/4 & 1/4 & 1/4 & 1/3 & 1 \end{Bmatrix}$$

3.2 Calculate the eigenvector and maximum eigenvalue

3.2.1 The eigenvector of judgment matrix $J_{6\times6}$ is calculated as follows:

Step 1: calculate the product of every row element of the judgment matrix $J_{6\times6}$ using equation (1)

$$M_m=\prod_{n=1}^{6} J_{mn} \quad (m=1, 2, 3, 4, 5, 6) \tag{1}$$

The result is $M_m=(1200, 8, 6, 0.667, 0.025, 0.001)^T$.

Step 2: calculate the 6th roots of M_m using equation (2)

$$\overline{W_m}=\sqrt[6]{M_m} \quad (m=1, 2, 3, 4, 5, 6) \tag{2}$$

The result is $\overline{W_m}$=(3.260, 1.414, 1.348, 0.935, 0.541, 0.316)^T

Step 3: Standardized $\overline{W_m}$ using equation (3)

$$W_m = \frac{\overline{W_m}}{\sum_{m=1}^{6} \overline{W_m}} \quad (m=1, 2, 3, 4, 5, 6) \tag{3}$$

The result is $W_m = (0.417, 0.181, 0.173, 0.120, 0.069, 0.040)^T$, and W_m is the weight matrix.

3.2.2 The maximum eigenvalue of judgment matrix $J_{6\times 6}$ is calculated by using equation (4)

$$\lambda_{\max} = \frac{1}{6}\sum_{m=1}^{6} \frac{\sum_{n=1}^{6} J_{mn} W_n}{W_m} \quad (m=1, 2, 3, 4, 5, 6) \tag{4}$$

The result is $\lambda_{\max} = 6.467$.

3.3 Check consistency

Whether the weight of index obtained is reasonable or not, it must be checked as follows:

Step 1: calculate consistency index (CI) using equation (5)

$$CI = \frac{\lambda_{\max} - n}{n-1} \tag{5}$$

The result is $CI = \frac{6.467-6}{6-1} = 0.093$

Step 2: calculate random index (RI)

According to the research result given by Saaty, the value of RI for 1-9 scale judgment matrixes is listed in Table 3.

Table 3. The value of RI

n	1	2	3	4	5	6	7	8	9
RI	0	0	0.58	0.94	1.12	1.24	1.32	1.41	1.45

Since n is 6 in this paper, the value of RI is 1.24. RI=1.24

Step 3: calculate consistency ratio (CR) using equation (6)

$$CR = \frac{CI}{RI} \tag{6}$$

The result is $CR = \frac{0.093}{1.24} = 0.075 < 0.1$. When CR is less than 0.1, the value of the weight of index is considered to be reasonable.

4. The application of Fuzzy clustering

4.1 The choose of research object

In order to study the safety culture, six chemical enterprises, that is, A_1, A_2, A_3, A_4, A_5 and A_6 in Nanjing, China are chosen in this paper. At the same time, fifteen safety experts are invited to give scores for the safety culture of six chemical enterprises according to the above safety culture assessment index system. Based on centesimal grade, the results of six chemical enterprises are listed in Table 4.

Table 4. The scores of six chemical enterprises

Sample	Score of index					
	U_1	U_2	U_3	U_4	U_5	U_6
A_1	95	81	89	84	91	90
A_2	87	79	88	83	82	87
A_3	93	81	89	82	88	83
A_4	78	91	74	80	85	83
A_5	95	78	89	83	91	84
A_6	90	79	83	81	81	84

4.2 Fuzzy clustering process

Based on the above scores of six chemical enterprises, an object matrix $A_{6\times 6}$ is obtained as follows:

$$A_{6\times 6}=\begin{Bmatrix} 95 & 81 & 89 & 84 & 91 & 90 \\ 87 & 79 & 88 & 83 & 82 & 87 \\ 93 & 81 & 89 & 82 & 88 & 83 \\ 78 & 91 & 74 & 80 & 85 & 83 \\ 95 & 78 & 89 & 83 & 91 & 84 \\ 90 & 79 & 83 & 81 & 81 & 84 \end{Bmatrix}$$

A kind of fuzzy clustering method based on fuzzy equivalence relation is applied to analyzing the object matrix $A_{6\times 6}$ [6, 7]. Steps are as follows:

Step 1: normalize data by shift and range transform using equation (7)-(10)

$$\overline{A_k}=\frac{1}{n}\sum_{i=1}^{n}A_{ik} \tag{7}$$

$$s_k=\sqrt{\frac{1}{n}\sum_{i=1}^{n}(A_{ik}-\overline{A_k})^2} \tag{8}$$

$$A'_{ik}=\frac{A_{ik}-\overline{A_k}}{s_k} \quad (i=1, 2, \cdots, 6;\ k=1, 2, \cdots, 6) \tag{9}$$

$$A''_{ik}=\frac{A'_{ik}-\min\limits_{1\le i\le n}\{A'_{ik}\}}{\max\limits_{1\le i\le n}\{A'_{ik}\}-\min\limits_{1\le i\le n}\{A'_{ik}\}} \quad (k=1, 2, \cdots, 6) \tag{10}$$

The average and variance of the data are computed respectively by using equation (7) and equation (8). In order to compress the data to the interval of [0, 1], the standard deviation equation (9) and range equation (10) are used. In the equation (10), $\max_{1\le i\le n}\{A'_{ik}\}$ and $\min_{1\le i\le n}\{A'_{ik}\}$ stand for the maximum and minimum data of A'_{ik} respectively.

Step 2: construct fuzzy similar matrix R_s using the equation (11)-(12).

$$r_{ij} = \begin{cases} 1, & i = j \\ \dfrac{c}{\sum_{k=1}^{m}\left|A_{ik} - A_{jk}\right|}, & i \neq j \end{cases} \tag{11}$$

$$R_s = [r_{ij}]_{n\times n} \tag{12}$$

The fuzzy similar matrix is the fundamental work for fuzzy cluster analysis, and more than thirteen kinds of methods are put forward to construct fuzzy similar matrix. Research shows that the best one is the reciprocal of absolute value method shown in the equation (11), which meets three principles including correctness, invariance and discernibility at the same time in the construction of the fuzzy similar matrix [8]. In the equation (11), r_{ij} is similar coefficient which represents the similar degree between A_i and A_j . C represents a constant.

Step 3: construct fuzzy equivalent matrix R_e using "transitive closure" method

Transitivity is required for matrix to fuzzy cluster analysis further, and transitive closure method is used to construct the fuzzy equivalent matrix. It is calculated as shown in equation (13).

$$R_s^2 = R_s \circ R_s, \quad R_s^4 = R_s^2 \circ R_s^2, \cdots \tag{13}$$

When $R_s^k = R_s^k \circ R_s^k$, the equivalent matrix $R_e = R_s^k$ is obtained.

Step 4: get a dynamic clustering figure

For the equivalent matrix R_e, when the intercept λ changes from 1 to 0, the dynamic cluster results $R_{e\lambda}$ are obtained.

4.3 The result of Fuzzy clustering

The steps from step 1 to step 4 are turned into program language based on MATLAB software platform. The result of the operation procedure is shown in Figure 2 and Table 5.

Figure 2: The result of dynamic clustering

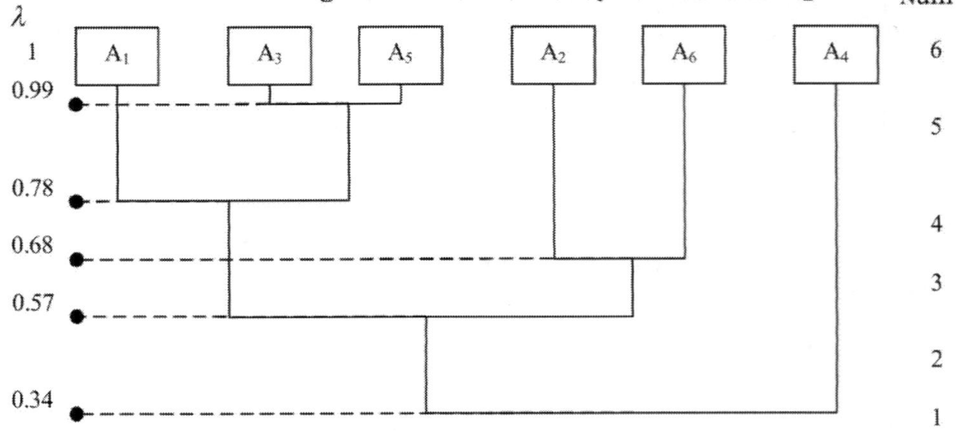

Table 5. The result of dynamic clustering

λ	0.34	$0.34<\lambda\leq 0.57$	$0.57<\lambda\leq 0.68$	$0.68<\lambda\leq 0.78$	$0.78<\lambda\leq 0.99$	$0.99<\lambda\leq 1$
num	1	2	3	4	5	6

From Figure 2 and Table 5, when the intercept λ changes from 1 to 0, six different results are obtained. When λ is 0.34 and 1, the six chemical enterprises are in the same category and completely different categories respectively. The above two situations are meaningless to the real situation. Taking into account the overall situation of safety culture for chemical enterprises in Nanjing, the best classification is 3, that is, $\{A_1, A_3, A_5\}$, $\{A_2, A_6\}$, and $\{A_4\}$, when $0.57<\lambda\leq 0.68$.

5. Confirming the level of safety culture

After the best classifications have been confirmed, the fifteen safety experts give the standard score of three levels which is listed in Table 6.

Table 6. The standard score of three levels

Level	Score of index					
	U_1	U_2	U_3	U_4	U_5	U_6
1	91	82	92	81	88	90
2	88	80	87	78	85	86
3	83	77	82	75	80	82

Comparing the scores of safety culture of six chemical enterprises (Table 4) with the standard scores (Table 6) one by one utilizing the Generalized Hamming distance equation (14), the result is listed in Table 7.

$$d(A_i, L_j) = \sum_{i=1}^{n} W_m(A_i - L_j) \quad (i=1,2,3,4,5,6;\ j=1,2,3;\ m=1,2,3,4,5,6) \tag{14}$$

Table 7. The value of Generalized hamming distance

Sample	Level		
	L1	L2	L3
A_1	1.535	4.740	9.098
A_2	-3.197	-0.008	4.366
A_3	-0.026	3.179	7.537
A_4	-7.513	-4.308	0.050
A_5	0.632	3.837	8.195
A_6	-3.240	-0.035	4.323

The smaller the absolute value of Table 7 is, the closer distance between samples and levels is. From the value of Table 7, the levels of safety culture for six chemical enterprises are respectively $\{A_1, A_3, A_5\} \in L1$, $\{A_2, A_6\} \in L2$ and $\{A_4\} \in L3$.

6. Conclusions

Generally, safety culture is hard to be measured in quantitative methods, so fuzzy mathematics is introduced to solve this problem in the paper. The safety culture target chart covers most of the safety culture indexes for six chemical enterprises, which is the basis of assessment. The application of AHP and Fuzzy clustering improves the accuracy of safety culture assessment. The results got by fuzzy clustering are consistent with the ones got by the Generalized Hamming distance. Moreover, the levels of safety culture for six chemical enterprises are obtained as well. As it turns out that the Method based on Fuzzy Clustering and AHP for Safety Culture Assessment of Chemical Enterprises is reasonable and feasible.

7. References

[1] Building Process Safety Culture: Tools to Enhance Process Safety Performance, New York: Center for Chemical Process Safety of the American Institute of Chemical Engineers, 2005.

[2] Gao Xin-Bo. "Fuzzy Cluster Analysis and its Applications". Xi'an: Xi'an University of Electronic Science and Technology Press, 2004, pg.4-5.

[3] F.W. Guldenmund. "The nature of safety culture: a review of theory and research". Safety science, Volume 34, No.1, 2000.

[4] Lei Wang, Ruishan Sun. "The Development of a New Safety Culture Evaluation Index System". Procedia Engineering, Volume 43, 2012.

[5] Xiao Ci-yuan. "Engineering Fuzzy systems". Beijing: Science Press, 2004, pg.181-182.

[6] Liang Bao-song, Cao Dian-li. "Fuzzy mathematics and its applications". Beijing: Science Press, 2007, pg.67-70.

[7] Gin-Shuh Liang, Tsung-Yu Chou, Tzeu-Chen Han. "Cluster analysis based on fuzzy equivalence relation". European Journal of Operational Research, Volume 166, 2005.

[8] Wang Xin-zhou, Shu Hai-chi. "Construction of Fuzzy Similar Matrix", Journal of Jishou University (Natural Science Edition), Volume 24, No.3, September 2003.

The Neural Network Control Algorithm Research of Straight Pull Single Crystal Furnace

Li, M.,
Meng, H.,

In straight pull single crystal furnace's pulling system, Traditional PID controller's parameter can't meet the changes of the DC motor system. In view of this defect that the traditional PID controller parameter can not turn in a timely manner with the DC motor speed change. This paper used a new control method PID neural network to the monocrystalline pulling speed control system and established PID neural network pulling system, at any time to adjust to the optimal PID parameters, to achieve an adaptive control effect and improve the quality of monocrystalline silicon.

1ST CCPS ASIA-PACIFIC CONFERENCE ON PROCESS SAFETY℠

The study of suppression effect of porous material for the flame spreading in the tube

Fenglei Han
China University of Petroleum
No.66 Changjiang Western Road, Huangdao District, Qingdao, Shandong Province, P.R.C.
andyhan0595@gmail.com

Jianliang Yu
Dalian University of Technology
No.2 Linggong Road, Ganjingzi District, Dalian City, Liaoning Province, P. R. C.
yjl63@21cn.com

Xi'e Chen
Liaoyang Petrochemical Branch Office, China National Petroleum Corporation
Liaoyao City, Liaoning Province, P.R.C.
chenxe@petrochina.com.cn

Prepared for Presentation at
American Institute of Chemical Engineers
1st CCPS Asia-Pacific Conference on Process Safety
Qingdao, China
September 4 – 5, 2013

UNPUBLISHED

The study of suppression effect of porous material for the flame spreading in the tube

Fenglei Han
China University of Petroleum
No.66 Changjiang Western Road, Huangdao District, Qingdao, Shandong Province, P.R.C.
andyhan0595@gmail.com

Jianliang Yu
Dalian University of Technology
No.2 Linggong Road, Ganjingzi District, Dalian City, Liaoning Province, P. R. C.
yjl63@21cn.com

Xi'e Chen
Liaoyang Petrochemical Branch Office, China National Petroleum Corporation
Liaoyao City, Liaoning Province, P.R.C.
chenxe@petrochina.com.cn

Keywords: porous; flame; experiment; simulation

Abstract

The suppression effect of porous material for flame spreading in the tubes was studied by the experiment. The experiment got the law of the suppression effect when the size of the porous material changed. The results of the experiments and simulation indicate that the porous material can suppress the blame wave effectively, and when the size of the material changed, included the length and the thickness, the suppression effect also changed. This paper can propose the theoretical foundation and an effective guidance of the fire explosion suppression.

1. Instruction

Combustible gas explosion is one of the most destructive disasters in the modern industrial accidents, it causes enormous losses, so the study of explosion suppression technology has been a concerned issue for a long time. In most cases, the combustible gas flows in the pipeline. Therefore, it has a great significance to study the flame spread in the pipeline and the suppression method. It can be found in the previous studies that, the mechanical failure of the pressure wave and the combustion due to the explosive flame were the two means of destruction[1,2]. Porous material can suppress the pressure of combustible

Supported by "the Fundamental Research Funds for the Central Universities" gas explosion in the pipeline and vessels[3,4,5]. It has an ideal suppression effect for explosion pressure, meanwhile it enhance the turbulence and combustion due to the porous surface [6,7]. Consequently, the combined action of the suppression and turbulence was taken into consideration to study the flame wave suppression effect of the porous material. And also, the quenching effect was taken into account. The results can propose a theoretical basis for the flame suppression studies.

2. Instruments

The schematic diagram of the instrument is shown in the figure 1. It consists of pipeline, process control ignition system, and dynamic data collecting and analyzing system. The inner diameter of the pipeline is 81mm, and the length is 2.2 m. There are 8 detecting point for the flame speed and pressure detecting. The ignition system consists of ignition circuit and control circuit. A program controls the spark plug to ignite by an electric relay. The data collecting system has 16 channels, and the sampling frequency is 1.25MHz. This system has the functions of auto collecting, storage, data processing, display, printing, and output. The programmer was compiled by C language.

The porous material used in this experiment is aluminum silicate board, which is made of proper fractions of aluminum silicate base, through the processes of pressing, stereotype in the solidification room, and vertical and horizontal cutting. It has heat stability and the resistance to shock. Besides that, it also has the merit of insensitive resistance to pressure, good absorptivity, anti-salt, and acid resistance. So it is a pollution-free, green product. The parameters of the material are shown in table 1.

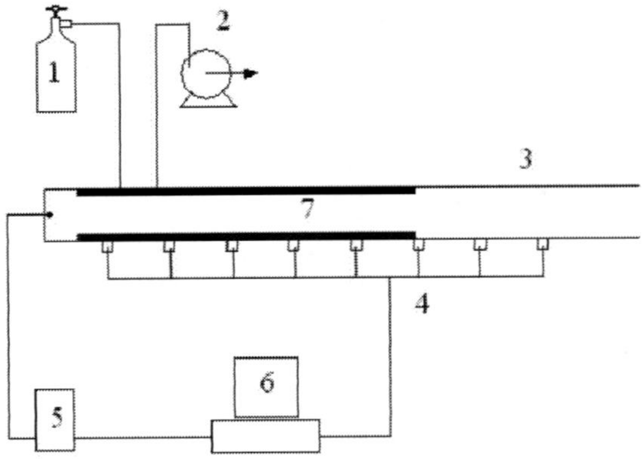

Fig. 1 Sketch map of experimental instrument: 1 flammable gas; 2 vacuum pump ; 3 Pipeline ; 4 pressure and flame sensor ; 5 program controlled ignition system; 6 computer; 7 aluminum silicate board.

When the experiment started, the porous material was putted in the pipeline 1 m from the exit end. As shown in the figure 2, the flame speed and maximum overpressure in the pipeline was measured by the photosensitive flame sensor and high frequency dynamic pressure transmitter, which is settled at the position a and b, respectively. Changing the length and thickness of the material, it can got different groups of experimental data.

Table 1 The parameters of aluminum silicate board

Color	Size(mm)	Thickness (mm)	Density (kg/m^3)	fiber fineness (μm)	shot content (%)Φ>0.25	Permanent linear shrinkage (24 小时)
White	400×600 600×800	10-150	170-600	0.315	0.95	-4 (900-1000℃)

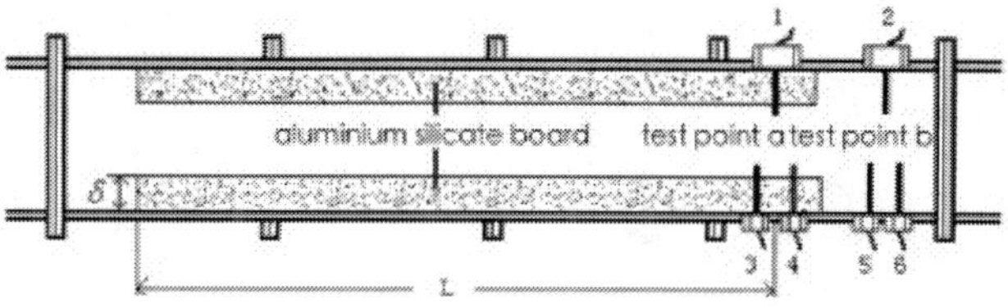

Fig. 2 Diagram of test points: 1~2, pressure sensor; 3~6, photosensitive flame sensor

3. Results and discussion

3.1 The effect of aluminum silicate board on the flame speed

For studying the suppression effect of pre-mixed combustible gas flame spread in the pipeline, the flame wave spreading in the pipeline was compared with and without the aluminum silicate board. Figure 3 and figure 4 showed the flame wave when the acetylene/ air exploded in the pipeline. Because of the aluminum silicate board inside the pipeline, the retention time of the flame in the tube increased from 8 ms to 30 ms, which means the flame speed decreased from 327 m/s to 118 m/s. That's because the aluminum silicate board delayed the spread of the flame, it decrease the flame speed. The results show that the aluminum silicate board has a good suppression effect on the flame spread in the pipeline.

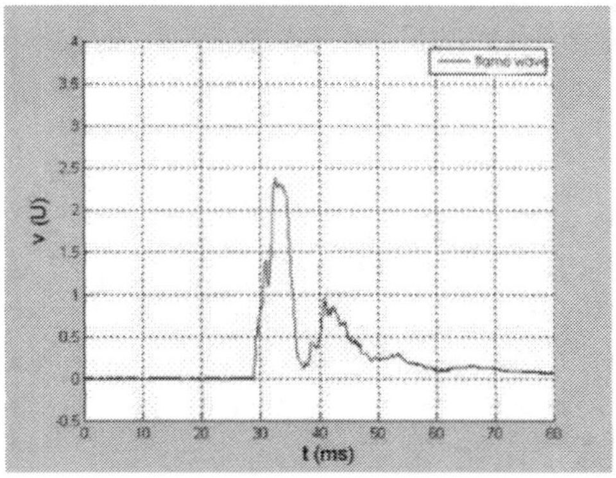

Fig. 3 Flame wave propagating in tube

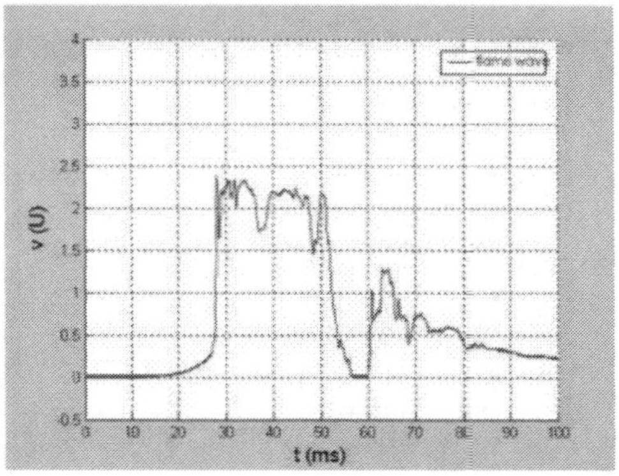

Fig.4 Flame wave propagating in tube with aluminum silicate board

3.2 The suppression effect of different length on the acetylene flame speed

A series of aluminum silicate board with the same thickness but different length were attached inside of the pipeline to study the suppression effect on the acetylene flame speed. Figure 5 shows the schematic of the flame speed on the test point a and b when the thickness is 10 mm. It can be seen clearly that, the flame speed on the same point decrease when there was aluminum silicate board attached inside the pipeline. And with the length increasing, the flame speed on the test point decreased gradually. However, when the length was close to the length of the accelerating section, the suppression effect changed little. That means there is an optimal length for the suppression effect.

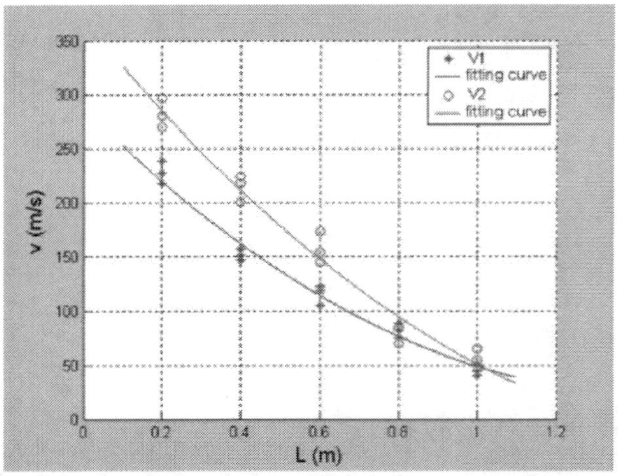

Fig.5 Effects of the length of aluminum silicate board on flame speed (δ=10 mm)

In addition, it can be found in the experiment that, when the length of the aluminum silicate board is equal or lesser than 0.8m, the flame speed on the test point b is bigger than point a, that means the flame speed get suddenly increased after getting out of the aluminum silicate board stage. That is because the diameter of section changed suddenly, which induced a turbulent boundary layer at the end of aluminum silicate board stage. The turbulence increased the transport effect of the system, and make the flame front drape transformed, which increased the flame speed. But due to the suppression effect of the aluminum silicate board, the speed was still smaller than the speed in the empty pipeline (377m/s).

3.3 The suppression effect of different thickness on the acetylene flame speed

A series of aluminum silicate board with the same length but different thickness were attached inside of the pipeline to study the suppression effect on the acetylene flame speed. Just like length, different thickness also has different suppression effect on the flame speed, as the experimental results showed. And with different length, the suppression trend is different, as showed in figure 6 and figure 7. As the length is 0.2 m, 0.4 m and 0.6m, with the increasing of the aluminum silicate board, the flame speed decreased firstly. When the thickness reached a certain value, the speed began increasing. This indicates that there is an optical thickness value for the suppression effect when the length is 0.2 m, 0.4 m and 0.6m. Moreover, when the length is 0.8 m and 1.0m, the flame speed will always show a downward trend. That's because if the aluminum silicate board is long enough, when the flame gets out of this section, most part of the energy has been absorbed by the board. There will not be enough energy for the turbulence to enhance the flame speed.

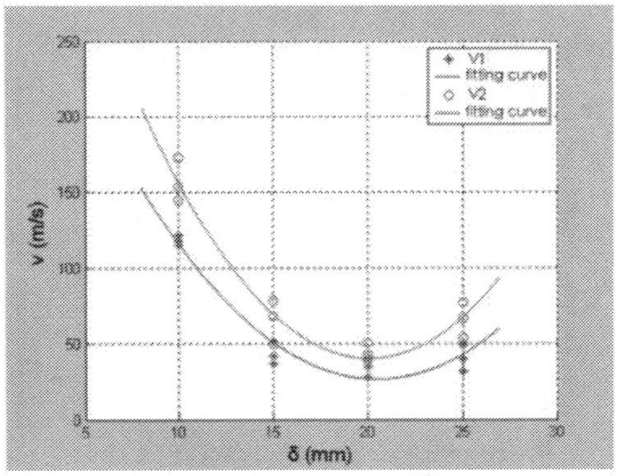

Fig.6 Flame wave propagating in tube

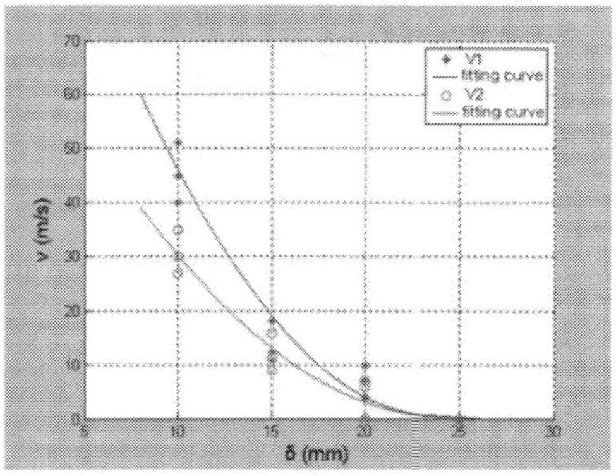

Fig.7 Effects of the thickness of aluminum silicate board on flame speed

4. Simulation results

The turbulence combustion flow theory model was settled up based on homogeneous reaction Reynolds-averaging equations, $\kappa-\varepsilon$ turbulent model and EBU-Arrhenius combustion model. The simulation chose acetylene and air mixture for instance, and the concentration ratio was chemical reaction equivalent ratio. Using SIMPLEC algorithm, the pre-mixture flame spread in pipeline with porous materials was simulated by the CFD software Fluent.

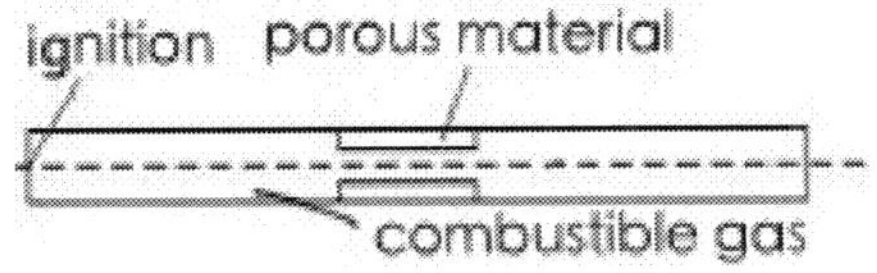

Fig.8 The sketch map of physical model

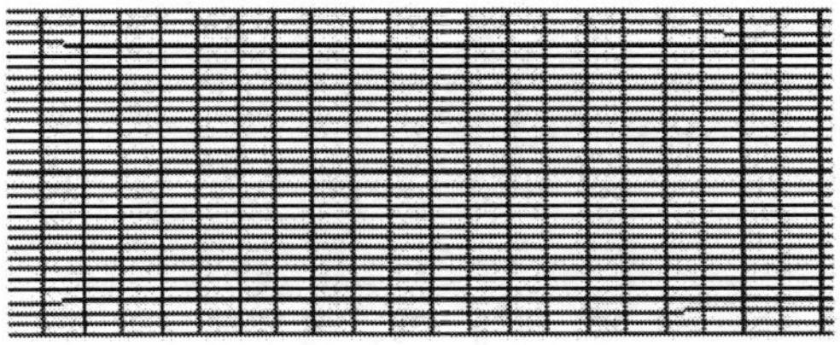

Fig.9 The gridding

The physical model is shown in figure 8, and figure 9 shows the gridding meshed for the model. Disperse un-steady solver, standard $\kappa-\varepsilon$ turbulent model and finite-rate model was chosen for the combustion simulation. At the beginning of the ignition, laminar model was used for the reaction model. When the combustion was sufficient, eddy disperse model was chosen for a more precise simulation.

The flame front at t=5ms, 5.5ms, 6ms, 6.5ms, 7ms was shown in figure 10. It is can be seen in the figure that the flame front has an obvious change when it gets the position of the porous materials, and the flame temperature declines evidently. If the porous material is long enough, the flame will fault, as show in the figure. And longer porous material has better suppression effect. The simulation results confirm the experiment well. And the experimental results are in good agreement with the calculated results.

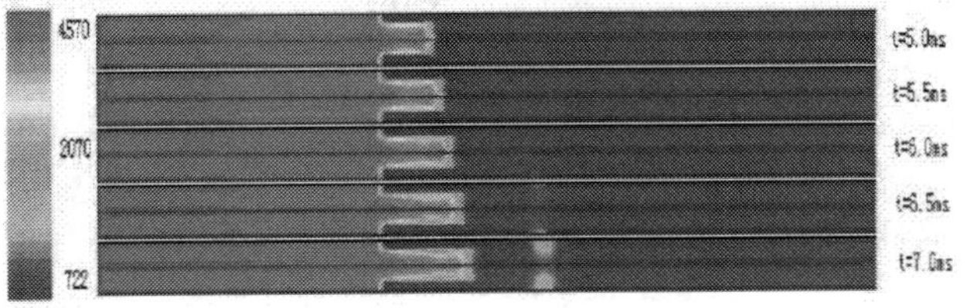

Fig.10 The shape of the flame at different time

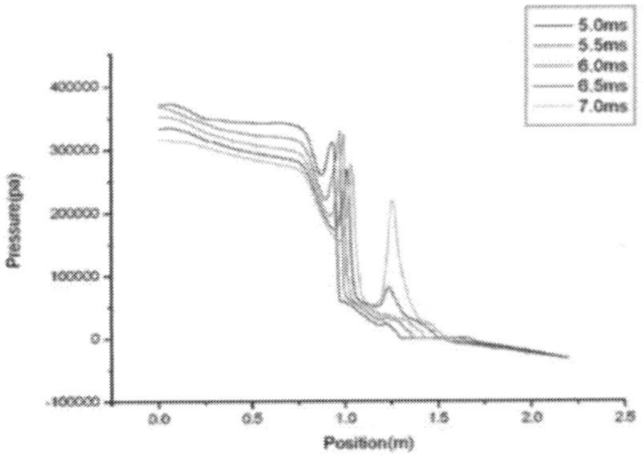

Fig.11 The sketch of maximal pressure at different

Figure 11 shows the sketch of maximal pressure at different time. And it shows in the figure that, the pressure has an obvious drop at the position of the porous material attached inside. This is in agreement with the analysis before, and we can see that the porous material has a good suppression and absorption effect for the flame wave.

5. Conclusion

The suppression effect of porous material for flame spreading in the tubes was studied by the experiment. The experiment and simulation results show that:

(1) The porous material has a good suppression and absorption effect for the flame wave, it is a good flame and explosion retardant material.

(2) The trend of overpressure and spread speed is synchronous. That is to say, they will decrease and increase simultaneously.

(3) When the porous material is relative short, there is an optimal thickness of the material for a best suppression effect.

6. Reference

[1] Teodoreczyk A,et al. Detonation Attenuation by Foams and Wire Meshes lining the Walls. Shock Waves,1994.

[2] Boquan L , Research on Flame Transmission and Acceleration Mechanism in Gas Explosion. Journal of China Coal Society,24(1):56-58,1999.

[3] Hengyang ZH, The explosion theory of gas and dust. Beijing institute of technology press,1996.
[4] Vanwingerden C J M, Zeeven J P. On the role of acoustically driven flame instabilities in vented gas explosions and their elimination. Combustion and Flame, 51:109-111,1983.
[5] Baoxing W, Zhenyan L. Study on the elimination measures of pressure peak driven by acoustically unstable combustion in vented gas explosions. Explosion and shock waves, 9(2)：130-136,1989.
[6] Baoxing W. Point for attention in the application of metal explosion reduction plate. Fire Technique and Products Information, 3:31-33,2004.
[7] Changing G, Zhigang C. Experimental Study on Gaseous Detonation Propagation in a Perforated Steel Plate Section. Journal of experimental mechanics, 15(4): 400-407, 2000.

Acknowledgements

All the research was supported by "the Fundamental Research Funds for the Central Universities"

1ST CCPS ASIA-PACIFIC CONFERENCE ON PROCESS SAFETY℠

CFD Based Two-Phase Dispersion Model of Flashing Liquefied Jet and Experiment Validation

Zhu Yuan
Centre for Offshore Engineering and Safety Technology
China University of Petroleum (Eastern China), No. 66 The Yangtze River West Road, Economic& Technological Development Zone, Qingdao, Shandong Province
zhy3323@163.com

Chen Guoming
Centre for Offshore Engineering and Safety Technology
China University of Petroleum (Eastern China), No. 66 The Yangtze River West Road, Economic& Technological Development Zone, Qingdao, Shandong Province
offshore@126.com

Prepared for Presentation at
American Institute of Chemical Engineers
2013 Spring Meeting
1st CCPS Asia-Pacific Conference on Process Safety
Qingdao, China
September 4 – 5, 2013

UNPUBLISHED

CFD Based Two-Phase Dispersion Model of Flashing Liquefied Jet and Experiment Validation

Zhu Yuan
Centre for Offshore Engineering and Safety Technology
China University of Petroleum (Eastern China), No. 66 The Yangtze River West Road, Economic& Technological Development Zone, Qingdao, Shandong Province
zhy3323@163.com

Chen Guoming
Centre for Offshore Engineering and Safety Technology
China University of Petroleum (Eastern China), No. 66 The Yangtze River West Road, Economic& Technological Development Zone, Qingdao, Shandong Province
offshore@126.com

Keywords: Two-Phase Dispersion, Flashing Liquefied Jet, CFD

Abstract

The prediction of the hazardous chemical release is important for the risk assessment on the process safety. Though there are various atmospheric dispersion models, it is still hard to analyze the toxic and/or flammable hazard of flashing liquefied jet due to the rapid phase transition from liquid to gas, which usually happens in the condition of the liquefied gas such as ammonia, LPG, LNG releasing from vessels and/or pipes. Based on the CFD method, we propose a two-phase dispersion model to analyze the flashing liquefied jet with the focus on the gas build-up, variation and distribution. The jet of flashing liquid is directly defined as flow inlet boundary according to the liquid discharge rate. Then the Lagrangian particles are introduced to simulate the movement of the liquid phase as discrete airborne spheres propelled through the gas, or as rectangular blocks that collectively form a thin liquid film on solid objects. Simultaneously, the gas phase deriving from the rapid phase transition is calculated based on the mass and energy equilibrium equations between the gas and the liquid. Then the gas dispersion in the atmosphere is determined by the commonly used species transport equations. And the large eddy simulation is worked for the treatment of turbulence influences. The model validation is conducted against the Lathen gas field experiments NO. EEC 86 - EEC 98, in which the liquefied propane is discharged with the rates of 170 – 770 g/s, wind speeds of 0.8 – 2.0 m/s, release duration of 105 - 330 s and nozzle diameter of 0.1 m in an open area. First of all, the atmospheric environment is simulated through the boundary conditions defined by the on-site temperature, wind speed/direction and atmospheric humidity. Then, the liquefied propane is introduced into the simulation and the model calculates the two-phase transition, liquid movement and gas dispersion automatically. The predicted downwind concentrations of propane are compared with the monitoring data recorded during the experiments. The maximum concentration, duration of gas existence, concentration trends are analyzed at the different monitoring points. Though the average wind speeds are substituted for the real speeds due to the lack of the on-site information. It is shows that the simulated results are quite in accordance with the observed values. The variation in the concentration fluctuations is quit the same. The simulated concentrations are within the acceptable range. The time when gas appears and departures at the monitoring points is satisfied with the experiments. In general, the model proposed is capable of determining the subsequent two-phase dispersion of flashing liquefied jet. And the next study will be concentrated on the model built-up and validation with the two-phase movement influenced by the obstacles which are not concerned in this study and experiments.

1. Introduction

The prediction of the hazardous chemical release is important for the risk assessment on the process safety. Though there are various atmospheric dispersion models including the plume mode, puff model, Pasquill-Gifford model et al.[1], it is still hard to analyze the toxic and/or flammable hazard of flashing liquefied jet due to the rapid phase transition from liquid to gas, which usually happens in the condition of the liquefied gas such as ammonia, LPG, LNG releasing from vessels and/or pipes.

According to the review of P. Bricard, L. Friedel, the hazardous dispersion with special attention to two-phase jet conditions have appeared only in the last decade in the literature. They could be divided into integral or multidimensional models and were insufficiently validated [2]. M.J. Pattison, R. Martini et. al built a model for the dispersion of two-phase releases[3, 4]. Recently, with the growing energy consumption, Liquefied Natural Gas (LNG), has become an important sources of energy. Lots of studies concerning the LNG release, which is a typical two-phase jet, were done in order to satisfy the risk assessment on the LNG fire or explosion. S.G. Giannissi, A.G. Venetsanos et al. conducted the numerical simulation of LNG dispersion over water surface under two-phase release conditions[5]. Filippo Gavelli, Edward Bullister et al. simulated the LNG dispersion after spilt into geometrically complex environments[6].

In this paper, we proposed a two-phase dispersion model to describe the flashing liquefied jet by using computational fluid dynamics (CFD). And the results of the model were verified by the field experiment.

2. The basis of the two-phase model

The two-phase model we proposed divides the flashing liquefied jet into the liquid movement, gas dispersion and the relationship between each other.

2.1 Liquid movement

Lagrangian particles are used to describe the movement of liquid droplets forming during the jet. The trajectory of an individual droplet is governed by the equation

$$\frac{d}{dt}(m_d \mathbf{u}_d) = m_d \mathbf{g} - \frac{1}{2}\rho C_D \pi r_d^2 (\mathbf{u}_d - \mathbf{u})|\mathbf{u}_d - \mathbf{u}| \qquad \text{[Eq. 1]}$$

where m_d is the mass of the droplet, $\mathbf{u}_d$ is the velocity of the droplet, $\mathbf{u}$ is the velocity of the gas, ρ is the density of the gas, C_D is the drag coefficient, r_d is the droplet radius, $\mathbf{g}$ is the acceleration of gravity, t is time.

The drag coefficient, C_D is a function of the local Reynolds number (based on droplet diameter)

$$C_D = \begin{cases} 24/\mathrm{Re}_D & \mathrm{Re}_D < 1 \\ 24(1+0.15\,\mathrm{Re}_D^{\ 0.687})/\mathrm{Re}_D & 1 < \mathrm{Re}_D < 1000 \\ 0.44 & 1000 < \mathrm{Re}_D \end{cases} \qquad \text{[Eq. 2]}$$

$$\mathrm{Re}_D = \frac{\rho\left|\mathbf{u}_d - \mathbf{u}\right| 2r_d}{\mu(T)}$$

where Re_D is the Reynolds number, and $\mu(T)$ is the dynamic viscosity of air at temperature T. When a liquid droplet hits a solid horizontal surface, it is assigned a random horizontal direction and moves at a fixed velocity until it reaches the edge. While attached to a surface, the droplet is assumed to form a thin film of liquid that transfers heat to the solid, and heat and mass to the gas.

2.2 Gas dispersion

The conservation equations for mass, momentum and energy for a Newtonian fluid are used to determine the gas dispersion:

Conservation of mass fractions of the individual gaseous species[7]

$$\frac{\partial}{\partial t}(\rho Y_i) + \nabla\cdot\rho Y_i \mathbf{u} = \nabla\cdot(\rho D_i \nabla Y_i) \qquad \text{[Eq. 3]}$$

$$\sum Y_i = 1 \qquad \text{[Eq. 4]}$$

Conservation of momentum

$$\frac{\partial}{\partial t}(\rho\mathbf{u}) + \nabla\cdot\rho\mathbf{u}\mathbf{u} + \nabla p = \rho\mathbf{g} + \nabla\cdot\tau_{ij} \qquad \text{[Eq. 5]}$$

Conservation of energy

$$\frac{\partial}{\partial t}(\rho h) + \nabla\cdot\rho h\mathbf{u} = \frac{\partial p}{\partial t} + \mathbf{u}\cdot\nabla p + \nabla\cdot k\nabla T + \sum_i \nabla\cdot h_i \rho D_i \nabla Y_i \qquad \text{[Eq. 6]}$$

Equation of state for a gas

$$p = \rho TR\sum_i \frac{Y_i}{M_i} \qquad \text{[Eq. 7]}$$

where Y_i is the mass fraction of species i, D_i is the diffusion coefficient for species i, p is pressure, τ_{ij} is stress tensor, h is the total enthalpy, k is thermal conductivity, T is temperature, R is the universal gas constant, and M_i is the molecular weight of species i.

Large eddy simulation (LES) associated with the mixture fraction transport equation is used as the theoretical basis for gas dispersion model. In LES, flow features that are larger than the grid size are resolved directly, whereas flow structures that are smaller than the grid size are modeled using a subgrid scale model.

2.3 Balance between the liquid and gas

The mass and energy transfer between the gas and the liquid can be described by the following equations

$$\frac{dm_l}{dt} = -Ah_m\rho(\mathrm{Y}_l - \mathrm{Y}_g) \qquad \text{[Eq. 8]}$$

$$m_l c_l \frac{dT_l}{dt} = Ah(\mathrm{T}_g - \mathrm{T}_l) = \mathrm{Ah}_s(\mathrm{T}_s - \mathrm{T}_l) + \dot{q}_r + \frac{dm_l}{dt}h_v \qquad \text{[Eq. 9]}$$

where, m_l is the mass of the liquid droplet, A is the surface area of the liquid droplet, h_m is the mass transfer coefficient, c_l is the liquid specific heat, h is the heat transfer coefficient between the liquid and the gas, h_s is the heat transfer coefficient between the liquid and the solid surface, $\dot{q}_r$ is the rate of radiative heating of the droplet, h_v is the latent heat of vaporization of the liquid, Y_g is the vapor mass fraction of the gas. The liquid equilibrium vapor mass fraction, Y_l is obtained from the Clausius-Clapeyron equation

$$X_l = \exp\left[\frac{h_v W_l}{R}\left(\frac{1}{T_b} - \frac{1}{T_l}\right)\right]$$
$$Y_l = \frac{X_l}{X_l(1 - \mathrm{W}_a / \mathrm{W}_l) + \mathrm{W}_a / W_l}$$ [Eq. 10]

where X_d is the equilibrium vapor volume fraction, W_l is the molecular weight of the evaporated liquid, W_a is the molecular weight of air, and T_b is the boiling temperature of the liquid.

The mass transfer coefficient, h_m is defined as

$$h_m = \frac{S_h D_{\mathrm{lg}}}{L}$$
$$S_h = \begin{cases} 2 + 0.6\,\mathrm{Re}_D^{\ 1/2} S_c^{\ 1/3} & \text{droplet} \\ 0.037\,\mathrm{Re}_L^{\ 4/5} S_c^{\ 1/3} & \text{film} \end{cases}$$ [Eq. 11]

where S_h is the Sherwood number, D_{lg} is the binary diffusion coefficient between the liquid vapor and the surrounding gas, L is a length scale equal to either the droplet diameter or 1 m for a surface film, Re_L is the Reynolds number based on the length scale L, and S_c is the Schmidt number.

The heat transfer coefficient, h is defined as:

$$h = \frac{N_u k}{L}$$
$$N_u = \begin{cases} 2 + 0.6\,\mathrm{Re}_D^{\ 1/2} P_r^{1/3} & \text{droplet} \\ 0.037\,\mathrm{Re}_L^{\ 4/5} P_r^{1/3} & \text{film} \end{cases}$$ [Eq. 12]

where N_u the Nusselt number, and P_r is the Prandtl number.

3. Description of Lathen experiments

The Lathen gas field experiments used a continuous source of liquefied propane, with the purpose of understanding the accidental gas release. The results of EEC 88, EEC 91, EEC 95, EEC 98 were chosen to do model validation. In these experiments, the liquefied propane was released under the condition of the nozzle diameter of 0.1 m, the height of 2.1 m, the wind speeds of 0.8 – 2.0 m/s, the exit temperature of -42 ℃ (the same as the boiling temperature of liquefied propane), the exit pressure of the atmospheric pressure, the release rates between about 170 – 770 g/s, and the release duration between 105 – 330 s.

All obstacles were removed during these experiments. The gas concentration distribution was measured with catalytic gas sensors arranged along the downwind direction. And the wind speeds and directions were recorded simultaneously. The experimental layout of the Lathen experiments is showed in Fig. 1.

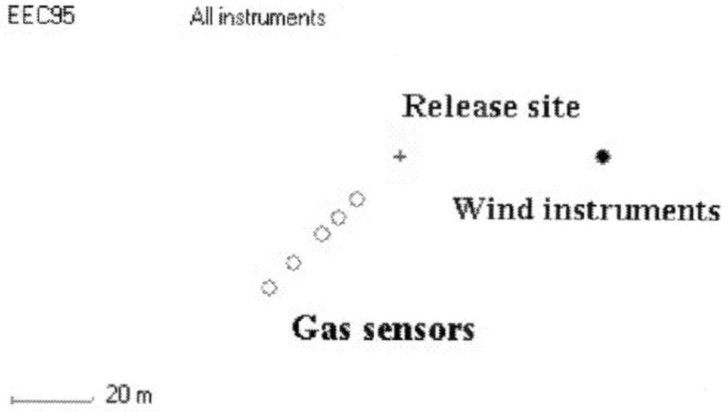

Fig. 1 The experimental layout of the Lathen experiments

4. Model validation

4.1 Simulation of the experiments

For every experiment, first of all, the geometry is built according to the actual conditions. Then, the atmospheric environment is simulated through the boundary conditions defined by the on-site temperature, wind speed/direction and atmospheric humidity. Due to the lack of the on-site information, the average values are substituted for the real values. Finally, the liquefied propane is introduced into the simulation and the model calculates the two-phase transition, liquid movement and gas dispersion automatically.

The simulation model is showed in Fig. 2.

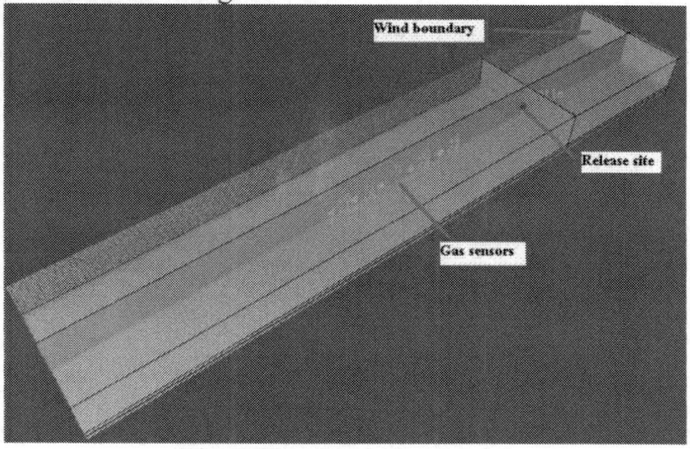

Fig. 2 The simulation model

4.2 The movement of the propane

Both the liquid and gaseous propane is predicted by the two-phase model. For the liquid propane, it is evaporated into gaseous propane immediately due to the environmental temperature that is much higher than the exit temperature and the boiling temperature of liquefied propane. The propane droplets fall under gravity, then hit the ground and spread out, and evaporate into gas eventually. For the gaseous propane, it is transited from the liquid propane and disperses along the wind direction.

The movement of the propane is showed in Fig. 3.

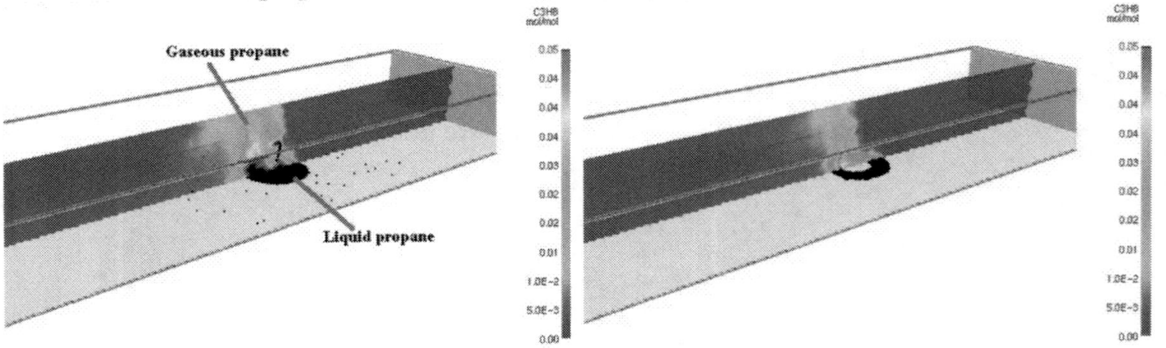

Fig. 3 The movement of the propane

4.3 The results analysis on the gaseous propane

Taking EEC 91 as example, the arrangement and labels of the sensors (according to the field experiment) are showed in Fig. 4.

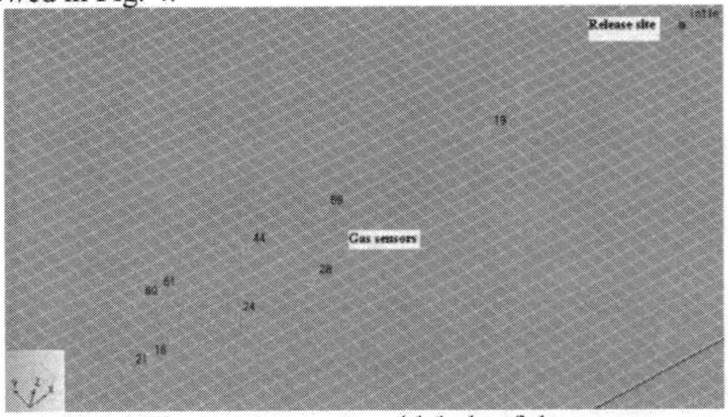

Fig. 4 The arrangement and labels of the sensors

The predicted downwind concentrations of propane against the monitoring data recorded during the experiments are showed in Fig. 5.

The simulated maximum concentration is lower than the experimental value of 13.24 %. Totally, the simulated time when gaseous propane appears at the monitoring points is earlier than that of the experiments, and for the departure time, the simulated time is almost the same as or a little earlier than the experimental values. Except for the No. 19 sensors, the changes of simulated concentrations against the time are in accordance with the trends of the experimental values. Considering that the average values of environmental parameters such as wind speed are substituted for the real values. And it is hard to exactly estimate the appearance and departure time of the gaseous propane due to the delay time and the accuracy of the sensors.

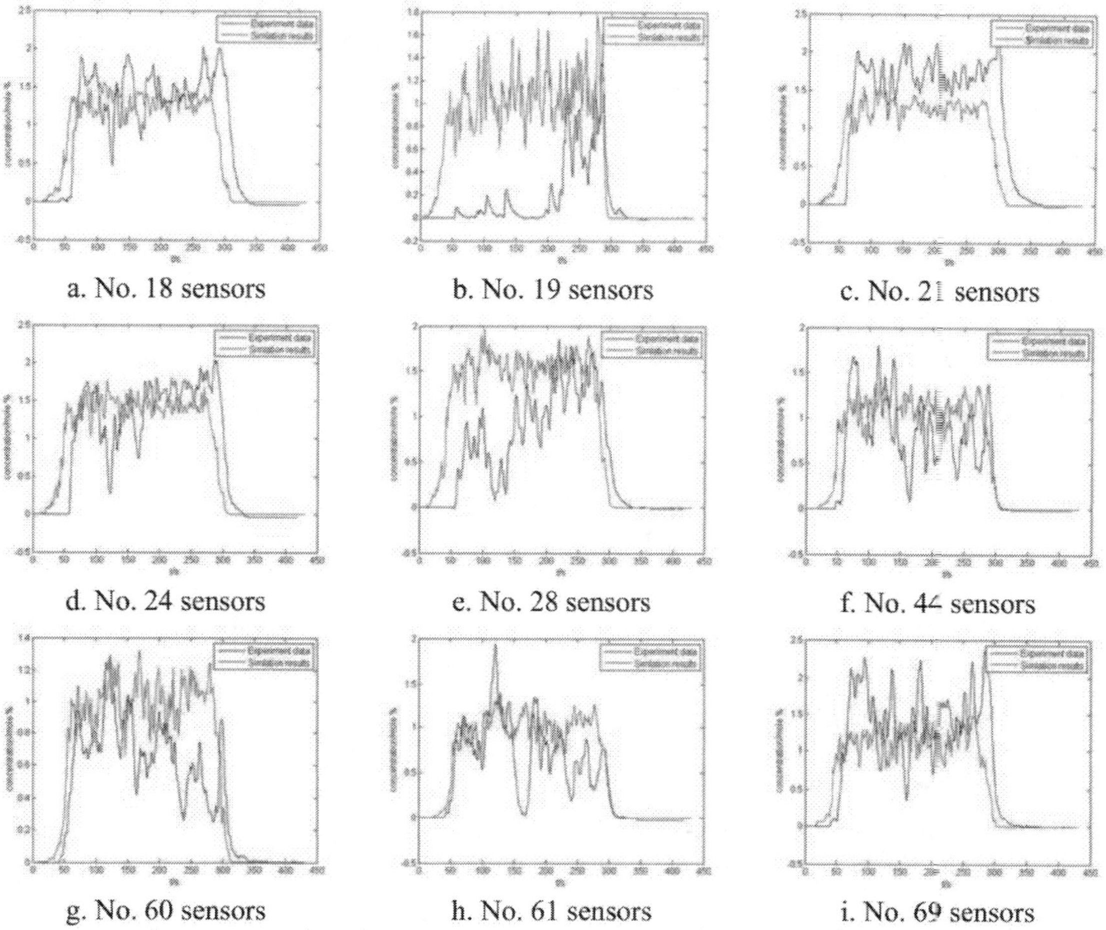

a. No. 18 sensors b. No. 19 sensors c. No. 21 sensors

d. No. 24 sensors e. No. 28 sensors f. No. 44 sensors

g. No. 60 sensors h. No. 61 sensors i. No. 69 sensors

Fig. 5 The predicted propane concentrations against the monitoring data

For other experiments, it can be concluded that the simulated concentrations are within the acceptable range. Some results are showed in Fig. 6.

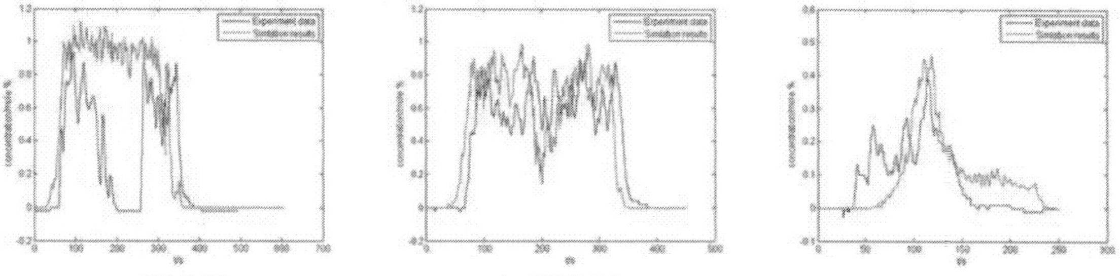

a. EEC 88 b. EEC 95 c. EEC 98

Fig. 6 The propane concentrations of the simulations and the experiments

5. Conclusion

A two-phase dispersion model based on CFD is built to simulate the jet of flashing liquid. By comparing with the Lathen experiments, the model is capable of analyzing the two-phase transition, liquid movement and gaseous dispersion. And the next study will be concentrated on the model built-up and validation with the two-phase movement influenced by the obstacles which are not concerned in this study and experiments.

Acknowledgement

This research is supported by "National Key Technology Research and Development Program of the Ministry of Science and Technology of China" (2011BAK03B08), "the Shandong Provincial Natural Science Foundation, China" (ZR2011EL048), "the Fundamental Research Funds for the Central Universities" (12CX04060A).

6. References

[1] Daniel A.Crowl, Joseph F.Louvar, "Chemical process safety: fundamental with applications," Prentice-Hall, 2002.

[2] P. Bricard, L. Friedel, "Two-phase jet dispersion," *Journal of Hazardous Materials*, Volume 59, Issues 2–3, 1998.

[3] M.J. Pattison, R. Martini, S. Banerjee, G.F. Hewitt, "Modelling of Dispersion of Two-Phase Releases: Part 1—Conservation Equations and Closure Relationships," *Process Safety and Environmental Protection*, Volume 76, Issue 1, 1998.

[4] M.J. Pattison, R. Martini, S. Banerjee, "Modelling of Dispersion of Two-Phase Releases: Part 2—Numerical Solution Scheme and Validation," *Process Safety and Environmental Protection*, Volume 76, Issue 1, 1998.

[5] S.G. Giannissi, A.G. Venetsanos, N. Markatos, J.G. Bartzisc, "Numerical simulation of LNG dispersion under two-phase release conditions," *Journal of Loss Prevention in the Process Industries*, Volume 26, Issue 1, 2013.

[6] Filippo Gavelli, Edward Bullister, Harri Kytomaa, "Application of CFD (Fluent) to LNG spills into geometrically complex environments," *Journal of Hazardous Materials*, Volume 159, Issue 1, 2008.

[7] LEI Ting, CHEN Guo-ming, ZHU Yuan, LIU De-xu, "Analysis on behavior of jet fire from separator in natural gas station based simulation by FDS," *Journal of Safety and Envirommnt*, Volume 10, Issue 4, 2010.

1ST CCPS
ASIA-PACIFIC
CONFERENCE
ON PROCESS SAFETY℠

Simulation and Full-scale Experiments of Tunnel Fire with Shafts on Roof

Yan Fu WANG[1]
China University of Petroleum
Qing Dao 266580, China
wangyanfu@upc.edu.cn

Li ZHANG[2], Jun Cheng JIANG[3]
[2]Jiangsu Province of safety science and technology; [3]Nanjing University of Technology, School of Urban Construction and Safety
[2]Nanjing 210042, China; [3]Nanjing 210009, China
[2]llizhang@163.com; [3]jcjiang@njut.edu.cn

[This work was supported by "the Fundamental Research Funds for the Central Universities" (27R1203008A) and Shandong Province Natural Science Fund Project (Number: ZR2012EEM023)]

Prepared for Presentation at
1st CCPS Asia-Pacific Conference on Process Safety
Qing Dao, China
September 4 –September 5, 2013

UNPUBLISHED

Simulation and Full-scale Experiments of Tunnel Fire with Shafts on Roof

Yan Fu WANG[1]
China University of Petroleum
Qing Dao 266580, China
wangyanfu@upc.edu.cn

Li ZHANG[2], Jun Cheng JIANG[3]
[2]Jiangsu Province of safety science and technology; [3]Nanjing University of Technology, School of Urban Construction and Safety

Keywords: Simulation; Fire experiment; Natural ventilation; Mathematical model

Abstract

Based on CFD theory and the knowledge gained from the full-scale experiments a mathematical model is built to study the characteristic of fire in tunnel which adopts natural ventilation systems. The system consists of groups of openings on tunnel roof, and vertical shafts on each opening. The smoke temperature, smoke concentration, smoke spread and sedimentation laws, are, firstly, simulated using FLUENT software under full-scale experiments' condition. Secondly, simulated results are compared with the data gathered from full-scale experiments. Lastly the conclusion is drawn that the presented mathematical model is applicable for researching the tunnel fire with natural ventilation systems. It provides theoretical support to adopt this innovative tunnel with vertical shaft on roof in future.

1. Physical model of the tunnel with shafts

Due to the symmetry of the tunnel framework, the single hole of this tunnel is researched to save time when the physical model is built [1-2]. The built physical model is 320m long, 12.3m wide and 5.75m high. The underground area is 240m long. All the shafts are identical. The dimensions are 3.6 m long, 2.6m wide and 4.65m high. Each shaft is separated by 0.8m thick concrete plate. Every four shafts connect and form a group. Every group has an interval of 8.8 meters [3-4]. The fire source is located below one of the openings. The detailed model is illustrated in Fig.1

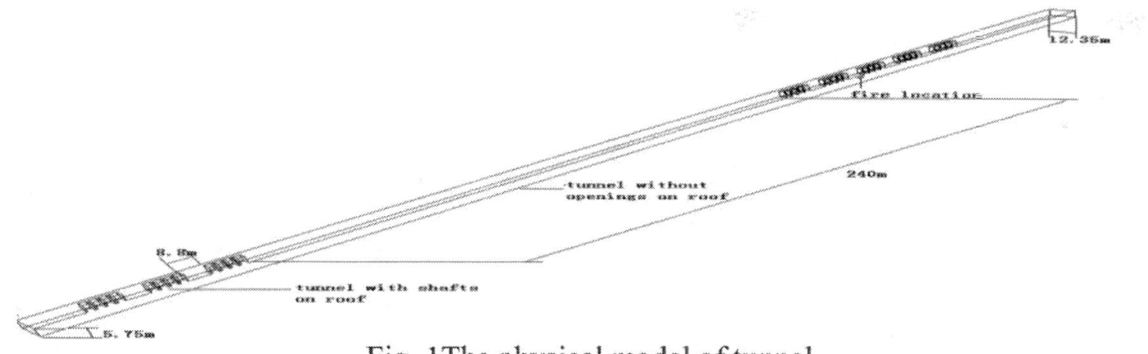

Fig. 1The physical model of tunnel

2. Mathematical model

The solution flowchart of field simulation is shown in Fig.2.

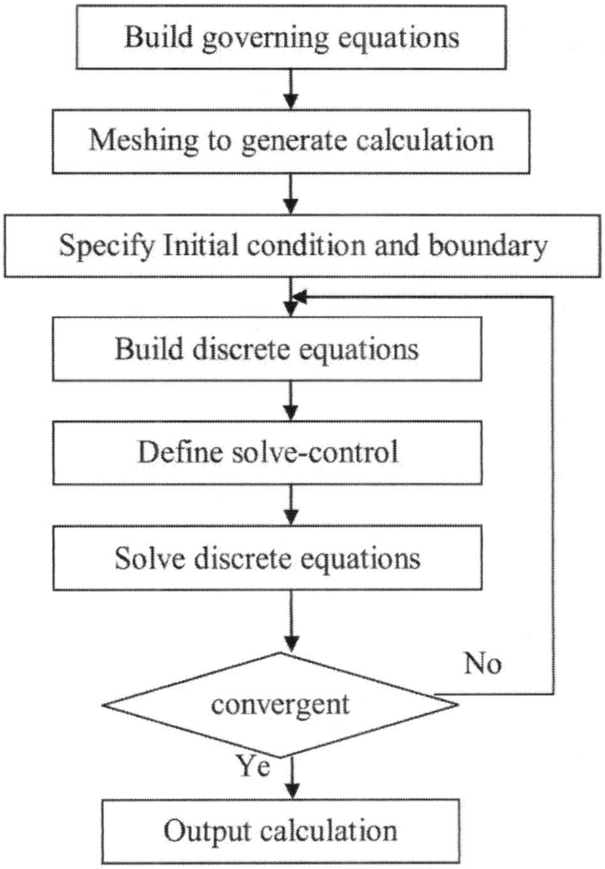

Fig.2 Flow diagram of solution

2.1 Build governing equations

In order to use the same program to analyze all control methods and solve all control equations, a general equation for every governing equation is built [6]:

$$\frac{\partial}{\partial t}(\rho\phi)+div(\rho u\phi)=div(\Gamma grad\phi)+S_{\phi} \tag{1}$$

Where, ϕ: General dependent variable, ρ: density, u: velocity vector, Γ: diffusion coefficient, S_ϕ: source term

In equation (1), from left to right, the four terms are separately: unsteady term, convection term, diffusion coefficient and source term. The detailed equations are shown in Table 1.

Table. 1 Coupled equations for control of 3D field model for smoke movement

Equation	ϕ	Γ	S_ϕ
Continuity equation	*1*	0	0
x- momentum equation	u	μ	$-\frac{\partial p}{\partial x}+\frac{\partial}{\partial x}(\mu\frac{\partial u}{\partial x})+\frac{\partial}{\partial y}(\mu\frac{\partial v}{\partial x})+\frac{\partial}{\partial z}(\mu\frac{\partial w}{\partial x})$
y- momentum equation	v	μ	$-\frac{\partial p}{\partial y}+\frac{\partial}{\partial x}(\mu\frac{\partial u}{\partial y})+\frac{\partial}{\partial y}(\mu\frac{\partial v}{\partial y})+\frac{\partial}{\partial z}(\mu\frac{\partial w}{\partial y})$
z- momentum equation	w	μ	$-\frac{\partial p}{\partial z}+\frac{\partial}{\partial x}(\mu\frac{\partial u}{\partial z})+\frac{\partial}{\partial y}(\mu\frac{\partial v}{\partial z})+\frac{\partial}{\partial z}(\mu\frac{\partial w}{\partial z})-\rho g$
enthalpy equation	h	$\frac{\mu}{\sigma_h}$	q_s-q_r
composition equation	C_s	$\frac{\mu}{\sigma_c}$	w_s
turbulent kinetic energy equation	k	$\frac{\mu}{\sigma_k}$	G- $\rho\varepsilon$
turbulent dissipation rate equation	ε	$\frac{\mu}{\sigma_\varepsilon}$	$\frac{\varepsilon}{k}(C_1G-C_2\rho\varepsilon)$
$\mu=\mu_l+\mu_T$		$\mu_T=C_\mu\rho k^2/\varepsilon$	$G=G_k+G_b=\mu_T(\frac{\partial u_i}{\partial x_j}+\frac{\partial u_j}{\partial x_i})\frac{\partial u_j}{\partial x_i}-\mu_t\frac{g_i}{\rho}\frac{\partial\rho}{\partial x_i}$
$C_u=0.09$ $C_1=1.44$	$C_2=1.92$	$\sigma_k=1.0$	$\sigma_\varepsilon=1.3$ $\sigma_h=0.9$ $\sigma_c=0.9$

q_s: Combustion heat flux, q_r: Radiation source term, w_s: Smoke quantity

2.2 Meshing

sub-section text Fire source adopts Hex mesh, type of which is Map. The tunnel adopts Tet/Hybrid mesh, type of which is TGrid. Taken into the consideration of strong turbulence flow of fire plume, the mesh density near fire source is refined to 10 times of it of other parts.

2.3 Specify boundary conditions and initial conditions

(1) Inlet conditions

The pressure inlet boundary condition is adopted for inlets and shaft openings. The parameters need to be inputted are Gauge Total Pressure, Supersonic/Initial Gauge Pressure, Direction Specification Method and Turbulence Specification Method. It is important to accurately describe the turbulence of the boundaries in the turbulent RNG k-ε model. The temperature of back-flow through shafts is environmental temperature (276K). The static pressure of incompressible fluid can be calculated as equation (2):

$$p_0 = p_s + \rho|v|^2 \qquad (2)$$

Where, P_0: General pressure, Ps: Static pressure

Turbulence intensity I can be calculated as:

$$I = \frac{u'}{u_{avg}} = 0.16(\mathrm{Re})^{-\frac{1}{8}} \qquad (3)$$

(2) Outlet conditions

The pressure outlet boundary condition is adopted for the outlet. The parameters need to be inputted are gauge pressure, backflow conditions, general temperature and turbulence parameters. Turbulence energy and energy dissipation rate can be calculated using equation:

$$\mathrm{k} = \frac{3}{2}(u_{avg}I)^2 \qquad (4)$$

$$\varepsilon = C_\mu^{\frac{3}{4}}\frac{k^{\frac{3}{2}}}{l} \qquad (5)$$

(3)Solid wall conditions

All velocity component (u, v, w) adopt no-slip boundary conditions. For energy equation, the conduction of wall is dealt with as one dimensional heat transfer and the heat flux of wall is 0, namely adiabatic wall. The roughness of wall influences the resistance, heat transfer and mass transportation. In order to reflect the influence, two parameters should be defined: roughness height K_s and roughness constant CK_s. In this paper, K_s equal to 0.003 and CK_s equals to 0.5.

(4)Fire source

An independent cube is set up as body fire when the physical model is built using Gambit. When the boundary conditions are defined in FLUENT software, the "Source" item is activated on the fluid region panel. The energy source item, mass flow and mass of smoke component are defined with User Defined Function for fire source. To simplify the computational processes and reduce the amount of calculation, VHS model is chosen as combustion model in this paper. VHS model can properly predict the diffusion range of smoke and the temperature field. The fire source is set at 7.5MW. The heat release rate of the fire can be described as [7]:

$$\begin{cases} \mathrm{t} < 326\text{时} & Q = \alpha t^2 \\ \mathrm{t} \geq 326\text{时} & Q = 7500\mathrm{kw} \end{cases} \qquad (6)$$

Where, α——Fire growth coefficient (kw/s^2), α=0.047;

t——Time after ignition(s)

(5)Radiation model

Heat transfer by radiation is evident in the process of heat transfer for tunnel fire, so the radiation model needs to be considered. Kown from the experiments, tunnel fires always emit heavy smoke, and absorption coefficient is larger than 0.5 (a>0.5). The characteristic length of tunnel main structure L is 7.65838 (aL>3.82919). So Rosseland model is chosen to calculate radiation heat transfer when simulating tunnel fire in this paper.

(6)Initial conditions

According to the full-scale experiments data and CFD related basic theory, the initial conditions are set: Operating pressure is one atmosphere; Gauge pressure is calculated according to equation (2) under different conditions; Temperature is the atmospheric temperature in tunnel before the fire; X velocity is the average speed of natural wind before the fire; the mass fraction of smoke is 0; Turbulence Kinetic Energy and kinetic energy dissipation rate can be approximately calculated according to equation (4) and equation (5).

3. Simulation results

3.1 Smoke temperature of vertical center plane of the tunnel

Smoke temperature of vertical center plane varies with time as shown from Fig.5 to Fig.14.

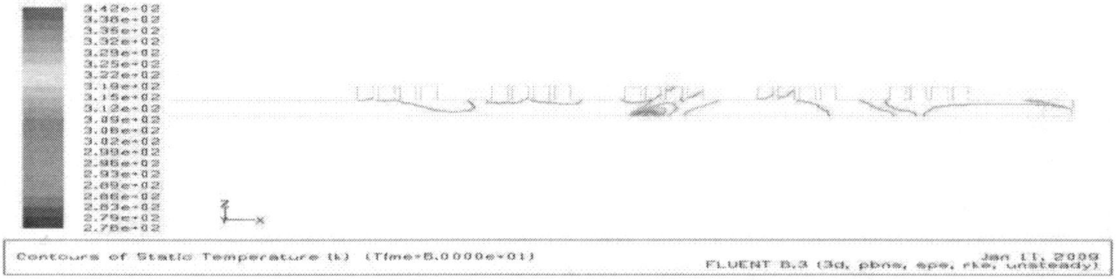

Fig. 5 Smoke temperature map (60s)

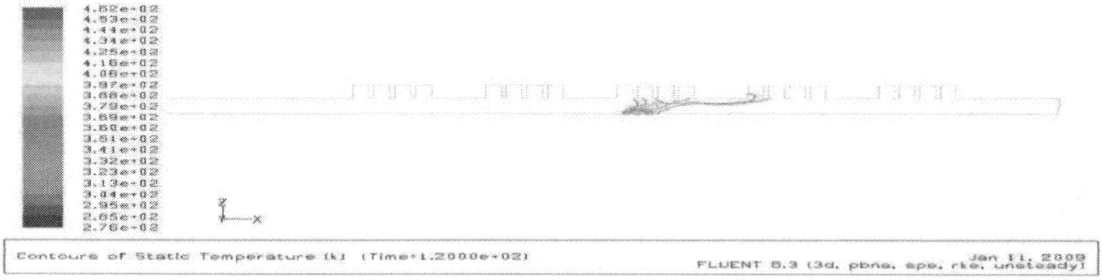

Fig. 6 Smoke temperature map (120s)

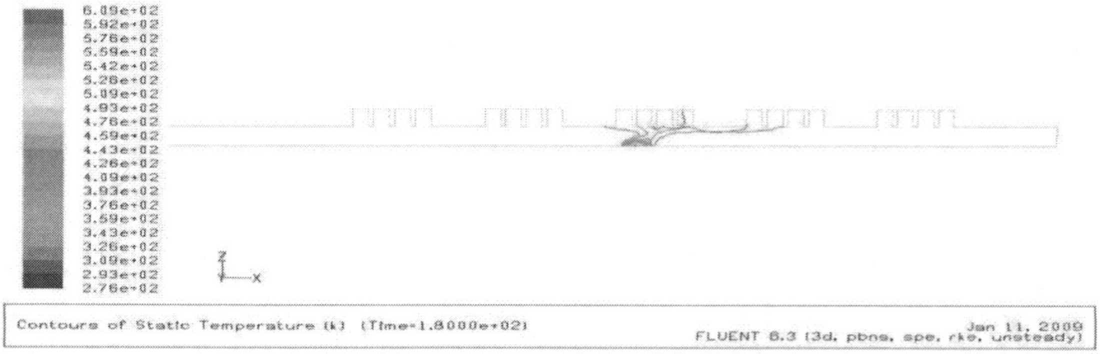

Fig. 7 Smoke temperature map (180s)

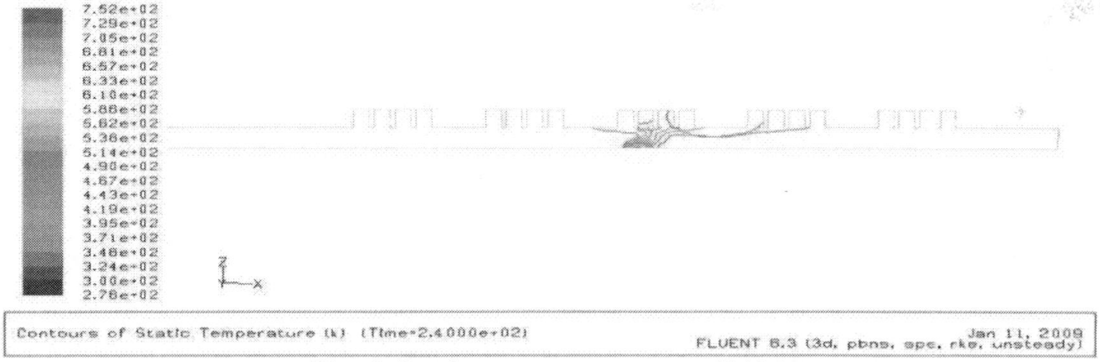

Fig. 8 Smoke temperature map (240s)

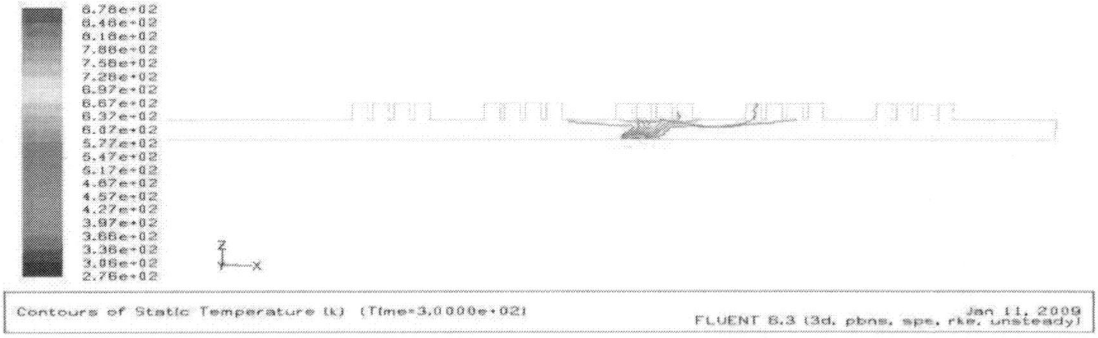

Fig. 9 Smoke temperature map (300s)

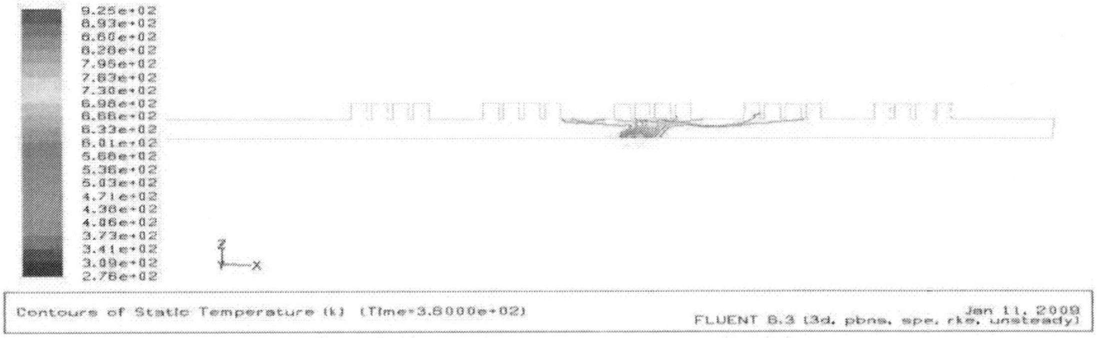

Fig. 10 Smoke temperature map (360s)

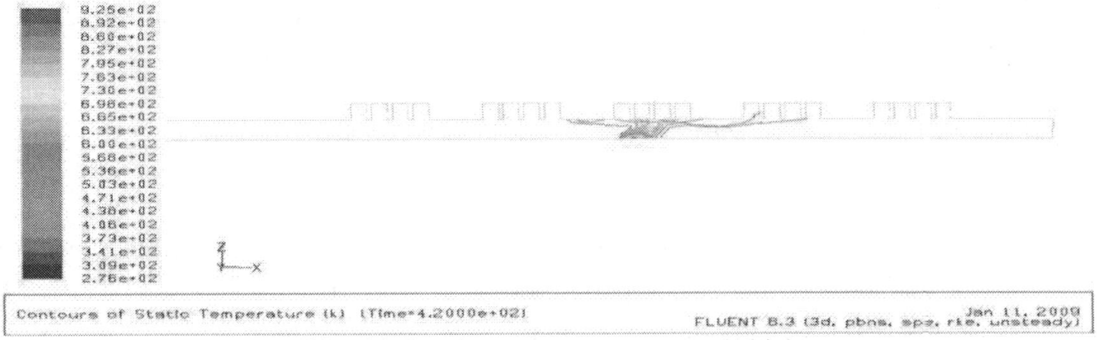

Fig. 11 Smoke temperature map (420s)

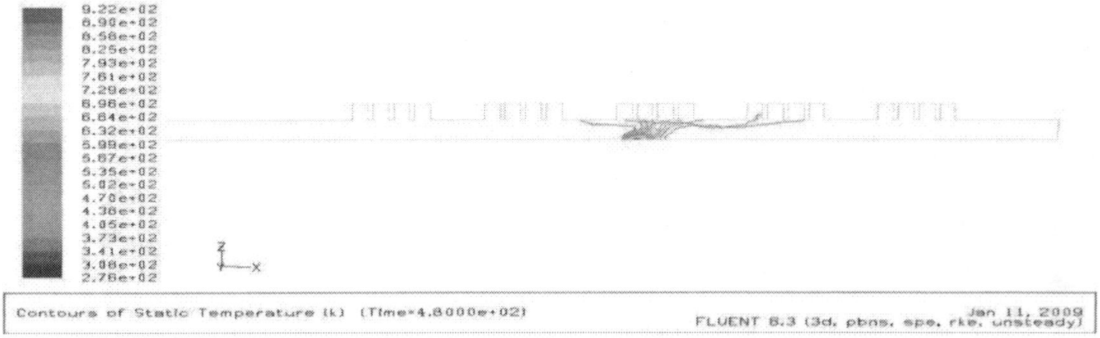

Fig. 12 Smoke temperature map (480s)

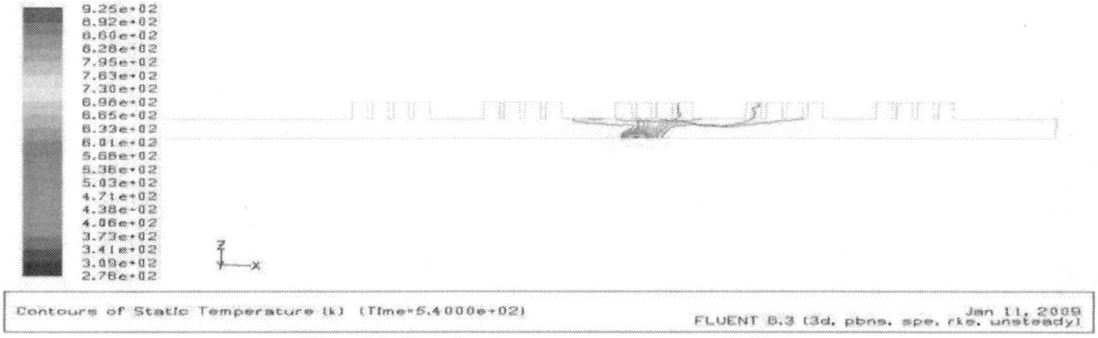

Fig. 13 Smoke temperature map (540s)

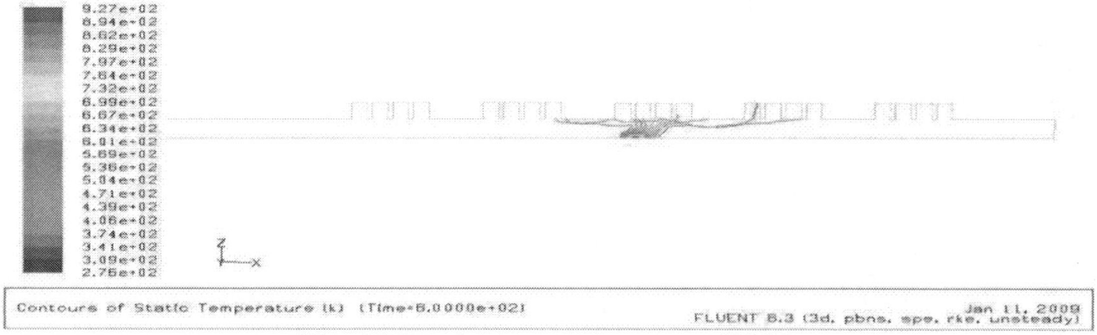

Fig. 14 Smoke temperature map (600s)

From the simulation results, it can be seen that

(1) The smoke temperature near fire source is higher: the highest temperature of fire source achieves 927K. But the smoke temperature drops rapidly with the increase of the distance away from fire source.

(2) The smoke temperature changes reflects the process of smoke diffusion: the smoke hits the tunnel ceiling and a small amount of smoke emits from the shafts at 60s; the smoke reaches the first shafts group downstream (the direction of natural wind is from left to right) at 120s and almost all smoke exhausts from the first group shafts; There are also some amount of smoke flows backwards at 240s; it reaches the first shafts group upstream at 360s.the smoke is well controlled between the first group shafts upstream and downstream in the first 10 minutes.

3.2 Cross-sectional concentration distribution of smoke near fire source

The inhalation of toxic gases in fire accidents is the main cause of death, so smoke concentration distribution laws are deserved to research. Cross-sectional concentration of smoke at fire center, upstream 5m,15m and downstream 5m,15m are distributed as Fig.15 to Fig.24.

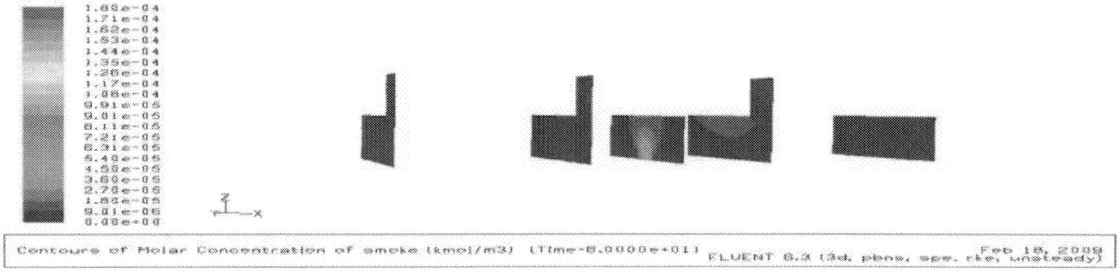

Fig. 15 Cross-sectional map of smoke concentration (60s)

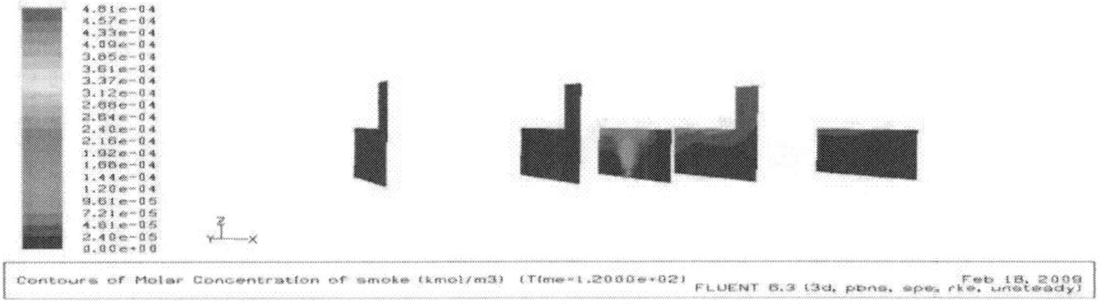

Fig. 16 Cross-sectional map of smoke concentration (120s)

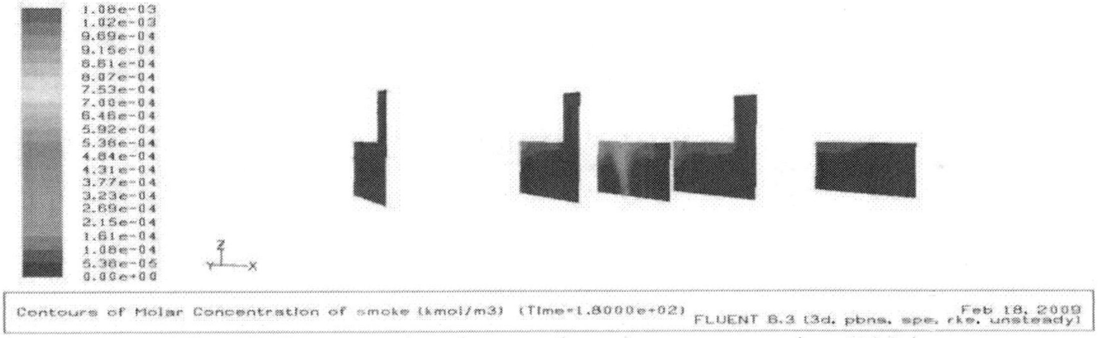

Fig. 17 Cross-sectional map of smoke concentration (180s)

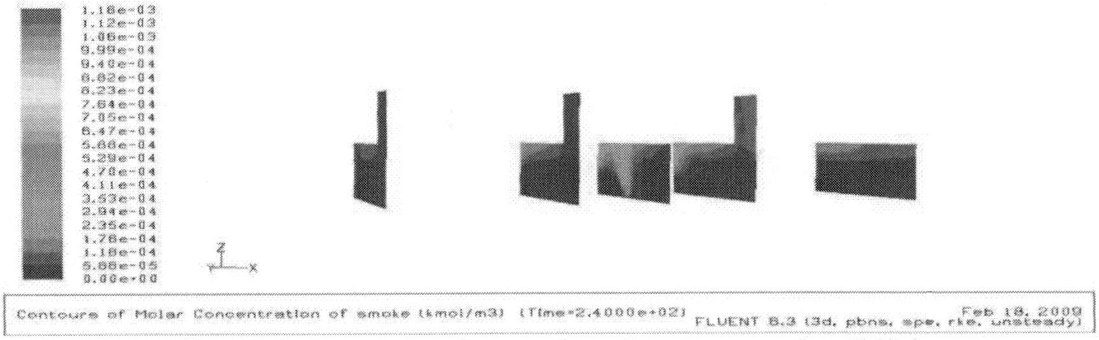

Fig. 18 Cross-sectional map of smoke concentration (240s)

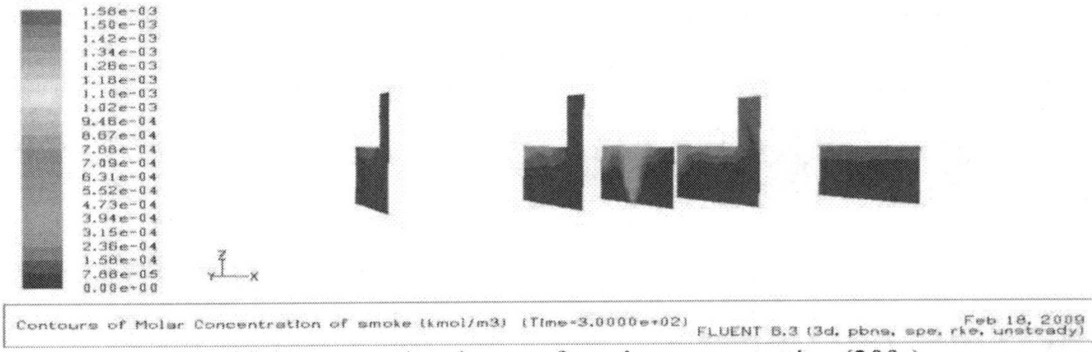

Fig.19 Cross-sectional map of smoke concentration (300s)

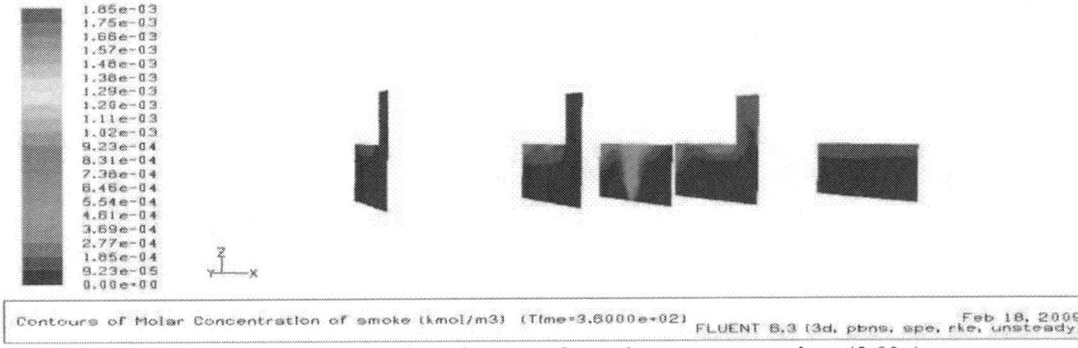

Fig. 20 Cross-sectional map of smoke concentration (360s)

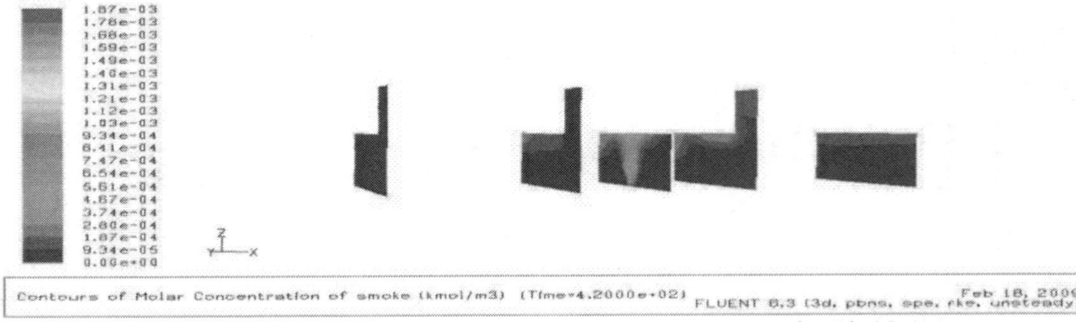

Fig. 21 Cross-sectional map of smoke concentration (420s)

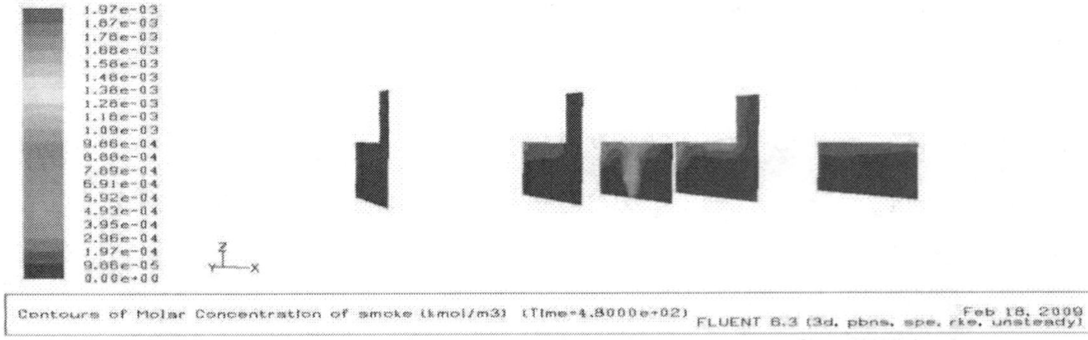

Fig. 22 Cross-sectional map of smoke concentration (480s)

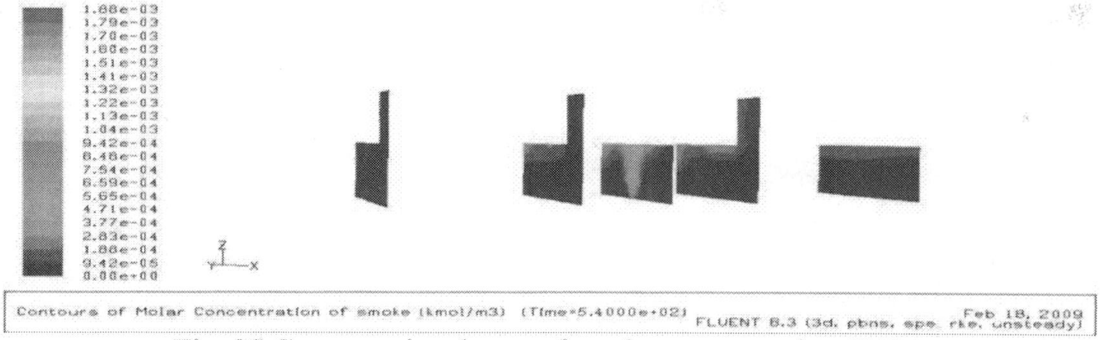

Fig. 23 Cross-sectional map of smoke concentration (540s)

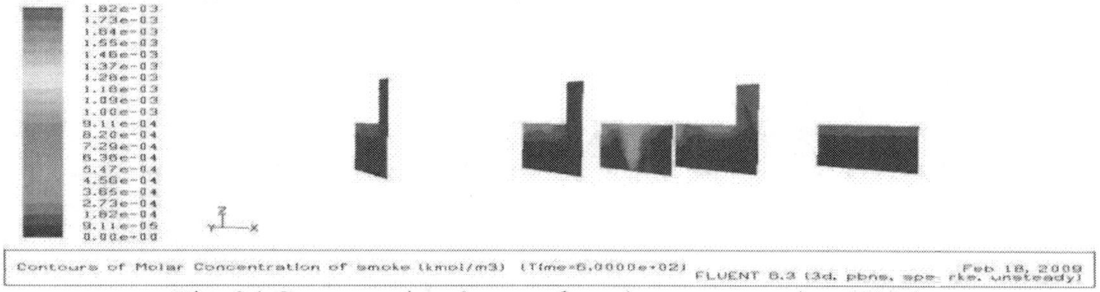

Fig. 24 Cross-sectional map of smoke concentration (600s)

From the simulation results, it can be seen that

(1) there is already a small amount of smoke which evacuates from the shafts at 5m downstream at 60s; there are considerable amount of smoke exhausts from these shafts when reach 120s; a small proportion of smoke reaches 5m and 15m upstream at 180s and 240s respectively; almost all the smoke exhausted from the shafts; the smoke layer keeps at around 4.5m high.

(2) Near fire source, the highest smoke concentration level reaches 1.97mol/m3, but the smoke concentration level lowers rapidly with the increase of distance away from fire source and height away from roof. There is no smoke in most space of tunnel.

(3) The smoke layer keeps at around 4to 5m high because a great amount of smoke evacuates from the shafts.

4. Comparison between simulation results and full-scale experimental data

In this section, the simulation results and full-scale experimental data are compared.

4.1 Longitudinal variation of smoke maximum temperature under tunnel ceiling with the distance away from fire source

Longitudinal variation laws of smoke maximum temperature under tunnel ceiling with the distance away from fire source are shown as Fig.25 and Fig.26.

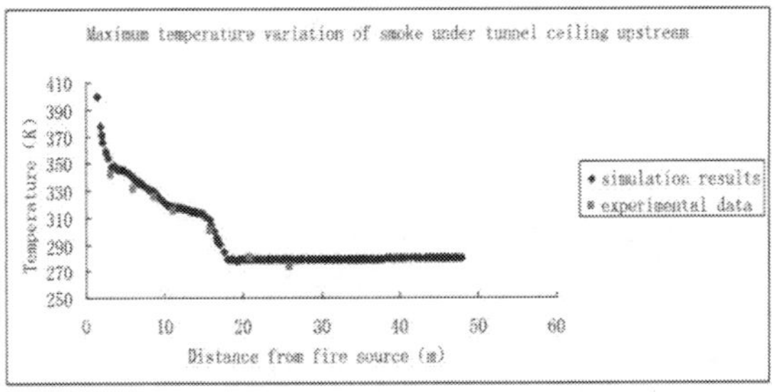

Fig.25 Maximum temperature variation of smoke under tunnel ceiling upstream

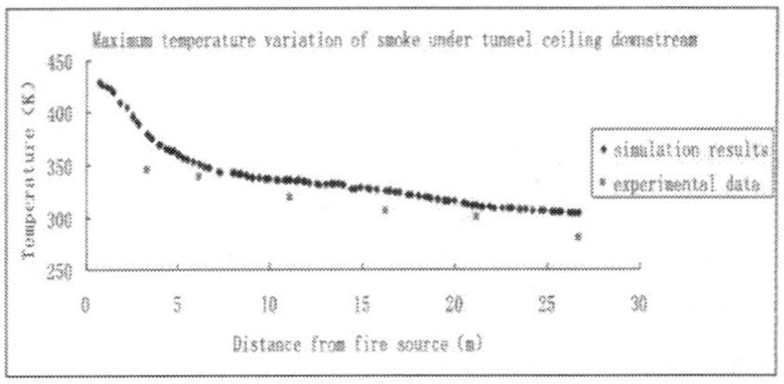

Fig.26 Maximum temperature variation of smoke under tunnel ceiling downstream

(1) Variation laws of simulation results and experimental data are identical: smoke temperature under tunnel ceiling near the fire source is higher but it decreases rapidly. The smoke temperature keeps at around 300K-350K, which cannot threaten the structure of tunnel.

(2) Simulation results agree well with experimental data, which shows that this numerical model can be used to simulate the maximum temperature under tunnel ceiling.

(3) Near fire source, simulation results are higher than experimental data, which is due to the simplification of the fire source. The fire model needs to be improved.

4.2 Longitudinal variation laws of smoke maximum temperature at 1.8m high

Longitudinal variation laws of smoke maximum temperature at 1.8m high are shown as Fig. 27 and Fig. 28.

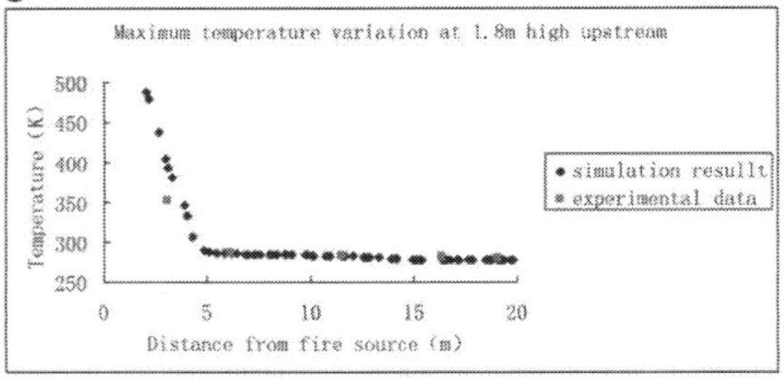

Fig. 27 Maximum temperature variation at 1.8m high upstream

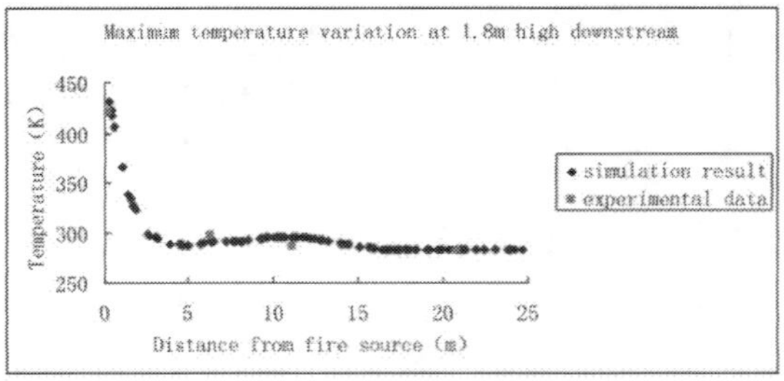

Fig. 28 Maximum temperature variation at 1.8m high downstream

(1) Variation laws of simulation results and experimental data are identical: smoke temperature at 1.8m high near fire source is higher, but it rapidly decreases to environmental temperature.

(2) Under the tunnel with shafts, simulation results agree well with experimental data, which illustrates that this numerical model can be used to simulate the maximum temperature at 1.8m high from the ground.

4.3 Longitudinal movement laws of smoke

The smoke longitudinal movements upstream and downstream are respectively shown in Fig. 29 and Fig. 30.

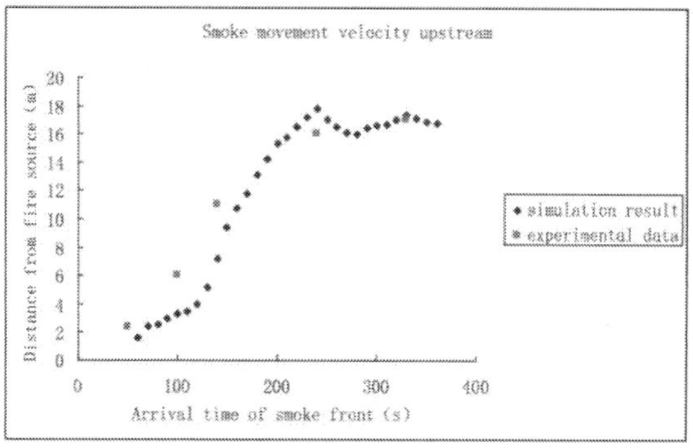

Fig. 29 Smoke movement velocity upstream

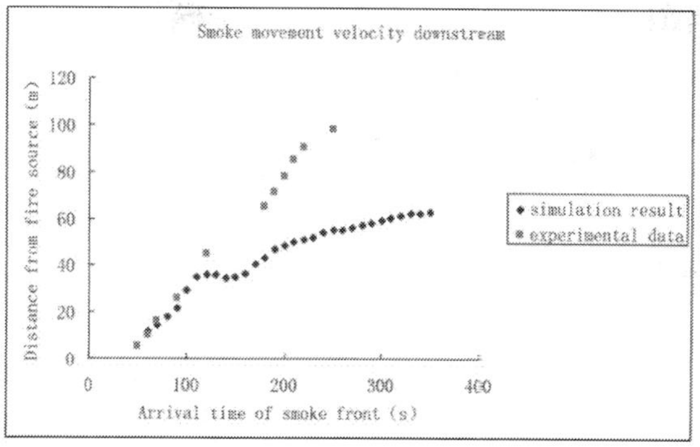

Fig. 30 Smoke movement velocity downstream

(1) Variation laws of simulation results and experimental data are identical: near the fire source, smoke at downstream spreads faster than it does at upstream. Therefore, persons at downstream need to shorten their response time and have to evacuate more rapidly.

(2) Simulation results of smoke spread velocity for upstream are very close to experimental data. For downstream, simulation results near the fire source are consistent with experimental data, but differ considerably after 40m away from the fire source. This is mainly because the temperature difference between the smoke and the air becomes smaller with the distance from fire source increases which results in a smaller thermal driving force in experiment; The smoke stops moving and even back flow because the disturbance of external natural wind from shafts become comparatively heavier; it is difficult for testers to catch exactly the arrival time of smoke.

4.4 Sedimentation laws of smoke layer

Sedimentation laws of smoke layer upstream and downstream are respectively shown in Fig. 31 and Fig. 32.

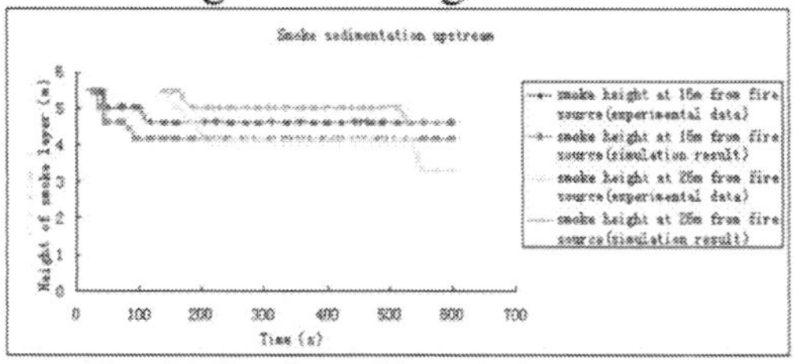

Fig. 31 Smoke sedimentation upstream

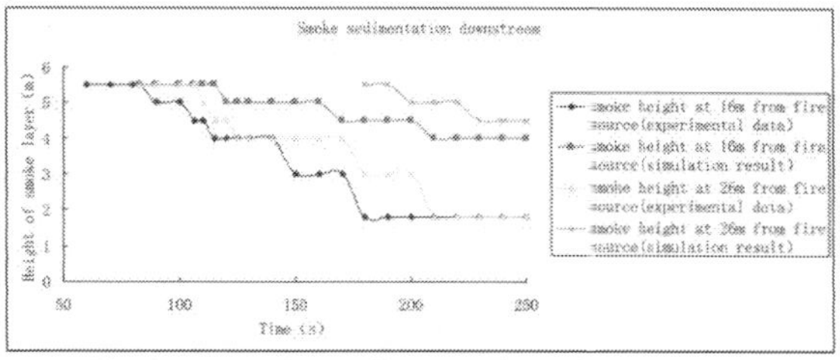

Fig. 32 Smoke sedimentation downstream

(1) Smoke layer from simulation are higher than it from experiment. This is mainly because it is difficult for testers to judge exactly the height of smoke layer before a noticeable amount of smoke reaches that level. Moreover, it is also subject to tester's judgment.

(2) From the simulation results, smoke down-draught and flow-backward are observed, which agree well with experimental data. It is resulted from the influence of shafts on roof. However, the smoke layering in simulation results is more obvious than that in experiment, which is mainly due to the external air turbulence in the experiment.

(3) From the experimental data, it can be seen that most smoke layer does not

subside to 1.8m high. It means the smoke will not threaten people while evacuating. However, there is some amount of smoke at downstream subsiding to 1.8m high under the effect of downdraught from shafts in the later stages of fire. This is bad for evacuation.

5. Conclusions

From the comparison and analysis, it can be seen that simulation results agree well with experimental data and the errors are acceptable. This illustrate that the built numerical model can be used to research the smoke temperature, smoke spreading and smoke sedimentation laws of fire in tunnels with shafts on roof. Smoke temperature near the fire source is higher, but it rapidly decreases to environmental temperature, which cannot threaten the safety of tunnel structure and personnel. Smoke spread figures illustrate that shafts are very useful for preventing smoke from spreading. From the simulation results, smoke down-draught and flow-backward under the influence of shafts can be observed, which agrees well with experimental data. If the model is chosen properly and boundary conditions are defined accurately, it is feasible to simulate tunnel fire with shafts using Computational Fluid Dynamics technique.

6. References

[1] Woodburn P J, Britter R E., "CFD simulation of a tunnel fire - Part 1," Fire Safety Journal, Volume 26, NO. 1, 1996

[2] Woodburn P J, Britter R E., "CFD simulation of a tunnel fire - Part 2," *Fire Safety Journal*, Volume 26, NO. 1, 1996

[3] WANG Yanfu, JIANG Juncheng, ZHU Dezhi, "Diesel oil pool fire characteristic under natural ventilation conditions in the tunnel with roof openings," *Journal of Hazardous Material*. NO. 166, 2009

[4] WANG Yanfu, JIANG Juncheng, ZHU Dezhi, "Full-scale experiment research and theoretical study for fires in tunnel with roof openings," *Fire Safety Journal*, NO. 44, 2009

[5] Besserre R, Delort P, "Recent studies prove that the main cause of death during urban fires is poisoning by smoke," *Urgencies Medicals*, NO. 16, 1997

[6] WANG Yanfu, JIANG Juncheng, "Real-scale experiment research and numerical simulation for fires in tunnels with roof openings," Nanjing University of Technology, 2009

[7] Babrauskas, V. in SFPE Handbook of Fire Protection Engineering, 2nd ed., National Fire Protection Association, Quincy, MA, 1995.

The numerical analysis of the cylindrical vessel wall response under different energy load

Wenling Guan[a]
College of Environmental Science and Safety Engineer, Tianjin University of Technology, NO.391 Binshui Xidao, Xiqing District Tianjin 300384, China
hjgwl@163.com

Juncheng Jiang[b], Xiao-li Wang[a], Xiangkai Meng[a]
a. College of Environmental Science and Safety Engineer, Tianjin University of Technology,
NO.391 Binshui Xidao, Xiqing District Tianjin 300384, China
b. Institute of Safety Engineering, Nanjing University of Technology
Mail box 186, No.5 Xinmofan Road, Nanjing University of Technology, Nanjing 210009, China
j_c_jiang@163.com, wangxiaoli_cbm@sina.com,436057616@QQ.com

Prepared for Presentation at
American Institute of Chemical Engineers
1st CCPS Asia-Pacific Conference on Process Safety
Qingdao, China
September 4 – 5, 2013

UNPUBLISHED

The numerical analysis of the cylindrical vessel wall response under different energy load

Wenling Guan[a]
College of Environmental Science and Safety Engineer, Tianjin University of Technology, NO.391 Binshui Xidao, Xiqing District Tianjin 300384, China
hjgwl@163.com

Juncheng Jiang[b], Xiao-li Wang[a], Xiangkai Meng[a]
a. College of Environmental Science and Safety Engineer, Tianjin University of Technology,
NO.391 Binshui Xidao, Xiqing District Tianjin 300384, China
b. Institute of Safety Engineering, Nanjing University of Technology
Mail box 186, No.5 Xinmofan Road, Nanjing University of Technology, Nanjing 210009, China
j_c_jiang@163.com, wangxiaoli_cbm@sina.com,436057616@QQ.com
m

Keywords: energy load, the cylindrical vessel, LS-DYNA, wall response

Abstract

Structural fragments are one of the important reasons to cause a domino effect in an industrial plant. The study of the vessel wall response law under different energy loading is conducive to analysis structural fragment formation mechanism and the cause of the industrial accident. This study selects the LS-DYNA to simulate the cylindrical vessel wall response law under different energy load. The results show that the minimum value of explosion load could be determined by the vessel wall response result and the kinetic energy of structural fragment forming process can be traced by means of energy conservation.

1. Introduction

In the chemical industry, a lot of vessel is filled with high pressure gas or liquid, these vessels may be invalid and produced fragments when overpressure, mechanical failure or under the action of external force. Fragment is one of the important reasons for the Domino effect [1]. This study about the vessel wall response law under different energy loading provides a method to analyze the cause of the accident and to research fragment formation mechanism. In recent years, the research about explosion vessel dynamic response focuses on the field of military and engineering blasting. Yuanyuan et al. (2009) studied the response of cylindrical explosion vessels with the elliptical head using LS-DYNA when TNT quality were 11.9g, 23.7g, 35.6g and 47.4g[2]. WangDing et al.(2011) numerically analyzed strain behavior of blasting center section parts on a cylindrical explosion vessel under different TNT dosage loading [3]. Bayi et al. (2001) simulated explosion of spherical vessel with the 1000g TNT equivalent, equivalent stress nephogram and stress variation about a few characteristic points are given [4]. Numerical analysis research is less about the chemical vessels wall response under different energy load. The numerical analysis based on LS-DYNA soft was used to simulate vessel wall response process under different energy load in the present study. The energy of a liquid or gaseous matter is replaced with condensed explosives. When LS-DYNA is used to analyze the large deformation dynamic response problems in the process of condensed explosive, simulate result consistent with experimental. Therefore, this study selects the LS-DYNA to simulate different energy response under different energy load.

2. Calculation model

The cylindrical is more common shape of chemical vessels. In this paper, the ideal cylindrical vessel is seen as the simulate object. The ideal structure refers that the explosion cylindrical vessel is composed of head and shell only.

The object vessel' inner diameter (R) is equal to 0.6m, the cylinder height (H) is 0.91m, and the wall thickness is 18mm. Two standard elliptical head is joined with cylinder. Two pieces of parallel steel plate whose thickness is 10mm is been used to support whole vessels. According to Vons Mises' Isotropic hardening law, the material chooses ideal elastic-plastic model, three-dimensional and eight-node hexahedral Solid45 structural unit and three translational degrees of freedom per node. Mess with a total of 198472 nodes is automatically generated by program. Mesh model shown in Figure 1.

Fig. 1: Mesh model of cylindrical

3. Material model's selection and construction

3.1 Meshing

Lagrange, Euler and ALE are widely used to treat continuum in numerical simulation. The Lagrange method is commonly used in the stress-strain analysis of solid structures; The Euler method that based on the spatial coordinates is used to analyze fluid. The ALE method with the advantage of the above two methods is suitable for analysis of large deformation problems.

This paper selects the ALE method. Explosives are defined as fluid in order to avoid excessive distortion of the grid during the explosion and the adverse effect on the results.

3.2 Vessel damage criterion

In 1913, R.Von Mises proposed Maximum deformation criterion. In 1926, Lode' experiments show that Maximum deformation criterion can better describe yield state of many ductile materials such as copper, nickel , aluminum, ductile, etc. Alloy steel often is used in the manufacture of chemical vessels, so material' constitutive model is been established in accordance with the above criterion in this paper.

3.3 Material constitutive model

3.3.1 Finite element simulation of explosives

In explosives unit, "Programmed+beta burn" technology is selected to simulate the detonation process. Model is built by Euler grid and ALE algorithm is to adopt. "Explosive Burn" model is used as material model. Combustion factor control the release of explosive chemical energy. Ignition time of each explosive unit is definite according to the distance that cell center to the detonation point and the detonation velocity [5]. The JWL state equation is used to simulate pressure and volume relationship in the process of detonation:

$$P_e = A(1-\frac{\omega}{R_1 V_e})e^{-R_1 V_e} + B(1-\frac{\omega}{R_2 V_e})e^{-R_2 V_e} + \frac{\omega E_{TNT}}{V_e} \quad \text{[Eq. 1]}$$

Where E_{TNT} is initial specific internal energy; P_e is pressure; Ve is relative volume; A, B, R_1, R_2 and ω are material constants.

TNT is used in this study and its related parameters are shown in table1:

Table 1. The state equation parameters of explosive [5]

Number	Project (unit)	Values
1	density (kg/m^3)	1600
2	Material constant A (GPa)	371.2
3	Material constant B (Gpa)	3.231
4	Material constant R_1	4.15
5	Material constant R_2	0.95
6	Material constant ω	0.3
7	initial internal energy E_0 (GJ/kg)	4.184

3.3.2 The constitutive model of steel materials

Lagrangian Algorithm is used to describe steel vessel response process. Fluid-solid coupling algorithm is used to describe interact with steel container and TNT. Steel container adopt 16MnR steel. Its parameters are shown in Table 2.

Table 2. Elastic-plastic model parameters of steel

Number	Project (unit)	Values
1	density (kg/m^3)	7800
2	Young modulus (MPa)	2.01×10^5
3	poisson ratio	0.3
4	yield stress (MPa)	325

When the blast loading put on the vessel, explosion may occur and shock wave is formed.

It is a very complex action process that the explosion shock wave propagate, diffract and reflect in explosion product, that collision and motion is happed between air, explosion products , vessel and the internal structure of the vessel deform[6]. Because the vessel wall deformation is relatively small on the flow field, it is ignored that the vessel wall effect to the fluid when the influence that explosive fluids put on the container wall is been studied. Above assumptions enables boundary conditions to simplify when solving load problem. This study uses the fluid-structure coupling algorithm to deal with the fluid move on the wall of the explosion. TNT and ALE space grid is respectively defined more material group, and is defined fluid-structure interaction with a solid grid. The initial conditions are that the initial displacement, velocity and acceleration are zero for all nodes.

4. Simulation results and analysis

4.1 vessel deformation analysis under different energy load

In calculating time within 80us, different blast loading cause the different vessel wall response. When blast energy is 220GJ, only small deformations occurs in Container wall(seen Fig2(1)) ; When blast energy is 384GJ, the wall of the container began to break(seen Fig2(2)) ; When Explosion energy is 529GJ, the wall of the container is broken into pieces(seen Fig2(3)) 。 So it is convenient to use numerical analysis techniques to inverse inference the minimum load of the vessels wall based on the deformation of the vessel in accident survey.

Fig. 2: The response consequences of the cylindrical vessel under different energy load

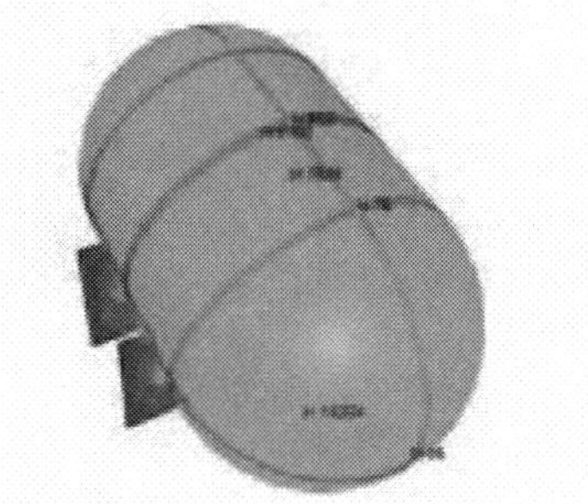

(1) E=220GJ, small deformation

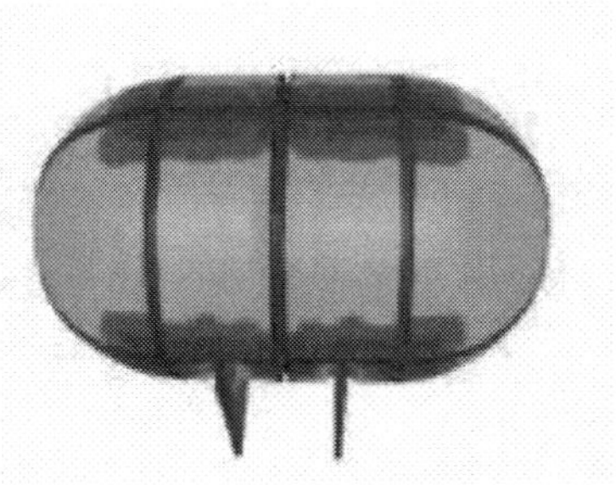

(2) E=384GJ cracking

(3) E=529GJ fracture

4.2 Energy analysis of vessel wall under different energy load

The total kinetic energy is analyzed and calculated when small deformation and the wall fragment formation, the results shown in Fig 3-4. In zero time, the kinetic energy of the vessel wall is zero. Kinetic energy is divided into two parts include weak area (or connection region) kinetic energy and wall kinetic energy. Comparison of the kinetic energy of the above two parts, connection parts have certain energy in the case of small deformation (seen Fig3.) When the vessel rupture, kinetic energy is zero in joints cracking(seen Fig4) . It can be seen the kinetic energy of vessel shell that formed fragments is much higher than that of small deformation. The above result is consistent with the actual situation. So numerical analysis techniques can simulate the total kinetic energy of explosion fragments with the proposed methodology in this paper.

Fig.3: The kinetic energy of cylindrical vessel when deformation

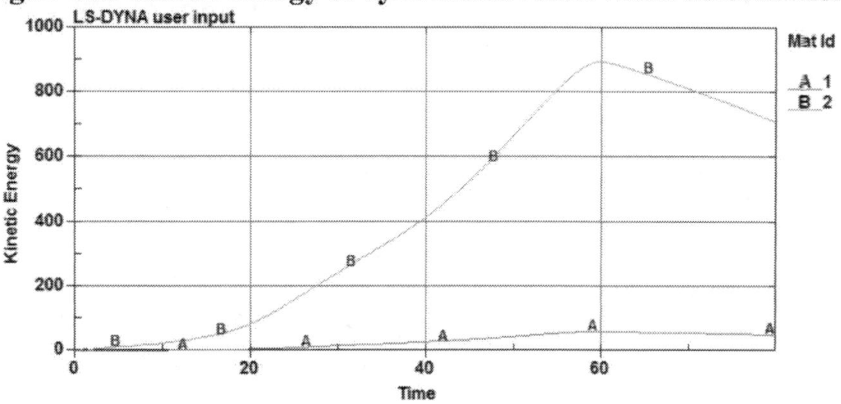

A line is kinetic energy in wall connections ;
B line is the kinetic energy of vessel wall in addition to the junction

Fig.4: The kinetic energy of cylindrical vessel when fracture

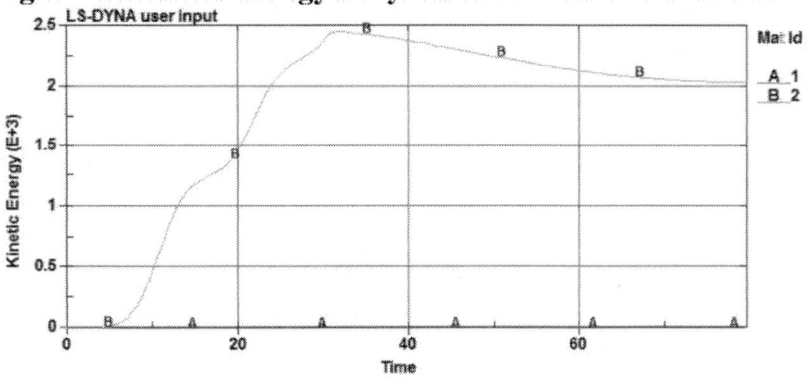

A line is kinetic energy in wall connections ;
B line is the kinetic energy of vessel wall in addition to the junction

4.3 Stress analysis of the vessel wall under different energy load

The stress nephogram of vessel wall with fracture is extract in different time periods (seen Fig5 (1), 5(2). By stress nephogram, it is seen visually that each unit' force in vessel wall is changing with time of explosion. When blast load acts on the vessel wall, its weak points gathered in the nodes' connecting parts such as head and center torus. Explosives are placed in the middle of the vessel. Center torus is nearest from the explosion center, first suffer blast attack and crack (seen Fig5(1)).

Fig.5 Equialent stress contour of cylindrical vessel at different time

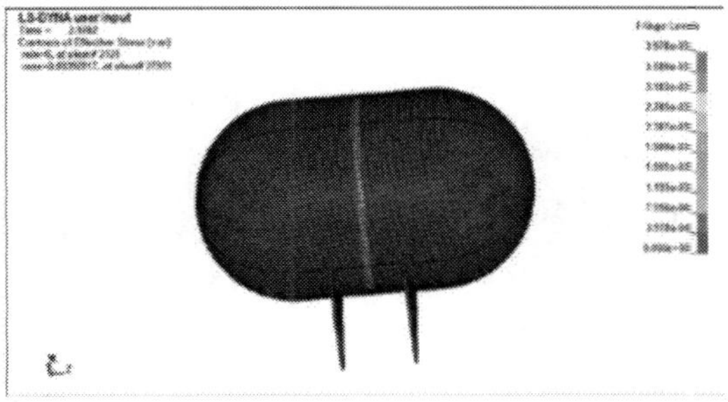

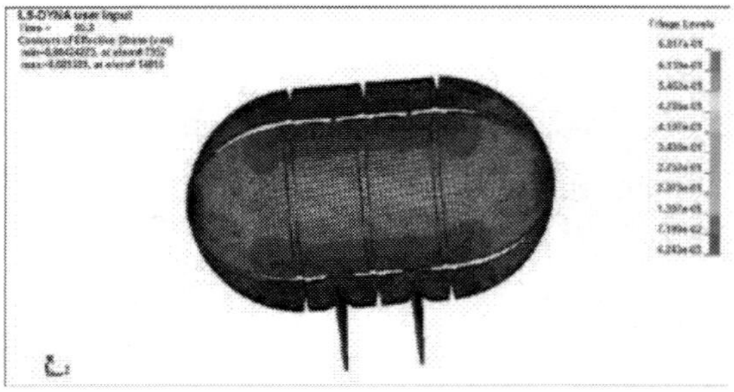

(2) t=80.3us

5 Conclusion

It is been analyzed that the cylindrical vessels wall response under different blast loading by element analysis software ANSYS/LS-DYNA. The results were as follows: under different blast loading, different responses happen in vessel wall. The minimum explosive load can be determined by vessel response consequences; the fracture site focused on vessel connections and weak point nearest from the detonation point is most likely to break; based on the conservation of energy, the kinetic energy at different times can be traced during fragments formation process.

6 Acknowledge

This research is financed by National Science Foundation (NO.21206040), College students' innovative training program of Tianjin university of technology (NO.X201210060043) and the Teaching Fund of Tianjin University of Technology (NO. Y1311-28).

7. References

[1] Wenling G. Juncheng J., "Estimation of explosion energy based on fragment characteristics," *Journal of Loss Prevention in the Process Industries*. Volume 24,2011: 69-704

[2] Yuanyuan M., Jinyang Zh., Yongjun Ch., "Numerical simulation on dynamic response of the cylindrical explosion containment vessel with an elliptical cover".*Explosion and Shock Waves*,Volume29, NO.3,2009: 249-254.

[3] Wangding X., Hao H., Wanpeng W., "Application of the Dynamical Coefficient Method in Explosion Vessel Design," *Journal of Sichuan Ordnance*. Volume 32, NO.4, 2011:24- 26.

[4] Bayi H., Jinsong B., Ming Zh., "The dynamic response analysis of spherical explosion-containment vessels," *Engineering mechanics*,Volume18,No.4, 2001: 136-139.

[5] Wangding X, Wanpeng W, Hao Hu., "Finite element simulation and experimental proving on the dynamics response of the cylindrical explosion vessel," *Pressure Vessel*. Volume 25, No.7, 2008:13 - 16.

[6] Xifeng Pu, Ming L, Liefeng Sh., "Numerical simulation of 3 kg explosion vessel structural response," the 8th impact dynamics academic discussion, 2007:312-317.

Analysis of Major Accidents in European Chemical Industries

Fu, J., State key Laboratory of Chemicals Safety, Qingdao Safety Engineering Institute, SINOPEC

After a series of major accidents in the 1970s and 1980s, such as Flixborough(1974), Seveso(1976), and Bhopal(1984), accident database systems were created in Europe. The European major accident reporting system (MARS) has been developed and operated since 1984 by the Major Accident Hazard Bureau (MAHB) within the framework of the SEVESO directives. Other European databases concerning chemical accidents also exit, such as Major Hazard Incident Data Service (MHIDAS) in UK, Failure and Accidents Technical information System (FACTS) in the Netherlands, Central Major Accident Notification System (ZEMA) in Germany, Analysis Research and Information on Accidents (ARIA) in France.

In the paper, European accident databases are reviewed and the characteristics of these databases are compared. Based on major accidents of chemical industries in MARS database, this paper presents an analysis of some characteristics of European chemical accidents. Bayesian theory is used to model the rate of occurrence of accidents and to formulate probability density distributions for their relevant causes (e.g. equipment failure, Human errors, etc.). Systematic analyses of the databases help to avoid similar accidents in the future.

The investment of integrate model of HSE management adopted at the HongKong-Zhuhai-Macao bridge during construction

Liukun[1] Cao hanjiang[2]
(1 Guangdong technology center of work safe, Guangzhou Guangdong 510060
2 HongKong-Zhuhai-Macao bridge authority, Zhuhai Guangdong 519015)

Keywords HongKong-Zhuhai-Macao bridge integrate management model HSE system

Abstract: This paper introduces the integrated HSE (Health, Safety and Environment) management model adopted by the owner of the Hong Kong-Zhuhai-Macao Bridge during its construction, including the preliminary work before this creative adoption, the main content as well as the performance of this model. Considering the satisfying operation status indicated by the monitoring data, the author advocates that the HSE management model being applied in major infrastructure construction.

1. Introduction

1.1 The Hong Kong-Zhuhai-Macao Bridge project

The Hong Kong-Zhuhai-Macao Bridge(HZMB)is the first large scale infrastructure project developed jointly by the three regional governments of Guangdong Province, Hong Kong Special Administrative Region, and Macao Special Administrative Region under the policy of "One country, Two systems". Based on the planning of the national highway network, the Bridge starts from Hong Kong, through Zhuhai in Guangdong Province, and ends in Macao as its west end. It will provide a significant part of the expressway in the Pearl River Delta regional transportation loop; meanwhile, it will be a key project crossing the Lingdingyang and connecting the east and west banks of the Pearl River. The construction of the HongKong-Zhuhai-Macao Bridge is of strategic importance to achieve our magnificent goals of reform and growth in the Pearl River Delta region, and to maintain sustainable prosperify and stability of Hong Kong and Macao regions, as well as to build a word-class city group.[1]

The HongKong-Zhuhai-Macao Bridge consists of three major parts: Offshore Bridge-cum- Tunnel. Boundary Crossing Facilitates (BCF) at Hong Kong, Zhuhai, and Macao, and Link Roads in these three regions. The Offshore Bridge-cum-Tunnel within mainland Chinese territory (from the borderline of Guangdong and Hong Kong to the BCF of Zhuhai and Macao) will be built jointly by the collaboration of three regional governments, and the remaining section in HongKong territory (from San Shek Wan in Hong Kong to the borderline of Guangdong and Hong Kong) will be completed by the Hong Kong Government. The BCF of the three regions and their regions and their link roads will be constructed independently by each jurisdiction.

1.2 The Integrative Management model

This paper analyzes, from the view of system integration and functional optimization, the HSE system based on the system of "six in one" in major infrastructure projects. The analysis also integrates process control and system management. The "six in one" model includes organization assurance systems, institutional guarantee systems, training education systems, precontrol protection systems, examination target systems, and emergency rescue systems. Based on this model, this study innovatively examines the HSE management system used by the proprietor for the Hong Kong–Zhuhai–Macaou Bridge, to create an integrated HSE system for infrastructure projects.

1.3 Safety problems with HZMB construction industry

According to the investigation, occupational safety problems are complex, multivariate, and dynamic process[2-4]. In this project, falls from height, falling objects, electronics, working near water, hit by site vehicles, short-term response(mental illness, changes in behaviour, etc.), long-term health outcomes(depression, hypertension, injuries, etc.) are all interrelated[5]. The safety problems mainly derive from six sources: (1)inherently dangerous activities(eg heavy materials being moved about a site); (2)harsh working environment(eg close to the sea, advers weather conditions); (3)constantly changing work environment(eg workers may be unfamiliar with the sage routes through the site, including emergency exit routes); (4)untrained workforce (who may not have a good understanding of the risks they are exposed to and a safe way of working);(5)stress related to career development, concluding under- or over- promotion and lack of job security;(6)long time off-shore construction works which specific to off-shore settings. Therefore, construction workers in this project are at increased risk for several safety problems. [5]

4. The aim to construct the integrated HSE management model .

Compared to the other industries, traffic construction safe work still have ill-defined procedures[6]. A collection of hard barriers, engineering and administrative controls is often required, this should include improved principles, initial safe design to reduce risks to operators and workers.[7]

With the improvement of our Chinese basic traffic construction, series of problems about occupational health, the Engineering safety and environmental protection are emerging. It is necessary to apply a greater user-centred design system that to balance the relationship between the HSE measure and the HZMB construction process. This paper take the HZMB as an example, to discuss if the HSE system is effective enough to discuss the effectiveness of the HSE for infrastructure projects .[8]

2. Methods

The key focus on this paper is the detailed exposition of the methodology used to generate the HSE system, rather than the background and actual question set of the system. A summary is given here: Fig.1 shows the main framework of the system.

to unpack Fig.1, the HSE management system is divided into three parts: (1)HSE management standard, (2)HSE technical implementation guide in construction,(3) HSE implement tools. The practitioner can make improvement and adjustment by the appraisal of the three parts mentioned above, and securing the workers` life. The three parts mentioned above are designed independently, however, they can integrate successfully without mutual interference.

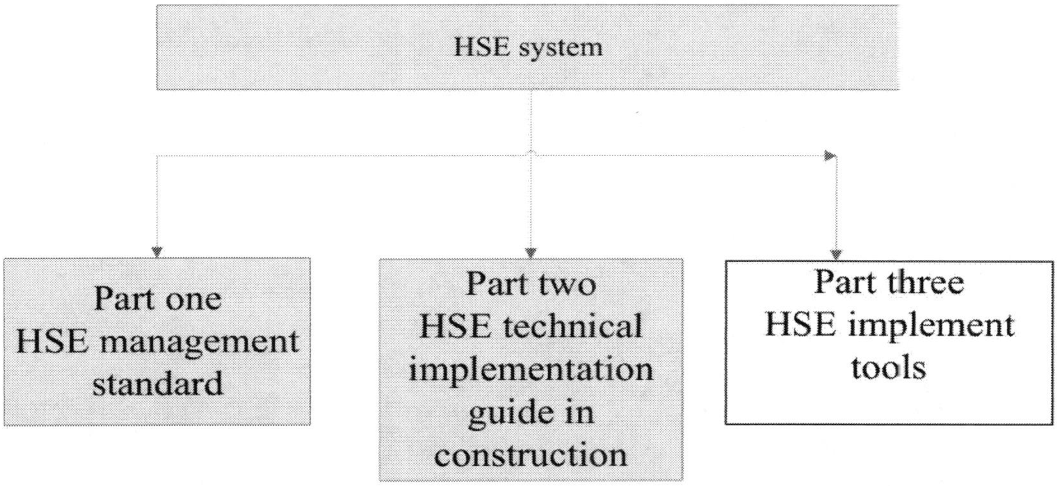

Fig.1 the main framework of the system

To create the HSE system meet the special requirements, two questions will be presented below, together with the research program.

(1)How to create a general and practical system completely during the technical challenges construction process.

(2)it must be a common practice that the operation and practice of the system have safety control within a safety assurance scheme, and the performance can be adjusted at any time.

It is necessary that to understand every part and the element of the system, throughout this exercise, it is important for the system creator to improve the huge system engineering.

2.1 HSE management standard

HZMB and construction unites developed the standard, which included selecting the essence of safety technical content based on the considering the national and international norms and standards. Besides, the experience of present and traditional HSE system used in other projects is also be studied. Fig.2 shows the main element(s) and framework of the HSE management standard.

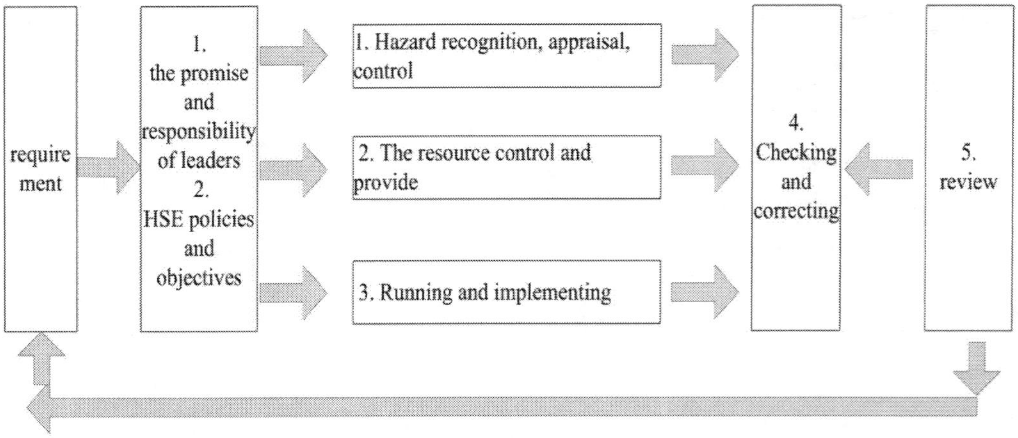

Fig.2 the main element and framework of the system

2.2 HSE technical implementation guideline in construction

For controls are focus on operator protection and the elimination of hazards, this part work are promoted based on the HSE management standards. The implementation guide directs the implementation procedure of constructions, which is the key point for creating the HSE management system. Fig.3 shows the main element and framework of the HSE technical implementation guide.

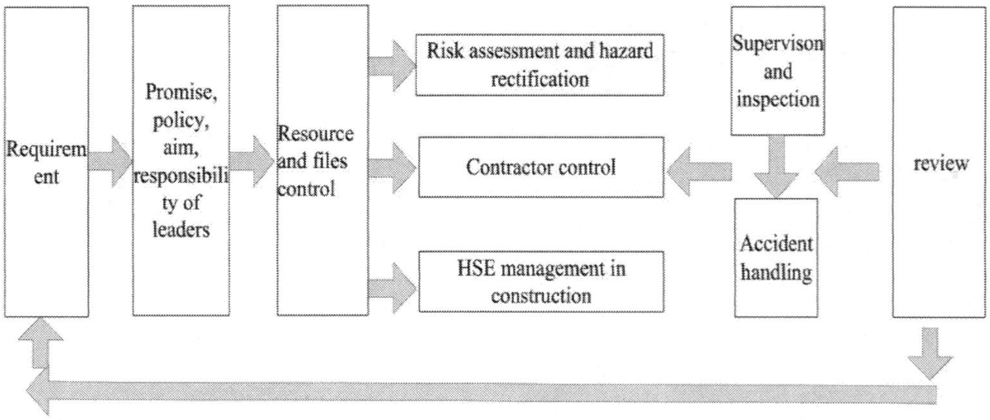

Fig.3 shows the main element and framework of the HSE technical implementation guide.

2.3 HSE implementation tool

As the technical support of the part 1and 2, HSE implementation tools nearly cover all the elements of HSE managements in construction enterprise. Elimination of hazards is controlled during the process of construction. It is presented in the follow stratgies:

Strategy 1: security evaluation;

Strategy 2: training;

Strategy 3: rules and regulations;

Strategy 4: operation management norms permit;

Strategy 5: change management process;

Strategy 6: operation instruction;

Strategy 7: operation specification;

Strategy 8: emergency preparedness and response

3. Result

1. The HSE system is first study to used in traffic construction, and according to the 3 years practice, the results shows that the system is efficient.
2. The HSE system of HZMB successfully integrates the responsibility, methods, procedure, process and the resource of a dynamic management model.
3. the good performance of HZMB has offered the others experience used for reference.

4.Discuss

Actually few projects can perfectly eliminate the hazards, especially in the area of construction, but we can confidently expect from our day-to-day experience the attention to every details in every respect can only help the system creators to achieve our objective. That means system creators need to design in every detail, however, experience leads us to anticipate that the good planning major construction projects will often go wrong unexpectedly, so the system creators can seek to learn from the past project experience to guarantee success-risk assessment and contingency planning.

Excellent design of system can ensure the success, but sometimes we must notice that the workforce are sometimes not adequately educated, experienced, controlled, motivated or trained and often that they have to work in harsh conditions. It is not common practice that the productions are regular factory type operation.

To make the HSE system perfect, the creator should： 1）fully identify the every role and fulfill it to the best of ability；2） Actively seek to arrange guidance and training for those who work for HSE aimed at enhancing their ability to understand the deliver quality； 3）No matter any time, the HSE manager should know that all others working on the project give priority to "safe" rather than "progress".

5.Inference

[1] 朱永灵，张劲文，港珠澳大桥主体工程建设项目管理规划，2011;12:12–25.

[2] Cooper CL, Marshall J. Occupational sources of stress: areview of the literature relating to coronary heart disease andill health. J Occup Psychol 1976;49:11–28.

[3] Williams S, Cooper CL. Measuring occupational stress:development of the pressure management indicator. J OccupHealth Psychol 1998;3:306–21.

[4] Sutherland VJ, Flin RH. Stress at sea: a review of workingconditions in the off-shore oil and fishing industries. WorkStress 1989;3:269–85.

[5] WEI-QING CHEN, TCZ-WAI WONG. TAK-SUN YU, Infuence of occupational

stress on mental health among Chinese off-shore oil workers. Scandinavian Journal of Public health 2009;37:766–773.
[6] Parkes KR. Psychosocial aspects of stress, health and safety on North Sea installations. Scand J Work Environ Health 1998;24:321–33.
[7] Tim Horberry, Safe design of mobile equipment traffic management systems.International Journal of Industrial Ergonomics ,41{2011} 351–560.
[8] J.S.Boschman, H.f.van der Molen,M.H.W.Frins-dresen,J.K.Sluiter, The impact of common mental disorders no work ability in mentally and physically demanding construction work. Int Arch Occup environ Heath DOI 10.1007/s00420-012-0837-6.

Study On Safety Evaluation Technology Of Plant Layout Based On Domino Effect

Meng, Y., China University of Petroleum
Zhao, D., China University of Petroleum

The domino effect is one of the most serious hazards in petrochemical plant. The study on safety evaluation technology of plant layout based on domino effect will be helpful to the optimal design of plant layout and thus reduce the domino risk of a plant. Firstly, the domino risk evaluation technologies that currently used were collected and analyzed in this work, and two shortages were concluded: (1) the mathematic models used in the evaluation process are very complicated, which decrease the effectiveness of operation; (2) the evaluation simply focus on the absolute value of domino risk, but ignores the rationality of the plant layout in consideration of domino effect, while the rationality is an important indicator of the gap between current layout and theoretical optimal layout. Secondly, in response to the above shortages, a new concept of '*domino effect irrational degree*' was proposed to evaluate whether a layout is rational in consideration of domino effect. Finally, based on the domino effect theories and the principle of 'easy to implement', the mathematic expression of '*domino effect irrational degree*' was established.

1ST CCPS ASIA-PACIFIC CONFERENCE ON PROCESS SAFETY℠

Application research on inherent safety method in chemical process safety design

Yu Wendi
Nanjing University of Technology
No. 200 Zhongshan North Road, Nanjing, Jiangsu
yuwd0123@163.com

Fang Jiajing, Pan Xuhai, Hua Min, Jiang Juncheng

[a]Safety Engineering Institute, Nanjing University of Technology, Nanjing 210009, Jiangsu, China;
[b]Jiangsu Key Laboratory of Dangerous Chemical Safety and Control, Nanjing 210009, Jiangsu, China;
xuhaipan@163.com

Prepared for Presentation at
American Institute of Chemical Engineers
1st CCPS Asia-Pacific Conference on Process Safety
Qingdao, China
September 4 – 5, 2013

UNPUBLISHED

Application research on inherent safety method in chemical process safety design*

Yu Wendi
Nanjing University of Technology
No. 200 Zhongshan North Road, Nanjing, Jiangsu
yuwd0123@163.com

Fang Jiajing, Pan Xuhai, Hua Min, Jiang Juncheng

[a] Safety Engineering Institute, Nanjing University of Technology, Nanjing 210009, Jiangsu, China;
[b] Jiangsu Key Laboratory of Dangerous Chemical Safety and Control, Nanjing 210009, Jiangsu, China;
xuhaipan@163.com

Keywords: Inherent safe; Inherent safer design; Hazardous reaction; Hazard assessment; Risk analysis

Abstract

Conventional process safety design techniques were used to identify potential hazards and suggested corresponding solutions. Nevertheless, the recommend risk reduction alternatives against application of inherent safety design (ISD) concept to prevent or minimize the inherent hazards. One of the main reasons could be the shortage of reasonable techniques or tools to support implementation of the guidewords in process safety design. Thus, a risk-based procedure was conducted to analyze the performance of various inherent safety principles in chemical process. The developed method namely Reformulate Inherently Safer Idea (RISI) could be used to quantitative analysis the hazard and inherent safety degree based on the hazard characteristics of the exothermic reaction and thermal runaway risk assessment theory. Under the condition of the thermal effect of the reaction were more considered, the potential of thermal runaway was represented for the trigger conditions and the power were represented the severity of the accident, then the potential hazards of process was obtained. If the hazard was unaccepted, the inherently safer design must be taken and the inherent safe impact value was calculated. RISI was applied to a nitric acid oxidation of alcohols combined with the experimental data by RC1. The results show that the potential hazard value was reduced by 60% after ISD. Meanwhile, it was found that the sensitivity of RISI to basic process parameters was higher than SWeHI and I2SI index.

1. Introduction

* Supported by the National Natural Science Foundation of China (No.71001051) and the Natural Science Foundation of Jiangsu Province, China (10KJA620013).

Barton [1] has carried on the statistical analysis of the chemical batch reaction process accident cases occurred in British, and results were classified based on the accident reasons. 29% of the accident was initiated by chemical process conditions, 19% of the accident was derived from the poor temperature control, 10% of the accident was triggered by agitation failure, and 21% of the accident was brought about by inadequate dosing procedures. Situations can upset the process such as poor temperature control, agitation failure, or incorrect recipes affiliated to the chemical process thermal safety field where an uncontrollable reaction system (i.e., heat generated is larger than the heat removed) would lead to the accident of fire explosion, leakage poisoning, etc. An [inherently safer design] ISD is one that avoids hazards instead of controlling them, particularly by reducing the amount of hazardous material or the opportunity of hazardous operations [2, 4]. So the essence of the inherently safer approach to plant design is the avoidance of hazards rather than their control by adding protective equipment to minimize the likelihood and severity of the accident [2, 3, 5].

Several relevance of the projects, such as the construction of inherent safety evaluation index, the performance analysis of IS strategies and the application research of ISD, have been implemented in order to apply inherent safety concepts into the hazard reactive chemicals [6-8]. A series of ISD tools have been developed and been fixed in timeline to meet the hazard challenges [9, 10]. DOW and ICI method are the pioneer index for evaluating process hazards and corresponding solutions [11]. The both index have been used prosperously for safety assessment and accident investigation. The prototype index of Inherent Safety (PIIS) is applied a hazard indexing score to identify the inherently safer route which including substance and reaction process at the research and development stage [12]. The revision index of Inherent Safety (ISI) completes the hazard-based indexing by reintroducing the equipment arose hazard while the essential features of PIIS is still retained [13]. Both of them largely refer to the classic DOW and ICI index for the parameter assignment, but the assignment and weight is uncertain. So Gentile et al. [14] propose a new safety index based on fuzzy logic to reduce the uncertainties in index scoring through implementation of fuzzy logic rules. Meanwhile, a further extended the index, called i-Safe by developing an expert system, automating the ISI for inherently safer route selection and flow-sheet value [13, 15]. Another further expanded evaluation tool of inherent safety, namely Integrated Inherent Safety Index (I2SI) which derived from Safety Weighted Hazard Index (SWeHI) [16], is produced by Khan to measure the potential consequence of hazards at preliminary design stage [17]. The index is used to manifesto the influence degree on guidewords to the process, although comprehensive considering the potential hazards and control measures on the usual situation, but it is for the low sensitivity to the specific process conditions [10, 13, 18]. Environmental, Health and Safety (EHS) assessment method, qualitatively and quantitatively, is used in the early phases of chemical process [19, 20]. The evaluation structure of model has been established relatively simple so that the material or process information can be managed flexibly. As the evaluation object mainly is reaction substances which cannot fully reflect the whole security of a process, thus it not suitable for a complex process development or optimization, such as thermal runaway reaction scale-up.

In this paper, a few attempts have been made to develop a risk-based inherent safety procedure (Reformulate Inherently Safer Idea, RISI) of dangerous exothermic process. The fundamental philosophy of RISI is inherent safety strategies that reduce the system energy and avoid thermal runaway to remove fire explosion accident scenario generation.

2. Framework of Inherently Safer Design (ISD) in Chemical Process

The ISD framework [21, 22] shown in Fig. 1 comprises of three key steps. First choose a process as the research object, quantify the damage index (DI) and the possibility of accident (PI) to establish the potential risk of the evaluation process (HI). Then, do an inherently safer design alternative if process potentially hazard is unacceptable, and calculate the inherent safety design effect index (ISDEI) that shows inherently safer design utility. Finally determine the rationality of the inherently safer design which decreases the potential hazard and maintains the risk within acceptable range.

Fig.1 Dangerous exothermic process inherent safety process framework

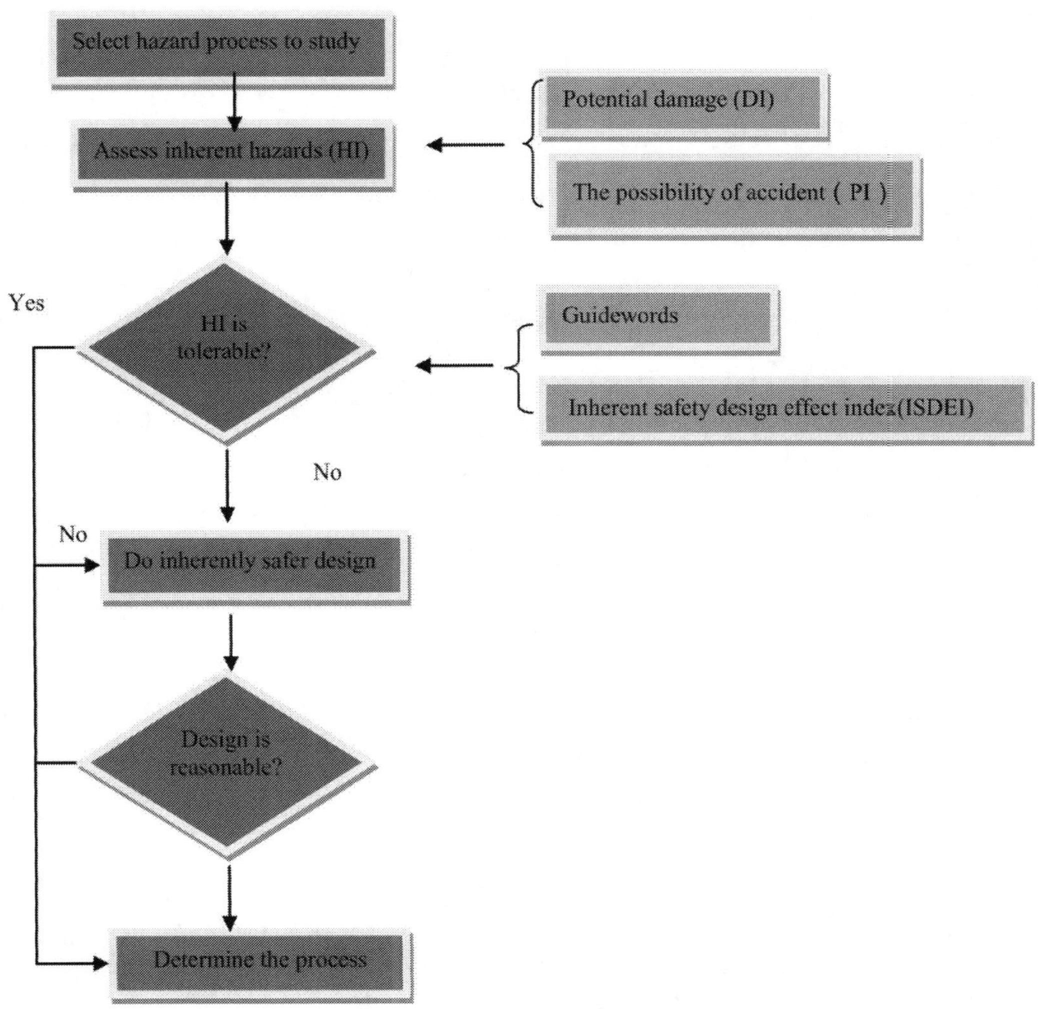

3. Quantify indicators

3.1 Quantify Potential Hazards Index (HI)

The risk-based process potential hazards composition by two parts that include the damage index (DI) and the probability index (PI). The potential hazard of thermal runaway is represented for the trigger conditions, and the energy or toxicity hazard on the surrounding environment is converted as the severity of the accident.

The severity of the hazards is obtained by calculating the potential energy or harm at the worst-case scenario. The quantification of hazards is developed based on the enhancement of Safety Weighted Hazard Index (SWeHI) [16, 25] and I2SI [17] tool to fulfill the current study, and the results in the range of 0 ~ 100. Corrosion and chronic poisoning is not instantaneously harm process relative to these transient occurred situations such as combustion, explosion and acute toxicity. Corrosion and chronic poisoning hazards are not considered here, therefore, only potential fire explosion hazard and acute toxicity hazard are regarded. Fire explosion hazard is converted as energy factors that can be refined into combustion energy, chemical reaction energy and physical damage for base case unit. The parameter of acute toxicity quantifies the toxic load over a circle area that is 50% probability of lethally affected by the toxic load [26, 27].

Quantification about the likelihood of an accident is implemented by identifying adiabatic temperature rise (Δ Tad) and process hazards control indicators (PHCI) based on the characteristics of runaway reaction. ΔTad is an index of runaway reaction severity, the greater Δ Tad , the more energy releasing in reaction process, the higher MTSR (the maximum temperature synthetic reaction) in certain material accumulation degrees, to the extent that system temperature uncontrolled capability is reinforced. The other sub-index, PHCI, is calculated for various add-on process or hazard control measures that are required in the system. This index is quantified based on the necessity of these control arrangements to maintain safe operation. The larger necessity of adding these control measures, the lower inherent safe degree of process so that a chance of an accident is easier grasped. ΔTad and PHCI are factored into the possibility indicators of accident and both the grading assessment criteria are integrated, the results as shown in Table 1.

Table1. The Possibility Index of Accident

PI	PHCI and ΔTad
0.5	PHCI=1,2 ; ΔTad< 50K
1	PHCI=3,4 ; ΔTad =50~200K
1.5	PHCI=5,6,7 ; ΔTad=200~400K
2	PHCI=8,9,10 ; ΔTad >400K

The estimation of potential risk value (HI) is based on Eq.(1) while Eq.(2) and Eq.(3) are the revised equations to estimate the damage index(DI) and probability index(PI). The energy and toxicity hazards, in Eq. (2), are computed by combining the initial energy and toxicity factors with their respective related compensation factors. The detail estimation of the compensations and damage factors for combustion, physical energy and toxicity factors can be obtained from Khan et al. [17]. These parameters, in Eq.(2), are derived from empirical models based on the amount of chemicals involved in the unit, the physical state of the chemicals, the process operating conditions and the toxicity of the chemicals. For reaction energy, moreover, the compensation coefficients about the reaction types and secondary decomposition reaction are introduced to make up for the reaction uncontrolled complexity. For the other revised sub-factors,

PI in Eq. (1), the criterion of △Tad can be obtained from Francis [28] and the assignment reference of PHCI is form Khan, F. I [17] combined with the DOW and ICI index. It must be kept in mind, however, that "*" proposed in this method is only subjected to which should be considered the correlation of parameters in calculation process, instead of specific mathematical algorithm.

$$HI = DI \times PI \qquad \text{[Eq.1]}$$
$$DI = p(energy, \; acutetoxicity) * f(\text{combustion}) \qquad \text{[Eq.2]}$$
$$PI = g(\text{thermal runaway}) * \Psi(\text{controllability}) \qquad \text{[Eq.3]}$$

HI is ranked using the lowest to upper limit outputs of the risk index attained from DI and PI to pronounce the potential performance of process [29]. The ranking references to a series of documents and guidelines to determine the maximum acceptable hazards degree after considering the assumption of worst-case scenario [23, 30]. HI at 76~200 is proposed as the "High" hazard situation where process is considered as inherent unsafe and redesign is highly demanded [24]. For "Medium" index, the HI range at 26~75, is proposed have a certain degree of risk where redesign or the corresponding safety protection measures are required. Then, the "Low" index, HI at 0~25, is informed an inherently safer process due to its low damage degree and small likelihood of hazard. At this time, the additional risk reduction measures are unnecessary and can just follow the standardization safety management in the process of production.

3.2 Inherently Safer Design (ISD) and ISDEI Quantification

The ISD option should be introduced for HI value in "Medium" and "High" levels, and the detail considering factors as Table 2 shows [30]. The performance of an ISD option could be conducted an investigation by observing two trade-off assessments. One is the applicability of the inherent safety principles (or guidewords) produced by ISD option in chemical process, and the other one is the ability of recommend ISD option for eliminating or reducing hazards. An index-assignment, Inherent safety design effect index (ISDEI), is developed as a guideline of decision- making to capture the performance of ISD option and to illustrate the degree of the hazards. ISDEI is quantified using Eq. (4), and the detail assessment criteria can be informed from empirical models of I2SI. The greater ISDEI, the stronger ability to improve process inherent safety degree by suggested ISD option.

$$ISDEI = Min\left(200, \left[(ISDEI_{min})^2 + (ISDEI_{sub})^2 + (ISDEI_{att})^2 + (ISDEI_{sim})^2\right]\right)^{\frac{1}{2}} \qquad \text{[Eq.4]}$$

Where *ISDEImin*, *ISDEIsub*, *ISDEIatt* and *ISDEIsim* are the impact factors of inherent safety principle generated by ISD adopted.

Table2. Factors Considered in ISD Procedure

guidewords	Inherent safety design factor	
	Thermal Runaway	**Fire and Explosion , Acute toxicity**

Minimization	Concentration, Heat of reaction , The material accumulation, Heat accumulation Toxicity for feed	Process inventory, Volume, Heat of reaction Toxicity
Substitution	Reactor, Solvent Stability, Reactivity , By-product;	Reactor, Solvent Flammability, Explosive; Toxicity;
	Semi-batch instead of batch	Reaction path
Attenuation	Temperature effect: Adiabatic temperature rise, Time to maximum rate of runaway	Flammability limits, Explosion limit
	Pressure effect: Vapor pressure, Amount of solvent evaporated	
Simplification	Feed-temperature interlocking	————

ISI value is a macroscopic assignment based on a applicability of the inherent safety principles into process safety design, however, that yielded to the low sensitivity of calculation results for the specific online process conditions [10]. For example, reaction might have been beyond controlled under a reaction condition of superior temperature, but the ISI value shows less dangerous under the same condition. The trade-off or conflict of hazards could have been relieved by introducing the reactor thermal safety technology [28] due to the upper sensitivity from thermodynamic parameters to reaction condition. For instance, a faint temperature modification will lead to obvious exothermic change since heat release rate changing with temperature exponentially. So in this work, the chemical reaction thermodynamics calculation is performed more to remit the conflict hazard assessment. Thus reaction thermodynamics effect, process hazards, process hazard controllability as well as the applicability of the inherent safety guidewords are regarded comprehensively in RISI procedure. RISI is expounded using:

$$\begin{aligned} RISI &= I(inherent - safe) * G(hazards, \text{thermo - runaway controllability}) \\ &= f(guidewords) * \mu(\text{thermal - effect}) * \eta(\text{process - control}) * \rho(\text{energy - destruction}) \\ &= f(ISI) * \mu(\Delta T_{ad}) * \eta(PHCI) * \rho(DI) \end{aligned} \quad \text{[Eq.5]}$$

Where similarly, "*" proposed in this method is only subjected to which should be considered the correlation of parameters in the process of calculation, instead of specific mathematical algorithm.

It must be kept clear that, however, HI and ISDEI proposed in this method cannot be integrated into a simple unified index to characterize the effect degree of ISD, because each index value represents different physical significance.

4. Application of RISI in Nitric Acid Oxidation 2-octyl Alcohols Process

4.1 Reactor Overview

A conventional batch reactor unit used in an oxidation process, nitric acid oxidation 2-octyl alcohols liquid-liquid heterogeneous reaction, is selected as the case study with a purpose of identifying the inherently safer process. Details of this process have been described by B.A.A. [31] and the typical process properties are described in Table 3. It shows there is a major potential thermal runaway hazard but no acute poisoning hazards in this process.

Table 3 Typical Properties of Nitric Acid Oxidation 2-octyl Alcohols in Batch Reactor

Process and safety properties	Typical values
Chemical substances	Reactant: nitric acid, 60~65% 2-octyl alcohols wt≥97% Product: carboxylic acid By-products: ketone, nitriles, nitro compounds, organic, etc
Reaction condition	Temperature: 50~60℃ Concentration: 65% excessive nitrate Phase: Liquid-liquid heterogeneous
The reactor	Bator
Heat of reaction	139kJ/kg, 178kJ/kg
Decomposition temperature	105~143℃
Flammability, reactivity	nitric acid , nitro compounds, organic, etc

4.2 The Hazard Assessment of Traditional Process

Table 4 shows the HI calculation results of nitric acid oxidation 2-octyl alcohols in a batch reactor using Eq. (1) ~ Eq. (3). The result declares a "medium" risk level in view of HI value is 23 and potential risk value is 50. The energy contained in the reactor is huge, especially the thermal energy from the reaction itself. Meanwhile, the reaction is highly exothermic and a dominated potential hazard in this process is thermal runaway since the ΔTad value reaches 382K. If an unexpected scenario of failure is introduced such as uncontrolled system temperature, the reaction would easily runaway. Thus, identifying the inherently safer reactor to avoid fire and explosion is crucial.

Table4. Assessment Results of Nitric Acid Oxidation 2-octyl Alcohols in Batch Reactor

Typical Index	Value
F_1	91.38
F_2	133.7
Dfe	905
DI	23
ΔTad	382K
PHCI	2
HI	50
Risk level	medium Have a certain degree of risk and inherent safety measures should be taken

4.3 Implement ISD Option

A detailed calorimetric experiment has been carried out by RC1 (Reaction Calorimeter) to make an ISD decision that has the lower potential damage and a considerably low likelihood of hazard. Table 5 shows the exothermic information that can be applied into the hazard assessment for nitric acid oxidation 2-octyl alcohols. The IDH tool developed by Risza R. [30] is used to generate a series of ISD options based on inherent safety principles. In current work, two ISD options are implemented to avoid or reduce thermal runaway hazard. Changing the type of reactor, ID1, is a promising alternative since it could minimize the reaction inventory and energy in the process. Another ISD option, ID2, is a good performance of the lower temperature in semi-batch process to moderate exothermic reaction.

Table5. The Adiabatic Temperature Rise (ΔTad) and The Maximum Temperature Synthetic Reaction (MTSR) under Different Reaction Conditions

Reaction temperature T/°C	stirring rate R/r/min	feeding time t/min	reaction heat ΔH/kJ	ΔT_{ad} /K	MTSR /K	Thermal runaway severity
20	100	30	286.82	91.82	304.5	Critical
30	100	30	337.92	103.62	312.5	Critical
40	100	30	394.03	121.06	329.8	Critical
50	100	30	489.1	143.39	331.9	Critical
20	50	30	201.63	68.16	296.4	Critical
20	125	30	301.78	96.89	301.9	Critical
20	150	30	319.28	99.05	309.5	Critical
20	100	10	256.92	78.19	298.7	Critical
20	100	45	266.33	83.01	301.5	Critical
20	100	60	185.79	61.38	288.1	Critical
40	150	30	497.7	146.65	325.3	Critical

4.4 Results and discussion

Table 6 shows the summary of data concerning evaluating ID1 and ID2, and the change of main parameters are exhibited in in Fig. 2. The ID1 and ID2 realize an expected result for minimizing the HI and ΔTad, since the mass of reactants and size of the reactor is highly smaller than the batch reactor. Moreover, the exothermic factors in semi-batch reactor are pulled the tolerable region from the intolerable condition, which indicates the ID1 option enable the severity of thermal runaway and the possibility of fire and explosion greatly reduce. The DI results of two options are in agreement with Dow F&EI and SWeH for this tool can also assess the hazard based on a consequence.

Process with a higher inherent safety degree (ISDEI = 124) after doing ISD options, where ID1 demonstrates a significant likelihood of hazard being removed by the guidewords of "minimization" and "substitution". For the "attenuation" principle, ID2 illustrates that an appropriate lower reaction temperature is decisive requirement for the vapor phase hazard. However, the trade-off and conflict must be compromise since excessive low-temperature not only is unfavorable to yield benefits, but also will contribute to material accumulation in reactor that leads to thermal runaway.

Fig.3 compares the change trends of three methods (SWeH, I2SI, RISI) with basic parameters, and the result is expected since three methods reveal a positive effect has been generated by ISD options. Furthermore, it is found that the sensitivity of RISI to basic process parameters is higher than SWeHI and I2SI index.

Fig.2 Distribution of ΔTad, ID and HD after ISD

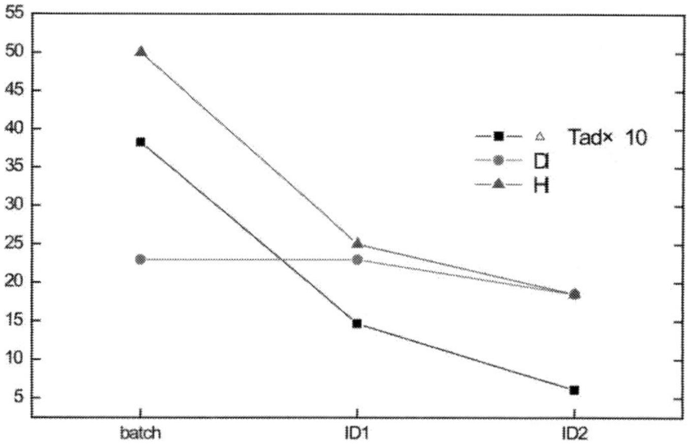

Table6. The Results of ISD Options

		ISDEI	ΔTad	DI	HI	Level
ID1	Minimization	60				
	Substitution	80				
	Simplification	20				
		102	146.65	23	25	low
ID2	Minimization	5				
	Moderation	60				
	Simplification	10				
		61	61.38	18.5	18.5	low
ID1+ ID2		124				

Fig.3 The comparison of different inherent safety methods

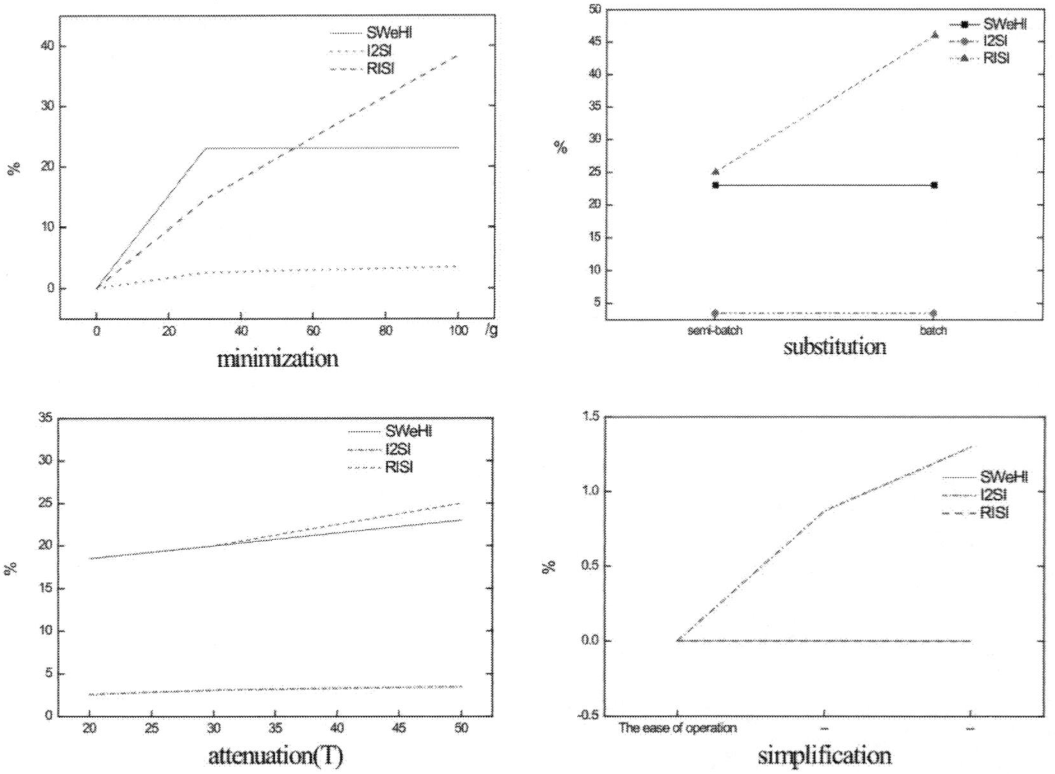

5. Conclusions

In this work, an ISD procedure namely RISI has been proposed based on risk approach and reaction thermal hazard analysis methodology. The procedure identifies the potential damage and evaluates the ISD level of performance after IS principles are considered. The case study shows that a semi-batch reactor option and an appropriate lower-temperature can be considered as an inherently safer process for nitric acid oxidation 2-octyl alcohols, supposing IS guidewords have been taken into account to reduce the risk of thermal runaway and fire or explosion. The result showed that the potential hazard value was reduced by 60% and the inherent safety degree was promoted to 124 after implementing ISD option for the process nitric acid oxidation 2-octyl alcohols. Meanwhile, RISI is proved to be an effective tool of inherently safer process by comparing with the other two methods.

Acknowledgements

The authors would like to thank Prof. X.H.PAN, for valuable suggestions on experiments and thesis supervision. Also, we wish to thank foe the financial support of the National Natural Science Foundation of China (No. 71001051) and the Natural Science Foundation of Jiangsu Province, China (10KJA620013).

References

[1] John Barton, Richard Rogers, Chemical Reaction Hazards, Institution of Chemical Engineers, 1993

[2] Kletz, T. A., "Inherently safer plants: an update," Plant/Operations Progress,10(2), 81-84,1996

[3] Kletz, Trevor, Taylor and Francis, Process Plants: A Handbook for Inherently Safer Design. Bristol, PA5-6: The essence of the inherently safer approach to plant design is the avoidance of hazards rather than their control by adding protective equipment, 1998.

[4] Kletz, Trevor, An [inherently safer] IS design is one that avoids hazards instead of controlling them, particularly by reducing the amount of hazardous material or the number of hazardous operations, personal communication , February, 2010.

[5] Anna-Mari Heikkila, Inherent safety in process plant design [Dissertation]. Espoo, Finland: Helsinki University of Technology, 1999.

[6] Tian Z., "Strategies for inherent safer design during chemical process development,". China Safety Science Journal, 16(2)：4-9, 2006.

[7] Wu Z.Z., "Industrial accident risk management method based on inherent safety,". Engineering Science, 9(5): 46-49, 2007.

[8] Markku Hurme, Stafizur Rahman, "Implementing inherent safety throughout process lifecycle," Journal of Loss Prevention in the Process Industries, 18(4-6): 238-244, 2005.

[9] Faisall. Khan, Rehan Sadiq, Paul R. Amyole, "Evaluation of Available Indices for Inherently Safer Design Options," Process Safety Progress, 2(22): 83-97, 2003.

[10] Wei J., Liu Q. & Xiang S.G., "The research progress of chemical process intrinsic safety evaluation index," Computer and Applied Chemistry, 26: 1084-1088, 2009.

[11] C.B. Etowa, P.R. Amyotte, M.J. Pegg, et al. "Quantification of inherent safety aspects of the Dow indices," Journal of Loss Prevention in the Process Industries, 15(6):477-487, 2002.

[12] Palaniappan, C., Srinivasan, R., Tan, R.B., "Expert system for the design of inherently safer processes in Flowsheet development stage," Industrial & Engineering Chemical Research, 41: 6711–6722, 2002.

[13] ParisaAbedi, Shahriari, "Inherent safety evaluation in process plants - a comparison of methodologies," Central European Journal of Chemistry, 3(4): 756-779, 2005

[14] Michela Gentile, William J. Rogers, M. Sam Mannan. Development of a Fuzzy Logic-based Inherent Safety Index[J]. Institute of Chemical Engineers. 2003, 81(6): 444-456.

[15] J.P. Gupta, David. W. Edwards, "A simple graphical method for measuring inherent safety," Journal of Hazardous Materials, 104(1-3): 15-30, 2003.

[16] Khan, F.I., Husain, T., Abbasi, S.A., "Safety Weighted Hazard Index (SWEHI): a new, user-friendly tool for swift yet comprehensive hazard identification and safety evaluation in chemical process industries." Process Safety and Environment Protection 79, 65–80, 2001.

[17] Khan, F. I., & Amyotte, P. R., "Integrated inherent safety index (I2SI): A tool for inherent safety evaluation," Proceedings of 37th Annual Loss Prevention Symposium, New Orleans, LA (pp. 263–291).New York: American Institute of Chemical Engineers, 263–291, (2003).

[18] Wei J., Liu Q. and Xiang S.G., "Reaction path synthesis considered," Computer and Applied Chemistry, 27(8), 2010.

[19] Guntram Koller, Ulrich Fischer, and Konrad Hungerbuhler, "Assessing Safety, Health, and Environmental Impact Early during Process Development," American Chemical Society, 39(4): 960-972, 2000.

[20] Isaac Kweku Adu, "Comparison of methods for assessing environmental, health and safety (EHS) hazards in early phases of chemical process design," process safety and environmental protection, 8(6):77– 93, 2008.

[21] CCPS, Guidelines for Risk Based Process Safety. Center for Chemical Process Safety/American Institute of Chemical Engineers, New York. 2007.

[22] Edwards, V. H., "Designing safer process plants," Chemical Engineering, 118(4),44-48, 2011.

[23] CCPS, Guidelines for Chemical Process Quantitative Risk Analysis, second ed. Center for Chemical Process Safety/American Institute of Chemical Engineers, New York, 2000.

[24] Hauptmanns, U., A risk-based approach to land-use planning. Journal of Hazardous Materials 12(5): 1-9, 2005.

[25] Piang, L.W., Arshad, A., "Mathematical modelling and simulation of heat dispersion due to fire and explosion," J. Teknol. F 37, 2002.

[26] Fawcett, H. H., and Wood, W. S. Safety and accident prevention in chemical operation(3rd ed.). New York: Wiley, 1993.

[27] Tyler, B. J., Thomas, A. R., Doran, P., and Greig, T. R.,"A toxicity hazard index," Chemical Health and Safety, 3, 19–25, 1996.

[28] Francis self with straw, Chen W H, Peng J H, Chen L P. Thermal safety of chemical process - risk assessment and process design . Beijing: Science Press, 2009.

[29] Li Q. , Chen J., "Study on risk assessment method based on inherent safety in chemical techniques," Journal of Safety Science and Technology, 5 (2) :45-50, 2009.

[30] RiszaRusli, AzmiMohdShariff, "Qualitative Assessment for Inherently Safer Design (QAISD) at preliminary design stage," Journal of Loss Prevention in the Process Industries. 23(1): 157-165, 2010.

[31] B.A.A. van Woezik, K.R. Westerterp, "The nitric acid oxidation of 2-octanol. A model reaction for multiple heterogeneous liquid–liquid reactions," Chemical Engineering and Processing, 39 (6): 521–537, 2000.

1ST CCPS ASIA-PACIFIC CONFERENCE ON PROCESS SAFETY℠

Modelling and simulation of major hazard installation dynamic classifications on storage tank area of wharf using LabVIEW

Xiao-li Wang[a]
College of Environmental Science and Safety Engineering, Tianjin Technology of University, Tianjin, China
wangxiaoli_cbm@sina.com

Yi-yuan Cui[a], Wen-jie Wang[a], Wen-ling Guan[a], Jia-qi Zhang[a], Shi-tao Peng[b]
a. College of Environmental Science and Safety Engineering, Tianjin Technology of University, Tianjin 300384, China
b. Laboratory of Environmental Protection in Water Transport Engineering, Tianjin Research Institute for Water Transport Engineering, Ministry of Transport, Tianjin 300456, China
cuiyiyuan0531@gmail.com 120417825@qq.com
hjgwl@163.com zhangjq@sina.com pengshitaotj@163.com

Prepared for Presentation at
American Institute of Chemical Engineers
1st CCPS Asia-Pacific Conference on Process Safety
Qingdao, China

September 4 – 5, 2013

UNPUBLISHED

Modelling and simulation of major hazard installation dynamic classifications on storage tank area of wharf using LabVIEW

Xiao-li Wang
College of Environmental Science and Safety Engineering, Tianjin Technology of University, Tianjin 300384, China
wangxiaoli_cbm@sina.com

Yi-yuan Cui[a], Wen-jie Wang[a], Wen-ling Guan[a], Jia-qi Zhang[a], Shi-tao Peng[b*]
a. College of Environmental Science and Safety Engineering, Tianjin Technology of University, Tianjin 300384, China
b. Laboratory of Environmental Protection in Water Transport Engineering, Tianjin Research Institute for Water Transport Engineering, Ministry of Transport, Tianjin 300456, China

Key words: major hazard installation, classification, storage tank area of wharf, LabVIEW

Abstract

SPA, which was proposed and applied to many fields including system engineering, artificial intelligence, forecasting, and multi-attribute assessment, can be chosen as a possible method with major hazard installation dynamic classifications on storage tank area of wharf. This paper presents a modeling and simulation system of major hazard installation dynamic classification on wharf storage tank area based on SPA using LabVIEW real-time interface. LabVIEW9.0, which is a platform and development environment for a visual programming language, was used to configure a computer program to model and simulate major hazard installation dynamic classifications. The principal functions of the system are represented in a main block diagram which consists of 16 sub-diagrams , which stimulate the site condition of wharf storage tank area and calculates 11 different index based on Pool fire accident model, VCE model and Gaussian model, respectively. The calculation and data recording are carried out constantly, and the results of major hazard installation dynamic classification are performed simultaneously. The developed computer program has been implemented in a certain chemical wharf located in China. The results showed the status of the dangerous chemicals in chemical wharf were

* Corresponding author. E-mail address: pengshitaotj@163.com(Shi-tao Peng)

constantly changed with the variation of the dangerous chemicals amount.

1. Introduction

The global petrochemical industry is now under industrial structure adjustment and trends to large-scale and the centralization of the installations. More and more chemical industry zones have been generated with the development of the petrochemical industry[1]. The petrochemical wharf is a complicated dynamic system including interaction process of transportation, storage and discharging[2]. Since the quantity of major hazard installation(MHI) real-time changes, traditional static classification of MHI with manual scoring calculation method could not figure out the grade of MHIs, stored in petrochemical wharf storage tank zone, quickly and accurately[3]. Thus, dynamic classification mode and computer calculation are urgent needed in petrochemical wharf to prevent significant accident and promote the healthy development of the petrochemical transportation industry.

LabVIEW is a kind of program-developed environment, developed by the U.S. national instrument (NI). Different from other computer language, LabVIEW is using a graphical editor language but not generating code based on text language, such as C and BASIC development environment, to write programs. Functions such as data input, model construction and graphics output, can be completed as long as user connects the flow chart of the sub-VI icons. Compared with conventional program language, LabVIEW greatly reduces the difficulty and development cycle[4-5].

This paper uses set pair analysis theory to a certain petrochemical wharf located in China for MHI dynamic classification. The pool fire model, vapor cloud explosion model and gauss smoke diffusion model were introduced to calculate accident consequences establishing the classification index system. Based on the real-time volume change of chemical substances in storage tank zone, a calculation program developed by LabVIEW, was constructed to obtain the risk level of MHIs. The realization of dynamic classification for dangerous liquid chemical substances in LabVIEW provides reference for safety precaution and reinforces the technical support for accident prevention.

2. Material and method

Set Pair Analysis(SPA) is a novel systemic analysis method dealing with uncertain problem[6]. With respect to a given problem, a set pair is formed by putting together two interrelated sets. Then, the connection degree formula of the two sets, including identity degree, discrepancy degree and contrary degree under certain circumstances, can be given as follow:

$$\mu_{A-B} = \frac{S}{N} + \frac{F}{N} I + \frac{P}{N} J \quad (1)$$

Where μ is the connection degree of the set pair, W is the given problem, N is the total number of characteristics of the set pair, S denotes the number of identity characteristics, P represents the number of contrary characteristics, $F = N S - P$ is the number of the characteristics of these two sets that are neither identity nor contrary. S/N, F/N, and P/N respectively represent the identity degree, discrepancy degree and contrary degree of the two sets. j is the coefficient of the contrary degree, and is specified as−1, i, the coefficient of the discrepancy degree, is an uncertain value between −1 and 1, i.e., i∈[−1,1].

Given a = S/N, b = F/N, c = P/N, Eq.(1) can be rewritten as:

$$\mu_{A\sim B} = a + bI + cJ \quad (2)$$

In Eq.(2), a, b and c are all greater than 0 and a+b+c=1。

The 2 equations above show the core concept of SPA that is using $\mu_{A\sim B}$, a function that fully reflects the identity, discrepancy and contrary characteristics, to describe the uncertain problem. A key to applying SPA is to figure out the quantitative criterion of the identity, discrepancy and contrary characteristics and then using connection degree to solve problems. From mentioned above, SPA, whose advantage is to simplify complicate problem, is an effective analysis method to achieve the function of analysis, calculation, evaluation, prediction and decision making[7].

3. Design of software

3.1 Structure of classification model

When the leakage accident happened in storage tank zone, pool fire, vapor cloud explosion and poisonous vapor diffusion were considered to the possible disastrous consequences. Based on above three consequences, pool fire model, VCE model and Gaussian model were introduced to obtain parameters of dangerous chemical MHI. The procedure of accident happened in storage tank zone are concluded as:

(1). The liquid pool [8] is formed when chemical substance stored in tank leaking in the condition of eroding hole or rip.

(2). When the liquid pool is ignited immediately, Pool fire accident was caused. According to the calculation of the thermal radiation rate impacted on staff, 3 parameters, Thermal radiation death probability(%),Thermal radiation seriously injury probability(%),Thermal radiation minor injury probability(%), are found to be the classification data in Pool fire accident model[9-11].

(3). If the ignition source doesn't occur immediately, the flammable and toxic chemical liquid will evaporate into steam forming explosive gas or toxic gas. The evaporation could be calculated by dynamic evaporation model. If explosive gas is ignited later, The maximum overpressure of target shock wave (MPa), the Blast radius of death (m), Blast seriously injury radius Rd0.5(m) and Blast minor injury radius Rd0.01(m) could be calculated by VCE model[9,11,12]. However, if toxic gas has not been lit, a toxic area will have been formed under the direction of the wind. Immediate threat to life and health distance(m), Immediate threat to life and health area (m^2), Short time allowable contact distance (m), Immediate threat to life and health area (m^2) can be determined by Gaussian model[13].

So the 11 parameters relevant to the characters of MHI are Thermal radiation death probability(%),Thermal radiation seriously injury probability(%),Thermal radiation minor injury probability(%), Immediate threat to life and health distance(m), Immediate threat to life and health area (m^2), Short time allowable contact distance(m), Immediate threat to life and health area (m^2), The maximum overpressure of target shock wave(MPa), Blast radius of death(m), Blast seriously injury radius $Rd_{0.05}$(m), Blast minor injury radius $Rd_{0.01}$(m). The flow chart of classification model is shown in Figure1.

Figure 1: The classification flow diagram of major hazard installations in wharf storage tank zone

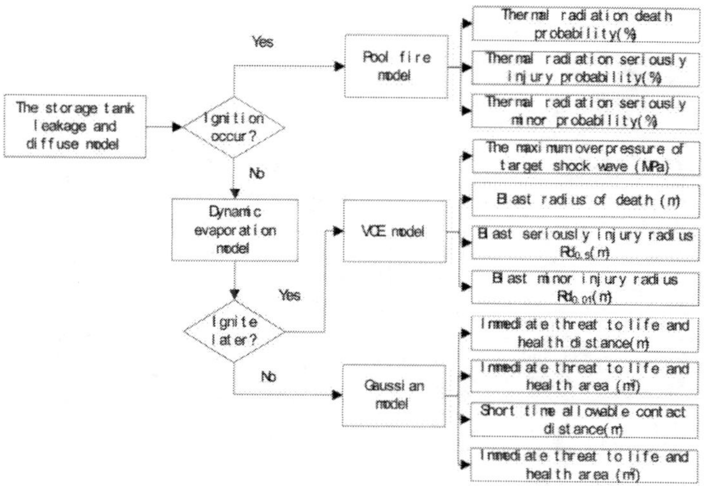

3.2 Realization of software

The main idea of software design is using the LabVIEW platform to compile models in Fig.1 and inputting 11 parameters, came from Pool fire model, VCE model and Gaussian model, into SPA calculation procedure. The calculation procedure, developed by LabVIEW, provides risk level to MHI with the variation of MHI real-time so that dynamic classification in petrochemical storage tank zone is realized.

Figure 2 shows the structure of software. In this software, data input interface, parameter calculation module, classification module and result display block should be the core parts of program.

Figure 2: Structure of software in designed system

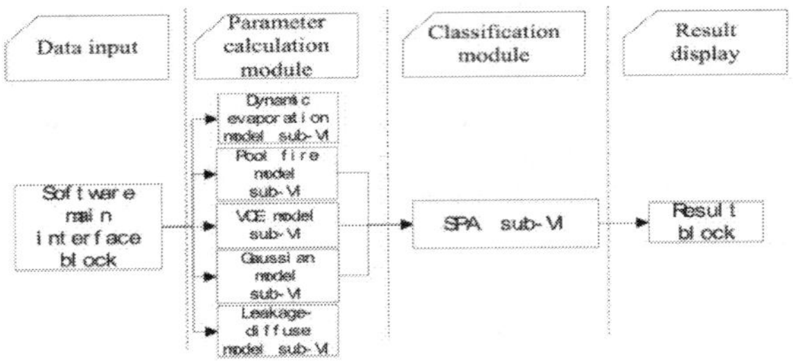

Data input module, complied by LabVIEW, should be fed in the accident simulation parameters. Based on accident model construction, the accident simulation parameters inputted in program include month, period, day-night, cloud cover, underlying surface, cracking shape, cracking measure, geographic coordinates, temperature, time of the accident, wind speed, the evaporation and leakage time, distance between fire spot and target and time of radiation. The design of data input module is demonstrated in Figure 3.

Figure 3: Design of data input module

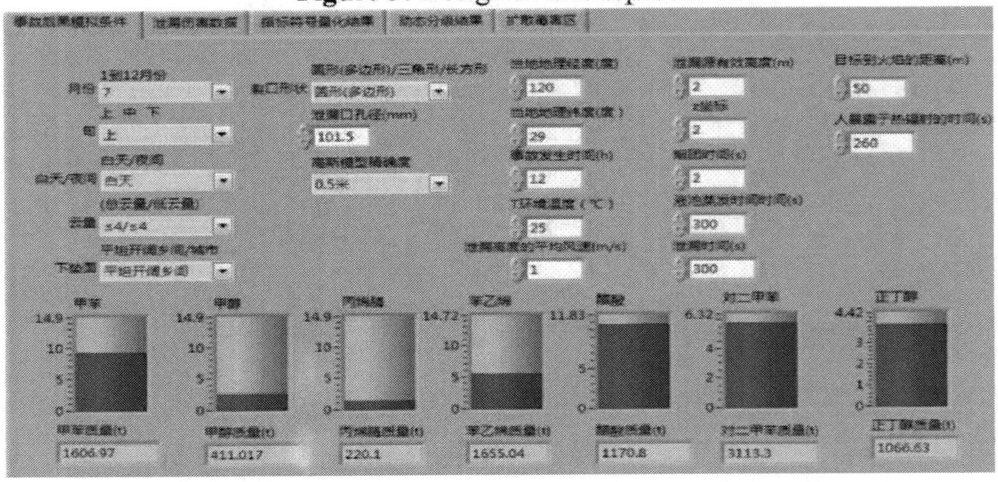

Parameter calculation module consists of five sub-VIs reflect the procedure of classification model, displayed in Fig.1. In this module, Pool fire model sub-VI, VEC model sub-VI and Gaussian model sub-VI figure out, in total, 11 parameters and transfer them to next module as initial data of classification. The design of parameter calculation module is shown in Figure 4

Figure: 4 Design of parameter calculation module

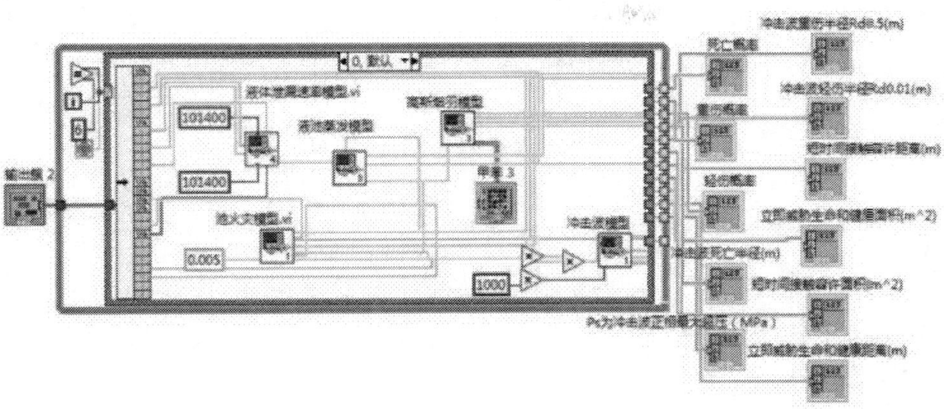

Receiving the data transferred from previous module, MHIs are divided into different categories and outputted in order of risk level in the classification module. In this module, SPA was realized as algorithm in LabVIEW. After calculating, the result should be demonstrated in result show module. The design of SPA sub-VI is shown in Figure 5. The design of result show module is shown in Figure 6.

Figure 5: Design of classification module and the realization of SPA in LabVIEW

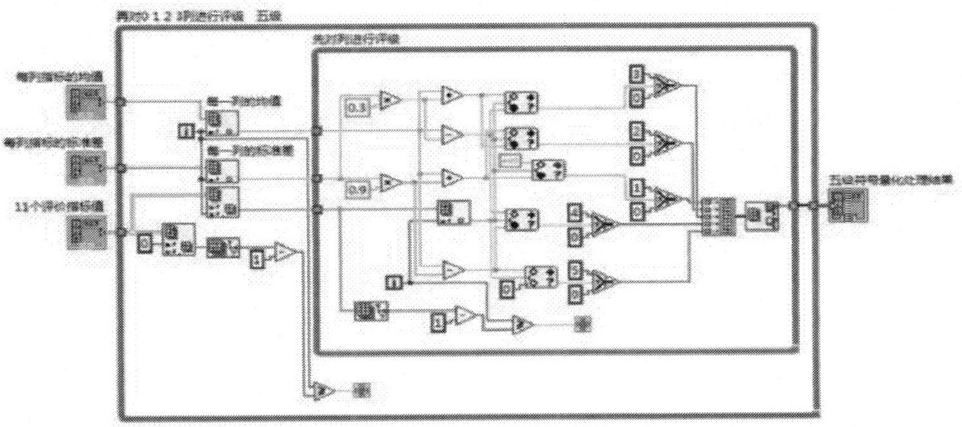

Figure 6: Design of result show module

4. Experiment

In this paper, a petrochemical wharf storage tank area located in China was selected to simulate dangerous chemical accidents based on the actual reserves of each chemical substance as the case study examples. These liquid chemical substances stored in these chemical wharf storage tanks are present as follow: toluene, methanol, acrylonitrile, styrene, acetic acid, paraxylene, butanol.

4.1 Parameter Setting

According to the record of local meteorological conditions and geography feature[14], the accident factors were simulated as follow: July, daylight, 1m/s wind speed, 250℃, sunny. In this paper, circle was selected to be the simulation eroding hole. The diameter of circle was set to 101.6mm, leaking time was 300sec based on API[15]. Table 1 is the change of volume in each chemical storage tank from 1:00 PM to 5:00 PM.

Table 1: The storage of substances for 3 different times

Material	1:00 PM	3:00 PM	5:00 PM
Toluene/t	1659	1649	1686
Methanol/t	411	411	2200
Acrylonitrile/t	195	195	195
Styrene/t	4183	1930	873
Acetic acid/t	159	1170	172
Paraxylene/t	734	3138	3148
Butanol/t	1017	1066	1091

4.2 Experiment Results Analyzing

Figure 7 shows the output from this program when simulation meteorological condition, accident factors and the volume of chemical substance for 3 periods from 1:00PM to 5:00PM are fed into it. In 1:00PM, acrylonitrile possesses the highest risk 1st level, Toluene in 2nd level, 3rd level is empty, styrene and paraxylene are in 4th level and the lowest risk, 5th level, is acetic acid and butanol. Situations about 3:00PM and 5:00PM can also be summarized like this based on figure 7. The result shows that the risk levels of MHIs are adjusting constantly with volume changes. Generally, MHI risk level will rise with the increasing of volume based on inherent risk of unit. But there are a few exceptions showing in this result. Methanol is changing from 2nd in 1:00PM to 3rd in 3:00PM even its volume doesn't change. However, styrene volume is keeping decrease but its level keeps steady in 4th. This is because the application of SPA in wharf storage tank area can reflect the interaction relations among MHIs. Each MHI's variation has effect on the classification interval that making MHIs conform to the actual situation changes.

Figure 7: The comparison of results for 3 periods from 1:00 PM to 5:00 PM

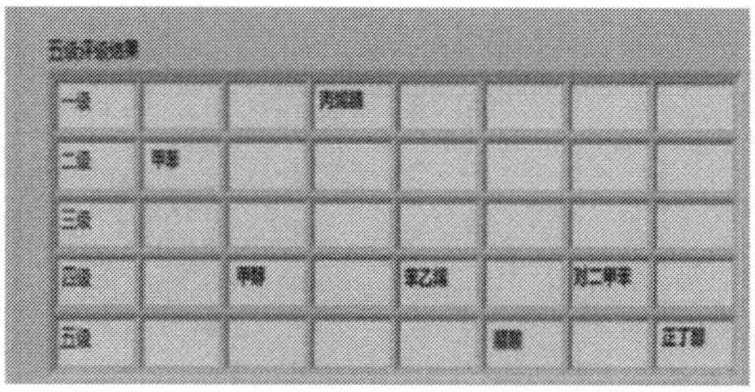

Result in 1:00PM

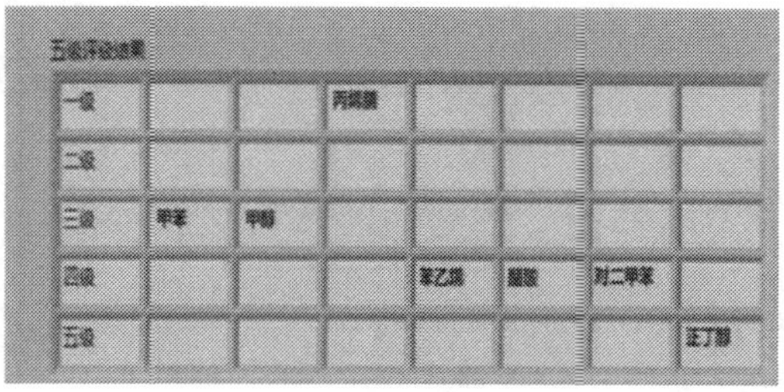

Result in 3:00PM

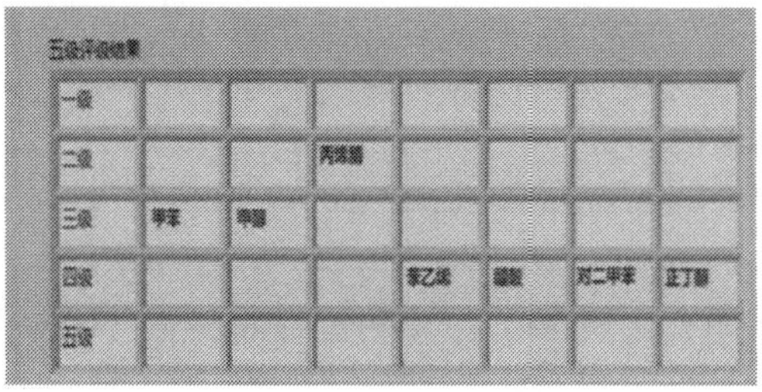

Result in 5:00PM

5. Conclusion

This paper proposes a method for dynamic classification on wharf storage tank area major hazard installations using SPA .Using this method, the disadvantages of static classification method, such as single form of accidents and lack of evaluation indicators, have been overcame. 11 parameters were proposed applying to wharf dynamic classification. LabVIEW that been used to simulate accident situation can give out MHIs level accurately and take less time than artificial calculation. The results of the simulation experiment on a practical problem show that this classification system is fit for the wharf storage zone MHIs.

6. Acknowledge

This work is supported by the National Natural Science Foundation of China (21206064) .

7. Reference

[1] Zhang Xin-mei, Chen Chen., "Mechanism analysis and risk assessment of escalation scenario in chemical industry zones," *Process Safety and Environmental N*, Volume 9, NO.1, 2012.

[2] Pula R, Khan F I, Veitch B, et al., "A grid based approach for fire and explosion consequence analysis," *Process Safety and Environmental Protection*, Volume 84, NO.2, 2006.

[3] Zhong Mao-hua., "The research of dynamic classification on major hazard based on neural network," *China Safety Science Journal*, Volume 7, NO.2, 1997.

[4] Samer M, Loebsin C, von Bobrutzki K, et al., "A computer program for monitoring and controlling ultrasonic anemometers for aerodynamic measurements in animal buildings," *Computers and Electronics in Agriculture*, Volume 79, NO.1, 2011.

[5] Chouder A, Silvestre S, Taghezouit B, et al. Monitoring, modelling and simulation of PV systems using LabVIEW. Solar Energy, 2013, 91:337-349.

[6] Zhang Jing-fei, Chen Na., "Application of SPA in Safety Assessment of Coal Mine," *Industrial safety and environmental protection*, Volume 37, NO.8, 2011

[7] Su M R, Yang Z F, Chen B, et al., "Urban ecosystem health assessment based on emergy and set pair analysis—A comparative study of typical Chinese cities," *Ecological modelling*, Volume 220, NO.18, 2009.

[8] Hu Yuan-yuan., "Modeling and engineering application of leakage process of hazardous liquid storage tank," *Chemical Engineering (China)*, Volume 38, NO.12, 2010.

[9] Liu Ji., "Probe in to the grading method for major hazard Installation," *China safety science journal*, Volume 18, NO.6, 2008.

[10] Phani K. R., "A physical model and improved experimental data correlation for wind induced flame drag in Pool Fires," *Fire Technology*, Volume 46, NO.3 2010

[11] Sha Xi-dong., "Study on classification of major hazard installation of hazardous chemicals," *Journal of Safety Science and Technology*, Volume 7, NO.3, 2011.

[12] S. M. Tauseef., "Risks of Fire and Explosion Associated With the Increasing Use of Liquefied Petroleum Gas," *Journal of Failure Analysis and Prevention*, Volume 10, NO.4, 2010.

[13] Zhang Qing-long., "Research of toxic gas diffusion simulation technology based on Arc Engine," *Lecture Notes in Electrical Engineering*, Volume 255, 2013

[14] Wang Yue-feng., "The safety analysis and research of chemical storage and transport in Qingsi chemical zone of Ningbo wharf," *Shanghai Maritime*

University, 2006.
[15] API581(2000),. "Risk-Based Inspection Base Resource Document." *American Petroleum Institute*, 2000.

The Effect of Particle Size on the Minimum Ignition Temperature of Flour Powder

SHAO Yue & YOU Fei Nanjing University of Technology

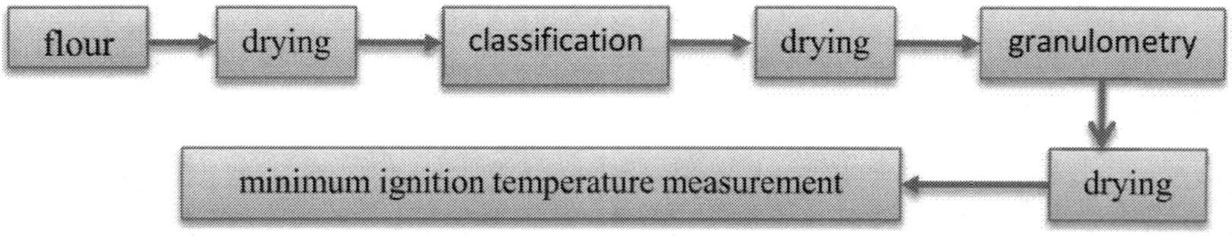

Experimental Flow Chart

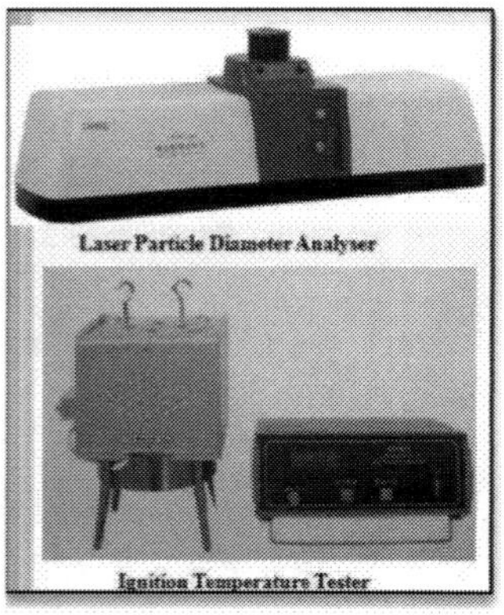

Experimental Equipments

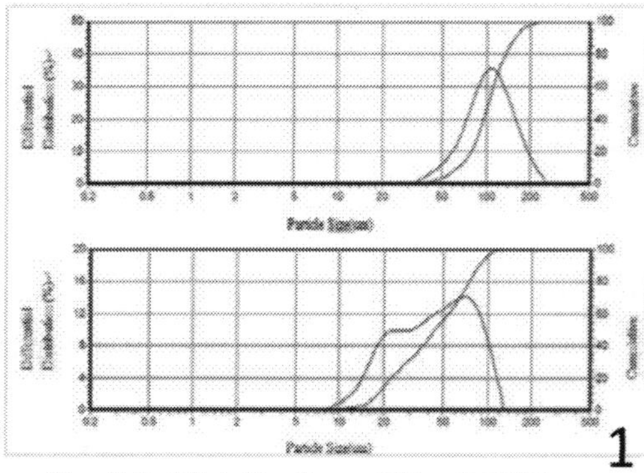

Particle Distribution of Graded Flour (up) and Unprocessed Flour (down)

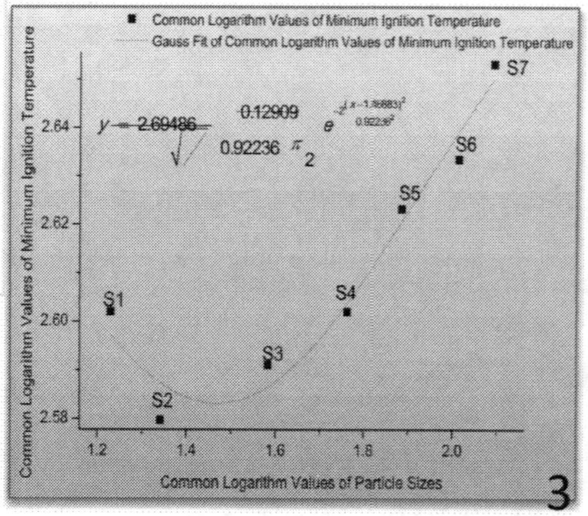

Fitting Curve of Common Logarithm Values of Minimum Ignition Temperature with Those of Particle Sizes

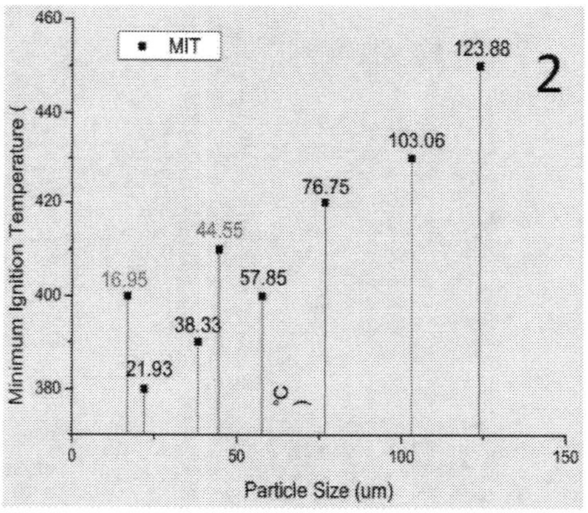

Vertical Drop Line of Minimum Ignition Temperature and Particle Size

For graded flour, when its particle size is more or less than 21.93um, its minimum ignition temperature will increase or reduce with the enhancement of particle sizes. Primary reason lies in that the coalescence strength is higher when the particle size is tinier. When the particle size is over 21.93um, the amount of the combustible gas released and the ability of oxygen absorption will decrease with the increase of particle size, specific surface area is reduced, which all makes the minimum ignition temperatures higher.

Considering the interaction of various particles, the minimum ignition temperature of particle mixtures does not follow the rule that
the minimum ignition temperature decreases with the reduction of particle size.

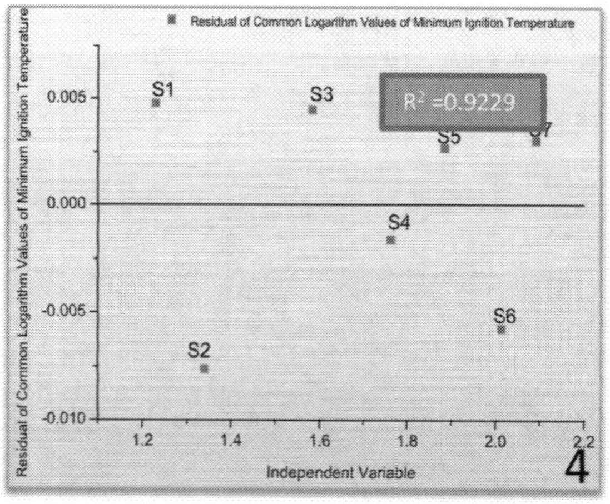

The Fitting Residue of Common Logarithm Values of Minimum Ignition Temperature with Those of Particle Sizes

The minimum ignition temperatures have positive relationship with average particle sizes. The corresponding fitting formula with R^2 of 0.9229 is as followed:

$$y = 2.69486 - \frac{0.12909}{0.92236\sqrt{\pi/2}} e^{-2\frac{(x-1.46883)^2}{0.92236^2}}$$

The Effect of Particle Size on the Minimum Ignition Temperature of Flour Powder

*SHAO Yue, YOU Fei**

(College of Urban Construction and Safety Engineering, Nanjing University of Technology, Nanjing 210009, China)

Abstract: The minimum ignition temperature is an important parameter to show the explosion risk of dust, which is affected by particle size, moisture content and ash etc. According to GBT19077.1-2008, the sizes and minimum-ignition-temperatures of 8 flour samples were determined by laser particle analyzer and ignition temperature meter, respectively. The relationship curve between the minimum ignition temperature and particle size was fitted based on the experimental data. Results show that the minimum-ignition-temperatures of flour particles reduce and increase with the reduction of particle sizes above 21.93 μm and below 21.93 μm, respectively. There is a positive correlation between the average size and minimum-ignition temperature based on the fitting curve.

Keywords: particle size; minimum ignition temperature; flour powder

* To whom correspondence should be addressed. Email: yfei@njut.edu.cn.

1 Introduction

With the development of economy, powder industry is also booming. However, some combustible dusts can cause fire and explosion accidents at times in their production, storage and transportation processes. Serious life and property losses have been brought about, and social stability has thus been destroyed badly. There have been about 150～300 local explosions and 1～3 systematic explosions involving combustible dusts in recent years. On February 24th in 2010, a dust explosion accident involving corn starch that happened in Qinhuangdao of Hebei province and regarded as the most serious dust explosion case in grain trade had made 20 people die and 49 people injure (1). Subsequently, it is urgent to deepen understanding about explosion characteristics and mechanism of combustible dusts. In reality, dust fires and explosions are influenced and restricted not only by self-properties (i.e. particle size, ash, water content, etc.) but also by some external factors (i.e. application environment, ventilation conditions, etc.).

Although there already exit some researches about the effect of inhibitors and inert agents on dust burning and explosion characteristics, the influence of particle sizes has been few studied. Zhang Maozeng et.al (2) tested the effect of different particle sizes on the minimum ignition temperature of coal dust layer and cloud. Results show that both minimum ignition temperatures decreased with the reduction of particle size, while the minimum ignition temperature of coal dust layer was lower than that of coal dust cloud. For food industries, the majority of combustible dusts have certain cohesion, so it is important to study those special dusts. Based on IEC standard, Li Gang et.al. (3) used self-designed testing rig to test minimum ignition temperatures of 8 type of grain layers and dusts sampled from industrial fields. Besides, the effects of ash, particle size and water contents on MIT were explored. The results showed the hazards decreased with the increasement of particle sizes. To sum up, in existing studies editable powders (flours and starches) have not been involved, and the burning and explosion characteristics of cohesive flours have not been focused on specially. Therefore, the minimum ignition temperatures of flours with 8 different particle sizes were tested based on gas phase combustion mechanism to discuss the effect of particle size and cohesion force on combustion properties, the fitting curves between the flour particle size and the minimum ignition temperature were gained numerically.

2 Experiment

2.1 Experimental Samples and Devices

Self-made wheat flours were adopted as experimental dust source. Standard test sieves were used to sift powders with different mesh numbers. The basic situations of samples and devices are as followed:

Table 1 Experimental Samples and Devices

Name	Type	Manufacturer
flour	wheat flour	self-made
standard test sieve	160 mesh	Xiongchen Screen Plant, Nanjing, Jiangsu
standard test sieve	180 mesh	
standard test sieve	200 mesh	
standard test sieve	320 mesh	
standard test sieve	400 mesh	
standard test sieve	500 mesh	
drying oven	electric blast	Shanghai Yiheng Scientific Apparatus Co., Ltd
laser particle analyser (as Fig.1)	LS-C(Ⅱ)	Zhuhai OMEC Scientific Co., Ltd
ignition temperature tester（as Fig.2）	DW-02	Nanjing Jiangning Analytic Instrumental Factory

The objective pictures of ignition temperature tester and laser particle analyser are presented as follows.

Temperature:10～30℃; humidity ＜85%; measuring range of particle sizes: 0.5～300um; repeatability error (D50) ＜±3%.

Fig. 1 Photo of Laser Particle Diameter Analyser (right) and Photo of Ignition Temperature Tester (left)

2.2 Experimental Processes

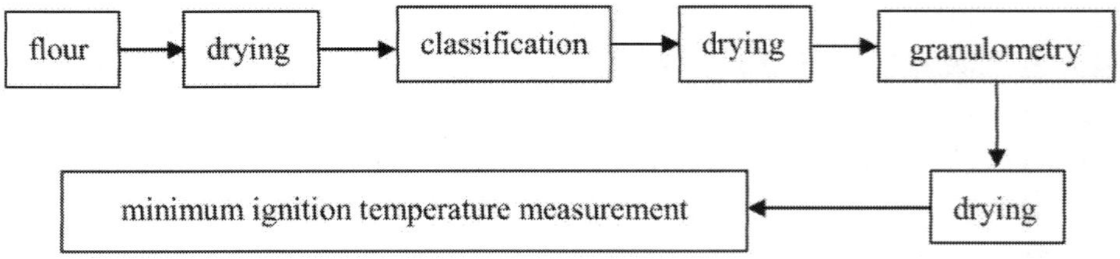

Fig.2 Experimental Flow Chart

Firstly, unprocessed flour was put into thermostatic electric blast drying oven to dry at a temperature of 70°C for 4 hours. Secondly, standard test sieves with different mesh numbers were used to classify the samples. Thirdly the sizes of samples were tested by laser particle analyser and then samples were put back to the drying oven again. Lastly, the minimum ignition temperatures of dried and classified samples with different sizes were measured by ignition temperature tester. The experimental process was schemed as Fig.2.

What happens to wheat flour in ignition temperature tester is as follows: flour particles are heated by external fire source, the surface temperature is raised until a critical value is reached to release volatile combustible gases. The mixture gas of volatile combustible components and air is ignited by external fire source to form flame (4).

3 Discussions and Analyses

3.1 The Relationship between Minimum Ignition Temperature and Flour Particle Size

Table 2 presented the average sizes and their corresponding minimum ignition temperatures of 8 wheat flour samples.

Table 2 Size Distribution and Corresponding Minimum Ignition Temperature

No.	Mesh Number	Average Size (μm)	Minimum Ignition Temperature (℃)
S1	< 500	16.95	400
S2	400～500	21.93	380
S3	320～400	38.33	390
S4	200～320	57.85	400
S5	180～200	76.75	420
S6	160～180	103.06	430
S7	>160	123.88	450
S8	unprocessed flour	44.55	410

Fig.3 is the vertical drop line between the minimum ignition temperatures and particle sizes based on the data in Table 2. The fourth line (S8) represents the minimum ignition temperature of unprocessed flour.

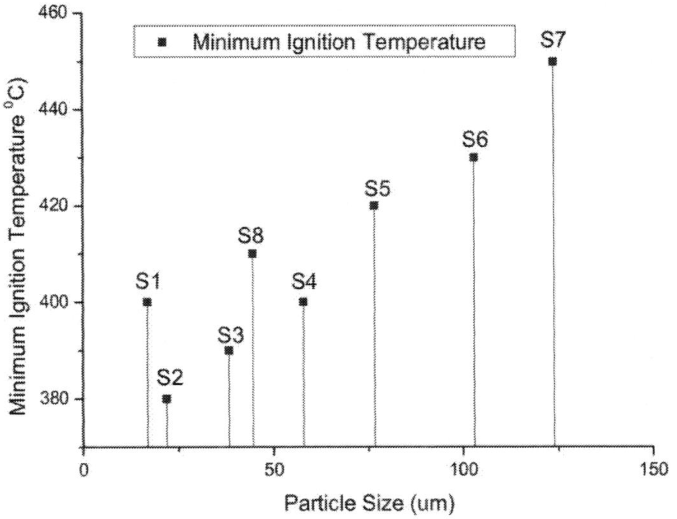

Fig.3 Vertical Drop Line of Minimum Ignition Temperature and Particle Size

From **Table 2** and **Fig.3**, the minimum ignition temperatures are between 380℃and 450℃, the average particle size ranges from 16.95μm to 123.88μm.

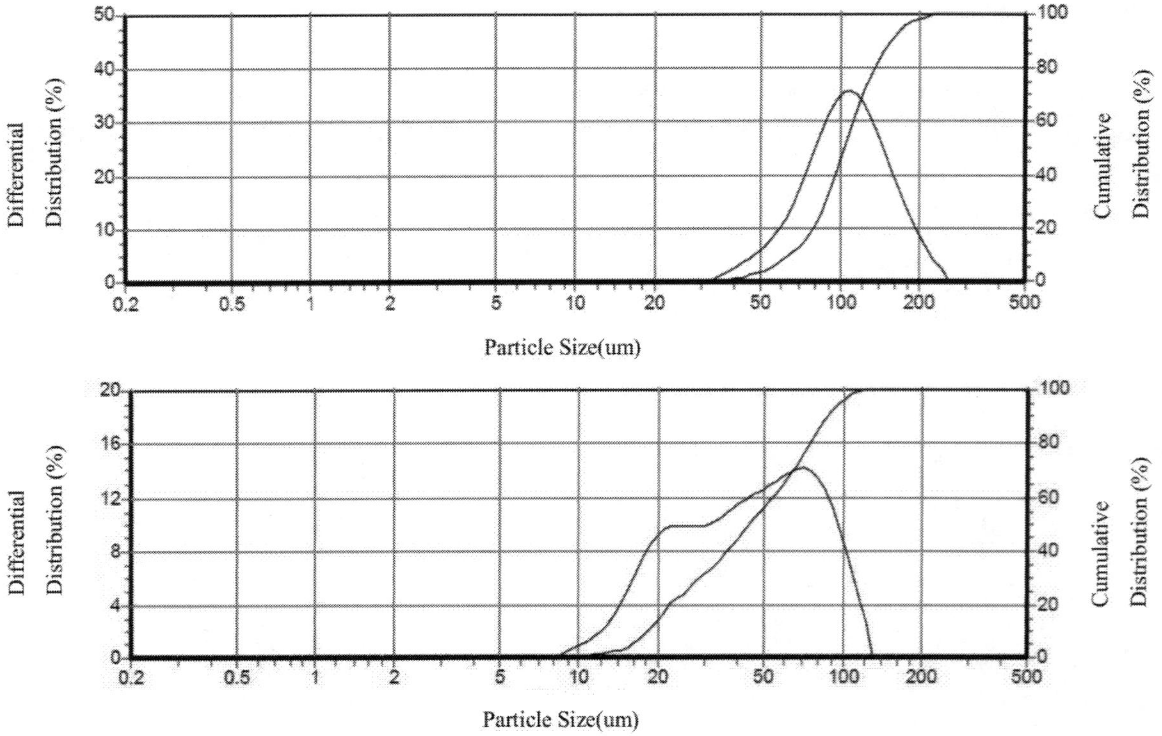

Fig.4 Particle Distribution of Graded Flour (up) and Unprocessed Flour (down)

The size distributions of unprocessed flour and graded flour are presented as **Fig.4**. through laser particle analyser. The upper figure with only one single peak is about the graded flour, that is to say, the majority of particles are well distributed and small particles and large particles are less. The lower one with two peaks is about the unprocessed flour, that is to say, the particles with various sizes are irregularly mixed.

3.2 The Variation Trend of Minimum Ignition Temperature with the Particle Size of Graded Samples

Fig.5 presented the variation trend curve that showed the variation trend of minimum ignition temperatures with particle sizes of graded flour.

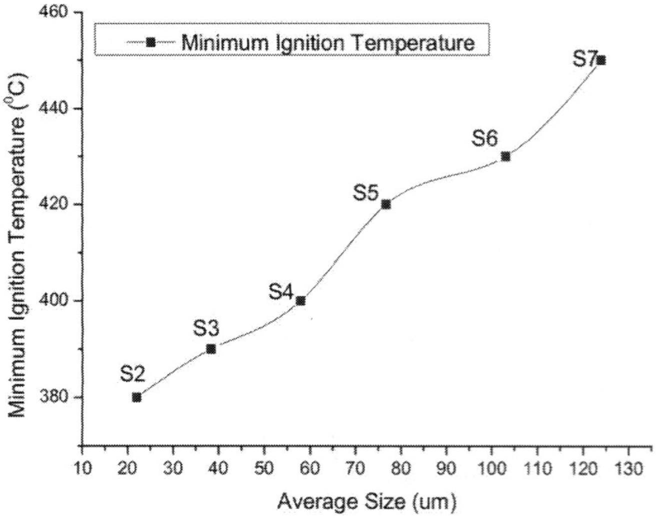

Fig.5 Variation Trend of Minimum Ignition Temperatures with Average Sizes of Graded Flour

From **Fig. 5**, the minimum ignition temperature reduced with the decrease of particle sizes. From right to left, ratios of two neighbor values decreased by 4.44%, 2.33%, 4.76%, 2.50% and 2.56%, respectively. One of the main reasons is: the specific surface area increases largely with decrease of the particle size, the flour particles can thus be heated more uniformly only to accelerate the auto thermal oxidation process.

In addition, the mass transfer coefficient α_{zl} in heat transfer is expressed as follows:

$$\alpha_{zl} = \frac{Nu_{zl} D}{\delta}$$

Where Nu_{zl} was nusser number; δ was the diameter of carbon sphere and D was oxygen diffusion coefficient.

As can be seen from the above formula, the smaller the particle size is, the larger the α_{zl} is, the more combustible gases will be released in unit time. The initial stage in the test of the minimum ignition temperature is an oxygen-enriched combustion. In other words, considering the more combustible gas released, flour powders with smaller particle sizes is prone to be ignited and the minimum ignition temperature of smaller particles is lower than that of larger particles (5).

3.3 The Comparison between Sample 1 (S1) and Sample 8 (S8)

Based on the basic combustion theory of solid particles, the minimum ignition temperature of sample 1 should be below 380℃, but its measured value reaches at 400℃ (See Table 2). It is mainly because the tinier dispersed flour particles show stronger water absorption capability and thus their residual moisture content is higher than that of larger particles in the same drying time. Besides, agglomeration happened for tinier particles because of collision and other impact action forces during the sample loading process.

The average particle size of S8 is between that of S3 and S4, while its minimum ignition temperature is between that of S4 and S5.The main reason for this is because the coalescence rate of the flour particles with larger size differences is higher than that of flour particles with similar sizes; the possibility to collide of the former is also higher than that of the latter; flour particles with different sizes bring various electrified capacity and make the mutual Coulombs force larger, agglomeration and amalgamation effects are then aggravated .

3.4 Numerical Fitting of the Minimum Ignition Temperature with the Particle Size of Graded Flour

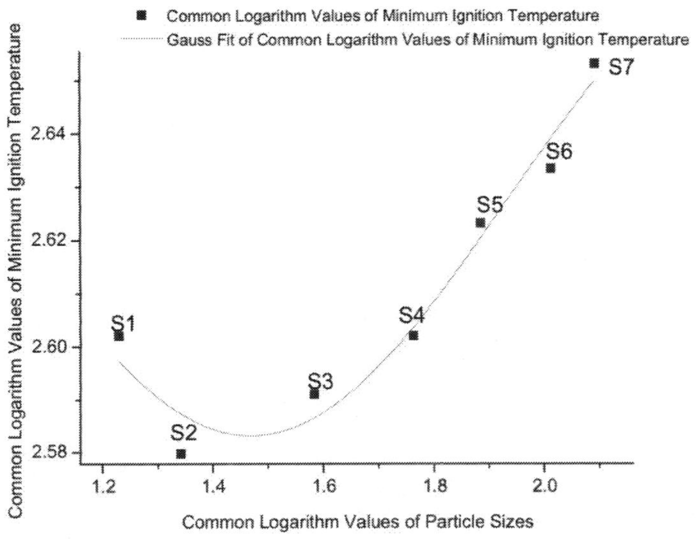

Fig.6 Fitting Curve of Common Logarithm Values of Minimum Ignition Temperature with Those of Particle Sizes

The common logarithm relationship between minimum ignition temperatures and average particle sizes was gained by origin8.5. The fitting curve was shown as **Fig.6**. A liner model was used to describe their relationship just as follows.

$$y = 2.69486 - \frac{0.12909}{0.92236\sqrt{\pi/2}} e^{-2\frac{(x-1.46883)^2}{0.92236^2}}$$

In the formula, x is common logarithm of average particle size and y is common logarithm of minimum ignition temperature.

The fitting residue is presented as **Fig.7**.

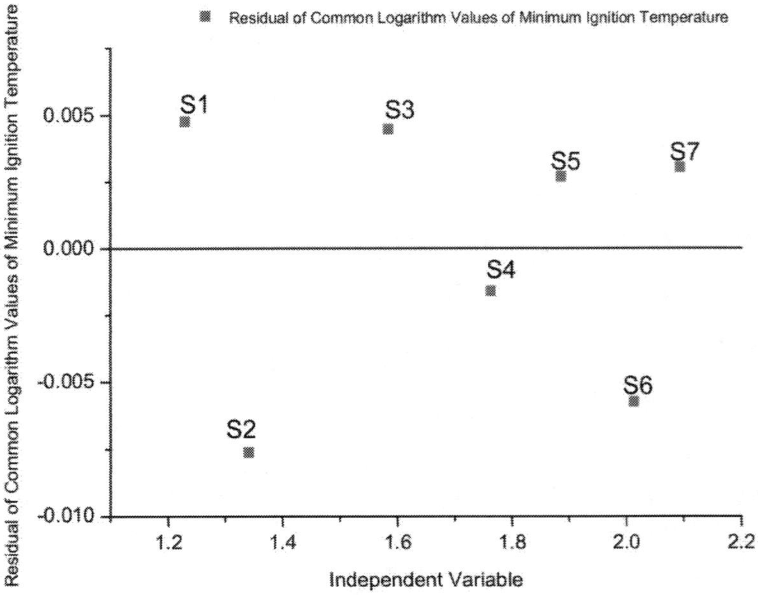

Fig. 7 The Fitting Residue of Common Logarithm Values of Minimum Ignition Temperature with Those of Particle Sizes

From **Fig.7**, the numerical distribution is irregular and its R^2 is 0.9229, which is in correct fitting range. The agreement is fairly good. And the logarithms of minimum ignition temperatures have positively correlated with that of average particle sizes. Therefore, it has sound guidance significance to predict the explosion limit of different graded flour in practical production.

4 Conclusions

1) For graded flour, when its particle size is more or less than 21.93um, its minimum ignition temperature will increase or reduce with the enhancement of particle sizes. Primary reason lies in that the coalescence strength is higher when the particle size is tinier. When the particle size is over 21.93um, the amount of the combustible gas released and the ability of oxygen absorption will decrease with the increase of particle size, specific surface area is reduced, which all makes the minimum ignition temperatures higher.

2) Considering the interaction of various particles, the minimum ignition temperature of particle mixtures does not follow the rule that the minimum ignition temperature decreases with the reduction of particle size.

3) The minimum ignition temperatures have positive relationship with average particle sizes. The corresponding fitting formula with R^2 of 0.9229 is as followed:

$$y = 2.69486 - \frac{0.12909}{0.92236\sqrt{\pi/2}} e^{-2\frac{(x-1.46883)^2}{0.92236^2}}$$

Acknowledgement

This study was financially supported by "Key Laboratory of Mine Thermo-motive Disaster and Prevention, Ministry of Education" and "National Natural Science Foundation of China"(No. 50906039).

References

1. PAN Feng, MA Chao, CAO Wei-guo, ZHANG Jian-xin and XU Sen (2011), "Research on Explosion Risk of Corn Starch Dust", China Safety Science Journal, July, pp.46-51.

2. ZHANG Mao-zeng, MA Shang-quan, DONG Hui and GU Xin-yu (2010), "Study on relationship between coal dust ignition temperature and particle size", Coal Engineering, January, pp. 83-85.

3. LI Gang, LIU Xiao-yan, ZHONG Sheng-jun and DANG Jun-xiang (2005). "Experimental Investigation on Minimum Ignition Temperature (MIT) of Dust Concomitant with Grain", JOURNAL OF NORTHEASTERN UNIVERSITY, February, pp.145-147.

4. HUO Ran (2007). "Fire and Explosion Prevention and Control Engineering", Beijing: China Machine Press, pp.91-100.

5. LIU Zhong, LIU Han-xiao, FENG Xin-xin, LI Huai-liang and XING Zhen-zhong (2012). "Comparative Study on the Different Coalescence Models of Ultrafine Particles". Journal of Combustion Science and Technology, March, pp.212-216.

1ST CCPS ASIA-PACIFIC CONFERENCE ON PROCESS SAFETY℠

A Discussion of the Grey Model Based Methodology in Work Safety Target Forecasting and Setting

ZHANG Hua-wen
College of Petrochemical and Energy Engineering, Zhejiang Ocean University
Wenhua Road 109, Dinghai District, Zhoushan City, Zhejiang 316004, China
hwzhang@zjou.edu.cn

CHEN Guo-hua
Institute of Safety Science and Engineering, South China University of Technology
Wushan Road 381, Tianhe District, Guangzhou 510640, China
mmghchen@scut.edu.cn

ZHU Bai-kang
College of Petrochemical and Energy Engineering, Zhejiang Ocean University
Wenhua Road 109, Dinghai District, Zhoushan City, Zhejiang 316004, China
zszbk@126.com

Prepared for Presentation at
American Institute of Chemical Engineers
1st CCPS Asia-Pacific Conference on Process Safety
Qingdao, China
September 4 – 5, 2013

UNPUBLISHED

A Discussion of the Grey Model Based Methodology in Work Safety Target Forecasting and Setting

ZHANG Hua-wen
College of Petrochemical and Energy Engineering, Zhejiang Ocean University
Wenhua Road 109, Dinghai District, Zhoushan City, Zhejiang 316004, China
hwzhang@zjou.edu.cn

CHEN Guo-hua
Institute of Safety Science and Engineering, South China University of Technology
Wushan Road 381, Tianhe District, Guangzhou 510640, China
mmghchen@scut.edu.cn

ZHU Bai-kang
College of Petrochemical and Energy Engineering, Zhejiang Ocean University
Wenhua Road 109, Dinghai District, Zhoushan City, Zhejiang 316004, China
zszbk@126.com

Keywords: safety target; target setting; Accident Reduction Factor; Multivariate Grey Model

Abstract

In managing occupational safety or work safety, it has been a common approach in many countries of forecasting and setting quantified and time-limited targets to reduce work-related accidents, fatalities and injuries, and this is usually included in a wide range of work safety plans and policies formulated and implemented by the government. In this paper, we present a new method for forecasting and setting safety targets by using a combination of Multivariate Grey Model and Linear Regression Model. To develop a better understanding of how to forecast the likely number of fatalities in the target year, we first perform a descriptive analysis to identify what factors within the broader economic, societal and technological backgrounds may contribute towards changes in the incidence of workplace fatalities over time, and then use seven indicators as the mark for these factors to generate a Multivariate Grey Forecast Model. Also, Accident Reduction Factors (ARFs) commonly used in road safety field are incorporated to provide a computationally simple and quick way of estimating fatality reductions, taking account the aggregated effectiveness performance of the work safety policy interventions. Finally, an application of the proposed function is made by using fatality records in several consecutive years to forecast and set safety targets for the target year. Our final objective is to explore the potential utility of the multivariate forecasting model to provide decision support for related agencies to forecast and set new safety targets and to estimate the effects of new safety measures in the target year.

1. Introduction

1.1 work safety in China

With the continual and rapid economic growth and social development during the last several decades, China has taken an enormous leap towards modern society with the increasing pace of globalization, industrialization, urbanization, and also, however, with coexistence of all kinds of natural and social risks. Among all these risks, work-related accidents obviously constitute a formidable public challenge in China, as they have claimed more than 100,000 lives yearly (as of 2006), and inflicted injuries upon at least one million people annually over the ten consecutive years. Undoubtedly, China, as well as other developing economies, is experiencing an industrial accident crisis of world-historical proportions. According to the latest data of the ILO report, accidents account for the biggest share of work-related mortality in China, which is a far cry from the level of developed countries, and of course impose a massive and unacceptable burden of death, injury and suffering on China. Fig. 1 summarizes the work safety statistical background, namely the annual national totals of work-related fatalities (all types and levels of severity) and major accidents between 1990 and 2008.

Figure 1: Trends of work-related fatalities and major accident in China: 1990-2008

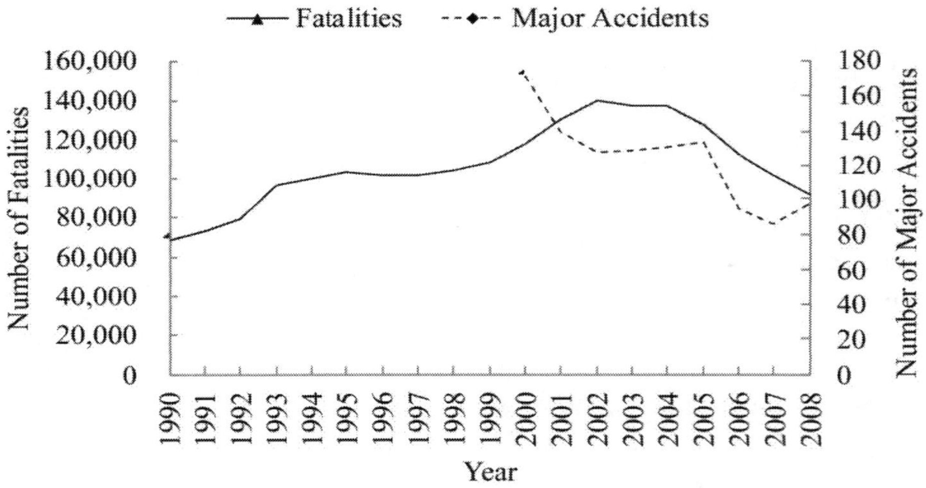

Note: major accidents are those that cause multiple fatalities (more than 10) and substantial property and environmental damages.

The terrible work safety situation was timely wake-up calls for firm and decisive action. The governments have become increasingly aware of this problem, and thus adopted safety intervention policies and strengthened relevant laws and regulations to attempt to deal with this issue to reduce this burden by making China's workplaces safer for everyone. On the threshold of the 21st century, China is continuously making considerable advances in work safety. In 2001, there was an important change in work safety supervision and management; this was the foundation of the State Administration of Work Safety (SAWS), a national safety administrative body which is responsible for work safety regulation and inspections, industrial hygiene, work safety training, accident investigation and analysis, and similar functions. The year 2002 is another important milestone in work safety management – China enacted the Law on Work Safety, aimed at enhancing accountability for the entire of stakeholders and stiffens penalties for poor work safety management. In 2004, the State Council defined the ambitious vision in its document, *Decision on Further Enhancing Work Safety*, to serve as an effective catalyst that motivates decision making to support the formulation of comprehensive work safety strategy, and calls for pronounced reductions in the number of workplace casualties and major accidents by 2010.

Also, a wide range of work safety plans and strategies has been formulated and implemented to promote work safety. As a result of the joint efforts of various government departments and organizations at central and local levels on the development and implementation of work safety plans and strategies, especially in the fields of safety education, legislation, enforcement, engineering, and safety assessment, work safety has continued to improve in recent years. The number of work-related accidents and fatalities has fallen from peak levels and exhibited a significant downward trend. In 2008, work-related fatalities toll was registered at 91,172, falling below the 100,000 mark for the first time since 1995. Although the progress is rather slow, there is a large room for improvement.

1.2 Work safety targets

For a work safety plan or strategy to be successfully implemented and accomplished, it is generally believed that realistic quantified work safety targets should be set/ necessary. In the area of work safety, quantitative targets represent the results that policy makers or public administrators wish to achieve for a certain geographic area (country, province, region, etc.) in a certain period of time. In this paper, a quantified work safety target is referred to as a permitted limit or index that is set by a national or local government to reduce the number of accidents or people killed or injured in accidents. For instance, China set a yearly target for reducing work-related fatalities (e.g., not more than 89,900 in 2009) since 2005. In many countries, such as the United States, UK, Japan, Australia, Singapore, and Viet Nam, quantified safety targets may be expressed as targeted percentage reductions in fatalities or fatality rates, or workplace injuries, and adopted as a key component in work safety programs in many sectors, e.g., mining, construction, hazard chemicals, and especially in road transportation (U.S. Department of Labor, 2006; National Occupational Health and Safety Commission, 2002; Health and Safety Commission, 2004; Ministry of Health, Labour and Welfare, 2008; Bureau for Safe Work, 2006; Workplace Safety and Health Advisory Committee, 2007).

In the broader arena of safety target (especially in road safety target), a number of studies conducted by researchers and organizations have attempted to address the issue of target setting, distribution and monitoring. Peter Marlow (2005) reviewed literatures and developed full discussions on the effects of target setting on performance, the characteristics of targets, the influences on target commitment, and the use of incentives to achieve targets, and finally proposed recommendations to the UK health and safety initiative 'Revitalising Health and Safety'. ETSC (2003) gave a research report to identify best practice in risk assessment and target setting in transport safety and the scope for their wider application and further useful development. LTSA (1998) described the process of setting road safety targets, including discussion of the factors which determine the target levels, and further discussed how to estimate the amount of resources required to achieve the targets. Elvik (1993) made an early attempt to study this problem using rigorous statistical analyses, and showed that Norwegian counties that had set quantified safety targets had a better performance in road safety than those that had set qualitative targets only, and that the best performance was achieved in counties with highly ambitious quantified targets. A more comprehensive study was again conducted by Elvik (2001) to examine the effects of setting road safety targets by national and local governments in different countries. The TRL (2000) generated a report to describe the flexible future methodology that has been developed to estimate the future number of casualties which could be taken into account in setting the new casualty reduction targets. Also Broughton (2010) summarizes the statistical analyses of accident data from 1983 to 1998 upon which the forecasts were based, and compared the approach with alternatives that had been adopted elsewhere. Robert Raeside (2004) developed an exploration of recent casualty time series in Great Britain, and used autoregressive models to forecast if targets for the number of those killed or seriously injured in vehicle accident can be met. By using a before-and-after analysis method, Wong et al. believed that the establishment of quantified road safety targets had an appreciable association with road fatality reduction during the period in which the targets were in effect (Wong et al., 2006, 2010; Allsop et al., 2011). Tolón-Becerra et al. (2009) proposed a simple method of distributing the total effort required to reach the EU target of halving fatalities in a territory in such a way that those areas with a higher proportion of fatalities have the highest targets and the sum of all the areas is the 50% reduction.

From reviews of past studies and documents, it seems that there are many discussions about the use and development of safety target setting model in road transportation, but few research in this field were structured and summarized in other industries, this may be partly because road accidents constitute such a major public safety problem world-wide that more attention was focused on this issue, or in respect that the process of setting and forecasting occupational safety targets based on scientific studies, if any, was not well researched and documented. So there is a need for effective assimilation of road safety target concepts into occupational safety study. Meanwhile, work safety targets for the reduction of accidents and fatalities should lie within a reasonable and achievable range in a theoretical setting, but most agencies appear to base safety targets on arbitrary standards or marginal adjustments to trends. This highlights the need of establishing scientific approaches for setting and forecasting safety targets to promote the implementation of work safety plan or policy on a national or regional scale.

2. Literature review and background

2.1 Model-based target setting

Marsden and Bonsall (2006) summarized the process of setting targets for transportation policy and grouped the approaches for setting performance targets into three categories: (1) model-based, (2) extrapolation and evidence-led/ based judgment, and (3) aspirational approaches. The model-based approach is the most rigorous in setting targets. It requires developing models of transportation users and system interaction, meaning that the relationships between the performance measures and influencing factors should be established. As for work safety, however, there are several influencing factors to be identified, and the relationship between work safety and its causal factors needs to be well understood. To generate a more rigorous model, we also need accurate time-series data about various influencing factors to forecast the future work safety scenarios and extrapolation of trends in the number of fatalities and accidents, and the potential safety effects of work safety programs that are likely to be implemented.

2.1 Influencing factors

As is known, a variety of factors (some are intangible factors) have been found to be crucial for determining work safety outcome (in terms of the number of occupational accidents, fatalities and injuries or the relative rates thereof), either directly or indirectly, including fluctuating economic trends, educational status, technologies used (low automation, discontinuous operating) (Bruno Fabiano et al., 2009), work conditions, safety training, and social factors (Jovica Jovanović, 2004; Asim SAHA et al., 2004). The first aim of this paper is to take a holistic approach to review the evidence for these factors as documented in the available economic, sociological and technological literatures and implications of work safety for whole workers, economies and societies, and to provide an assessment of which of the factors may most contribute towards changes in the incidence of workplace fatalities and injuries overtime.

2.2.1 Economic factors

Work is by definition an economic activity, and there is a sporadic but long tradition in the study of the relation between the economic cycle and occupational safety at work (Anna & Antonis, 2010; Eila Kankaanpää et al., 2008; Rhys Davies et al, 2009; Yoshitsugu Kitazawa, 2010; Liu, 2005; Rhys Davies & Paul Jones, 2005; Rhys Davies & Peter Elias, 2000; D Loomis et al., 2004). Fernando et al. (2004) reviewed two theories relevant to the impact of economic fluctuations or cycles over the effect indicators of safety, and put forward an integrative theory to demonstrate the relationship. The statistical analysis made by Alfred Barth et al. (2007) showed that a growing economy is associated with declining rates (fatal and non-fatal), and two mechanisms were discussed: 1) rising GDP is accompanied by greater investment in safer technologies and occupational safety measures; and 2) booming economies are associated with a reduced risk of unemployment, which is already known to be a risk factor for occupational injuries. However, this pattern may not clearly applies in the developing world for which it will take many years, if following historic trends, to achieve work safety situations of high-income economies (Elizabeth Kopits & Maureen Cropper, 2003, 2005, 2008; A. Saloniemi & H. Oksanen, 1998). Research suggests that the fatality risk level caused by workplace accidents at first increases as economic and social development and industrialization increases, but eventually declines as incomes grow and fatalities per unit GDP decrease due primarily to improved work safety. (The fatality risk level relates in a non-linear way to the income level and the long-term risk decay to the long-term increase in the income level of many countries (Koornstra, 2007; see Péter Holló, Vojtech Eksler & Joanna Zukowska, 2010; Rhys Davies and Paul Jones, 2005).)

We had considered that the economic status would be a strong confounder of the relationship in question based on the following excerpt from an ILO Report (2007): 'A heavier toll of occupational fatalities and injuries was still occurring in developing countries whilst industrialized countries have been a clear reduction in the number of serious injuries as a result of structural changes in the nature of work and real improvements in making the workplace healthier and safer.'

2.2.2 Sociological factors

It is believed that economic development will lead to profound changes in social structure and organization. Also the application of sociological theories in organizational safety is not new. Some studies have discussed the association between work safety and its contributing factors in the broader social and cultural context (Britkov, V. & Sergeev, G, 1998; Suman Rao, 2007; Pouliakas and Theodossiou, 2010; see Dembe et al, 2004). Many factors of social nature, including political-economic characteristics (Dana Loomis et al., 2009), regulatory functions and mechanisms, legal aspects and standard setting, safety culture or climate (Gordon S. Smith et al., 2006), as well as inspection, investigation and enforcement measures, safety policy interventions (A.R. Hale et al., 2009), can create preconditions that encourage proactive accident prevention (Fitzgerald, 2005; Roxane L. Gervais et al., 2009; Rhys Davies & Paul Jones, 2005; Roni Factor et al., 2007; Joanne M. Nigg et al., 1993). The significance of sociological factors and their impact on the situation both for work safety and for the dynamics and development of a society as a whole can be best seen by looking at what happened in China over the last few decades on the tails of an economic boom, and cultural and societal transformation.

The social and economic changes have had a significant impact on unemployment (rate), labor market, urbanization, public consciousness of safety, which were often accompanied by a change in the coverage of labor protection and protection systems, causing new challenges for the protection of workers at work (Brigitte Froneberg, 2005; Pouliakas and Theodossiou, 2010).

Level of education, which is generally considered as one important proxy of socioeconomic status of people, was strongly associated with the risk for being injured in an occupational accident (Swaen et al., 2003). It is believed that those, especially among male workers, with a lower level of education have higher rates of accidental injury or fatality at work (in traffic accidents, occupational accidents, home and leisure accidents, et al) than the well-educated group, mainly because of their 'accident-proneness', more feasibility of exposure to danger (for example, less ability to use protective devices, or to process information rapidly and accurately in a emergency), and more susceptibility (Johan P. Mackenbach, 2006; C. Borrell et al, 2005).

2.2.3 Technological factors

Over the last several decades dramatic advances have been seen in the development of new techniques and methods in every industry. Technological improvements do not only reduce or eliminate some types of accidents (Nancy et al., 1998); they may also transfer accident risks to different types of tasks, groups of workers or production phases, as well as create new risks, and make injuries more severe (Laflamme,1993; Vera, 1996; Hovden, Albrechtsen & Herrera, 2009). Lewchuk (1991), for France, and Aldrich (1997), for the United States, suggest that mechanization and new factory work practices – as two of the main factors – could have increased the level of risk. Also shifts in demand between sectors and technological advances within sectors can contribute to changes in the occupational composition of employment that will influence rates of workplace injury (Rhys Davies & Paul Jones, 2005). Robert B. Noland (2004) reviewed that medical care and technology improvements, in terms of high-tech emergency medical response systems, advanced trauma care procedures, specific surgical intervention techniques, have been seen dramatically effective in fatality reductions in traffic accidents. In particular, technological advances in industrial activities can give rise to improvement in productivity and in occupational health and safety, but not necessarily simultaneously (Bruno Fabiano et al., 2009).

3. Methodology

3.1 Modeling thoughtway

The target setting model is essentially a forecasting model in that what a safety target is built on and whether it will be reached should depend not only on the net effect of new measures of work safety policy intervention programmes, but also on the safety outcome forecasts (in terms of work-related fatalities) that are strongly based upon analysis of past changes, present status and ongoing trends (Stefan Siegrist, 2010). Therefore a forecast representing the continuation of recent trends throughout the forecasting period is necessary.

To be better understood, the Grey Model based forecasting methodology shown in Figure 2 is structured in two major parts; they are forecasting work-related fatalities and preparing reduction targets. The former part is a key element in the whole model because it provides a sound theoretical prediction foundation for the latter, taking account of any factors which may have a substantial influence upon the target without additional intervention of safety policy. From a time series forecasting standpoint, many consider a multivariate approach may shade more light on influencing factors, and to be the better choice over the univariate model. In the second part, the before-mentioned forecasts are adjusted by considering the likely effects of work safety policy intervention which have been reckoned as potential determinants of geographic or temporal variation in occupational fatality and injury rates. So the methodology relies on an assumption that the safety policy intervention works on the safety outcome in a different way from other influencing factors, which may make it possible to assess the feasibility of target setting model and the likely individual effect of each safety measure.

Figure 2: Illustration of the target setting and forecasting method

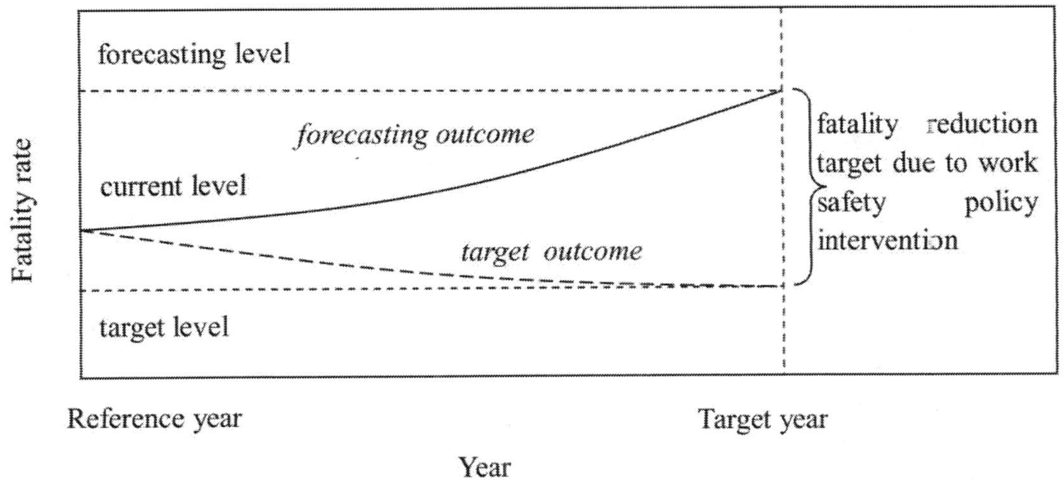

3.2 Accident / fatality forecast model

3.2.1 Grey forecast theory

Among all of more than 300 forecasting tools, many qualitative & quantitative methods, including, but not limited to, expert system, Delphi method, multivariate or univariate regression analysis, exponential smoothing, neural networks, trend prediction, Grey theory, Markov Switching model, and some empirical rules (Csaba KOREN & Attila BORSOS, 2009; El-Basyouny and Sayed, 2009; Rune Elvik, 2011), are used for the purposes of forecasting work-related accidents and fatalities, from a time series perspective, to set safety targets, or estimate the effect of a safety policy, programme or measure, (Chun-I Chen et al., 2010; Zheng, 2008; Ezra Hauer, 2010), and attempt to study all the time series components in order to identify a reasonably good fit to make specific forecasts into the future. Forecasting the future trends of work-related safety, however, is generally an uncertain and difficult process, for the outcome will depend on the future change of causal factors, and is often characterized by complex and multiple criteria. Thus traditional forecasting methods are probably constrained by many underlying hypotheses, and hindered by insufficient amounts of sample data which may perform fluctuations in time series that do not meet with a growth trend or seasonal and periodic behavior of fatality rates.

Grey theory, originally proposed by Deng in 1979, is a truly multi-disciplinary and systematic analysis methodology that deals with grey systems that are characterized by poor information, and has drawn increased attention from academic society since its inception mostly because of its special capability of rectifying the above disadvantages. The fields covered by grey theory include systems analysis, data processing, modeling, forecasting, decision making and control. Recently, the two hot topics of grey theory, grey relational analysis (GRA) and grey forecasting model, have been successfully employed in various fields and have proven good performance (Yen-Tseng Hsu et al., 2009). Grey forecasting models have been widely used and demonstrated promising results in various applications, such as environmental engineering, agriculture, communications, construction, transportation, economics, medical treatments, education, geology, management (Chien-Hua Wang, 2010; Yonghong Hao et al. 2006), and other microcosmic or macroscopic systems (Zou Lihua et al., 2009), and especially in cases where limited samples are required (Gou-sheng Liu & Jian-guo Yu, 2007), and have been constantly improved in performance by researchers (Tzu-Li Tien, 2009; Che-Tsung Tung & Yu-Je Lee, 2010), so many hybrid grey forecasting models were developed to increase the grey model accuracy of forecast, such as Grey-Fuzzy, Grey-Taguchi, Grey-Markov etc.(Chun-I Chen et al., 2010).

As mentioned above, the grey theory has proven to be more effective at obtaining reasonable and good predictive results with minimal data (e.g. only four observations are sufficient to characterize an unknown system in the simplest model (Chin-Tsai Lin & In-Fun Lee, 2009)), compared with traditional regression statistical methods which require mode data and time, and can be applied without consideration of assumptions regarding the statistical distribution of data. In addition, one of the most important features of grey theory is the accumulated generation operation (AGO) whose main purpose is to reduce the randomness of data to a monotonic increasing function. In fact, functions derived from AGO formulations of the original series are always well fitted to exponential functions.

3.2.2 Multivariate Grey forecasting model

Generally, the grey model is expressed as GM(n,m) where n is the order of the difference equation and m is the number of variables. The most simple and commonly used Grey Model, called GM (1,1), is a single variable first-order linear model, whose major weakness is its inability of identifying multiple factor interactions among the model parameters, and another form of Grey Model, written as GM(1,n), focuses mainly not upon forecasting but on the interrelationships of multiple independent variables, as well as the impact thereof upon the dependent variable (Baohong GUAN et al, 2010). To increase the accuracy and application potential of the Grey forecasting model, some studies have made attempts to develop improved versions, namely, multivariate grey forecasting models, represented as MGM(1,n), which are essentially an multivariate extension, but not a simple combination of GM(1,1)(s), and also excel GM(1,n)s in the ability to relate each variable to past values of itself as well as to past values of the other variables. Many applications have demonstrated its better performance than GM(1,1) and GM(1,n) (LOU et al., 2005; WANG et al., 2006; ZHENG, MA & GUAN, 2006; SU, CAO & WANG, 2007; LIU, WANG & LIU, 2007).

The MGM(1,n) model is processed in 5 main steps in the following.

- Obtaining original data series: Assume that $X(0)$ stands for the observed non-negative data series of n independent variables, namely,

$$X^{(0)}=\{X^{(0)}(1), X^{(0)}(2),\ldots, X^{(0)}(n)\}$$

- Taking the first order accumulated generating operation (1-AGO): generate the 1-AGO sequence $X(1)$ based on the initial sequence X(0),

$$X^{(1)}=\{X^{(1)}(1), X^{(1)}(2),\ldots, X^{(1)}(n)\}$$

where $X^{(1)}(k)=\sum_{i=1}^{k} X^{(0)}(i)$, for k=1, 2, …, n.

In the two $\boldsymbol{m}$-dimensional line vectors,

$X^{(0)}(k)=\left(x_1^{(0)}(k),x_2^{(0)}(k),\cdots,x_m^{(0)}(k)\right)^{\mathrm{T}}$ and $X^{(1)}(k)=\left(x_1^{(1)}(k),x_2^{(1)}(k),\cdots,x_m^{(1)}(k)\right)^{\mathrm{T}}$, $\boldsymbol{m}$ indicates the sample size.

- Putting forward the whitening differential equation: Thus, from grey system theory the first-order grey differential equation of MGM(1,n) and its whitening equation are obtained, respectively, as follows:

$$\begin{cases} \dfrac{dx_1^{(1)}}{dt} = a_{11}x_1^{(1)} + a_{12}x_2^{(1)} + \cdots + a_{1n}x_n^{(1)} + b_1, \\ \dfrac{dx_2^{(1)}}{dt} = a_{21}x_1^{(1)} + a_{22}x_2^{(1)} + \cdots + a_{2n}x_n^{(1)} + b_2, \\ \vdots \\ \dfrac{dx_n^{(1)}}{dt} = a_{n1}x_1^{(1)} + a_{n2}x_2^{(1)} + \cdots + a_{nn}x_n^{(1)} + b_n \end{cases}$$

Let A and B denote, respectively, two regression coefficient matrices:

$$A=\begin{bmatrix} a_{11} & a_{12} & \cdots & a_{1m} \\ a_{21} & a_{22} & \cdots & a_{2m} \\ \vdots & \vdots & \vdots & \vdots \\ a_{m1} & a_{m2} & \cdots & a_{mm} \end{bmatrix}, \text{ and } B=\left(b_1,b_2,\cdots,b_n\right)^{\mathrm{T}}$$

Then the dynamic differential equation of MGM(1,n) can be represented by/ rewritten as

$$\frac{dX^{(1)}(t)}{dt}=AX^{(1)}(t)+B$$

where t denotes the independent variables in the system. The vectors A and B becomes by the least-square method as

$$\hat{D}=\left[\hat{A},B\right]^{\mathrm{T}}=\left(L^{\mathrm{T}}L\right)^{-1}L^{\mathrm{T}}Y$$

where

$$\boldsymbol{L}=\begin{bmatrix} \left(x_1^{(1)}(1)+x_1^{(1)}(2)\right)/2 & \cdots & \left(x_m^{(1)}(1)+x_m^{(1)}(2)\right)/2 & 1 \\ \left(x_1^{(1)}(2)+x_1^{(1)}(3)\right)/2 & \cdots & \left(x_m^{(1)}(2)+x_m^{(1)}(3)\right)/2 & 1 \\ \vdots & \ddots & \vdots & \vdots \\ \left(x_1^{(1)}(n-1)+x_1^{(1)}(n)\right)/2 & \cdots & \left(x_m^{(1)}(n-1)+x_m^{(1)}(n)\right)/2 & 1 \end{bmatrix}$$

and

$$\boldsymbol{Y}=\begin{bmatrix} x_1^{(0)}(2) & x_2^{(0)}(2) & \cdots & x_m^{(0)}(2) \\ x_1^{(0)}(3) & x_2^{(0)}(3) & \cdots & x_m^{(0)}(3) \\ \vdots & \vdots & \ddots & \vdots \\ x_1^{(0)}(n) & x_2^{(0)}(n) & \cdots & x_m^{(0)}(n) \end{bmatrix}$$

From formula (6), the estimated value of the two parameter matrixes, $\hat{A}$ and $\hat{B}$, can be obtained by using least square method.

Thus the solution of grey differential equation, is given by

$$\hat{X}^{(1)}(k)=e^{\hat{A}(k-1)}X^{(1)}(1)+\hat{A}^{-1}(e^{A(k-1)}-I)B$$

- Calculating the forecasting value: Get the predicted result from inverse accumulated generating operation (Inverse IGO) by

  $\hat{X}^{(0)}(k)=X^{(1)}(k)-X^{(1)}(k-1)$, for k=2, 3, …

  where $\hat{X}^{(0)}(1)=X^{(1)}(1)$.

  But, using the original GM(1,1) model may face problems in that this model cannot reflect real system growth trends among different periods and it is not suitable for long-term forecasting.

3.2.3 Multivariate linear regression model based on MGM(1,n)

Several mathematical models are available in the literature to describe the fuzzy relationship between dependent variable and multiple independent variables thereof, and quantify the various influences of independent variables on dependent variable. In this work, the multivariate linear regression model based on MGM(1,n) presented by Liu *et al* (2007) is used to represent the linear association between the independent variables and the dependent variable. A form of multivariate linear regression model based on MGM(1,n) can be shown/ taken as

$$y(k) = a_0 + a_1 x_1^{(0)}(k) + a_2 x_2^{(0)}(k) + \cdots + a_m x_m^{(0)}(k)$$

where $y(k)$ is response matrix, a_i (for i=1,2,…,m) is regression coefficient matrix and a_0 is fitting error matrix.

This ordinary model assumes that the data observed over time, which usually have strong correlations among neighboring observations, are independent, and therefore is not appropriate for this analysis in the paper. Here we bring forward an improvement of Multivariate linear regression model expressed as

$$y(k) = a_0 + \hat{\alpha}\hat{X}^{(0)}(k)$$

where $\hat{X}^{(0)}(k) = (\hat{x}_1^{(0)}(k), x_2^{(0)}(k), \cdots, x_m^{(0)}(k))$ and $\hat{\alpha} = (\hat{a}_1, a_2, \cdots, a_m)$, in which $\hat{x}_i^{(0)}(k)$ is the fitted value of $x_i^{(0)}(k)$ by using Multivariate Grey forecasting model, and $\hat{a}_i$ is the estimated value of a_i by substitforuting the values of $\hat{x}_i^{(0)}(k)$ and $y(k)$ into the equation. Finally we could get the fitted value of $y(k)$ denoted as $\hat{y}(k)$.

As is a combination of Multivariate Grey Forecasting Model and Regression Model, the developed model should be tested for its forecast accuracy through the standard deviation and mean square deviation of original data series by using statistical analysis and posterior-variance-test techniques.

The standard deviation of original data series and residual errors, denoted, respectively, by S_1, S_2, are defined as follows:

$$S_1 = \sqrt{\frac{\sum_{k=1}^{n}\left[\hat{y}(k) - \bar{y}(k)\right]^2}{n-1}}$$

$$S_2 = \sqrt{\frac{\sum_{k=1}^{n}\left[\varepsilon(k) - \bar{\varepsilon}\right]}{n-1}}$$

where $\varepsilon(k) = y(k) - \hat{y}(k)$ and $\bar{\varepsilon} = \frac{\sum_{k=1}^{n}\varepsilon(k)}{n}$.

Then we can calculate the posterior variance ratio and infinitesimal error probability by the formulas

$$\begin{cases} c = S_2/S_1 \\ p = \mathrm{P}\{|\varepsilon(k) - \bar{\varepsilon}| < 0.6745 S_1\} \end{cases}$$

and estimate the forecast accuracy of this framework according to Table 1 where p is the probability of small error, and c is ratio of posterior error.

Table1. the accuracy test degree of grey forecasting model

Degree	Test Results	p	c
I	Good	p≥0.95	c≤0.35
II	Better	0.80≤p < 0.95	0.35 < c≤0.50
III	Medium	0.70≤p < 0.80	0.50 < c≤0.65
IV	Poor	p < 0.70	c > 0.65

3.3 Preparing safety target using Accident Reduction Factor

Accident Reduction Factor (ARF), also known as Accident Modification Factor (AMF), is a piece of terminology initially brought forward in road transportation, and used i.e. in cost-optimization procedure to rank safety improvements at locations identified as high accident locations using traffic accident data (Kenneth R. Agent et al., 1996; Gary A. Davis, 2000). An ARF is the percentage of road accident reduction that might be expected after implementing a given countermeasure; that is to say, in some cases, the ARF is negative, i.e. the implementation of a countermeasure is expected to lead to a percentage increase in road accidents. Therefore, many existing AMFs are derived from **before-after** analysis of actual countermeasure implementation. Indeed, such **before-after** studies, as opposed to cross-sectional/regression-type analysis, will produce the best AMF estimates, but only if conducted properly.

As above-mentioned, ARFs provide a computationally simple and quick way of estimating accident reductions.

By using the ARF model, this paper presents an algorithm for estimating the aggregated effectiveness performance of the work safety policy interventions that have been implemented for a period, i.e. one year, in a region. The algorithm is made on the assumption that the safety policy or program, including safety legislation, regulation, inspection and other intervention by authorities at all levels, is introduced and implemented at the beginning of a year, and the number of fatalities after the improvement is divided by the number of fatalities before the improvement to calculate the ARF, which is expressed as follows

$$ARF = 100 \times \left[1 - \frac{a_B}{(E_A/E_B) \times a_A}\right] = 100 \times \left[1 - 100 \times \frac{a_B/a_A}{I_{GDP}}\right]$$

where a_B and a_A denote, respectively, the annual number of fatalities before and after the safety policies are implemented, E_A and E_B, respectively, the annual GDP before and after the safety policies are implemented, and I_{GDP} represents the GDP growth rate.

The change of work-related injuries and fatalities in a period is the result of various factors, among which economic activity is believed to be a dominant one, thus the parameter I_{GDP} is introduced to estimate the actual effectiveness of the safety measure by eliminating the influence of economy growth (marked by GDP Index).

However, it is not an easy process to estimate precisely the safety outcomes of intervention programmes that consist of several measures that interact with each other when they are implemented at the same time; even the estimation of effects of a single measure should take into account five parameters, namely, A) potential number of (severe and fatal) injuries that could be influenced by the measure, B) area of theoretical impact, C) effectiveness, D) degree of implementation, and E) degree of compliance (Stefan Siegrist, 2010). The introduction of multiple safety measures leads to synergy effects, which can increase the effects of the single measure. In this work we utilized a model of aging intervention degree (*AID*) proposed by Ren et al. (2008) to estimate the combined effect of safety measures in the target year, and thence to adjust the base forecasting model predictions, accounting for the safety effects of various measures, to obtain the fatality reduction targets.

4. Materials and data

4.1 data sources

With the aim of determining which of the economic, sociological and technological factors could forecast the work safety situation in subsequent years, we utilized a multiple indicator approach to measure these factors by using various items or variables. This process for generating items included successive stages. First, we conducted an exhaustive review of related literatures in public journals and practices in government documents, which provided a pool of items that were mostly possible to theoretically measure some of the factors considered in this research. But we observed that some items were unquantifiable or unavailable of data, thus we opted here to build an initial structure of 20 items, adapted to the theoretical model developed in this study, and then a draft of the questionnaire to be used to collect the information was elaborated.

Next the questionnaire together with a presentation letter was sent by mail or E-mail to several members of the Guangdong Provincial Work Safety Expert Panel which was made up of a number of specialists from government agencies and other regulators, NGOs, education and research organizations as well as industrial corporations and aimed at bringing together their experiences, perspectives and expertise to advise the Bureau and other agencies on matters related to work safety.

Finally, we received a total of 18 questionnaires. Some of these had been incorrectly completed in some respect (some data was either missing or erroneous, or some questionnaires had serious internal inconsistencies), so we decided to eliminate them, following the complete case approach described by Liang (2007). In this way, we obtained a total of 12 valid responses, and ultimately identified out 7 critical factors (Table 2) that are easily expressed by available observations as independent variables, and utilized 1 indicator which can accommodate the outcome of work safety, e.g. the number of fatalities or fatality rate, as dependent variable.

Table 2. fatality influencing factors and variables

Characteristics	Factors	Variables/Notation
Economic	GDP (trillion Yuan RMB)	*X1*
	GDP per capita (thousand Yuan RMB)	*X2*
Sociological	Percentage of non-agricultural product to GDP (%)	*X3*
	Percentage of urban population to total (%)	*X4*
	Number of high school students per 10 thousand people	*X5*
Technological	Number of doctors per 10 thousand people	*X6*
	Percentage of expenditure on R&D to GDP (%)	*X7*
Outcome	Number of work-related fatalities	*Y*

Note: GDP – gross domestic product; R&D – research and development

4.2 Data analysis

To illustrate our methodology, we tabulated eligible records data of above variables by calendar year. The information was extracted from the yearbook, gazette and official reports, covering the time period from 1990 to 2009 (in Table 3). The variables/data values observed in 1990-2007 (sample size n=18) were/are used for forecasting/ to predict the values for the target year and those in 2008 for accuracy validation. We applied the Multivariate Grey Forecasting Model in all sample analyses, which we carried out with the Matlab procedure in version 7.0 for Windows system, and obtained the fitted values of variables for the years 1990-2009 (in Table 4). Then we performed multivariate regression analysis by using the fitted values of explanatory variables to predict the response variable, y(k), (in Table 5) and get regression coefficients and variance (in Table 6) through SPSS analysis.

Table 3. Original data of variables for 1990 to 2007

Year	*Y*	*X1*	*X2*	*X3*	*X4*	*X5*	*X6*	*X7*
1990	68342	1.87	1.644	72.9	26.41	447	12.89	0.71
1991	72618	2.18	1.893	75.5	26.94	451	12.87	0.72
1992	78568	2.69	2.311	78.2	27.46	457	12.96	0.70
1993	96298	3.53	2.998	80.3	27.99	454	13.02	0.62
1994	99672	4.82	4.044	80.2	28.51	476	13.29	0.50
1995	103543	6.08	5.046	80.1	29.04	511	13.45	0.60
1996	101600	7.12	5.846	80.3	30.48	542	13.50	0.60
1997	101037	7.90	6.420	81.7	31.91	566	13.72	0.64
1998	104126	8.44	6.796	82.4	33.35	588	13.71	0.69
1999	108086	8.97	7.159	83.5	34.78	621	13.96	0.83
2000	117718	9.92	7.858	84.9	36.22	660	14.10	1.00
2001	130491	10.97	8.622	85.6	37.66	697	14.19	1.07
2002	139393	12.03	9.398	86.3	39.09	733	14.36	1.23
2003	137070	13.58	10.542	87.2	40.53	763	14.46	1.13
2004	136755	15.99	12.336	86.6	41.76	785	14.66	1.23
2005	127089	18.32	14.053	87.8	42.99	785	14.82	1.34
2006	112822	21.19	16.165	88.5	43.90	788	15.18	1.42
2007	101480	24.95	18.934	88.7	44.94	777	15.24	1.49

Table 4. Fitted value of variables for 1990-2009

Year	relative error	residual error ε(k)	$\hat{y}(k)$	$y(k)$	*X1*	*X2*	*X3*	*X4*	*X5*	*X6*	*X7*
1990	1.36	927	67415	68342	1.87	1.644	72.90	26.41	12.89	0.71	447
1991	-2.35	-1710	74328	72618	2.12	1.841	75.64	26.97	12.86	0.72	450
1992	-0.64	-499	79067	78568	2.60	2.238	78.67	27.45	12.95	0.70	449
1993	3.90	3756	92542	96298	3.67	3.120	79.73	27.82	13.09	0.62	463
1994	-2.47	-2459	102131	99672	4.95	4.153	79.92	28.40	13.25	0.55	485
1995	-0.28	-294	103837	103543	6.13	5.086	80.05	29.31	13.40	0.52	509
1996	-0.41	-412	102012	101600	7.07	5.800	80.55	30.50	13.54	0.56	534
1997	0.71	719	100318	101037	7.77	6.311	81.47	31.88	13.67	0.65	561
1998	1.93	2010	102116	104126	8.35	6.721	82.65	33.34	13.80	0.76	591
1999	-1.47	-1587	109673	108086	8.97	7.158	83.84	34.80	13.93	0.88	626
2000	-1.68	-1983	119701	117718	9.75	7.728	84.84	36.25	14.06	0.98	663
2001	0.24	307	130184	130491	10.79	8.490	85.60	37.67	14.20	1.06	700
2002	1.35	1877	137516	139393	12.10	9.463	86.17	39.09	14.34	1.13	735
2003	-1.55	-2122	139192	137070	13.71	10.652	86.68	40.50	14.49	1.19	764
2004	1.23	1687	135068	136755	15.65	12.076	87.22	41.85	14.66	1.25	784
2005	0.93	1176	125913	127089	17.98	13.791	87.84	43.09	14.86	1.32	794
2006	-1.32	-1485	114307	112822	20.86	15.906	88.49	44.15	15.07	1.40	792
2007	-0.27	-275	101755	101480	24.48	18.583	89.03	44.96	15.31	1.49	778
2008	0.40	927	90803	91172	29.13	22.034	89.23	45.44	15.56	1.58	750
2009	—	—	83731	—	35.15	26.519	88.83	45.54	15.81	1.68	708

Table 5. Significance test of partial regression coefficients

Coefficients [a]

Model	Variables	Statistics				
		Unstandardized Coefficients		Standardized Coefficients	t	Sig.
		B	Std. Error	Beta		
1	(Constant)	215662.264	182630.709	—	1.181	0.261
	*X*1	8705.999	1333.087	3.194	6.531	0.000
	*X*3	7515.940	626.039	1.669	12.006	0.000
	*X*4	-36352.903	4009.268	-11.625	-9.067	0.000
	*X*5	-26983.264	16709.865	-1.074	-1.615	0.132
	*X*6	59851.766	14064.267	0.979	4.256	0.001
	*X*7	1237.047	99.936	7.912	12.378	0.000

a: Dependent Variable: *Y*

Table 6. Variance analysis (ANOVA) of regression coefficients

ANOVA [b]

Model		Sum of Squares	df	Mean Square	F	Sig.
1	Regression	7911718475.300	6	1318619745.883	315.103	0.000 [a]
	Residual	50216777.437	12	4184731.453	—	—
	Total	7961935252.737	18	—	—	—

a: Predictors: (Constant), *X*7, *X*1, *X*6, *X*3, *X*5, *X*4

b: Dependent Variable: *Y*

5. Results and Discussion

The output of SPSS provides information concerned with the parameters of predictors, and an assessment of overall significance of the model.

Table 6 reports on an analysis of variance (ANOVA) that tests whether the model is significantly good at forecasting the outcome. The F-ratio represents the ratio of the improvement in prediction that results from fitting the model (labeled 'Regression' in the table), relative to the inaccuracy that still exists in the model (labeled 'Residual' in the table). If the improvement due to fitting the regression model is much greater than the inaccuracy within the model then the value of F will be greater than 1 and SPSS calculates the exact probability of obtaining the value of F by chance. For this model the F-ratio is 315.103, which is very unlikely to have happened by chance as Sig. $(p)<.001$, then we can interpret these results as meaning that the developed model is statistically significant and can improve our ability to predict the outcome variable.

Table 5 gives us estimates for the coefficients (labeled 'B') contained in the multiple regression equation for corresponding predictors, the confidence with which we can support the estimate for each such estimate (see the column "Sig."), and the t value (and 2 tailed p value) used in testing the null hypothesis that the coefficient/ parameter is 0.

From the b values, which imply the relationship between dependent variable and each predictor, and to what degree each predictor affects the outcome if the effects of all predictors are held constant (and the individual contribution of each predictor to the model), we can tell that the predictors, $X1$, $X3$, $X6$ and $X7$, indicate positive relationships with the dependent variable, and other ones, $X4$ and $X5$, show negative relationships. The results, however, were a little opposite to what we had expected that the predictors, $X6$ and $X7$, respectively denoting the number of doctors per 10 thousand people and the percentage of expenditure on R&D to GDP, would exhibit theoretically inverse relationships with the dependent variable, namely the number of work-related fatalities. This may be because the records were observed on an irregular basis.

Each of these beta values in the table has an associated standard error indicating to what extent these values would vary across different samples, and these standard errors are used to determine whether or not the b values differs significantly from zero (using the t-statistic that you come across last year). Therefore, if the t-test associated with a b value is significant (if the value in the column Sig. is less than 0.05) then that predictor is making a significant contribution to the model. The smaller the value of sig. (and the larger the value of t), the greater the contribution of that predictor. For this model, *X*1, *X*3, *X*4, *X*6 and *X*7, whose *p* values are less than 0.05, are significant predictors of the dependent variable. From the magnitude of the t-statistics we can see that both *X*3 and *X*7 had slightly more impact than the other predictors. It can be also seen in Table 5 that *X*2 has not entered into the model.

The Standardized Beta Coefficients give a measure of the contribution of each variable to the model. A large value indicates that a unit change in this predictor variable has a large effect on the criterion variable. The t and Sig. (*p*) values give a rough indication of the impact of each predictor variable - a big absolute t value and small *p* value suggests that a predictor variable is having a large impact on the criterion variable.

The *b* values and their significance are important to look at; however, the standardized versions of the b values are easier to interpret because they are not dependent on the units of measurement of the variables. The standardized beta values are provided by SPSS and they tell us the number of standard deviations that the outcome will change as a result of one standard deviation change in the predictor. The standardized beta values (β) are all measured in standard deviation units and so are directly comparable: therefore, they provide a better insight into the 'importance' of a predictor in the model. Among the above predictors, *X*1, *X*4 and *X*7 are higher than other ones in terms of standardized beta values, which tell us that these predictors have more impact in the model.

As analyzed above, all the explanatory variables (predictors) are said to be statistically significant, and useful in forecasting the dependent variable (*Y*). With 6 predictors (+ a constant) and coefficients thereof, the estimated multivariate regression model here is:

$$y(k)=215662.264+8705.999x_1^{(0)}(k)+7515.940x_3^{(0)}(k)-36352.903x_4^{(0)}(k)$$
$$-26983.264x_5^{(0)}(k)+59851.766x_6^{(0)}(k)+1237.047x_7^{(0)}(k)$$

By substituting the fitted values in 2009 for the predictors accordingly in the above equation, we can forecast the number of fatalities in that year.

Then we can calculate the posterior variance ratio and infinitesimal error probability by

$$\begin{cases} c=S_2/S_1=1670.28/21031.64=0.0794<0.35 \\ p=\mathrm{P}\{|\varepsilon(k)-\bar{\varepsilon}|<0.6745S_1\}=1>0.95 \end{cases}$$

where $S_1=21031.64$, $S_2=1670.28$, $\bar{\varepsilon}=0.11$, and $\max|\varepsilon(k)-\bar{\varepsilon}|=3755.90$.

According to table 1, we know that the model accuracy degree ranks at I level, which demonstrate a good predicting performance of this model.

By using the data of national work fatalities and GDP, we could figure out the value of *F* separately for the years 1999-2008 (Table 7).

Table 7. Time series data of *AID* and *F*

Year	Fatalities	GDP Index	Aging Intervention Degree *AID*	Accident Reduction Factor *F*

1999	108086	107.6	283.6	10.48
2000	117718	108.4	274.5	15.32
2001	130491	108.3	345.4	16.70
2002	139393	109.1	445.9	14.18
2003	137070	110.0	592.9	7.57
2004	136755	110.1	669.9	8.95
2005	127089	110.4	788.0	2.56
2006	112822	111.6	1040.0	-0.94
2007	101480	111.9	1189.0	0.65
2008	91172	109.0	1327.1	-2.12
2009	—	—	1396.1	—

Then we can fit a regression equation/ model that included indicator variables for *AID* and *F*, and graph the regression equation and the data points.

$$F = 3.791\text{E}-08AID^3 - 8.215E-05AID^2 + 3.378\text{E}-02AID + 1.044\text{E}+01$$

Figure 3: The trajectory of *F* and *AID*, over the period 1990-2008

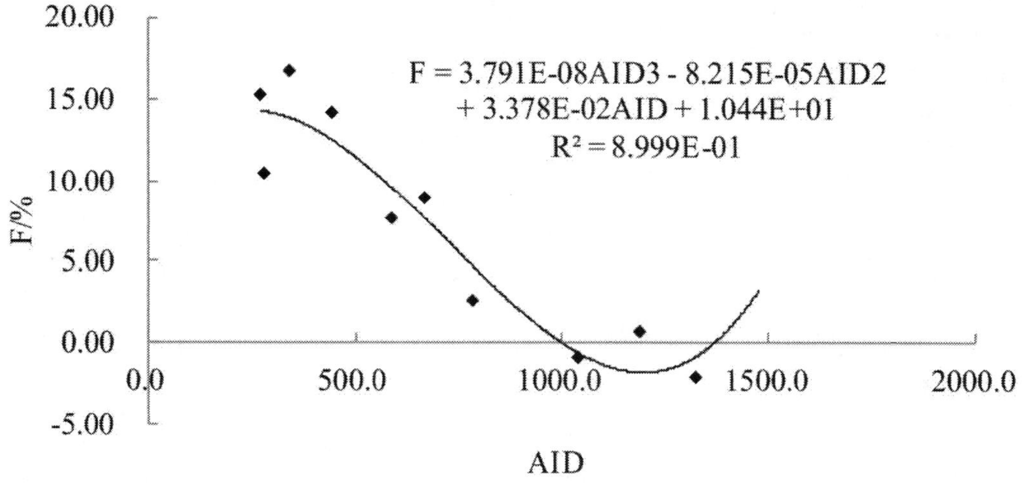

The non-monotone curve fitted to the data in Figure 3 suggests that *F* has an unclear correlation with work safety policy intervention, as long as the economic activity variables remain constant. When the value of *AID* is less than about 1000, *F* declines as *AID* grows (negative correlation), which means that the increase of safety policy intervention has not achieved its intended results. This is maybe because that the impact of any intensification of safety enforcement measures can occur with a lag due to the time involved in making the capital investments or organizational changes required for compliance with safety standards (Viscusi, 1986), that is, the safety intervention did not take effect immediately when implemented. On the other hand, it is one of the biggest challenges for further reduction in the number of fatalities and injuries because the total number is still at an unacceptable level. While the value of *AID* exceeds 1200, safety intervention exhibits its due effect.

Through the value of *AID* in 2009 which is estimated / obtained by extrapolating from the data in previous years, we can estimate the *ARF* is 0.64%, and set the safety target at 83195 (83731×(1-0.64%)=83195), taking account the predicted fatalities in 2009.

6. Conclusion

This paper has reached a threefold purpose. It first engaged in an interdisciplinary review of the current state of knowledge related to the theory, determinants and consequences of work safety, and managed to provide an overview and information/ new insights into how work safety outcome (in terms of the number or rate(s) of accidents, fatalities and injuries) are influenced by the economic, societal and technical characteristics. Secondly, it proposed a Grey theory based multivariate regression model, which may cast more light on factors that achieve the safety outcome than univariate methods, to be used for work safety target forecasting and setting. Finally this paper illustrated with an empirical case how to use the model as a technical tool for related agencies to judge of the work safety target and to estimate the effects of new safety policy interventions in the target year.

Despite proving to be a good technique for forecasting and demonstrating promising results, the proposed model is still limited by some drawbacks, one of which is the steep regularity of data needed; this would make it impossible to fully take account all the factors influencing work safety outcomes. On the other hand, quantitative data, though indisputably important, are hardly captured with regard to administrative factors, e.g., the effectiveness of safety policy intervention, so we had to either eliminate such items, or make an estimate by extrapolating or interpolating from other incomplete data series. The model case was illustrated at national level, and may not accurately applies in other individual cases, so further research is needed to explore an 'almighty' model that has higher performances not only on model fitting but also on forecasting, and to deeply investigate, with greater depth and precision, the correlation between the risk of work injury and the variety of causal factors thereof.

7. References

[1] U.S. Department of Labor, "Strategic Plan Fiscal Years 2006-2011", Washington, D.C., 2006, http://www.dol.gov/_sec/stratplan/

[2] National Occupational Health and Safety Commission, "National OHS Strategy 2002-2012", Canberra, 2002, http://www.nohsc.gov.au/nationalstrategy/

[3] Health and Safety Commission, "A Strategy for Workplace Health and Safety in Great Britain to 2010 and Beyond", London, 2004, http://www.hse.gov.uk/aboutus/strategiesandplans/

[4] Ministry of Health, Labour and Welfare, "11th Industrial Accident Prevention Plan 2008 to 2012", Tokyo, 2008, http://www.jisha.or.jp/english/

[5] Bureau for Safe Work, "National Programme on Labour Protection, Occupational Safety and Occupational Health up to 2010", Hanoi, Ministry of Labour, Invalids and Social Affairs, 2006, http://www.antoanlaodong.gov.vn

[6] Workplace Safety and Health Advisory Committee, Ministry of Manpower, "WSH 2015 - A Strategy for Workplace Safety and Health in Singapore", Singapore, 2007, http://www.wsh.sg

[7] Peter Marlow, "Literature Review on the Value of Target Setting, Buxton", Health & Safety Laboratory, 2005

[8] European Transport Safety Council, "Assessing Risk and Setting Targets in Transport Safety Programmes", Brussels, 2003

[9] Tony Bliss, Jagadish Guria, Wayne Jones, et al., "Setting Road Safety Targets: Working Paper 4", Wellington, Land Transport Safety Authority, 1998

[10] Rune Elvik, "Quantified Road Safety targets: a useful tool for policy making", *Accident Analysis & Prevention*, 25(5), 1993

[11] Rune Elvik, Quantified Road Safety Targets: An Assessment of Evaluation Methodology, Oslo, Institute of Transport Economics, 2001

[12] J. Broughton, R.E. Allop, D.A. Lynam, et al, The Numerical Context for Setting National Casualty Reduction Targets, Crowthorne, Transport Research Laboratory, 2000

[13] Jeremy Broughton, Jackie Knowles, "Providing the numerical context for British casualty reduction targets", *Safety Science*, 48(9), 2010

[14] Robert Raeside, "Predicting and monitoring Casualty Numbers in Great Britain", *Journal of Transportation and Statistics*, 7(1), 2004

[15] Wong, S.C., et al., "Association between setting quantified road safety targets and road fatality reduction", *Accident Analysis & Prevention*, 38(5), 2006

[16] S.C. Wong, N.N. Sze, "Is the effect of quantfied road safety targets sustainable ", *Safety Science*, 48(9), 2010

[17] Allsop, R.E., et al., "An update on the association between setting quantified road safety targets and road fatality reduction", *Accident Analysis & Prevention*, 43(3), 2011

[18] A. Tolón-Becerra, X. Lastra-Bravo, F. Bienvenido-Bárcena, "Proposal for territorial distribution of the 2010 EU road safety target", *Accident Analysis & Prevention*, 41(5), 2009

[19] Marsden, G., Bonsall, P., "Performance targets in transport policy", *Transport Policy*, 13(3), 2006

[20] Young-Jun Kweon, "Data-driven reduction targets for a highway safety plan", *Transport Policy*, 17(4), 2010

[21] Bruno Fabiano et al., "Port safety and the container revolution: A statistical study on human factor and occupational accidents over the long period", *Safety Science.*, 48(8), 2010

[22] Jovica Jovanović, Mirjana Aranđelović, Milan Jovanović, "Multidisciplinary aspects of occupational accidents and injuries", *Working and Living Environmental Protection*, 2(4), 2004

[23] Asim SAHA et al., "An Accident-Risk Assessment Study of Temporary Piece Rated Workers", *Industrial Health*, 42(2), 2004

[24] Anna Maria MOUZA, Antonis TARGOUTZIDIS, "The Effect of the Economic Cycle on Workplace Accidents in Six European Countries", *Ege Academic Review*, 10(1), 2010

[25] Eila Kankaanpää et al, "Economics for occupational safety and health", *SJWEH Supplement*, (5), 2008
[26] Rhys Davies, Paul Jones & Imanol Nuñez, "The impact of the business cycle on occupational injuries in the UK", *Social Science & Medicine*, 69(2), 2009
[27] Yoshitsugu Kitazawa, "Size of economic activity and occurrence of fatal traffic accidents: a count panel data analysis on Fukuoka prefecture in Japan", Discussion Paper Series, Faculty of Economics, Kyushu Sangyo University, Discussion Paper, March 2010, No. 41, http://www.ip.kyusan-u.ac.jp/J/kitazawa/ERRATA/errata_fp.html
[28] LIU Tie-min, "Orange GDP and its evolvement regularity", *Journal of Safety Science and Technology*, 1(2), 2005 (*in Chinese*)
[29] Rhys Davies and Paul Jones, "Trends and context to rates of workplace injury", HSE, 2005
[30] Rhys Davies, Peter Elias, "An Analysis of Temporal and National Variations in Reported Workplace Injury Rates", HSE, 2000
[31] D Loomis, D B Richardson, J F Bena, A J Bailer, "Deindustrialisation and the long term decline in fatal occupational injuries", *Occupational Environmental Medicine*, 61(7), 2004
[32] Fernando Terrés de Ercilla et al., "Economic fluctuations affecting occupational safety: the Spanish case", *Occupational Ergonomics*, 4(4), 2004
[33] Alfred Barth et al., "Economic growth and the incidence of occupational injuries in Austria", *Wien Klin Wochenschr*, 119(5–6), 2007
[34] Elizabeth Kopits, Maureen Cropper, "Traffic fatalities and economic growth", Washington, DC, The World Bank, 2003
[35] Elizabeth Kopits, Maureen Cropper, "Traffic fatalities and economic growth", *Accident Analysis & Prevention*, 37(1), 2005
[36] Elizabeth Kopits, Maureen Cropper, "Why Have Traffic Fatalities Declined in Industrialised Countries", *Journal of Transport Economics and Policy*, 42(1), 2008
[37] A. Saloniemi, H. Oksanen, "Accidents and fatal accidents - some paradoxes", *Safety Science*, 29(1), 1998
[38] Koornstra, M., "Prediction of traffic fatalities and prospects for mobility becoming sustainable-safe", *Sadhana*, 32(4), 2007
[39] Péter Holló, Vojtech Eksler & Joanna Zukowska, "Road safety performance indicators and their explanatory value: A critical view based on the experience of Central European countries", *Safety Science*, 48(9), 2010
[40] Rhys Davies & Paul Jones, "Trends and context to rates of workplace injury", HSE, 2005
[41] Donald J Wilson et al., "The Ratification Status of ILO Conventions Related to Occupational Safety and Health and Its Relationship with Reported Occupational Fatality Rates", *Journal of Occupational Health*, 49(1), 2007
[42] Britkov, V. & Sergeev, G., "Risk Management: Role of Social Factors in Major Industrial Accidents", *Safety Science*, 30(1-2), 1998
[43] Suman Rao, "Safety culture and accident analysis—A socio-management approach based on organizational safety social capital", *Journal of Hazardous Materials*, 142(3), 2007

[44] Pouliakas, K., Theodossiou, I., "An Inquiry into the Theory, Causes and Consequences of Monitoring Indicators of Health and Safety at Work", IZA Discussion Paper No. 4734, Bonn, 2010
[45] Dembe, A.E., Erickson, J.B. & Delbos, R., "Predictors of Work-Related Injuries and Illnesses: National Survey Findings", *Journal of Occupational and Environmental Hygiene*, 1(8), 2004
[46] Dana Loomis et al., "Political Economy of US States and Rates of Fatal Occupational Injury", *American Journal of Public Health*, 99(8), 2009
[47] Gordon S. Smith et al., "The relationship between safety climate and injury rates across industries: The need to adjust for injury hazards", *Accident Analysis & Prevention*, 38(3), 2006
[48] A.R. Hale et al., "Evaluating safety management and culture interventions to improve safety: Effective intervention strategies", *Safety Science*, 48(8), 2009
[49] Fitzgerald, M. K., "Safety performance improvement through culture change", *Process Safety and Environmental Protection*, 83(4), 2005
[50] Roxane L. Gervais et al., "Occupational safety and health and economic performance in small and medium-sized enterprises: a review", European Agency for Safety and Health at Work, 2009
[51] Roni Factor et al., "The social accident: A theoretical model and a research agenda for studying the influence of social and cultural characteristics on motor vehicle accidents". *Accident Analysis & Prevention*, 39(5), 2007
[52] Joanne M. Nigg et al., "Disasters and social change: consequences for community construct and affect", Paper presented at the annual meeting of the American Sociological Association, Miami, Florida, 1993
[53] Froneberg B., "Challenges in occupational safety and health from the global market economy and from demographic change - facts, trends, policy response and actual need for preventive occupational health services in Europe", *SJWEH Supplements*, (1), 2005
[54] Johan P. Mackenbach, "Health Inequalities: Europe in Profile", ROTTERDAM, the Netherlands, 2006
[55] C. Borrell et al., "Education level inequalities and transportation injury mortality in the middle aged and elderly in European settings", *Injury Prevention*, 11(3), 2005
[56] Nancy Stout, et al., "Traumatic Occupational Injury Research Needs and Priorities", A Report by the NORA Traumatic Injury Team, Cincinnati, National Institute for Occupational Safety and Health, 1998
[57] Vera L. G. Blank, et al., "Technological Development and Occupational Accidents as a Conditional Relationship: A Study of Over Eighty Years in the Swedish Mining Industry", *Journal of Safety Research*, 27(3), 1996
[58] Jan Hovden, Eirik Albrechtsen, Ivonne A. Herrera, "Is there a need for new theories, models and approaches to occupational accident prevention", *Safety Science*, 48(8), 2009
[59] Lewchuk, W., "Industrialization and Occupational Mortality in France prior to 1914", *Explorations in Economic History*, 28(3), 1991
[60] Aldrich, M. Safety First, "Technology, Labor and Business in the Building of American Work Safety, 1870-1939", The John Hopkins University Press,

Baltimore, 1997
[61] Robert B. Noland, "A Review of the Impact of Medical Care and Technology in Reducing Traffic Fatalities", IATSS Research, 28(2), 2004
[62] Stefan Siegrist, "Towards a method to forecast the effectiveness of national road safety programmes", *Safety Science*, 48(9), 2010
[63] Csaba KOREN & Attila BORSOS, "GDP, vehicle ownership and fatality rate: similarities and differences among countries", 4th IRTAD Conference 16-17 Sept., 2009, Seoul, Korea
[64] El-Basyouny, K., Sayed, T., "Accident prediction models with random corridor parameters", *Accident Analysis & Prevention*, 41(5), 2009
[65] Rune Elvik, "Assessing causality in multivariate accident models", *Accident Analysis and Prevention*, 43(1), 2011
[66] Chun-I Chen et al., "Forecasting Taiwan's major stock indices by the Nash nonlinear grey Bernoulli model", *Expert Systems with Applications*, 37(12), 2010
[67] ZHENG Xiao-ping, "Review on the accident-forecasting research methods", *Journal of Safety and Environment*, 8(3), 2008 (*in Chinese*)
[68] Ezra Hauer, "On prediction in road safety", *Safety Science*, 48(9), 2010
[69] Yen-Tseng Hsu et al., "Forecasting the turning time of stock market based on Markov–Fourier grey model", *Expert Systems with Applications*, 36(4), 2009
[70] Chien-Hua Wang, "A Study of the Relationship between Schooling and Enterprises' Demands for Talents". *Journal of Social Sciences*, 6 (3), 393-398, 2010
[71] Yonghong Hao et al., "A gray system model for studying the response to climatic change - The Liulin karst springs", China Journal of Hydrology, 328(3-4), 2006
[72] Zou Lihua et al., "Grey forecasting model for active vibration control systems", *Journal of Sound and Vibration*, 322(4-5), 2009
[73] Gou-sheng Liu & Jian-guo, "Gray correlation analysis and prediction models of living refuse generation in Shanghai city", *Waste Management*, 27(3), 2007
[74] Tzu-Li Tien, "The deterministic grey dynamic model with convolution integral DGDMC(1, n)", *Applied Mathematical Modeling*, 33(8), 2009
[75] Che-Tsung Tung & Yu-Je Lee, "The innovative performance evaluation model of grey factor analysis: A case study of listed biotechnology corporations in Taiwan", *Expert Systems with Applications*, 2010, 37(12), 7844-7851
[76] Chun-I Chen, Hong Long Chen, Shuo-Pei Chen, "Forecasting of foreign exchange rates of Taiwan's major trading partners by novel nonlinear Grey Bernoulli model NGBM(1,1)", *Communications in Nonlinear Science and Numerical Simulation*, 13(6), 2008
[77] Chin-Tsai Lin & In-Fun Lee, "Artificial intelligence diagnosis algorithm for expanding a precision expert forecasting system", *Expert Systems with Applications*, 36(4), 2009
[78] Baohong GUAN *et al.*, "Analysis of the relationship between particle size distribution of α-calcium sulfate hemihydrate and compressive strength of set plaster—Using grey model", *Powder Technology*, 200(3), 2010
[79] LOU Yu et al., "Application of grey multi-variable forecasting model in predicting urban water consumption", *Water Resources Protection*, 21(1), 2005 (*in Chinese*)
[80] WANG Wu-xiang et al., "Application of Grey Multi-variable Model MGM (1, n)

in the R&D Investment Forecast", *Research & Development Management*, 18(2), 2006 (*in Chinese*)

[81] ZHENG Shu-qing, MA Jing-zhong, GUAN Jun, "Apply of a Multi-variable Grey Model in Prediction", Journal of Hebei University (Natural Science Edition), 26(4), 2006 (*in Chinese*)

[82] SU Bian-ping, CAO Yan-ping, WANG Ting, "Research on grey forecasting model of multi-factor time series", *Journal of Xi'an University of Architecture & Technology (Natural Science Edition)*, 39(2), 2007 (*in Chinese*)

[83] LIU, WANG & LIU, "Multiple Linear Regression Model Based upon Multi-variables Grey Forecast Model", *Science Technology and Engineering*, 24(7), 2007

[84] LIU Wen-dian, WANG Feng-xiao, LIU You-run, "Multiple linear regression model based upon multi-variables grey forecast model", *Science Technology and Engineering*, 7(24), 2007 (*in Chinese*)

[85] Kenneth R. Agent et al., "Development of Accident Reduction Factors", Lexington, Kentucky Transportation Center, 1996

[86] Gary A. Davis, "Accident reduction factors and causal inference in traffic safety studies: a review", *Accident Analysis & Prevention*, 32, 2000

[87] Ren Zhi-gang, Liu Tie-min, Liu Gong-zhi, "The quantitative analysis of the relationship between the trend of industrial accidents and the intervention of the safety policy in China", *Journal of Safety Science and Technology*, 4(4), 2008 (*in Chinese*)

[88] LIANG Tao, "Study on the Assessment System of the Carrying Capability of Emergency Accidents and Disasters in Regional Area", Guangzhou, South China University of Technology, 2007 (*in Chinese*)

[89] Viscusi W. Kip, "The Impact of Occupational Safety and Health Regulation, 1973-1983", *RAND Journal of Economics*, 17(4), 1986

Application of the Bow-tie Risk Analysis Method on Methanol Tank Leakage

Xin Wang
College of Urban Construction and Safety Engineering, Nanjing University of Technology
Xin Mofan road 5# Nanjing China
wangxin_88@126.com

Lijing Zhang
Gang Tao
College of Urban Construction and Safety Engineering, Nanjing University of Technology
Xin Mofan road 5# Nanjing China
zhanglj@njut.edu.cn
taogangs@163.com

Hui Han
Jiangsu Academy of Safety Science and Technology
Hua Yuan road 9# Nanjing China
njhh66@163.com

Prepared for Presentation at
American Institute of Chemical Engineers
1st CCPS Asia-Pacific Conference on Process Safety
Qingdao, China
September 4–5, 2013

UNPUBLISHED

Application of the Bow-tie Risk Analysis Method on Methanol Tank Leakage

Xin Wang
College of Urban Construction and Safety Engineering, Nanjing University of Technology
Nanjing, China
wangxin_88@126.com

Lijing Zhang*
Gang Tao
College of Urban Construction and Safety Engineering, Nanjing University of Technology
Xin Mofan road 5# Nanjing China
zhanglj@njut.edu.cn
taogangs@163.com

Hui Han
Jiangsu Academy of Safety Science and Technology
Hua Yuan road 9# Nanjing China
njhh66@163.com

Keywords: Bow-tie, Tank leakage, Critical event, Safety barriers, Methanol tank

Abstract

The leakage is a common accident for the tank and pipeline in chemical industry, of which the consequence is usually severe. In view of the fact that a kind of practical risk analysis technology, the Bow-tie method, is introduced, and its history, background and definitions are mentioned. Then according to the research on the Bow-tie, five steps are described in detail to build the Bow-tie diagram. Finally, the application of the Bow-tie risk analysis technology on methanol tank leakage in Zhangjiagang city is given, including how to identify the causes and possible outcomes of methanol tank leakage, how to select the preventive and mitigative measures. This paper is intended to be of interest to those who are new to the Bow-tie method, and provides a guidance for the risk analysis of related enterprise.

1. Introduction

With the economic development, various fuel gas and chemical products are used widely in China. And both chemical industries and urban gas pipelines have become the dangerous source which threatens the people's life and possession. There are several examples as follow: A gas fire and explosion accident happened in xi'an gas company in 1998, which was caused by the leakage.

* * Corresponding author. Tel.: +86 13851810118.

The accident left 11 dead, 1 missing and nearly 33 injured. The company suffered from a great economic loss [1]. On July 28, 2012, a shuttered plastics factory in Nanjing exploded, killing at least 13 people, because the propylene pipeline was excavated to break [2]. On December 31, 2012, a large aniline spill was happened in a chemical company in Shanxi Province. Affected area was range of 80 kilometers, and about 20 thousand people were affected. As we can see from these accidents, they were all caused by the leakage. They not only caused the economic loss and casualties, also had a negative impact on society.

The causes of the leakage are complex, and some hazardous factors are difficult to measure. A number of qualitative and quantitative techniques are used for the risk assessment, such as event tree analysis (ETA), fault tree analysis (FTA), hazard and operability study (HAZOP), failure mode and effect analysis (FMEA), etc. Among them, event tree analysis and fault tree analysis are widely used. Fault tree analysis assesses the risk by providing qualitative analysis of hazards identification, and event tree analysis assists a detail quantitative assessment of likelihood for the undesired events. But they are used individually. In this paper, a kind of more practical risk analysis technology, the Bow-tie method, is introduced. It couples FTA and ETA and makes remedies to the disadvantages of both methods, and is a better risk management model [3].

This paper aims to demonstrate the practical uses and benefits of the Bow-tie method through a reference to the case studies of methanol tank leakage, and demonstrate that the Bow-tie method can be used in many industries and business sectors.

2. Bow-tie method

2.1 The history and background of the Bow-tie

The exact origins of the bow-tie method may never know for sure. The earliest mention appears to the Imperial Chemical Industries(ICI) plc Hazan Course Notes which were presented by University of Queensland, Australia in 1979 [3,4]. But the methodology which is widely in use today should be owed to the Royal Dutch/Shell Group. It integrated fully the Bow-tie methodology into its business practices, and was the first major company [5, 6, 7].

The methodology is used to identify potential hazards and assess risks. There are numbers of examples of Bow-tie analysis which have been published in international standards by the UK Health and Safety Executive [8,9], the UK defence industry [10], the French government, etc[11].

Recent years, use of Bow-tie methodology has spread around the countries, industries and companies. Most companies involved in hazardous activities combine the Bow-tie methodology with a health, safety and environment (HSE) management system [12, 13, 14].

2.2 Proposed framework of the Bow-tie

The Bow-tie method analyzes the accident seniors by assessing the probabilities and pathways of occurrences and it can be seen as an integrated probabilistic technique [15]. It is used to prevent, control and mitigate undesired events, and the way is to develop a logical relationship between causes and consequences of an undesired event [16].

Bow-tie diagram can be divided into six basic elements: critical event, event tree, fault tree, causes, consequences and safety barriers [17]. The general scheme of the Bow-tie is shown in Figure 1.The critical event is generally defined as a loss of containment (LOC) or a loss of physical integrity (LPI), and it is also called top-event. The fault tree is placed on the left part of the bow-tie and it begins from the critical event. According to the critical event, the basic or intermediate causes should be described by using logic gates and basic events, such as AND and OR gates. The event tree is placed on the right part of the Bow-tie diagram, and it also starts with the critical event as initiating event. The possible outcome events also called consequences will be finalized by following the sequences of events. If so, the causes and consequences related to the top-event can be identified in the Bow-tie diagram. Finally, the safety barriers, such as "avoid" barriers, "prevent", "control" barriers or "limit" barriers, must be identified and exported to the Bow-tie diagram. If the barriers avoid or prevent a critical event, it should be placed upstream of the event. If it controls the event, it will be placed downstream [15, 18].

Figure 1: General scheme of the Bow-tie

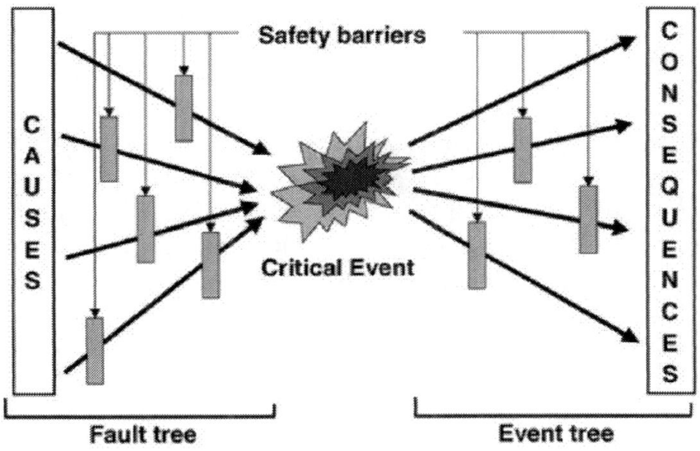

2.3 Methodology for building the Bow-tie

The whole proposed framework of the Bow-tie diagram is introduced above. But the key here is how to build the Bow-tie diagram coherently and exhaustively. According to the feature of the Bow-tie and relevant experience, it can be divided into five steps to build the Bow-tie diagram.

Step 1: Collect necessary information and associate critical events

Before identifying the critical event, data must be collected. The way to get is various. One part is from general data, and the other is from the analysis. General data includes description of processes and storage tank, plant layout, etc. Also, the information about the substances stored or handled, and substances' hazardous properties need to be obtained. Moreover, the experience of industrialists, related industry and academic research is a really important source. After that, to identify the critical event is the center of the Bow-tie diagram. For different industries, processes, substances, etc, the critical event is different. Also there may be several critical events for the same industry, process or substance. It depends on what you research [18].

Step 2: Build a fault tree according to each critical event

A fault tree for the critical event identified should be obtained in this step. The method is analogous to build the generic fault tree, and the detailed steps on how to build a fault tree will not be discussed here. At the same time, the list of possible causes should be considered. The fault tree is linked by AND or OR gates. Because the causes can be different, it may have different fault trees, regardless of the same or different critical event. But the fault tree should not be in opposition with other risk analysis methods. So, the possible causes identified by other methods should be included.

Step 3: Build an event tree according to each critical event

The event tree is placed on the right part of the Bow-tie diagram, and used to identify the possible consequences of the critical event. The critical event can be divided to secondary critical events, tertiary critical events,etc. The final critical events usually lead to dangerous phenomena. For example, if there is a tank leakage accident, the dangerous phenomena contains explosion, toxic effects, boilover, overpressure, tankfire, poolfire, jetfire, thermal radiation, environmental damage, etc. Also the method on how to build an event tree will not go into details.

Step 4: Identify safety barriers on the fault tree and event tree

The concept of safety barriers is introduced to prevent or reduce the influence on the occurrence of accident. Both the fault tree and event tree have safety barriers which have many forms, such as human actions, engineered systems, physical protecting, etc. The safety barrier directly guarantees the safety function. They can be passive barriers (retention wall, corrosion prevention systems), activated barriers (a sequence of detection-diagnosis-action), human actions (the action of the operator) and symbolic barriers (passive warnings). Occasionally, these safety barriers can be interchangeable and work together in order to increase the effectiveness of the safety function. In short, the safety barriers mean the "how" to implement safety functions [19, 20].

Step 5: Build and review the complete Bow-tie

Once the steps above finished, the construction of the Bow-tie is complete. It is composed of the critical event, the event tree on the right, the fault tree on the left and safety barriers. It is worth noting that the accident scenarios are assumed, also the safety system are assumed to be no installed or ineffective. It is the basis for the application of the Bow-tie method. Finally, don't forget to review the Bow-tie diagram.

3. Application of the Bow- tie method on methanol tank leakage

3.1 The profile of methanol storage tank farm

The case which is a hazardous chemical storage company in Zhangjiagang city, in China, located in the north of Yangtze river is presented. The company had been operating the transportation and storage business of methanol. Its tank farm covers an area of 135.1*56.6 m^2, and contains 12 different tanks which are divided into two platoons. The tanks are a 14.1 or 15.4m diameter, 18m high and 6mm thick. There are many supporting safety facilities, such as a 1.2m high firewall, the electric camera, the liquid level monitoring system, the electric lighting, etc. The sketch map is shown in Figure 2.

Figure 2: The sketch map of methanol tank farm

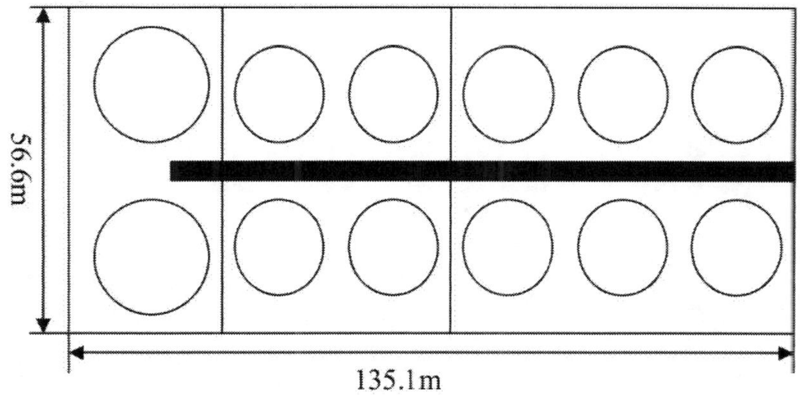

3.2 Establishing the Bow- tie of methanol tank leakage

In this case, the author is just to present the application of the Bow-tie method. The analysis for methanol tank leakage is simple. The leakage tank is anyone of 12 methanol tanks, and other tanks whether be affected won't take into account [21].

3.2.1 Identifying the critical event

There are many choices for the critical event on methanol tank leakage, such as leakage, overpressure, fire, etc. But according to data, in China, the leakage accidents are not only numerous but also quite harmful. So the methanol tank leakage is chosen as the critical event.

3.2.2 Build the fault tree

The fault tree should be built according to the critical event that is tank leakage. The leakage can occur due to various mechanisms. For example, the environmental situation, design factors and daily operation all can increase failure rate. Failure mode for the tank can be rupture or puncture, while each mode can happen in multiple ways. Based on the cases and literature, the causes have been identified, namely corrosion (external or internal), natural hazards (typhoon, flood, earthquake, etc.), structural defects (design, material quality, weld, accessory, etc.), operational defects (system failure and human failure errors including lack of adequate inspection and maintenance), third party activities (malicious, incidental, sabotage, accidental, etc.) [22, 23]. Due to space constraints, a few representative causes are shown in Figure 3.

Figure 3: Bow-tie diagram of methanol tank leakage

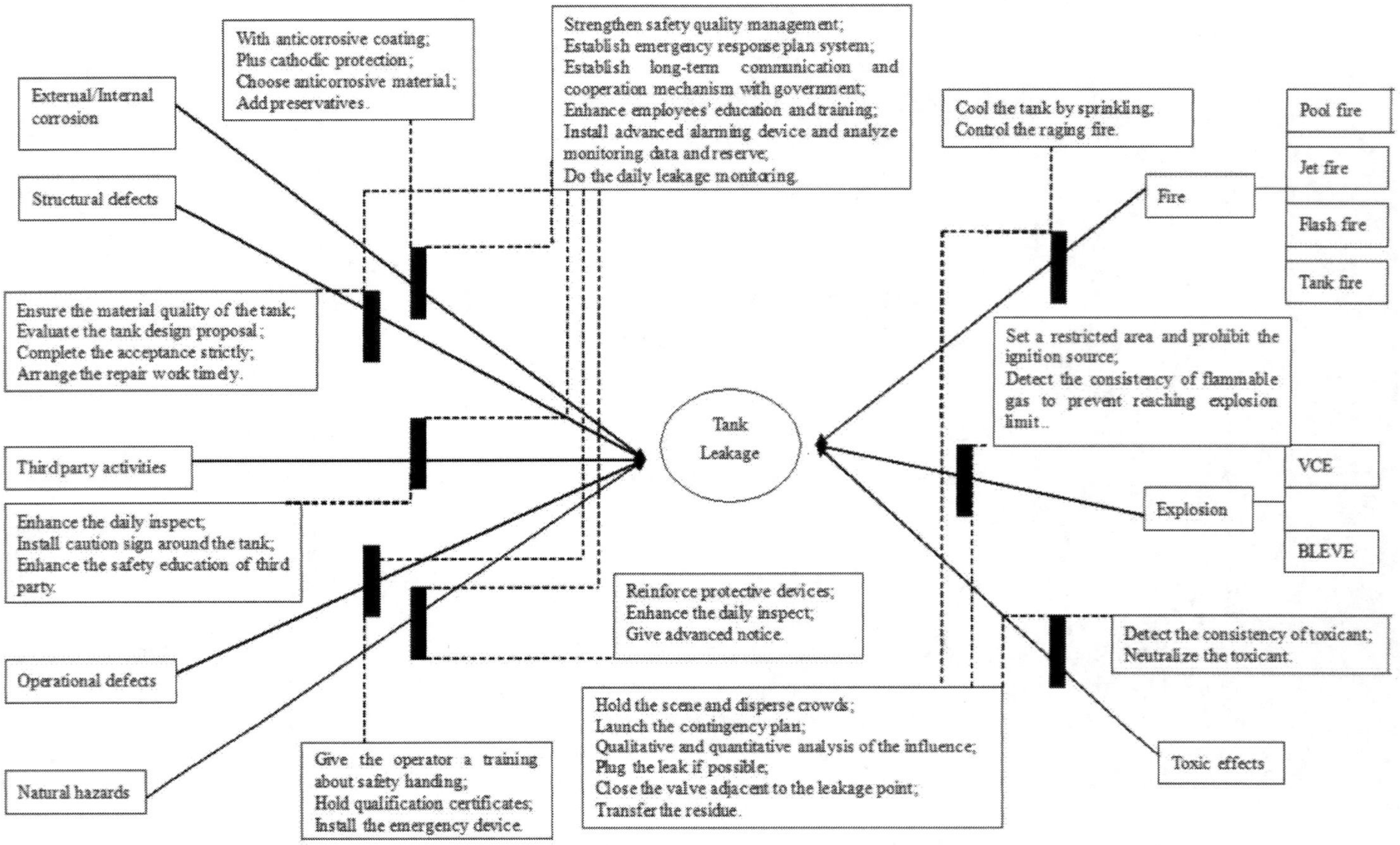
With anticorrosive coating;
Plus cathodic protection;
Choose anticorrosive material;
Add preservatives.
Strengthen safety quality management;
Establish emergency response plan system;
Establish long-term communication and cooperation mechanism with government;
Enhance employees' education and training;
Install advanced alarming device and analyze monitoring data and reserve;
Do the daily leakage monitoring.
External/Internal corrosion
Structural defects
Ensure the material quality of the tank;
Evaluate the tank design proposal;
Complete the acceptance strictly;
Arrange the repair work timely.
Third party activities
Enhance the daily inspect;
Install caution sign around the tank;
Enhance the safety education of third party.
Operational defects
Natural hazards
Give the operator a training about safety handing;
Hold qualification certificates;
Install the emergency device.
Reinforce protective devices;
Enhance the daily inspect;
Give advanced notice.
Tank Leakage
Hold the scene and disperse crowds;
Launch the contingency plan;
Qualitative and quantitative analysis of the influence;
Plug the leak if possible;
Close the valve adjacent to the leakage point;
Transfer the residue.
Cool the tank by sprinkling;
Control the raging fire.
Fire
Pool fire
Jet fire
Flash fire
Tank fire
Set a restricted area and prohibit the ignition source;
Detect the consistency of flammable gas to prevent reaching explosion limit..
Explosion
VCE
BLEVE
Detect the consistency of toxicant;
Neutralize the toxicant.
Toxic effects

3.2.3 Build the event tree

If methanol tank leaks, methanol will go from a liquid form to a gaseous form. And due to the different physical and chemical properties (diffusible, toxic, inflammable, explosive, etc.) of the methanol, the degree of space confinement produced by the surroundings and the time gap between the leakage and possible ignition, different accident scenarios (pool fire, jet fire, flash fire, BLEVE, VCE, etc.) may be happened and can lead to an obvious threat to people and property. The detailed event tree is shown on the right part of the Bow-tie diagram in Figure 3.

3.2.4 Identification of safety barriers

For each cause and consequence created in the Bow-tie diagram of methanol tank leakage, the safety barriers are attempted to be found to prevent the critical event happened, or after it happened, to eliminate or reduce its consequences. As pointed out above, safety barriers have many forms, such as human actions, engineered systems, physical protecting, etc. For the methanol tank leakage, the safety barriers shown in Figure 3 contain the organizational measures (establish long-term communication and cooperation mechanism with government), management measures (enhance the employees' education and training, strengthen safety quality management, enhance the safety education of third-party, etc.) and physical protecting (install advanced alarming device, reinforce protective devices, choose anticorrosive material, etc.).

After completing the steps described above, the complete Bow-tie diagram which is shown in Figure 3 is finished. And something omission or wrong can be perfect by reviewing it. From the case studies of methanol tank leakage, the Bow-tie method is proved that it is fairly easy to grasp, and can be adapted to different critical event and applied to different fields.

4. Conclusion

According to the application of the Bow-tie method on methanol tank leakage, the Bow-tie method shows its advantages and benefits than other risk analysis methods.

Clear communication: All elements, such as the critical event, consequences, causes and the safety barriers, are visually displayed. Besides, according to the Bow-tie diagram, the meaning people want to convey can be easily understood by all levels, such as senior managers, regulators, third parties, operations personnel, etc. A picture will often be understood more quickly than a thousand words.

High logicality: With the combining of fault tree analysis and event tree analysis, all aspects of the management of risk is considered, from initial cause to final consequence, from safety barriers upstream to the barriers downstream in sequence. Each step is reasonable and fit tightly.

Endless possibilities: Based on the case above, the method is not limited to assess one risk. It has broader use. In recent years, the Bow-tie has been developed from assessment of HSE risks to management of security, from assessment of project risks to management of business interruption, etc.

So the Bow-tie is an ideal method and can be used to qualitatively assess and demonstrate control of all types of risk, and has strong practical application and promoting values. It is intended to be of interest to those who are new to the technique and experienced users alike.

5. Acknowledgments

The authors would like to thank all the persons for providing valuable support and feedback in the course of the research, and we also thank Jiangsu ordinary university postgraduate research and innovation project (Grant No. CXLX12_0458). We are very grateful to the unknown referees for valuable remarks.

6. References

[1] ZHANG Jingjing, ZHANG Lijing, TAO Gang. "Study on Quantitative Risk Assessment for Storage Devices of Petrochemical Industry." Industrial Safety and Environmental Protection, September 2011, vol37, 7-8.

[2] Zhu Chengsi. "Thinking about prevention of industrial disasters in cities and towns. Enhance city's management", 2011, 60-64.

[3] Hurst, S., Lewis, S.. "Lessons Learned from Real World Application of the Bow-tie Method." Risk World, 2004.

[4] University of Queensland, Australia, Minerals Industry Safety and Health Centre, National Minerals Industry Safety and Health Risk Assessment Guide. www.mishc.uq.edu.au/NMIRAG/NMISHRAG_Chapter4_4.1.5.asp.

[5] Primrose, M. J., Bentley, P. D., van der Graaf, G. C., Sykes, R. M. Shell International Exploration and Production B.V.. "The HSE Management System in Practice-Implementation," SPE 35826, 1996.

[6] Gower-Jones, A. D., van der Graaf, G. C., Milne, D. J. Shell International Exploration and Production B.V.. "Application of Hazard and Effects Management Tools and Links to the HSE Case", SPE 36031, 1996.

[7] "Bow-Tie an Elegant Solution". Strategic Risk, 2005.

[8] Marine Risk Assessment, Prepared by Det Norske Veritas Ltd. for the Health and Safety Executive, Offshore Technology Report 2001/063, 2002.

[9] Application of QRA in Operational Safety Issues, Prepared by Det Norske Veritas Ltd. for the Health and Safety Executive, Research Report 025, 2002.

[10] Gifford, M. J., Gilbert, S. M.. "Atkins Defence Systems, Barnes, I. Ministry of Defence", The Use of Bow-tie Analysis in OME Safety Cases, ESAS 03, 2003.

[11] Couronneau, J. C., Tripathi, A, Transoft International, Implementation of the New Approach of Risk Analysis in France, 41st International Petroleum Conference, Bratislava, 2003.

[12] BS EN ISO 14001, Environmental Management Systems, 1996.

[13] BSI-OHSAS 18001, Occupational Health and Safety Management Systems-Specification, 1999.

[14] E&P Forum, Guidelines for the Development and Application of Health, Safety and Environmental Management Systems, Report No. 6.36/210, July 1994.

[15] Duijm, N. J.. "Safety-barrier diagrams as a safety management tool. Reliability Engineering and System Safety", 94(2) 2009, 332-341.

[16] Dianous, V., & Fiévez, C.. "ARAMIS project: a more explicit demonstration of risk control through the use of bow-tie diagrams and the evaluation of safety barrier performance." Journal of Hazardous Materials, 130(3) (2006)220-233.

[17] Shahriar, A, Sadiq, R, Tesfamariam,S.. "Risk analysis for oil & gas pipelines: A sustainability assessment approach using fuzzy based bow-tie analysis." Journal of Loss Prevention in the Process Industries 25 (2012) 505-523

[18] C. Delvosalle, C. Fievez, A. Pipart, B. Debray, ARAMIS project: A comprehensive methodology for the identification of reference accident scenarios in process industries, J. Hazard. Mater. 130 (3) (2006)200–219.

[19] B. Debray, C. Delvosalle, C. Fievez, A. Pipart, H. Londiche, E.Hubert, Defining safety functions and safety barriers from fault and event trees analysis of major industrial hazards, in: Proceedings ESREL, Berlin, Germany, 14–18 June 2004.

[20] F. Guldenmund, A. Hale, L. Goosens, J. Betten, N. J. Duijm,The development of an audit technique to assess the quality of safety barrier management, J. Hazard. Mater. 130 (3) (2006) 234–241.

[21] Kandola, B., Sullivan, M, ABS Consulting, The Use of Bow-tie Model and THESIS in the Control of Major Accident Hazards, FABIG Technical Meeting, October 2003.

[22] Guénana, T. L., Manceaua, J. C., Bouc, O., Rohmer, J., Ledouxa, A. GERICO: A database for CO2 geological storage risk management.Energy Procedia 4 (2011) 4124–4131.

[23] Ruihua Yuan. Pre-study of Leaking Fault of the Oil Pipelines in Maowusu Desert Areas.Proceedings of the International Conference on Pipelines and Trenchless Technology, 2009, 210-216.

Optimization analysis of Fire consequence of BOG compressor room based on FDS

Xiaoping Sun
College of Mechanical and Electronic Engineering
China University of Petroleum
sunxiaoping87@163.com

Yuan Zhu
Guoming Chen
Changhang Xu
Xi Chen
College of Mechanical and Electronic Engineering
China University of Petroleum
zhy3323@163.com
offshore@126.com
changhangxu@126.com
chenxi0831@126.com

Prepared for Presentation at
American Institute of Chemical Engineers
1st CCPS Asia-Pacific Conference on Process Safety
Qingdao, China
September 4 – 5, 2013

UNPUBLISHED

Optimization analysis of Fire consequence of BOG compressor room based on FDS

Sun Xiaoping
College of Mechanical and Electronic Engineering
China University of Petroleum
sunxiaoping87@163.com

Yuan Zhu
Guoming Chen
Changhang Xu
Xi Chen
College of Mechanical and Electronic Engineering
China University of Petroleum

Keywords: fire accidents; BOG compressor room; LNG tank farm; FDS software; fire dynamic model

Abstract

To study the fire accidents happening in BOG compressor room as result of compressor discharge, a fire dynamic model of LNG tank farm is developed by FDS software. Water sprinklers are added to the model in order to analyze the effect of water spray system. Two criteria including temperature and radiation intensity are used to assess the fire suppression of water sprinklers on objects such as equipment both beside and away from the BOG compressor room, BOG compressors and the whole tank farm. It is showed that water sprinklers can observably reduce the height of the flame, but increase the amount of smoke. The effects of water spray system for the farther equipment is more effective than the nearer. The water spray system can reduce the range of the radiation and decrease the hazardous areas from north-south 180m and east-west 270m to north-south 140m and east-west 160m. Thus, water spray system can play a significant role in fire rescue and the protection of equipment and personnel.

1. Introduction

In order to reduce the waste of resources, BOG compressor is used for the recovery of BOG (Boil off Gas). BOG is pressurized to 0.55MPa by BOG compressor, then transfer heat with superheated liquid from LNG low pressure pump, and condensed into LNG by affer-condenser at last. BOG compressor plays an important role to ensure the normal pressure of LNG tank. To study the fire accidents happening in BOG compressor room as result of compressor discharge by FDS software, and verify the effect of water spray system to fire consequence, then provide guidance for fire rescue.

2. FDS software system

FDS (Fire Dynamics Simulator) which is developed by NIST (National Institute of Standards and Technology), sets fire fluid motion as the main simulation object and analyses fire smoke and heat transfer process in the fire. FDS provides two numerical simulation methods which are DNS (Direct Numerical Simulation) and LES (Large Eddy Simulation). LES is used more often than DNS, for its high accuracy and small workload. The basic equations of LES are as follows[1, 2]:

(1) Mass conservation equation:

$$\frac{\partial \rho}{\partial t} + \nabla \cdot \rho \overline{\mu} = 0$$

(2) Component conservation equation:

$$\frac{\partial}{\partial t}(\rho Y_i) + \nabla \cdot (\rho Y_i \overline{\mu}) = \nabla\left(\rho D_i \nabla Y_i\right) + \dot{m}_i$$

(3) Momentum conservation equation:

$$\rho\left[\frac{\partial \overline{u}}{\partial t} + \left(\overline{u} \cdot \nabla\right)\overline{u}\right] = \rho g + \overline{f} + \nabla \cdot \tau$$

(4) Energy conservation equation:

$$\frac{\partial}{\partial_t}(\rho h) + \nabla \cdot \left(\rho h \overline{u}\right) = \frac{\partial p}{\partial t} + \overline{u} \cdot \nabla p - \nabla \cdot q_r + \nabla \cdot (k \nabla T) + \sum_i \nabla \cdot \left(h_i \rho D_i \nabla Y_i\right)$$

(5) Equation of state :

$$P_0 = \rho TR \sum_i \frac{Y_i}{M_i}$$

Where: ρ is density (kg/m^3) , $\overline{\mu}$ is velocity vector(m/s), Y_i is the mass fraction of the i-th component, D_i is the diffusion coefficient of the i-th component (m^2/s), $\dot{m}_i$ is the mass generation rate of the i-th component in unit volume [kg/(m3.s)], g is acceleration of gravity (m/s^2), $\overline{f}$ is the external force acting on the gas besides gravity (N), τ is viscous stress tensor (N), h is enthalpy (J/kg), k is thermal conductivity [W/(m·K)], T is thermodynamic temperature (K), q_r is radiant heat flux (W/m^2), P_0 is background pressure (Pa), R is gas constant [J/(mol.K)], M_i is the molar mass of the i-th component(kg/mol).

3. The setting of simulated conditions

3.1 Leak Size

Many reasons can cause spill accident of BOG compressor, which is belongs to pressure equipment, such as insufficient strength of pipeline, poor weld quality, no pressure test before operation, damage of lubrication system, cooling system, exhaust system and so on[3, 4]. The leak size of BOG compressor is summarized in Table 1[5].

Table 1. Reference size of compressor leak-caliber

Leakage form	Leak size

Enclosure damaged.	20% -100% of connecting pipe diameter
Gland leakage	20% of the diameter

3.2 *Amount of leakage*

The way of gas leakage in pressure vessel or piping is usually jet. The main application model of gas leakage is R.H.Perry modle. When calculating the amount of leakage, the speed of leakage which can be sound or subsonic should be determined. Sound means critical flow and subsonic means subcritical flow. Assuming that leakage diffusion is an adiabatic and irreversible process and the leakage gas is ideal gas.

If $P \geq P_0\left(\frac{r+1}{2}\right)^{\frac{r}{r-1}}$, the speed of leakage is sound, leaking formula:

$$Q=\delta_g A_0\sqrt{P\rho_g r\left(\frac{2}{r+1}\right)^{\frac{r+1}{r-1}}}$$

If $P < P_0\left(\frac{r+1}{2}\right)^{\frac{r}{r-1}}$, the speed of leakage is subsonic, leaking formula:

$$Q=\delta_g A_0\left\{\left[2P\rho_g\left(\frac{r}{r-1}\right)\right]\times\left[\left(\frac{P_0}{P}\right)^{2/r}-\left(\frac{P_0}{P}\right)^{(r+1)/r}\right]\right\}^{1/2}$$

$$\rho_g=\frac{MP}{8314T}$$

Where: Q is leak rate (kg/s), A_0 is leaking port area (m^2), P is medium pressure in the vessel (Pa), P_0 is environmental pressures (Pa), r is adiabatic index,δ_g is gas leakage coefficient, whose value depends on the shape of the leaking mouth, Ex: round take 1.0, triangles take 0.95, rectangles take 0.9, ρ_g is medium density in the container (kg/m^3), M is medium molecular weight (kg/mol), T is medium temperature inside the container(K).

3.3 *Model parameter setting*

Meshing the tank farm, "composite calculation region" function is used[6, 7]. Considering devices of the tank farm is not very high, grid of Z axis was set to 58m. Specific mesh generation was shown in Table 2 and Figure 1.

Table 2 Calculation region and mesh generation of BOG compressor fire

Calculation area	Coordinate dimension (m)	Grid accuracy (m*m*m)	Number of grid	Total
Fire mesh	80≤X≤240;135≤Y≤215;-2≤Z≤58	1.00*1.00*1.00	975,750	975,750
Mesh 1	80≤X≤240;15≤Y≤135;-2≤Z≤58	2.00*2.00*2.00	144,000	
Mesh 2	240≤X≤500;15≤Y≤215;-2≤Z≤58	4.00*4.00*4.00	48,750	
Mesh 3	0≤X≤80;15≤Y≤215;-2≤Z≤58	4.00*4.00*4.00	15,000	

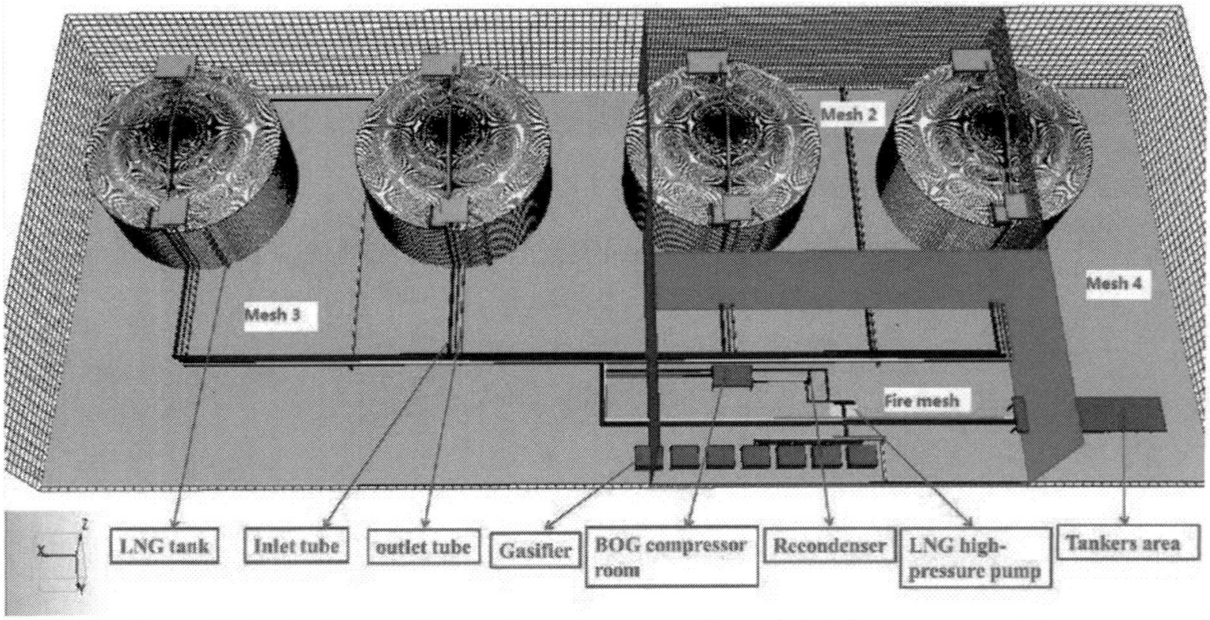

Fig. 1 Calculation region and mesh generation of BOG compressor fire

3.4 Water spray system parameter setting

FDS model does not need to establish sprinkler piping, it only need to set a sprinkler device in the place where should install sprinklers. The function of fire suppression for water spray system can achieve through controlling nozzle by temperature and time [8, 9]. Actual arrange of sprinklers in BOG compressor room is shown in Figure 2. Arrange of sprinkler device in FDS model is shown in figure 3.

Fig. 2 Nozzle arrangement of BOG compressor room

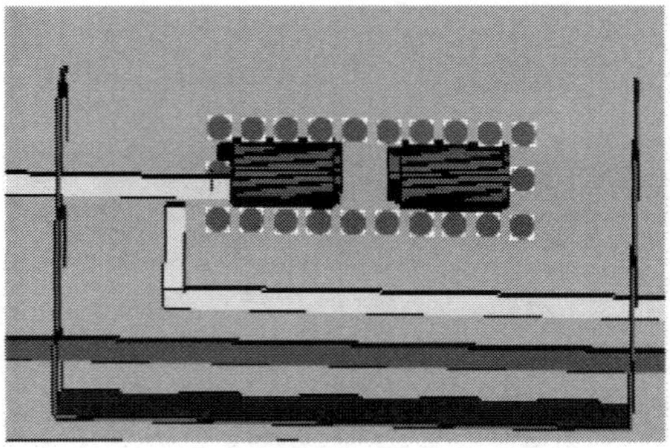

Fig. 3 Sprinker arrangement in FDS

The standard diameter of sprinkler is 15mm, calculation formula of sprinkler flow and speed is shown below:

$$Q=K\sqrt{10H}$$

Where: Q is sprinkler flow (L/min), K is characteristic coefficient of sprinkler, for standard sprinkler K=80, H is work pressure of sprinkler (MPa).

$$\upsilon=K_cQ$$

Where: v is speed (m/s), K_c is flow coefficient, when diameter is 15mm K_c takes 5.85, Q is sprinkler flow (L/s).
The parameters of water spray system in BOG compressor room are shown in Table 3.

Table 3 Parameters of water spray system in BOG compressor room

Location	Spray intensity	Work pressure	flow	speed	Spacing
BOG compressor room	20.4L/min.m^2	0.1MPa	80L/min	7.8m/s	1m

4. Analysis of simulation results

4.1 Temperature impact

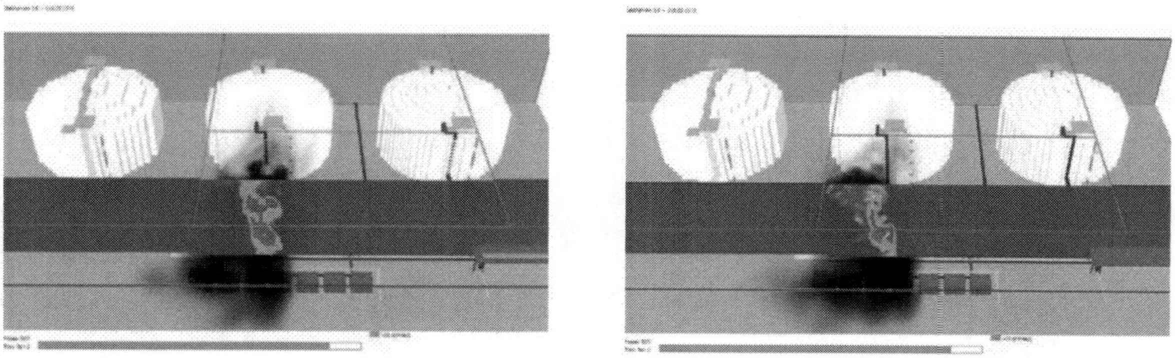

(a) No water spray system **(b) With water spray system**

Fig. 4 Temperature section and flue gas figure

Slice and flue gas temperature diagram is shown in Figure 4. The slice, whose maximum temperature is 1020℃, is located at the opening position of the BOG compressor room. The high temperature region in diagram (a) is significantly larger than the range in diagram (b). In diagram (a), we can see flame behind the slice but can't see in diagram (b). Affected by the water spray system, flame is inhibited and more gas is generated in diagram (b). The smoke spread over 60 meters downwind from outside under the wind.

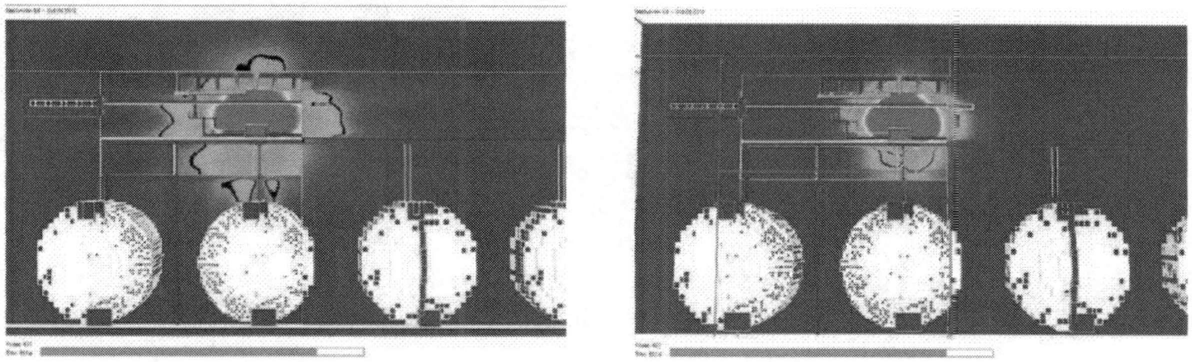

(a) No water spray system **(b) With water spray system**

Fig. 5 Water spray system's effect on overall temperature range of LNG tank

Overall temperature range of LNG tank farm is shown in Figure 5. Black range is 60℃ temperature limit and the relatively safe area is beyond the limit range. In diagram (a), high-temperature region is widely scattered. The range of hazardous areas expands to north-south

125m and east-west 160m. The majority of process equipment are covered in the hazardous areas such as recondenser, LNG high-pressure pump, 6 sets of gasification units. Hazardous area extends to the bottom of the tank. In diagram (b), high-temperature region is relative centralized. The range of hazardous areas expands to north-south 55m and east-west 90m. Recondenser and 4 sets of gasification units are covered in the hazardous areas. The maximum temperature can't be cut down but the range of hazardous area can be reduced by automatic water spray system.

4.2 Radiation analysis

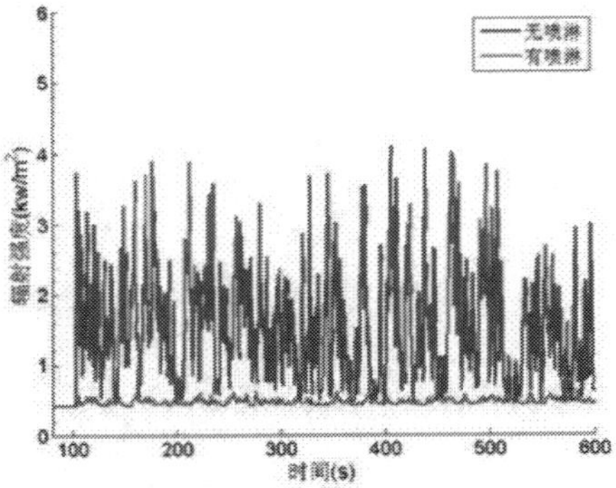

(a)radiation contrast in tankers area

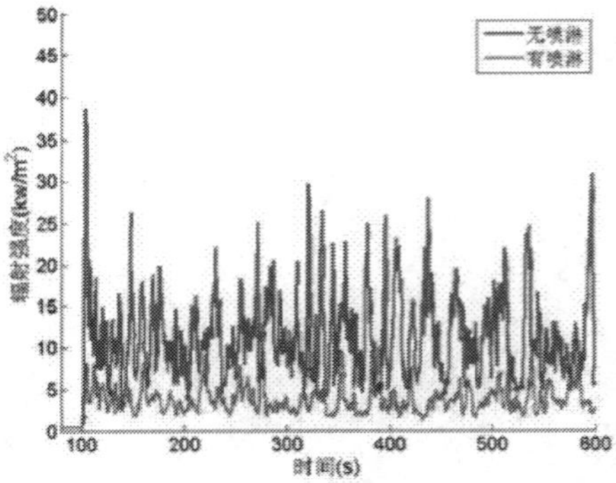

(b) radiation contrast of LNG high-pressure pump

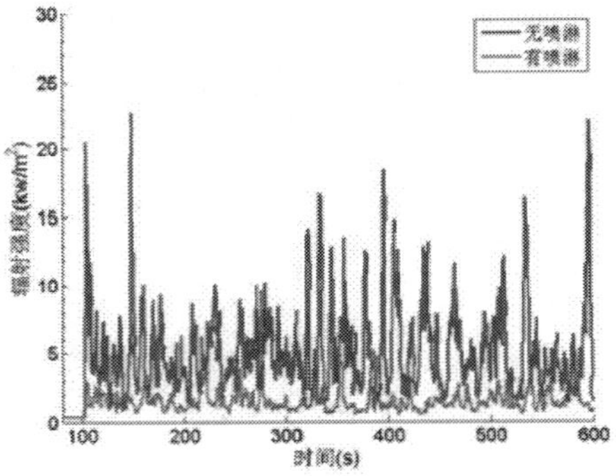

(c) radiation contrast of gasification units

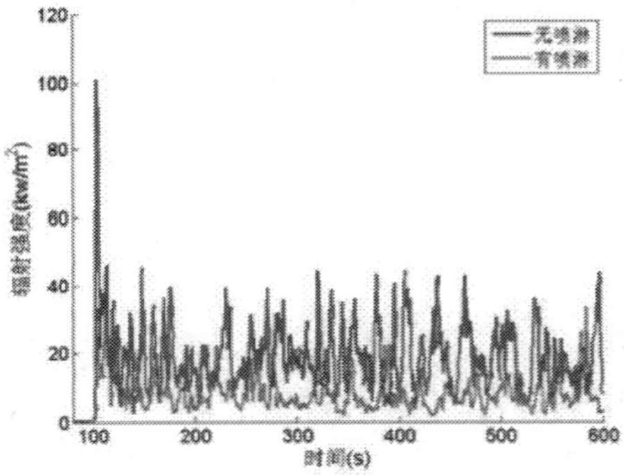

(d) radiation contrast of recondenser

Fig.6 Water spray system's effect on device radiation intensity

Water spray system's effect on device radiation intensity is shown in Figure 6. Contrast the four diagrams, we can see: No water spray system's radiation intensity reaches to the peak in a few seconds when the fire started, and is with strong fluctuations. Water spray system's radiation intensity doesn't have obvious radiation peak and is with weak fluctuations.

Contrast of mean radiation of water spray system's effect on device is shown in table 4.

Table 4 Contrast of mean radiation of water spray system's effect on device

Device	Distance away from the fire	No water spray system (kW/m^2)	With water spray system (kW/m^2)	reduction ratio of radiation intensity
Tankers area	124m	2.67	0.93	65.2%
Gasification	62m	8.47	2.67	68.4%
LNG high-pressure pump	45m	10.68	3.88	63.68%

recondenser	30m	37.23	16.6	55.4%

In diagram (a), tankers area has the farthest distance from the BOG compressor room and has a low radiation. The risk of personnel and equipment is relatively small.
In diagram (b), the radiation is 10.68kW/m^2 without water spray system. Staff in this area will have life-threatening and minor damage will cause to the equipment. When the water spray system is on, the radiation is reduced to 3.88kW/m^2 and no damage will appear to the equipment and staff.
In diagram (c), the radiation is 8.47kW/m^2 without water spray system. Staff who stays for a long time in this area will have different degrees of burns. When the water spray system is on, radiation value falls below human acceptable range.
In diagram (d), recondenser has the nearest distance from the BOG compressor room and has a high radiation. The radiation is 37.23kW/m^2 without water spray system. Prolonged baking can cause serious damage to recondenser. When the water spray system is on, radiation value falls to 16.6kW/m^2, equipment can be well protected.
Compared Figure 6 and Table 4, radiation intensity of device can have a decrease of more than 50%. Devices farther away from the scene have the larger decrease of radiation intensity.

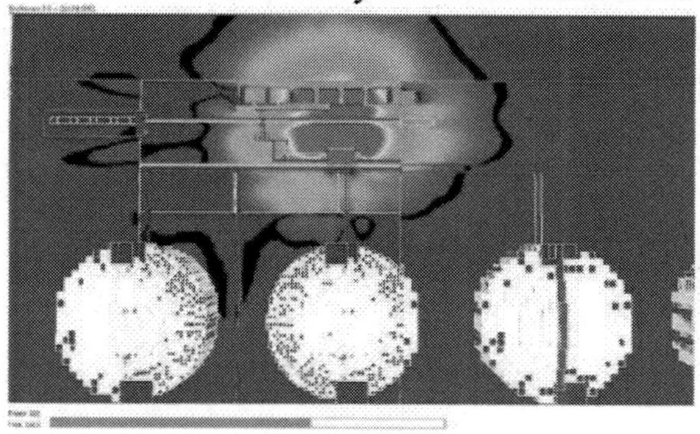

(a) No water spray system

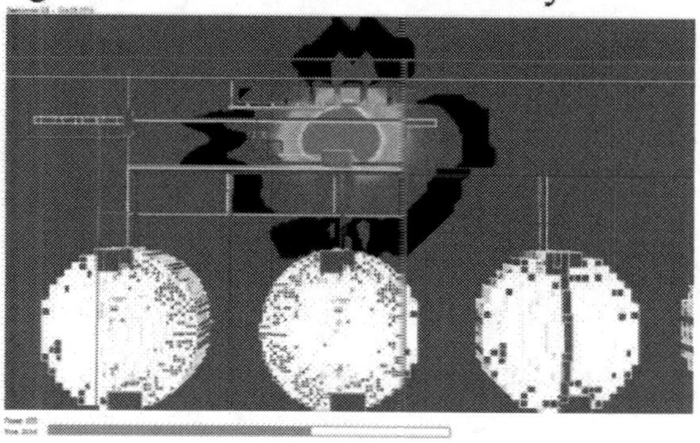

(b) With water spray system

Fig. 7 Water spray system's effect on overall radiation range of LNG tank

Radiation range of the whole tank farm is shown in Figure 7. Black range is 4kW/m^2 radiation limit and the relatively safe area is beyond the limit range. In diagram (a), hazardous area is relatively large and instabilities. The range of hazardous areas expands to north-south 180m and east-west 270m. The tank, who is 48m away from the fire, will be affected. In diagram (b), the range of hazardous areas expands to north-south 140m and east-west 167m. Hazardous area is more concentrated and is very favorable for the rescue.

5. Conclusion

To study the fire accidents happening in BOG compressor room as result of compressor discharge, a fire dynamic model of LNG tank farm is developed by FDS software. Water sprinklers are added to the model in order to analyze the effect of water spray system. Two criteria including temperature and radiation intensity are used to assess the fire suppression. The conclusions are shown below:
(1) Affected by the water spray system, flame is inhibited and more gas is generated. The smoke spread over 60 meters downwind from outside under the wind.
(2) Affected by the water spray system, radiation intensity of device can have a decrease of more than 50%. Devices farther away from the scene have the larger decrease of radiation intensity.

(3) The water spray system can reduce the range of the radiation and decrease the hazardous areas from north-south 180m and east-west 270m to north-south 140m and east-west 160m and can play a significant role in fire rescue and the protection of equipment and personnel.

Acknowledgement

This research is supported by "National Key Technology Research and Development Program of the Ministry of Science and Technology of China" (2011BAK03B08), "the Shandong Provincial Natural Science Foundation, China" (ZR2011EL048), "the Fundamental Research Funds for the Central Universities" (12CX04060A).

References

[1] Jun L, Min L, "Basic Theory and Applications of FDS fire simulation techniques," *Safety Science Cnology*, NO. 1, 2006

[2] Mcgrattan K, Forney G, "Fire Dynamic Simulator User's Guide: U.S governing printing office," Washington, 2004

[3] Ruilian S, "Fire and Explosion Accident Fault Tree Analysis of Natural gas refueling Stations Compressor Room," *Safety & health & environment*, No. 10, 2004

[4] Xiaoping S, Yuan Z, "An Analysis of Foreign and Domestic Explosion in LNG Tank Fields andProposals for Preventing Measures," *Nature gas industry*, Volume 33, NO. 5, 2013

[5] Juncheng J, "Accident Investigation and Analysis Technology," Chemical Industry Press, 2004

[6] Jie G, Yuan Z, "Fire Hazard Analysis on Oil & Gas Processing System on The Offshore Platform Based on FDS," *China offshore Oil and Gas*, Volume 23, NO. 2, 2011

[7] Ting L, Guoming C, "Characteristics of Injection Fire based on FDS Simulation in separator of natural gas station," *Journal of Safety & Environment*, 2010

[8] Wenhua Z, Shengli T, "Study on The Performance of The Sprinkler System," *Fire Technique and Products Information*, 2007

[9] Peihua L, "Fire Characteristics of Offshore Platform Confined Space and Early Control Strategy Research," *China University of petroleum*, 2011

[10]Yan S, Juxia W, Arrangement of Automatic Sprinkler System Nozzle," *Science & Technology of Baotou Steel(Group) Corporation*, 2001

1ST CCPS ASIA-PACIFIC CONFERENCE ON PROCESS SAFETY℠

Design of Monitoring and Early-Warning System in Chemical Industry Parks Based On Risk Assessment

Zhang Yong
Nanjing University of Technology
No.200 Zhongshan Road, Nanjing, JiangSu
zmgnj198173@163.com

Zhang Mingguang, Qian chengjiang, Chen yinting
Nanjing University of Technology
No.200 Zhongshan Road,Nanjing, JiangSu
Mingguang_zhang@163.com

Prepared for Presentation at
American Institute of Chemical Engineers
1st CCPS Asia-Pacific Conference on Process Safety
Qingdao, China
September 4 – 5, 2013

UNPUBLISHED

Design of Monitoring and Early-Warning System in Chemical Industry Parks Based On Risk Assessment*

Zhang Yong
Nanjing University of Technology
No.200 Zhongshan Road, Nanjing, JiangSu
zmgnj198173@163.com

Zhang Mingguang, Qian chengjiang, Chen yinting
Nanjing University of Technology
No.200 Zhongshan Road, Nanjing, JiangSu

Keywords: risk assessment, fixed hazard, transportable hazard, domino effect, monitoring and early-warning system

Abstract

The hazards in chemical industry parks were classed into two kinds: fixed hazards and transportable hazards, whose definition and content were discussed and distinguished. Statistics of accidents were analyzed to achieve accident process datas, such as material phase, accident damage type and threshold value which initated the accident. The domino effect between fixed hazard- fixed hazard, transportable hazard- fixed hazard and transportable hazard- transportable hazard were studied, thus corresponding domino accident risk assessment models were built. A set of early-warning indices were designed for risk calculation of single hazard accident and domino accidents among different hazards, finally a risk coupling technology was developed to get overall risk of the chemical industry park.Based on that a method was brought to assist construction and lay-out of monitoring hardware system, and gather information as input of risk calculation. Based on the design, a accident monitoring and early-warning system for chemical industry park was established.

1. Introduction

Currently, chemical industry parks and industry parks with independent chemical region has exceeded 1200[1]. The most significant problem in developing chemical industry parks is to secure the production and environment. There are large quantities of flammable, explosive, toxic chemicals in the production, storage, usage and transportation parts. Furthermore, since the existence of a large number of major hazards, once incidents like fires, explosives and toxic leakages take place, it is easily to result in major accidents which causes lots of fatalities and injuries. Meanwhile, due to the land utilization and economic reasons, distances between installations were getting smaller and smaller. When a primary accident occurs, it is likely to damage nearby equipment and even cause catastrophic domino effects. Therefore, in order to prevent major accidents and decrease accident loss, it is necessary to develop monitoring and early-warning systems in chemical industry park.

Britain is the first country to research the technology of major hazards monitoring systematically. Thereafter, Netherlands, Germany, France, Italy, Belgium and other EU

** Corresponding author: Zhang Mingguang Tel.: +86 13815892792.

(European Union) member states and the United States, Australia and some other countries have started to research major hazards monitoring system and its related technologies successively [2]. EU has developed a major hazards monitoring system which collected geographic information of major hazards and their surroundings from variety of countries and used GIS (Geographic information system) to achieve particular managements of major hazards [3]. H.Jeong proposed a method that multi-transducer monitoring system can be used in the process of chemical industry automation [4]. S.Bagavathiappan studied the infrared imaging technology and believed that the technology has significant advantages in equipment defects detection and system condition monitoring [5]. However, due to the limitations like excessive development of chemical industry parks in China, basic data shortages, insufficient capital investment in monitoring equipment, inadequate management mechanisms and so on, the risk assessment technology is backward and the safety planning of chemical industry parks are imperfect. These situations lead to the design of monitoring and early-warning system in chemical industry parks is not adequate and major accidents had taken place frequently.

2. Risk assessment

Risk assessment is a method that is based on the analysis of the probability and consequences of accidents. Risk assessment can be used to classify production safety accidents and to guide accidents preventing and early-warning.

2.1 Characteristics of chemical industrial accident risk

In order to figure out characteristics of risk in chemical industry park, typical accident cases have been analyzed. Accident details were listed in Table 1.

Table 1. Typical accident cases

Date	Location	Event	Consequence	Cause
2012.2	Keer Chemical Plant, Shijiazhuang, Hebei	Ammonium nitrate explosion	25 fatalities, 46 injuries	Fixed major hazard, domino effect
2010.7	International Logistics Co., Ltd. Of China National Petroleum Corporation, Dalian, Liaoning	Pipeline explosion	1 fatalities, 2 injuries	Fixed major hazard, domino effect
2010.1	Lanzhou Petrochemical Co., Ltd., Gansu	Tank leakage and explosion	6 fatalities, 1injuries	Fixed major hazard, domino effect
2008.8	Guangxi Vinylon Chemical Co., Ltd.	plant explosion	20 fatalities, 50 injuries	Fixed major hazard, domino effect
2010.8	Taizhou, Jiangsu	Ammonia Leakage (Traffic accident)	1 fatalities, 10 injuries	Transportable hazard, environment uncertainty
2005.3	Huaian, Jiangsu	Chlorine Leakage (Traffic accident)	28 fatalities, more than 300 people were poisoned	Transportable hazard, environment uncertainty

From the table 1, it can be seen that the accident risk were divided into two categories: fixed hazard risk and Transportable hazard risk. The definitions are listed as follow

Fixed Hazard

Places and facilities which product, use and store hazardous chemicals that have destructive energy[6].

Transportable hazard

Conveyance which convey hazardous chemicals by land, waterway and airway that constitute hazards [7].

Figure 1: Risk assessment system of chemical industry parks

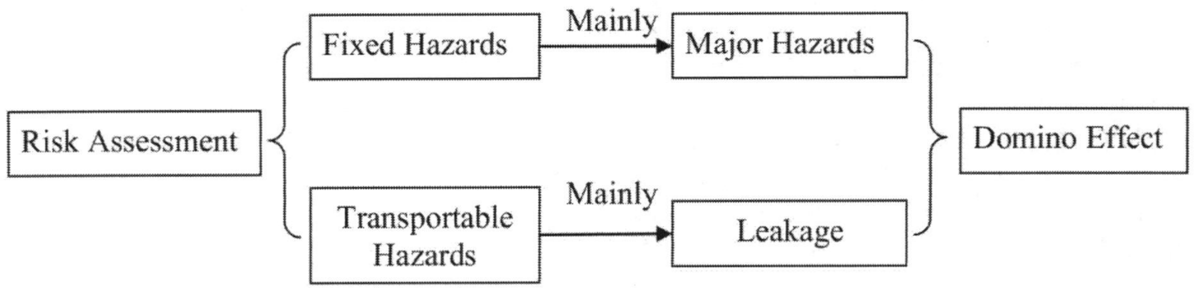

At present, individual risk and societal risk have been widely used in the domestic and international risk assessment of fixed hazards and Transportable hazards. As for the risk assessment of fixed hazards, analysis models of fire, explosion and toxic leakage have been developed to calculate changes of individual risk and societal risk in chemical industry parks. While for the risk assessment of Transportable hazards, some other researchers have developed a lot of transportation risk assessment model. These models should be applied according to the actual situation.

2.2 Risk Assessment for Fixed Major Hazard Installations

1. The identification of fixed major hazard installations

The identification of fixed major hazard installations is mainly based on the *Identification of major hazard installations for dangerous chemicals (GB18218-2009)*. Fixed major hazard installations can be judged by that whether the quantity of dangerous chemicals in the region exceeds the threshold quantity.

One hazard installation can be determined as a major hazard installation when the hazard installation satisfies one of the following two conditions[8].

① The quality of any one of the existing hazardous chemicals in the region reaches or exceeds the threshold quantity.

② The quality of any one of the existing hazardous chemicals in the region didn't reaches or exceeds the threshold quantity but satisfied the Eq. 1.

$$\frac{q_1}{Q_1} + \frac{q_2}{Q_2} + \cdots + \frac{q_n}{Q_n} \geq 1 \quad \text{[Eq. 1]}$$

where $q_1, q_2, \ldots, q_n$ are the storage contents of each one of the existing hazardous chemicals and $Q_1, Q_2, \ldots, Q_n$ are threshold quantities of corresponding hazardous chemicals.

2. Risk Assessment of Major Hazard Installations

Individual risk at the location (x, y) in hazardous area of a major hazard can be calculated by the Eq. 2 [9].

$$R_g = \sum_{s=1}^{S} \sum_{w=1}^{W} \sum_{i=1}^{I} F_{s,o} F_E F_M P_i V_s(x, y) \quad \text{[Eq. 2]}$$

where $R(x, y)$ is individual risk at the location (x, y) caused by the major hazard; $F_{s,o}$ is original frequency of leaking accident for each container equipment; F_E is equipment correction factor; F_M is safety management and personnel correction factor; P_w is the probability of weather conditions; P_i is the probability of ignition source; $V_s(x,y)$ is frequency of death at the location (x, y) caused by the accident s; S is the number of leaking accident; W is the number of weather conditions and I is the number of ignition source.

2.3 Risk Assessment of transportable hazards

Hazardous transportable risk mainly refers to the combination of the consequence and frequency of unwanted events when transport the hazardous chemicals [10]. Total transportable risk is equal to the algebraic sum of risks on various sections.

$$R_t = \sum_{i=1}^{m} p_i \cdot c_i \quad \text{[Eq. 3]}$$

where R_t is total transportable risk, p_i is frequency of hazardous chemicals leaking accident in section i, c_i is the comprehensive consequence of the unwanted events and m is the number of road sections. The risk in any section can be calculated by Eq. 4, if the number of sections is infinity.

$$R_y = p_i \cdot c_i \quad \text{[Eq. 4]}$$

The casualty area should be the corresponding evacuation distance in practical engineering applications. Individual risk caused by the transportable hazards can be assumed in the casualty area. This individual risk is the risk at the point i. The individual risk excessing casualty area could be ignored.

2.4 Superposition of Fixed and transportable hazards risk

There are two prerequisites, one is that transportable hazards move on the roads in the chemical industry parks and the other is that the risk of chemical industry parks can be represented by individual risk.

2.4.1 Case without domino effect

The equation for calculating the superposition of fixed and transportable hazards risk would be the following one considering the Eq. 2 and Eq. 4.

$$R = \begin{cases} R_g + R_y \\ R_g \text{ or } R_y & \text{The risk could be ignored when the individual risk is less than } 10^{-8}. \end{cases}$$

2.4.2 Case with domino effect

There are three main forms of domino effect: fixed hazards—fixed hazards, transportable hazards—fixed hazards and transportable hazards—transportable hazards. Among the three forms, probability of occurrence is from high to low.

(1) Fixed hazards—fixed hazards

The total individual risk is initial accident's individual risk plus chain accident's individual risk. So the individual risk R_o can be calculated by Eq. 5 [11].

$$R_o = \sum_{m=1}^{h}\sum_{i=1}^{n}\left(R_i^m + \sum_{k=1}^{w}\sum_{j=1}^{s} R_{i,j}^{m,k}\right) = \sum_{m=1}^{h}\sum_{i=1}^{n}\left(P_i^m V_i^m + \sum_{k=1}^{w}\sum_{j=1}^{s} P_{i,j}^{m,k} V_{i,j}^{m,k}\right) \quad \text{[Eq. 5]}$$

where R_i^m is the individual risk caused by the initial accident scenario i of device m . $R_{i,j}^{m,k}$ is the individual risk caused by the secondary accident scenario j of device k which is destroyed by initial accident scenario i of device m . h is the total number of devices have initial accidents. n is the total number of initial accident scenarios of device m . w is the total number of devices which are destroyed by all the initial accident scenarios of device m .s is the total number of the secondary accident scenarios when device k is destroyed. V_i^m is the probability of death which is determined by the initial accident scenario i of device m . $V_{i,j}^{m,k}$ is the probability of death that is determined by the secondary accident scenario j of device k which is destroyed by initial accident scenario i of device m .

(2) Transportable hazards—fixed hazards

Assuming that transportable hazards are fixed hazards with probability, we can calculate the distance that transportable hazards accidents will cause fixed hazards accidents. Then we can obtain the probability P_c of domino effect between transportable hazards and fixed hazards. Through counting the time when transportable hazards will appear and cause domino effect, we can obtain the probability P_a of that transportable hazards appear in the scope cause domino effect. At this point, the personal risk of chemical industry parks can be calculated by the following equation under domino effects.

$$R = \begin{cases} P_c \cdot P_a \cdot (R_g + R_y) \\ P_c P_a R_g \text{ or } P_c P_a R_y \end{cases}$$

The risk could be ignored when the individual risk is less than 10^{-8}.

(3) Transportable hazards—transportable hazards

In reality, this situation is very rare. From both the accident statistics and news reports, the probability is very low. Therefore, litter attention is paid on this situation.

3. Design of monitoring and early-warning system in chemical industry parks

Design of monitoring and early-warning system in chemical industrial parks is always based on the risk assessment of major hazard installations, transportable hazards and domino effect. Applicability, quantity, installation points, layout of the monitoring and early-warning devices must be established on the basis of reasonable risk assessment of chemical industrial parks.

3.1 Monitoring and early-warning system for fixed hazards

Monitoring and early-warning system of fixed major hazard installations consists of indices management subsystem, information gathering subsystem and decision making subsystem.

Due to the difference of the major hazard installations in storage tank farm, storage farm and production area, the indices of the monitoring and early-warning system were divided into 6 groups.

(1) the key process parameters of storage tanks and production installations, such as temperature, pressure, liquid level, flow, etc

(2) concentration of the flammable and toxic gases

(3) environmental parameters, such as temperature, humidity, wind speed, wind direction, etc

(4) audio and video signals and the condition of staff in and out

(5) fire and smoke

(6) grounding resistance of lightning rod and antistatic devices and power supply conditions

According to the risk assessment of chemical industry parks, fixed major hazard installations must be under the control of monitoring and early-warning system. At the same time, reasonable selection, design, installation, commissioning and maintenance of monitoring equipment and facilities can be done depend on the site condition and characteristics of the monitoring object.

For the domino effect, we can calculate the comprehensive personal risk of chemical industrial parks with the probability of initial accidents and chain accidents and the consequences of accidents. At last, the layout of monitoring and early-warning system was research based on the results of personal risk distribution in chemical industrial park.

Among the 6 groups mentioned above, devices for monitoring the indices in group 1 and 6 are usually installed on the main body structure of the target; devices for monitoring the indices in group 2, 3, 4 and 5 can be installed surrounding the monitoring target in the chemical industrial park, and the monitoring parameters which were obtained through the monitoring equipment are the characteristic parameters of accident. Therefore, the layout of these four types monitoring devices should be based on the personal risk in chemical industry parks.

Once the monitoring target is determined, we will calculate the personal risk which is caused by the hazards of the target and obtained the corresponding risk contour distribution of personal risk. At the same time, considering the risk superposition of transportable hazards and domino effect, we choose the high risk value points as the final candidates for the monitoring devices installation. Then with the other relevant requirements, the installation positions of monitoring devices are determined ultimately.

3.2 Monitoring and early-warning system for transportable hazards

Dynamic of transportable hazards brings dynamic individual risk. Therefore, the main monitoring devices should be installed on the transportable hazards for monitoring and early-warning. In this paper, two types of monitoring and early-warning systems are put forward. First, the technology like mobile wireless communication network is used to establish the monitoring and early-warning systems. Second, the two-dimensional codes are used to establish this systems for their dynamic store information characteristics, and monitoring and early warning information can be transmitted intermittently.

3.2.1 The application of wireless communication network in the monitoring and early- warning system

Before the transport vehicles start, an information terminal launcher similar to SIM card is installed on the car, using wireless communication network terminal and remote platform to guarantee the information unobstructed. The temperature sensor, pressure sensor, concentration sensor, speed monitor or other such devices will practice the real-time collection of monitoring information and have the information stored in the SIM card information terminals. The stored information will be sent in set intervals via wireless communication network to the remote monitoring and early-warning system platform. Finally, it will realize the monitoring and early warning of road hazards. The whole process is shown in figure 2.

Figure 2: Communication Theory of wireless monitoring and early-warning systems

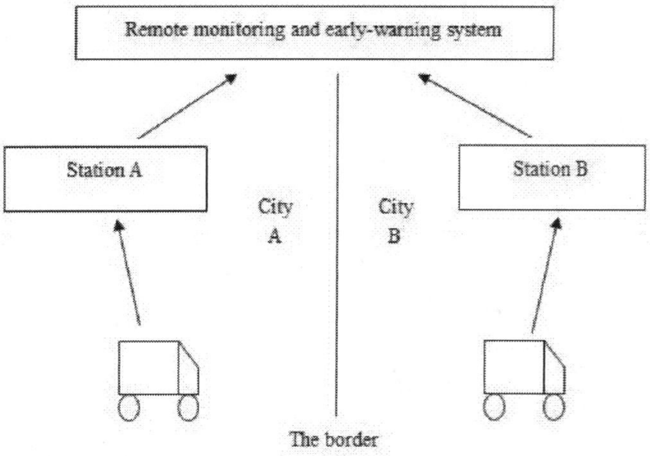

3.2.2 The application of two-dimensional codes in monitoring and early-warning systems

For some areas and roads wireless communication network is not covered, the hazardous chemical transportation vehicles can use two-dimensional codes to store large amounts of information and implement dynamic storage. And the state of the running vehicle and hazardous chemicals will be intermittently sent to the remote monitoring and early-warning platform through a dedicated network transmission. During the time of hazardous chemical transportation vehicles from provenance to destination, vehicle-mounted two-dimensional code compilation devices compile the sensor collected information every once in a while (recommended time interval not more than ten minutes), then store the information in the onboard storage terminals, when through the setting sites, the vehicles can transmit the stored information to the remote monitoring and early-warning platform through the two-dimensional transmission port established in these sites. Implementation process is shown in figure 3.

Figure 3: Communication Theory of monitoring and early-warning systems using two-dimensional codes

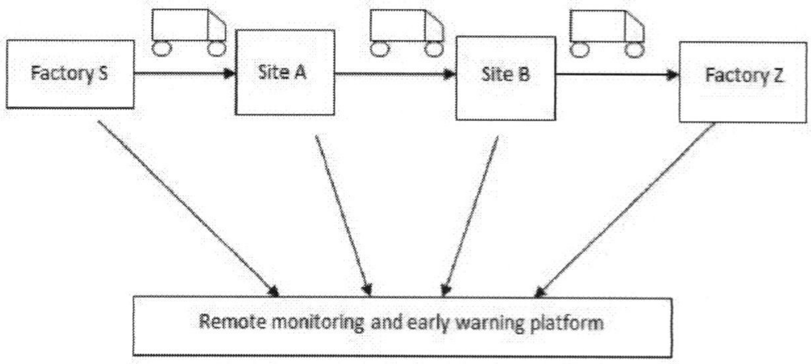

4. Conclusion

1. Through the analysis of accident cases, we can summarize that the risk assessment in chemical industrial park must include fixed hazards, transportable hazards and domino effects.
2. Based on the previous calculation of individual risk caused by fixed hazards and transportable hazards in chemical industry parks, the comprehensive risk assessment model is established between these two hazards with the influence of domino effects.
3. According to the risk contour which is calculate by risk assessment of chemical industry parks, a monitoring and early-warning system for chemical industry parks is designed. For the superposition of risk that is caused by domino effects, monitoring devices for

fixed hazards should be chose to install at the high risk value points. And monitoring devices for transportable hazards are commonly installed on their own body. So the dynamic monitoring and early warning can be realized.

5. References

[1] Sun Nannan. Construction of emergency evaluating system on incidents occurred in chemical industrial parks based on process management[C]. 2011 International Conference on Engineering and Business Management,2011:3509-3512.

[2] Sun Jinfeng. Research on Application of Major Hazard Installation Safety Monitoring and Early-Warning System[D].South China University of Technology,2010.

[3] Sergio Contini, Furio Bellezza, Michalis D.Christou. The Use of Geographic InformationSystems in Major Accidents Risk Assessment and Management[J]. Journal of Hazardous Material, 2000, 78(2): 223-245.

[4] H. Jeong, H. Kim, S. Lee, D. Dornfeld. Multi-Sensor Monitoring System in Chemical Mechanical Planarization (CMP) for Correlations with Process Issues[J]. CIRP Annals-Manufacturing Technology,2006,55(1):325-328.

[5] S. Bagavathiappan, B.B. Lahiri, T. Saravanan, John Philip, T. Jayakumar. Infrared thermography for condition monitoring – A review[J]. Infrared Physics & Technology,2013,60:35-55.

[6] Code for construction of fixed hazard safety monitoring and early-warning system Part 1: General[S]. DB32/1321.1—2009,2009.

[7] Yao Feng. Study of Supervision and Emergent Rescue of Transportable Hazards on Road Transportation in a province[D].Jiangsu University, 2007.

[8] Identification of major hazard installations for dangerous chemicals[S]. GB18218-2009,2009.

[9] Chen Xiaodong, Shi Lichen, Liu Ji, Zeng Mingrong, Duo Yingquan. Study on the Appropriate Risk Capacity of Chemical Industry Park [J].China Safety Science Journal,2009,19(3):132-137.

[10] Tan Xiaoqun. Research on Evaluation Model for Safety Capacity of Chemical Industrial Park Based on Regional Risk and its Application in Chemical Industrial Park[D]. South China University of Technology, 2011.

[11] Zhang Mingguang, Jiang Juncheng, Sun Dongliang, Ye Junle. Research on the Layout of Safety Monitoring and Control System in Chemical Plant Based on Risk[J].Industrial Safety and Environmental Protection,2010,36(12):49-51.

Study on Typical Chemical Accidents' Emergency Rescue Decision Making Based on GIS

Shanshan Guo
College of Urban Construction and Safety Engineering, Nanjing University of Technology
Xin mo fan road 5#, Nanjing, China
shanshanguo33@163.com

Huihui Song
Administration of Work Safety of Linzi District
Xue gong road 309#, Zibo, China
187359790@qq.com

Gang Tao
Lijing Zhang
College of Urban Construction and Safety Engineering, Nanjing University of Technology
Xin mo fan road 5#, Nanjing, China
taogangs@163.com
zhanglj@njut.edu.cn

Prepared for Presentation at American Institute of Chemical Engineers

1st CCPS Asia-Pacific Conference on Process Safety

Qingdao, China

September 4 – 5, 2013

UNPUBLISHED

Study on Typical Chemical Accidents' Emergency Rescue Decision Making Based on GIS

Shanshan Guo
College of Urban Construction and Safety Engineering, Nanjing University of Technology
Xin mo fan road 5#, Nanjing, China
shanshanguo33@163.com

Huihui Song
Administration of Work Safety of Linzi District

Gang Tao*
Lijing Zhang
College of Urban Construction and Safety Engineering, Nanjing University of Technology

Keywords: Emergency rescue; Decision analysis; Typical Chemical accident; Geographical Information System (GIS)

Abstract

With the development of the chemical industry, fires, explosions and poisoning is probable to happen. Effective emergency responses would be of great importance for reducing the consequences of accident. The typical leakage accident is chosen firstly. Impact scale is determined by accident consequence models and divided into different zones at the same time. Then, according to detailed consequence, a rule system which would be built on GIS platform was established by consulting resources. Finally, after judgments, comparing and calculation, decision result is obtained by the data of GIS platform.

1. Introduction

Along with the densification of modern enterprise device, the complication of process, the large-scale of production and the diversification of products, once the chemical accident occurs in large-scale chemical enterprise, the consequences will far beyond the acceptable degree. Therefor, the importance of the emergency rescue work is becoming more and more prominent.

GIS always integrates geography, geometry, computer science and all the application objects, CAD technology, remote sensing, GPS technology, internet, multimedia, virtual reality technology, utilizing computer graphics and data base technology to collect, store, edit, display, transform, analyze and deliver geographical graphics and attribute data. So that the information can be delivered to the users for their analysis and decision making purpose. Now GIS is widely used in transportation, environment protection, mining and other economic activities [1].

2. Research method

Analytic Hierarchy Process (AHP) in multi-objective decision-making is used to make the risk degree of different areas affected by the accident be comprehensive and quantitative [12]. Combined with the basic information for decision-making that given by GIS system, we can judge and solve the decision-making model and sort the risk degree of all areas affected by the accident. And then the most severe affected unit is determined. Finally which area is rescued on priority can be decided. The ideological line of this research is shown in Figure 1.:

Figure 1:Research procedure

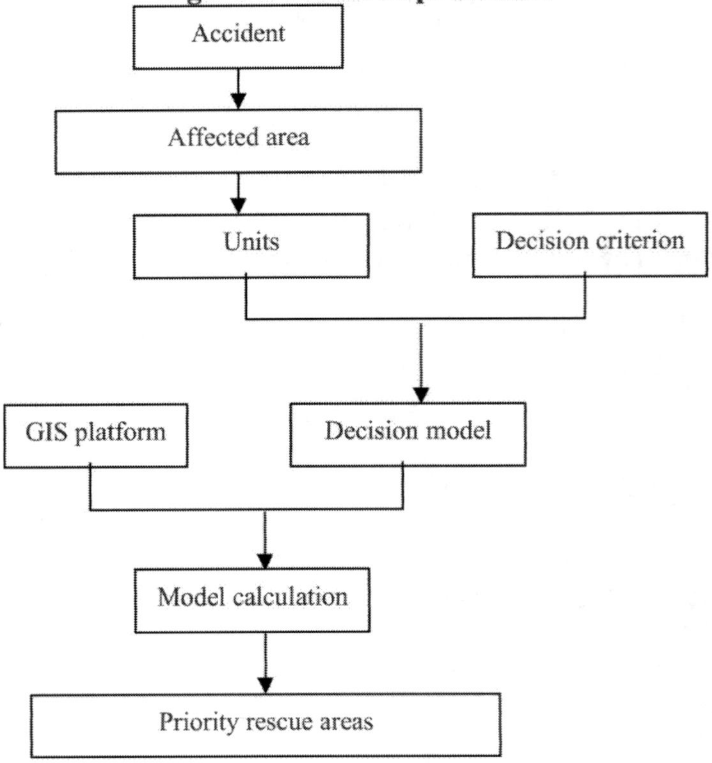

3. Decision criterion

The area affected by accident can be obtained via analyzing the model of accident consequence. And then priority rescue areas in the area affected by accident are determined. These priority rescue areas are mainly determined by some factors as the criterion. Through literature and the analysis of related accident cases found that, for a certain area affected by the accident of dangerous degree, we can consider from several aspects. During the decision-making process, according to the incident and the area to identify a specific decision criterion.

- the severity of accident consequences;
- the size of population factors;
- the impact on the environment;
- the loss of property;
- the key systems in affected area ;
- social influence and political influence;

- the emergency resources in affected scope and the quality of public emergency [2-6].

Criteria information database and spatial database is the key basis of work. The information is reflected by GIS data.

4. The analysis of decision-making procedure

For all kinds of accidents, the analysis of corresponding accident consequences model can give influence scope and extent affected by the accident. It provides certain guiding significance for emergency rescue.

4.1. Consequence analysis of case accident

The common liquid chlorine leakage accident and diffusion model for leakage diffusion simulation in the chemical production are chosen in the paper. Damage radius can be obtained after a specified period according to the poison of the toxic dose [6]. Table 1. shows the danger concentration of chlorine under different contact time.

Table 1. Standard the toxic for dose of chlorine

Toxic	Dose standards(g.min/m3)		
	Severe damage	Moderate damage	Mitigated injury
Chlorine	6.40	3.20	0.48

Assume the following scenario, the liquid chlorine tank ruptures. The leakage is continuous and the rate is $5\text{kg}\cdot\text{s}^{-1}$. The temperature in the storage tank is 298K before the leakage. And the liquid chlorine leak out of the tank becomes gasification instantly. It happens in daytime and atmospheric stability for the medium is not stable and the wind speed is $2.5\text{m}\cdot\text{s}^{-1}$.

The density of chlorine is about 2.48 times that of the air. After the rupture the chlorine forms heavy gas clouds. Due to the leakage is continuous, steam cloud forms continuously and mixes with air dilution constantly. It turns into the neutral cloud and spreads immediately. Because of the heavy gas stage has little impact on calculation of the whole spread range, the neutral cloud model simulation and the gaussian model are used directly in this paper [7-9].

Through the preliminary calculation, dangerous substances diffused to 1882 m under the direction of the wind, longitudinal maximum distance is 243 m and isoline concentration is 10 mg m^{-3}. The range of influence is large and the damage consequence is very serious. By using the c# program to simulate the diffusion, GIS calls diffusion calculation and draws to get scope influenced by accident.

4.2. Emergency rescue decision model

The accident consequence analysis and numerical simulation indicate the area influenced by the accident can extend for miles. It was divided several regions for further evaluation. These regions would be as object of decision-making. The importance degree of the region is the decision-making general objective G. Considering workload of the basic data investigation and according to the influence degree of accident consequences, the area influenced by accident is divided into eight regional units, which are A1, A2, A3, A4, A5, A6, A7, A8. After the diffusion simulation calculation, the diffusion range and divided areas can be got. The information is shown through the GIS platform.

On the basis of the specific accident types, the nature of the dangerous substances and operability in practice, hierarchy analysis diagram is constructed as Figure 2 shows:

Figure 2: Regional hierarchy diagram of emergency rescue

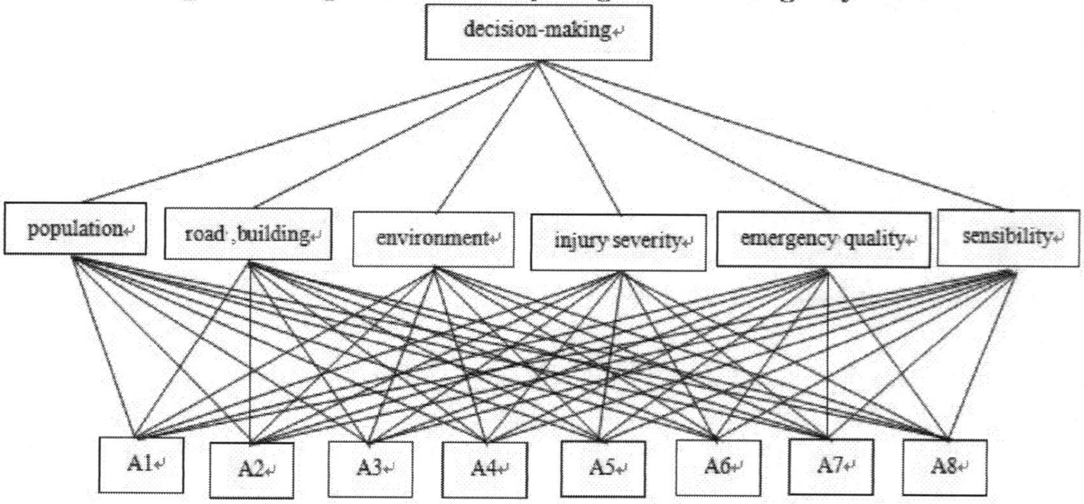

In GIS maps, different types of data based on different layers, all data is displayed through the grid view. At the same time, for some information needed can be obtained by using the GIS map file editing interface. Selecting the data associated with each criterion, standard system data is got and shown in Table 2.

Table 2. The data information Based on GIS

area	population	road network(m)	building(m^2)	river system	land types	sensitive unit
A1	62	182	326	——	——	——
A2	24	356	521	——	——	——
A3	83	504	536	lake	vegetation	hospital
A4	255	176	333	——	——	——
A5	423	81	105	river	——	——
A6	651	199	158	river	——	——
A7	406	374	544	river	vegetation	school
A8	169	305	469	river	vegetation	——

4.3. Solution of decision-making model

For the general goal of decision-making, according to expert's judgment, related literature and the analysis of accident case, a judgment matrix B(G-C) for the importance degree of each criterion can be made up:

G	C1	C2	C3	C4	C5	C6

C1	1	5	3	1	4	3
C2	1/5	1	1/3	1/5	1/2	1/2
C3	1/3	3	1	1/3	2	1
C4	1	5	3	1	4	3
C5	1/4	2	1/2	1/4	1	1/2
C6	1/3	3	1	1/3	2	1

Another judgment matrix B' is obtained by normalizing each column of the judgment matrix B(G-C).

$$B' = \begin{bmatrix} 0.339 & 0.263 & 0.360 & 0.296 & 0.321 & 0.340 \\ 0.068 & 0.052 & 0.040 & 0.037 & 0.064 & 0.038 \\ 0.113 & 0.157 & 0.120 & 0.148 & 0.107 & 0.113 \\ 0.339 & 0.262 & 0.360 & 0.296 & 0.321 & 0.340 \\ 0.085 & 0.105 & 0.060 & 0.074 & 0.080 & 0.157 \\ 0.113 & 0.157 & 0.120 & 0.148 & 0.107 & 0.113 \end{bmatrix}$$

Vector quantity M is obtained by adding together judgment matrix B' according to the line: $M = (1.191, 0.299, 0.758, 1.918, 0.561, 0.758)$. W_c is obtained by normalizing Vector quantity M. $W_c = (0.308, 0.048, 0.122, 0.308, 0.09, 0.122)^T$ is the eigenvector of matrix G-C, W_{c1}, W_{c2}, W_{c3} are weighted value of the relative important degree of each criterion.

By the type $\lambda_{max} = \frac{1}{n}\sum_{i=1}^{n}\frac{(BW)_i}{W_i}$, for judgment matrices B(G-C):

$$BW_c = \begin{bmatrix} 1 & 5 & 3 & 1 & 4 & 3 \\ 1/5 & 1 & 1/3 & 1/5 & 1/2 & 1/3 \\ 1/3 & 3 & 1 & 1/3 & 2 & 1 \\ 1 & 5 & 3 & 1 & 4 & 3 \\ 1/4 & 2 & 1/2 & 1/4 & 1 & 1/2 \\ 1/3 & 3 & 1 & 1/3 & 2 & 1 \end{bmatrix} \bullet \begin{bmatrix} 0.308 \\ 0.048 \\ 0.122 \\ 0.308 \\ 0.09 \\ 0.122 \end{bmatrix} = \begin{bmatrix} 1.948 \\ 0.298 \\ 0.773 \\ 1.948 \\ 0.462 \\ 0.773 \end{bmatrix}$$ Then,

$$\lambda_{max} = \frac{1}{n}\sum_{i=1}^{n}\frac{(BW_c)_i}{W_{ci}} = \frac{1}{6}\left[\frac{1.948}{0.308} + \frac{0.298}{0.048} + \frac{0.773}{0.122} + \frac{1.948}{0.308} + \frac{0.462}{0.09} + \frac{0.773}{0.122}\right] = 6.11$$

$$CI = \frac{\lambda_{max} - n}{n-1} = \frac{6.11-6}{5} = 0.022$$

Standard values of CR can be found by looking up the literature data. Because of $CR = \frac{CI}{RI} = \frac{0.022}{1.26} = 0.018 < 0.1$, judgment matrices B(G-C)has the satisfactory consistency.

Similarly, pairwise comparison is made to attribute values under the various decision criteria of a few units affected by the accident. Then the valuation is carried according to the principle. Six judgment matrixes of 8×8 is made, thereinto:

The criterion C4, according to the calculation of accident consequences, is combined with the Table 1. The size of damage people suffered in the scope is achieved and death or serious injury area is shown by GIS. As shown in Figure 3.

According to the area size of each regions within the scope of the different degree of damage, judgment matrix B(C4) is established :

C4	A1	A2	A3	A4	A5	A6	A7	A8
A1	1	1	4	4	6	6	8	8
A2	1	1	4	4	6	6	8	8
A3	1/4	1/4	1	1	3	3	5	5
A4	1/4	1/4	1	1	3	3	5	5
A5	1/6	1/6	1/3	1/3	1	1	3	3
A6	1/6	1/6	1/3	1/3	1	1	3	3
A7	1/8	1/8	1/5	1/5	1/2	1/2	1	1
A8	1/8	1/8	1/5	1/5	1/2	1/2	1	1

Another judgment matrix B_4' is obtained by normalizing each column of the judgment matrix B(C4), then vector quantity W_4 is obtained:

$$W_4 = (0.302, 0.302, 0.115, 0.115, 0.055, 0.055, 0.028, 0.028)^{\mathrm{T}}$$

$$B_4 W_4 = (2.634, 2.634, 0.991, 0.991, 0.455, 0.455, 0.232, 0.232)^{\mathrm{T}}$$

$$\lambda_{\max} = 8.479,\quad CI = 0.068,\quad CR = 0.048 < 0.1$$

Figure 3: The spread of serious injury and death

(a) Seriously injured concentration range (b) Lethal concentration range

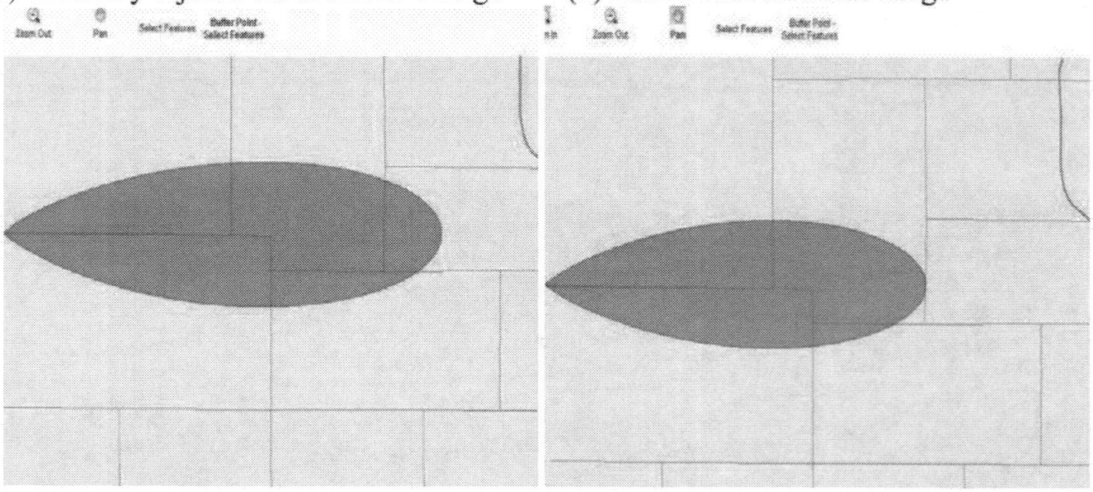

At the same time, for other 5 criteria, the similar calculation is made and finally make general calculation of the Weight.

The calculation above indicates that all judgment matrixes are accord with satisfactory consistency.

$$W = W_c \cdot W_i = \begin{bmatrix} 0.064 & 0.189 & 0.051 & 0.302 & 0.063 & 0.056 \\ 0.025 & 0.051 & 0.051 & 0.302 & 0.040 & 0.056 \\ 0.064 & 0.029 & 0.324 & 0.115 & 0.124 & 0.172 \\ 0.111 & 0.188 & 0.051 & 0.115 & 0.124 & 0.101 \\ 0.177 & 0.324 & 0.087 & 0.055 & 0.063 & 0.101 \\ 0.316 & 0.116 & 0.086 & 0.055 & 0.211 & 0.056 \\ 0.177 & 0.051 & 0.205 & 0.028 & 0.335 & 0.357 \\ 0.067 & 0.051 & 0.143 & 0.028 & 0.040 & 0.101 \end{bmatrix} \bullet \begin{bmatrix} 0.308 \\ 0.048 \\ 0.122 \\ 0.308 \\ 0.09 \\ 0.122 \end{bmatrix} = \begin{bmatrix} 0.140 \\ 0.120 \\ 0.128 \\ 0.108 \\ 0.115 \\ 0.156 \\ 0.164 \\ 0.065 \end{bmatrix}$$

By comparing the other four, areas 1, 3, 6 and 7 should be protected preferentially. So investment in rescue personnel and resources should be increased [10-12].

5. Conclusions

The Analytic Hierarchy Process in multi-objective decision-making is used in this paper to study decision-making of emergency rescue in each area. Risk assessment can be accomplished quickly by GIS system. For some people in danger and other valuable resources can be marked by GIS intuitively, and then homologous action should be adopted.

6. References

[1] Yao Haowei, D. Wenli, L. Dong, "Application of GIS on Emergency Rescue", Procedia Engineering, Volume 11, pp:185-188, 2011

[2] Zongzhi Wu , Mao Li. "An introduction to major accident emergency rescue system and preparedness", Beijing, Metallurgical Industry Press, 2003

[3] Zongzhi Wu, Mao Li, "Overview of emergency rescue — The system and plan of emergency rescue", Beijing, Chemical Industry Press, 2004

[4] Cui W Z,Liu C. "Study on establishing emergency mechanism for handing major sudden accidents of water contamination", Pearl River, pp:1-6, 2005

[5] Jihong Guo, "The Accidents Forecast Simulation and Safety Programming of City MHI", Huaian:Anhui University of Science and Technology, 2006

[6] Wang Ziji, Zhao Jinyuan, "Chemical accidents and emergency rescue", Beijing, Chemical Industry Press, 1997

[7] Zhang Lijing, Ding Li, Tao Gang, "Simulation of accident consequences of liquefied chlorine pipeline leakage, Journal of Chemical Industry and Engineering," Volume 59, NO 5, 2008

[8] Venkata G,Ashok K, "An evaluation of four box models for Instantaneous dense gas releases, Journal of Hazardous Materials," pp: 237-255, 1990

[9] Mohan M, Panwar T S, Singh M P, "Development of dense gas dispersion model for emergency preparedness", Atmospheric Environment, Volume 29, NO 16, pp: 2076-2087, 1995

[10] Xu Jiuping, Wu Wei, "Theory and methods of multiple attribute decision making" , Beijing, Tsinghua University Press, 2006

[11] Wang Lianfen, Xu Shubo, "An introduction to the analytic hierarchy process", Beijing, The Chinese People's University Press, 1990

[12]Saaty T L,Vargas L G., "Ucertainty and rank order in the analytic hierarchy process", European Journal of Operational Research," Volume 32, NO. 1, pp: 107-117, 1987

1ST CCPS ASIA-PACIFIC CONFERENCE ON PROCESS SAFETY℠

Investigation on the Safety Performance of the Organic Insulation Materials in Petrochemical Enterprises

Yanfei CHEN
Shaung ZHUANG

Tianjin Fire Research Institute of MPS,
Tianjin 300381, China

chenyanfei@tfri.com.cn
zhuangshuang@tfri.com.cn

Prepared for Presentation at
American Institute of Chemical Engineers
2013 Spring Meeting
9th Global Congress on Process Safety
San Antonio, Texas
April 28 – May 1, 2013

UNPUBLISHED

Investigation on the Safety Performance of the Organic Insulation Materials in Petrochemical Enterprises

Yanfei CHEN
Shaung ZHUANG

Tianjin Fire Research Institute of MPS,
Tianjin 300381, China

chenyanfei@tfri.com.cn
zhuangshuang@tfri.com.cn

Keywords: petrochemical enterprise organic insulation materials, calorific value, non-combustibility

Abstract

Most work environment of the petrochemical production plants is above or below ambient temperatures, the chemical equipment for thermal insulation is to reduce heat loss and an important measure is to achieve thermal insulation. The use of organic insulation materials in modern buildings has made a huge contribution to energy saving. Common organic insulation materials are sprayed polyurethane borad and polystyrene particles. During the complex producing progress, in the event of an accident, fire explosion caused a high probability; while the organic insulation materials burn with releasing lat of heat and toxic fumes for their bad heat resistance and which will lead the spilled materials fire again and explosion, causing more serious accident of life and property. The calorific value of the materials is the important parameters to characterize their potential fire hazard and it's also the essential basic data to calculate its combustion releases heat and fire load; non-combustibility test can determine whether the material directly contributes to fire. The methods mentioned above were used in this paper to study the typical organic insulation materials in China, and further combustion performance analysis and proposal were made to give recommendations for improvement.

1. Introduction

Petrochemical production devices usually work during high temperature, high pressure or low temperature, high vacuum, so using insulation materials an important measure to reduce heat loss and achieve thermal insulation [1]. Because the typical organic insulation materials, such as polyurethane foam, polystyrene board, phenolic foam. The advantages of organic insulation materials are lightweight, a high density, good thermal insulation, and simple installation, good mechanical properties [2]. Hence, organic materials are common insulation materials used in petrochemical equipments. In addition, as petrochemical plants often handle flammable, corrosive materials, so a lot of combustible and flammable materials will leak in case of accidents, which will lead to fire or explosion, and the materials with high temperature will directly get fire if the temperature exceeds its ignition point. Many fires were caused by the equipments problems in China.

As we know, the shortcomings of the organic insulation materials are poor heat resistance and flammable. Once on fire, it releases a large amounts of heat and produces large amounts of carbon monoxide, hydrogen cyanide, hydrogen chloride and other toxic, highly toxic gas, so it only takes tens of seconds to be fatal in the event of a fire, which do great harm to people's lives and property. Thus, if the insulation materials get fire first, petrochemical equipments will go wrong easily and lead to further fire explosion. On the opposite, if the fire is caused by device leakage, the insulation materials will accelerate the process of deterioration of the fire and explosion losses. Therefore, the insulation materials used in petrochemical enterprises should be in strict accordance with relevant regulations, try to use flame retardant or flame-retardant materials. The thermal insulation performance and security of use are both very important to be considered.

Flammability of the insulation materials are all the physical and chemical changes that occur in a fire. The performance is measured by the material surface fire, flame propagation, heat, smoke, charring, weight loss, as well as toxic product. The calorific value of insulation materials is the important parameters to characterize the potential fire hazard of the materials and it's also the essential basic data to calculate its combustion releases heat and fire load [3]; non-combustibility test can determine whether the material directly contributes to fire. This study is made on the three typical kinds of organic materials by oxygen bomb calorimeter and non-combustibility test methods to give improvement propose on the combustibility performance.

2. Combustion performance test

2.1 test samples

The three types of organic insulation materials used in China were tested in this study. Phenolic material, polyurethane material, and pollyphenylene materials are included.

2.2 test instruments and conditions

Oxygen bomb calorimeter was used test the calorific value of the organic insulation materials. The determination of the heat of shells is in the case of the excess oxygen in an oxygen bomb, the heat generated by predetermined conditions of combustion per unit weight of the sample, shown as the bomb calorific value [J/g or kJ/kg].

Non-combustibility test method of insulation materials were made by the FTT non-combustibility apparatus. The furnace temperature was kept 750±5°Cduring the test for 30min ~ 60min. Cylindrical sample with (76±8) cubic centimeter in volume, 45^{+0}_{-2} in diameter, (50±3)mm in height. Moreover, evaluation indexes are mass lose (%), continuous flame time (s) and furnace average temperature rise (°C).

In addition, to avoid the residue dripping during the test, metal meshes were put on the bottom of the sample holder.

The two test instruments are shown in Figure 1.and Figure 2.

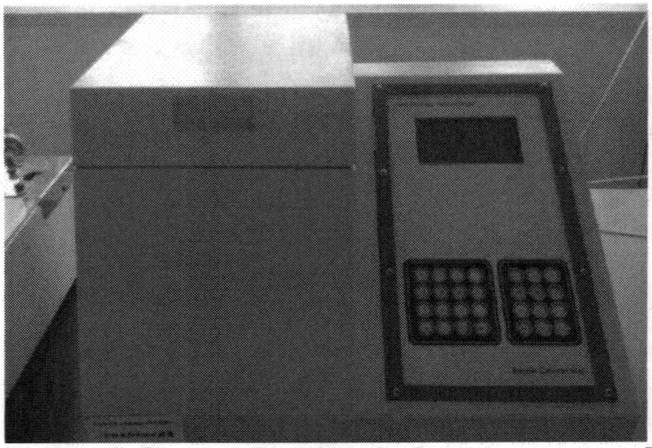
Figure 1. Oxygen bomb calorimeter

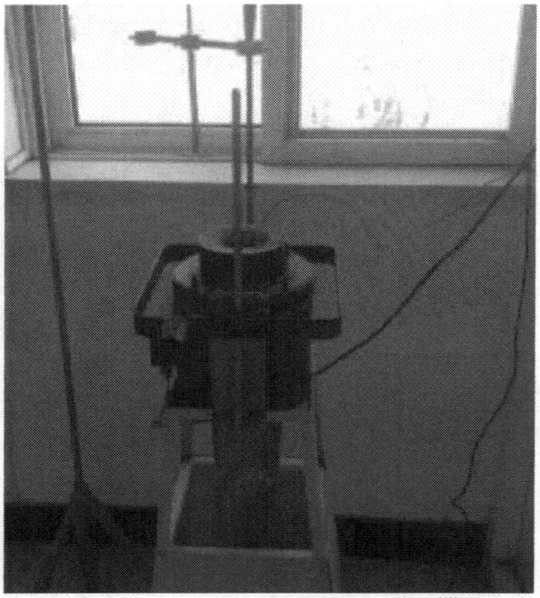
Figure 2. Non-combustibility test instrument

3. Test phenomenon and conclusions

3.1 phenomenon of the oxygen bomb test

Three types, equivalent to fifteen samples were tested in this study and the phenomenon is concluded as follows.

(1) After the tests of polyurethane, phenolic, extruded board materials, irritant gases ran out when the oxygen bomb was opened, which was produced during the burning in oxygen and not absorbed by the water in it.

(2) After the test, only black charcoal-like scaling on the wall of the crucible pot. This is for these types of organic materials, completely combustion happened and the black was the combustion solid product.

3.2 data of the oxygen bomb test

Bomb calorific value ranges of the typical organic insulation materials are shown in Table 1.

Table 1 Bomb calorific value of the organic insulation materials

Type	Bomb calorific value ranges[MJ/kg]
Phenolic	20 ~ 30
Polyurethane	25 ~ 30
Pollyphenylene	20 ~ 35

3.3 phenomenon of the non-combustibility test

The three types marerials selected in the test were from five different plants in China, and three groups of each type were tested in this study.

(1) During the test of two common phenolic materials, the samples burned immediately, generating much irritant gases and a very small amount of residual gray wreckage after the test. But during the test of the modified phenolic materials, the condition was much improved, three parameters were all the lowest in the table.

(2) During the tests of polyurethane material, much thick black smoke released as the samples were put into the furnace immediately, although the time of the continuous flame is the shortest, the temperature rising is the longest in the three types materials. After the test, nothing was left on the metal mesh.

During the tests of pollyphenylene materials, no flame at the start when they are put into the furnace for a few seconds, and then bright flame appeared with strong pungent odor, this was the combustion products of benzene. The temperature rising is the shortest in the three types materials. After the test, only a little coke-like residue left on the metal mesh.

3.4 data of the non-combustibility test

Data ranges of the non-combustibility test on the typical organic insulation materials are shown in Table 2.

Table 2 Bomb calorific value of the organic insulation materials

Type	Temperature rising[°C]	Mass lose[%]	Time of continuous flame[s]
Phenolic	60 ~ 100	90 ~ 95	100 ~ 400
Polyurethane	100 ~ 150	100	20 ~ 50
Pollyphenylene	30 ~ 50	98	30 ~ 100

4. Test phenomenon and conclusions

Petrochemical plant once on fire, burning speed is quite high and fire spread rapidly. One reason is for the reason of devices explosion, a high temperature combustion zone formed and the fire spread quickly in all directions, or even cause explosion again, and further to form a large area of the combustion. The second reason is the petrochemical materials calorific value is too large, and combustion heat generated by radiation rapidly heat the surrounding adjoining container equipment and the insulation materials, new fire and explosion happens, resulting in the materials in the adjacent vessel rapidly pressurized, volatilization or decomposition, to create the conditions for the spread of fire[4]. The third is the flow and diffusion of the materials, especially the combustible gas diffusion, increasing the risk of fire expanding instantly. Therefore, petrochemical fire can affect a large range of combustion area in a short period of time, big heat release and burning speed.

In recent years, related departments in China have paid great attention on the fire supervision and management of the insulation materials. The fire performance of the organic insulation materials used in petrochemical became more and more important. Fire resistance of the organic insulation materials are in great demand to be improved, which can reduce the harm in the petrochemical fire and explosion.

5. Conclusion

Above all, conclusions and recommendations about the organic insulation materials used in petrochemical equipments can be made as follows:

(1) On condition of the same volume, ordinary phenolic material release the most heat and toxic gases when burning, and the parameters of polyurethane is in the middle, pollyphenylene material shows better heat resistance performance in the test.

(2) Modified phenolic materia showed great advantage in the test.

(3) More modified organic insulation materials with similar other physical characteristics are in great demand to be used in petrochemical industry.

6. References

[1] Qian YE, The Insulation Materials' application in Chemical Equipments, Chemical Engineering and equipment, Vol.8 (2012),p 161-162,166.

[2] Information on http://info.china.alibaba.com/

[3] W.Ye. in: *The Characteristics of the Building's Exterior Wall Material Fire and Prevention*. Journal of New Technology and New Products, Vol. 8(2011), p. 171.

[4] X.B. Wan,The chemical equipment fire safety in chemical industy,Chemical Engineering & Equipment, Vol.7(2012),174-176.

1ST CCPS ASIA-PACIFIC CONFERENCE ON PROCESS SAFETY℠

Thermal stress analysis of hydrogenation reactor using the finite element software

Xinmeng Sun
College of Urban Construction and Safety Engineering, Nanjing University of Technology
Xin Mofan Road 5#, Nanjing, China
sunxinmeng@126.com

Wan Xu
Nanjing College of Chemical Technology
Ge Guan Road 625#, Nanjing, China
12764685@qq.com

Gang Tao
Lijing Zhang
College of Urban Construction and Safety Engineering, Nanjing University of Technology
Xin Mofan Road 5#, Nanjing, China
taogangs@163.com
zhanglj@njut.edu.cn

Prepared for Presentation at American Institute of Chemical Engineers
1st CCPS Asia-Pacific Conference on Process Safety
Qingdao, China
September 4 – 5, 2013

UNPUBLISHED

Thermal stress analysis of hydrogenation reactor using the finite element software

Xinmeng Sun
College of Urban Construction and Safety Engineering, Nanjing University of Technology
Xin Mofan Road 5#, Nanjing, China
sunxinmeng@126.com

Wan Xu
Nanjing College of Chemical Technology
Ge Guan Road 625#, Nanjing, China

Gang Tao
Lijing Zhang*
College of Urban Construction and Safety Engineering, Nanjing University of Technology
Xin Mofan Road 5#, Nanjing, China

Keywords: Thermal stress analysis, Hydrogenation reactor, PTA plant

Abstract

A hydrogenation reactor of PTA plant is selected in this paper as the object. By using finite element software, the steady-state temperature field, transient-state temperature field under shutdown and the corresponding thermal stress of the key section T-connector are analyzed. The analysis results show that the thermal stress can be ignored in the operation for the temperature field is steady. But large thermal stress is generated in the process of shutdown for the temperature is changing over time. Therefore during the period of reactor cooling the unevenness of temperature field lead the corrosion cracking easily.

1. Introduction

Hydrogenation reactor is the key equipment in modern oil refining device. The equipment is mainly used to convert the intermediate products such as crude terephthalic acid (TA) to the fiber grade purified terephthalic acid (PTA) under hydrogenation reduction. The operation condition is very critical because of high temperature, high pressure and hydrogen environment. Still, there will be serious corrosion which can reduce the service life of equipment such as crack, temper brittleness, hydrogen brittleness and hydrogen stripping of surfacing layer, while using high strength materials such as low carbon stainless steel in the reactor [1].

According to the previous inspected results of hydrogenation reactor of refined section in PTA plants, the cracks are mainly distributed in several places such as the head, the cylinder liner, the inner wall of T-connector and the weld between T-connector and head [1]. The weld in the hydrogenation inlet pipe is more severe and long transverse crack is generated in the weld

* Corresponding author. Tel.:+86 13851810118.

between pipe and cylinder liner. Therefore the key section T-connector of hydrogenation reactor is selected as the object of thermal stress analysis in this paper in order to find out the main cause of corrosion failure. Providing reliable basis in researching similar corrosion fracture mechanism and providing technical support in avoiding and reducing the corrosion are also expected [2-3].

2. Model of temperature field analysis

2.1 Finite element software ANSYS

The temperature field analysis on the T-connector of hydrogenation reactor using finite element software ANSYS is divided into three main processing stages [4]:

1. The preprocessor, in which the calculation model is established, the material parameters are set and the form of unit is chosen, etc.
2. The solution, in which the load is applied to the model
3. The postprocessor, in which the results is analyzed and outputted in several forms [2], [5].

2.2 The temperature field analysis model of steady-state heat conduction

Understanding the situation of temperature field is the premise of analyzing thermal stress. The structure which connects the half of T-connector with head was selected as the analysis model because of its symmetry. The structural parameters of hydrogenation reactor are shown in Figure 1, and characteristic parameters of materials at 281℃ are shown in Table 1.

Figure 1: Framework sketch of connector

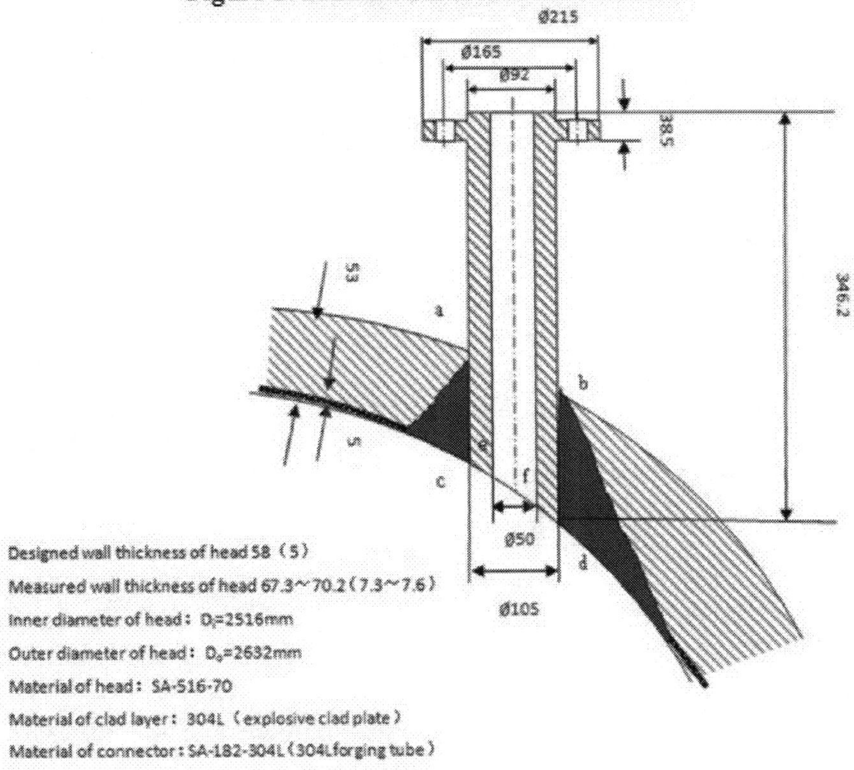

Table 1. The characteristic parameters of the materials

NO.	Material	Thermal conductivity W/(m·℃)	Specific heat J/(kg·℃)	Density Kg/m^3
# 1	Carbon steel WSTE36(281℃)	36.6	515	7.85×10^3
# 2	Stainless steel 304L(281℃)	18.4	500	1.79×10^4

Figure 2: Calculation model of thermal conduction

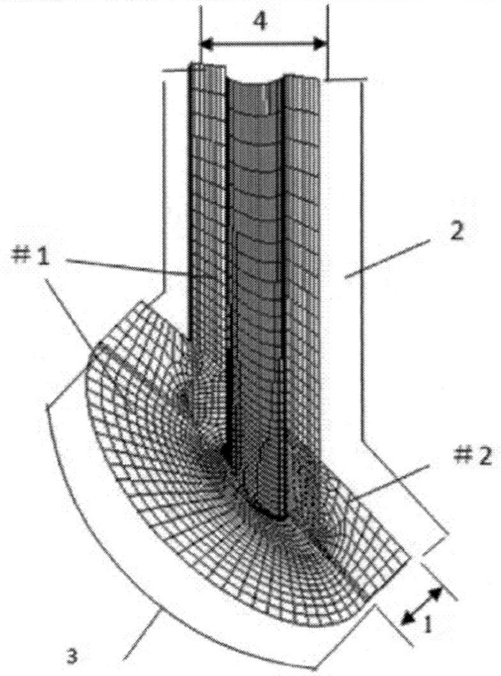

The specific steps of steady-state temperature field analysis adopted in this study are summarized below:

1. The finite element model of temperature field analysis are established based on above given structural parameters and characteristic parameters, as shown in Figure 2. In this figure, # 1, # 2 represent different materials, 1, 2, 3, 4 show that the boundary is divided into four zones in a steady-state temperature field analysis, and each zone has different boundary condition. Zone1and 2 are convective heat transfer boundary conditions which are not exactly same considering the internal and external temperature of actual cylinder walls are different. Zone3 is set as adiabatic boundary and zone4 is also regarded as adiabatic boundary considering the upper air layer.
2. The SOLID70 three-dimensional entity eight node hexahedral elements are chosen according to the characteristics of the ANSYS and the structure of analysis model. The model is divided into 11130 units and 13113 nodes.
3. Applied load is the internal and external temperature of connector. And the convective heat transfer coefficients between wall and medium or environment are also applied to the model.

4. Finally the results are analyzed by using the processing functions included in the ANSYS software. Figure 3 shows the distribution of steady-state temperature field of hydrogen connector.

Figure 3: Distribution of steady-state temperature field

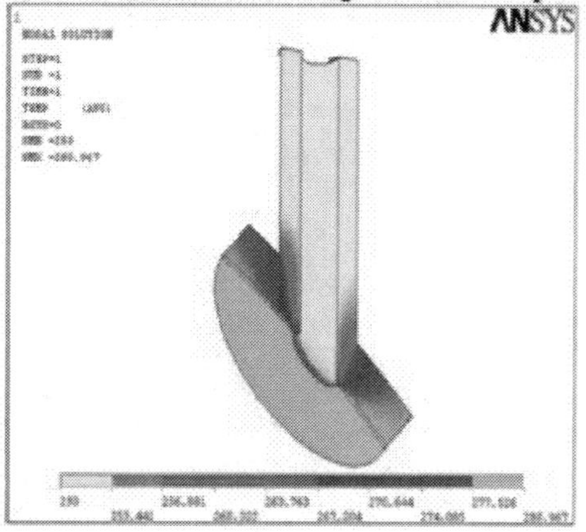

Figure 4: The curve of temperature from a to c on the connector

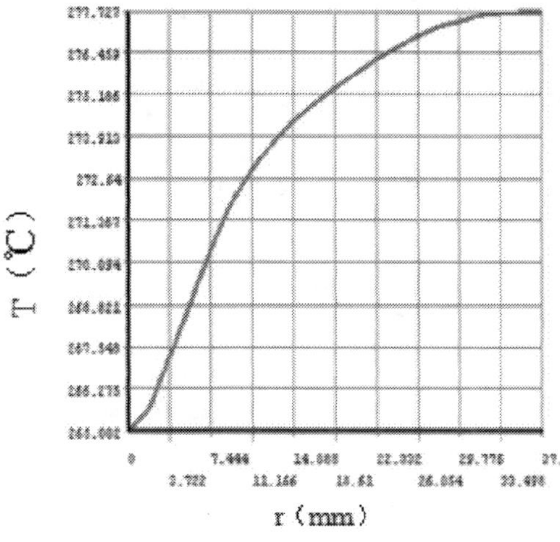

Figure 5: The curve of temperature from e to c on the connector

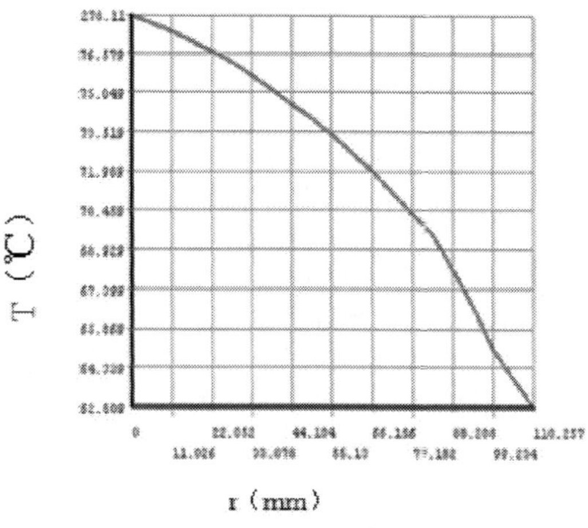

According to the above outputs, the distribution of steady-state temperature field of the hydrogen connector is basically radial distribution on the cylinder, which can be calculated using T = T (r). The temperature from the inside to outside of the connector is at constant because the wall is thin. But the thermal stress cannot be ignored because the temperature gradient is generated in the weld between connector and the head. Furthermore, the finite element software is used to get the thermal stress distribution, for the temperature of the position between connector and head has no unified form and cannot be calculated by formula [6].

2.3 The temperature field analysis model of transient-state heat conduction

Transient-state temperature field will be produced in some accidents of hydrogenation reactor such as a fault or the medium cooling too quickly [7]. The analysis of transient-state temperature field under shutdown is simulated in this paper. The transient-state boundary condition can be obtained assuming that the change rules of boundary condition is consistent with the change rules of temperature [8-9].

The analysis model of transient-state temperature field including geometry model and physical model, in which the geometric model is same with the steady-state temperature field but the physical model is different. The boundary condition and the temperature and the convective heat transfer coefficient at the border of transient-state temperature field are a function of time while the boundary condition of steady-state temperature field and load does not change over time [10]. Table 1 shows the characteristic parameters of the materials, and the Table 2 shows the heat capability parameters of the transient-state temperature field analysis.

Table 2 the heat capability parameters of the transient-state temperature field analysis

parameter	convective heat transfer coefficient W/(m²·℃)	T ℃
Boundary 1	25(t=0min),43(t=125min)	280（t=0min），30（t=125min）
Boundary 2	33	30
Boundary 3	adiabatic boundary	
Boundary 4	adiabatic boundary	

The specific steps of transient-state temperature field analysis adopted in this study are summarized below:

1. Appropriate PLANE55 thermal analysis unit is adopted and the model is divided into 11130 units and 13113 nodes according to the physical model defined by above conditions.
2. The type of analysis is defined as transient (ANTYPE, TRANSIENT) for the default type of finite element is steady-state.
3. The initial conditions of transient-state thermal analysis must be determined by steady-state thermal analysis because the initial temperature is unknown.
4. Then the loads need to be applied to the model. In the process of analysis, a load-step per minute is defined. There have 125 load-steps and each step is divided into four time-steps. In the finite element analysis, the load equal to the boundary conditions. Boundary1 is changing over time, its temperature and convection coefficient are a function of time. Boundary2 is the convective heat transfer boundary conditions. Boundary 3 and 4 are adiabatic boundary conditions.

Through the above analysis and calculation, the transient-state temperature field situation under shutdown of hydrogen connector in the hydrogenation reactor is obtained in this paper. And the temperature distribution at any moments is displayed using the postprocessor of ANSYS finite element software. The Figure 6 to 9 show the temperature distribution after shutdown.

Figure 6: The temperature distribution of connector after stopping for 30min

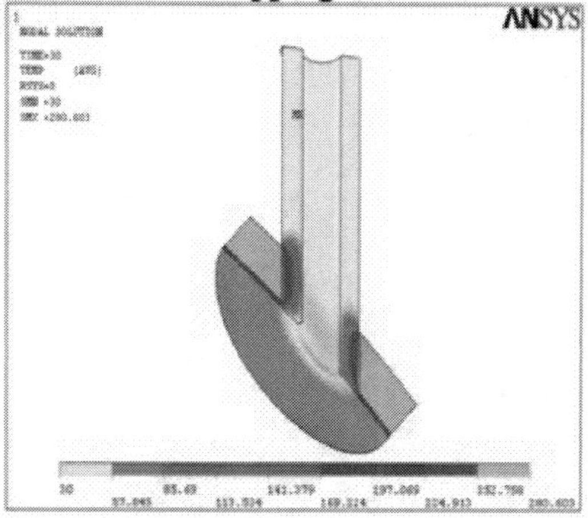

Figure 7: The temperature distribution of connector after stopping for 125min

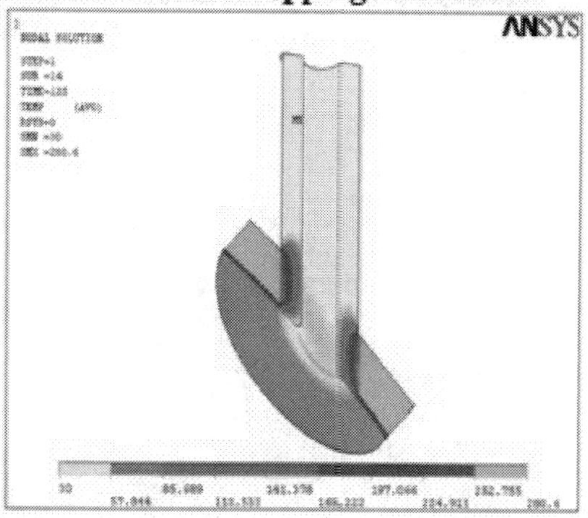

Figure 8: The curve of temperature from c to a on the connector

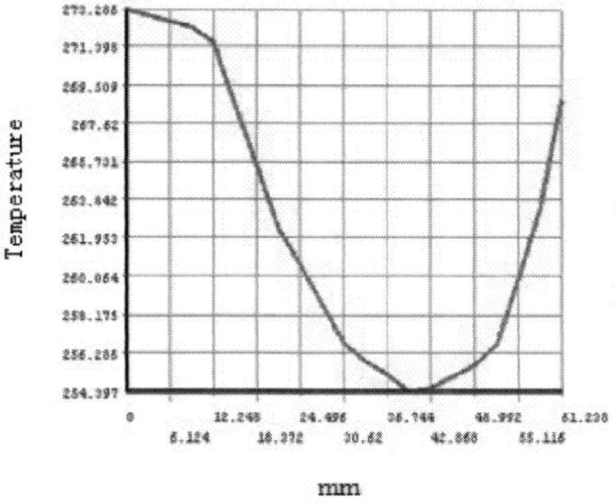

Figure 9: The curve of temperature from e to c on the connector

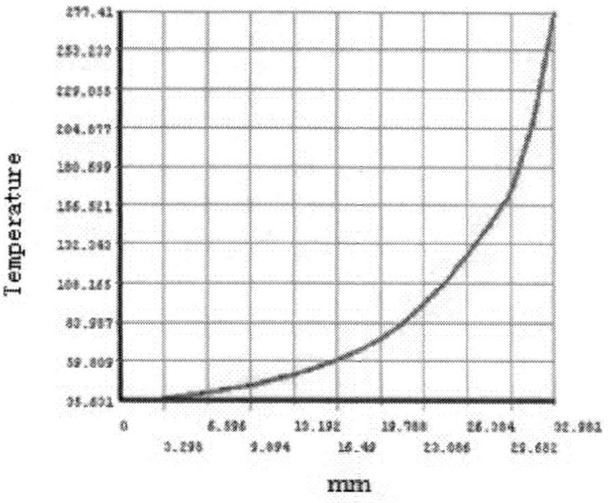

3. Thermal stress analysis using the results of temperature field analysis

Linear strain $\alpha_m(T - T_0)$ will be generated when the temperature of object changes, where α_m is the linear expansion coefficient of material, T is the current temperature, T_0 is the initial temperature. Furthermore, the thermal stress of the elastomer can be obtained by using the temperature field.

The above analysis indicates that the thermal stress cannot be obtained by analytic calculation due to the irregular distribution of temperature. Therefore the indirect method of ANSYS is adopted in this paper [11]. If the thermal analysis is transient-state, the temperature at the moment of the largest temperature gradient are set as apply load which is only applicable to simple structure. For the complex structure involved in this paper, the moment of the largest temperature gradient is different from the moment of the largest internal stress, so the thermal stress needs to be analyzed by using the temperature field results of transient-state [12-14]. The indirect method of ANSYS is adopted in this paper as follows:

1. The thermal analysis unit is converted into the corresponding structure unit.
2. The material properties and the temperature of thermal analysis are set.
3. Finally the post-processing is done.

The distribution of thermal stress of transient-state is obtained through analysis, as shown in the Figure10 to 12.

Figure 10: Thermal stress distribution of connector

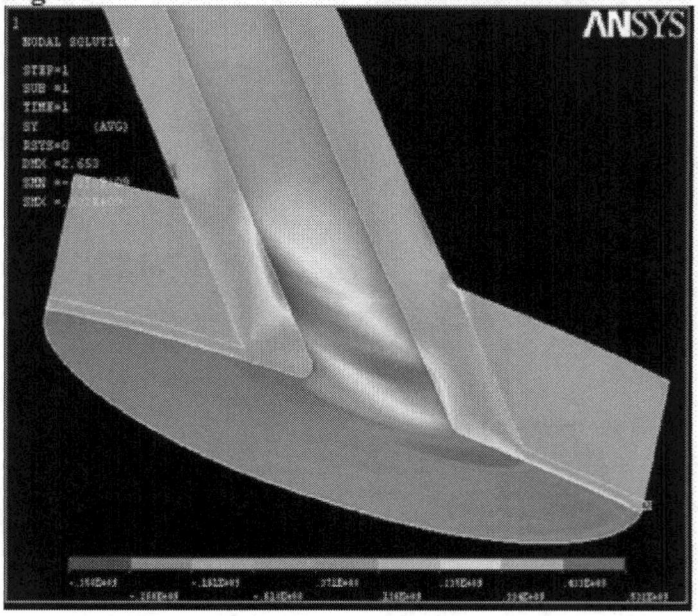

Figure 11: the curve of thermal stress from c to a on the connector

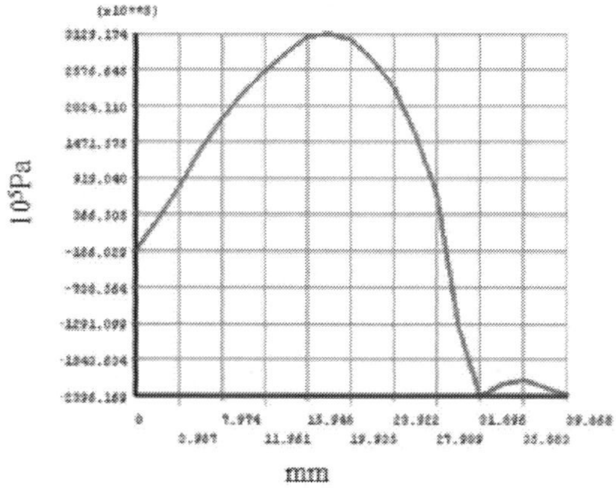

Figure 12: the curve of thermal stress from e to c on the connector

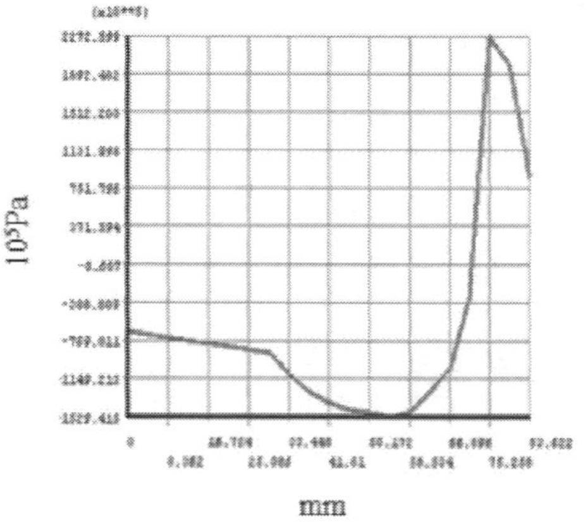

The outputs show the uneven distribution of thermal stress and local stress concentration at the T-connector. The thermal stress at the part of connector outside the head remains unchanged basically. But the part of connector inside the head is more complex which has tensile stress and compressive stress. The outermost stress is tensile stress at 239MPa. From the medial to lateral of the connector which connected to the weld, the tensile stress increase from 76.9 MPa to152.9 MPa and then convert to compressive stress. The results show that the thermal stress generated at the connector in the process of shutdown is near to the allowable stress of material (the yield strength of 304 L is 290 MPa). Therefore, the generation and level of thermal stress have a direct impact on the distribution and size of crack.

4. Conclusion

In this paper, the steady-state temperature field, transient-state temperature field under shutdown and the corresponding thermal stress of the key section T-connector are analyzed by using finite element software. The results show that the thermal stress can be ignored in the operation but the large thermal stress is generated in the process of shutdown. At the instant of the reactor cools, the uneven temperature field led the cracks easily. Therefore, in the process of shutdown, special attention should be paid to the cooling speed of reactors in order to avoid the thermal stress corrosion cracking.

5. References

[1] Chang Xiaojun, Cui Wenyong, "The fracture analysis and safety evaluation of wall cracks of hot wall hydrogenation reactor," Petro-Chemical Equipment Technology, Volume 20, NO. 4(1992): 5-12

[2] XU Wan, ZHANG Lijing, MEMG Yifei, ZHENG Weining, TIAN Jianpin, "Stress Corrosion Analysis of the Hydrogeneration Reactors of the PTA Plants in Alkali Environments and Their Prevention," Chemical Engineering & Machinery, Volume 33, NO. 4(2006): 248-250

[3] YANG Huosheng, DONG Shaoping, CAO Shuiquan, LIN Jianhong, CHEN Jin, "Safety analyses on operating condition for hydrogenation reactor" Petro-Chemical Equipment, Volume 29, NO. 2(2000): 10-12

[4] Wang Maocheng., Shao Min, Basic principles and numerical methods of the finite element method, Pergamon Press, Beijing, 1997

[5] Wang Guoqiang, Practical engineering simulation technology and its application on ANSYS, Northwest industrial university press, Xian, 1999

[6] PENG Cuiling, AI Huaning, LIU Qingshong, XIANG Wenyuan, "Reliable Analysis for Pressure Vessel Based on ANSYS," Nuclear Power Engineering, Volume 30, No. 1, 2009

[7] Wang Honghai, Gaobingjun, Lichunli, Yanshengzan, "Numeric simulation of transient temperature field on hydrogenation reactor," China Petroleum Machinery, Volume 35, No. 6(2007): 21-24

[8] LI Shiyun, LI Shiping, “Analysis of Transient Temperature Distribion and Thermal Stress Distribution in a Casting During Solidification,” Journal of Kunming University of Science and Technology（Science and Technology）, Volume 25, No. 1(2000): 47-50

[9] Fu Dongmei, Li Xiaogang, “A Software System to Estimate the Remaining Life for Hydrocracking Tower on Line with Infrared Image,” Laser & Infrared, Volume 31, No. 4(2001): 230-233

[10] Heming, C., Shuhong, Z., Honggang, W., & Jianyun, L., “Finite element analysis of temperature field with phase transformation and non-linear surface heat-transfer coefficient during quenching,” Applied Mathematics and Mechanics, Volume 19, No. 1(1998): 15-20

[11] Segall, A. E., “Thermoelastic analysis of thick-walled vessels subjected to transient thermal loading,” Journal of pressure vessel technology, Volume 123, No. 1(2001):146-149

[12] Liu Bin, Shen Shiming, “Finite element analysis of circumferential thermal stress on cylinder of the hot-wall hydrogenation reactor,” Petroleum Refinery Engineering, Volume 12, No. 5, 2007

[13] Cai Changquan, Guo Jing, Cheng Junming, Wei Yuting, “Finite Element Analysis of Thermal Stress on Cylinder of the Hydrogenation Reactor,” Chemical Equipment Technology, Volume 32, No. 3, 2011

[14] Transient and residual thermal stresses in quenched cylindrical bodies
Sen, S., Aksakal, B., Ozel, A., “Transient and residual thermal stresses in quenched cylindrical bodies”. International Journal of Mechanical Sciences, Volume 42, No. 10(2000):2013-2029

A correlation method of quality problems and process safety for improving security status of chemical production

Yaqi Gao
College of Urban Construction and Safety Engineering, Nanjing University of Technology
Xin Mofan road 5# Nanjing China
gaoyaqi90@126.com

Bin Sun
Tangshan Caofeidian Shihua Crude Oil & Terminal Co., Ltd
Tangshan, China
sinopecsunb@sina.com

Gang Tao
Lijing Zhang
College of Urban Construction and Safety Engineering, Nanjing University of Technology
Xin Mofan road 5# Nanjing China
taogangs@163.com
zhanglj@njut.edu.cn

Prepared for Presentation at
American Institute of Chemical Engineers
1st CCPS Asia-Pacific Conference on Process Safety
Qingdao, China
September 4 – 5, 2013

UNPUBLISHED

A correlation method of quality problems and process safety for improving security status of chemical production

Yaqi Gao
College of Urban Construction and Safety Engineering, Nanjing University of Technology
Xin Mofan road 5# Nanjing China
gaoyaqi90@126.com

Bin Sun
Tangshan Caofeidian Shihua Crude Oil & Terminal Co., Ltd
Tangshan, China
sinopecsunb@sina.com

Gang Tao*
Lijing Zhang
College of Urban Construction and Safety Engineering, Nanjing University of Technology
Xin Mofan road 5# Nanjing China
taogangs@163.com
zhanglj@njut.edu.cn

Keywords: quality problems; process safety; method; chemical production

Abstract

The chemical process is always accompanied by a series of quality problems which are easy to cause accidents. According to probabilistic principle and Swiss Cheese Model, the number of quality problems is consistent with the accident probability. A large number of quality problems are more likely to lead to an accident. An improved theory based on Process Safety Pyramid is presented in the paper. On the basis of the theory, quality problems and process safety correlation method is put forward in the paper. In this method a checklist is used to investigate quality problems and determine the risk level of units. The intent of the method is to improve the security status of chemical production and prevent occurrence of accidents. When the method is applied to the process of actual chemical production, some problems are discussed such as the division of risk level, table preparation, the determining of time interval and so on. The security status of chemical production can be improved effectively by using the method.

1. Introduction

The chemical process is always accompanied by a series of problems, such as physiological and psychological status of the person, the operation of workers, the condition of equipments and devices, surrounding environment and so on. If a single problem occurs in the chemical process, it rarely leads to an accident. However, near misses and process safety incidents may happen with the accumulation of problems in the future. These elementary process safety factors which

* Corresponding author. Tel.: +86 13451919654.

are presented at the bottom of the process safety pyramid are defined as quality problems in this paper. Quality problems also mean man, machine, environment and management beyond the security status or the hidden troubles of safety in the process of chemical production.

Compared with process safety incidents, quality problems did not receive strenuous investigation in some organization. Only casualties and property losses occur, the investigation may be executed. At most time, they may not apply sufficient rigor to find the root causes. Corrective actions may not solve the real problems and could create the probability of a more serious incident in the future [1]. Most of process safety incidents are caused by a series of quality problems. If we take some measures and pay more attention to quality problems in the daily process of chemical production, many process safety incidents may be avoided. Quality problems play a very important role in the process of chemical production. By analysis of quality problems, a method is put forward to improve process safety in the latter part of the paper.

The intent of this paper is to explore how quality problems can be an accident and what should be done to improve the security status of chemical process by quality problems. Accidents can be controlled effectively by the identification and correction of quality problems.

2. Theoretical Background

2.1. Swiss Cheese Model

Process safety incidents are rarely caused by a single quality problem, but rather by multiple problems or failures that coincide and collectively result in an accident. This relationship between simultaneous and sequential failures of multiple systems is illustrated by the "Swiss Cheese Model," as shown in Figure 1 [2].

Figure 1: Swiss Cheese Model

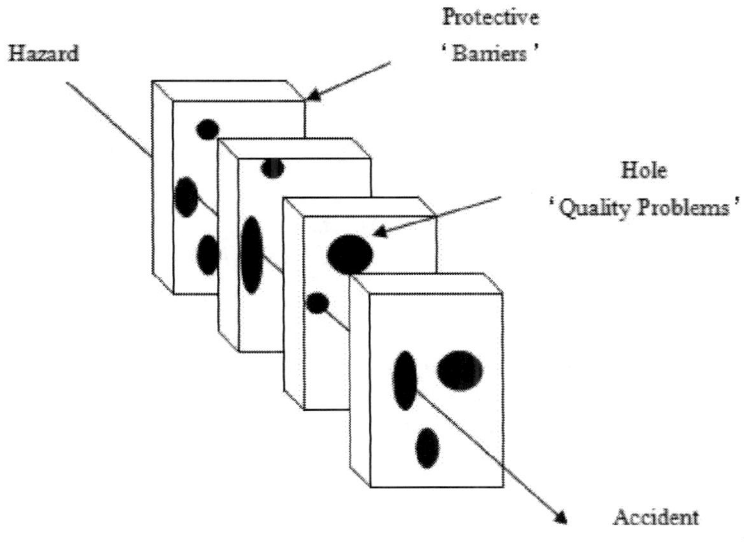

The barriers are elements of a process safety management system or some other protective layers designed to prevent an accident from occurring. The holes in the slice of cheese represent quality problems and they may allow an event to escalate into an accident. When holes align, an accident may occur [3]. Events that occur by passing through holes in a process safety management system or protective system can be described as process safety incidents.

There are many layers that exist in the process of chemical production. The point, however, is that nothing is foolproof and each piece of cheese has holes. When these holes line up, an error occurs. If there are just a few holes, they are difficult to line up. But if the number of holes increases, the accident probability will increase too. That means if the number of quality problems can be controlled, the accident probability may be controlled effectively. A method which how to control the number of quality problems would be presented in the latter part of this paper.

At the same time, holes are different in each piece of cheese and the potential impact of them is also discrepant. That means different quality problems play different roles in the chemical process. The influence of every quality problems is different, some of them may be easier to result in an accident and play a more important role than other quality problems. However, each of them should not be neglected. If we pay no attention to them, they may also lead to serious events in the future. The diversity of quality problems is not the emphasis of this paper. We would not discuss it specifically.

2.2. Process Safety Pyramid

Serious process safety incidents may be engendered by a number of less-severe related incidents. Such related events may be low-consequence incidents such as near misses, insufficient operating discipline or unsafe behaviors. This relationship between no-impact or low-impact events and actual process safety incidents is described in the Process Safety Pyramid [3].

Figure 2: Process Safety Pyramid

- 1. Incidents which meet the threshold of severity which should be reported as part of the process safety metric.
- 2. Incidents which didn't meet the definition of PS incident metric (e.g., all other Loss of Primary Containment or fire).

- 3. System failures which could have led to an incident.
- 4. Other Process Safety Factors, e.g., Equipment Selection, Engineering Design, Specification of Inspection Frequency and Technique, Unsafe Behaviors and Insufficient Operating Discipline.

2.3. An Improved theory on the relationship between accident probability and quality problems

As Swiss Cheese Model and Process Safety Pyramid said, process safety incidents are caused by a series of quality problems like unsafe behaviors, insufficient operating discipline or equipment failure and so on [4]. According to probabilistic principle and Swiss Cheese Model, the number of quality problems is consistent with the accident probability. That means if just a few quality problems occur, the probability of accident is low. But if many quality problems occur, the probability of accident may be very high. In order to describe the relationship between quality problems and the risk level of no-impact or low-impact events and actual process safety incidents, we define a process safety quadrangle which is presented in Figure 3.

Figure 3: Process Safety Quadrangle

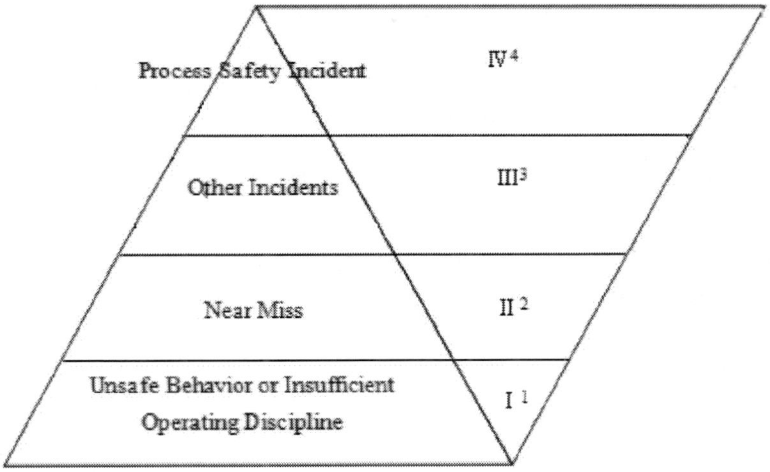

- I means the risk of the chemical production is low. In this part of the quadrangle, only a few quality problems happen. Accidents may not occur easily.
- II means the risk of the chemical production is general. In this part of the quadrangle, several quality problems happen. Near miss may be happen.
- III means the risk of the chemical production is moderate. In this part of the quadrangle, some quality problems happen. Some incidents may arise.
- IV means the risk of the chemical production is high. In this part of the quadrangle, many quality problems happen. Accidents may occur easily.

Based on probabilistic principle, the number of quality problems in the process of chemical production has an obvious impact on the security status of chemical process, a large number of quality problems are more likely to lead to an incident. Therefore, four risk levels which have different degree of risk are defined by the number of quality problems. Security status of chemical process can be obtained by the different risk levels. The division of grade is presented in Figure 3.

However, the boundary between adjacent levels is vague. A clear figure can't be determined to divide different levels because the definition of risk levels is based on the probabilistic principle. The boundary is an indeterminate figure. In the process of actual production, the boundary has to be determined according to work experience and a lot of accident investigations. After investigation and research, an approximate figure can be determined. Then risk levels can be divided according to the finger. The security status of units is also determined by the risk level.

After the definition of security status of units by quality problems, timely and proper corrective measures are also very important. If quality problems can not be solved properly, an accident may arise in the future. However, if all problems can be completed properly, the security status of chemical process can be improved effectively.

In the process of actual production, what we need to do is not only to solve quality problems but also to reflect on why they happen. Everything happens for a reason. If the level of risk is just Ⅰ or Ⅱ, it may be some normal reasons like the aging of equipment or the normal reaction phenomenon and so on. It just needs to fix them and pay an attention to them. However, if the level of risk reaches Ⅲ or Ⅳ, we have to reflect on it. It represents not only quality problems on the surface but also deeper problems such as loopholes in management or manufacturing defect and so on[5,6]. Root causes have to be found out, otherwise the same quality problems may occur frequently and it may lead to a serious accident at last.

3. The Application of Quality Problems and Process Safety Correlation Method to Natural Gas Storage Area

The method is used to the process of production. The whole chemical production process is very complicated. If the method is applied to the whole chemical production process, the pertinence is not clear and the security status of various regions can not be determined. The entire chemical area may be divided into cells so that the risk level of different cells can be analyzed and determined. We can acquaint the risk levels of different cells better by using this method. Then we can solve quality problems duly and improve the security status effectively. In this paper, the method will be applied to natural gas storage area.

The process of chemical production is a very complicated system. Specific details may not be illustrated in the paper. We just presented approximate procedure to show how the method is carried out.

In order to improve the security status effectively by the method, several steps are included:

- Firstly, specialized staffs are needed to survey and analyze the unit so that all quality problems can be found out. All quality problems have to be summarized into a table at last. With the change of production, some devices may be added or surrounding environment may be changed and so on. At the same time, some new quality problems will arise. We need to add them to the table and update it timely.

- In order to simplify the analysis of the natural gas storage area in the paper, management factors are assumed to meet the requirements. Additionally, some steady factors are also not considered like the location of unit in the chemical plant or the space between equipments and so on. All quality problems should be summarized into a table. The style of table is impressed in Table 1.

Table 1. Quality Problems Checklist

	Potential Quality Problems[1]	True（T）[2]	False（F）[3]	Description of Quality Problems[4]
1				
2				
…				

- 1. Potential Quality Problems, e.g., worker training, operating technique of worker, operating according to rules and regulations, safety protection articles (e.g., protective clothes, safety helmet, safety boots and so on), lightning protection system, open flame(e.g., cigarette butts, high temperature surface and so on), electric spark(e.g., power up/down the electrical equipment, fuse burn-out, short circuit and so on), safe sign, stacking of goods, the aging and damage of pipeline, tightness of valves, tightness of natural gas storage tanks, gas leakage alarm device, the ventilation system and so on [4,5].
- 2. If the potential quality problem does not occur, T will be filled in the table.
- 3. If the potential quality problem occurs, F will be filled in the table.
- 4. If the potential quality problem occurs, Specialized staffs must descript it in detail in the table so that workers can understand the problem and solve it easily.

- A part of quality problems are listed. There are many other problems in the process of actual production. The paper just submits the method and does not introduce quality problems one by one. The process of chemical production is variable and some new quality problems may appear. Specialized staffs have to add these problems into the table and update it timely.

- Secondly, specialized staffs need to be arranged to inspect quality problems according to the table regularly. After all quality problems are checked out, the level of risk can be determined according to the number of quality problems. It represents the security status of the unit and the urgency of rectification. The division of the level is impressed in Figure 3. The level of risk table is impressed in Table 2.

- The determination of time interval is also a problem. With the accumulating of quality problems, accident will happen if time interval is too long. However, if time interval is too short, the workload and cost will be improved. Different units also should have different time interval because different units have different devices, equipments and environment, which have different time of the accumulating of quality problems. In the process of actual production, time interval should be determined on the basis of actual situation [6].

Table 2. Level of risk

Level of risk	I	II	III	IV
Number of quality problems				

- Thirdly, timely and proper correction should be implemented. If the level of risk is just I or II, workers just need to solve the problems. If the level of risk reaches III or IV, it represents not only quality problems on the surface but also deeper problems. Workers must find root causes and solve problems properly.

- At last, when the entire process is completed, it must be documented to provide a tangible record of the quality problems, its causes, and corrective actions. A date base should be set up to store the entire process. It will have a great impact on the chemical process safety with the accumulation of information in the future.

If all steps are completed greatly, the security status of unit can be improved effective and most of accidents can be prevented.

4. Difficulties in application of method to the process of actual production

The security status of units can be improved effectively in the process of actual production by using the method. However, several difficulties are existed:

- In the process of actual production, the table will be very complicated. Some quality problems are difficult to be found. Experienced staffs and experts are needed to complete the table and update it timely, the application of the method will cost a lot of time and money. It may be a problem to some factories but it is worthy.

- The division of the level of risk is difficult. It is a probability problem and hard to get a deterministic finger. The definition of the finger has to be based on experience and a large number of accident investigations. It is a difficult point in the method.

- Staffs can check the unit according to the table. However, the problem is how to determine a appropriate time interval. Time interval can not be too long or too short, which have to be based on the actual situation of the unit.

5. References

[1] John W. Herber. Quality Problems as Process Safety Warning Signs. Process Safety Progress, 2013, 32(2): 175-178.

[2] Center for Chemical Process Safety (CCPS), “Process Safety Leading and Lagging Metrics– You Don’t Improve What You Don’t Measure”, January 2011.

[3] Center for Chemical Process Safety (CCPS), “Guidelines for Process Safety Metrics”, May 2010.

[4] Crowl D A, Louvar J F. Chemical process safety: fundamentals with applications. 3nd ed. USA: Prentice Hall PTR, 2011.

[5] Keqing Cui, Lijing Zhang, Gang Tao. Introduction to Safety Engineering and Science. Beijing: Chemical Industry Press. 2004.

[6] Keqing Cui. Chemical Process Safety Engineering. Beijing: Chemical Industry Press. 2002.

A new collection system of reliability data for equipment applied to Mechanical Integrity management

(Paper number: 305999)

Hang Wang
Hai-qing Wang
Yu-tao Zhang

China University of Petroleum (Huadong)
Department of Safety Science and Engineering, Yangtze River road 66, economic and technological development zone, Qingdao, Shandong province, China
wanghaiqing@upc.edu.cn

Prepared for Presentation at
American Institute of Chemical Engineers
1st CCPS Asia-Pacific Conference on Process Safety
Qingdao, China
September 4 – 5, 2013

UNPUBLISHED

A new collection system of reliability data for equipment applied to Mechanical Integrity management

(Paper number: 305999)

Hang Wang
Hai-qing Wang*
Yu-tao Zhang

China University of Petroleum (Huadong)
Department of Safety Science and Engineering, Yangtze River road 66, economic and technological development zone, Qingdao, Shandong province, China
wanghaiqing@upc.edu.cn

Keywords: Reliability; Data collection; Failure; Maintenance; Mechanical Integrity (MI)

Abstract

A new collection and management system of reliability data was proposed for petrochemical industry users. Based on Android or IOS platform, the software of data collection was developed with Smartphone and table PC as its terminals, with the goals of equipment reliability prediction, equipment maintenance/change, spare parts and suppliers selection, workload and task execution efficiency optimization. Then, by building enterprise internal Local Area Network (LAN), the real-time information of reliability such as equipment failure, repair, and replacement could be shared.

1. Introduction

Petrochemical craft process is of high risk and often causes severe accidents. According to a global chemical accidents investigation report, in August 2011, only between July 30 and August 5, forty-five chemical accidents took place, among them, 17 were oil/gas/petrochemical/chemical industry accidents, and killed seven people [1]. On the basis of a statistical report on 100 largest chemicals, oil and gas accident losses of the United States, which was released by An American enterprise called Marsh, the damage caused by MI system failure accounted for 46%. Through the analysis of the past petrochemical production accidents, a conclusion was drew that most of the petrochemical enterprises in the field of MI were lacking of systematic management. However, MI management must be supported by enough historical data and test data. But the present situation is that reliability data have not been systematically collected and sorted, also, the existing reliability data is relatively fragmented, which makes it sluggish to process service requests, major failures and maintenance work. Meanwhile, Management of Change (MOC) is not effectively implemented, and the hidden dangers of equipment are not found in time, which can result in a lot of unforeseen losses.

---

* Corresponding author. +86-0532-86983500 Ext. 8516; E-mail address: wanghaiqing@upc.edu.cn; this research is supported partially by the Excellent Scholar Project of China university of petroleum (Huadong) (Grant No. SD2013211153) and the Fundamental Research Funds for the Central Universities

In order to solve the above mentioned situations, the equipment reliability data acquisition and the closed-loop management system of MI were integrated. With the objectives of equipment reliability prediction, equipment maintenance/change, spare parts and suppliers selection, workload and task execution efficiency optimization, an idea of petrochemical equipment reliability data collection procedures was proposed, by making the reliability data collection forms electronization which was treated as a part of MI management system.

2. New collection system of reliability data for equipment

MI management requires plenty of reliability data, which can be obtained mainly from two aspects—equipment tested in laboratory and devices used on spot [2]. The data acquisition mentioned in this article is the latter one.

The information obtained by traditional reliability data collection method was interpreted manually, and it was typed into the target database which needed a certain type and format, and the data entry was also by people. Therefore, it was a waste of lots of manpower and material resources procedure, the accuracy of data entry could not be guaranteed yet, and data analysis could only be offline.

On the basis of Guidelines for process equipment reliability data [3], Petroleum and natural gas industries — Collection and exchange of reliability and maintenance data for equipment (ISO 14224-2006) [4] and Offshore Reliability Data [5], the new reliability data collection process for equipment was established. By developing suitable data acquisition software for Smartphone and table PC, reliability data could be easily collected and summarized when using mobile terminals. Through online data analysis system, real-time change information of devices could be sent to computing unit for decision-making, finally, petrochemical equipment/devices were adjusted.

2.1 list of key equipment and subdivision level

First, the research object must be confirmed. Taking oil and gas pipelines transport stations as an example, their main work area should be included, namely as oil pumping room, heating system, the main valve chamber, pigging distribution room, measuring room, oil tank area, the station control room, oil pretreatment facilities, power supply systems, heating supply systems, water supply and drainage systems, communication systems, air supply systems, cathodic protection facilities and fire-fighting equipment. All working area could be divided into two main categories: use/location, equipment/subdivision, as shown in table 1

Table 1 Equipment list of transport station of oil and gas pipelines

Main categories	Taxonomy level	Taxonomy hierarchy	Examples of use or location
Use/location	1	Transport medium	Crude oil, refined oil, natural gas
	2	Types of the station	Compression/decompression station, injection station
	3	Process area	Pigging, metering, receiver &launcher
Equipment/subdivision	4	Equipment	Heat exchanger, compressor,

			pipeline, pump
	5	Subunit	The lubrication unit, cooling unit, control monitoring unit
	6	Parts/repair parts	Cooler, coupler, the transmission, lubrication pump
	7	components	Sealed, catheter, casing, impeller

2.2 boundary definition of devices

The purpose of boundary definition is to ensure a common understanding of the "subunit/component" and "maintainable item/part" included within the boundary of a particular equipment unit and, hence, which failure and maintenance events to record. For definition of the boundaries, the following rules are recommended:

- Do not include items of unique design or configuration-dependant items. Include only those items that are considered to be generic for the equipment class being considered in order to compare "like with like."
- Exclude connected items from the equipment-class boundary, unless specifically included by the boundary specification. Failures that occur in a connection (e.g. leak), and that cannot be solely related to the connected item should be included within the boundary definition.
- If a driver and the driven unit use a common subunit (e.g. lubrication system, relate failure and maintenance events on this subunit, as a general rule, to the driven unit.
- Include instrumentation only where it has a specific control and/or monitoring function for the equipment unit in question and/or is locally mounted on the equipment unit. Control and supervisory instrumentation of more general use should not, as a rule, be included.

2.3 Equipment code and failure mode

Coding equipment and failure mode is of great significance and very helpful to the later stage of software design. Code is not unique, but must be consistent when used.

Still taking oil and gas pipelines transport stations as an example, according to the hierarchical coding method, rules of code design could be determined as follows: location area/name of the station/process/equipment/child element/component code. As codes NW/XY/PIG/PIP/ LUB/VA-GA represents northwest /Xian Yang/pigging/pipe/lubrication unit/gate valves.

Meanwhile, the rules of designing failure mode code could be determined as follows: name of a device/code of failure mode/definition of failure mode/description of failure mode. Code of failure mode and corresponding description of valves were listed in table 2.

Table 2 Code of failure mode and corresponding description of valves

Equipment	Code	Failure mode	Description of failure mode
Valves	FTC	Fail to close on demand	Normally open or not fully close
	FTO	Fail to open on demand	Often close or not fully open
	FTR	Fail to regulate	cannot adjust, only for control valve
	DOP	Delayed operation	Switching time is different from specification
	HIO	High output	Error adjust, only for control valve
	LOO	Low output	Error adjust, only for control valve
	ELP	External leakage-Process medium	Process medium fled to environment
	ELU	External leakage-Utility medium	Incentive fluids, lubricants, etc
	INL	Internal leakage	Incentive of fluid leakage or valve actuator is connected
	LCP	Valve leakage in closed position	Leakage through the closed position
	PLU	Plugged/Choked	Some or all of the liquid are limited
	STD	Structural deficiency	Integrity decrease due to impact, cracks
	AIR	Abnormal instrument reading	Wrong position indicator, etc
	SER	Minor in-service problems	Loose parts, discoloration, dirt, etc
	OTH	Other	Stipulated in the annotations
	UNK	Unknown	Not enough information or omission

2.4 The data collection forms and electronization

Data collection forms design was a major task of data acquisition system. The forms should have unified contents and standardized formats according to the requirements.

With Smartphone and tablet PC as mobile terminals, the basic data collection table, failure data collection table (as shown in Figure 1), maintenance data collection table (as shown in Figure 2) and special data collection table were all created as software which was based on Android or IOS platform. By using mobile terminals, on which the software had a function of data entry, query, classification, summary, statistics, etc, reliability data could be processed. Then through enterprise internal LAN, the real-time information of reliability such as equipment failure, repair, and replacement could be easily shared.

Category	Data to be recorded	Description	Data (examples)
Identification	Failure record	Unique maintenance identification	FGA12345x1001
	Equipment identification/location	Tag number	PT0101
Maintenance data	Failure date	Date of failure detection (year/month/day)	2011/12/18
	Failure mode	What reason caused a failure(e.g. valve)	LOO
	intendance impact on plant	Usually zero, partial or total	
	Failure impact an equipment	Usually zero, partial or total	
	Failure mechanism	The physical, chemical or other processes which have led to a failure	physical

DOP
HIO
LOO
ELP
ELU
INL
LCP
PLU

Figure 1 Failure data collection table (partial)

Category	Data to be recorded	Description	Data (examples)
Identification	Failure record	Unique maintenance identification	FGA12345x1001
	Equipment identification/location	Tag number	
	Failure record	Corresponding failure identification record	
Maintenance data	Date of maintenance	Date when maintenance action was undertaken	2011/12/18
	Maintenance category	Corrective, preventive,predictive	Preventive
	Maintenance priority	High, medium or low priority	Medium
	Intendance impact on plant operations	Zero partial or total	Partial
		Name of subunit	Control of

FGA12345x1001
FGA12345x1002
FGA12345x1003
FGA12345x1004
FGA12345x1005

Figure 2 Maintenance data collection table (partial)

3. FRACAS integrated with the collection system of reliability data

3.1 The ideology of closed-loop in FRACAS

FRACAS's full name is "Failure Report Analysis and Corrective Action System". Its main goal is to achieve a closed-loop management system. Through FRACAS software, all kinds of information and data flow in each process are available, and the fault information database of enterprise will be established gradually. Therefore, the formulation of maintenance strategy, spare parts strategy and reliability design and analysis are supported by the reliability database.

There are fourteen steps for a standard FRACAS procedure. But, with the target of MI, a simplified FRACAS closed-loop system for petrochemical industry users was put forward as follows, as shown in Figure 3.

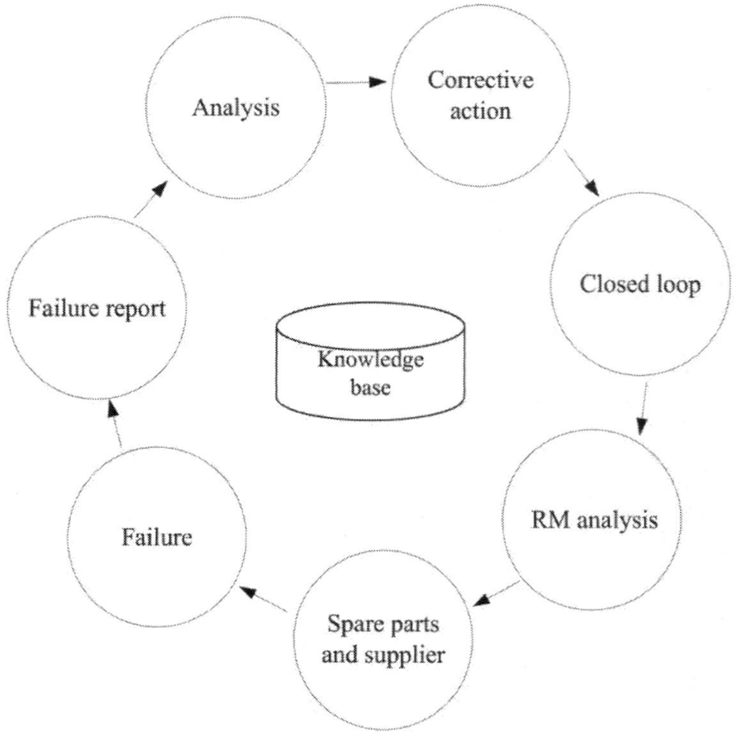

Figure 3 Simplified closed-loop systems for petrochemical industry users

3.2 Mobile terminals applied in FRACAS closed-loop system

With the reliability data increasing, equipment integrity management could be made from four aspects, namely as equipment reliability prediction, equipment maintenance/change, spare parts and suppliers selection, workload and task execution efficiency optimization. For example, when equipment reliability prediction was taken into consideration, the mean time between failure(MTBF) of a certain key equipment was obtained immediately by the classification and statistics function of the software, so as to the life change curve and estimated timing of a repair or replacement, etc. Equipment maintenance or replacement/change real-time information could also be shared through the closed-loop system, which could avoid the information asymmetry and disoperation accidents. Another example, only a comprehensive assessment of reliability and maintenance cost of the same kind equipment was finished, could a better equipment supplier be selected. The framework of data collection system for petrochemical equipment used in MI closed-loop system is shown in Figure 4.

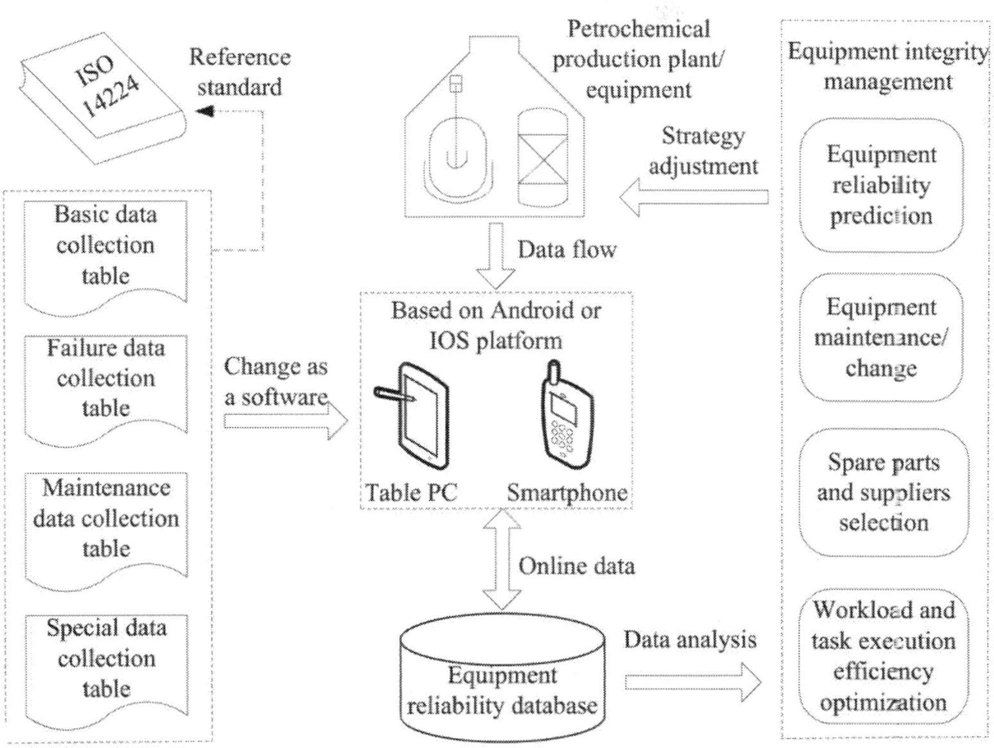

Figure .4 Data collection system for petrochemical equipment used in MI closed-loop system

4. Conclusion

The new equipment reliability data collection and management system seems more convenient and efficiency prior to the tedious work that should be done manually to type data into the database. Such a system may enables the equipment to form a closed loop from malfunction, analysis, maintenance and spare parts management, and finally forms the equipment reliability database which has more practical value than general database (like OREDA). What's more, this system also has a multi-level and broad application value in risk assessment, safety critical systems integrity analysis, designing maintenance and spare parts strategy and other fields.

Nevertheless, the Petrochemical system is so complex that the idea data acquisition system integrated with MI still requires further practice and validation, and a lot of work also has to be done, and finally a perfect system will be turned out.

5. References

[1] “Global chemical accident reporting and investigation trends in August” From NRCC, August 2011, http://info.nrcc.com.cn/document(accessed June. 24, 2013) (In Chinese)

[2] Zhao Yu. “Data analysis of reliability”, National Defense Industry Press, 2011(In Chinese)

[3] Center for Chemical Process Safety (CCPS). “Guidelines for Process Equipment Reliability Data with Data Tables”, 1989

[4] International Organization for Standardization (ISO 14224), “Petroleum and natural gas industries-Collection and exchange of reliability and maintenance data for equipment”, 2006

[5] Det Norske Veritas (DNV), “Offshore Reliability Data Handbook”, 4th Edition, 2002

Dynamic risk analysis to improve process safety using Bayesian network

Jianfeng Zhu
State Key Laboratory of Chemical Engineering, Department of Chemical Engineering, Tsinghua University, Beijing 100084, China
zhujf07@mails.tsinghua.edu.cn

Jinsong Zhao
State Key Laboratory of Chemical Engineering, Department of Chemical Engineering, Tsinghua University, Beijing 100084, China
jinsongzhao@tsinghua.edu.cn

Fan Yang
Tsinghua National Laboratory for Information Science and Technology (TNList) Department of Automation, Tsinghua University, Beijing 100084, China
yangfan@tsinghua.edu.cn

Prepared for Presentation at
American Institute of Chemical Engineers
2013
The 1st CCPS Asia-Pacific Conference on Process Safety
Qingdao, China
Sep 4 – Sep 5, 2013

UNPUBLISHED

Dynamic risk analysis to improve process safety using Bayesian network

Jianfeng Zhu
State Key Laboratory of Chemical Engineering, Department of Chemical Engineering, Tsinghua University, Beijing 100084, China
zhujf07@mails.tsinghua.edu.cn

Jinsong Zhao
State Key Laboratory of Chemical Engineering, Department of Chemical Engineering, Tsinghua University, Beijing 100084, China
jinsongzhao@tsinghua.edu.cn

Fan Yang
Tsinghua National Laboratory for Information Science and Technology (TNList) Department of Automation, Tsinghua University, Beijing 100084, China
yangfan@tsinghua.edu.cn

Keywords: dynamic risk analysis; Bayesian network; event tree; dynamic Bayesian network

Abstract

Chemical process industries often suffer from abnormal events of varying magnitudes, which may lead to different consequences, including incipient faults, near-miss, incidents, and accidents. Typically, when an abnormal event occurs, various safety systems, such as alarm systems and safety instrumented systems, come into play to prevent the event from propagating. To improve process safety, estimation of the failure probabilities of safety systems and probabilities of corresponding consequences is extremely important and should receive more attention. Risk analysis techniques, such as event tree analysis, have been used popularly for this purpose. However they suffer limitations of static structures and uncertainty handling, which are of great significance in safety system analysis. In this paper, Bayesian network (BN) for dynamic risk analysis is proposed. Bayesian networks are a probabilistic approach to modeling and representing influences between safety systems. First, a mapping algorithm which translates an event tree to a Bayesian network is introduced. In a BN model, causal relationships reflect the dependencies among safety systems which are represented as nodes. After building the BN causal structure, we estimate the parameters of the network, namely, the failure probabilities of safety systems and probabilities of corresponding consequences, using accident precursor data. Then reasoning under uncertainty in BN is performed, including predictive and updating analysis. In addition, other advantages of BN are discussed. First, it is easy to extend the model when the structure of BN is modified because of functions required. Second, dynamic Bayesian network is introduced. Finally dynamic risk analysis using Bayesian networks is applied to a case study to show its performance.

1. Introduction

In chemical process industries, process safety is monitored and ensured by Distributed Control Systems (DCSs) and Emergency Shutdown (ESD) Systems, which interact to take actions to nullify the influence of abnormal events. Typically, involving human behavior, the DCS and ESD system can be decomposed into different components, i.e. safety systems or SQOSs (safety, quality, and operability systems), which respond to every abnormal events and prevent them from propagating to upset consequences [1].

In most situations, event tree analysis is used to perform risk analysis [2, 3, 12]. However, how to model/reflect the dependencies among safe systems is an important issue to be considered. In the probabilistic risk analysis method (PRA), the accident frequency is estimated by the failure probability of individual components. Typically, independence relationships between components are assumed. Some types of dependency could be modeled, but depend on the analyst's ability to identify and what's more, some are difficult to model.

Dependencies among these safety systems are significant for dynamic risk analysis. The reasons why dependencies exist are as follows [4, 5, 6]. First, independence relationship between variables in the plant leads to the interactions among those safety systems. For example, with the variable of root cause returning to normal operating range, the impact will get channeled to its causal variables, which return to normal situations gradually. Second, if we take operator factors into consideration, failure of human-operated safety systems will be influenced by other safety systems or stressor performance-shaping factors. For example, the increases in the failure rates of basic control system are likely to increase the failure probability of human related safety systems because of operator's nervous mind or under stress. Thus, to obtain accurate and reliable dynamic risk results, it is important to account for dependencies among those safety systems. Method of copulas, which computes the joint distribution with each marginal distribution, is used to represent dependencies among safety systems in [4]. In [6], causal relationships between safety systems are expressed with a dependence matrix, which consists of pairwise interaction parameters and then several copulas types are used. However, before the copulas method is used, some parameters in copulas functions need to be described. For example, measures of dependences are typically represented by linear correlations or rank-correlations and experts will determine these measures.

Generally, we just know there is causal relationship between two variables in qualitative analysis but have no idea of quantitative measure. Therefore Bayesian network is introduced to represent dependencies among safety systems instead of method of copulas. BN has the advantage of modeling and analysis aspects. Expert knowledge can construct the qualitative representation of causal relationships between variables through local conditional dependences. Then learning parameters of BN structure with data, quantitative representation of causal relationships is obtained. Meanwhile, because of the feature of flexible modeling in BN, we can modify the graph structure easily, for example, adjusting, adding or removing causal related nodes. In the structure learning stage, Bayesian estimation is used and expert quantitative knowledge will be prior information of Bayesian method, which could deal with rare data information of one node.

The rest of the paper is organized as follows. In sections 2 and 3, event tree analysis and Bayesian networks are introduced, respectively. Then a mapping algorithm from an event tree to a Bayesian network is described briefly in section 4. In section 5, such dynamic risk analysis using Bayesian networks is applied to a case study compared with event tree analysis, followed by concluding remarks in Section 6.

2. Event Tree Analysis (ETA)

Event Tree Analysis (ETA) [2, 3] is one of quantified risk assessment techniques and is used for identifying and evaluating the sequence of events in a potential accident scenario following the occurrence of an initialing event, for example, a component failure, increase in temperature or a release of a hazardous substance. The initialing event will be prevented from resulting in accidents by sequential safety systems. ETA is based on binary logic, in which a safety system succeeds or fails. Therefore the consequences of the initialing event are followed through a series of possible paths and multiple outcomes and outcome probabilities are yielded. Event trees are popular for modeling accident scenarios in risk analysis. The consequence probabilities of each scenario can be estimated given estimates of failure rate and probabilities of the sequence safety systems, which can be obtained from a FTA of the event typically. A simple event tree of modeling safety systems is illustrated in Figure 1 [12]. Suppose that there are two safety systems, the first and second safety system, responding to the initial abnormal event. Considering the performance of the binary state, success or failure, of each safety system, three different severe consequences are obtained. When the abnormal event occurs, the path leads to continued operation if the first safety system succeeds to deal with the abnormal event. On the contrary, if the first safety system fails and the second one takes action successfully, the consequence is minor. If both two safety systems fail, the end state of the abnormal event will be severe consequence. Given failure probabilities of each safety system under each path of the event tree, the occurrence of every consequence can be estimated. In Figure 1, assume that the failure rate of the first safety system is 0.01 and the failure rate of the second safety system is 0.05 given the failure of the first safety system. Then estimates of occurrences of three consequences are 0.99, 0.0095 and 0.0005, respectively. As discussed above, this is the forward analysis, namely, predicting the occurrences of every consequence given failure probabilities of those safety systems. However, in practice, backward analysis is needed, which updates failure probabilities of those safety systems given specific consequence occurring. BN can satisfy this requirement and this is one reason why we chose BN to model dynamic risk.

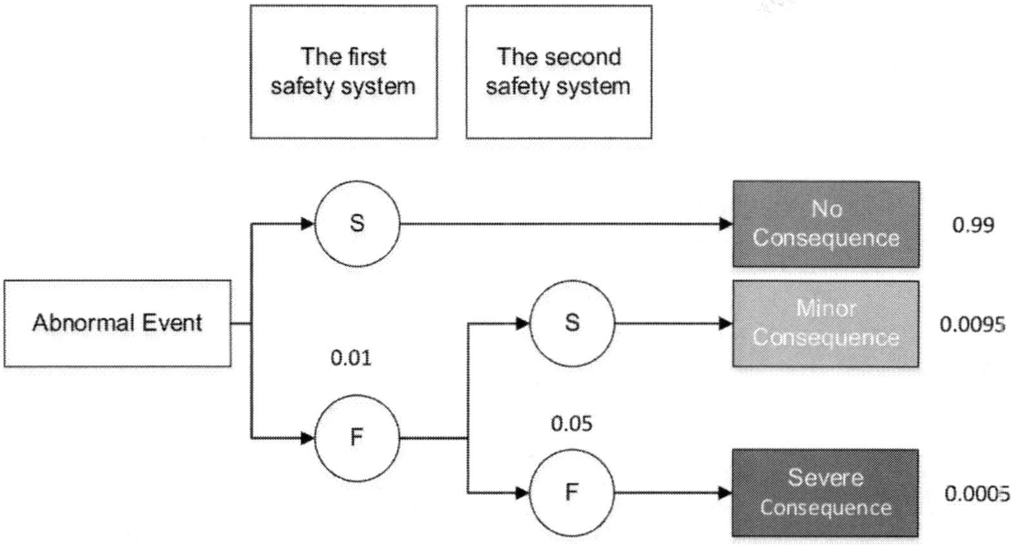

Figure 1. A simple event tree of modeling safety systems

ETA is a structured approach and is relatively easy to learn. However, the ETA technique has the following disadvantages: An event tree can only model one initialing event, and therefore if there are multiple initialing events, multiple ETAs should be constructed. ETA can express simple cause/effect relationships but overlook subtle system dependencies when modeling the events. Besides, partial successes or failures are not distinguishable [8].

3. Bayesian networks

3.1 Introduction to Bayesian networks

Bayesian networks have received considerable attention from scientists and engineers across lots of fields, including reliability analysis, risk management and safety analysis [10].
A Bayesian network is a directed acyclic graph (DAG) and consists of a set of nodes and a set of directed edges between nodes [9, 11]. Each node corresponds to a variable, such as a safety system. The edges represent the casual relationship between the linked nodes. For every node, a conditional probability distribution (CPT) should be included, which defines the probability distribution of a node given its parents. For instance, let us consider a simple Bayesian network structure shown in Figure 2. The BN contains three discrete nodes (A_1, A_2 and A_3) and each node has two states (0 and 1). The conditional probability tables, the simplest kind of CPT for discrete nodes, of three nodes are also included in the Figure 2.

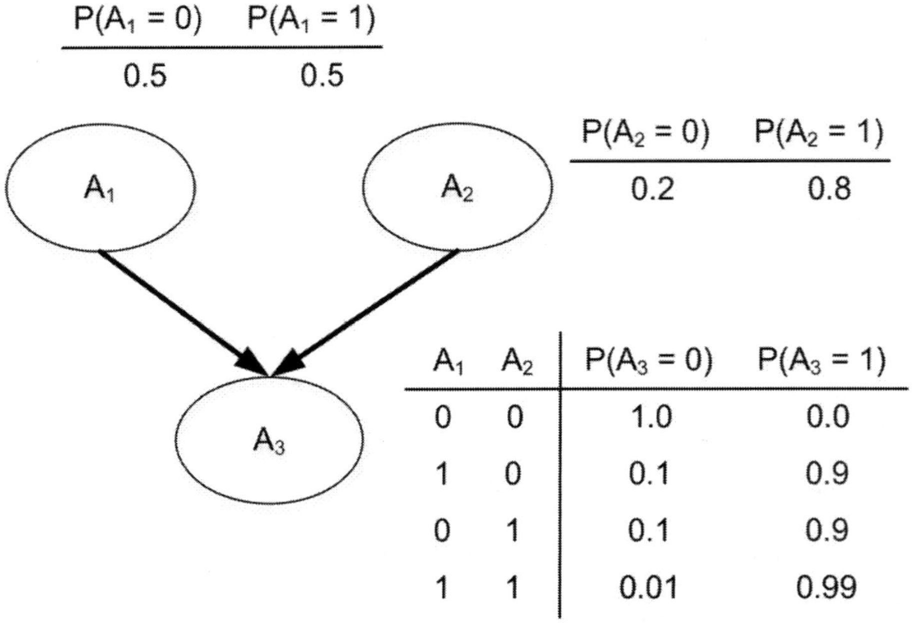

$P(A_1 = 0)$	$P(A_1 = 1)$
0.5	0.5

$P(A_2 = 0)$	$P(A_2 = 1)$
0.2	0.8

A_1	A_2	$P(A_3 = 0)$	$P(A_3 = 1)$
0	0	1.0	0.0
1	0	0.1	0.9
0	1	0.1	0.9
1	1	0.01	0.99

Figure 2. A simple Bayesian network

Typically, there are three main ways to construct Bayesian networks. In the first method, casual Influences depend largely on experts' knowledge. The second method to construct Bayesian networks is based on another type of knowledge, such as Fault tree. Also learning from data is a way to model Bayesian networks. Here, a mapping algorithm is used to construct the structure of Bayesian networks from event trees. The conditional probability distribution of each node is learned from data based on Bayesian approach with prior information.

The main feature of BN is the conditional independence and the chain rule, which makes the joint probability distribution P(U) of variables $U=\{A_1,\ldots A_n\}$ included in the network easy to compute. The computation process is as follows:

$$P(U)=\prod_{i=1}^{n} P(A_i \mid pa(A_i))$$

where $pa(A_i)$ are the parents of A_i in BN, and $P(U)$ is the joint probability distribution.

3.2 Advantages of Bayesian networks

For the risk analysis, the main advantage of using Bayesian networks is its ability for abductive reasoning, namely, the inference for updating the failure probability of safety systems given the occurrence of accident precursors [9]. The process of updating the failure probability cannot be done in the method of event tree analysis, which can solve the deductive reasoning, namely, predicting the occurrence of consequences. Besides, the structure of BN expresses the dependencies among the safety systems. As discussed above, the dependencies among the safety systems are important in the dynamic risk analysis.

For the modeling aspect, BN is more flexible than event tree [11]. New nodes can be added just with new arcs linking to relative nodes and modifying relative conditional probability distributions. Similarly, it is easy to remove a node. Any analysis that can be performed on the event tree can be performed by means of BN inference.

As we know, dynamic risk analysis not only needs to take account of severity of consequence that will follow when safety systems do not respond appropriately, but also depends on the time available compared with the time required for the corrective action to be performed and to have the desired effect. Then dynamic Bayesian network (DBN) can be introduced to dynamic risk analysis when modeling a dynamic system is taken account for. A general DBN structure of Bayesian network in Figure 2 is illustrated in Figure 3, where a two-slice temporal Bayesian network (2TBN) is considered for modelling the system temporal evolution [13]. Denote that A_i^t is the i's node at time t. The conditional probability distribution of A_i^t is as follows:

$$P(A_i^t \mid pa(A_i^t)) = P(A_i^t \mid pa^{t-1}(A_i), pa^t(A_i))$$

where $pa^{t-1}(A_i)$ and $pa^t(A_i)$ represent the parents from the first and the second slice of the 2TBN, respectively. For example, at time t, the node A_3 is the child of nodes A_1 and A_2 at time t, and the node A_3 at time t-1, i.e., $pa^{t-1}(A_i) = \{A_3^{t-1}\}$ and $pa^t(A_i) = \{A_1^t, A_2^t\}$.

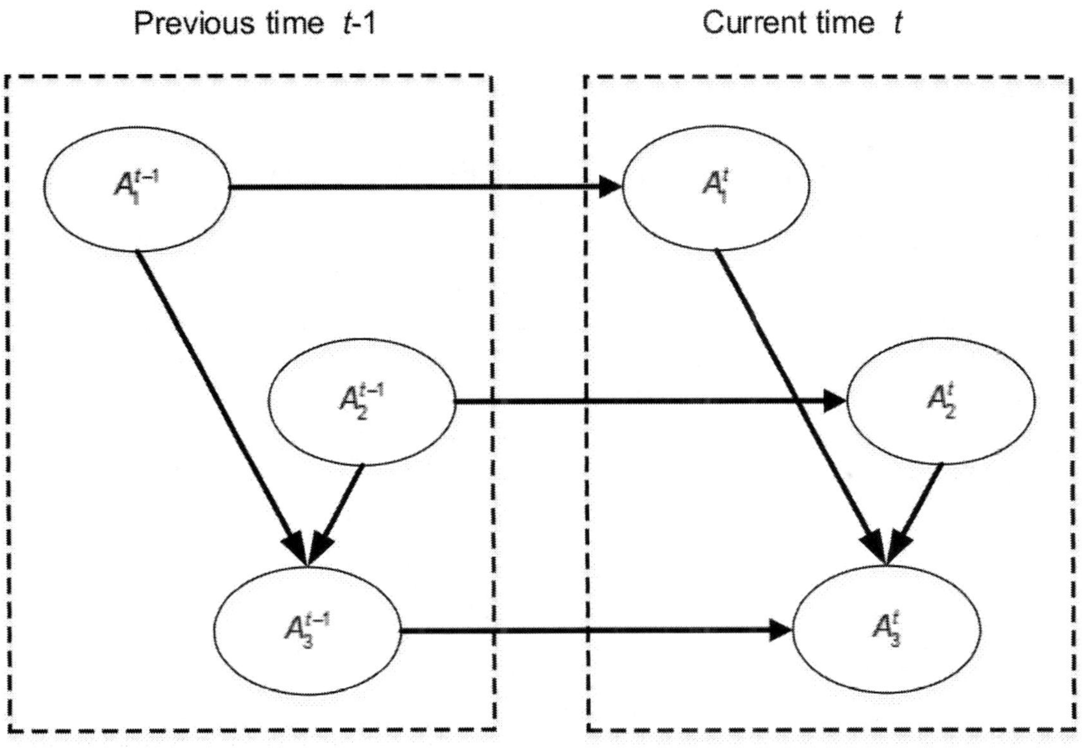

Figure 3. A general DBN structure of Bayesian network in Figure 2

4. The mapping algorithm from an event tree to a Bayesian network

In [8], a mapping algorithm from an event tree to a Bayesian network has been presented. Herein, a brief description of the algorithm is introduced.

Any event tree can be translated into a BN through a mapping algorithm. First, construct the graph structure of BN. The nodes are every single safety system. The consequence event is also as a leaf node. The arcs between the nodes are divided into two types: consequence and causal arcs. Consequence arcs connect each safety system node to the consequence node because success or failure of every safety system impacts the final consequence. Causal arcs connect each former safety system node to all later safety system nodes because of dependencies among those safety systems as discussed above. This is the generic BN representation of an event tree. A generic BN represents a complete event tree in which every path splits into two branches for every safety system, that is, if the number of safety systems is n, there are 2^n final consequences for a complete event tree. However, we then need to check the model's d-separation properties and ensure that they correspond to our perception of conditional independence properties. However, not all arcs may be needed and there exists a condition where not all paths need to split into two branches for every safety system. Consider the event tree of Figure 1. If the first safety system successes to prevent the abnormal event, then the path leads to continued operation no matter how other safety systems perform. In this situation, the arcs between the first safety system and other safety systems can be canceled. This method can simplify the Bayesian network and meanwhile does not influence the computation results.

5. A case study

In order to verify the method proposed in this paper, a case study used in [1, 6] is presented. The case study is an industrial, fluidized-catalytic-cracker at a major petroleum refinery that processes over 250,000 barrels of oil per day. There are nearly 150-200 alarmed variables and as many as 5,000-10,000 alarms occurring per day. Therefore, the alarm database is large and it is perfect to do the dynamic risk analysis. For the details of the FCCU please refer to the paper of [1].

In [1, 6], a dynamic risk analysis is proposed based on event-tree and set-theoretic formulation and Bayesian analysis to calculate the likelihood of the occurrence of incidents. In this case study, five safety systems are introduced, namely, basic process control system (BPCS), two levels of operator (machine + human) corrective actions, override controller and ESD system. The historical alarm data are summarized and represented by the branches of the event tree in that article. The event tree can be found in Figure 4. Herein, the primary variable, which is associated with ESD system, *P1*, is considered. The total failure and success counts, denoted as K_T^j and L_T^j (j = 2, ... 5), respectively, for safety systems in response to the abnormal events of the primary variable, *P1*, are shown in [1]. Then, Bayesian analysis is used to calculate the failure rates and probabilities of safety systems. It accounts for the dependencies among the safety systems using copulas. Two types of copula functions, multivariate normal and Cuadras–Auge´ copula, are used. Table 1 shows the results for P1 which is a comparison of the means using two copulas, Bayesian analysis with independent safety systems, and classic statistics. Parameters $\theta_i (i = 1,2,3,4,5)$ represent the mean of failure frequency of the first safety system, BPCS, and failure probabilities from the second safety system to the fifth safety system.

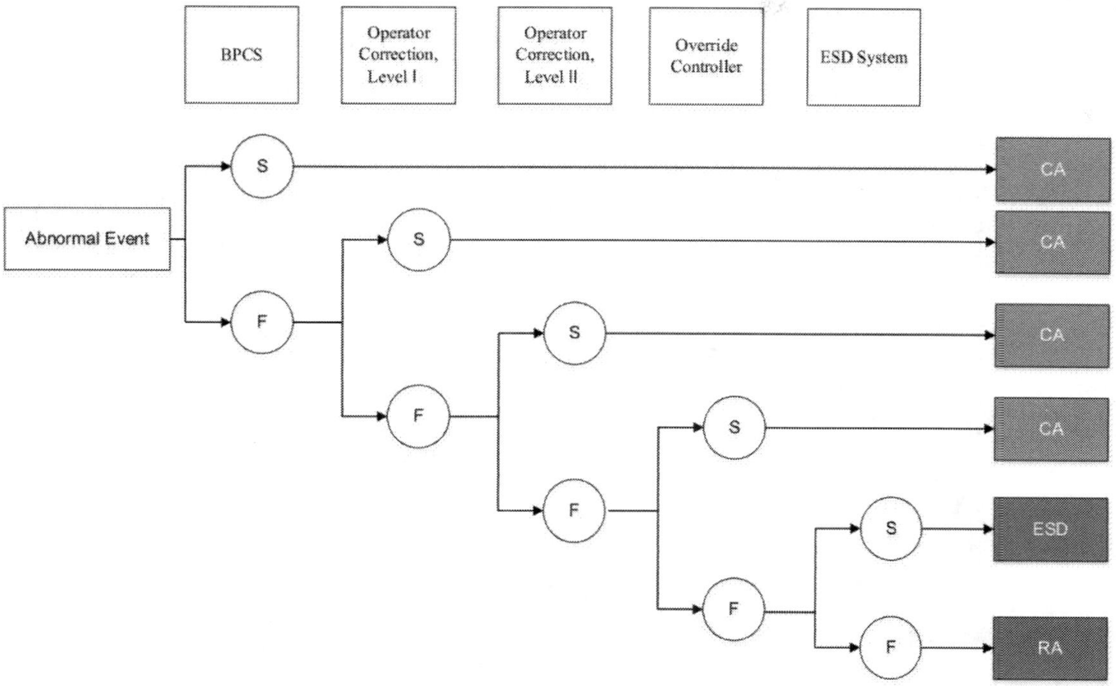

Figure 4. The event tree of the case study

Table 1. Comparison of Mean Failure Probabilities

	Bayesian network	Mean (Cuadras–Auge′ copula)	Mean (Multivariate normal copula)	Means (Independent safety systems)	MLE (Classic)
θ_1	142.6	142.5	142.6	142.7	142.8
θ_2	0.074	0.072	0.074	0.074	0.074
θ_3	0.844	0.838	0.851	0.844	0.847
θ_4	0.022	0.024	0.017	0.021	0.017
θ_5	0.259	0.247	0.215	0.250	0

Then, according to the mapping algorithm, the event tree shown in Figure 4 can be translated to a Bayesian network, which is presented in Figure 5. There are six nodes, including five safety systems and the final consequence node. The arcs describe dependent relationship among the nodes. The data which were used in [1] are inputted to the Bayesian network and the means of failure probabilities of safety systems are shown in Table 1. The marginal posterior distributions for using Bayesian network are presented in Figure 6. After the comparison with results of [6], little difference with methods of copulas can be found.

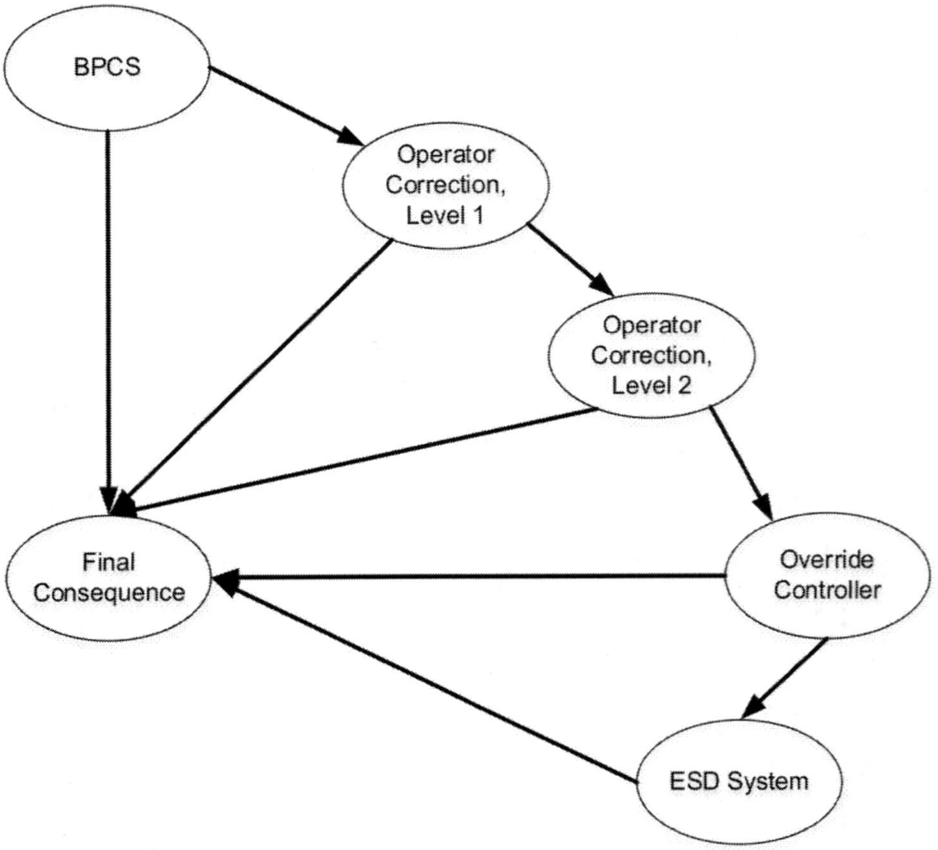

Figure 5. The Bayesian network of the case study

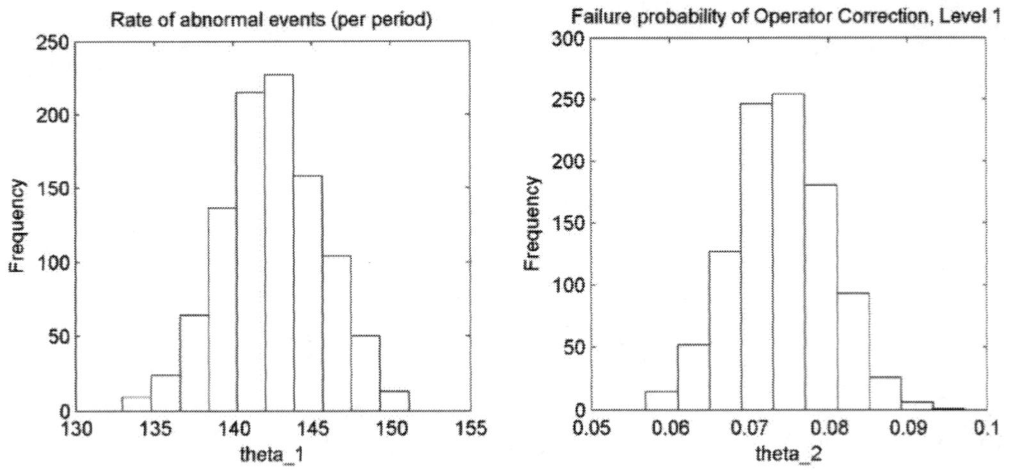

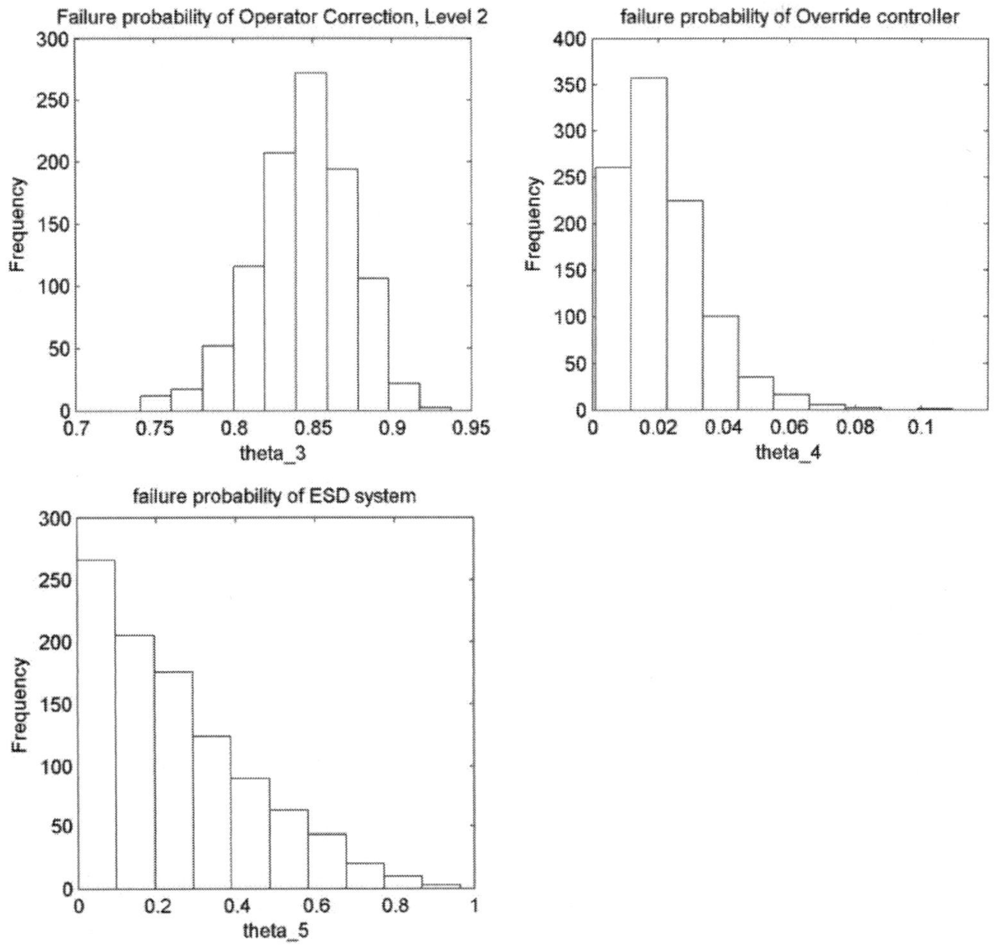

Figure 6. The marginal posterior distributions for using Bayesian network

The following part discusses the advantages of Bayesian network. Management actions and human behaviors have a significant impact on process reliability [14]. For example, the impacts of poor operator training, hard stress of operators, operator inabilities, and wrong decision-making of managers are involved. In event tree analysis, these factors are difficult to be involved to study, while it is easy to consider these in Bayesian network. The safety systems of two levels of operator correction are dependent on human behavior and management level, so the nodes of the corresponding factors can be added, which can make the dynamic risk analysis more accurate. Similarly, if the failure rate of final consequences is known, the economic and environmental loss can be estimated. For example, it is estimated that the financial loss of an emergency shutdown is $10,000 for a special plant. Then the estimated loss can be calculated if one economy loss node is extended after the final consequence node in Bayesian network. In case study, these two scenarios are shown in Figures 7 and 8, respectively.

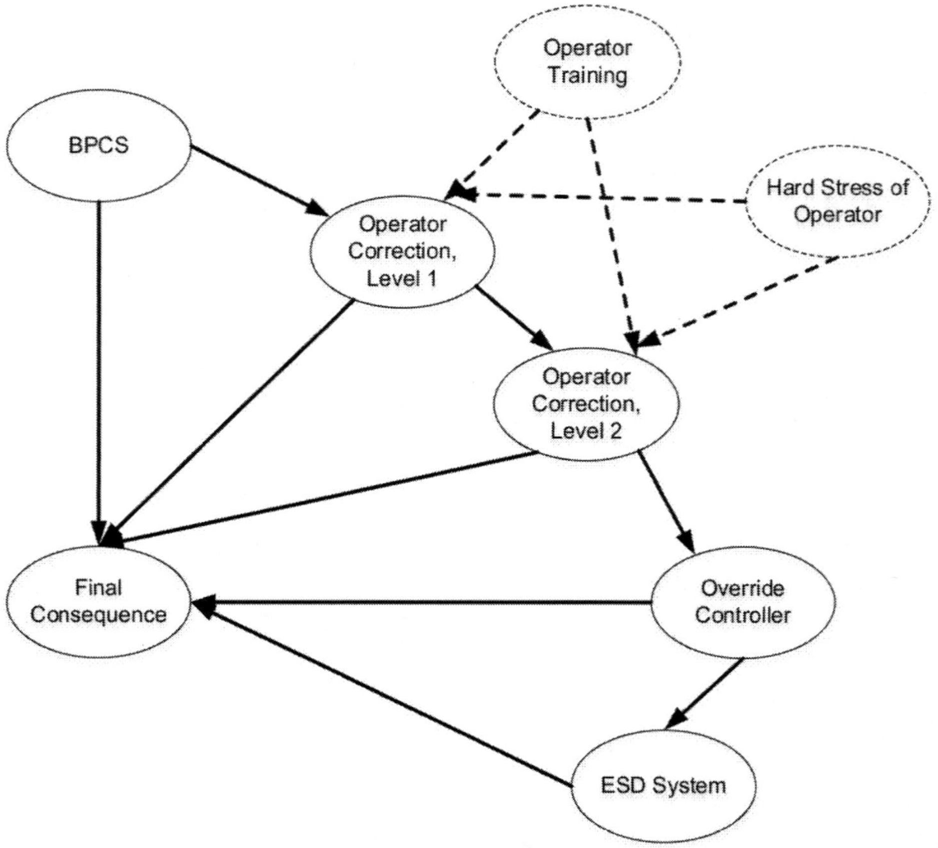

Figure 7. The extended Bayesian network with poor operator training and hard stress of operator

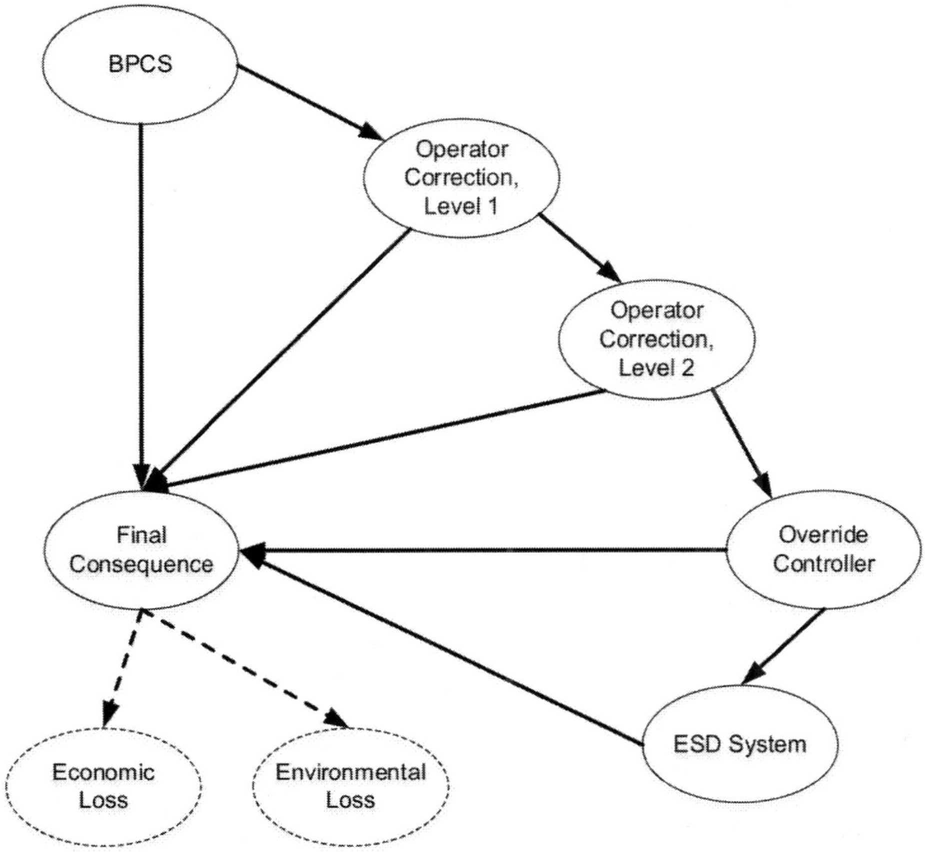

Figure 8. The extended Bayesian network with the loss of economic and environment

It is necessary to model the dynamic Bayesian network in dynamic risk analysis, because of temporal evolution of safety systems. For instance, consider the safety systems of operator correction levels, which highly depend on human behaviors. At time *t*-1, an abnormal event propagates to the safety system of operator correction level 1 and one operator, operator A, takes action to prevent the abnormal event. However, the action is wrong, that is, this safety system fails, but it will not trigger the next safety system. At time *t*, another operator, operator B, finds the problem, takes the right action, and then this safety system succeeds to lead the abnormal event to the consequence of CO. In static Bayesian network and event tree analysis, this process will be recognized and a path of success of safety system of operator correction level 1 is considered. However, the important details of two operators' behavior will be ignored. If dynamic Bayesian network is introduced, the state transition of operator behavior between time *t*-1 and time *t* will be analyzed. Bad performance by operator A will be concluded, while static Bayesian network and event tree analysis cannot achieve it.

6 . Conclusion

For the normal operation of chemical process industries, safety systems play a significant role. Event tree analysis is used for risk analysis mostly. However, Bayesian networks have advantages over event tree analysis, for example, extended modelling ability and dynamic Bayesian network which can be introduced easily. Through the paper, event tree analysis and

Bayesian networks are introduced. Meanwhile, a mapping algorithm from event tree to Bayesian network is described briefly. Finally, the proposed method is applied to a case study.

Acknowledgement

The authors gratefully acknowledge financial support from the National High Technology Research and Development Program of China (863 Program, Grant No. 2013AA040702) and Tsinghua National Laboratory for Information Science and Technology (TNList) Cross-discipline Foundation.

References

[1] Pariyani, Ankur, et al. "Dynamic risk analysis using alarm databases to improve process safety and product quality: Part I—Data compaction," AIChE Journal 58.3 (2012): 812-825.

[2] Crowl, Daniel A., and Joseph F. Louvar. Chemical Process Safety: Fundamentals with Applications. Pearson Education, 2001.

[3] Ericson, Clifton A. Hazard analysis techniques for system safety. Wiley.com, 2005.

[4] Yi, Woojune, and Vicki M. Bier. "An application of copulas to accident precursor analysis," Management Science 44.12-Part-2 (1998): S257-S270.

[5] Meel, Anjana, and Warren D. Seider. "Plant-specific dynamic failure assessment using Bayesian theory," Chemical engineering science 61.21 (2006): 7036-7056.

[6] Pariyani, Ankur, et al. "Dynamic risk analysis using alarm databases to improve process safety and product quality: Part II—Bayesian analysis," AIChE Journal 58.3 (2012): 826-841.

[7] Bobbio, Andrea, et al. "Improving the analysis of dependable systems by mapping fault trees into Bayesian networks," Reliability Engineering & System Safety 71.3 (2001): 249-260.

[8] Bearfield, George, and William Marsh. "Generalising event trees using bayesian networks with a case study of train derailment," Computer Safety, Reliability, and Security. Springer Berlin Heidelberg, 2005. 52-66.

[9] Khakzad, Nima, Faisal Khan, and Paul Amyotte. "Safety analysis in process facilities: Comparison of fault tree and Bayesian network approaches," Reliability Engineering & System Safety 96.8 (2011): 925-932.

[10] Weber, Philippe, et al. "Overview on Bayesian networks applications for dependability, risk analysis and maintenance areas," Engineering Applications of Artificial Intelligence 25.4 (2012): 671-682.

[11] Adnan Darwiche. "Bayesian networks," Communications of the ACM 53.12 (2010): 80-90.

[12] Jun, Chi-Hyuck, et al. "A Bayesian approach to prediction of system failure rates by criticalities under event trees," International Journal of Production Economics 60 (1999): 623-628.

[13] Murphy, Kevin P. "Dynamic Bayesian networks," Probabilistic Graphical Models, M. Jordan (2002).

[14] Meel, Anjana, Warren D. Seider, and Ulku Oktem. "Analysis of management actions, human behavior, and process reliability in chemical plants. I. impact of management actions," Process Safety Progress 27.1 (2008): 7-14.

1ST CCPS ASIA-PACIFIC CONFERENCE ON PROCESS SAFETY℠

A new SIL conception for emergency system and its application in LNG terminals

Jun-fang LIU, Hai-qing WANG, Fang LIU
China University of Petroleum (East China), Department of Safety Science and Engineering
Qingdao, Shandong province, 266580. CHINA
wanghaiqing@upc.edu.cn

Prepared for Presentation at
American Institute of Chemical Engineers
1st CCPS Asia-Pacific Conference on Process Safety
Qingdao, China
September 4 – 5, 2013

UNPUBLISHED

A new SIL conception for emergency system and its application in LNG terminals

NO. 306033

Jun-fang LIU, Hai-qing WANG *, Fang LIU
China University of Petroleum (Huadong), Department of Safety Science and Engineering

Keywords: emergency system; SIL; PFD; LNG terminal

Abstract

A new safety integrity level (SIL) conception introduced in this paper to analyze the risk reduction requirement of the emergency system (e.g., Fire and Gas System, flare system etc.) in oil & gas industry, meanwhile analyze the corresponding framework integrated with the safety function's qualitative description for such passive protection system, with emphasis and application to LNG terminals. In this paper, human factor and the actual mitigation function of emergency system are explored to establish the proposed SIL concept, and the numeric case study shows this new SIL conception provides an efficient approach to characterize the effectiveness of mitigation, and thus more practicable than the traditional E/E/PE integrate based SIL concept.

1. Introduction

Since the development of the standard IEC61508, the safety integrity level (SIL) conception has been widely used for the safety instrumented system (SIS) to quantify its reliability which corresponding to a risk reduction degree in process safety society[1]. But at the same time, some potential problems arise. To the industry users, a higher SIL generally means much safer (while relatively expensive), but to the emergency system, another kind of SIS which is different from that of the prevention system, it may means something else. This is the initiative of this paper, and a new SIL conception will be introduced to analyze the risk reduction requirement of the emergency system (e.g., Fire and Gas System, flare system etc.) in oil & gas industry.

2. The SIS and the emergency system

Safety instrumented systems (SIS) are widely used in different industrial areas such as the oil & gas industry so as to protect humans, the environment and assets. A SIS is usually composed of three parts: input element (various sensors, transmits), logic unit (PLC, relay based logic, etc.) and final element (alarm, valve, sprinkle, etc.), see Figure 1. The mandatory requirement of being defined as a SIS is that the safety system must include at least one element utilizing E/E/PE technology [2].

* Corresponding author. +86-0532-86983500 Ext. 8516; E-mail address: wanghaiqing@upc.edu.cn; This research is supported by the Excellent Scholar Project of China university of petroleum (Huadong) with No. SD2013211153

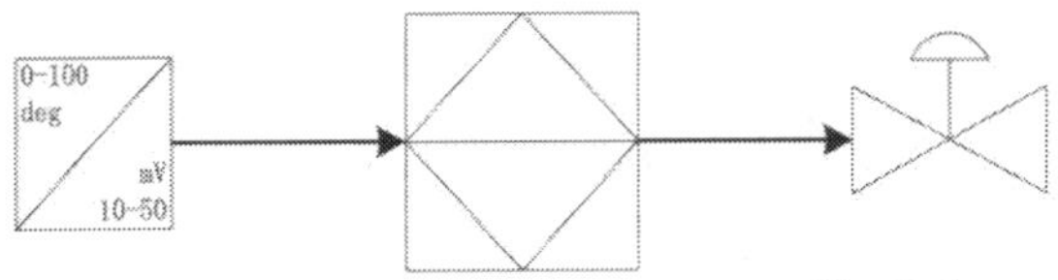

Figure1. The schematic of a SIS loop

The safety related systems can be categorized as two types: The prevention system and the emergency system [3]. The former is mainly utilized to prevent the accidents of loss of containment which may further result in severe fire and explosion, while the latter is to mitigate the severity of consequence after those accidents happens. These systems are designed to keep the plant in or back to a safe state, and both of them apply the same standard IEC 61508 in the design, manufacturing, installation and follow up, as shown in Figure 2.
When accidents occur, it will soon spread out and threaten people's life and cause property loss if there is no emergency system. However, with the use of the emergency system, the resulting fire or explosion can be controlled or mitigated, for those people in dangerous areas will have more time for evacuation.

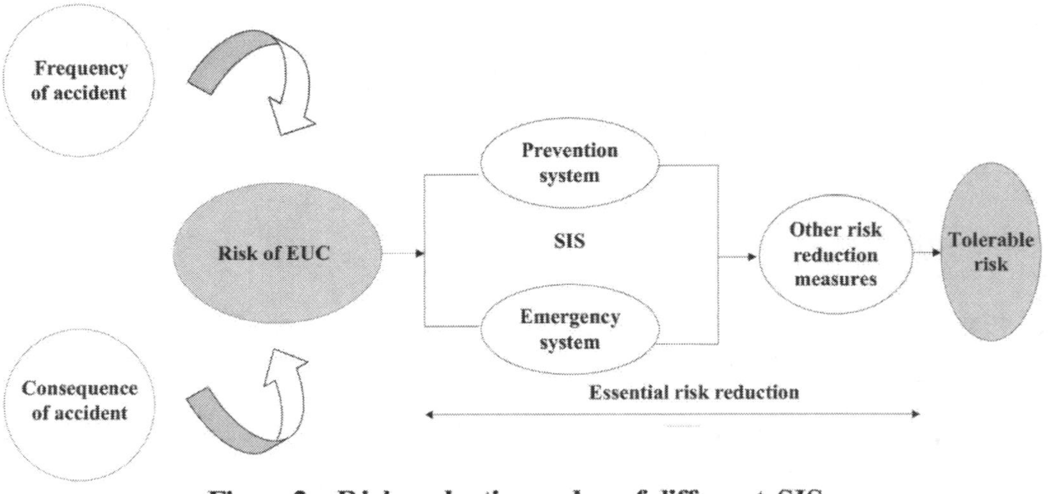

Figure2. Risk reduction roles of different SISs

As it known, the risk reduction degree of a safety function loop is referred as a SIL number, which is actually a measure of unavailability of the corresponding loop, and specifically the SIL is calculated as the average probability of failure on demand (PFD_{avg}) in process industry [4]. This concept of SIL is appropriate for the prevention system, but for the emergency system, as a special kind of SIS, may has a high SIL number while not meeting its safety function target in practical application because of the human factor or unreasonable layout of the detectors. As it may not start up because of the failure of the system or because the operators do something wrong, and even after it started it seems difficult to stop the accidents. Thus, the traditional SIL concept is not appropriate to measure the risk reduction requirement of the emergency system.

3. SIL conception based on human factor and mitigation analysis

The proposed SIL concept, as well as its analysis approach, is shown in Figure3. Contrary to the traditional SIL concept used to describe the safety function of safety instrumented system, the new SIL conception will take human factor and actual mitigation into consideration, as operators usually play a role in mitigation systems[5] and the ability of SIS is measured by risk reduction concept according to the international standards, such as IEC and ISA. In summary, the actual risk reduction function of the emergency system is related to the following factors:

- The probability of failure on demand (PFD)
- The human factor
- The actual risk reduction

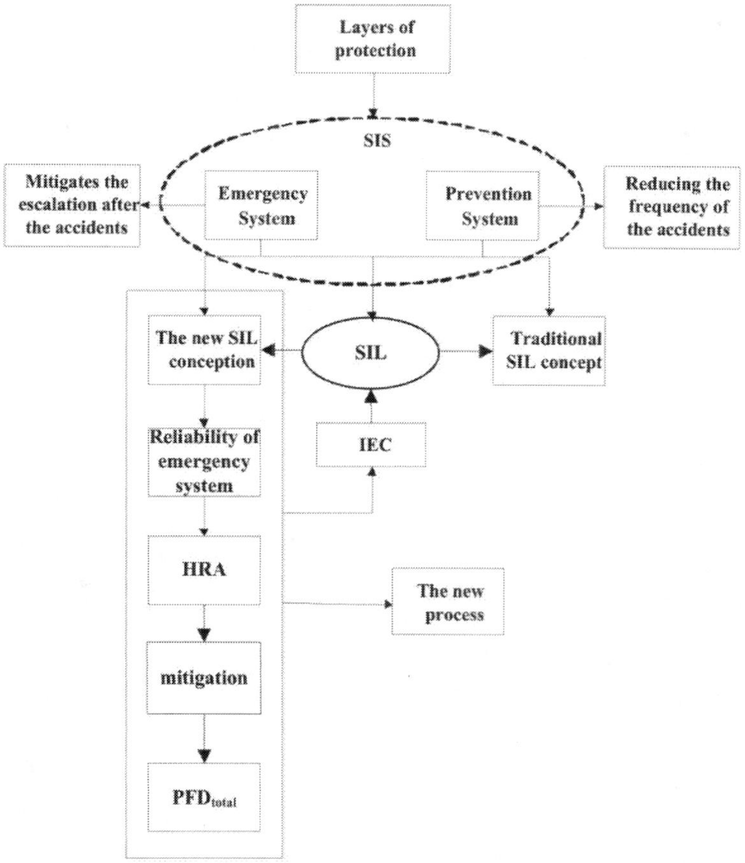

Figure3. Framework and application process of the proposed SIL concept

For the traditional SIL concept, PFD is an important parameter in determining SIL, since PFD represents the unreliability of a safety function loop of the SIS.
As be shown in Figure1, a SIS loop covers three parts. Thus the system PFD can be calculated by taking the sum of contributions of PFDs from three different elements[2], as Eq. (1):

$$PFD_0 = PFD(IE) + PFD(LU) + PFD(PE) \quad (1)$$

Where:

PFD (IE) is the contribution of PFD from the input element;
PFD (LU) is the contribution of PFD from the logic unit;
PFD (FE) is the contribution of PFD from the final element.

3.1 The human factor in mitigation system

Although the emergency system is designed as automatic, in practice it's not uncommon the operators or the site manager play the decisive role to ascertain the arising hazard and avoid spurious trip. For example, when a pool fire occurs, the emergency system such as Fire & Gas System will detect flame or the temperature change and thus trigger an F&G alarm. The Caretakers or Security officers may ignore this alarm signal, e.g., as nuisance alarm and fail to start the rescue measures and mitigates system such as ESD and Water spray system. Furthermore, it should be noted that operators are likely to make a mistake; it's possible for them to choose a wrong button when starting the ESD, especially in the emergency situation, so the human factor really has influence on the reliability of the emergency system.
When the operators receive the alarm signal, he or she has to make a judgment and take further action, i.e., the output of the sensor is the input of the operators, the PFD of operators must be a part of the total PFD, that is:

$$PFD_1 = PFD(IE) + PFD(LU) + PFD(PE) + PFD(OP) \tag{2}$$

Where:

PFD (OP) is the contribution of PFD from the operators.

Here the point is to determine the PFD (OP) since the former three parameters can be obtained from e.g., the OREDA handbook. To calculate PFD (OP), the Human Cognitive Reliability (HRA) model is used.

HEP of the control room operators							
1.Determining the allowed time	$t(s)=$	80					
2.Choosing the cognitive processing		Types of cognitive processing					
		Rule based					
Determining the parameters	$\alpha=$	0.601	$\beta=$	0.9	$\gamma=$	0.6	
3.Choosing the influencing factor	The experience of the operator	1.Expert or well trained	the stress level	1.High	The level of man-machine interface	2.Good	
	K_1	-0.22	K_2	0.44	K_3	0	
4.Determining the median response time	$T_{1/2,\ normal}(s)=$	15					
5.Adjusting the median response time	$T_{1/2}(s)=$	16.848					
6.the probability of non-response	$e^{-(\frac{t/T_{1/2}-\gamma}{\alpha})^{\beta}}$	0.003380149					

Figure 4 Interface of developed HRA software

Human cognitive reliability (HCR) model is one of widely adopted HRA models, which quantifies the probability of human being unable to fulfill cognitive judgment and actions within a limited time in nuclear industry and common large-scale complicated man–machine systems. It is based on Rasmussen's SRK three-level action model, which divides all the human actions in a system into three categories of skill-base (S), rule-base (R) and knowledge-base (K)[6]. Equipped by this HCR model with consideration of Chinese operator performance data[5], a dedicated software is developed to get a more accurate PFD (OP), it's easily and convenient for users as the users can get the result within just 6 steps choosing the right option. Figure4 shows the main interface of this HCR based software.

3.2 Effectiveness of risk reduction

The accidents may still cause lots of damage and loss even though the emergency system started, because the fire or explosion can't be stop immediately. To the same emergency scenario, the longer time it takes to recover to the safe state, the lower of emergency system's ability to mitigate the risk. In another word, if there is no emergency system, the fire or explosion will escalate to an unacceptable level, but with the function of the emergency system, it will take a longer time to reach the same tragic result, the added time ΔT, called additional time to act (ATTA), contributed to emergency system, relates directly to the effectiveness of the mitigation to the accidents.

To formulate the effectiveness of emergency mitigation system, some notations and assumptions are provided as below:

T_0: the time it takes to drive the accidents to a severe level of severity without emergency system, it can be obtained by the software.

T: the time it takes to drive the accidents to a severe level of severity with emergency system, can be obtained by the use of the same software.

PFD (MF): PFD of the mitigating function of the emergency system and can be derived from Eq. (3) on the base of practical condition.

It can be seen that quantities of T0 and T are another two important parameters to measure the performance of the emergency system. Apparently T is larger than T_0 and according to the above, PFD (MF) is related to T_0 and T, the relationship between them can be expressed as below.

$$PFD(MF) = 0.1 \times \frac{T_0}{T} \tag{3}$$

3.3 Calculation of new SIL.

Finally, the new SIL can be calculated by the three parts of PFD, which is denoted as PFD_{total} by Eq. (4):

$$PFD_{total} = PFD(IE) + PFD(LU) + PFD(PE) + PFD(OP) + PFD(MF) \tag{4}$$

PFD_{total} is obtained; the corresponding new SIL can be determined simply by categorizing into four levels, which is same with the traditional SIL concept.

4. Case study

The case study aims to verify the proposed SIL conception for the emergency system using actual Liquefied Natural Gas (LNG) plant field data.

4.1 LNG terminal description

The function of LNG terminal is to receive liquefied natural gas from the LNG carrier, store the LNG with tanks and supply nature gas to the clients after gasification [7]. A LNG terminal usually covers LNG storage zone, LNG unloading zone, BOG compression and re-condensation zone, LNG vaporization and sent out zone, and office zone. According to the different ways in disposing BOG, LNG terminal can be divided into direct output one and re-condensation type, usually the large-scale LNG terminals adopt re-condensation process [8].

4.2 Calculation of SIL of the emergency system.

During the transmission, storage, re-condensation and gasification process, the nature gas is likely to leak to the air resulting in possible fire and explosion hazards. The main areas suffering from leakage include LNG storage zone, LNG unloading zone, BOG compression and re-condensation zone, LNG vaporization and sent out zone[7]. BOG compression and re-condensation zone is considered in this study because it's so near from both the office and LNG tanks, when accidents occur in this area, it will be more dangerous than any other area.
The emergency system in this area consists of: fire and gas system, Caretakers or Security officers, water spray system, ESD.
In order to get the value of the SIL, three parameters must be calculated first, that is, PFD_0, PFD (OP) and PFD (MF).

- PFD_0: FFD_0 is the initial PFD of F&G; F&G is usually designed as high reliability of traditional SIL3 with Triple redundant controller, so the value of PFD_0 is between 0.0001 and 0.001, for this study $PFD_0=6.11*10^{-4}$, determined from OREDA handbook.

- PFD (OP): PFD (OP) can be calculated by the HRA software mentioned above. The process and result is shown in Figure5. Choose the type of cognitive process as rule based, the experience and the stress level and the interface level is respectively general (not so bad but far to expert), high (emergency situation) and good (not so bad and not excellent). Finally, PFD (OP) $=2.388*10^{-4}$.

t(s)=	150				
	Types of cognitive processing				
	Rule based				
α=	0.601	β=	0.9	γ=	0.6
The experience of the operator	2.General	the stress level	1.High	The level of man-machine interface	2.Good
K_1	0	K_2	0.44	K_3	0
$T_{1/2,\ normal}(s)=$	15				
$T_{1/2}(s)=$	21.6				
$e^{-(\frac{t/T_{1/2}-\gamma}{\alpha})^{\beta}}$	0.000238767				

Figure 5 the calculating process and result of PFD (OP)

- PFD (MF):To validate the proposed approach and to identify the T_0 and T, software is needed to simulate the real chemical plants and the accidents, from what T_0 and T can be precisely calculated, therefore the SIL of the emergency system can be identified. FDS is used in this paper to simulate the fire and explosion scenario in order to get T and T_0.

Data used in this process:
(1) According to NFPA the thermal radiation must be under 4KW/m2 in office area, 32KW/m2 for the skin of LNG tank, thus this paper define thermal radiation higher than 4KW/m2 in office area and 32KW/m2 in LNG tank skin area[7] as the higher level of severity.
(2) To prevent fire and explosion, fixed water spray system is installed in BOG compression and re-condensation zone with intensity of $10.2L/min{\cdot}m^2$ [7].

By using the data above and combining with the field model [9, 10], FDS model is built to calculate PFD (MF), as a result, PFD (MF)=0.1*13/3600=3.61*10-4.

- PFD $_{total}$:

Take PFD_0, PFD (OP) and PFD (MF) into formula (4):

$$PFD_{total} = PFD_0 + PFD(MF) + PFD(OP) = 6.11*10^{-4} + 2.388*10^{-4} + 3.611*10^{-4} = 1.211*10^{-3}$$

So the reliability of the SIF for emergency system is just SIL2, rather than the designed SIL3, and it seems tally with the practical use. At the same time, the result shows many differences between the new SIL level and the tradition SIL level which determined by the logic structure only. Since the new SIL conception take human factor and the practical mitigation function into consideration, it's more accurate and reliable than the traditional SIL concept for the emergency system.

5. Conclusions

For the emergency system a new SIL conception is proposed, and the approach is validated using a LNG terminal field data. The result shows the SIL of emergency system in LNG terminal considered is SIL2 instead of SIL3 which is designed as in logic structure. The following are some strategies to increase the reliability of the emergency system in actual operation:

- Improving the reliability of the operator: operators should me well trained and know the operating procedures very well, so they can shorten the response time; man-machine inter face should be improved for it can provide operators a nice work condition.
- Improving the mitigate function of the water spray system: the layout of the spray-head should be optimized to mitigate the accidents.
- Mitigating the layout of the detectors: the gas or smog or temperature may not be detected if the detectors are not installed correctly which can lead to no alarm.

References

[1] IEC61508 Part 1-7. (2000). Requirement for electrical/electronic/programmable electronic safety-related systems. Geneva: International Electrotechnical Commission[S].

[2] Chen Yanghou. Reliability Analysis of a Fire Alarm System[J]. Procedia Engineering, 2011, 24: 731-736.

[3] Masayuki Tanabe, Atsumi Miyake. Risk reduction concept to provide design criteria for Emergency Systems for onshore LNG plants[J]. Journal of Loss Prevention in the Process Industries, 2011(24):383-390.

[4] YANG Xianhui, GUO Haitao. Function safety of the safety instrument system [M]. Beijing, Tsinghua University Press, 2007.10(in Chinese).

[5] HE Xuhong, HUANG Xiangrui. Human reliability analysis in engineering system: theory,methods and applications [M]. Tsinghua University Press, 2007(in Chinese).

[6] Zhang L, He X, Dai L C, et al. The simulator experimental study on the operator reliability of Qinshan nuclear power plant[J]. Reliability Engineering & System Safety, 2007, 92(2): 252-259.

[7] GUO Kuichang. Liquefied natural gas (LNG) applications and security [M]. China Petrochemical Press, 2007(in Chinese).

[8] Li Ning-ning, Liu Yu-bao, Liu Kai. Primary Discussion on LNG Terminal Fire Prevention System Design [J]. Procedia Engineering, 2013, 52: 181-185.

[9] Tanabe M, Miyake A. Approach enhancing inherent safety application in onshore LNG plant design[J]. Journal of Loss Prevention in the Process Industries, 2012, 25(5): 809-819.

[10] ZHU Jianlu, LI Yuxing. Dynamic simulation of LNG import terminal process[J]. CIESC Journal, 2013, 64(3): 1000-1007(in Chinese).

Feature Extraction of Acoustic Emission for Offshore Platform Structural Health Monitoring

Guo-rui Li
China University of petroleum (East China)
Qingdao 266580, China
guorui_li@126.com

Chang-hang Xu
Guo-ming Chen
Su-ping Ai
China University of petroleum (East China)
Qingdao 266580, China
changhangxu@126.com
offshore@126.com
aisuping123456@163.com

Prepared for Presentation at
American Institute of Chemical Engineers
1st CCPS Asia-Pacific Conference on Process Safety
Qingdao, China
September 4 – 5, 2013

UNPUBLISHED

Feature Extraction of Acoustic Emission for Offshore Platform Structural Health Monitoring

Guo-rui Li
China University of petroleum (East China)
Qingdao 266580, China
guorui_li@126.com

Chang-hang Xu
Guo-ming Chen
Su-ping Ai
China University of petroleum (East China)
Qingdao 266580, China

Keywords: acoustic emission; offshore platform; feature extraction; HHT

Abstract

Offshore platforms are the basis of offshore oil exploration and development. Living in harsh environment, fatigue damage and processing defects will gradually come out, which may lead to the platform collapse. Acoustic emission (AE) technology is potentially a promising method for crack detection in the offshore structures. The AE device could detect defects' changes timely and effectively with exquisite sensitivity and online real-time monitoring features. It has been successfully used in foreign offshore platform monitoring, but seldom in domestic ones. It is well known that AE is proficient in data acquisition and monitoring data changes, while not in signal processing. Signal processing plays an extremely important part in structural monitoring assessment. Without a good signal processing, the monitoring process is a failure to a certain extent.

In this paper, we address an application of the Hilbert-Huang transform (HHT) and wave packet transform (WPT) to characterize the AE signals released from the offshore structure model. Without artificially determining the decomposition level, original signals are adaptively decomposed into several Intrinsic Mode Functions (IMFs) by HHT, while the wavelet (packet) transform could not. Filter out non-acoustic emission frequency band components for signal de-noising and reserve the larger IMFs by Fourier transform and cross-correlation analysis between each IMF and original signal. By HHT analysis, 3D joint time-frequency distribution and wavelet packet transform, the differences between structure normal signal and crack AE are clear and which highlight the AE characteristics. It can be concluded that the joint of HHT and wave packet transform is an effective tool to extract the features in offshore structures, which have a certain sense in the application of acoustic emission technique.

1. Introduction

At present, offshore platform construction costs a lot. The service year of a fairly large numbers of platforms are much long in offshore oil development of our country, and its environment is harsh, this aged platforms will bring risks to our safety in production.

As a global non-destructive testing (NDT) method, acoustic emission (AE) has been successfully applied for the structural integrity assessment of large-scale structures, such as metallic pressure vessels, tanks and offshore structures. Building up early warning system for offshore platform based on AE technology can ensure working safety on offshore platform which can monitor offshore platform in all weather and find the hurts of the platform, so the early warning system has immense research value. In recent years, much attention has been paid to AE technique development by a large number of research scholars from Northeast Petroleum University, Dalian University of Technology, China University of Petroleum and so on.

Acoustic emission (AE), known as micro-seismic activity, is the energy released when a flaw, crack initiation or propagation, happens. The technology makes it possible to monitor the offshore platforms online and real-timely. However, in most instances, AE signal is very faint, and it is easily covered by noise. Conventional signal processing method is difficult to extract AE signal features, which restricts the application and promotion of this technology.

Signal processing is an important step in AE monitoring process. Therefore, AE signal processing has fascinated many workers in this field, such as research scholars from Northeast Petroleum University, Dalian University of Technology, China University of Petroleum and so on. Knowledge of AE signal processing has to be further enlarged.

This paper addresses an idea based on the Hilbert-Huang transform (HHT) and time-frequency analysis methods to process AE signal released from the offshore structure model. The processing flowchart is shown in Figure 1.

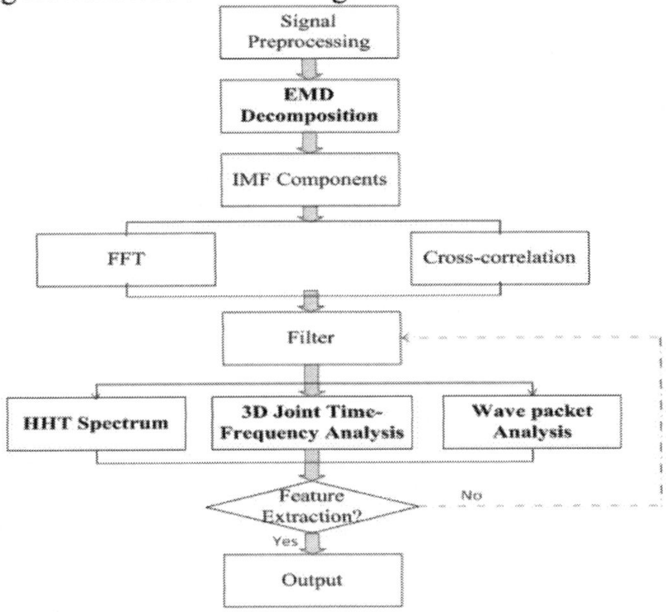

Figure 1: Flow diagram of signal processing

2. Signal Processing methods

2.1 Hilbert-Huang Transform

The Hilbert-Huang transform (HHT) is an empirical method to analyze nonlinear and non-stationary signal. The AE data, including crack growth signal and environment noises, may have many different coexisting modes of oscillation at the same time. Each of these modes is represented by an Intrinsic Mode Function (IMF). The definition of IMF is:

1) The number of extreme (maximum and minimum points of dataset) and the number of the zero crossing of function must be equal or differ at most by one.

2) The mean value of the envelopes defined by local extreme must be zero.

The first stage is to calculate and compute the IMF's of the signal (The EMD process). The EMD process itself can be done in several steps:

(1) Find the location of all maxima, $x_{\max}(t)$, and minima $x_{\min}(t)$ of $x(t)$.

(2) Connect all $x_{\max}(t)$ points with cubic spline (upper envelope) and do the same procedure to $x_{\min}(t)$ to produce lower envelope.

(3) Compute the mean of spline curves at each point:

$$m(t) = \frac{x_{\max}(t) + x_{\min}(t)}{2} \qquad \text{[Eq. 1]}$$

(4) Remove the trend, $m(t)$. Let

$$d(t) = x(t) - m(t) \qquad \text{[Eq. 2]}$$

(5) Now if d(t) is an IMF according the definition of IMF. Let

$$c_i(t) = d(t) \qquad \text{[Eq. 3]}$$

and advance i by 1 .Find the residual

$$r(t) = x(t) - d(t) \qquad \text{[Eq. 4]}$$

If $d(t)$ is not an IMF (according to definition of IMF), Further sifting is needed.

(6) Repeat steps 1 to 5.

2.2 Wave packet Transform

Assuming $g_j^n(t) \in U_j^n$, so $g_j^n(t)$ can be expressed as

$$g_j^n(t) = \sum_l d_l^{j,n} u_n(2^j t - l) \qquad \text{[Eq. 5]}$$

Wavelet packet decomposition algorithm: $\{d_l^{j,2n}\}$ and $\{d_l^{j,2n+1}\}$ are obtained by the $\{d_l^{j+1,n}\}$

$$\begin{cases} d_l^{j,2n} = \sum_k a_{k-2l} d_k^{j+1,n} \\ d_l^{j,2n+1} = \sum_k b_{k-2l} d_k^{j+1,n} \end{cases} \qquad \text{[Eq. 6]}$$

Wavelet packet reconstruction algorithm: $\{d_l^{j+1,n}\}$ are obtained by the $\{d_l^{j,2n}\}$ and $\{d_l^{j,2n+1}\}$

$$d_l^{j+1,n} = \sum_k \left[h_{l-2k} d_k^{j,2n} + g_{l-2k} d_k^{j,2n+1} \right] \quad \text{[Eq. 7]}$$

3. Experiment Apparatuses

This experiment model is a fixed offshore jacket platform, which is built in scale (Figure 2). The model is a four-legged jacket with a combination of pipes and deck. The main legs are fixed on the ground with bolts, which are made of 45 steel.

The maximum length of the platform model is 846mm, the maximum width is 852mm and the total height is 1980mm. The model consists of four layers, including 10mm thick steel plate of the top deck and pipe welded structures of the remaining three.

Sensor Highway II system (Figure 2) from Physical Acoustics Corporation (USA) is used to acquire AE signal for structural health monitoring (SHM). The sensors are R series resonance probes, with silica gel as the coupling agent. B&K vibration equipment from Denmark is used to motivate AE signal in simulated Marine vibration environment, and the relevant signals propagate along the platform model components are received by the sensors ultimately.

Figure 2: Offshore platform model

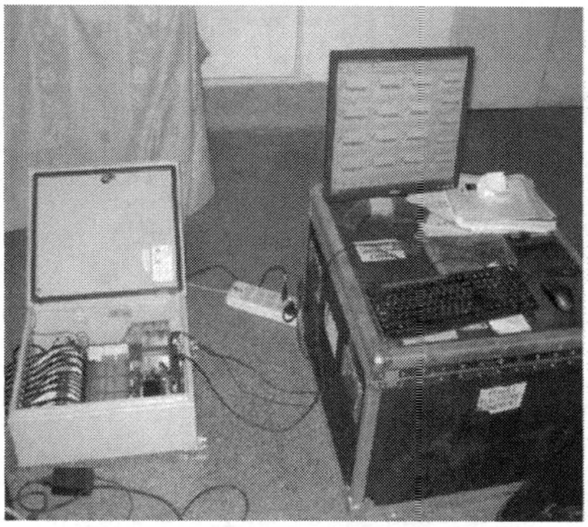

Figure 3: Sensor Highway II

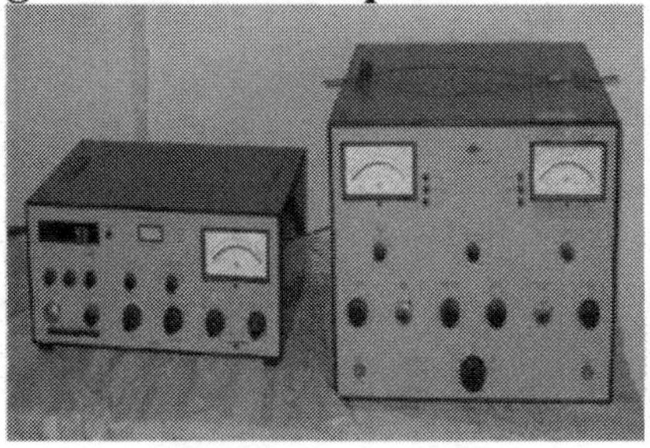

Figure 4: vibration control equipment

Figure 5: vibration loading equipment

The Sensor Highway II hardware environment parameter settings are shown in Table 1.

Table 1. AE hardware setting parameters

Threshold	Floating gate	Sampling frequency	PDT /us	HDT /us	HLT /us	Maximum duration/us
28dB	18dB	1MSPS	300	600	1000	1000

4. AE waveforms analysis based on HHT

In this part, signals will be analyzed in the time domain, the frequency domain and the joint time-frequency domain. The methods include Fast Fourier Transform (FFT), cross-correlation analysis, HHT and wave packet energy spectrum analysis. The signals include two types:

(1) Normal Signal: signals under vibration load in normal state (Figure 6);

(2) Crack AE: signals under vibration loading in crack status (Figure 7).

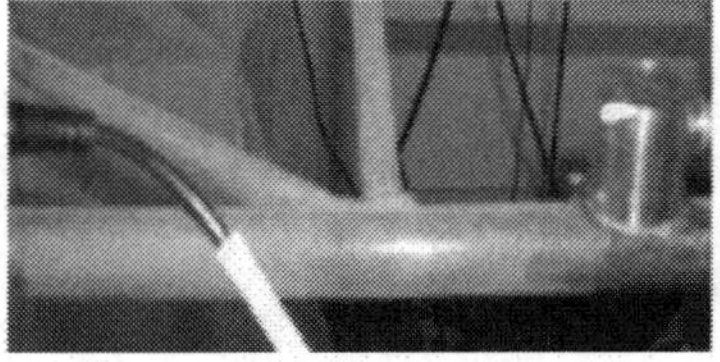

Figure 6: Normal state

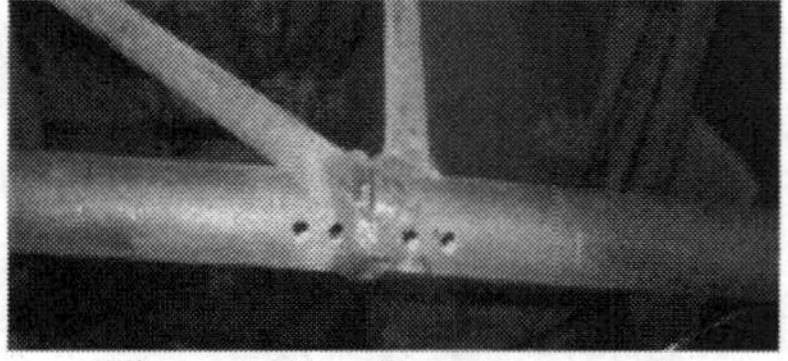

Figure 7: Crack status

4.1 Time Domain Waveform

The two types of signal are acquired in the same conditions by Sensor Highway II system. Comparing Figure 6 with Figure 7, it can be seen that:

(1) The amplitudes of normal signal are generally low, and the amplitude differences between signals are very small; the whole signal shows randomness and only several higher amplitude signals arise occasionally in the acquisition process.

(2) The high amplitudes of crack AE are much higher than ones of normal signal, but the low ones are similar. The change of crack AE amplitudes shows a cyclical fluctuation law, and some signal features like lead-broken signals: the amplitude increases rapidly and eventually drops quickly.

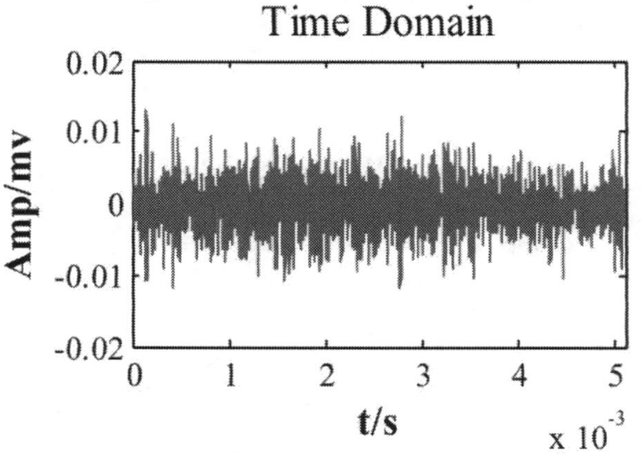

Figure 8: Normal Signal waveform

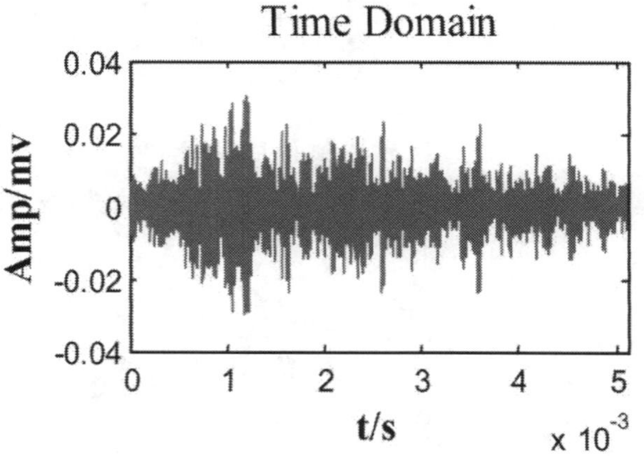

Figure 9: Crack AE waveform

4.2 Empirical Mode Decomposition (EMD) Analysis

Empirical Mode Decomposition (EMD) is the core of Hilbert-huang transform. It is a signal adaptive decomposition method, and particularly suitable for processing non-stationary signals. EMD method yields ten IMF components (shown in Figure 10 or Figure 11) from each signal. They are similar in some feature and different in many areas.

(1) The average amplitudes of the first five IMF components, IMF1-IMF5, are much higher than the residual IMF components in the two figures. The five IMFs constitute the main amplitude of signal and they present the non-stationary features of AE. While the last several IMF components, IMF6-IMF10, show a similar fluctuation in the two figures, which could not characterize the signal features.

(2) Though the total amplitudes of normal signal are lower, its IMF components, IMF2-IMF5, are higher than the corresponding IMFs of crack AE, while IMF1 of normal signal is much lower than the crack AE ones. Thus, the first five IMFs, IMF1-IMF5, are the key objects in the next analysis.

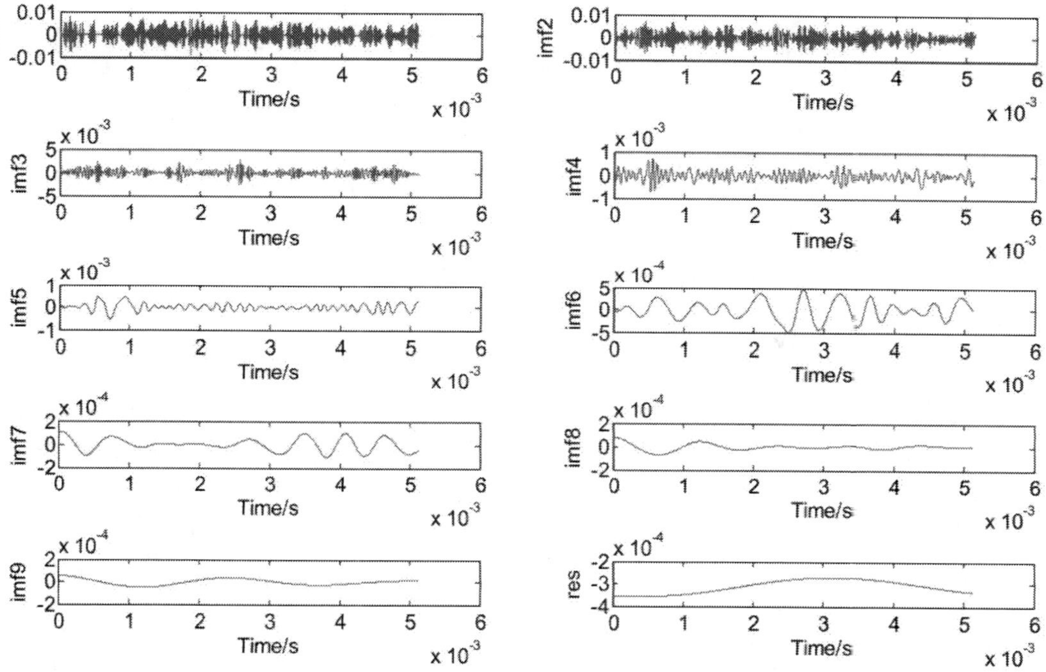

Figure 10: EMD of Normal Signal

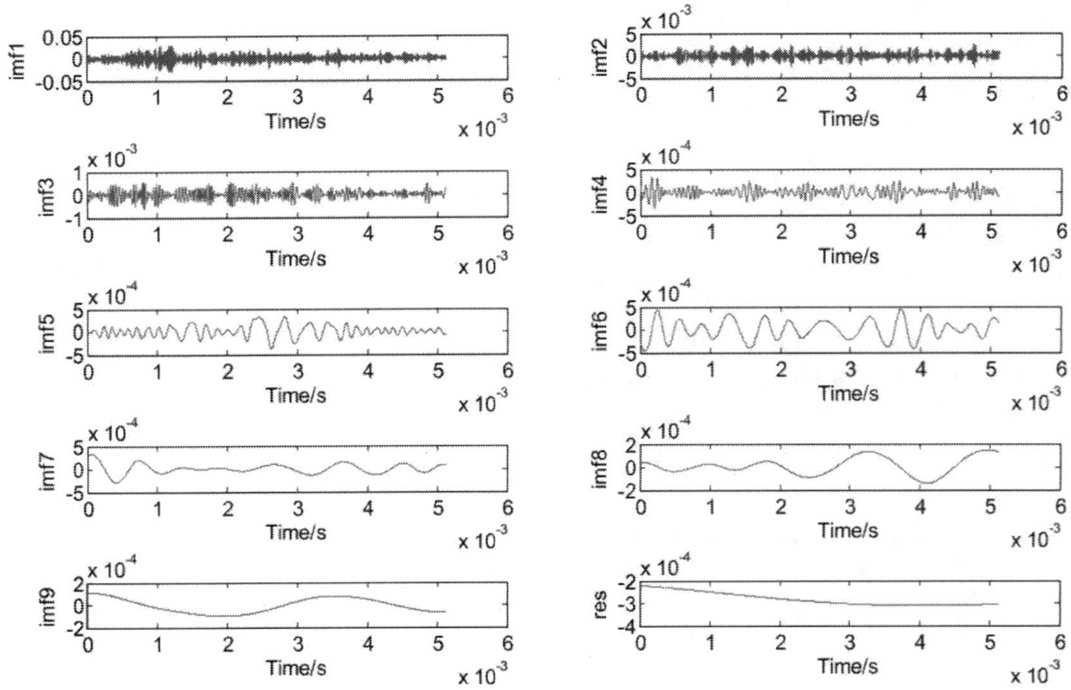

Figure 11: EMD of Crack AE

4.3 FFT & Cross-correlation Analysis

Taking the results of EMD into account, Fast Fourier transform (FFT) of each IMF and cross-correlation analysis between each IMF and original signal were carried out. The FFT center frequencies and the cross-correlation coefficients (shown in Table 2) are the most important parameters to filter the unimportant signal components.

The two parameters were plotted in the same graphs (Figure 12 and Figure 13) and we can find more intuitive features that

(1) In Figure 12, the cross-correlation coefficients between IMF4-IMF10 and normal signal are very small, and they are almost no contribution. These IMFs are concentrated in 0-15K Hz, which is not the frequency range of structural AE.

(2) In Figure 13, the cross-correlation coefficients between IMF3-IMF10 and crack AE are also very small, and their center frequency is less than 25K Hz, which should be filtered.

Therefore, the IMF4-IMF10 of normal signal and IMF3-IMF10 of crack AE are neglected, the signals are de-noised, which is very important for AE feature extraction. Then the remaining IMFs reconstruct new signals, which are used as the input of the post processing.

Table 2. FFT & Cross-correlation data

Normal Signal			Crack AE		
IMF	**FFT Center Frequency (K Hz)**	**Cross-correlation coefficient**	**IMF**	**FFT Center Frequency (K Hz)**	**Cross-correlation coefficient**
1	79	0.7472	1	83	0.9932

2	42	0.661	2	43	0.1653
3	27	0.1537	3	21	0.0087
4	12	0.0031	4	12	0.0012
5	4	0.0252	5	3	0.0098
6	2	0.0466	6	2	0.0231
7	1.5	0.0122	7	1	0.0119
8	0.9	0.001	8	0.6	0.0081
9	0.3	-0.0004	9	0.2	0.0059
10	0.01	0.0003	10	0	0.0015

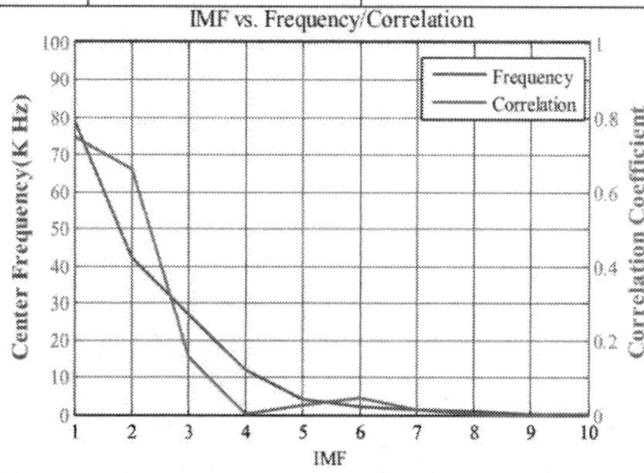

Figure 12: Frequency & Correlation of Normal Signal IMFs

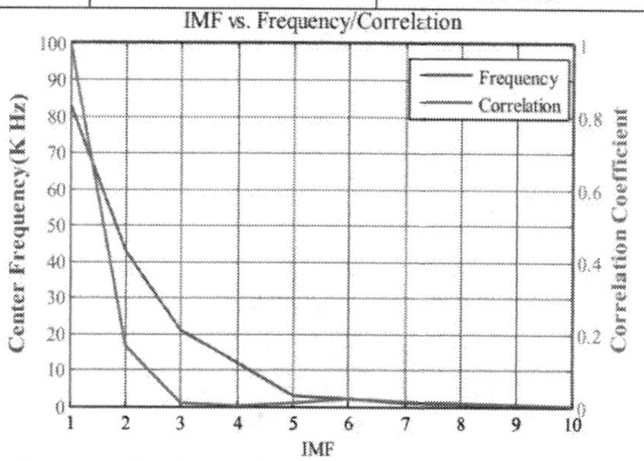

Figure 13: Frequency & Correlation of Crack AE IMFs

4. 4 Hilbert-Huang Spectrum

The reconstructed new signals are decomposed by EMD, and the new IMF components are used for the spectral analysis. The Hilbert spectrum is given in Figure 14 and Figure 15. From the Hilbert spectrum, we can see that:

(1) The major energies of two signals are concentrated in a frequency range of 30-100K Hz; the normal signal energies in time domain changes little, and they are much lower than the crack ones. From the two figures, the differences between normal signal and crack one is clear.

(2) In Figure 15, the several highest energy signals occur around 1.3ms, and they are very different from the remaining ones. At these times, platform crack is most likely to grow. At other times, the structure may change, but the change is little.

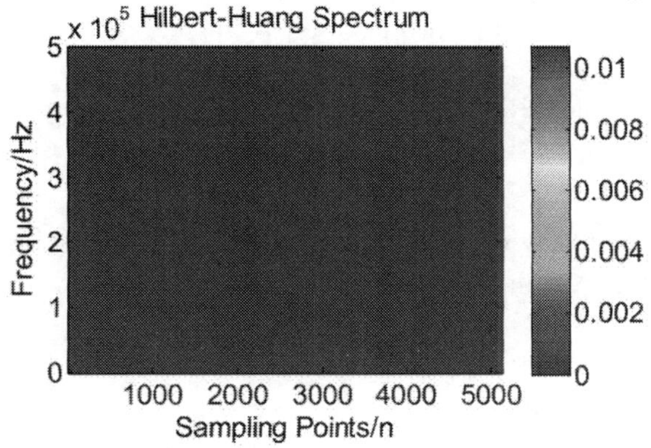

Figure 14: HHT of Normal Signal

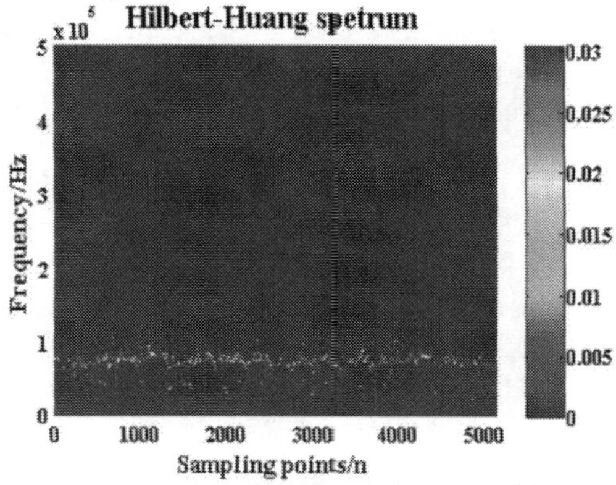

Figure 15: HHT of Crack AE

4.5 *3D Joint Time-Frequency Analysis*

Though the colors of Figure 15 present some changes, it is not intuitive enough. By calculating the signal instantaneous frequency of each moment, two three-dimensional (3D) images (Figure 16, Figure 17) of the time-frequency results is drawn, which present signal changes or features better:

(1) The amplitudes of the whole normal signal change little in time domain, and the signals are uniformly distributed in a frequency range of 30-100K Hz.

(2) The amplitudes of crack AE change a lot. The signal mutation means some new signals occur. The amplitudes raise high at 1.3ms, 1.8ms, 2.2ms and 3.1ms in Figure 17, but none in Figure 16. These times is the key time of crack AE occurring. In addition, these times may be attained in Figure 15, but it is not easy for you.

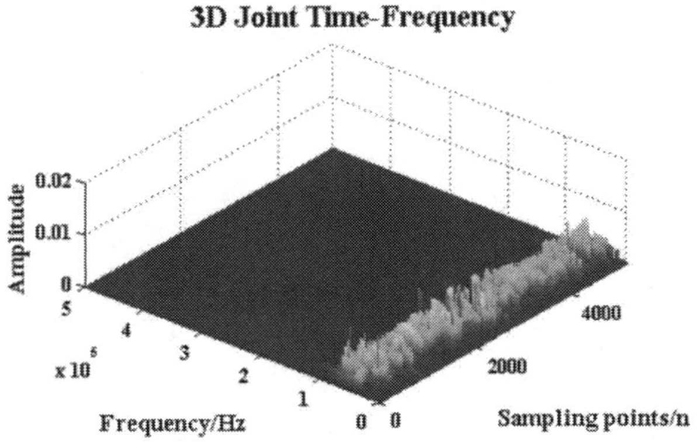

Figure 16: 3D Joint Time-Frequency spectrum of Normal Signal

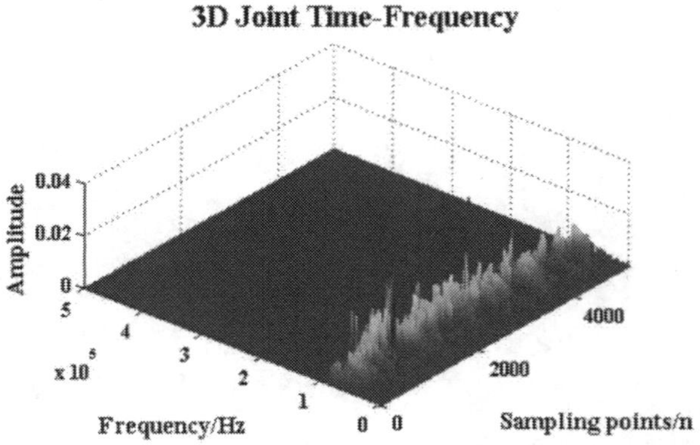

Figure 17: 3D Joint Time-Frequency spectrum of Crack AE

4.6 *Wave packet Energy Spectrum Analysis*

To get more information, wave packet energy spectrum is used to extract AE features. Wavelet and decomposition scale must be selected. As Daubechies (db*N*) wavelet is compact and orthonormal, it is suitable for processing AE signals. However, the scale of 5 is confirmed by a series of tests (decomposition of 2-5 layers). Compare the fourth segment (Dec5) in Figure 18 with Figure 19, the difference between normal and crack AE is very clear.

(1) The highest energy in normal signal is the fourth frequency order, the total energy of the first four orders is more than 60%, and the sum energy of the seventh and eighth order is about 32%.

(2) The highest energy in crack one is the eighth frequency order, the total energy of the first four orders is only 23%, and the sum energy of the seventh and eighth order is about 65%.

(3) The differences of highest energy frequency order are one of the most important points, and the great differences in percentage are much useful in AE feature extraction.

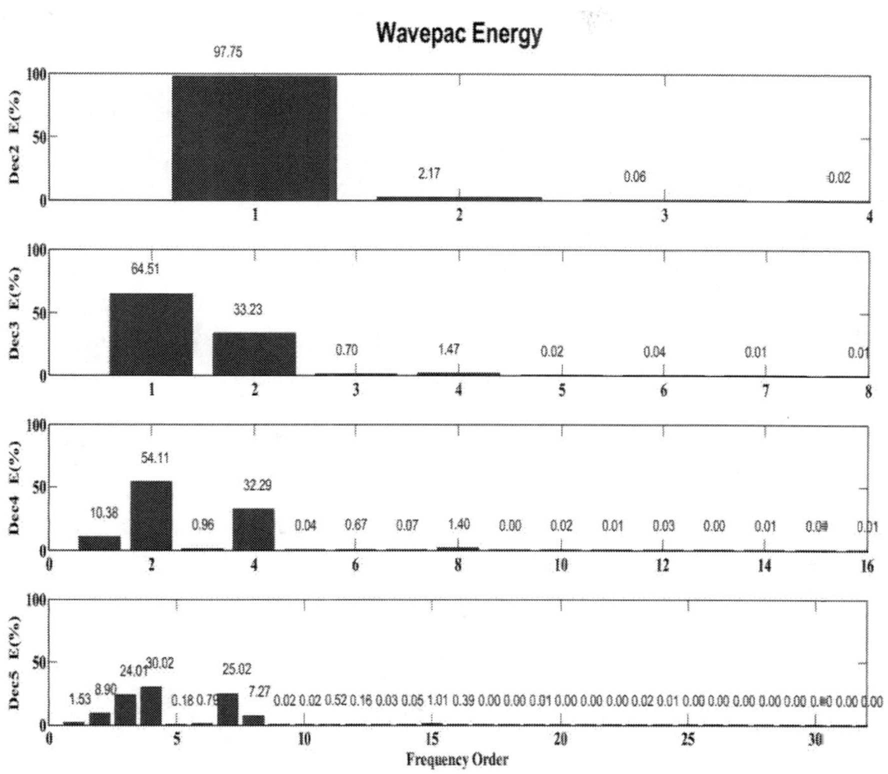

Figure 18: Wave packet Energy Spectrum of Normal Signal

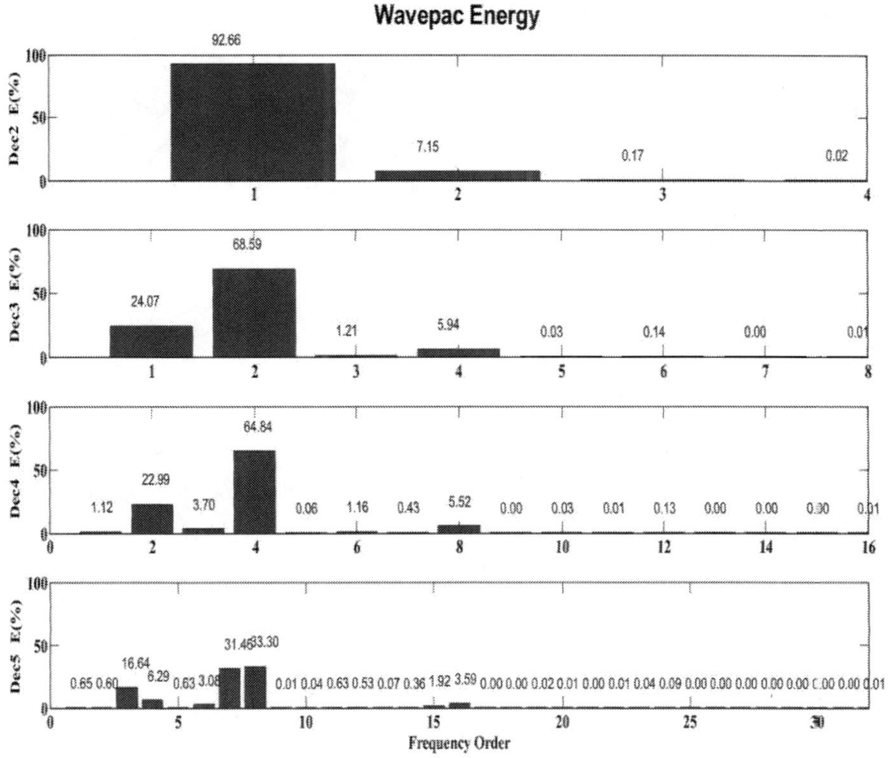

Figure 19: Wave packet Energy Spectrum of Crack AE

5. Conclusion

In this paper, different methods are used to extract AE features. By experiment and data processing, it can be concluded that:

(1) Without artificial control of basic functions and decomposition level, Empirical Mode Decomposition can adaptively decompose signal.

(2) By FFT and cross-correlation analysis, we can effectively find characteristics of various frequency components and filter out the irrelevant components in signal.

(3) Hilbert-Huang spectrum presents the signal feature by different colors in time and frequency domain. 3D Joint Time-Frequency Analysis reflects signal features more intuitively over time and frequency change.

(4) At the same time, the wave packet energy spectrum shows the difference between AE signal and noise by histograms. The differences of highest energy frequency order are one of the most important points, and the great differences in percentage are much useful in AE feature extraction.

All in all, the joint of HHT and time-frequency analysis is an effective tool to extract the features in offshore structures. It may make a difference in the future application of acoustic emission technique.

6. References

[1] Non-Destructive Qualification Book Editorial Board for Industry of national defense science and technology, "Acoustic Emission Testing," Beijing: China Machine Press, 2009.

[2] Changhua Hu, Guohua Li, Tao Zhou, "MATLAB7.x based System Analysis and Design - wavelet analysis," Xi'an: Xi'an University of Electronic Science and Technology Press, 2008.

[3] L. Lin, F. Chu, "Feature extraction of AE characteristics in offshore structure model using Hilbert-Huang transform," *Measurement*, Volume 2011, NO. 44, 2011.

[4] J. eisenblatter et al, "Survey of Acoustic Emission Research in Europe and investigation Conducted at Battelle Frankfurt," *2nd Acoustic Emission Symposium*, Tokyo, 1987, Session 7-1/99.

[5] "Standard Practice for Acoustic Emission Monitoring of Structures during Controlled Stimulation," ASTME 569-85, *American Society for Testing Materials.*

[6] "Acoustic Emission Testing of Spherical Pressure Vessels and Classification of Test Results," NDIS 2410-80, *Japanese Society For Non–Destructive Inspection*, Tokyo, 1980.

Experimental Study on Fan Water Curtain Diluting Water-Insoluble Heavy Gas in Open Space

Pi Xiaoyue
Institute of Safety Engineering, Nanjing University of Technology
No. 200 Zhongshan North Road, Nanjing, Jiangsu
pxyphys@gmail.com

Wang Ruizhi, Guo Weiwei, Sun Cheng, Hua Min, Pan Xuhai
[a.] Institute of Safety Engineering, Nanjing University of Technology
No. 200 Zhongshan North Road, Nanjing, Jiangsu
[b.] Jiangsu Key Laboratory of Dangerous Chemical Safety and Control
Nanjing 210009, Jiangsu, China;
xuhaipan@163.com

Prepared for Presentation at
American Institute of Chemical Engineers
1st CCPS Asia-Pacific Conference on Process Safety
Qingdao, China
September 4 – 5, 2013

UNPUBLISHED

Experimental Study on Fan Water Curtain Diluting Water-Insoluble Heavy Gas in Open Space*

Pi Xiaoyue
Institute of Safety Engineering, Nanjing University of Technology
No. 200 Zhongshan North Road, Nanjing, Jiangsu
pxyphys@gmail.com

Wang Ruizhi, Guo Weiwei, Sun Cheng, Hua Min, Pan Xuhai
[a] Institute of Safety Engineering, Nanjing University of Technology
No. 200 Zhongshan North Road, Nanjing, Jiangsu
[b] Jiangsu Key Laboratory of Dangerous Chemical Safety and Control,
Nanjing 210009, Jiangsu, China;

Keywords: Fan water curtain, Heavy gas, Carbon dioxide, Dispersion

Abstract

Experiments were done to research fan water curtain diluting water-insoluble heavy gas (carbon dioxide as an example) dispersion in open space in this paper. In experiments, the influence of water flow pressure, distance between leakage source and water curtain, height of leakage source and squirt angle of fan water curtain on dilution effect were done. Experimental results show that the higher the water pressure of fan water curtain is, the lower the concentration of carbon dioxide behind the water curtain is, and thus, more efficient the dilution effect is. However, when water pressure is no lower than 0.25MPa, the concentration of carbon dioxide behind the water curtain will not continue decreasing. Also, when leakage source height is higher, the concentration of carbon dioxide is higher, and the dilution effect is worse. When water curtain is closer to leakage source, the concentration of carbon dioxide downwind is higher. When the ratio of K, which is leakage source height to the distance of water curtain away from leakage source, is 0.6 and dilution effect is the best. Furthermore, when the squirt angle is 60°, the dilution effect is comparatively good..

1. Introduction

Chemicals have been widely used in all walks of life, along with the vigorous development of the modern chemical industry. Once the dangerous ones leak, they will release a large number of poisonous and harmful gas into the air. Since considerable proportions of the water-insoluble heavy gas's density are greater than that of the air, they will accumulate and diffuse along the earth's surface [1-2], which may cause chain reaction in a very short period, and bring huge

* Supported by Doctoral Fund of Ministry of Education of China (Grant No. 20113221120010); the Major Program of Natural Science Foundation of the Jiangsu Higher Education Institutions of China(Grant No. 09KJA620001); the Science and Technology Support Program of Jiangsu Province, China (Grant No. BE2010740); University Students' Innovative Undertaking Training Program of Jiangsu Province (the key project).

losses. Water curtain can dilute the leakage of toxic gas, separate toxic gas diffusion to the sensitive zones and avoid secondary chain reaction safely and effectively, by diluting and separating the spread of heavy gas clouds in the atmosphere. However, the process of water curtain diluting and separating heavy gas clouds' spread is very complex; thus, people lack exhaustive understanding of its mechanism.

2. Research status and the dilution mechanism

Hansen [3], Morse [4], Spyros [5], etc. find that water curtain reduce the concentration of the gas cloud mainly by mechanical dilution, air entrainment and heat exchange through numerical simulation and experimental study on water curtain inhibiting LNG's (Liquefied Natural Gas) diffusion. Aurelia [6], etc. has carried on field experiments on the small-scale liquid chlorine continuous diffusion, they point out the feasibility and disadvantages of water curtain diluting and separating the spread of heavy gas clouds. Bara[7], etc. find that the water curtain which squirts upward is better at diluting gas's diffusion, while the one squirts downward is better at separating them through the experiment on water curtain diluting and separating ammonia's diffusion. Gavellia [8], Morshed [9], etc. studied on the experiment on diluting and separating LNG's diffusion, test and verify its efficiency according to experimental data and obtain that the air disturbance efficiency made by cone water curtain is higher than that of fan water curtain. Thus, cone water curtain is better at diluting heavy gas clouds near the water curtain, while on the horizontal ground, fan water curtain is better at separating them. Dimbour [10, 11], etc. studied three different types of water curtain (hollow cone, solid cone and fan) which squirting downward, compared the effect of diluting and separating the chlorine's diffusion of these three, find that fan water curtain is best at diluting and separating the chlorine's diffusion.

The mechanism that water curtain diluting and separating gas diffusion mainly includes:

- Barrier and mechanical action: due to the continuity of droplets, water curtain forms barrier so that it can separate the leaked gas from going through the water curtain effectively, disperse clouds diffuse, and increase the mechanical effect of clouds' vertical movements.
- Air entrainment: the momentum transfer between droplets and air will cause them to mix and dilute, and therefore reduces the concentration of gas in downwind position of the water curtain.
- Heat exchange: heat transfer between droplets and cold gas clouds also plays an important role in dispersing the cold gas. It is mainly achieved by improving the gas buoyancy.
- Water curtain's physical and chemical absorption: studies have shown that water curtain can more effectively control the unconstrained leakage of high-solubility gas.
-

3. Experimental study

3.1 The experimental platform

3.1.1 Experimental instruments

Main testing instruments are shown in Tab.1

Table 1. Main apparatus in the experiment

Testing instrument	Manufacturer (model)	Range	Precision
Infrared CO_2 concentration sensor	Beijing Collihigh Sensor Technology Co., LTD (JQAW-4AC)	$0 \sim 5000\times10^{-6}$	$\pm4\times10^{-6}$
Data collecting instrument	Beijing Collihigh Sensor Technology Co., LTD (1-7017)	4 ~ 20mA	±0.02%
Hot-bulb anemometer	Beijing Environmental Protection Instrument Factory (AVM-07)	0 ~ 5m/s	±0.01m/s
Rotameter (gas)	Shanghai Tianchuan Instrument Factory (LZB-25)	$1 \sim 10m^3/h$	Level 2.5
Rotameter (liquid)	Shanghai Tianchuan Instrument Factory (LZB-32S)	600 ~ 6000L/h	Level 4
Pressure gauge	Shanghai Tianchuan Instrument Factory (YTZ-150)	0 ~ 0.6MPa	±0.01MPa

3.1.2 Water curtain system

The water curtain system mainly consists of rotameters, gate valves, pressure gauges and drencher nozzles. It supplies water through the fire pipe. Its flow and pressure is controlled by gate valves. And the nozzle of water curtain is fan-shaped.

3.1.3 Gas release system

The gas release system uses CO_2 as the water-insoluble gas for the experiment, the purity is 99%, the pressure is 10.0 MPa. It connects to a pressure-reducing valve in the outlets for gas cylinder, and the pressure-reducing valve connects to a gas meter externally. The diameter of the pipe's outlet is 6 mm, and the direction of the pipe is horizontal.

3.1.4 Data collection system

The data collection system consists of infrared CO_2 gas concentration sensors, transmitters, data collection instruments and computer data processing systems.

3.2 Experimental schemes and results

We set benchmark condition as follows: the pressure of water curtain was 0.3 MPa; CO_2 was kept continuously and horizontally through a pipe, with the leakage flow of 6 m/h, the height of 1.2 m, and the distance of 2 m from the water curtain; nozzles of fan-shaped. The experiment temperature is 20±5 ℃, sunny,and the wind speed is 0 ~ 0.5 m/s.

3.2.1 Influences of the height of leakage source

When the experiment began, the valve of water curtain was opened, then the gas was kept releasing for 5-10 minutes. Immediately after the experiment was over, the air valve and the water curtain system were shut simultaneously.
The distribution of CO_2 concentration sensor is shown in Fig. 1.

According to Fig. 1, the coordinates of the 6 testing sites are: No. 1 (X = 6 m, Y = 0, Z = 0.3 m); No. 2 (6,0,0.6); No. 3 (6,0,0.8); No. 4 (8,0,0.3); No. 5 (8,0,0.6) and No. 6 (8,0,0.8). (X represents the distance from the testing sites to leakage source, Y represents the deviation distance perpendicular to the XY plane, Z represents the height that the testing sites from the ground). The height of leakage source is 60 cm, 120 cm and 180 cm.

Figure 1: The distribution of CO2 concentration sensors

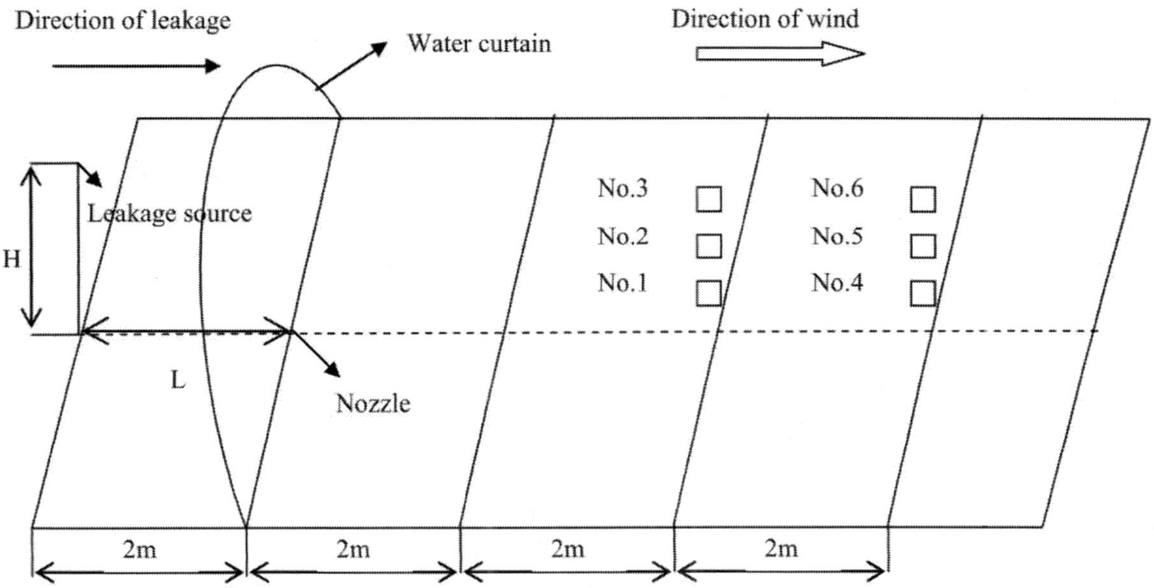

Fig. 2 shows the distribution figure of CO_2 concentration in accordance with time behind the fan water curtain (the solid curve and the hollow one represent results with and without the water curtain respectively.), when the height of the testing sites 1 ~ 3 is different. (Since the regularities of testing sites 4, 5, 6 and 1, 2, 3 are consistent, limited to the paper's space, the former one is no longer illustrated in detail, similarly hereinafter.) Tab. 2 is the average value of CO_2 concentration (ppm) at testing sites No. 1 ~ 3 after complete dispersion. As it is shown in Fig. 2 and Tab. 2, when these testing sites are as the same, the higher the leakage source is, the lower the downwind CO_2 concentration is without water curtain down. But when water curtain is open, the result displays the opposite. This is because the vertical mechanical action of fan water curtain's droplets upward drive gas diffusion upward, and the more close to the top of water curtain, the more momentum droplets lose, the worse of the vertical mechanical action is, the worse the effect of air entrainment is, and the higher the downwind CO_2 concentration is. In addition, fan water curtain's droplets squirt to top from bottom, because gaps between the upper droplets are larger, and it's easy for CO_2 to penetrate. Therefore, to set up water curtain on scene, we should set the height of water curtain close to set that of leakage source.

Table 2. Average value of CO_2 concentration (ppm) at testing sites 1 ~ 3 after complete dispersion

The height of leakage source (m) / Testing sites	0.6 (with/ without water curtain)	1.2 (with/ without water curtain)	1.8 (with/ without water curtain)
No. 1	24.34/3460.91	131.93/3094.03	237.81/2463.42
No. 2	31.53/3326.82	165.06/2867.05	278.81/2269.36
No. 3	8.12/3258.75	92.81/2918.17	117.81/2539.56

Figure 2: Effects of the heights of leakage sources on CO2 concentration (ppm)

Figure 2(a): Testing sites 1

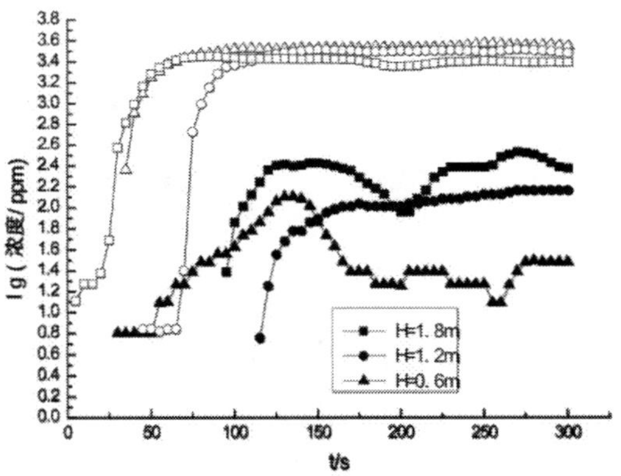

Figure 2(b): Testing sites 2

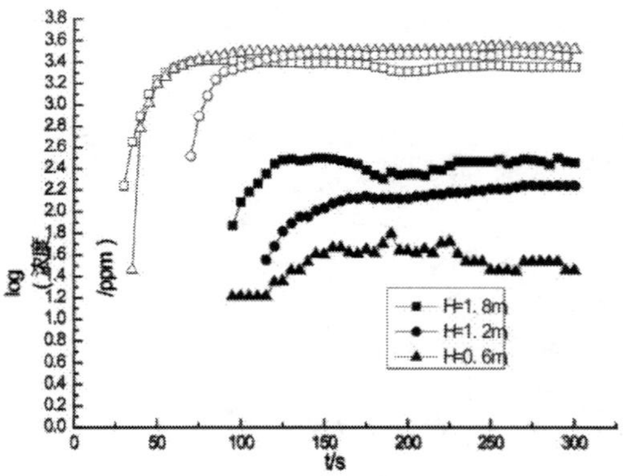

Figure 2(c): Testing sites 3

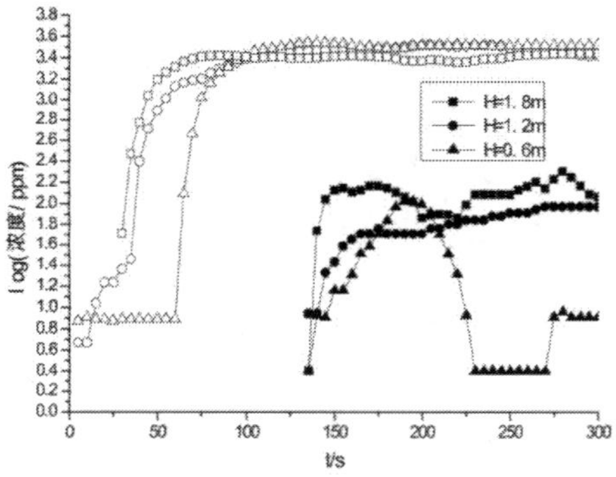

3.2.2 Influences of the pressure of fan water curtain

The benchmark condition, settings of water curtain system and steps of experiment mainly remained the same throughout this section; and the only variable is the pressure of water curtain, which is set to be 0.1 MPa, 0.15 MPa and 0.2 MPa, 0.25 MPa and 0.3 MPa respectively.

Fig.3 illustrates the CO_2 concentration (ppm) distribution of testing sites No.1, 2, 3 in accordance with time when the pressure of water curtain is different. (Y coordinate represents for the exponential of the difference between the value of the CO_2 concentration (ppm) in the experiment and air). Fig. 4 displays Variation diagram of dilution effectiveness in accordance with water curtain pressure. Tab. 3 is average value of CO_2 concentration (ppm) at testing sites 1 ~ 3 after complete dispersion. As it is shown in Fig. 3 and Tab. 3, after open the water curtain system, the CO_2 concentration downwind significantly decreases, which proves that water curtain can effectively inhibit and separate CO_2's diffusion. As time goes on, the CO_2 concentration behind water curtain tends to be stable after a period of increase. Influenced by environmental factors, the CO_2 concentration behind water curtain fluctuates. This requires us to take the average value of CO_2 concentration after complete dispersion at testing sites 1 ~ 3 to achieve a more accurate comparison result. And the result shows an inversely proportional relationship between the pressure of water curtain and the CO_2 concentration for each testing site; the greater the pressure of water curtain is, the lower the concentration of CO_2 downwind tends to be. However, when the pressure of water curtain reaches 0.25 MPa, the CO_2 concentration seems to decline slowly or even remain flat as the pressure of water curtain increases. The reason behind is that increased pressure of water curtain would trigger a chain effect; larger water curtain's droplets, increased air turbulence around water curtain and the air entrainment as well as stronger effect on gas to disperse, and lastly lower the CO_2 concentration downwind. However, when water curtain pressure increases to a certain degree, the dilution that droplets to CO_2 reaches saturation and therefore increasing the water curtain's pressure no longer causes the CO_2 concentration behind water curtain to drop. When there is no water curtain, the lower the testing sites are, the higher the concentration of CO_2 downwind is, which reflects the general law of heavy gas's free diffusion. After opening water curtain, the lower the testing sites are, the lower the CO_2 concentration is behind the water curtain when the pressure of water curtain remains the same. This is due to mechanical actions of droplets themselves can drive gas move upward, and make the CO_2 downwind in relatively high denser. As it is shown in Fig.4, water curtain's dilution efficiency ($\eta = 1 - \frac{C_{forced\_disersion}}{C_{free\_disersion}}$) is within 91% ~ 99%. The dilution efficiency increases in accordance with the water curtain's pressure; however, when the pressure of water curtain increases over 0.25 MPa, it is nearly still; and the lower the testing site's position is, the better the dilution efficiency is. In conclusion, higher pressure of fan water curtain is more likely to cause lower CO_2 concentration behind the water curtain and more efficient dilution; however, when the pressure exceeds the level of 0.25 MPa, its increase no longer function the same way; instead, the CO_2 concentration declines slowly or almost remains the same.

Table 3. Average value of CO_2 concentration (ppm) at testing sites No.1 ~ 3 after complete dispersion

Water curtain's pressure (MPa) / Testing sites	0.1	0.15	0.2	0.25	0.3	Without water curtain
No. 1	167.81	166.76	59.22	32.33	30.66	3262.51
No. 2	188.43	183.67	92.18	49.45	47.54	2866.88
No. 3	228.41	192.43	101.87	74.61	63.68	2573.41

Figure 3: Effect of the water curtain pressure on CO_2 concentration

Figure 3(a): Testing sites 1

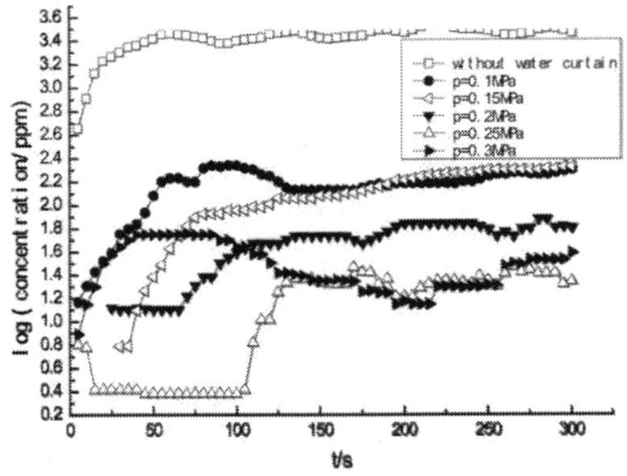

Figure 3(b): Testing sites 2

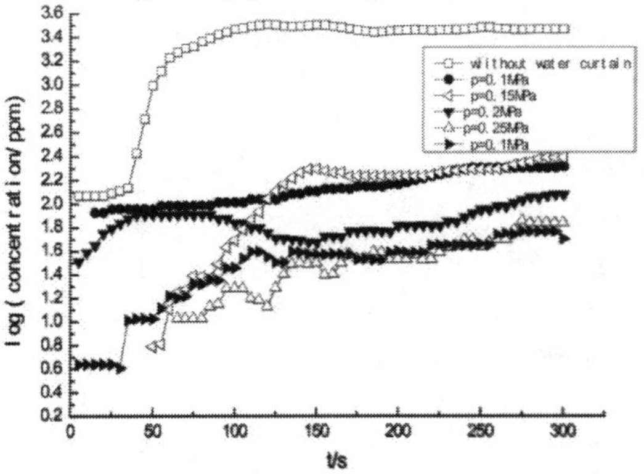

Figure 3(c): Testing sites 3

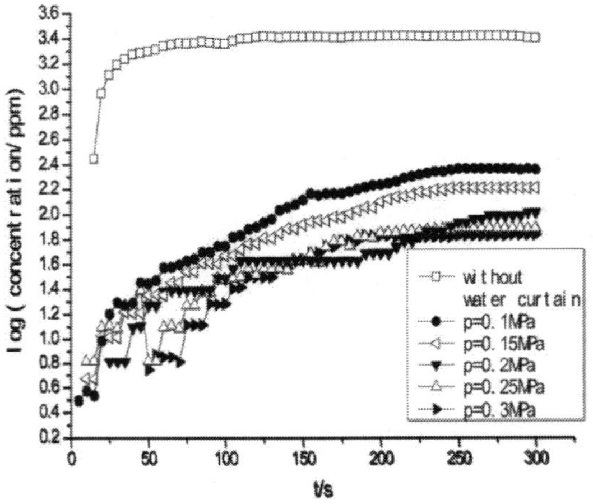

Figure 4: Variation diagram of dilution effectiveness vs. fan water curtain pressure

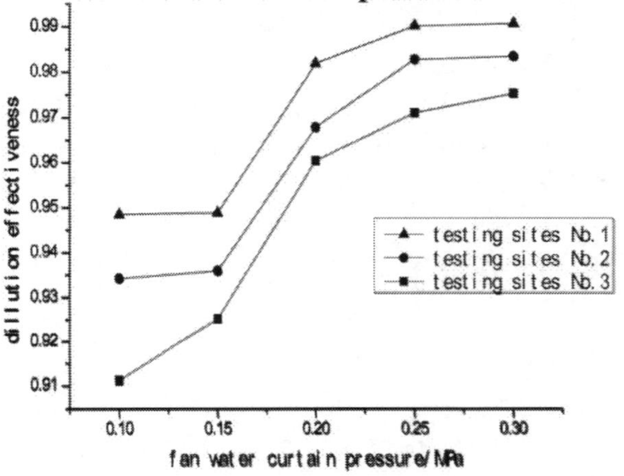

3.2.3 Influences of the distance between leakage source and fan water curtain

① Experiment

The benchmark condition, settings of water curtain system and steps of experiment mainly remained the same throughout this section; and the only variable is the distance between leakage source and water curtain, which is set to be 100cm, 200cm and 300cm respectively.

Fig. 5 illustrates the effect of distance between water curtain and leakage sources on CO_2 concentration (ppm) in accordance with time at testing sites No.1. (Since the laws of results in No.2 and No.3 are the same, they won't be discussed respectively in this paper. Fig. 6 shows the variation diagram of dilution effectiveness distance between water curtain and leakage source under the same circumstance as it is in Fig.5. As it is shown in Fig. 5 and Fig. 6, when the distance is 1 m, 2 m, the average value of CO_2 concentration after complete dispersion is 203.96 ppm, 153.50 respectively; however, in the case of 3 m, the CO2 concentration curve is significantly lower than that of the previous two cases. Its average value is only 19.71 ppm, which shows an obvious improvement on dilution efficiency. This is because the closer water curtain from leakage sources are, the more momentum of CO_2 that reaches the water curtain is, and the droplets cannot disperse CO_2 upwards through their own mechanical actions; thus, CO_2 can penetrate water curtain to reach its rear, making the CO_2 concentration downwind higher.

Figure 5: Effect of distance between water curtain and leakage sources on CO_2 concentration vs. time at testing sites No.1

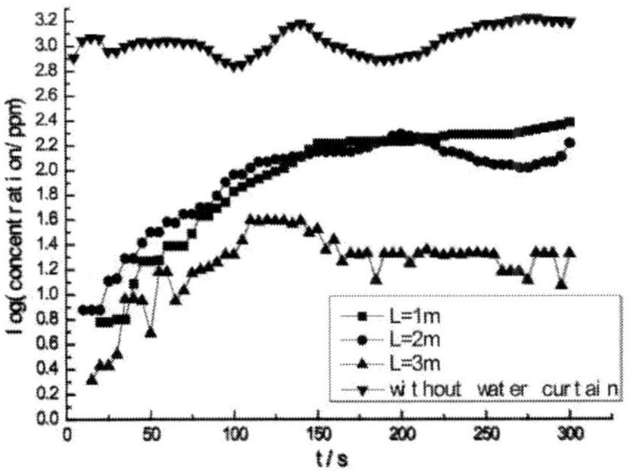

Figure 6: Variation diagram of dilution effectiveness vs. the distance between water curtain and leakage sources

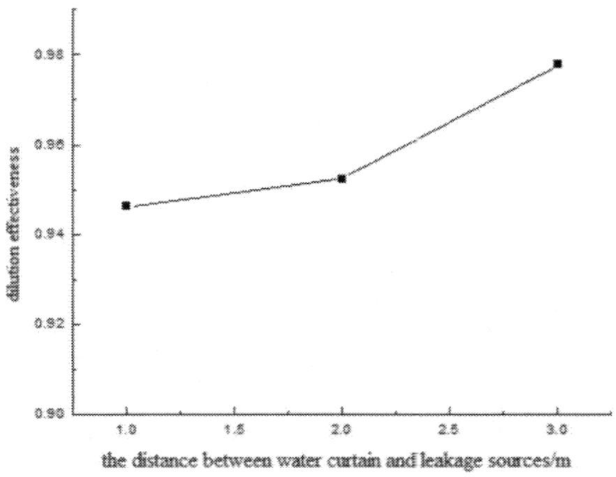

② Experimental study on fan water curtain diluting water-insoluble heavy gas of the dimensionless distance

In this part, we simply nondimensionalize one the factor of fan water curtain , and define the dimensionless distance K as [Eq. 1]:

$$K = \frac{H}{L} \quad \text{[Eq. 1]}$$

Q is the water curtain flow (m^3/h) , q is CO_2's leakage flow (m^3/h), H is the height of leakage source (m), L is the distance between leakage source and water curtain (m).

Table 4. the value of K and corresponding height of leakage source (m) H

Value of K	0.2	0.4	0.6	0.9	1.2	1.8
H(m)	0.6	1.2	0.6	1.8	1.2	1.8

Figure 7: Distribution diagram of CO2 concentration (ppm) vs. leakage flow volume under the different K values

Figure 8: Distribution diagram of CO2 concentration (ppm) vs. K

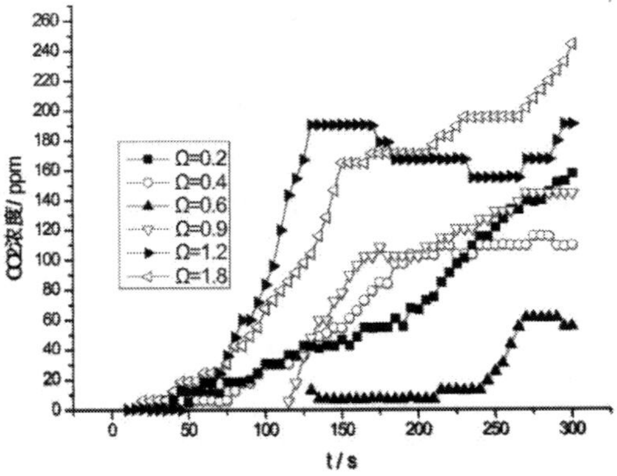

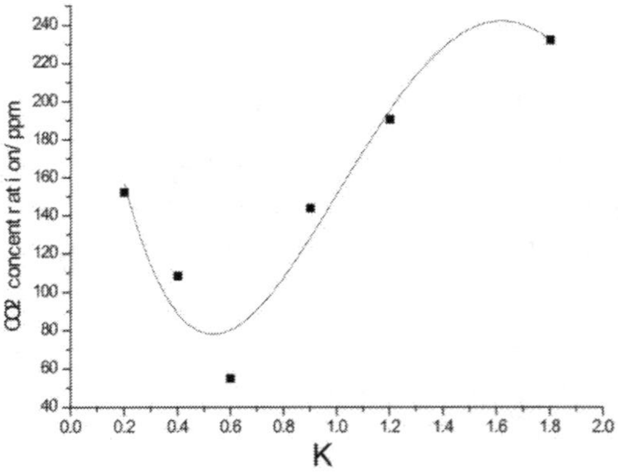

This experiment mainly studies the effects of value K on the CO_2 concentration behind water curtain, and the set of values used in this experiment is shown in Tab. 4.
Fig. 7 displays distribution diagram of CO_2 concentration (ppm) in accordance with time as K varies. The CO_2 concentration behind the water curtain decreases as K increases from 0.2 to 0.6 and increases in the case when K increases from 0.6 to 1.8. Thus, when K = 0.6, the CO_2 concentration behind water curtain remains the lowest, and effects of water curtain dilution is the best.

Table 5. Average value of CO_2 concentration after complete dispersion (ppm)

K	0.2	0.4	0.6	0.9	1.2	1.8
CO_2 concentration (ppm)	152.25	108.37	54.87	143.81	190.40	231.88

Tab. 5 shows the average values of CO_2 concentration (ppm) after complete dispersion taken from Fig. 7. And Fig. 8 displays the distribution diagram of CO_2 concentration (ppm) in accordance with K, fitted by Oringe 8.5, using data in Tab. 5. As it is shown in Tab. 5 and Fig. 8, the closer the K is to 0.6, the smaller the CO_2 concentration is behind water curtain, the better the dilution effect is. Then, when the CO_2 concentration and height of leakage source are known, we can determine the appropriate distance between water curtain and leakage source by referencing Fig. 8, considering the landform, and making K close to 0.6.

3.2.4 Influences of the squirt angle of fan water curtain

We kept the parameters of the benchmark condition unchanged. At the beginning of the experiment, the CO_2 valve was opened to let the gas diffuse free when the began; at the 2nd minute, we opened the water curtain and the software to collect data; and at the 8th minute, we close air supply valve and water curtain.

6 CO_2 concentration sensors were set respectively in the downwind area of the leakage sources. The coordinates of the 6 testing sites are: No.1 (X = 4 m, m, Y = 0 Z = 0.2 m); No.2 (4,0,0.8); No.3 (4,0,1.2); No.4 (6,0,0.2); No.5 (6,0,0.8) and No.6 (6,0,1.2) while the squirt angles of water curtain were 45°, 60°, 75°, 90°, 105°, and 135°. (We define the angle between the section that water curtain squirts and the section that water curtain points to leakage source as the squirt angle. That is, the squirt angle of water curtain that parallels the ground is 0°.)

① **Fig. 9 shows the numeric curves of CO_2 concentration that sensors collected at each testing site under different squirt angles. (Because of the problems that appeared on senor N0.5 in the experiment, its data won't be discussed here. Similarly hereinafter)**

Figure 9: Numeric curves of CO2 concentration (ppm) that sensors collected at each testing sites under different squirt angles

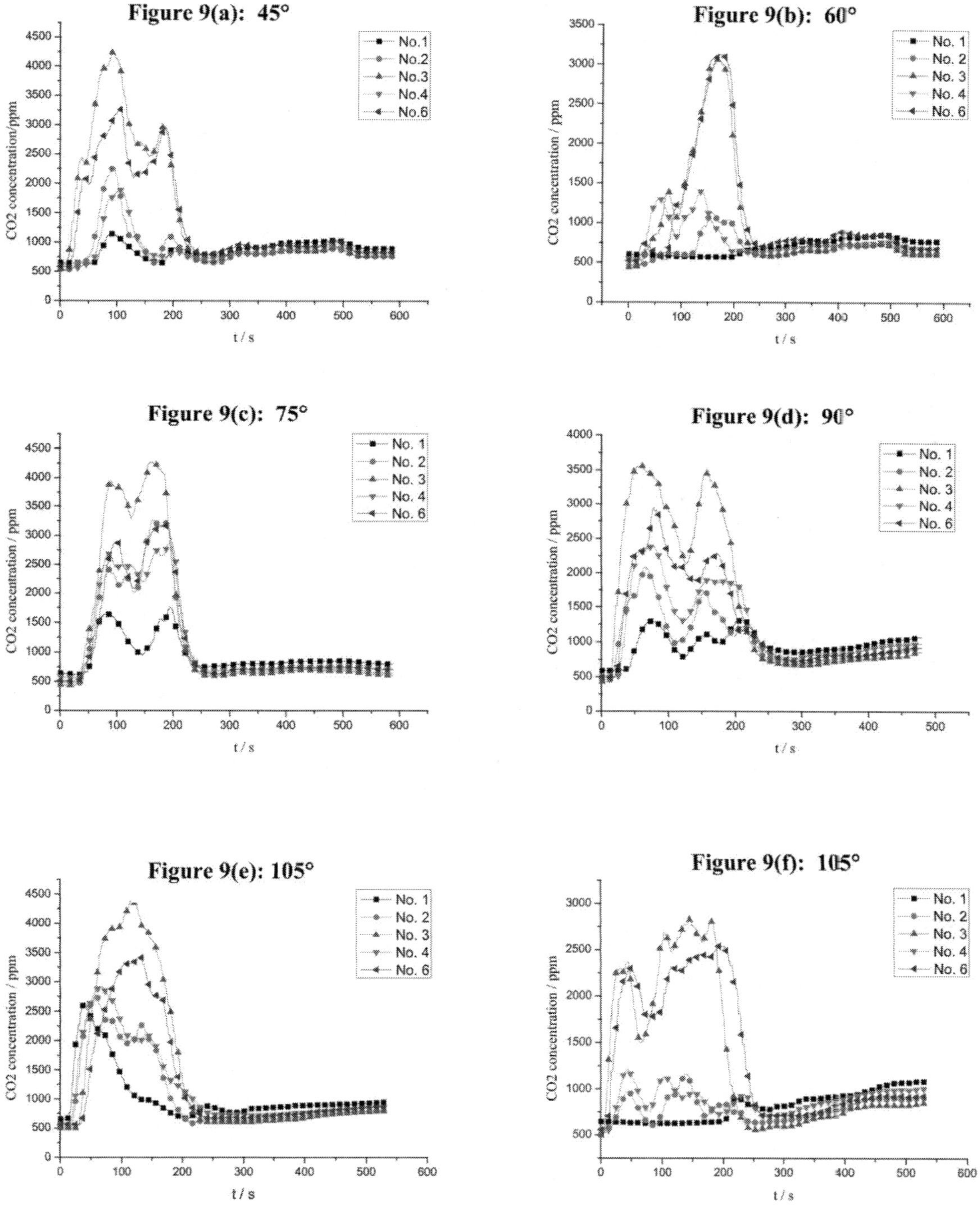

As it is shown in the initial stage in Fig. 9, the CO_2 concentration (ppm) at testing site No.1 is higher than 500ppm, the height of No.2, 3's curve diminishes; that is, the closer the position to the ground is, the higher the concentration of heavy gas is. This is because the density of CO_2 is greater than air, and it is easy to move down to the earth's surface. Also, as it is illustrated that the majority of the height of curve No. 1 and No. 4 is higher than that of No. 3 and No. 6 relatively before 2 minutes, we know that, the closer the position to leakage source is, the higher the concentration of heavy gas is. This shows that when CO_2 diffuses to the surroundings free, the CO_2 concentrations measured in testing sites accords with the general law of heavy gas diffusion. But the sequence changes after opening fan water curtain. This is because of the mechanical action of fan water curtain's droplets themselves, which makes the farther the position to leakage source is, the higher the concentration of heavy gas is in a certain time. After about 200 s from the beginning, all values of CO_2 concentration on sensors' screens falls, and they are stability in the interval $[600\text{ppm}, 1200\text{ppm}]$ after complete dispersion. It is the separate action of fan water system at that time makes the CO_2 concentration gradually reducing, and ultimately achieves a steady state.

From Fig. 9, we can also find that most curves have two peaks. This is because, in the first 2 minutes, CO_2 released and its concentration at each observation site increased uniformly; and after we opened the water curtain at 2 minutes, the mechanical action of droplets themselves may drive gas move upward, which leads to the second rise test point of CO_2 concentration, and gradually formed the second wave. These do contain great potential dangers. If and only if the squirt angle of fan water curtain is 60° and 105 °, there is sole vertex, and the rests all have two or more. Consequently, under the same conditions, compared with other angles' dilutions, these two make the possibility of dangers, which occur after the mixture of heavy gas and air, to be the minimum. However, according to Fig. 9 (e), when the squirt angle is 105°, the maximum value of No. 3 reaches around 4500ppm, which is much greater than the peaks in other angles. As it is illustrated in Fig. 9 (d), when the spray angle is 90°, the CO_2 concentration in some parts of testing sites can finally be greater than 1000ppm after complete diffusion, which is also greater than in the rest Fig.9. This is because the water curtain droplets' mechanical actions will make the surrounding air turbulent, which disperse the CO_2 behind water curtain away. When the squirt angle of water curtain is 90°, the system lacks the velocity of horizontal dispersed flow, and it is not easy for heavy gas to disperse faraway. Thus, different squirt angles have various effects on diffusion of heavy gas, and the water curtain of 90° is not efficient for separating heavy gas's diffusion.

② Fig. 10 shows the numeric curves of CO_2 concentration that sensors collected at each squirt under different testing sites.

Figure 10: Numeric curves of CO2 concentration (ppm) that sensors collected a each squirt angle under different testing sites

Figure 10(a): Testing sites No. 1

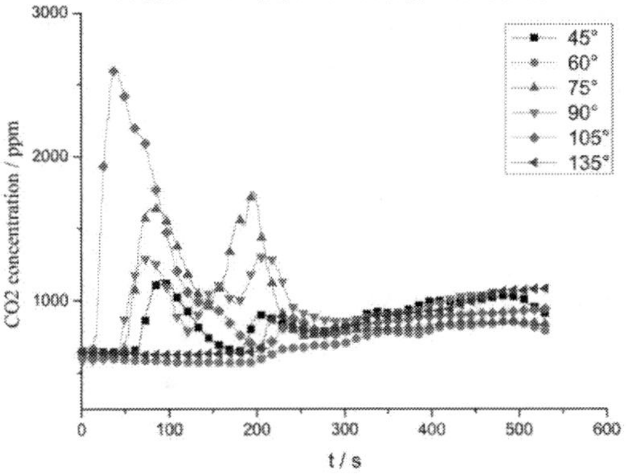

Figure 10(b): Testing sites No. 2

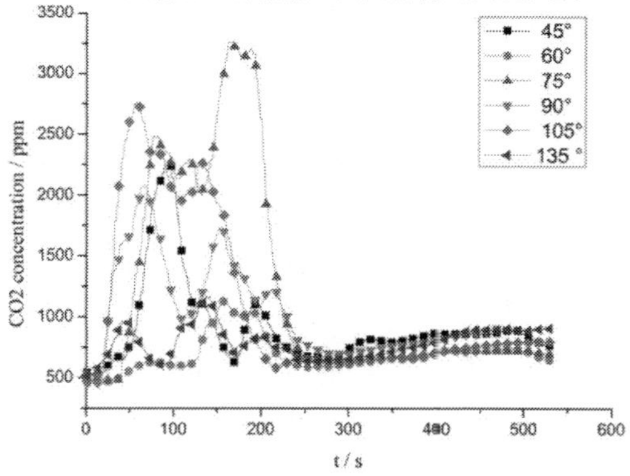

Figure 10(c): Testing sites No. 3

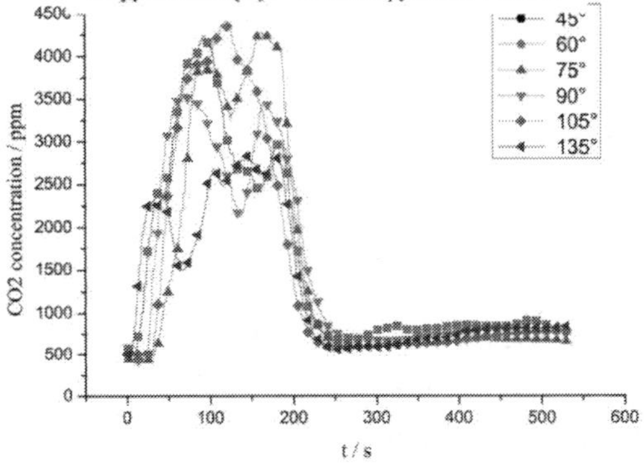

Figure 10(d): Testing sites No. 4

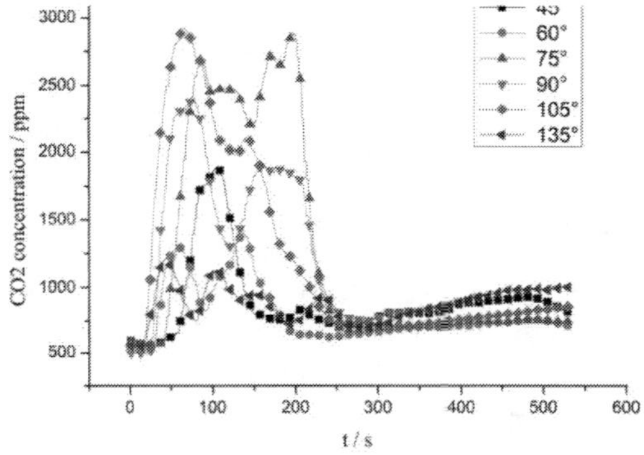

Figure 10(e): Testing sites No. 6

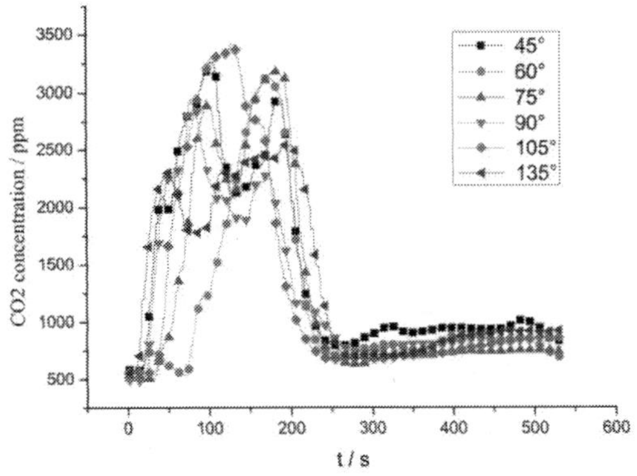

As it is shown in Fig. 10, the first peak values of 105° and the second peak values of 75° at various testing sites are great, that is, the maximum CO_2 concentration of 75° and 105° may easily induce dangers under the same parameter settings. Thus, it is better to choose other angles for the dilution.

Judging from Fig. 10(a), only when the squirt angle is set to 60°, the peak value of CO_2 concentration (ppm) is less than 1000ppm, thus it is the most efficient. From Fig. 10(a), we can find the dilution effect of 60° and 135° the best, since their peak values of CO_2 concentration (ppm) is approximately equal to 1000ppm, which are much smaller than the rest. Fig. 10 (c) illustrates that, when the squirt angle is 135°, the peak value is the lowest, and the time CO_2 used to diffuse completely is the least; thus, the dilution effect is the best at this angle. The effect of dilution is similar when the squirt angle is set to 60° and 135° at testing sites No. 4. As it is shown in Fig. 10(e), water curtain with the squirt angle 60°, 90° and 135° is relatively more efficient in diluting and separating CO_2 after it opens.

Overall, in most cases, the squirt angle of 60° is the most efficient in dilution.

4. Conclusion

- The higher a leakage source is, the higher the concentration of CO_2 (carbon dioxide) downwind is, and the worse a dilution effect is.
- As pressure of fan water curtain increases, the concentration of CO_2 decreases and the efficiency of water curtain dilution goes up. This is because the fan water curtain that squirts upward pushes heavy gas up through vertically mechanical actions, air entrainment, etc for its dilution and separation. However, when water pressure exceeds 0.25MPa, the concentration of carbon dioxide behind the water curtain would stop decreasing.
- When water curtain is closer to leakage sources, the more momentum of CO_2 that reaches the water curtain, CO_2 can penetrate water curtain to reach its rear. And since the droplets cannot disperse CO_2 upwards through their own mechanical actions, the CO_2 concentration downwind becomes higher. Especially when the ratio of K, which is the height of leakage source to the distance of water curtain away from leakage source, is 0.6, the concentration of carbon dioxide behind water curtain is the lowest and dilution effect is the best.
- 60° is the optimal squirt angle when parameters are set similarly to those in the experiment above, considering the scenes' safety and the fan water curtain system's efficiency. Meanwhile, other angles are to be considered to make some auxiliary dilution in case of any specific requirement for the process of dilution.

5. References

[1] PAN Xuhai, JIANG Juncheng. Research on discharging dispersion of chemical dangerous gases and its influence factors [J]. Journal of Nanjing University of Chemical Technology, 2001, 1(1): 19-221

[2] PAN Xuhai, JIANG Juncheng. Numerical simulation research on instantaneous release dispersion of heavy gas [J]. Chemical Engineering (China), 2003, 31(1): 35-391

[3] HANSEN O R, GAVELLI F. Validation of FLACS against experimental data sets from the model evaluation database for LNG vapor dispersion [J].Journal of Loss Prevention in the Process Industries, 2010, 23(6): 57-671

[4] MORSETL, KYT&OMAAHK. The effect of turbulence on the rate of evaporation of LNG on water [J].Journal of Loss Prevention in the Process Industries, 2011, 19 (3): 1-71

[5] SPYROS S, FOTIS R. Validation of turbulence models in heavy gas dispersion over obstacles [J]. Journal of Hazardous Materials, 2004, 108(1/2): 9-201

[6] Aurelia Dandrieux, Gilles Dusserre, James Ollivier, Small scale field experiments of chlorine dispersion [J]. Journal of Loss Prevention in the Process Industries, 2002, 15(1): 5-10

[7] A. Bara, G. Dusserre, The use of water curtains to protect firemen in case of heavy gas dispersion [J]. Journal of Loss Prevention in the Process Industries, 1997, 3(10): 179-183

[8] Gavellia F, Chernovskya M K Modeling of LNG spills into trenches[J]. Journal of Hazardous Materials, 2010, 180(1/2/3): 332-3391

[9] Morshed A. Rana, Benjamin R. Cormier, Jaffee A. Suardin, Experimental Study of Effective Water Spray Curtain Application in Dispersing Liquefied Natural Gas Vapor Clouds[J]. Process Safety Progress, 2008, 27(4): 345-353

[10] J. P. Dimbour, D. Gillbert, A. Dandrieux, G. Dusserre, Assessment of the effectiveness of downward water sprays for mitigating gaseous chlorine releases in partially confined spaces [J], Journal of Hazardous Materials, 2003, A96(2-3): 127-141

[11] J. P. Dimbour, A. Dandrieux, D. Gibert, G. Dusserre, The use of water sprays for mitigating chlorine gaseous releases escaping from a storage shed [J], Journal of Loss Prevention in the Process Industries , 2003, 16 (4): 259-269.

1ST CCPS ASIA-PACIFIC CONFERENCE ON PROCESS SAFETY℠

Research on emergency response system of chemical industrial park on the AHP-Fuzzy comprehensive evaluation method

Xue'e ZHAO
Xin CAI
Han-yu SU
Hong-qi YAO
School of Chemical Engineering,
Zhengzhou University,
Zhengzhou Henan 450001,P.R.China
zhao.xue.e@163.com
caixin1909@163.com
853527360@qq.com
hongqiyao@126.com

Prepared for Presentation at
American Institute of Chemical Engineers
1st CCPS Asia-Pacific Conference on Process Safety
Qingdao, China
September 4 – 5, 2013

UNPUBLISHED

Research on emergency response system of chemical industrial park on the AHP-Fuzzy comprehensive evaluation method

Xue'e ZHAO
Xin CAI
Han-yu SU
Hong-qi YAO
School of Chemical Engineering,
Zhengzhou University,
Zhengzhou Henan 450001,P.R.China
zhao.xue.e@163.com
caixin1909@163.com
853527360@qq.com
hongqiyao@126.com

Keywords: emergency response ,index system,fuzzy-AHP,multilevel valuation,weight

Abstract

The emergency response system of chemical industrial park is giant with complex interaction. This paper aims to establish a model of emergency response system for chemical industrial park through an analysis of influential factors from three perspectives: human resource, material support and information acquiring & dissemination. By applying the Fuzzy-AHP mathematics theory which incorporates qualitative and quantitative method , the model works out weights to determine the importance of each factor, which helps make a judgment of the important factors and weak links. The outcome of multistage fuzzy comprehensive evaluation is *'equal'*. Index figures illustrate degree of influence of the single factor. The paper proposes some targeted suggestions to the improvement of the emergency response of the park.

1. Introduction

Due to the chemical industrial park is involved with frequent accidents by large amount of dangerous chemicals combined with various complicated chemical processes leading to an incalculable damage to the society, economy and environment. The Bhopal chemical leak accident on December 13th, 1984 resulted in 3150 death and millions of people affected. On October 13th, 2005, the Aniline set of dual benzene plant in Jilin Petrochemical Company of CNPC had an enormous explosion caused by dangerous chemicals, 8 deaths,60 casualty and 10,000 emergency evacuations. The economic loss is enormous . Therefore, there has been an urgent need to improve the emergency response ability of chemical industrial park.

The emergency response system of chemical industrial park is a giant system. Factors of this complex system are fuzzy and random through connection and interaction. Fuzzy comprehensive evaluation method[1], incorporated the quantitative result of analytical hierarchy process and comprehensive evaluation of fuzzy mathematics[2], can not only avoid the subjective arbitrariness produced by some expert's personal opinion, but also get rid of the errors caused by the uncertainty of fuzzy mathematics.

Based on the AHP - Fuzzy mathematical theory, the paper studies the emergency response system of chemical industrial park. According to the actual situation of the chemical industrial park, it is established that multi-hierarchy index system for emergency response. The paper calculates weight of key elements by using the AHP which is a combination of qualitative and quantitative method .

In this paper,it is proposed that the model of two-stage fuzzy multifactorial evaluations which is used to assess the emergency response system of chemical park. It is shown that this study can get grasp affiliation of the pivotal elements and judge the weaknesses in the emergency response process timely . We can make preparation for the further improvement of the emergency response system and enhance the efficiency of emergency rescue action.

2. Establishment of emergency response hierarchy model

Multi-hierarchy index system is a set of measurable data to describe and evaluate an event. There exists a comprehensive and systematic multi-hierarchy index system in favor of describing the content and characteristics of the chemical emergency response. The model considers multi-hierarchy index system as its base. Meanwhile, The function of model aims to dissolve complex system into some functional components and visualize each component systematically.

Referred to the Capability Assessment for Readiness (CAR) and evaluating programs of managing capability on disasters of Japan, the conventional evaluated system includes three consideration: whether it covers all key factors affecting emergence response action, whether it can measure and evaluate performance in advance from existing database and whether it can conduce to solve problems in reality. Combined with relative safety statistic analysis material [3-5] and experts opinions, the hierarchy model is established from three perspectives: human resource (H), material support (M) and information acquisition &dissemination (I). Base on these perspectives above, 21 sub-factors are considered to build the safety evaluating index system for chemical industrial park, which is shown in blank of Table-1.

Table1. Elements& sub-factors for emergency response system

Elements	Sub-factors
Human resource(H)	Emergency decision- making capacity (H1) Evolution assessment capacity (H2) Command and coordination capacity (H3) Ability to dealing with the aftermath (H4) Medical assistance capability (H5)

	Fire disposal capacity (H6) Emergency drill capability (H7) Self-rescue capability (H8) Evacuation decision-making capacity (H9)
Material support(M)	Monitoring and warning system (M1)
	Communication system (M2)
	Medical care devices (M3)
	Fire fighting equipment (M4)
	Park shared system of relief supplies (M5)
	Emergency Special funds (M6)
	Park road network of emergency evacuation (M7)
	Personal protective equipment (M8)
Information acquiring & dissemination(I)	Hazardous information database (I1)
	Feedback of Emergency drill (I2) Dangerous information dissemination (I3) Information sharing at all levels of the emergency center (I4)

Human resource indicator includes all personnel-related influential factors[6]. The performance capabilities of discriminating, physiological and psychological in emergency are seriously considered. For example, the organization and coordination ability of commanders , the site disposal capacity of rescuers, the self-rescue ability for operations personnel in dangerous area,etc. This section is classified as nine sub-indicators according to critical capabilities of all the major functional departments which are closely related with the emergency operations.

Material support indicator refers to the necessary supplies in the emergency rescue[7], such as Special funds, communication equipment, fire equipment, medical devices and so on. It includes eight sub-indicators.

Information acquiring and dissemination indicator is used to measure the processing capabilities of all types of information in hazardous disposal process and the information sharing of hazardous substances . It contains four sub-indicators.

These factors are distinguished and ranked according to predefined performance. Figure 1 explains the AHP model.

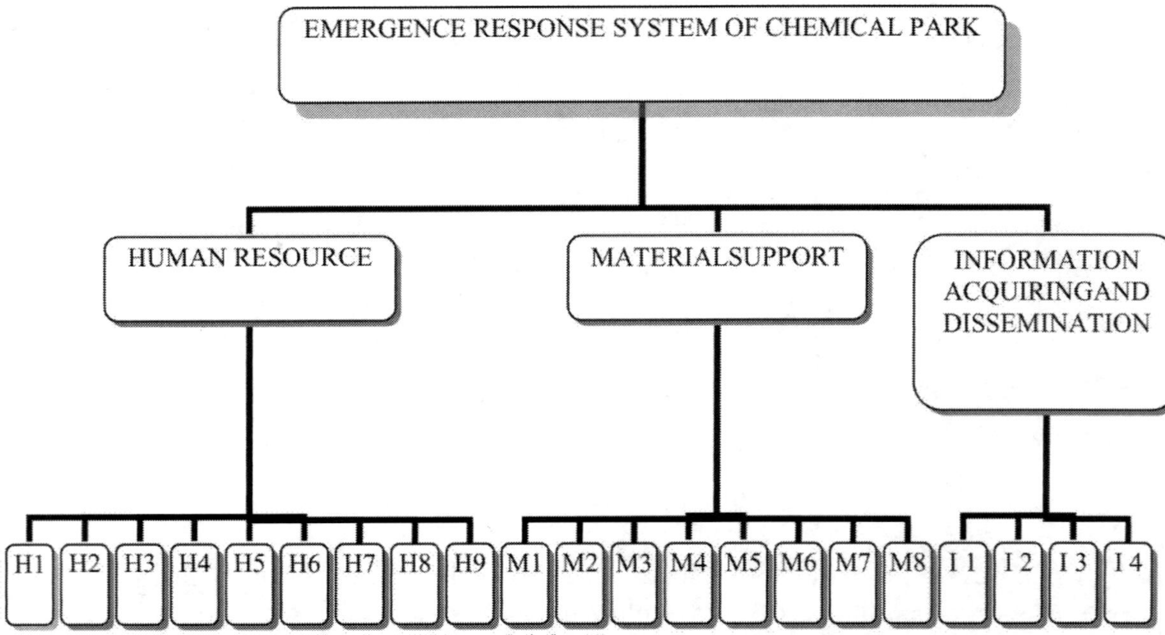

Fig. 1.AHP model for Emergence response

Figure 1 is comprised of three levels for emergence response system. Level 1 stands for the whole. Level 2 represents the three elements: human resource (H), material support (M) and information acquisition &dissemination (I)and level 3represents 21 sub-factors.

3 AHP-Fuzzy comprehensive Evaluation Method theory

The AHP-Fuzzy comprehensive evaluation method aims to calculate the importance of factors in the emergency response system .The mechanism can be described with three steps. Firstly, fuzzy pair-wise comparison judgments are obtained from specialistic team. Secondly, the qualitative descriptions and subjective judgments is quantified. Finally, memberships are counted to distinguish factors on different levels in the system[8].

3.1 Establishment of baseline set and review set

All major factors in emergency response system are classified into collection. The first grade indicators are belongs to baseline set ,namely ,$U=\{U_1, U_2, \ldots, U_n\}$.Each baseline set can be divided into m subsets ,that is to say, $u_i=\{u_{i1}, u_{i2}, \ldots, u_{im}\}$,(i=1,2,…,n).

After decision making team discusses, which includes experts and emergence managers of chemical industrial. The most prevalent and important criteria for emergence response are selected. [9]. Factors are designed as a questionnaire for survey. The team will evaluate the survey and reach a conclusion as review. In general, review set is divided into five levels to judge. It is shown in Table2.

Table2. Wise comparison between two elements(adapted from Saaty [10])

Preference Weights	Definition	Explanation
1	Perfect	Two activities contribute equally to the objective

3	Moderately perfect	Experience and judgments slightly favor one activity over another
5	Equal	Experience and judgments strongly or essentially favor one activity over another
7	Moderately imperfect	An activity is strongly favored over another and its dominance demonstrated in practice
9	Imperfect	The evidence favoring one activity over another is of the highest degree possible of affirmation
2,4,6,8	Intermediate values	Used to represent compromise between the preferences listed above

3.2 Establishment of comparative judgments

According to the results of pair-wise comparison, the values generate the fuzzy judgments matrix.Supposing u_{ij}(i=1,2…,n; j=1,2,…,m) means the *i-th* element and the *j-th* grade and corresponding weight is a_{ij}. So the weight set of u_{ij} is expressed as A={a_{i1},a_{i2},… ,a_{im}}. With the Analytic Hierarchy Process , the factors' weights are calculated. Moreover, they should meet the following conditions: *a*.0<Ai≤1; *b*. $\sum_{i=1}^{n} A_i = 1$ (i=1,2,…,n).

3.3 Hierarchical fuzzy evaluation

Single-stage fuzzy comprehensive evaluation method measures the impact of various factors of the emergency response system. B_i=A_i°R_i, [Eq. 1]the sign of '°' is equal to the synthetic operators .The computational procedure is considered that the factors' weight set matrix multiply membership matrix. The expression can be written as $b_{ik} = \sum a_{ij} r_{ijk}$.

3.4 Measurement of consistency

The consistency ratio is an important validating parameter in AHP. In this paper, the acceptable CR is 0.1 because of n ≥ 5[12]. If the value of CR is equal to, or less than that value, it represents that the evaluation within the matrix is acceptable or indicates a good level of consistency in the comparative judgments. These steps to calculate the consistency ratio are shown as follow [13]:

1. Compute the eigenvector or the relative weights and λmax for each matrix of order *n* ;

2. Calculate the consistency index for each matrix of order n by the formulate:

CI = (λmax− n) /(n − 1) [Eq. 2]

3. The consistency ratio is then calculated by the formulate:

CR = CI/RI [Eq. 3]

4 Application of the model for emergency response system

4.1 Conclusion of factors' weight

This step allows the pair-wise comparisons from the decision team to enter into data matrices. It is shown in Table3.Weight of all 21criteria for emergence response have been obtained by using FAHP and the results are tabulated in the 2nd and 4th column in the Table4.

Table 3. Pair -wise comparison matrix

Elements form level Π	H	M	I
H	1	2	3
M	1/2	1	1
I	1/3	1	1

Sub-elements of H	H1	H2	H2	H4	H5	H6	H7	H8	H9
H1	1	6	1	3	4	9	5	7	8
H2	1/6	1	1/5	1/3	1/2	3	1	1	2
H3	1	5	1	2	3	7	4	6	7
H4	1/3	3	1/2	1	1	6	2	4	5
H5	1/4	2	1/3	1	1	5	1	3	4
H6	1/9	1/3	1/7	1/6	1/5	1	1/4	1/2	1
H7	1/5	1	1/4	1/2	1	4	1	2	3
H8	1/7	1	1/6	1/4	1/3	2	1/2	1	1
H9	1/8	1/2	1/7	1/5	1/4	1	1/3	1	1

Sub-elements of M	M1	M2	M3	M4	M5	M6	M7	M8
M1	1	1/2	1/4	1/5	1/2	1/4	1/2	1
M2	2	1	1/2	1/3	1	1/2	1	2
M3	4	2	1	1/2	2	1	2	5
M4	2	3	1	1	2	2	2	6
M5	3	1	2	1/2	1	1/2	1	2
M6	2	2	2	1/2	2	1	2	4
M7	1	1	1/2	1/2	1	1/2	1	2
M8	1	1/2	1/5	1/6	1/2	1/4	1/2	1

Sub-elements of I	I1	I2	I3	I5
I1	1	3	9	3
I2	1/3	1	7	1
I3	1/9	1/7	1	1/5
I4	1/3	1	5	1

In theory, the matrix have a unique maximum eigenvalue.However, eigenvalues and eigenvectors are difficult to be computed precisely [14]. Using the split-root method, it is feasible to obtain the approximate solution .W represents the value of eigenvectors of all elements and W_i (i=1,2,…,n) stands for factors' weight . The relationship between W and W_i follows as $W=(W_1, W_2, ..., W_n)$.

In conclusion,the multiply factors' weight are assigned as follows: W={0.5473 , 0.2409 , 0.2118};W_H={0.2327 , 0.0458 , 0.1179 , 0.0656 , 0.1463 , 0.2248 , 0.0949 , 0.0325 , 0.0231};W_M={0.0487 , 0.0976 , 0.2011 , 0.0978 , 0.0923 , 0.1733 , 0.0449};W_I={0.5166 , 0.3021 , 0.0417 , 0.1396}.

4.2 Multilevel fuzzy comprehensive valuation

After evaluation of special-team was obtained ,Table4 summarizes the weights and memberships of emergence response system factors. The results of memberships are tabulated in the 5th to 9th column in Table4.

Table4. Weights and memberships of factors of emergence response plan

Elements	Weights	Factors	Weights	Membership				
				m_5	m_4	m_3	m_2	m_1
Human resource	0.5473	H1	0.2327	0.2	0.4	0.3	0.1	0.0
		H2	0.0458	0.3	0.2	0.3	0.2	0.0
		H3	0.1179	0.1	0.4	0.3	0.1	0.1
		H4	0.0656	0.1	0.4	0.4	0.3	0.1
		H5	0.1463	0.2	0.4	0.3	0.1	0.0
		H6	0.2248	0.0	0.5	0.3	0.2	0.0
		H7	0.0949	0.2	0.1	0.4	0.2	0.1
		H8	0.0325	0.3	0.2	0.3	0.2	0.0
		H9	0.0231	0.1	0.2	0.4	0.1	0.2
Material support	0.2409	M1	0.0487	0.2	0.3	0.2	0.3	0.0
		M2	0.0976	0.3	0.2	0.5	0.0	0.0
		M3	0.0978	0.1	0.4	0.4	0.0	0.1
		M4	0.2443	0.3	0.5	0.2	0.0	0.0
		M5	0.2011	0.0	0.1	0.5	0.2	0.2
		M6	0.0923	0.2	0.3	0.4	0.0	0.1
		M7	0.1733	0.2	0.1	0.4	0.3	0.0
		M8	0.0449	0.1	0.1	0.5	0.2	0.1
Information acquiring and dissemination	0.2118	I1	0.5166	0.1	0.2	0.3	0.3	0.1
		I2	0.3021	0.1	0.3	0.2	0.4	0.0
		I3	0.0417	0.3	0.4	0.3	0.0	0.0
		I4	0.1396	0.2	0.4	0.4	0.0	0.0

Comment:m_1-m_5 stand for Perfect, Moderately perfect, Equal, Moderately imperfect, Imperfect

The value of CR is 0.06271,less than 0.1. So the comparative judgments should be considered consistency.The membership data of sub-factors of levels II are entered into fuzzy-assemble matrix. It is shown as follows.

$$R_A = \begin{bmatrix} 0.2 & 0.4 & 0.3 & 0.1 & 0.0 \\ 0.3 & 0.2 & 0.3 & 0.2 & 0.0 \\ 0.1 & 0.4 & 0.3 & 0.1 & 0.1 \\ 0.1 & 0.4 & 0.4 & 0.3 & 0.1 \\ 0.2 & 0.4 & 0.3 & 0.1 & 0.0 \\ 0.0 & 0.5 & 0.3 & 0.2 & 0.0 \\ 0.2 & 0.1 & 0.4 & 0.2 & 0.1 \\ 0.3 & 0.2 & 0.3 & 0.2 & 0.0 \\ 0.1 & 0.2 & 0.4 & 0.1 & 0.2 \end{bmatrix}$$

$$R_B = \begin{bmatrix} 0.2 & 0.3 & 0.2 & 0.3 & 0.0 \\ 0.3 & 0.2 & 0.5 & 0.0 & 0.0 \\ 0.1 & 0.4 & 0.4 & 0.0 & 0.1 \\ 0.3 & 0.5 & 0.2 & 0.0 & 0.0 \\ 0.0 & 0.1 & 0.5 & 0.2 & 0.2 \\ 0.2 & 0.3 & 0.4 & 0.0 & 0.1 \\ 0.2 & 0.1 & 0.4 & 0.3 & 0.0 \\ 0.1 & 0.1 & 0.5 & 0.2 & 0.1 \end{bmatrix}$$

$$R_C = \begin{bmatrix} 0.1 & 0.2 & 0.3 & 0.3 & 0.1 \\ 0.1 & 0.3 & 0.2 & 0.4 & 0.0 \\ 0.3 & 0.4 & 0.3 & 0.0 & 0.0 \\ 0.2 & 0.4 & 0.4 & 0.0 & 0.0 \end{bmatrix}$$

The single-stage fuzzy evaluation result is matrix R_1:

$$R_1 = \begin{bmatrix} 0.13893 & 0.36717 & 0.31344 & 0.15128 & 0.03246 \\ 0.1797 & 0.26502 & 0.37576 & 0.1158 & 0.06372 \\ 0.1223 & 0.26647 & 0.27582 & 0.28375 & 0.05166 \end{bmatrix}$$

Then, the first-index weight matrix multiply the matrix R_1. The result of two-stage fuzzy evaluation is B={0.145229,0.321234,0.320485,0.17079,0.044057} .Finally, the scores are calculated as 0.7324181 by weighted method. In accordance with Table5,the membership levels is *equal.*

Table5. Weight value and standard score

Levels	1	2	3	4	5
Weighted value	1.0	0.85	0.70	0.50	0.10
Standard score	0.900~1	0.80~0.899	0.60~0.799	0.40~0.599	<0.400
Preference	Perfect	Moderately perfect	Equal	Moderately imperfect	Imperfect
Comment	Weighted value and standard score depend on experience.				

In order to improve the performance of emergency response , factors whose weight are greater must be noticed. In table4, the result of weights implies that the emergency decision- making capacity, the fire disposal ability of participants , the fire fighting equipment and hazardous information database maybe play an important role.

At the same time, attention should be paid to the weak links which are caused by smaller values in two-stages fuzzy valuation. Therefore , the emergency drills should be taken into account prudently. The skill of self-rescue and the ability of risk decision-making should be trained. In addition,the monitoring and warning system should be improved.

5. Conclusion

This paper establishes a emergency response hierarchy model and proposes a systematic approach using Fuzzy-AHP ,which is used in identifying all key factors from the qualitative and quantitative angle. Pair-wise comparison is applied to measure the dependency of element. Considering weights in factors of all levels, the conclusion of multilevel fuzzy comprehensive valuation is *equal* . The results show that the model has the capability to choose important issue in the flexible emergency response. The approach could help in proposing some targeted suggestions to the improvement of the emergency response of the park.

References

[1] Saaty T L. "A Scaling Method for Priorities in Hierarchical Structures." *Journal of Math. Psychology*, 1997;

[2] Sheu J-B. "A hybrid fuzzy-based approach for identifying global logistics Strategies."*Transp Res* 40,2004;

[3] DENG Yun-feng, ZHENG Shuang-zhong, LIU Gong-zhi; " Study on city emergency capability assessment system."*China Academy of Safety Science and Technology.*Vol1,No. 6, Dec, 2005;

[4] Wu Zong-zhi ,LIU Mao; " Important accident emergency rescue system and preparedness Introduction." Metallurgical industrial press, 2003;

[5] Chen G H, Zhang X M; " Fuzzy-based methodology for performance assessment of emergency planning and its application."*Journal of Loss Prevention in the Process Industries,* 2009,22(2);

[6] Markowski A S, Mannan M S, Kotynia A,et al. "Uncertainty aspects in process safety Analysis".*Journal of Loss Prevention in the Process Industries*, 2010,23(3);

[7] Green L V, Kolesar P J; "Improving emergency responsiveness with management Science." *Management Science*,2004, 50(8);

[8]LIMei-xia. "A NewMethod for Improving the Consistency of the Comparison Matrix in AHP."*Systems Engineering-Theory& Practice*, No. 2, 2002;

[9]Tam MCY,Tummala VMR, "An application of the AHP in vendor selection of a telecommunications system."*Omega, Int J Manage Sci* ,2001,29(2);

[10] Saaty TL, "Fundamentals of decision making and priority theory."2nd edu.RWS, PA ,2000;

[11] Laarhoven PJM, Pedrycz W "A fuzzy extension of Saaty's spriority theory.*"Fuzzy Sets Syst* 11,1983;

[12]Cheng EWL, Li H, " Information priority-setting for better resource allocation using analytic hierarchy process (AHP)." *Inf Manage Compute Secure* 9(2),2001;

[13]A. Noorul Haq · G. Kannan , "Fuzzy analytical hierarchy process for evaluating and selecting a vendor in a supply chain model."*Int J Adv Manuf Technol* ,29,2006) ;

[14]Chang D-Y "Application of the extent analysis method on fuzzy AHP."*Eur J Oper Res* 95(3),1996;

1ST CCPS ASIA-PACIFIC CONFERENCE ON PROCESS SAFETY℠

Integrated Alarm Management and Human Factor Analysis for Pipeline Industry

(Paper number: 306040)

Yu-tao ZHANG
Hai-qing WANG
Hang WANG

China University of Petroleum （Huadong）
Department of Safety Science and Engineering, Qingdao, 266580, China
wanghaiqing@upc.edu.cn

Prepared for Presentation at
American Institute of Chemical Engineers
1st CCPS Asia-Pacific Conference on Process Safety
Qingdao, China
September 4 – 5, 2013

UNPUBLISHED

Integrated Alarm Management and Human Factor Analysis for Pipeline Industry

(Paper number: 306040)

Yu-tao ZHANG, Hai-qing WANG*, Hang WANG

China University of Petroleum （Huadong）, Department of Safety Science and Engineering, Qingdao, 266580, China

wanghaiqing@upc.edu.cn

Keywords: alarm management; pipeline industry; control room; human factors

Abstract

To improve the operation excellence of oil & gas pipeline, a new design scheme of integrated alarm management is proposed, which includes a plan for control room management, a project for alarm rationalization and a solution for several typical alarms in pipeline industry, according to the API-1167. Alarm rationalization is an important step in alarm program. HCR model is applied in priority assignment to get a reasonable priority. An idea is discussed that control room was classified three levels to help pipeline operator to do an efficient and effective job.

Background

In 2010, American Petroleum Institute (API) published API 1167 - Recommended Practice for Pipeline SCADA Alarm Management, which is the first standard for alarm management in pipeline industry. This document is intended to provide pipeline Operators with recommended industry practices in the development, implementation, and maintenance of an alarm management program [1].

Before this document, there are other referenced standards for Alarm Systems and Alarm Management, as is showed in Table1.

Table1. Referenced Standards for Alarm Systems and Alarm Management

Document	Institution	Year	Main Content
CRR166	**Health and Safety Executive**	**1998**	**Describe the current situation and measures of the alarm management for process industries**
EEMUA-191	**Electrical Engineering Manufacturers Users Association**	**1999 first edition 2007 second edition**	**Make the alarm management for process industries standardization Propose to manage the human factor firstly**
The Alarm Management Handbook	**PAS Corporation**	**2006**	**A simple method of alarm management in engineering practice**
ISA-18.2	**International Society of Automation**	**2008 2009 ANSI Adoption**	**make a comprehensive description of the life circle and detailed function of alarm management**
API-RP-1167	**American Petroleum Institute**	**2010**	**Recommended Practice for Pipeline SCADA Alarm Management**

We find the EEMUA-191-Alarm Systems-A Guide to Design, Management and Procurement to

* Corresponding author. +86-0532-86983500 Ext. 8516; E-mail address: wanghaiqing@upc.edu.cn; This research is supported by the Excellent Scholar Project of China university of petroleum (Huadong) with No. SD2013211153

be the facto world standard for a long time before ISA-18.2-Management of Alarm Systems for the Process Industries is published. Then API-RP-1167 is published for pipeline industry, and it is more detailed, more comprehensive, and more professional.

1. Introduction

1.1 Definition

An alarm is a visible or audible means of indicating to the controller an equipment malfunction, process deviation, or other abnormal condition requiring a response [2]. The alarm has three basic functions:

- Notifying the controller that an abnormal change has occurred;
- Communicating to the controller the nature of the change and possible causes;
- Directing the controller to take the appropriate corrective action.

Alarm management is the processes and practices for determining, documenting, designing, operating, monitoring, and maintaining alarm systems [2]. The alarm management has three basic targets:

- In normal situation, provide enough information of the alarm;
- In abnormal situation, provide the most important distinguished information of the alarm;

Try best to avoid the excessively alarm, nuisance alarm, chattering alarm, non-annunciated alarm, and alarm flood.

1.2 Alarm Philosophy

An alarm philosophy is a comprehensive document that covers the proper ways to define, design, implement, maintain, monitor, and test an alarm system. Typically the alarm philosophy will cover the topics of the alarm management review process, the management of change control, roles and responsibilities, SCADA(Supervisory Control and Data Acquisition) specifics, alarm setting guidelines, alarm handling criteria, system monitoring, testing, documentation control, and training. Successful organizations have found that the process of creating the alarm philosophy can be as valuable as the document itself, as it brings all the views, opinions, and methodologies of all the employee resources into a common framework. It becomes the corner stone of the alarm management program as it develops into a guideline, rule book, interpretation guide, and the center for rationalization.

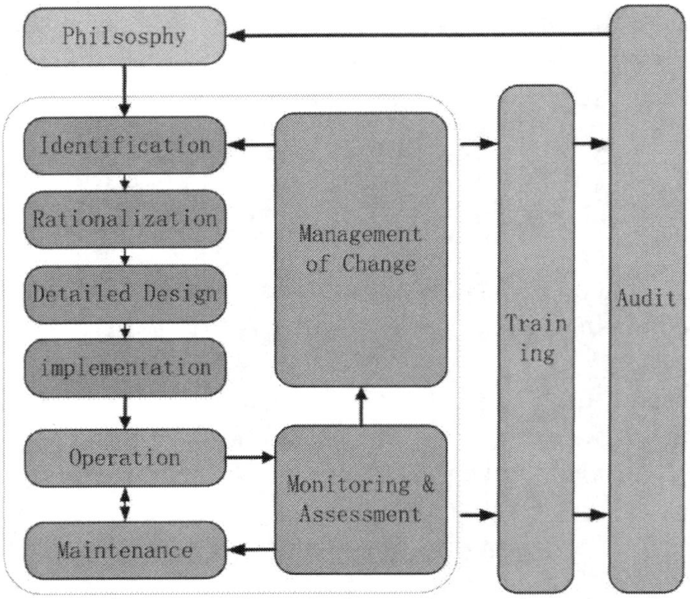

Fig1. The Life Circle of Alarm Management

2. Alarm Rationalization

Rationalization is the method by which existing or proposed alarms are reviewed and altered in order to be consistent with an operator's alarm philosophy or with recommended practices [3]. Rationalization should be implemented for the following purposes:

- To identify and reduce duplicate alarms;
- To ensure proper and meaningful alarm setpoints and priorities;
- To provide detailed alarm information for use by the controllers;
- To create the master alarm database.

2.1 A Method of Priority Determination

The process of rationalization is a process of bad actor cleanup, by analyzing and determining the alarm priority. An advanced method of priority assignment is presented, which combines the available response time and the severity of consequence.

2.1.1 The Available Response Time

The maximum response time of an alarm is the time between the alarm occurrence and the unavailable consequence happen, during which the controller action, the operator action, and sometimes they are one person. As response need person action, it should take human factors into consideration.

The Human Cognizance Reliability (HCR) model classified the human cognitive behavior into three types: knowledge based, rule based, and knowledge based [4]. Each types of cognitive behavior have different parameters. As the HCR model defines, the probability of human error could be fitted by Weibull distribution function in this formula:

$$P = \exp\left\{\frac{-(t/T_{1/2}) - \gamma}{\alpha}\right\}^{\beta}, \quad \frac{t}{T_{1/2}} \geq \gamma \tag{1}$$

P - The probability of non-response;
$T_{1/2}$ - Adjusting the median response time;
α – Feature response time;
β – Shape parameter;
γ – Normalized minimum response time

The formula can figure out the probability of non-response, which is not we want. Turn the formula back, we get this formula:

$$t = -T_{1/2} * \left(\gamma + \frac{\alpha}{\beta} * \ln P \right) \tag{2}$$

So there is an available probability of non-response, there comes an available response time. What is more, the HCR model think the Performance Shaping Factor (PSF) will have influence on the median response time, and provide three important PSF: the experience of the operator, the stress level, the level of human-machine interface. Through the PSFs, the median response time $T_{1/2}$ can be amended close to field condition. The formula is:

$$(T_{1/2})_{amend} = T_{1/2} * (1 + K_1) * (1 + K_2) * (1 + K_3) \tag{3}$$

2.1.2 The Severity of Consequence

The six categories of the severity of accident consequence are: Personnel Safety, Equipment Loss, Public Impact, Environmental Impact, Financial Loss and Down-time. The weight is given to each severity of each impact category as the Table2 showed:

Table2. Recommended Weight of the Severity of Consequence

Severity / Impact Category	NONE	MINOR	MAJOR	SEVERE
Personnel Safety	0	0.5	50	100
Equipment Loss	0	0.5	5	30
Public Impact	0	0.5	1	50
Environmental Impact	0	0.5	1	50
Financial Loss	0	0.5	5	30
Down-time	0	0.5	5	30

2.1.2 Priority Assignment

How to combines the available response time and the severity of consequence? There is weight stand for the severity of consequence, then what stand for the available response time? This is Mean Time to Response (MTTR). The relation between MTTR Factor and the available response time is showed in Table3. The score to assignment alarm priority is figured out by this formula:

$$\text{Score} = (\text{MTTR Factor}) * \Sigma (\text{Weight}) \tag{4}$$

Table3. The Relation between MTTR Factor and Response Time

Response Time	MTTR Factor
>30min	0.64
15—30min	0.8
5—15min	1
<5min	1.2

Table4. The Priority Assignment Principle and Recommended Distributions

Priority	Score	Recommended Distributions	Alarm Display

Critical priority	>300	<1%	Connect to The ESD
Priority 1(P1)	50-300	5%	Shrill Sound, Flashing Display Record
Priority 2(P2)	5-50	15%	Audible Sound, Display, Record
Priority 3(P3)	1-5	80%	Display, Record
Diagnostic priority	0-1	<1%	Record

2.2 Dynamic and State Based Alarming

The method of priority assignment is applied to the static alarm, but the dynamic and state based alarming should be investigated individually. This is not a simple activity and takes a solid understanding of the operational and control philosophies of the facilities during all relevant states [5]. During this stage you:

- Identify time of alarm floods
- Determine how to sense the initiating event
- Identify tags that alarmed during the flood
- Determine what is normal during the event
- List problems
- List available tags
- Engineer new alarms
- Build new alarms
- Remove old alarms
- Suppress non-relevant alarms

Care must be taken to test the logic prior to implementation to ensure that unsafe conditions are not generated as a result of alarm suppression.

2.3 Key Performance Indicators （KPIs）

Alarm management has a life cycle and is not a one-time project. Continuous performance monitoring helps to identify new opportunities for improvement, such as dynamic alarm strategies. Some Key Performance Indicators is applied during the operating, for example, how busy the controller is with respect to alarms, the burst rate or the maximum rate that alarms are received by the operator [2]. The following table is taken from the standard API-RP-1167.

Table5. Alarm KPI Summary

Alarm Performance Metrics per Controller Position Based upon at least 30 days of data		
Metric	**Target Value**	
Annunciated Alarms per Time:	**Target Value: Very Likely to be**	**Acceptable Target Value: Maximum**
Manageable Annunciated Alarms Per Day per Controller Position	**~150 alarms per day**	**~300 alarms per day**
Annunciated Alarms Per Hour per Controller Position	**~6 (average)**	**~12 (average)**
Annunciated Alarms Per 10 Minutes per Controller Position	**~1 (average)**	**~2 (average)**
Metric	**Target Value**	

Percentage of hours containing more than 30 alarms	**~<1%**
Percentage of 10-minute periods containing more than 5 alarms	**~<1%**
Maximum number of alarms in a 10 minute period	**10 or less**
Percentage of time the alarm system is in a flood condition	**~<1%**
Percentage contribution of the top 10 most frequent alarms to the overall alarm load	**~<1% to 5% maximum, with action plans to address deficiencies.**
Quantity of chattering and fleeting alarms	**Zero, action plans to correct any that occur.**
Stale Alarms	**Less than 5 present on any day, with action plans to address**

At the start of any alarm management program, chattering alarms tend to contribute heavily to the controllers alarm load. These are alarms that recur with a significant frequency and have the potential to mask important alarms from the operator. A KPI to measure the number of chattering alarms may be useful at the beginning of an alarm management program, but less useful when the program has been established. KPI's should be specific to the application being improved.

3. The Three-Level-Control Room Alarm Management system

3.1 Similarities and Differences between the Process Industry and the Pipeline Industry

For the most part both process industry and pipeline industry have centralized control rooms that are used by operators/controllers to run the facilities. This may seem somewhat trivial, but both industries measure the raw process variables of pressure, flow, temperature, level, and status to provide the controller with information about the operational condition of their facilities. Both have piping, pressure relief, safety valves, rotating equipment, processing facilities (yes, the natural gas transmission industry has refrigeration plants, straddle plants, joule Thomson cooling plants, compression, salt cavern storage, propane-air, dehydration, and LNG plants) [5].
One of the basic differences between the process industry and the pipeline industry is that there is no fence around the pipeline. Pipeline systems are long, linear, facilities that tend to cover long distances. This basic difference introduces several nuances. The remote locations associated with pipelines, often mean unmanned facilities. The distance factor results in the need to travel and increases time to respond when a field operator's attention is required.

3.2 The Three-Level-Control Room Alarm Management System in Pipeline Industry

As the result of the peculiarity of pipeline, we start to see a sharing of role and responsibilities among central control room, area control room and station control room. The framework and application process of three-level-control room is shown in Fig2.
The architecture of the control systems is different with a Distributed Control System (DCS) tending to exist in process facilities and SCADA systems in the pipeline control center. Remote locations in the pipeline industry tend to have a HMI application coupled with either Programmable Logic Controller (PLC) or Remote Terminal Unit (RTU) applications. That is called station control room. There are great majority of alarms that should be responded without delay by the controllers in station and should not be represented to the senior controllers. However, some alarms happened between station and station, for example, overpressure or low

pressure before getting into station and water hammer should be represented to the controllers in area control room. If the alarm isn't responded by both station and area control room, it should be sent to central controllers. Central controllers also handle significant alarms, such as shut down along the whole line.

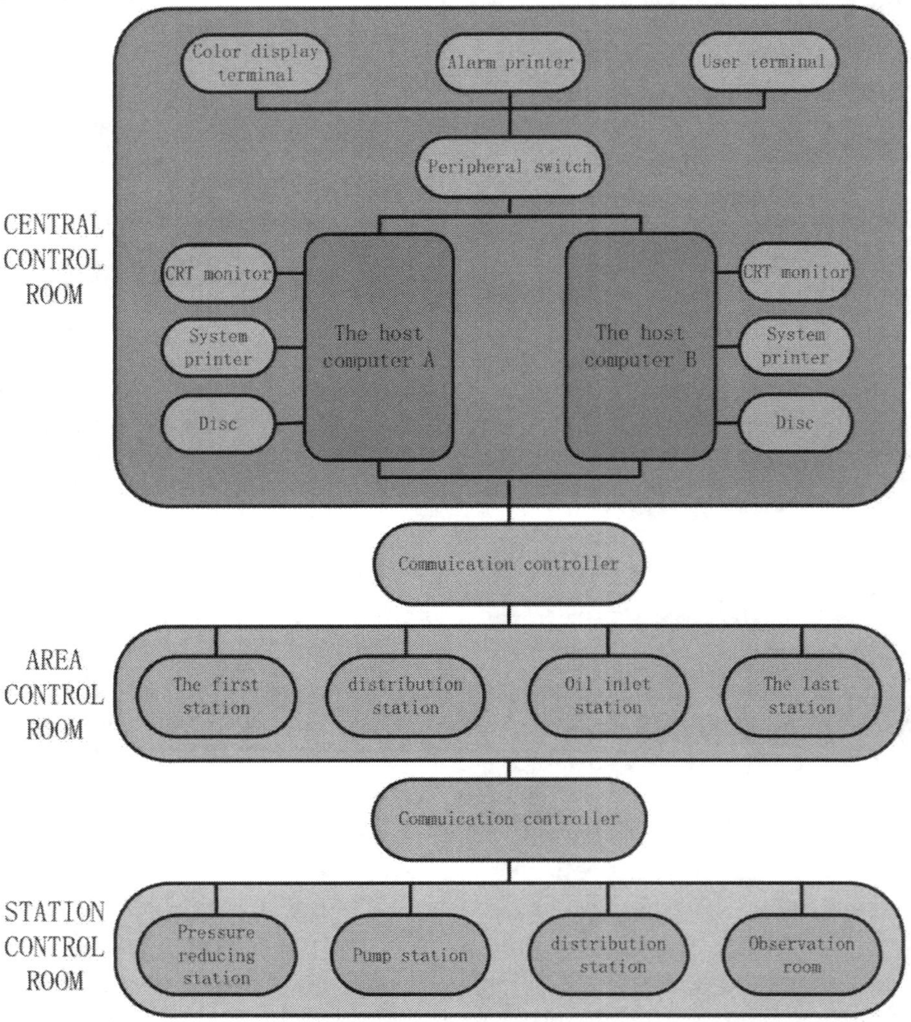

Fig2. The Framework and Application Process of Three-Level-Control Room

3.3 Water Hammer Alarm in Three-Level-Control Room

Water hammer alarm is common in pipeline control room, which caused by power cut, the pump stop by mistake, the storage tanks switched, the valves switched, and do serious harm to the pipeline by sending pressure wave. At present, there are some common protect measures like automatic protection (including regulating valve, pump and relief valve), pre-protection (intercept the pressure wave) pump stopping and so on [6] [7].

The three-level-control rooms have different authority to implement a protect measures. For example, an oil pipe is 150km long, Φ273 diameter, and the maximal output is $200m^3/h$, the design pressure is 6.4MPa.

Table6. Water Hammer Alarm in Three-Level-Control Room

Monitor Site	Alarm Parameter	Response Measure	Control Room
	>5.5MPa	**Regulating valve adjust**	**Station Control Room**

Out of station	**>5.6MPa**	**Relief valve turn on**	**Automatic**
	>5.8MPa	**Pre-protection turn on**	**Area Control Room**
	>5.85MPa	**Pump stopping controlled by PLC**	**Area / Central Control Room**
	>5.9MPa	**Pump stopping on the whole pipeline**	**Central Control Room**
	<0.2MPa	**Regulating valve adjust**	**Station Control Room**
Out of pump	**>6.2MPa**	**Pre-protection turn on**	**Area Control Room**
	>6.25MPa	**Pump stopping controlled by PLC**	**Area / Central Control Room**
	>6.3MPa	**Pump stopping on the whole pipeline**	**Central Control Room**
Get in station	**<0.15MPa**	**Pre-protection turn on**	**Area Control Room**
	<0.12MPa	**Pump stopping controlled by PLC**	**Area / Central Control Room**
	<0.1MPa	**Pump stopping on the whole pipeline**	**Central Control Room**

4. Conclusion

This paper is contributed to build an integrated alarm management, referring to the life circle of alarm management in API-1167. An advanced method of priority assignment is presented, which combines the available response time which figured out from the HCR model and the severity of consequence. A system is discussed that control room was classified three levels, which help the pipeline operate more efficient and effective. Alarm management is a durable processing, the performance of alarm system should be measured and audited, then understand your starting point and what needs to be improved upon.

5. References

[1] Yang Fan, Xiao De-yun. Research Topics of Intelligent Alarm Management [J].Computers and Applied Chemistry, 2011(12) (In Chinese)

[2] API-1167-2010.Recommended Practice for Pipeline SCADA Alarm Management. American Petroleum Institute, 2010

[3] Rothenberg D H. Alarm Management for Process Control. New York: Momentum Press, 2009

[4] He Xu-hong, Huang Rui-xiang. Human reliability analysis in engineering system: theory, methods and applications [M].Beijing: Qinghua university publishing house, 2007(In Chinese)

[5] Mark K McTavish. Alarm Management for Pipelines.7th International Pipeline Conference.IPC2008-64450

[6] Zhang Lin-cheng, Ma Bo-tao. Water Hammer and Pre-protection in Oil Pipeline between Zhongyuan and Luoyang [J].Technology and Life, 2010(3) (In Chinese)

[7] Zhao Yun-lei, Zhang Guang-bin. Analysis and Control of Water Hammer in Long Distance Crude Oil Pipeline [J].Technology and Life, 2012(12) (In Chinese)

1ST CCPS ASIA-PACIFIC CONFERENCE ON PROCESS SAFETY℠

Study on Advanced Petrochemical Safety Technology with Optimal Design of Fire & Gas System (F&G)

(Paper number: 306041)

Shengmei Shang
Haiqing Wang
Fang Liu

China University of Petroleum (Huadong)
Safety Science and Engineering Department
Qingdao, Shandong, 266580, China
wanghaiqing@upc.edu.cn

Prepared for Presentation at
American Institute of Chemical Engineers
1st CCPS Asia-Pacific Conference on Process Safety
Qingdao, China
September 4 – 5, 2013

UNPUBLISHED

Study on Advanced Petrochemical Safety Technology with Optimal Design of Fire & Gas System (F&G)

(Paper number: 306041)

Shengmei Shang
Haiqing Wang*
Fang Liu

China University of Petroleum (Huadong), Safety Science and Engineering Department
Qingdao, Shandong, 266580, CHINA
wanghaiqing@upc.edu.cn

Keywords: Petrochemical safety technology; F&G; Optimal design; Detectors layout

Abstract

For petrochemical facility with high fire and explosion risk, the paper proposes a new optimal design solution of F&G that belongs to safety instrumented systems (SIS), and the proposed solution is explored from both system design and detectors layout sides. The integrated F&G of petrochemical industry can reduce management and maintenance costs; whilst the optimized layout of detectors can strengthen the prevention of fire and explosion accidents by improving the F&G coverage rates and thus has a strong impact on safety, environment or business of petrochemical industry. This advanced technical scheme will reduce the serious consequences of accidents and make people take emergency responses measures in time. The optimal design caters to the modern concept of win-win and the demand of production safety and economic benefits.

1. Introduction

In the petrochemical industry, both a large amount flammable liquids , superheated liquids and other dangerous chemicals are stored, and complex high-risk chemical production facilities and process , such as high temperature , high pressure etc. are contained, and thus large area fires are easily spread. Besides, greater risks of device collapse and explosion are resulted because of complex fire and explosion accidents forms, disasters evolution and dissemination process. F&G belong to "mitigate safety layer" of petrochemical process industry for monitoring fire explosion and leakage accident of fire and combustible gas, and it also can carry out alarm and certain fire extinguishing function as safety control system. As a new system that integrates fire with combustible gas detection, F&G is not widely used in the country currently [1].

2. Optimal design of F&G integration

* Corresponding author. +86-0532-86983500 Ext. 8516; E-mail address: wanghaiqing@upc.edu.cn .This research is supported by the Excellent Scholar Project of China university of petroleum (Huadong) with No. SD2013211153

Currently the relevant mandatory requirements have not yet been put forward clearly by state departments, so the general designing institutes and the user don't understand clearly the concept of F&G and usually use ordinary civilian fire alarm combined with DCS (Distribution Control System) for gas detection and alarm instead of designing and setting the F&G as a safety system independently. However, in big international projects, the independent F&G is generally designed by the equipment of manufacturers with dedicated safety system, and thus the integration plant with comprehensive safety system is constituted by the SIS and F&G [1].

2.1 Safety layer of process industry

As an independent safety layer of process industry, the petrochemical F&G belongs to the active protection layer and is defined as safety instrumented category in the IEC [2]. Layers of protective measures are set in the process industry in order to prevent the occurrence of accidents, and the typical protective measures is composed by safety layers as shown in Figure 1, while the F&G is used for fire and gas detection and safety control of SIS.

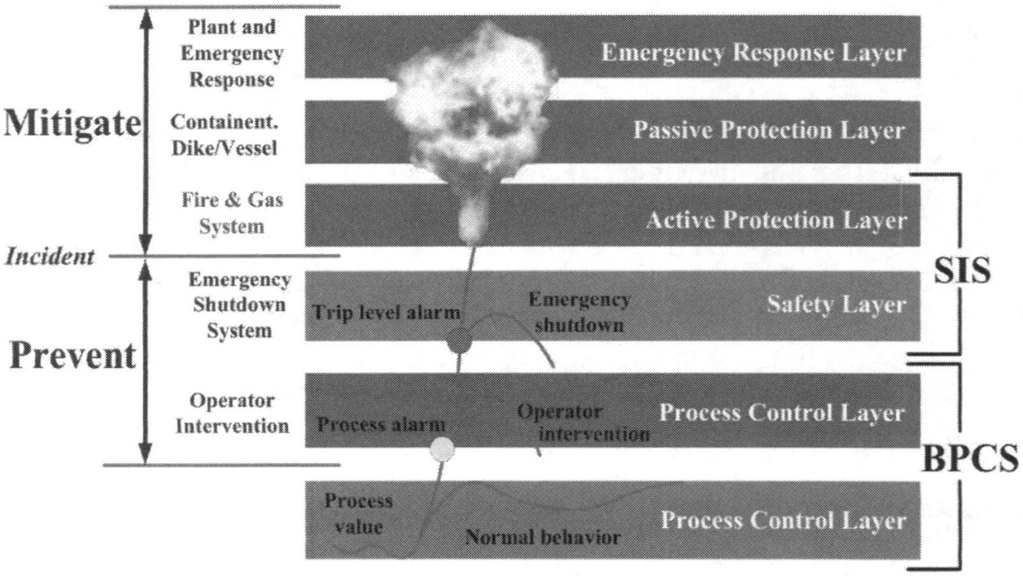

Figure 1 Safety layer of process industry

Under normal conditions, by the DCS, the process is controlled and operated reposefully with the design parameters, and the production need to be adjusted to meet the regular production control, and thus the basic process control system (BPCS) is formed. Through the emergency shutdown system (ESD), the process is always protected from exceeding the acceptable safety limit, that is a limit line of safe operation of the BPCS, and it will be an unacceptable abnormality situation if the process is beyond the safety limit. However, when the process is out of control or a disaster occurs, the F&G will prevent or delay the situations development to make the loss to lowest [2]. The F&G not only has very direct relationship with the DCS, ESD and automatic fire alarm system, but also has close association with other ancillary facilities, as shown in Figure 2.

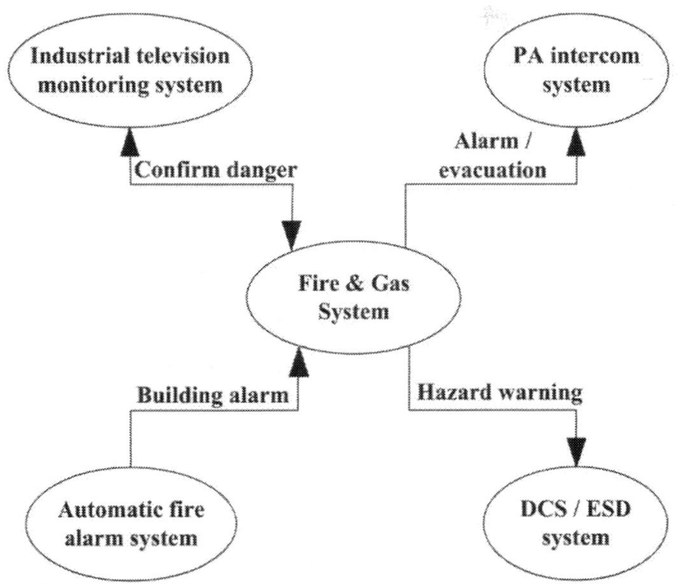

Figure 2 Relationship between the F&G and each other system

2.2 Traditional F&G

Traditional F&G is usually independent outside of Distribution Control System (DCS) and connects the alarm panel by hard-wiring. Furthermore, the alarm location and fire pump are displayed through the signal light on the skeleton map of protective area. At present the fire alarm and the detection alarm of toxic or combustible gas are set on different alarm panels in most petrochemical plants, and thus there is more than one panel that needs to be monitored. For joint device, the traditional alarm panels of combustible gas are dispersed in different control rooms, which are not convenient to manage and maintain.

As shown in Figure 3, the traditional F&G was set in every independent production unit, where respective DCS was corresponded through redundant serial communication, and alarm signal had to be connected to the alarm panel of respective control room by hard-wiring, and thus alarm panels in every integrated building need to be monitored. It is more inconvenient that redundant Ethernet interface must be configured in each F&G and should be linked to redundant central switch via fiber optic communication. HMI between emergency response center and fire brigade communicate with various F&G via Ethernet, therefore the alarm status of each detector must be always monitored [1].

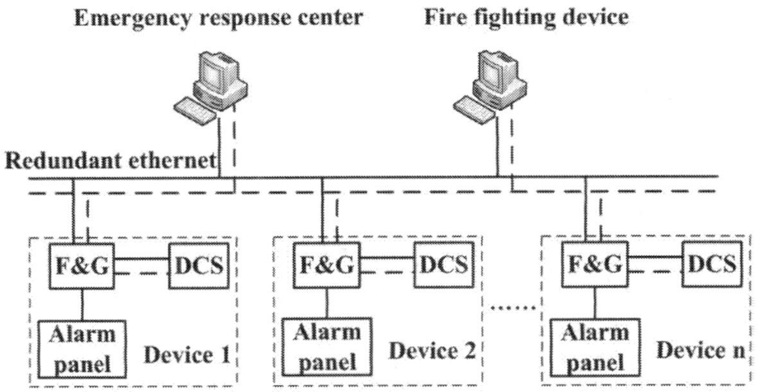

Figure 3 Traditional F&G

2.3 Integration of the F&G

The integrated control and safety protection strategies are carried out by the modern F&G. Through FTE public network, the controller of DCS, safety management system and other devices are seamlessly connected into the integrated control and safety technology solutions by F&G platform, as shown in Figure 4. In this way, the independent system such as DCS, Emergency Shutdown System (ESD), and F&G etc. will use a unified operating station, which will make the operational information, system diagnostics and alarm / SOE list display in the same window. Therefore, the modern F&G can reduce the operation and maintenance costs as well as improve safety and reliability of the petrochemical plant.

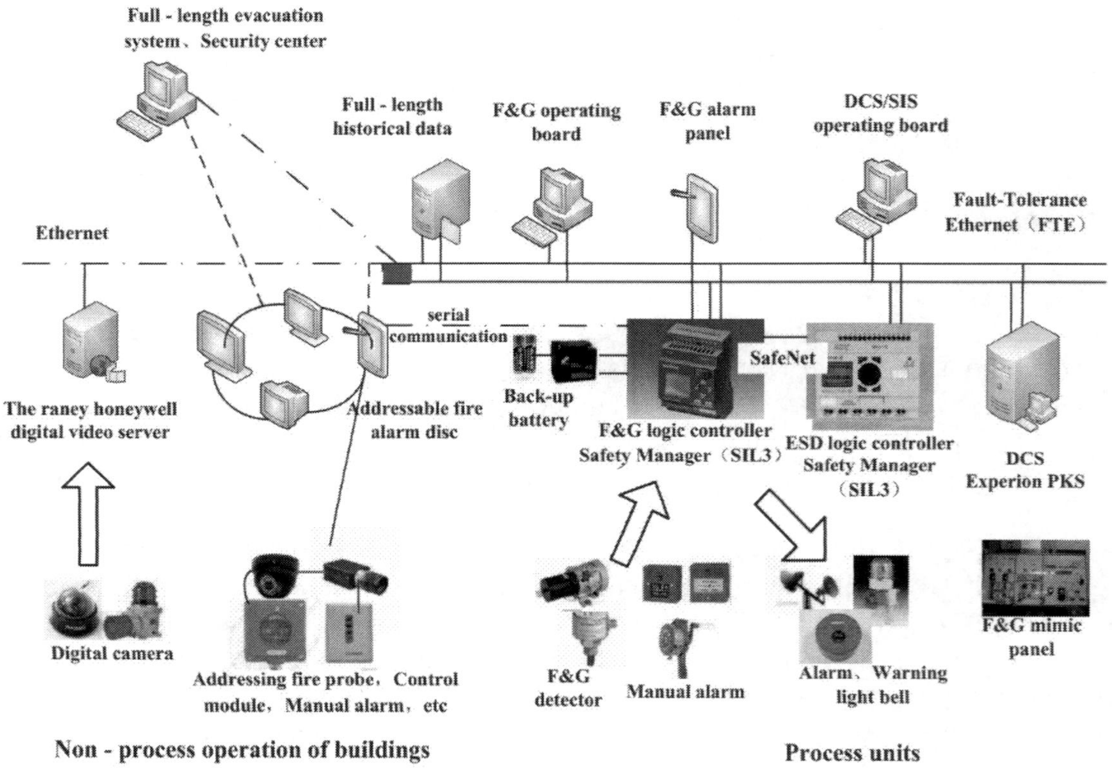

Figure 4 Integrated F&G

3. Optimal layout of the flame detectors

As shown in Figure 4 , the whole system is divided into three parts: Local detectors(combustible gas detectors , thermal detectors, flame detectors, etc.) ; logic controllers of local cabinet interior; local executive component (deluge valve , fire hydrant , alarm , etc.) ; alarm panel of the central control room, fire alarm panel ,public alarm, etc. The local detectors are most important parts of the F&G, but the staff of petrochemical industry always cannot discover process anomalies in time, and a large fire or even explosion occurs because of inadequate numbers, unreasonable layout and false alarms of detectors. Therefore, it is very important to optimize the layout of detectors to ensure the function of the F&G in petrochemical plants. But the concept of optimized detectors layout has not been proposed in the existing “standards”, furthermore, the method of lookup limit curve is multi-valued and very complicated to determine detectors layout spacing. Besides, the lookup table method and traditional experience method lay out detectors at a fixed spacing, which are rigid and not flexible enough to insure the hazardous zone achieves detectors coverage rate. Therefore, the above methods are all difficult to optimize the layout of detectors [3]. In this paper, the layout of flame detectors were optimized in the detection zone that is 18m in length, 10m in width and 5m in height.

3.1 Coverage indicators

Article 2.2.1 in the “Design Standard on Electrical Device in Explosion and Fire Hazard Zone” GB50058-92: Explosive atmospheres zoning should be divided according to the following provisions based on the frequency and duration of appearance of explosive gas mixture [4]:
Zone0: The environment where explosive gas mixtures appear continuously or chronically;
Zone1: The environment where explosive gas mixtures may appear during normal operation;
Zone2: The environment where explosive gas mixtures never appear or even appear for only a short time during normal operation.
Computational formula of the detector coverage rate V is defined as follows:

$$Vi = \frac{Si}{S} \quad (i=1, 2, 3) \tag{1}$$

Vi: $V1$ indicates that the detectors coverage arrived two or more layers; $V2$ indicates that the detectors coverage arrived only one layer, $V3$ indicates the dead zone of detectors (i.e. zero coverage layer, undetectable area);
Si: $S1$, $S2$ and $S3$ indicate the area of two or more coverage layers, the area of only one coverage layer and the dead zone area, respectively;
S: The total area of detection zone.
The detector coverage indicators should be determined according to hazardous zone classification as shown in Table 1.

Table 1 Coverage indicators of detectors in different hazardous zones

Hazardous zone classification	Total coverage rate($V1+V2$)	Coverage rate of two or more layers $V1$
Class 0	≥90%	≥50%
Class 1	≥80%	≥40%
Class 2	≥70%	≥30%

3.2 Calculation of detectors numbers

The selection and setting for flame detectors are detailedly formulated in the “Automatic fire alarm system design” [5].Table 2 shows protected area and radius of smoke and thermal detectors.

Table 2 Protected area and radius of smoke and thermal detectors

Types of flame detectors	Floor area S/m^2	Room height h/ m	Protected area A and radius R of one detector					
			Room slope θ					
			$\theta \le 15°$		$15° < \theta \le 30°$		$\theta > 30°$	
			A/m^2	R/m	A/m^2	R/m	A/m^2	R/m

Smoke detectors	$S \le 80$	$h \le 12$	80	6.7	80	7.2	80	8.0
	$S > 80$	$6 < h \le 12$	80	6.7	100	8.0	120	9.9
		$h \le 6$	60	5.8	80	7.2	100	9.0
Thermal detectors	$S \le 30$	$h \le 8$	30	4.4	30	4.9	30	5.5
	$S > 30$	$h \le 8$	20	3.6	30	4.9	40	6.3

The numbers of detectors in a detection zone can be calculated according to the following formula [6]:

$$N \geqslant \frac{S}{K \ A} \tag{2}$$

N: The detector numbers needed in a detection zone;
S: The floor area of a detection zone (m^2);
A: The protected area of a detection zone (m^2);
K: Safety correction factor. The value of K is 0.7 ~ 0.9 in important protected areas and is 1 in generally protected areas.

The detection zone: $L = 18m; M = 10m; h = 5m$. Floor area: $S = 10 \times 18 \approx 180m^2$.

Table 2 shows that protection area of smoke detectors in the detection zone is 60 m^2, and K is selected to be 0.8. By formula (2), the smoke detector numbers needed in the detection zone: $N = \frac{S}{K \ A} = \frac{180}{0.8 \times 60} \approx 4$.

3.3 Optimal layout of detectors in the detection zone

The detection zone model was created with CAD software, then detectors are decorated in four corners of the detection zone, and the decoration angles can be adjusted to reach the maximum coverage rate. The experiments and calculations indicate when the decoration angles of four detectors are equal, the coverage rate reaches the maximum. Figure 5 shows decoration angles of four detectors in the detection zone, and Figure 6 shows vertical view of the detection zone.

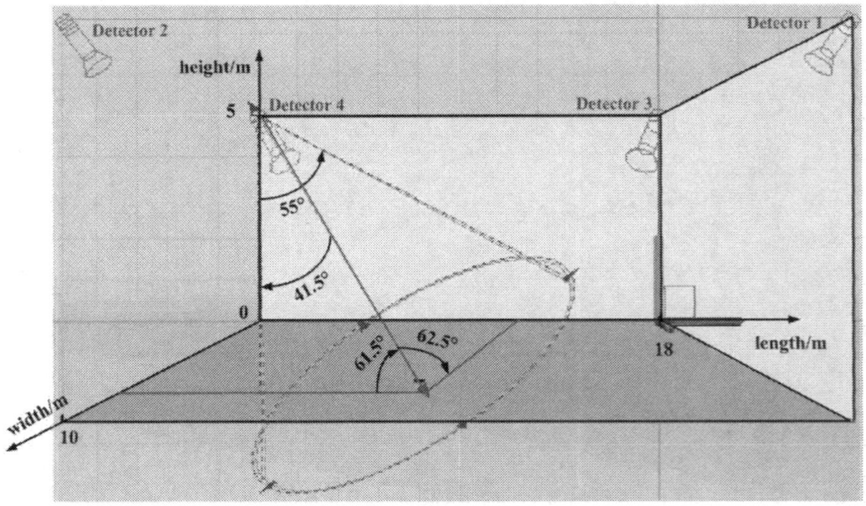

Figure 5 Left view of the detection zone

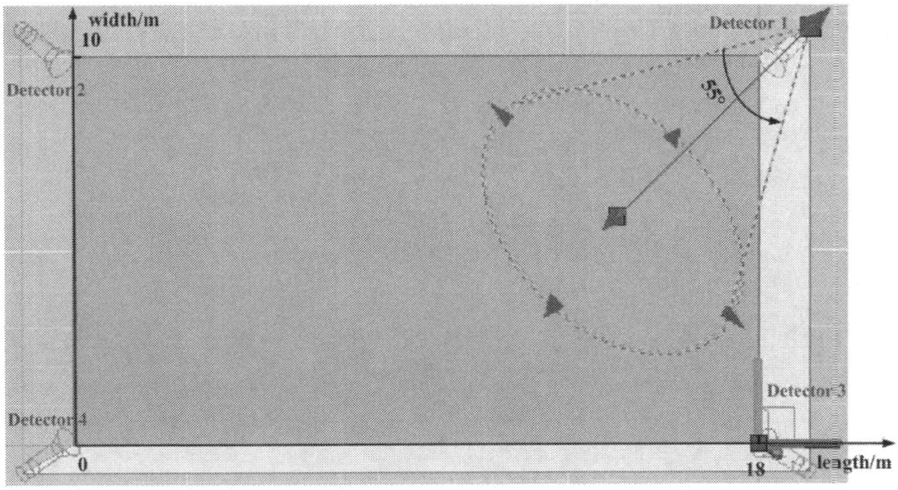

Figure 6 Vertical view of the detection zone

The detection zone model can be rendered with the CAD software as shown in Figure 7.

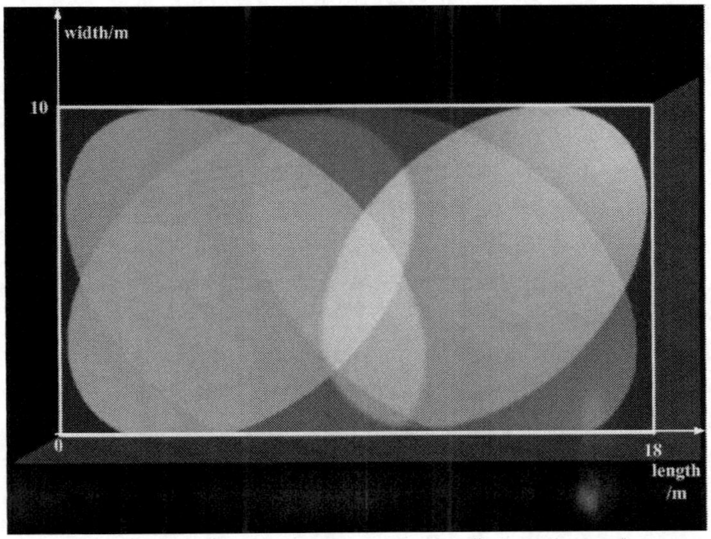

Figure 7 Vertical renderings view of the detection zone

Based on the decoration angles of four detectors showed in Figure 5, the detectors coverage diagram in the detection zone should be plotted by MATLAB software as shown in Figure 8.

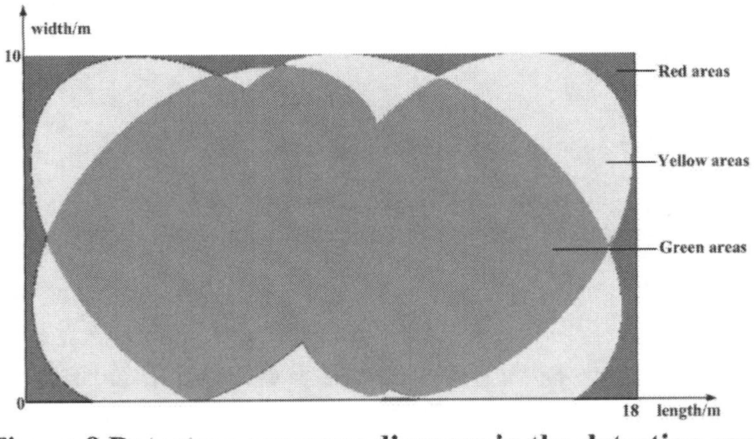

Figure 8 Detectors coverage diagram in the detection zone

NOTE: The green areas indicate that the detectors coverage arrived two or more layers and the yellow areas indicate that there is only one coverage layer, while the red areas indicate the dead zone of detectors (i.e. zero coverage layer, undetectable area).
The total area of the detection zone: S=180m^2. The computed result of coverage area by MATLAB software: $S1$=116.8m^2, $S2$=48.67m^2, $S3$=14.53m^2.
Through formula (1),the computed result of coverage rate: $V1$= 64.89%,$V2$=27.04%,$V3$=8.07%, so $V1$+ $V2$=91.93%.As shown in table3,through optimal layout of detectors, the coverage rates absolutely meet the coverage indicators of class 0 hazardous zone.

Table 3 Compare between class 0 hazardous zone and optimal layout of detectors

Detection zone	Total coverage rate($V1$+$V2$)	Coverage rate of two or more layers $V1$
Class 0 hazardous zone	⩾90%	⩾50%
Optimal layout of detectors	91.93%	64.89%

4. Conclusions

(1) Through the comparison with the traditional F&G ,the integration of the F&G proposed in this paper not only has a high degree of automation that detection, transmission, display , alarm, fire started can be completed almost at the same time, but also has a simple structure for easy maintenance and accurate positioning to improve maintenance efficiency.
(2) For coverage indicators needed in hazardous zone of different class of the petrochemical industry, the F&G was optimized to better ensure its reliability and safety. This optimized layout method of detectors can not only improve coverage rates of detectors, but also enhance accuracy of fire detection, and thus reduce frequency of false alarms in the petrochemical plant. Once an unusual condition occurs in hazardous zone, the alarm of F&G will be aroused faster and more accurate to get more emergency rescue time for eliminating a hidden danger and avoiding accidents, and thus the loss of accident can reach minimum.

5. References

[1] Yilin Zhang. Application of F&G in Petrochemical Equipment [J].Petrochemical Automation, 2012(4), 48(2):11-14. (In Chinese)

[2] Junfang Qi. F&G Design in Petrochemical Industry [J].Science and Technology Communication, 2010(9):85-89. (In Chinese)

[3] Guoshu Zhang, Shujie Yuan. New Ways about Determine Layout Spacing of Fire Detectors [J]. Fire Science and Technology, 2008(6), 27(6):431-432. (In Chinese)

[4] GB50058-92. Design Standard on Electrical Device in Explosion and Fire Hazard Zone [S].Beijing: Chemical Ministry of the People's Republic of China, 1992. (In Chinese)

[5] Zhi Yang, Renming Deng, Qiguo Zhou. Intelligent Systems and Engineering Applications of Architectural [J] .Beijing: Chemical Industry Press.2002. (In Chinese)

[6] Ganhua Zhang. Study on Smoke Device and Fire Monitoring of Infrared Fiber for Ships [D] .Dalian: Dalian University of Technology.2005. (In Chinese)

Effect of Heat Treatments On Stability of Penicillin G in Waste Penicillium Chrysogenum

Wang, P.,
Liu, H.,
Fu, H., Harbin Institute of Technology
Cheng, X.,
Wang, B.,
Cheng, Q.,

Abstract

Penicillin G (PENG) is one of the most widely used bacteriostatic antibiotics in human and veterinary medicine. . It may cause biological drug resistance and environmental toxicity. In this study, we investigate the degradation efficiency of PENG by heat treatment. A kinetic study was performed under different temperature conditions (60°C, 70°C, 80°C, 90°C and 100°C). The first-order rate constants for the direct reaction of PENG with heat treatment were measured to be 0.008 to 0.027 under the temperature conditions tested. The rate constants were greater at higher temperature. These results suggest that higher temperature conditions and short term are needed for the removal of PENG by heat treatment in order to mitigate the residual toxicity that can remain in waste *penicillium chrysogenum* manufactured during the penicillin pharmaceutical process

Theoretical flammability diagrams

Tingguang Ma

Fire Protection and Safety Technology

Oklahoma State University

492 Cordell South, Stillwater, OK 74078

Tingguang.ma@gmail.com

Abstract

Flammability diagrams play an important role on guiding dilution and purging operations. Recently, a theoretical flammability diagram, generic to Coward?s diluted flammability diagram, is proposed to check the dilution requirement under various diluent/oxygen/temperature conditions. This method is based on the enthalpy conservation at the critical flame temperature, still a hand-calculation method with simple treatment of fuel/oxygen/diluent. Any gaseous mixtures involving flammable gases, will be presented in the flammability diagram, with quadrant information on flammable state, dilution requirement to non-flammable state, and variations due to oxygen or temperature. It will find more applications in gas-handling industry for safety-related operations.

The role of flammability limits on determining the fire and explosion potential of a flammable gas mixture has long been recognized since Sir Humphrey Davy?s classical experiments (1816). Starting from Davy?s work, Britton (2002) reviewed 200 years of experimental work, especially identified some debatable topics on flammability measurements, such as ignition strength, wall heat loss, ignition criteria, incompleteness of reactions, etc. This fact calls for a theoretical view on flammability, without any ambiguity from experimental methods.

Flammability theory started by Le Chatelier?s (1898) ?mixture law?, while his work was mostly empirical and its derivation was given almost a century later (Mashuga et al. 2000). Burgess and Wheeler (1911) first proposed that the lower limit of flammability should vary inversely as the calorific value of the gas, which is adopted by many correlations (Ma et al. 2003). They also proposed the temperature dependence of flammability limits, which is termed as Burgess-Wheeler?s law. Thornton (1917) proposed that the upper limit bears a direct relationship to the amount of oxygen needed for perfect combustion, which lay the foundation for the concept of heat of oxidation (Britton, 2002). White (1925) confirmed the net calorific value of the vapor, and further proposed that the critical adiabatic flame temperature is unchanged by the presence of raised ambient temperature or added fuels. Jones (1938) noticed the ratio of the lower limit to the amount of combustible needed for complete combustion is approximately constant. This is termed as the Jone?s rule or Lloyd?s rule (1948). Coward et al. (1952) first proposed the diluted flammability diagram, which used the diluent/fuel ratio as the primary variable. Zabetakis (1965) instead proposed a ternary diagram, which gained wide acceptance in industry. The last contribution is Huggett?s oxygen calorimetry (1978), which is not directly related to flammability but useful to understand the role of oxygen at ignition.

Based on previous experience on Le Chatelier's Rule, Burgess Wheeler's law, Jone's (or Lloyd) rule, White's principle, Coward Diagram, Zabetakis? ternary diagram, Huggett?s oxygen calorimetry, a theoretical flammability diagram is proposed based on the thermal balance at the critical flammable limits (Ma, 2013). This thermal balance method has been applied to solve the flammability limits of a mixture (Ma, 2011), check and propose correlations (Ma et al. 2013), determine the dilution requirements to safety (Ma, 2013). It will be summarized here with several solved example to show the utility of flammability diagram in solving flammable mixture problems.

Reference:

1. H. Davy, On the Fire-Damp of Coal Mines, and on Methods of Lighting the Mines So as to Prevent Its Explosion, Philosophical Transactions of the Royal Society of London, Vol. 106 (1816), pp. 1-22

2. L.G. Britton, Two-hundred years of flammable limits, Prog. Safe. Prog. 21 (1) (2002) 1?11.

3. H. Le Chatelier, O. Boudouard, On the Flammable Limits of Gas Mixtures, *Bull Soc Chim*, 1898

4. C.V. Mashuga, D.A. Crowl, Derivation of Le Chatelier?s Mixing Rule for Flammable Limits, Process Safety. Progress. 19(2) (2000)112-117.

5. MJ., Burgess, RV., Wheeler, The lower limit of Inflammation of Mixtures of the Paraffin Hydrocarbons with air, Journal of Chemical society, 1911

6. W.M. Thornton XV. The relation of oxygen to the heat of combustion of organic compounds, Philosophical Magazine Series 6, Vol. 33, Iss. 194, 1917

7. A.G. White, Limits for the propagation of flame in inflammable gas?air mixtures. III. The effect of temperature on the limits, J. Chem. Soc. 127 (1925) 672?684.

8. GW. Jones, Inflammation Limits and Their Practical Application in Hazardous Industrial Operations. Chem. Rev., v. 22, February 1938, pp. 1-26.

9. Lloyd, P. The Fuel Problem in Gas Turbines. Inst. Mech. Eng. Proc., War Emergency, Issue No. 41, v. 159, 1948, p. 220.

10. H.F. Coward, G.W. Jones, Limits of Flammability of Gases and Vapors, Bulletin No. 503, US Bureau of Mines, 1952.

11. M.G. Zabetakis, Flammability characteristics of combustible gases and vapors, Bulletin 627, Bureau of Mines, Washington, 1965.

12. B. Huggett, Estimation of rate of heat release by means of oxygen consumption measurements, Fire Mater. 4 (2) (1980).

13. V. Babrauskas, Ignition Handbook, Fire Science Publishers, 2003, Appendix.

14. T. Ma, Q. Wang, M. Larrañaga, Correlations for estimating flammability limits of pure fuels and fuel-inert mixtures, Fire Safety Journal 56 (2013) 9?19

15. T. Ma, Q. Wang, M. Larrañaga, A thermal balance method for estimating the flammability limits of a mixture, FW-World, Fall 2012.

16. T. Ma, A thermal Theory for Estimating the Flammability limits of a Mixture, Fire Safety Journal, 46 (2011) 558-567

Study on safety capacity and layout of chemical industrial park based on overall regional risk analysis

Chen Yinting
Nanjing University of Technology
No. 200 Zhongshan North Road, Nanjing, Jiangsu
chenyintingnjut@163.com

ZhangMingguang [a], Wang Rui [a], Qian Chengjiang [a], Jiang Juncheng [a], Ahmed Marbarki [b].
[a] Nanjing University of Technology
[b] University Paris-Est Marne-la-Vallée
[a] No. 200 Zhongshan North Road, Nanjing, Jiangsu
[b] Marne la Vallée Cedex 2 , France
Mingguang_zhang@163.com

Prepared for Presentation at
American Institute of Chemical Engineers
2013 Spring Meeting
1st CCPS Asia-Pacific Conference on Process Safety
Qingdao, China
September 4 – 5, 2013

UNPUBLISHED

Study on safety capacity and layout of chemical industrial park based on overall regional risk analysis*

Chen Yinting
Nanjing University of Technology
No. 200 Zhongshan North Road, Nanjing, Jiangsu
chenyintingnjut@163.com

ZhangMingguang [a], Wang Rui [a], Qian Chengjiang [a], Jiang Juncheng [a], Ahmed Marbarki [b].
[a] Nanjing University of Technology
[b] University Paris-Est Marne-la-Vallée

Keywords: risk analysis, safety capacity, layout design, domino effect

Abstract

As chemical industry parks developed fast in recent decades, the risks of chemical plants had been increased dramatically and major accidents had taken place frequently. In order to implement inherent safety, avoid and reduce accidents loss, the concept of safety capacity of chemical industry park was proposed. At the same time, suitable layout design was used to prevent the domino effect and decrease the overall regional risk. In the present paper, the definition of safety capacity was studied. Based on risk capacity calculation model which used individual risk, societal risk and potential loss of life as indicators, influence factors of safety capacity (such as safety management, safety technique and emergency system) were took into consideration and an adaptable safety capacity calculation model was constructed. The new calculation model was applied to evaluate both overall and existing safety capacity of a partially completed chemical industry park and predict the scales of chemical plants which may settle down in the future. On this basis, for the purpose of preventing major accidents and external domino effect accidents, regional risks of different layout design options were analyzed and the best suitable layout design was determined. The paper proposed a new safety capacity calculation model and a case study was presented to illustrate the applicability of this methodology in planning and layout design of the chemical industrial park.

1. Introduction

With the rapid economic development, the requirement of chemical industrial products of our nation is continuously growing. In this case, a great number of chemical industry parks have been constructed. According to statistics, there have been 1568 provincial and national chemical industry parks in China. A lot of large-sized complex facilities were used in chemical plants, which resulted in the great increase in storage of hazardous chemicals. Thus, the risks of chemical plants had been increased dramatically and major accidents had taken place frequently. Furthermore, when a primary accident occurs in a unit, it is likely to damage equipment surrounded and even cause secondary incidents, which called domino effect.

* Supported by the National Natural Science Foundation of China (No. 71001051) and the Natural Science Foundation of Jiangsu Province, China (BK 2012824).
Corresponding author: Zhang Mingguang Tel.: +86 13815892792.

In recent years, the concept of safety capacity of chemical industry park has been proposed to control the scale of hazardous chemicals and reduce the occurrence probability of major accidents.

Yang Yusheng and Wu Zongzhi et al. [1] (2008) believed that the fundamental reason of risks was the presence of hazardous chemicals and the accidents investigation statistics showed that most major accidents are related to the excessive amount of hazardous chemicals. Chen Xiaodong and Shi Lichen et al. [2] (2009) proposed a new concept: risk capacity of chemical industry park. The risk capacity was the maximum capacity of hazardous substances under the conditions that risks of facilities in a chemical industry park were at an acceptable level. On the basis of researching risk capacity, Chen Xiaodong and Duo Yingquan [3] (2009) defined safety capacity of chemical industry park as a risk-associated threshold which is the amount of hazardous chemicals determined by the production risks, storage risks, transportation risks, usage risks and other risks jointly within the affordable range of overall risk. Tan Xiaoqun [4] (2011) defined safety capacity of chemical industry park as the total amount of hazardous chemicals determined by both maximum stock amount and maximum transportation amount. The maximum stock and transportation amount of hazardous chemicals were assessed under the conditions that risks could be accepted by people.

The purpose of study the safety capacity of chemical industry park is to solve the problem that what is the reasonable scale of hazardous chemicals. Safety capacity can be divided into two parts: fixed safety capacity and moved safety capacity. Fixed safety capacity can be used in following aspects: land-use planning, controlling scales of companies, layout design and so on. The authors hold the point that fixed safety capacity is a determinant of overall safety capacity while moved safety capacity is a restriction. Although it is difficult to reduce transportation risks, traffic restrictions and reasonable arrangements for transportation routes can be used to minimize the impact on overall safety capacity generated by moved safety capacity. Therefore, in order to facilitate the method for practical application, this paper just analyzed the fixed safety capacity. In the following text, the words "safety capacity" all means "fixed safety capacity".

Suitable layout design can be used to implement inherent safety, prevent domino effects and reduce accidents loss. The present paper intends to use the following steps to determine the optimal layout design. First of all, based on the calculations of safety capacity and area of the land, the maximum safety capacity of enterprises that may settle down is estimated. Then according to some factors like traffic, surroundings and public works, layout design alternatives can be proposed. At last, indicators (individual risk, societal risk and potential loss of life) of regional quantitative risk assessment were analyzed to determine the most suitable layout design.

2. Quantitative Risk Assessment (QRA)

2.1 Indicators of Quantitative Risk Assessment

2.1.1 Definition

Individual Risk

The individual risk for a point-location around a hazardous activity is defined as the probability that an average unprotected person permanently present at that point location, would get killed due to an accident at the hazardous activity [5] (P. H. Bottelberghs, 2000).

Societal Risk

The societal risk for a hazardous activity is defined as the probability that a group of more than N persons would get killed due to an accident at the hazardous activity. The societal risk is given as a function of the group size in the form of a plot of frequency (F=probability per year) vs. group size N [5] (P. H. Bottelberghs, 2000).Where individual risk gives the probability of dying on a certain location, the societal risk gives a number for a whole area, no matter precisely where the harm occurs within that area [6] (S. N. Jonkman et al.).

Potential Loss of Life (PLL)

The potential loss of life is defined as the expected number of fatalities per year in a certain region. It is a simple measure for societal risk.

2.1.2 Calculation Model

Calculation Model of Individual Risk

For a certain risk source inside an area, the individual risk generated by the risk source on location (x, y) can be calculated by the following equation:

$$R(x, y) = \sum_{S=1}^{N} f_s v_s(x, y) \quad \text{[Eq. 1]}$$

Where R(x, y) is the individual risk generated by a certain risk source on location (x, y); f_s is the frequency of an accident scene; $v_s(x, y)$ is the probability of dying of an individual in the accident scene on location (x, y); N is the number of accident scenes.

Calculation Model of Societal Risk

Expected number of fatalities (group size N) could be obtained from multiplying the individual risk calculation results by population figures. Then the societal risk curve plot (F , N plot) can be plotted by using parameters of expected number of fatalities (group size N) and frequency (F=frequency per year).

Calculation Model of PLL

In quantitative risk assessment, PLL can be calculated by the following equation [7] (J. E. Vinnem, 2007):

$$PLL = \sum_{s} \sum_{j} \left(f_{sj} \times C_{sj} \right) \quad \text{[Eq. 2]}$$

Where f_{sj} is the frequency of a consequence (j) generated by an accident scene (s); C_{sj} is the expected number of fatalities.

2.1.3 Acceptability Risk Criteria

Currently, the most widely used acceptability risk criteria is ALARP i.e. As Low As Reasonable Practice. In the criteria, risk is divided by two limits into three regions: unacceptable region, ALARP region and acceptable region. The two limits are acceptable risk limit and negligible risk limit.

The core of ALAEP is that all measures have to be taken to reduce the risk, as long as these measures are reasonable for costs or other aspects. If safety measures would be extremely costly, this would not be considered reasonable in the sense of ALARP.

On the premise of criteria proposed by *Major Hazards of Hazardous Chemicals Supervision Interim Provisions (The 40th order of State Administration of Work Safety)*, the authors referenced to the relevant domestic and international research achievements and decided to use the individual risk of 1×10^{-4} as the maximum acceptable risk criteria.

As the societal risk criteria had been elaborated clearly, the authors referenced the societal risk criteria proposed by *Major Hazards of Hazardous Chemicals Supervision Interim Provisions (The 40th order of State Administration of Work Safety)* directly.

3. Safety Capacity of Chemical Industry Park Analysis

3.1 The Correlation of Risk Capacity and Safety Capacity

The main purpose of study the safety capacity of chemical industry park is to control the scale of hazardous chemicals. Safety capacity could not be calculated directly by using amount of hazardous chemicals, since there are a great number of hazardous chemicals with different physical and chemical properties. Risk is a recognized measure of a chemical industry park`s safety status. Furthermore, there are generally accepted criteria for risk assessment. Therefore, safety capacity calculation in this paper was on the basis of the regional quantitative risk assessment and risk capacity.

3.2 Risk Capacity Assessment

The definition of risk capacity is that the sum of PLLs generated by accidents (fire, explosion and poisoning) of all hazardous sources in a certain region. Risk capacity can be calculated by following formula:

$$RC = \sum_{i=1}^{M} PLL_i \qquad \text{[Eq. 3]}$$

Where RC is risk capacity; PLL_i is potential loss of life generated by a certain hazardous source (i); M is the number of hazardous sources.

3.3 Safety Capacity Assessment

The authors have analyzed the existing safety capacity calculation methods and found the following shortages:

1 Did not take into account of the uneven distribution of hazardous chemicals.
2 Did not consider the risks generated by different hazardous chemicals sufficiently.
3 It is too complicated for the calculation of safety capacity.

The new safety capacity calculation model presented in following text had overcome these shortages.

Although PLL can be used to calculate risk capacity of chemical industry park, as the lack of appropriate acceptable risk criteria, PLL can be hardly applied for the further assessment of safety capacity of chemical industry park. Therefore, a new parameter called average individual risk has been introduced.

$$R_{average} = \frac{RC}{N_{onduty}} \qquad \text{[Eq. 4]}$$

Where $R_{average}$ is the average individual risk of workers on duty; N_{onduty} is the number of workers on duty.

According to the foregoing analysis, $R_{average}$ means the average individual risk of workers on duty generated by hazardous chemicals in total. Through collecting hazardous chemicals data of every company in the chemical industry park, the total mass of hazardous chemicals can be estimated and used to calculate the average individual risk of workers on duty generated by unit mass of hazardous chemicals ($\bar{R}$).

For the reason that there are numerous kinds of hazardous chemicals, furthermore, different types of hazardous chemicals are with different hazards. It is difficult to classify hazardous chemicals in detail and calculate their safety capacity separately. In order to ensure the accuracy of the results, this paper referred to the contents of *Major Hazards of Hazardous Chemicals Supervision Interim Provisions (The 40th order of State Administration of Work Safety)* and proposed a method to calculate the hazardous chemicals equivalent (C).

$$C = \beta_1 C_1 + \beta_2 C_2 + \cdots + \beta_n C_n \qquad \text{[Eq. 5]}$$

Where β_1, β_2... β_n is correction coefficient; C_1, C_2, ... C_n is existing inventory of each kind of hazardous chemical.

In the 40th order, hazardous chemicals have been classified into four kinds: toxic gases, explosives, flammable gases and others. Except toxic gases, corresponding coefficients of the three kinds of chemicals are 2, 1.5 and 1. Particularly, chemicals from toxic gases have been assigned correction coefficient specifically.

The $\bar{R}$ can be calculated by following equation:

$$\bar{R} = \frac{R_{average}}{C_{existing}} \quad \text{[Eq. 6]}$$

Where $C_{existing}$ is the equivalent of total mass of hazardous chemicals which are existing in the chemical industry park.

According to the maximum acceptable individual risk criteria, the maximum safety capacity of existing companies can be calculated.

$$C_{maxexisting} = \frac{R_{threahold}}{\bar{R}} \quad \text{[Eq. 7]}$$

Where $R_{threshold}$ is the maximum acceptable individual risk criteria ($R_{threshold} = 1\times10^{-4}$).

In order to calculate the maximum expanding safety capacity of new chemical plants, the maximum existing safety capacity of per unit storage area should be calculated.

$$\bar{C}_{maxexisting} = \frac{C_{maxexisting}}{S_{existing}} \quad \text{[Eq. 8]}$$

Where $\bar{C}_{maxexisting}$ is the maximum existing safety capacity of per unit storage area; $S_{existing}$ is the sum of the existing storage area

Then the maximum existing safety capacity of per unit storage area ($\bar{C}_{maxexisting}$) and the sum of expanding storage area ($S_{expanding}$) can be used to calculated the maximum expanding safety capacity of new chemical plants.

$$C_{max\exp anding} = \bar{C}_{maxexisting} \times S_{\exp anding} \quad \text{[Eq. 9]}$$

This paper improved the calculation model proposed by Tan Xiaoqun [4] (2011). The new method used storage area instead of total area of chemical plants to estimate the maximum expanding safety capacity. This improvement emphasized the uneven distribution of hazardous chemicals.

3.4 Influence Factors of Safety Capacity

The results of analyses above are theoretical safety capacity. In practical calculation process, influence factors of safety capacity should be considered. This paper focused on 4 first level indicators (location, planning, safety management and emergency system) and 16 second level indicators. Then the Analytic Hierarchy Process (AHP) was used to calculate weight coefficients of indicators. The detailed influence factors and corresponding weight coefficients are listed in Table 1.

Table 1. Influence Factors of Safety Capacity

First Level Indicator	Weight Coefficient	Second Level Indicator	Weight Coefficient
Location	0.154	Natural conditions	0.464
		Transportation	0.333
		Environmental conditions	0.203

Planning	0.179	Industry chain	0.104
		Regional division	0.321
		Road planning	0.217
		Utilities planning	0.093
		Firefighting planning	0.265
Safety Management	0.354	Safety management system	0.296
		Major hazards management	0.372
		Key companies management	0.332
Emergency System	0.313	Emergency organization system	0.322
		Emergency monitoring and warning	0.172
		Rescue capability	0.133
		Emergency plan	0.171
		Emergency management	0.202

Experts, civil servants, EHS managers of chemical companies and ordinary people should be invited to evaluate all influence factors and give out scores. By collecting scores provided by respondents, the composite score have been drawn by the following equation:

$$P = \frac{P_1 + P_2 + \cdots + P_n}{n} \quad \text{[Eq. 10]}$$

Where P is the composite score of the influence factor system; P_1, P_2, ... P_n is the score given by each respondent; n is the number of respondents.

Therefore, the practical safety capacity can be calculated by the following equation:

$$C_{practical} = C_{theoretical} \times P \quad \text{[Eq. 11]}$$

Where $C_{practical}$ is the practical safety capacity; $C_{theoretical}$ is the theoretical safety capacity; P is the composite score.

4. Case Studies

4. 1 Application

A fine chemical industry park in Suzhou has been presented as the example of case studies. The total area of the park is approximate 1.5 square kilometers. There are 9 non-chemical companies and 37 chemical companies. According to statistics, the scale of hazardous chemicals reached 5289 tons, in which are mostly organic solvent such as aromatics and resins. The sum of employees in the park is 2951. Generally, the number of employees on duty is about 2300.

Based on the above studies on safety capacity, regional quantitative risk assessment software which was developed by Nanjing University of Technology and An Yuan Technology Co., Ltd. jointly was used to calculate parameters. The individual risk contour map (Figure 1) and societal risk curve plot (Figure 2) are shown below.

Figure 1: Individual Risk Contour Map

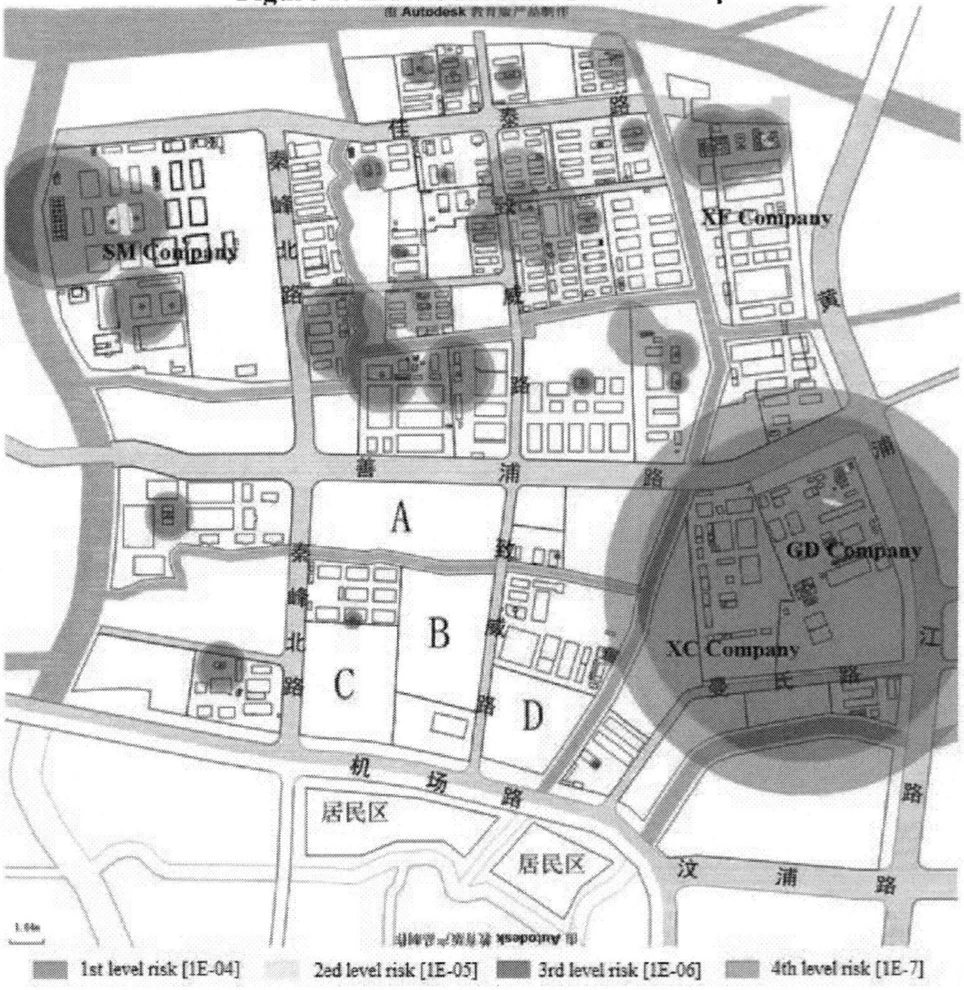

Figure 2: Societal Risk Curve Plot

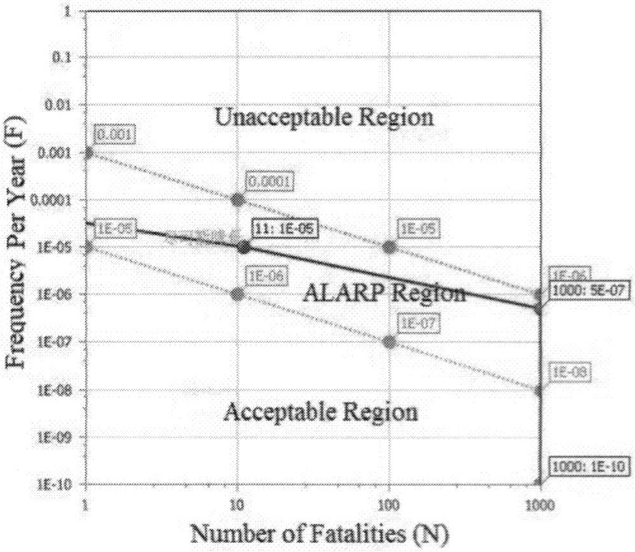

According to the safety capacity calculation model and results drawn from the software, RC i.e. risk capacity is 0.00992. Once the parameters substituted into the [Eq. 4], $R_{average}$ is 4.313×

10^{-6} which is at the level of medium personnel density areas. After substituting the data of hazardous chemicals and corresponding correction coefficients referenced from the 40th order of State Administration of Work Safety into [Eq. 5], we got the existing hazardous chemicals equivalent $C_{existing}$=6772.7 tons. According to [Eq. 6] and [Eq. 7], $\bar{R}$ is 6.368×10^{-10} and $C_{max, existing}$ is 1.57×10^{5} tons. Statistics of the fine chemical industry park show that the total area of storage regions, chemical companies and undeveloped regions is 56145 m^2, 924289.3 m^2 and 349168 m^2. The ratio between the storage area and the chemical companies' area is 0.03. Therefore, according to [Eq. 8] and [Eq. 9], $\bar{C}_{max,existing}$ =2.796 tons and $C_{max, expanding}$ is 29291.7 tons.

31 respondents were invited to give out their scores for the influence factor system of the fine chemical industry park. Finally, the composite score calculated by [Eq. 10] is 0.72. According to the [Eq. 11], the practical maximum safety capacity of existing companies and new companies are 1.13×10^{5} tons and 2.10×10^{4} tons.

There are four companies that will be settled in the fine chemical industry park. These four companies are engaged in ink manufacturing (No.1), paint manufacturing (No.2), chemical raw materials production (No.3) and wastewater treatment (No.4). The information of four companies has been listed in Table 2.

Table 2. Information of Four Companies

No.	Company Category	Planning Area/Storage Area (m^2)	Hazardous Chemicals	Estimated Amount of Hazardous Chemicals Equivalent (t)
1	Ink manufacturing	33333 / 3460	Organic solvent	1057
2	Paint manufacturing	56000 / 6224	Organic solvent	1568
3	Chemical raw materials production	33333 / 1554	Ammonia and hydrogen fluoride	255
4	Wastewater treatment	40000 / 2616	Chlorine	40

According to the calculation results and equations mentioned above, the practical maximum expanding safety capacities of the companies are 6965.4 tons, 12529.7 tons, 3128.4 tons and 5226.3 tons. Therefore, expanding safety capacities all meet the requirements. Considering some factors like traffic, surroundings and public works, especially the land area, the No. 1company can be only located in land A. On this basis, four alternative layout designs have been presented. Authors used the regional quantitative risk assessment software to assess all four layout designs and results have been shown in Figure 3 and Table 3.

Figure 3: Regional Risks of Alternative Layout Designs

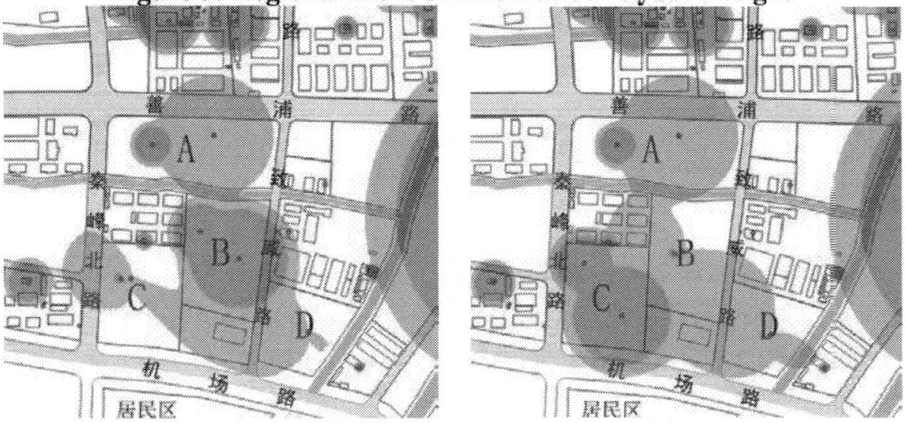

a b

c d

Table 3. Additional PLL of Alternative Layout Designs

No.	Additional PLL
Ⅰ	0.00120
Ⅱ	0.00128
Ⅲ	0.00128
Ⅳ	0.00121

From the four pictures presented in Figure 3, we have found that No. 1 company had the 1×10^{-6} risk contour and when it was located in land C or D, the hazardous area may affected residential areas in south side. As listed in Table 3, the PLL of layout design Ⅰ had the minimum value. Therefore, according to the calculation results and analyses mentioned above, the optimal layout design is Design Ⅰ.

4.2 Discussions

The Individual Risk Contour Map has shown that 1×10^{-5} risk contour appeared in three companies (SM Company, XF Company and GD Company). The causes of this phenomenon are that these companies had a great amount of hazardous chemicals and warehouses / tanks had concentrated layout. Particularly, risks generated by GD Company and XC Company covered a large area and even affected companies outside the park. Authors argued that, major hazards in these three companies should be monitored and supervised carefully.

According to the Societal Risk Curve Plot, the societal risk curve lay in the ALARP region. Therefore, some measures like strengthening the safety management, investigating and eliminating hidden dangers, implementing risk control technologies and so on.

5. Conclusions

1. Hazardous chemicals in chemical industry park are the sources of risks and the fundamental reason to cause accidents. The purpose of researching the safety capacity is to control the scale of hazardous chemicals and reduce the occurrence probability of major accidents.
2. Based on the regional quantitative risk assessment, a new safety capacity calculation model has been proposed. The calculation model has considered the uneven distribution

of hazardous chemicals and the different risks generated by different hazardous chemicals.

3. A regional quantitative risk assessment software developed by Nanjing University of Technology and An Yuan Technology Co., Ltd. Has been used to calculate indicators like individual risk, societal risk and PLL and draw result figures.
4. An example has been studied to illustrate the applicability of the safety capacity calculation model and to elaborate the steps of selecting the optimal layout design.

6. Acknowledgements

The author thanks the strict guidance of tutor Zhang Mingguang and the financial support of the National Natural Science Foundation of China (No. 71001051) and the Natural Science Foundation of Jiangsu Province, China (BK 2012824).

7. References

[1] Yang Yusheng and Wu Zongzhi et al. Analysis on the causes of typical petrochemical industry accidents based on safety planning [J]. Journal of Safety Science and Technology, 2008, 4(1):120-123.

[2] Chen Xiaodong and Shi Lichen et al. Study on the Appropriate Risk Capacity of Chemical Industry Park [J]. China Safety Science Journal, 2009, 19(3):132-137.

[3] Chen Xiaodong and Duo Yingquan. Researches on safety capacity of chemical industry park [J]. Journal of Safety Science and Technology, 2009, 5(2):10-13.

[4] Tan Xiaoqun. Research on Evaluation Model for Safety Capacity of Chemical Industry Park Based on Regional Risk and its Application in Chemical Industrial Park [D]. South China University of Technology, 2011.

[5] P. H. Bottelberghs. Risk Analysis and Safety Policy Developments in the Netherlands [J]. Journal of Hazardous Materials, 2000, 71(1-3): 59–84.

[6] S.N. Jonkman, P.H.A.J.M. van Gelder, J.K. Vrijling. An overview of quantitative risk measures for loss of life and economic damage [J]. Journal of Hazardous Materials, 2003, A99(1-3): 1–30.

[7] Jan Erik Vinnem. Offshore Risk Assessment: Principles, Modeling and Applications of QRA Studies (2nd ed) [M]. London: Springer-Verlag London Limited, 2007.

1ST CCPS ASIA-PACIFIC CONFERENCE ON PROCESS SAFETY℠

Design of a management of change platform based on Web and RFID

Guoliang Cao
College of Electromechanical Engineering
China University of Petroleum (East China)
No.66, West Changjiang Road, Qingdao
824145817@qq.com

Changhang Xu[1]
Huandi Shi[1]
Lefeng Ren[1,2]
1. **College of Electromechanical Engineering**
China University of Petroleum (East China)
2. **Gudong oil production plant of Shengli Oil company**
changhangxu@126.com
hualishitou@126.com
xiaoc2008@126.com

Prepared for Presentation at
American Institute of Chemical Engineers
1st CCPS Asia-Pacific Conference on Process Safety
Qingdao, China
September 4 – 5, 2013

UNPUBLISHED

Design of a management of change platform based on Web and RFID

Guoliang Cao
College of Electromechanical Engineering
China University of Petroleum (East China)
No.66, West Changjiang Road, Qingdao
824145817@qq.com

Changhang Xu[1]
Huandi Shi[1]
Lefeng Ren[1,2]
1.College of Electromechanical Engineering
China University of Petroleum (East China)
2.Gudong oil production plant of Shengli Oil company

Keywords: management of change(MOC), Web, Radio Frequency Identification(RFID)

Abstract

Taking the demand of the enterprise change management into consideration, a management of change platform based on web and RFID is designed and developed. It possesses change process management module which contains change application, review and examination, database about change history, risk analysis, change notice and real-time monitoring and other functions. Based on this method, a MOC system for a joint station of oil-gas gathering is developed and put into practice. The application results demonstrate that the system could extend its field of application immensely since it is designed based on web technology and RFID. It becomes more flexible for MOC procedure with the help of MOC system. It contributes to manage the update of regulations, process engineering and equipments, which will finally promote the improvement of the support capability in work safety.

1. Introduction

With the rapid development of petrochemical industry, grave worksite accidents frequently occur. In order to lower the prevalence of accident and raise the safety level, people has taken stringent steps to control risk of accident, such as building organization of safety management, increasing labor protection product, formulating fire control safety system and operation rules .etc. Practice has shown that strictly standing by the rules and regulations definitely lower accident rates. However, numerous activities such as process change, new facility installation, unstable operational staffing levels are executed in the chemical plant by the process technology, facility management and economical environment. Inappropriate changes can affect employee safety, damage the equipments, or result in significant process interruptions.[1,2]

Management of change (MOC) is a process for evaluating and controlling modifications to facility design, operation, organization, or activities-prior to implementation-to make certain that no new hazards are introduced and that the risk of existing hazards to employees, the public, or the environment is not unknowingly increased.[3] As an important part of chemical industry process safety management, management of change (MOC) has been approved by great majority enterprises. However, in practice, more and more problems are exposed in traditional MOC methods.

First, it is very difficult to manage the change information which includes the management of change file and the issue of change notice. The delivery, store and locating of change files is a time-consuming work, which seriously lowered the management efficiency of change activity. Second, in the process of MOC including change application, risk analysis, audit, implement, examination, staff error is inevitable. For examples, risk analysis is left out before change application. Or there's no safety warning equipment in the process of implement. Third, real-time monitoring and management is a critical part in MOC but we lack of effective tools. A new change activity may have great influence on the already existing one. Therefore, the manager should quickly find out all the affected change activities and take responding measures. Forth, it is difficult to achieve the quick query of change history for the existing information management method. But it is very important for assisting site operation and risk analysis.

As the most extensive distributed application structure, Web can share all the interconnected messages spreading every web site. RFID can connect entity with the internet, and realize the informatization of entity . RFID has been a necessary and practical tool for secure traceability management system. These advantages can be effective methods to solve problems exposed in traditional MOC methods.[4,5]

2. Design of the MOC system based on web and RFID

2.1 System Requirements Analysis

The main requirements of the change management system include two parts, one part is the search of change history, and another is the management of change process. It is meaningful to achieve the search of change history, which is the main ground to carry on the change management work. On the one hand, we can learn from the change history to guide the management work which is similar to the current change project. On the other hand, it is convenient to get the latest information of the equipment state, avoiding the inaccurate risk analysis results due to the delayed information. The change management system should be able to ensure that the history change records database is updated in time to maintain the timeliness. Change process management is the main demand of the users, but also is an important part of the system. The system should achieve electronic operations including change application, review and examination.etc. It can supervise the change application process and implementation process in the scene on the basis of web and RFID, which is effective to improve the change efficiency, reduce human errors, and avoid employees carrying out unauthorized change activity without referrals.

2.2 Design of the MOC system module

Based on the above analysis, the paper proposed the MOC system structure based on Web and radio frequency identification devices (RFID). The modules of the system are as follows:

1. Change process management module

The module functions for change application, review and examination. According to the time characteristic, change activity can be classified into two categories: emergency change and general change. And general change can be divided into permanent change and temporary change. For different change type, the change process and files are different. Every change application is regard as a project, and can be numbered automatically as its identification code. In the change file, there is the detail (time, site, equipment characteristic, etc.) about the change activity, the adopted plan and safety precautions. All in all, the module contains the important part of a change activity, and can provide original data for change notice, real-time monitoring and database update. What's more, it also provides the radio frequency identification interface, which can identify data about file, equipment, and process variable message imported from RFID.

2. Database about change history

The database records the detail about the change activities in the enterprise, which can be queried by project numbers, equipment name, station name, responsible person or identifier to get the historical change records.

3. Risk analysis module

The applicants and auditors can evaluate the change risk using the Change checklist risk analysis method. This module provides basic risk information for HOZOP conference and other advanced risk evaluation methods.

4. Change notice and real-time monitoring module

Managerial personnel released the change notice in the area, and then when the staff entered, he can get the change notice by wireless network. First, we can integrate the current and historical information about staff, equipment, and process into RFID, and update by wireless network, then, the staff can get information by the RFID reading device. To achieve effective supervision in change procedure, the system includes video surveillance and regular reports to the manager every one or two hours. The former is efficient for sample survey and detail query, while the latter is achieved by the manager's real-time record, which is aimed at quick understanding about change implement for the manager.

According to the enterprise change management requirements, the system includes four roles, including staff, department management, security management and system management. Every role has its specific module permission. The system module function and user permission are shown in the following table.

Table 1. Function and authority of system modules

Module	Submodule	Authority
Change Request module	New Change Request	Department managements
	Change Acceptance Request	
	Revocation Request	
Risk analysis module	Primary Part	Department managements, security management
	Advanced Part	
Change Approval module	Emergency Change Approval	security management
	Temporary Change Approval	
	Permanent Change Approval	

Change Notification module	Implementing Change	All staff
	Completed Change	
Change implementation and monitoring module	RFID connector	Department managements, security management
	Monitor record	
Change Acceptance module	Emergency Change Acceptance	security management
	Temporary Change Acceptance	
	Permanent Change Acceptance	
Change history search module	Precise search	All staff
	Advanced search	
Authority and system management module	vacant	system management

2.3 Design of application terminal

Based on the sites of the change management system, we designed the application terminal. Considering that the system aims at the change activities management of technology, equipment, management of different positions in the enterprise, the system is applied in the following categories: every production position, the responsible departments and security department. The production positions submit change application by the RFID read device and website, and then the responsible department audits the information of the application on the internet; the security department verifies it and supervises the working scene by the RFID technology. What's more, currently, the mobile equipment transmit from personal computer, pad to smart phone, which is more intelligent and portable. Therefore the change management can be achieved on-site office in the future. Finally, it will promote the convenient and efficient management of change activity.

3. Development example of the MOC system

3.1 System development platform

The MOC system based on B/S architecture is designed on the open source MVC framework and developed with JavaBean. What's more, the system view is developed by JSP, and the database is manipulated by Hibernate technology to maintain data persistence. RFID middleware service module is used to filter, logical judgment, transmit the RFID tag information and keep touch with change history database and RFID reader.

3.2 Introductions of system modules

When different roles login the system, the left part displays its corresponding module. To understand the system, we will extract all the modules to introduce its brief function in the following part, rather than based on the user roles.

- We can have a view of the latest information of the system according to "change notification" module, including the being implemented and recently completed change project. As shown in Figure 1, we can access the PDF files about the change project when click the “display” button.

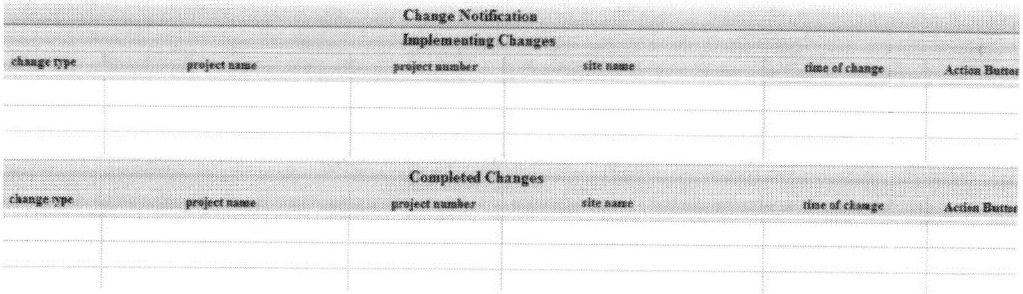

Figure 1 "change notification" module

- "Change Request" module consists of the input, display and modification of new change request, change acceptance request, revocation request. Meanwhile, there are three kinds of change project: permanent change, temporary change and emergency change. Click the button to access the responding change application form. Take permanent change as an example, click” change project application--permanent change” menu, and then enter the permanent change module. This kind of application includes basic information of change application, technology change form, facility change form, staff change form and training program application form as shown in the Figure 2.

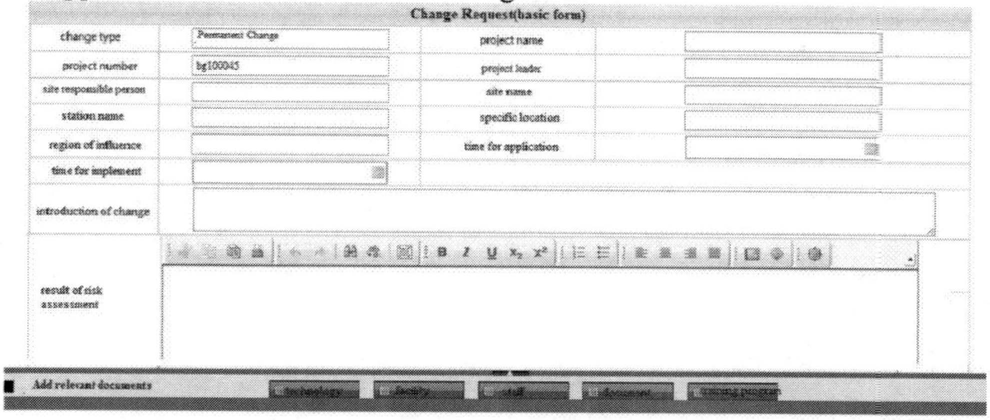

Figure 2 "Change Request" module

- In the "change approval" module, we can receive or handle different kind of change applications. Risks caused by change activities are determined by advanced analysis, and then risk prevention measures are modified. At last, the approval personnel reply the application, and meanwhile, release the change notification. Figure 3 shows the main interface of "change approval" module.

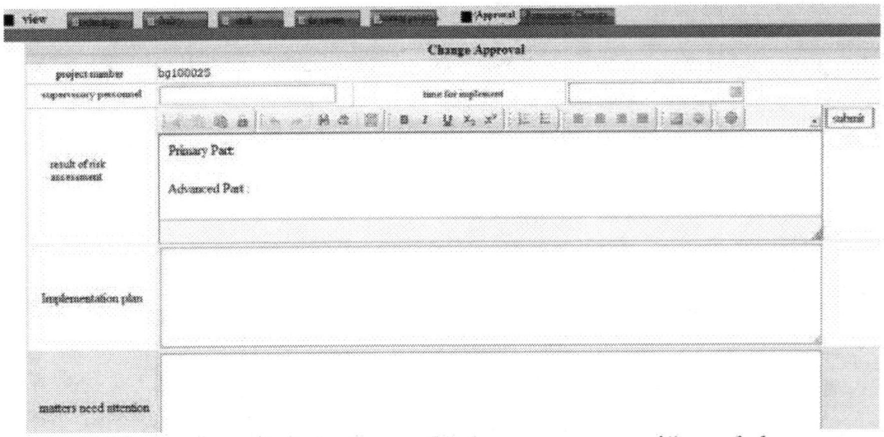

Figure 3 main interface of "change approval" module

- The properties in the "change implementation and monitoring" module includes time, the implementation content and situation, existing problems, proposed solutions, phase responsible persons and so on. The equipment with electronic tags provided with status information is convenient for real-time scanning monitoring.
- "Change acceptance" module checks the submitted change acceptance request. It is to ensure that the implementation content and announcements in the application have been carried out. If qualified, the application is accepted. What's more, it can collect all the information about the change project and update the database. Otherwise, if unqualified, the application will be given advice to inform department management for rectification.
- "Risk analysis" module is divided into two parts--primary and advanced parts. Primary part, consisting of a checklist for the conventional change risk, is relatively simple. Advanced part provides the guidance of two common risk analysis methods -- HAZOP, LOPA, which are applied to assess the underlying risk of change. Risk analysis results automatically aggregate to "Change Request" module and "Change Approval" module. Figure 4 shows the main interface of advanced analysis.

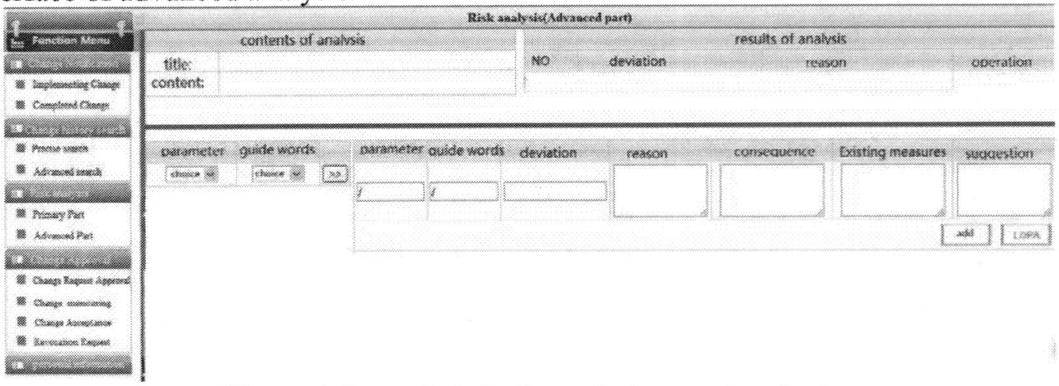

Figure 4 the main interface of advanced analysis

- "Change history search" module includes precise search and advanced search. Item number of change is used for precise search. A combination of certain conditions or keywords is used for advanced search. Click on the "History Search - Advanced Search" menu to enter the interface shown in Figure 5.

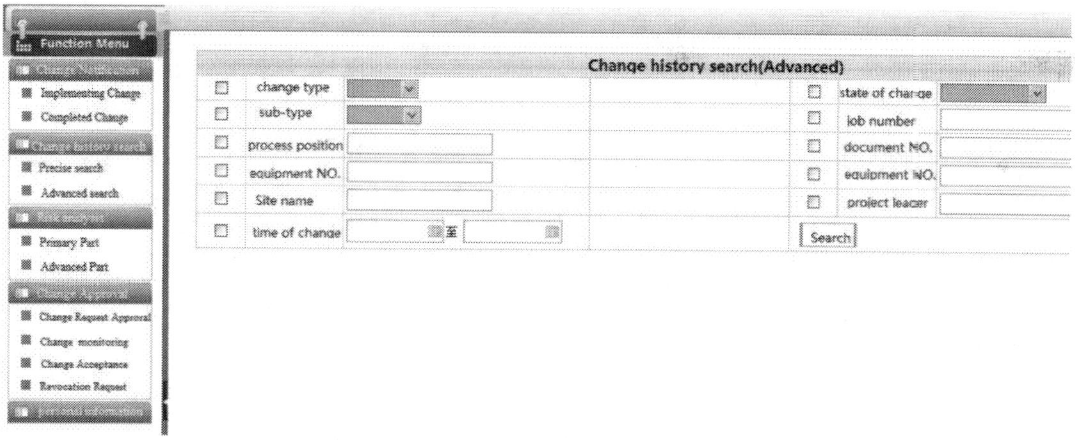

Figure 5 "Advanced Search" menu

- "Authority and system management" module achieves the following functions: assign roles to the appropriate users, assign permissions to users, manage user roles and permissions. When the user selects System Maintenance identity and enters user name and password information, he will enter the main interface of System Maintenance.

3.3 Implementation effects of the MOC system

The system releases on the information center server in a joint station enterprise. It realizes change application, review and examination, database about change history, risk analysis, change notice and real-time monitoring and other functions. Practice has proved that the system has a significant role in safeguarding process safety during transformation and maintenance work. It contributes to solve a series of problems existed in former system caused by lagged updating. It lessens the labor intensity in the process of MOC and changes the event execution and update efficiency.

4 Conclusion

The MOC system based on web and RFID is very effective for reducing the incorrect operation, at the same time, it improves the management efficiency to ensure safety in process. With the continuous development of the mobile terminal client technology and internet of things technology, management of change platform based on web and RFID will have a broader opportunity for application.

Acknowledgement

This work was financially supported by "the Fundamental Research Funds for the Central Universities" (14CX06056A), National science and technology support project (2011BAK03B08), Shandong Province Natural Science Foundation (ZR2011EL048) and Graduate Innovation Fund of China University of Petroleum (East China) (JD12-04).

5. References

[1] Yoo, J.H., H.S. Lee and J.W. Ko, "Risk assessment and drawing information system based change management," Korean Journal of Chemical Engineering, Volume 28, No. 2, 2011.

[2] Reniers, G., "A Management of change approach for assessing and Evaluating operational Staffing Levels (MocESL) in chemical plants," Safety Science, Volume 48, No. 7,2010.

[3] Grossel, S.S., "Guidelines for Management of Change for Process Safety (2008)," Process Safety Progress, Volume 28, No. 2, 2009.

[4] Ma Zhaoding, Zhang Pengcheng, "Research on the application of IOT in fixed-assets management of oilfield enterprises," value engineering, Volume 36, 2011.

[5] XU Lingling, LIU Bingwu, LI Juntao, "The application of RFID technology in goods' safety trace ability management," Logistics engineering and management, Volume 36, 2011.

1ST CCPS ASIA-PACIFIC CONFERENCE ON PROCESS SAFETY℠

Fire Consequence Analysis for Two Parallel-running Pipelines

Tianming Ge
Scandpower China Inc.
Rm. 1002, Tower E, Global Trade Centre, 36 North Third Ring Road East Dong Cheng District, Beijing, 100013, P.R. China
Tianming.Ge@lr.org

Lin Hong
Scandpower AS
Malmskriverveien 4, N-1337 Sandvika, Norway
Lin.Hong@lr.org

Zhe Yao
Petrochina Beijing Natural Gas Pipeline Inc.
Datun Road, Chaoyang District, Beijing, China
yaozhe@petrochina.com.cn

Prepared for Presentation at
American Institute of Chemical Engineers
1st CCPS Asia-Pacific Conference on Process Safety
Qingdao, China
September 4 – 5, 2013

UNPUBLISHED

Fire Consequence Analysis for Two Parallel-running Pipelines

Tianming Ge
Scandpower China Inc.
Rm. 1002, Tower E, Global Trade Centre, 36 North Third Ring Road East Dong Cheng District, Beijing, 100013, P.R. China
Tianming.Ge@lr.org

Lin Hong
Scandpower AS

Zhe Yao
Petrochina Beijing Natural Gas Pipeline Inc.

Keywords: Fire, Pipeline, Parallel, CFD

Abstract

One natural gas pipeline and product oil pipeline is designed to be routed through the same pipe gallery to minimize land use. The horizontal distances of two parallel pipelines are 1.5m, 1.1m and 0.7m at buried sector, tunnel sector and box culvert when crossing railway, respectively. How to prevent fire on one pipeline from affecting the other becomes a major concern for authority and operator.

Scandpower was invited to evaluate risk that failure of one pipeline to affecting the integrity of another. The most severe scenario, i.e. jet fire, was simulated by Kameleon FireEx KFX®, a commercial CFD tool, to estimate the fire accidental load; while VessFire®, a commercial simulation program for time-dependent non-linear analysis of thermo-mechanical response exposed to fire, was used to evaluate the integrity of the exposed pipeline. The results show that exposed natural gas line will lose its integrity but the product oil line will not. Sensitivity cases were studied as well to verify the recommendations.

1. Introduction

To feed China's burgeoning economy, more and more pipelines are under construction to increase the length of pipeline network, which means pipeline routes will be through some developed area, such as the more densely populated area. The planning department with the principle of reducing occupied land, one or more pipelines are designed to be routed through the same pipe gallery. Especially when transporting different media, how to manage the safety is a big challenge for both authority and operator.

For such problems, especially oil pipeline and gas pipeline lying in same gallery, there are not only industry standards in the blank area, but also the lack of proven engineering experience. Given this scenario, Scandpower was invited to evaluate risk that failure of one pipeline to affecting the integrity of another.

Research achievement successfully passed professionals review meeting on June, 2011, organized by Pre-assessment Center of Petroleum Industry Construction Project. They recognized that "technique, modeling and consequence assessment standard are scientific and reasonable, and research achievement can be used as technique support for engineering design". The achievement also won full approval of professionals review organized by Beijing Supervision and Administration of Production Safety in Dec. 2011, with the conclusion of "rich content, appropriate approach and authentic conclusion".

This paper will focus on risk research achievement on two parallel-running pipelines.

2. Pipelines Specification

Two pipelines design specification is listed in Table 1.

Table 1. Pipelines Specification

Parameter	Product Oil Pipeline	Natural Gas Pipeline
Welding types	Seamless steel tube	SAWH
Material	L360MB	X70
Diameter	D323.9mm	D1016mm
Wall Thickness	8.8mm	14.6/17.5/26.2mm
Pressure	10MPa	10MPa
Temperature	Ambient	12°C
Flow Rate	180×10^4 t/a	150×10^8 Nm^3/a
Velocity	0.95 m/s	8.5 m/s (6.1MPa)

3. Accidental Fire Load

3.1 Fire Scenario

As per conclusion in *Key Technique Research on Two Parallel-running Pipelines of the Second West-east Gas Line*[1] and *Pipeline Rupture and Parallel Pipe Spacing Study* [2] published by PRCI, failure effects on adjacent parallel-running pipelines mainly are:

- Impact pressure resulting from natural gas pipeline rupture led to unbalanced radial of and damage to parallel-running pipelines (breach facing parallel-running pipelines);

- Jet fire resulting from natural gas pipeline rupture led to decreased intensity of and damage to parallel-running pipelines.

This paper will only focus on failure impact of jet fire while impact pressure influence will be further discussed in another paper. Meanwhile, to guarantee analysis integrity, impact analysis of production oil pipeline failure will also be demonstrated besides natural gas pipeline failure. Therefore, the finalized fire scenario is as:

- Jet fire resulting from leak will cover adjacent pipeline

3.2 Crater Size Analysis

Empirical formula on size of crater resulting from high pressurized natural gas pipeline leak is demonstrated in *Pipeline Rupture and Parallel Pipe Spacing Study* [2]. This empirical formula will be adopted in this paper. For production oil pipeline leak, the same formula will be used to calculate bullet size. Refer to Table 2 for calculation result:

Table 2. Crater Size

Ruptured pipeline	Crater diameter	
	Pipeline center	Bottom of pipeline
Natural gas pipeline	3.33	3.00
Production oil pipeline	2.24	2.13

Based on current design condition, crater diameter resulting from natural gas pipeline or production oil pipeline leak will exceed current design parallel design distance of 1.5 m; therefore, there is no more overburden impact on parallel pipelines upon leak.

3.3 Jet Fire Analysis

Internationally recognized and widely used KFX is adopted in thermal load intensity analysis. KFX can take into consideration of pipe diameter impact on fire.

3.3.1 Natural Gas Pipeline Leak

KFX is adopted in modeling of parallel distance of 1.5m, 1.1m and 0.7m respectively. As per modeling result, distance has small impact on thermal load. Production oil pipeline does not locate in the place with the maximum heat radiation intensity. When jet fire reaches production oil pipeline, it will extend to each end of the pipeline.

Figure 1. Thermal Radiation Intensity Distribution on Production Oil Pipeline

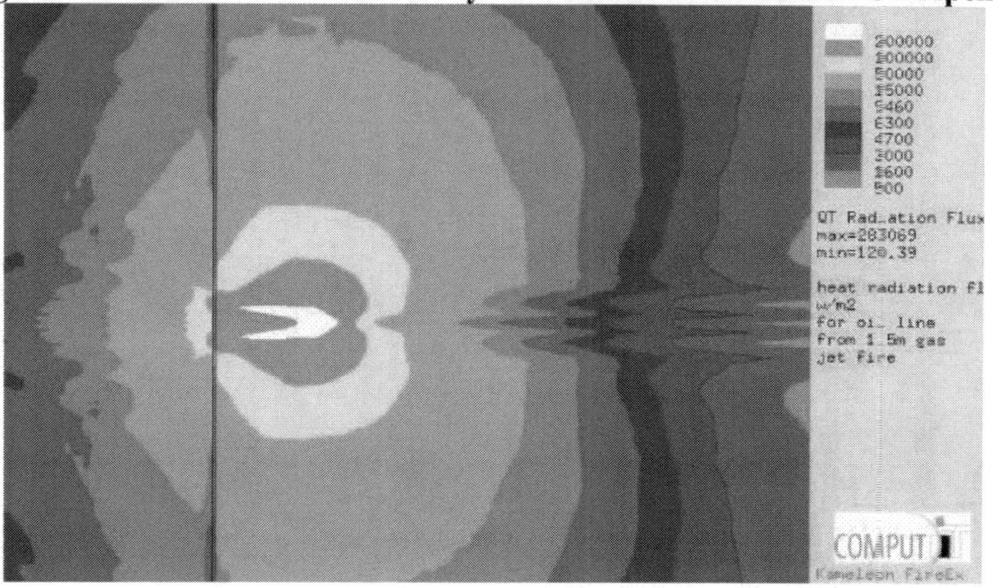

Impact sphere of thermal load on production oil pipeline is set as 10 m with heat radiation intensity of 150KW/m^2. As the pipeline will be surrounded by the fire, heat convection strength is set as 100KW/m^2. Therefore, the gross heat load intensity is set as 250KW/m^2.

3.3.2 Production Oil Pipeline Leak

KFX is adopted in modeling of parallel distance of 1.5m, 1.1m and 0.7m respectively. As per modeling result, distance has small impact on thermal load. Natural gas pipeline is close to but does not locate in the place with the maximum heat radiation intensity.

Impact sphere of thermal load on natural gas pipeline is set as 25 m with heat radiation intensity of 150KW/m^2. As the pipeline will be surrounded by the fire, heat convection strength is set as 100KW/m^2. Therefore, the gross heat load intensity is set as 250KW/m^2.

Figure 2. Thermal Radiation Intensity Distribution on Natural Oil Pipeline

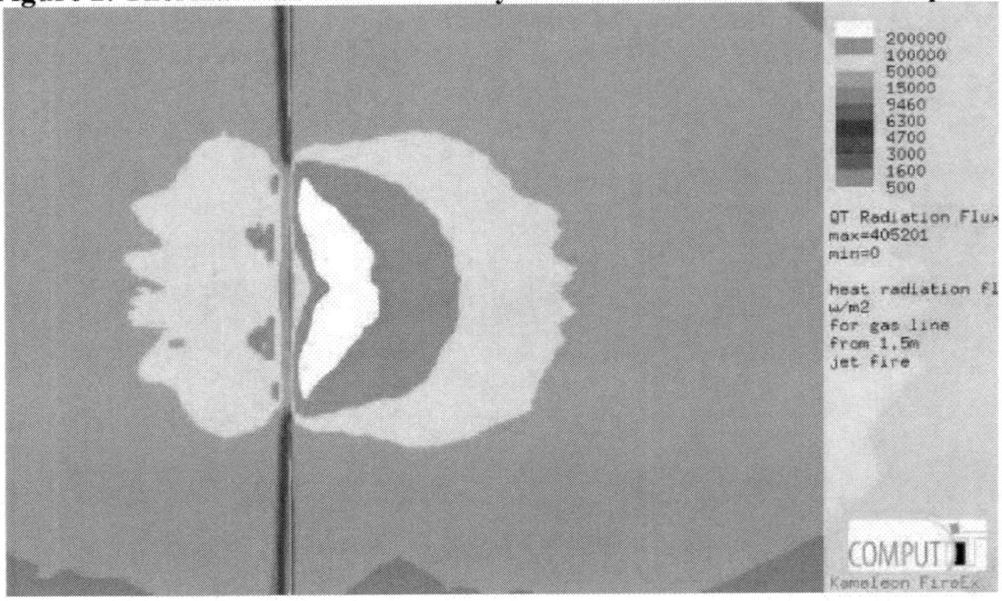

4. Pipeline Thermal Load Response Analysis

4.1 Elementary Case

Vessfire will be adopted in pipeline structure response calculation upon impact of thermal load. Following factors will be considered:

- Internal pipeline pressure
- Gas or liquid flow within pipeline
- Pipeline material property subject to temperature variation

Figure 3. The Stress and UTS at Elevated Temperature

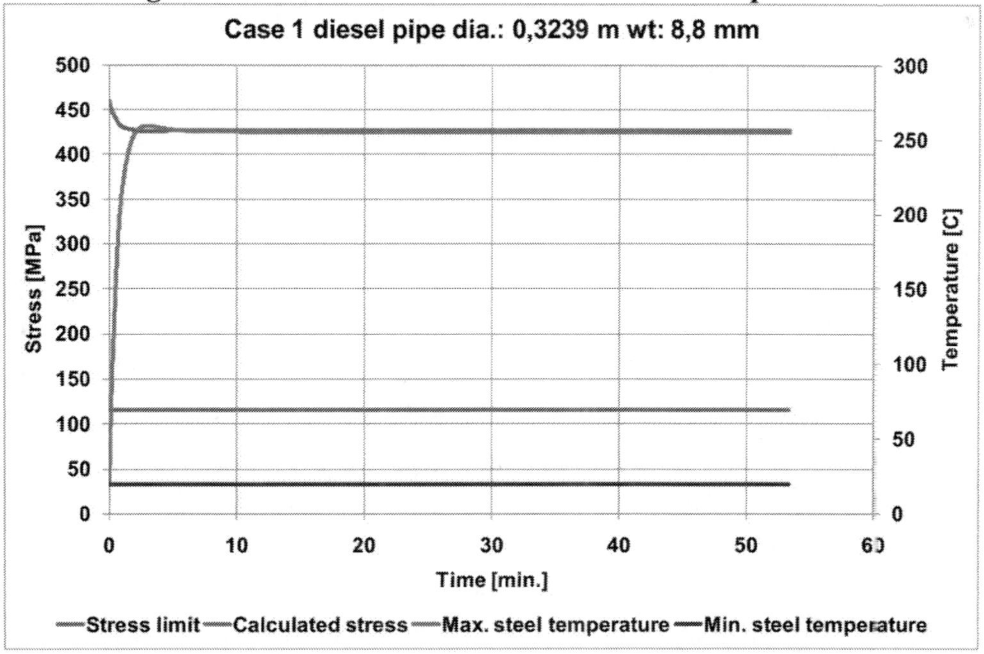

Figure 3 demonstrates calculation result of production oil pipeline, in which:

- Light green curve represents material ultimate strength
- Dark green curve represents pipeline structural stress
- Red curve represents the maximum temperature on the pipeline
- Purple curve represents the minimum temperature on the pipeline

Final conclusion is arrived as:

- Production oil pipeline will not fail because its material strength is far above structural stress;
- No big changes on pipeline structural stress, which means internal pressure does not increase due to heated pipeline;
- Material strength decreases in the first 2 minutes upon heating due to increased pipeline temperature, but no more decrease in strength once pipeline temperature is stable without any increasing;
- The highest pipeline temperature, that is , the average temperature during heating phase, climbs to 250°C in the first 2 minutes upon heating. Due to continuous internal fluid flow that takes away heat, pipeline temperature is gradually stabilized.

Figure 4. The Stress and UTS at Elevated Temperature (14.6 mm)

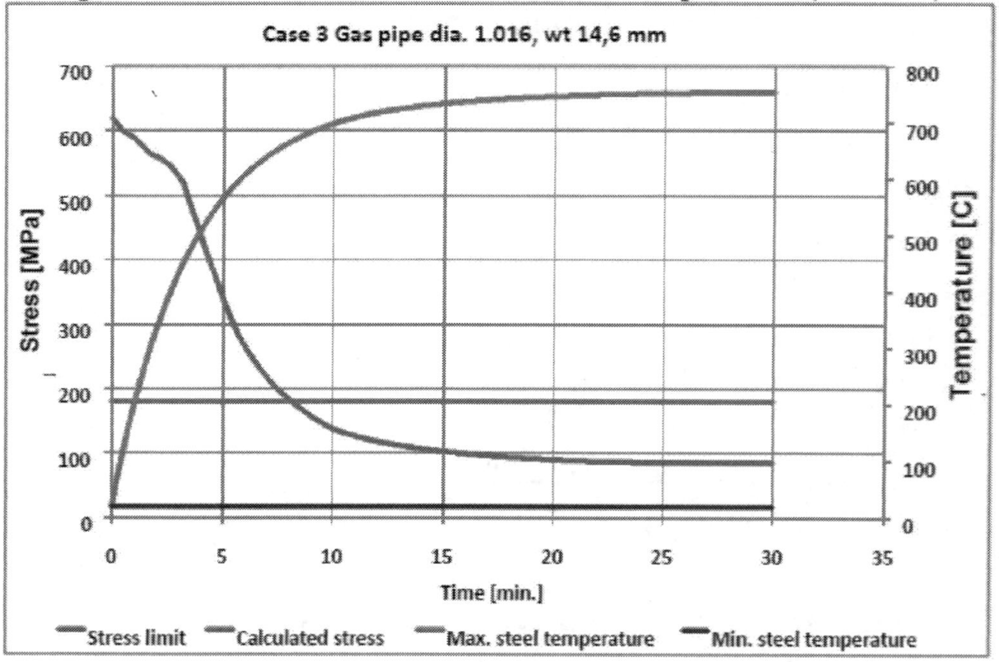

Figure 5. The Stress and UTS at Elevated Temperature (17.5 mm)

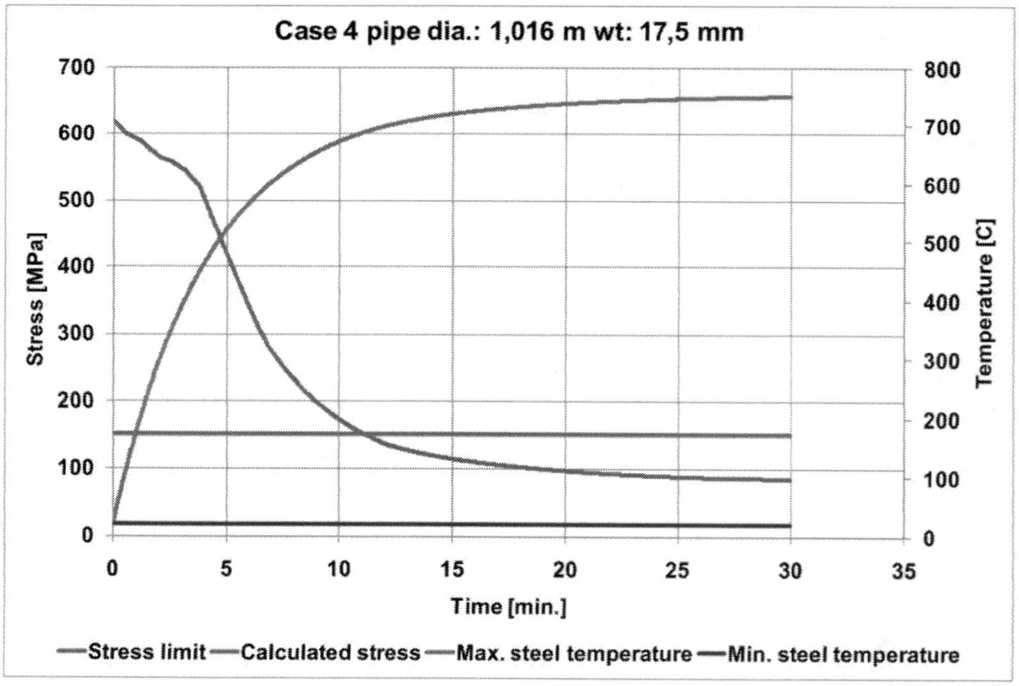

Figure 6. The Stress and UTS at Elevated Temperature (26.2 mm)

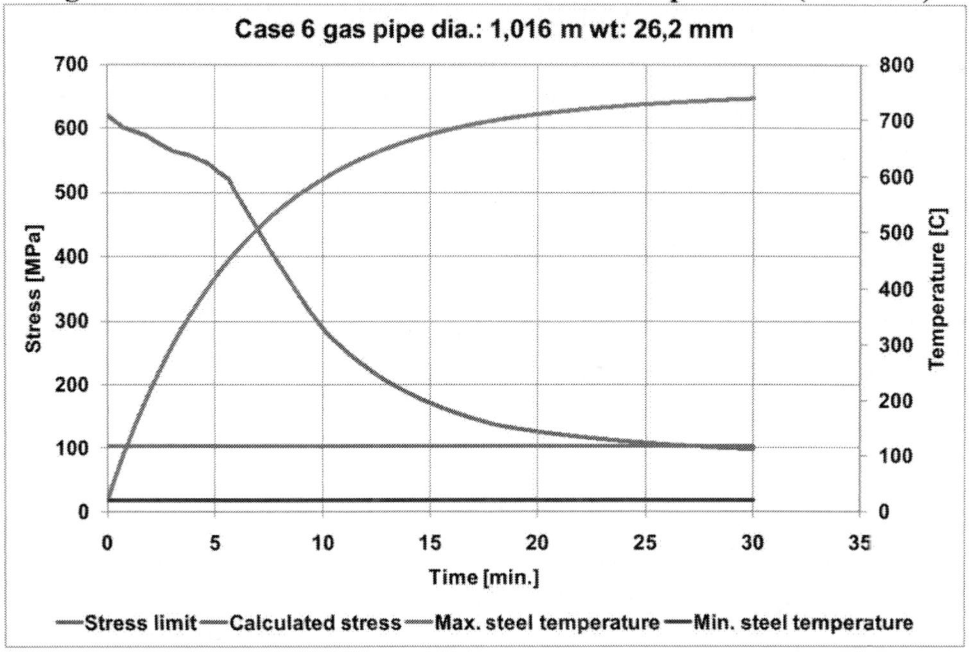

Figure 4 demonstrates that natual gas pipeline (with thickness of 14.6mm) fails at the 8th minutes upon heating, Figure 5 demonstrates that natual gas pipeline (with thickness of 17.6mm) fails at the 11th minutes upon heating while Figure 6 demonstrates that natual gas pipeline (with thickness of 26.2mm) fails at the 26th minutes upon heating

4.2 Sensitivity Analysis

Sensitivity analysis on pipelines with possible thickness of 30mm, 32mm and 36mm are perfromed, with conclusion of natural gas pipeline failing at the 40th, 45th and 60th minutes respectively.

Following conclusion is drawn:

- Drastic decrease in material strength of natural gas pipeline less than structural stress will result in pipeline failure;
- Compared to production oil pipeline, internal gas flow can take away less heat due to small gas specific heat capacity.

5. Discussion

To guarantee natural gas pipeline does not fail due to production oil jet fire, we recommend to reduce buried depth of production oil pipeline or increase that of natural gas pipeline to make sure enough overbarden available for natural gas pipeline as heat insulating material upon jet fire generated from leak of production oil pipeline, and to gurantee integrity of natural gas pipeline. Pipeline buried depth estimation shall follow priciple below:

- For jet fire accident scenario, consider that natural gas pipeline locates out of crater formed by production oil pipeline;
- Buried depth of production oil pipeline is determined as 1.2m as the minimum engineering requirement. Production oil pipeline crater size is estimated as per data below;
- Overbarden thickness is generated based on the condition that buried depth of natural gas pipeline is 1.2 m, and Vessfire is adopted for verification.

Overbarden thickness is finalized as 509mm, as Figure 7 below:

Figure 7. Sectional View of Two Pipelines

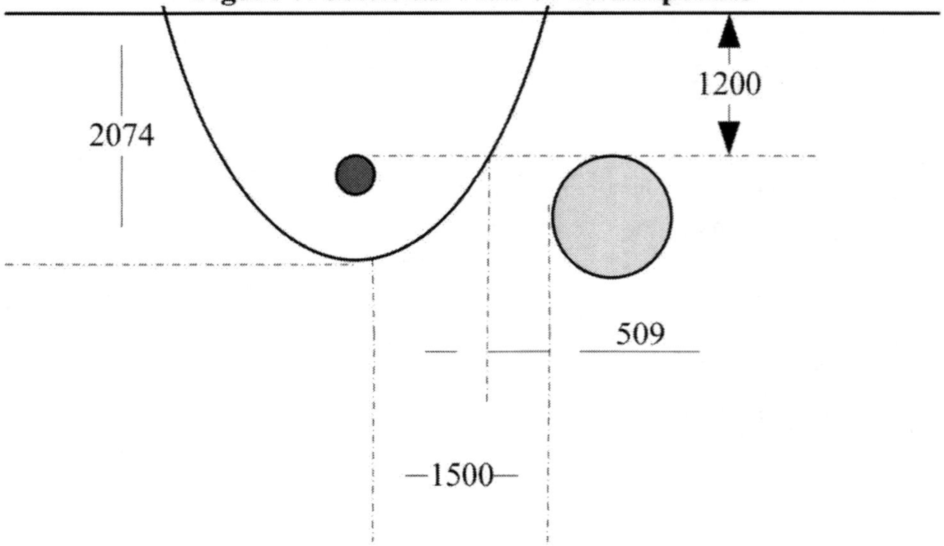

Verification analysis is performed through Vessfire, with soil property below:

- Heat capacity: 890 J/kg*k
- Heat conductivity coefficient: 0.25 W/K*m2
- Density: 1600 Kg/m3
- Thickness: 50mm

The result demonstrates that natural gas pipeline can maintain quite low temperture beyong 6 hours, as shown in Figure 8.

Since soil property is set as constant in analysis, that is, it does not change over time, but in practical condition, soil property will increase over temperature, as a result, Vessfire analysis result is optimistic to some extent.

Verifiction with soil of heat conductivity coefficient of 1 W/K*m^2 is performed, which demonstrates that temperature remains quite low within 6 hours. Meanwhile, since actual overbarden thickness is over 50mm, assumption is arrived that natural gas pipeline will not fail due to jet fire generated due to production oil pipeline leak unde such parallel-running condition and soil cover can provide good heat insulation effect.

Figure 8. Verification Sensitivity Analysis on Overbarden of 50mm

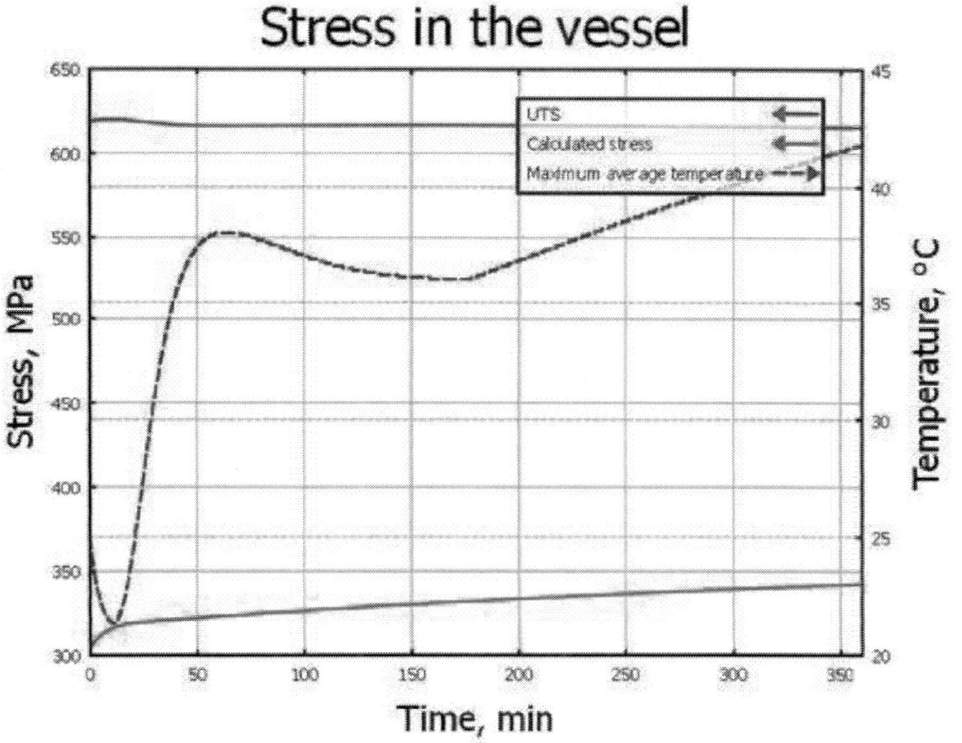

6. Conclusion

Fire consequence analysis is performed on production oil pipeline and natural gas pipeline with parallel-running distance of 1.5m, 1.1m and 0.7m respectively, mainly considering effect on one pipeline if jet fire was generated from failure of the other pipeline.
We find that natual gas pipeline will fail once jet fire is generated due to production oil pipeling leak.

Sesitivity analysis is performed by maintaining parallel distance of 1.5m, with the conclusion that buried depth of natural gas and production oil pipeline is determined as 1.2m.

Under the condition of maintaining parallel distance of 1.1m (with shared tunnel), use composite soil to cover production oil pipeline. Top of upper overbarden is the same as that of top of natural gas pipeline. Depth of soil covered on top of production oil pipeline is 734mm while soil bedding depth is 250mm.

For shared box culverts with parallel distance of 0.7m, lay seperately into two dependent box culverts.

7. References

[1] The Second West-east Gas Line Project Team, "Key Technique Research on Two Parallel-running Pipelines of the Second West-east Gas Line" [R], *Langfang: The Second West-east Gas Line Project Team*, 2009

[2] Pipeline Research Council International (PRCI), "Pipeline Rupture and Parallel Pipe Spacing Study" [R], *US: PRCI*, 2002

1ST CCPS ASIA-PACIFIC CONFERENCE ON PROCESS SAFETY℠

Offshore Rig Kitchen Facility Change Risk

Zufeng Zhang
Scandpower Inc.
Rm.1002, Tower E, Global Trade Center, 36 North
Third Ring Road East, Beijing, 100013, P.R.China
zufeng.zhang@ scandpower.com.cn

Prepared for Presentation at
1st CCPS Asia-Pacific Conference on Process Safety
Kempinski Hotel, Qingdao, China
September 3rd -5th 2013

UNPUBLISHED

Offshore Rig Kitchen Facility Change Risk

Zufeng Zhang
Scandpower Inc.
Rm.1002, Tower E, Global Trade Center, 36 North
Third Ring Road East, Beijing, 100013, P.R.China
zufeng.zhang@ scandpower.com.cn

Keywords: MOC, offshore, rig kitchen facility, gas

Abstract

The kitchen on offshore platform is mostly equipped with electric cooking stove in original design, and then changed to gas stove after start-up since Chinese chef prefers that. The natural gas, as fuel gas for gas stove, is normally introduced from equipment/pipeline in production processing area to kitchen; however, the hazard provoked by this kind of modification has been ignored in most of time.

1. MOC of offshore rig kitchen facility

This platform has been initially designed with electric stove; foam extinguishing system is installed in the kitchen, as well as fire/gas detectors.

Some while after platform start-up, the electric stoves are changed to gas ones, where gas is supplied through a new pipe connected to fuel gas (natural gas) line of hot media furnace. In order to minimize the length, this new pipe goes across the support frame of turbine stack gas flue, along the building wall of living quarter and then passes below living quarter, finally enters kitchen through a hole next to kitchen window in the wall(lifeboat side). Valves are set at gas picking point and also in the kitchen, behind gas stove.

2. Problems raised with the modification

2.1 Inappropriate arrangement of fuel gas pipe

Pipe going below living quarter will cause accumulation of fuel gas once leakage occurs.

2.2 Improper fire/gas detector

Fire and smoke detectors are initially installed, whereas gas stoves are added without combustible gas detector.

2.3 Unmatched Fire extinguishing system

Foam system is set for grease fire, not suitable for fuel gas fire.

2.4 Affecting operation of lifeboat

Fuel gas pipe goes through safety area at lifeboat side, if jet fire happens due to pipe leakage, emergency evacuation will be affected.

2.5 Inadequate valve location

Located behind gas stove, valve is difficult to be approached in case of fire.

2.6 Lack of training for operators

People working in the kitchen are short of training after the modification. Fire from gas stove triggers alarm of fire detector, which is not bypassed, and subsequently activates foam extinguishing system. As a result, spray nozzle of foam system is purposefully blocked.

3. Hazard emerged from the modification

3.1 Modification management

This kind of modification should strictly follow modification management procedures (referring to process safety focus, issue 006), modification scheme and hazard analysis should be made before implementation. Protection and safety equipment should be provided accordingly. Welding inspection and pressure test should be made before operating new fuel gas pipe, safety review as well. Related documents and drawings need to be updated after modification; trainings should be scheduled so that operators could appropriately react in case of fuel gas leaking or fire. Hazard and response for fuel gas leaking/fire need to be added in platform emergency plan.

3.2 Stability of fuel gas

Composition and pressure of fuel gas could change since it's directly from process side, which will be potential risks for gas stove and pipe material.

3.3 Fuel gas pipe

Gas picking point should be on the non-critical equipment/pipeline and could be cut off once ESD is activated; pipe should be well fixed and protected, its path should avoid important safety areas such as firewall, living quarter or any evacuation facilities. Fuel gas pipe should be fireproof and equipped with pressure meter and dual block valves, which are accessible for operators.

3.4 Firefighting system in kitchen

Facilities are able to handle gas fire, grease fire and electrical fire.

3.5 Detection and alarm system

Kitchen and dining room shall install combustible gas detector; existing flame detector should be bypassed in DCS so as to avoid activation of firefighting system due to fake alarm.

3.6 Others

Flue gas discharge path shall be fire-resistant because of possible incomplete combustion and fuel gas leaking; Influence of fuel gas discharge to surrounding facilities such as helicopter operation should be considered; besides, kitchen ventilation, explosion-proof on electrical facilities etc. shall also be set if necessary after hazard analysis.

1ST CCPS ASIA-PACIFIC CONFERENCE ON PROCESS SAFETY℠

Quality Assurance of HAZOP (Hazard and Operability) Study

Zhaohui Wang
Scandpower Inc.
Rm.1002, Tower E, Global Trade Center, 36 North
Third Ring Road East, Beijing, 100013, P.R.China
Zhaohui.wang@scandpower.com.cn

Prepared for Presentation at
American Institute of Chemical Engineers
1st CCPS Asia-Pacific Conference on Process Safety
Qingdao, China
September 4 – 5, 2013

UNPUBLISHED

Quality Assurance of HAZOP (Hazard and Operability) Study

Zhaohui Wang
Scandpower Inc.
Rm.1002, Tower E, Global Trade Center, 36 North
Third Ring Road East, Beijing, 100013, P.R.China
Zhaohui.wang@scandpower.com.cn

Keywords: HAZOP analysis, quality control, PSM management system, evaluate

Abstract

Importance of HAZOP analysis quality is illustrated in a real accident case resulting from failed HAZOP analysis. Firstly, instructions the necessary resources ensure of HAZOP analysis by technology, timing, personnel, organization and management, etc. Secondly, discourses the important part which need quality control during the process of HAZOP analysis through the HAZOP methodological, PSM management system, technical standards and norms, practical work experience, etc. Finally, discusses how to evaluate the quality of HAZOP analysis. The purpose is to fully guarantee the quality of HAZOP analysis, making the analysis work to be properly and accurately carried out.

1. Introduction

Many process safety accidents still happened over the years. One key reason is the quality of process hazard analysis is not assured. This paper introduces an accident due to poor HAZOP study and hence quality assurance of HAZOP study has been discussed and highlighted.

Lessons of process safety accident have not been well learned and the major accidents repeated over the years have raised our concern and thoughts in China. These are due to high but coarse development, production expanding of petrochemical industry, complicated and critical requirements of process safety control and operation, and poor management of process safety. Therefore, the application of process safety management has been promoted from the government, through the industry to operators, in order to prevent the process incidents due to loss of containment. In particular, the hazard and operability (HAZOP) study, as one method of process hazard analysis (PHA) has being gradually recognized, highlighted and applied extensively.

However, there are a couple of issues not properly tackled during the application and spreading of HAZOP, in particular the quality assurance of HAZOP. If this issue were not properly resolved, it may impact the effectiveness of PHA and the performance of process safety management of the company. Furthermore, the establishment of process safety culture, value and the even the development and implementation of Chinese safety policy, regulations and codes may be negatively affected.

2. Lesson from one Process Safety Accident

A process accident happened in the new building hydrogenation unit of a chemical plant during commissioning stage in China. The catalyst for hydrocarbon vapor at reforming furnace were totally damaged, see Figure 1. As a result, the startup of the plant had to be postponed by 6 months and significant economic loss occurred. Methane feed inlet flow meter gave wrong indication (showed low flow), and hence led the production technician and operators wrong judge. Consequently, raw material of steam and methane were fed with wrong ratio. Massive carbon solids formed on the bed of catalyst, furthermore, they penetrated and caused the catalyst entirely damage and useless. This is the primary cause of the process accident, in addition, a couple of failures of the HAZOP study for this process were observed.

Figure 1: Damaged Catalyst

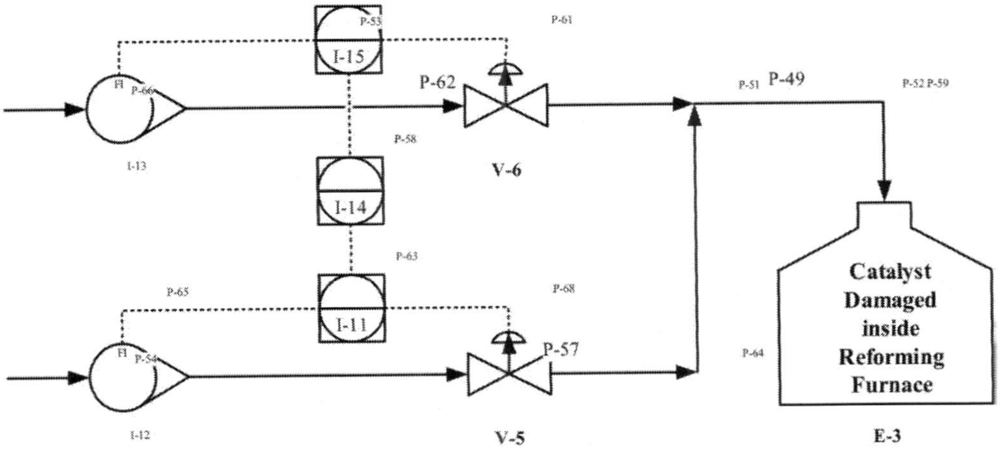

Figure 2: Simplified Process Flow Diagram for Hydrogenation Unit

The hydrogenation process unit was engineered, procured and constructed (EPC) by one engineering company. HAZOP of this process unit was performed by this EPC Company during design stage. See Table 1 below for the record sheet of HAZOP study.

The following minutes of HAZOP analysis were observed:

(1) Deviation: fluctuation of the ratio of steam and methane;
(2) Possible causes: fluctuation of methane feed, steam pressure and total hydrocarbon vapor;
(3) Possible consequences: increased the energy consumption of reforming furnace, carbon solid formed and catalyst carbonized;
(4) Safeguards in place: stabilize the methane feed, regulate the steam supply as per the methane feed, and stabilize the steam pressure;
(5) Recommendations: no need for further actions.

Table 1: Worksheet of HAZOP Study for Reforming System

Node 2: Reforming system		Version: A		
Description: Prior to feed into Reforming Furnace (F24002), refined methane mixes with steam of 3.5 MPa at ratio of 3.5. The mix stream is preheated by 600°C via convection section of F24002 and it enters the heat radiation section of F24002 via upper collecting pipes. Reforming reaction occurs when mixed gas is exposed at high temperature.				
Deviation	**Causes**	**Consequences**	**Safeguards**	**Recommendations**
2.2 fluctuation of the ratio of steam and methane feed	fluctuation of methane feed, steam pressure and total hydrocarbon vapor	energy consumption increased for reforming furnace, carbon solid formed and catalyst carbonized	methane feed stabilized, the steam supply monitored as per the methane feed, the steam pressure stabilized	No
Conclusion: no further actions required				

The following failures of HAZOP analysis were observed:

(1) Guidewords and deviations were used inappropriately. Fluctuation of the ratio of mix feed actually is neither guidewords nor deviation actually, but a consequence that deviation of process control occurs. The failure of applying appropriate guidewords had cause failure of identifying right causes of process deviation.
(2) Causes analyses were logically unclear. The real causes of abnormal process were not identified, such as the causes of more/less flow of methane feed, more/less concentration of total carbon and more/less pressure of steam were failed to identify.
(3) Consequences analyses were not adequate. The consequences analyses of catalyst damage due to flow deviation of gas feed were not clear. Energy consumption increased for reforming furnace, carbon solid formed and catalyst carbonized i.e. these were not real and eventual consequences of accident loss but abnormal process.
(4) Safeguards were not specified clearly. Methane feed stabilized, the steam supply monitored as per the methane feed, the steam pressure stabilized, i.e. these poor operability measures are not engineering control but management measures. Furthermore, these measures did not make sense against the failure of flow meter and control of gas feed.
(5) Recommendations were failure. No useful recommendations/control measures were proposed due to failure of the above analysis. The system was running in a hazardous situation and accident would happen sooner or later!

The analysis above indicated that the HAZOP leader and the team failed to correctly apply the HAZOP technique. The analysis of causes, consequences, safeguards and recommendations were not logically and adequately conducted during the workshop. As a result, this analysis has not followed the requirements of IEC 61508 and wrong conclusion was made. Despite manpower, physical and financial resources had been taken for the HAZOP study of this unit, the result was undesired event due to poor quality assurance of the work.

3. Resource Assurance of HAZOP Study

Adequate resources shall be assured, in order to ensure the quality of HAZOP study. The resources include the following four requirements in terms of organizational management, technical, schedule and competence.

3.1 Management Resource

HAZOP shall be included in the requirement of process safety management, as it is a significant method of process hazard analysis. The responsibility, organization, method, competence evaluation of facilitator, follow-up of recommendations, quality evaluation of HAZOP study report, etc, shall be included in the management procedure of process hazard analysis. In other words, the management procedure shall require the manpower, physical and financial resources for the HAZOP study.

3.2 Technical Resource

HAZOP analysis is one important method of process hazard analysis (PHA), which is closely linked with other elements of process safety management system. It means for the execution of HAZOP study, extensive technical inputs are required from other elements of the system. For example, prior to the study, unified safety criteria of a company shall be developed. Complete and accurate process technology information, i.e. P&ID, PFD and operating procedure, etc. are normally required, and it is also depending on the quality of process safety information (PSI). The element of mechanical integrity shall be required to provide the operating status/feedback of the equipments/ components of the process, which will be also considered/ discussed during the workshop study. In order to improve the quality of HAZOP study, it may also require the inputs/results of other process hazard analysis, i.e. the outcome of HAZID, QRA, etc. In addition, the limitation of HAZOP method shall be considered, in particular, HAZOP and PHA; one element of PSM shall be not confused.

3.3 Time Resource

Time and schedule requirements of HAZOP study are normally affected by the complexity and criticality of the process, competence and experience of the team, scope of work and specific method applied. High quality HAZOP study is depending on the maximum efforts of the team given proper time and schedule. The time fitting selection for HAZOP studies of process life cycle, preparation of the study, control of the schedule, reporting and acceptance review, etc. all these factors affect the quality of HAZOP study in terms of time and schedule. For example, the definition of the scope of the node to be studied, which may impact the quality, i.e. if the node is determined too small, it is normally unable to appreciate the whole system and consuming more time; on the contrary, if the node is determined too big, the detailed level of the analysis cannot be balanced and quality may be affected. The determination of the scope of the node is not only depending on the amount of the equipments, pipes of node, but also depending on the process technology information, equipment and component information, instrument control requirements, criticality of the process material, the competence of the team to appreciate the system and the experience of facilitator. All these may affect the time and schedule requirement for a HAZOP study.

3.4 Competence Resource

Competence requirement of a HAZOP leader is more important than anything else, for assurance of the quality of the study. Most companies require qualification and ability of the facilitators as she/he is the key person of the team, which may determine the quality of the workshop. First of all, the leader shall have ethics of integrity, honesty, objective, impartial and highly responsible. In addition, relevant experiences of process design, operation and safety management together with good communication skill and team management are preferred. The quality is also relying on the inputs and collective wisdom of the team members and hence it also requires the ethics of honesty, open, share and cooperative for team members. Knowledge of various disciplines and operating experience and feedbacks are also key factors for a success study. The quality is doubted if the study were facilitated and performed by a team without appropriate knowledge of discipline, process management experience and site operating experiences. Therefore, the knowledge of process and disciplines for the study should be fully covered, and this is requirement of the competence for the team and quality assurance.

4. Quality Control during Study

It requires correctly understand the technical standards of HAZOP study IEC61882, and accurately apply it during HAZOP study. There are some common mistakes during HAZOP study, including:

(1) Node divided inappropriately;
(2) Design intent of the node being analyzed not well understood;
(3) Deviations not selected properly and/or incompletely;
(4) Causes of deviation and its consequences not analyzed clearly and/or accurately;
(5) Logically wrong from deviations, through causes to consequences and safeguards;
(6) Risk analysis and judgment not reasonable;
(7) Proposed measures not specific and/or without good operability, and so on.

A couple of important elements for HAZOP study shall be properly managed, in order to avoid the problems as mentioned above and hence assure the quality. The requirements to the company and HAZOP study team include:

(1) All members should stick to prudent and objective attitude together with facts to be identified;
(2) All members should have sufficient knowledge of PSM management system and be aware of requirements related;
(3) Some team members should understand the process design and be familiar with the technical specifications;
(4) Some team members should have sufficient operating experience of process unit;
(5) Company should provide solid support for HAZOP organization, management and technical requirements;
(6) HAZOP facilitator should have sufficient knowledge of HAZOP related standards/codes and adequate working experience.

For example, the following technical standards extensively applied should be familiar to HAZOP study team members:

(1) API RP 14 C Recommended Practice for Analysis, Design, Installation and Testing of Basic Surface Safety Systems for Offshore Production Platforms;
(2) API RP 520 Sizing, Selection and Installation of Pressure-Relieving Devices in refineries;
(3) API RP 521 Guide for Pressure-Relieving and Depressuring Systems;
(4) ISO 10418 Petroleum and natural gas industries - analysis, design, installation and testing of basic surface process safety systems on offshore production, installations - Requirements and guidelines;
(5) IEC 61508/61511 Functional safety of electrical / electronic / programmable electronic safety related systems;
(6) Etc, etc.

Furthermore, in order to justify the adequacy of the safeguards for process system in terms of risk management, some principles for process safety should be adhered, as follows:

(1) No single failure (be it technical or operational) shall be sufficient to trigger an unwanted incident;
(2) For critical operations / systems / substances two independent barriers shall be established against unwanted incidents.

The following missions have to be fulfilled by company, so as to improve the quality of HAZOP study:

(1) First of all, the company should establish a right safety value, and continuously provide support for this work. The job conducted with high quality will not only help improve the process safety level of plants, but also enable to effectively promote the organizational safety culture.
(2) Secondly, effective implementation of HAZOP study within process management system provides support and inputs for better implementation of other elements of process safety management. For example, improve the integrity and accuracy for process safety information; enhance the technical knowledge and competence for team members through training; establish and improve the management procedure and technical requirements for

the element of process hazard analysis. In addition, HAZOP quality could be evaluated through the implementation of element of compliance audit of process safety management of the company.

5. Conclusion

Currently, it is quite inspiring that most national enterprises are gradually carrying out HAZOP study work, following the requirements of China policies, regulations and standards. However, the quality of HAZOP study varies from companies to companies, even from plants to plants at one company. Despite HAZOP study has been applied for few decades in foreign countries, there are still many bottlenecks for Chinese enterprises at an early stage of development, such as, gaps process safety management system, lack of sufficient qualified HAZOP study leader, etc. As known to all, Rome was not built in a day, thus we are not able to acquire high quality HAZOP study overnight.

However, the quality of HAZOP study should continuously improved through Plan, Do, Check and Action (PDCA) cycle of HAZOP management procedure within the process safety management system. The responsibility, resources required, and criteria could be defined so as to ensure effective implementation of HAZOP and hence the quality could be assured. In other words, the quality HAZOP study could be assured only relying on effective process, procedure, competent people and implementation.

6. References

[1] API RP 14 C Recommended Practice for Analysis, Design, Installation and Testing of Basic Surface Safety Systems for Offshore Production Platforms

[2] API RP 520 Sizing, Selection and Installation of Pressure-Relieving Devices in refineries

[3] API RP 521 Guide for Pressure-Relieving and Depressuring Systems

[4] ISO 10418 Petroleum and natural gas industries - analysis, design, installation and testing of basic surface process safety systems on offshore production, installations - Requirements and guidelines

[5] IEC 61508/61511 Functional safety of electrical / electronic / programmable electronic safety related systems

[6] IEC 61882 Hazard and operability studies (HAZOP studies) —Application guide

1ST CCPS ASIA-PACIFIC CONFERENCE ON PROCESS SAFETY℠

Thermal Hazards Evaluation on Three Initiators in Solvent under Adiabatic Condition

PENG Min-Jun
15261894065
No. 200, Xiaolingwei Street, Xuanwu Distract, Nanjing City, Jiangsu Province, China
pmj_sun@163.com

LV Jia-Yu
15150514002
No. 200, Xiaolingwei Street, Xuanwu Distract, Nanjing City, Jiangsu Province, China
jiayu-lv@163.com

CHEN Wang-Hua*
13851670846
No. 200, Xiaolingwei Street, Xuanwu Distract, Nanjing City, Jiangsu Province, China
chenwh_nust@163.com

CHEN Li-Ping
13913897044
No. 200, Xiaolingwei Street, Xuanwu Distract, Nanjing City, Jiangsu Province, China
clp2005@hotmail.com

Prepared for Presentation at
American Institute of Chemical Engineers
1st CCPS Asia-Pacific Conference on Process Safety
Qingdao, China
September 4 – 5, 2013

Thermal Hazards Evaluation on Three Initiators in Solvent under Adiabatic Condition

PENG Min-Jun
15261894065
No. 200, Xiaolingwei Street, Xuanwu Distract, Nanjing City, Jiangsu Province, China
pmj_sun@163.com

LV Jia-Yu
15150514002

CHEN Wang-Hua *
13851670846

CHEN Li-Ping
13913897044

Keywords: Initiator; Thermal hazard; Solvent; Accelerating rate calorimeter; Kinetic parameter; Time to maximum rate

Abstract

2,2'-Azobis(isobutyronitrile) (AIBN), 1,1'-Azobis(cyclohexane-1-carbonitrile) (ACCN) and Lauroyl peroxide (LPO) are commonly used initiators in solution radical polymerization. To study thermal hazards of the three initiators in solvent contrastively, accelerating rate calorimeter (ARC) was used to study the adiabatic decomposition processes of AIBN, ACCN, LPO with the same mass in xylene. The characteristic curves and thermal hazard parameters of adiabatic decomposition were obtained, and the relative kinetic parameters were calculated. The correction and analysis results of the measured data show that the onset temperature of AIBN, ACCN, LPO are 61.75℃, 85.08℃, 55.19℃ respectively, the activation energy are 148.45kJ/mol, 154.14kJ/mol, 140.41kJ/mol based on first-order law, and their respective temperature at which the time to the maximum rate under adiabatic condition equal to 24 hours is 60.26℃, 82.91℃, 52.58℃. These all show that the decompositions of the three initiators in xylene have highly thermal hazards (LPO>AIBN>ACCN). So in the relevant processes, we must strengthen temperature monitoring and take effective cooling measures to prevent fire and explosion.

1. Introduction

Radical solution polymerization has been widely applied for more easily avoiding local overheating due to the smaller viscosity than other polymerization processes. The common radical polymerization initiators are azo initiators, such as 2,2'-Azobis(isobutyronitrile) (AIBN), 1,1'-Azobis(cyclohexane-1-carbonitrile) (ACCN), etc., and organic peroxide initiators such as lauroyl peroxide (LPO), etc. But the azo group and peroxy group of such initiators' molecular structures are unstable, so it's highly easy for them to cause fire and explosion in the process of production, using, storage and transportation, and lead great losses of life and property. For example, in November 2003, an AIBN explosion happened in a chemical company in Tongzhou city Jiangsu province, which caused two killed, five injured, and the losses of several million dollars.

The studies of thermal decomposition of initiator in solvent have obtained some achievements both home and abroad. Scholars used the UV/visible spectroscopy[1-4], different scanning calorimeter (DSC)[5,6], dilatometer method[7], nitrogen gas volume measurement[1] to study the decomposition rate of AIBN in different media systems under isothermal conditions. Provder, T *et al* [8] used dynamic DSC to study the thermal decomposition characteristics of AIBN in DBP solvent, calculated the activation energy and pre-exponential factor. Li, X R *et al* [9] dissolved AIBN in acetone and toluene to characterize its pure thermal decomposition behavior by C80 micro-calorimeter. Liu, Y F *et al* [10] found that acetone increased the onset temperature and the activation energy of methyl ethyl ketone peroxide (MEKPO) by dynamic DSC. Wang, C X *et al* [11] found the alcohols and aldehydes decreased the initial decomposition temperature and self accelerating decomposition temperature (SADT) of Di-t-butylperoxide (DTBP), Tert-butyl peroxy-2-ethylhexanoate (TBPEH), Tert-butyl perbenzoate (TBPB) by accelerating rate calorimeter (ARC). To sum up, the thermal hazard experimental studies of AIBN, ACCN, LPO in solvent under adiabatic condition are few.

In order to obtain the thermal hazard parameters of the three initiators above under adiabatic conditions in xylene, ARC was used to study their thermal decomposition processes and the thermal decomposition characteristics, and then their risk evaluations were carried out. These efforts would give some useful information to prevent thermal runaway in the relevant chemical process.

2. Experimental

2.1 Reagent

AIBN, CP, white crystal, was purchased from Shanghai No.4 Reagent and H.V. Chemical Co., Ltd. ACCN, which has a purity of 98%, white crystal, was from American Sigma-Alorich Co. LPO, which has a purity of 98%, white powder, was from Sino-pharm Chemical Reagent Co., Ltd. Xylene, AR, colorless transparent liquid, was from Sino-pharm Chemical Reagent Co., Ltd. All of them were stored in a refrigerator at 6-8℃.

2.2 Apparatus and test conditions

Accelerating rate calorimeter (ARC), produced by the British THT Co., is a common apparatus used to evaluate the thermal hazard potential of energetic materials and measurements are usually carried out under adiabatic conditions. The samples tested by ARC were the AIBN, ACCN, LPO in xylene solvent respectively. The samples were placed in a Ti-LCQ bomb, sealed in air atmosphere, had a 10 min seek for an exothermic signal, which can be detected if the self heaing rate > 0.02℃/min. If the exothermic signal wasn't detected, then the temperature was increased by 5 ℃. The other specific relative experimental conditions were listed in Table 1.

Table 1. Experimental conditions

Sample Number	Sample Name	Sample Mass (Mass Fraction)/g	Temperature Range /℃	Bomb Mass /g
1#	AIBN+Xylene	3.04(16.53%)	40~150	6.256
2#	ACCN+Xylene	3.43(14.67%)	70~180	6.347
3#	LPO+Xylene	2.16(24.76%)	30~140	6.258

3. Results and discussion

3.1 Data analysis

The adiabatic decomposition curves of 1#, 2#, 3# samples are shown in Figures 1 to 4. And the corresponding adiabatic decomposition characteristic parameters are shown in Table 2.

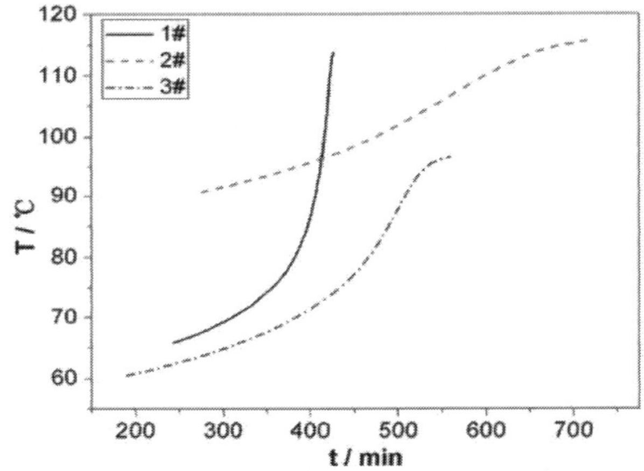

Figure.1 Curves of temperature vs. time for sample 1# to 3#

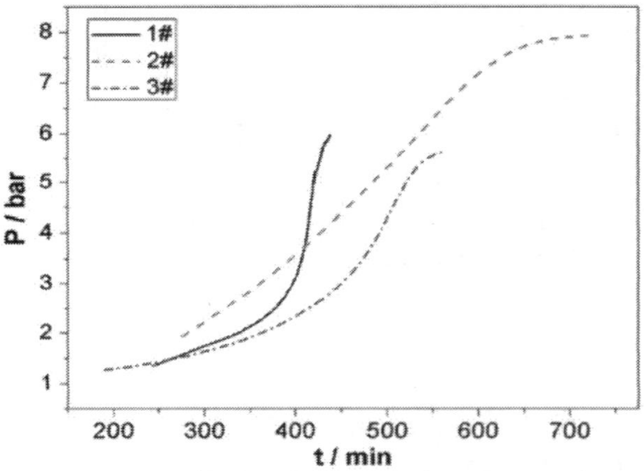

Figure.2 Curves of pressure vs. time for sample 1# to 3#

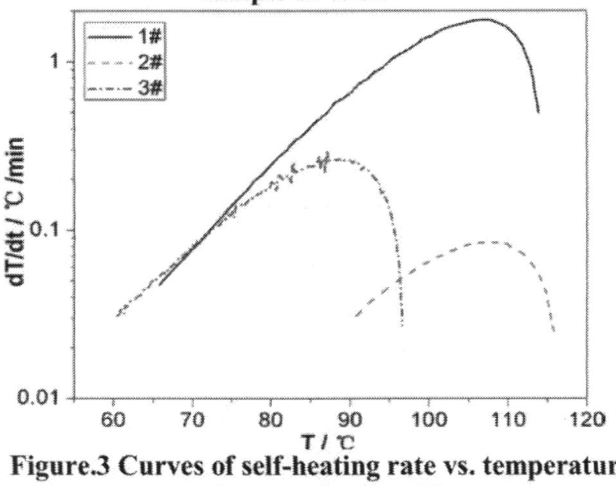

Figure.3 Curves of self-heating rate vs. temperature for sample 1# to 3#

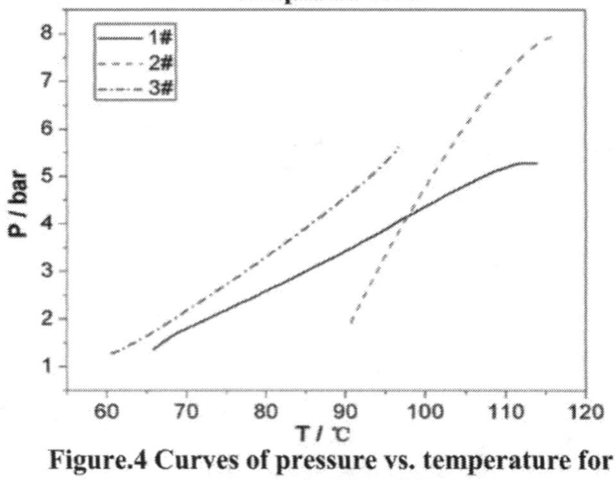

Figure.4 Curves of pressure vs. temperature for sample 1# to 3#

Table.2 Adiabatic thermal decomposition characteristic parameters of 1#、2#、3# samples

Sample Number	T_0 /°C	m_0/°C·min^{-1}	θ/min	m_m/°C·min^{-1}	dP/dt/bar.min^{-1}	T_f/°C	P_{max}/bar	ΔT_{ad}/°C
1#	65.8	0.047	176.7	1.782	0.157	112.9	6.2	47.1
2#	90.7	0.031	297.3	0.085	0.021	114.8	7.9	24.1
3#	60.5	0.031	306.9	0.302	0.057	96.4	5.6	35.9

As shown in Figure 1 to 3 and Table 2, the self decomposition of sample 2 takes place at a about 25°C higher temperature than sample 1# and 3#, and the initial decomposition temperature of sample 1# and 3# has little difference. The initial temperature rise rate of sample 1# is the highest. In the adiabatic decomposition processes of three samples, the temperature and pressure increase with the time. The maximum values of self heating rate of three samples follow the order: 1#>3#>2#, and the time to maximum rate of sample 3# is the shortest. The values of final decomposition temperature of samples 1# and 2# are basically same, but sample 3# decreases about 20°C. Figure 4 shows that the three samples' pressure increases linearly with temperature. When the adiabatic decompositions are over, the pressures of per unit mass are 2.0 bar, 2.3 bar, 2.6 bar, respectively.

3.2 Data correction

Under the real experimental condition of ARC, heat from the exothermic reaction of the sample rises the temperature of both sample and reaction bomb. For accurately comparative analysis of thermal hazards of the above samples, it's necessary to correct relative thermal hazard parameters with the thermal inertia factor.

For an *n*th-order reaction with a single reaction, the self -heat rate of adiabatic system can be expressed as the follows [12]:

$$m_T = dT/dt = \Delta T_{ad} \cdot k \cdot ((T_f - T)/(T_f - T_0))^n \cdot C_0^{n-1} \quad \text{[Eq. 1]}$$

Where m_T is the self heat rate at arbitrary temperature T of adiabatic system; k is the rate constant; A is the pre-exponential factor; E is the apparent activation energy; R is the gas constant; ΔT_{ad} is the measured adiabatic temperature rise; T_f is the measured final decomposition temperature; T_0 is the measured initial decomposition temperature; n is the reaction order ; C_0 is initial concentration.

From the equation, the rate constant can be expressed as the follows:

$$k = m_T/(\Delta T_{ad} \cdot C_0^{n-1} \cdot ((T_f - T)/(T_f - T_0))^n) \quad \text{[Eq. 2]}$$

According to the Arrhenius equation:

$$\ln k = \ln A - E/RT \quad \text{[Eq. 3]}$$

If the order of reaction is chosen properly, the plot of lnk vs. 1/T is expected to be a straight line, and then E and A can be calculated from the plot.

Figure 5 shows the fitting diagram of lnk vs. 1/T based on first-order law. As expected, the fairly linear correlations are obtained. The calculated kinetic parameters of the three initiators are shown in Table 3.

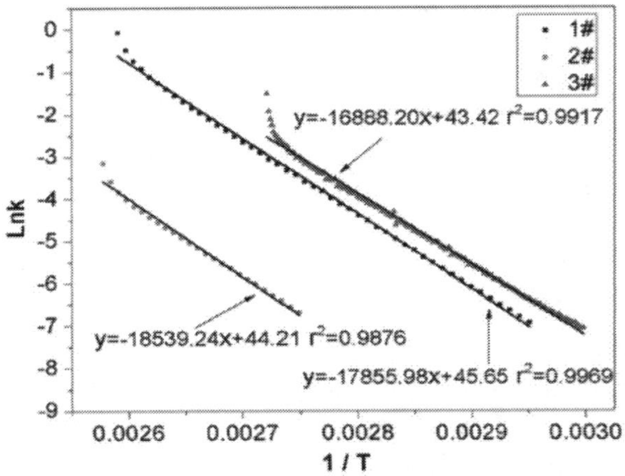

Figure5. Curves of pseudo rate constant vs. temperature for sample 1# to 3#

Table3 Calculated kinetic parameters

Sample Number	E /kJ.mol^{-1}	A/min^{-1}	n	γ^2
1#	148.45	6.69×10^{19}	1	0.9968
2#	154.14	1.59×10^{19}	1	0.9876
3#	140.41	7.20×10^{18}	1	0.9917

After the activation energy calculation, the thermal inertia factor is used to correct the tested parameters, which is defined as follows:

$$\Phi=1+(M_b \cdot C_{v,b})/(M_s \cdot C_{v,s}) \quad \text{[Eq. 4]}$$

where M_s is the mass of the reactive sample, $C_{v,s}$ is the heat capacity of reaction sample, M_b is the mass of the reaction bomb, $C_{v,b}$ is the heat capacity of reaction bomb. The corrected equations for some temperature parameters are as follows [13, 14]:

$$T_{0,s}=[1/T_0+R \cdot \ln\Phi/E]^{-1} \quad \text{[Eq. 5]}$$

$$m_{m,s}=\Phi \cdot m_m \quad \text{[Eq. 6]}$$

$$\Delta T_{ad,s}=\Phi \cdot \Delta T_{ad} \quad \text{[Eq. 7]}$$

$$T_{f,s}=\Phi \cdot \Delta T_{ad}+[1/T_0+R \cdot \ln\Phi/E]^{-1} \quad \text{[Eq. 8]}$$

$$\theta_{m,s}=\theta_m/\Phi \quad \text{[Eq. 9]}$$

where $T_{0,s}$ is the corrected initial decomposition temperature, T_0 is the measured initial decomposition temperature; $m_{m,s}$ is the corrected maximum temperature rise rate, m_m is the measured maximum temperature rise rate; $\Delta T_{ad,s}$ is the corrected adiabatic temperature rise, ΔT_{ad} is the measured adiabatic temperature rise; $T_{f,s}$ is corrected final decomposition temperature, T_f is measured final decomposition temperature; $\theta_{m,s}$ is the corrected time to maximum rate, θ_m is the measured time to maximum rate. And the time required to reach the maximum self-heating rate under adiabatic conditions is called the time to maximum rate (TMR_{ad})[15].

Table4. Corrected parameters

Sample Number	Φ	$T_{0,s}$/°C	$\theta_{m,s}$ /min	$\Delta T_{ad,s}$ /°C	$m_{m,s}$/°C·min^{-1}	$T_{f,s}$/°C
1#	1.88	61.8	94.0	88.6	3.23	150.3
2#	1.80	86.5	165.2	43.4	0.15	126.9
3#	2.24	55.2	137.0	80.4	0.68	135.6

correction, the initial decomposition temperatures of sample 1# and 3# are low and basically same, but the $T_{0,s}$ of sample 2# is still about 25 °C higher than sample 1# and 3#. The maximum reaction rate and final decomposition temperature of three samples follow the order: 1#>3#>2#, and the values of AIBN are significantly larger than ACCN and LPO. The TMR_{ad} of the three samples are low, and the value of sample 1# is the smallest. The adiabatic temperature rise of sample 1# and sample 3# are about 85 °C, while sample 2# is about 40 °C lower. And as shown in Table 2, we can see that three samples all have great maximum pressures and follow the order: 3#>2#>1#, so thermal explosion risk of sample 2# is the lowest in a sealed container. From the above results, it's not enough to comparative analysis the thermal hazard of AIBN and ACCN in xylene from the initial decomposition temperature, adiabatic temperature rise, time to maximum rate and the final decomposition temperature.

3.3 TD_{24}

In industry, the temperature at *TMRad*=24h is an important concept in the thermal hazard evaluation, and it is defined as TD_{24}[16]. Townsend, D etc.[17] formulated the expression of time to maximum rate under adiabatic conditions:

$$\mathbf{Ln\theta=E/RT-LnA} \quad \text{[Eq. 10]}$$

According to Equation 10, the TD_{24} of sample 1#, 2#, 3# are 60.26°C, 82.91°C, 52.58°C, respectively. Generally, the lower value TD_{24} indicates the higher hazard. On this basis, the thermal runaway possibilities of three kinds of initiators in the xylene follow the order: LPO >AIBN >ACCN. Combined with the analysis in 3.2, it can be found that ACCN in xylene is of the best thermal stability.

4. Conclusion

In this paper, the thermal hazards of AIBN, ACCN, LPO in xylene solvent were contrastively studied by ARC. The conclusions are as follows:

(1) Comparing to the thermal hazard parameters of the three samples above, such as $T_{0,s}$, θ, $\Delta T_{ad,s}$, $m_{m,s}$, and $T_{f,s}$, AIBN has the shortest decomposition time, and the thermal explosion severity of AIBN and LPO are basically same.

(2) The decomposition kinetic parameters have been calculated from the measured self-heating rate data based on first-order law. The activation energy of the three samples above are

148.45kJ/mol, 154.14kJ/mol, 140.41kJ/mol respectively, and the pre-exponential factor are 45.65 min^{-1}, 44.72 min^{-1}, 43.42 min^{-1}, respectively.

(3) TD_{24} of the three samples are 60.26℃, 82.91℃, 52.58℃ respectively. These all indicate that the three initiators have great possibilities to cause fire and explosion due to thermal decomposition runaway in xylene, the possibilities of thermal runaway follow the following order: LPO>AIBN>ACCN, and ACCN has the best thermal stability in xylene.

5. References

[1] Tabka, M. T., Chenal, J. M., Widmainer J. M., "Effect of Stannous Octoate on the Thermal Decomposition of 2,2-Azobis(isobutyronitrile)," *Polymer International*, Volume 49, NO. 4, 2000

[2] Overberger, C. G., O'Shaughnessy, M. T., Shalit, H., "The Preparation of Some Aliphatic Azonitriles and Their Decomposition in Solution," *Journal of the American Chemical Society*, Volume 71, NO. 8, 1999

[3] Xu, G., Nambiar, R. R., Blum, F. D., "Room-temperature Decomposition of 2,2-Azobis (isobutyronitrile) in Emulsion Gels with and without Silica," *Journal of Colloid and Interface Science*, Volume 302, 2006

[4] Li, H. P., Liu, J., Zhang, H. F., Han, B. X., "Effect of Co-solvent and Pressure on the Thermal Decomposition of 2,2′-Azobis-(isobutyronitrile) in Supercritical CO_2," *Chinese Chemical Letters*, Volume 12, NO. 8, 2001

[5] Bessiere, J. M., Boutevin, B., Loubet, O., "Determination of Kinetic Parameters for Isothermal Decomposition of Azo Initiators of Polymerization by Differential Scanning Calorimeter," *Polymer Bulletin*, Volume 30, NO. 5, 1993

[6] Lazar, M., Ambrovic, P., Mikovic, J., "Thermal Decomposition of Azobisisobutyro Nitrile in the Solid Phase," *Journal of Thermal Analysis*, Volume 36, 1973

[7] Huang, H. J., Yan, J., Zhang, W. Y., Wang, X. F., "Comparative Study and Theoretical Analysis of Characteristics of Polystyrene Aggregation Initiated by Three Kinds of Initiators," *Chemistry*, NO. 8, 2002

[8] Provder, T., Holsworth, R. M., Grentzer, T. H., "Use of the Single Dynamic Temperature Scan Method in Differential Scanning Calorimeter for Quantitative Reaction Kinetics," *Advances in Chemistry Series*, Volume 203, NO. 13, 1983

[9] Li, X. R., Wang, X. L., Koseki, H.,"Study on Thermal Decomposition Characteristics of AIBN," *Journal of Hazardous Materials*, NO. 159, 2008

[10] Liu, Y. F., Tseng, J. M., Wu, T. C., Shu, C. M., "Effects of Acetone on Methyl Ethyl Ketone Peroxide Runaway Reaction," *Journal of hazardous materials*, Volume 153, 2008

[11] Wang, C. X., Xie, C. X., Chen, Q. F., Xu, J., Wu, Y. M., "Effect of Impurities on Self-accelerating Decomposition Temperature(SADT) of Organic Peroxides," *Journal of Qingdao University of Science and Technology(Natural Science Edition)*, Volume 28, NO. 2, 2007

[12] Liu, R.H., Chen, W. H., Hu, Y. T., Safety principle and assessment technology for dangerous chemicals, Beijing: Chemical Industry Press, 2004

[13] Wilberforce J. K., "Comparison of Method of Determination of Adiabatic Times to Maximum Rate of Exothermic Reactions," *Journal of thermal analysis*, Volume 25, 1982

[14] Zhang, Z., Jiang H., Huang, P., "Study of Hydrogen Peroxide Explosion Mechanism and its Prevention Measures," *Chemical Safety and Environment (in Chinese)*, Volume 7, NO. 4, 2007

[15] Stoessel, F., Thermal Safety of Chemical Processes: Risk Assessment and Process Design , Chen, W. H., Peng, J. H., Chen, L. P, Translate. Beijing: Science Press, 2009.

[16] Grewer, T., Fruip, D. J., Harrison, B. K., "Prediction of Thermal Hazard of Chemical Reactions," *Journal of Loss Prevention in the Process Industries*, Volume 12, 1999.

[17] Townsend, D. I., Tou, J. C., "Thermal Hazard Evaluation by an accelerating rate calorimeter," *Thermochimica Acta*, Volume 37, 1980.

1ST CCPS ASIA-PACIFIC CONFERENCE ON PROCESS SAFETY℠

The Experimental Study on Explosion-proof Property of HAN Explosion Suppression Material to Premixed Methane-air Flame Propagating

Yi Liu
National Key Laboratory of Fire Science, University of Science and Technology of China
Anhui Hefei, China
Chemical engineering collage, China University of Petroleum
Shandong Qingdao, China
liuyi@upc.edu.cn

Tongshan Yin
Mechanical and electrical engineering college, China University of Petroleum
Shandong Qingdao, China
ytsupc@163.com

Prepared for Presentation at
American Institute of Chemical Engineers
1st CCPS Asia-Pacific Conference on Process Safety
Qingdao, China
September 4 – 5, 2013

UNPUBLISHED

The Experimental Study on Explosion-proof Property of HAN Explosion Suppression Material to Premixed Methane-air Flame Propagating

Yi Liu
National Key Laboratory of Fire Science, University of Science and Technology of China
Anhui Hefei, China
Chemical engineering collage, China University of Petroleum
Shandong Qingdao, China
liuyi@upc.edu.cn

Tongshan Yin
Mechanical and electrical engineering college, China University of Petroleum

Keywords: HAN explosion suppression material, premixed flame, explosion suppression, the speed of flame propagation, explosion overpressure

Abstract

Inflammable and explosive dangerous chemical accidents occurred frequently in recent years, and new HAN explosion suppression materials have been developed with their good suppression to fire and explosion. The investigation into premixed flame propagation and pressure wave in pipes installed porous materials of aluminum is great important for explosion suppression in pipe and design of flame arresters.

The experimental apparatus for suppressing the combustion and explosion of flammable gases were set up in the paper. The premixed methane-air flame propagating and suppression with different volume of methane were studied experimentally. The explosion suppression performance of porous materials of aluminum on the premixed flame propagation was analyzed, which includes the effects of the packed density, blank rate and installing position of HAN material on flame propagation speed and explosion overpressure. Based on the influence of each parameter on explosion suppression performance, the best filling of HAN explosion suppression material was presented. The results show that the best packed density is 15.4 g/L, the smaller blank rate is better, and the installing position shall be close to the ignition.

1. Introduction

HAN is the abbreviation of "Hypostasis Anchor-hold No-explosion". HAN explosion-proof technology effectively prevents the flammable, explosive, gaseous, liquid dangerous chemicals exploding in storage and transportation caused by an accident (static electricity, welding, shot, collision, faulty operation, etc.), and fundamentally guarantees the flammable liquid and gas in storage and transportation be in the nature of the security[1,2]. America successfully developed suppression material of polyurethane foaming plastic in the 1960s, and was immediately used in the United States air force fighter jet[3]. In the 1970s, they developed the resistance to hydrolysis of polyether foam plastics. Canada developed the reticular aluminum alloy explosion suppression materials later. Foreign industrial and commercial has used gasoline containers with the reticular aluminum alloy explosion suppression material for many years, and has been successfully applied to various kinds of the vehicle's fuel tank, in daily life can also be used for liquefied petroleum gas (LPG) storage container.

Research on explosion suppression materials in China started later, began to develop in the late 1980s, and in 1992 got the appraisal of ministry technology. In 1992, the development explosion suppression performance of the reticular aluminum alloy explosion suppression material which was developed by original weapons industry in our country reached international advanced level of similar products. The research on reticular aluminum alloy explosion suppression material was mainly from the selection of raw materials, different structure size, installation, use, explosion suppression mechanism, test methods, etc. Yuan Tian, etc.[4-6] had studied the packed density, blank rate affecting the performance of mesh aluminum alloy explosion suppression, and found that the pressure wave transmission is suppressed by the changed container inner cavity shape which is filled with the reticular aluminum alloy material; At the same time they can absorb the most of the heat energy released by combustion, thus reach explosion suppression effect. Shuyou Wang[7] found that aluminum alloy explosion suppression material is able to suppress the explosion pressure of 1.40MPa to 0.14MPa, and the packing density in 28-32kg/m3 is best. ZhiXiang Xing[8] chose different blank rate and packed density to design test, got the explosion pressure of liquefied petroleum gas and air mixture, and recorded the flame propagation process with a high-speed camera, when filled with materials. Then explosion suppression performance of the material were studied comparing with air explosive. This paper established the combustible gas explosion and explosion suppression experimental device, and the suppression of HAN explosion-proof material was studied in methane/air premixed flame propagation.

2. Experiment system and process

2.1 experiment system

Figure 1 is the self-designed combustion and explosion system of combustible gas. Experiment platform including combustion tube, gas distribution system, ignition system, logical time control system, data (pressure and flame signal) acquisition system, data recording system.

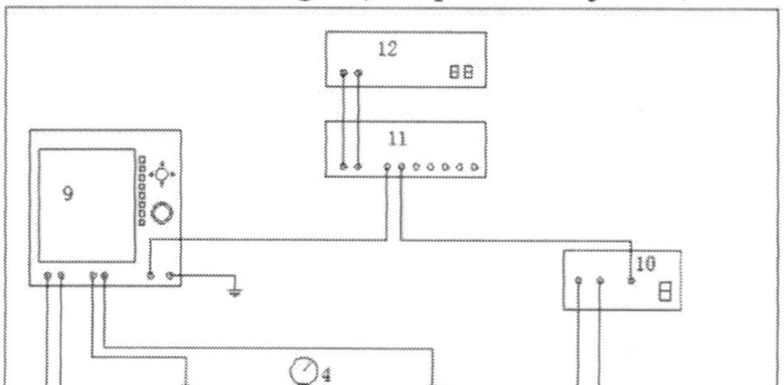

Fig.1 Experiment apparatus schematic diagram

Note: 1 Combustion tube; 2 Pressure sensor; 3 Flame sensor; 4 Vacuum pressure gauge; 5 Spark plug; 6 Explosion venting membrane; 7 Vacuum pump; 8 Distribution system; 9 HIOKI8860 type data acquisition instrument; 10 Ignition; 11 PLC; 12 Power supply

The explosion container is a 2m long carbon steel pipe, whose inner diameter is 50mm and outer diameter is 60mm. The vacuum pump and the inlet pipe are 200mm far from ignition electrode. A pressure sensor and a flame sensor are installed at 600 mm and 1400 mm. The vacuum gauge is in the central of the combustion tube, and guarantee the whole tube complete sealed. System can realize premixed methane air flame temperature and pressure monitoring, and calculate the flame velocity changes before and after the loading explosion-proof material, explosion pressure change and temperature change, etc. In the end the explosion-proof property of porous aluminum alloy HAN explosion suppression material can be evaluated on this basis.

2.2 Experiment material and explosive media

The porous aluminum alloy HAN explosion-proof material adopted by this experiment is produced by Jiangsu Ampute Explosion Prevention Technology Co. Ltd. The materials are made by hot rolling way, and change the special mechanical properties of the aluminum foil cutting form a reticular structure material.

The explosive media is the mixture of methane and air, and the fraction of methane is respectively 8%, 10%and12%.

2.3 Experiment process

1. Check the tightness of device, during the experiment the pipe often be pumped into vacuum state and removed, which greatly increases the possibility of pipeline leak, so before each experiment the tightness of the pipeline must be checked. Open all device into the ready state.
2. Open the air mass flow controller, preheat 20 min, and adjust the given velocity.
3. Open the combustion tube inlet valve, open the vacuum pump valve and start the vacuum pump, close all valves and vacuum pump until the precision pressure vacuum gauge is -0.1MPa.
4. Open the two cylinder valve, open the two pressure reducing valve adjust the pressure of reducing valve to 0.3MPa, and open the combustion tube inlet valve. When precision pressure vacuum gauge shows 0MPa, close all valves.
5. Charge 25kv high voltage to the igniter, press the reset button. Start the PLC, and record the pressure and temperature of the pipe.
6. Record the experimental results.

3. Results and Discussion

3.1 The influence of different packed density on flame

Fill different quality explosion-proof material between two points in the circular pipe the pipeline, choose the mixed gas of methane and air as the research object which volume ratio are 8%, 10% and 12%, study the changing rule of the flame speed and overpressure explosion. The speed change and the explosion overpressure of the flame passed the explosion-proof material is the important basis for judging explosion-proof property.

Different quality of HAN explosion-proof material (respectively30g, 40g, 50g, 60g, 70g or 7.69 g/L, 10.26 g/L, 12.82 g/L, 15.39 g/L, 17.95 g/L) is filled in a certain position, record the maximum explosion overpressure through the second pressure sensors. The results are shown in table 1 and figure 2.

Table 1. The max explosion overpressure with different packed density

Packed quantity g	Packed density g/L	Pmax (MPa) 8%	10%	12%
0	0	0.0595	0.085	0.081
30	7.69	0.0394	0.083	0.081
40	10.26	0.037	0.05625	0.0557
50	12.82	0.032	0.0485	0.049
60	15.38	0.0295	0.048	0.041
70	17.95	0.03	0.048	0.041

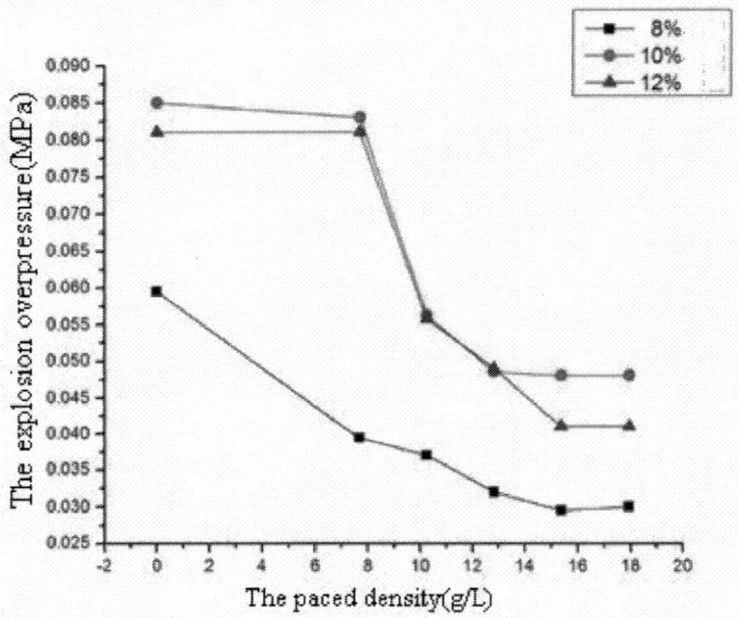

Fig.2 The max explosion overpressure with different packed density

From the diagram, explosion overpressure has obvious decline before and after HAN explosion-proof material is filled in the pipe. When the packed density is less than 7.7 g/L, explosion suppression effect is not obvious for 10% and 12% volume ratio of methane and air mixture, and the explosion suppression effect appears only when the packed density is greater than this value. Overpressure decreases with the increase of filling density for 8% and 12% methane and air mixture, and the explosion overpressure reaches the lowest when packed density is 15.4 g/L. For 10% methane air mixture, while packed density reduces to 12.8 g/L explosion overpressure is close to the minimum.

The above comprehensive results indicate, when the packed density is 15.4 g/L, the impact on the explosion overpressure is almost close to the maximum. If packed density increases, the suppression of explosion overpressure will be not obvious, and only greatly increase the project expenses. So the best packed density of HAN explosion-proof material is 15.4 g/L under this experimental condition.

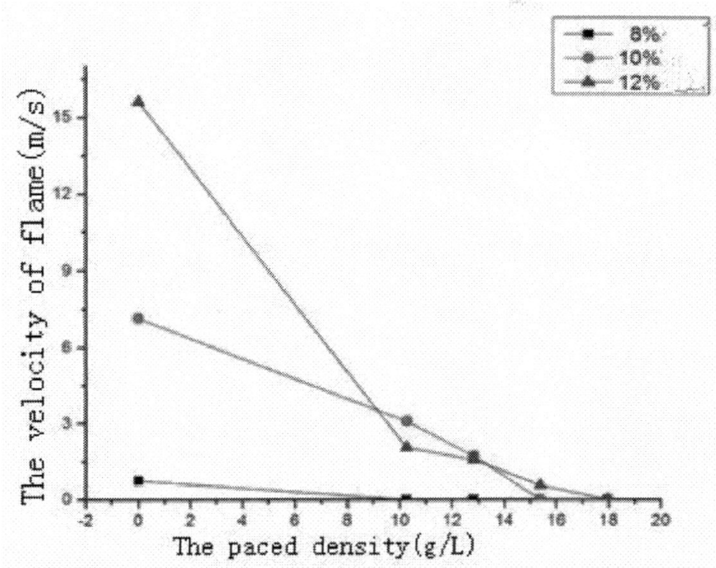

Fig.3 The velocity of flame with different paced density

As can be seen from the above, the speed of the flame will be significantly lower after HAN explosion-proof material is filled in the pipe, and the flame may be extinguished with the increase of packed density. When the packed density is 10.3g/L, the flame will be extinguished for 8% methane air mixture. For 10% methane air mixture, flame speed significantly decrease between two points with the increase of packed density, and when the packed density is 15.4 g/L the flame will be extinguished. When the packing density is 17.9 g/L, the flame will be extinguished for 12% methane air mixture.

Because the 10% methane air mixture has a larger explosion overpressure, destructive power is stronger, the flame is extinguished with the 15.4 g/L packed density. To increase packing density will increase the cost. So the best packed density of HAN explosion-proof material is 15.4 g/L under this experimental condition.

3.2 The influence of different blank rate on flame

By adjusting the explosion-proof material accounted for the entire container volume, to verify the influence of different blank rate for explosion-proof material performance. Its influence on the explosion overpressure and flame propagation is shown in table 2.

Table 2. The influence of different blank rate

Blank rate%	Pmax (MPa)					
	8%		10%		12%	
60	0.032	extinguished	0.0585	passed	0.049	passed
50	0.025	extinguished	0.044	extinguished	0.032	extinguished
40	0.019	extinguished	0.0485	passed	0.044	passed

Since the blank rate is reduced, the density of explosion-proof material itself also gradually decreases, and thus the phenomenon "pass-extinguish-pass" will be appear. For 8% methane air mixture, the flames are not through HAN explosion-proof material, and the explosion overpressure also gradually reduces with the decrease of the blank rate. For 10% and 12% methane air mixture: under the condition of the flame extinguished, the explosion overpressure is

gradually decreased with the decrease of the blank rate; the explosion overpressure of passed flame with a smaller blank rate is higher than extinguished flame with a larger blank rate.
It can be seen that the explosion overpressure will be reduced accordingly with the decrease of the blank rate whether the flame pass or not. But whether the flame extinguish or not, which is also quite important factor influence on explosion overpressure.

3.3 The influence of different installing position on flame

A certain quality of the explosion-proof material is filled in the pipe, and verify the different installing position affecting the performance of the explosion-proof material by adjusting the distance between explosion-proof material and the ignition. Due to the flame resistance between two points is different, the speed of the flame won't be discussed.

Table 3. The influence of different installing position

The distance to ignition/m	Pmax (MPa)					
	8%		10%		12%	
0.6	0.032	extinguished	0.0485	passed	0.049	passed
0.4	0.024	extinguished	0.042	extinguished	0.07	passed
0.2	0.012	extinguished	0.026	extinguished	0.062	passed

When the HAN explosion-proof material close to the ignition, the explosion overpressure will decrease, and the flame may also extinguish with the distance decreasing. The flame cannot pass explosion-proof material for the 8% methane air mixture, and explosion overpressure will decreases with the decrease of the distance from the ignition. For the 10% methane air mixture, when the distance between HAN explosion-proof material and the ignition is less than 0.4 m, the flame will be extinguished, and the explosion overpressure will decrease as the distance decreases. For the 12% methane and air mixture, flame fails to pass. When HAN explosion-proof material is between two points, due to the blocking of the explosion-proof material to the stress wave, the explosion overpressure is small. But when explosion-proof material close to the ignition, explosion overpressure will increase somewhat where no explosion-proof material is under the measure point.
This shows: when the flame passes, HAN explosion-proof material has obvious inhibiting function for explosion overpressure; the flame will be easier extinguished, and the explosion overpressure will be smaller, when explosion-proof material closes the ignition position.

3.4 Analysis of the flame and pressure wave

3.4.1 Pressure analysis in empty circular pipe

When explosion-proof material is not filled, the flame propagation speed is accelerated, and the pipe pressure is gradually increased with the spread of flame. The overpressure and flame changes of 12% methane air premixed flame are shown in figure 4.

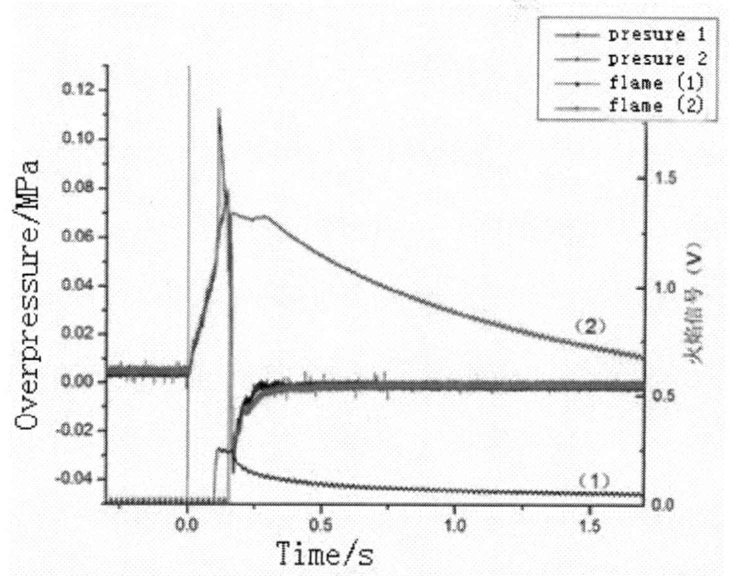

Fig.4 The pressure and flame signal without explosion-proof material

The changes of explosion overpressure can be divided into four stages from the figure. Take the second point for example: 0-0.1s is the precursor shock wave stage, 0.1-0.15s is the booster stage, 0.15-0.2s is the depressurization stage, and it is the aftermath stage after 0.2s. The light intensity changes of flame is measured by the voltage value, the changes of flame combustion can be divided into three stages from the flame signal: 0-0.1s is the no burning stage, 0.1-0.15s is a burning stage, and it is the combustion phase after 0.15s.

Gas begins to burn after the spark plug ignites, burning releases heat to make the temperature of the gas soar, the volume rapidly expand and lead to form the precursor shock wave. The overpressure of measuring point gradually rises with the arrival of precursor shock wave. When the flame go through measuring point will produce a small pressure abruptly rise process, this is caused by high temperature on the surface of the flame, and the explosion overpressure peaks at this time. The pressure suddenly drops after flame front passes the measuring point, because as the spread of flame the unburned gas is pushed forward, a negative pressure zone will appear after the flame front.

3.4.2 Pressure analysis filling with explosion-proof material

The flame propagation speed and explosion overpressure won't increase all the time after pipe is filled with a certain amount of explosion-proof material. The overpressure and flame changes of 12% methane air premixed flame with 60g HAN explosion-proof material are shown in figure 5.

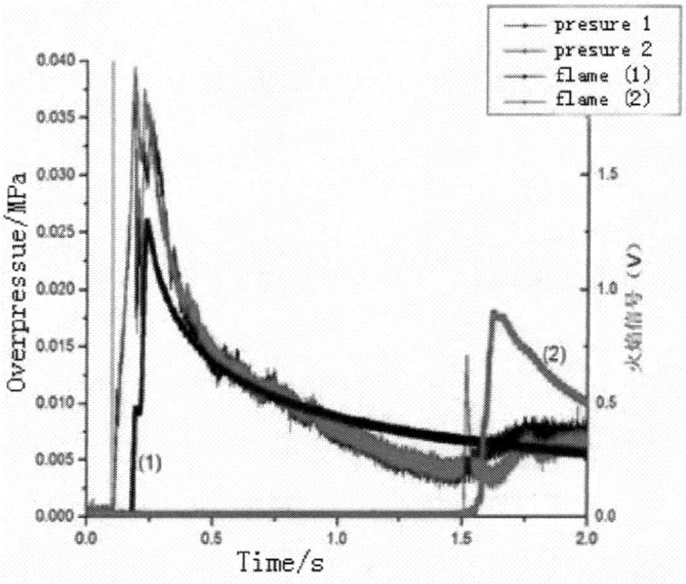

Fig.5 The pressure and flame signal with explosion-proof material

After the pipe is filled with HAN explosion-proof material, the overpressure changes basically similar to not filling.

At the first point, the precursor pressure wave rises faster than not filling. When precursor pressure wave push the unburned gas to move to the other end, which is blocked by the explosion-proof material, then the gas accumulate between the flame front and explosion-proof material, and pressure rise abruptly.

Flame front passes the measuring point, the pressure can produce a peak in a moment, this is never appear when there is no explosion-proof material. Flame front goes through measuring point, the temperature is higher, resulting in gas expansion, but gas cannot be timely to spread around because of explosion-proof material blocking, the pressure instantly increases and the peak appears.

4. Conclusion

1. Establish the combustible gas explosion and explosion suppression experimental system. System realizes the programmable ignition and automatic collection of data, the pressure and speed of flame collected by the pressure and flame signal sampling system can reflect the development process of premixed flame propagation in the combustion pipe.

2. When the packed density of HAN explosion-proof material is 15.4 g/L, the explosion-proof property will be the best under the condition of this experiment and the density of explosion-proof material itself is 38.5 g/L. When the packed density continues to increase, the increase of explosion suppression performance will not be obvious. The relationship between the explosion suppression performance and the blank rate is acquired. When the flame is extinguished, the blank rate is smaller and the explosion suppression performance is better. If the blank rate and the packed density are the same, and the flame is also extinguished, explosion-proof material is closer to the ignition position and the explosion-proof performance is better. At the same time the paper verifies it is an important index of explosion-proof property whether the explosion-proof material makes the flame extinguished or not.

3. Through the comparative analysis of the pressure wave and the explosion overpressure with the HAN explosion-proof material or not, acquire the process of overpressure change, and verifies the HAN explosion-proof material good explosion suppression effect.
Due to the explosion wave number is the result of the interaction between flame and pressure wave, the study on explosion disaster prevention and control technique preventing the spread of flame with lower explosion overpressure is not comprehensive enough. In order to establish more perfect theory of explosion suppression structure through the theoretical exploration of the deeper fire explosion suppression mechanism, a large amount of experimental study and theoretical analysis need to do.

5. Acknowledgement

This study was supported by the open subject of national key laboratory of fire science (No : HZ2009-KF14).

6. References

[1] RU Chengyou, Wang Dexian, "Explosion-proof Principle and Application of Han Explosion-proof Technology", *China Science and Technology Information*, Volume 18, NO. 5, 2006

[2] "The United States Military standard MIL-B-87162, The Mesh Aluminum Protective Materials Used in Explosion-proof of Fuel Tank of Plane", 1982

[3] CHEN Sining, "Application of Inherent Safety Explosion-proof Technology in Oil Storage & Transportation Device", *Procedia Engineering*, Volume 15, 2011

[4] TIAN Yuan, GU Wei-fang, TIAN Hong, "The Performance of New Reticulate Aluminum Alloy Fire and Explosion Suppression Material and Its Application", *Industrial Safety and Environmental Protection*, Volume 33, NO 3, 2007

[5] LU Xu-jie, DING Bang-qin, "Research on Aluminum-magnesium Alloy Explosion Suppression Material and Its Performance", *Journal of Shenyang Institute of Engineering: Natural Science*, Volume 2, NO 2, 2006

[6] HE Hong-wen, CHENG Jin-yuan, "Experimental Verification on The Repressive Function to Explosion for Aluminum Alloyreticular Material", *Electric Explosion Protection*, Volume 14, NO 2, 2005

[7] WANG Shu-you,ZHENG Ying-min,GU Xiao-hui, "Experimental Research on Explosion Suppression of Aluminum Alloy Material", *Explosive Materials*, Volume 34, NO 3, 2005

[8] XING Zhi-xiang, ZHANG Yi-guo, MA Guo-liang, "Research on Explosion Suppression Performance of Reticulate Aluminum Alloy Explosion Suppression Material", *China Safety Science Journal*, Volume 22, NO. 2, 2012

1ST CCPS ASIA-PACIFIC CONFERENCE ON PROCESS SAFETY℠

A Technical Solution for the Optimization of Gas Detectors' Placement in High-sulfur Crude Oil Processing Installations

ZHANG Bo
Department of Safety Science and Engineering, Centre for Offshore Engineering and Safety Technology, China University of Petroleum (East China)
No. 66 Changjiang West Road, Qingdao, P.R.China
zhangbo@upc.edu.cn

JIA Yan
Department of Safety Science and Engineering, China University of Petroleum (East China)
No. 66 Changjiang West Road, Qingdao, P.R.China
jyan716@163.com

XIONG Wei
Engineering Training Center， Shandong University
No. 12550 Second Ring East Road, Engineering Training Center, Jinan, P.R.China
xiongwei@sdu.edu.cn (corresponding author)

WANG Lei, WANG Zhi-gang
Centre for Offshore Engineering and Safety Technology, China University of Petroleum (East China)
No. 66 Changjiang West Road, Qingdao, P.R.China
13256891019@163.com, sdwangzhigang@foxmail.com

1st CCPS Asia-Pacific Conference on Process Safety
Kempinski Hotel, Qingdao, China
September 4-5, 2013

UNPUBLISHED

A Technical Solution for the Optimization of Gas Detectors' Placement in High-sulfur Crude Oil Processing Installations

ZHANG Bo
Department of Safety Science and Engineering, Centre for Offshore Engineering and Safety Technology, China University of Petroleum (East China)
No. 66 Changjiang West Road, Qingdao, P.R.China
zhangbo@upc.edu.cn

JIA Yan
Department of Safety Science and Engineering, China University of Petroleum (East China)

XIONG Wei
Engineering Training Center , Shandong University

WANG Lei, WANG Zhi-gang
Centre for Offshore Engineering and Safety Technology, China University of Petroleum (East China)

Keywords: high-sulfur crude oil processing; gas detector; optimization of placement; hydrogen sulfide leakage; CFD

Abstract

With the fast development of petroleum and chemical industry in China, the leakage accidents of hazardous gases of petrochemical plant are increasing rapidly. Once the hazardous gases leaked, it will result in serious human casualty, property loss and environmental pollution. High-sulfur crude oil processing is prone to cause significant risk of hydrogen sulfide leakage. This paper tried to find out the possible leakage sources, studied its leak probabilities and put forward a method to enhance the detection efficiency and accuracy of hydrogen sulfide leakage.

The accident scenarios most likely to happen could be acquired by the leak probability analysis of typical installations. In order to do this, a three-dimensional CFD (Computational Fluid Dynamics) model for predicting hydrogen sulfide leakage and dispersion consequences in typical installation was set up. Gas detector positions were set as monitor points in the model, and the model parameter sensitivity analysis were carried out. What's more, a downscaling installation model was used to test correctness of the model. Then, the validated CFD model could be used to predict real-time gas concentration fields for the leakage scenarios most likely to happen. The horizontal spacing, vertical height, selection of flammable or toxic gas detectors and other aspects of optimization analysis could be decided according to gas leakage and dispersion variation. The optimization scheme should be suitable for the safety investment benefits relationship by the safety economics decision analysis. The results could obtain the best leakage detection optimization scheme of high-sulfur crude oil processing.

Based on the above research, a technical scheme for the optimization of dangerous gas detection and alarm instrument layout in high sulfur crude oil process installations based on CFD and fuzz probability analysis was proposed. The scheme is expected to enhance the efficiency of gas leakage detection, prevent fire explosion and personnel poisoning accident. It's an important way to improve inherent safety degree in the design stage of refinery plant. The results of the project can reduce accident risks in view of hydrogen sulfide leaks, enhance the prevention and emergency rescue abilities for such accidents, as well as improve the level of work safety in high-sulfur crude oil processing.

1. Introduction

With the rapid development of national economy, the demand for oil and gas resources is increasing gradually in China. According to the prediction provided by some research institutions, by the end of "13th Five Year Plan" of China, the import volume of crude oil will stay at relatively high level. Statistics demonstrate that over the past decade the proportion of high-sulfur crude oil in world's oil outputs has been continuously increasing, and the average growth is about 100 million tons each year [1]. So the high sulfur crude oil processing ability of oil refinery industry has to be enhanced in the future for a very long stage.

However, the in-service oil refining facilities in China were mainly designed for processing of low-sulfur crude oil. Most cannot fulfill the requirement of high-sulfur crude oil processing. Due to the sulphur content increased in the crude oil processing, the number of acute poisoning accidents caused by hydrogen sulfide leakage has been increasing. For example, between 2006 and 2010, several accidents of hydrogen sulfide poisoning in oil refining installations happened in Shandong province every year. As the high-sulfur crude oil processing capacity of refinery plants increasing, the safety issues of hydrogen sulfide leakage and detection will become more critical. Thus, the studies on identifying hydrogen sulfide leakage risks and enhancing leakage detection efficiency and accuracy are of great importance.

Gas leakage detection optimization means, on the premise of gas leakage and dispersion rules, hazardous gas detectors are arranged scientifically so as to achieve the objective of efficient and reliable detection, accurately and timely provide alarm information, technical support for emergency responses. Effective gas detector networks allow a refinery plant to address leaks before they become serious, by automatically alarming, initiating isolation and safe venting. With ineffective gas detection, there is very limited defense against a flammable gas leakage developing into a fire or explosion that could cause loss of life or escalate to cascading failures of nearby vessels, piping, and equipment [3]. The hydrocarbon leakage database of the UK Health and Safety Executive (HSE) agency shows: a total of 873 gas releases occurred in the period between1992 and 1999, only 540 (62 %) were initially detected by gas detection systems. [2]. Therefore, there still has a large ascension space for enhancing of the efficiency of gas leakage detection system.

2. The gas leakage detection optimization overview in related fields and the influence factors of gas detection efficiency

It is widely known that we need to set up gas detectors in the petrochemical installations or the places where combustible and poisonous gases exist. But the specific number of detectors has not been specified. Factors such as site-specific, arrangement of plant equipment and structures, chemical composition, process conditions, piping design standards number of potential leak sites, design of isolation and depressurization systems and frequency of detector testing usually have influences on the gas detection. When determining the optimization method for the position of gas detectors, these factors should be comprehensively considered. And also the consideration has to be combined with safety economics theories, which could guarantees the best return on investment. Currently, the study on the optimization of placement for gas detectors mainly has the following aspects.

2.1 The optimization of gas leakage detection in some related areas

At present, the study of systematic optimization for gas leakage detection system in petrochemical installations is rarely reported. The placement optimization of gas detectors must first predict a gas dispersion concentration field. The method of Computational Fluid Dynamics (CFD) has been widely used for its low cost and high precision characteristics in the area of gas dispersion consequences prediction. In marine and offshore facilities, researchers have already used CFD modeling to determine the size of the combustible gas cloud that would result from a specific leak scenario. The results from CFD analyses on several leak scenarios can be plotted to determine the size of a flammable cloud that could result in an explosion and damage the structure of plant. Gas detection system of optimization design must be capable of detecting a gas release or vapor cloud, and transport alarm information [3].

Based on the idea of optimization of sensor placement, some research work has been carried out on the study of placement optimization of the gas detectors [4-7].Stephen proposed to design a combustible gas detection system based on the combined risk from a representative selection of leakage scenarios, rather than a worst-case evaluation, and it can hopefully improve the accuracy of the gas leakage detection. This method determines the distance of detectors by a risk array method [3]. Kelsey studied the relationship among gas detectors and gas cloud size, vertically higher or lower placement, detector type (point and beam), alarm detection levels and response time [2]. The comparative study of the infrared point contact detectors spacing indicated that half spacing can contribute to reduce detection time slightly , but need more additional investment of detectors. On the contrary, double-spaced extends the detection time greatly [2]. Based on the principle that the combustible gas should be accurately detected before a flammable cloud reaches 5 m diameter, researchers examined the performance of detector networks based on placing gas detectors on a 5 m × 5 m grid as current guidance [8]. Studies have shown that it can reduce response time by changing alarm limits of detectors, especially for small mass flow rate releases. Compared single response time of the detector with its alarm logic type, the latter is more obviously impact on the detection rate of the entire detection network. Infrared point detectors need less response time and more reliable compared with contacted detectors [8].

2.2 The alarm response time of gas detectors

There are many technical standards in the aspect of the response time. According to some standards, the response time of combustible gas detectors should fulfill the following provisions: diffusing type is less than 30 s, and suction type is less than 20 s [9]. The standard of the response time for hydrogen sulfide gas detectors: diffusing type is less than 60 s, and pump suction type is less than 30 s [9-10]. According to the provision of the lag time for detectors, the detection and alarm time of combustible gas detectors must be less than 30 s, and that of toxic gas detectors must be within 60 s [11]. The China industry standard "Technical conditions of toxic gas detectors and test methods" (HG23006) stipulated the response time of toxic gas detectors: diffusing type is 60 s, and pump suction type is within 30s.

While based on manufacturer's information, the range of response times is about 9–12 s for catalytic point being detectors; 3–6 s for infrared point detectors and 1–5 s infrared beam detectors. And the response time is shorter than the requirement of standards during the industrial practice [12].

2.3 The horizontal and vertical position of detectors

Some China industry standards such as "Specifications for the design of combustible gas and toxic gas detection and alarm for petrochemical enterprises" (SH3063) make certain provisions for the combustible and the design of toxic gas detectors [13]. In terms of the effective coverage of detectors, the horizontal plane radius of the effective coverage of the detector is about 7.5 m in indoor environments, and about 15 m in outdoor environments. The distance between toxic gas detectors and the release source should be less than 2 m in outdoor environments, and that should be less than 1 m in indoor environments. The alarm setting value should be determined in accordance with the following provisions: the primary alarm (high limit) setting value of the combustible gas is less than or equal to 25% LEL (Low Explosion Limit), and its secondary alarm (high high limit) setting value is less than or equal to 50% LEL. And the alarm setting value of the toxic gas should be less than or equal to 1 time of the TLV (Threshold Limit Value) [14,15].

The vertical installation position of the combustible and toxic gas detectors should be decided according to density of target gases. Detectors of the gases which density are larger than air should be higher than the floor about 0.3 to 0.6 m, and detectors of the gases which density are less than air should be higher than the release source about 0.5 to 2 m. When detecting the hydrogen leakage in the specified position, the detector should be installed around the release source or above of the release source within 1 m. Otherwise, the detection effect cannot be obtained. Moreover, when detecting toxic gas (such as carbon monoxide or hydrogen cyanide) which molecular weight is close to air and can easily mixed with air, the detectors should be installed away from the release source within the range of 1 m. As the density of toxic gas is slightly less than air, the detector should be installed above of the release source. On the contrary, the detector should be installed below the release source, and the horizontal distance between detector and of the release source should be within 1 m [14,15].

2.4 Influence factors for the placement of gas detectors

The influence factors of the gas detectors spacing also include the alarm logic design of the detection system. 1-out-of-2 logic is different from 2-out-of-2 logic or 2-out-of-3 voting logic in aspects of the reliability and the false alarm rate. Since the reliability of the detection system is not the same, it also affects the design of detectors spacing. The maintenance situation of the detector is also the factor affects its placement spacing. For example, the gas detector maintenance frequency of an installation is once every 3 months, and that of another installation is twice a week. According to the theory of the optimization placement for gas detectors based on risk, with a repair interval of 3 months, the required number of detectors is 1.74 times of the facility where detectors are repaired twice a week [3].

To sum up, we can see the study of monitoring point optimization of toxic and combustible gas leakage is still in its primary stage, so it's necessary to carry out the hydrogen sulfide leakage detection optimization research in high-sulfur crude oil processing installations. The objective is to study the optimization of detectors by using CFD technology, to obtain the fundamental principles of horizontal and vertical placement of combustible/toxic gas detectors, type selection, response time and other optimization issues [16].

3 . Technical scheme of the Optimization for Gas Detectors' Placement

3.1 Analysis of the gas leakage probability of a typical high-sulfur crude oil processing installation

Taking high-sulfur crude oil processing installations as a typical research object, we can identify potential leakage sources and analyze possible leakage conditions. Combined with the equipment leakage database established by petroleum and petrochemical companies and organizations (e.g. Oil & Gas Producer, DNV, HSE, etc.), set up a typical quantitative assessment model to calculate the leak probability. Based on this, several possible leak scenarios based on the leakage probability analysis could be obtained.

3.2 Set up, sensitivity and correctness analysis of the CFD model for the gas leakage and dispersion simulation of a typical high-sulfur crude processing installation

After considering the structure of installations, the wind velocity and direction, ground surface roughness, air temperature and other factors, establish a three-dimensional CFD model of gas leakage and dispersion of a typical high-sulfur crude oil processing installations. Also, according to the gas leakage condition, correct boundary conditions should be calculated. The positions of gas detectors should be set as monitoring points in CFD model.

3.2.1 Construction of three-dimensional CFD model for typical installations

Combined with the leak scenario of typical installations and the terrain conditions, complex terrain grid is built by Gambit, and CFD model is built by Fluent. Under the consideration of the factors such as wind direction, wind velocity and the ground roughness, the appropriate boundary conditions and the gas leakage monitoring points are set. Based on the above conditions, the CFD model is set up.

3.2.2 The sensitivity analysis of the CFD model

The second step of establish a CFD model is the sensitivity analysis. The sensitivity analysis of Computational Fluid Dynamics (CFD) modeling for gas dispersion is an uncertainty analysis method during the modeling process which firstly identifies the most sensitive ones from a number of uncertain factors, and then monitors and analyzes their impacts on the simulation results, and lastly selects the most suitable modeling parameters. The analysis orders have been advised to be grid dependency, boundary conditions, and then turbulence models and solve controls analysis.

Grid dependency analysis could obtain a balance between model prediction accuracy and computation cost. Sensitivity analysis is important for establishing proper computation models, enhancing prediction accuracy and descending computation costs, and it is an important step for gas dispersion CFD modeling [17].There is a positive correlation between the accuracy of the calculation and the intensive degree of the grid in computational domain. The greater the density of the grid, the greater the calculation accuracy is. But the intensive degree of the grid is restricted by the cost of calculation. So the dependency analysis of the grid should be carried out before CFD calculation. Let the intensity degree of the grid in computational domain achieve a balance of calculation accuracy and time cost.

There is an important premise is that the computational domain wind field can reach the wind velocity of set in advance to determine the dangerous gas leakage occurs in the set of basic wind field, when the dangerous gas leakage takes place. Therefore, there also is need to validate the consistency of the basic wind field before the dangerous gas leakage occurs.

3.2.3 The effective validation of the CFD model

The correctness of CFD model could also be verified to make sure that the CFD prediction is correct. Although CFD method has the characteristics of predicting leakage and dispersion concentration field economically and efficiently, the calculation results also have certain uncertainty because of some factors such as grid, simplified boundary conditions and turbulence model. It is necessary to test and verify the validity of the CFD model.

The common validation methods are experiment and the contrast test of mathematical dispersion model. One of the most convenient ways is using empirical formula of gas dispersion. According to the similarity theory, we can build a scaled-down model of a typical device to detect gas leakage and verify the correctness of the CFD model.

3.3 The CFD predictions of each leak scenario consequences based on the leak probability

Then we use validated CFD model to predict the consequences of the maximum credible leak scenarios which obtained by the leak probability analysis. The process of dispersion of hazardous gas, trends and patterns could be simulated. The information of real-time hydrogen sulfide concentration at important monitoring points and gas concentration field after the leakage of hydrogen sulfide in typical installation can be obtained.

3.4 The optimization analysis for gas detectors

Based on the information of gas concentration field after the leakage of hydrogen sulfide for typical device, the position of hydrogen sulfide detectors could be optimized. It includes optimization of vertical position, horizontal position optimization, hydrogen sulfide and combustible gas detectors type selection, and alarm logic design, etc.

According to the gas concentration predicted by CFD modeling, we could find out the actual alarm time which could be defined as time needed for the concentration reaches the desired level in CFD modeling of each monitoring point in each optimization scheme. The alarm time analysis is an important reference for the optimization of gas detector networks. One of the principles is to choose the fastest detection and alarm scheme.

According to the comparison study for hydrogen sulfide and combustible gas detectors of each optimization scheme, we could choose the correct detector type. Through the comparison of the sensitivity of combustible gas detectors for the high and low level placement, the decision of vertical position could be made. According to the above conclusion of systemic analysis, we can get the scheme and principle of optimal placement of the leakage detection in high sulphur crude oil processing production.

3.5 The optimization scheme of Detectors based on safety economics decisions

Another important factor we should concern about is the investment cost of detectors. So various optimization schemes that have been considered should be comprehensively analyzed using safety economics methods. Under the premise of the cost benefit analysis decision and the safety investment risk decision, considering the factors of accidents prevention benefits and the investment cost ratio, the safety equipment depreciation, the best gas optimization scheme of detectors network for high sulphur crude oil processing could finally be chosen.

4. Conclusions

The optimization scheme of dangerous gas detectors' network in oil refinery installations based on CFD and fuzzy probability analysis can hopefully to improve gas leakage detection efficiency, and prevent fire, explosion and personnel poisoning accidents, which is an important way to enhance the degree of intrinsic safety in the design phrase of petrochemical installations. This research can reduce the risk of malignant accidents caused by hydrogen sulfide leakage; improve the ability of prevention and emergency rescue for this accident.

5 . Acknowledgement

This research is supported by "Natural science foundation of Shandong province" (ZR2012EEQ026), "2012 Shandong science development project for work safety" (LAK2012-6), "Fundamental Research Funds for the Central Universities" (13CX02079A), and "2012 key technology science project for major accident prevention in work safety" (2012-487). We appreciate those who have assisted us in the process of study, and the anonymous reviewers and the editor for their constructive comments and suggestions.

6. References

[1] DU Donghua. The study of sour crude oil processing of Guangxi Petrochemical enterprise [D]: East China University of Science and Technology , 2011

[2] A. Kelsey, M. A. Hemingway, P. T. Walshet al. Evaluation of flammable gas detector networks based on experimental simulations of offshore, high pressure gas releases [J]. Process Safety and Environmental Protection , 2002 , 80 (2) : 78-86

[3] Stephen DeFriend, Mark Dejmek, Leisa Porteret al. A risk-based approach to flammable gas detector spacing [J]. Journal of Hazardous Materials , 2008 , 159 (1) : 142-151

[4] M. Abbaspour, N. Mansouri. City hazardous gas monitoring network [J]. Journal of Loss Prevention in the Process Industries , 2005 , 18 (4-6) : 481-487

[5] Carsten Niebuhr. Aging effects in gas detectors [J]. Nuclear Instruments and Methods in Physics Research Section A: Accelerators, Spectrometers, Detectors and Associated Equipment , 2006 , 566 (1) : 118-122

[6] Xin-Rui Li, Hiroshi Koseki, Yusaku Iwata. Risk assessment on processing facility of raw organic garbage[J]. Journal of Hazardous Materials , 2008 , 154 (1-3) : 38-43

[7] Jan Stawczyk. Experimental evaluation of LPG tank explosion hazards[J]. Journal of Hazardous Materials , 2003 , 96 (2-3) : 189-200

[8] Health and Safety Executive. Offshore detector siting criterion investigation of detector spacing by Lloyds register, offshore technology report OTO 93002[R]. Croydon , 1993

[9] General Administration of Quality Supervision, Inspection and Quarantine of the People's Republic of China (AQSIQ), Standardization Administration of the People's Republic of China. GB12358-2006 Gas monitors and alarms for workplace-general technical requirement[S].Beijing: Standards Press of China, 2006.

[10] State Administration of Petroleum and Chemical Industry. SH3063-1999 Specification for the design of combustible gas and toxic gas detection and alarm for petrochemical industry[S].Beijing: China Petrochemical Press, 1999.

[11] General Administration of Quality Supervision, Inspection and Quarantine of the People's Republic of China (AQSIQ), Standardization Administration of the People's Republic of China. GB12358-2006 Gas monitors and alarms for workplace-general technical requirement[S].Beijing: Standards Press of China, 2006

[12] A. Kelsey, M. A. Hemingway, P. T. Walshet al. Evaluation of flammable gas detector networks based on experimental simulations of offshore, high pressure gas releases [J]. Process Safety and Environmental Protection , 2002 , 80 (2) : 78-86

[13] WANG Huan-qiang , ZHU Qiu-hong , YU Chen. Lows and technological requirements for toxic and harmful gases detection alarm in workplace[J]. China Safety Science Journal , 2010 (02) : 121-129

[14] State Development and Reform Commission. SY6503-2008 Safety technical specification of combustible gas detection and alarm system for petroleum and natural gas engineering[S]. Beijing: Petroleum Industry Press , 2008.

[15] State Development and Reform Commission. SY/T6277-2005 Specification of hydrogen sulfide monitoring and protecting for operation personnel in oil and gas field involving hydrogen sulfide[S]. Beijing: Petroleum Industry Press, 2005.

[16] ZHANG Bo. Analysis on Risk due to Corrosion and Leakage of High Sulfide Natural Gas Gathering Pipeline [D]: China University Of Petroleum , 2010

[17] ZHANG Bo, CHEN Guo-ming. Sensitivity analysis for Computational Fluid Dynamics modeling of gas dispersion [J]. Science & Technology Review. 2011, 29(13): 57-61.

1ST CCPS ASIA-PACIFIC CONFERENCE ON PROCESS SAFETY℠

H_2S Leakage Grading Warning and Protective Measures in High-Sulfur Crude Oil Processing

Ming Chen
College of Mechanical and Electronic Engineering, China University of Petroleum
Shandong Qingdao, China
a54cm@163.com

Dongfeng Zhao
College of Chemical Engineering, China University of Petroleum
Shandong Qingdao, China
zhaodf@vip.sina.com

Prepared for Presentation at
American Institute of Chemical Engineers
1st CCPS Asia-Pacific Conference on Process Safety
Qingdao, China
September 4 – 5, 2013

UNPUBLISHED

H_2S Leakage Grading Warning and Protective Measures in High-Sulfur Crude Oil Processing

Ming Chen
College of Mechanical and Electronic Engineering, China University of Petroleum
a54cm@163.com

Dongfeng Zhao
College of Chemical Engineering, China University of Petroleum
zhaodf@vip.sina.com

Keywords: High-sulfur Crude Oil, H_2S, Grading Warning, Protection

Abstract

As one of the major risks in high-sulfur crude oil processing, the H_2S exposure could be reduced by early warning of H_2S leakage. Based on the source and distribution of H_2S, the four-level warning mechanisms of H_2S leakage is developed by using HSE risk identification system, which makes the H_2S concentration values as the grading index. The appropriate emergency measures are formed. To provide guidance for H_2S poisoning protection during high-sulfur crude oil processing, the safeguards to prevent H_2S poisoning are presented according to the cause of H_2S poisoning and "Human - Machine - Material - Regulation - Environment".

1 . Introduction

At present, with the rapid development of petroleum industry, conventional energy is drying up due to the consumption of word petroleum resource. According to related investigation, more than 75% of the total oil production is high-sulfur crude oil. Therefore, the development trend of world oil is heavy, poor quality and high-sulfur content, etc.
In recent years, more and more domestic oil refineries have changed to refine the poor quality crude oil because of the low-cost and expanding import of high-sulfur crude oil [1]. However, most of the domestic oil refining devices were originally designed for low-sulfur crude oil, processing high-sulfur crude oil in refinery will bring a series of problems to the process units, such as equipment corrosion, hydrogen sulfide poisoning, and the spontaneous combustion of ferrous sulfide, especially the damage from hydrogen sulfide poisoning can not be ignored. Hydrogen sulfide leakage can be resulted from various reasons and is a mainly threat to life.

2 . The source, distribution and harm of H_2S

2.1 The source of H_2S

The common oil refining devices mainly includes Atmospheric and Vacuum Distillation , Catalytic Cracking , Catalytic Reforming , Hydrofining , Hydrocracking , Delayed Coking ,

Hydrogen Production, Stripping and Sulfur Recovery. Under corresponding high temperature, pressure and catalyst, cracking will bring a series of proton acid reaction, hydrogen transfer process, isomerization reaction, alkylation reaction and cracking reaction, which make the combined sulfur in the oil separated out. Hydrogen sulfide can be generated by the following reaction in the process of hydrogenation: Reaction with thiol: $RSH+H_2 \rightarrow RH+H_2S\uparrow$; Reaction with thioether: $R\text{-}S\text{-}R+2H_2 \rightarrow 2RH+H_2S\uparrow$; Reaction with disulphide: $(RS)_2+3H_2 \rightarrow 2RH+2H_2S\uparrow$. These show that it will produce a large number of hydrogen sulfide in the process of Catalytic Cracking , Catalytic Reforming and Hydrofining Cracking [3].

According to the varieties of refinery processing crude oil and various forms of sulfide, the source of hydrogen sulfide in machining process mainly includes the following three situations [4]:

(1) The hydrogen sulfide existed in crude oil.

(2) Sulfide in oil after thermal decomposition produces H_2S.

(3) For removing the sulfur in oil or improve the quality of product, sulfur element will be transformed to H_2S with hydrogenation process.

2.2 The distribution of H_2S

Hydrogen sulfide exists in most refinery devices, the distribution of hydrogen sulfide in the refining unit is closely related to various factors such as the crude oil types, the sulfur content of crude oil, and process scheme. In the process of refining crude oil, hydrogen sulfide, which is produced by sulfide after thermal decomposition and hydrogenation process, mainly concentrate in the overhead gas, light component oil and dissolved in sulfur-containing wastewater on top of the tower. The distribution characteristics of hydrogen sulfide in typical refining device are summarized in Table 1[5].

Table 1 Key location of hydrogen sulfide in common refinery devices

Typical devices	Key location of hydrogen sulfide
Atmospheric-Vacuum Distillation Unit	Overhead gas compressor of atmospheric tower, Overhead gas compressor of decompression tower, Overhead reflux and production tank of atmospheric tower, Water separator of decompression tower, Overhead reflux and production tank of stabilization tower, Overhead air cooler heat exchanger, Sulfur-containing sewage pump, Sulfur medium sampling port, Dehydration port, etc.
Delayed Coking Unit	Rich gas compressor, Coking furnace, Coke drum, Coking fractional distillation column, Blowdown tower, Gas-Liquid seperation tank at the top of fractionating tower, Blending and coking dry gas seperation tank, Coking rich gas balance tank, Absorption tower, Desorption tower, Stabilization tower, Overhead air cooler heat exchanger, Sulfur medium sampling port, Dehydration port, etc.
Catalytic Cracking Unit	Rich Gas Compressor, Sulfur-containing sewage tank, Sewage pump, Oil-Gas separator, Gas compressor condensate, Condensate pump, Overhead reflux tank of stabilization tower, Reflux pump, Absorption tower, Desorption tower, Stabilization tower, Overhead air cooler heat exchanger, Sulfur medium sampling port, Dehydration port, etc.
Hydrofining Unit	Recycle hydrogen compressor, Hydrofining reactor, Production fractionator, H_2S Stripping tower, Recycle hydrogen desulfurization tower, Recycle hydrogen

	separation tank, Overhead reflux pump of reflux tank, Sulfur medium sampling port , Dehydration port, etc.
Continuous Catalytic Reforming Unit	Pre-hydrogenating reactor, Pre-hydrogenating production separation tank, Pre-hydrogenating stripping tower, Reflux tank at the top of stripping tower, Reflux tank of stripping tower, Reflux pump of stripping tower, Overhead air cooler heat exchanger, Sulfur medium sampling port, Dehydration port, etc.
Sour Water Stripping Unit	Sour water tank, Sour water degassing tank, Sour gas separation tank, Sulfur medium sampling port, Dehydration port, etc.
Desulfurization and solvent regeneration unit	Dry gas separation tank, Dry gas desulfurizing tower, Flooded tank, Solvent regenerator, Flooded pump, Regeneration tower water-cooled air cooler, Sulfur medium sampling port, Dehydration port, etc.
Sulfur Recovery Unit	Combustion furnace of sulfur manufacturing, Waste heat boiler, Sour gas accumulator tank, Sulphur condenser, Heat exchanger, First and second claus reactor, Tail gas treatment combustion furnace, Liquefied sulfur pool, Liquefied sulfur tank, Sulfur medium sampling port, Dehydration port, etc.
Gas Holder and the Torch System	The compressor input or output seperation tank, Gasometer and relevant manual sampling port, Water-Cutting off-take point, Sulfur medium sampling port, Dehydration port, etc.

2.3 The Harm of H_2S

Hydrogen sulfide, which is a kind of nerve agents, is easy to cause poisoning and also has strong stimulation to mucous membrane. High concentrations of hydrogen sulfide can restrain respiratory center directly and then cause death by suffocation. Table 2 shows its harm to people in different concentrations of hydrogen sulfide.

Table 2 The harm to people in different concentrations of hydrogen sulfide

Concentration /(mg·/m^3)	**The harm to people**
15	Obvious bad smell, long-term exposure is allowed
30	In the open air, safe exposure work for 8h
70~150	After 2~5min, olfactory fatigue, no longer smell the odor. Respiratory and eye irritation symptoms will occur in 1~2h
150~300	Inhibition of the sense of smell, coughs, eye inflammation in 2~5mins. Eye pain and breathing disorder in 8h and it will lead to the deaths of bleeding in 8~48h
450~600	Collapse, loss of consciousness, and make artificial respiration immediately in 2mins, eye severe inflammation and cardiac arrhythmias in 15~30mins, Eyes, headache, arms and legs tremble, and even death in 30~60mins
900~1000	Loss of consciousness and leads to the death in 2mins
1200	In seconds, respiratory arrest suddenly, and then sudden cardiac arrest, people will be rescued with implementation of artificial CPR in time
1500	In seconds, electric shock death
Over 3000	Immediately kill people and can not to be rescued

From 1984 to 1994, about 80 cases of hydrogen sulfide poisoning death accident occurred in the United States, 20 cases happened in oil industry. 86.3 percent of H_2S poisoning accidents

happened in a limited space. Many workers, which were not aware of being in dangerous working area and were also trying to rescue companions, died of poisoning [6].
Hydrogen sulfide mainly enters the body through the respiratory tract, in which a small percentage of hydrogen sulfide by respiratory tract is expelled from the body and will not be accumulated. The most part of hydrogen sulfide in the body will be oxidized under the action of metabolism into non-toxic sulfide, sulfate or thiosulfate, and then excreted into urine by the kidneys. According to "Classification for Hazards of Occupational Exposure to Toxicant" (GB5044-85) , hydrogen sulfide occupational hazards classification as follows: high hazard (II).
The body reactions to hydrogen sulfide are different. The main influence factors are individual differences, concentration and contact time, etc. With the combination of concentration and contact time, hydrogen sulfide poisoning symptoms can be divided into three types: mild poisoning, moderate poisoning, and severe poisoning [7].

3 . Grading warning system of H_2S

Grading warning system is widely applied in all kinds of emergency plans and is an important basis and reference of direct emergency rescue. Grading warning is usually based on the likelihood and consequences of the incident as a premise for classification, after the incident it directs workers to enter a state of grading warning and timely to be prepared for possible emergency action in advance.
Accurate monitoring of the hydrogen sulfide in refining system is the basis of the H_2S poisoning prevention job. In this paper, grading warning system is mainly carried out in accordance with the density of H_2S. As a result, grading warning is closely connected with the concentration of hydrogen sulfide.

3.1 HSE risk identification system

Rapid effective risk identification is the key to the HSE management system. So, how to accurately identify risks is the premise to prevent hydrogen sulfide poisoning. First of all, setting testing alarming device and necessary protective equipment in the place where H_2S is easy to leak and the concentration is high; Secondly, establishing special, complete emergency contingency plans, establish risk identification group and accident treatment group; At last, work site should be divided into safe area, emergency assembly area, evacuation area and alert area, and set up the escape route. When the concentration of hydrogen sulfide is more than 15mg/m^3, all passageways should order the rules of warning signs, set appropriate warning signs (Each letter and blank are black print over yellow) to remind workers of potential hazardous situations around. When the concentration of hydrogen sulfide is more than 30mg/m^3, all workers should be taken away from the site [8].

3.2 Grading warning system of H_2S

The system makes the H_2S concentration values as the grading index. According to "Recommended Practice for Safe Drilling Operations involving Hydrogen Sulfide" (SY/T5087—2005) and "Specification of Hydrogen Sulfide Monitoring and Protecting for Operation Personnel in Oil and Gas Field involving Hydrogen Sulfide" (SY/T6277—2005), set up the concentration threshold of H_2S monitor and instruments can automatically give alarm when the concentration of hydrogen sulfide reaches 15mg/m^3.

According to the concentration of hydrogen sulfide measured by monitoring devices, grading warning is divided into blue, yellow, orange and red alert, which respectively indicates workers the situation in order to take corresponding measures. Table 3 shows the situation of grading warning.

Table 3 Grading warning system of H_2S

H_2S concentration / (mg·/m^3)	Warning Situation	Warning Level	Countermeasures
C < 15	Allowable concentration in the air	blue	Fixed-point real-time monitoring
15≤C < 30	Little H_2S leakage	yellow	Risk monitoring and warning
30≤C < 150	More serious H_2S leakage	orange	Use breathing apparatus, real-time monitor the wind and the H_2S content
C≥150	H_2S Serious leakage	red	Wearing gas masks, evacuate the site

H_2S blue indication: No H_2S in the air or the concentration of H_2S below 15mg/m^3, there is still a potential risk of H_2S leakage, fixed-point real-time monitoring is required.

H_2S yellow indication: There is a small amount of H_2S leakage but have not reached 30 mg/m3 safe critical concentration, the respirator is not necessary and also do not need immediately evacuate the site, but it must be to monitor and warn of potential danger and be ready to enter the orange alert.

H_2S orange indication: Serious H_2S leakage accident happened, the H_2S concentration exceeds the safe critical concentration. The respirator is necessary in leakage danger zone. Continuous monitoring of wind direction and concentration of H_2S in air around are required, if there are any changes need to be further processing.

H_2S red indication: Sever H_2S leakage accident happened, the H_2S concentration has greatly exceeded the safe critical concentration. The field operating persons have been threatened and they should immediately evacuate the site with gas marks.

According to warning system , set up the corresponding grading treatment measures, Table 4 shows the detail treatment measures.

Table 4 Grading warning treatment measures

H_2S concentration /(mg/m^3)	Grading Treatment Measures
C < 15	Check, monitor and protect devices, real-time monitoring of the concentration of H_2S and be prepared for emergency.
15≤C < 30	Wind direction, wind speed should be monitored and be charged by designated employee, wear positive pressure breathing apparatus in the danger zone to check leakage and dispose emergency.
30≤C < 150	Wear positive pressure respirator, report to the superior, monitor H_2S concentration change at the downwind distance of 100m,initiate emergency response plans, control the leakage source, count and evacuees, cut off potential ignition source and inform the aid agencies.
C≥150	Evacuate leakage danger zone with wear positive pressure respirator, Set the alert zone on the spot, workers wear protective clothing enter into the leakage field to control

	leakage source and report to the superior or request support。

When H_2S leakage occurs, starting grading warning model can deal with it efficiently and orderly and make full use of existing emergency resources and protective equipment to minimize or control the harm.

4 . Safeguards of H_2S poisoning

Based on the cause analysis of H_2S poisoning and combined with refinery field condition, the safeguards to prevent H_2S poisoning are presented according to "Human - Machine - Material - Regulation - Environment"[2].

1. Strengthen ideological education, raise safety awareness

In high-sulfur Crude Oil processing, H_2S leakage has become the focus of safety management and risk prevention and control. According to statistics, 50% of H_2S poisoning accidents were caused by poor safety management and poor safety awareness. Therefore, the basic safeguard to prevent H_2S poisoning accident is to strengthen the promotion of H_2S hazards and the safety awareness of staffs or leaders.

2. Ensure the integrity of the equipment, realize intrinsically safe of the equipment

The unplanned interface between pipeline materials with the surrounding environment is the essential reason of H_2S leakage. As a result, to ensure the maximum intrinsically safe and the integrity of the equipment in the entire life cycle can reduce H_2S poisoning accidents from the source. Especially the integrity detection of acidic gas pipeline and the overhead pipeline is the key of the H_2S corrosion leakage monitoring.

3. Master the H_2S dynamic distribution, the key control

To establish refinery H_2S dynamic map according to the technological process and H_2S generation area, and set up intensive monitoring and management in H_2S possible leakage area such as hydrotreating unit and diesel hydrogenation unit, in which setting H_2S alarm device and providing related personnel protective equipment. H_2S poisoning accidents can be reduced by knowing well about hazardous substances and improving safety awareness during the normal operation of equipment.

4. Set relevant safety management regulations and H_2S standard operations

Against H_2S process operation, the refinery should formulate corresponding safety operation procedures and eliminate violations of rules to reduce H_2S leakage or H_2S poisoning accidents caused by anthropogenic factors.

5. Consider the surrounding environment, reduce secondary accidents caused by H_2S leakage.

Since the crowded around large refineries and natural environmental factors such as topography and wind direction, H_2S leakage is easy to spread out of the factory bound. Therefore, preliminary design and process equipment arrangement should fully consider the effect of environment factors to reduce secondary accidents caused by H_2S leakage. Meanwhile, when making contingency plans of H_2S, the natural environment and the surrounding personnel concentration areas should be taken into account to avoid accidents because of improper emergency preparedness.

5 . Conclusion

At present, there is still no very perfect grading warning system for H_2S leakage. Based on the source and distribution of H_2S, grading warning system is established according to the possibility and maximum concentration of hydrogen sulfide leakage. The disposal measures of different risk levels are formed on the basis of grading warning system. This paper mainly discusses the prevention of H_2S poisoning accidents according to "Human - Machine - Material - Regulation - Environment", and recommends that through reducing H_2S leakage probability, strengthening the personnel and equipment safety management and completing the H_2S poisoning emergency measures, the security of people's lives and properties can be guaranteed.

6. References：

[1] Zhitao Cao. Technical status quo and development trend of China's oil refining industry [J]. Refining and Chemical Industry, 2010, 21(2):1-3.

[2] Hongsheng Zhang, Yanhui Zhang. Discussion on the safety countermeasures of processing high-sulfur crude oil[J]. China Petroleum and Chemical Standard and Quality, 2011(6):225.

[3] Shumei Jia, Mingying Ma, Xuecheng Liu. Potential hazards of hydrogen sulfide and control measures in a petrochemical factory[J]. Occupational Health and Emergency Rescue,2006, 24(2):72-74.

[4] Jinqiu Ma, Dongfeng Zhao, Jiangbo Jiu, Yuqi Shen. Safety Analysis and Countermeasures of Typical Oil Refinery after the Change of Raw Material Changed to High-sulfur Crude Oil[J].2010, 17(4):55-59.

[5] Jinqiu Ma. Safety Analysis and Countermeasures Study of Typical Refinery Units with Raw Material Replaced by High-sulfur Crude Oil[D]. China University Of Petroleum. 2010.

[6] Jianzhong Yang. Countermeasures of the Hidden Danger of Accidents for processing high-sulfur crude oil[J]. Chemical Industry Occupational Safety & Health. 2001, 22(9):308-310.

[7] Hongsheng Zhang, Yanhui Zhang. The safety countermeasures of processing high-sulfur crude oil[J]. China Petroleum and Chemical Standard and Quality, 2010, 6:225.

[8] Wanping Li, Qiuping Hu. Environment monitoring and protecting for operators in petroleum and nature gas field with hydrogen sulfide[J]. The world's oil technology review, 2004(01):234-238.40(11):32-34.

The response to the leak accidents of long-distance natural gas pipeline

Cui-ping WANG
Dong-feng ZHAO, Feng-lei HAN
China University of Petroleum (East China), Department of Safety Science and Engineering
Qingdao, Shandong province, 266580. CHINA
wangcuiping0215@163.com

Prepared for Presentation at
American Institute of Chemical Engineers
1st CCPS Asia-Pacific Conference on Process Safety
Qingdao, China
September 4-5, 2013

UNPUBLISHED

The response to the leak accidents of long-distance natural gas pipeline

Cui-ping WANG, Dong-feng ZHAO, Feng-lei HAN
China University of Petroleum (East China), Department of Safety Science and Engineering

Keywords: Long-distance natural gas pipeline; Leak accident; Counter measures

Abstract

The long-distance natural gas pipeline has characteristics of large diameter, long distance, high pressure and large amount of output. There is flammable, explosive, volatile gas in the pipeline. If pipeline leaked, it may lead to fire or explosion accident .So there are huge losses to people's lives and property, and prolonged negative effects to the public health and the environment. In this article, pipeline lifecycle was taken as the timeline of the accident causes. For each stage of the lifecycle, the hidden dangers of accidents and the counter measures were expounded in chronological order. Meanwhile, the causes of pipeline leak accident were supplemented, and more detailed security measures were formulated.

1. Introduction

The long-distance natural gas pipeline has characteristics of large diameter, long distance, high pressure and large amount of output [1]. Because there is flammable, explosive, volatile gas in the pipeline. Once pipeline leaked, it may lead to fire or explosion accident .Even worse there will be huge losses to people's lives and property, and prolonged negative effects to the public health and the environment [2].Since 1971, the cumulative failure probability of European natural gas pipeline was 0.347 times / (1000 km·a) in the last 40 years. After the implementation of integrity management, over the past decade, this value had been reduced to 0.167times / (1000 km·a)[3].In China, Petro China Pipelines Company has made some attempts and achieved initial results in the pipeline integrity management in recent years, effectively improved the safety of pipeline operation[4]. However, considering that the original analysis of the accidents was incomprehensive, it was necessary to continue studying the causes and countermeasures of the pipeline leak accidents, to further enhance the security of the natural gas pipeline system, and to promote the safety of gas pipeline construction and operation. In this article, pipeline lifecycle was taken as the timeline of the accident causes. For each stage of the lifecycle, the hidden dangers of accidents and the counter measures were expounded in chronological order. Meanwhile, the causes of pipeline leak accident were supplemented, and more detailed security measures were developed.

2. Overview of the causes for long-distance natural gas pipeline leak accidents

At each stage of the pipeline life cycle, including design, construction, operation, maintenance and emergency, there are a variety of causes that may lead to leak accidents. According to the pipeline life cycle timeline, the causes of long-distance natural gas pipeline leak accident were summarized as follows.

Table 1. Causes of long-distance natural gas pipeline leak accidents

<table>
<tr><th>Stage</th><th colspan="2">Accident causes</th></tr>
<tr><td>Design</td><td>Improper connection of pipe fittings, unreasonable pipeline routing, poor qualification of the construction unit, pipe material and auxiliary facilities defects, misuse of substitute material</td><td rowspan="5">Improper command, mismanagement and others</td></tr>
<tr><td>Construction</td><td>Cracks, slag inclusion, incomplete fusion, overlap, air hole, undercut</td></tr>
<tr><td>Operation</td><td>Natural factors, human factors, corrosion factors</td></tr>
<tr><td>Maintenance</td><td>Improper maintenance</td></tr>
<tr><td>Emergency</td><td>Incomplete emergency measures</td></tr>
</table>

3. Causes and prevention of long-distance natural gas pipeline leak accidents

3.1 Design stage

3.1.1 Improper connection of pipe fittings

At the design stage, the pipeline structure, the connection of pipe fittings and valve and the screw type should be drawed high attention. Without appropriate fixation, the equipment was prone to vibration, which may lead to crack, fatigue damage and fulcrum deformation. And at last, the pipeline broke.

Preventive measures [5]:

(1)Pipeline should be laid straight, and the thermal expansion problem should be considered when parallel pipe was connected.
(2)Design defects should be avoided. For example, for inert gas and combustible gas pipe, it was necessary to install two valves. Meanwhile, an exhaust valve should be mounted between them.

3.1.2 Unreasonable pipeline routing

Pipeline routing is a guiding and basic work at the design stage. It determines not only the project cost but also the safety of pipeline operation[6]. It's necessary to understand their hydrological data, geological conditions and regional development planning where the pipeline will cross. Otherwise it may cause such consequences that the pipeline was laid in the unstable region, and leak accident may occur.

In order to enhance the rationality, the work of pipeline routing should be in accordance with the relevant provisions strictly. The selected line should have good stability, and should be laid away from the crowd as far as possible. Works that need to be done are prospecting the geological structure, the stratigraphic lithology, the hydrogeology. Possible natural disasters and man-made harmful factors should be determined, so as to take corresponding measures. At last, the selected line should be submitted to the experts for examination.

3.1.3 Poor qualification of the construction unit

The natural gas is flammable and explosive, so the construction of the pipeline is very dangerous. The construction unit should have qualification recognized by state or industrial departments. The units without qualification generally don't have enough construction experience, special tools, and qualified welding personnel. And they may even have no professional builders, so the construction quality cannot be guaranteed.

Preventive measures: In the construction process of the long-distance natural gas pipeline, the offer-bid must be in accordance with the relevant provisions of the state for the project, and only the unit who has corresponding qualification could be permitted to undertake the task.

3.1.4 Pipe material and auxiliary facilities defects

Defects of pipe material and affiliated facilities could lead to pipeline accidents. Material defects will cause pipeline leak, and poor welding quality will lead to crack, dislocation, burn through, incomplete fusion, overlap and undercut, leave hidden dangers of pipeline accidents.

Preventive measures[7]:

(1)Check the qualification of the suppliers, and strictly carry out non-destructive testing on the pipe material and welding quality. Once defects were found, especially in castings, forgings, and high pressure pipe, the material should not be used.
(2)Carry out appearance inspection and non-destructive testing to the pipe welds to confirm the welding quality. Check the qualification of the construction unit, and strictly implement the system of supervision.

3.1.5 Misuse of substitute material

Misuse of materials may occur in each stage, including design, material classification, and processing. The corrosion resistance and pressure resistance of different materials are disparate, so it's easy to cause accidents.

To avoid this problem, non-destructive testing must be carried out strictly on the materials of castings, forgings, and high pressure pipeline. In addition, defective materials shall not be allowed to put into use. And after installation, hydrostatic test is required for the pipeline. The testing pressure is 1.5 times of working pressure. The selection of the form and material of the pipeline should be based on the technological condition, instead of free replacement.

3.2 Construction stage

The accident potentials during the construction stage weld defects, including crack, slag inclusion, incomplete fusion, overlap, air hole and undercut, etc.

Preventive measures: Implement welding procedure qualification before welding, and choose qualified welders. Welders should strictly follow the operation procedures. After the construction, carry out nondestructive testing on the pipeline in accordance with the specifications and design requirements. In order to ensure the reliability and integrity of the test results, the nondestructive testing personnel must belong to a third party, and the inspection fees should be paid directly by the owner.

In addition, even if the design of the long-distance natural gas pipeline conforms to the specifications, and the construction complies with the safety requirements, there are still many factors that may cause accidents. For example, the pressure fluctuations[8], the high velocity impact, the medium temperature change and pipeline deformation, etc. All of these factors may cause localized stress concentration, and lead to fatigue cracking. Thereby, at the construction stage, the operation of the pipeline should be taken into account. And the stress-strain detection must be strengthened during the operation stage.

3.3 Operation stage

3.3.1 Natural factors

There are many natural factors, such as rain-wash, flood and debris flow, which may cause pipeline accidents. Earthquake and river water are also potential factors. If the pipeline cross the river, it may be broken by the fast-flowing river water.

For the destruction of the rain-wash, wall protection, slope protection, retaining wall and gutter should be used to protect the pipeline where it is prone to landslides.

Choose the region carefully where the pipeline will cross the river. The natural condition and the climatic condition must be taken into account when making the pipeline routing. Try to avoid laying pipeline in areas where extreme weather and earthquake are frequent. And corresponding protective measures should be taken to ensure the safety of the pipeline[7].

In addition, frost heaving must be taken into account if the temperature is low in winter [9].

3.3.2 Human factors

Human factors, such as stealing gas by drilling and construction near the pipeline, are likely to lead to pipeline accidents. And for pipelines under ground, driving and planting of deep-root plant up the pipeline are also two factors that should not be ignored.

To avoid accident caused by human factors, public propaganda work must be carried out to improve people's safety awareness. The patrol work of the pipeline should be strengthened. Combine inspection and key protection to find problem and prevent man-made damage in time. Strengthen the propaganda work of security and law on the residents along the pipeline, and employ them as pipeline patrol. Confirm that problem can be found, reported and disposed in time. When constructing near the pipeline, appoint specially-assigned person to monitor it. Under the condition of driving up the pipeline, soil and boards must be used for protection [7].

3.3.3 Corrosion factors

Pipeline corrosion includes outer wall corrosion, internal wall corrosion and stress corrosion. When corrosion occurs, spill accidents may occur too. Because the environment of pipeline is complex, and moisture, H_2S, CO_2 and other impurities are contained in the pipeline, so corrosion of inner and outer walls are prone to happen[10].

Currently the pipeline protection methods include[11]: coating protection , electrochemical protection, stray current protection. The specific measures are as follows:

(1)The coating used must be stable, it has features of anticorrosive, insulation, thermal insulation, increasing strength and so on.
(2)Strengthen the cathodic protection of pipeline, and ensure that the protection of pipes is in good condition. Electrical protection is divided into three categories[12]: plus cathodic protection, sacrificial anode protection and exhaust flow protection. Since different protection methods are suitable for different situations, appropriate protection methods must be chosen.
(3)Generally both coating protection and electrochemical protection are used to protect buried pipeline to achieve the desired preservative effect [12].
(4)Periodic inspection and surveys should be carried out on the pipe sections are prone to corrosion. And modification of coating aging, perforation also should be carried out timely.
(5)For the process of dehydration without desulfurization in the transportation of natural gas, appropriate measures must be taken to prevent hydrogen sulfide corrosion occurring[13].
Stress corrosion cracking refers to the cracking caused by the joint action of the metal in a fixed tensile stress and the specific media. This form of damage of corrosion is brittle fracture, which occurs without warning. So it has great dangers and destructibility for oil and gas pipeline[14].

In order to avoid pipeline stress corrosion, an anticorrosion layer with good adhesion must be adopted. In the new design of pipeline, it is emphasized adopting anticorrosion layer with good adhesion to steel pipeline, such as three layer polyethylene coating, epoxy powder coated, polyethylene, liquid epoxy, etc. And a joint approach of coatings and catholic protection should be taken. Regular testing should be conducted, so problems could be detected and disposed early.

3.3.4 Others

During the construction and pipeline pressure test, a large number of free water invaded and left in the lower-lying areas .If the water are not clear , the ice block may occur in the pipeline .It seriously affects the safe operation of pipelines. So the regular gas pipeline pigging is recommended. And once hydrates are formed, measures of reducing the pressure or elevating the temperatures must be taken as soon as possible, in order to destroy the formation process of hydrates. Alcohols are needed to be filled to the pipeline regularly, to ensure safe operation of pipelines[15].

Safety measures of heating, soil replacement and insulation should be taken to avoid frost heave[9].

3.4 Maintenance stage

In the maintenance stage, formulas of "early detection, early treatment" should be achieved. Regularly inspect the equipment wear; take reasonable pipe routing and proper reinforcement measures; the cushioning device should be settled nearby the chemical machinery and equipment with severe vibration to reduce the interference on the pipeline; Check pressure gauge regularly; re-adjust the opening pressure of security valve; once pressure gauge and safety valve fail, they should be promptly repaired or replaced ; when the largest crack depth is greater than 10% of the wall thickness ,or the total crack length exceeds the crack length of the pressure test with the specified minimum yield strength of 110%, the pipe segment should be repaired[16] .

3.5 Emergency stage

3.5.1 Emergency measures

The severity of the accident depends on the emergency measures when the accident happens. The emergency measures for the natural gas pipeline should be set in advance, and the rescue exercises should be held on a regular basis. Generally, the emergency response of the gas pipeline has the following problems[17]: The basic data of the pipeline is incomplete. The emergency rescue forces and emergency resources lack unified management. The operability of the emergency plan is poor so that the rescue work is difficult to carry out orderly. There is no scientific prediction to the development of the accident. The condition of the road along the pipeline is very complex, it’s hard to find a best route for rescuing and evacuation when accident occur.

A good emergency rescue measure should meet the following conditions [18]: It should have a scientific organization system for the rescuer. The system framework of the emergency plan should be graded explicitly. All kinds of information can be transmitted, published and collected timely in a rescue. The reserve mechanism of the emergency supplies should be appropriate. The daily training and exercises of the rescuer should be carried out strictly. Once accident happens, communicate with the public timely, to enhance their confidence in dealing with emergency. And at last, emergency information should be recorded completely.

3.5.2 Strengthen the research of security technology

Main aspects:

(1)Establish accident database for the gas pipeline. It's a necessary and basic work to collect the accident data and study the characteristics of the pipeline accident [19].
(2)Improve the level of detection of the gas pipeline. Special intelligent detection system must be developed on the basis of the pipeline corrosion detector. And a comprehensive inspection and safety assessment are needed for the pipelines in service [19].
(3) The emergency rescue system should be developped by coupling the data of GIS and SCADA[20] on the principle of high cohesion and low coupling. The risks management of the long-distance pipeline should be carried out according to the method of quantitative analysis. In order to reduce operational errors, staff should be provided security education to safety awareness.

4. Conclusions

In this article, pipeline lifecycle was taken as the timeline of the accident causes. For each stage of the lifecycle, the hidden dangers of accidents and the counter measures were expounded in chronological order. Meanwhile, the causes of pipeline leak accident were supplemented, and more detailed security measures were formulated. Main conclusions:

(1)The causes of the leak accident were described under the condition that factors of all sides in the pipeline's lifecycle were taken together, which include design, construction, operation, maintenance and emergency of the long-distance natural gas pipeline. This analytical method is very clear so that people can understand it at a glance.
(2)When formulating safety measures for the long-distance natural gas pipeline, it's better to make that according to the causes of each accident, by which more factors could be taken into account and the operability could be improved.

5. References

[1] Li Helin, Zhao Xinwei, Ji Lingkang. The failure analysis and integrity management to oil and gas pipelines. Physical and chemical inspection: physical volumes, 2005,41:24-31 .
[2] Che Fuli,Dai Qianmin,Huang Kun,ect. The safety problems and counter measures for Long-distance natural gas pipeline [J]. Petro & Chemical Equipment, 2012(3):83-85.
[3] Liu Guozhi, Peng Yingwei, WuDong, ect. Natural Gas Pipeline Failure Analysis and Precautions[J]. Safety and Environmental Engineering, 2012(3):100-104.

[4] For the first time PetroChina "pulse" integrity and repair defects of oil pipeline [N / OL]. China Energy Report, http://paper.people.com.cn/zgnyb/html/2010-09/27/content_632432.htm
[5] Ni Jianfeng,Yang Ronghua. The causes and prevention to the rupture of pressure pipeline [J]. Chemical Equipment Technology, 2005(5):46-47.
[6] Zhao Zhicheng,ZhaoYuluo,Ren Hongqi. Risk Factors and Prevention Measures for Safety Operation of Natural Gas Pipeline[J]. GAS & HEAT, 2011(9):41-44.
[7] He Xiaoqiong. The accidents and countermeasures of Long gas pipeline [J].SHANXI METALLURGY, 2007(6):44-45.
[8] Chang Xing. The causes of weld cracking for Long gas pipeline [J]. Oil & Gas Storage and Transportation, 2010(12):910-912.
[9] Zheng Xiaoming,Wang Xiaoyan,Yan Jie,ect. Freezing bulge analysis and countermeasures for gas pipeline[J]. Oil & Gas Storage and Transportation, 2010(12):910-912.
[10] Luo Peng,Zhang Yiling,Cai Peiei,ect. Analysis and Countermeasures of Natural Gas Transmission Pipeline Internal Corrosion Accidents[J]. Total Corrosion Control, 2010(6):16-21.
[11] Li Haichuan. Analysis of Crude Oil and Corrosion and Protection of Natural Gas Pipeline [J].China Petroleum and Chemical Standard and Quality, 2011(7):120-121.
[12] Wang Xiaoming,Feng Yanyan. The corrosion and protection technology for the natural gas pipeline [J]. ZHONGGUO JITI JINGJI, 2008(13):166-167.
[13] Liu Liewei， Li Guomin,Zheng Jiashen,ect. Investigation of the Corrosion of Collection Pipe of Dehydrated natural Gas Transport [J]. Journal of Chinese Society for Corrosion and Protection, 2000(6):374-378.
[14] Lu Qimin. The causes and protection about stress corrosion cracking for Oil and gas pipeline [J]. Petroleum Planning & Engineering , 11 (2) :42- 43.
[15] Wang Wei,Chen Chen,Li Yingqin,ect. Analysis on typical natural Gas Pipeline Accidents and The Countermeasures[J]. Petrochemical Safety and Environmental Protection Technology, 2010(6):25-30.
[16] Stress Corrosion Cracking on Canadian Oil and Gas Pipelines. National Energy Board.Canda,2006.
[17] Zhu Benting,Wu Ming,Chen Jun. Problems and Solution in Oil and Gas Transmission Pipeline Accident Emergency Response[J]. Computer Applications of Petroleum, 2010(3):28-31.
[18] Zhu Xiaohua,Yang Jun. Development and Research on Emergency Rescue System for Long-distance Natural Gas Pipeline[J]. Natural Gas and Oil, 2009(2):38-41.
[19] Zhu Yuan. Leakage Control and Emergency Response for Gas Gathering System of Highly Sour Gas Field[D]. 2010,College of Mechanical & Electronic Engineering, China University of Petroleum (East China),Qing Dao.
[20] Lang Xuqing,Zhao Zhiyong,Gong Hong,ect. Statistical analysis of oil and gas pipeline accidents and safety operation countermeasures [J]. Safety, health and environment, 2006 (10) :15 -17 .

1ST CCPS ASIA-PACIFIC CONFERENCE ON PROCESS SAFETY℠

The Safety Distance Study of the LPG Storage Field Based on the Numerical Simulation of Vapor Cloud Explosion

ZHANG Chao
College of Mechanical and Electronic Engineering, China University of Petroleum
Qingdao 266580, China
zhchaoupc@126.com

ZHAO Dongfeng[a], MENG Yifei[a], QIU Long[b]
[a] College of Chemical Engineering, China University of Petroleum
[b] College of Mechanical and Electronic Engineering, China University of Petroleum
Qingdao 266580, China
zhaodf@vip.sina.com, myf213@163.com, qiulong2013@yeah.net

Prepared for Presentation at
American Institute of Chemical Engineers
1st CCPS Asia-Pacific Conference on Process Safety
Qingdao, China
September 4 – 5, 2013

UNPUBLISHED

The Safety Distance Study of the LPG Storage Field Based on the Numerical Simulation of Vapor Cloud Explosion

ZHANG Chao
College of Mechanical and Electronic Engineering, China University of Petroleum
Qingdao 266580, China
zhchaoupc@126.com

ZHAO Dongfeng[a], MENG Yifei[a], QIU Long[b]
[a]College of Chemical Engineering, China University of Petroleum
[b]College of Mechanical and Electronic Engineering, China University of Petroleum
Qingdao 266580, China

Keywords: safety distance; vapor cloud explosion; numerical simulation; Layout Optimization

Abstract

Abstract: The safety distance among the petrochemical units is usually based on the fireproof distance, and rarely takes the impacts of the explosion into account. As to the zones which is high-risk of explosion, such as liquefied petroleum gas (LPG) storage field, it is quite necessary to involve the effects of the explosive blast wave when design the safety distance . In this paper, the computational fluid dynamics tool named AutoReaGas is used to simulate the vapor cloud explosion (VCE) in the LPG storage field. The influences of the obstacles, which consist of fire dike, storage structure, on the vapor cloud explosion of the LPG have been analyzed, and the damage probability analyses are conducted to the devices. On basis of the analyses, the safety distance in the LPG storage field has been discussed combined with the fireproof distance. The results show that in the high danger zones of VCE, the safety distance based only on the fire prevention has not meet the inherent safety goal of the plant layout. The reliable numerical simulation is an effective method to analyze the explosion hazards and can offer persuasive data to the appropriate safety distance design between the units. The method that integrates the fire hazards with the explosion influences to determine the safety distance can contribute to the safety design and layout optimization, thus improves the inherent safety level of the plants to a large extent.

1. Introduction

Petrochemical tank farms are the important areas intended to store a large amount of chemical raw materials, and intermediate or finished products. Most of substances belong to the hazardous chemicals, with toxic, harmful, flammable and explosive properties, and have a high risk of fire and explosion. Due to the intensive equipment and the large amount of the chemicals, the accidents in the tank farms tend to trigger more events, which cause domino effect in the whole area. According to the statistics both at home and abroad, storage areas such as tank farms and oil depots come to be the most dangerous areas that cause the most domino effect[1-3]. In the study of R.M.Darbra and CHEN Yinting, the domino accidents happened in the storage areas account for about 41% or more of all the domino accidents[2, 3]. One typical accident is that the fires and explosions occurred in Buncefield Oil Depot on December 11 2005, which leaded to the European biggest blast in peacetime. The investigate report showed that the cause was that a large amount of gasoline overflowed from the 912 tank, then mixed into the explosive mixture with the air. The explosive mixture ignited by a car and lead to a serious of fires and explosions. The biggest blast wave reached the range from 70 to 100KPa. The fires and explosions involved more than 20 tanks and destroyed large parts of the Buncefield oil storage and transfer depot, Hemel Hempstead, and caused 43 persons injured and widespread damage to neighboring properties.

To effectively prevent and control the consequence of the accident, especially the domino effect, a good practice is usually to keep certain safety distance between the equipment and the targets when the plant is at the stage of design. Once the accidents happen, as an important safety measure, safety distance can prevent the heat radiation, blast wave and projective fragments form causing significant damage to the target equipment, thus decrease the possibility of the domino accidents.

At present, three kinds of ways are widely recognized to determine the safety distance: consequence-based method, risk-based method, and experienced distance[4]. The first method focuses on analyzing the probably major hazardous accidents. The second one quantifies the risk of the dangerous incidents from the perspective of the consequence severity and the probability. The experienced method is widely accepted in the engineering practice, and the distance is usually from the experienced data of the experts and research institutions.

In China, the safety distance among the petrochemical installations is determined by the experienced method, and refers to the standards and regulations from advanced countries. The fact is that the safety distance is mainly used to prevent fire damage. However, it is unreasonable to only consider the fire consequence, because the range of explosive blast influence is wider than that of fire, and the tanks are easier damaged by the explosion. The study[1, 3] shows that in the storage areas which have high risk of domino accident, fire and explosion are the main initiate events, and explosion is in the first place, then the fire. Furthermore, fire events are often followed by the explosion.

Thus, the liquefied hydrocarbon tank farms such as that stores liquefied petroleum gas (LPG) usually belong to the high risk of explosion areas. The safety distance only from the perspective of fire prevention would simplified the plant layout, and also create major potential risk to the tank farms and even the whole plant. The paper applies the consequence based method and makes the LPG tank farms as the object, uses the computational fluid dynamics software to analyze the consequence of the vapor cloud explosion in the LPG tank farms, and aims to get the reasonable safety distance.

2. Fire prevention distance in the LPG tank farms

As a kind of fuel, LPG has a high calorific value, easy to operate and store, and has been widely used in people's daily life. What's more, it is also an important basic chemical raw material. In chemical production, LPG can be used to make the ethylene, propylene, etc., which are the essential substances to produce synthetic plastics, synthetic rubber, synthetic fibers and other chemical products. According to the statistics, there have been over 200 LPG storage tank farms in China by 2012, and most of the LPG is stored in the sphere tanks over the atmosphere pressure. The safety problem becomes the focus of the government and the common people.

2.1 Safety distance in LPG tank farms

At present, the layout of the LPG tank farms in the petrochemical plants follows the principle of the regulations, such as 'Fire prevention code of petrochemical enterprise design', 'Specification for design of tank farms for petrochemical storage and transportation system' and 'Code for design of fire-dike in storage tank farm', etc. To a lager extent, safety distance in the LPG tank farms is based on the fire prevention.

In the same tank group, fireproof distance among the liquefied hydrocarbon tank such as LPG sphere tank is half of the diameter of the larger tank[5].

Table 1. Fireproof distance among the liquefied hydrocarbon tanks

substance			Sphere tank	Horizontal tank
Liquefied hydrocarbon	Pressure or semi-refrigerated storage tank	Have measures to discharge substance to the flare system	0.5D	1.0D
		No measures to discharge substance to the flare system	1.0D	

Note: D is the diameter of the larger tank.

The distance between the tank groups is also determined in the regulation. The fireproof distance between the liquefied hydrocarbon tanks is 60 meters.

Table 2. Fireproof distance between the petrochemical enterprises

Matter	Liquefied hydrocarbon tank group(outer wall of tank)	Tank group stored class A and B liquid(outer wall of tank)	Class A and B process equipment or facilities	Overall or regional major facility
Liquefied hydrocarbon tank group(outer wall of tank)	60 meters	60 meters	70 meters	90 meters
Tank group stored class A and B liquid(outer wall of tank)	60 meters	1.5 times of diameter	50 meters	60 meters

Protective walls are needed in the LPG tank farms. When the LPG with a high pressure leaks from the tank, it can vapor quickly. The protective wall intends to restrain the LPG form overflowing from the tank farm. In the regulation, the height of the wall is 0.6 meter[6].

2.2 Hazard characteristics of LPG tank farm

The main components of LPG are C3 and C4, and include a small amount of ethylene, propylene, etc. In general, LPG is colorless and transparent, and has the special smell of hydro carbons. At atmospheric temperature, LPG is gaseous, and can be compressed or refrigerated to liquid. LPG is a typical substance with a property of flammable and explosive, and belongs to class A fire hazard. Under normal temperature and pressure, LPG that is stored in the tank is easily to volatilize, and the volume can rapidly expand 250-300 times than before[7]. Once the LPG leaks from the tank, it can form a continuous and instantaneous leaking source near the ground. If the LPG is ignited immediately near the leakage hole, jet fire may occur. If not, due to the heavy effect, the LPG will mixed with air into the explosive cloud. At this time, if the mixture is ignited, the vapor cloud explosion happens. When the LPG tank is heated by the external flame and internal pressure of tank increases, LPG will be gasified sharply. In this case, if the LPG is immediately ignited, boiling liquid expanding vapor explosion (BLEVE) will occur.

LPG has a wide range of explosion, and the upper explosive limitation is 9.5%, and the lower explosive limitation is 1.5%. The rate of explosion is about 2000-3000 m/s, and the calorific value reaches 104.67Mj. Thus, the chemical explosion can occur in a twinkling. In LPG tank farms, compared with the fire, vapor cloud explosion occurs more likely, and consequence is more serious. Overall, safety distance based on fire prevention can control the fire hazard effectively to some extent, but it is not enough to prevent the blast from damaging the surrounding tanks.

3. Methods to evaluate the vapor cloud explosion

Due to the mechanical, chemical or thermal impact to the LPG tank, the LPG tank ruptures. Once LPG leaks from the tank and forms the explosive cloud, the vapor cloud explosion may occur. The destructive effects of VCE depend on the damage from the blast, thermal radiation on the surrounding personnel, buildings, storage tanks and other equipment. However, the blast causes the biggest damage, and has the largest destroying regions. It comes to be a key indicator to quantify the vapor cloud explosion hazard[8].

Currently, several methods are used to evaluate the vapor cloud explosion blast wave. They are TNT equivalent method, TNO method, and multi-energy method, etc.

3.1 TNT equivalent method

TNT equivalent method intends to transform the damaging effect of the VCE into that of TNT explosion. The key point is figure out the mass of the cloud and changes it into the TNT equivalent amount[9]. The formulas are as following.

$$W_{TNT} = \alpha W_{Hf} / H_{TNT} \;\ldots\ldots [\text{Eq. }1]$$

Once the TNT equivalent amount is got, the blast value at any distance from the center of the explosion can be calculated by the formulas:

$$z = R/(W_{TNT})^{1/3} \;\ldots\ldots [\text{Eq. }2]$$

$$p_i = 3.9/z^{1.85} + 0.5/z \;\ldots\ldots [\text{Eq. }3]$$

Where W_{TNT} is TNT equivalent amount (kg); W is mass of the flammable cloud, kg; Hf is the heat of combustion, MJ/kg; η is equivalent coefficient of TNT; HTNT is explosion heat of the TNT, 4.5MJ/Kg; p_i is blast value at the distance of R, MPa。

However, the method has obvious shortcomings on the VCE evaluation. First, gas explosion belongs to non-ideal source explosion, and has a significant difference with the TNT explosion. What's more, the combustion of gas explosion usually is not sufficient. According to statistics, the actual release energy is just a little fraction of the gas cloud explosion potential energy, typically 10% or less, and in most cases, the value does not exceed 2%. Therefore, the TNT equivalent coefficient is difficult to make it accurately.

3.2 Multi-energy method

Multi-energy method is created and gradually improved by Dutch TNO (The Netherlands Organization) laboratory based on the TNO model and the combination of a large number of experimental and numerical studies. The model takes into account many factors such as the turbulent acceleration, local constraints and gas deflagration activity, which impact the strength of the deflagration. The method is currently commonly used in the industry to simulate and predict the gas cloud explosion in the open space.

The method divides the gas combustion intensity into 10 levels. The first 10 is detonation, and class decreases gradually with the combustion intensity[10].

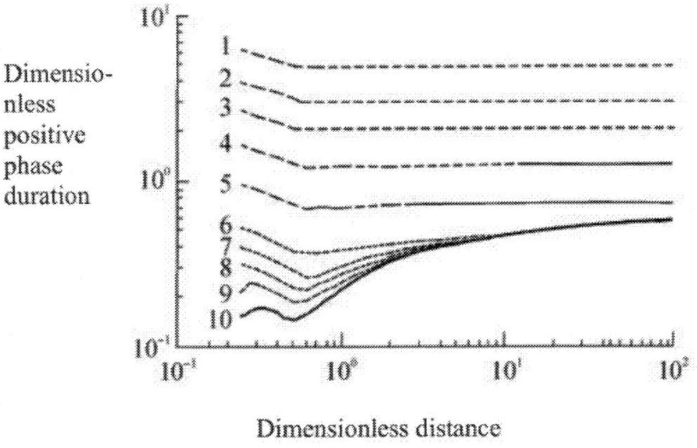

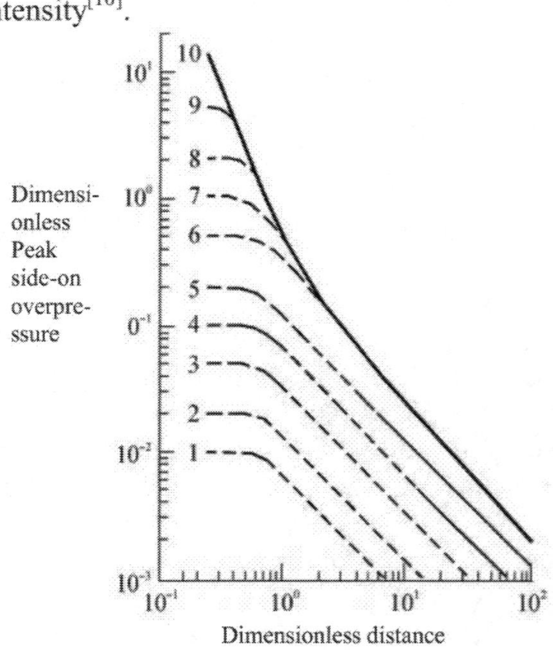

Figure1-Change curves of dimensionless positive phase duration versus dimensionless distance

Figure2-Change curves of dimensionless peak side-on overpressure versus dimensionless distance

After determining the burst intensity and the released energy of fuel - air mixture, the Sachs distance can get by the equation4, then the Sachs lateral blast overpressure and the positive phase duration time at the distance of R can be obtained from Figure 1 and Figure 2.

$$\overline{R} = \frac{R}{(E/P_0)^{1/3}} \quad ......[Eq. 4]$$

In the equation:

$\overline{R}$ —Sachs distance(dimensionless); R—distance, m;

E—heat of combustion, J; P0—Ambient Pressure, Pa。

The Lateral explosion overpressure peak and negative phase duration can be calculated according to the Sachs overpressure and the duration of the positive phase calculation. The overpressure peak can be obtained by the following formula:

$$p_0 = \Delta\overline{P}_s P_a ......[Eq. 5]$$

The negative phase duration can be obtained by the following formula:

$$t_d = \overline{t}_d \left[\frac{(E/p_0)^{1/3}}{C_0} \right] \quad ......[Eq. 6]$$

In the formula:

p_0--Lateral explosion overpressure, Pa; $\Delta\overline{P}_s$ --Sachs explosion overpressure (dimensionless)

P_a--Ambient Pressure, Pa; E--heat of combustion, J;

t_d --negative phase duration, s; C_0--velocity of sound, m/s;

$\bar{t}_d$ --Sachs negative phase duration, (dimensionless).

The model is verified and corrected by a large number of experimental results, and is closer to the reality. Due to the clearly concept, easily solved, the method is widely used. However, the user must select an explosive level according to their own experiences when use the method. Some researchers have discussed how to select explosion level by experimental data. Because the volume of the actual gas cloud explosion open space is relatively large, and the shape of the obstacle is various, the way to use a small-scale experiment to predict large-scale open space vapor cloud explosion is impossible.

3.3 CFD method

Compared with TNT equivalent method and Multi-energy method that need much experimental data and verification, the numerical simulation is simple and convenient and costs less. By now, there have developed a variety of simulation methods, and one used in gas explosion is CFD model. It is based on the computational fluid dynamics, combines the basic equations of fluid dynamics and chemical reaction kinetics equation, and uses the finite element or finite difference method for differential equations. Given the appropriate boundary and initial conditions, the gas explosion is calculated and solved[11].

At present, several CFD softwares have been developed to solve the gas explosion, such as FLACS, AutoReaGas and EXISM. Above all, the three-dimensional fluid dynamics software, AutoReaGas, builds the turbulent combustion model taking the advantage of empirical equations and theoretical analysis. It can simulate the turbulence caused by obstructions within the each cell and the any ignition position. Also, it can calculate the gas explosion with different fuel-air concentration ratio in the open space. So, the model has certain advantages in the simulation of turbulence caused by obstructions.

AutoReaGas is a dedicated software package to simulate the gas explosion by three-dimensional modeling, and developed by the British Century Dynamics company and the Dutch TNO Laboratory in the 1990s. The software includes two calculation parts: gas explosion calculation and blast wave calculation. Gas deflagration is considered as a viscous perfect gas thermal expansion. Combustion aerodynamics is expressed by mass, momentum and energy conservation equation[12].

Mass conservation equation:

$$\frac{\partial \rho}{\partial t} + \frac{\partial(\rho u_i)}{x_i} = 0 \cdots\cdots [\text{Eq.7}]$$

Momentum conservation equation:

$$\frac{\partial}{\partial t}(\rho u_i) + \frac{\partial}{\partial x_i}(\rho u_i u_j) = -\frac{\partial P}{\partial x_i} + \frac{\partial \tau_{ij}}{\partial x_j} \cdots\cdots [\text{Eq. 8}]$$

Energy conservation equation:

$$\frac{\partial}{\partial t}(\rho E) + \frac{\partial}{\partial x_i}(\rho u_j E) = \frac{\partial}{\partial x_i}\left(\Gamma_E \frac{\partial E}{\partial x}\right) - \frac{\partial}{\partial x_i}(\rho u_j) + \tau_{ij}\frac{\partial u_j}{\partial x_j} \cdots\cdots [\text{Eq. 9}]$$

Fuel component conservation equation:

$$\frac{\partial}{\partial t}(\rho m_{fu}) + \frac{\partial}{\partial x_i}(\rho u_j m_{fu}) = \frac{\partial}{\partial x_i}\left(\Gamma_{m_{fu}} \frac{\partial m_{fu}}{\partial x_j}\right) + R_{fu} \cdots\cdots [\text{Eq. 10}]$$

In the equations:

$$\tau_{ij} = u_t\left(\frac{\partial u_i}{\partial x_i} + \frac{\partial u_i}{\partial x_j}\right) - \frac{2}{3}\delta_{ij}\left(\rho k + u_t \frac{\partial u_i}{\partial x_i}\right) \cdots\cdots [\text{Eq. 11}]$$

ρ --Density, kg/m^3; p --static pressure, pa; E --energy, J; m_{fu} --fuel mass fraction, %;
R_{fu} --volume combustion speed; Γ --turbulent dissipation factor in transportation;

I, j--coordinate direction, m^3/s; u_i --fluid velocity in the i direction, m/s;
μ_t --turbulent viscosity coefficient, m2/s; δ_{ij} --Kronecker operator.

4. Simulation and assessment of VCE in LPG tank farm

4.1 LPG tank farm layout

In one tank farm of a petrochemical industry, a list of LPG spherical tanks have been built sine in 2008. The problem is that the spheres are adjacent to the ethylene production installations, other tanks and two control rooms. As shown in the figure3, because of the large amount of the LPG and the probability of domino accident, the layout is a big potential danger to whole the plant.

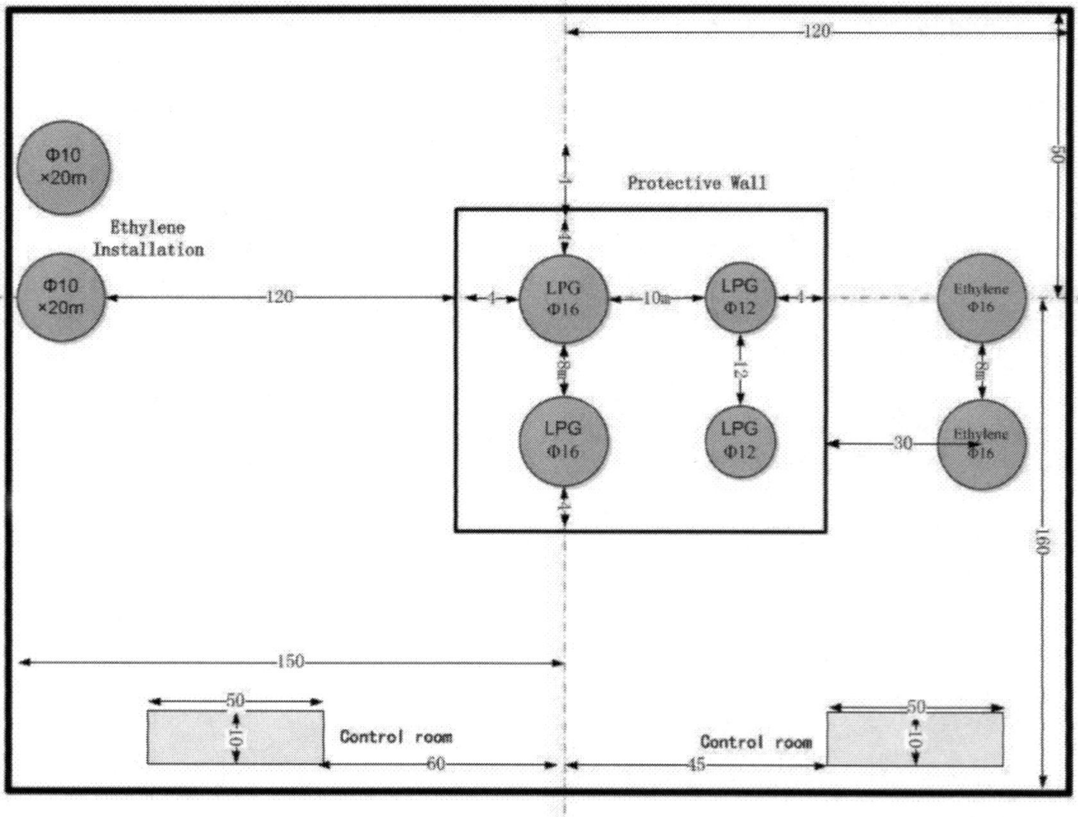

Figure3-the layout of the LPG tank farm

In the LPG tank farm, the protective wall with 0.6 meter high and 48 meters long circles the tanks. Because the density of the LPG is higher than that of air, the LPG is prone to gather in the protective wall. When the VCE happens, the confined space makes the consequence of the explosion more severe.

4.2 VCE simulation and analysis

In Autoreagas, there is no sphere construction, thus the cylinder whose diameter equals to the length is used to replace the LPG spherical tank. The model shows in the figure-4.

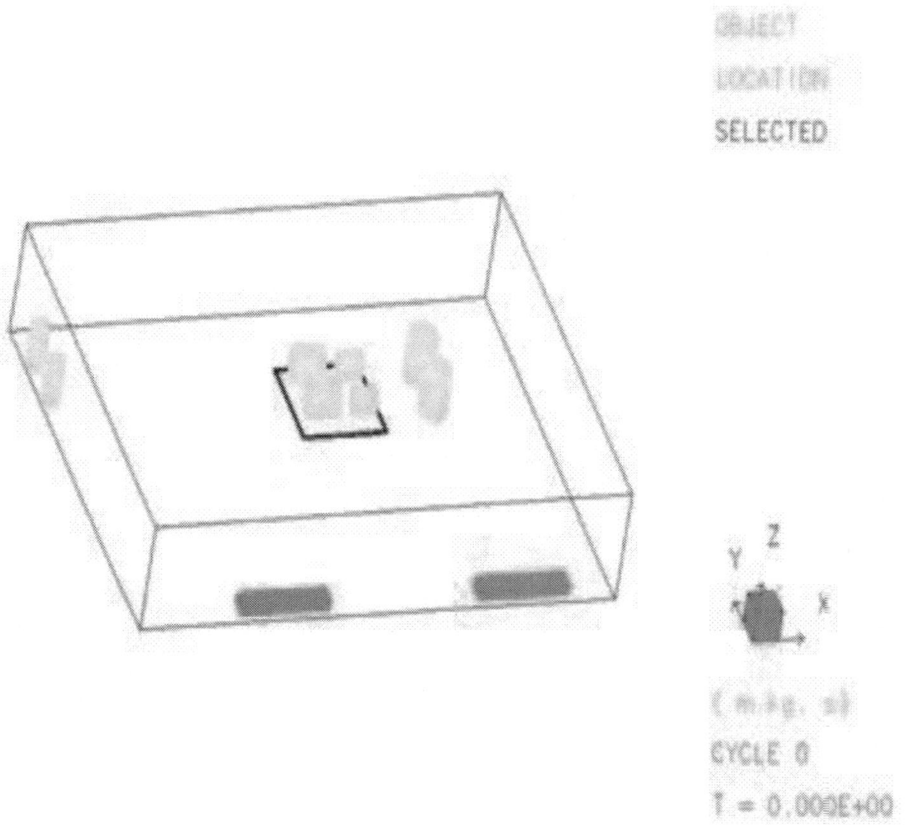

Figure4-the 3D model

The scene is assumed that LPG leaks from the bottom of the spherical tank. The leakage hole is 25mm in diameter and the leakage rate is constant in 10min.

According to the Bernoulli equation, the leakage of rate follows the below formula:

$$Q_0 = \rho A C_0 \sqrt{2(P - P_0)/\rho + 2gh} \ldots\ldots [\text{Eq. } 12]$$

In the equation 7, Q_0--leakage rate, kg/s; C_0--leakage coefficient, dimensionless， 0.81；

A--area of the leakage hole,m2; ρ --density of the substance,0.58g/cm^3；

h--height of the hole, 15m； g--gravity acceleration， 9.8kg/m^2；

P--pressure of the tank, 680000Pa； P_0--ambient pressure ,101×10^{3}Pa；

The leakage rate of the LPG is 10.9kg/s. The amount of the LPG leak form the tank in 10min is 6.54×10^{3}kg. Here we assumed that the proportion of the butane and the propane is 1. When the LPG leaks from the tank, only a certain amount of the LPG evaporate into the air and form the explosive mixture. The flash factor is different between the butane and propane. Through the calculation, the mass of the propane in the air is 981kg, and that of butane is 327kg. In the simulation, there are 11 gauges selected to measure the overpressure at different distance. The results are shown in the figure 5 and 6.

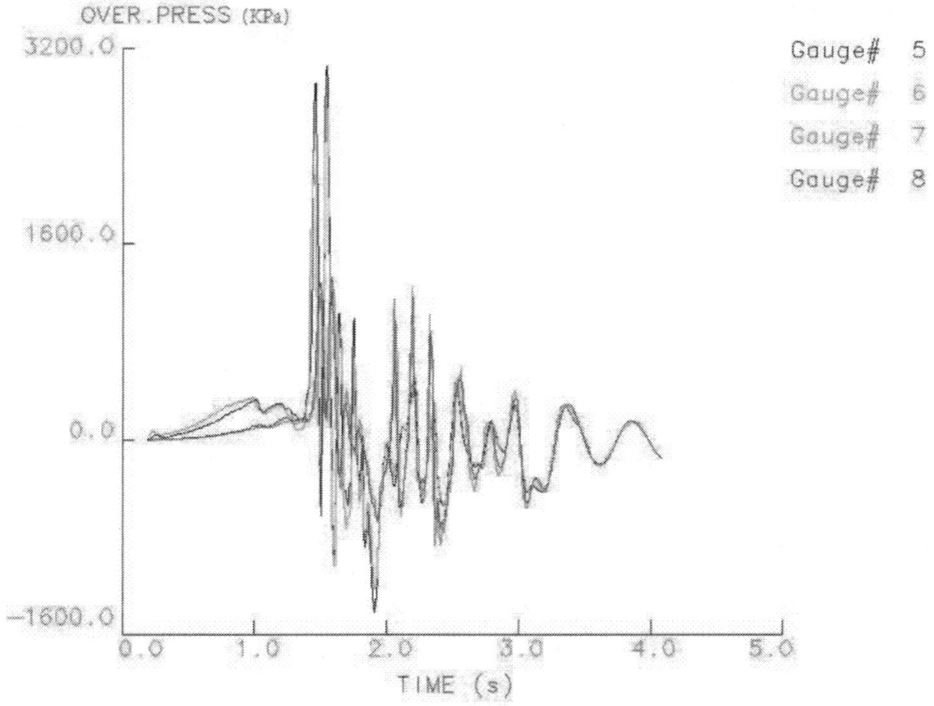

Figure 5 –overpressure around the LPG tanks

The result shows that the overpressure in the tank farm can reach about 3.0MPa, which can almost destroy all the devices. It means that the VCE in the LPG tank can damage the LPG tank next to it, and lead to another VCE or fire accident, thus may probable trigger domino effect in the whole plant.

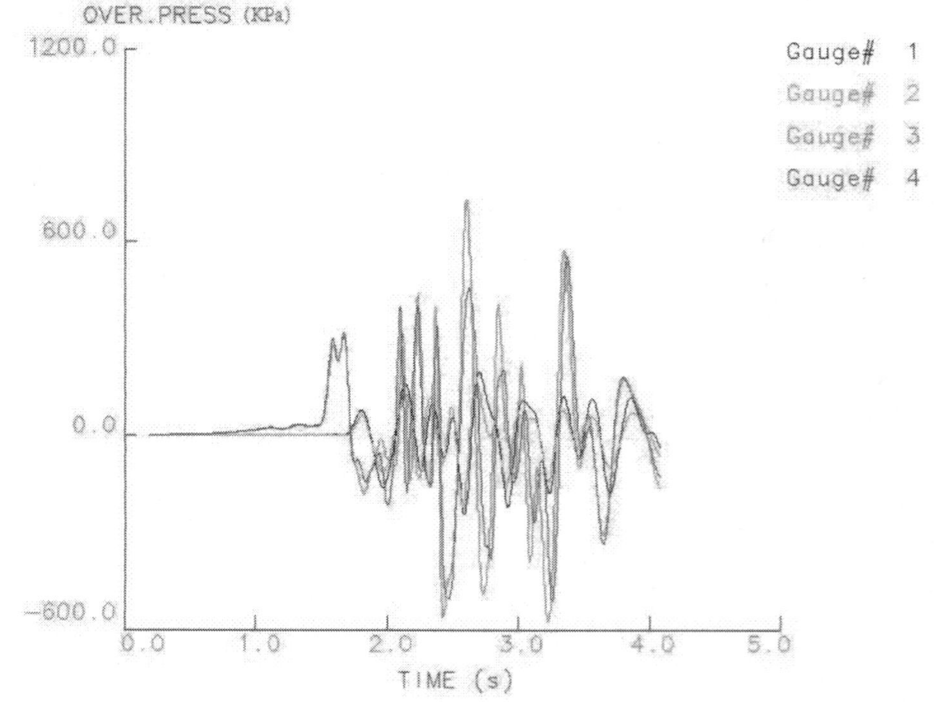

Figure 6-overpressure in the ethylene installations

The ethylene production also suffers a high overpressure. As shown in the figure 6 (Gauge 3), the overpressure can reach 600KPa, which can destroy the ethylene production to a large degree.

Currently, the threshold value is commonly used to assess the damage of the equipment. The threshold value indicates the minimum overpressure at which damage is expected at 'target' equipment. There are different threshold value criterions of overpressure due to different understanding of damage to equipment. Here the values proposed by Zhang Ming guang[13] are used. Based on damage probability ranges discussed in Zhang, 30% damage probability is the critical point of minor loss, light damage state and intense loss and damage state; 70% damage probability is the critical point of intense or catastrophic loss, intense damage state. 30% damage probability value means the state of short time (about 10 min) leakage of the inventory of dangerous materials contained in equipment. 70% damage probability value means the state of instantaneous (about 1 min) leakage of the inventory of dangerous materials contained in equipment. So, threshold values of overpressure which cause 30% and 70% damage probability are used to assess the chemical process equipment.

In terms of pressurized vessels, the threshold value of 30% damage is 32KPa, and that of 70% damage is 58KPa.

According to the threshold values, the damage of the equipment can be estimated and it shows in the table 3.

Table 3-The damage of different equipment

Equipment	Distance (m)	Overpressure (KPa)	Damage
LPG spherical tank	Less than 20	More than 100	At least 70%
Ethylene spherical tank	60	About 90	At least 70%
Ethylene devices	120	About 55	At least 30%
Power distribution Room	170	About 35	About 30%
control room	250	About 21	Less than 30%

4.3 Layout optimization discussion

Through the simulation and assessment of the VCE that happens in the LPG tank farm, the LPG tanks in the farm are almost destroyed by the blast wave of the VCE, and the ethylene devices and adjacent tanks are damaged more than 30% by the overpressure. However, the distances among the LPG tank are based on the fireproof distance and all the layout is in accordance with the current standards and regulations. Thus, the result indicates that the fireproof distance is not enough to decrease the damage form the explosion. The layout based on the fireproof distance cannot meet the inherent safety goal.

To guarantee the safety, the layout should not ignore the explosion consequence. According to the study, given the VCE, the safety distance between the LPG tank and the ethylene production should be at least 170 meters to assure the damage of the production units less than 30%. The distance away from the adjacent LPG tank farm is 60 meters according to the regulations. However, at this distance, the damage caused by the overpressure is more than 70% damage, so there should be measures to protect the adjacent tank farms. It is worth mentioning that keeping a certain distance is not always the best choice, and the protective walls also are an effective method in many cases. No matter what measures are taken, the credible consequence assessment is important.

The layout optimization should take the fire and explosion into consideration and the assessment of the consequence is necessary to make reasonable decisions. Because the consequence assessment needs more detail information and credible data, and the assessment is usually carried out by the professional persons and cost much time, the assessment is not be widely used in the present.

5. Conclusion

The layout of the LPG tank farms is essential to the inherent safety of the whole plant. Compared to the fire accidents in the LPG tank farms, the vapor cloud explosion is prone to cause serious damage to the surrounding devices directly and to trigger domino effect.

Due to the drawbacks of the TNT equivalent method and multi-energy method in assessing the consequence of the VCE, the VCE simulation was conducted using the 3-Dimension software autoreagas. The results show that the traditional safety distance based on the fireproof has limitation in controlling the consequence of the VCE effectively, and should not be used to prevent the domino effect as well. The explosion, fire, land planning and surroundings all have influence on the Safety distance, however, not all of them can become the leading factor, which dominate the safety distance. Numerical simulation can be an ideal way to confirm the leading factors and determine the safety distance. As to the VCE event, the autoreagas has unique advantages on analyzing overpressure of the explosion.

The layout of the tank farms was also discussed. To optimize the layout, the consequence assessment is necessary and a certain number of credible data and information is the precondition. The numerical simulation is proved to be an effective ways to analyze the VCE consequence and make the layout optimization more reasonable.

References

[1] Bahman Abdolhamidzadeh, Abbasi, Domino effect in process-industry accidents – An inventory of past events and identification of some patterns [J]. *Journal of Loss Prevention in the Process Industries* **24**, 575 (2011).

[2] Chen Yinting, Mingguang Zhang, Investigation and Analysis of Historical Domino Effects Statistic [J]. *Procedia Engineering* **45**, 152 (2012).

[3] R. M. Darbra, A. Palacios, J. Casal, Domino effect in chemical accidents: main features and accident sequences [J]. *Joural of Hazardous Materials* **183**, 565 (Nov 15, 2010).

[4] Liu Ji, Gao Jian-Ming, Determination of Safety Protection Distancefor Installations of Dangerous Chemicals [J]. *China Safety Science Journal* **22**, (2012).

[5] Ministry of Housing and Urban-Rural Development of the People's Republic Of China, GB 50160-2008 Fire prevention code of petrochemical enterprise design [S]. (2008).

[6] Ching State Construction, GB50351-2005 Code for design of fire-dike in storage tank farm [S]. (2005).

[7] Sun Wen-Dong Fang Jiang-Min, Research on the Simulant Methods of the Fire and Explosion Accidents of the LPG Tank [J]. *Industry Safety and Environmental Protection* **35**, (2009).

[8] Jiang Jun-Cheng Wang Ruo-Jun, Parameter uncertainty analysis of LPG vapour cloud explosion risk assessment [J]. *Joural of Nanjing University of Technology* **27**, (2005).

[9] China Petroleum & Chemical Co. Ltd. Qingdao Safety Engineering Institute, A Guidance for Quantitative Risk Assessment in the etrochemical Plant [M]. (China Petrochemical Press 2007).

[10] Zhang Rui Hua, Chen Guo Hua, Engineering Applications of TNO Multi Energy Method in VCE Simulation Assessment [J]. *Journal of South China University of Technology* **34**, (2006).

[11] Zhao Xiang Di, Zhai Liang Yun, Study on Explosion Risk Assessment Method for Control Room in Petrochemical Industry [J]. China Safety Science Journal **19**, (2009).

[12] Chai Jian She, Chen Wen Ying, Numerical Simulationof Explosion Overpressure Field of Flammable Vapor with Barriers in Open Space [J]. *China Safety Science Journal* **19**, (2009).

[13] Zhang Ming Guang, Jiang Juncheng. An improved probit method for assessment of domino effect to chemical process equipment caused by overpressure [J]. *Joural of Hazardous Materials* **158**, 280 (Oct 30, 2008).

The Basic Characteristics and Hazard Assessment Of Particular Risk Of Dust Explosion As Well As Security Prevention and Control Countermeasures Analysis
Ding, H., China University of Petroleum

Abstract: China is currently in a period of accelerating industrialization and urbanization. During the economic development process, the increasing emissions of industrial waste gas especially dust have damaged the ecological environment so badly. As a special risk of process industry, dust explosion accidents occur each year in the chemical, coal, metallurgy, wood processing industries which were too numerous to mention. People exert their efforts for up to hundred years in order to prevent dust explosion accidents, but so far, dust explosion is still a serious threat to the safety of process industry, causing casualties and significant property damage.For that reason, put forward some available security measures based on the analysis of dust explosion hazards, which has become a serious problem demanding prompt solution. For this particular risk of dust explosion, firstly, basic characteristics such as the conditions, procedures and mechanisms were introduced so that we can have a basic understanding of dust explosion; Secondly, combined with significant dust explosion accidents occurred in recent years, analyze three main explosion hazards: strong destruction, prone to secondary explosions, produce toxic gases and then raise hazard assessment method, with using FLACS for accurate simulation and quantitative assessment of the whole explosion process. Next, focus on analyzing the factors of dust explosion such as the dust characteristics, latent solvent strength, potential ignition sources, ambient temperature, flammable gases and inert substances, presenting risk identification methods. Simultaneously, have a experimental analysis for a variety of common dust explosion parameters which have important reference value for dust safety protection through the results of experimental determination. Finally, the latest safety protection methods of dust explosion are summarized, mainly from two aspects that the dust explosion prevention and dust explosion control were discussed. Prevention measures of dust explosion were achieved by drawing up well-established and rational hazard assessment. The highlights of this measure include: rationalization in the industrial layout; prevent combustible dust suspend, reduce dust accumulation and spread range; eliminate dust ignition sources and suppress dust combustion by using inert gases. Control strategies of dust explosion include these following measures: (1) Pressure-proof method. Have a design for silos, pressure vessels and equipment, enable it to withstand the maximum explosion pressure which is a basic and most effective control method. (2) Flame-proof method. Between the two processes interconnecting with each other inside the pipe material, cut off by a certain means to prevent the spread of dust explosion. This method is divided into active isolation and passive isolation. (3) explosion suppression method. Using detectors to monitor the danger signals of dust explosion at manufacturing equipments or work areas, immediately trigger explosion suppression processing systems, rapidly jet inhibitors aiming to prevent the spread of dust explosion and rebirth. All that can reach the purpose of eliminating explosion hazards as a result. (4) vented explosion method. When a dust explosion occurs, a timely manner through a fixed pressure relief opening for release is required, meanwhile, exhaust the high temperature-pressure dust and combustion gases to outside, limit the pressure of dust explosion to a certain numerical value that the container material strength can sustain, reduce explosion damage to facilities and personnel. It is the most widely used explosion-proof method. Safety prevention measures of dust can greatly reduce explosion hazard, provide recommendations for safety and civilized production.

Study on Emergency Management Capability Evaluation of Our Oil Companies' Overseas Emergency Based on the Catastrophe Progression Method

Hao Ding, Haodong Shi and Ming Liu, China University of Petroleum (East China)

Abstract

According to the fact that the overseas emergencies of our oil companies occurred frequently and emergency management capability is improving, and with the index system construction method, the evaluation index system is established starting from the whole process of emergency management and complying with the process of index construction, which can reflect the emergency management capability level of our oil companies roundly. This index system includes the capabilities of risk pre-warning and control, emergency command, emergency control, incompletion and restoring, and conclusion analysis. Primary index risk pre-warning and control including emergency plan, information collection of potential accidents and information analysis and predication; Primary index emergency command including personnel training of overseas emergency and emergency decision-making; Primary index emergency control including dynamic assessment of damage, communication and collaboration with the resource countries, regional materials linkage and information acquisition and transmission; Primary index incompletion and restoring including survey evaluation and treatment and psychological intervention and guidance of responsibility; Primary index conclusion analysis including event reason summary analysis ability and disposal process and experience learning. With this index system, the CPM was used to assess the emergency management of our oil companies 'overseas accidents. Growing out of the catastrophe theory, the CPM is a kind of comprehensive evaluation method with the advantages of considering the importance of every index relatively, combining the qualitative and quantitative, reducing the subjectivity, scientific and reasonable, and accuracy. The expert scoring method was used to assign the emergency management capability of four types of emergency including natural disaster, kidnap, variation of resource policy and coup. Then the questionnaire was made and the secondary indexes were explained through one to three questions. The workers with the overseas emergency management experience and related experts were invited to grade those questions, and then took the average to acquire the score of each indicator respectively. Based on this and the CPM, the grades of emergency management capability of our oil companies in different accidents were given. The result indicates that the emergency management capability of variation of resource policy is the strongest and that of the coup is the weakest.

Key words: Oil Companies; Emergency; Capability of Emergency; Catastrophe Progression Method

1. Introduction

According to our country's strategic deployment of "making the most use of domestic international markets and two kinds of resources", China's oil enterprises established the development strategy of "based on domestic, developing overseas and implementing the international operation". Recently, overseas business has formed a scale and better benefit was gained. With the development of overseas business and the expansion of scope of business, the risk of health, safety and environment have increased relatively. With the characteristics of multi-project, large cover rage, large staff mobility and complex social situation, the overseas business of oil companies may face various kinds of business no matter whether the domestic business may face or not. In order to understand the basic situation of emergency management capability and improve the emergency reaction capability of overseas emergency of our oil companies, it is necessary to evaluate its emergency management capability to ensure the emergency measures can be taken timely and effectively. It can prevent and reduce the damage of the emergency to the uttermost.

2. The basic theory and procedure of catastrophe progression method

2.1Basic Theory

Based on singularity theory, stability theory and other mathematical theory, catastrophe theories as a kind of mathematical theory was constructed by Thom, a France scientist, to describe how a series of successional variables evolve into salutatory quantitative change in 1970s. Although the mathematical foundation is rather profound, the application model of catastrophe theory is simple. Therefore, the application of catastrophe theory has extended to many fields of natural science and social science.

2.2 Procedure

(1) Decide catastrophe model.

In catastrophe theory, analyzing the stability or instability of system is to analyze the variation minimal value of functions, potential function for short. According to the classification theorem of catastrophe theory, there are a large number of discontinuous phenomena existing in the nature and society, which can be showed in some specific geometrical shape. In summary, there are seven basic shapes, including fold catastrophe, cusp catastrophe, swallowtail catastrophe, butterfly catastrophe, hyperbolic umbilical point catastrophe, oval umbilical point catastrophe, parabolic umbilical point catastrophe. Each of those catastrophes corresponds to different potential function, as shown in the following table.

Table 1. The Basic Catastrophe Shapes Table

Catastroph e shapes	**Control variable dimension**	**State variable dimension**	**Potential function**
Fold catastrophe	1	1	$V(x)=x^3+ux$
Cusp catastrophe	2	1	$V(x)=x^4+ux^2+vx$

Continued Table 1. The Basic Catastrophe Shapes Table

Swallowtail catastrophe	3	1	$V(x)=x^5+ux^3+vx^2+wx$
Butterfly catastrophe	4	1	$V(x)=x^6+tx^4+ux^3+vx^2+wx$
Hyperbolic umbilical point catastrophe	3	2	$V_{abc}(x,y)=x^3+y^3+wxy+ux+vy$
Oval umbilical point catastrophe	3	2	$V_{abc}(x,y)=x^3-xy^2+w(x^2+y^2)+ux+vy$
Parabolic umbilical point catastrophe	4	2	$V(x,y)=y^4-x^2y+wx^2+ty^2+ux+vy$

Growing out of the catastrophe theory, the CPM is a kind of comprehensive evaluation method with the advantages of considering the importance of every index relatively, combining the qualitative and quantitative, reducing the subjectivity, while with scientific nature and reasonability and accuracy.

(2) The unitary formula is derived from bifurcation equation of catastrophe model. In catastrophe theory, potential function includes state variable and control variable. State variable indicates behavioral of the system, while the control variable shows the each aspect of internal contradictions of the system. Its final unified result is state variable, and depending on the different effect on the state variable, the internal contradiction can be divided into primary and secondary contradictions. For example, cusp catastrophe theory is a three-dimensional phase space, so the profile of equilibrium M was drawn and the equation is as follows:

$$4x^3+2ux+v=0$$

Its singularity set S, which is the subset of M.

$$12x^2+2u=0$$

Then was eliminated by simultaneous above two equations, and the bifurcation equation was acquired.

$$8u^3+27v^2=0$$

This bifurcation equation can also be showed in decomposed form:

$$\begin{cases} u = -6x^2 \\ v = 8x^3 \end{cases}$$

In evaluating the decision, as an evaluation decision variable x was elected as a goal, and as the main objective, u reflect the main aspect of the contradiction, and the change of u will result in the qualitative change of the evaluation decision. As the secondary objective, v is the minor aspect of the contradiction, which defined the primary and secondary relationships of u and v. In the evaluation decision, the control variation corresponding to each objective is well-founded. And the relationship between control variation and state variation in cusp catastrophe is showed as follows:

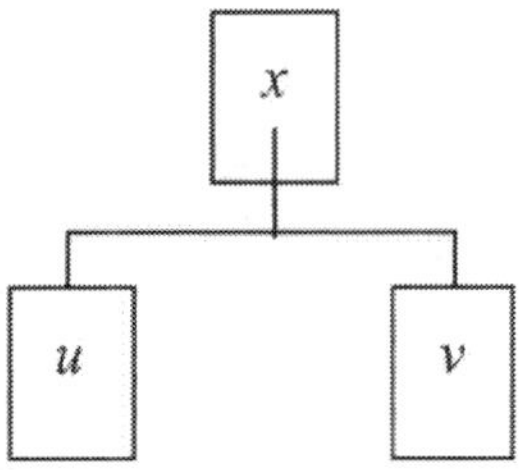

The following is the normalization formula of catastrophe progression method, which acquired from the bifurcation set of the cusp catastrophe decomposed form:

$$x_u = \sqrt{\frac{u}{-6}}$$

$$x_v = \sqrt[3]{\frac{v}{8}}$$

x_u and x_v corresponds to the x value of u and the x value of v. But in order to calculate conveniently, in practice application the unitary formula of cusp catastrophe can be showed as follows:

$$\begin{cases} x_u = \sqrt{|u|} \\ x_v = \sqrt[3]{|v|} \end{cases}$$

Catastrophe progression method is the method to calculate the contradictions, which can unify different control vitiations into one qualitative state. In catastrophe progression method, the amount of the elements of each level of the constructed hierarchical structure should be less than four because we live in four-dimensional space.

Based on the above analysis, the system chart and bifurcation point set equation of other catastrophe shapes can be acquired. x_u、x_v、x_w and x_t are the x's value of u, v, w and t respectively.

Table 2. The System Chart and Point Set Equation of Typical Catastrophes

Catastrophe shapes	**System chart**	**Bifurcation point set equation**	**Unitary formula**
Cusp catastrophe	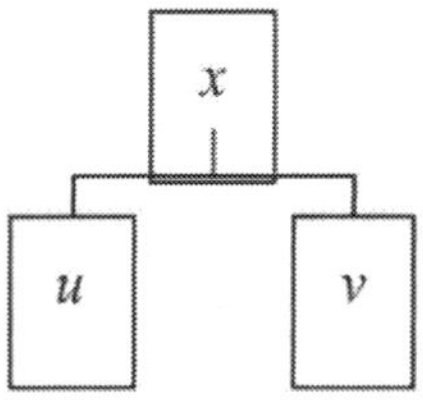	$u=-6x^2$ $v=8x^3$	$x_u=\sqrt{\lvert u\rvert}$ $x_v=\sqrt[3]{\lvert v\rvert}$
Swallowtail catastrophe	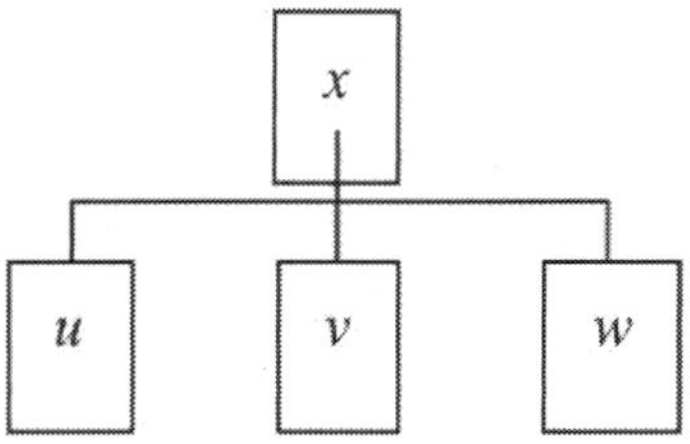	$u=-6x^2$ $v=8x^3$ $w=-3x^4$	$x_u=\sqrt{\lvert u\rvert}$ $x_v=\sqrt[3]{\lvert v\rvert}$ $x_w=\sqrt[4]{\lvert w\rvert}$
Butterfly catastrophe	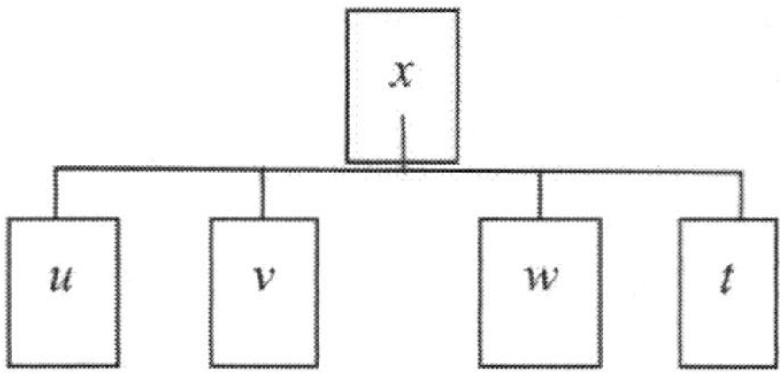	$u=-10x^2$ $v=20x^3$ $w=-15x^4$ $t=4x^5$	$x_u=\sqrt{\lvert u\rvert}$ $x_v=\sqrt[3]{\lvert v\rvert}$ $x_w=\sqrt[4]{\lvert w\rvert}$ $x_t=\sqrt[5]{\lvert t\rvert}$

(3) Chose normalization formula to evaluate. Using the unitary formula the x of each control variation for the same control target is calculated with the principle of "complementary" and "non-complementary", and the catastrophic affiliated functional values is acquired. The principle of "complementary" means that the control variations (u, v, w and t) of the system can complement each other perfectly to reach the higher average value of x value accordingly, that is $x = (x_u + x_v + x_w + x_t)/4$. The principle of "non-complementary" means the control variations of the system cannot complement each other. Therefore, the x value of the whole system is the smallest of the x_u, x_v, x_w and x_t corresponding to each control variation when solving the stare variation x according to the unitary formula. Then the affiliated functional value can be acquired with the step by step computation.

3. The application of the catastrophe progression method in the emergency management capability evaluation of our overseas oil companies' emergency

3.1 The construction of indicator system of emergency management capability evaluation of our overseas oil companies' emergency.

With the specific conditions of our oil companies, the constructed indicator system of emergency management capability evaluation of our oil companies' overseas emergency is showing in table 3.

Table 3. The Indicator System of Emergency Management Capability Evaluation of Our Oil Companies' Overseas Emergency

Assessment projective	**First level index**	**Second level index**
The emergency management of our overseas oil companies	Risk pre-warning and control *C*1	Contingency plan *D*1
		Information collection of potential accident *D*2
		Analysis and prediction *D*3
	Emergency command *C*2	The training degree of overseas emergency personal *D*4
		Emergency decision *D*5
	Emergency control *C*3	Dynamic damage assessment *D*6
		Communication and cooperation with the resources country *D*7
		Regional materials linkage *D*8
		Information collection and transmission *D*9
	Aftermath recovery *C*4	Investigation, evaluation and dispose *D*10
		Mental intervention and coaching *D*11
	Summarize ability *C*5	Reason and the summarize of disposal process *D*12
		Experiencing learning *D*13

3.2 The emergency management capability evaluation of our overseas oil companies' emergency based on catastrophe progression method.

With the catastrophe progression method, the emergency capability of our oil companies was analyzed in the overseas emergencies including natural disaster, kidnap, policy change of resource country and coup. And four kinds of representative emergencies were selected from the above emergencies to carry on the instance analysis.

The expert scoring method was used to assign the emergency management capability of four types of emergency including, natural disaster, kidnap, variation of resource policy and coup. Then the questionnaire was made and the secondary indexes were explained through one to three questions. The workers with the overseas emergency management experience and related experts were invited to grade those questions, and then took the average to acquire the score of each indicator respectively. As the single influential factor, one to five assignment was applied. If the effect of emergency is the best, the factor was assigned 5. If the effect of emergency is the worst, it was assigned 1. The initial data of all influential factors that experts scored is showing in table 4.

Table 4. The Experts rating of Emergency Capability of Oil Companies' Overseas Emergency

Incident	D_1	D_2	D_3	D_4	D_5	D_6	D_7	D_8	D_9	D_{10}	D_{11}	D_{12}	D_{13}
1	4.652	3.988	2.023	4.325	4.855	4.546	3.255	3.052	4.762	4.575	2.188	4.248	4.236
2	4.658	4.124	3.223	4.274	4.756	4.457	3.672	2.235	4.45	4.25	2.218	4.227	3.88
3	4.521	4.223	2.262	4.336	4.764	4.762	3.573	2.336	3.25	3.494	3.625	4.258	3.67
4	4.541	3.978	3.177	4.289	4.768	4.527	3.627	4.286	3.762	3.176	3.579	4.226	3.587

Normalized handing of source data with the following formula:

$$x_i' = \frac{x_i - \bar{x}}{s}$$

$s = \sqrt{\frac{1}{n}\sum(x_i - \bar{x})^2}$, x_i is the source data, x_i is the normalized data.

Table 5. Normalized Third Level Indicator Data

Incident	D_1	D_2	D_3	D_4	D_5	D_6	D_7	D_8	D_9	D_{10}	D_{11}	D_{12}	D_{13}
1	0.818	-0.770	-1.048	0.649	1.491	-0.205	-1.465	0.079	1.037	1.080	-0.884	0.521	1.357
2	0.902	0.390	0.892	-1.092	-0.641	-0.881	0.743	-0.785	0.579	0.580	-0.847	-0.805	0.127
3	-0.999	1.234	-0.662	1.024	-0.468	1.435	0.218	-0.678	-1.184	-0.585	0.894	1.152	-0.599
4	-0.721	-0.855	0.818	-0.580	-0.382	-0.349	0.504	1.385	-0.432	-1.075	0.837	-0.868	-0.885

In the process of decomposing secondary indicator, it was decomposed into three factors, two factors, four factors, two factors and two factors respectively, and using the swallowtail type, cusp type, butterfly type, cusp type and cusp type respectively, these factors were calculated with unitary formula. With the unitary formula of every catastrophe models, it was integrated step by step to get the catastrophic affiliated functional values of the highest level. The bifurcation set x_{ij} of the four incidents were calculated, showing in table 6.

Table 6. The Bifurcation Set of Catastrophe Samples

Incident	D_1	D_2	D_3	D_4	D_5	D_6	D_7	D_8	D_9	D_{10}	D_{11}	D_{12}	D_{13}
1	0.905	0.916	1.012	0.805	1.143	0.453	1.136	0.530	1.007	1.039	0.960	0.722	1.107
2	0.950	0.731	0.972	1.045	0.862	0.959	0.906	0.941	0.896	0.761	0.946	0.897	0.503
3	0.999	1.073	0.902	1.012	0.777	1.198	0.602	0.908	1.034	0.765	0.963	1.073	0.843
4	0.849	0.949	0.951	0.762	0.726	0.591	0.796	1.085	0.845	1.037	0.943	0.932	0.960

With the principle of "complement", the catastrophic affiliated functional values of the C level were calculated, showing in table 7.

Table 7. The Catastrophic Affiliated Functional Values of the C Level

Incident	C_1	C_2	C_3	C_4	C_5	B
1	0.97174	0.99124	0.94024	0.99991	0.9852	0.977668
2	0.94022	0.9843	0.98083	0.96887	0.94226	0.963297
3	0.99565	0.96345	0.98348	0.97121	0.99289	0.981337
4	0.9573	0.90604	0.95428	0.99791	0.99078	0.961266

3.3 The result modification of emergency capability evaluation of our oil companies' overseas emergency.

We can find from the above evaluation that with the catastrophe evaluation method, the evaluation values of the emergency management capability for the four typical incidents are generally larger (more than 0.09); In addition, the difference of the comprehensive evaluation value between those four incidents is rather small. The emergency management capability of the incident 3(0.981337) is the strongest only 0.020071 shy of the weakest emergency management capability of the incident 4 (0.961266), which goes beyond people's behavior using evaluation value to make an intuitive judgment for the evaluation object and leads to misunderstanding.

In people's behavior, the comprehensive evaluation value of "0.9" belongs to the superior, but in fact, it belongs to the inferior in emergency capability. Therefore, there is a shortcoming in the comprehensive evaluation values calculated from the catastrophe evaluation method: the generic values are a little higher, which makes it difficult to evaluate the evaluation object intuitively. In order to overcome the above shortcomings of catastrophe evaluation method, the score transformation method was used to correct it.

This paper carried out in the following thoughts: with the given catastrophe evaluation index system, the comprehensive evaluation value of the global evaluation should also be x in theoretical significance, assumed membership values corresponding to the underlying indicators are all x. Assumed the membership value corresponding to the underlying indicators are $x_i (i = 1,2,\cdots, n)$, the comprehensive evaluation values are $y_i (i = 1,2,\cdots, n)$ based on the catastrophe evaluation method. When n big enough, the correspondences are tables for y_i and x_i can be built. For the membership value of underlying indicators has the difference of superior and inferior behaviorally, the transformed catastrophe evaluation value also has that kind of difference behaviorally. And the transform relationship between x and y can also be built respectively. The table 8 compares the transformed evaluation value based on the catastrophe evaluation method with the evaluation value based on the analytic hierarchy process.

Table 8. $x-y$ Correspondence Table

x	y	x	y	x	y	x	y
1.00	1.0000	0.88	0.9939	0.70	0.9833	0.45	0.9632
0.98	0.9991	0.86	0.9929	0.66	0.9805	0.40	0.9580
0.96	0.9981	0.83	0.9912	0.63	0.9784	0.30	0.9456
0.94	0.9971	0.80	0.9895	0.60	0.9762	0.20	0.9285
0.92	0.9960	0.77	0.9877	0.55	0.9722	0.10	0.9006
0.90	0.9950	0.73	0.9852	0.50	0.9680	0.00	0.0000

We can see from the above table that the transformed scores overcome the drawback that the original values are too high to differentiate them, and the values have the concept of "superior" and "inferior" behaviorally, which makes the catastrophe evaluation method practical.

The scores of transformed companies' emergency management were acquired, showing in table 9.

Table 9. The Scores of Emergency Management Capability of Our Oil Companies' Overseas Emergency

Incident	B	A
1	0.977668	0.61
2	0.963297	0.45
3	0.981337	0.68
4	0.961266	0.43

The emergency management capability of the incident 3 is the strongest, while the emergency management capability of the incident 4 is the weakest. The incident 3 is the policy variation of resource-rich country. Our country is always facing the policy variation of resources country since the overseas business was conducted. In order to limit the investment enterprise, many resource countries' policy changed frequently to limit the benefit of investment enterprise. Our oil companies have rich coping strategy to protect themselves. Incident 4 is the coup. Coups always go with the introduction of new government and new laws. Most countries with coups are poor and the negotiation effect is not good. The oil companies have no choice but to quietly to protect self-interest, so the emergency management capability of those incidents is modest.

4. Conclusion

Recently, oil companies' overseas emergency occurred frequently, and oil companies are lack of the capacity of emergency management. Therefore, it is meaningful to strength the evaluation of emergency management capability of our oil companies' overseas emergency improve the emergency reaction capability. Based on the catastrophe progression method, this paper evaluated the emergency management capability of our oil companies' overseas emergency, the result indicates that the emergency management capability of the variation of the resource policy is the strongest and the emergency management capability of the coup is the weakest in our oil companies. In the future, our oil companies need to strength the propaganda of emergencies, establish the emergency mechanism and reserve the emergency material.

References

1. Zhao Laijun and Cheng Jingjing(2010), "Study of Individual Behavior State in Unconventional Emergencies Based on Theory of Catastrophe," *China Safety Science Journal*, 20（12）:14-19.
2. Li Yan, Chen Xiaohong and Zhang Pengfei(2007), "Application of Catastrophe Progression Method to Evaluation of Regional Ecosystem Health," *CHINA POPULATION, RESOURCES AND ENVIRONMENT*,17（3）: 50-54.
3. LiRun-Qiu, ShiShi-Liang and Peng Xin(2008), "Catastrophe Model of Accident Evolvement of Gas Explosion in Coal Mines," *China Safety Science Journal*,18（3）: 22-27.
4. Chen weike and Zhangliying(2011), "Research on Evaluation of Emergency Management Capability in University Public Incident," *Journal of Safety Science and Technology*,7(11): 92-95
5. Tong Shujiao, Zhang Peihong, Zhong Maohua and Chen Baozhi(2012), "Assessment to Capability of Emergency Management of Enterprise Based on AHP," *Journal of Northeastern University(Natural Science)*,33（6）: 900-903

1ST CCPS ASIA-PACIFIC CONFERENCE ON PROCESS SAFETY℠

Evaluation System and Application of the Integral Safety in Chemical Industry Zone

Chen Chen
College of Economics and Management, China University of Petroleum
Qingdao 266580, Shandong, China
chenchen@upc.edu.cn

Zhang Xin-mei
College of Mechanical and Electronic Engineering, China University of Petroleum,
Qingdao 266580, Shandong, China
zhangxm@upc.edu.cn

Prepared for Presentation at
American Institute of Chemical Engineers
1st CCPS Asia-Pacific Conference on Process Safety
Qingdao, China
September 4 – 5, 2013

UNPUBLISHED

Evaluation System and Application of the Integral Safety in Chemical Industry Zone

Chen Chen
College of Economics and Management, China University of Petroleum
Qingdao 266580, Shandong, China
chenchen@upc.edu.cn

Zhang Xin-mei
College of Mechanical and Electronic Engineering, China University of Petroleum,
Qingdao 266580, Shandong, China
zhangxm@upc.edu.cn

Keywords: chemical industry zone; integral safety; index system; AHP

Abstract

Cluster regional development has become the main mode of the petrochemical industry development. How to evaluate and optimize the integral safety has become the key issue in the running of the chemical industry zone. Many factors impact the integral safety including zone location, integral planning, safety conditions, safety supervision, emergency system and social effect. 62 indexes were proposed in detail based of the six effect factors. By the introduction of AHP (Analytical Hierarchy Process), the integral safety level performance of each index will be obtained and the shortcomings of the safety works will be revealed. Based on the results, long term effective work safety mechanism will be adopted to improve the level of integral safety more pertinently.

1. Introduction

With the construction and development of the zone, the issues of integral safety and management become prominent increasingly. The various chemical companies are more concentrated in the chemical industry zone. The companies of chemicals production, use, storage, operation, transportation and other aspects are gathered in the zone. The hazardous chemicals are various and the danger of leaks, fires, explosions and other major industrial accidents are in potential. Relatively the dense of installations in the zone makes that the major accidents will spread to the neighbor if they occur on one. It is possible to lead to catastrophic domino effect, and make the safety production situation of zone more serious.

Developed countries began to study the regional risk assessment issues from the 1970s, and had achieved good application effect. The Collvey Island planning projects in England, the Netherlands Rijnmond regional planning projects in Holland and the regional risk research plan of the Ravenna Company in Italy, they all apply quantitative risk assessment methods to integral regional risk assessment[1-3]. In China, Chen Guohua studied the index determined problem of regional risk superposition and risk compensation composite function from two sides of inherent risk and actual risk[4]. Wu Zhongzhi proposed the quantitative evaluation method of regional risk based on the accident consequences model[5]. Shi Peijun revealed the dynamics mechanism of disaster formation during disaster through the risk assessment of hazard factors [6]. Luo Yun presented the concept of regional vulnerability analysis[7]. Zhu Zhengwei established an index system to evaluate Chinese regional public safety level[8].

Current research surrounding the safety issues of chemical industry zone is extensive, but most is focused research for a particular aspect of the problem. Chemical industry zone is an operating entity possessing combined functions, and its integral safety conditions can be affected by multiple influences from multiple factors. Establishing a method becomes the critical issue required to solve from a broader and more systematic level to explain the safety issues in chemical industry zone, which is used to systematically characterize the problem of the integral safety and quantitatively solving the assessment of the integral safety in chemical industry zone.

Factors which affect the integral safety of chemical industry zone deeply are analyzed and the index system is used to evaluate the integral safety of chemical industry zone. Quantitative evaluation of the integral safety of chemical industry zone can be completed relying on theory of Analytic Hierarchy Process. Factors which affect the integral safety of chemical industry zone are sorted based on the evaluation results and then the weaknesses which affect the safe operation of chemical industry zone are cleared. The analysis can provide the suggestions to improve the safety management more directly.

2. Building Index System of integral safety for chemical industry zone

Index system is a set to describe and evaluate the measurable parameters of something. Index system for chemical industry zone is a set for measurable parameters which is describing and identifying the content, scale and scope of chemicals business practitioners, and reflecting the direction and focus of safety. The safety objectives of safe level for chemical industry zone should base on index system, to determine individual indicators' level of controlling and degree can be controlled according to the planning period. Safety evaluation system should allocate basing on the weights of safety index. The evaluation indexes of Safe level for chemical industry zone can serve as a regional evaluation system.

Currently, the planning and layout of chemical industry zone belong to regional planning policy areas. And some of the targets and factors which planning involves have no quantitative index, particularly ecological safety and social environment, etc. Therefore, it is required that the index system should apply to all situations. And it is required to carry out quantitative evaluation, so as to avoid subjective evaluation.

For the elements to characterize the integral safety of the chemical industry zone are complex, and index selection plays an important role to the evaluation results, the principle of completeness should be followed when evaluation index system was established. The index will be set according to the actual situation and data of the study area in practical applications. But to chemical industry zone, the production and operation features of chemical companies should be more focused on. Therefore, to establish evaluation index system for chemical industry zone should combine status of resources and environment, socio-economic development level and the views of experts of specific regions, in accordance with the general principles identified by indexes such as science, operability, the relative completeness, the relative independence and direction.

The indexes affecting integrated safety of chemical industry zone are mainly based on countries and the chemical industry regulations, standards or norms. Six effect factors including zone location, integral planning, safety conditions, safety supervision, emergency system and social effect are concerned and specific assessment indexes are shown in Fig. 1.

Fig. 1 Integrated safety evaluation index system for chemical industry zone

1. Zone location

Chemical industry zone and the surrounding social environment are interdependent and interacted. Production requires a lot of raw materials, so good traffic condition is required. And the waste generated in the production process will impact the surrounding environment, thus the surrounding natural environment should also be considered. Meanwhile, the construction planning of zone and other aspects must meet the relevant national standards. In normal development stage, the zone can bring economic growth of surrounding area. However, if accidents happen, surrounding area will be affected. Therefore, the location of zone should be considered from three aspects involving natural conditions, traffic conditions and national standards.

2. Integral planning of zone

Zone integrated planning is to regulate the activities of human beings, to reduce risk, and to reduce the probability of accidents. Thus the normal production and stable development of zone can be ensured. The long-term arrangements of satisfying safe manufacture needs and safety measures and facilities construction should be taken into account. To conduct zone integrated planning should adhere to a comprehensive planning, distribute rationally, stress focal point, balance the general, accord to the core principles to improve the economic, social and safety benefits and adapt to local condition. Zone integrated planning and economic and social development influence and restrict to each other. It is also related to the control and management staff exposure risk. So the assessment criteria can be established from five aspects including industry related degree, reasonableness of zone integrated layout, road and traffic conditions in zone, rationality safety distance between devices, temporary shelters.

3. Safety conditions of zone

Work safety manufacture is all measures which to make labor process conducting in the material conditions and working in order to meet safety requirements, to prevent accidents in production such as personal injury or property damage and so on, to eliminate or control hazardous and harmful factors, and to protect the health and safety of workers, equipment and facilities and environment from damage. Safety condition in an enterprise is related to the hazardous and harmful factors including production process, materials, equipment, etc. So the assessment criteria of safety conditions of chemical industry zone can be constituted by the index of risk of production process, risk of material, risk of the device, potential occupational disease hazards.

4. Safety supervision conditions of zone

Safety supervision refers to carry out the decision making, planning, organizing, monitoring and to controlling measures on the purpose of safety. It is combined with accident prevention, emergency measures and insurance compensation closely together in order to reach the goal of safety by using supervision and management activities. Therefore, the assessment criteria can be established from the completeness of safety production supervision and management agencies and systems.

5. Emergency system of zone

Emergency rescue is an organized, fast, breaking the routine and effective act to rescue personnel and to support material when accidents happen, and is the generic name of countermeasure, control measures and rescue operations in a variety of emergency situations. Emergency rescue capability reflects the ability to respond to emergencies, also characterizing the completeness of public safety in the available rescue resources. So the emergency rescue includes six indexes: fire emergency organizations, medical assistance organizations, police emergency organizations, experts database resources, environmental monitoring organizations, perfection of emergency mechanism.

6. Comprehensive social effect index of zone

As a large-scale gathering place of chemical companies, chemical industry zone has some commonness necessarily with the whole society in certain aspects. The commonness affects normal production and life in the zone, and restricts sustainable development process of the chemical industry zone. It is can be considered from several aspects including per capita GDP in this region, eco-environment construction indexes, safety investment situations of zone, proportion of science and technology investment.

3. Integral Safety case analysis of chemical industry zone

3.1 Basic profiles of zone

The zone has good conditions to construct deep-water port. More than ten petrochemical storage and transportation companies constructing large-scale petrochemical storage and transportation projects have been located here. The zone is divided into port terminal operation area, storage and logistics area, industrial area, administrative center area and facilities support services area. The schematic figure of the zone is shown in Fig. 2.

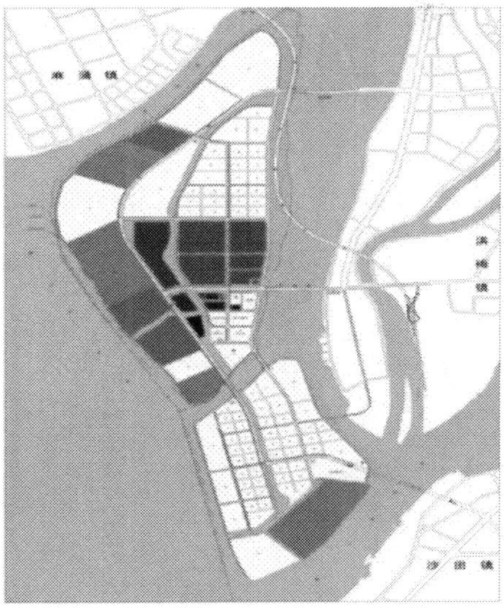

Fig. 2 Schematic figure of the application zone

3.2 Construct judgment matrix

The problem has been transformed into sequence and computation problem in hierarchy after the hierarchy structure index system has been completed. Sorting of each layer is paired judgment comparison of factors in this layer. According to a certain percentage scale, pair comparisons can be quantified and judgment matrix in quantity will be achieved.

The percentage scale of comparing paired factors is detailed in Table 1.

Table 1 Scale meaning

Scale	1	3	5	7	9	2, 4, 6, 8
Meaning	Equally important	Slightly important	Obviously important	Highly important	Extremely important	Intermediate values of aforementioned adjacent judgments

Aimed at the goal of the integral safety in chemical industry zone, O-A judgment matrix can be achieved by the relative importance comparison between indexes in standard layer (A) (see Table 2). Similarly, A-B judgment matrix, B-C judgment matrix can be achieved.

Table 2 O-A judgment matrix

O	**A1**	**A2**	**A3**	**A4**	**A5**	**A6**
A1	1	0.5	0.2	0.25	0.33	0.33
A2	2	1	0.25	0.33	0.5	0.5
A3	5	4	1	2	3	3
A4	4	3	0.5	1	2	2
A5	3	2	0.33	0.5	1	1
A6	3	2	0.33	0.5	1	1

3.3 Calculation index weight

To each judgment matrix, consistency test is required after the solving of the largest eigenvalue and corresponding eigenvector. The consistency ratio(*CR*) of every judgment matrix satisfy the condition that *CR*<0.10, it is concluded that the construction of judgment matrix meets the consistency requirements.

Combining with the specific circumstances of the zone, the index weight can be achieved according to the process of AHP (detailed in Fig. 3).

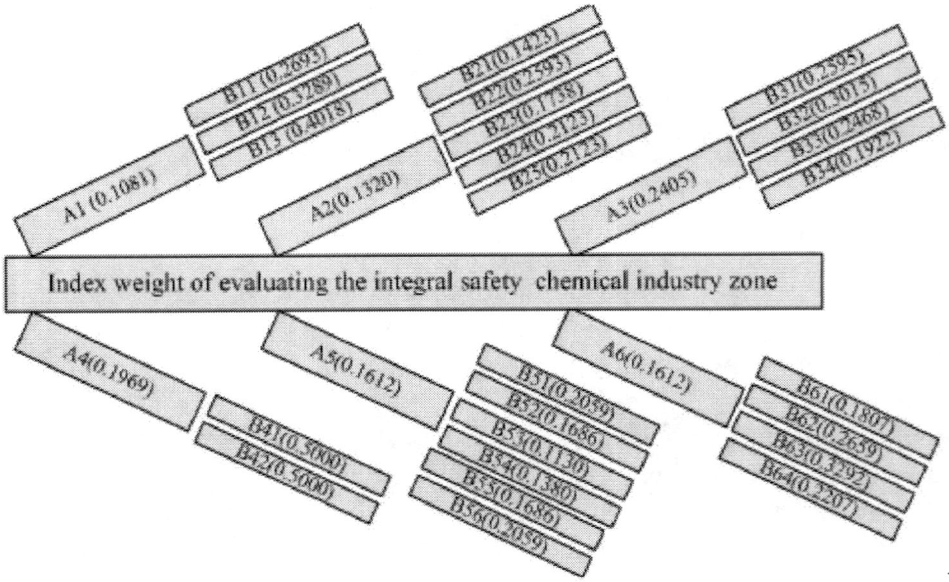

Fig. 3 Calculation results of index weight

From the calculation results of Fig. 3, it can be seen that safety conditions of zone has the biggest weight of 0.2405 among the six indexes in the criteria layer. So it is the foundation of ensuring safety throughout the chemical industry zone to do a good job within the safety conditions of all the companies in zone. However, it should also be seen that safety planning, capability response to emergency, surrounding production environment and social factors of zone is closely effect the integral safety not only the safety conditions.

3.4 Analysis results of the integral safety in chemical industry zone

By the combination of weight results and actual rating of the application, the analysis result of the integral safety in chemical industry zone is 72.52. The integral safety of zone can satisfy the demand of normal safety production basically. In order to find the weak index in the zone, achievement rate was calculated to each index and the results can be seen in Fig.4.

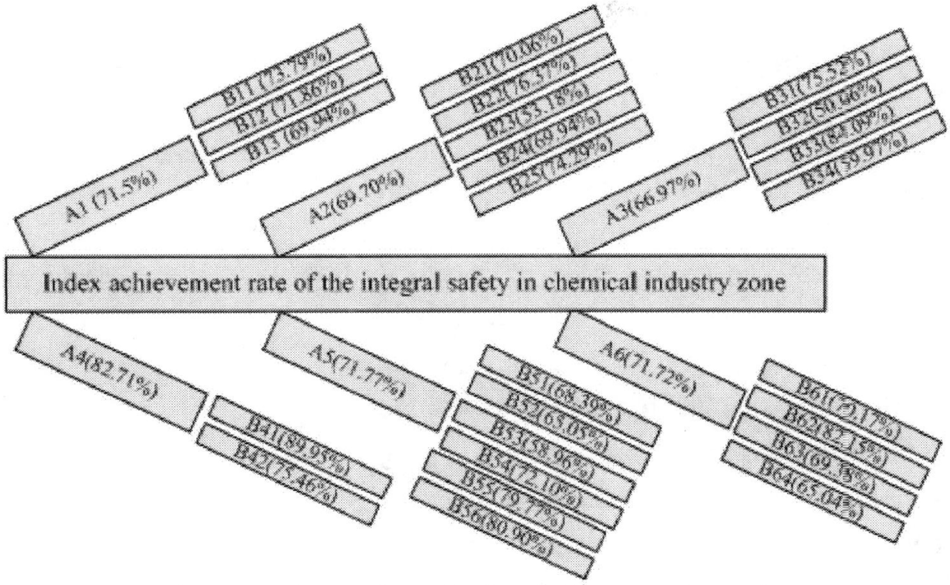

Fig.4 The analysis results of index achievement rate in zone

From achievement rate of each index shown in Fig.4, it is clear that several aspects including road conditions around zone, road width in zone and special passage of dangerous goods, quality of personnel and research and development investment have the poor achievement rate, more efforts are required to improve the performance in these indexes.

4. Conclusions

According to the deeply analysis of key factors to the integral safety, index system of evaluating the integral safety in chemical industry zone has been established. The analysis results of the integral safety in chemical industry zone have been analyzed quantitatively through introduction of AHP. And the weaknesses of affecting integral safety in chemical industry zone have been cleared through the analysis method of index system's achievement rate of integral safety in chemical industry zone. It can provide important policy decision reference to improve the integral safety in chemical industry zone.

5. Acknowledgements

The paper was supported by National Natural Science and Foundation of China (51104174) and the Fundamental Research Funds for the Central Universities (14CX04024B, 14CX02067A). We also appreciate those who assisted us in the process of observation, the anonymous reviewers and the editor for their constructive comments and suggestions.

6. References

[1] NormanCR. An Assessment of Accident Risk in US Commercial Nuclear Power Plants Draft, WASH-1400, 1974, 10

[2] Papazoglou I A, Nivolianitou Z S, Bonanos G S. "Land use planning policies stemming from the implementation of the SEVESO-II directive in the EU". *Journal of Hazardous Materials*, Volume 61, NO.1-3, 1998

[3] Brazier A M, Greenwood R L. "Geographic information systems: A consistent approach to land use planning decisions around hazardous installations". *Journal of Hazardous Materials*, Volume 61,NO 1-3, 1998

[4] Chen G H, Zhang J, Zhang H, et.al. "Study on regional risk assessment methodology". *China Safety Science Journal*, Volume 16, NO. 6, 2006.

[5] Wu Z Z, Gao J D, Zhang X K. Identification and Assessment of Industry Hazard Installations. Beijing: China Meteorological Press, 2000

[6] Shi P J. "Theory and practice on disaster system research in a fourth time". *Journal of Natural Disasters*, Volume 14, NO. 6, 2005

[7] Luo Y, Fan Y X, Ma X C. Risk Analysis and Safety Assessment. Beijing: Chemical Industry Press, 2004.

[8] Zhu Z W, Zhang R, Zhou B. "Regional public safety assessment and its correlative factors analysis in China". *Chinese Public Administration*, NO. 1, 2006.

1ST CCPS ASIA-PACIFIC CONFERENCE ON PROCESS SAFETY℠

ARAMIS: powerful methodology in quantitative risk assessment of chemical industry zones

Zhang Xin-mei
College of Mechanical and Electronic Engineering, China University of Petroleum, Qingdao 266580, Shandong, China
zhangxm@upc.edu.cn

Chen Chen
College of Economics and Management, China University of Petroleum Qingdao 266580, Shandong, China
chenchen@upc.edu.cn

ZENG Yue-mei
College of Mechanical and Electronic Engineering, China University of Petroleum, Qingdao 266580, Shandong, China

Prepared for Presentation at
American Institute of Chemical Engineers
1st CCPS Asia-Pacific Conference on Process Safety
Qingdao, China
September 4 – 5, 2013

UNPUBLISHED

ARAMIS: powerful methodology in quantitative risk assessment of chemical industry zones

Zhang Xin-mei
College of Mechanical and Electronic Engineering, China University of Petroleum,
Qingdao 266580, Shandong, China
ZhangXM@upc.edu.cn

Chen Chen
College of Economics and Management, China University of Petroleum
Qingdao 266580, Shandong, China

ZENG Yue-mei
College of Mechanical and Electronic Engineering, China University of Petroleum,
Qingdao 266580, Shandong, China

Keywords: ARAMIS; quantitative risk assessment; vulnerability; accident scenario

Abstract

Many chemical industry installations exist in Chemical Industry Zone(CIZ) and catastrophic accidents may be caused if the primary accident expanded. ARAMIS(Accidental Risk Assessment Methodology for Industries System) is a systematic methodology to the process industry risk analysis and was proposed in the fifth Framework Program of the European Commission. The methodology is consisted of several sections. The first is the building of a methodology for the identification of major accident hazards (MIMAH), which is help to find the worst accidents likely to occur on the equipment. The second section is called methodology for the identification of reference accident scenarios (MIRAS) takes into account the influence of safety systems on both the frequencies and possible consequences of accidents. This section helps us to identify the more realistic accident scenarios. Vulnerability and risk distribution figure in the whole CIZ will be achieved visually at last. The application of the methodology indicates that it offers an alternative to purely deterministic and probabilistic approaches to risk assessment of process plants and the results will provide the decision-making basis of land-using and emergency planning.

1 Introduction

As a carrier of chemical companies, CIZ has become a successful model for the development of the petrochemical industry. According to the statistics, above-scale enterprises in the CIZs have accounted for the total number of petrochemical enterprises 1/3 nation widely in 2011. In the development planning published in February 2012, it required that hazardous chemical production enterprises must move into the CIZs. Not only China, many CIZs has occurs all over the world, such as USA, EU etc.

Many chemical industry enterprises exist in CIZ, which have large quantities of flammable, explosive or toxic chemicals, mostly in the production, storage, use, transport and other sectors. The event of fire, explosion or dangerous chemical leak may easily result in serious consequences. Ensuring safety is the primary problem in the sustainable development of CIZ.
Developed countries began to study the regional risk assessment issues from the 1970s, and had achieved good application effect[1-6]. ARAMIS methodology was proposed in the fifth Framework Program of the European Commission with the objective to answer the specific requirements of the SEVESO II directive. ARAMIS overall objective was to build up a new accidental risk assessment methodology for industries that combines the strengths of both deterministic and risk-based approaches. The potential users interested by the results of ARAMIS are numerous but the most concerned are the plant operators, the competent authorities and the local authorities.

ARAMIS methodology was applied in the to quantitative risk analysis to a chemical tank. Likely accident scenario was analyzed by the construction of risk matrix. Severity and vulnerability analysis was conducted to the likely accident scenario. According to the definition of risk in ARAMIS methodology, the escalation risk distribution in the subject chemical tank area can be demonstrated and weakness of the risk control was revealed. Furthermore, the chemical rank area will achieve the most optimal and safe operation.

2. Framework of ARAMIS methodology

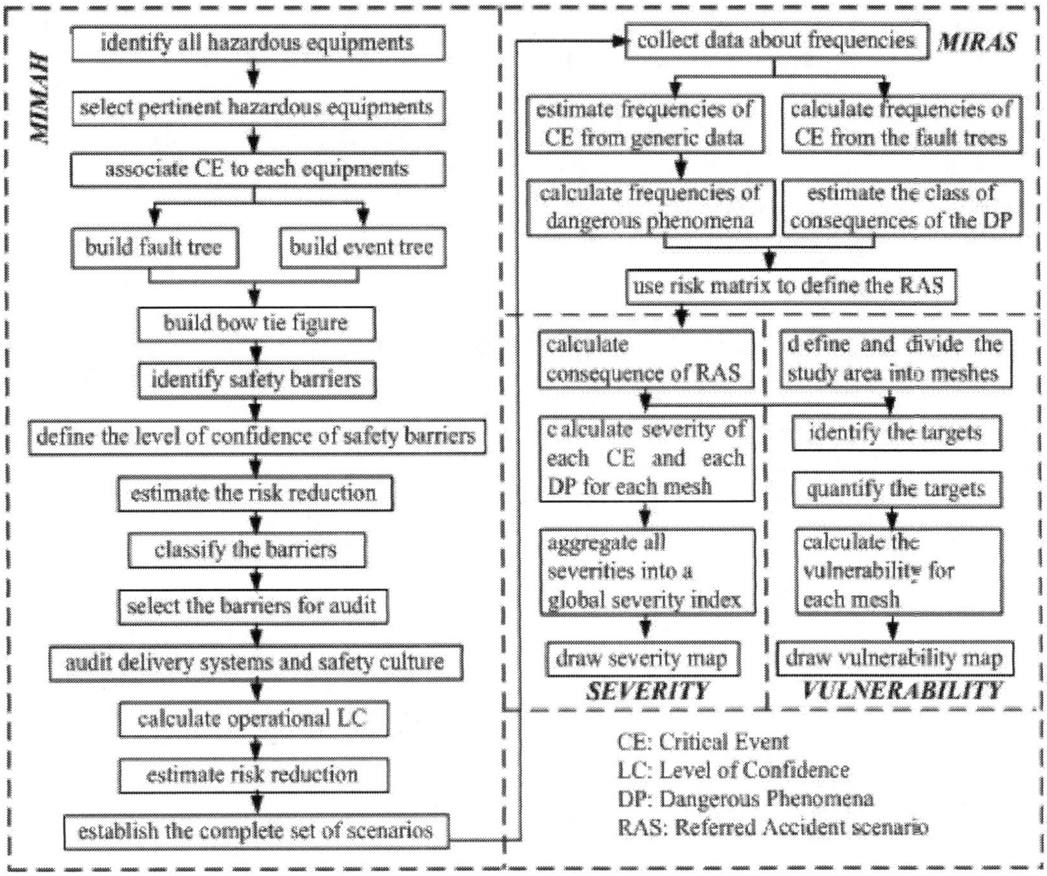

Figure.1 Structure diagram of the ARAMIS methodology

In risk assessment of process industry, identify potential accident scenarios is a critical step. General methods mainly focus on deterministic methods considering the worst case, regardless of the safety equipment and the role of policy, and these approaches may lead to an over-estimation of the risk-level, and does not promote the implementation of safety systems. ARAMIS methodology has many advantages to avoid such problem and it is divided into the following major steps[7]:

(1) Identification of major accident hazards (MIMAH).
(2) Identification of the safety barriers and assessment of their performances.
(3) Evaluation of safety management efficiency to barrier reliability.
(4) Identification of reference accident scenarios (MIRAS).
(5) Assessment and mapping of the risk severity of reference scenarios.
(6) Evaluation and mapping of the vulnerability of the plant's surroundings.

The framework of ARAMIS is demonstrated in Figure.1.

3. Application of ARAMIS methodology

3.1 Identify potentially hazardous equipment in the plant

According to the classification of the equipment, the equipment in the applied depot can be mainly classify as oil storage facilities, oil loading and unloading facilities, electrical equipment, power machinery, fire fighting equipment, equipment, machine repair equipment, the main production building, etc.[8]. But the specific dangerous equipments are tanks and pipelines.
ARAMIS proposes 16 equipment categories in the four classification including storage equipment, transport equipment, pipes networks and process equipment[8]. In the following analysis, the storage equipment is selected as the hazardous equipment.

3.2 For selected equipment, associate critical events

MIMAH considers 12 different critical events which are specified hereunder: decomposition (CE1), explosion (CE2), materials set in motion (entrainment by air) (CE3), materials set in motion (entrainment by a liquid) (CE4), start of fire (LPI) (CE5), breach on the shell in vapor phase (CE6), breach on the shell in liquid phase (CE7), leak from liquid pipe (CE8), leak from gas pipe (CE9), catastrophic rupture (CE10), vessel collapse (CE11), collapse of the roof (CE12).

The critical events of start of fire (LPI) (CE5) and breach on the shell in liquid phase (CE7) are considered as the critical events to the selected hazardous equipment[8].

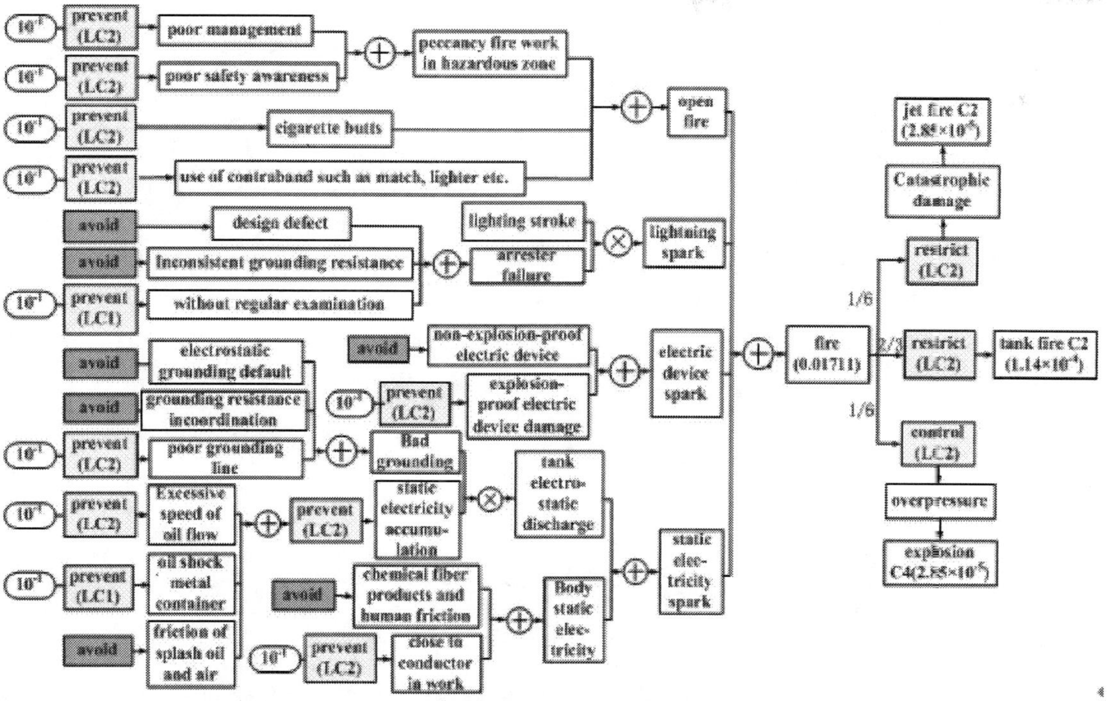

Figure.3 Start of fire of oil tank of the bow-tie

3.3 Identification of reference accident scenarios

(1)Demonstration of risk control through the use of bow-tie diagrams

The bow tie diagrams turn out to be a very powerful tool to communicate about risks. Combined with the literatures[9, 10] and the characteristics of the device, key event of fire bow-tie diagram shown in Figure.3 was established.

Since the oil depot was in operation for a short time, the initial incident probability was estimated according to the initial incident qualitative table[8]. In Figure.3, the operators "+" and "×" represent the operation "and gate" and "or gate". According calculation rule of fault tree and event tree, the frequency of each dangerous phenomenon can be calculated.

(2) Estimate the class of consequences of dangerous phenomena

The selection of reference accident scenarios is based on the evaluation of the frequency of dangerous phenomena together with their potential consequences. At this stage, it is thus necessary to evaluate roughly the consequences of each dangerous phenomenon. For each dangerous phenomenon obtained during the development of the event trees, a class of consequence must be chosen according to the definitions given in Table 1[11].

Table 1. Class of consequences

Domino effect	Effect on human target	Effect on environment	Class ranking
To take into account domino effects, the class of consequences attributed to the studied dangerous phenomenon will be increased to the class of the secondary dangerous phenomenon that the first	No injury or slight injury with no stoppage of work	No action necessary, just watching	C1
	Injury leading to an hospitalisation >24 hours	Serious effects on environment, requiring local means of intervention	C2
	Irreversible injuries or death inside the site, reversible injuries	Effects on environment outside the site, requiring national means	C3

can bring about by domino effects	outside the site		
	Irreversible injuries or death outside the site	Irreversible effects on environment outside the site, requiring national means	C4

(3)Use the risk matrix to select reference accident scenarios

The objective of this step is to select the reference accident scenarios which will be modeled in the calculation of the severity. The tool used here is a risk matrix (Figure.4)[8]. The X-axis corresponds to the four consequence classes, and the Y-axis corresponds to the frequency of the dangerous phenomena. Three zones are defined in this matrix.

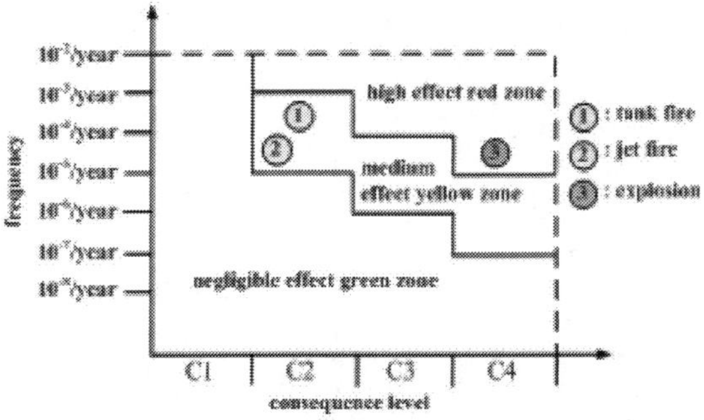

Figure.4 Risk matrix

According to the Figure.4, it is clear that the risk of three accident scenarios is relatively high, including oil tank fire, jet fire and explosion. The reference accident scenarios can be got and they are needed to be analyzed quantitatively in depth.

3.4 Calculate of the severity

Each reference accident scenario (RAS) is defined by an initiating event that leads to a critical event, which can potentially lead to different dangerous phenomena. For each phenomenon, a specific severity index has been defined.

For each RAS, a risk severity map is then produced. Risk severity is defined for one scenario as the combination of the level of frequency with the intensity of the effects. The combination of risk severity with the vulnerability of the targets produces the actual risk. Risk severity can be represented for each scenario in a geographical way.

The three reference accident scenarios of the tank named PC0304 was simulated by the software PHAST, their respective principal effect arising from the relationship was shown in Table 2.

Table 2. Characteristic distances of heat radiation caused by tank fire

Dangerous phenomena	**Probability**	**Influence**	**d0**	**d1**	**d2**	**d3**	**d4**
		S_{DP}	**0**	**25**	**50**	**75**	**100**
Pool fire	1.14×10^{-4}	Heat radiation	70(1)①	60(1.8)	46(3)	15(5)	6(8)
Jet fire	2.85×10^{-5}	Heat radiation	105(1)	85(1.8)	63(3)	49(5)	40(8)
Explosion	2.85×10^{-5}	overpressure	150(1)②	105(30)	80(50)	37(140)	28(250)

① the characteristic distance is 70m when the intensity of heat radiation is 1 KWm^{-2}.

② the characteristic distance is 150m when the overpressure of explosion blast is 1mbar.

3.5 Calculate the vulnerability

The last innovative attempt from ARAMIS is to address the vulnerability of the environment independently of the hazardous site[11-13]. The vulnerability is calculated on the basis of a multi-criteria decision-aiding approach. According to the layout characteristics of, the oil depot can be divided into six regions and the vulnerability analysis result was distributed in Figure.5.

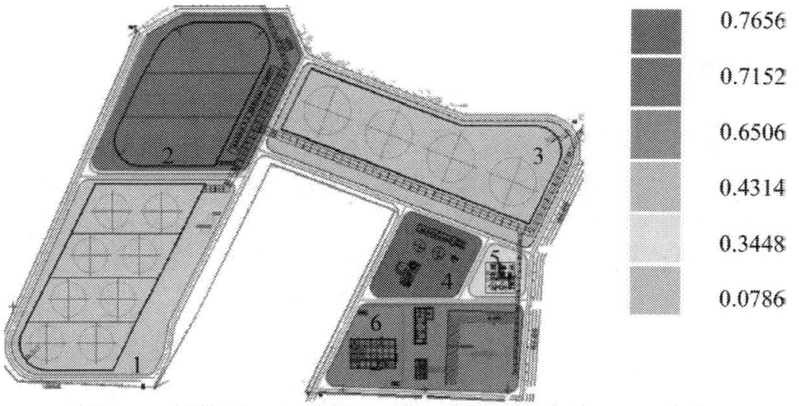

Figure.5 Vulnerability distribution of materials

3.6 Draw the risk cloud figure

According ARAMIS method definition: Risk = frequency × severity × vulnerability, the risk distribution cloud figure when and fire or explosion occurs on the tank named PC0304 was drawn in Figure.6.

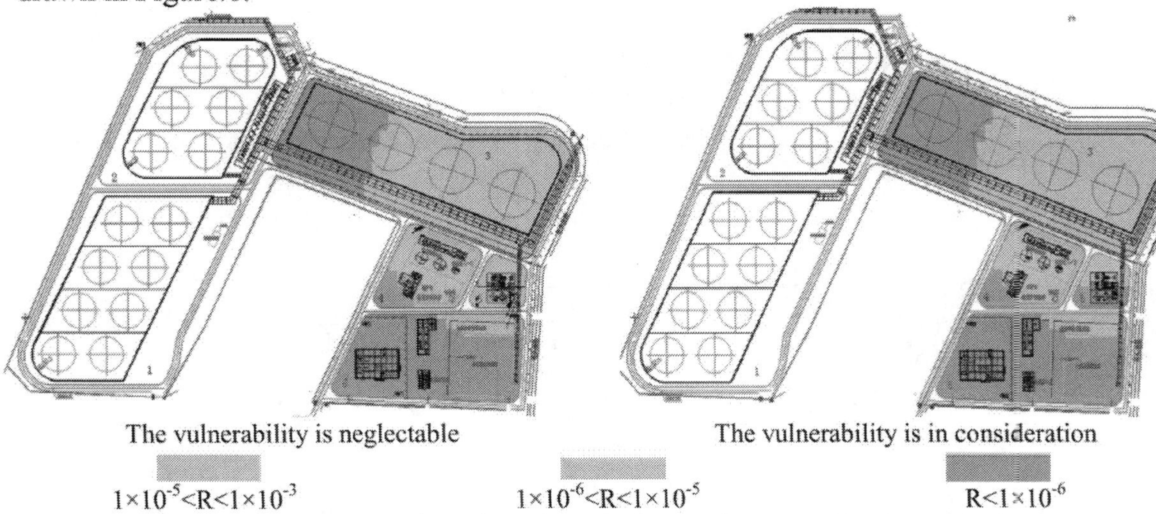

Figure.6 Escalation risk distribution of the application tank zone

Taking the vulnerability effect into account, the risk result can reflect the risk distribution under the special layout of the chemicals installations. Further more, the result can provide suggestions to the functional layout and safety risk prevention in the CIZ.

4. Conclusions

ARAMIS methodology is a systematic method of risk analysis device, the application of the methodology shows that it has the good applicability in petrochemical hazardous installations risk analysis. The core of ARAMIS methodology is the established of bow-tie figure. It can achieve the completely and systematic analysis to the potential risk and unsafe factor existing in the petrochemical enterprises and the probability calculation with the lack of data was solved. Vulnerability was introduced into the characteristic of risk. It can better reflect the risk of the spatial expansion. It can help to optimize the distribution of land-use planning and improve the risk prevention in the decision making.

5. Acknowledgements

The paper was supported by National Natural Science and Foundation of China (51104174) and the Fundamental Research Funds for the Central Universities (14CX02067A, 14CX04024B). We also appreciate those who assisted us in the process of observation, the anonymous reviewers and the editor for their constructive comments and suggestions.

6. References

[1] NormanCR. An Assessment of Accident Risk in US Commercial Nuclear Power Plants Draft, WASH-1400, 1974, 10

[2] Papazoglou I A, Nivolianitou Z S, Bonanos G S. "Land use planning policies stemming from the implementation of the SEVESO-II directive in the EU". *Journal of Hazardous Materials*, Volume 61, NO.1-3, 1998

[3] Brazier A M, Greenwood R L. "Geographic information systems: A consistent approach to land use planning decisions around hazardous installations". *Journal of Hazardous Materials*, Volume 61,NO 1-3, 1998

[4] Chen G H, Zhang J, Zhang H, et.al. "Study on regional risk assessment methodology". *China Safety Science Journal*, Volume 16, NO. 6, 2006.

[5] Wu Z Z, Gao J D, Zhang X K. Identification and Assessment of Industry Hazard Installations. Beijing: China Meteorological Press, 2000

[6] Luo Y, Fan Y X, Ma X C. Risk Analysis and Safety Assessment. Beijing: Chemical Industry Press, 2004.

[7] O. Salvi, B. Debray. A global view on ARAMIS, a risk assessment methodology for industries in the framework of the SEVESO II directive. Journal of Hazardous Materials, 2006, 130(3): 187-199.

[8] C. Delvosalle, C. Fievez, A. Pipart, et al. "ARAMIS project: A comprehensive methodology for the idenfication of reference accident scenarios in process industries". *Journal of Hazardous Materials*, Volume 130, NO. 3, 2006.

[9] LI Xin-zhan, HUANG Kun , CHEN Jian-xia , et al. "Fault Tree Analysis in Oil Depot Fire and Explosion". *Oil and Gas Transportation and Storage*, Volume 29, NO. 3, 2011.

[10]ZHANG Yuan-xiu, WANG Shu-li. "Analyses on the causes of oil tank fires and the control measures". *Industrial Safety and Environmental Protection*, Volume 33, NO. 4, 2007.

[11]V. de Dianous, C. Fievez. "ARAMIS project: A more explicit demonstration of risk control through the use of bow–tie diagrams and the evaluation of safety barrier performance". *Journal of Hazardous Materials*, Volume 130, NO. 3, 2006.

[12]J. Tixier, A. Dandrieux, G. Dusserre, et al. "Environmental vulnerability assessment in the vicinity of an industrial site in the frame of ARAMIS European project". *Journal of Hazardous Materials*, Volume 130, NO. 3, 2006.

[13]X. D. Wang, X. H. Zhong, S.Z. Liu, et al. "Regional assessment of environmental

vulnerability in the Tibetan Plateau: Development and application of a new method". *Journal of Arid Environments*, Volume 72, NO. 10, 2008.

Loss Functions and Their Applications in Process Safety Assessment

Faisal Khan[a,b]
fikhan@mun.ca

S. Javad Hashemi[a]
j.hashemi@mun.ca

Salim Ahmed[a]
sahmed@mun.ca

(a) Safety and Risk Engineering Group, Faculty of Engineering & Applied Science, Memorial University, St. John's, Canada
(b) Australian Maritime College, University of Tasmania, Launceston, TAS 7450, Australia

Prepared for Presentation at
American Institute of Chemical Engineers
1st CCPS Asia-Pacific Conference on Process Safety
Ingdao, China
September 4-5, 2013

UNPUBLISHED

Loss Functions and Their Applications in Process Safety Assessment

Faisal Khan[a,b]
fikhan@mun.ca

S. Javad Hashemi[a]
j.hashemi@mun.ca

Salim Ahmed[a]
sahmed@mun.ca

(a) Safety and Risk Engineering Group, Faculty of Engineering & Applied Science, Memorial University, St. John's, Canada
(b) Australian Maritime College, University of Tasmania, Launceston, TAS 7450, Australia

Keywords: process safety, loss function, risk assessment, early warning, alarm management

Abstract

Process deviations, along with failure of control systems and protection layers, result in safety and quality loss in plant operations. This paper proposes an operational risk-based warning system design methodology based on overall system loss. Loss functions are used to define the relationship between process deviations and system loss. For this purpose, properties associated with quadratic loss function and a set of inverted probability loss functions are investigated and compared. The results suggest that loss functions can be used in a novel way to assess operational stability and system safety. The proposed safety assessment methodology using loss functions is then incorporated into a risk-based warning system design model. This methodology provides practitioners a real-time operational risk-based tool to analyze warnings associated with process deviations. A simulated case study is presented to demonstrate potential application of the proposed methodology; the study examines the response to a temperature surge for a reactor system. The risks associated with two alerts caused by a low and a high temperature are analyzed for warning generation and prioritization.

1. Introduction

An important aspect of any industrial operation is conformance to standards. It relates to how closely the operational performance and process safety, as well as quality of the operation and final products, match the design specifications. In particular to the process industry, meeting the target performance is of great importance due to increasing demands for higher efficiency and strict environmental regulations. Achieving conformance requires investment on appraisal, prevention, and improvement techniques. Evaluating the loss incurred due to deviation from the expected level of conformance provides a leading process performance indicator to justify both the cost of investment on continuous improvement of operations and a renewed commitment to run a safer and greener process.

Taguchi [1] proposed a quadratic loss function to illustrate losses to society associated with deviations of quality characteristics from their operational targets in industrial applications. Taguchi's quality philosophy states that there is a cost for any finite deviation. Commonly referred to as Taguchi's loss function, this loss imposes maintenance and repair costs to consumers, warranty and scrap costs to manufacturers, and pollution and environmental costs to the society [2].

The loss function (LF) approach is widely used to quantify losses associated with deviation from target value for various purposes such as economic design of specification region [3], obtaining optimal inventory and investment policies [4], business decision-making [5], quality assurance [6], risk assessment [7], marine and offshore safety [8], supplier selection [9] and reliability settings [10]. There are many other reported applications of LFs to quantify unobservable variability costs. These are mostly in the area of quality engineering; interested readers are referred to [4], [7], [9], [11], [12].

Despite successful application of the LFs, an obvious difficulty regarding their use is to determine their exact form. Various LFs have been discussed in the literature and in some applications different types of LFs are used for similar purposes. While many researchers agree in principle with the premise of using LFs in the context of managing continuous improvement, there is no general agreement on the specific form of the LF to be used for this purpose [13]. As a result, one of the most critical issues encountered in any application of LFs is the selection of a proper LF to relate a key characteristic of a system to its performance. Using inappropriate LFs may lead to inaccurate results that either underestimate or overestimate the loss [14].

The fundamental objective of this work is to study the advantageous application of LFs in process safety management. In this paper, the general basis of LFs and a review of relevant literature on LFs are discussed in Section 2. Use of LFs in different industrial applications is analyzed in Section 3. The potential applications in the context of operational risk-based process safety management and associated challenges are discussed in Section 4. A methodology for warning system design by integrating the loss functions in risk determination is proposed in Section 5. Numerical comparisons of the LFs by examining the response to a temperature surge for a reactor system and a sensitivity analysis to study the impact of changing the shape of loss functions on the results are provided in Section 6. Conclusions and recommendations are drawn in Section 7.

2. The loss functions

2.1 Univariate loss functions

Taguchi [1] defined the quadratic loss function (QLF) as $L(y)=B(y-T)^2$ where y denotes the quality characteristic, $L(y)$ is the actual loss at y, T is the target value, and B is a constant. With its infinite maximum loss and symmetric shape, QLF is inadequate for some applications in describing the loss [15]. Joseph [16] proposed a set of LFs based on Taguchi's loss concept to address problems requiring asymmetric shapes . In practice, the maximum loss is generally finite. Truncating the QLF at the points where the function intersects the maximum loss addresses this concern [6]. To implement this idea, the concept of inverted normal loss function (INLF) was proposed by Spiring [6] which has a bounded value from above and its supremum can be specified by the user. Sun et al. [17] refined the INLF further and developed the modified INLF (MINLF). This LF has a shape parameter γ that is specified by the user and its value determines the slope of the function in the neighborhood of the target value. Later, Spiring and Leung [5] extended the concept of INLF further to other inverted probability density functions including inverted beta loss functions (IBLF) which not only provide the traditional properties of LFs but also include the asymmetrical loss cases [5], [11], [18].

Table 1 provides the formulations for the most widely used univariate LFs in the literature. In Table 1, y denotes the quality characteristic, $L(y)$ is the actual loss at y, T is the target value, EML is the estimated maximum loss, Δ is the distance from the target to the point where the maximum loss EML first occurs, and α and γ are shape parameters and need to be determined from additional information for MINLF and IBLF.

Table 1. Univariate loss functions formulations.

Type of LF	Reference	Formulation of Loss Function
QLF	[1]	$L(y)=B(y-T)^2$ where $B=EML/\Delta^2$
INLF	[6]	$L(y)=EML\{1-exp(-(y-T)^2/2\gamma^2)\}$ where $\gamma=\Delta/4$
MINLF	[17]	$L(y)=\dfrac{EML_\Delta}{1-exp\{-0.5(\Delta/\gamma)^2\}}\{1-exp(-(y-T)^2/2\gamma^2)\}$
IBLF	[11]	$L(y)=EML\{1-C[y(1-y)^{(1-T)/T}]^{(\alpha-1)}\}$ where $C=[T(1-T)^{1-T/T}]^{1-\alpha}$

2.2 *Multivariate loss functions*

While the existing literature widely covers the univariate LFs, there are some attempts, mainly in the area of quality engineering, to extend the LF concept to multivariate problems. A review of some multivariate LFs can be found in [19], [20], [21]. Traditionally, assigning a weight for each response has been considered to be an effective way to a multivariate loss function [20]. The main drawback of these approaches is that in real-life industrial applications, quality characteristics are often not independent or additive [19].

Pignatiello [22] and Artiles-Leon [23] are among the early researchers reporting the application of multivariate LFs and their works are the bases for many later research [21]. Pignatiello [22] presented a quadratic LF for multi-response problems and established a predictive regression model by using controllable variables. As argued in [20], in Pignatiello's approach, it is difficult to determine the cost matrix and additional experimental observations may be required. Artiles-Leon [23] included specification limits in the LF itself. Later, Ma [24] improved the Artiles-Leon's work by applying principal component analysis (PCA) to consider correlation structures among the various responses. Hsu [20] and Su and Tong [25] have also used PCA to use Taguchi's method for multiple system characteristics.

Spiring [6] extended the univariate INLF to the case where two characteristics are of interest in assessing loss. However, the dependency between the parameters is not considered in [6].

Extensions of the concept of QLF for considering three main categories of multiple quality characteristics including nominal-the-best, smaller-the-better, and larger-the-better cases are reported in [14], [19], and [26] where it is assumed that the quality characteristics follow multivariate distributions. To consider the dependency among quality characteristics, the Bessel function and a correlation coefficient were used to obtain the joint probability distribution of multiple quality characteristics when calculating the expected loss.

3. Use of the loss functions

In the past two decades the use of loss functions for measuring quality costs for the purpose of process quality improvement, has rekindled interest for researchers and practitioners alike [4]. One of the active areas of research has been the application of LFs for process improvement and production planning. For a review of the literature on application of LFs in production planning, economic lot-sizing, and inspection planning interested readers are referred to [4] and [27].

LFs have also been used to optimize the expected loss to the customer as well as to the manufacturer. Abdul-Kader [27] presented a review of the literature on product and process optimization and loss minimization models and proposed an integrated cost model composed of a tolerance model and an investment model using Taguchi's QLF. However, only the single variable optimization problems are considered in [27]. Raiman [28] emphasized the importance of the multivariate quality characteristic case and concluded that the total loss for a given process can be obtained by adding the losses resulting from each quality characteristic with each loss based on an independent quality characteristic and its associated loss. References [3], [15], [29], [30] have also studied and proposed process optimization methods and loss minimization models for multiple quality characteristics problems. To relax the assumption of independency, references [20] and [25] presented an approach based on PCA to optimize the multi-response problems.

Pan [31] proposed a LF-based risk assessment method by linking the process capability indices and LFs. In [31] the univariate capability indices of processes and LFs describe the likelihood and consequences of non-conformities in the system, respectively. More recently, Pan and Chen [7] extended the methodology in [31] to multivariate cases and developed a correlated risk assessment technique for manufacturing and environmental systems. This approach is based on Pignatiello's quadratic multivariate LF, where in addition to restrictions of QLF in modeling loss, it requires to determine the cost matrix which could be difficult to obtain. However, this work is among the early efforts to apply LFs in quantitative risk assessment.

4. Application of loss functions in process safety assessment

The work by Sii et al. [8] to apply the Taguchi concept in maritime safety engineering systems is a prime effort to use LF for improvement of system safety. More recently, a work by Pan and Chen [7] to find the relationship between process capability indices and LFs to develop a LF based risk assessment approach has been reported. However, there appears to be virtually no study that uses LFs for risk-based safety assessment and management for process industries.

The limited application of LFs in process systems can be related, in part, to the original intention of their application to optimize engineering processes [1]. Compared with manufacturing industries, deviation of the key process characteristic from the target markedly increases the probability of process safety loss. This is mainly due to hazardous operations under high pressure and temperature in process industries and involvement of dangerous substances. In addition to the original intention of LFs to optimize manufacturing settings, when comparing process industries with its manufacturing counterparts, there are some reasons why the LF approach have not been employed in process safety studies. The main reasons include the difficulty in measuring the safety of a system precisely and relating it to monetary values. In addition, process systems comprise of more unknown, or imprecisely known, parameters.

In traditional process safety assessment, no loss with respect to product quality and system safety is assumed until the key process characteristic deviates from the operational boundary. However, in the LF approach a process imparts "no loss" only if the process characteristic of interest equals its target. In terms of system safety, as the deviation from the target increases there is a larger possible demand for activation and functionality of control and safety systems. Due to inevitable uncertainty in the successful operation of control and safety systems, the deviation of the process characteristic increases the failure possibility of safety systems. Application of the LF approach in process safety assessment and consideration of zero loss only for on-target operation can lead to the direction of continuous safety improvement of process operations.

Process safety management is widely recognized and credited for accident risk reduction and improved operational performance. However, process safety management appear to have stagnated within many organizations due to application of inadequate management system performance indicators such as incident investigation [32]. Key to achieving a safe and economic process operation is the application of leading indicators that focus on the process safety and quality elements that matter, thereby providing a true measure of how an asset is performing. A LF approach is used in this work to quantify loss due to deviation of process characteristics from their target operation. The measured loss is then integrated in a risk estimation model which provides a real-time leading indicator of process safety performance. The estimated risk is finally used as a criterion to annunciate, prioritize, and analyze warnings which are generated as a result of process deviations.

5. A Risk-based warning system design methodology

Integration of loss function based consequence analysis into an operational risk assessment model is proposed to be used in alarm system design. An alarm system is one of the critical layers of protection in a process safety system. Alarms prevent the escalation of abnormal situations and minimize the demand for activation of safety instrumented systems (SIS) such as a trip or an emergency shutdown device (ESD) [33]. However, the appearance of too many alarms, also known as alarm flooding, when the process approaches an abnormal situation has been reported as a contributor to abnormal events [34]. The discussion on the shortcoming of the current alarm management systems can be found in [35], [36].

To address the alarm flooding problem in process facilities, Chang et al. [33] proposed a quantitative risk-based alarm design approach for annunciation and prioritization of alarms. An integrated model consisting of the probability and the impact of the potential hazards, and the process safety time is used to estimate the risk according to Eq. 1 [33].

$$R = P \times I \times 100^{\left(1-t/60\right)} \quad \text{[Eq. 1]}$$

where R is the final risk of a hazard, P is the probability of the potential hazard, I is the severity or impact of the consequences, and t is the process safety time between the initiating event and the occurrence of a hazardous event in the unit of minutes [33], [34]. Scaling factors based on expert's judgment and process knowledge to determine the $P \times I$ score for each hazard is used in [34]. Although this approach to assign $P \times I$ score has the advantage of simplicity and consistency when analyzing multiple hazards, it adds uncertainty to the calculated risk.

We propose to integrate the loss function approach to estimate system loss as a function of process deviation into risk equation. The formulation of the proposed model is in accordance with Eq. 2:

$$R = P \times L(x,T) \times 100^{\left(1-t/60\right)} \quad \text{[Eq. 2]}$$

where x is the key process characteristic, T is the target operating value of the process characteristic of interest, P is the probability of the potential hazard, t is the process safety time, and $L(x,T)$ is the overall loss calculated by using loss functions (listed in Table 1). The method continuously evaluates the risk associated with the current state of a process and issues warnings based on the estimated risk. By comparing the estimated risk with a predefined threshold risk, warnings are issued and categorized into alerts and alarms which will be annunciated on different displays. In addition to affecting the system loss, process deviations affect the probabilities of failure of safety systems which is the subject for another ongoing research.

The proposed approach determines risk in dollar value instead of a risk ranking based on the $P \times I$ score, thereby providing a more sensible criterion to make operational risk-based decisions. Choosing the threshold risk can be based on the tolerable risk limit of the system which is a challenging task in industrial applications. Wrong assignment of the threshold risk may have adverse effect on process safety. It may result in too many alarms or, on the other hand, may downgrade an alarm to an alert. Another advantage of the proposed method in representing the risk in dollar values is to provide a convenient tool for business managers to assign a threshold risk for a particular process system.

Despite several advantages of using the LF approach in risk-based warning system design, as discussed above, the choice of the LF could be a practical difficulty. The effect of changing the shape of the LFs on the estimated risk and a comparison of different LFs are discussed in the next section. The procedure is demonstrated by application of the methodology to a reactor system to examine the response to a temperature surge. This comparison will help industrial practitioners in choosing the appropriate LF for their applications.

6. Application and analysis of results

6.1 Application of loss functions to a reactor system

To compare the performance of different loss functions, namely QLF, INLF, MINLF, and IBLF (see Table 1), system loss associated with temperature deviations in a reactor system is calculated using the above mentioned LFs. This example is taken from [33]. The hazards related to the temperature surge in the reactor are identified as overheating and under-heating of the reactor liquid. The nomenclature of different parameters used in the calculations and the user-defined values are listed in Table 2.

As the reactor temperature deviates more from the desired target reaction temperature (120°C) and upon failure of the control system and other protection layers, the system experiences more loss. As shown in Table 2, the amount of loss associated with overheating of the liquid (temperature above the target) is considered to be more compared to the loss encountered when reactor temperature decreases (temperature below the target). Therefore, a combined function is used for each of the LFs allowing the LFs to have asymmetric shapes. For example, for QLF, the piecewise function is used as $L(T_{reactor}, T_T) = B_1\ (T_{reactor} - T_T)^2$ for $T_T \leq T_{reactor} < T_{max}$ and $L(T_{reactor}, T_T) = B_2\ (T_{reactor} - T_T)^2$ for $T_{min} < T_{reactor} < T_T$, respectively. T_T, T_{min}, and T_{max} are defined in Table 2. Similar piecewise functions are used for the other three LFs.

To apply each of the four LFs, the parameters of the LFs (estimated maximum loss and shape parameter) should be determined first. Based on the formulations given in Table 1, the QLF and the INLF require only the primary information which is the target reaction temperature and estimated maximum loss values; i.e. T_T, EML_1, and EML_2 in Table 2. For the MINLF and IBLF, auxiliary information given in Table 2 for the loss values at upper and lower specified reactor temperature limits are used to obtain the shape parameters. The least-squares estimation method is used for this purpose in accordance with Eq. 3:

$$\min_{SP>0} = \sum_{i=1}^{n} \{L_i - LF_{T_{reactor\_i}}\}^2 \quad \text{[Eq. 3]}$$

where $LF_{T_{reactor\_i}}$ is the value of any of MINLF or IBLF at a given $T_{reactor}$, L_i is the assumed loss values mentioned in Table 2, and SP is the shape parameter γ for MINLF and the shape parameter α for IBLF.

All four loss functions are plotted in Figure 1, along with the two auxiliary data points which are the assumed amount of system loss at upper specified reactor temperature (set point for high temperature alarm) and lower specified reactor temperature (set point for low temperature alarm); see Table 2 for the description and the value of the defined process information. Figure 1 shows the flexibility of MINLF and IBLF as it is possible to estimate their shape parameters using the available process information. If more information is available, more accurate shape parameters can be estimated. As shown in Figure 1, QLF and INLF are not flexible as it is not possible to change the shapes of the functions. For this case, QLF underestimates and the INLF overestimates the reactor system loss due to temperature deviations.

Table 2. Reactor process information used to develop the LFs.

Symbol	Description	Value
T_T	Target reaction temperature of the reactor	120 °C
T_{HA}	Temperature set point for high temperature alarm	130 °C
USL	Upper specified limit for liquid temperature associated with high-high temperature alarm (set point for ESD system)	150 °C
LSL	Lower specified limit for liquid temperature associated with low temperature alarm	100 °C
T_{max}	Maximum tolerable liquid temperature; it is assumed that at this temperature the reactor fails catastrophically	200 °C
T_{min}	Minimum tolerable liquid temperature; it is assumed that at this temperature ESD system will be activated	80 °C
EML_{USL}	Loss associated with liquid temperature at USL	2,000,000 US$
EML_{LSL}	Loss associated with liquid temperature at LSL	1,000,000 US$
EML_1	Estimated maximum loss due to high temperature scenario in the reactor	10,000,000 US$
EML_2	Estimated maximum loss due to low temperature scenario in the reactor	2,000,000 US$

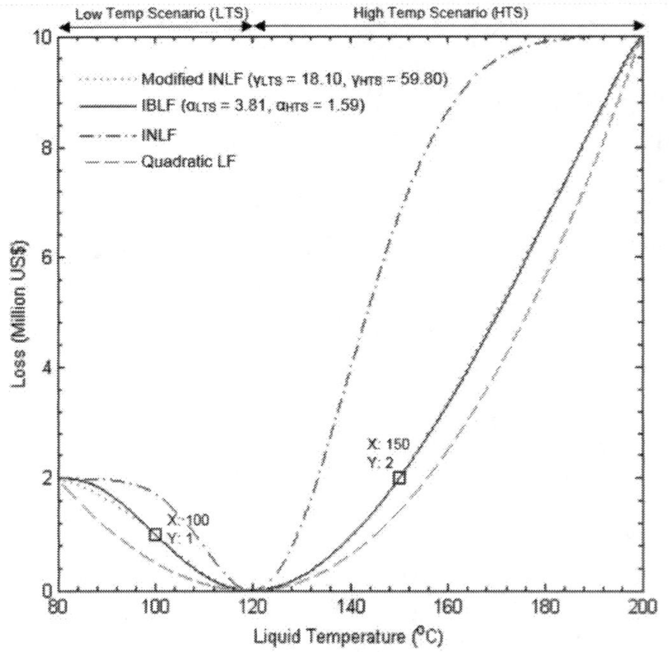

Figure 1. Illustration of the reactor risk using different loss functions

6.2 Risk-based warning analysis of the reactor system

The calculated system loss using LFs is then used in Eq. 1 to determine the operational risk of the reactor. The proposed operational risk-based method is then used to design the alarm system. To make the case simple, it is assumed that there is only one alert generated from the temperature control system. The probability of occurrence of both high temperature and low temperature scenarios is assumed as 0.01 per year. The process safety response time for both temperature deviation scenarios is considered as three minutes. These values are obtained from process history and expert knowledge. Figure 2 shows the final risk of the alert determined by application of the four LFs in Eq. 2.

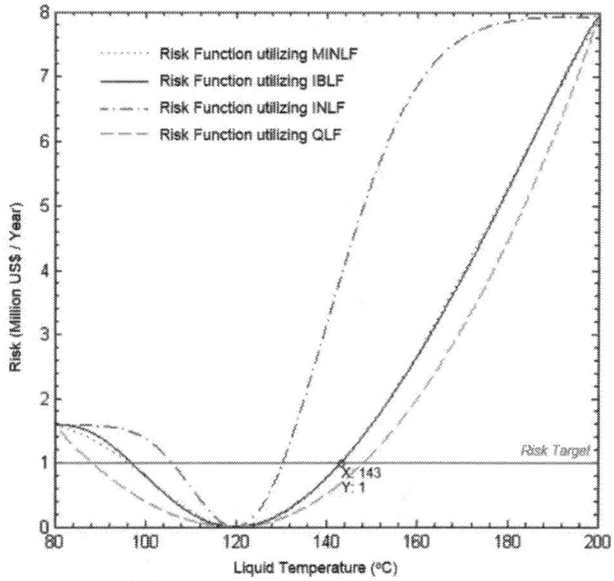

Figure 2. Reactor operational risk for different loss functions

When the reactor temperature deviation exceeds the set point for the high alarm (120°C) a warning will be issued. Based on the deviation, the proposed model calculates the real-time risk for the current reactor temperature. Then the estimated risk is compared with the risk target. The operational risk target for the reactor system is considered as 1 million dollars per year; see Figure 2. If the estimated risk is less than the target risk, the warning will be classified as alert. If the estimated risk exceeds the target risk, the warning will be upgraded to alarm and will be annunciated on a different display. The target risk may need to be adjusted for the alarm management system to satisfy the plant safety goals [33].

6.3 Sensitivity analysis

As discussed in Section 6.1, the application of the LFs requires determination of the parameters, i.e. the estimated maximum loss (EML) and the shape factor. To study the effect of utilizing different LFs and the effect of uncertainty in the parameters of the LFs on the estimated risk values, a sensitivity analysis is conducted on four LFs. These are: IBLF with $\alpha = 1.59$, INLF with $\gamma = 20$, MINLF with $\gamma = 59.80$, and QLF with $B = 0.0016$. See Table 1 and Section 2.1 for the list of formulations of each LF and the definition of the parameters, respectively. The selected shape parameters are associated with the LFs shown in Figure 1.

The sensitivity analysis results are presented in Figure 3. For the case study conducted in this work, it is evident from Figure 3(a) that the maximum effect on the estimated risk occurs when the INLF is used followed by QLF. A small deviation in the shape of these two LFs results in a significant change in the estimated risk. The sensitivity of the estimated risk is less for both the MINLF and IBLF. However, the estimated risk is more sensitive for the shape parameter when using IBLF as compared to MINLF. A positive error of about 20% in the shape parameter of

MINLF may underestimate the risk by 9.6%, whereas an error of the same magnitude in the shape parameter for the IBLF will overestimate the risk by 45%.

MINLF and IBLF have different behavior regarding their impact on the estimated risk as a result of an error in the shape parameter. When using IBLF, positive errors in the shape parameter (higher than the base value) causes positive error in the estimated risk and negative errors (lower than the base value) causes negative error of the same magnitude in the estimated risk. However, as can be seen in Figure 3(a), MINLF has opposite behavior as compared to IBLF. Also, when comparing MINLF with IBLF, the estimated risk is more sensitive when there is a negative change in the shape factor as compared to a positive change. A negative error of 20% in the shape factor may overestimate the risk by 18.5%, whereas a positive error of the same magnitude may underestimate the risk by just 9.6%.

Figure 3(b) shows the sensitivity analysis results related to estimated maximum loss change when using MINLF and IBLF. It is evident from Figure 3(b) that the sensitivity of the estimated risk is the same for both IBLF and MINLF. An error of 20% results in an error of the same magnitude in the estimated risk, i.e. a linear dependence.

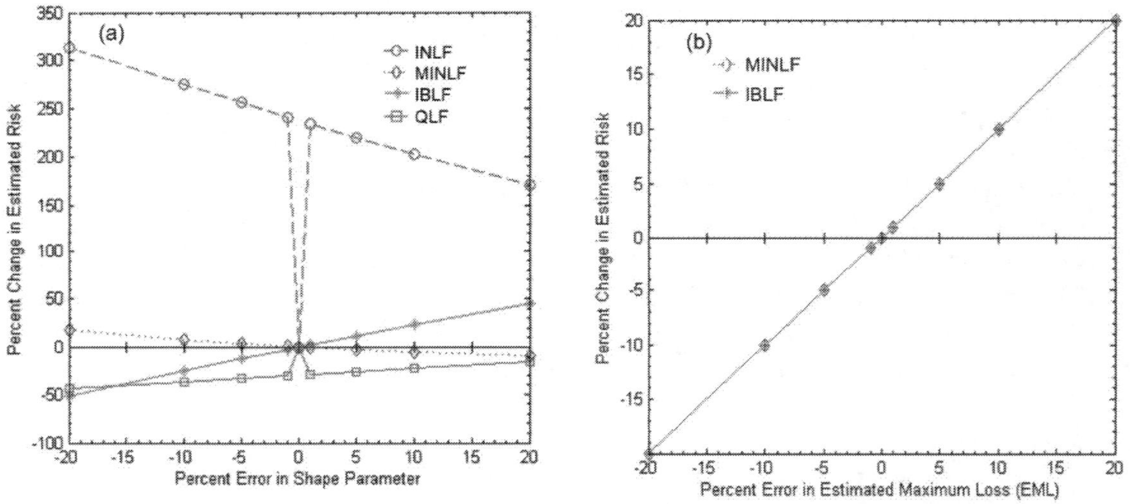

Figure 3. Sensitivity analysis: (a) for different estimated maximum loss (EML) values, (b) for different shape parameters

Overall, it can be concluded from this study that the proper selection of the LF is very important. From the four LFs studied, the MINLF shows comparatively more robust results followed by IBLF. The results of the sensitivity analysis show that a conservative value of the estimated maximum loss should be avoided for both MINLF and IBLF. Considering the results of the sensitivity analysis for the shape factor, it is evident that a conservative value of the shape factor for IBLF should also be avoided. However, it is advisable to use a conservative value of the shape factor for the MINLF.

6.4. Comparison of loss functions

MINLF and IBLF are found to demonstrate better performance than QLF and INLF. Compared to MINLF, IBLF has an additional advantage of covering asymmetrical cases, i.e. when the target is not at the middle of upper and lower specification limits. However, for the cases when the loss functions should approach different maximum loss values at both sides of the target, like the case study conducted in this work, a piecewise function should be used. Therefore, for asymmetric maximum losses at both sides of the target, which is the applicable case when analyzing most process systems, IBLF and MINLF have the same restriction.

By considering the overall performance of the LFs studied in this work, the MINLF and IBLF could be the optimum choice for depicting the system loss associated with process variations. Choice of LF should be based on the process behavior and availability of loss data. The shape of both MINLF and IBLF can be modified to suit the practitioner's needs for both symmetric and asymmetric problems. However, construction of the MINLF is easier when compared to IBLF due to simplicity of the formulation and also there is no need for transformation of the scales. Considering the sensitivity analysis results, it is also shown that the MINLF has comparatively more robust performance.

7. Concluding remarks

Loss functions are used in this work to design an operational risk-based warning system. This is a paradigm shift that is expected to benefit the process industry in continuous improvement of process safety. Instead of relying on the safety levels which are set during the design stage, the utilization of the proposed approach integrates the safety assessment into the day to day activities at operational level. By applying the operational risk-based warning system design, the warnings will be differentiated into alerts and alarms. This helps in significant reduction of the generated alarms to meet the recommended alarm regulations. Using the proposed approach, operators can prioritize the alarms and will have more time to make informed decisions. This prevents unnecessary activation of the safety instrumented systems and minimizes the system loss by reducing operational and maintenance costs.

Use of loss functions in risk estimation provides a real-time risk assessment tool which continuously updates the operational risk based on the current state of the process. The proposed approach to system loss modeling helps to foster a zero-loss safety culture and provides the infrastructure for continuous safety improvement. Extending the proposed methodology to multivariate problems is a topic of future research.

8. References

[1] G. Taguchi, *Quality Engineering In Production Systems*, First ed. New York: McGraw-Hill, 1989.

[2] G. Taguchi and D. Clausing, "Robust quality," *Harvard Business Review*, vol. 68, no. 1, pp. 65–75, 1990.

[3] K. C. Kapur and B. R. Cho, "Economic design of the specification region for multiple quality characteristics," *IIE Transactions*, vol. 28, no. 3, pp. 237–248, Mar. 1996.

[4] S. S. Kulkarni, "Loss-based quality costs and inventory planning: General models and insights," *European Journal of Operational Research*, vol. 188, no. 2, pp. 428–449, Jul. 2008.

[5] B. P. K. Leung and F. A. Spiring, "Some Properties of the Family of Inverted Probability Loss Functions," *Quality Technology & Quantitative Management*, vol. 1, no. 1, pp. 125–147, 2004.

[6] F. A. Spiring, "The reflected normal loss function," *The Canadian Journal of Statistics*, vol. 21, no. 3, pp. 321–330, 1993.

[7] J. Pan and S. Chen, "A new approach for assessing the correlated risk," *Industrial Management & Data Systems*, vol. 112, no. 9, pp. 1348–1365, 2012.

[8] H. S. Sii, T. Ruxton, and J. Wang, "Taguchi concepts and their applications in marine and offshore," *Journal of Engineering Design*, vol. 12, no. 4, pp. 331–358, 2001.

[9] S. Ordoobadi, "Application of Taguchi loss functions for supplier selection," *Supply Chain Management: An International Journal*, vol. 14, no. 1, pp. 22–30, 2009.

[10] S. Dey, "Bayesian Estimation of the Parameter and Reliability Function of an Inverse Rayleigh Distribution," *Malaysian Journal of Mathematical Sciences*, vol. 6, no. 1, pp. 113–124, 2012.

[11] B. P. K. Leung and F. A. Spiring, "The inverted beta loss function: properties and applications," *IIE Transactions*, vol. 34, no. 12, pp. 1101–1109, Dec. 2002.

[12] T. E. Murphy, K.-L. Tsui, and J. K. Allen, "A review of robust design methods for multiple responses," *Research in Engineering Design*, vol. 15, no. 4, pp. 201–215, Sep. 2004.

[13] Y. Fathi and C. Poonthanomsook, "A Quartic Quality Loss Function and Its Properties," *Journal of Industrial and Systems Engineering*, vol. 1, no. 1, pp. 8–22, 2007.

[14] W. M. Chan, R. N. Ibrahim, and P. B. Lochert, "Quality evaluation model using loss function for multiple S-type quality characteristics," *The International Journal of Advanced Manufacturing Technology*, vol. 26, no. 1–2, pp. 98–101, Apr. 2004.

[15] J. Pan and J. Pan, "A Comparative Study of Various Loss Functions in the Economic Tolerance Design," *2006 IEEE International Conference on Management of Innovation and Technology*, vol. 2, pp. 783–787, Jun. 2006.

[16] V. R. Joseph, "Quality Loss Functions for Nonnegative Variables and Their Applications," *Journal of Quality Technology*, vol. 36, no. 2, pp. 129–138, 2004.

[17] F.-B. Sun, J.-Y. Laramee, and J. S. Ramber, "On Spiring 's normal loss function," *The Canadian Journal of Statistics*, vol. 24, no. 2, pp. 241–249, 1996.

[18] F. A. Spiring and A. S. Yeung, "A general class of loss functions with industrial applications," *Journal of Quality Technology*, vol. 30, no. 2, pp. 152–160, 1998.

[19] W. M. Chan and R. N. Ibrahim, "Evaluating the quality level of a product with multiple quality characterisitcs," *The Int J of Advanced Manufacturing Technology*, vol. 24, no. 9–10, pp. 738–742, Apr. 2004.

[20] C. M. Hsu, "Solving Multi-Response Problems Through Neural Networks and Principal Component," *Journal of the Chinese Institute of Industrial Engineers*, vol. 18, no. 5, pp. 47–54, 2001.

[21] R. Suhr and R. G. Batson, "Constrained Multivariate Loss Function Minimization," *Quality Engineering*, vol. 13, no. 3, pp. 475–483, 2001.

[22] J. J. Pignatiello, "Strategies for robust multiresponse quality engineering," *IIE Transactions*, vol. 25, no. 3, pp. 5–15, 1993.

[23] N. Artiles-León, "A pragmatic approach to multi-response problems using loss functions," *Quality Engineering*, vol. 9, no. 2, pp. 213–220, 1996.

[24] Y. Ma and F. Zhao, "An Improved Multivariate Loss Function Approach to Optimization," *Journal of Systems Science and Systems Engineering*, vol. 13, no. 3, pp. 318–325, 2004.

[25] C. Su and L. Tong, "Multi-response robust design by principal component analysis," *Total Quality Management*, vol. 8, no. 6, pp. 409– 416, 1997.

[26] W. M. Chan, R. N. Ibrahim, and P. B. Lochert, "Evaluating the product quality level under multiple L-type quality characteristics," *The International Journal of Advanced Manufacturing Technology*, vol. 27, no. 1–2, pp. 90–95, Jun. 2004.

[27] W. Abdul-Kader, O. Ganjavi, and A. Solaiman, "An integrated model for optimisation of production and quality costs," *International Journal of Production Research*, vol. 48, no. 24, pp. 7357–7370, Dec. 2010.

[28] L. Raiman and K. Case, "Monitoring continuous improvement using the Taguchi loss function," *Int Conf Industr Eng*, pp. 391–396, 1990.

[29] L.-I. Tong, C.-T. Su, and C.-H. Wang, "The optimization of multiresponse problems in the Taguchi method," *International Journal of Quality and Reliability Management*, vol. 14, pp. 367– 380, 1997.

[30] L.-I. Tong and C.-T. Su, "Optimizing multi-response problems in the Taguchi method by Fuzzy multiple attribute decision making," *Quality and Reliability Engineering International*, vol. 13, pp. 25–34, 1997.

[31] J.-N. Pan, "A new loss function-based method for evaluating manufacturing and environmental risks," *International Journal of Quality & Reliability Management*, vol. 24, no. 8, pp. 861–887, 2007.

[32] AIChE, *Guidelines for Risk Based Process Safety*, First. New York; Hoboken, NJ.: Wiley, 2007.

[33] Y. Chang, F. Khan, and S. Ahmed, "A risk-based approach to design warning system for processing facilities," *Process Safety and Environmental Protection*, vol. 89, no. 5, pp. 310–316, Sep. 2011.

[34] S. Ahmed, H. A. Gabbar, Y. Chang, and F. I. Khan, "Risk based alarm design: A systems approach," in *2011 International Symposium on Advanced Control of Industrial Processes, ADCONIP 2011*, 2011, pp. 42–47.

[35] D. D. Woods, "The alarm problem and directed attention in dynamic fault management," *Ergonomics*, vol. 23, no. 11, pp. 2371–2393, 1995.

[36] EEMUA, *Alarm Systems: A Guide to Design, Management and Procurement*, 2nd ed. EEMUA, 2007.

1ST CCPS ASIA-PACIFIC CONFERENCE ON PROCESS SAFETY℠

The risk analysis and control research of oil shale processing

Jing Lan
Mechanical and Electrical Engineering College, China University of Petroleum
Shandong Qingdao, China
lanjing06@126.com

Prepared for Presentation at
American Institute of Chemical Engineers
1st CCPS Asia-Pacific Conference on Process Safety
Qingdao, China
September 4 – 5, 2013

UNPUBLISHED

The risk analysis and control research of oil shale processing

Jing Lan
Mechanical and Electrical Engineering College, China University of Petroleum ,Shandong Qingdao, China
lanjing06@126.com

Keyword: Oil shale, Processing and utilization, Risk analysis

Abstract

The oil shale resources is abundant in China, shale oil can be obtained by retorting the oil shale, which is an important supplement to the future energy, it can directly be used as fuel or produce high-quality fuel and chemical products through further processing, With the accelerating utilization of oil shale development, its safety and environmental issues attract more and more people's attention. In the side of safety, oil shale processing involves a lot of flammable, explosive, toxic and hazardous substances, and due to human and equipment factors, there exists many dangers. such as the pile of oil shale, semicoke and incomplete combustion shale ash will self-ignite or cause fire. retort furnace will explode if the air excess. Environmental aspects, the good and bad are intermingled in enterprises, the comprehensive utilization of oil shale in large enterprises is higher, but some small enterprises use the old technology and equipment because of financial and technical problems, not only low utilization of the resource, there is a larger environmental pollution problems.

Oil shale processing risk analysis focus primarily on environmental risks, less on safety risk analysis, which primarily focus on retort furnace safety risk analysis, not consider the entire process. Otherwise, safety risk analysis and environmental risk analysis contents are repetitive and confusing. Therefore, combing their characteristics, conduct a comprehensive risk analysis of the oil shale processing. identify shale oil, ammonia, hydrogen sulfide, pyridine, phenols and other poisonous and harmful substances, analyze their harmfulness and the risk of devices, Finally, put forward the corresponding preventive measures.

The processing and utilization of oil shale brings the gospel to the world, but its impact on society is enormous. at the time of our country encourage and accelerate the comprehensive utilization of oil shale, enterprises should strengthen the oil shale processing safety and environmental management, so that the development and utilization of the oil shale can achieve safety , environmental and economic benefits, and it plays an important role to achieve sustainable development of unconventional oil and gas resources.

1 Introduction

China began to carry out the processing and utilization of oil shale in the early 1928, the comprehensive utilization of oil shale in our country mainly concentrated in Fushun, Huadian, Maoming and other places.

Owing to the relatively backward technology, some small businesses use the old technology and equipment and the environmental protection measures are not perfect, not only the shale oil recovery rate is low, but exist safety risks and environmental problems. Chinese scholars on the oil shale industry risk analysis focuses on environmental risk assessment [1,2,3], mainly for the study of impact of toxic and harmful chemical substances. Han Fang [4] has been analyzed and evaluate Fushun retort furnace, but the safety evaluation did not consider the danger of crushing, screening device, condensate recovery system. Therefore, conduct a comprehensive analysis of the safety and environmental risk of oil shale process, providing guarantee for the sustainable development of the industry of oil shale.

2 Oil shale processing technology

Oil shale retort is a low temperature (450°C~600°C) retort method for generating shale oil, refining gas, sewage, shale residue under the condition of isolation from the air . The process is shown in figure 1.

Figure 1 Flow chart of oil shale retort process

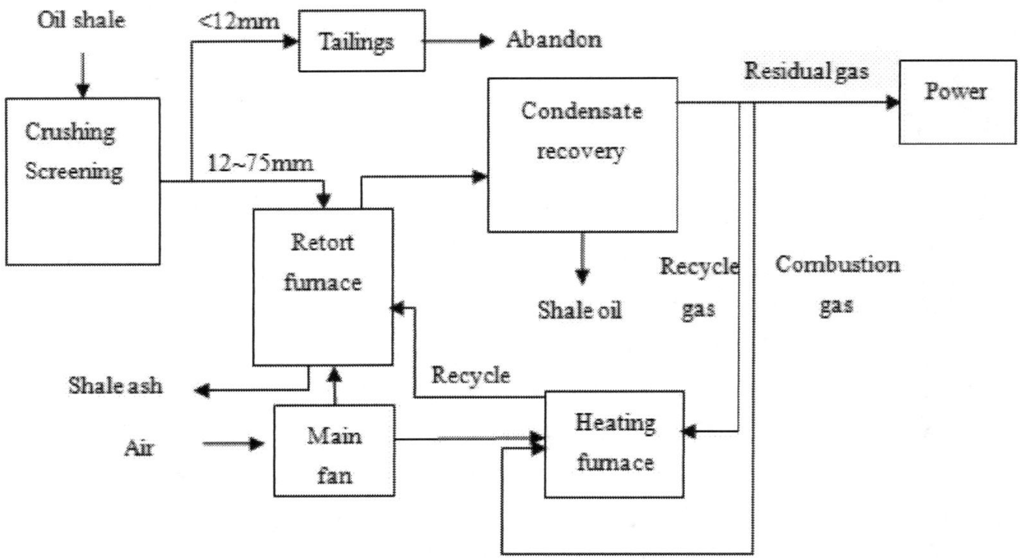

After crushing, screening ,12 ~ 75mm shale particles are added from furnace roof into the furnace, the majority of shale oil is released after drying, preheating, retorting in the retorting zone, the remaining semi-coke go into the gasification zone , with the rise of the main air oxidation reaction into shale ash, the ash are discharged from oil shale ash pan ,with the tailings transported to the waste heap. Gasification and retorting oil and gas generated from the furnace top into condensate recovery system to get the shale oil and refining gas, one part of the gas for heating fuel, one for furnace recycle gas , and the remaining used for power generation .

3 Risk analysis of oil shale processing

3.1 Material hazard analysis

The major oil shale processing hazardous material is shale oil, gas gasoline, ammonia, hydrogen sulfide, pyridine, refining gas , etc.

Table 1 Characteristics of hazardous material

Material	Flash point (℃)	Ignition point (℃)	Normal temperature status	Explosive limit (%)	Fire hazard category	Hazardous characteristics
Shale oil	127	140	liquid	—	C	flammable
Gasoline	-50	415 ~ 530	gas	1.3~6.0	A	flammable explosive
NH_3	—	652	gas	15.7~27.4	B	flammable explosive toxic
H_2S	—	260	gas	4.0~46.0	A	flammable explosive toxic
Pyridine	17	482	liquid	1.7~12.4	A	flammable explosive

						toxic
CO	<-50	610	gas	12.5~74.2	B	flammable explosive toxic

Table 2 Environmental hazard of toxic and harmful material

Material	Maximum concentrations in residential atmosphere [5] mg/m³	Maximum concentration in ground water [5] mg/m³	Chemical substances allowable concentration in the workplace air[6] mg/m³	Acute toxicity[8]	
				Lethal concentration LC50 (mg/m³)	Half lethal dose LD50 (mg/kg)
Gasoline	—	0.3	—	103000	67000
NH_3	0.20	—	30 (PC-STEL)	1390	350
H_2S	0.01	—	10(MAC)	618	—
Pyridine	0.08	0.2	4(PC-TWA)	—	1580
CO	3.00	—	30 (PC-STEL)	2069	—

3.2 Analysis the cause of accidents

3.2.1 Cause analysis of fire and explosion in production operations

(1)Shale oil leakage happened in flange connection or welding of retort furnace top and retort export pipe, outer wall forming oil stain and not cleared in time, when the furnace air is excessive , inner fire rising causes the oil furnace outlet temperature rising to ignite oil and then cause fire. (2)The bottom basin is short of water, furnace gas get out of the framework ,and then ignite the oil stain. (3) Improper retort capacity and air volume control causes fire layer increase ,forming coke, the retort may explode. (4)Rain and snow season, wet shale will block equipment, make the charging extend, causing the air enter the furnace, it may trigger an explosion.(5)Shale chunk is too big or mixed with others, Shale was stuck in the upper part of the umbrella , resulting in uneven distribution of shale , furnace coking occurs , it may cause an explosion.(6)The operator did not timely charge the shale, air enter the furnace causing the retort furnace temperature increases, and the gas explosion limit increases.(7)Heating furnace manual operation is not coordinate, cause the pressure increase, it may result in heating furnace explosion accident.

3.2.2 Leak and repair process hazard analysis

The main substance of retort furnace and condensate recovery system are refining gas, pyridine, ammonia and hydrogen sulfide, most are inflammable ,explosive and toxic ,when happens leak, it will cause fire and explosion accidents. hazards of leakage: (1)Retort flange, weld joint is not tight, and gate control loading equipment failure cause leakage, when operators at the top of retort ,it will easy happen poisoning accidents ,fire and explosion. (2)The condensate recovery equipment corrosion, valves, pumps and so on seal is not close, gas happens leak, and then mixed with air to form explosive mixture gas, causing fire and explosion.(3)When parking overhaul, if equipment, pipelines, valves, etc substitution or replacement is not clear, the flammable and explosive gases in hot work areas, it can lead to fire, explosion and poisoning accidents. (4)The traditional probe material is prone to leak gas, cause operator poisoning.

In addition , in summer ,oil shale , shale semi-coke , shale ash and other large piled up will cause spontaneous combustion or ignition; crushing and screening process will produce shale dust, if ventilation is inadequate , when encounter high temperature or sources of ignition ,it has dust explosion hazard , these are required to take appropriate measures to prevent.

3.2.3 Other hazards

(1)Using crushing ,screening equipment, conveyor and other kinds of machinery equipment during the preparation of shale ,which may cause mechanical damage during operation.(2)When the furnace temperature doesn't fall to optimum temperature , the shale coke on furnace wall doesn't completely fall off, or there remains residual gas, the workers will be easy exposed to high temperature and poison gas during the retort decoking operations.(3)The high falling accidents may occur working on each operating platform which is 2m higher than the assignments datum. (4)Electric shock accidents may be caused by electrical equipment leakage, improper grounding and insulation damage, etc.

3.2.4 Environmental Impact

The oil shale retorting process of "three wastes" emissions has certain pollution to the environment.

The production of 1t shale oil will produce 10~20t waste residue, through weathering, oxidation, rain soaked, the sulfur nitrogen compounds, PAHs, trace toxic elements leaching from waste and oil shale pile impact on the surrounding ecological environment.,

Retort furnace, cooling towers, chimneys and other equipment emit gas smell, and other harmful ingredients containing hydrogen sulfide, pyridine, aromatic hydrocarbons, has a certain impact on the health of personnel, and pollute the surrounding environment, harm the biological growth.

Sewage containing oil substances, volatile phenols, organic sulfur, organic nitrogen, polycyclic aromatic hydrocarbons and other pollutants produced by retorting process.

Oil shale crushing and screening process produces a large number of particles, adsorbing harmful metal elements and organic compounds , when diffused, they will settle in certain conditions, which has impact on the surrounding environment

3.3 Accident probability calculation

Retort leakage generally occurs in pipeline, valve and flange joints, typical leakage probability is 1×10^{-5}, at present, the fire and explosion accident proportion of China's petrochemical enterprises device is 30.8%, leak accident proportion Is 59.4%. [9],Figure 2 uses ETA method analysis retort gas leak accident consequence.

Figure 2 ETA of gas leak

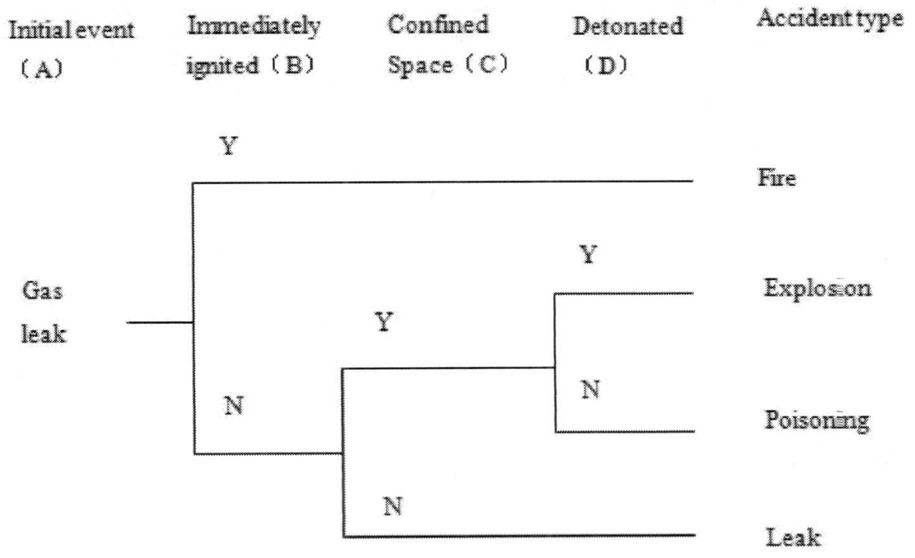

P(fire、explosion)=P(A)P(B)+P(A)(1-P(B))P(C)P(D)

P(poisoning)=P(A)(1-P(B))P(C)(1-P(D))
P(leakage)=P(A)(1-P(B))(1-P(C))

Fire explosion accident probability is 1.09×10^{-6}, poisoning accident probability is 8.1×10^{-7}, leak accident probability is 8.1×10^{-6}.

3.4 Consequence calculations

3.4.1 Gas leakage flow calculation

Diameter of small hole leakage is 10 mm, the gas composition as follows in table 3.

Table 3 Refining gas composition

Composition	CO_2(%)	CO(%)	H_2(%)	CH_4(%)	O_2(%)	C_nH_m(%)	N_2(%)
Proportion	20.5	4.0	10.5	7.5	0.4	0.8	56.3

The content of H_2S is 3~8g/m^3, NH_3 is 5~7g/m^3, pyridine is 0.28~1.21g/m^3, gas gasoline is 20~30g/m^3, The density is 1.266kg/m^3.

Gas leak rate calculation equation[7] as following:

$$Q = \delta_g A_0 \sqrt{P\rho_g r[\frac{2}{r+1}]^{\frac{r+1}{r-1}}} \qquad \text{[Eq. 1]}$$

In the equation , Q—Gas leak rate , kg/s; δ_g—Gas leak coefficient; A_0—Gap area , m^2; P—Medium pressure in container , Pa; ρ_g—gas density , kg/m^3; r—Gas adiabatic index.

Calculated gas leak rate is 19.46g/s, according to the composition , CO leak rate is 0.769g/s, H_2S leak rate is 0.123g/s, NH_3 leak rate is 0.11g/s, pyridine leak rate is 0.018g/s.

3.4.2 Diffusion effect calculation

Assume the leakage is continuous leakage, uses the Gaussian plume model [8] to simulate, usually consider to calculate the maximum concentration of ground axis formed by ground -point source leakage, calculation equation is:

$$C(x,0,0) = \frac{Q}{\pi\sigma_y\sigma_z\mu} \qquad \text{[Eq. 2]}$$

In the equation , $C(x,0,0)$—maximum concentration on ground (kg/m^3) ; Q—strength of the leakage source (kg/s) ; σ_y、 σ_z—The diffusion coefficient of y and z axis, x, y, z is prediction point coordinates ; μ—Environmental average wind speed (m/s) ;

Parameter: wind speed is 2.6m/s, atmospheric stability is D.The diffusion consequence as follows.

Table 4 CO Leakage and diffusion results

X (m)	10	20	50	100	150	250	1000
C (mg/m^3)	196.23	49.059	7.850	1.962	0.872	0.314	0.02

Table 5 H_2S leakage and diffusion results

X (m)	10	20	50	100	150	250	1000
C (mg/m^3)	31.38	7.845	1.255	0.314	0.14	0.05	0.003

Table 6 NH_3 leakage and diffusion results

X (m)	10	20	50	100	150	250	1000
C (mg/m^3)	28.07	7.017	1.122	0.281	0.125	0.045	0.0028

Table 7 Pyridine leakage and diffusion results

X (m)	10	20	50	100	150	250	1000
C (mg/m^3)	4.59	1.148	0.184	0.046	0.020	0.007	0.0005

Table 8 Harmful substances leakage hazardous areas

Substance	Concentration threshold (mg/m3)	Maximum diffusion distance in downwind (m)
CO	30 (PC-STEL)	25.58
H_2S	10(MAC)	17.7
NH_3	30 (PC-STEL)	9.67
Pyridine	4(PC-TWA)	10.7

Table 4-7 show that after the gas leak, along with the diffusion distance of each substance increase, their concentration gradually reduce , when spread to neighboring residential areas, the concentration is very small, which have no effect on the health of residents . Table 8 shows the workplace CO, H_2S, NH_3, pyridine maximum diffusion distance of 25.58m, 17.7m, 9.67m, 10.7m. The diffusion influence area is small

4 Conclusion

Oil shale retorting process involves shale oil, refining gas and other harmful and toxic substances, retort furnace, condensate recovery systems are the key risk prevention area , which prone to cause fire , explosion , leak and other accidents .

"Three wastes" of retort process has certain influence to the environment, which can be discharged after treatment

After retort gas leak, harmful substances diffusion influence area is small ,and there has no influence on the surrounding residents

5 References

[1] Wang songyun , Liang jing, "Oil Shale Refining Environmental Risk Assessment" , Oil and Gas Field of Environmental Protection， Volume 19, NO. 3, 2009

[2] Zhang liping, Zeng rongshu,Xu wending,Jin fujie, "Impact of oil shale utilization on the surrounding environment", Coal Geology of China, Volume 18, NO. 2, 2006

[3] Ye jiwen, Yang yang ,Xu mingzhu, "Oil shale resource utilization and prospects",China Resources Comprehensive Utilization, Volume 28, NO. 6, 2010

[4] Han fang,Bao mingfu,Gao jian, "Fushun oil shale retort furnace safety analysis and evaluation", China Occupational Safety and Health Association 2009 Annual Conference Proceedings， 2009

[5] "Standard Design of Industrial Enterprises TJ36-1979"

[6] "Workplace Occupational Exposure Limits for Hazardous Chemicals Harmful Factors GBZ2.1-2007"

[7] Chen Guohua, "Risk Engineering", 2007

[8] Hu erbang, "Environmental Risk Assessment Practical Technology, Mmethods and Case", China Environmental Science Press, 2009

1ST CCPS ASIA-PACIFIC CONFERENCE ON PROCESS SAFETY℠

Identification and Thermokinetics of Autocatalytic Exothermic Decomposition of β-HMX

Shilong Bao
Department of safety engineering, Institute of Chemical Technology, Nanjing University of Technology and Science
Xiaolingwei Street No. 200 , Xuanwu District, Nanjing City, Jiangsu Province, China
Baoshilongnjust@163.com

Liping Chen* Wanghua Chen Jiayu Lv Haisu Gao
Department of safety engineering, Institute of Chemical Technology, Nanjing University of Technology and Science
Xiaolingwei Street No. 200 , Xuanwu District, Nanjing City, Jiangsu Province, China
clp2005@hotmail.com

Prepared for Presentation at
American Institute of Chemical Engineers
1st CCPS Asia-Pacific Conference on Process Safety
Qingdao, China
September 4 – 5, 2013

UNPUBLISHED

Identification and Thermokinetics of Autocatalytic Exothermic Decomposition of β-HMX

Shilong Bao
Department of safety engineering, Institute of Chemical Technology, Nanjing University of Technology and Science
Xiaolingwei Street No. 200 , Xuanwu District, Nanjing City, Jiangsu Province, China
Baoshilongnjust@163.com

Liping Chen* Wanghua Chen Jiayu Lv Haisu Gao
Department of safety engineering, Institute of Chemical Technology, Nanjing University of Technology and Science
clp2005@hotmail.com

Keywords: Autocatalytic decomposition; DSC; Rapid identification; Malek method; Decomposition kinetics; AKTS

Abstract

β-HMX is investigated under dynamic conditions using differential scanning calorimetry (DSC). An identification method based on numerical simulation technique is used to determine the characteristic parameters of the decomposition reaction, revealing that the decomposition of β-HMX is autocatalytic. Isothermal DSC experiments confirmed that the decomposition of β-HMX is autocatalytic, and Security parameter of adiabatic condition is tested with accelerating rate calorimeter (ARC).

1. Introduction

Cyclotetramethylene tetranitramine(HMX) is a kind of nitrobenzene explosives, it has been widely used in civil and military fields since its excellent detonation performance. In order to improve its performance and estimate the risk, the research on thermal stability and pyrolysis mechanism of HMX is always the focus. Zhongxing Jing and Mulan Bai applied time of flight mass spectrography (TOFMS) to research the decomposition mechanism of HMX [1]. Huiping Jia and Mulan Bai concentrated on investigating the autocatalysis of HMX by DSC& TOFMS [2]. Zhi Jiang and Shufen Li studied the effects of ordinary and nano nickel, aluminum and bronze powder for HMX decomposition [3]. TG was used to study the influence of crystal defect on sensitivity and thermal stability of HMX [4]. The influences of contaminants on HMX are investigated by DengJr Peng, ChengMing Chang and Miin Chiu[5]. Ziru Liu and her colleagues studied thermal decomposition behavior, thermal decomposition kinetics and its decomposition mechanism of HMX systematically with TG & DSC-TG-FTIR [6-9]. To investigate the properties of HMX, molecular simulation was implemented by Yiding Shi and Fenglei Huang to calculate the isotherms of HMX at different temperatures using COMPASS force field [10]. In 2011, Li Yang and Xiaojing Qiao gave out the most probable mechanism function of HMX [11]. In the same year, Rui Liu, Yanli Zhang and Tonglai

Zhang studied the stability of HMX with Dynamic Vacuum Stability Test (DVST), and gave out a relationship between decomposition temperature and time [12]. The thermal decomposition mechanisms of condensed phase HMX at various densities and at 2500K were studied using ReaxFF reactive molecular dynamics simulations by Tingting Zhou and Yiding Shi [13].

In conclusion, a great many studies have been done on the decomposition mechanism and thermal behavior of HMX, but there is little research on its autocatalysis of decomposition. In order to accurately describe the decomposition of HMX, this article studies its autocatalysis identification, and give out the kinetic equations and T_{D24}.

2. Experimental

2.1 Reagent

The experiment sample β-HMX is white crystalline powder manufactured by Gansu Yinguang Chemical Industry Group co., LTD, which satisfies China's GJB2335-95 standard.

2.2 Measurements by DSC

Tests were conducted using Mettler Toledo DSC-1 differential scanning calorimetry. In the DSC tests, samples were sealed in the high-pressure crucibles in the atmosphere and heated from 25°C to 350°C. The heating rates used in the dynamic scanning experiments were 1, 2, 4 and 8 $°C \cdot min^{-1}$. The sample mass were 0.96, 0.84, 0.83 and 0.88 mg, respectively. The temperatures selected in isothermal experiments were 240°C and 260°C, and the corresponding sample mass were 0.72 and 0.73mg.

2.3 Adiabatic measurements by ARC

The ARC is a common apparatus used to evaluate the thermal stability of energetic materials and measurements are usually carried out under adiabatic conditions. Several tests have been taken. All samples in the Ti-LCQ bomb were measured from 150 °C to 400 °C with a heat step of 5 °C, while the temperature rate sensitivity was 0.02 $°C \cdot min^{-1}$.

3. Results and discussions

3.1 Non-Isothermal DSC experiment

The dynamic DSC curves of β-HMX are shown in Figure 1, and the tested results are provided in Table 1. DSC tests show that there is only one exothermic peak, and the onset temperature is high than room temperature, which indicated a good thermal stability of β-HMX. The onset temperature is close to peak temperature, which shows that β-HMX decomposes quickly. The exothermic peak is very sharp, and peak value is high. According to criterions from reference [14], the severity of β-HMX decomposition is high. It is judged by average enthalpy with the value of 2050.9 $J \cdot g^{-1}$,which far exceed the valve of 800 $J \cdot g^{-1}$.

Table 1 Results of dynamic DSC tests

Sample	β(°C·min^{-1})	T_0 (°C)	T_p (°C)	ΔH_d (J·g^{-1})
β-HMX	1	259.03	260.79	2000.4
	2	268.43	271.53	2232.5
	4	275.74	280.03	2109.9
	8	282.63	288.48	1860.9

β: Heating rate; T_0: Onset temperature; T_p: Peak temperature; ΔH_d: Enthalpy

Figure1:Dynamic DSC experimental curves of HMX

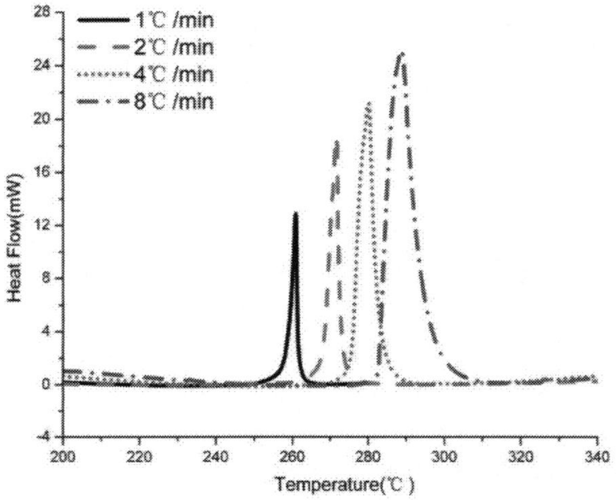

3.2 Identification of autocatalytic exothermic decomposition of β-HMX

Here, a numerical simulation technique from the Swiss Institute for the Promotion of Safety and Security (Swiss method) as the means for identification of autocatalytic exothermic decomposition of β-HMX was used. Swiss method is on the base of a first order model: in fact the beginning of the reaction determines the behavior of the reaction mass under adiabatic conditions. It can be shown that at small conversion any reaction kinetics can be described by a first order reaction [14]. Therefore a simple first order model is fitted to the beginning of the peak by adjusting three parameters of the model: the decomposition energy, the activation energy and a heat production rate at a certain temperature. The latter can be extracted from the DSC curve (the baseline is known). The resulting activation energy from the fit is an apparent activation energy and the higher is its value the greater is the probability that the reaction is autocatalytic. For further details, please look at the reference [15-16].

Swiss method is based on Eq. 1.

$$q(t+\Delta t) = q_0 \exp\left(\frac{-E}{R}\left(\frac{1}{T(t+\Delta t)} - \frac{1}{T_0}\right)\right)\left(1 - \frac{\Delta H(t+\Delta t)}{\Delta H_d}\right) \quad \text{[Eq. 1]}$$

In Eq. 1, β(°C ·min^{-1}) is heating rate, α is reaction progress, R is ideal gas constant, E(kJ·mol^{-1}) is activation energy, q(W·kg^{-1}) is heat release rate, ΔH_d(J·g^{-1}) is Enthalpy. T_0(°C) is the temperature which deviated from the baseline, and q_0(W·kg^{-1}) is the heat release rate at T_0. $\Delta H(t+\Delta t)$ is the decomposition heat at time $t+\Delta t$, and its approximate equation is Eq. 2.

$$\Delta H(t+\Delta t) = \Delta H(t) + q(t)\Delta t \quad \text{[Eq. 2]}$$

Eq. 2 is plugged into Eq. 1, and Eq. 3 is obtained:

$$q(t+\Delta t) = q_0 \exp(\frac{-E}{R}(\frac{1}{T(t+\Delta t)} - \frac{1}{T_0}))(1 - \frac{\Delta H(t) + q(t)\Delta t}{\Delta H_d}) \quad \text{[Eq. 3]}$$

In Eq. 3, q_0 and ΔH_d are measured value, E、t、T、$\Delta H(t)$、$q(t)$ are variables. Except E, other variables can be calculated with t.

In this article, activation energy E with minimum deviation is fitted by computer program. Definition of deviation:

$$\varepsilon = \frac{1}{n}\sum_{i=1}^{n}\left|p_{si} - p_{ei}\right| \quad \text{[Eq. 4]}$$

Unit of ε is mW, n is the number of data, p_{si} and p_{ei} is the simulating value and experimental value of q at the same temperature. Step size of t and E are 1s and 0.1kJ·mol^{-1}.The reaction is probable autocatalytic if obtained E greater than 220 kJ·mol^{-1}.

The dynamic DSC thermal curves were fitted. The fitting results are shown in Figure2 and Table 2. The deviation ε of 8°C·min^{-1} curve is large, it might because the beginning stages of decomposition was triggered strongly. All obtained activation energy E are much more than 220 kJ·mol^{-1}, it illustrates decomposition of HMX is probable autocatalytic.

Figure 2:Fitted and experimental curves of HMX at different heating rates

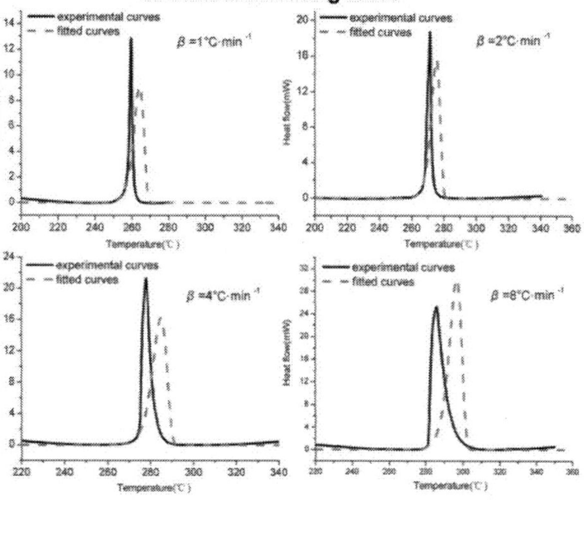

Table 2 Fitting results of dynamic DSC tests

Sample	β(°C·min^{-1})	E (kJ·mol^{-1})	ε(mW)
β-HMX	1	891.0	0.01120
	2	933.8	0.02434

	4	779.2	0.01723
	8	810.7	0.06835

β: Heating rate; E: Activation energy; ε: Deviations

3.3 Isothermal DSC experiment

Isothermal DSC experiments were performed to verify the autocatalysis of β-HMX [14]. Experimental results are shown in Figure 3. In the two experiments, a classic bell-shaped curve has been detected. It illustrates decomposition of β-HMX is autocatalytic. The induction period of 260°C isothermal DSC experiment is about 10min, and that of 240°C experiment is about 50min.

Figure 3:Isothermal DSC curves Of HMX

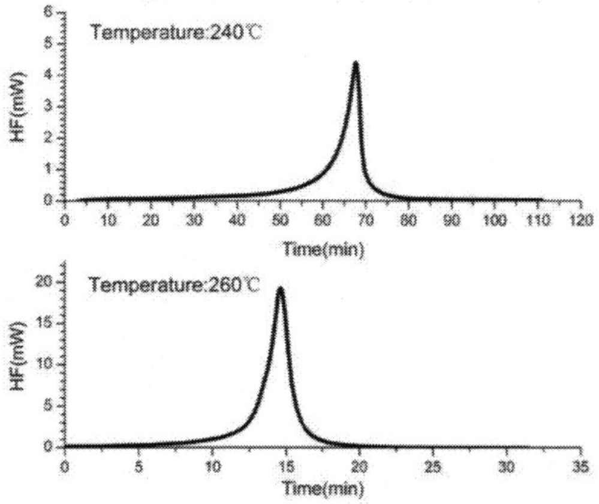

3.4 Decomposition kinetics of β-HMX

In order to determine the activation energy E, Ozawa method was used. Malek method was used to determine the most probable mechanism function, reaction order and pre-exponential factor in thermal analysis kinetics.

The activation energy from Ozawa method is 178.883 kJ·mol^{-1}, and correlation is 0.9916. For further research of the decomposition Kinetics of β-HMX, Malek method is used to calculate kinetic equation. Normalized functions $y(\alpha)$ and $Z(\alpha)$ are applied to determine most probable mechanism function $f(\alpha)$ and $G(\alpha)$ in Malek method, and this method has been proved to be effective and accurate[16].

$$y(\alpha) = \frac{d\alpha}{dt}\exp(\frac{-E}{R}) = Af(\alpha) \quad \text{[Eq. 5]}$$

$$y(\alpha) = \left(\frac{T}{T_{0.5}}\right)^2 (\frac{d\alpha}{dt}) \bigg/ \left(\frac{d\alpha}{dt}\right)_{0.5} \quad \text{[Eq.6]}$$

$$Z(\alpha) = \pi(u)(\frac{d\alpha}{dt})T \Big/ \beta = f(\alpha)G(\alpha) \quad \text{[Eq. 7]}$$

The definition formulas and calculation formulas for $y(\alpha)$ are shown in Eq. 5and Eq. 6. The expression of $Z(\alpha)$ is shown in Eq. 7. In the previously formulas, $T=T_0+\beta t$, $u=E/RT$, and the value of E is 178.883 kJ·mol^{-1}, $\pi(u)$ is Luke second approximation formula. And obtained characteristic values (α_M, α_P^∞ & α_P) are shown in Table 3, based on these characteristic values the most probable mechanism can be distinguished.

Table 3 The value of α_M、 α_P^∞ and α_P of HMX

β(°C·min^{-1})	α_M	α_P^∞	α_P
1	0.609	0.609	0.593
2	0.527	0.527	0.515
4	0.464	0.464	0.45
8	0.342	0.342	0.477

α_M, Fraction conversion at the peak of $y(\alpha)$-α curve; α_P^∞, Fraction conversion at the peak of $Z(\alpha)$-α curve; α_P, Fraction conversion at the peak of DSC curve

Judging criteria of determine most probable mechanism is proposed in reference [16], and we find the Sestal-Berggren model ($f(\alpha)= \alpha^m(1-\alpha)^n$) is most appropriate. It is the same to previous conclusion: decomposition of β-HMX is autocatalytic. Then reaction order m, n and pre-exponential factor A is obtained, and the results are listed in Table 4.

Table 4 Calculated kinetic parameters of HMX at different heating rates

β(°C·min^{-1})	n	m	lgA(s^{-1})
1	1.212	1.887	32.857
2	1.758	1.959	33.044
4	1.430	1.238	32.714
8	1.559	0.810	32.707
Average	1.490	1.474	32.841

n, m, reaction order of autocatalytic decomposition; A, pre-exponential factor

3.5 ARC experiment

The experimental results of β-HMX in ARC are listed in Table 5. Figure 4 shows the curves of temperature and pressure changing with operating time. From Figure 4 we can see that the initial reaction is mild, but become very violent in later stage, and it can be conclude from the pressure curve that during thermal decomposition a large number of gas has been released. To avoid the chance of explosion, the small sample mass is needed. Through data reduction and analysis, we find that its T_{D24} is about 211.5°C.

Table 5 ARC results of β-HMX

Thermal inertia, Φ	37.1
Sample mass(mg)	0.06
Onset temperature (°C)	225.489
End temperature (°C)	236.791
Adiabatic temperature rise (°C)	11.302
Maximal temperature rate (°C·min^{-1})	35.515
Temperature of maximal rate (°C)	236.791
T_{D24}(°C)	211.5

Figure 4:ARC curve of HMX

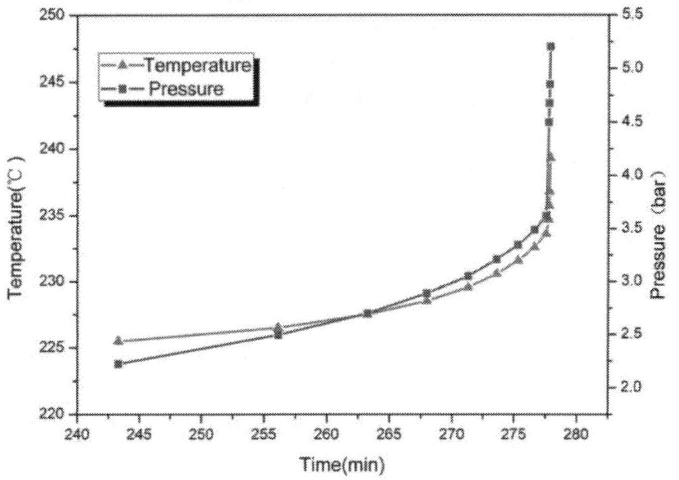

4. Conclusion

1. From dynamic DSC experiment, we know that the decomposition of β-HMX is violent, once decomposition has occurred, it could cause an accident with high explosive severity.

2. Both the results from Swiss method and Malek method show that β-HMX decomposes autocatalytic, which is vilified by the isothermal DSC experiments. The most probable mechanism function is Sestal-Berggren model ($f(\alpha)= \alpha^m(1-\alpha)^n$), and its kinetic equation is $d\alpha / dt = 10^{32.841} \times \exp(-178883 / RT)\alpha^{1.474}(1-\alpha)^{1.490}$ (these numbers are averages).

3. ARC is applied to determine the onset temperature under the adiabatic states, and onset temperature of β-HMX is 225.5°C, T_{D24} of β-HMX is 216.75°C, so the stability of HMX is good.

5. References

[1] Zhongxing Jing, Mulan Bai, "The research on the thermal decomposition of HMX with time-of-flight mass spectrometry ," *Transactions of Beijing Institute of Technology*, Volume 36, NO. 3, 1988

[2] Huiping Jia, Mulan Bai, "Thermal decomposition and autocatalysis of HMX," *Acta Armamentarii*, NO. 4, 1993

[3] Zhi Jiang, Shufen Li, Fengqi Zhao, etc., "Effect of nano metal powder on the thermal decomposition characteristics of HMX," *Environmental Science & Technology*, Volume 123, NO.13, 2002

[4] Hengjian Huang, Haishan Dong, Yuanjie Shu, etc., "Get crystal defects and its effect to thermal sensitivity and stability," *Chinese journal of energetic material*, Volume 11, NO. 3, 2003

[5] DengJr Peng, ChengMing Chang, Miin Chiu, "Thermal reactive hazards of HMX with contaminants," *Journal of Hazardous Materials*, Volume 114, NO. 1-3, 2004

[6] Ziru Liu, Yan Liu, Xiping Fan, etc., "Thermal Decomposition of RDX and HMX art I: Characteristic Values of Thermal Analysis," *Chinese journal of Explosives & Propellants*, Volume 27, NO. 2, 2004

[7] Ziru Liu, Cuimei Yin, Yan Liu, etc., "Thermal Decomposition of RDX and HMX Part II: Kinetic Parameters and Kinetic Compensation Effects," *Chinese journal of Explosives & Propellants*, Volume 27, NO. 4, 2004

[8] Ziru Liu, Yan Liu, Xiping Fan, etc., "Thermal Decomposition of RDX and HMX Explosives Part III: Mechanism of thermal Decomposition," *Chinese journal of Explosives & Propellants*, Volume 29, NO. 4, 2006

[9] Ziru Liu, Zhenhao Shi,Cuimei Yin, etc., "Investigation on Thermal Decomposition of Mixed Systems of AP with RDX and HMX by DSC-TG-FTIR," *Chinese journal of Explosives & Propellants*, Volume 30, NO. 5, 2007

[10] Yiding Shi, Fenglei Huang, "Molecular Dynamics Simulation of Isothermal Equations of State of β -HMX at Different Temperatures," *Transactions of Beijing Institute of Technology*, Volume 30, NO. 7, 2010

[11] Zhan Tang, Li Yang, Xiaojing Qiao, etc., "Kinetics of thermal decomposition of HMX and thermal safety studies," *Chinese journal of energetic material*, Volume19, NO. 4, 2011

[12] Rui Liu,Yanli Yin,Tonglai Zhang, etc., "Dynamic vacuum stability test method research (IV) : HMX thermal decomposition," *Chinese journal of energetic material*, Volume19, NO. 6, 2011

[13] Tingting Zhou, Yiding Shi, Fenglei Huang, "Thermal Decomposition Mechanism of β-HMX under High Pressures via ReaxFF Reactive Molecular Dynamics Simulations," *Acta Phys. -Chim. Sin.*, Volume28, NO. 11, 2012

[14] Francis Stoessel. Thermal Safety of Chemical Process: Risk Assessment and Process Design; Science Press: Beijing, 2009; translated by Chen, W. H, Peng, J. H., Chen, L. P.

[15] Shilong Bao,Wanghua Chen,Liping Chen, "Identification and Thermokinetics of Autocatalytic Exothermic Decomposition of 2,4-Dinitrotoluene," *Acta Phys. -Chim. Sin.*, Volume 29, NO. 3, 2013

[16] Rongzu Hu, Shengli Gao, Fengqi Zhao, Thermal analysis kinetics; Science Press: Beijing, 2008

基于 ISM 确定 HAZOP 分析顺序软件的开发和应用

王峰*，杨欧，陈存银，金旭

（北京化工大学 国家级危险化学品生产系统故障预防及监控基础研究实验室，北京市 北京 100029）

摘 要： 论文将解释结构模型和 HAZOP 分析方法相结合，开发了基于解释结构模型的 HAZOP 分析序列确定软件。该软件用于复杂化工过程中 HAZOP 分析顺序的确定以及提高了 HAZOP 分析的效率。软件的程序包括：输入设备名称、子系统和变量信息，获得子系统和过程变量关系索引值，建立子系统和过程变量的影响关系矩阵，生成子系统和过程变量的层次结构，确定复杂化工过程的分析序列。利用这个软件获得了加氢裂化反应部分的 HAZOP 分析顺序。同时，以 9 号子系统为例，系统中 12 个过程变量的 HAZOP 分析顺序也是由该软件决定的。可以肯定，该软件可以提高分析效率和使 HAZOP 分析过程更加合理。

关键词： 解释结构模型；HAZOP(危险与可操作性分析)；复杂化工过程；HAZOP 分析顺序；加氢裂化过程

[中图分类号] T **[文献标识码]** A **[文章编号]**

Development and Implementation of Software of HAZOP Analysis Sequence Determination Method Based on the Interpretive Structural Modeling

Wang Feng*, Yang Ou, Cheng Cunyin, Jin Xu

(National Foundation Research Laboratory of Fault Prevention and Hazardous

Chemicals Production System, Beijing University of Chemical Technology, Beijing 100029)

Abstract: This paper combined the interpretive structural modeling and the HAZOP analysis method. Then software of HAZOP analysis sequence determination method based on the interpretive structural modeling was developed, which is used for the HAZOP analysis sequence determination of a complex chemical process and improve the efficiency of HAZOP analysis. The procedure of this software includes: inputting plants names, subsystems and variables information, acquiring the values of the relationship indexes of the subsystems and process variables, establishing influence relation matrix of subsystems and process variables, generating a hierarchy of the subsystems and process variables, and determining the analysis sequence of the complex chemical process. Furthermore, the HAZOP analysis sequence of the reaction section of the VGO unicracking process unit is determined by using the software. In contrast with the customary analysis, this software acquired the HAZOP analysis sequence of thirteen subsystems in the reaction section of the VGO unicracking process unit. Meanwhile, the number nine subsystem was taken as an example, the HAZOP analysis sequence of twelve process variables in this subsystem was also determined by this software. It is confirmed that this software can enhance the analysis efficiency and make the HAZOP analysis process more reasonable.

Key words: Interpretive Structural Modeling; HAZOP (hazard and operability); Complex chemical process; HAZOP analysis sequence; VGO unicracking process

1 Instruction

Chemical process plants are often operated at pressure and temperature extremes to achieve optimal performance, making them more vulnerable to equipment failures. HAZOP is the most widely used method and recognized as the preferred process hazard analysis approach in chemical industry. In this analysis, P&ID can be examined by a multi-disciplinary team of experts systematically, and all conceivable deviations far from design intent in the plant can be identified and all the possible abnormal causes and the adverse consequences of those deviations can be determined (Liao, X. P. 2004)[1]. And experts will propose safeguards according to the existing protective measures, to perfect the process design.

Nowadays, the HAZOP method and its computer aided software, which are discussed in detailed in a large number of literatures and research reports, are used to identify the dangerous events and to help experts focus on the urgent and serious problems in the first time (Cui, L., Zhao,

J. S., & Zhang, R. Q. 2010; Kim, M. C., & Seong, P. H. 2006; Manuel, R., & Jose Luis, D. L. M. 2012; Wang, F., Gao, J. J., & Wang, H. Q. 2012)[2-5]. However, there are some disadvantages remaining in the practical chemical engineering application of HAZOP analysis. First of all, a complicated chemical process system consists of interconnected or interwoven parts, thus it is difficult for HAZOP analyzers to grasp and understand elaborate relationship of parts and to reach the primary reasons and the deepest dangerous consequences of an abnormal situation by direct reasoning. So subsequence determination between subsystems or nodes is avail to HAZOP meeting. Secondly, a process variable analysis may provide the basis for other variables analysis in a subsystem or node of a complex process system. Thus identification of the basic process variables is very important. Thirdly, although the above two problems are conquered, determining the sequence of HAZOP analysis by human will consume more time and resources. Therefore, it is important for HAZOP meeting that acquire sequence of HAZOP analysis fleetly.

The Interpretive Structural Modeling(ISM) approach is an analysis method which has been most widely used in modern system engineering (Huang, L. J., Yu, J., & Huang, X. W. 2012; Liu, W., &Liu, L. 2006)[6,7]. ISM divides complicated system into several subsystems, constituting a multilevel hierarchical structure model by utilizing experts' experience and computer. In this paper, ISM is combined with HAZOP and a software of HAZOP analysis sequence determination method based on the interpretive structural modeling (HAZOPISM) is introduced in detail which determines the sequence of HAZOP analysis of complicated chemical system. HAZOPISM has been developed according to the HAZOP analysis sequence determination method based on the ISM. The procedure of HAZOPISM is composed of three parts which are inputting plants names, subsystems and variables information, determining the subsystems analysis sequences and determining the variable analysis sequences. The application of HAZOPISM improves the analysis efficiency and ensures the analysis accuracy.

2 The procedure of the HAZOP analysis sequence determination method based on the ISM

The ISM procedure mainly includes procedures (1) acquiring the internal relationship of the factors of a complex system, (2) dividing the complex system into some subsystems, (3) ascertaining the process variables of each subsystem, (4) respectively acquiring the values of influence relationship indexes of the subsystems and the values of the indexes of the process variables, (5) establishing the matrixes of the subsystems and the process variables, (6) generating a hierarchy of the subsystems and process variables, (7) determining the analysis sequence of the complex process and so on. After the analysis sequence determination of a complex system, meetings of basic HAZOP analysis can be held.

In order to determine the values of the influence relation indexes, the subsystems in a system or the process variables in a subsystem should be compared with each other. There are two factors and the factors may be two subsystems or two process variables. One factor can be named the key factor and the other can be named the influence factor if a factor influences the other factor. The value of the influence relation index which is an important indicator can be used to reflect the influence extent of the influence factor on the key factor. And the influence relation indexes values, which are 0 or 1, can be estimated by experts according to their experience and the following rules.

1) The value of the influence relation index will be 0 if there is no influence relation between these two factors;

2) The value of the influence relation index will be 1 if the key factor is influenced by the influence factor.

3 Framework of HAZOPISM

The application procedure of HAZOPISM consists of inputting plants names, subsystems and variables information, determining the subsystems analysis sequences and determining the variable analysis sequences. Fig. 1 shows the analysis procedure of the HAZOPISM program.

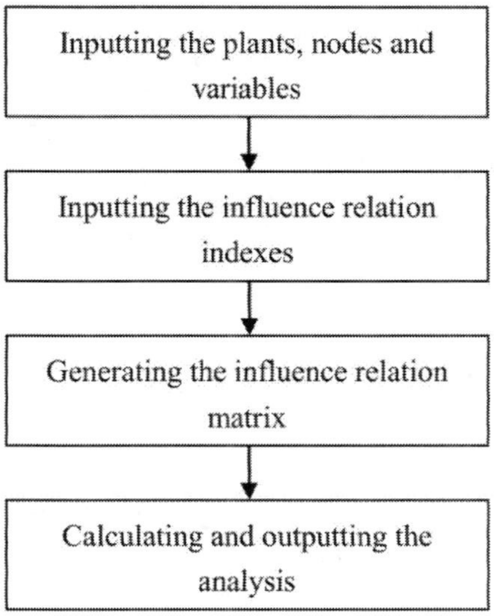

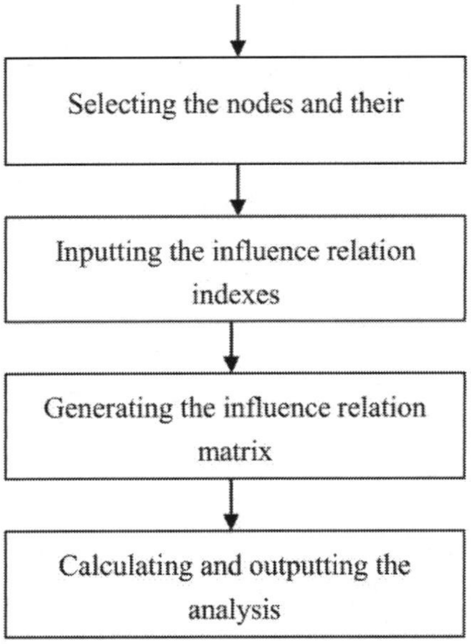

Fig. 1-The analysis procedure of $HAZOP_{ISM}$ program

4 HAZOPISM application instruction

The application of HAZOPISM mainly includes four stages.

1) Inputting the plant, subsystem and variable information

Analyzers can select a plant and decide subsystems and variables of each subsystem. The information of the plant, subsystems and variables can be inputted, deleted and modified.

2) Determining the analysis sequences of the plant subsystems

a. Inputting the influence relation indexes values (0 or 1) of the subsystems, and establishing the influence relation matrix.

b. Calculating and outputting the analysis sequence of the subsystems.

3) Determining of the analysis sequences of the subsystem variables

a. Selecting a subsystem;

b. Inputting the influence relation indexes values (0 or 1) of the variables, and establishing the influence relation matrix.

c. Calculating and outputting the analysis sequence of the variables.

4) Determining the final analysis sequences of the plant.

Combining the results of 2) and 3), the final analysis sequences of the selected plant can be determined.

5 Case study

The plant of the VGO unicracking process is one of the most dangerous plants in chemical industry. Firstly, many types of equipments are often operated at the pressure and temperature extremes, making them more vulnerable to equipment failures. Secondly, the operations are so complex that it is almost impossible for operators to identify and analyze all the abnormal situations in the first time. (Jin, D. H., Liu, J. H.,& Shen, T. 2006)[8]. Therefore, the HAZOP analysis and risk identification of the complex process are worth of being done for the accident preventions and the HAZOP analysis based on the ISM is a necessary method for them. Fig.2 shows the flowchart of the reaction section of this plant. Raw oil is mixed with hydrogen from PSA and raw materials for fractionation are prepared after a series of heat transfer and reactions.

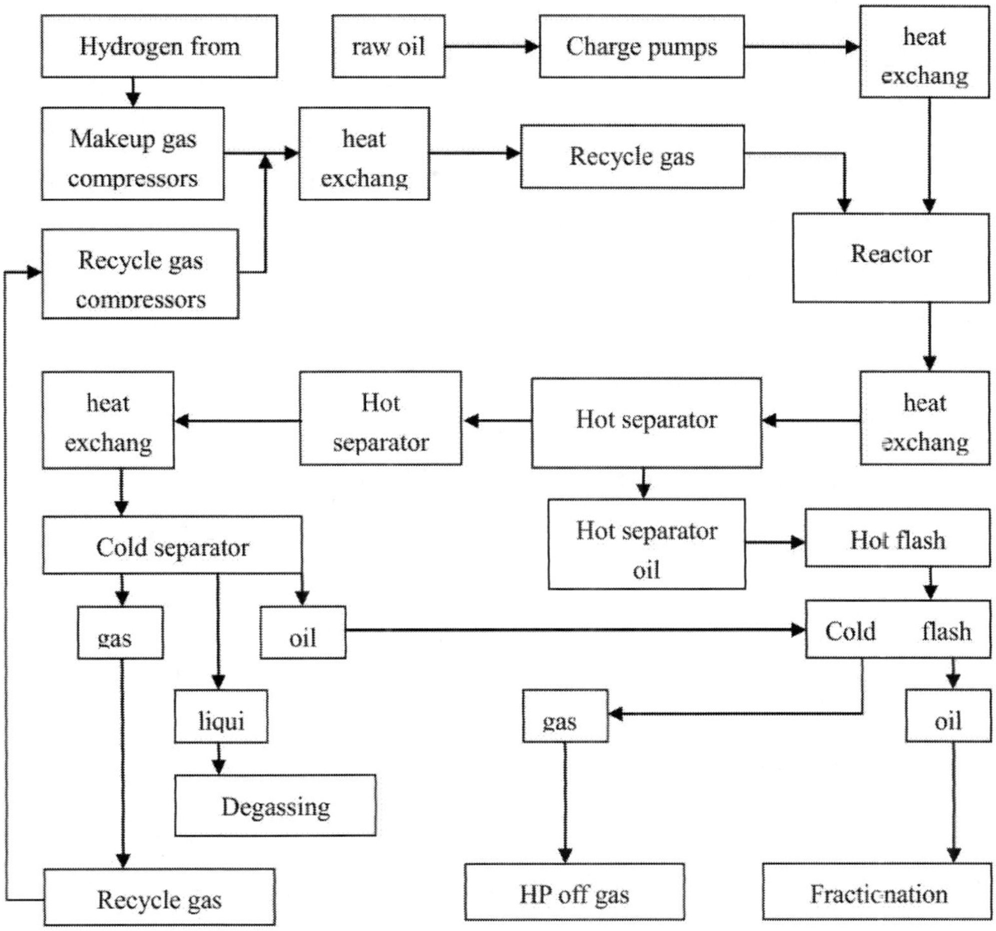

Fig.2- the flowchart of the reaction section of VGO unicracking

5.1 Subsystem determination

As shown in Table 1 and Fig.3, the reaction section of the VGO unicracking process can be divided into 13 subsystems and the subsystem contents can be written down into the database of the program. The arrows in table stand for material flow.

Subsystem order	Contents of subsystems
1	Feed filter S201□Feed surge drum V201□Charge pumps P201
2	Hot separator vapor-feed cold exchanger E209□Hot separator vapor-feed hot exchanger E206□Effluent-feed cold exchanger E204□Effluent-feed hot exchanger E202
3	Guard reactor R201□Reactor No.1 R202 □Reactor No.2 R203
4	Effluent-recycle gas hot exchanger E201□Effluent-feed hot exchanger E202□ Effluent-recycle gas cold exchanger E203□Effluent-feed cold exchanger E204□Effluent-stripper bottoms exchanger E205
5	Hot separator V202: hot gas□Hot separator vapor-feed hot exchanger E206□ Hot separator vapor-recycle gas hot exchanger E207□Hot separator vapor-stripper feed hot exchanger E208□Hot separator vapor-recycle gas cold exchanger E210□Hot separator vapor-stripper feed hot exchanger E211□ Hot separator vapor condenser EA202
6	Hot separator V202: Hot oil□Hot flash drum V203：Gas□Hot flash vapor condenser EA201□Cold flash drum V205: Oil□Fractionation section
7	Cold flash drum V205：Oil□Hot separator vapor-stripper feed hot exchanger E211□Hot separator vapor-stripper feed hot exchanger E208□Fractionation section
8	Cold flash drum V205 Hot separator vapor-feed hot exchanger E206: Gas□HP off gas scrubber C201：Gas□HP treated offgas knockout drum V207: Liquid□ Rich amine flash drum V217
9	Cold separator V204：Gas□Recycle gas scrubber C202□Recycle gas knockout drum V208: Oil□Cold flash drum V205: Water□Degassing Drum V206
10	Recycle gas compressors K201
11	Makeup gas compressors K202

12	Hot separator vapor-recycle gas cold exchanger E210□Hot separator vapor-recycle gas hot exchanger E207□Effluent-recycle gas cold exchanger E203□Effluent-recycle gas hot exchanger E201
13	Recycle gas heater F201

Table 1-Contents of the subsystems

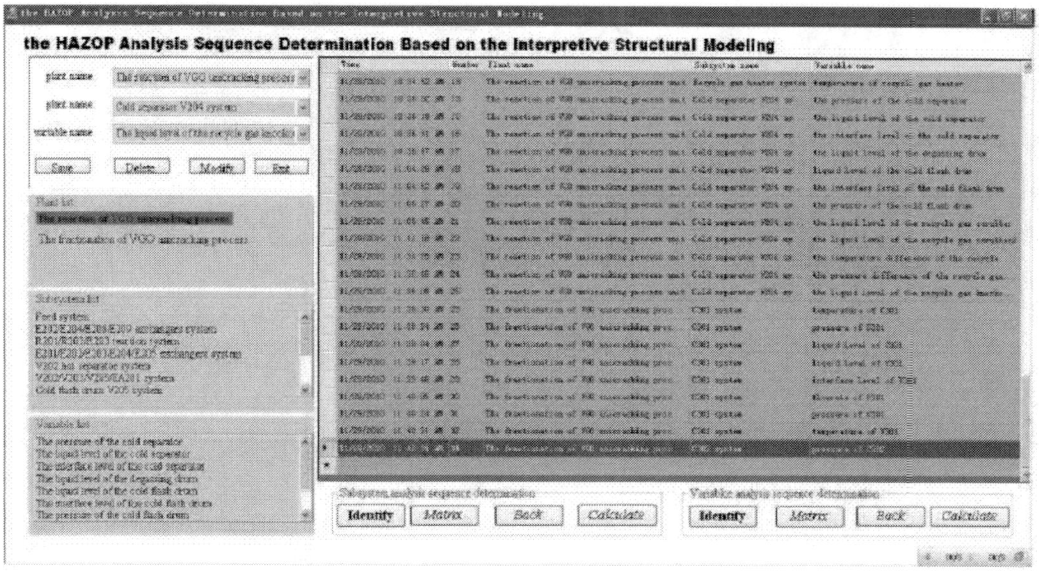

Fig.3-The program interface

The subsystem names can be listed in the “Subsystem list” space, and the subsystem names that belong to the plant will be shown in the “Subsystem list” space after a plant name in the “Plant list” space is selected and clicked.

5.2 The influence relation matrix of the subsystems

As shown in Fig.4, the values of the influence relation indexes can be determined according to the experts’ experiences, and then an influence relation matrix can be generated. The flowchart of the HAZOP analysis sequence of the subsystems can be formulated. The flowchart of the HAZOP analysis sequence needs modification if more than two subsystems influence their key subsystem. The values of the influence extent can be compared with each other. The modification can be done according to the following rules.

1) The influence relation index values will be more than 0 and less than 1 if more than two subsystems influence their key subsystems.

2) The more influence extent the subsystems are, the greater the values of the influence relation indexes of the subsystems will be.

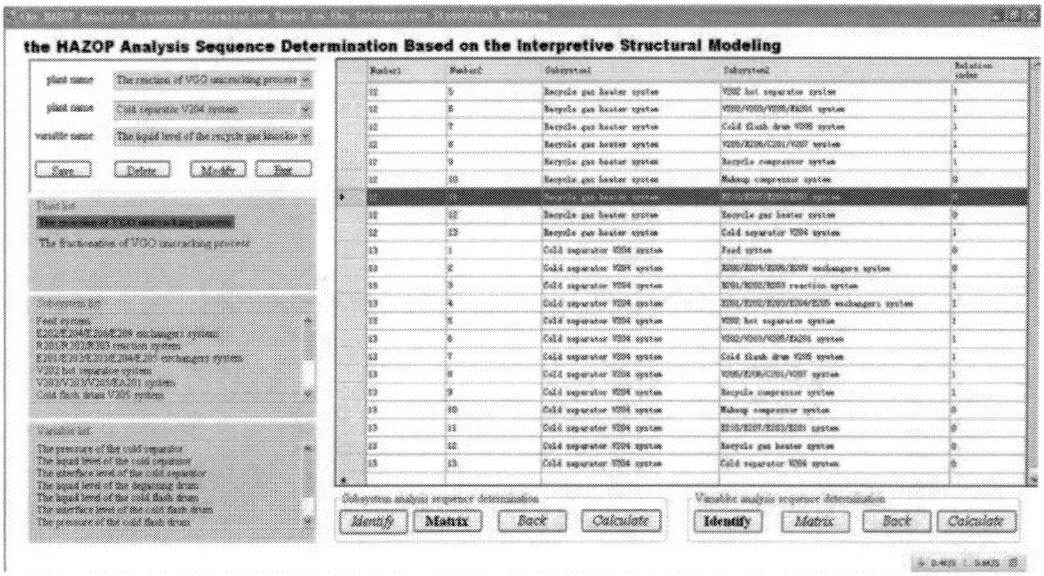

Fig.4-Value input of the relation indexes of the subsystems

According to Fig.4, the HAZOP analysis sequence of the subsystems can be determined: Subsystem10→Subsystem11→Subsystem12→Subsystem13→Subsystem1→Subsystem2→Subsystem3→Subsystem4→Subsystem5→Subsystem9→Subsystem6→Subsystem7→Subsystem8.

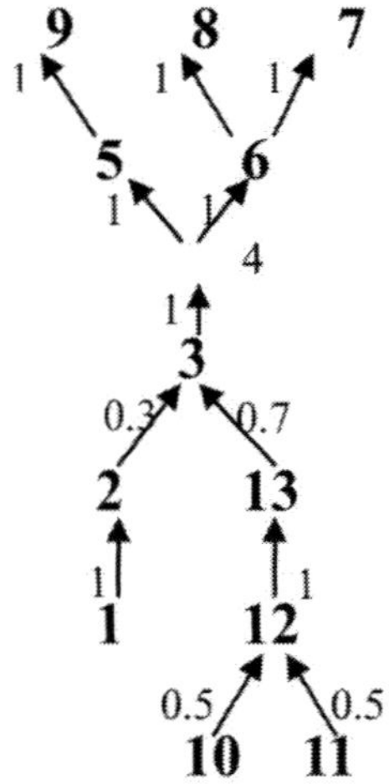

Fig.5-Analysis sequence of the subsystems

5.3 Determination of the analysis sequence of the variables of the subsystems

The 9th subsystem is taken as an example to express the analysis sequence determination of the process variables in a subsystem, and the analysis sequence of other subsystems can be determined according to the following example.

It can be seen from Fig.5 that the equipments of the 9th subsystem mainly includes cold separator, recycle gas scrubber, recycle gas knockout drum, cold flash drum and degassing drum. And the variables of the subsystem include (1) the pressure of the cold separator; (2) the liquid level of the cold separator; (3) the interface level of the cold separator; (4) the liquid level of the degassing drum; (5) liquid level of the cold flash drum; (6) the interface level of the cold flash drum; (7) the pressure of the cold flash drum; (8) the liquid level of the recycle gas scrubber; (9) the liquid level of the recycle gas scrubber2; (10) the temperature difference of the recycle gas scrubber; (11) the pressure difference of the recycle gas scrubber; (12) the liquid level of the recycle gas knockout drum.

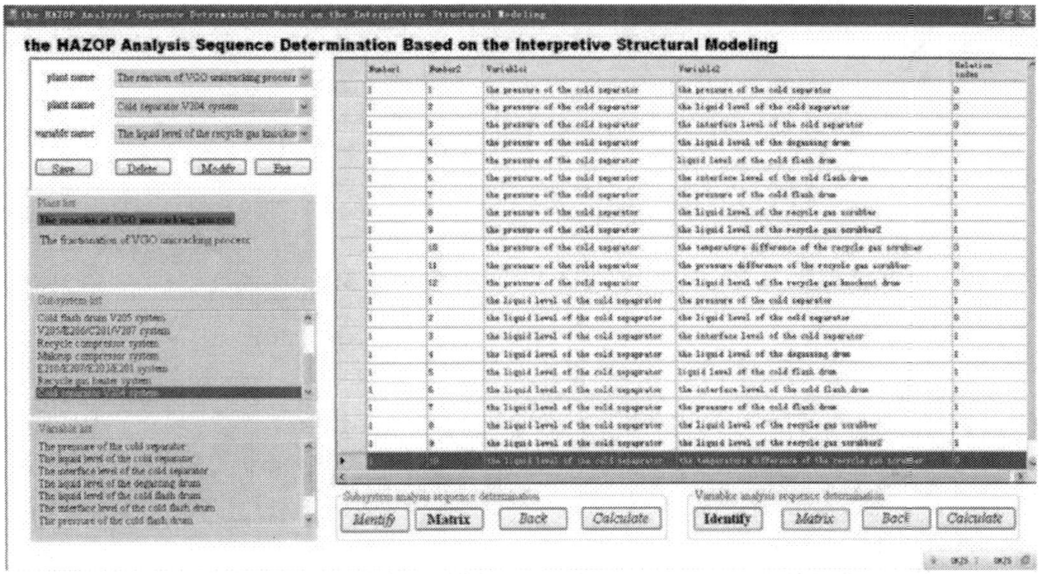

Fig.6-Value input of the relation indexes of the variables

According to Fig.6, the HAZOP analysis sequence of the 9th subsystem can be determined:variable2→variable3→variable1→variable8→variable4→variable5→variable6→variable7→variable92→variable10→variable11→variable12.

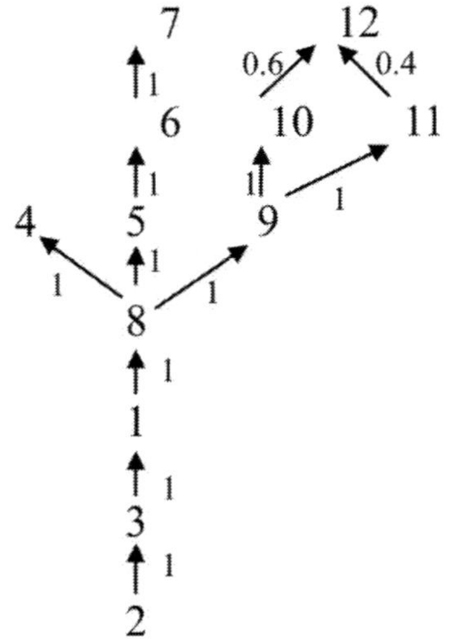

Fig.7-Analysis sequence of the variables

After the sequence determination of the subsystems and all the process variables, the final HAZOP analysis sequence can be acquired.

6. Conclusions

The HAZOP analysis sequence determination based on the ISM which provides a reference method for the analyzers, not only improves the analysis efficiency, but also ensures the analysis accuracy. The program plays an important role in guiding analysis, calculating matrixes and reducing the workload of the analyzers. However, the HAZOP analysis refers to the process, operations, environment, instrument and control subject, and there are too many unknown factors exiting in the complex system, so that the analyzers should widely listen to the opinions of the experts in the method application. It is necessary for certain process variables or subsystem analysis to be postponed to get more information later in the analysis.

Acknowledgment

This work was supported by the National Key Technologies R&D Program, Grant No. 2011BAK06B02 & the Major State Basic Research Development Program of China (973 Program) Projects, Grant No.2012CB026000 & the Chinese universities scientific fund, Grant No. ZZ1309.

References

[1] Liao, X. P. *Chemical Process Hazard Analysis*, Chemical Industry Press, city: Beijing,2004:12-63.

[2] Cui, L., Zhao, J. S. and Zhang, R. Q.The integration of HAZOP expert system and piping and instrumentation diagrams. *Process Safety & Environmental Protection*,2010, (5): 327-334.

[3] Kim, M. C. and Seong, P. H. (2006). A computational method for probabilistic safety assessment of I&C systems and human operators in nuclear power plants, *Reliability Engineering & System Safety*,2006, (5), : 580–593.

[4] Manuel, R. and Jose Luis, D. L. M. Automating HAZOP studies using D-higraphs, *Computers & Chemical Engineering*, 2012:102-113.

[5] Wang, F., Gao, J. J. and Wang, H. Q. A new intelligent assistant system for HAZOP analysis of complex process plant. *Journal of Loss Prevention in the Process Industries*, 2012,(3): 636-642.

[6] Huang, L. J., Yu, J. and Huang, X. W. (2012). Modeling agricultural logistics distribution center location based on ISM, *Journal of Software*,2012, (3), :638-643

[7] Liu, W. and Liu, L. Risk analysis of knowledge system based on ISM, *Computer Engineering and application*, 2006,(14):134-136.

[8] Jin, D. H., Liu, J.H. and Shen, T. (2006). *Technologies ask and answer for the VGO unicracking process unit.* Chemical Industry Press, city: Beijing ,2006,103-110.

An Overview of Modern Calorimetric Techniques for Energetic Materials and Runaway Reaction Studies

J. Zhou
Center for Hazardous Chemical Safety
College of Chemistry and Chemical Engineering
Central South University
Hunan Changsha, China

flwu@mail.csu.edu.cn

Prepared for Presentation at
American Institute of Chemical Engineers
1st CCPS Asia-Pacific Conference on Process Safety
Qingdao, China
September 4 – 5, 2013

UNPUBLISHED

An Overview of Modern Calorimetric Techniques for Energetic Materials and Runaway Reaction Studies

J. Zhou, P. Fu, Z. Ming and F. Wu*
Center for Hazardous Chemical Safety
College of Chemistry and Chemical Engineering
Central South University, Changsha, Hunan 410083 P.R.China
flwu@mail.csu.edu.cn

Keywords: Process Safety, Thermal Hazard, Reactive Chemical, Microcalorimetry, Reaction Calorimeter, Adiabatic Calorimeter

Abstract

Accurate knowledge of modern calorimetric techniques and proper calorimetric testing strategies are essential for the successful operation of energetic materials and process safety testing laboratories. Although many tools and techniques have been introduced in open literature, most safety scientists are not knowledgeable about modern calorimetry. This presentation will review the most recent calorimetric techniques and scenario-based testing protocols for energetic materials/process safety assessment. Key issues in early-stage process safety evaluation will be discussed, but emphasis will be given to the issue of calorimetric data scalability.
Since instrument developments in calorimetry science are usually not reported in primary literature, and calorimetric experiments are usually performed with whatever instruments that are available, a guide for choosing the best calorimeter for a given task will be given. In this guide, the data scalability and applicability of various calorimetric methods (RC1, SuperCRC, C80, DSC, ARC, DARC, VSP, etc.) are reviewed. Examples of the use and misuse of calorimetric tools and data will be demonstrated.

1. Introduction

Runaway reactions in chemical operations always involve energetic but unstable chemical materials, such as organic synthesis in reaction vessels, unit operation, and chemical storage and transportation. Dealing with process safety pressure is a major challenge for most chemical companies, but the right safety testing strategy may aid them. There has been an increase in recent years of the use of modern calorimetry on complex, often catalyzed organic reaction systems and lithium-ion battery systems.

Accurate knowledge of modern calorimetric techniques and proper calorimetric testing strategies are necessary for evaluating the risk of thermal runaway. One of the important aspects in thermal runaway reaction assessment is to understand the limits of the applicability of the thermoanalytical techniques. Although a large number of literatures on reaction calorimetric testing exists, key issues, such as calorimeter applicability and data scalability, still remain unclear to many testing laboratories.

In this paper, we will explain why most near-adiabatic and isothermal calorimeters and temperature programmed methods may not be suitable for obtaining sound kinetic information. In accordance with current laboratory practices, we propose some effective thermal hazard evaluation methodologies.

2. The Scenario-based thermal testing protocols

Generally, the scenario-based thermal hazard testing categorizes into two distinct, but related fields, a) reaction processes, and 2) reactive chemicals. Regarding the first field, the reaction process related runaway scenario can be described as a progressive temperature rise of the reaction mixture due to a heat generation rate (proportional to the reaction rate) which exceeds the cooling capacity of the reactor. And, if the energy release is large enough, gaseous products may form, which can lead to over-pressurization and possible mechanical destruction of the vessel (thermal explosion). The maximum energy release may come from the desired organic synthesis reaction, reaction mass, and/or thermodynamically unstable solvent system. Regarding the second field, the reactive chemical related runaway scenario, mostly happens during the hot-stacking, storage, and transportation of chemically unstable compounds. Severity and probability of this type of the runaway reactions can be described by both adiabatic temperature rise, ΔTad, and the time-to-maximum rate under adiabatic conditions, TMRad.

Depending on various stages of the chemical process, different testing protocols/schemes have been introduced[1,2]. Here, we propose a revised Gygax runaway reaction model with proper calorimetry testing, as illustrated in Figure 1. The revised Gygax model represents a two-stage runaway reaction encountered mostly in manufacturing safety incidents. The key issue of safety testing based on the revised Gygax model is to define the critical region between the two stages, i.e. how high the temperature that the uncontrolled target reaction can go (MTSR), and how low the temperature where the secondary runaway decomposition will kick in (Tonset).

To define the first stage of the model, isothermal heat-flow reaction calorimetry at the process temperatures are recommended in order to obtain the so-called Maximum Temperature of the Synthesis Reaction, MTSR (= Tproc + Xc*CpΔH). To define the second stage of the model, adiabatic calorimetry is recommended. To accurately measure TMRad, the most advanced, differential power-compensation accelerating rate calorimeter (dARC by OmniCal) is highly recommended. Since a typical process acceptable TMRad of 8~24 h provides a critical time for taking appropriate measures to prevent a runaway reaction, estimated TMRad values by either the dynamic DSC method or the simple model-based ARC phi-correction are not reliable for most complex runaway chemical systems.

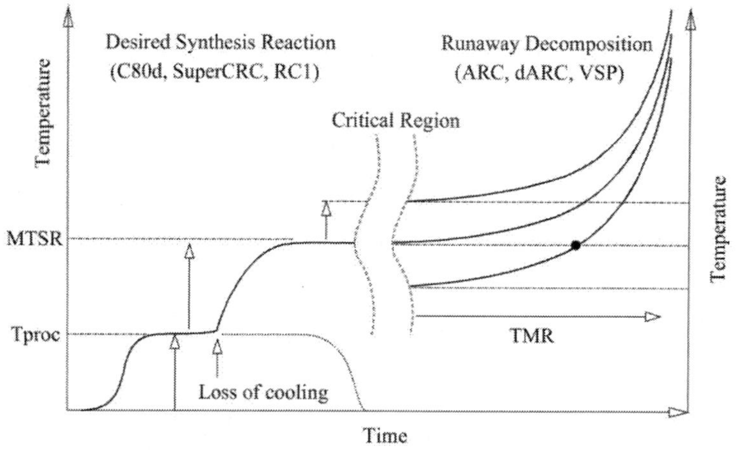

Figure 1 Scenario-based calorimetric testing based on the revised Gygax runaway reaction model.

3. Modern calorimetric techniques

In the current safety testing community, three main categories of calorimeters are mostly used to determine the severity (max. energy release) and probability (TMRad) of a thermal runaway event: temperature scanning calorimeters, isothermal reaction calorimeters, and pseudo-adiabatic calorimeters. Since instrument developments in calorimetry science are usually not reported in primary literature, calorimetric experiments are usually performed by investigators with limited knowledge on how to choose a right calorimeter for a specific task. As a result, only commercially available calorimeters, but not the best instruments, are used. Here, we provide an overview of various commercially available calorimetric devices with their pros and cons in safety testing.

The temperature scanning calorimeter is a temperature-programmed calorimeter among which the differential scanning calorimeter (DSC) is the most popular. However, DSCs have been designed mainly to study thermophysical or thermochemical properties of materials with a typical sample size in the range of milligrams. The major disadvantage of a DSC, from a thermal hazard screening perspective, is its inability to provide pressure rise information.

There is a new proprietary differential scanning calorimeter (pDSC by OmniCal) that overcomes DSC limitations as a gram-scale DSC with sealed metal 2ml vessels and pressure-tracking up to 2000psi. This type of DSC device has an apparent advantage in identifying runaway pressure excursion that may cause catastrophic plant damage. To evaluate this pDSC, an organic peroxide decomposition experiment was performed, as shown in Figure 2.

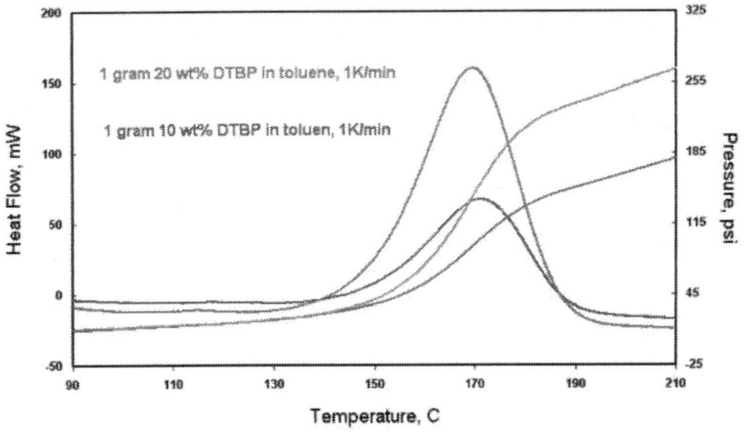

Figure 2 Di-tert-butyl-peroxide in toluene using OmniCal pDSC with simultaneous pressure-tracking

Isothermal mixing/reaction calorimeters are mainly used for monitoring time-resolved reaction processes. While liter-scale reactions are normally tested on single channel reaction calorimeters, gram-scale reaction studies are performed on isothermal microcalorimeters based on the differential calorimetry principle. For the desired organic synthesis that requires a reaction calorimeter evaluation, the liter-scale heat-flow reaction calorimeter (Mettler RC1) and the ml-scale heat-flow reaction microcalorimeters (OmniCal SuperCRC and Setaram C80d) are recommended.

The liter-scale reaction calorimeter RC1 is actually a calorimetric stirred-tank reactor, which can be operated in batch or semi-batch mode under isothermal and pseudo-adiabatic conditions. The heat flow measurement on RC1 is based on the Newtonian heat-transfer principle $d/Q/dT = U \cdot A \cdot (Ti - Tj)$ where U is the heat transfer coefficient and A is the heat exchange area. To correctly apply this calorimetric principle to a jacket reactor, the jacket fluid temperature must be homogeneous all the way through the inner jacket from the fluid inlet to outlet, which requires a high-power jet stream of the heat transfer fluid. While there are a number of similar reaction calorimeters, none of these calorimeters truly meet the jet stream criteria to apply the Newtonian cooling law. Although it can be used to construct the first part of the Revised Gygax model, the liter-scale reaction calorimeter RC1 is not suitable for measuring energy release rate in reactions under runaway conditions to construct the second part of the model. The RC1 operation, itself, imposes a serious safety issue to its operator since reactions at liter-scale have their runaway hazard potential.

The gram-scale reaction microcalorimeters, SuperCRC and C80d (15ml vessel size in a confined calorimeter chamber), are more suitable for both desired synthesis and runaway reaction hazard evaluation. Although the C80d can reach higher temperatures (30~300C), the SuperCRC (-80~200C) is more versatile for mixing and reaction studies due to its internal magnetic stirring and multiple dosing capabilities. In regards to user convenience and operator safety, the gram-scale reaction microcalorimeter SuperCRC is more useful than other microcalorimeters.

When choosing between various types of isothermal microcalorimeters for reaction kinetics investigation, attention must be given to the instrument cell-time constant or thermal-lag constant, a critical parameter that determines if the experimental heat flow data is eligible for kinetics calculation and scalable for reaction engineering. As shown in Figure 3, the three experimental heat flow curves (solid) at the three temperatures indicate significant distortion due to the thermal lag of the calorimeter system, in comparison to the dynamically-corrected heat flow curves (dotted). Kinetic evaluation from peak rates influenced by this thermal lag results in a much smaller and erratic Ea value, also shown in the small graph of Figure 3. The SuperCRC is the only microcalorimeter device capable of eliminating the thermal lag effect and dynamically correcting experimental heat flow curves (thermograms) back to its origin (thermogenesis). This mixing/reaction microcalorimeter has recently gained its worldwide reputation in tracking fast reaction kinetics[3-5].

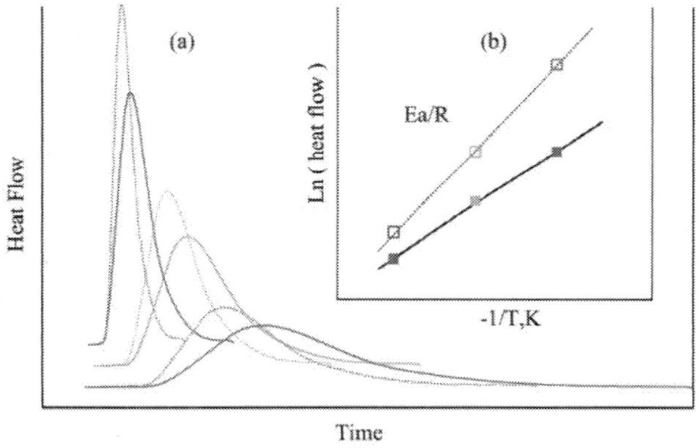

Figure 3 Thermogenesis (dotted lines) vs. thermograms (solid lines) by isothermal microcalorimetry

Adiabatic calorimeters are used to assess the severity (ΔTmax) and probability (TMRad) of self-perpetuated thermal runaway reactions. Conventionally, an adiabatic device is a calorimeter that can eliminate heat leak between the sample (including container) and the environment (adiabatic jacket). However, even if the heat leak can be completely eliminated, a large amount of reaction heat will sink into the sample container. Before 2010, no adiabatic calorimeter was a 100% true adiabatic calorimeter, because the sample holder always absorbs the reaction heat, causing a smaller temperature rise and a much longer TMR.

To study self-perpetuating (self-heating) reactions, an adiabatic reaction calorimeter, called the accelerating rate calorimeter (ARC), was developed by Dow Chemical in late 1970s[6]. The Dow ARC is actually a pseudo-adiabatic device, since a large portion of the heat released by sample is absorbed by or sinks into ARC sample container. Because of this sample container heat-sink effect, the final temperature rise is reduced and possible reactions at higher temperatures may have been missed. Also, because of this heat-sink or "adiabatic thermal lag" effect, the rate of temperature rise is also significantly reduced, which seriously slows the reaction progress in the ARC calorimeter and causes an erratic TMR.

From the mid 1980s, a few other near-adiabatic devices were developed, such as VSP and similar devices with their thinner vessel and large-volume sample to reduce the heat sink effect (phi-factor) to compensate for the loss of thermochemical sensitivity. This hardware approach does partially mitigate the thermal dilution effect, but further caused a host of other problems, such as: 1) loss of thermophysical sensitivity due to inadequate adiabatic temperature control, and 2) an addition of a costly and complex pressure-balancing system, as seen in the VSP and other similar devices.

In addition, there are a number of quick and easy screening devices (TSu by HEL, BPC by THT, etc.) which produced flawed thermograms quickly, and the VSP and dARC, which produces a better or true thermogram at a greater cost, having little in common in terms of calorimetry science and reaction kinetics. In these quick and easy devices, one can easily find violations of physiochemical laws in the heat-compensation calorimeters, as they place the internal cartridge compensation heaters (VariPhi by Netzsch) or coiled compensation heaters (Simular by HEL) in the mid of sample where their high heat density surfaces provide an ideal hot spot to catalyze reaction acceleration. Such heater placement muddles the kinetics and reaction time-clock, and demonstrates a general lack of calorimetry fundamentals.

A new development, the OmniCal dARC, a differential accelerating rate calorimeter, obtains a true adiabatic state by a differential compensation feedback circuit, and therefore, the theoretical maximum temperature rise and TMRad can all be experimentally measured without thermal dilution and unsubstantiated phi-correction. While both the ARC and the dARC have a similar detection limits (thermophysical sensitivity), the thermal dilution effect of the ARC phi-factor causes the preceding reaction to be almost undetectable and loses its thermochemical sensitivity. Significance of the emerging dARC technology is its realization of 100% adiabaticity. In theory, if a sample system becomes 100% adiabatic, i.e. neither heat-leak from nor heat-sink into the sample, its adiabatic temperature rise becomes both sample mass- and container-independent (i.e. $\Delta T = (m*Q)/(m*Cp) = Q/Cp$), which is clearly demonstrated in Figure 4. This mass-independent adiabatic calorimeter is incredibly simple for operation and data reduction. It allows even the most unskilled chemists to reliably produce specialist-caliber runaway kinetic results without the need of complicated phi-corrections, because sample weight and container weight are both no longer irrelevant to the testing.

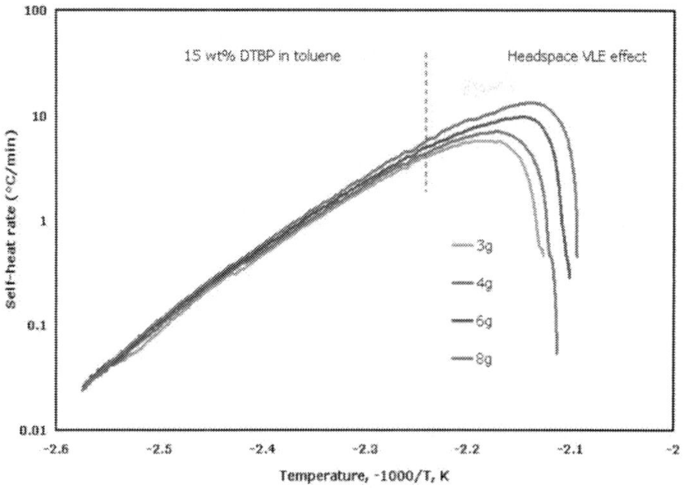

Figure 4 Size of the sample becomes irrelevant to the dARC operation

Note that the upper of the curves was influenced by the VLE-driven phase change in the different sizable head-spaces

Also, the dARC, using its differential compensation heaters, eliminates the thermal dilution of the sample container and allows much quicker reaction acceleration that result in a true TMRad, as shown in Figure 5.

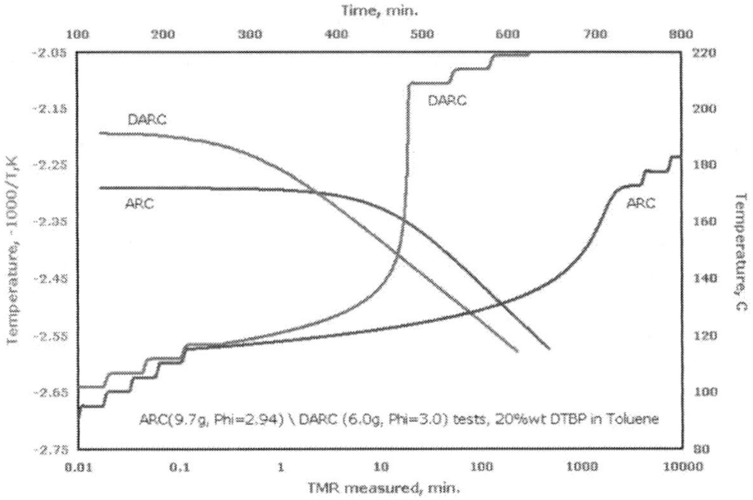

Figure 5 Organic peroxide decomposition on ARC and dARC

Note that the original phi value of the DARC sample plus cell system is 3.0. After the compensating heat is applied to the DARC cells, the phi becomes exactly equal to 1.00.

Regarding the onset temperature detection limits and maximum temperature rises, a results comparison of the dARC and a Round-Robin test[7] of 15%wt di-tert-butyl-peroxide (DTBP) in toluene on various quasi-adiabatic and near-adiabatic calorimeters are shown in Table 1. All near-adiabatic (Phi-Tec and VSP) and quasi-adiabatic (ARC) tests were performed by AEN (Adiabatic Experience Network) member organizations. Results of the dARC runs on heat-soak-search™ (HSS) modes with 10%wt and 20%wt DTBP in toluene are also shown in Table 1. Note that the values of TMR24 are all experimentally measured on dARC and no phi-correction is ever needed.

Table 1 Results comparison of the dARC and Round-Robin tests[7] and the darker numbers indicate the values were experimentally measured.

Calorimeter	in toluene	Tonset	Ea	ΔT	TMR24	Pmax	# of expts
	20%	**109.5**	155.9	**116.7**	**97.9**	**495.8**	6
2 dARC	15%	**114.5**	154.8	**88.3**	**100.1**	**308.1**	7
3 CSI ARC	15%	**117.0**	157.0	88.0			3
3 Phi-TEC	15%	**128.0**	158.0	93.0			3
2 Dewar	15%	**120.0**	155.0	103.0			3
	10%	**115.5**	153.6	**58.7**	**103.8**	**168.2**	11

4. A Combined isothermal microcalorimeter and adiabatic reaction calorimeter approach

Regardless which types of temperature sensors are used, none of the current pure-adiabatic and near-adiabatic calorimeters have reliable thermophysical sensitivity or heat rate detection limits better than 0.005C/min. As far as thermophysical sensitivity is concerned, this detection limit can be easily met or surpassed by using heat flow microcalorimeters. To determine a heat flow at or below 0.005C/min on a microcalorimeter, however, the conventional drop-in-and-measure microcalorimetry approach should not be used. At such a small heat rate of a thermal runaway during its initial stage (low conversion), no one can easily separate the baseline from its pseudo-zero order heat flow curve. Instead, a so-called switching microcalorimetry approach should be used, a niche technique with virtually no penetration into most safety testing laboratories, as shown in Figure 6.

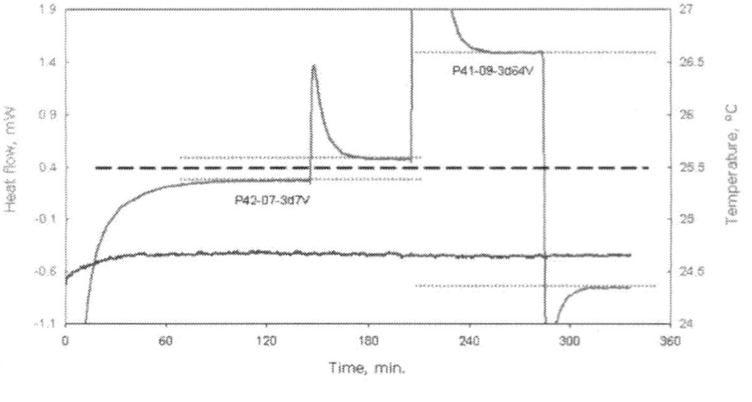

Figure 6 Self-discharge study of lithium-ion battery cells using the switching technique on SuperCRC microcalorimeter

In this switching microcalorimetric experiment, a dummy battery cell was used for the reference side of the OmniCal SuperCRC microcalorimeter for heat flow microcalorimetric tests. To convert the SuperCRC heat flow data (microwatt) to dARC self-heat rate data (K/min), the heat capacity of the battery cell was determined using an auxiliary drop calorimetric device of the SuperCRC microcalorimeter.

5. Summary

Chemical process situations are often complicated, nonetheless, the recommended calorimetric strategy based on the Revised Gygax model can be successfully applied to exclude or reduce the occurrence of thermal runaway incidents. Although no specific type of calorimeter can perform all aspects of thermal hazards at all stages, it must be emphasized that the issue of data scalability should be considered regardless of which device is chosen. Thermal dilution in near-adiabatic calorimetry and thermal lag in isothermal calorimetry must be removed before the obtained data can be utilized for sound kinetics and engineering calculations.

To define a two-stage Gygax runaway scenario, a minimum of two calorimeters are needed, i.e. an isothermal reaction calorimeter for stage-1 desired reaction process and an adiabatic calorimeter for stage-2 unwanted runaway reaction. In the view of the authors, the following main features should be considered prior to acquiring any calorimeter for academic or corporate safety testing laboratory:

<u>Differential scanning calorimeter</u>

Consideration: gram-scale pressure-tracking vs. mg-scale

Choice: DSC, pDSC

<u>Isothermal reaction calorimeter</u>

Consideration: ml-scale magnetic stirring vs. liter-scale agitation

Choice: RC1, SuperCRC

<u>Adiabatic reaction calorimeter</u>

Consideration: 100% adiabatic with no sample prep vs. near-adiabatic with phi-correction

Choice: dARC, VSP

Safety testing communities need to establish calorimetric testing protocols based on sound calorimetric principles to study reaction processes and reactive chemicals. Process safety is an area where many chemical companies feel there is much room for improvement, yet many companies place this opportunity second to other concerns. When all companies are thinking about emerging technologies, there's a danger that you will miss the boat without sufficient investment. Whatever happens, though, one thing is sure, investing the right calorimetry is costly, but companies that fail to look forward will almost certainly be left behind in an increasingly competitive, globalized world.

6. Acknowledgement

The authors would like to thank Prof. Wu's previous Dow Chemical colleague, Dr. James C. Tou for helpful discussions.

7. References

[1] Gygax, R. "Chemical Reaction Engineering for Safety", Chem. Eng. Sci., 1988, 43, 8, 1759-1771.

[2] Frurip, D. "Computational Thermochemistry in Support of Calorimetry for Reactive Chemicals Assessment", Chim. Ind. 2006, 88, 3, 140.

[3] Klingensmith, L.M.; Strieter, E,R.; Barder, T.E.; Buchwald, S.L. "New Insights into Xantphos/Pd-Catalyzed C-N Bond Forming Reactions: A Structural and Kinetic Study", Organometallics, 2006, 25, 82-91.

[4] Duan, W.L.; Iwamura, H.; Shintani, R.; Hayashi, T. "Chiral Phosphine-Olefin Ligands in the Rhodium-Catalyzed Asymmetric 1, 4-Addition Reactions", J. Am. Chem. Soc. 2007, 129, 2130-2138.

[5] Hansen, K.B.; Rosner, T.; Kubryk, M.; Dormer, P.G.; Armstrong, J.D. "Detection and Elimination of Product Inhibition from the Asymmetric Catalytic Hydrogenation of Enamines", Org. Lett. 2005, 7, 22, 4935-4938.

[6] Townsend, D.I.; Tou, J.C. "Thermal Hazard Evaluation by an Accelerating Rate Calorimeter", Thermochimica Acta, 1980, 37, 1-30.

[7] Kersten, R.J.A.; Boers, M.N.; Stork, M.M.; Visser, C. "Results of a Round-Robin with di-tertiary-butyl peroxide in various adiabatic equipment for assessment of runaway reaction hazards", J. Loss. Prev., 2005, 18, 145-151.

Risk Factor Identification of the Risk Evaluation for Industrial parks along the Yangtze Rive

Li ZHANG[1]
Jiangsu Province academy of safety science and technology
Nan Jing 210042, China
llizhang@163.com

Ping BAI [2]
[2]**Jiangsu Province academy of safety science and technology**
[2]**NanJing 210042, China**
[2]**bpnihao@163.com**

Prepared for Presentation at
American Institute of Chemical Engineers
1st CCPS Asia-Pacific Conference on Process Safety
Qingdao, China
September 4 – 5, 2013

UNPUBLISHED

Risk Factor Identification of the Risk Evaluation for Industrial Parks along the Yangtze River

Li ZHANG[1]
Jiangsu Province academy of safety science and technology
Nan Jing 210042, China
llizhang@163.com

Ping BAI [2]
[2]Jiangsu Province academy of safety science and technology
[2]NanJing 210042, China
[2]bpnihao@163.com

Abstract

A large number of industrial parks are distributed along the Yangtze River of China. The safety and environmental problems of the industrial park are increasingly prominent. The structure, identification method and identification basis of risk factor in risk evaluation for industrial parks along the Yangtze River are advanced. From three fields, the framework of identification is put forward, which are risk factor of production activity accident, risk factor of environment and ecology, risk factor of social response and rescue. The identification process and detailed composition of three fields are described briefly. The study of author has proper reference significance for similar evaluation.

1. Introduction

According to statistics, seven provinces and two cities along the Yangtze River take up more than 41% of the total economy in our country. Among them, the steel production accounted for 36% of the country, the petrochemical production 28%, the output of cars 47%, and the thermal power installed capacity 16%. The Yangtze River water transport took for the 80% of iron ore's transportation, 72% of the crude oil's transportation, and 83% of the thermal coal's transportation required by the production of the enterprise along the Yangtze River[1]. As China's economy continues to grow, its increasingly powerful transport capacity made a batch of emerging industries economic zone on both sides along the Yangtze River formed contiguously in a riverside development boom. Along the Yangtze River, ports and terminals densely covered, production device stand in great numbers. In Jiangsu province as an example, up to August 2007, there are 50 industrial parks at or above the provincial level along the Yangtze River in Jiangsu province, including three national economic and technological development zones and four national high-tech industrial development zones[2]. All kinds of industrial parks approved by city, county and township governments have more than 1200[3-4].

In recent years, frequent serious accidents occur frequently in industrial parks along the Yangtze River. For example, in November 3, 2005, Aniline unit in Sinopec Jilin Petrochemical

Co double-benzene factory exploded. This accident causes dozens of casualties, tens of thousands of people evacuated. Chemicals leakage polluted the water of Songhua River. The pollution led directly to downstream Harbin cutting off the water for a few days. The pollution even caused neighboring Russia's panic and caused a very bad social influence. Thus it can be seen, in industrial parks along the Yangtze Rive, because concentrated industrial production enterprises, all kinds of dangerous and harmful factors, major hazards, and major hidden dangers relatively concentrated. Once an accident happened and loss of effective control, the consequences are often unimaginable. With the development along the Yangtze River and the irreversible high-speed economic growth in area along the Yangtze River, the damage and influence to the society from all kinds of accident risk factors is unavoidable. Therefore, to strengthen the risk management of industrial parks along the Yangtze Rive and the whole riverside area are imperative.

2. The purpose and significance of risk evaluation for industrial parks along the Yangtze River

Risk evaluation is a work that quantifies evaluation for the influence and loss of people's life, property, etc. which are caused by the risk events. Risk evaluation will help the decision makers to choose optimal technical and policy specifically in the construction and management. At present, the risk evaluation of industrial parks just started in our country. And the risk evaluation of industrial parks along the Yangtze Rive or coast, or other waterside emphatically from the angle of safety, there is no ripe experience to draw lessons from.

Risk evaluation studied in this paper objectively analyses the safety situation of key industries group and social or economic risks to the regional development of industrial parks along the Yangtze River. Through a comprehensive understanding of the major hazards installations existing in the regional economic development and all kinds of risk factors derived from production and economic activity, reference should be provided to the establishment of long-term effective mechanism of safety production, to sustainable development of social and economic for industrial parks along the Yangtze River.

3. Risk factor identification of the risk evaluation for industrial parks along the Yangtze River

Finding risk factors accurately and completely and establishing reasonable evaluation index system are the important foundation of risk evaluation for the industrial parks along the Yangtze River. Through the field investigation and research of the industrial parks along the Yangtze River, risk factors are identified from three aspects in this paper.

3.1 Risk factor identification and analysis of production accidents

3.1.1 Risk analysis of major industries in the park

At present, the proportional number's difference of industrial injuries death, serious injury, minor injuries and occupational diseases in various industries in China is very big. Safety production situation of high-risk industry in China is serious. According to individual dead risk statistical data of worker in all industries from 2000 to 2005 in our country, the industrial injury accident risk of extractive industries, manufacturing, construction industry, power industry, gas industry, water production and supply industry is higher. Sampling survey of industrial injury accident countrywide in 2005 by China Academy of Safety Science and Technology shows the ratio of death, injuries and minor injuries is1:5.34:36.57, the average direct economic loss are 22.44million, 11.32 million, and 1.48 million. Calculation according to the survey sampling results, the ratio of direct economic loss is 1:2.69:2.4[5]. Based on the statistic and computation of employment and occupational disease number in various industries in 1996 ~ 2006, we know the industry distribution of occupational disease hazards industry widely distributed in our country and the damage is heavy in small and medium-sized enterprises. From the coal, metallurgical, chemical, construction and other traditional industries to automobile manufacturing, medicine, computer, biological engineering and other emerging industries, different degree occupation hazards exist. Among them, the mining, manufacturing, construction, geological exploration industry and other high accident risk industries also have a high incidence of occupational diseases[5].

There are many enterprises in the evaluation park. Raw materials, products, technology, layout, process, equipment, personnel quality and management level of the enterprises differ in thousands of ways. The risk existing objectively in each industry largely determines the accident risk of the assessment park, so we will first analyze the macro and innate danger of each enterprise from the industry which enterprise belongs and industry risk.

Industry classification in assessment should be based on standard *Industrial classification for national economic activities* (GB/T 4754-2011). Usually, industries in the industrial parks along the Yangtze River which included the largest number of enterprises are water transportation industry, warehousing industry, chemical raw materials and chemical products manufacturing industry, ferrous metal smelting and rolling processing industry, fabricated metal products, electricity or heat production and supply industry, ship building industry in general machinery manufacturing, etc.. In practical evaluation, many large companies may be a chemical production enterprise and a steel smelting enterprise, but also has a self-provided power plant or port of loading and unloading of raw materials and products. Then in the analysis of the industry risk, the risk of various industries should be considered. Industry risk analysis should consider the threat to whole regional environment when the most common accidents happened and the response should be taken by the government.

(1) Water transportation industry

Water transportation industry includes water transport of passengers, water transport of goods, passenger transportation port, and freight port. The major accident risk of this industry are accident and casualties caused by the reason such as poor condition of car and ship , weak management , overload , bad weather and so on. Transport materials through port cargo are always dangerous chemicals, crude and product oil with a huge number or various kinds. The hazard of fire, explosion or leak is enormous. Once some accidents happen, the dangerous chemicals , oils, fire fighting water will leak into the river and bring a huge damage to the water, environment and ecology. This damage could also cause huge social panic.

(2) Warehousing industry

Warehousing industry specializes in goods storage, storage transit in freightage and logistics activities which give priority to with warehouse. The serious consequence of warehousing industry is domino effect destruction caused by the peripheral storage container which affected by the fire or explosion.

(3) Chemical raw material and chemical products manufacturing

Chemical raw material and chemical products manufacturing includes basic chemical raw materials manufacturing, fertilizer manufacturing, pesticide manufacturing, coatings and ink pigment and similar products manufacture and so on. The major accident risk of this industry includes fire, explosion, poisoning and asphyxia. Group die and group of injury caused by fire or explosion, large-scale evacuation caused by leakage, water bodies contaminated by fire fighting water which inadequately treated should be attached great importance by the government.

3.1.2 Identification and analysis of dangerous substance in the park

Dangerous chemical is a special type of chemical whose characteristics are flammable, explosive, toxic, harmful and corrosive. As a special commodity, dangerous chemical greatly improves people's lives, but its inherent risk brings threat to the survival of mankind. According to statistics, with the development of economic, production of dangerous chemicals increased significantly each year and dangerous chemicals accident in our country show an obvious rising trend in recent years.

In our country, there are strict laws and regulations for production, management, storage, transportation, use and scrap of dangerous chemicals. A strict regulatory scrutiny exists in each link. In recent years, major dangerous chemicals accidents occurred frequently. In the risk evaluation for industrial parks along the Yangtze River, we should first make statistics and analysis on hazardous substance especially the dangerous chemicals.

Hazardous substance used for production, use and storage of each enterprise in the park should be listed according to the eight categories of *Catalog of hazardous chemicals* (2012 edition. A Huge number of flammable liquid, flammable solid or other hazardous substances which are not in the dangerous chemicals list should be listed as a separate category. During the identification and analysis, the storing site, maximum storage capacity and storage method of each dangerous chemical should be analyzed statistically. This work will provide the basis for future qualitative and quantitative evaluation.

3.1.3 Identification and analysis of highly toxic chemicals, precursor chemicals and monitored and control chemicals in the park

In the evaluation process, we should in addition to considering of flammable and explosive characters of dangerous chemicals, also from avoiding collective poisoning, preventing crime, protect public safety and other way, focus on the existence of the following 3 types of chemicals.

(1) Highly toxic chemicals.

They refer to chemicals with highly toxic hazards, including synthetic chemicals and their mixtures (including pesticides) and natural toxins. With respect to the flammable and explosive hazardous chemicals, toxic hazards to human and animal more. Relative to the flammable and explosive of dangerous chemicals, highly toxic chemicals play a bigger threat to humans and animals.

(2) Precursor chemicals.

They refer to the raw materials, dispensing and other chemicals used in illegal production, manufacturing or synthetic drug, including the raw precursor, reagents, solvents and thinners, additive and so on used in the manufacture of drugs. Precursor chemicals itself is not a drug. But it has a dual nature, precursor chemicals is both general raw materials of medicine and chemical industry, and essential chemicals to production, manufacture or synthetic drugs.

(3) Monitored and Control Chemicals (MCCS)

Four types of chemicals are defined as MCCS in China in *the People's Republic of China Regulations on Monitored and Controlled Chemicals*. The first category is chemicals can be used as chemical weapons. The second type is chemicals can be used to produce chemical weapons precursor. The third class is chemicals can be used as major raw materials to produce chemical weapons. The fourth class is certain organic chemicals besides explosives and pure hydrocarbon compounds.

In identification and analysis, we will be screening and classifying the dangerous chemicals in assessment area according to *the Catalog of Highly Toxic Chemicals* (2002 Edition), *the Classification and Varieties Catalogue of the Precursor Chemicals*, *the People's Republic of China Regulations on Monitored and Controlled Chemicals* (the State Council Decree No. 190, 1995) , *the List of All Kinds of MCCS* (chemical industry ministry Decree No.11, 1996), *New Varieties Listed Included in Category III MCCS* (the national bureau of petrochemical Decree No1, 1998). The storing site, maximum storage capacity, storage method and transportation mode of every substance should be explained.

3.1.4 Identification and analysis of major hazard installations in the park

An important link to prevent industrial accidents which happened in the assessment park is to identify or confirm the high risk of hazards (Dangerous substances or plant facilities). Investigation and assessment for major hazard installations is an important work of safety administration department and fire protection department in recent years. The investigation and assessment for major hazard installations can determine their distribution, type and quantity. The investigation and assessment not only can make the safety supervision department, emergency rescue department, medical and other department to more clearly understand and grasp the important enterprises and site condition, the present situation of rescue facility, the rescue ability level, etc., but also can more perfect safety design planning for the area and provide scientific basis for the strengthening of the safety facilities investment, the construction of security team, the improvement of the emergency plan, the emergency drill, the safety establishment and approval and so on.

Identification of major hazard installations is mainly based on *Identification of major hazard installations for dangerous chemicals* (GB 18218-2009) and *Guidance on the development of major hazard installations supervision and management* issued by State Administration of Work Safety.

In the industrial parks along the Yangtze River, many large chemical enterprises, storage enterprises and terminal are located, so the consumption, reserves and throughput of petrochemical products is huge and the energy consumption is huge too. In the actual evaluation process, according to the two identification of substance classes often appear contradictory or deficiency. In the actual evaluation process, when reference to these two identification bases, conflict or lack of data often occurs in the identification of the material class. Some substances have a high flash point when listed separately, but because they are a huge number, once combustion or explosion, they will pose a huge threat to the surrounding equipments and substances in the equipments. Some substances storage sites are adjacent to important public network, once accidents happen, they will cause widespread blackout or water failure resulting in regional production paralysis. Some chemicals are less toxic, if leakage, it will produce serious pollution towards water or required huge funds to disposal and recycling. In the identification of major hazard installations, should not be only based on the above two basis. When industrial parks along the Yangtze River located in different cities, these cities' economic level is different, the geographical environment and humanities environment are not the same either, so the definition of major hazard installations should not be the same. In addition to the two major hazard installations identification bases, assessment person can develop major hazards installations grading standards containing different weighting factor and weight coefficient and so on according to the characteristics of the evaluation area. Quantitative comparison after calculation according to these grading standards, assessment person can tell the government supervision departments what need supervision, why to supervision and how to regulate.

3.2 Identification and analysis of risk factor of environment and ecology

In identification of dangerous and harmful factor during the assessment, relatively speaking, many accidents themselves are not terrible. But the reason such as unreasonable safety planning, lack of safety distance, emergency rescue force insufficiency will expand the accident consequences and then cause the inevitable environmental and ecological impact. In the risk evaluation for industrial parks along the Yangtze River, we can choose the following accidents or disasters which happened in the park as the risk factor of environment and zoology.

(1) Environment impact analysis for chain reaction of major accident

After the accident, whether timely and effectively control, response, rescue and recovery is very important. We can consider the possible chain reaction and its consequences from the following aspects: clean water conditions, emergency rescue capabilities of enterprise, emergency rescue capabilities of society, emergency rescue capabilities, effects of the leakage of chemicals on the environment, effect of products of chemical combustion on the environment, cost needed of accident recovery and other fields. So chain reaction of major accident should be chosen as the primary risk factors which put threat to environment and ecology.

(2) Impact analysis for environment and ecology of the leakage of toxic material

Leakage of toxic material will first put threat to life safety of people near the leakage point. Along with the spread of the poison, it will pose a threat to the surrounding personnel, facilities and environment. So, we should choose it as a risk factor. The nature of poisons, amount of leakage, leakage location, surrounding emergency rescue forces, emergency rescue measures and so on should be analyzed carefully.

(3) Impact analysis for environment and ecology of major transportation accidents

Traffic accidents occur frequently in our country. The traffic situation in the industrial parks along the Yangtze River such as road, ferry, bridge and so on is complex. Many ports and wharfs distribute along the Yangtze River area, so the transport of hazardous materials is wide. Therefore, the possibility of transportation accident in the park is big, choosing it as one of the environmental and ecology risk factor for analysis is very necessary

3.3 Risk factor identification and analysis of social response and rescue

In industrial parks along the Yangtze River, in general, there are developed industries, intensive workers and advanced equipments. Once an accident occurs, it not only causes casualties and property losses, but also can produce serious social influence. Therefore, the establishment of a more perfect emergency rescue system to meet the needs of the industrial park is particularly important. In this emergency rescue system, multilateral rescue forces are needed such as fire, health and environment department.

(1) Analysis of personal emergency rescue capabilities of production and business units

Enterprise is the main body of safe production. After the accident, the production and business units to carry out a reasonable and effective self-help in the first time is an important means to control the harm, insure the safety of personnel and reduce accident loss. It requests us to analyze the emergency rescue ability of the enterprise through many forms and many aspects such as configuration of emergency rescue equipment, emergency rescue plan, training education, rescue teams, response time, and continued response ability and so on.

(2) Analysis of emergency rescue capabilities of government regulators and other relevant departments

The government plays a leading role of command and coordination in emergency rescue. In recent years, some problems exist in emergency rescue of major security accident, such as accident expanded and losses increased caused by ineffective command, improper disposal and lagged disposal. In the risk evaluation for industrial parks, the emergency rescue capabilities of government regulators and other relevant departments should be analyzed from regional emergency response basic conditions, setting up of medical institutions and fire department, cooperation ability of regional aid agencies.

4. Prospect

To conduct the risk evaluation of industrial parks or other functional areas and to implement the corresponding safety precautions has got more and more attention from the government and the production safety supervision department. At present, the risk evaluation of industrial parks in China has just started. For the risk evaluation along the Yangtze River, coastal, or other waterside overall planning region, there is no fixed or standard mode to refer to. The structure, identification method and identification basis of risk factor in risk evaluation for industrial parks along the Yangtze Rive are advanced. The study of author provides some thoughts for the identification and determination of risk factors in other industrial parks, areas along the Yangtze River region, coastal areas and similar waterside areas. The study can also provide reference for similar risk evaluation study.

5. References

[1] LIN Hongmei, MA Jie, "The golden waterway of the Yangtze River in the golden period of development," *Xinhua Daily*, Nov. 22, 2006

[2] Notice- Carrying out investigates of industry development situation of development zones above the provincial level: Annex 1 development zones above the provincial level, www.jsdpc.gov.cn/pub/jsdpc/sylm/tzgg (accessed Aug. 21, 2007)

[3] LU Rui, "Industrial characteristics research of industrial parks along the Yangtze Rive in Jiangsu province," *Shanghai Enterprise*, NO. 3, 2004

[4] Research group of development of chemical industry along the Yangtze River in Jiangsu Province, "Development situation and prospects of chemical industry along the Yangtze River in Jiangsu Province," *Jiangsu chemical industry*, NO. 1, 2004

[5] ZHOU Jianxin, LIU Gongzhi, Sample survey report of the direct economy loss caused by work safety accidents in 2005, State Administration of Work Safety, Beijing, China Academy of Safety Science and Technology, Apr. 2007

1ST CCPS
ASIA-PACIFIC
CONFERENCE
ON PROCESS SAFETY℠

Formation and Characteristics of Tulip Flame propagating Through the Premixed Methane/air Mixture in a Duct

D. L. Chen
College of Mechanical and Electrical Engineering
Beijing University of Chemical Technology
Beijing, P. R. China

chendl@mail.buct.edu.cn

Prepared for Presentation at
American Institute of Chemical Engineers
1st CCPS Asia-Pacific Conference on Process Safety
Qingdao, China
September 4 – 5, 2013

UNPUBLISHED

Formation and Characteristics of Tulip Flame propagating Through the Premixed Methane/air Mixture in a Duct

D. L.CHEN[1,3,*], D. S. ZHANG[1], J.H. Sun[3], Y. LIU[2,3], K. S.WANG[1], J. F.YANG[1]
1.College of Mechanical and Electrical Engineering, Beijing University of Chemical Technology ,Beijing 100029, P. R. China
2.College of Chemistry and Chemical Engineering, China University of Petroleum, Dongying, Shandong 257061, P. R. China
3. State Key Laboratory of Fire Science, University of Science and Technology of China, Hefei 230026, China
Chendl@mail.buct.edu.cn
Keywords: Tulip Flame, Methane, Laminar Flame, Turbulent Combustion, Rarefaction Wave

Abstract

In order to explore the formation process and characteristics of tulip flame propagating through methane/air premixed gases in a duct, the techniques of Schlieren, high-speed video camera and micro-thermocouples are used to study the propagating flame in the closed duct. Based on the experimental results, Combustion behaviors and characteristics of the premixed gases are analyzed. The results show that, the period of the tulip flame structure formation is about 20ms and the zone of that is about 6cm in width. The temperature of the tulip flame is higher than that of the laminar flame, which is mainly due to the turbulent combustion in the tulip flame. The turbulent intensity increases greatly after the rarefaction wave interferences the flame front, which make the combustion process in the duct more comprehensive.

1. Introduction

Many calamities have been caused by methane and coal dust explosion in coal mines in recent years. Once methane-coal dust explosion happens in the coal mine processes, serious harms often spread far and wide. There were 2945 calamities in coal mine in 2006, caused 4746 coal miner deaths in China (State Administration of Work Safety; Jan. 11,2007) [1]. Above 80 percentages of these calamities were caused by methane explosions or methane/coal dust hybrid explosions.

The first photographs of the shape of a flame of the premixed gases propagating irregularly in tubes were published by Ellis [2] in 1928. The flame propagation experiment of premixed CO/O_2 was carried out in a closed cylinder duct with 5cm in radius and 19.5cm in length. He observed that the shape of the flame changed suddenly from a forward pointing finger to a backward pointing cusp. After this inversion, the cusped shape remained stable. In 1959 Salamandra et al[3] coined the name "tulip flame" to describe this curious phenomenon.

After that, many studies have been performed for many years on this fascinating flame structure experimentally and numerically. Dunn-Rankin et al[4] compared experimental results with inviscid numerical simulations, observed that the tulip formation coincides with the flame arrival, after a semi-spherical propagation, at the vessel side walls. Matalon and Metzener [5] thought the actual formation of the tulip flame results from the vortical motion created in the burned gas which is a consequence of the vorticity produced at the flame front.Guenoche[6] thought that the tulip flame is caused by interaction between the pressure wave and the flame front. Biao Zhou et al[7] found that, the interaction between the flame and the flame-induced flows results in higher dynamic pressure acting forward on the corner portions of the flame front than the middle and therefore a tulip-shape flame could be formed. C. Clanet et al[8] found the formation of tulip flame structure of premixed propane/air mixture in a cylinder tube with 2.5-5cm in radius and 0.6-6m in length. L. Kagan, G. Sivashinsky [9] performed a numerical study of a two-dimensional model for premixed gas combustion in a thin, semi-infinite and thermally-insulated channel. They found that the detonation first develops in the boundary layer where the effect of hydraulic resistance is stronger, and thereupon spreads over the channel's interior. However, the second stage of the transition does not proceed gradually but rather through a localized auto-ignition within the tulip.

As mentioned above, during the past about 80 years, many researchers have studied the tulip flame structure, and many various mechanisms have been proposed to explain the formation of the tulip flame. Unfortunately, a recognized explanation is not got so far, and the formation process and characteristics of the tulip flame structure have not been sufficiently explored yet. In this work, the techniques of Schlieren, high-speed video camera and micro-thermocouples are used to explore the flame propagation characteristics and tulip flame structure formation process of premixed methane/air mixture.

2 Experimental Apparatus and Procedure

2.1 Experimental Apparatus

The experimental system is schematically shown in Fig.1, which is composed of a closed combustion duct, an ignition system, a Schlieren image system, a high-speed video camera, a data recorder and a controller.

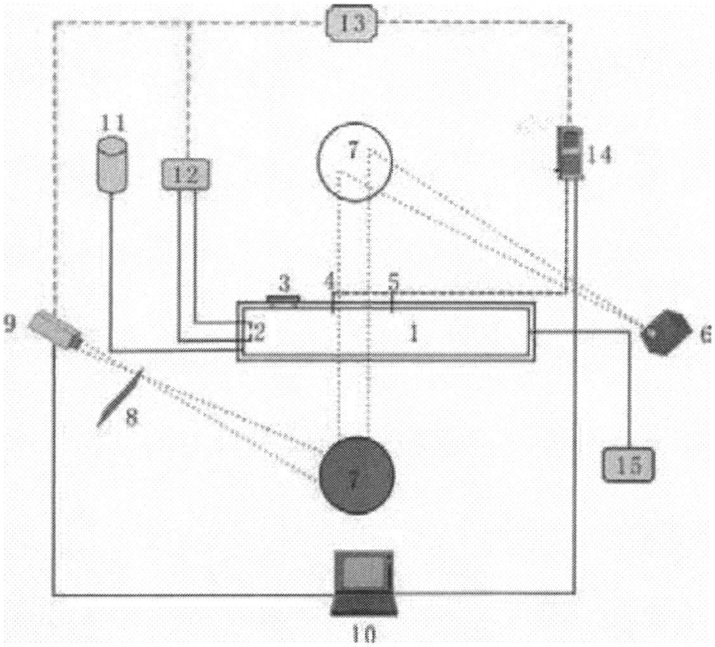

Fig.1 Experimental system

1-combustion duct; 2-ignition electrodes; 3-pressure vent; 4-thermocouple 1; 5-thermocouple 2; 6-halogen tungsten lamp; 7-concave mirrors; 8-knife edge; 9-high-speed video camera; 10-computer; 11-mixed gases tank; 12-ignition system; 13-controller; 14-data recorder; 15-vacuum pump.

The volume of the combustion duct L×W×H is 580mm×80mm, which is a rectangular tube. To observe flame propagation characteristics and flame acceleration evolution process conveniently, two sides of the chamber were made of glass and the other two sides were made of stainless steel. To explore the flame propagation characteristics of the premixed flame, two thermocouples were fixed at different position in the upper wall (16.5cm and 26.5cm to the left end of the duct respectively). The thermocouples are made from 25 μm diameter wires of Pt-Pt/Rh13%. The Schlieren image system shown in Fig. 1 consists of a 50W halogen tungsten lamp, two concave mirrors and a knife edge. A pair of ignition electrodes was set in the section which is 2cm to the left end of the duct, and a pressure vent was set in the top wall of combustion duct for safety in the process of experiment.

2.2 Experimental Procedure and Conditions

Connect the high-speed video camera, ignition system and data recorder to the controller as Fig.1 shows. Close the vent with a membrane and turn on the vacuum pump to vacuum the duct. After the duct was vacuumed, the premixed methane/air mixture (methane: air=1:9) was filled into it to 1 atmospheric pressure. When all the experimental apparatuses were ready, the controller was started and the apparatuses would be started at the scheduled time. The startup time of the high- speed video camera, the ignition time and discharge period of the spark were controlled by the controller. The detailed experimental conditions are listed as follows:

ignition voltage: 30,000V; discharge period: 0.01s; recording speed of high-speed video camera: 2000 frames/s; sampling rate of data recorder: 200,000Hz; membrane material: terylene membrane of 0.04mm in thickness.

3 Results and Discussion

Fig. 2 is a series of typical high-speed Schlieren photographs of the premixed methane/air mixture in the closed duct, depicting the pullulating, developing and propagating processes of the tulip flame.

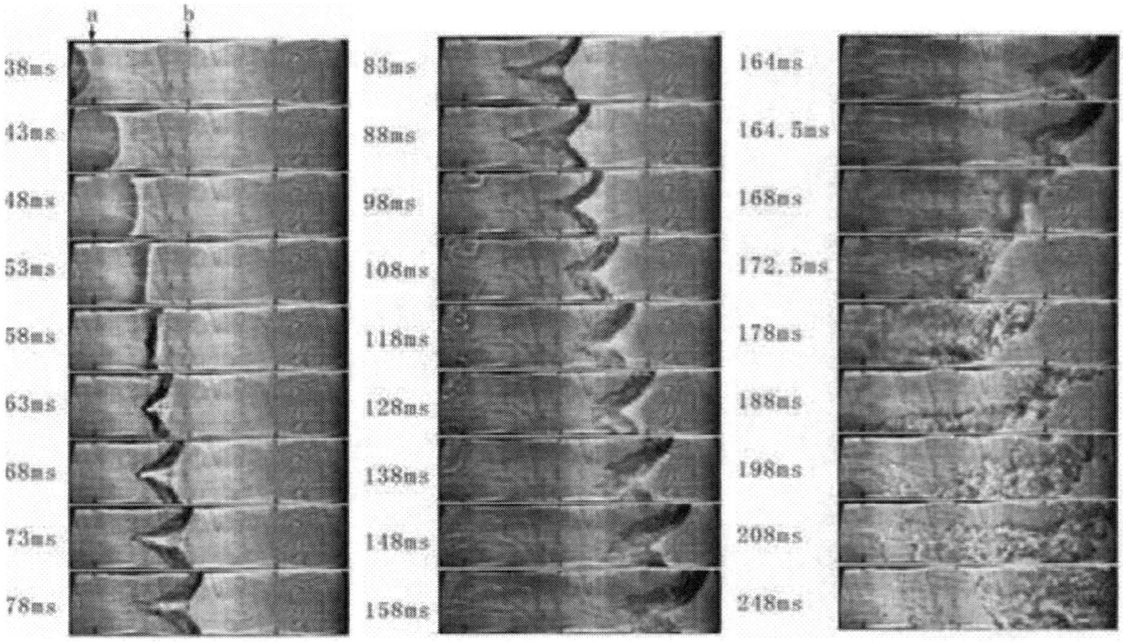

Fig.2 High-speed Schlieren photographs of premixed methane/air mixture in closed duct. Methane concentration: 10%; framing rate: 2000 frames/s

It is clear that the flame propagates with a spherical flame front in the initial stage, and it shows laminar combustion characteristics. The flame front begins to change at 43ms as the Fig.2 shows. The curvature radius of the flame front begins to increase at 43s. The flame front almost turns into a plane at 53ms, and it forms a "spindle" structure at about 58ms. The flame front begins to bend from unburned zone to burned zone at 63ms, after that the distance from the flame front to the back cusp increases gradually. The tulip flame structure changes greatly after 98ms and it shows that the turbulent combustion exists in the flame front. The rarefaction wave produced by the breaken of the membrane in the pressure vent begins to affect the flame front at 164.5ms, and the rightward propagating flame begins to back off. The turbulent combustion intensity of the flame front increases greatly because of the effect of the rarefaction wave. The backward flame front turns into propagating rightward again at 172.5ms. According to the photographs during 168ms to 248ms, it shows that, the turbulent quantity in the flame front increase greatly after the rarefaction wave interferences the propagating flame. The upper part of the flame front is before the lower part due to the buoyancy.

Based on the above analysis, the combustion of the premixed gas mixture in closed duct can be considered as laminar combustion at the initial stage, and the turbulent in the flame front is still very small during the formation process of the tulip flame. But, the instability of the flame front increases with the development of the tulip flame. After the tulip flame structure forms, the turbulent combustion characteristics of the propagating flame increases continuously and it shows turbulent combustion characteristics.

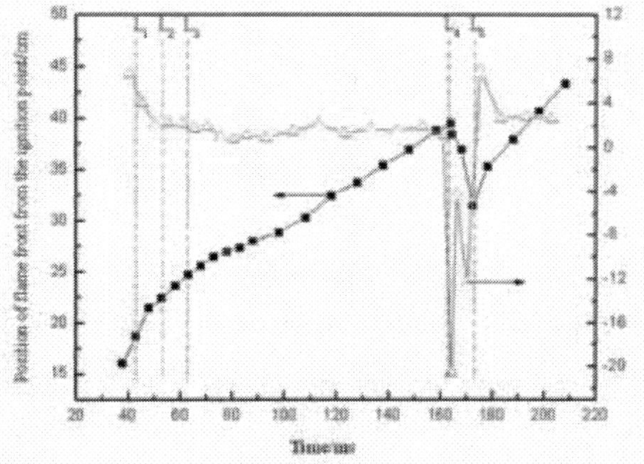

Fig.3 Relationships between flame propagation velocity and flame front position with time.

Fig.3 shows relationships between flame propagation velocity and flame front position with time of the premixed methane/air mixture in closed duct. It show that the propagating velocity of the flame decreases continually from 6.7m/s to 2.0m/s during the period of 38ms to 53ms. The 53ms is the transition time of the flame, after that the flame velocity shows fluctuant during 164.5ms to172.5ms. The flame propagates rightwards with a velocity of about 3.03m/s after 172.5ms.

According to the analysis above, the tulip flame structure of the premixed methane/air mixture begins to form at L_1 as Fig.3 shows. The L_2 is its transition point at which the flame front begins to bend from unburned zone to burned zone. The tulip flame structure formation process completes at L_3. So the region from L_1 to L_3 is the tulip flame structure formation region. The width of it is about 6cm and the period is about 20ms. The region from L_3 to L_4 is the tulip structure development region which is about 14.5cm in width, and the period is about 100ms. The region from L_4 to L_5 is leftwards propagation region which is about 8cm, and its period is 8ms. After L_5, the turbulent in the flame front increases with the flame propagating.

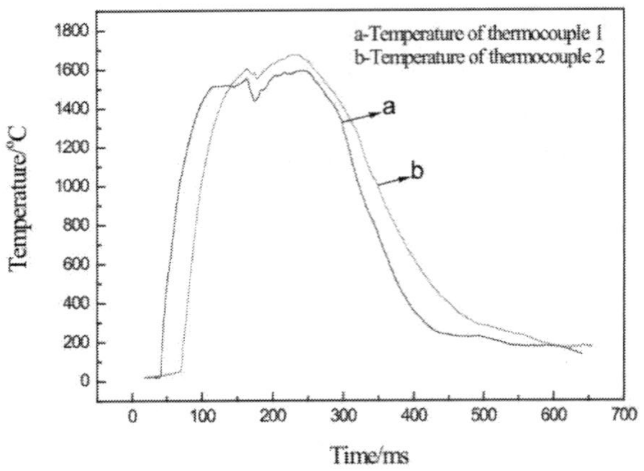

Fig.4 Temperature profiles of the premixed methane/air flame

Fig.4 shows temperature profiles of the premixed methane/air flame by the two thermocouples on the duct. It is clear that the temperatures increase to the maximum in several ten milliseconds while the flame propagates through, which indicates the combustion reaction is very intensive. The temperature of the thermocouple 1 begins to increase at 40.07ms and the temperature increases quickly from 20°Cto 1513°Cduring 40.07ms to 118.1ms. During the period from 118.1ms to 163.5ms the temperature increase gradually from1513°Cto 1555°CThen it begins to decrease from 1555°Cto 1438.2°Cduring 163.5ms to 174.9ms. The temperature increases again at 174.9ms, and it reaches to the maximum volume of 1592.1°Cat 242.8ms. After that, it decreases gradually.

Comparing the temperature of thermocouple 1 with the high-speed Schlieren photographs, the flame propagates through the thermocouple 1 at 39.26ms, and at the same time the temperature increases quickly. It begins to decrease at 163.5ms and increases again at 174.9ms. According to the Schlieren photographs, the rarefaction wave begin to interference the flame front at 164.5ms, then the flame propagation turn into back off. So the backward gas flow passes through thermocouple 1, which bring the second temperature increase of the thermocouple 1. And the second increase of the temperature stops at 242.8ms, then decreases.

According to the Fig. 4, the temperature of the thermocouple 2 begins to increase at 69ms. It increases quickly from 50°Cto 1603.3°Cduring 69ms to 164.4ms, and then it begin to decrease to 1552.8°Ctill 178.4ms. The temperature decrease stops at 178.4ms, and then increases again to the maximum volume of 1674.1°Cat 232.5ms. After that, it decreases gradually. According to the analysis above, the temperature of thermocouple 2 is higher than that of the thermocouple 1 while the flame propagates through, which is maybe resulted from the turbulent combustion in the tulip flame.

Comparing the temperature of thermocouple 2 with the High-speed Schlieren photographs, the flame propagates through the thermocouple 2 at 69ms. It is clear that the tulip flame structure has been formed before it propagates through the thermocouple 2. So the flame structure pass by the thermocouple 2 is the tulip flame structure. The temperature increases quickly while the flame pass by, it increases to 1603.3°Cduring 95ms, and then decreases gradually. The temperature increases again at 178.4ms and reaches to the maximum volume at 232.5ms. According to the Schlieren photographs, the rarefaction wave begins to interference the flame front at 164.5ms, then the flame propagation turn into back off. The backward gas flow passes through thermocouple 2, which bring the second temperatures increase of the thermocouple 2. And the second increase of the temperature stops at 232.5ms, then decreases.

4 Conclusions

In this study, the Schlieren, high-speed video camera, micro-thermocouples and other test methods are used to study the premixed methane/air mixture flame propagating in a horizontal rectangular duct. Based on experimental results, Combustion behaviors and flame structure of methane/coal dust hybrid propagating flame are analyzed. The following conclusions are drawn:

The tulip flame structure of the premixed methane/air mixture forms during about 20ms in the experimental condition, and the formation region is about 6cm in width. The temperature of the tulip flame is higher than that of the laminar flame, which is mainly due to the turbulent combustion in the tulip flame. The turbulent intensity increases greatly after the rarefaction wave interferences the flame front, which make the combustion process in the duct more comprehensive.

5 Acknowledgments

This work was supported financially by the Open Project of State Key laboraroty of Fire science(HZ2010- KF09) , "Youth Talent Project" of Beijing Higher Education Collages and Universities.

6.References

[1] http://www.chinasafety.gov.cn/anquanfenxi/2007-01/11/content_214963.htm

[2] Ellis C de C O, Flame movement in gaseous explosive mixtures,J. Fuel Sci. 1928,7,502-508.

[3] Salamandra G D, Bazhenova T V, Naboko I M, Formation of detonation wave during combustion of gas in combustion tube, Seventh Symposium（International）on combustion, Butterworths, London, 1959. 851-855.

[4] Dunn-Rankin D, Barr P K, Sawyer R F, Numerical and experimental study of "tulip" flame formation in a closed vessel, in: Twenty-First Symposium (International) on Combustion, The Combustion Institute, Pittsburgh, 1986. 1291-1301.

[5] Matalon M, Metzener P, The propagation of premixed flames in closed tubes, J. Fluid Mech. 1997, 336,331-350.

[6] Guenoche H, in: Markstein G H (Ed.), Nonsteady Flame Propagation, Pergamon, New York, 1964.

[7] Zhou B, Sobiesiak A and Quan P，Flame behavior and flame-induced flow in a closed rectangular duct with a 90° bend， International Journal of Thermal Sciences, 2006, 45 (5) : 457-474.
[8] Clanet C, Geoffrey S, On the "tulip flame" phenomenon, Combust. Flame , 1996,105,225 —238.
[9] Kagan L, Sivashinsky G, The transition from deflagration to detonation in thin channels, Combustion and Flame, 2003,134,389–397.

1ST CCPS ASIA-PACIFIC CONFERENCE ON PROCESS SAFETY℠

Study on Establishment and Analysis of Fault Tree Model of Buried Natural Gas Pipelines Failure

X. Y. Liu
School of Urban Construction and Safety Engineering, Nanjing University of Technology
200 North Zhongshan Rd, Nanjing, China 210009
781022045@qq.com

Y. N. Wang
Jiangsu Province Special Equipment Safety Supervision Inspection Institute
107 Caochangmen Street, Nanjing, China 210000
njwyn@126.com

Z. R. Wang
School of Urban Construction and Safety Engineering, Nanjing University of Technology
200 North Zhongshan Rd, Nanjing, China 210009
wangzr@njut.edu.cn

Prepared for Presentation at
American Institute of Chemical Engineers
1st CCPS Asia-Pacific Conference on Process Safety
Qingdao, China
September 4 – 5, 2013

UNPUBLISHED

Study on Establishment and Analysis of Fault Tree Model of Buried Natural Gas Pipelines Failure[1]

X. Y. Liu
School of Urban Construction and Safety Engineering, Nanjing University of Technology 200 North Zhongshan Rd, Nanjing, China 210009
781022045@qq.com

Y. N. Wang
Jiangsu Province Special Equipment Safety Supervision Inspection Institute
Z. R. Wang *
School of Urban Construction and Safety Engineering, Nanjing University of Technology

Keywords: Natural gas; Buried pipelines; Fault tree model; Analysis

Abstract

Abstract: Buried natural gas pipelines in using process, due to external influence, corrosion and material defects, such as pipeline failure, threaten people's life and property safety and the environment. According to the characteristics of the buried natural gas pipelines, gas pipelines failure fault tree model is established; The minimal cut sets and minimal path sets are found out by the analysis of pipelines failure fault tree, based on which countermeasures and suggestions are put forward for preventing accidents of the buried natural gas pipelines; Then the application methods of the fault tree model in industrial enterprise or urban buried gas pipelines risk assessment are discussed and suggestions are put forward for the application and popularization of the fault tree analysis model.

1. Introduction

Leakage, puncture or rupture is of common occurrence in using process of buried natural gas pipelines. As a result, serious fire and explosion accidents may happen. The natural pipelines may fail caused by many facors. Because they are mainly installed underground, the cost of laying, repair or replacement is generally high. Thus reliability of buried pipeline system should be taken into consideration. On the basis of it, the main fault modes and failure mechanism can be found out. And the relevant safety measures of improving the reliability and available life of the buried natural gas pipelines can be proposed.

Fault tree analysis (FTA) is a logical and diagrammatic method for evaluating failure probability of an accident, which has the advantages of concision and visuality[1-2]. An unexpected event is considered as a top event of a fault tree. All kinds of reasons and effect relationship which can cause the top event are analyzed by logical reasoning and the logic diagram. At last through the structure of the fault tree, the failure of top event is evaluated. In this paper, various influencing factors are considered and a fault tree of buried natural gas pipelines is built and analyzed.

2. Establishment of fault tree

The failure of buried natural gas pipelines is considered as the top event of the fault tree. Three kinds of failure modes, such as puncture, rupture and accessory equipments failure are considered as sub-top

* Corresponding author. Tel.: +86 25 83587423; fax: +86 25 83587423; *E-mail address*: iseofnjut@gmail.com

events and all basic events are achieved by fault tree analysis. The fault tree of buried natural gas pipelines is shown in Fig. 1[3-8]. All basic events are shown in Table 1.

3. FTA of buried natural gas pipelines

3.1 Minimum cut sets

The minimum basic events sets which can cause the top event are called minimum cut sets. The more the minimum cut sets fault tree contains, the more possibility the top event happens and the more dangerous the system is.

In this paper, by using Boolean algebraic operation methods, 167 minimum cut sets of above mentioned for buried pipelines fault tree are obtained. 87 first-order minimum cut sets include: $\{X_1\}$, $\{X_2\}$,$\{X_3\}$,…$\{X_{113}\}$; 77 second-order minimum cut sets include: $\{X_{65},X_{72}\}$, $\{X_{66},X_{77}\}$, $\{X_{70},X_{73}\}$,…$\{X_{71},X_{82}\}$; 3 third-order minimum cut sets include: $\{X_{57},X_{58},X_{61}\}$, $\{X_{57},X_{59},X_{61}\}$, $\{X_{57},X_{60},X_{61}\}$.

Generally speaking, the less the order of minimum cut sets is, the higher the occurrence frequence is. Therefore , the first-order minimum cut sets should be considered in advance in FTA.

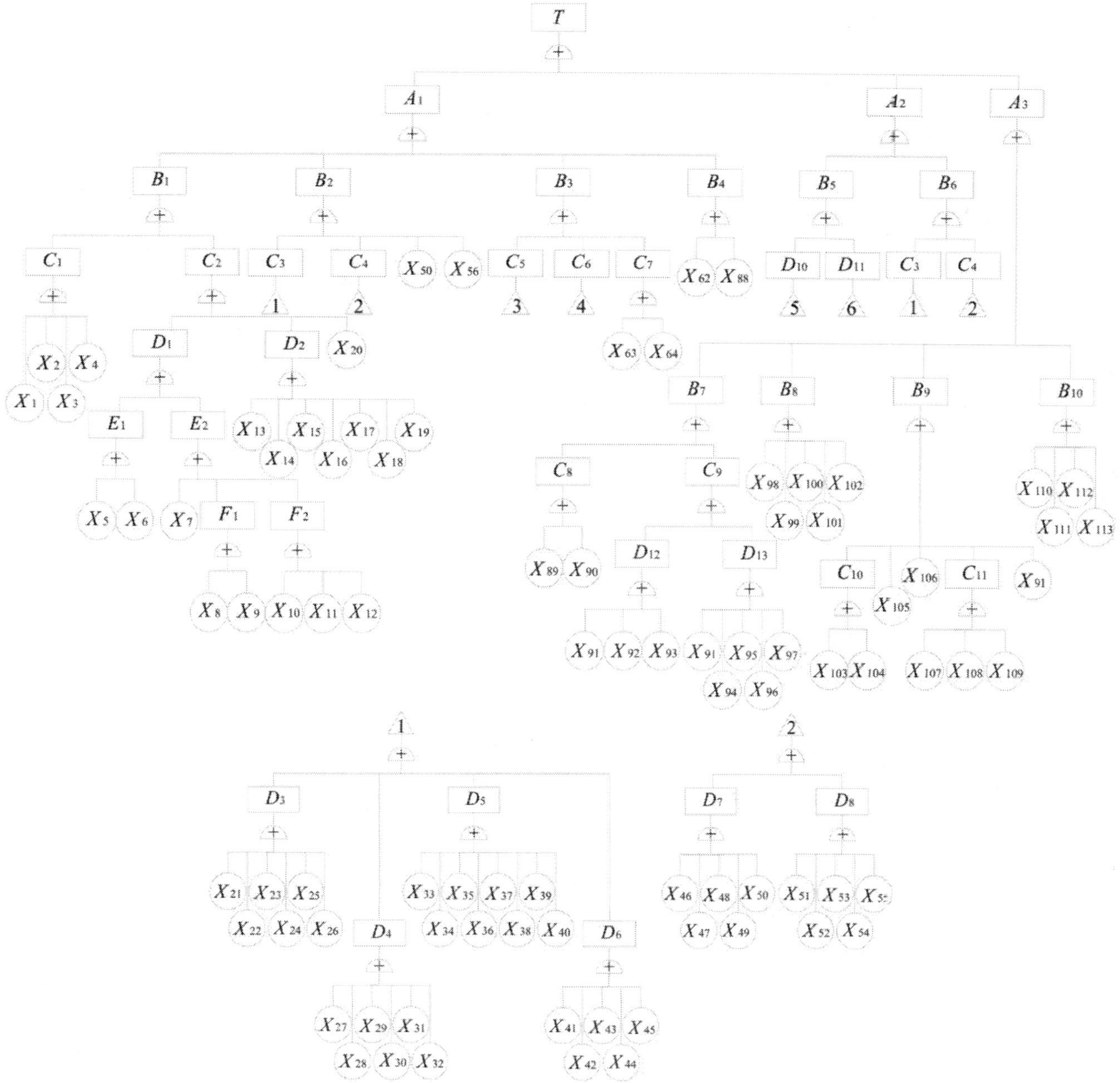

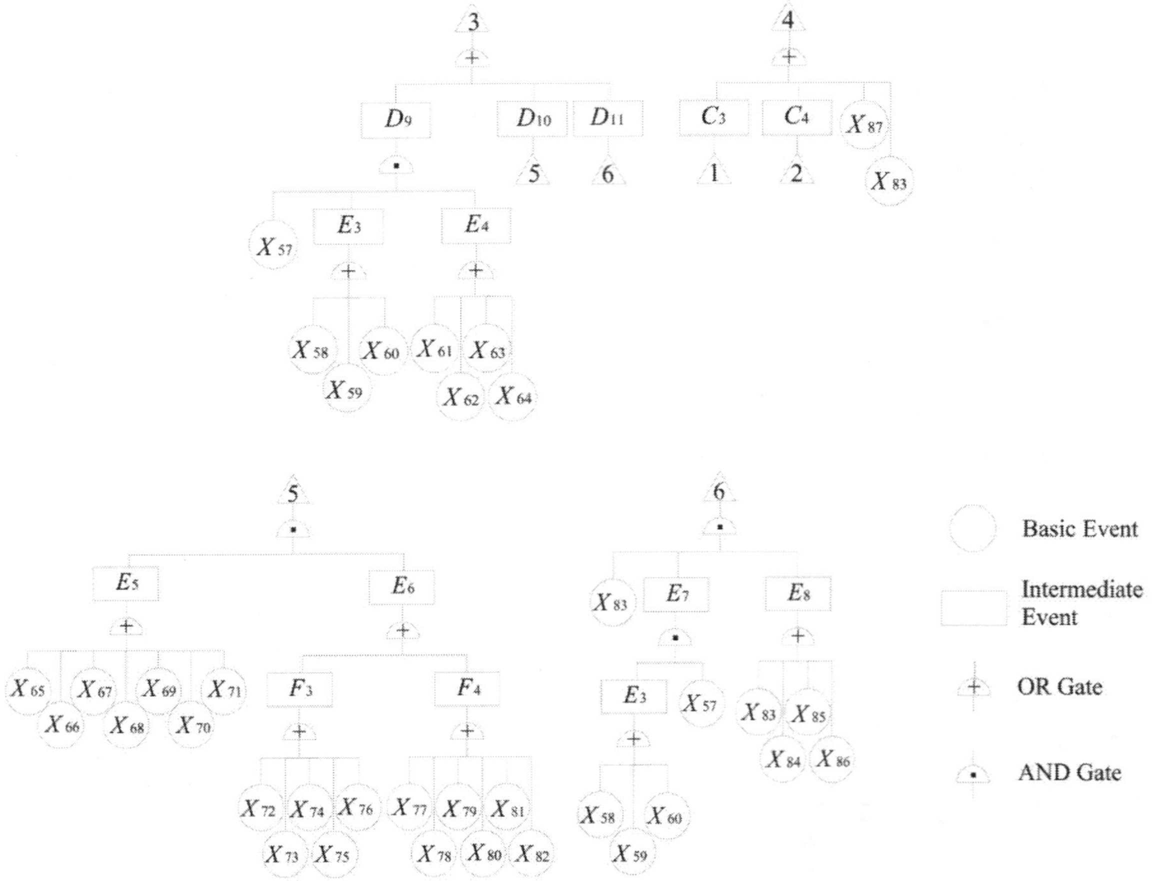

Fig.1 Fault tree model of buried natural gas pipelines failure

Table 1 All basic events in the fault tree

Code	Event	Code	Event	Code	Event
T	Pipeline failure	X_6	Malicious blasting	X_{61}	Residual stress
A_1	Pipeline rupture	X_7	Incorrect operation	X_{62}	Stress concentration
A_2	Pipeline puncture	X_8	No coordinating with pipeline company	X_{63}	External force
A_3	Accessory equipments failure	X_9	Pipeline company information is not complete	X_{64}	Internal stress
B_1	Third party interference	X_{10}	Geologic explosion	X_{65}	Plant rhizomes penetrate
B_2	Low bearing capacity	X_{11}	Geological subsidence caused by construction	X_{66}	Sulfide of soil
B_3	Corrosion	X_{12}	Digging and pile driving	X_{67}	High salt of soil
B_4	Stress	X_{13}	Heavy vehicles peccancy	X_{68}	Low pH of soil
B_5	Serious corrosion	X_{14}	Traffic changes	X_{69}	Soil with SRB
B_6	Pipeline defects	X_{15}	Illegal structures over pipeline	X_{70}	High redox potential of soil
B_7	Valve leakage	X_{16}	Illegal stacking above pipeline	X_{71}	High water ratio of soil
B_8	Flange leakage	X_{17}	Deficient buried depth	X_{72}	Short distance of cathodic protection
B_9	Compensator leakage	X_{18}	Illegal planting trees above pipeline	X_{73}	Small protection potential
B_{10}	Condenser cylinder leakage	X_{19}	Secondary disasters caused by explosion	X_{74}	Stray current of ground bed
C_1	Natural factor	X_{20}	Implicit signage	X_{75}	Improper protection
C_2	Unnatural factor	X_{21}	Poor quality coating	X_{76}	Protective materials failure

Code	Event	Code	Event	Code	Event
C_3	Construction defect	X_{22}	Uneven coating	X_{77}	Coating thinning
C_4	Initial defect	X_{23}	Unsuitable kind of coating	X_{78}	Coating stickiness decreased
C_5	Corrosion crack	X_{24}	Coating is not timely	X_{79}	Coating brittleness increases
C_6	Poor performance of pipe	X_{25}	Incorrect coating method	X_{80}	Coating is damaged
C_7	Large stress	X_{26}	Coating containing impurities	X_{81}	Coating aging stripping
C_8	Packing room leakage	X_{27}	Deficient depth of pipe trench	X_{82}	Coating lower hydrops
C_9	Valve body leakage	X_{28}	Poor slope stability	X_{83}	Poor corrosion resistance of pipe
C_{10}	Poor quality of construction	X_{29}	Coarse grain of backfill soil	X_{84}	Corrosion inhibitor failure
C_{11}	Compensator breakage	X_{30}	High water ratio of backfill soil	X_{85}	Internal coating thinning
D_1	Illegal construction	X_{31}	Poor drainage performance of pipe trench	X_{86}	Pipeline lining exfoliates
D_2	Occupied pressure	X_{32}	Corrosion medium of backfill soil	X_{87}	Poor mechanical properties of pipeline
D_3	Bad anti-corrosion insulation coating	X_{33}	Improper pipeline welding method	X_{88}	Large static stress
D_4	Pipe trench construction	X_{34}	Unqualified welding material	X_{89}	Valve body and cover not seal strictly
D_5	Pipeline welding	X_{35}	Poor quality of surface pretreatment	X_{90}	Valve stem contacts adversely
D_6	Pipeline installation	X_{36}	Stomata of weld surface	X_{91}	Third party damage
D_7	Material defects	X_{37}	Lack od penetration part is large	X_{92}	Incorrect selection of valve body mateeeerial
D_8	Processing technology	X_{38}	Serious carburizing	X_{93}	Pipeline defects
D_9	Stress corrosion	X_{39}	Unclear weiding slag	X_{94}	Poor quality of valve body material
D_{10}	External corrosion	X_{40}	Weld surface with slag	X_{95}	Valve body is not match pipeline
D_{11}	Internal corrosion	X_{41}	Unsmooth surface of elbow	X_{96}	Serious corrosion of valve body
D_{12}	Cast iron valve body leakage	X_{42}	Crack of elbow surface	X_{97}	Stress corrosion cracking of valve body
D_{13}	Cast steel valve body leakage	X_{43}	Large dislocation of pipe section	X_{98}	Nut looses
E_1	Malicious damage	X_{44}	Cracks of flange	X_{99}	Uneven bolt pretightening
E_2	Accidental damage by construction	X_{45}	Bolt material is different from pipeline	X_{100}	Deficient bolt pretightening
E_3	With acid medium	X_{46}	Much inclusion	X_{101}	Uncorrect gasket installation
E_4	Acting stress	X_{47}	Bad microstructure	X_{102}	Gasket aging
E_5	Soil corrosion	X_{48}	Coarse grain	X_{103}	Pipeline coaxiality error
E_6	External anti-corrosion failure	X_{49}	Unsuitable pipeline material	X_{104}	Forced assembly
E_7	Internal corrosive environment	X_{50}	Poor corrosion resistance of pipeline	X_{105}	Matal compensator leakage
E_8	Internal anti-corrosion failure	X_{51}	Heat treatment measures	X_{106}	Aging and cracking of other materials compensator
F_1	Unclear pipeline position	X_{52}	Ovality of pipeline	X_{107}	Low compensation quantity
F_2	External construction unit	X_{53}	Cold working technology	X_{108}	Not consider the heat-expansion and cold-contraction of pipe
F_3	Cathode protect	X_{54}	Uneven pipeline wall thickness	X_{109}	Other causes of rupture
F_4	Anti-corrosion insulation coating	X_{55}	Mechanical scar of pipe wall	X_{110}	Bad anti-corrosion of pumping pipe
X_1	Natural hazard	X_{56}	Unreasonable strength design	X_{111}	damage of pumping pipe

Code	Event	Code	Event	Code	Event
					anti-corrosion
X_2	Soil movement	X_{57}	With water	X_{112}	Pumping pipe has no regular replacement
X_3	Plant sabotage	X_{58}	With H_2S	X_{113}	Naked patr of pumping pipe has no protection
X_4	Water impact	X_{59}	With SO_2		
X_5	Enforced digging and pile driving	X_{60}	With CO_2		

3.2 Minimal path sets

Contrary to the cut sets, the minimum basic events sets which can not cause the top event are called minimum path sets. The more minimum path sets are, the safer the system is.

Through the success tree of the fault tree, by using Boolean algebraic operation methods, 6 minimum path sets of buried pipelines fault tree are obtained:

$$J_1=\{X_1',X_2',X_3',X_4',X_5',X_6',X_7',X_8',X_9',X_{10}',X_{11}',X_{12}',X_{13}',X_{14}',X_{15}',X_{16}',X_{17}',X_{18}',X_{19}',X_{20}',X_{21}',X_{22}',X_{23}',X_{24}',X_{25}',X_{26}',X_{27}',X_{28}',X_{29}',X_{30}',X_{31}',X_{32}',X_{33}',X_{34}',X_{35}',X_{36}',X_{37}',X_{38}',X_{39}',X_{40}',X_{41}',X_{42}',X_{43}',X_{44}',X_{45}',X_{46}',X_{47}',X_{48}',X_{49}',X_{50}',X_{51}',X_{52}',X_{53}',X_{54}',X_{55}',X_{56}',X_{61}',X_{62}',X_{63}',X_{64}',X_{65}',X_{66}',X_{67}',X_{68}',X_{69}',X_{70}',X_{71}',X_{83}',X_{87}',X_{88}',X_{89}',X_{90}',X_{91}',X_{92}',X_{93}',X_{94}',X_{95}',X_{96}',X_{97}',X_{98}',X_{99}',X_{100}',X_{101}',X_{102}',X_{103}',X_{104}',X_{105}',X_{106}',X_{107}',X_{108}',X_{109}',X_{110}',X_{111}',X_{112}',X_{113}'\}$$

$$J_2=\{X_1',X_2',X_3',X_4',X_5',X_6',X_7',X_8',X_9',X_{10}',X_{11}',X_{12}',X_{13}',X_{14}',X_{15}',X_{16}',X_{17}',X_{18}',X_{19}',X_{20}',X_{21}',X_{22}',X_{23}',X_{24}',X_{25}',X_{26}',X_{27}',X_{28}',X_{29}',X_{30}',X_{31}',X_{32}',X_{33}',X_{34}',X_{35}',X_{36}',X_{37}',X_{38}',X_{39}',X_{40}',X_{41}',X_{42}',X_{43}',X_{44}',X_{45}',X_{46}',X_{47}',X_{48}',X_{49}',X_{50}',X_{51}',X_{52}',X_{53}',X_{54}',X_{55}',X_{56}',X_{61}',X_{62}',X_{63}',X_{64}',X_{72}',X_{73}',X_{74}',X_{75}',X_{76}',X_{77}',X_{78}',X_{79}',X_{80}',X_{81}',X_{82}',X_{83}',X_{87}',X_{88}',X_{89}',X_{90}',X_{91}',X_{92}',X_{93}',X_{94}',X_{95}',X_{96}',X_{97}',X_{98}',X_{99}',X_{100}',X_{101}',X_{102}',X_{103}',X_{104}',X_{105}',X_{106}',X_{107}',X_{108}',X_{109}',X_{110}',X_{111}',X_{112}',X_{113}'\}$$

$$J_3=\{X_1',X_2',X_3',X_4',X_5',X_6',X_7',X_8',X_9',X_{10}',X_{11}',X_{12}',X_{13}',X_{14}',X_{15}',X_{16}',X_{17}',X_{18}',X_{19}',X_{20}',X_{21}',X_{22}',X_{23}',X_{24}',X_{25}',X_{26}',X_{27}',X_{28}',X_{29}',X_{30}',X_{31}',X_{32}',X_{33}',X_{34}',X_{35}',X_{36}',X_{37}',X_{38}',X_{39}',X_{40}',X_{41}',X_{42}',X_{43}',X_{44}',X_{45}',X_{46}',X_{47}',X_{48}',X_{49}',X_{50}',X_{51}',X_{52}',X_{53}',X_{54}',X_{55}',X_{56}',X_{58}',X_{59}',X_{60}',X_{62}',X_{63}',X_{64}',X_{65}',X_{66}',X_{67}',X_{68}',X_{69}',X_{70}',X_{71}',X_{83}',X_{87}',X_{88}',X_{89}',X_{90}',X_{91}',X_{92}',X_{93}',X_{94}',X_{95}',X_{96}',X_{97}',X_{98}',X_{99}',X_{100}',X_{101}',X_{102}',X_{103}',X_{104}',X_{105}',X_{106}',X_{107}',X_{108}',X_{109}',X_{110}',X_{111}',X_{112}',X_{113}'\}$$

$$J_4=\{X_1',X_2',X_3',X_4',X_5',X_6',X_7',X_8',X_9',X_{10}',X_{11}',X_{12}',X_{13}',X_{14}',X_{15}',X_{16}',X_{17}',X_{18}',X_{19}',X_{20}',X_{21}',X_{22}',X_{23}',X_{24}',X_{25}',X_{26}',X_{27}',X_{28}',X_{29}',X_{30}',X_{31}',X_{32}',X_{33}',X_{34}',X_{35}',X_{36}',X_{37}',X_{38}',X_{39}',X_{40}',X_{41}',X_{42}',X_{43}',X_{44}',X_{45}',X_{46}',X_{47}',X_{48}',X_{49}',X_{50}',X_{51}',X_{52}',X_{53}',X_{54}',X_{55}',X_{56}',X_{58}',X_{59}',X_{60}',X_{62}',X_{63}',X_{64}',X_{72}',X_{73}',X_{74}',X_{75}',X_{76}',X_{77}',X_{78}',X_{79}',X_{80}',X_{81}',X_{82}',X_{83}',X_{87}',X_{88}',X_{89}',X_{90}',X_{91}',X_{92}',X_{93}',X_{94}',X_{95}',X_{96}',X_{97}',X_{98}',X_{99}',X_{100}',X_{101}',X_{102}',X_{103}',X_{104}',X_{105}',X_{106}',X_{107}',X_{108}',X_{109}',X_{110}',X_{111}',X_{112}',X_{113}'\}$$

$$J_5=\{X_1',X_2',X_3',X_4',X_5',X_6',X_7',X_8',X_9',X_{10}',X_{11}',X_{12}',X_{13}',X_{14}',X_{15}',X_{16}',X_{17}',X_{18}',X_{19}',X_{20}',X_{21}',X_{22}',X_{23}',X_{24}',X_{25}',X_{26}',X_{27}',X_{28}',X_{29}',X_{30}',X_{31}',X_{32}',X_{33}',X_{34}',X_{35}',X_{36}',X_{37}',X_{38}',X_{39}',X_{40}',X_{41}',X_{42}',X_{43}',X_{44}',X_{45}',X_{46}',X_{47}',X_{48}',X_{49}',X_{50}',X_{51}',X_{52}',X_{53}',X_{54}',X_{55}',X_{56}',X_{57}',X_{62}',X_{63}',X_{64}',X_{65}',X_{66}',X_{67}',X_{68}',X_{69}',X_{70}',X_{71}',X_{83}',X_{87}',X_{88}',X_{89}',X_{90}',X_{91}',X_{92}',X_{93}',X_{94}',X_{95}',X_{96}',X_{97}',X_{98}',X_{99}',X_{100}',X_{101}',X_{102}',X_{103}',X_{104}',X_{105}',X_{106}',X_{107}',X_{108}',X_{109}',X_{110}',X_{111}',X_{112}',X_{113}'\}$$

$$J_6=\{X_1',X_2',X_3',X_4',X_5',X_6',X_7',X_8',X_9',X_{10}',X_{11}',X_{12}',X_{13}',X_{14}',X_{15}',X_{16}',X_{17}',X_{18}',X_{19}',X_{20}',X_{21}',X_{22}',X_{23}',X_{24}',X_{25}',X_{26}',X_{27}',X_{28}',X_{29}',X_{30}',X_{31}',X_{32}',X_{33}',X_{34}',X_{35}',X_{36}',X_{37}',X_{38}',X_{39}',X_{40}',X_{41}',X_{42}',X_{43}',X_{44}',X_{45}',X_{46}',X_{47}',X_{48}',X_{49}',X_{50}',X_{51}',X_{52}',X_{53}',X_{54}',X_{55}',X_{56}',X_{57}',X_{62}',X_{63}',X_{64}',X_{72}',X_{73}',X_{74}',X_{75}',X_{76}',X_{77}',X_{78}',X_{79}',X_{80}',X_{81}',$$

$X_{82}{}',X_{83}{}',X_{87}{}',X_{88}{}',X_{89}{}',X_{90}{}',X_{91}{}',X_{92}{}',X_{93}{}',X_{94}{}',X_{95}{}',X_{96}{}',X_{97}{}',X_{98}{}',X_{99}{}',X_{100}{}',X_{101}{}',X_{102}{}',X_{103}{}',X_{104}{}',X_{105}{}',X_{106}{}'$
$, X_{107}{}',X_{108}{}',X_{109}{}',X_{110}{}',X_{111}{}',X_{112}{}',X_{113}{}'\}$

Each minimum path set contains many events, and the failure probability and risk of pipelines system are high. Therefore, the reliability of pipeline system should be improved by reasonable and accurate design, manufacture, installation, operation and maintenance.

3.3 Structural importance

The structural importance I_i, in the fault tree shows the effect of basic events on pipelines failure[9]. The larger structural importance of basic event is, the greater the influence of basic event on the top event is. The structural importance is obtained by using the minimum cut sets.

Order of structural importance:

$I_1=I_2=I_3=I_4=I_5=I_6=I_7=I_8=I_9=I_{10}=I_{11}=I_{12}=I_{13}=I_{14}=I_{15}=I_{16}=I_{17}=I_{18}=I_{19}=I_{20}=I_{21}=I_{22}=I_{23}=I_{24}=$
$I_{25}=I_{26}=I_{27}=I_{28}=I_{29}=I_{30}=I_{31}=I_{32}=I_{33}=I_{34}=I_{35}=I_{36}=I_{37}=I_{38}=I_{39}=I_{40}=I_{41}=I_{42}=I_{43}=I_{44}=I_{45}=I_{46}=$
$I_{47}=I_{48}=I_{49}=I_{50}=I_{51}=I_{52}=I_{53}=I_{54}=I_{55}=I_{56}=I_{62}=I_{63}=I_{64}=I_{83}=I_{87}=I_{88}=I_{89}=I_{90}=I_{91}=I_{92}=I_{93}=I_{94}=$
$I_{95}=I_{96}=I_{97}=I_{98}=I_{99}=I_{100}=I_{101}=I_{102}=I_{103}=I_{104}=I_{105}=I_{106}=I_{107}=I_{108}=I_{109}=I_{110}=I_{111}=I_{112}=I_{113}$

$I_{57}=I_{61}$

$I_{58}=I_{59}=I_{60}$

$I_{65}=I_{66}=I_{67}=I_{68}=I_{69}=I_{70}=I_{71}$

$I_{72}=I_{73}=I_{74}=I_{75}=I_{76}=I_{77}=I_{78}=I_{79}=I_{80}=I_{81}=I_{82}$

$I_{84}=I_{85}=I_{86}$

And $I_1>I_{57}>I_{58}>I_{65}>I_{72}>I_{84}$

By the analysis of structural importance, 87 basic events with higher value have the greatest influence on the top event. While the corrosion environment, external anti-corrosion failure and internal anti-corrosion failure have a relatively smaller effect.

4. Main influencing factors analysis and suggestion

By the analysis of minimum cut sets, minimum path sets and structural importance of fault tree, the major causes of pipelines failure can be obtained and the safety precautions as follows:

4.1 Third-party interference

Third-party interference includes natural and human factors. Natural disasters also make serious damage to pipelines and consequent leakage of natural gas makes a worse situation. At the design stage, pipelines construction lines should be investigated overall. Related data on geology, meteorology, surrounding environment are collected as comprehensively as possible to make the best design scheme. Artificial damage is mainly illegal construction and occupied pressure, such as incorrect operation, illegal structures over pipeline, enforced digging and pile driving. Construction side should get in touch with pipe company and familiar with piping drawings and other related data before construction near the natural gas pipelines. Pipeline operating companies must strengthen the protection and propaganda of pipes, such as establishing pipeline regular patrol system, perfecting the warning marks of buried pipelines, etc.

4.2 Corrosion

Serious corrosion includes external and internal corrosion which may lead to wall thinning, puncture and even rupture of pipe.

External corrosion is mainly caused by soil corrosion and external anti-corrosion failure. Therefore, the anti-corrosion level should be improved when pipelines cross rivers, lowland and high salt zone. Pipelines detection and assessment must be done on a regular basis to ensure the normal operation of cathodic protection system.

Special medium, such as water, sulfide, acid medium, has the most effect to internal corrosion. Therefore, the quality of natural gas in pipelines should be controled strictly. Termly detecting and analyzing of dehydration, anti-corrosive coating and corrosion inhibitor could reduce unexpected failure. Reasonable pigging period and suitable pipe cleaner are also preferred.

4.3 Pipeline defects

Pipeline defects include initial and construction defect. Initial defect is mainly due to the unqualified pipes, such as containing impurities and uneven wall thickness. Improper operation, such as insufficient pipe trench depth and welding, is the mainly factor of construction defect. In order to reduce the affect and ensure the safety of pipelines, quality of manufacturing pipe, quality of materials and quality of installation process should be inspected carefully.

5. Applicaton discussion of fault tree

As mentioned above, through site investigation and theoretical analysis, a more complete buried natural gas pipelines fault tree model considering various events has been established..It can be applied for risk analysis of buried natural gas pipelines in various environmental conditions. In the case of specific application, basic events are selected from Fig. 1 and Table 1, which should be in accordance with the actual situation first. Then Boolean algebraic operation methods are used to find out the minimum cut sets of the fault tree. The minimum path sets can be obtained through the success tree of the fault tree. Based on minimum cut sets analysis, the structural importance of basic events is analyzed. And then the main influencing factors of pipelines failure are analyzed comprehensively. At last, safety precautions are presented.

FTA is applicable for the reliability and safety analysis of complex system. Due to lack of adequate experimental data of the natural gas pipelines system, fault tree can be used to analyze the main influencing factors. And the qualitative and quantitative evaluation can also be carried out on the buried pipelines system. And then targeted prevention measures are put forward to improve the level of safe operation, which can provide reference for natural gas pipeline integrity management. FTA model established in this paper has been applied for risk analysis and safety evaluation of buried nantural gas pipelines in 30 projects in Jiangsu province. Practice proves that the FTA model built in this paper has universal practicability.

6. Conclusions

The research shows that FTA provides an effective method to assess buried pipelines reliability and risk analysis. A complete buried natural gas pipelines fault tree has been constructed, including 113 basic events which can result in natural gas pipelines failure. Through the FTA, the main factors which cause pipelines failure can be determined. And preventive measures and suggestions can also be put forward. Practice has proved that the FTA model of natural gas pipelines can be widely used in buried pipelines risk analysis and safety evaluation.

7. Acknowledgements

The authors are grateful for the support given by Jiangsu Province Special Equipment Safety Supervision Inspection Institute.

8. References

[1] Wang jianyou, Accident statistical analysis and countermeasures for gas transmission and distuibution system[J]. Gas and Heat, 2001, 21(2): 178-179.

[2] He shujing, Zhou weiguo, Yan mingqing, Accident statistical analysis and countermeasures for urban gas transmission and distribution system[J]. Gas and Heat, 2003, 23(12): 753-755.

[3] Liao kexi, Yao anlin, Zhag huaixin, Fault tree analysis of natural gas pipeline failure[J]. Natural Gas Industry, 2001, 22(2): 94-96.

[4] Liu yang, Jia fengrui, Duan changgui, Study on fault tree of urban gas buried steel pipeline[J]. Gas and Heat, 2009, 29(7): 35-38.

[5] Dong yuhua, Yu dabao, Estimation of failure probability of oil and gas transmission pipeline by fuzzy fault tree analysis[J]. Journal of Loss Prevention in the Process Industries, 2005(18): 83-88.

[6] Liang wei, Hu jinqiu, Zhang laibin, Guo cunjie, Lin weipeng, Assessing and classifying risk of pipeline third-party interference based on fault tree and SOM[J]. Engineering Applications of Artificial Intelligence, 2012(25): 594-608.

[7] Liao kexi, Yao anlin, Zhang huaixin, Fault tree analysis on the long distance pipelines[J]. Oil and Gas Transportation, 2001, 20(1): 27-30.

[8] Dong yuhua, Yu dabao, Fault tree analysis of oil and gas pipeline[J]. Oil and Gas Transportation, 2002, 21(6): 15-17.

[9] Wang yuanhui, Safety system engineering[M].Tianjin: Tianjin University Press, 1999.

1ST CCPS ASIA-PACIFIC CONFERENCE ON PROCESS SAFETY℠

A Performance Evaluation System of Safety Culture Construction in Petroleum Enterprise Based on the Balanced Scorecard

Daizhi Tang
China University of Petroleum(East China)
Shandong Qingdao, China
tangdzh@cnooc.com.cn

Prepared for Presentation at
American Institute of Chemical Engineers
1st CCPS Asia-Pacific Conference on Process Safety
Qingdao, China
September 4 – 5, 2013

UNPUBLISHED

A Performance Evaluation System of Safety Culture Construction in Petroleum Enterprise Based on the Balanced Scorecard

Daizhi Tang
China University of Petroleum(East China)
Shandong Qingdao, China
tangdzh@cnooc.com.cn

Keywords: Petroleum Enterprise, Performance Proposal, Safety Culture, Balanced Scorecard

Abstract

Facing the severe security situation in the oil industry, it has been becoming scholars' research priority about how to strengthen the safety culture construction of petroleum enterprise, particularly how to conduct effective performance management of safety culture construction. As a performance proposal method which can give a comprehensive examination of various factors affecting the performance, the balanced scorecard became an effective tool of the performance management of safety culture construction in petroleum enterprises. The main contents of the thesis include (1) Summarizing the current situation of performance management of safety culture construction in petroleum enterprises.(2) Introducing the basic theory of the balanced scorecard and analyzing its applicability in performance evaluation of safety culture construction of petroleum enterprises. (3) According to the characteristics and requirements of the performance management of safety culture construction in petroleum enterprises, modulating the four dimensions of the balanced scorecard into the security dimension, the social satisfaction dimension, the internal management dimension, and the learning and growth dimension. The paper also sets the corresponding indicators for each dimension. The security dimension indicators include production accident rate, personnel casualty rate, and violating operation rate. The social satisfaction dimension indicators include government satisfaction, investment party satisfaction, and employees' families' satisfaction. The internal management indicators include security concept identity rate, security activities participation rate, and security system implementation force. The learning and growth dimension degrees indicators include training expenditures, training time, and supportive learning environment. (4) Introducing the calculation and application method of performance results of petroleum enterprises safety culture construction. Based on the practice of performance management of safety culture construction in petroleum enterprises and the balanced scorecard, the performance measurement system of safety culture construction in petroleum enterprises is of strong pertinence and operability.

1. Introduction

The characteristics of Oil Company including large scale, complex system, and high risk in oil industry make the health, safety and environment management (HSE) become a long-term and complicated project. In recent years, the accidents of petroleum enterprises happened continually, which exacerbated environmental pollution and endangered people's health and lives all over the world. Facing the severe security situation, it has become scholars' research priority about how to strengthen the safety culture construction of petroleum enterprise, in particular how to conduct effective performance management of safety culture construction. As a performance proposal method which can give a comprehensive examination of various factors affecting the performance, the balanced scorecard became an effective tool of the performance management of petroleum enterprises safety culture construction [1].

2. Current situation of performance evaluation of safety culture construction in petroleum enterprises

Many academics did some related research on performance evaluation of safety culture construction in oil enterprises. Guo Chen analyzed the problem of HSE performance management in pipeline engineering project of the oil company in China combining performance management theory and integration mode of PMT (owners) +PMC (supervision) +EPC (construction total contract party) and the practice of West pipeline engineering. A performance management system construction thought was came up with, which used target for oriented, unified performance management standard, covered all the contractors in performance evaluation, following PDCA principles, combined positive and negative incentive phase, and standardized managed contract forms[2]; Mu Bo used performance measurement method of security management system based on event investigation combined with performance index parameters including organization and personnel, dangerous identification and risk evaluation, operation control, change management and emergency plans to evaluate the whole performance. Liu Shuanxiao put forward that HSE effect should be evaluated from three aspects including wide publicity, and fortified training, and activities orienting[3]; Ding Hongtao deemed that performance management system of safety construction should be constructed in three aspects including composite security performance target, decomposition and implementation of security performance target, communicating safety performance; Dong Zhengliang built a balanced scorecard from the dimension of values, objectives, indicators, targets, action, task, and built performance management system of safety construction[4].

To sum up, security-building performance management is an integral and important part of safety culture construction, which occupied an important position in this long-term project of security culture. The construction of petroleum enterprises safety culture requires a new approach and new thought. Based on a balanced scorecard, this article builds a system of performance management of safety culture construction in petroleum enterprises.

3. Advantages of Using Balanced Scorecard for Performance Evaluation of Petroleum Enterprises Safety Culture Construction

In the early 1990s, Robot and David Norton created the balanced scorecard. Based on the financial evaluation system of which focused on financial index previously, it conducted a comprehensive evaluation of a corporation's performance considering the other non-financial factors such as customers, internal processes, learning and growth. The balanced scorecard linked to the strategic objectives of a company, overcame the insufficient of traditional single financial indicator. Using balanced scorecard in the performance evaluation of safety culture construction in oil companies provides a new way of thinking for the performance evaluation of safety culture construction. The core framework of the balanced scorecard is shown in Figure 1.

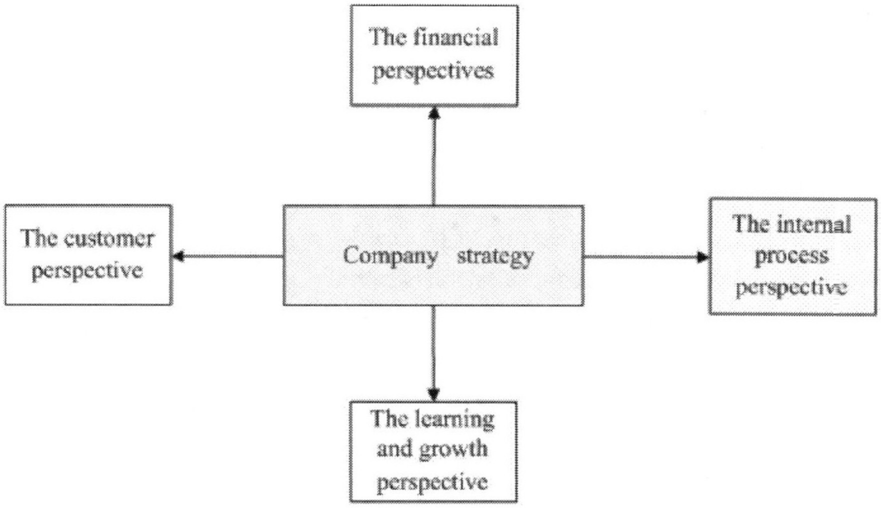

Figure 1 Core framework of the balanced scorecard

The Balanced scorecard provides a comprehensive method to survey enterprise value creation from financial, customer, internal processes, learning and growth. Firstly, the financial perspective: from the perspective of shareholders, including business growth, profitability and risk strategy. Secondly, the customer perspective: from the perspective of customers, including the strategy of creating value for customers and providing differentiated products and services. Thirdly, the internal process perspective: from the perspective of internal operation processes, business operation processes to meet the needs of customers and strategies of shareholder demands. IV, learning and growth perspective: from the corporate and employee learning and growth perspective, giving priority to the creation of a climate to support the organizational change and growth.

The Balanced scorecard is an indicator evaluation tool and a strategic management system combined with financial indicators and non-financial indicators, internal management and social oversight, long-term target and short-term target, objective measurement and subjective wishes, enterprise acts and personal efforts. It related strategy, process, management personnel, and operation workers together to reflect the real situation of enterprises 'safety culture systematically and dynamically, and point out the problems in security culture construction intuitively.

Security culture is a long-term, comprehensive, and strategic problem. Its performance results are always time-lagged and invisible. It is hard to fully evaluate in other methods. However, based on indicators' information of recent period, combined with the past information and long-term strategy of an enterprise, the balance scorecard can evaluate the performance of security culture construction more scientifically and accurately avoiding short-term acts of company effectively.

BSC is as much a safety culture assessment process as a process of strategic task decomposition and management of safety production. It is not traditional plan and control but creating a learning organization to foster and enhance the core competitiveness of enterprises. And this is the operating environment that healthy development of safety culture depends on.

4. Performance Evaluation of Safety Culture Construction in Petroleum Enterprises Based on the Balanced Scorecard

1. Determination of performance evaluation dimensions

The power of the balanced scorecard focused on strategic management. The balanced scorecard can also be used for other corporate management only when it aligns with corporate strategy. Safety culture is an integral part of business strategy, but its strategic manifestations and strategy implementation paths are different from the other strategic goals of the enterprise. Therefore, the balanced scorecard dimensions of safety culture evaluation must meet the specific connotation of safety culture.

(1) Security is the common goal of the enterprise and society, the major perspective of safety culture performance evaluation, which can be identified as the core dimension of the balanced scorecard;

(2) Internal management level of an enterprise is the foundation of security. It is also an important dimension in evaluating safety culture construction performance;

(3) Social satisfaction is the external evaluation of enterprise security and an indispensable evaluation dimension;

(4) Learning and growth represents the development potential of the enterprise. It is another basic dimension to evaluate the safety culture performance comprehensively.

Therefore, the balanced scorecard performance measurement system of safety culture construction performance should apply four-dimensional structure including security dimension, the social satisfaction dimension, the internal management dimension, and the learning and growth dimension. Its structure model is shown in Figure 2.

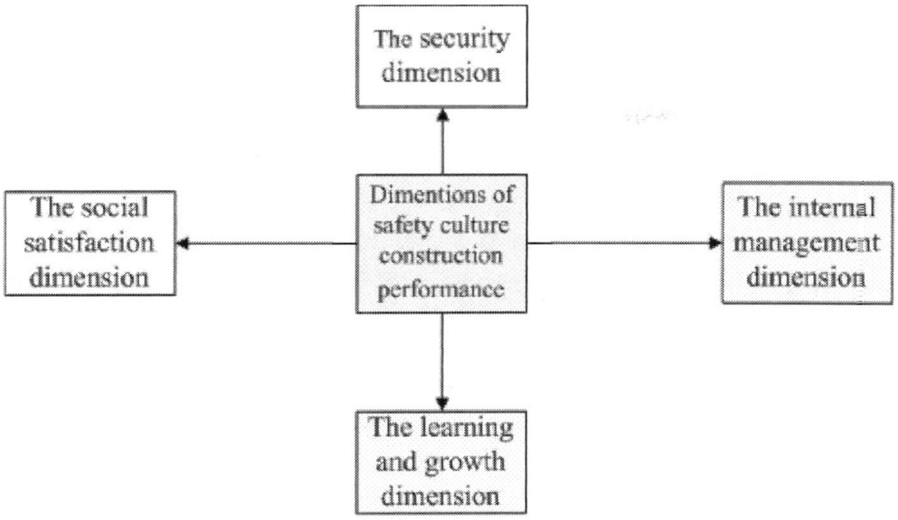

Figure 2 Dimensions of safety culture construction performance

2. Setting the performance evaluation index

According to the characteristics of enterprise security, the relevant, representative, quantifiable parameters were selected as evaluation indicators.

(1) Security dimension

Security is the foundation of enterprises' survival and the value proposition of safety culture. Safety is benefits including social benefits and economic benefits. Therefore, security is both a strategic indicator and a financial indicator. The evaluation indicators of security dimension include industrial accident rates, casualty rates, unprofessional operation rate and security cycles. Those four main indicators reflect the recognition degree of safety values in the minds of employees and self-consciousness in the action of employees. The deeper the identity of safety culture is, the higher the self-consciousness of security operation is. Besides the safety incident rate, the casualty rate and the unprofessional operation rate are lower and the safety period is longer.

(2) Social satisfaction dimension

Social satisfaction dimension is the starting point for security, and also is the destination of security. We can consider community satisfaction from three dimensions including government satisfaction, shareholder satisfaction and employees' family members' satisfaction. On the one hand, government satisfaction reflects humanism and respect for human rights. On the other hand, shareholder satisfaction reflects the financial return of investment; Moreover, family satisfaction reflects the social harmony and happiness. These three main indicators reflect the effective role of safety culture. The better the effect is, the higher the satisfaction is.

(3) Internal management dimension

Internal management should meet the need of safety requirements and safety culture construction. Internal management indicators include the safety recognition rate, safety activities participation rates, security system executive, security capital inputs, and quality standardization. Safety recognition rate, safety activities participation rates, security system executive reflect the behavior of employees, while safety quality standardization level, and safety capital input reflect the construction level in material level of safety culture.

(4) Learning and growth dimension

Learning and growth dimension is one of the dimensions of the balanced scorecard used to assess a company's strategy. It is also in the basic position of all the dimensions, which ensures the great significance of learning and growth. In continuing security enterprise organizations, proper techniques and methods used to motivate employees are the key factors to improve the management processes, create security value, satisfy social needs, and ultimately get the financial returns. Learning and growth dimension indicators include training expenditure, training time, supportive learning environment, creative product and technological innovation, employee incentives. Among them, the training expenditure, training time, supportive learning environment, staff incentive level reflects the extent of input and emphasis of safety culture construction. Creative product and technological innovation is the result of safety culture construction.

3. Calculation of the performance score

(1) Determination of the index weight

The weight of index reflects the relative importance of the indicator, and is an accurate scale to measure safety culture construction performance. Expert scoring method, AHP can be used to determine the weights.

(2) Determines the directivity of index

The index can be divided into positive indicators and negative indicators. Positive indicators include the safe period, investor satisfaction and innovations. These indexes should be kept as large as possible, while negative indicators such as casualty rates as low as possible.

(3) Standardize different unit of measurement

The unit of measurement of each standard is different from others. It is necessary to standardize different unit of measurement, and convert them into dimensionless value. The formula is: $Yi=(Xi-x)/SD$, where, xi is the actual value of evaluation indicators, x is average value of the indicators; SD is the standard deviation of the samples.

(4) Standardized score

The score of positive indicators is the result of standardize results and the score of negative indicators is the result of standardize results multiplied by -1.

(5) Scoring the performance

After adding the score of each indicator, it comes the score of each dimension, and the 4 dimensions of scoring phase may get a performance assessment of enterprise safety culture construction.

4. Integrated evaluation of oil enterprise security culture construction performance
According to the score of 4 dimension of BSC and the total score, through horizontal comparison we can find the weak project of security culture construction recently, and through vertical comparison we can s pros and cons in different periods, gains and losses in security culture construction, thus to take targeted improved measures, help to do further security work.

5. Conclusion

This article analyses the application of BSC in the performance evaluation of petroleum enterprises safety culture construction and we can mainly draw the following conclusions:
First, the balanced scorecard is a performance evaluation method based on enterprise strategy and safety culture construction is an important part of the business strategy of petroleum enterprises. Thus BSC has a strong applicability in the performance evaluation of the petroleum enterprises safety culture construction.
Second, the performance evaluation of the petroleum enterprises safety culture construction has its own characteristics, therefore it cannot fully copying the four dimensions of the traditional balanced scorecard, the modulation the four dimensions of the balanced scorecard of the performance evaluation of the petroleum enterprises safety culture construction into the security dimension, the social satisfaction dimension, the internal management dimension, and the learning and growth dimension.

6. References

[1] whir (1999), "Analyzing the Competitive Advantaged Disadvantage so Germany with the TOWS Matrix an Alternative to Porter Model". European Business Review, MCB University Press,September,22.

[2] Guo Chen(2010), "Balanced on the Balanced Scorecard-A Critical Analysis of Some of It Assumptions", Management Accounting Research, November.

[3] Mu Bo H(2008), "The Balanced on the Balanced Scorecard-A Critical Analysis of Some of Its Assumptions". Management Accounting Research, November, 65-88.

[4] Norrelit H(2007), "The Balanced on the Balanced Scorecard-A Critical Analysis of Some of It Assumptions", Management Accounting Research, November.

Study on Establishment and Analysis of Fault Tree Model of Buried Natural Gas Pipelines Failure

X. Y. Liu
School of Urban Construction and Safety Engineering, Nanjing University of Technology
200 North Zhongshan Rd, Nanjing, China 210009
781022045@qq.com

Y. N. Wang
Jiangsu Province Special Equipment Safety Supervision Inspection Institute
107 Caochangmen Street, Nanjing, China 210000
njwyn@126.com

Z. R. Wang
School of Urban Construction and Safety Engineering, Nanjing University of Technology
200 North Zhongshan Rd, Nanjing, China 210009
wangzr@njut.edu.cn

Prepared for Presentation at
American Institute of Chemical Engineers
1st CCPS Asia-Pacific Conference on Process Safety
Qingdao, China
September 4 – 5, 2013

UNPUBLISHED

Study on Establishment and Analysis of Fault Tree Model of Buried Natural Gas Pipelines Failure[1]

X. Y. Liu
School of Urban Construction and Safety Engineering, Nanjing University of Technology 200 North Zhongshan Rd, Nanjing, China 210009
781022045@qq.com

Y. N. Wang
Jiangsu Province Special Equipment Safety Supervision Inspection Institute
Z. R. Wang *
School of Urban Construction and Safety Engineering, Nanjing University of Technology

Keywords: Natural gas; Buried pipelines; Fault tree model; Analysis

Abstract

Abstract: Buried natural gas pipelines in using process, due to external influence, corrosion and material defects, such as pipeline failure, threaten people's life and property safety and the environment. According to the characteristics of the buried natural gas pipelines, gas pipelines failure fault tree model is established; The minimal cut sets and minimal path sets are found out by the analysis of pipelines failure fault tree, based on which countermeasures and suggestions are put forward for preventing accidents of the buried natural gas pipelines; Then the application methods of the fault tree model in industrial enterprise or urban buried gas pipelines risk assessment are discussed and suggestions are put forward for the application and popularization of the fault tree analysis model.

1. Introduction

Leakage, puncture or rupture is of common occurrence in using process of buried natural gas pipelines. As a result, serious fire and explosion accidents may happen. The natural pipelines may fail caused by many facors. Because they are mainly installed underground, the cost of laying, repair or replacement is generally high. Thus reliability of buried pipeline system should be taken into consideration. On the basis of it, the main fault modes and failure mechanism can be found out. And the relevant safety measures of improving the reliability and available life of the buried natural gas pipelines can be proposed.

Fault tree analysis (FTA) is a logical and diagrammatic method for evaluating failure probability of an accident, which has the advantages of concision and visuality[1-2]. An unexpected event is considered as a top event of a fault tree. All kinds of reasons and effect relationship which can cause the top event are analyzed by logical reasoning and the logic diagram. At last through the structure of the fault tree, the failure of top event is evaluated. In this paper, various influencing factors are considered and a fault tree of buried natural gas pipelines is built and analyzed.

2. Establishment of fault tree

The failure of buried natural gas pipelines is considered as the top event of the fault tree. Three kinds of failure modes, such as puncture, rupture and accessory equipments failure are considered as sub-top

* Corresponding author. Tel.: +86 25 83587423; fax: +86 25 83587423; *E-mail address*: iseofnjut@gmail.com

events and all basic events are achieved by fault tree analysis. The fault tree of buried natural gas pipelines is shown in Fig. 1[3-8]. All basic events are shown in Table 1.

3. FTA of buried natural gas pipelines

3.1 Minimum cut sets

The minimum basic events sets which can cause the top event are called minimum cut sets. The more the minimum cut sets fault tree contains, the more possibility the top event happens and the more dangerous the system is.

In this paper, by using Boolean algebraic operation methods, 167 minimum cut sets of above mentioned for buried pipelines fault tree are obtained. 87 first-order minimum cut sets include: $\{X_1\}$, $\{X_2\}$,$\{X_3\}$,...$\{X_{113}\}$; 77 second-order minimum cut sets include: $\{X_{65},X_{72}\}$, $\{X_{66},X_{77}\}$, $\{X_{70},X_{73}\}$,...$\{X_{71},X_{82}\}$; 3 third-order minimum cut sets include: $\{X_{57},X_{58},X_{61}\}$, $\{X_{57},X_{59},X_{61}\}$, $\{X_{57},X_{60},X_{61}\}$.

Generally speaking, the less the order of minimum cut sets is, the higher the occurrence frequence is. Therefore , the first-order minimum cut sets should be considered in advance in FTA.

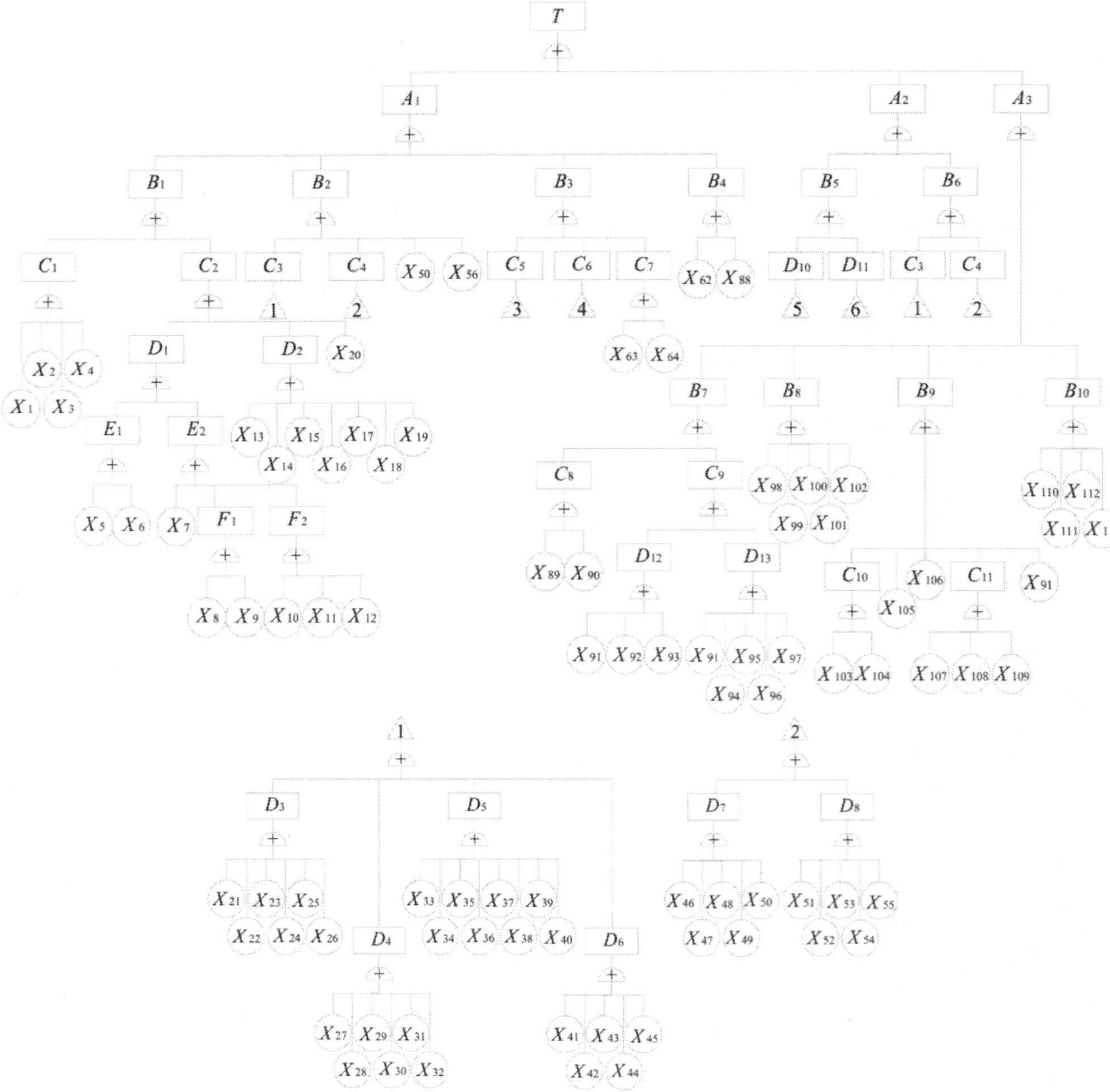

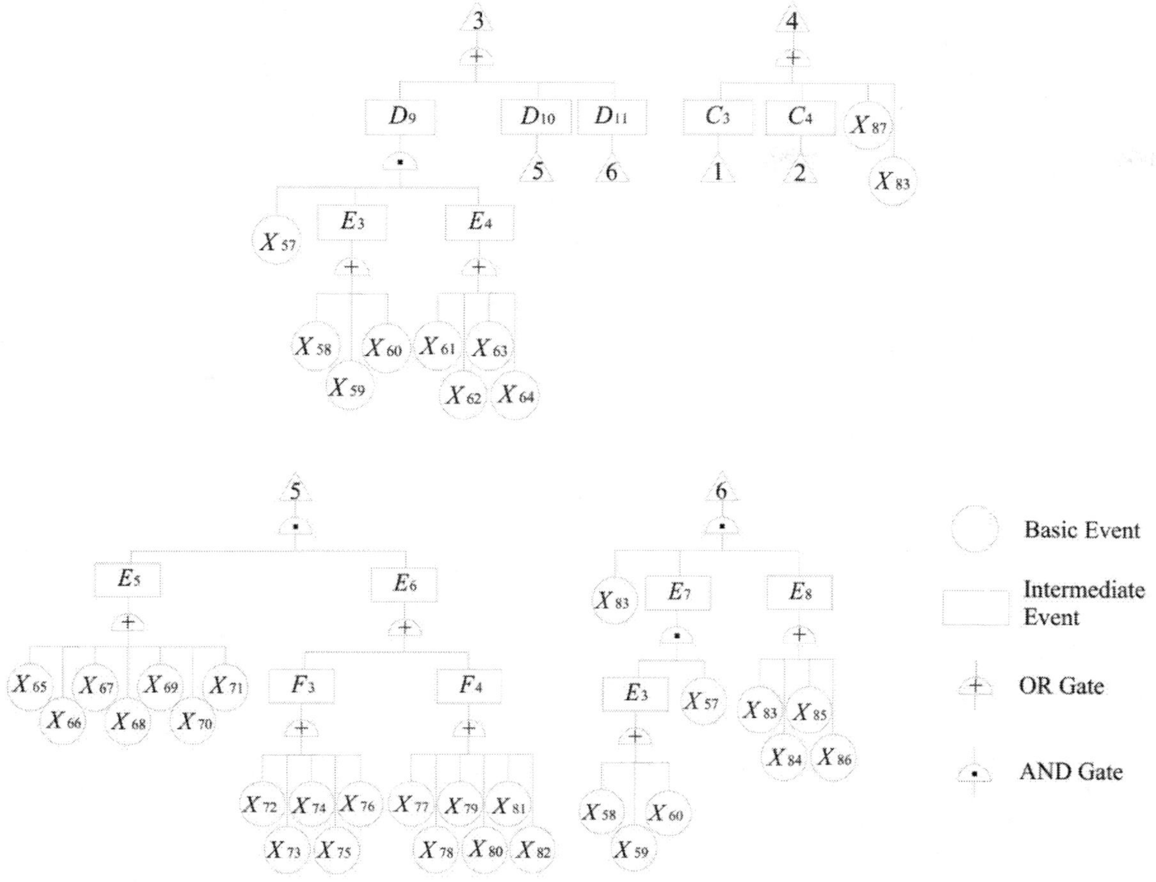

Fig.1 Fault tree model of buried natural gas pipelines failure

Table 1 All basic events in the fault tree

Code	Event	Code	Event	Code	Event
T	Pipeline failure	X_{6}	Malicious blasting	X_{61}	Residual stress
A_{1}	Pipeline rupture	X_{7}	Incorrect operation	X_{62}	Stress concentration
A_{2}	Pipeline puncture	X_{8}	No coordinating with pipeline company	X_{63}	External force
A_{3}	Accessory equipments failure	X_{9}	Pipeline company information is not complete	X_{64}	Internal stress
B_{1}	Third party interference	X_{10}	Geologic explosion	X_{65}	Plant rhizomes penetrate
B_{2}	Low bearing capacity	X_{11}	Geological subsidence caused by construction	X_{66}	Sulfide of soil
B_{3}	Corrosion	X_{12}	Digging and pile driving	X_{67}	High salt of soil
B_{4}	Stress	X_{13}	Heavy vehicles peccancy	X_{68}	Low pH of soil
B_{5}	Serious corrosion	X_{14}	Traffic changes	X_{69}	Soil with SRB
B_{6}	Pipeline defects	X_{15}	Illegal structures over pipeline	X_{70}	High redox potential of soil
B_{7}	Valve leakage	X_{16}	Illegal stacking above pipeline	X_{71}	High water ratio of soil
B_{8}	Flange leakage	X_{17}	Deficient buried depth	X_{72}	Short distance of cathodic protection
B_{9}	Compensator leakage	X_{18}	Illegal planting trees above pipeline	X_{73}	Small protection potential
B_{10}	Condenser cylinder leakage	X_{19}	Secondary disasters caused by explosion	X_{74}	Stray current of ground bed
C_{1}	Natural factor	X_{20}	Implicit signage	X_{75}	Improper protection
C_{2}	Unnatural factor	X_{21}	Poor quality coating	X_{76}	Protective materials failure

Code	Event	Code	Event	Code	Event
C_3	Construction defect	X_{22}	Uneven coating	X_{77}	Coating thinning
C_4	Initial defect	X_{23}	Unsuitable kind of coating	X_{78}	Coating stickiness decreased
C_5	Corrosion crack	X_{24}	Coating is not timely	X_{79}	Coating brittleness increases
C_6	Poor performance of pipe	X_{25}	Incorrect coating method	X_{80}	Coating is damaged
C_7	Large stress	X_{26}	Coating containing impurities	X_{81}	Coating aging stripping
C_8	Packing room leakage	X_{27}	Deficient depth of pipe trench	X_{82}	Coating lower hydrops
C_9	Valve body leakage	X_{28}	Poor slope stability	X_{83}	Poor corrosion resistance of pipe
C_{10}	Poor quality of construction	X_{29}	Coarse grain of backfill soil	X_{84}	Corrosion inhibitor failure
C_{11}	Compensator breakage	X_{30}	High water ratio of backfill soil	X_{85}	Internal coating thinning
D_1	Illegal construction	X_{31}	Poor drainage performance of pipe trench	X_{86}	Pipeline lining exfoliates
D_2	Occupied pressure	X_{32}	Corrosion medium of backfill soil	X_{87}	Poor mechanical properties of pipeline
D_3	Bad anti-corrosion insulation coating	X_{33}	Improper pipeline welding method	X_{88}	Large static stress
D_4	Pipe trench construction	X_{34}	Unqualified welding material	X_{89}	Valve body and cover not seal strictly
D_5	Pipeline welding	X_{35}	Poor quality of surface pretreatment	X_{90}	Valve stem contacts adversely
D_6	Pipeline installation	X_{36}	Stomata of weld surface	X_{91}	Third party damage
D_7	Material defects	X_{37}	Lack od penetration part is large	X_{92}	Incorrect selection of valve body mateeeerial
D_8	Processing technology	X_{38}	Serious carburizing	X_{93}	Pipeline defects
D_9	Stress corrosion	X_{39}	Unclear weiding slag	X_{94}	Poor quality of valve body material
D_{10}	External corrosion	X_{40}	Weld surface with slag	X_{95}	Valve body is not match pipeline
D_{11}	Internal corrosion	X_{41}	Unsmooth surface of elbow	X_{96}	Serious corrosion of valve body
D_{12}	Cast iron valve body leakage	X_{42}	Crack of elbow surface	X_{97}	Stress corrosion cracking of valve body
D_{13}	Cast steel valve body leakage	X_{43}	Large dislocation of pipe section	X_{98}	Nut looses
E_1	Malicious damage	X_{44}	Cracks of flange	X_{99}	Uneven bolt pretightening
E_2	Accidental damage by construction	X_{45}	Bolt material is different from pipeline	X_{100}	Deficient bolt pretightening
E_3	With acid medium	X_{46}	Much inclusion	X_{101}	Uncorrect gasket installation
E_4	Acting stress	X_{47}	Bad microstructure	X_{102}	Gasket aging
E_5	Soil corrosion	X_{48}	Coarse grain	X_{103}	Pipeline coaxiality error
E_6	External anti-corrosion failure	X_{49}	Unsuitable pipeline material	X_{104}	Forced assembly
E_7	Internal corrosive environment	X_{50}	Poor corrosion resistance of pipeline	X_{105}	Matal compensator leakage
E_8	Internal anti-corrosion failure	X_{51}	Heat treatment measures	X_{106}	Aging and cracking of other materials compensator
F_1	Unclear pipeline position	X_{52}	Ovality of pipeline	X_{107}	Low compensation quantity
F_2	External construction unit	X_{53}	Cold working technology	X_{108}	Not consider the heat-expansion and cold-contraction of pipe
F_3	Cathode protect	X_{54}	Uneven pipeline wall thickness	X_{109}	Other causes of rupture
F_4	Anti-corrosion insulation coating	X_{55}	Mechanical scar of pipe wall	X_{110}	Bad anti-corrosion of pumping pipe
X_1	Natural hazard	X_{56}	Unreasonable strength design	X_{111}	damage of pumping pipe

Code	Event	Code	Event	Code	Event
					anti-corrosion
X_2	Soil movement	X_{57}	With water	X_{112}	Pumping pipe has no regular replacement
X_3	Plant sabotage	X_{58}	With H_2S	X_{113}	Naked patr of pumping pipe has no protection
X_4	Water impact	X_{59}	With SO_2		
X_5	Enforced digging and pile driving	X_{60}	With CO_2		

3.2 Minimal path sets

Contrary to the cut sets, the minimum basic events sets which can not cause the top event are called minimum path sets. The more minimum path sets are, the safer the system is.

Through the success tree of the fault tree, by using Boolean algebraic operation methods, 6 minimum path sets of buried pipelines fault tree are obtained:

$J_1=\{X_1',X_2',X_3',X_4',X_5',X_6',X_7',X_8',X_9',X_{10}',X_{11}',X_{12}',X_{13}',X_{14}',X_{15}',X_{16}',X_{17}',X_{18}',X_{19}',X_{20}',X_{21}',X_{22}',X_{23}',X_{24}',X_{25}',X_{26}',X_{27}',X_{28}',X_{29}',X_{30}',X_{31}',X_{32}',X_{33}',X_{34}',X_{35}',X_{36}',X_{37}',X_{38}',X_{39}',X_{40}',X_{41}',X_{42}',X_{43}',X_{44}',X_{45}',X_{46}',X_{47}',X_{48}',X_{49}',X_{50}',X_{51}',X_{52}',X_{53}',X_{54}',X_{55}',X_{56}',X_{61}',X_{62}',X_{63}',X_{64}',X_{65}',X_{66}',X_{67}',X_{68}',X_{69}',X_{70}',X_{71}',X_{83}',X_{87}',X_{88}',X_{89}',X_{90}',X_{91}',X_{92}',X_{93}',X_{94}',X_{95}',X_{96}',X_{97}',X_{98}',X_{99}',X_{100}',X_{101}',X_{102}',X_{103}',X_{104}',X_{105}',X_{106}',X_{107}',X_{108}',X_{109}',X_{110}',X_{111}',X_{112}',X_{113}'\}$

$J_2=\{X_1',X_2',X_3',X_4',X_5',X_6',X_7',X_8',X_9',X_{10}',X_{11}',X_{12}',X_{13}',X_{14}',X_{15}',X_{16}',X_{17}',X_{18}',X_{19}',X_{20}',X_{21}',X_{22}',X_{23}',X_{24}',X_{25}',X_{26}',X_{27}',X_{28}',X_{29}',X_{30}',X_{31}',X_{32}',X_{33}',X_{34}',X_{35}',X_{36}',X_{37}',X_{38}',X_{39}',X_{40}',X_{41}',X_{42}',X_{43}',X_{44}',X_{45}',X_{46}',X_{47}',X_{48}',X_{49}',X_{50}',X_{51}',X_{52}',X_{53}',X_{54}',X_{55}',X_{56}',X_{61}',X_{62}',X_{63}',X_{64}',X_{72}',X_{73}',X_{74}',X_{75}',X_{76}',X_{77}',X_{78}',X_{79}',X_{80}',X_{81}',X_{82}',X_{83}',X_{87}',X_{88}',X_{89}',X_{90}',X_{91}',X_{92}',X_{93}',X_{94}',X_{95}',X_{96}',X_{97}',X_{98}',X_{99}',X_{100}',X_{101}',X_{102}',X_{103}',X_{104}',X_{105}',X_{106}',X_{107}',X_{108}',X_{109}',X_{110}',X_{111}',X_{112}',X_{113}'\}$

$J_3=\{X_1',X_2',X_3',X_4',X_5',X_6',X_7',X_8',X_9',X_{10}',X_{11}',X_{12}',X_{13}',X_{14}',X_{15}',X_{16}',X_{17}',X_{18}',X_{19}',X_{20}',X_{21}',X_{22}',X_{23}',X_{24}',X_{25}',X_{26}',X_{27}',X_{28}',X_{29}',X_{30}',X_{31}',X_{32}',X_{33}',X_{34}',X_{35}',X_{36}',X_{37}',X_{38}',X_{39}',X_{40}',X_{41}',X_{42}',X_{43}',X_{44}',X_{45}',X_{46}',X_{47}',X_{48}',X_{49}',X_{50}',X_{51}',X_{52}',X_{53}',X_{54}',X_{55}',X_{56}',X_{58}',X_{59}',X_{60}',X_{62}',X_{63}',X_{64}',X_{65}',X_{66}',X_{67}',X_{68}',X_{69}',X_{70}',X_{71}',X_{83}',X_{87}',X_{88}',X_{89}',X_{90}',X_{91}',X_{92}',X_{93}',X_{94}',X_{95}',X_{96}',X_{97}',X_{98}',X_{99}',X_{100}',X_{101}',X_{102}',X_{103}',X_{104}',X_{105}',X_{106}',X_{107}',X_{108}',X_{109}',X_{110}',X_{111}',X_{112}',X_{113}'\}$

$J_4=\{X_1',X_2',X_3',X_4',X_5',X_6',X_7',X_8',X_9',X_{10}',X_{11}',X_{12}',X_{13}',X_{14}',X_{15}',X_{16}',X_{17}',X_{18}',X_{19}',X_{20}',X_{21}',X_{22}',X_{23}',X_{24}',X_{25}',X_{26}',X_{27}',X_{28}',X_{29}',X_{30}',X_{31}',X_{32}',X_{33}',X_{34}',X_{35}',X_{36}',X_{37}',X_{38}',X_{39}',X_{40}',X_{41}',X_{42}',X_{43}',X_{44}',X_{45}',X_{46}',X_{47}',X_{48}',X_{49}',X_{50}',X_{51}',X_{52}',X_{53}',X_{54}',X_{55}',X_{56}',X_{58}',X_{59}',X_{60}',X_{62}',X_{63}',X_{64}',X_{72}',X_{73}',X_{74}',X_{75}',X_{76}',X_{77}',X_{78}',X_{79}',X_{80}',X_{81}',X_{82}',X_{83}',X_{87}',X_{88}',X_{89}',X_{90}',X_{91}',X_{92}',X_{93}',X_{94}',X_{95}',X_{96}',X_{97}',X_{98}',X_{99}',X_{100}',X_{101}',X_{102}',X_{103}',X_{104}',X_{105}',X_{106}',X_{107}',X_{108}',X_{109}',X_{110}',X_{111}',X_{112}',X_{113}'\}$

$J_5=\{X_1',X_2',X_3',X_4',X_5',X_6',X_7',X_8',X_9',X_{10}',X_{11}',X_{12}',X_{13}',X_{14}',X_{15}',X_{16}',X_{17}',X_{18}',X_{19}',X_{20}',X_{21}',X_{22}',X_{23}',X_{24}',X_{25}',X_{26}',X_{27}',X_{28}',X_{29}',X_{30}',X_{31}',X_{32}',X_{33}',X_{34}',X_{35}',X_{36}',X_{37}',X_{38}',X_{39}',X_{40}',X_{41}',X_{42}',X_{43}',X_{44}',X_{45}',X_{46}',X_{47}',X_{48}',X_{49}',X_{50}',X_{51}',X_{52}',X_{53}',X_{54}',X_{55}',X_{56}',X_{57}',X_{62}',X_{63}',X_{64}',X_{65}',X_{66}',X_{67}',X_{68}',X_{69}',X_{70}',X_{71}',X_{83}',X_{87}',X_{88}',X_{89}',X_{90}',X_{91}',X_{92}',X_{93}',X_{94}',X_{95}',X_{96}',X_{97}',X_{98}',X_{99}',X_{100}',X_{101}',X_{102}',X_{103}',X_{104}',X_{105}',X_{106}',X_{107}',X_{108}',X_{109}',X_{110}',X_{111}',X_{112}',X_{113}'\}$

$J_6=\{X_1',X_2',X_3',X_4',X_5',X_6',X_7',X_8',X_9',X_{10}',X_{11}',X_{12}',X_{13}',X_{14}',X_{15}',X_{16}',X_{17}',X_{18}',X_{19}',X_{20}',X_{21}',X_{22}',X_{23}',X_{24}',X_{25}',X_{26}',X_{27}',X_{28}',X_{29}',X_{30}',X_{31}',X_{32}',X_{33}',X_{34}',X_{35}',X_{36}',X_{37}',X_{38}',X_{39}',X_{40}',X_{41}',X_{42}',X_{43}',X_{44}',X_{45}',X_{46}',X_{47}',X_{48}',X_{49}',X_{50}',X_{51}',X_{52}',X_{53}',X_{54}',X_{55}',X_{56}',X_{57}',X_{62}',X_{63}',X_{64}',X_{72}',X_{73}',X_{74}',X_{75}',X_{76}',X_{77}',X_{78}',X_{79}',X_{80}',X_{81}',$

$X_{82}', X_{83}', X_{87}', X_{88}', X_{89}', X_{90}', X_{91}', X_{92}', X_{93}', X_{94}', X_{95}', X_{96}', X_{97}', X_{98}', X_{99}', X_{100}', X_{101}', X_{102}', X_{103}', X_{104}', X_{105}', X_{106}', X_{107}', X_{108}', X_{109}', X_{110}', X_{111}', X_{112}', X_{113}'\}$

Each minimum path set contains many events, and the failure probability and risk of pipelines system are high. Therefore, the reliability of pipeline system should be improved by reasonable and accurate design, manufacture, installation, operation and maintenance.

3.3 Structural importance

The structural importance I_i, in the fault tree shows the effect of basic events on pipelines failure[9]. The larger structural importance of basic event is, the greater the influence of basic event on the top event is. The structural importance is obtained by using the minimum cut sets.

Order of structural importance:

$I_1=I_2=I_3=I_4=I_5=I_6=I_7=I_8=I_9=I_{10}=I_{11}=I_{12}=I_{13}=I_{14}=I_{15}=I_{16}=I_{17}=I_{18}=I_{19}=I_{20}=I_{21}=I_{22}=I_{23}=I_{24}=$
$I_{25}=I_{26}=I_{27}=I_{28}=I_{29}=I_{30}=I_{31}=I_{32}=I_{33}=I_{34}=I_{35}=I_{36}=I_{37}=I_{38}=I_{39}=I_{40}=I_{41}=I_{42}=I_{43}=I_{44}=I_{45}=I_{46}=$
$I_{47}=I_{48}=I_{49}=I_{50}=I_{51}=I_{52}=I_{53}=I_{54}=I_{55}=I_{56}=I_{62}=I_{63}=I_{64}=I_{83}=I_{87}=I_{88}=I_{89}=I_{90}=I_{91}=I_{92}=I_{93}=I_{94}=$
$I_{95}=I_{96}=I_{97}=I_{98}=I_{99}=I_{100}=I_{101}=I_{102}=I_{103}=I_{104}=I_{105}=I_{106}=I_{107}=I_{108}=I_{109}=I_{110}=I_{111}=I_{112}=I_{113}$

$I_{57}=I_{61}$

$I_{58}=I_{59}=I_{60}$

$I_{65}=I_{66}=I_{67}=I_{68}=I_{69}=I_{70}=I_{71}$

$I_{72}=I_{73}=I_{74}=I_{75}=I_{76}=I_{77}=I_{78}=I_{79}=I_{80}=I_{81}=I_{82}$

$I_{84}=I_{85}=I_{86}$

And $I_1>I_{57}>I_{58}>I_{65}>I_{72}>I_{84}$

By the analysis of structural importance, 87 basic events with higher value have the greatest influence on the top event. While the corrosion environment, external anti-corrosion failure and internal anti-corrosion failure have a relatively smaller effect.

4. Main influencing factors analysis and suggestion

By the analysis of minimum cut sets, minimum path sets and structural importance of fault tree, the major causes of pipelines failure can be obtained and the safety precautions as follows:

4.1 Third-party interference

Third-party interference includes natural and human factors. Natural disasters also make serious damage to pipelines and consequent leakage of natural gas makes a worse situation. At the design stage, pipelines construction lines should be investigated overall. Related data on geology, meteorology, surrounding environment are collected as comprehensively as possible to make the best design scheme. Artificial damage is mainly illegal construction and occupied pressure, such as incorrect operation, illegal structures over pipeline, enforced digging and pile driving. Construction side should get in touch with pipe company and familiar with piping drawings and other related data before construction near the natural gas pipelines. Pipeline operating companies must strengthen the protection and propaganda of pipes, such as establishing pipeline regular patrol system, perfecting the warning marks of buried pipelines, etc.

4.2 Corrosion

Serious corrosion includes external and internal corrosion which may lead to wall thinning, puncture and even rupture of pipe.

External corrosion is mainly caused by soil corrosion and external anti-corrosion failure. Therefore, the anti-corrosion level should be improved when pipelines cross rivers, lowland and high salt zone. Pipelines detection and assessment must be done on a regular basis to ensure the normal operation of cathodic protection system.

Special medium, such as water, sulfide, acid medium, has the most effect to internal corrosion. Therefore, the quality of natural gas in pipelines should be controled strictly. Termly detecting and analyzing of dehydration, anti-corrosive coating and corrosion inhibitor could reduce unexpected failure. Reasonable pigging period and suitable pipe cleaner are also preferred.

4.3 Pipeline defects

Pipeline defects include initial and construction defect. Initial defect is mainly due to the unqualified pipes, such as containing impurities and uneven wall thickness. Improper operation, such as insufficient pipe trench depth and welding, is the mainly factor of construction defect. In order to reduce the affect and ensure the safety of pipelines, quality of manufacturing pipe, quality of materials and quality of installation process should be inspected carefully.

5. Applicaton discussion of fault tree

As mentioned above, through site investigation and theoretical analysis, a more complete buried natural gas pipelines fault tree model considering various events has been established..It can be applied for risk analysis of buried natural gas pipelines in various environmental conditions. In the case of specific application, basic events are selected from Fig. 1 and Table 1, which should be in accordance with the actual situation first. Then Boolean algebraic operation methods are used to find out the minimum cut sets of the fault tree. The minimum path sets can be obtained through the success tree of the fault tree. Based on minimum cut sets analysis, the structural importance of basic events is analyzed. And then the main influencing factors of pipelines failure are analyzed comprehensively. At last, safety precautions are presented.

FTA is applicable for the reliability and safety analysis of complex system. Due to lack of adequate experimental data of the natural gas pipelines system, fault tree can be used to analyze the main influencing factors. And the qualitative and quantitative evaluation can also be carried out on the buried pipelines system. And then targeted prevention measures are put forward to improve the level of safe operation, which can provide reference for natural gas pipeline integrity management. FTA model established in this paper has been applied for risk analysis and safety evaluation of buried nantural gas pipelines in 30 projects in Jiangsu province. Practice proves that the FTA model built in this paper has universal practicability.

6. Conclusions

The research shows that FTA provides an effective method to assess buried pipelines reliability and risk analysis. A complete buried natural gas pipelines fault tree has been constructed, including 113 basic events which can result in natural gas pipelines failure. Through the FTA, the main factors which cause pipelines failure can be determined. And preventive measures and suggestions can also be put forward. Practice has proved that the FTA model of natural gas pipelines can be widely used in buried pipelines risk analysis and safety evaluation.

7. Acknowledgements

The authors are grateful for the support given by Jiangsu Province Special Equipment Safety Supervision Inspection Institute.

8. References

[1] Wang jianyou, Accident statistical analysis and countermeasures for gas transmission and distuibution system[J]. Gas and Heat, 2001, 21(2): 178-179.

[2] He shujing, Zhou weiguo, Yan mingqing, Accident statistical analysis and countermeasures for urban gas transmission and distribution system[J]. Gas and Heat, 2003, 23(12): 753-755.

[3] Liao kexi, Yao anlin, Zhag huaixin, Fault tree analysis of natural gas pipeline failure[J]. Natural Gas Industry, 2001, 22(2): 94-96.

[4] Liu yang, Jia fengrui, Duan changgui, Study on fault tree of urban gas buried steel pipeline[J]. Gas and Heat, 2009, 29(7): 35-38.

[5] Dong yuhua, Yu dabao, Estimation of failure probability of oil and gas transmission pipeline by fuzzy fault tree analysis[J]. Journal of Loss Prevention in the Process Industries, 2005(18): 83-88.

[6] Liang wei, Hu jinqiu, Zhang laibin, Guo cunjie, Lin weipeng, Assessing and classifying risk of pipeline third-party interference based on fault tree and SOM[J]. Engineering Applications of Artificial Intelligence, 2012(25): 594-608.

[7] Liao kexi, Yao anlin, Zhang huaixin, Fault tree analysis on the long distance pipelines[J]. Oil and Gas Transportation, 2001, 20(1): 27-30.

[8] Dong yuhua, Yu dabao, Fault tree analysis of oil and gas pipeline[J]. Oil and Gas Transportation, 2002, 21(6): 15-17.

[9] Wang yuanhui, Safety system engineering[M].Tianjin: Tianjin University Press, 1999.

AUTHOR INDEX

Ahmed, Salim 450, 906
Arendt, Steve 361, 440
Bao, Shilong 931
Baybutt, Paul 39, 265
Bearrow, Mike 1
Cai, Tianxing 483
Cai, Xin 740
Cao, Guoliang 165, 784
Cheam, Ernest 392
Chen, Chen 888, 897
Chen, Dong-Liang 972
Chen, Guohua 594
Chen, Guoming 165, 628
Chen, Guo-Ming 713
Chen, Jianhong 104
Chen, Liping 817
Chen, Ming 846
Chen, Wanghua 817
Chen, Xi 165, 628
Chen, Yanfei 655
Chen, Yinting 773
Cheng, Qinghua 769
Cheng, Xiuwen 769
Cheng, Zonghua 194
Chu, Shengli 194
Ding, Hao 876, 878
Ebadat, Vahid 402, 423
Feng, Wenxing 204
Fu, Huan 769
Fu, Jingchun 554
Gao, Xiaodan 412
Gao, Yaqi 671
Garaniya, Vikram 227
Ge, Tianming 793
Guan, Wenling 545
Guang, Zhang Ming 321
Guo, Shanshan 646
Guoming, Chen 522
Han, Fenglei 514, 854
Han, Hui 618
Han, Wenhui 441
Hashemi, Javad 450, 906
Hassim, Mimi H. 54, 311
Hazzan, Michael 348
Hickey, Colin 65
Hong, Lin 793
Horng, Jaojia 104
Hurme, Markku 311
Ji, Xu 91
Jia, Yan 836
Jiang, Jiaojun 412
Jiang, Jun Cheng 531, 545
Jiang, Shu 474
Karthikeyan, S 2
Kasmani, Rafiziana M. 54, 311
Khan, Faisal I. 227, 450, 906
Kiah, Miss Hasimawaty Mat 54
Kong, Linghai 290
Lan, Jing 921
Lehmann, Scott 238
Li, Dong 91
Li, Guo-Rui 713
Li, Mingyu 513
Li, Zhang 963
Liao, Kuangyu 104
Liao, Yuanchun 104
Liou, Kehun 104
Liu, Huiling 769
Liu, Jun-Fang 704
Liu, Kun 555
Liu, Ming 878
Liu, Tony 83
Liu, X. Y. 998
Liu, Xinyong 981
Liu, Xuanya 184
Liu, Yi 83, 863
Long, Fei 194
Lu, Jianming 134
Lv, Jiayu 817
Ma, Tingguang 770
Mannan, M Sam 412
Meng, Hua 513
Meng, Yifei 174, 561, 863
Moore, David A. 348
Nara, Louisa 22, 215
Pandian, Santha 311
Peng, Minjun 817
Peterson, Eric 83
Pi, Xiaoyue 725
Ping, Bai 963
Pridy, Cindie 22
Qiu, Shaolin 339
Ramabhadran, Srinivasan 13, 382
Ramalingam, Ramesh 2
Reed, Morgan 83
Ren, Lefeng 784
Ru, Apeng 339
Shaba, Kehinde 65
Shang, Shengmei 760
Shao, Yue 586
Shi, Haodong 878
Shi, Huandi 784
Song, Huihui 646
Su, Hanyu 740
Subbarao, N V 2
Suikkanen, Johanna 460
Sun, Bin 671
Sun, Xiaoping 165, 628
Sun, Xinmeng 661
Tang, Daizhi 990

AUTHOR INDEX

Tao, Gang 91, 474, 503, 618, 646, 661, 671
Tian, Lianjun...290
Tong, Bing 503
Wang, Bing 769
Wang, Cuiping 854
Wang, Feng..940
Wang, Hai-Qing 680, 704, 760
Wang, Hang..680
Wang, Jenny...301
Wang, Kui-Sheng 972
Wang, Lei...836
Wang, Pu 769
Wang, Xiao-Li 545, 575
Wang, Xiaoyuan..290
Wang, Xin..618
Wang, Y. N. 981, 998
Wang, Yan Fu 531
Wang, Z. R...981, 998
Wang, Zhaohui..809
Wang, Zhigang..836
Wang, Zhihe 280
Wang, Zhongyu..474
Wenhe, Wang 155
Wu, Chaopeng...204
Xin, Chun-Ling 972
Xiong, Wei 836
Xu, Chang-Hang 165, 628, 713, 784
Xu, Wan...661
Xu, Xiaoyuan...184
Xu, Yongfu...119
Yang, Jianfeng...972
Yang, Ming 450
Yao, Hongqi...740
Yao, Zhe...793
Yi, Liu 826
Yin, Tongshan 826
Yong, Zhang 637
You, Fei..586
Yu, Ricky 330
Yu, Wendi...562
Yuan, Jing...174
Yuan, Zhu...522
Zeng, Xiaoming 301
Zhang, Bo...836
Zhang, Chao 863
Zhang, Dong-Sheng...972
Zhang, Hua 503
Zhang, Huawen 594
Zhang, Laibin...339
Zhang, Lijing............... 91, 474, 503, 618, 646, 661, 671
Zhang, Long 216
Zhang, Mingguang...773
Zhang, Qinlin 184
Zhang, Xinmei..888, 897
Zhang, Yu-Tao 680, 751
Zhang, Zufeng 805
Zhao, Dongfeng 561, 846, 854, 863
Zhao, Jinsong.. 460, 689
Zhao, John Jianmin... 249
Zhao, Xue'E 740
Zhou, J 952
Zhu, Baikang 594
Zhu, Jianfeng.. 689
Zhu, Yuan 165, 628

AIChE
120 Wall Street, FL 23
New York, NY 10005-4020

ISBN 978-1-5108-1715-9

CURRAN ASSOCIATES INC.
proceedings.com